T0390572

AGGREGATION-INDUCED EMISSION (AIE)

AGGREGATION-INDUCED EMISSION (AIE)

A Practical Guide

Edited by

JIANWEI XU

*Institute of Materials Research and Engineering, Agency for Science, Technology and Research (A*STAR), Singapore*

MING HUI CHUA

*Institute of Materials Research and Engineering, Agency for Science, Technology and Research (A*STAR), Singapore*

BEN ZHONG TANG

School of Science and Engineering, Shenzhen Key Laboratory of Functional Aggregate Materials, The Chinese University of Hong Kong, Shenzhen, Guangdong, China

ELSEVIER

Elsevier
Radarweg 29, PO Box 211, 1000 AE Amsterdam, Netherlands
The Boulevard, Langford Lane, Kidlington, Oxford OX5 1GB, United Kingdom
50 Hampshire Street, 5th Floor, Cambridge, MA 02139, United States

ISBN: 978-0-12-824335-0

For information on all Elsevier publications
visit our website at https://www.elsevier.com/books-and-journals

Publisher: Matthew Deans
Acquisitions Editor: Kayla Dos Santos
Editorial Project Manager: Isabella C. Silva
Production Project Manager: Anitha Sivaraj
Cover Designer: Mark Rogers

Typeset by STRAIVE, India

Contents

7. Photochromic and thermochromic luminescence in AIE luminogens

Ju Mei and He Tian

8. AIE-active rare-metal-free phosphorescent materials

Masaki Shimizu

9. AIE luminogens exhibiting thermally activated delayed fluorescence

Xiaojie Chen, Xiangyu Ge, Zhan Yang, Juan Zhao, Zhiyong Yang, Yi Zhang, and Zhenguo Chi

10. Aggregation-induced emission luminogens for organic light-emitting diodes

Suraj Kumar Pathak and Chuluo Yang

11. Liquid crystalline aggregation-induced emission luminogens for optical displays

Kyohei Hisano, Osamu Tsutsumi, and Supattra Panthai

12. Electrofluorochromism in AIE luminogens

Guey-Sheng Liou and Hung-Ju Yen

13. AIE-active materials for photovoltaics

Andrea Pucci

14. AIE molecular probes for biomedical applications

Alex Y.H. Wong, Fei Wang, Chuen Kam, and Sijie Chen

15. Recent advances of aggregation-induced emission nanoparticles (AIE-NPs) in biomedical applications

Soheila Sabouri, Bicheng Yao, and Yuning Hong

16. AIE bio-conjugates for biomedical applications

Zhiyuan Gao and Dan Ding

17. AIE-active polymers for explosive detection

Hui Zhou, Ming Hui Chua, Qiang Zhu, and Jianwei Xu

18. AIE-based chemosensors for vapor sensing

Meng Li, Dong Wang, and Ben Zhong Tang

19. AIEgen applications in rapid and portable sensing of foodstuff hazards

Qi Wang, Youheng Zhang, Yanting Lyu, Xiangyu Li, and Wei-Hong Zhu

20. Computational modeling of AIE luminogens

Qian Peng, Zhigang Shuai, and Qi Ou

Contributors

Massimo Cametti Department of Chemistry, Materials and Chemical Engineering "Giulio Natta", Politecnico di Milano, Milano, Italy

Sijie Chen Ming Wai Lau Centre for Reparative Medicine, Karolinska Institutet, Hong Kong, China

Xiaojie Chen PCFM Lab, GDHPPC Lab, Guangdong Engineering Technology Research Center for High-performance Organic and Polymer Photoelectric Functional Films, State Key Laboratory of OEMT, School of Chemistry, Sun Yat-sen University, Guangzhou, China

Zhenguo Chi PCFM Lab, GDHPPC Lab, Guangdong Engineering Technology Research Center for High-performance Organic and Polymer Photoelectric Functional Films, State Key Laboratory of OEMT, School of Chemistry; School of Materials Science and Engineering, Sun Yat-sen University, Guangzhou, China

Ming Hui Chua Institute of Materials Research and Engineering, A*STAR (Agency for Science, Technology and Research), Singapore, Singapore

Yoshiki Chujo Department of Polymer Chemistry, Graduate School of Engineering, Kyoto University, Kyoto, Japan

Dan Ding State Key Laboratory of Medicinal Chemical Biology, Key Laboratory of Bioactive Materials, Ministry of Education, and College of Life Sciences, Nankai University, Tianjin, China

Yong Qiang Dong Beijing Key Laboratory of Energy Conversion and Storage Materials, College of Chemistry, Beijing Normal University, Beijing, China

Zoran Džolić Ruđer Bošković Institute, Zagreb, Croatia

Zhiyuan Gao State Key Laboratory of Medicinal Chemical Biology, Key Laboratory of Bioactive Materials, Ministry of Education, and College of Life Sciences, Nankai University, Tianjin, China

Xiangyu Ge PCFM Lab, GDHPPC Lab, Guangdong Engineering Technology Research Center for High-performance Organic and Polymer Photoelectric Functional Films, State Key Laboratory of OEMT, School of Chemistry, Sun Yat-sen University, Guangzhou, China

Masayuki Gon Department of Polymer Chemistry, Graduate School of Engineering, Kyoto University, Kyoto, Japan

Pengbo Han State Key Laboratory of Luminescent Materials and Devices, Guangdong Provincial Key Laboratory of Luminescence from Molecular Aggregates, AIE Institute, Center for Aggregation-Induced Emission, South China University of Technology (SCUT), Guangzhou, China

Kyohei Hisano Department of Applied Chemistry, Ritsumeikan University, Kusatsu, Japan

Yuning Hong Department of Chemistry and Physics, La Trobe Institute for Molecular Science, La Trobe University, Melbourne, VIC, Australia

Rongrong Hu State Key Laboratory of Luminescent Materials and Devices, Guangdong Provincial Key Laboratory of Luminescence from Molecular Aggregates, AIE Institute, Center for Aggregation-Induced Emission, South China University of Technology (SCUT), Guangzhou, China

Yang Hu State Key Laboratory of Luminescent Materials and Devices, Guangdong Provincial

Key Laboratory of Luminescence from Molecular Aggregates, AIE Institute, Center for Aggregation-Induced Emission, South China University of Technology (SCUT), Guangzhou, China

Shunichiro Ito Department of Polymer Chemistry, Graduate School of Engineering, Kyoto University, Kyoto, Japan

Chuen Kam Ming Wai Lau Centre for Reparative Medicine, Karolinska Institutet, Hong Kong, China

Bing Shi Li Key Laboratory of New Lithium-Ion Battery and Mesoporous Material, College of Chemistry and Environmental Engineering, Shenzhen University, Shenzhen, China

Hongkun Li Laboratory of Advanced Optoelectronic Materials, College of Chemistry, Chemical Engineering and Materials Science, Soochow University, Suzhou, China

Meng Li Center for AIE Research, Shenzhen University, Shenzhen; Shenzhen Institute of Aggregate Science and Technology, School of Science and Engineering, The Chinese University of Hong Kong, Shenzhen, Guangdong, China

Xiangyu Li Institute of Fine Chemicals, East China University of Science and Technology, Shanghai, China

Guey-Sheng Liou Institute of Polymer Science and Engineering, National Taiwan University, Taipei, Taiwan

Yanting Lyu Institute of Fine Chemicals, East China University of Science and Technology, Shanghai, China

Ju Mei Key Laboratory for Advanced Materials, Feringa Nobel Prize Scientist Joint Research Center, Frontiers Science Center for Materiobiology and Dynamic Chemistry, Joint International Research Laboratory for Precision Chemistry and Molecular Engineering, Institute of Fine Chemicals, School of Chemistry & Molecular Engineering, East China University of Science & Technology, Shanghai, P. R. China

Qi Ou MOE Key Laboratory of Organic OptoElectronics and Molecular Engineering, Department of Chemistry, Tsinghua University, Beijing, China

Supattra Panthai Department of Applied Chemistry, Ritsumeikan University, Kusatsu, Japan

Suraj Kumar Pathak College of Materials Science and Engineering, Shenzhen University, Shenzhen, China

Qian Peng School of Chemical Sciences, University of Chinese Academy of Sciences, Beijing, China

Andrea Pucci Department of Chemistry and Industrial Chemistry of the University of Pisa, Pisa, Italy

Anjun Qin State Key Laboratory of Luminescent Materials and Devices, Guangdong Provincial Key Laboratory of Luminescence from Molecular Aggregates, AIE Institute, Center for Aggregation-Induced Emission, South China University of Technology (SCUT), Guangzhou, China

Soheila Sabouri Department of Chemistry and Physics, La Trobe Institute for Molecular Science, La Trobe University, Melbourne, VIC, Australia

Masaki Shimizu Faculty of Molecular Chemistry and Engineering, Kyoto Institute of Technology, Kyoto, Japan

Zhigang Shuai MOE Key Laboratory of Organic OptoElectronics and Molecular Engineering, Department of Chemistry, Tsinghua University, Beijing, China

Yue Si Beijing Key Laboratory of Energy Conversion and Storage Materials, College of Chemistry, Beijing Normal University, Beijing, China

Kazuo Tanaka Department of Polymer Chemistry, Graduate School of Engineering, Kyoto University, Kyoto, Japan

Ben Zhong Tang School of Science and Engineering, Shenzhen Key Laboratory of Functional Aggregate Materials, The Chinese University of Hong Kong, Shenzhen, Guangdong, China

He Tian Key Laboratory for Advanced Materials, Feringa Nobel Prize Scientist Joint

Research Center, Frontiers Science Center for Materiobiology and Dynamic Chemistry, Joint International Research Laboratory for Precision Chemistry and Molecular Engineering, Institute of Fine Chemicals, School of Chemistry & Molecular Engineering, East China University of Science & Technology, Shanghai, P. R. China

Osamu Tsutsumi Department of Applied Chemistry, Ritsumeikan University, Kusatsu, Japan

Dong Wang Center for AIE Research, Shenzhen University, Shenzhen, China

Fei Wang Ming Wai Lau Centre for Reparative Medicine, Karolinska Institutet, Hong Kong, China

Jia Wang State Key Laboratory of Luminescent Materials and Devices, Guangdong Provincial Key Laboratory of Luminescence from Molecular Aggregates, AIE Institute, Center for Aggregation-Induced Emission, South China University of Technology (SCUT), Guangzhou, China

Qi Wang Institute of Fine Chemicals, East China University of Science and Technology, Shanghai, China

Alex Y.H. Wong Ming Wai Lau Centre for Reparative Medicine, Karolinska Institutet, Hong Kong, China

Jianwei Xu Institute of Materials Research and Engineering, A*STAR (Agency for Science, Technology and Research), Singapore, Singapore

Chuluo Yang College of Materials Science and Engineering, Shenzhen University, Shenzhen, China

Zhan Yang PCFM Lab, GDHPPC Lab, Guangdong Engineering Technology Research Center for High-performance Organic and Polymer Photoelectric Functional Films, State Key Laboratory of OEMT, School of Chemistry, Sun Yat-sen University, Guangzhou, China

Zhiyong Yang PCFM Lab, GDHPPC Lab, Guangdong Engineering Technology Research Center for High-performance Organic and Polymer Photoelectric Functional Films, State Key Laboratory of OEMT, School of Chemistry, Sun Yat-sen University, Guangzhou, China

Bicheng Yao Department of Chemistry and Physics, La Trobe Institute for Molecular Science, La Trobe University, Melbourne, VIC, Australia

Hung-Ju Yen Institute of Chemistry, Academia Sinica, Taipei, Taiwan

Lihui Zhang State Key Laboratory of Luminescent Materials and Devices, Guangdong Provincial Key Laboratory of Luminescence from Molecular Aggregates, AIE Institute, Center for Aggregation-Induced Emission, South China University of Technology (SCUT), Guangzhou, China

Yi Zhang PCFM Lab, GDHPPC Lab, Guangdong Engineering Technology Research Center for High-performance Organic and Polymer Photoelectric Functional Films, State Key Laboratory of OEMT, School of Chemistry, Sun Yat-sen University, Guangzhou, China

Youheng Zhang Institute of Fine Chemicals, East China University of Science and Technology, Shanghai, China

Yucong Zhang Beijing Key Laboratory of Energy Conversion and Storage Materials, College of Chemistry, Beijing Normal University, Beijing, China

Juan Zhao School of Materials Science and Engineering, Sun Yat-sen University, Guangzhou, China

Hui Zhou Institute of Materials Research and Engineering, A*STAR (Agency for Science, Technology and Research), Singapore, Singapore

Qiang Zhu Institute of Materials Research and Engineering, A*STAR (Agency for Science, Technology and Research), Singapore, Singapore

Wei-Hong Zhu Institute of Fine Chemicals, East China University of Science and Technology, Shanghai, China

Preface

The discovery of aggregation-induced emission (AIE) in 2001 has served as a game changer in the development and application of luminogenic functional materials. Fundamental understanding of photophysical processes and properties in organic luminogens has been reshaped, creating vast opportunities for a wide range of applications of AIE luminogens (AIEgens). AIE has effectively overcome the limitations of aggregation-caused quenching (ACQ) commonly found in traditional luminogens. This not only contributes to improvement in material performance for existing applications but also leads to the emergence of new applications. Therefore, research interests and efforts in the area of AIE have soared over the past two decades, with an exponential growth of scientific publications and citations, serving as a testament to the usefulness of AIEgens. This book, therefore, aims to provide a holistic coverage of both fundamental principles and applications of AIEgens, covering the key scientific progresses in the AIE topic at an introductory-to-intermediate level, suitable for a wide range of scientific and academic audiences, both within and outside the AIE research fraternity.

This book begins with the introduction of fundamental concepts, principles, and mechanisms of AIE in Chapter 1. To date, a large number of novel AIEgens have been designed and synthesized, and therefore the first part of the book seeks to display the structural diversities of AIEgens. Chapter 2 highlights novel AIEgens with boron complexes while Chapter 3 summarizes numerous AIE-active polymers developed thus far for various applications. Thereafter, chiral AIEgens with circularly polarized luminescence and helical self-assembly properties are described in Chapter 4, followed by AIE supramolecular gel systems in Chapter 5.

The second part of the book seeks to showcase the application diversity and usefulness of AIEgens. AIEgens can be broadly categorized into four key domains in terms of their main applications: (i) stimuli-responsive AIE systems, (ii) optoelectronics, (iii) biomedical sensing, and (iv) chemical sensing. The subsequent chapters are therefore arranged in accordance with their applications in the respective domains. For stimuli-responsive AIE systems, Chapter 6 discusses mechanochromic AIEgens, whereas Chapter 7 reviews AIEgens with photochromic and thermochromic properties.

AIEgens are excellent candidates for optoelectronic applications, notably organic light-emitting diodes (OLEDs) and liquid crystal (LC) optical displays, due to their amplification of luminescence intensities in the solid states. Chapters 8 and 9 introduce pure organic AIEgens that exhibit (aggregation and crystallization-induced) phosphorescence and thermally activated delayed fluorescence (TADF) properties, respectively, both of which exhibit important solid-state luminescence properties particularly for the development of OLEDs. This leads us to Chapter 10, in which the application of AIEgens in OLEDs is summarized. The use of AIEgens in other notable optoelectronic applications such as LC optical displays, electrofluorochromic devices, and photovoltaics is discussed separately in Chapters 11–13.

Next, the book concentrates on AIEgens for biomedical applications, including biosensing and bioimaging, diagnostics, therapy, and drug delivery. AIEgens may be directly used as molecular fluorescent probes (mainly for biosensing and imaging) or be further prepared as fluorescent nanoparticles (FNPs) and bioconjugates to enhance luminescence intensity, sensitivity, and biocompatibility. The use of AIE-active molecular probes, FNPs, and bioconjugates for a spectrum of different biomedical applications is collectively discussed in Chapters 14–16, respectively.

In addition to biosensing, chemical sensing encompasses environmental monitoring of harmful substances to safeguard public health and the detection of trace chemicals for product quality assurance. Chapter 17 therefore summarizes AIE-based fluorescent chemosensors for explosive detection while the development of an AIE-based vapor sensor to detect analytes in the vapor state is discussed in Chapter 18. Thereafter, Chapter 19 highlights the use of AIEgens in sensing foodstuff hazards as well as a fluorescent thermometer. Finally, Chapter 20 summarizes the luminescence mechanism of AIEgens utilizing computational simulation and modeling methods, which provide another way to have in-depth and intrinsic understanding of the nature of AIE.

Through this book, it is envisaged that readers will gain knowledge and understanding about not only the fundamental principles of AIE but more importantly how AIE luminogens can be incorporated in various molecular and polymeric materials to specifically cater to targeted needs and applications.

Finally, the editors wish to express their immense appreciation to all authors for their dedicated efforts in contributing high-quality work to this book.

CHAPTER

1

Fundamental principles of AIE

Pengbo Han[a], *Jia Wang*[a], *Anjun Qin*[a], *and Ben Zhong Tang*[b]

[a]State Key Laboratory of Luminescent Materials and Devices, Guangdong Provincial Key Laboratory of Luminescence from Molecular Aggregates, AIE Institute, Center for Aggregation-Induced Emission, South China University of Technology (SCUT), Guangzhou, China [b]School of Science and Engineering, Shenzhen Key Laboratory of Functional Aggregate Materials, The Chinese University of Hong Kong, Shenzhen, Guangdong, China

1 Introduction

Luminescent materials with aggregation-induced emission (AIE) features have attracted tremendous attention for their potential practical applications. The concept of AIE was coined in 2001 by Tang et al. when they observed a unique phenomenon in a silole derivative, which is nonemissive in dilute solution but emits brightly when forming aggregates [1].

Deciphering the underneath mechanisms of AIE is crucial for the enrichment of fundamental photophysical knowledge, construction of new luminogens, and exploration of practical applications [2–4]. In principle, a matter absorbing light energy will be promoted to the excited state, which will fall back to lower energy states through photophysical or photochemical processes [5]. These photophysical processes mainly include radiative transition and nonradiative transition pathways, whereas the photochemical pathway mainly includes a chemical reaction. In the solution state, the excited-state decay of AIE luminogens (AIEgens) is mainly through nonradiative photophysical or photochemical processes. Meanwhile, in aggregate states, the nonradiative decay pathways are blocked and the radiative ones are opened. The combination effects readily result in the unique AIE feature.

Numerous efforts have been devoted to deciphering the AIE working principle and a number of possible mechanisms have been proposed, such as J-aggregation, conformational planarization, E/Z isomerization, twisted intramolecular charge transfer (TICT), and excited-state intramolecular proton transfer (ESIPT), but most of them were only applicable to limited AIE systems.

Aggregation-Induced Emission (AIE)
https://doi.org/10.1016/B978-0-12-824335-0.00010-6

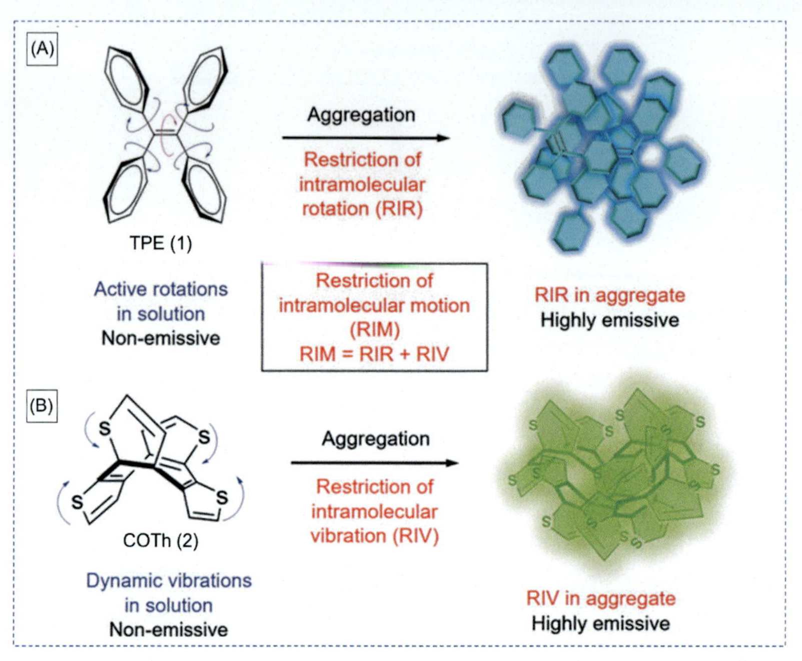

FIG. 1 (A) Tetraphenylethene (TPE) is nonemissive when molecularly dissolved but becomes emissive when aggregated due to the restriction of intramolecular rotations (RIR). (B) Cyclooctatetrathiophene (COTh) shows AIE activity due to the restriction of intramolecular vibration (RIV) in the aggregate state. *Reproduced with permission from Z. Zhao, H. Zhang, J.W.Y. Lam, B.Z. Tang, Aggregation-induced emission: new vistas at the aggregate level, Angew. Chem. Int. Edit. 59 (2020) 2. Copyright 2020 Wiley-VCH Verlag GmbH & Co. KGaA.*

With great and persistent efforts, the restriction of intramolecular rotation (RIR) mechanism has been proposed. However, as the family of AIEactive molecules grows, some AIE systems with no rotatable units cannot fully be explained by the RIR mode. Therefore, the restriction of intramolecular vibrations (RIV) was raised to explain these AIE cases. It has become clear that RIR and RIV have been rationalized as the main cause for the AIE effect. Therefore, they are integrated into a more comprehensive AIE mechanism, i.e., restriction of intramolecular motion (RIM) (Fig. 1) [6].

2 Restriction of intramolecular rotations

The RIR mechanism was proposed based on a careful and systematic study of an archetype of AIEgen of hexaphenylsilole (HPS, **3**) [7]. HPS is soluble in organic solvents, such as dichloromethane, acetone, THF, and methanol, but insoluble in water. Therefore, the aggregation of HPS molecules can be induced by adding water in HPS acetone solution, and the photoluminescence (PL) quantum yield ($[Fcy]_F$) was explored in acetone/water mixtures with different water fractions (f_w). As shown in Fig. 2A, HPS is nonemissive in dilute acetone

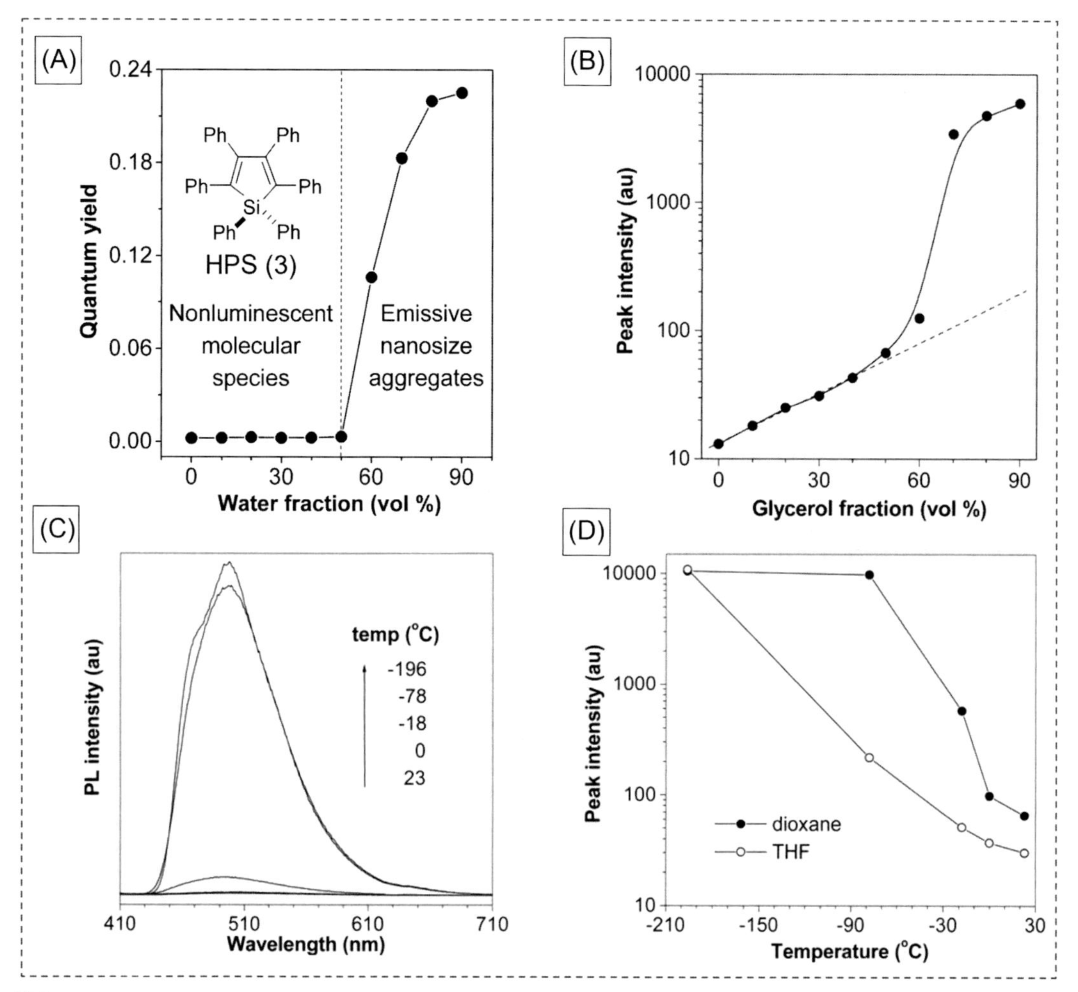

FIG. 2 Plots of (A) PL quantum yield of HPS vs water fraction in acetone/water mixtures and (B) its PL peak intensity vs glycerol fraction in glycerol/methanol mixtures. (C) PL spectra of HPS in 1,4-dioxane at different temperatures. (D) Effect of temperature on the peak intensity of the PL of HPS in dioxane and THF. Concentration = 10 μM. *Reproduced with permission from J. Chen, C.C.W. Law, J.W.Y. Lam, Y. Dong, S.M.F. Lo, I.D. Williams, et al., Synthesis, light emission, nanoaggregation, and restricted intramolecular rotation of 1,1-substituted 2,3,4,5-tetraphenylsiloles, Chem. Mater. 15 (7) (2003) 1535–1546. Copyright 2003 American Chemical Society.*

solution with a low $[Fcy]_F$ (~0.1%), which remains almost unchanged until the f_w reaches 50vol% but starts to increase swiftly afterward. When the f_w increases to 90%, the $[Fcy]_F$ value is boosted to 22%, which is ~200 times higher than that of the acetone solution. The higher $[Fcy]_F$ value of HPS in the aggregate state than that in dilute solutions demonstrates its AIE effect, which is attributed to the RIR mechanism. In solution, the peripheral phenyl rings of HPS can dynamically rotate around the central silole ring, which may effectively consume the energy of the excited state, making the HPS nonemissive, while in the aggregate state the intramolecular rotations are restricted because of the physical constraint. Therefore, the nonradiative channel of deexcitation is blocked and the radiative decay is activated, making the HPS emit strongly.

2.1 External physical control experiments

2.1.1 *Viscosity effect*

To further verify the rationality of the RIR mechanism, several control experiments were designed and conducted [7–9]. Strong emission is envisioned for HPS in the more viscous media because the high viscosity would retard the intramolecular rotations. The viscosity of glycerol (934 cP at 25°C) is 1720 times higher than that of methanol (0.544 cP at 25°C), and the viscosity of media will be enhanced by increasing the glycerol percentage in methanol. Therefore, the PL measurements of HPS were performed in such mixtures. As shown in Fig. 2B, the PL intensity increased linearly as the glycerol fraction (f_G) increased in the range of 0–50 vol% at 25°C. The emission enhancement in this region should be predominantly ascribed to the viscosity effect. When f_G is further increased, the peak intensity increased sharply due to the formation of nanoaggregates.

2.1.2 *Temperature effect*

Since decreasing the solution temperature can also hamper the intramolecular rotations, the temperature effects on the HPS emission were thus studied. When the dioxane solution of HPS was cooled, its PL intensity increased accordingly (Fig. 2C). This is because the dioxane solution changed to a glassy state when the temperature cooled below its melting point (11.8°C). Therefore, the intramolecular rotation of the phenyl rings of HPS would be restricted by the rigid environments. In addition, the emission of HPS decreased drastically when the solution was heated above the melting point of dioxane (Fig. 2D).

To further verify the RIR process restricted at low temperatures, dynamic NMR experiments were also carried out. The very fast conformational exchanges caused by the strong intramolecular rotations gave sharp NMR signals at room temperature. However, the NMR peaks were broadened at a lower temperature because the slow rotations led to the slower exchanges. Therefore, both increasing the solution viscosity and decreasing the solution temperature would hamper the intramolecular rotations of phenyl rings of HPS. As a result, the radiative transitions of AIEgens would be opened and the emission intensity would be boosted.

2.1.3 *Pressure effect*

In addition to viscosity and temperature, pressure will also influence the emission of HPS [8,10]. A traditional luminophore, tris(8-hydroxyquinolinato)-aluminum(III) (AlQ_3) was used as a contrast. AlQ_3 is highly emissive in dilute solution but less luminescent when aggregated (Fig. 3A). When applying different pressures, the emission of HPS becomes more complicated (Fig. 3B). The PL intensity of HPS increased swiftly by increasing its pressure. However, further increase of the pressure led to the decrease in PL intensity. Theoretically, the external pressurization decreases the intermolecular distance of HPS, thus imposing antagonistic

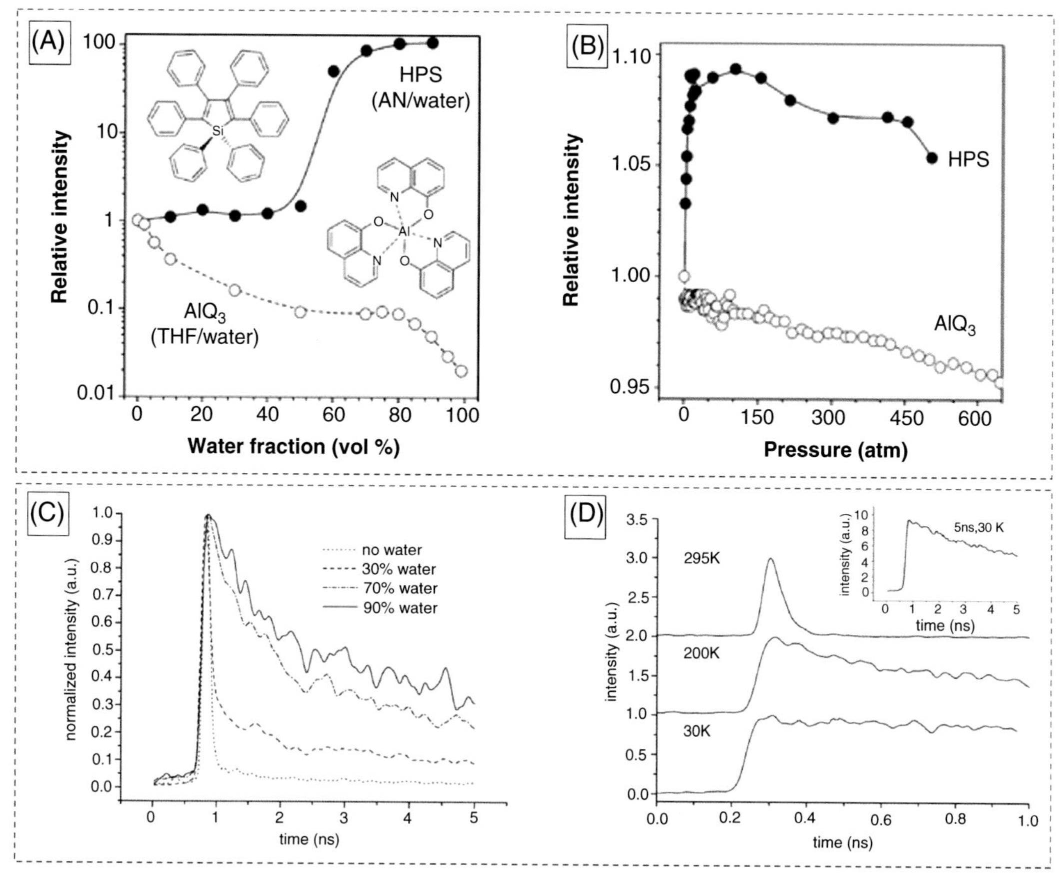

FIG. 3 (A) Changes in PL intensities of HPS and AlQ_3 solutions with water fractions of aqueous mixtures. Solution concentration: 10mM. (B) Effects of pressure on the PL intensities of HPS and AlQ_3 films. (C) Time-resolved fluorescence of HPS in solution with different fractions of water and DMF. The identical concentration for the mixtures is 1.3×10^{-5} mol/L. (D) Time-resolved fluorescence of DMF solution of HPS (2wt%) at different temperatures. Inset: PL decay at 30K at a 5ns timescale, which shows the slow decay component of PL at the low temperature. *(B) Reproduced with permission from X. Fan, J. Sun, F. Wang, Z. Chu, P. Wang, Y. Dong, et al., Photoluminescence and electroluminescence of hexaphenylsilole are enhanced by pressurization in the solid state, Chem. Commun. 26 (2008) 2989. Copyright 2008 Royal Society of Chemistry. (D) Reproduced with permission from Y. Ren, J.W.Y. Lam, Y. Dong, B.Z. Tang, K.S. Wong, Enhanced emission efficiency and excited state lifetime due to restricted intramolecular motion in silole aggregates, J. Phys. Chem. B 109 (2005) 1135. Copyright 2005 American Chemical Society.*

effects. On the one hand, the applied low pressure increases the intermolecular interaction but has little effect on the intermolecular distance. As a result, the freedom of the molecular rotations is inhibited and the emission is enhanced. On the other hand, the distances between the groups within the HPS molecule would be shortened, and the formation of excimers, etc. would be promoted at high pressures, thus weakening the emission. The quenching effect was found in AlQ_3 and its PL intensity was weakened monotonously within the pressure range of 1–650atm, indicating that the pressurization enhanced the unfavorable mutual interference between molecules.

The aggregation of a molecule can not only enhance its emission but also influence its PL lifetime [9]. Therefore, the time-resolved fluorescence of HPS was further measured to explore the AIE mechanism. As shown in Fig. 3C, the relaxation of the excited state of HPS is a single-exponential decay in dilute solution and its PL lifetime is only 40 ps. The low emission efficiency and short PL lifetime indicate a strong nonradiative recombination process. Increasing f_w causes two relaxation pathways of decay due to the formation of nanoaggregates, which results in the decay of more molecules radiatively by a slower channel. In the mixture solution containing 90% water, the excited state mainly decays through the slow pathway and the PL lifetime of the slow component rises to ~7 ns. The rapid rotation of the phenyl rings greatly consumes the energy of the excited state in dilute solutions, resulting in a ps-scale lifetime. However, the rotations of phenyl rings are largely restricted when aggregates are formed, and the radiative decay channels are activated with an ns lifetime. In addition, decreasing the temperature and increasing the medium viscosity can also enhance the PL lifetime of HPS (Fig. 3D). These results suggest that (a) the rotation of phenyl rings consumes the excited state energy and increases the nonradiative decay rates, resulting in a nonemission state of HPS in dilute solution, and (b) the restriction of rotational motions activates the radiative decay process, thus intensifying their emission in an aggregate state.

2.2 Chemical modification

All the control experiments described above have greatly proven the RIR mechanism. Moreover, these results also confirm that the luminescence of compounds can be controlled by physical or engineering encapsulation. These observations imply that the emission of compounds might also be influenced by controlling their intramolecular rotation processes at the molecular level [11–15].

2.2.1 Steric effect

The isopropyl (iPr) groups were attached to the phenyl rings of HPS to study how the steric effect would affect its AIE behavior [11]. A series of HPS derivatives have been designed and synthesized (Fig. 4A). Unlike HPS, regioisomers **4–6** are luminescent in dilute solutions, although their emission intensity varies dramatically. In acetone, the regioisomers **4–6** exhibit a blue-green emission with increased $[Fcy]_F$ values in the order of **6** > **5** > **4** (Fig. 4B). Similar results were also obtained in other solvents such as THF, which further confirmed the conclusion of the order of observed emission intensity in acetone. The Φ_F values of **4–6** are higher than that of HPS because the higher rotation barriers inhibit the intramolecular rotation process of the phenyl rings. It was well understood that a more rigid chromophore emits a stronger emission. Therefore, the structural rigidification plays a crucial role in making the regioisomers more emissive than HPS in dilute solutions.

To further verify that the RIR process of AIEgens can be activated at the molecular level via facile chemical modification, analogous works have also been done in TPE systems [12,13]. The multiple methyl groups were introduced at the ortho positions of TPE to check how the intramolecular steric effects would influence its photophysical properties [12]. TPE, a typical AIEgen, shows weak emission in THF solution due to the active intramolecular rotations of peripheral phenyl rings. When the sterically hindered methyl group was introduced, the

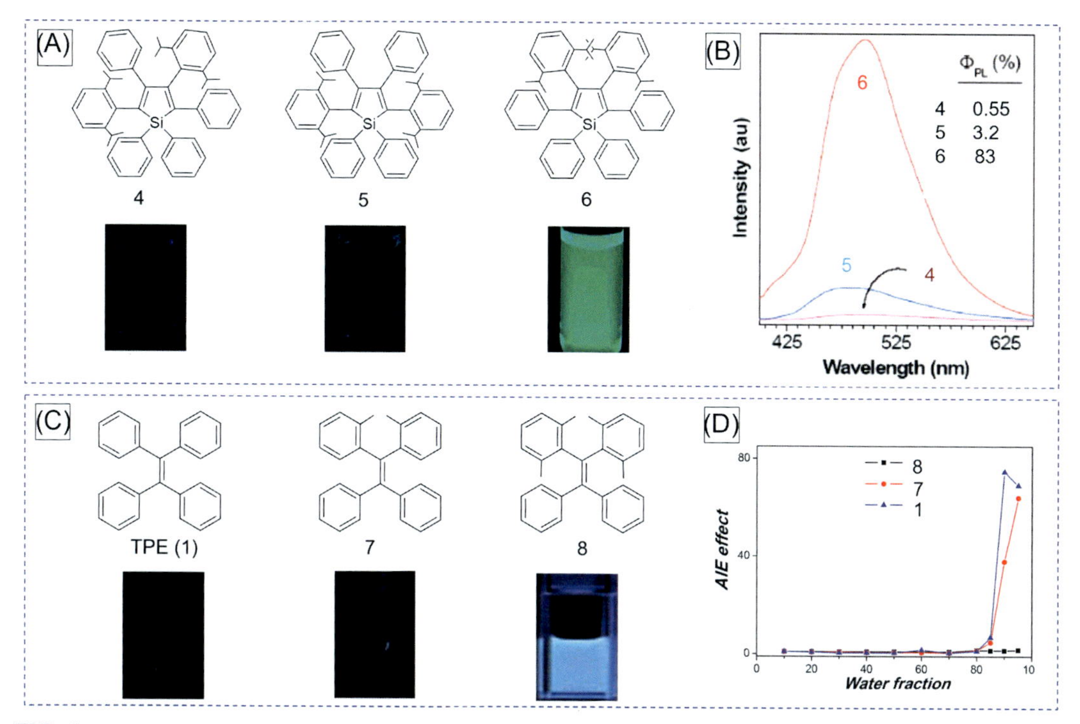

FIG. 4 (A) Chemical structures and fluorescent photographs and (B) PL spectra of solutions of siloles **4–6** in acetone (10 μM). (C) Chemical structures and fluorescent photographs. (D) Plots of I/I_0 of TPE, **7** and **8** versus water fractions in THF/water mixtures (10 μM), where I_0 and I are the PL intensities in THF solution and a THF/water mixture, respectively. Inset: fluorescence photographs of TPE, **7** and **8** in THF solutions. *(A and B) Reproduced with permission from Z. Li, Y. Dong, B. Mi, Y. Tang, M. Haussler, H. Tong, et al., Structural control of the photoluminescence of silole regioisomers and their utility as sensitive regiodiscriminating chemosensors and efficient electroluminescent materials, J. Phys. Chem. B 109 (2005) 10061. Copyright 2005 American Chemical Society. (C) Reproduced with permission from G. Zhang, Z. Chen, M.P. Aldred, Z. Hu, T. Chen, Z. Huang, et al., Direct validation of the restriction of intramolecular rotation hypothesis via the synthesis of novel ortho-methyl substituted tetraphenylethenes and their application in cell imaging, Chem. Commun. 50 (2014) 12058. Copyright 2014 Royal Society of Chemistry.*

luminescence of the TPE derivatives was further enhanced. The Φ_F of compound **8** in THF solution is higher than that of TPE and compound **7** (Fig. 3C). The compound **8** showed a bright cyan luminescence in THF solution, while weak fluorescence was observed by the naked eye for both TPE and **7**. Similar to TPE, compound **7** exhibits obvious AIE characteristics. In contrast, as the steric hindrance is further enhanced by increasing the number of methyl groups, the sterically crowded compound **8** loses its AIE feature. The introduction of tetra(ortho-methyl) groups in TPE greatly suppressed the rotational motion of intramolecular phenyl rings, and thus inhibited the nonradiative decay, further verifying the RIR mechanism.

2.2.2 Electronic conjugation effect

Besides the steric effect, the electronic interaction can make a contribution to the RIR process of AIEgens [14,15]. To study how electronic conjugation affects the AIE behavior

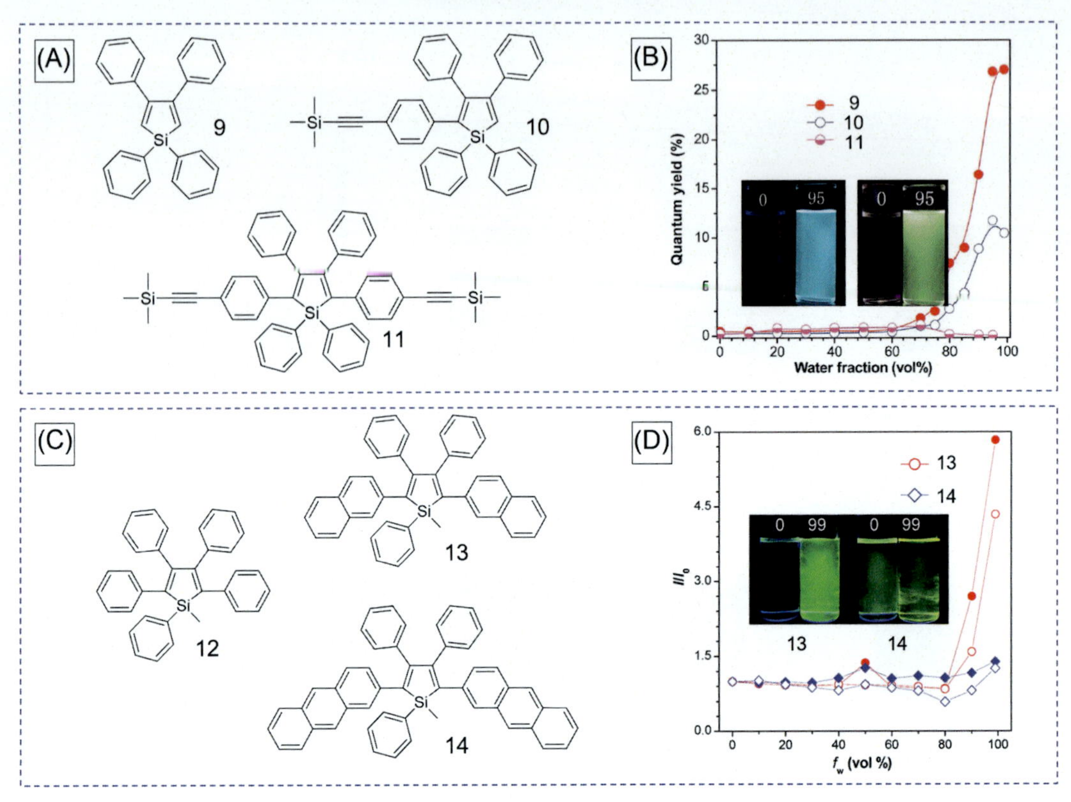

FIG. 5 (A) Chemical structures and (B) plots of I/I_0 of **9**, **10,** and **11** versus water fractions in THF/water mixtures (10 μM), where I_0 and I are the PL intensities in THF solutions and THF/water mixtures, respectively. Inset: fluorescence photographs of **10** and **11** in THF solutions. (C) Chemical structures of silole derivatives **12**, **13**, and **14** and (D) plots of their fluorescence intensity versus water fractions in THF/water mixtures. Inset: Fluorescent photographs of **13** and **14** in THF/water mixtures (f_w = 0, 99 vol%). *(A and B) Reproduced with permission from E. Zhao, J.W.Y. Lam, Y. Hong, J. Liu, Q. Peng, J. Hao, et al., How do substituents affect silole emission? J. Mater. Chem. C 1 (2013): 5661. Copyright 2013 Royal Society of Chemistry. (C and D) Reproduced with permission from B. Chen, H. Nie, P. Lu, J. Zhou, A. Qin, H. Qiu, et al., Conjugation versus rotation: good conjugation weakens the aggregation-induced emission effect of siloles, Chem. Commun. 50 (2014) 4500. Copyright 2014 Royal Society of Chemistry.*

of 1,1,3,4-tetraphenylsilole (TPS, **9**), compounds **10** and **11** with the trimethylsilylethynylphenyl (TMSEP) group on its 2,5 positions were designed and prepared (Fig. 5A) [14]. Weak emission was observed in both solution and aggregate states due to the poor conjugation of TPS. In contrast, compounds **10** and **11** showed distinctly different emission behaviors. They are nonemissive in dilute solutions but exhibit a bright blue-green emission at aggregate states, clearly demonstrating the AIE property (Fig. 5B). The Φ_F of **10** in solid state was measured to be 54.80%, while greatly intensified luminescence was found in two substituent(s) at the 2,5 positions of TPS (**11**) with a Φ_F of 90.88% in its solid state. The steric effect influences the emission behavior of TPS for the former (**10**), while the electronic effect affects the emission efficiency and wavelength of TPS for the latter (**11**).

In addition, similar work has also been performed to examine the contribution of electronic effect [15]. Polycyclic aromatic groups were attached at the 2,5-positions of silole **12** and naphthyl- and anthracyl-substituted silole derivatives **13** and **14**, respectively (Fig. 5C). Compound **13** exhibited weak emission in the solution state with a Φ_F of 2.4%, while strong luminescence was observed for **14** in the solution state with a Φ_F of 11.0%. These results suggest that the emission of **12** was enhanced in the solution state with an increased conjugation effect. Luminescence of **13** ($\Phi_F = 37.0\%$) in the solid thin film state was greatly intensified, while that of **14** ($\Phi_F = 14.0\%$) was barely enhanced relative to its solution state. Compound **14** suffered from quenching in solid state due to π–π stacking, thus weakening its AIE effect (Fig. 5D). These results also indicate that there is also a competitive relationship between the electronic conjugation effect and the intramolecular rotation processes.

2.2.3 *Effect of locking the phenyl rings*

Locking the phenyl rings of AIEgens and keeping its twisted conformation are effective ways to obtain high emission efficiency through the activation of the RIR process [16–19]. Related studies were performed with compounds 1OTPE (**15**) and 2OTPE (**16**) where the phenyl rings of TPE are locked with the "O" atom bridge (Fig. 5A) [16] TPE is nonemissive in dilute solution with a negligible Φ_F, while the Φ_F of **15** and **16** in solutions were increased to 4.6% and 30.1%, respectively. This implies that locking the phenyl ring with O bridges can restrict the intramolecular rotation and further block the nonradiative decay process, thus making the TPE derivatives emissive in solution states. The emission efficiency of luminogens in solution gradually increased with the stepwise locking of phenyl rings of TPE, thus verifying the RIR mechanism again.

As shown in Fig. 6B, AIEgen **17** can also be changed to an ACQ fluorophore (ACQphore) by blocking its phenyl ring [18]. Diphenyldibenzofulvene (**17**) is AIEactive, giving a very weak emission in dilute solution but bright blue light in solid state. The derivative of **17**, i.e., **18**, in which a methoxyl group was attached in a phenyl ring is also AIEactive. Unlike **17** and **18**, its locked form of **19** is an ACQphore. Bright emission was observed for **19** in solution ($\Phi_F = 38.0\%$) but weak emission in solid state ($\Phi_F = 5.5\%$). In the solution state, the intramolecular rotation of **19** is an invalid way to consume the excited state energy. Meanwhile, the strong π–π interaction between the flat benzo[e]acephenanthrylene stators was observed in **19**, which effectively quenched its emission in the solid state.

2.3 Supramolecular interaction

Intramolecular rotations can also be restricted through supramolecular interactions [20–22]. For example, a series of TPE derivatives **20–22** was used for detecting protein and DNA (Fig. 7A) [21]. Their increased Φ_F values in their aggregate states demonstrated their AIE feature (Fig. 7B). Compounds **20–22** are nonemissive in dilute solutions but show bright fluorescence upon addition of the DNA and BSA (Fig. 7C). Similar work had been done for the coordination interaction of another TPE derivative **23** and cations. Nearly no emission was observed for compound **23** in dilute solution, while the fluorescence was turned on after the addition of Hg^{2+} cations (Fig. 7D). Moreover, the emission of compound **23** can also be enhanced by the subsequent addition of HSO_4^- into the mixture of **23** and Hg^{2+} (Fig. 7E) [22].

FIG. 6 (A) Structures of TPE and its derivatives with "O" bridges (**15** and **16**), and fluorescent photographs of their solutions and crystals. (B) Structures of TPE derivatives (**17, 18,** and **19**), and fluorescent photographs of their solutions and crystals. *(A) Reproduced with permission from J. Shi, N. Chang, C. Li, J. Mei, C. Deng, X. Luo, et al., Locking the phenyl rings of tetraphenylethene step by step: understanding the mechanism of aggregation-induced emission, Chem. Commun. 48 (2012) 10675. Copyright 2012 Royal Society of Chemistry. (B) Reproduced with permission from H. Tong, Y. Dong, Y. Hong, M. Häussler, J.W.Y. Lam, H.H.Y. Sung, et al., Aggregation-induced emission: effects of molecular structure, solid-state conformation, and morphological packing arrangement on light-emitting behaviors of diphenyldibenzofulvene derivatives, J. Phys. Chem. C 111 (2017) 2287. Copyright 2007 American Chemical Society.*

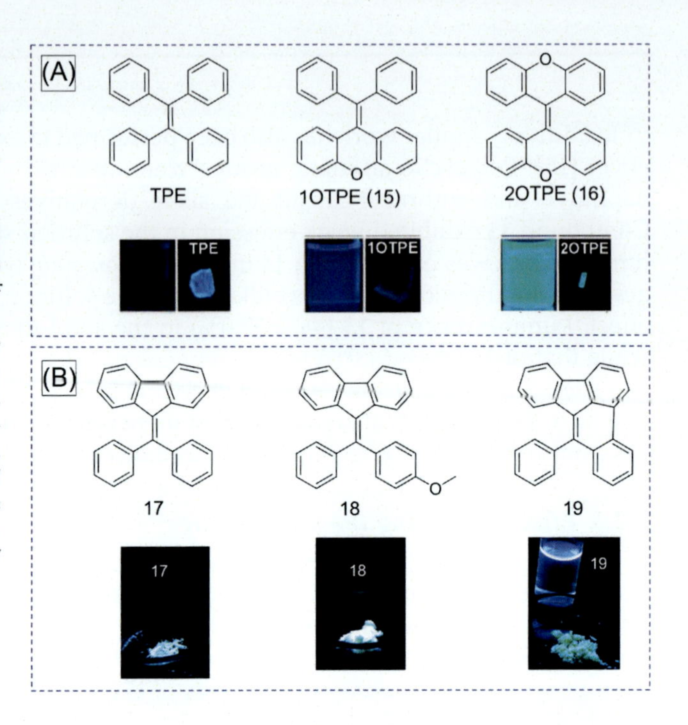

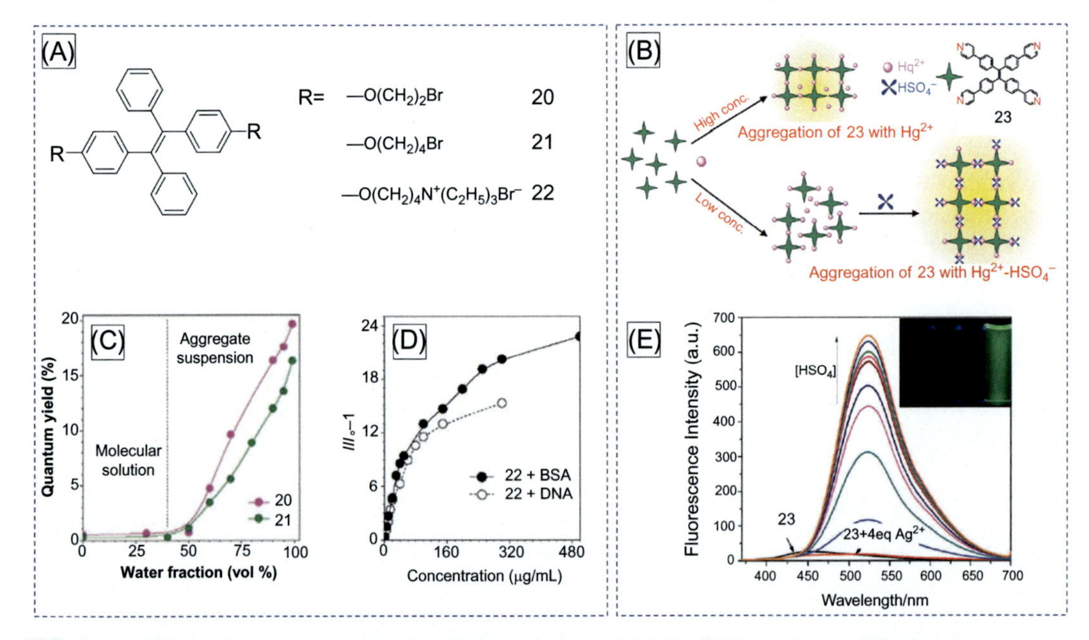

FIG. 7 (A) Chemical structures of tetraphenylethene derivatives **20–22**. (B) Dependence of fluorescence quantum yields of solutions of **20** and **21** on the solvent composition of acetonitrile (AN)-water mixtures. (C) Plots of fluorescence intensities of buffer solutions of **22** at 463 nm versus concentrations of ctDNA and BSA. (D) Schematic illustrations of coordination-induced restriction of intramolecular rotations based on luminogen **23**. (E) Fluorescence spectra of 1 (10 μM in DMF) with Hg^{2+} and after addition of 1.0–16.0 equiv. HSO_4^{-1}. Inset: from left to right: photos of solutions of 1, 1 + 4.0 equiv. Hg^{2+}, 1 + 8.0 equiv. HSO_4^{-1}, 1 + 4.0 equiv. Hg^{2+} + 8.0 equiv. HSO_4^{-1} under UV light illumination. *(B and C) Reproduced with permission from H. Tong, Y. Hong, Y. Dong, M. Häußler, J.W.Y. Lam, Z. Li, et al., Fluorescent "light-up" bioprobes based on tetraphenylethylene derivatives with aggregation-induced emission characteristics, Chem. Commun. 35 (2006) 3705. Copyright 2006 Royal Society of Chemistry. (D and E) Reproduced with permission from G. Huang, G. Zhang, D. Zhang, Turn-on of the fluorescence of tetra(4-pyridylphenyl)ethylene by the synergistic interactions of mercury(II) cation and hydrogen sulfate anion, Chem. Commun. 48 (2012) 7504. Copyright 2012 Royal Society of Chemistry.*

In these systems, before the addition of cations, the excited state energy can be consumed through the intramolecular rotations, making it nonemissive in the solution state. However, the addition of cations forms coordination complexes, generating a higher rotational barrier for the rotatory groups in the ligands, thus leading to the distinct light emission.

2.4 Theoretical studies

Besides experimental approaches, theoretical studies of the AIE mechanism were performed [23–33]. It is documented that DCDPP (**24**) is an AIEgen, while its locked-form DCPP (**25**) is an ACQphore. Shuai et al. modeled the AIEgen of **24** and its locked-form **25** by using quantum mechanics and molecular mechanics (QM/MM) approaches [26,27,29]. This provides crucial insights into their photophysical behaviors in different states. Huang-Rhys (HR) factors at different normal modes are shown in Fig. 8A and B. Three modes with large HR factors of isolated **24** were found in the low frequency region, and thus these normal modes are the main channel to consume its excited state energy (Fig. 8A). However, smaller HR factors of **24** clusters were observed in the higher frequency region, signifying that excited-state energy is reduced substantially by the low-frequency vibrations, such as the twisting of phenyl rings. In contrast, **25** in isolated and cluster states did not show low-frequency normal modes, but much smaller HR factors were found in a much higher frequency region (Fig. 8B). These results provide a clear support to RIR.

Furthermore, resonance Raman spectroscopy (RRS) was also used to study the microscopic mechanism of AIE in combination with the computational studies [34]. Taking HPDMCb (**26**) as an example and non-AIEactive DCPP for comparison, the intensities of low-frequency peaks of **26** cluster in RRS obviously decreased compared with the high-frequency peaks. However, the RRS of DCPP remained unaffected after aggregation. These results can be attributed to the RIR of AIEgen, such as the low-frequency phenyl ring twisting. Therefore, the RRS was also a direct approach to explain the above theoretical studies on the RIR mechanism.

All the results above, from experimental measurements to theoretical calculations, from external physical and engineering controls to internal structural and chemical modification, provide strong evidence to confirm that RIR is the working mechanism of AIEgens with rotatable aromatic rings.

3 Restriction of intramolecular vibrations

Under the guidance of RIR, many AIEgens have been designed, prepared, and applied in diverse areas. However, part of the molecules with the AIE feature cannot be fully explained by the RIR mechanism, such as BDBA (**27**) and THBDBA (**28**) displayed in Fig. 9 [35,36]. They possess no rotatable units because the two pairs of phenyl rings of BDBA and THBDBA are locked through vinyl linkages and ethylene tethers, respectively. These molecules should be emissive even in dilute solutions based on the mechanism of RIR because the excited-state energy is no longer consumed by nonradiative pathways. However, they show AIE property. Nearly no florescence was observed in solution but bright emission was recorded in

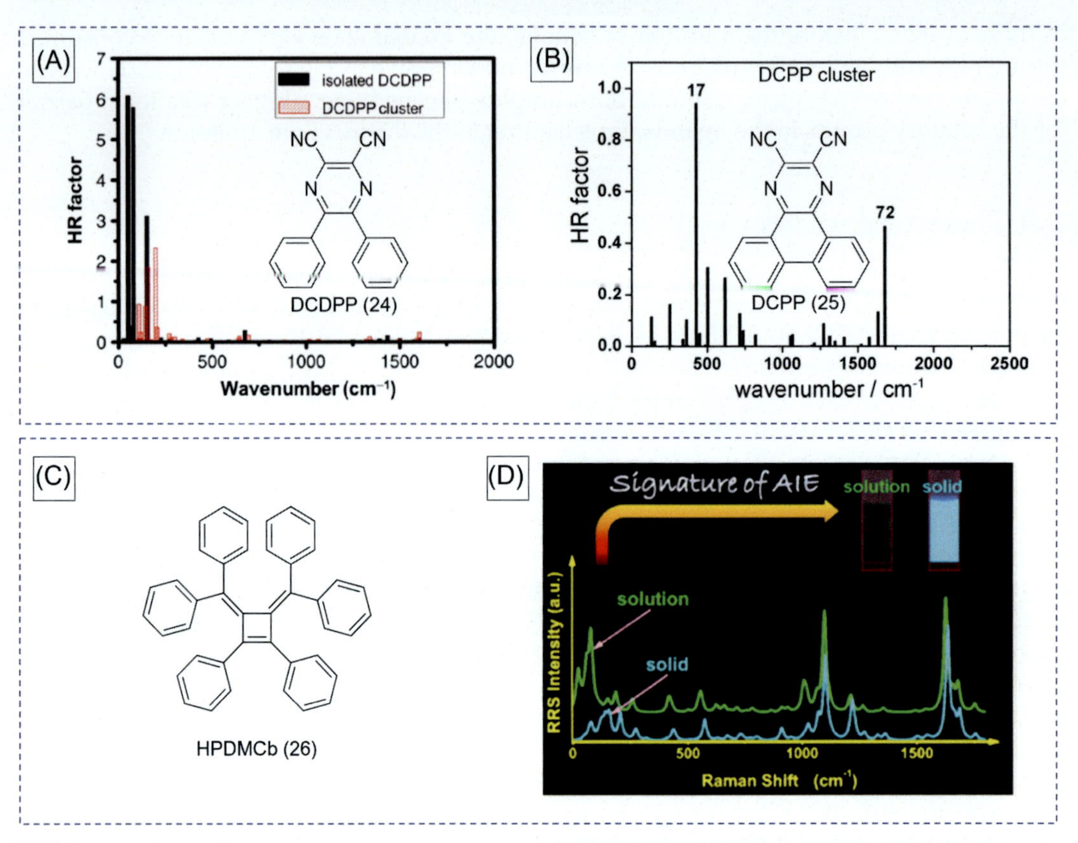

FIG. 8 Calculated Huang-Rhys (HR) factors versus normal mode wave numbers for (A) isolated DCDPP (**24**) and cluster and (B) DCPP (**25**) cluster. (C) Chemical Structure of HPDMCb (**26**). (D) Resonance Raman spectroscopy (RRS) intensity in both solution and solid phases for HPDMCb. *(A and B) Reproduced with permission from Q. Wu, C. Deng, Q. Peng, Y. Niu, Z. Shuai, Quantum chemical insights into the aggregation induced emission phenomena: a QM/MM study for pyrazine derivatives, J. Comput. Chem. 33 (2012) 1862. Copyright 2012 Wiley. Copyright 2013 Elsevier B.V. (D) Reproduced with permission from T. Zhang, H. Ma, Y. Niu, W. Li, D. Wang, Q. Peng, et al., Spectroscopic signature of the aggregation-induced emission phenomena caused by restricted nonradiative decay: a theoretical proposal, J. Phys. Chem. C 119 (2015) 5040. Copyright 2015 American Chemical Society.*

aggregate states (Fig. 9A and B). Similar to the rotation of phenyl rings, the vibrational motions should also consume exciton energy. In other words, restriction of intramolecular vibrations (RIV) is the cause of the AIE effect of BDBA and THBDBA. Computational analyses were also employed to explain the AIE phenomenon. As depicted in Fig. 9C, QM/MM modeling results show that there are mainly six normal modes for THBDBA in a single molecule consuming the energy of the excited state. Among these, every reorganization energy exceeded $200\,cm^{-1}$, resulting in a total energy of $5679\,cm^{-1}$. In contrast, THBDBA clusters only have three significant normal mode frequencies in the vicinity of the low-frequency range ($\sim 4016\,cm^{-1}$; Fig. 9D). In the clusters, the combination of a decrease in the total reorganization energy and vibrational channels was likely the cause of the observed AIE effect of THBDBA.

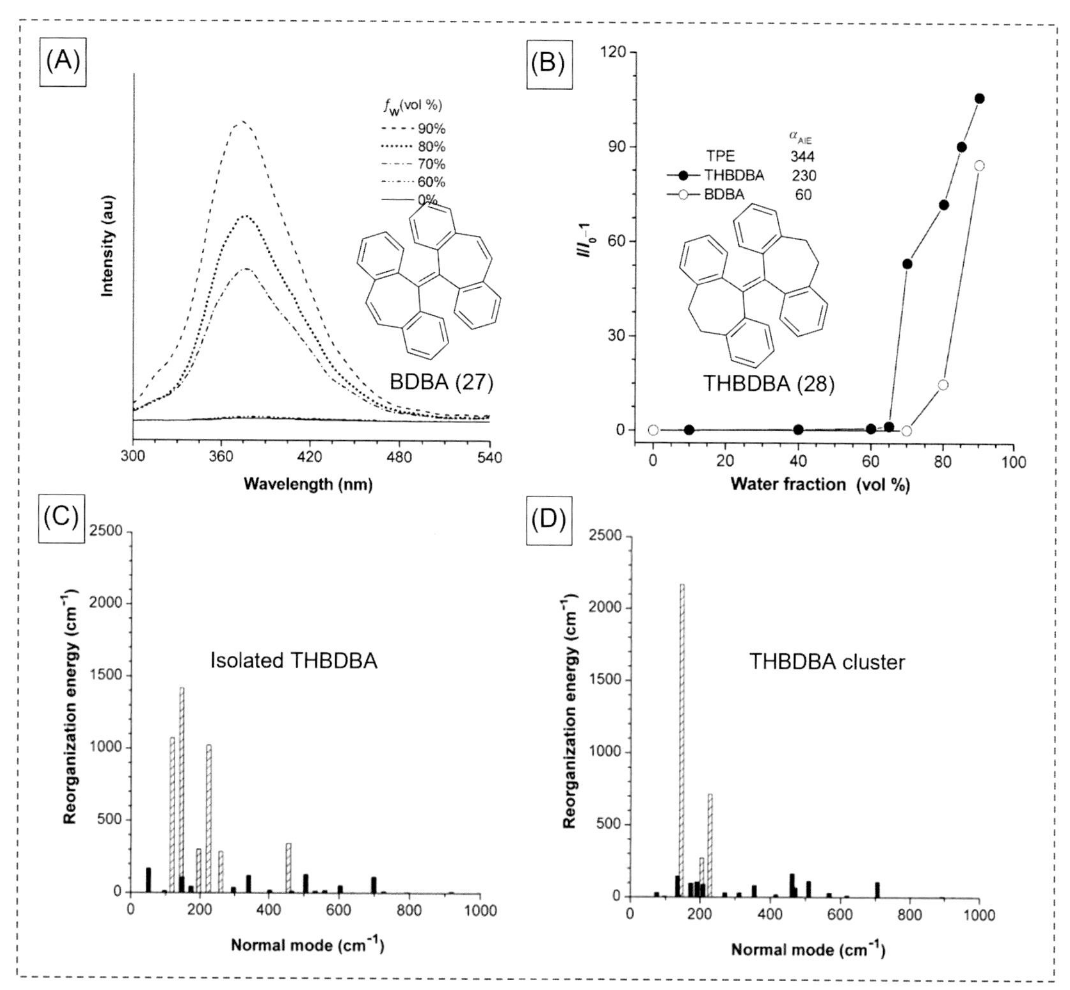

FIG. 9 (A) PL spectra of BDBA (**27**) in THF/water mixtures with different f_w and (B) change in the PL intensity of BDBA and THBDBA with f_w (20 μM). Plots of reorganization energy versus normal mode wave numbers for excited states of (C) molecular and (D) clustered species of **28**. *Reproduced with permission from N.L. C. Leung, N. Xie, W. Yuan, Y. Liu, Q. Wu, Q. Peng, et al., Restriction of intramolecular motions: the general mechanism behind aggregation-induced emission, Chem. Eur. J. 20 (2014) 15349. Copyright 2014 Wiley-VCH Verlag GmbH & Co. KGaA.*

The RIV mechanism has been further confirmed in other AIEgen systems [37–42]. Goel et al. reported a 5,6-dihydro-2H-pyrano[3,2-g]indolizine (DPI) derivative **29** which exhibits a unique solution/solid dual emission behavior with a stronger emission in the solid state (Fig. 10A) [41]. To check the AIE feature of **29**, the PL was recorded in THF/water mixtures with varying f_w. As shown in Fig. 10B, the PL intensity of **29** was remarkably enhanced up to 20-fold higher than that of its solution when the f_w reached 99%. Considering the structural characteristic of **29**, the RIV of C2-flexure might be responsible for its AIE feature. From the perspective of the crystal structure, two strong noncovalent C—H⋯N and C—H⋯O

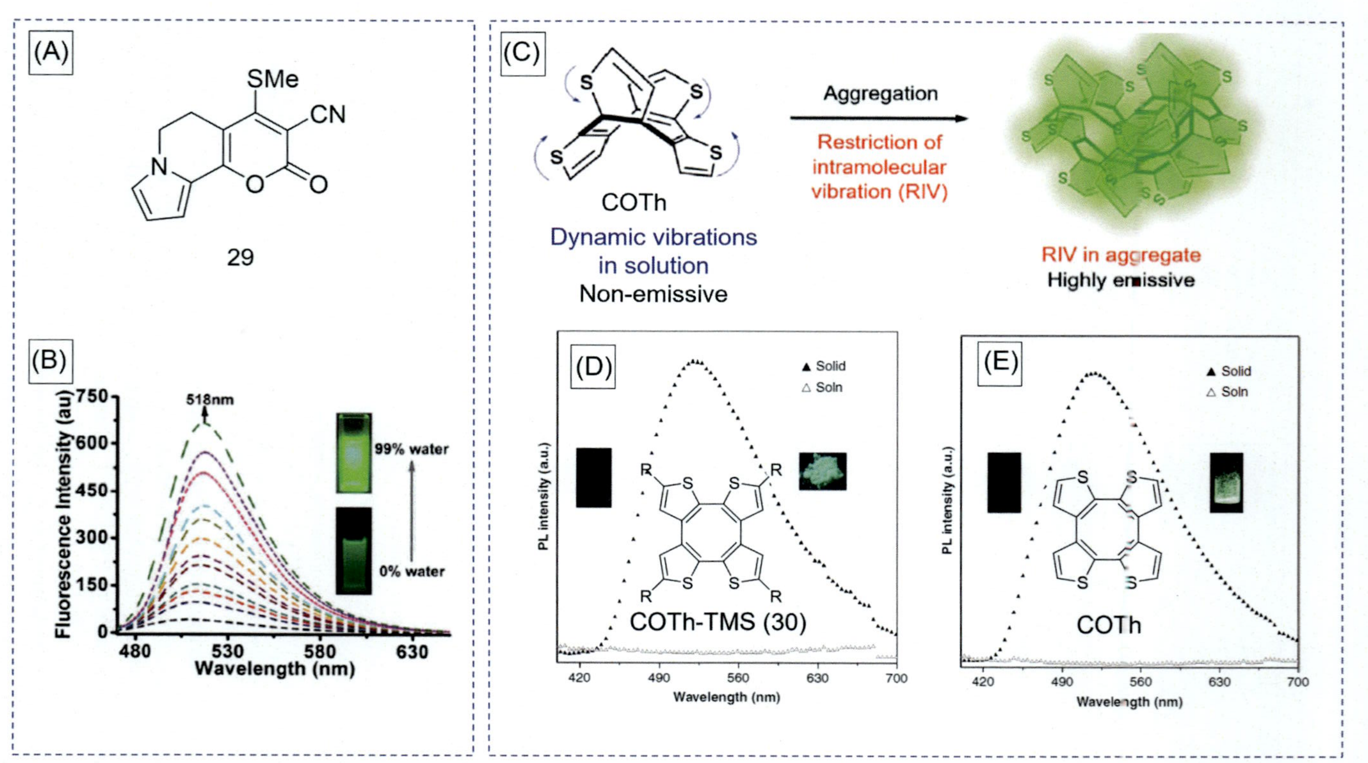

FIG. 10 (A) Chemical structure of **29**. (B) PL spectra of **29** in THF/water mixtures. Inset: Fluorescent photographs of **29** in THF/water mixtures ($f_w = 0$, 99 vol%). (C) Illustration of cyclooctatetrathiophene COTh that shows the AIE activity due to the restriction of intramolecular vibration (RIV) in the aggregate state. (D) PL spectra of COTh-TMS in THF solution *(empty triangles)* and solid states *(full triangles)*. Concentration: 10 μM, excitation wavelength: 365 nm. Inset: the fluorescence pictures of **30** in THF solution (left, *dark*) and solid states (right, *green fluorescence*) taken under an excitation wavelength of 365 nm by a portable UV lamp. (E) PL spectra of COTh in THF solution *(empty triangles)* and solid states *(full triangles)*. Concentration: 10 μM, excitation wavelength: 350 nm. Inset: the fluorescence pictures of COTh in THF solution (left, *dark*) and solid states (right, *green fluorescence*) taken under an excitation wavelength of 365 nm by a portable UV lamp. *(B) Reproduced with permission from A. Raghuvanshi, A.K. Jha, A. Sharma, S. Umar, S. Mishra, R. Kant, et al., A nonarchetypal 5,6-dihydro-2H-pyrano[3,2-g]indolizine-based solution-solid dual emissive AIEgen with multicolor tunability, Chem. Eur. J. 23 (2017) 4527. Copyright 2017 Wiley-VCH Verlag GmbH & Co. KGaA. (E) Reproduced with permission from Z. Zhao, X. Zheng, L. Du, Y. Xiong, W. He, X. Gao, et al., Non-aromatic annulene-based aggregation-induced emission system via aromaticity reversal process, Nat. Commun. 10 (2019) 1. Copyright 2019 Nature Publishing Group.*

interactions were found between the adjacent molecules. These interactions effectively restricted the molecular vibrational motion of C2-flexure, thus leading to strong emission in solid state.

Cyclooctotetraene (COT) derivatives such as COTh and COTh-TMS (**30**) were also found to be AIE-active (Fig. 10C) [42]. Weak emission with a Φ_F of 0.7% and bright green luminescence with a Φ_F of 10% were recorded for COTh-TMS in dilute solution and solid state, respectively (Fig. 10D). Initially, the TMS groups were proposed to possibly act as rotors to consume the exciton energy. However, a further study indicated that COTh without TMS groups also exhibit an AIE feature, in which the Φ_F in solution was only 0.4% but enhanced to 11% in its solid state (Fig. 10E). These results indicate that the AIE activity of the COTh system is caused through the molecular vibration rather than the rotation of TMS groups. To gain a deeper insight into the AIE activity of COTh derivatives, their crystal structures were analyzed. No obvious intermolecular π–π interaction was found due to their noncoplanar and saddle-type conformations. Furthermore, multiple intermolecular C—H⋯π and S⋯π interactions were observed in the crystals, which can rigidify the molecular conformation resulting in solid-state emission. In the solution state, the vibrational molecular motion would occur for the COTh derivatives due to the lack of aromaticity stabilization, making them nonemissive. However, the aromaticity of the COTh derivatives is stabilized because of the RIV process in solid state and the activation of the radiative channel, making the molecules emit brightly. The reversal from the ground state to the excited state of aromaticity serves as a driving force to induce the excited-state intramolecular vibration and leads to the AIE phenomenon.

The examples above therefore indicate that molecular vibrational motions, including in-plane/out-of-plane bending, flapping, stretching, scissoring, wagging, twisting, rocking, etc., can also consume the exciton energy. Similar to RIR, restriction of intramolecular vibration can also turn on the fluorescence of the molecule in the aggregate states. The proposal of the RIV mechanism can not only offer new perspectives for photophysical fundamentals but also open new ways for the design and development of new AIEgen systems, which will further broaden the scope of AIE research.

4 Restriction of intramolecular motions

According to the above discussion, it becomes clear that RIR and RIV have been rationalized as the main cause for the AIE effect. Thus, RIR and RIV could be unified as restriction of intramolecular motion (RIM).

Interestingly, in some AIE systems, both RIR and RIV are involved, and the working mechanism can only be ascribed to RIM [43–46]. Two representative examples of such AIEgens are shown in Fig. 11. Ma et al. reported a butterfly-shaped phenothiazine derivative **31** (Fig. 11A), which is not emissive in solution, but bright red fluorescence in the THF/water mixtures with $f_w \geq 70$ vol% was observed, indicating its AIE effect [43]. The theoretical optimized ground-state geometry showed that the phenothiazine unit exhibits a nonplanar butterfly-like structure with a large C—S—N—C dihedral angle of 142 degrees. Moreover, there is a large twist-linked distorted angle of 145 degrees between the phenothiazine and benzothiadiazole groups. In the solution state, its excited-state energy is mainly consumed through the

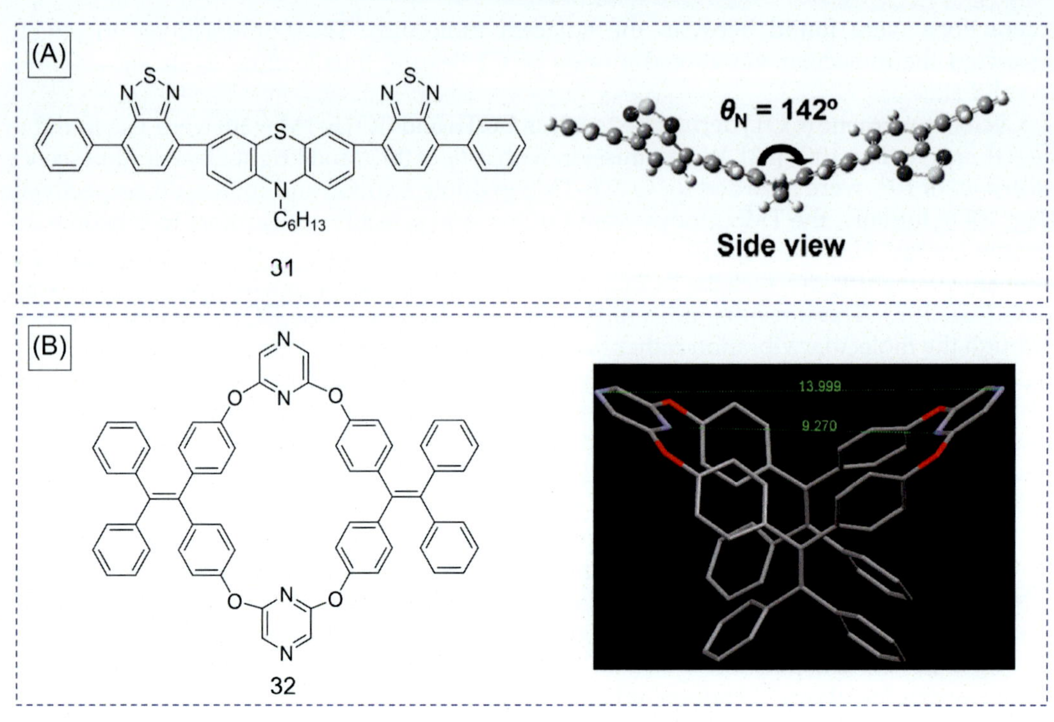

FIG. 11 (A) Chemical structure and theoretically optimized geometry of **31**. To simplify the calculation, the hexyl group was replaced with a methyl substituent. KGaA. (B) Chemical and crystal structure of **32**. Hydrogen atoms were omitted for clarity. *(A) Reproduced with permission from L. Yao, S. Zhang, R. Wang, W. Li, F. Shen, B. Yang, et al., Highly efficient near-infrared organic light-emitting diode based on a butterfly-shaped donor-acceptor chromophore with strong solid-state fluorescence and a large proportion of radiative excitons, Angew. Chem. Int. Ed. 53 (2014) 2119. Copyright 2014 Wiley-VCH Verlag GmbH & Co. (B) Reproduced with permission from C. Zhang, Z. Wang, S. Song, X. Meng, Y. Zheng, X. Yang, et al., Tetraphenylethylene-based expanded oxacalixarene: synthesis, structure, and its supramolecular grid assemblies directed by guests in the solid state, J. Org. Chem. 79 (2014) 2729. Copyright 2014 American Chemical Society.*

vibrational motions of the phenothiazine core and the rotational motions of the benzothiadiazole and phenyl rings. These intramolecular motions were, however, confined in the aggregate state. As a result, the RIM process is the cause of the AIE effect of **31**.

In addition to the typical luminogens including rotatable periphery phenyl rings and vibratable cores, many macrocycles also exhibit the AIE feature controlled by the RIM process [47–50]. For example, a TPE-based oxacalixarene (**32**) exhibits the typical AIE effect. It is almost nonemissive in dilute solution, while emits bright emission after aggregation (Fig. 11B) [47]. The rotations of phenyl rings and diphenylmethylene units in TPE units consume the exciton energy. In addition, the asymmetrically linked pyrazine units in the crystal structure of **32** adopt slanted conformations, which allow flap-like vibratory motions occurring to further dissipate the excited-state energy. As a result, the rotation and vibration consume the exciton energy, which in turn leads to emission quenching in dilute solution. However, upon the formation of aggregates, the intramolecular motion is restricted, and hence its luminescence is turned on.

5 New perspective: Quantum chemistry calculation

The proposal of the RIM mechanism encourages researchers to design and synthesize new AIEgens by introducing some rotors/vibrators. However, not all molecular motions will cause fluorescence quenching. Recently, some theoretical calculations were also performed to explore the crucial molecular motion related to the nonradiative transition and the decay pathways of the excited state of AIEgens. Four models have been established to explain the connotation of RIM mechanism (Fig. 12A).

5.1 Restriction of vibronic coupling

For the AIEgens with strong molecular motion, the internal conversion between S_1 and S_0 consumes the excited state energy, thereby inhibiting the generation of fluorescence. For example, AIEgen **33** absorbs light, forming an excited state, and then undergoes double-bond twisting and phenyl-ring torsion, which allow strong vibrational coupling between S_1 and S_0 (Fig. 13A) [51]. However, the potential energy surface of **33** becomes steeper because of RIM in the aggregate state. A small nuclear displacement might lead to a large potential energy

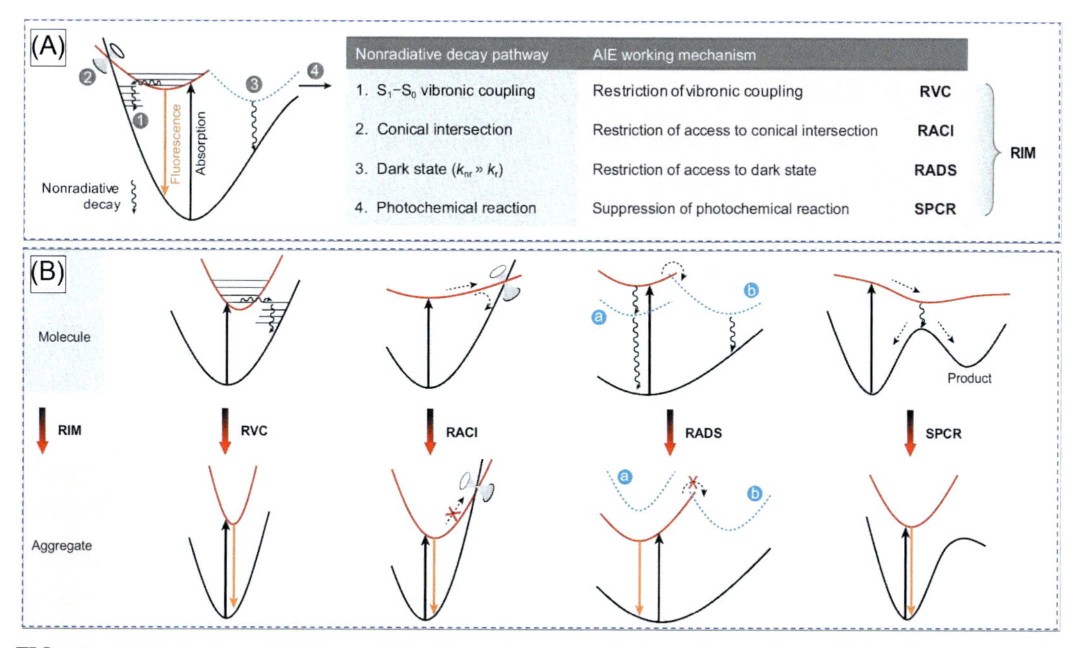

FIG. 12 (A) Activation of RIM through blocking various nonradiative pathways. k_r = radiative decay constant, k_{nr} = nonradiative decay constant. (B) Potential energy surfaces for the nonradiative and radiative pathways at molecular and aggregate levels, respectively. *Reproduced with permission from Y. Tu, Z. Zhao, J.W. Lam, B.Z. Tang, Mechanistic connotations of restriction of intramolecular motions (RIM). Natl. Sci. Rev. (2020). https://doi.org/10.1093/nsr/nwaa260. Copyright 2020 Oxford University Press.*

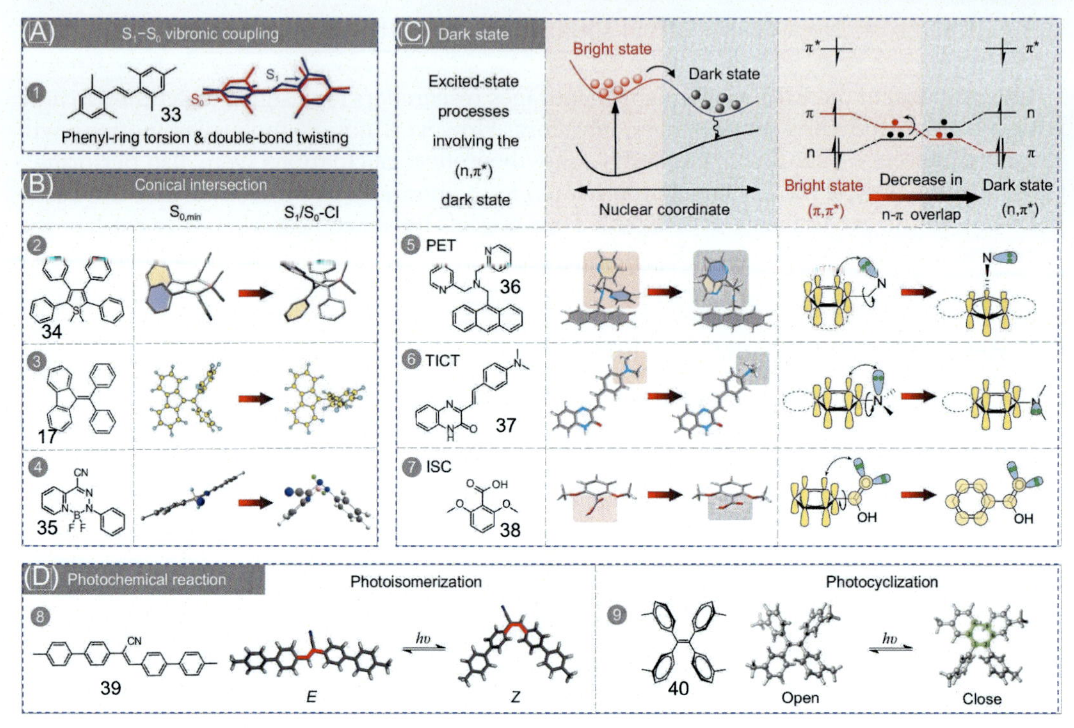

FIG. 13 Examples of excited-state molecular motions leading to different nonradiative pathways including (A) S_1–S_0 vibronic coupling, (B) conical intersection, (C) dark state, and (D) photochemical reaction. Abbreviations: *CI*, conical intersection; *PET*, photo-induced electron transfer; *TICT*, twisted intramolecular charge transfer; *ISC*, intersystem crossing. *Reproduced with permission from Y. Tu, Z. Zhao, J.W. Lam, B.Z. Tang, Mechanistic connotations of restriction of intramolecular motions (RIM). Natl. Sci. Rev. (2020). https://doi.org/10.1093/nsr/nwaa260. Copyright 2020 Oxford University Press.*

increase. As a result, the vibrational coupling between S_1 and S_0 is weak, and the coverage between the two wave functions is less effective [52]. Therefore, the fluorescence of AIEgen is turned on because of the restriction of vibration coupling between S_1 and S_0 (Fig. 12C).

5.2 Restriction of access to conical intersection

Many flexible AIEgens (e.g., **17**, **34**, and **35**) in the excited state rapidly relax to the conical intersection (CI), where two electronic states of the same spin multiplicity have the same energy (degenerate) [53]. Moreover, the magnitude of the vibration coupling of the point is close to infinity, and thus the excitons decay through the nonradiative channel. In the aggregate state, large amplitude motions that are required to arrive at the conical intersection are restricted [23,54,55]. Therefore, the fluorescence is turned on because of the restriction of access to the conical intersection (Fig. 12C).

5.3 Restriction of access to dark state

Due to the differences of the transition origin, such as (π, π*), (n, π*), (n, σ*), and (π, σ*), the spatial overlap of the transition orbits, e.g., local excited state and charge transfer state, orbital multiplicity (singlet state and triplet state), and the symmetry of the transition, the excited states also have different characteristics. Some excited states show small molar absorption coefficients and small oscillator strengths, which result in low transition probability and large nonradiative transition rates. Therefore, this type of excited state is defined as a dark state. When compared with the (π, π*) state/LE state/single state/symmetry-allowed transition, the (n, π*) state/CT state/triple state/symmetry-forbidden transitions are all fluorescent dark states. For example, some heteroatom-containing molecules as examples (Fig. 13C) emit weakly in solutions because of the light-induced electron transfer, twisted intramolecular charge transfer, and intersystem crossing [56–58]. In solutions, its weak fluorescence is attributed to the easy access from the bright (π, π*) state to the close-lying dark (n, π*) state. However, in the aggregate state, the molecular motions that cause the dark state is restricted or its energy is elevated, and in turn making the dark kinetically or thermodynamically inaccessible. Therefore, the emission is restored due to the restriction of access to the dark state.

5.4 Restriction of the molecular motions that lead to new product formation

In addition to photophysical channels, some AIEgens undergo photochemical reactions such as photoisomerization and photocyclization upon photoexcitation [30]. For example, the excited-state AIEgens **39** and **40** undergo configuration changes along the reaction coordinate to reach an intersection point, where there is strong vibration coupling or a conical intersection [32,59,60]. As a result, AIEgens consume excitons energy mainly through the nonradiative transition channel, thus making them nonemissive in solution. However, this photochemical reaction is inhibited, and the emission is turned on upon the formation of aggregates.

6 Conclusions and outlook

In this chapter, various working mechanisms of AIE are elaborated. Among the proposed mechanisms, the RIM mechanism has been widely studied and well accepted. The RIM mechanism offers the fundamental and comprehensive understanding of the AIE effect and provides a platform for the creation of new AIEgen systems. In principle, the intramolecular motions of isolated molecules can enhance its nonradiative decay rates. However, after aggregation, the nonradiative decay rates are greatly inhibited because of the restriction of intramolecular motions. Therefore, the radiative relaxation becomes favorable. In general, the flexible AIEgen promotes an intramolecular motion and decreases the structural rigidity to enhance the nonradiative decay, and thus the molecule is nonemissive in the solution state. However, the aggregation enhances the molecular rigidification and restricts the nonradiative decay channels, and in turn the fluorescence is turned on. Therefore, the structural rigidification of flexible molecules is crucial for the design of AIE systems.

However, there are also some mechanistic studies of molecular motions to explore in further researches, such as the frequency and amplitude of molecular motions and the solid-state molecular motion. Meanwhile other mechanistic explanations are also worth exploring for the AIE systems, such as clusteroluminescence [61,62] and room temperature phosphorescence [63,64]. In addition, according to the RIM mechanism, the intramolecular motions will quench the emission of AIEgens. In other words, the nonradiative transition is activated, in which the exciton energy will be mainly transferred to heat. Thus, inspired by the RIM mechanism, the molecules with active intramolecular motions could be used for photothermal therapy, photoacoustic (PA) imaging, laser resurfacing, desalination of seawater, etc.

In summary, the development of the AIE mechanism is of great significance for the understanding of the science in mesoscopic aggregates and designing diverse and effective AIE materials for intriguing application, which will eventually accelerate the establishment of aggregology [65].

References

[1] J. Luo, Z. Xie, J.W.Y. Lam, L. Cheng, H. Chen, C. Qiu, et al., Aggregation-induced emission of 1-methyl-1,2,3,4,5-pentaphenylsilole, Chem. Commun. 18 (2001) 1740.

[2] J. Mei, N.L.C. Leung, R.T.K. Kwok, J.W.Y. Lam, B.Z. Tang, Aggregation-induced emission: together we shine, united we soar! Chem. Rev. 115 (2015) 11718.

[3] H. Zhang, Z. Zhao, A.T. Turley, L. Wang, P.R. McGonigal, Y. Tu, et al., Aggregate science: from structures to properties, Adv. Mater. 32 (2020) 2001457.

[4] Z. He, C. Ke, B.Z. Tang, Journey of aggregation-induced emission research, ACS Omega 3 (2018) 3267.

[5] Z. He, E. Zhao, J.W.Y. Lam, B.Z. Tang, New mechanistic insights into the AIE phenomenon, in: Aggregation-Induced Emission: Materials and Applications, vol. 1, American Chemical Society, 2016, p. 5.

[6] Z. Zhao, H. Zhang, J.W.Y. Lam, B.Z. Tang, Aggregation-induced emission: new vistas at the aggregate level, Angew. Chem. Int. Edit. 59 (2020) 2.

[7] J. Chen, C.C.W. Law, J.W.Y. Lam, Y. Dong, S.M.F. Lo, I.D. Williams, et al., Synthesis, light emission, nanoaggregation, and restricted intramolecular rotation of 1,1-substituted 2,3,4,5-tetraphenylsiloles, Chem. Mater. 15 (7) (2003) 1535–1546.

[8] X. Fan, J. Sun, F. Wang, Z. Chu, P. Wang, Y. Dong, et al., Photoluminescence and electroluminescence of hexaphenylsilole are enhanced by pressurization in the solid state, Chem. Commun. 26 (2008) 2989.

[9] Y. Ren, J.W.Y. Lam, Y. Dong, B.Z. Tang, K.S. Wong, Enhanced emission efficiency and excited state lifetime due to restricted intramolecular motion in silole aggregates, J. Phys. Chem. B 109 (2005) 1135.

[10] S. Zhang, Y. Dai, S. Luo, Y. Gao, N. Gao, K. Wang, et al., Rehybridization of nitrogen atom induced photoluminescence enhancement under pressure stimulation, Adv. Funct. Mater. 27 (2017) 1602276.

[11] Z. Li, Y. Dong, B. Mi, Y. Tang, M. Haussler, H. Tong, et al., Structural control of the photoluminescence of silole regioisomers and their utility as sensitive regiodiscriminating chemosensors and efficient electroluminescent materials, J. Phys. Chem. B 109 (2005) 10061.

[12] G. Zhang, Z. Chen, M.P. Aldred, Z. Hu, T. Chen, Z. Huang, et al., Direct validation of the restriction of intramolecular rotation hypothesis via the synthesis of novel ortho-methyl substituted tetraphenylethenes and their application in cell imaging, Chem. Commun. 50 (2014) 12058.

[13] Z. Zhao, B. He, H. Nie, B. Chen, P. Lu, A. Qin, et al., Stereoselective synthesis of folded luminogens with arene-arene stacking interactions and aggregation-enhanced emission, Chem. Commun. 50 (2014) 1131.

[14] E. Zhao, J.W.Y. Lam, Y. Hong, J. Liu, Q. Peng, J. Hao, et al., How do substituents affect silole emission? J. Mater. Chem. C 1 (2013) 5661.

[15] B. Chen, H. Nie, P. Lu, J. Zhou, A. Qin, H. Qiu, et al., Conjugation versus rotation: good conjugation weakens the aggregation-induced emission effect of siloles, Chem. Commun. 50 (2014) 4500.

[16] J. Shi, N. Chang, C. Li, J. Mei, C. Deng, X. Luo, et al., Locking the phenyl rings of tetraphenylethene step by step: understanding the mechanism of aggregation-induced emission, Chem. Commun. 48 (2012) 10675.

[17] B. Jiang, D. Guo, Y. Liu, K. Wang, Y. Liu, Photomodulated fluorescence of supramolecular assemblies of sulfonatocalixarenes and tetraphenylethene, ACS Nano 8 (2014) 1609.
[18] H. Tong, Y. Dong, Y. Hong, M. Häussler, J.W.Y. Lam, H.H.Y. Sung, et al., Aggregation-induced emission: effects of molecular structure, solid-state conformation, and morphological packing arrangement on light-emitting behaviors of diphenyldibenzofulvene derivatives, J. Phys. Chem. C 111 (2017) 2287.
[19] J. Xiong, Y. Yuan, L. Wang, J. Sun, W. Qiao, H. Zhang, et al., Evidence for aggregation-induced emission from free rotation restriction of double bond at excited state, Org. Lett. 20 (2018) 373.
[20] J. Zhao, D. Yang, Y. Zhao, X. Yang, Y. Wang, B. Wu, Anion-coordination-induced turn-on fluorescence of an oligourea-functionalized tetraphenylethene in a wide concentration range, Angew. Chem. Int. Ed. 53 (2014) 6632.
[21] H. Tong, Y. Hong, Y. Dong, M. Häußler, J.W.Y. Lam, Z. Li, et al., Fluorescent "light-up" bioprobes based on tetraphenylethylene derivatives with aggregation-induced emission characteristics, Chem. Commun. 35 (2006) 3705.
[22] G. Huang, G. Zhang, D. Zhang, Turn-on of the fluorescence of tetra(4-pyridylphenyl)ethylene by the synergistic interactions of mercury(II) cation and hydrogen sulfate anion, Chem. Commun. 48 (2012) 7504.
[23] Q. Li, L. Blancafort, A conical intersection model to explain aggregation induced emission in diphenyl dibenzofulvene, Chem. Commun. 49 (2013) 5966.
[24] Z. Shuai, Q. Peng, Excited states structure and processes: understanding organic light-emitting diodes at the molecular level, Phys. Rep. 537 (2014) 123.
[25] Q. Wu, T. Zhang, Q. Peng, D. Wang, Z. Shuai, Aggregation induced blue-shifted emission—the molecular picture from a QM/MM study, Phys. Chem. Chem. Phys. 16 (2014) 5545.
[26] C. Deng, Y. Niu, Q. Peng, A. Qin, Z. Shuai, B.Z. Tang, Theoretical study of radiative and non-radiative decay processes in pyrazine derivatives, J. Chem. Phys. 135 (2011) 14304.
[27] Q. Wu, C. Deng, Q. Peng, Y. Niu, Z. Shuai, Quantum chemical insights into the aggregation induced emission phenomena: a QM/MM study for pyrazine derivatives, J. Comput. Chem. 33 (2012) 1862.
[28] S. Yin, Q. Peng, Z. Shuai, W. Fang, Y. Wang, Y. Luo, Aggregation-enhanced luminescence and vibronic coupling of silole molecules from first principles, Phys. Rev. B 73 (2006), 205409.
[29] A. Qin, J.W.Y. Lam, F. Mahtab, C.K.W. Jim, L. Tang, J. Sun, et al., Pyrazine luminogens with "free" and "locked" phenyl rings: Understanding of restriction of intramolecular rotation as a cause for aggregation-induced emission, Appl. Phys. Lett. 94 (2009), 253308.
[30] K. Kokado, K. Sada, Consideration of molecular structure in the excited state to design new luminogens with aggregation-induced emission, Angew. Chem. Int. Ed. 131 (2019) 8724.
[31] W. Zhang, J. Liu, X. Jin, X. Gu, X.C. Zeng, X. He, et al., Quantitative prediction of aggregation-induced emission: a full quantum mechanical approach to the optical spectra, Angew. Chem. Int. Ed. 59 (2020) 11550.
[32] Y. Gao, X. Chang, X. Liu, Q. Li, G. Cui, W. Thiel, Excited-state decay paths in tetraphenylethene derivatives, J. Phys. Chem. A 121 (2017) 2572.
[33] D. Presti, L. Wilbraham, C. Targa, F. Labat, A. Pedone, M.C. Menziani, et al., Understanding aggregation-induced emission in molecular crystals: insights from theory, J. Phys. Chem. C 121 (2017) 5747.
[34] T. Zhang, H. Ma, Y. Niu, W. Li, D. Wang, Q. Peng, et al., Spectroscopic signature of the aggregation-induced emission phenomena caused by restricted nonradiative decay: a theoretical proposal, J. Phys. Chem. C 119 (2015) 5040.
[35] J. Luo, K. Song, F.L. Gu, Q. Miao, Switching of non-helical overcrowded tetrabenzoheptafulvalene derivatives, Chem. Sci. 2 (2011) 2029.
[36] N.L.C. Leung, N. Xie, W. Yuan, Y. Liu, Q. Wu, Q. Peng, et al., Restriction of intramolecular motions: the general mechanism behind aggregation-induced emission, Chem. Eur. J. 20 (2014) 15349.
[37] S. Kumar, P. Singh, A. Mahajan, S. Kumar, Aggregation induced emission enhancement in ionic self-assembled aggregates of benzimidazolium based cyclophane and sodium dodecylbenzenesulfonate, Org. Lett. 15 (2013) 3400.
[38] T. Nishiuchi, K. Tanaka, Y. Kuwatani, J. Sung, T. Nishinaga, D. Kim, et al., Solvent-induced crystalline-state emission and multichromism of a bent p-surface system composed of dibenzocyclooctatetraene units, Chem. Eur. J. 19 (2013) 4110.
[39] C. Yuan, S. Saito, C. Camacho, T. Kowalczyk, S. Irle, S. Yamaguchi, Hybridization of a flexible cyclooctatetraene core and rigid aceneimide wings for multiluminescent flapping π systems, Chem. Eur. J. 20 (2014) 2193.
[40] C. Yuan, X. Tao, Y. Ren, Y. Li, J. Yang, W. Yu, et al., Synthesis, structure, and aggregation-induced emission of a novel lambda (Λ)-shaped pyridinium salt based on tröger's base, J. Phys. Chem. C 111 (34) (2007) 12811.

[41] A. Raghuvanshi, A.K. Jha, A. Sharma, S. Umar, S. Mishra, R. Kant, et al., A nonarchetypal 5,6-dihydro-2H-pyrano[3,2-g]indolizine-based solution-solid dual emissive AIEgen with multicolor tunability, Chem. Eur. J. 23 (2017) 4527.

[42] Z. Zhao, X. Zheng, L. Du, Y. Xiong, W. He, X. Gao, et al., Non-aromatic annulene-based aggregation-induced emission system via aromaticity reversal process, Nat. Commun. 10 (2019) 1.

[43] L. Yao, S. Zhang, R. Wang, W. Li, F. Shen, B. Yang, et al., Highly efficient near-infrared organic light-emitting diode based on a butterfly-shaped donor-acceptor chromophore with strong solid-state fluorescence and a large proportion of radiative excitons, Angew. Chem. Int. Ed. 53 (2014) 2119.

[44] J. Liu, Q. Meng, X. Zhang, X. Lu, P. He, L. Jiang, et al., Aggregation-induced emission enhancement based on 11,11,12,12,-tetracyano-9,10-anthraquinodimethane, Chem. Commun. 49 (2013) 1199.

[45] K.S.N. Kamaldeep, S. Kaur, V. Bhalla, M. Kumar, A. Gupta, Pentacenequinone derivatives for preparation of gold nanoparticles: facile synthesis and catalytic application, J. Mater. Chem. A 2 (2014) 8369.

[46] J.L. Banal, J.M. White, K.P. Ghiggino, W.W.H. Wong, Concentrating aggregation-induced fluorescence in planar waveguides: a proof-of-principle, Sci. Rep. 4 (2015) 1.

[47] C. Zhang, Z. Wang, S. Song, X. Meng, Y. Zheng, X. Yang, et al., Tetraphenylethylene-based expanded oxacalixarene: synthesis, structure, and its supramolecular grid assemblies directed by guests in the solid state, J. Org. Chem. 79 (2014) 2729.

[48] J. Wang, H. Feng, J. Luo, Y. Zheng, Monomer emission and aggregate emission of an imidazolium macrocycle based on bridged tetraphenylethylene and their quenching by C60, J. Org. Chem. 79 (2014) 5746.

[49] G. Karthik, P.V. Krushna, A. Srinivasan, T.K. Chandrashekar, Calix[2]thia[4]phyrin: an expanded calixphyrin with aggregation-induced enhanced emission and anion receptor properties, J. Org. Chem. 78 (2013) 8496.

[50] P.S. Salini, A.P. Thomas, R. Sabarinathan, S. Ramakrishnan, K.C. Sreedevi, M.L. Reddy, et al., Calix[2]-m-benzo [4]phyrin with aggregation-induced enhanced-emission characteristics: application as a HgII chemosensor, Chemistry 17 (2011) 6598.

[51] H. Zhang, J. Liu, L. Du, C. Ma, N.L.C. Leung, Y. Niu, et al., Drawing a clear mechanistic picture for the aggregation-induced emission process, Mater. Chem. Front. 3 (2019) 1143.

[52] F. Bu, R. Duan, Y. Xie, Y. Yi, Q. Peng, R. Hu, et al., Unusual aggregation-induced emission of a coumarin derivative as a result of the restriction of an intramolecular twisting motion, Angew. Chem. Int. Ed. 254 (2015) 14492.

[53] S. Sasaki, S. Suzuki, W.M.C. Sameera, K. Igawa, K. Morokuma, G. Konishi, Highly twisted N,N-dialkylamines as a design strategy to tune simple aromatic hydrocarbons as steric environment-sensitive fluorophores, J. Am. Chem. Soc. 138 (2016) 8194.

[54] X. Peng, S. Ruiz-Barragan, Z. Li, Q. Li, L. Blancafort, Restricted access to a conical intersection to explain aggregation induced emission in dimethyl tetraphenylsilole, J. Mater. Chem. C 4 (2016) 2802.

[55] P. Zhou, P. Li, Y. Zhao, K. Han, Restriction of flip-flop motion as a mechanism for aggregation-induced emission, J. Phys. Chem. Lett. 10 (2019) 6929.

[56] Y. Tu, Y. Yu, D. Xiao, J. Liu, Z. Zhao, Z. Liu, et al., An intelligent AIEgen with nonmonotonic multiresponses to multistimuli, Adv. Sci. 7 (2020) 2001845.

[57] Y. Tu, J. Liu, H. Zhang, Q. Peng, J.W.Y. Lam, B.Z. Tang, Restriction of access to the dark state: a new mechanistic model for heteroatom-containing AIE systems, Angew. Chem. Int. Ed. 58 (2019) 14911.

[58] Y. Tu, J. Liu, X. Zhang, T. Cheung, X. He, J. Guo, et al., The Importance of Solid-state Molecular Motion to Room temperature phosphorescence, 2020, https://doi.org/10.26434/chemrxiv.12629903.

[59] J.W. Chung, S. Yoon, B. An, S.Y. Park, High-contrast on/off fluorescence switching via reversible E-Z isomerization of diphenylstilbene containing the α-cyanostilbenic moiety, J. Phys. Chem. C 117 (2013) 11285.

[60] Y. Tu, Z. Zhao, J.W. Lam, B.Z. Tang, Mechanistic connotations of restriction of intramolecular motions (RIM), Natl. Sci. Rev. (2020), https://doi.org/10.1093/nsr/nwaa260.

[61] L. Viglianti, N.L.C. Leung, N. Xie, X. Gu, H.H.Y. Sung, Q. Miao, et al., Aggregation-induced emission: mechanistic study of the clusteroluminescence of tetrathienylethene, Chem. Sci. 8 (2017) 2629.

[62] H. Zhang, Z. Zhao, P.R. McGonigal, R. Ye, S. Liu, J.W.Y. Lam, et al., Clusterization-triggered emission: uncommon luminescence from common materials, Mater. Today 32 (2020) 275.

[63] Y. Xiong, Z. Zhao, W. Zhao, H. Ma, Q. Peng, Z. He, et al., Designing efficient and ultralong pure organic room-temperature phosphorescent materials by structural isomerism, Angew. Chem. Int. Ed. 57 (2018) 7997.

[64] D. Li, F. Lu, J. Wang, W. Hu, X. Cao, X. Ma, et al., Amorphous metal-free room-temperature phosphorescent small molecules with multicolor photoluminescence via a host-guest and dual-emission strategy, J. Am. Chem. Soc. 140 (2018) 1916.

[65] B.Z. Tang, Aggregology: exploration and innovation at aggregate level, Aggregate 1 (2020) 4.

CHAPTER

2

Fundamental chemistry and applications of boron complexes having aggregation-induced emission properties

Shunichiro Ito, Masayuki Gon, Kazuo Tanaka, and Yoshiki Chujo

Department of Polymer Chemistry, Graduate School of Engineering, Kyoto University, Kyoto, Japan

1 Introduction

Luminescent materials are vital to modern societies and are pervasive these days. Particularly, luminescent molecules including artificial and biological macromolecules have become ubiquitous not only in the scientific field but also in our daily lives. This is because these molecules possess advantageous features such as lightweight, superb designability, high scalability and processability, biocompatibility, and so on. These significant merits have been encouraging us to develop advanced materials based on these compounds. In this context, there has been a severe problem awaiting solution for typical luminescent molecules, named as concentration quenching or aggregation-caused quenching (ACQ) [1,2]. Although normal π-conjugated compounds emit efficiently under diluted conditions, they lose their emission in a concentrated state due to intermolecular interactions, such as π–π stacking, enhanced by the planar structure.

In recent years, a class of luminescent materials exhibiting the opposite property, namely, aggregation-induced emission (AIE), has attracted tremendous attention. AIE-active molecules emit weakly in dilute solutions, while they provide far more efficient luminescence in condensed states [1,2]. Such luminescent characteristics have been paid much attention

https://doi.org/10.1016/B978-0-12-824335-0.00017-9

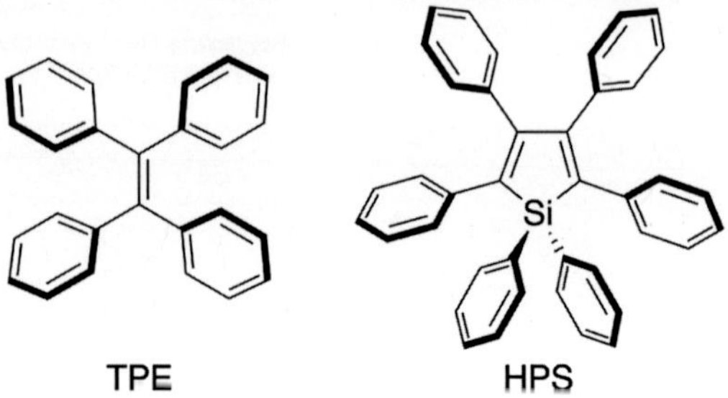

FIG. 1 Chemical structures of examples of AIE-active molecules: Left, tetraphenylethene (TPE); right, hexaphenylsilole (HPS).

because of their potential application in the field of organic light-emitting diodes, fluorescent sensors, bioimaging, and organic laser amplifiers. The incipient examples of AIE-active molecules, tetraphenylethene (TPE) and hexaphenylsilole (HPS), are shown in Fig. 1. The bulky peripheral aromatic substituents in these molecules consume the excited energy via molecular vibrations and motions in the solution. In contrast, aggregation and crystallization restrict the intramolecular vibrations and motions, and the hindered groups expel the intermolecular interactions, which lead to severe concentration quenching in the condensed state for typical luminescent molecules. As a result, intense solid-state emission is obtained from these systems. Furthermore, it has been found that the luminescence intensity and colors could be varied, due to high sensitivity toward environmental changes, by modulating the degree of molecular motions and morphology [3–6]. Thus, these complexes are expected to be a promising scaffold for developing stimulus-responsive optical materials as well as solid-state luminescent dyes.

Heteroatoms, including nonmetals, semimetals, and typical and transition metals, often serve as a key to functionalize both small molecules and polymers thanks to the promising electronic nature of these atoms. In this context, advanced functional materials can be achieved in a bottom-up manner by elaborately designing a minimal functional unit using heteroatoms. We have regarded such a minimal functional unit as an "element block" [7–9]. Based on this idea, various types of boron complexes and cluster compounds have been discovered to work as versatile element blocks for realizing the AIE behavior. This chapter describes the typical examples of the AIE- or crystallization-induced emission (CIE)-active four-coordinate boron complexes and discusses the proposed origins of the AIE nature of these classes of materials. It is worth noting that carboranes, which are cluster compounds composed of 10 boron and 2 carbon atoms, are also known to be a versatile scaffold for AIE-gens as well as the coordination compounds. Carborane-based systems have been covered in the recent comprehensive reviews [10,11] and will not be mentioned here.

Four-coordinate boron complexes are usually more stable toward air and water than three-coordinate ones because the 2p orbitals of four-coordinated boron are no longer vacant [12]. Consequently, four-coordinate boron complexes have been most extensively investigated in the chemistry of the AIE behavior of boron-containing molecules. Most studies mainly focus on the development of new chelate ligands for obtaining the AIE property, and the effects of substituents on the photophysical properties such as AIE activity, luminescent color, and stimuli responsivity.

2 Scaffolds for construction of AIE-active boron complexes

2.1 β-Diketonates

β-Diketone, such as acetylacetone (2,4-pentanedione), are one of the most significant scaffolds of ligands for typical and transition metal complexes. In recent years, photoluminescent boron complexes were based on β-diketonates with extended π-conjugation systems. However, most of these boron complexes exhibit ACQ due to their high structural planarity [13]. Nevertheless, an interesting example of a difluoroboron complex of substituted β-diketonate (**DKMeO**) with the AIE property was reported [14]. The luminescence from this compound showed not only AIE property but also dependence on polymorphs, mechanochromism, thermochromism, and solvatochromism. **DKMeO** exhibited slight emission in acetonitrile solution ($\Phi_f < 0.001$), but emitted much more strongly after the addition of an excess amount of water to form aggregates. Interestingly, the luminescent color was dependent on the way of adding water even if the final volume contents of water were the same. When water was added dropwise to the vigorously stirred acetonitrile solutions, the obtained suspensions showed blue emission (Fig. 2B). On the other hand, the emission color was dramatically changed from yellow to green when water was added all at once to prepare mixtures (Fig. 2C). From the results of the analyses of powder X-ray diffraction patterns and photophysical measurements, the blue and green emissions were assigned to the luminescence from the two distinct polymorphs, which are derived from the *syn-anti* conformers of the methoxy groups. The transient yellow-emissive suspension might be attributed to the metastable amorphous state. In addition, it should be noted that **DKMeO** showed relatively efficient emission in the less polar solvents (hexane, toluene, and tetrahydrofuran). Therefore, the mechanism of the AIE phenomenon might be not only the restriction of the molecular motion but also the difference of polarity between the solution

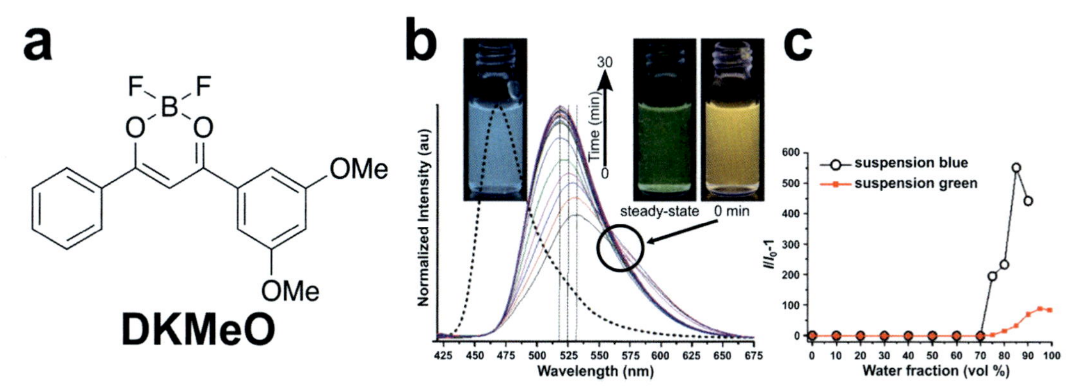

FIG. 2 (A) A chemical structure of an AIE-active boron β-diketonate complex **DKMeO**. (B) Normalized photoluminescence spectra of blue-emissive suspension *(dashed line)*, and time-dependent photoluminescence spectra from yellow- to green-emissive suspensions. (C) Emission intensity changes by altering water contents. *Reproduced from P. Galer, R.C. Korošec, M. Vidmar, B. Šket, Crystal structures and emission properties of the BF2 complex 1-phenyl-3-(3,5-dimethoxyphenyl)-propane-1,3-dione: multiple chromisms, aggregation- or crystallization-induced emission, and the self-assembly effect, J. Am. Chem. Soc. 136 (2014) 7383–7394. Copyright 2014 American Chemical Society.*

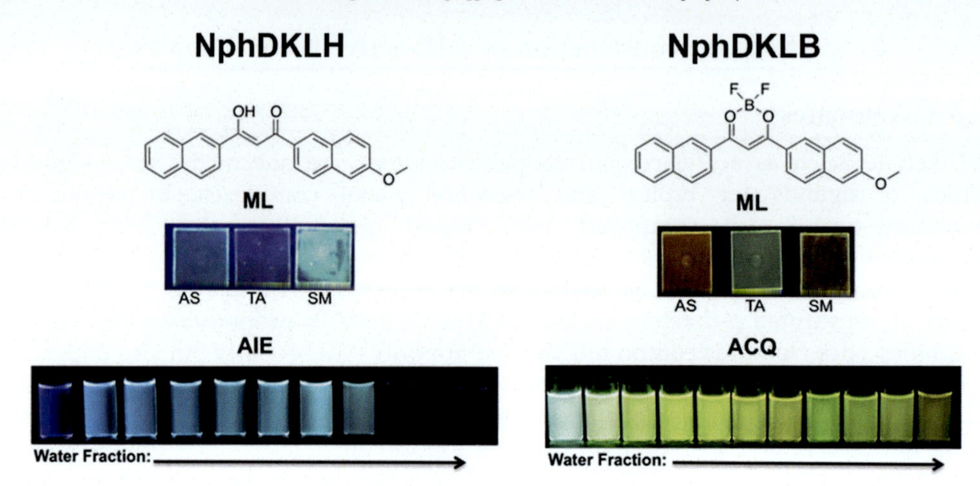

FIG. 3 Chemical structures of an AIE-active metal-free β-diketone **NphDKLH** and an AIE-inactive boron complex **NphDKLB**. Photographic images of thin films and THF/water mixtures of the compounds under UV irradiation. ML, AS, TA, and SM stand for mechanochromic luminescence, as spun, thermally annealed, and smeared, respectively. *Reproduced from T. Butler, W.A. Morris, J. Samonina-Kosicka, C.L. Fraser, Mechanochromic luminescence and aggregation induced emission of dinaphthoylmethane β-diketones and their boronated counterparts, ACS Appl. Mater. Interfaces 8 (2016) 1242–1251. Copyright 2016 American Chemical Society.*

and aggregate states. Mechanistic study of the AIE property of DKMeO can pave the way for designing AIE-active compounds based on β-diketonate complexes.

β-Diketones are inherently AIE-active and do not necessarily require boron complexation to trigger AIE. For example, Fraser et al. revealed that a metal-free β-diketone shows the AIE property and evaluated the effects of boron complexation on the photophysical properties [15,16]. The β-diketone with two naphthalene units, **NphDKLH**, shows AIE, while the spin-coated films of the corresponding boron complexes, **NphDKLB**, emit with weaker quantum yields than their solutions (Fig. 3). The boron complexation, on the other hand, leads to the increase in quantum yields and bathochromic shifts of the emission bands in their solutions. Importantly, the emission quantum yields of the boronated compounds are still significantly greater than those of the β-diketones even in the thin films. These results suggest that inherent nonradiative quenching processes of β-diketones in the solutions would be suppressed by the boron complexation, but the planar structures of the complexes bring about other nonradiative processes in the concentrated states probably due to intermolecular interactions. Therefore, the molecular design of AIE-active boron complexes based on β-diketonate ligands must need a proper balance among the suppression of molecularly intrinsic radiationless quenching paths and the planarization resulting in ACQ.

2.2 β-Ketoiminates

As mentioned above, β-diketonate may not be a sufficiently robust structural scaffold for the design of AIE-active luminophores. In contrast, it has been shown that β-ketoiminate ligands, also denoted as β-iminoenolates or β-enaminoketonates, serve as a versatile scaffold for the construction of AIE-gens [17–21]. The β-ketoiminate skeleton is analogous to the β-diketonate ligand, but one of the oxygen atoms is replaced with nitrogen-based moieties such as NR (R = aryl, alkyl, silyl). In general, B—N bonds (~445.6 kJ mol^{-1} or less) are weaker than B—O bonds (~536 kJ mol^{-1}). The weaker bond energy is likely to result in the larger molecular motions both at the ground and excited states. Therefore, nonradiative quenching processes of the corresponding boron complexes in the solution states are probably accelerated by the replacement of oxygen by nitrogen. In the solid states, such molecular motions would be restricted, and the substituent at the nitrogen could hinder intermolecular interaction because of the steric hindrance. As a result, it could be hypothesized that boron β-ketoiminate complexes are likely to emit more efficiently in the solid states than in the solution states.

AIE properties were observed and reported in the several boron β-ketoiminates synthesized. The structures of a few examples, **BKIa-BKIc**, are shown in Fig. 4 [18]. The solid of the β-diketonate complex, **BDK**, exhibited a decrease in the photoluminescence intensity and the quantum yield (Φ_{PL} = 0.36) compared to those of the solution (Φ_{PL} = 0.91). On the other hand, the corresponding boron β-ketoiminates, **BKIa**, **BKIb**, and **BKIc**, showed drastic emission enhancement caused by aggregation (Φ_{PL}^{solid}: 0.76 for **BKIa**; 0.42 for **BKIb**; 0.30 for **BKIc**), while the emissions from their solution samples were very weak ($\Phi_{PL}^{solution}$ ~ 0.01). Under frozen and viscous conditions, the emission intensity of these complexes increased compared to that in the solutions. These results therefore support the hypothesis that molecular motions play a critical role in emission annihilation in solution.

Since AIE-active boron β-ketoiminates have relatively high planarity, it is expected that the extension of the π-electronic planes could effectively modulate their photophysical properties. Mechanical stimuli, such as pressing, crushing, and grinding, occasionally change the luminescent color of crystalline samples of solid-state emissive molecules. This property is called mechano- or piezochromic luminescence and is expected to be utilized for pressure sensors and optical recording/memory devices [13,22–29]. A series of triads, **BKM**, containing two boron β-ketoiminates and the bithiophene linker with or without substituents at both ends were synthesized and studied for AIE and mechanochromic properties (Fig. 5A) [30]. All the compounds showed similar electronic absorption and photoluminescence spectra in the solution states (Fig. 5B and C). In the crystalline powder, the triads containing the smaller substituents, **BKM-H(a)** and **BKM-F**, exhibit yellow fluorescence, while those with relatively larger substituents, **BKM-Cl**, **BKM-Br**, and **BKM-I**, exhibited red fluorescence (Fig. 5D). Interestingly, mechanical grinding leads to the hypsochromic and the bathochromic shifts of the emission bands for the former group (**BKM-H(a)** and **BKM-F**) and the latter group (**BKM-Cl**, **BKM-Br**, and **BKM-I**), respectively (Fig. 5E). The initial emission colors were recovered by heating the ground samples. From the results of powder X-ray diffraction and differential scanning calorimetry before and after the mechanical treatment, the grinding process increased the content of the amorphous domain of the samples. These observations suggested the following plausible mechanism of the contrary mechanochromic behavior depending on the kind of the substituents (Fig. 5G): In the case of triads having the smaller substituents, the molecules could interact with a face-to-face motif and realize tight packing structures in the crystalline states. Consequently, emission bands appeared in the relatively longer wavelength regions in the initial crystalline states. By collapsing such tight packing, random molecular distributions should be obtained, resulting in the hypsochromic shifts. On the other hand, the steric hindrance of the larger substituents could disturb such a face-to-face stacking structure in the crystals. In the amorphous domains, π–π interactions predominate, leading to the bathochromic shifts.

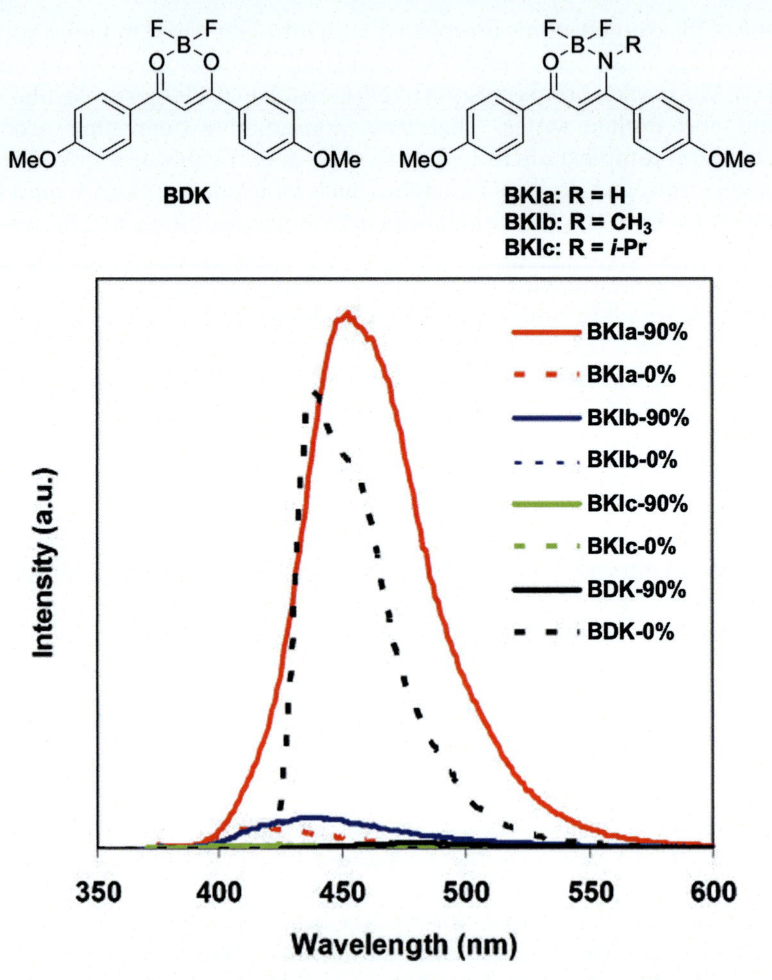

FIG. 4 Chemical structures of a series of β-diketonate and β-ketoiminate complexes. Photoluminescence spectra of the compounds (5.0×10^{-5} M) in THF *(solid line)* and THF/H_2O (1:9) mixed solvent *(dashed line)* upon excitation at each absorption maximum. *Reproduced from R. Yoshii, A. Nagai, K. Tanaka, Y. Chujo, Highly emissive boron ketoiminate derivatives as a new class of aggregation-induced emission fluorophores, Chem. Eur. J. 19 (2013) 4506–4512. Copyright 2013 WILEY-VCH Verlag GmbH & Co. KGaA, Weinheim*

2.3 β-Diketiminates

β-Diketiminate ligands, also known as β-diiminates, are the aza analogs of β-diketonate ligands as well as β-ketoiminate ligands. "Ketimine" is the name of an imine analogous to a ketone. "β-Ketoimine" is the name of an imine possessing a "keto" group at the β position of its imino group. Since both of the oxygen atoms in β-diketonates are replaced with aryl-substituted nitrogen atoms, the boron complexes of β-diketiminates are also likely to

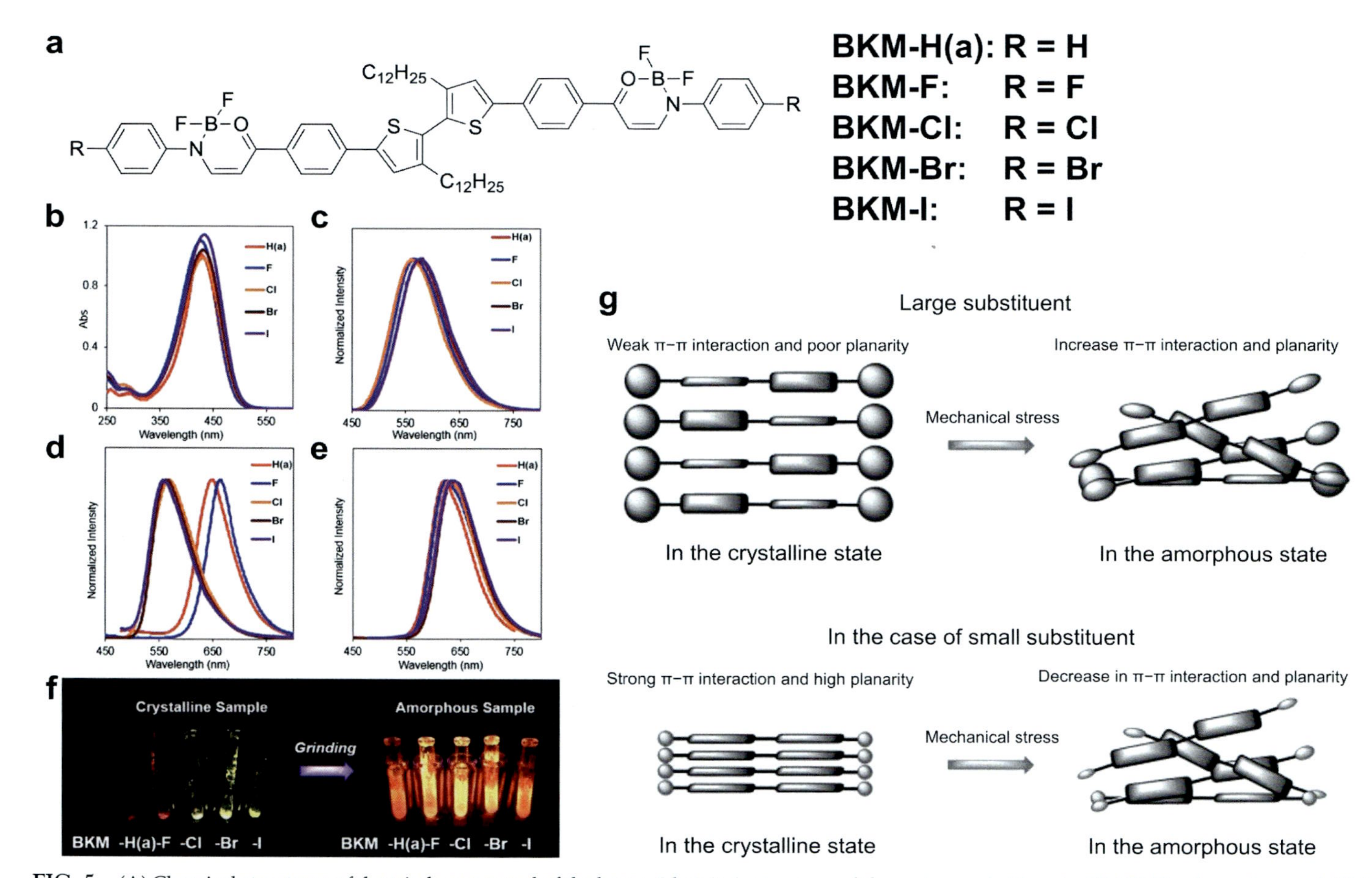

FIG. 5 (A) Chemical structures of the triads composed of the boron β-ketoiminate core and the various substituents. (B) UV-Vis absorption, and (C) PL spectra of **BKM-H(a)** and halogen-substituted boron β-ketoiminates in THF (1×10^{-5} M). PL spectra of **BKM-H(a)** and halogen-substituted boron β-ketoiminates in (D) crystalline, and (E) ground states. (F) Photographs of the boron β-ketoiminates in the crystalline and amorphous states under UV (365 nm) irradiation. (G) Speculated mechanisms of the mechanofluorochromism with different end groups. *Reproduced from R. Yoshii, K. Suenaga, K. Tanaka, Y. Chujo, Mechanofluorochromic materials based on aggregation-induced emission-active boron ketoiminates: regulation of the direction of the emission color changes, Chem. Eur. J. 21 (2015) 7231–7237. Copyright 2015 WILEY-VCH Verlag GmbH & Co. KGaA, Weinheim.*

be AIE active. Additionally, the steric and electronic properties of the β-diketiminate ligands are more easily tuned by modulating the substituents on the nitrogen and the carbon atoms than those of the β-diketonate and β-ketoiminate ones. Therefore, these ligands have been utilized in order to isolate the classes of unstable main-group and transition metal complexes composed of the group 13 elements, which had been not accessible using other bidentate ligands [31–37]. On the other hand, the optical and electronic properties of these complexes have not been extensively studied, probably because they are hardly colored due to the limited π-conjugation lengths of the β-diketiminate ligands. However, it could be envisioned that β-diketiminate ligands would enable us to access functional optoelectronic materials and possible applications of many kinds of metal complexes that have not been obtained by other ligands.

In the early study in 2008, it was clarified that the extension of the π-conjugated system involving β-diketiminate ligands makes the corresponding boron complexes luminescent in the visible region in the solid state. Importantly, it was reported that the emission intensity was enhanced by aggregation [38]. In the later reports, it was stated that the luminescence quantum yields of the boron β-diketiminate complexes, **BDKIa** and **BDKIb** (Fig. 6A), were higher in the crystalline states ($\Phi_{PL}=0.23$ for **BDKIa** and 0.11 for **BDKIb**) than those in the amorphous ($\Phi_{PL}=0.02$ for both compounds) and solution states ($\Phi_{PL}<0.01$ for both compounds) [39]. The crystalline aggregates were obtained by reprecipitation from the mixture of acetonitrile solutions and water (Fig. 6B and C). The amorphous solids were prepared by rapid quenching with the melt samples. The crystalline and amorphous states of these complexes were interconverted repeatedly by means of the fuming-cooling or heating-cooling cycles (Fig. 6D and E). It was proposed that the packing structures in their crystalline states play a part in suppressing molecular vibrations and motions leading to nonradiative quenching paths. Since amorphous states are generally sparser than crystalline states, the radiationless processes may be accessible enough to completely quench the excited energy without strong emission in the amorphous states. Therefore, the CIE property was observed from the boron β-diketiminates. In addition to these nonfused β-diketiminate complexes, Perumal et al. also reported the AIE behavior of the boron complexes having fused β-diketiminate [40], leading to the emergence of the chemistry of the emission from boron β-diketiminates [39,41–47].

Suzuki-Miyaura cross-coupling polycondensation of dihalogenated boron β-diketiminates and the corresponding comonomers successfully afforded AIE-active π-conjugated polymers [41]. The electron-donating and -withdrawing substituents at the complex moiety effectively change the emission colors of their films (Fig. 7). The complexes possessing an electron-donating substituent showed a bathochromic shift of their emission band compared to nonsubstituted analogs. On the other hand, the presence of electron-withdrawing substituents was found to hypsochromically shift the luminescence spectra.

Based on the above observations, film-type luminescent sensors based on acid-base [41] and redox [42] reactions were developed using the conjugated polymers composed of boron β-diketiminates. In the redox sensing system (Fig. 8) [42], for example, the emission properties were altered by the oxidation of sulfide-functionalized polymer films. The spin-coated films of the sulfide polymer showed weak yellow emission. When the films were treated with H_2O_2 aq., the emission intensity was gradually strengthened as the exposure time increased.

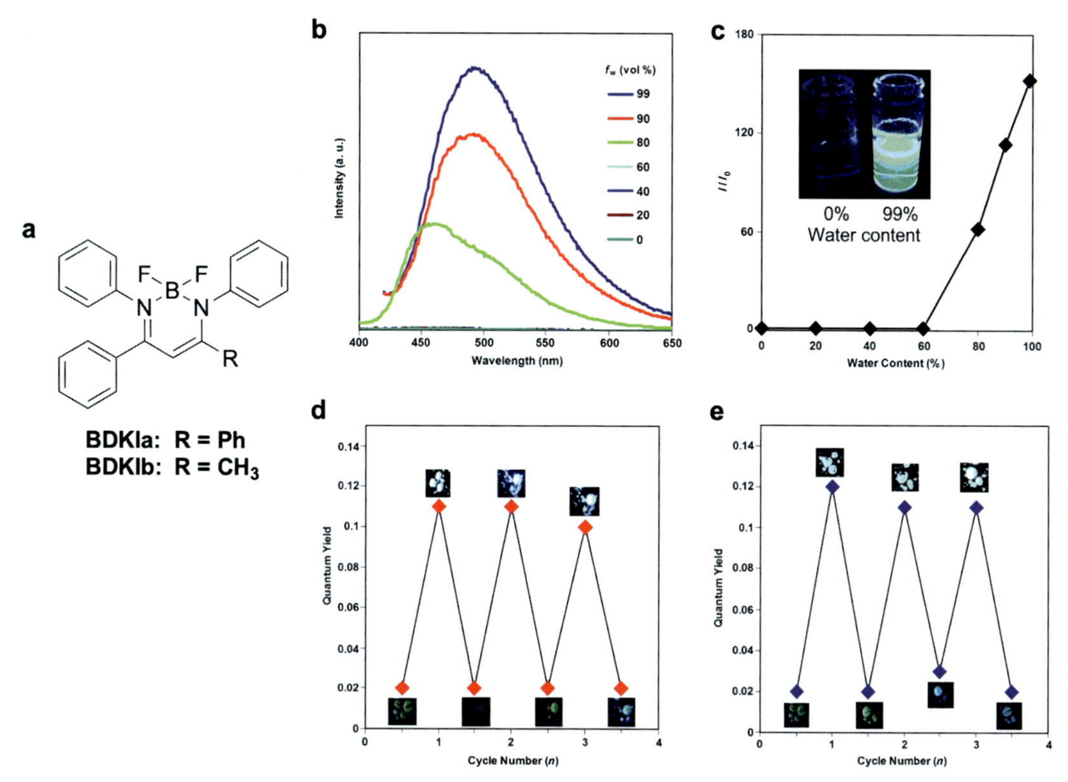

FIG. 6 (A) Chemical structures of boron β-diketiminate complexes. Dependence of (B) photoluminescence spectra and (C) emission-intensity ratio of **BDKIa** on solvent compositions of the acetonitrile/H_2O mixture (5×10^{-5} M) upon excitation at the peak position in the absorption spectra. Repeated switching between amorphous and crystalline states by (D) fuming-heating and (E) heating-cooling cycles. *Reproduced from R. Yoshii, A. Hirose, K. Tanaka, Y. Chujo, Boron diiminate with aggregation-induced emission and crystallization-induced emission-enhancement characteristics, Chem. Eur. J. 20 (2014) 8320–8324. Copyright 2014 WILEY-VCH Verlag GmbH & Co. KGaA, Weinheim.*

It was confirmed by ^{1}H NMR analyses that the sulfide groups in the polymer were oxidized to sulfoxide groups, which resulted in the enhancement of fluorescence.

2.4 Formazanates

Formazans have been utilized as the chromophore for the MTT (3-(4,5-dimethylthiazol-2-yl)-2,5-diphenyltetrazolium bromide) assay, in which cell viability can be estimated according to the amount of digested products from MTT, because of their redox activities with color change (Fig. 9A) [48–50]. In the recent years, formazanates, which can be regarded as the congeners of β-diketiminates, have attracted growing attention as versatile ligands for typical and transition metal ions [51]. In this context, AIE-active boron formazanate complexes were developed (Fig. 9B and C) [52]. The relatively planar structure of **CF1** contributes to its ACQ property and thus the quenching of solid-state emission. On the other hand, the

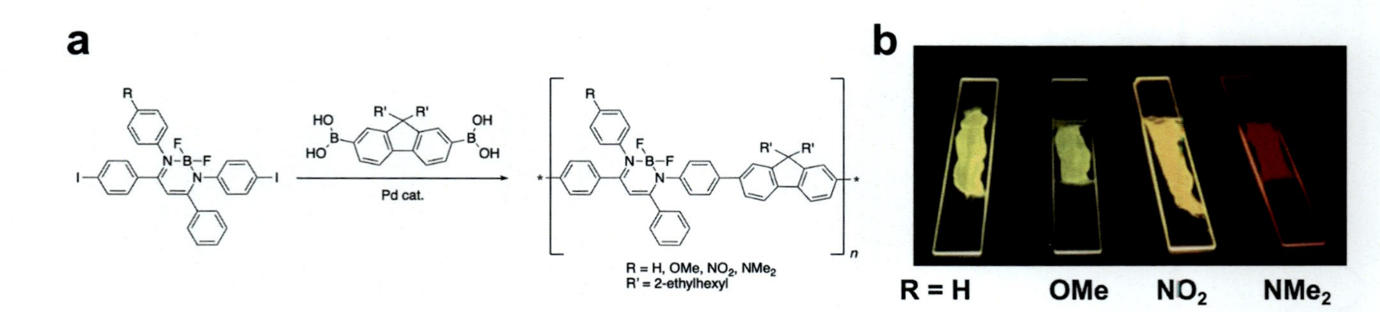

FIG. 7 (A) Synthetic scheme of conjugated polymers composed of β-diketiminate complexes with various substituents. (B) Photographic images of thin films of the polymers on quartz substrates under UV irradiation. *Reproduced from R. Yoshii, A. Hirose, K. Tanaka, Y. Chujo, Functionalization of boron diiminates with unique optical properties: multicolor tuning of crystallization-induced emission and introduction into the main chain of conjugated polymers, J. Am. Chem. Soc. 136 (2014) 18131–18139. Copyright 2014 American Chemical Society.*

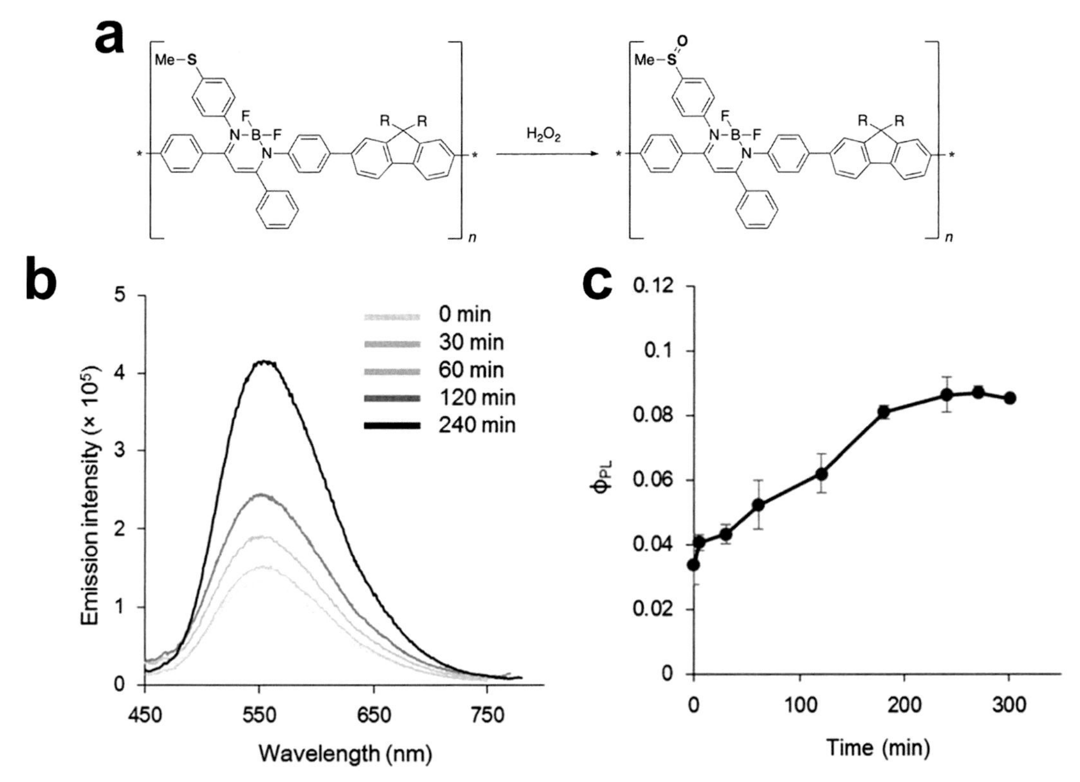

FIG. 8 (A) An oxidation reaction of a sulfide-containing polymer (R=2-ethylhexyl). Time dependence of (B) photoluminescence spectra and (C) emission quantum yield of the sulfide polymer film upon exposure to 30 wt% H_2O_2 aq. *Reproduced from A. Hirose, K. Tanaka, R. Yoshii, Y. Chujo, Film-type chemosensors based on boron diiminate polymers having oxidation-induced emission properties, Polym. Chem. 6 (2015) 5590–5595. Copyright 2015 The Royal Society of Chemistry.*

steric hindrance in **CF2** and **CF3** derived from the *ortho* substituents disturbs strong π-interactions in the solid state, resulting in the AIE property.

2.5 Azomethine complexes

Most AIE-active molecules possess peripheral phenyl groups and rotatable bonds, which accelerate nonradiative processes in the solution states, leading to emission annihilation. On the other hand, several boron complexes without such structural features, such as the azomethine complex, **Az** (Fig. 10A), have shown AIE behavior [53]. Optimized geometries calculated using the density functional theory (DFT) show that highly fused **Az** derivatives have high molecular planarity in the ground (S_0) state (Fig. 10B), which is in agreement with the structure determined from single crystal X-ray diffraction. Interestingly, the optimized calculated structure in the first singlet excited (S_1) state suggested drastic structural bending during the structural relaxation (Fig. 10C). This structural bending should give rise to the localization of π-electrons and radiationless decay of the excited energy. Hence, it could be

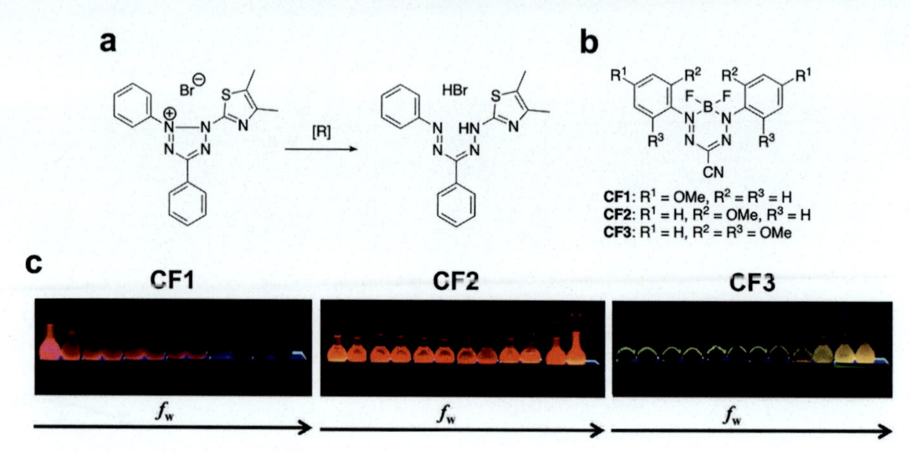

FIG. 9 (A) Reduction reaction of MTT to the corresponding formazan. (B) Chemical structures of AIE-active (**CF2** and **CF3**) and -inactive boron formazanate complexes. (C) Dependence of fluorescence of the formazanate complexes in the THF/water mixtures on the water content (f_w). *Reproduced from R.R. Maar, J.B. Gilroy, Aggregation-induced emission enhancement in boron difluoride complexes of 3-cyanoformazanates, J. Mater. Chem. C 4 (2016) 6478–6482. Copyright 2016 The Royal Society of Chemistry.*

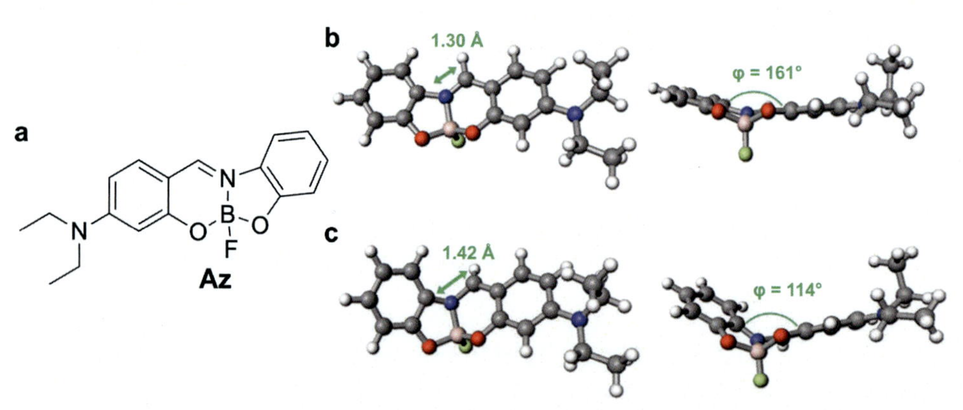

FIG. 10 (A) Chemical structure of the azomethine complex **Az**. The optimized geometric structures of **Az** in (B) the ground and (C) the excited states calculated at the B3LYP/6-311G** level and TD-B3LYP/6-311+G** level, respectively. *Reproduced from S. Ohtani, M. Gon, K. Tanaka, Y. Chujo, A flexible, fused, azomethine–boron complex: thermochromic luminescence and thermosalient behavior in structural transitions between crystalline polymorphs, Chem. Eur. J. 23 (2017) 11827–11833. Copyright 2017 Wiley-VCH Verlag GmbH & Co. KGaA, Weinheim.*

hypothesized that **Az** is likely to show the AIE property because aggregation would restrict the bending-flattening process.

To evaluate the justification of our presumption, we synthesized **Az**. Optical measurements suggested that **Az** showed not only AIE but also CIE properties. A large Stokes shift of **Az** was also detected in the solution state, implying the large structural relaxation in the excited state. The fluorine atom protruding from the molecular plane seems to repel strong

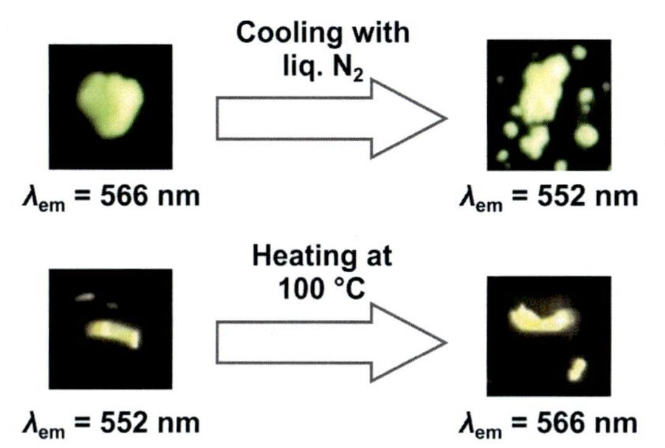

FIG. 11 Thermosalient effects of **Az** with thermochromic luminescence. *Reproduced from Supporting Movies of S. Ohtani, M. Gon, K. Tanaka, Y. Chujo, A flexible, fused, azomethine–boron complex: thermochromic luminescence and thermosalient behavior in structural transitions between crystalline polymorphs, Chem. Eur. J. 23 (2017) 11827–11833. Copyright 2017 Wiley-VCH Verlag GmbH & Co. KGaA.*

π–π interactions. These data therefore strongly encourage us to design AIE-active boron complexes with such flexible skeletons.

Additionally, the **Az** crystals showed unique physical and luminochromic behavior with temperature alterations (Fig. 11). This complex was crystallized in the two distinct polymorphs, which exhibited green and yellow emission. Differential scanning calorimetry (DSC) clearly indicated that these two polymorphs thermally transform each other. Surprisingly, the crystals showed an unusual physical motion such as hopping or fragmentation upon heating or cooling along with the crystal-crystal transition. Such motion induced by thermal stimuli is known as thermosalient effect. Moreover, thermochromism in the crystal appearance and luminescence was observed corresponding to the transition. It is supposed that the "flexible" boron structure of **Az** might provide room for physical motions during structural transition in the crystalline state. After this report, several other luminescent azomethine complexes have been developed [54–56].

2.6 Pyridine-based ligands

Pyridine plays a critical role as a robust scaffold for a wide variety of ligands [57–62], including the development of AIE-active boron complexes. For instance, pyridyl-enolates [63], whose structure is analogous to β-ketoiminates, were found to be valuable luminophores for constructing stimuli-responsive luminescent systems [64–66]. The first examples of AIE-active boron pyridyl-enolate complexes were reported in 2017 (Fig. 12). The synthesized complexes are decently emissive in their solutions (Φ_{PL} = 0.17–0.80), while the quantum yields become larger by crystallization. Importantly, these complexes exhibit drastic mechanochromic fluorescence (Fig. 12B and C). By scratching the crystals of each complex, their fluorescence spectra could be shifted to the shorter wavelength regions. **FBKIc** was crystallized in two distinct polymorphs with green and blue emissions. Single-crystal X-ray analysis suggested that the intermolecular interactions were stronger in the green-emissive crystal than those in the blue-emissive crystal. Therefore, the mechanical stimuli might unfasten the strong intermolecular interactions, resulting in the hypsochromic shift of the luminescence.

Hydrazones are one of the important functional groups in the wide range of chemistry [67,68]. Since the first series of AIE from boron difluorohydrazone complexes (BODIHYs,

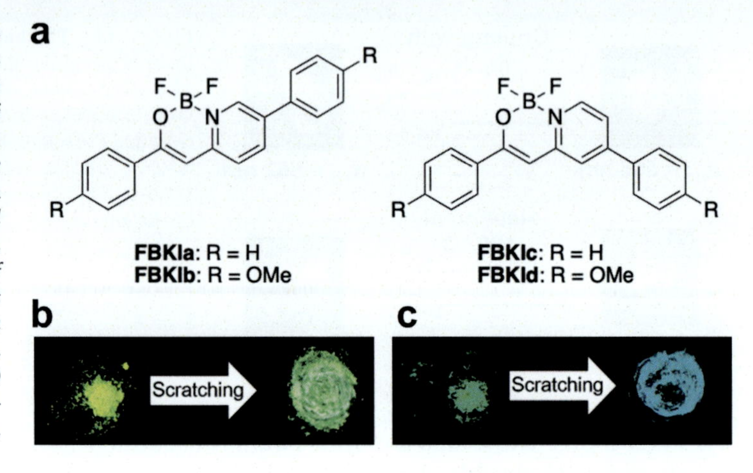

FIG. 12 (A) Chemical structures of pyridyl-enolate complexes (**FBKIa–d**). (B and C) Mechanochromic luminescence of **FBKIa** and **FBKIc**, respectively: Left, before scratching; right, after scratching. Photographs were taken under UV irradiation. *Reproduced from K. Suenaga, K. Tanaka, Y. Chujo, Design and luminescence chromism of fused boron complexes having constant emission efficiencies in solution and in the amorphous and crystalline states, Eur. J. Org. Chem. 2017 (2017) 5191–5196. Copyright 2017 WILEY-VCH Verlag GmbH & Co. KGaA, Weinheim.*

Fig. 13) were reported in 2012 [69], this class of molecules have been utilized as visible to near infrared emitters based on pyridine-containing ligands [70–76]. **BODIHY1**, for instance, shows fluorescence at 512 nm with $\Phi_{PL} = 0.06$ and at 497 nm with $\Phi_{PL} = 0.52$ in CH_2Cl_2 solution and the crystalline states, respectively. Comprehensive understanding of the mechanism of the AIE properties of BODIHYs is still not achieved. Two distinct mechanisms of the AIE properties have been proposed so far: (i) suppression of Kasha's rule (SOKR) [73] and (ii) restriction of flip-flop motion [77]. The rotational movements of the peripheral aromatic ring are responsible for the SOKR mechanism (Fig. 13B). The oscillator strengths of the first excited singlet state S_1 of BODIHYs are quite low. Therefore, the S_1 states should be accessible only through the structural relaxation at the higher excited states (S_2 or S_3) accompanied by the rotation of the aromatic rings. In the solutions, the molecules excited to the higher states would relax nonradiatively through S_1. The rotational relaxation should be restricted in the viscous media or solid states. Consequently, the excited molecules could emit from the optically accessible higher excited states. In the second mechanism (Fig. 13C), the bending motions of boron difluorohydrazones upon electronic excitation have been estimated. A series of DFT calculations suggested that the bending motion of these molecules, named "flip-flop," could lead the excited molecules to a conical intersection through which excitons nonradiatively quench. Interestingly, such a bending motion is similar to the structural relaxation in the azomethine complex **Az**, as mentioned above.

2.7 BODIPYs

Much effort has been also made for constructing AIE-active systems based on boron dipyrromethene (BODIPY) dyes, which are usually considered to be ineffective emitters in the solid states due to nonspecific intermolecular interactions in their planar structures [78–84]. One of the incipient examples was reported in 2009 for donor-acceptor type luminogens (Fig. 14a) [78]. These molecules are composed of a diphenylamino group as an electron donor and a BODIPY moiety as an electron acceptor with one or two phenylene spacers. These compounds were found to show twisted intramolecular charge transfer

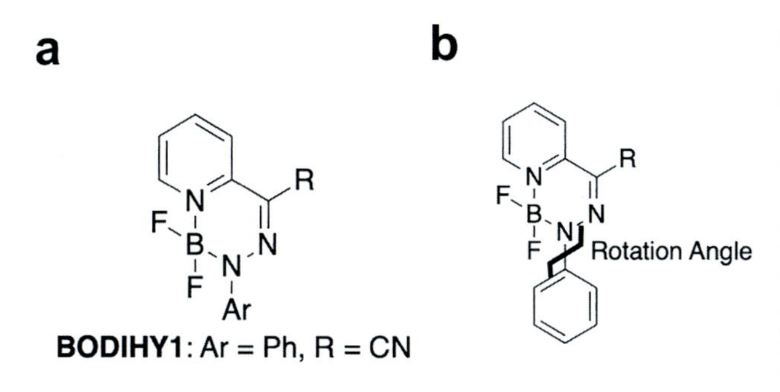

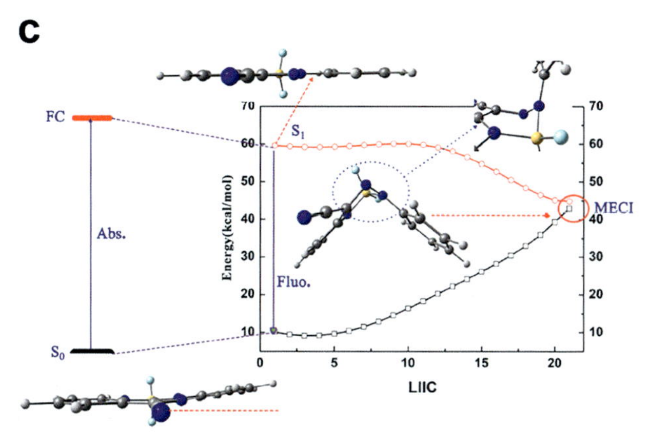

FIG. 13 (A) Chemical structure of **BODIHY1**. (B) The rotation angle responsible for the SOKR mechanism. (C) Energy diagram for the explanation of restriction of flip-flop motion mechanism. LIIC and MECI stand for linearly interpolated internal coordinate and a S_0/S_1 minimal energy conical intersection, respectively. *Reproduced from P. Zhou, P. Li, Y. Zhao, K. Han, Restriction of flip-flop motion as a mechanism for aggregation-induced emission, J. Phys. Chem. Lett. 10 (2019) 6929–6935. Copyright 2019 American Chemical Society.*

(TICT) property. They exhibit green emission derived from the locally excited states in the low-polarity solvents, while emission color changes to red originating from the TICT state when the solvent polarity increases (Fig. 14b and c). The intensities of the red TICT emissions in THF/water mixtures are strongly dependent to the contents of water (f_w) as shown in Fig. 14d for **BOD3**. When f_w is low (<60 vol%), the red emissions from the TICT state become negligible due to the very high polarity of the THF/water mixture. In contrast, the emission prevails when a large amount of water is added (f_w >70%) because nanoaggregates are formed and the polarity of the surroundings of the chromopores decreased. In the case of **BOD3** and **BOD4**, it is of interest to note that the emissions are enhanced compared from those in the pure THF solutions. Their intramolecular motions are likely to be restricted by the solidification.

Another exciting example was constructed with *meso*-trifluoromethyl-substituted BODIPY **BOD-CF3** (Fig. 15A) [79]. The emission intensity of **BOD-CF3** in the solution states is significantly lower (Φ_{PL} = 0.003 in acetonitrile) than that of the methyl counterpart **BOD-CH3** (Fig. 15B). Such weak luminescence may be originated from large structural difference between the S_0 and S_1 states resulting in the small Franck-Condon factor for the electronic transition. Surprisingly, the emission intensity of **BOD-CF3** is enhanced as the volume fraction of water (f_w) of the acetonitrile/water mixtures increases. The colloidal suspension of the **BOD-CF3** with 99% of f_w shows sharpened absorption and emission spectra with a

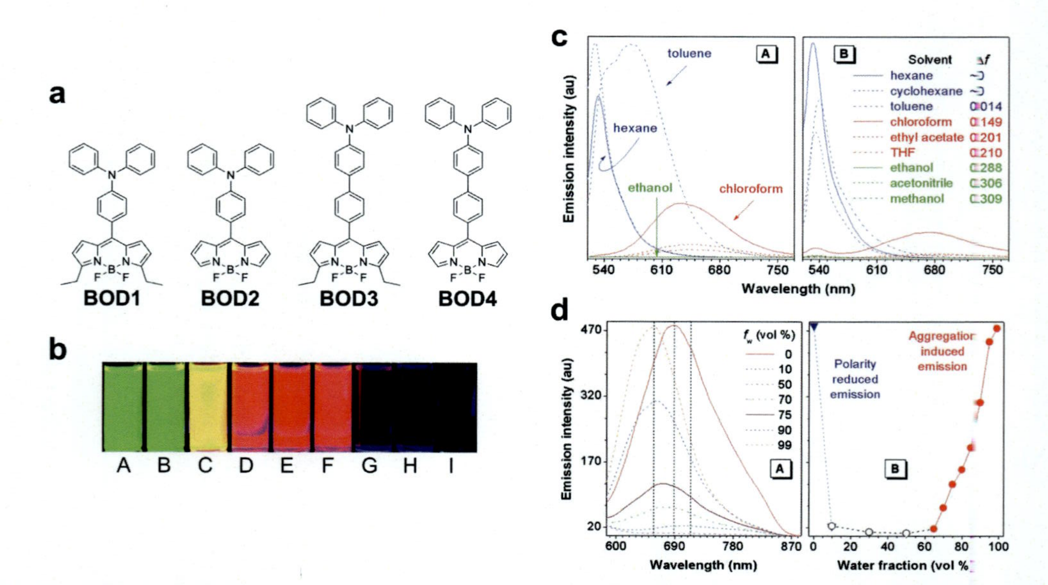

FIG. 14 (a) Chemical structures of the BODIPY-based donor-acceptor luminogens. (b) Photographs of **BOD1** under UV irradiation in the various solvents (A: hexane; B: cyclohexane; C: toluene; D: chloroform; E: ethyl acetate; F: THF; G: ethanol; H: acetonitrile; I: methanol.) (c) Photoluminescence spectra of (A) **BOD1** and (B) **BOD2** in the various solvents. (d) (A) Photoluminescence spectra and (B) emission intensities of **BOD3** in the THF/water mixtures with the different water contents. *Reproduced from R. Hu, E. Lager, A. Aguilar-Aguilar, J. Liu, J.W.Y. Lam, H.H.Y. Sung, I.D. Williams, Y. Zhong, K.S. Wong, E. Peña-Cabrera, B.Z. Tang, Twisted intramolecular charge transfer and aggregation-induced emission of BODIPY derivatives, J. Phys. Chem. C 113 (2009) 15845–15853. Copyright 2009 American Chemical Society.*

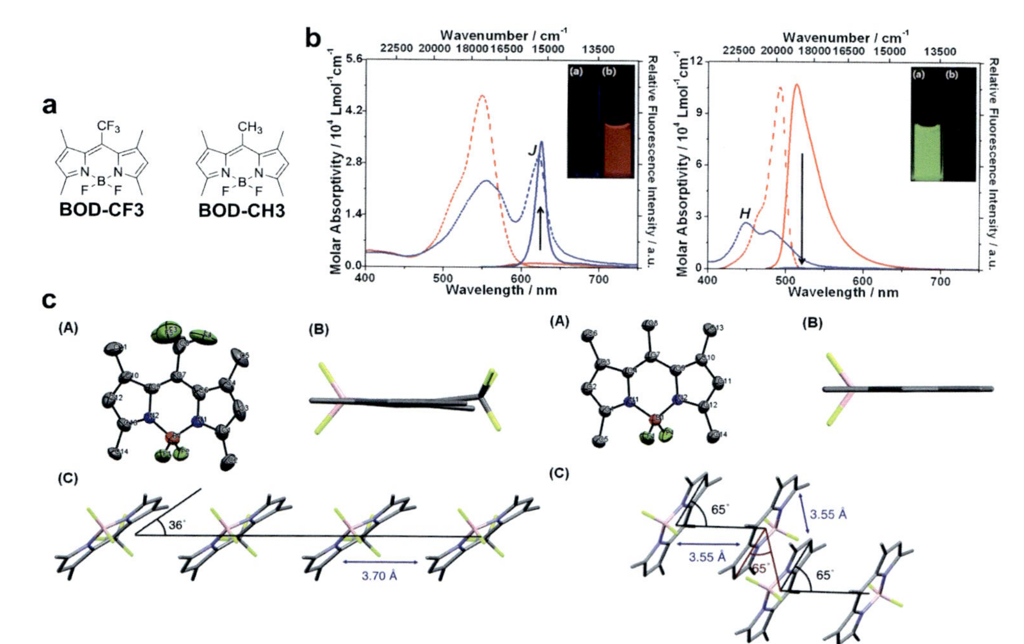

FIG. 15 (a) Chemical structures of the *meso*-substituted BODIPYs. (b) Absorption and emission spectra of **BOD-CF3** (left) and **BOD-CH3** (right). Single-crystal structures of (c) **BOD-CF3** and (d) **BOD-CH3**. (A, B, and C) represent a single molecule structure, side view, and packing structure for each compound, respectively. *Reproduced from S. Choi, J. Bouffard, Y. Kim, Aggregation-induced emission enhancement of a meso-trifluoromethyl BODIPY via J-aggregation, Chem. Sci. 5 (2013) 751–755. Copyright 2013 The Royal Society of Chemistry.*

significantly small Stokes shift (51 cm^{-1}; 2 nm) compared to the acetonitrile solution (Fig. 15B). In sharp contrast, the aggregation causes the annihilation of the emission of **BOD-CH3**. The results of the single-crystal X-ray analysis revealed that the transition dipole moments of **BOD-CF3** are aligned at an angle of 36 degrees, while those of **BOD-CH3** are with 65 degrees (Fig. 15C and D). These results clearly indicate that the formation of J-aggregation makes an efficiently emissive path in the **BOD-CF3** aggregates, leading to the strong AIE effect. **BOD-CH3**, on the other hand, exhibits the H-type aggregation resulting in the optically dark states. Solid-state luminescent materials based on BODIPY dyes with such a J-aggregation character have attracted broad attention [80–82,85,86].

3 Conclusions

Recent advances in the development of AIE-active four-coordinated boron complexes were briefly summarized in this chapter. A wide range of strategies to achieve AIE-active boron complexes have been developed in the recent years. Traditional approaches, such as the introduction of rotatable and hindered peripheral aromatics, have proven successful. It is also anticipated that "flexibility" of boron complexes at electronically excited states is likely to be one of the key factors in the AIE property of the complexes. Furthermore, intermolecular interactions which have been regarded as detrimental are now useful as shown in AIE-active BODIPYs. Insights into the origins of the AIE properties of boron complexes have been provided using DFT calculations and temperature- or viscosity-dependent measurements. However, much more efforts must be required for the comprehensive understanding of the mechanisms.

In order to detect tiny stimuli and slight environmental changes, we still require enhancement of sensitivity in optical sensors. In this regard, robust scaffolds for the construction of AIE-active complexes, such as β-ketoiminates and β-diketiminates, have been blazing a trail in a new brilliant chemistry of solid-state emissive materials. Therefore, AIE-active element blocks based on boron with novel chemical structures would be a scaffold for designing and constructing advanced sensing materials to meet these demands.

References

[1] Y. Hong, J.W.Y. Lam, B.Z. Tang, Aggregation-induced emission, Chem. Soc. Rev. 40 (2011) 5361–5388.
[2] J. Mei, N.L.C. Leung, R.T.K. Kwok, J.W.Y. Lam, B.Z. Tang, Aggregation-induced emission: together we shine, united we soar! Chem. Rev. 115 (2015) 11718–11940.
[3] G. Zhang, J. Lu, M. Sabat, C.L. Fraser, Polymorphism and reversible mechanochromic luminescence for solid-state difluoroboron avobenzone, J. Am. Chem. Soc. 132 (2010) 2160–2162.
[4] H. Ito, T. Saito, N. Oshima, N. Kitamura, S. Ishizaka, Y. Hinatsu, M. Wakeshima, M. Kato, K. Tsuge, M. Sawamura, Reversible mechanochromic luminescence of [$(C_6F_5Au)_2(\mu$-1,4-diisocyanobenzene)], J. Am. Chem. Soc. 130 (2008) 10044–10045.
[5] D. Zhao, G. Li, D. Wu, X. Qin, P. Neuhaus, Y. Cheng, S. Yang, Z. Lu, X. Pu, C. Long, J. You, Regiospecific *N*-heteroarylation of amidines for full-color-tunable boron difluoride dyes with mechanochromic luminescence, Angew. Chem. Int. Ed. 52 (2013) 13676–13680.
[6] M. Gon, K. Tanaka, Y. Chujo, Creative synthesis of organic–inorganic molecular hybrid materials, Bull. Chem. Soc. Jpn. 90 (2017) 463–474.
[7] Y. Chujo, K. Tanaka, New polymeric materials based on element-blocks, Bull. Chem. Soc. Jpn. 88 (2015) 633–643.

[8] M. Gon, K. Tanaka, Y. Chujo, Recent progress in the development of advanced element-block materials, Polym. J. 50 (2018) 109–126.

[9] M. Gon, K. Tanaka, Y. Chujo, Concept of excitation-driven boron complexes and their applications for functional luminescent materials, Bull. Chem. Soc. Jpn. 92 (2019) 7–18.

[10] S. Mukherjee, P. Thilagar, Boron clusters in luminescent materials, Chem. Commun. 52 (2015) 1070–1093.

[11] J. Ochi, K. Tanaka, Y. Chujo, Recent progress in the development of solid-state luminescent o -carboranes with stimuli responsivity, Angew. Chem. Int. Ed. 59 (2020) 9841–9855.

[12] D. Frath, J. Massue, G. Ulrich, R. Ziessel, Luminescent materials: locking π-conjugated and heterocyclic ligands with boron(III), Angew. Chem. Int. Ed. 53 (2014) 2290–2310.

[13] P.-Z. Chen, L.-Y. Niu, Y.-Z. Chen, Q.-Z. Yang, Difluoroboron β-diketonate dyes: spectroscopic properties and applications, Coord. Chem. Rev. 350 (2017) 196–216.

[14] P. Galer, R.C. Korošec, M. Vidmar, B. Šket, Crystal structures and emission properties of the BF_2 complex 1-phenyl-3-(3,5-dimethoxyphenyl)-propane-1,3-dione: multiple chromisms, aggregation- or crystallization-induced emission, and the self-assembly effect, J. Am. Chem. Soc. 136 (2014) 7383–7394.

[15] T. Butler, W.A. Morris, J. Samonina-Kosicka, C.L. Fraser, Mechanochromic luminescence and aggregation induced emission for a metal-free β-diketone, Chem. Commun. 51 (2015) 3359–3362.

[16] T. Butler, W.A. Morris, J. Samonina-Kosicka, C.L. Fraser, Mechanochromic luminescence and aggregation induced emission of dinaphthoylmethane β-diketones and their boronated counterparts, ACS Appl. Mater. Interfaces 8 (2016) 1242–1251.

[17] Y. Kubota, S. Tanaka, K. Funabiki, M. Matsui, Synthesis and fluorescence properties of thiazole–boron complexes bearing a β-ketoiminate ligand, Org. Lett. 14 (2012) 4682–4685.

[18] R. Yoshii, A. Nagai, K. Tanaka, Y. Chujo, Highly emissive boron ketoiminate derivatives as a new class of aggregation-induced emission fluorophores, Chem. Eur. J. 19 (2013) 4506–4512.

[19] R. Yoshii, A. Nagai, K. Tanaka, Y. Chujo, Boron-ketoiminate-based polymers: Fine-tuning of the emission color and expression of strong emission both in the solution and film states, Macromol. Rapid Commun. 35 (2014) 1315–1319.

[20] R. Yoshii, K. Tanaka, Y. Chujo, Conjugated polymers based on tautomeric units: regulation of main-chain conjugation and expression of aggregation induced emission property via boron-complexation, Macromolecules 47 (2014) 2268–2278.

[21] E.V. Fedorenko, A.G. Mirochnik, A.Y. Beloliptsev, I.V. Svistunova, G.O. Tretyakova, Design, synthesis, and crystallization-induced emission of boron difluorides β-ketoiminates, ChemPlusChem 83 (2018) 117–127.

[22] Z. Chi, X. Zhang, B. Xu, X. Zhou, C. Ma, Y. Zhang, S. Liu, J. Xu, Recent advances in organic mechanofluorochromic materials, Chem. Soc. Rev. 41 (2012) 3878–3896.

[23] K.M. Wiggins, J.N. Brantley, C.W. Bielawski, Methods for activating and characterizing mechanically responsive polymers, Chem. Soc. Rev. 42 (2013) 7130–7147.

[24] A.J. McConnell, C.S. Wood, P.P. Neelakandan, J.R. Nitschke, Stimuli-responsive metal–ligand assemblies, Chem. Rev. 115 (2015) 7729–7793.

[25] P. Xue, J. Ding, P. Wang, R. Lu, Recent progress in the mechanochromism of phosphorescent organic molecules and metal complexes, J. Mater. Chem. C 4 (2016) 6688–6706.

[26] S. Mukherjee, P. Thilagar, Stimuli and shape responsive 'boron-containing' luminescent organic materials, J. Mater. Chem. C 4 (2015) 2647–2662.

[27] C. Wang, Z. Li, Molecular conformation and packing: their critical roles in the emission performance of mechanochromic fluorescence materials, Mater. Chem. Front. 1 (2017) 2174–2194.

[28] Z. Yang, Z. Chi, Z. Mao, Y. Zhang, S. Liu, J. Zhao, M.P. Aldred, Z. Chi, Recent advances in mechano-responsive luminescence of tetraphenylethylene derivatives with aggregation-induced emission properties, Mater. Chem. Front. 2 (2018) 861–890.

[29] S.K. Mellerup, S. Wang, Boron-based stimuli responsive materials, Chem. Soc. Rev. 48 (2019) 3537–3549.

[30] R. Yoshii, K. Suenaga, K. Tanaka, Y. Chujo, Mechanofluorochromic materials based on aggregation-induced emission-active boron ketoiminates: regulation of the direction of the emission color changes, Chem. Eur. J. 21 (2015) 7231–7237.

[31] L. Bourget-Merle, M.F. Lappert, J.R. Severn, The chemistry of β-diketiminatometal complexes, Chem. Rev. 102 (2002) 3031–3066.

[32] C. Camp, J. Arnold, On the non-innocence of "Nacnacs": ligand-based reactivity in β-diketiminate supported coordination compounds, Dalton Trans. 45 (2016) 14462–14498.

[33] W.E. Piers, D.J.H. Emslie, Non-cyclopentadienyl ancillaries in organogroup 3 metal chemistry: a fine balance in ligand design, Coord. Chem. Rev. 233 (2002) 131–155.

[34] H.W. Roesky, S. Singh, V. Jancik, V. Chandrasekhar, A paradigm change in assembling OH functionalities on metal centers, Acc. Chem. Res. 37 (2004) 969–981.

[35] D.J. Mindiola, Oxidatively induced abstraction reactions. A synthetic approach to low-coordinate and reactive early transition metal complexes containing metal – ligand multiple bonds, Acc. Chem. Res. 39 (2006) 813–821.

[36] C.J. Cramer, W.B. Tolman, Mononuclear Cu–O_2 complexes: geometries, spectroscopic properties, electronic structures, and reactivity, Acc. Chem. Res. 40 (2007) 601–608.

[37] C. Chen, S.M. Bellows, P.L. Holland, Tuning steric and electronic effects in transition-metal β-diketiminate complexes, Dalton Trans. 44 (2015) 16654–16670.

[38] F.P. Macedo, C. Gwengo, S.V. Lindeman, M.D. Smith, J.R. Gardinier, β-Diketonate, β-ketoiminate, and β-diiminate complexes of difluoroboron, Eur. J. Inorg. Chem. 2008 (2008) 3200–3211.

[39] R. Yoshii, A. Hirose, K. Tanaka, Y. Chujo, Boron diiminate with aggregation-induced emission and crystallization-induced emission-enhancement characteristics, Chem. Eur. J. 20 (2014) 8320–8324.

[40] K. Perumal, J.A. Garg, O. Blacque, R. Saiganesh, S. Kabilan, K.K. Balasubramanian, K. Venkatesan, β-Iminoenamine-BF_2 complexes: aggregation-induced emission and pronounced effects of aliphatic rings on radiationless deactivation, Chem. Asian J. 7 (2012) 2670–2677.

[41] R. Yoshii, A. Hirose, K. Tanaka, Y. Chujo, Functionalization of boron diiminates with unique optical properties: multicolor tuning of crystallization-induced emission and introduction into the main chain of conjugated polymers, J. Am. Chem. Soc. 136 (2014) 18131–18139.

[42] A. Hirose, K. Tanaka, R. Yoshii, Y. Chujo, Film-type chemosensors based on boron diiminate polymers having oxidation-induced emission properties, Polym. Chem. 6 (2015) 5590–5595.

[43] K. Tanaka, T. Yanagida, A. Hirose, H. Yamane, R. Yoshii, Y. Chujo, Synthesis and color tuning of boron diiminate conjugated polymers with aggregation-induced scintillation properties, RSC Adv. 5 (2015) 96653–96659.

[44] M. Yamaguchi, S. Ito, A. Hirose, K. Tanaka, Y. Chujo, Modulation of sensitivity to mechanical stimulus in mechanofluorochromic properties by altering substituent positions in solid-state emissive diiodo boron diiminates, J. Mater. Chem. C 4 (2016) 5314–5319.

[45] M. Yamaguchi, S. Ito, A. Hirose, K. Tanaka, Y. Chujo, Control of aggregation-induced emission versus fluorescence aggregation-caused quenching by bond existence at a single site in boron pyridinoiminate complexes, Mater. Chem. Front. 1 (2017) 1573–1579.

[46] S. Ito, A. Hirose, M. Yamaguchi, K. Tanaka, Y. Chujo, Synthesis of aggregation-induced emission-active conjugated polymers composed of group 13 diiminate complexes with tunable energy levels via alteration of central element, Polymers 9 (2017) 68.

[47] M. Yamaguchi, S. Ito, A. Hirose, K. Tanaka, Y. Chujo, Luminescent color tuning with polymer films composed of boron diiminate conjugated copolymers by changing the connection points to comonomers, Polym. Chem. 9 (2018) 1942–1946.

[48] A.W. Nineham, The chemistry of formazans and tetrazolium salts, Chem. Rev. 55 (1955) 355–483.

[49] T. Mosmann, Rapid colorimetric assay for cellular growth and survival: application to proliferation and cytotoxicity assays, J. Immunol. Methods 65 (1983) 55–63.

[50] E. Grela, J. Kozłowska, A. Grabowiecka, Current methodology of MTT assay in bacteria—a review, Acta Histochem. 120 (2018) 303–311.

[51] J.B. Gilroy, E. Otten, Formazanate coordination compounds: synthesis, reactivity, and applications, Chem. Soc. Rev. 49 (2019) 85–113.

[52] R.R. Maar, J.B. Gilroy, Aggregation-induced emission enhancement in boron difluoride complexes of 3-cyanoformazanates, J. Mater. Chem. C 4 (2016) 6478–6482.

[53] S. Ohtani, M. Gon, K. Tanaka, Y. Chujo, A flexible, Fused, azomethine–boron complex: thermochromic luminescence and thermosalient behavior in structural transitions between crystalline polymorphs, Chem. Eur. J. 23 (2017) 11827–11833.

[54] S. Ohtani, Y. Takeda, M. Gon, K. Tanaka, Y. Chujo, Facile strategy for obtaining luminescent polymorphs based on the chirality of a boron-fused azomethine complex, Chem. Commun. 56 (2020) 15305–15308.

[55] S. Ohtani, M. Nakamura, M. Gon, K. Tanaka, Y. Chujo, Synthesis of fully-fused bisboron azomethine complexes and their conjugated polymers with solid-state near-infrared emission, Chem. Commun. 56 (2020) 6575–6578.

[56] S. Ohtani, M. Gon, K. Tanaka, Y. Chujo, The design strategy for an aggregation- and crystallization-induced emission-active molecule based on the introduction of skeletal distortion by boron complexation with a tridentate ligand, Crystals 10 (2020) 615.

[57] T. Yamamoto, T. Maruyama, Z.-H. Zhou, T. Ito, T. Fukuda, Y. Yoneda, F. Begum, T. Ikeda, S. Sasaki, π-Conjugated poly(pyridine-2,5-diyl), poly(2,2′-bipyridine-5,5′-diyl), and their alkyl derivatives. Preparation, linear structure, function as a ligand to form their transition metal complexes, catalytic reactions, n-type electrically conducting properties, optical properties, and alignment on substrates, J. Am. Chem. Soc. 116 (1994) 4832–4845.

[58] C. Gunanathan, Y. Ben-David, D. Milstein, Direct synthesis of amides from alcohols and amines with liberation of H_2, Science 317 (2007) 790–792.

[59] C. Gunanathan, D. Milstein, Metal–ligand cooperation by aromatization–dearomatization: a new paradigm in bond activation and "green" catalysis, Acc. Chem. Res. 44 (2011) 588–602.

[60] S.R. Neufeldt, M.S. Sanford, Controlling site selectivity in palladium-catalyzed C–H bond functionalization, Acc. Chem. Res. 45 (2012) 936–946.

[61] Z. Wang, G.A. Solan, W. Zhang, W.-H. Sun, Carbocyclic-fused *N,N,N*-pincer ligands as ring-strain adjustable supports for iron and cobalt catalysts in ethylene oligo-/polymerization, Coord. Chem. Rev. 363 (2018) 92–108.

[62] E. Peris, R.H. Crabtree, Key factors in pincer ligand design, Chem. Soc. Rev. 47 (2018) 1959–1968.

[63] V.C. Gibson, C. Newton, C. Redshaw, G.A. Solan, A.J.P. White, D.J. Williams, Low valent chromium complexes bearing *N,O*-chelating pyridyl-enolate ligands $[OC(Bu^t)(=2\text{-}CHN_5H_3Me\text{-}x)]^-$ (x = 3–6), Dalton Trans. (2003) 4612–4617.

[64] K. Suenaga, K. Tanaka, Y. Chujo, Design and luminescence chromism of fused boron complexes having constant emission efficiencies in solution and in the amorphous and crystalline states, Eur. J. Org. Chem. 2017 (2017) 5191–5196.

[65] K. Suenaga, K. Uemura, K. Tanaka, Y. Chujo, Stimuli-responsive luminochromic polymers consisting of multi-state emissive fused boron ketoiminate, Polym. Chem. 11 (2020) 1127–1133.

[66] S. Saotome, K. Suenaga, K. Tanaka, Y. Chujo, Design for multi-step mechanochromic luminescence property by enhancement of environmental sensitivity in a solid-state emissive boron complex, Mater. Chem. Front. 4 (2020) 1781–1788.

[67] X. Su, I. Aprahamian, Hydrazone-based switches, metallo-assemblies and sensors, Chem. Soc. Rev. 43 (2014) 1963–1981.

[68] L.A. Tatum, X. Su, I. Aprahamian, Simple hydrazone building blocks for complicated functional materials, Acc. Chem. Res. 47 (2014) 2141–2149.

[69] Y. Yang, X. Su, C.N. Carroll, I. Aprahamian, Aggregation-induced emission in BF_2–hydrazone (BODIHY) complexes, Chem. Sci. 3 (2011) 610–613.

[70] Y. Yang, R.P. Hughes, I. Aprahamian, Visible light switching of a BF_2-coordinated azo compound, J. Am. Chem. Soc. 134 (2012) 15221–15224.

[71] Y. Yang, R.P. Hughes, I. Aprahamian, Near-infrared light activated azo-BF_2 switches, J. Am. Chem. Soc. 136 (2014) 13190–13193.

[72] H. Qian, Y.-Y. Wang, D.-S. Guo, I. Aprahamian, Controlling the isomerization rate of an azo-BF_2 switch using aggregation, J. Am. Chem. Soc. 139 (2017) 1037–1040.

[73] H. Qian, M.E. Cousins, E.H. Horak, A. Wakefield, M.D. Liptak, I. Aprahamian, Suppression of Kasha's rule as a mechanism for fluorescent molecular rotors and aggregation-induced emission, Nat. Chem. 9 (2017) 83–87.

[74] D. Cappello, D.A.B. Therien, V.N. Staroverov, F. Lagugné-Labarthet, J.B. Gilroy, Optoelectronic, aggregation, and redox properties of double-rotor boron difluoride hydrazone dyes, Chem. Eur. J. 25 (2019) 5994–6006.

[75] D. Cappello, R.R. Maar, V.N. Staroverov, J.B. Gilroy, Optoelectronic properties of carbon-bound boron difluoride hydrazone dimers, Chem. Eur. J. 26 (2020) 5522–5529.

[76] D. Cappello, A.E.R. Watson, J.B. Gilroy, A boron difluoride hydrazone (BODIHY) polymer exhibits aggregation-induced emission, Macromol. Rapid Commun. (2021) 2000553.

[77] P. Zhou, P. Li, Y. Zhao, K. Han, Restriction of flip-flop motion as a mechanism for aggregation-induced emission, J. Phys. Chem. Lett. 10 (2019) 6929–6935.

[78] R. Hu, E. Lager, A. Aguilar-Aguilar, J. Liu, J.W.Y. Lam, H.H.Y. Sung, I.D. Williams, Y. Zhong, K.S. Wong, E. Peña-Cabrera, B.Z. Tang, Twisted intramolecular charge transfer and aggregation-induced emission of BODIPY derivatives, J. Phys. Chem. C 113 (2009) 15845–15853.

[79] S. Choi, J. Bouffard, Y. Kim, Aggregation-induced emission enhancement of a meso-trifluoromethyl BODIPY via J-aggregation, Chem. Sci. 5 (2013) 751–755.

[80] M.H. Chua, Y. Ni, M. Garai, B. Zheng, K. Huang, Q. Xu, J. Xu, J. Wu, Towards meso-ester BODIPYs with aggregation-induced emission properties: The effect of substitution positions, Chem. Asian J. 10 (2015) 1631–1634.

[81] S. Mukherjee, P. Thilagar, Fine-tuning dual emission and aggregation-induced emission switching in NPI–BODIPY dyads, Chem. Eur. J. 20 (2014) 9052–9062.

[82] P.P.P. Kumar, P. Yadav, A. Shanavas, P.P. Neelakandan, Aggregation enhances luminescence and photosensitization properties of a hexaiodo-BODIPY, Mater. Chem. Front. 4 (2020) 965–972.

[83] S. Kim, J. Bouffard, Y. Kim, Tailoring the solid-state fluorescence emission of BODIPY dyes by meso substitution, Chem. Eur. J. 21 (2015) 17459–17465.

[84] C. Spies, A.-M. Huynh, V. Huch, G. Jung, Correlation between crystal habit and luminescence properties of 4,4-difluoro-1,3-dimethyl-4-bora-3*a*,4*a*-diaza-*s*-indacene, an asymmetric BODIPY dye, J. Phys. Chem. C 117 (2013) 18163–18169.

[85] S.K. Sarkar, S. Mukherjee, A. Garai, P. Thilagar, A complementary aggregation induced emission pair for generating white light and four-colour (RGB and Near-IR) cell imaging, ChemPhotoChem 1 (2017) 84–88.

[86] D. Tian, F. Qi, H. Ma, X. Wang, Y. Pan, R. Chen, Z. Shen, Z. Liu, L. Huang, W. Huang, Domino-like multi-emissions across red and near infrared from solid-state 2-/2,6-aryl substituted BODIPY dyes, Nat. Commun. 9 (2018) 2688.

CHAPTER

3

Aggregation-induced emission polymers

Yang Hu[a], Lihui Zhang[a], Rongrong Hu[a], and Ben Zhong Tang[b]

[a]State Key Laboratory of Luminescent Materials and Devices, Guangdong Provincial Key Laboratory of Luminescence from Molecular Aggregates, AIE Institute, Center for Aggregation-Induced Emission, South China University of Technology (SCUT), Guangzhou, China [b]School of Science and Engineering, Shenzhen Key Laboratory of Functional Aggregate Materials, The Chinese University of Hong Kong, Shenzhen, Guangdong, China

1 Introduction

The development of luminescent materials has shown great influence on health, environment, energy, and safety [1–4]. Compared with inorganic materials, organic luminescent materials possess advantages such as low cost, low toxicity, facile structural modification, structural diversity, and so on [5–9]. To solve the aggregation-caused quenching (ACQ) problem of traditional organic luminescent materials when utilized in solid state or aggregated aqueous state, aggregation-induced emission (AIE) materials [10,11], including tetraphenylethylene (TPE) [12,13], hexaphenylsilole (HPS) [14,15], distyreneanthracene (DSA) [16,17], tetraphenylpyrazine (TPP) [18], and tetraphenylbenzene (TPB) [19], were developed. These compounds generally possess propeller-shaped structures, which show weak or no emission in dilute solution, but emit brightly in the aggregated or solid states. The widely accepted mechanism for AIE phenomenon is restriction of intramolecular motions (RIMs) in the aggregated states which rigidify the molecular structure and inhibit nonradiative decay of the excited state energy [20,21].

Compared with AIE-active small molecules, AIE polymers usually possess AIE-active functional units (AIEgens, AIE luminogens) in the polymer structure, enjoy higher thermal stability, good processability, high mechanical property, signal amplification effect,

https://doi.org/10.1016/B978-0-12-824335-0.00001-5

multifunctionality inherent from polymer structures, etc. [22,23]. When AIEgen is incorporated into the polymer chain, the intramolecular rotation is hindered by polymer chain to some extent even in the solution, and hence unlike AIE-active small molecules, the solution of AIE polymers normally possess weak emission. With their potential applications in fluorescence sensing, intelligent materials, biomedicine, and optoelectronic devices [23–26], AIE polymers are extensively studied recently, with many reviews and book chapters published to summarize the development of AIE polymers. For example, Tang's group has summarized the structures and luminescence properties of TPE- or silole containing AIE polymers such as polyacetylenes, polyphenylenes, polytriazoles, and poly(phenyleneethynylene)s in 2012 [22]. A comprehensive review on the design principle, synthetic approaches, topological structures, functionalities, and applications of AIE polymers was published in 2014 [23]. Later, Wei, Zhang, and coworkers summarized the design strategies and biomedical applications of AIE polymer-based luminescent nanomaterials [27]. Seixas de Melo et al. summarized the background of AIE phenomena, AIE working mechanism, and the photophysical studies of AIE polymers [28]. With the rapid development of this field, the recent progress in the synthesis, properties, and applications of AIE polymers during the past few years are summarized in some of the recent reviews [22–29]. Herein, we mainly focus on the most recent progress about AIE polymers in the past 3 years, from the aspects of new polymerization synthetic methodology for the preparation of AIE polymers, some typical structures of AIE polymers such as polyelectrolytes, chiral polymers, hyperbranched polymers, porous polymers, and nonconventional luminescent polymers, as well as the applications of AIE polymers, in the hope of providing the most up-to-date advances of this popular field.

2 Synthesis of AIE polymers

The common strategies for the preparation of AIE polymers include the incorporation of AIEgens in the monomer or initiator structures and the in situ generation of AIEgens directly from the polymerizations [21]. With the rapid development of polymerization methodologies, a large number of polymerization reactions such as free radical polymerization [30], ring-opening polymerization [31], metathesis polymerization [32], transition metal-catalyzed polycouplings [33,34], and other polycondensations are utilized to synthesize AIE polymers. Herein, the most recent advances of AIE polymer synthesis through living radical polymerizations and one-component, two-component, or multicomponent polycondensations are introduced as examples.

2.1 Living radical polymerizations

Living radical polymerizations enjoy the advantages of both living polymerization and free radical polymerization, which enable the synthesis of block copolymers or graft copolymers with controllable molecular weights, functionalized terminal groups, and various topological structures. Among them, atom transfer radical polymerization (ATRP) and reversible addition fragmentation chain transfer (RAFT) polymerization are the most widely used methods for the synthesis of AIE polymers. AIE-active initiators or monomers are designed for polymerization, and polymer products generated from these methods generally possess AIEgens as the terminal/central group or attached on the polymer main chain as the side chain.

For example, Lu, Xu, and coworkers have designed a red-emissive bifunctional ATRP initiator **1** and used *tert*-butyl methacrylate as the monomer to conduct ATRP in *N,N*-dimethylformamide (DMF) with CuBr and 1,1,4,7,7-pentamethyldiethylenetriamine (PMDETA) as the catalyst, generating P**1** with an M_w of 22,400 g/mol, which was then hydrolyzed by trifluoroacetic acid to obtain a hydrophilic polymer P**2** (Scheme 1A) [35]. In DMF/H_2O mixtures with 90 vol% water, P**2** emitted at 617 nm with a fluorescence quantum efficiency (Φ_F) of 1.71%, which was increased compared to that in DMF solution, suggesting its AIE characteristics. Similarly, a TPE-functionalized polyethylene (PE)-based initiator P**3** was designed and used for the ATRP of *tert*-butyl acrylate to afford the AIE-active block copolymer P**4** as shown in Scheme 1B [36]. In DMF solution, P**4** could self-assemble to form micelle, and the intramolecular rotation of TPE moieties were restricted in the DMF-phobic PE core region, hence enhancing the photoluminescence. Furthermore, to avoid the fluorescence quenching caused by the transition-metal catalyst residue in the polymer product, metal-free photo-mediated ATRP was adopted by Wei, Zhang, Ouyang, and coworkers using AIE-active initiator and photocatalyst [37]. The polymerization was carried out at room temperature without metal catalyst, generating AIE-active fluorescent polymeric nanoparticles P**5** with emission maximum located at 552 nm and a high Φ_F value of 41.2% (Scheme 1C).

RAFT polymerization is also extensively used for the preparation of AIE polymers. For example, Zhu, Deng, and coworkers reported a macromolecular chain transfer agent P**6** that was synthesized via RAFT polymerization of *N*-isopropylacrylamide **8** as shown in Scheme 1D. P**6** was then utilized as chain transfer agent for the RAFT polymerization of 2-(diisopropyl amino) ethyl methacrylate **9** and TPE-containing vinyl monomer **10** to afford the temperature and pH dual-responsive tri-block copolymer P**7** with AIE property [38].

Photo-initiated RAFT polymerization without involving peroxide or azo compound as initiator was also utilized to synthesize AIE polymers. As shown in Scheme 1E, Zhang, Cao, Tian, and coworkers synthesized a TPE-containing photoiniferter **11** and carried out the photo-initiated RAFT polymerization of itaconic acid and styrene at room temperature with UV light irradiation in the absence of catalyst, generating the fluorescent copolymer P**8** with a narrow molecular weight polydispersity of 1.15 [39]. The strong greenish yellow fluorescence at 516 nm was observed from the nanoparticles formed in DMF/H_2O mixtures with 90 vol% water. With the amphiphilic feature of the block copolymer, P**8** can self-assemble in aqueous medium and be fabricated to AIE-active fluorescent organic nanoparticles (FONs) with potential applications in cell imaging and drug delivery.

2.2 Condensation polymerizations

2.2.1 Single-component polymerization

The polymerization of single component monomer enjoys the advantages of simple system, easy operation, excellent stoichiometric balance, and so on, which has also been used in the preparation of AIE polymers. For example, Tang, Qin, and coworkers have developed a polycondensation reaction of diisocyanide monomers to prepare polyimidazoles with

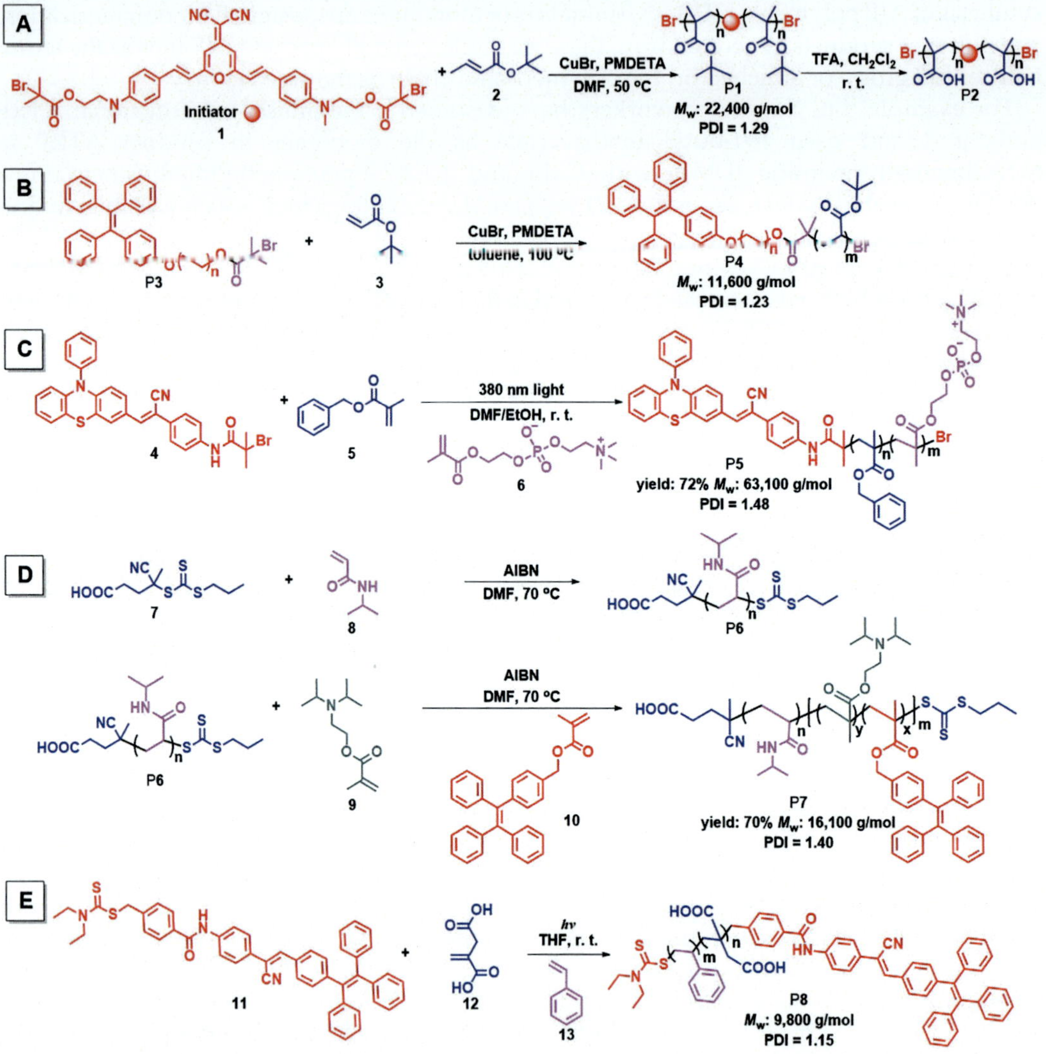

SCHEME 1 Synthesis of AIE polymers through (A, B) ATRP [35,36], (C) photo-mediated ATRP [37], (D) RAFT polymerization [38], and (E) photo-initiated RAFT polymerization [39].

potential postfunctionalization to form polyelectrolytes. Two isocyanide moieties from monomer **14** can undergo cyclodimerization under the catalysis of silver acetate to form an imidazole ring embedded in the polymer main chain of P**9** (Scheme 2A) [40]. Upon the excitation of 310 nm, the strongest fluorescence with the fluorescence quantum efficiency of 31.8% was observed in the DMF/H_2O mixed solution of P**9** with 50 vol% water.

Except for polycondensation, TPE-containing isocyanide monomers can also undergo cationic homo-polymerization to form AIE-active poly(isocyanide)s. For example, Li, Dong, and coworkers used silylium cationic initiator for the cationic homo-polymerization of TPE-

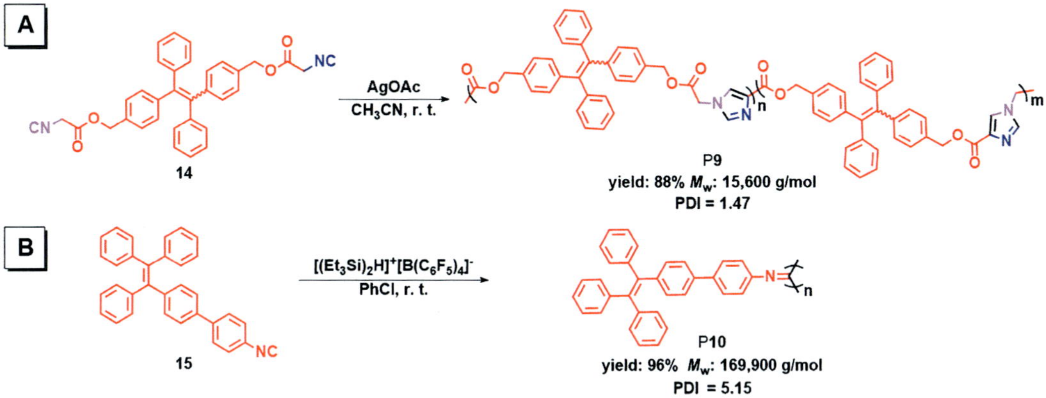

SCHEME 2 Synthesis of AIE polymers through (A) polycondensation of diisocyanide [40] and (B) cationic homopolymerization of isocyanide [41].

containing isocyanide monomer **15** and obtained AIE-active poly(isocyanide) P**10** with a high M_w of 169,900 g/mol (Scheme 2B) [41]. The polymerization enjoys fast reaction speed, high efficiency with a 96% conversion in 1 min.

2.2.2 Two-component polymerization

Click polymerization with mild reaction conditions, high efficiency, high atom economy, and high regioselectivity has attracted much attention recently and has been developed rapidly for the synthesis of AIE polymers, because of its potential of preparing functional polymers with heteroatoms and diversified structures such as linear and hyperbranched polymers [42–46]. One recent trend is to use activated alkyne monomers to realize efficient metal-free click polymerization. For example, a series of metal-free azide-alkyne [42], hydroxyl-yne [43], amino-yne [44], thiol-yne [45], and thiol-ene [46] click polymerizations incorporating ester group- or sulfonyl group-activated alkyne monomers have been developed for the synthesis of AIE polymers.

Tang, Zhang, Li, and coworkers have prepared an AIE-active polytriazole P**11** with 85.4% yield, an M_w of 10,500 g/mol, good regioregularities (up to 71.4%), good solubility, and thermal stability through a solvent- and catalyst-free butynoate-azide click polymerization at 100° C (Scheme 3A) [47]. The emission intensity of P**11** in THF/H_2O mixtures with 90 vol% water content at 485 nm was 220-fold higher compared with that of its pure THF solution. Later, Qin and coworkers have developed an efficient hydroxyl-yne click polymerization using ester active alkynes and TPE-containing aliphatic diols as monomers in the presence of bicyclo[2.2.2]-1,4-diazaoctane (DABCO) through the polyhydroalkoxylation of alkynes to afford poly(vinyl ether)s with high M_w of up to 71,000 g/mol in excellent yields of up to 99% in air in 1 h [48]. The phenol groups from the TPE-containing monomer **19** can also undergo this polymerization smoothly to afford AIE-active polymer P**12** (Scheme 3B), whose emission intensity in THF/H_2O mixtures with 90 vol% water content was 125-fold higher than that of its pure THF solution.

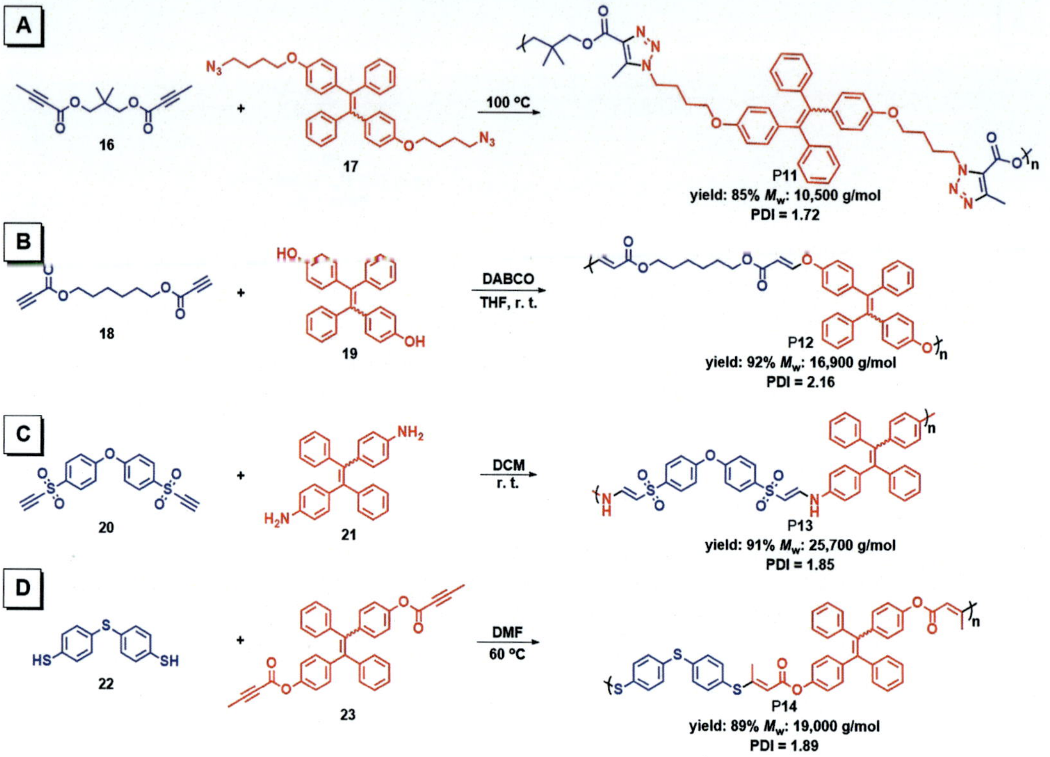

SCHEME 3 Synthesis of AIE polymers through (A) azide-yne [47], (B) hydroxyl-yne [48], (C) amino-yne [49], and (D) thiol-yne [50] click polymerizations.

Sulfonyl group-activated terminal alkynes are also very reactive monomers, which could undergo hydroamination of the alkyne group even with aromatic amines. Tang, Qin, and coworkers have developed an amino-yne click polymerization of sulfonyl-activated terminal alkynes and diamines, affording a regio- and stereoregular poly(β-aminovinylsulfone) P**13** with excellent stereoregularity (Scheme 3C) [49]. Because of the neighboring strong electron-withdrawing sulfonyl groups, the reactivity of the sulfonyl-activated alkynes is higher than that of the ester-activated alkynes, which could efficiently undergo polymerization with various aliphatic or aromatic primary/secondary amines. Besides amines and alcohols, thiols are also effective nucleophiles that could react with activated alkynes. Li's group developed a facile catalyst-free click polymerization of 4,4′-thiodibenzenethiol and ester group-activated internal alkynes based on the addition reaction of thiol and internal alkyne. Sulfur-containing functional poly(β-thioacrylate) P**14** with high *Z*-stereoregularities up to 81%, good solubility and thermal stability, high optical transparency, and high refractive indices were obtained at 60°C, despite that the polymerization involved relatively long reaction time of 24 h (Scheme 3D) [50]. Typical AIE property was observed for P**14**, whose PL intensity in THF/H_2O mixtures with 80 vol% water content was 39-fold higher compared with that of its THF solution.

Besides alkyne-based click polymerizations, there are many two-component polycondensations developed for the construction of linear and hyperbranched AIE polymers. As a representative example of transition metal-catalyzed polycouplings, Suzuki coupling polymerization of AIE-active monomer is recently used to synthesize an AIE-active conjugated polymer **P15** with excellent M_w of 205,800 g/mol under the catalysis of $Pd(PPh_3)_4$ and K_2CO_3 in toluene/water mixed solution (Scheme 4A) [51]. In 90 vol% aqueous solution of **P15**, the emission at 490 nm reached the maximum with the Φ_F value of 47%.

In addition, transition metal-catalyzed annulations involving C—H activation process have also been developed for the preparation of AIE polymers with unique fused cycle

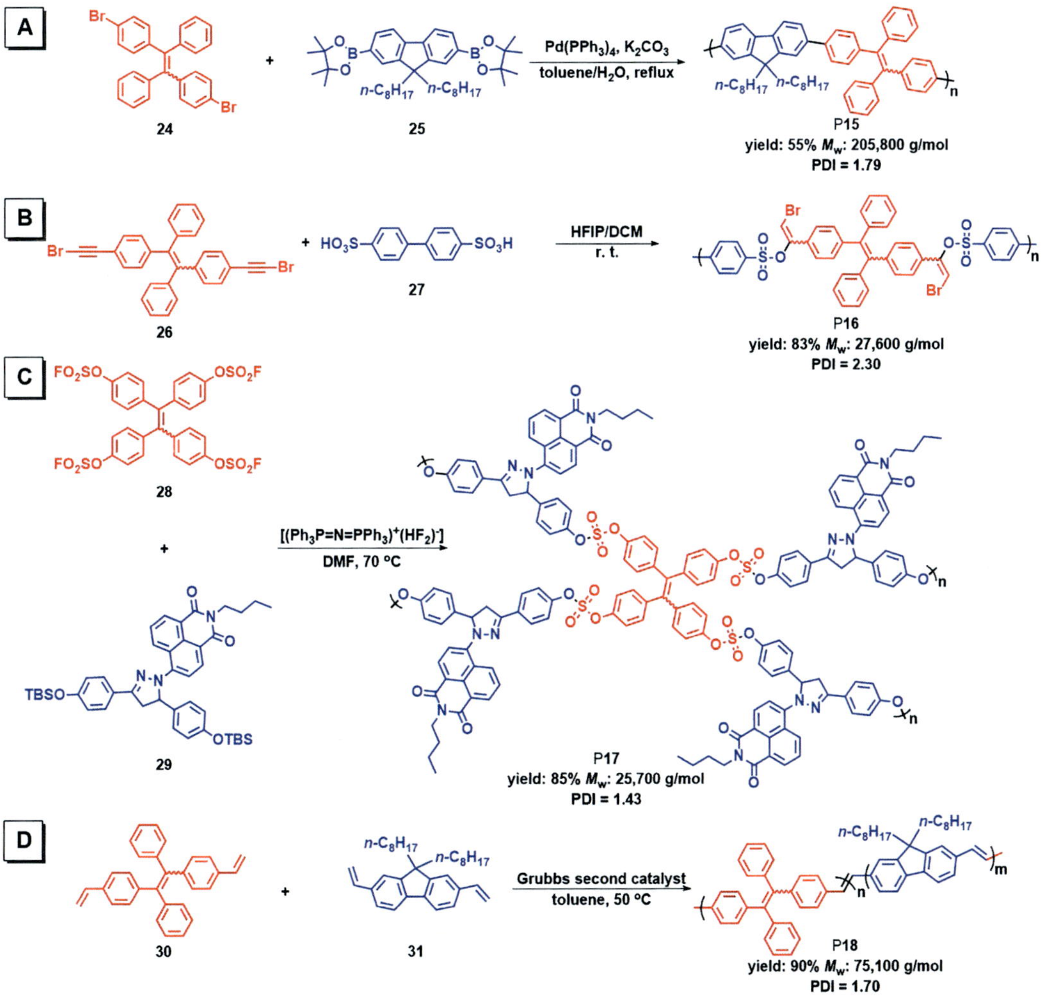

SCHEME 4 Synthesis of AIE polymers through (A) Suzuki coupling polymerization [51], (B) polyaddition of dihaloalkyne and disulfonic acid [52], (C) sulfur (VI) fluoride exchange click reaction [53], and (D) metathesis polymerization [54].

structures. For example, AIE-active polymers with multisubstituted fused ring were obtained by transition metal-catalyzed oxidative annulation of TPE-containing internal diynes with aryl iodide [55] or 1-metylpyrazole [56]. In these polymerizations, the fused rings embedded in the polymer backbone were generated in situ from the polymerization. Interestingly, charged fused ring could also be generated directly from one-pot polymerizations to furnish polyelectrolytes with active oxygen generation capabilities for biological applications [57]. In these transition metal-catalyzed polymerizations, monomer imbalance was found to promote polymerization. The polymeric products with unique multisubstituted polarized aromatic rings or heterocycles generally enjoy excellent thermal stability, high refractive indices, and low chromatic dispersion.

Moreover, a catalyst-free spontaneous polyaddition of dihaloalkyne and disulfonic acid was also developed by Tang, Lam, Han, and coworkers to proceed in hexafluoro-2-propanol (HFIP) and dichloromethane (DCM) mixed solution at room temperature in air to prepare an AIE-active halogen-rich polysulfate P**16** without generating any by-product (Scheme 4B) [52]. The polymer almost did not emit in its good solvent DCM, however, with the addition of its poor solvent, nonpolar *n*-hexane, P**16** emitted strong fluorescence at 487 nm in DCM/hexane mixtures with 80 vol% hexane content, demonstrating AIE characteristics. A hyperbranched AIE-active polysulfate P**17** with two types of AIEgens, TPE and naphthylamide, was also synthesized through sulfur (VI) fluoride exchange click reaction of fluorosulfate and silyl ethers in DMF at 70°C (Scheme 4C) [53]. The solid powder of P**17** emitted at 564 nm with the Φ_F value of 12.9%.

In addition to polycondensations, the acyclic diene metathesis polymerization was also used to construct AIE-active conjugated random copolymer P**18** from diyne monomers **30** and **31** in the presence of the second-generation Grubbs catalyst (Scheme 4D) [54]. P**18** possessed excellent M_w of 75,100 g/mol, good solubility, and exhibited multiple emission bands at 475, 508, and 540 nm with a Φ_F of 37% in THF/H_2O mixtures with 90 vol% water. Wu's group has also built a well-defined AIE-active diblock copolymer through room temperature living ring-opening metathesis polymerization of norbornene-based monomer using the third-generation Grubbs catalyst [58]. The polymer product can be obtained efficiently and rapidly with high molecular weight in a few minutes, and AIE polymers synthesized from these polymerizations usually possessed controlled molecular weight and narrow molecular weight distribution.

2.2.3 *Multicomponent polymerization*

Multicomponent polymerization (MCP), with three or more types of monomers polymerize in a one-pot manner to afford single-polymer product without isolation of intermediates, has been recently developed into a powerful tool to construct various functional polymers and has attracted much attention from polymer scientists owing to their unique advantages such as simple operation, high atom economy, high efficiency, and great structural diversity of products [59]. A large number of MCPs have been developed based on multicomponent reactions such as Passerini reaction [60], Ugi reaction [60], Hantzsch reaction [61], Biginelli reaction [62], A^3-coupling reaction [63], Debus-Radziszewski reaction [64], and so on. A series of AIE polymers with unique structures or even in situ built new AIEgens are synthesized from MCPs [65].

For example, Tang, Hu, and coworkers have developed a series of Cu(I)-catalyzed alkynes and sulfonyl azide-based MCPs, including (1) the MCP of alkynes, sulfonyl azides, and iminophosphorane **34** to afford poly(phosphorus amidine) P**19** (Scheme 5A), (2) the MCP

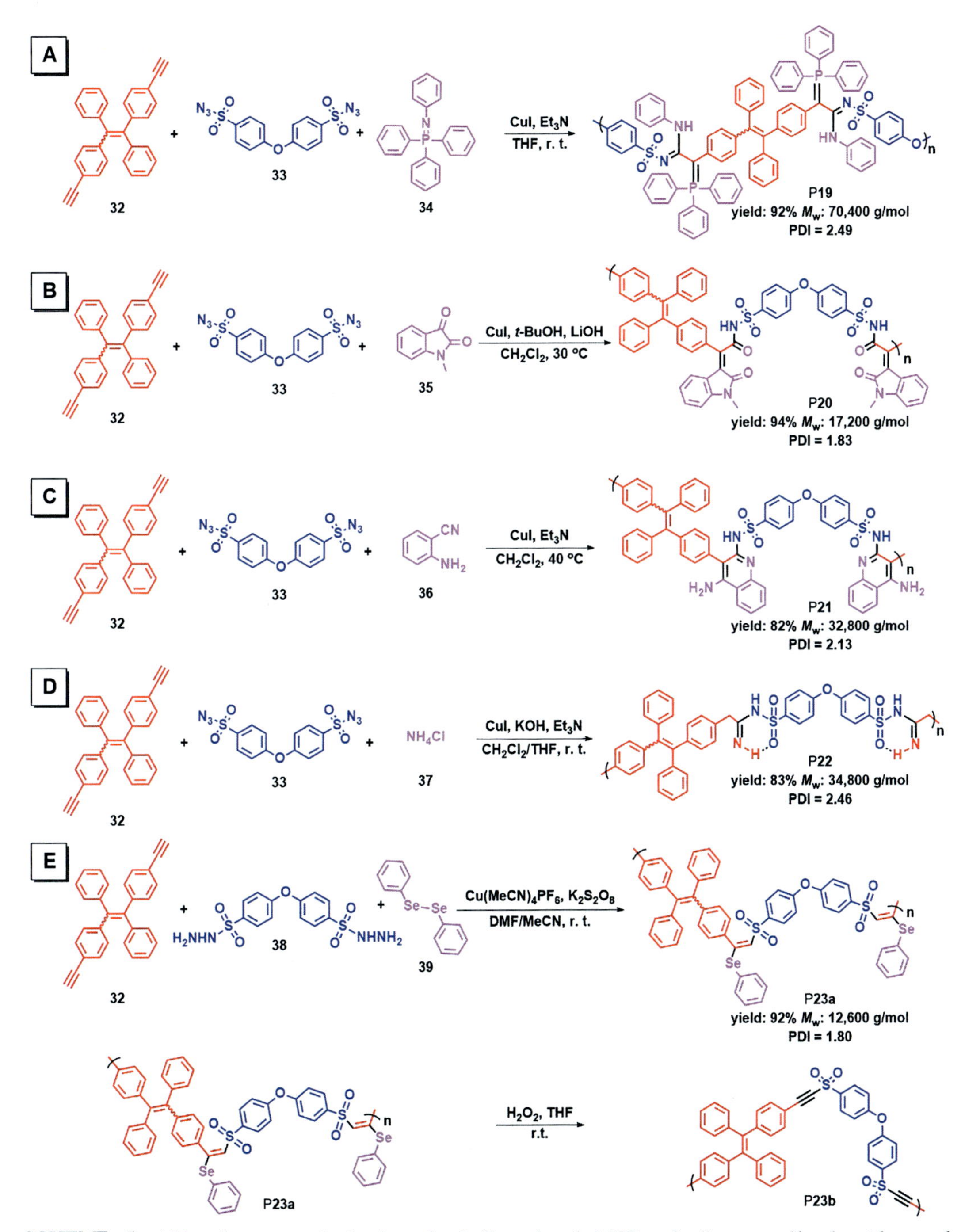

SCHEME 5 AIE polymer synthesis through Cu(I)-catalyzed MCPs of alkynes, sulfonyl azides, and (A) iminophosphorane [66], (B) *N*-protected isatin [67], (C) 2-aminobenzonitrile [68], (D) NH_4Cl [69], and (E) the MCP of alkyne, arylsulfonohydrazide, and diphenyl diselenide [70].

of alkynes, sulfonyl azides, and *N*-protected isatin **35** to afford poly(*N*-acylsulfonamide) P**20** (Scheme 5B), (3) the MCP of alkynes, sulfonyl azides, and 2-aminobenzonitrile **36** to afford poly(*N*-sulfonylimine) P**21** (Scheme 5C), and (4) the MCP of alkynes, sulfonyl azides, and NH_4Cl to afford poly(sulfonyl amidine) P**22** (Scheme 5D). These Cu(I)-catalyzed MCPs are generally highly efficient, possessed high atom economy, can be conducted at mild condition such as room temperature, and generally releasing N_2 gas as the only by-product or releasing N_2 and HCl in the case of NH_4Cl-based MCP [66–69]. The polymer products generally possessed rich heteroatoms or heterocycles, which showed unique AIE characteristics. Moreover, a copper catalyzed multicomponent polymerization of diyne, 4,4′-oxybis(benzenesulfonyl hydrazide), and diphenyl diselenide was developed based on the three-component selenosulfonation of alkynes, arylsulfonohydrazides, and diphenyl diselenide (Scheme 5E) [70], delivering chalcogen-rich functional polymer P**23a** with high regio- and stereoselectivity, good photostability, outstanding refractive indices, good film-forming ability, and unique redox properties. After postmodification with oxidant such as H_2O_2, the Se—C bond could break to form C≡C bond in the polymer backbone, and the resultant polymer P**23b** possessed a Φ_F value of 13.4% in the solid state, showing AIE property.

Metal-free multicomponent polymerizations are also developed for the synthesis of AIE polymers to avoid the influence on photophysical property from transition metal catalyst residue in the polymer. For instance, a metal-free A^3-polymerization of diynes, dialdehydes, and ureas was developed using trifluoroacetic acid and acetic acid as catalysts in acetonitrile at 90° C (Scheme 6A) [71]. The dehydration condensation of acetylene, aldehyde, and urea took place forming polyheterocycles, and hence AIE-active *N*-heterocycle polymer P**24** with emission at 528 nm with the Φ_F value of 5.6% in the solid state and fluorescence responses toward protonation and deprotonation was obtained. The three-component polycoupling of diynes, monoaldehydes, and BCl_3 was developed in dichloromethane, and BCl_3 served not only as a monomer but also as a mediator. Stereoregular chlorine-containing polymer P**25** with excellent M_w of 106,300 g/mol and the ratio of *E*-isomeric units of the newly formed vinyl group to be 95% (Scheme 6B). The fluorescence intensity of P**25** in THF/H_2O mixed solution with an f_w of 90 vol% was 37 times higher than that in pure THF, suggesting its AIE property [72].

Sulfur-containing polymers have attracted extensive attention recently owing to their good metal coordination ability, high refractive index, self-healing ability, and redox properties. From the perspective of green chemistry, it is of great significance to transform simple monomer from nature into functional polymers, and green monomers such as sulfur [73], water [77], carbon dioxide [74,78], and oxygen [79] have been adopted in multicomponent polymerizations. For instance, a catalyst-free MCP of elemental sulfur, aliphatic diamine, and diisocyanate was developed at room temperature to directly convert elemental sulfur into functional polythioureas with various structures. This MCP enjoyed unique advantages such as rapid reaction, high efficiency, mild conditions, high atom economy, and wide monomer scope including primary and secondary amines. AIE-active polythiourea P**26** was synthesized in toluene/DMF (v/v, 1/2) (Scheme 6C) [73], and its Φ_F value in DMF/water mixtures with 1:1 volume ratio was 13.3% with emission maximum at 493 nm. CO_2 could also be used to construct AIE polymers as monomer. For example, an MCP of dialkyne, 1,4-dibromobutane, and CO_2 was developed using Ag_2WO_4 as catalyst and Cs_2CO_3 as inorganic base in *N*,*N*-dimethylacetamide (Scheme 6D). WO_4^{2-} was used to activate CO_2 by forming $[WO_4^{2-}]/CO_2$ adduct, and Ag^+ was used to restrain the cyclization reaction between CO_2

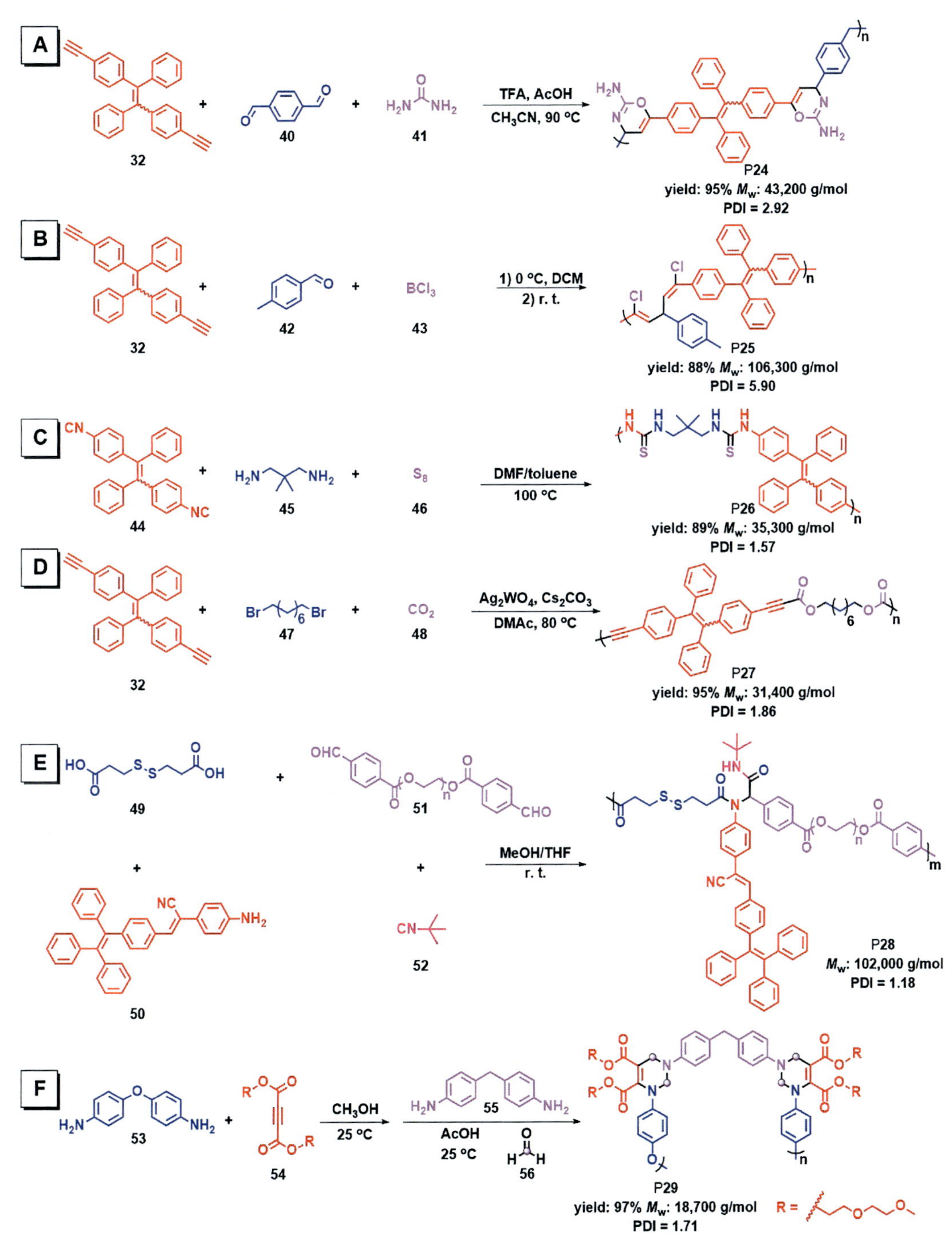

SCHEME 6 AIE polymer synthesis through (A) the MCP of alkyne, aldehyde, and urea [71], (B) the MCP of alkyne, aldehyde, and BCl_3 [72], (C) the MCP of isocyanide, amine, and elemental sulfur [73], (D) the MCP of alkyne, bromide, and CO_2 [74], (E) Ugi polymerization [75], and (F) metal-free four-component tandem polymerization [76].

and alkyne to ensure the formation of liner structures. Significantly, the absolute Φ_F of the polymer film of P**27** prepared from this MCP is as high as 61% with emission maximum at 499nm. It is worth mentioning that with the activated alkyne bonds in the polymer main chain, postmodification could be facilely realized [74].

Four-component polymerizations such as Ugi polymerization of aldehydes, amines, carboxylic acids, and isocyanides are used to synthesize AIE-active poly(α-acylamino amide) s. Wei, Zhang, Liu, and coworkers have synthesized AIE-active fluorescent polymeric nanoparticle P**28** with excellent M_w of 102,000g/mol via Ugi four-component polymerization of 3,3′-dithiodipropionic acid, dialdehyde-terminated polyethylene glycol, amino-containing AIEgen, and *tert*-butyl isocyanide in THF-methanol (v/v=1:1) at room temperature without catalysts (Scheme 6E) [75]. The fluorescent polymeric nanoparticles emitted at 536nm with a large Stokes shift of 142nm, and P**28** possessed reduction-responsive property, biocompatibility, low cytotoxicity, and excellent dispersibility.

Multicomponent tandem polymerization (MCTP), where the intermediate generated from the first step is directly converted to final product with the freshly added other monomer without further isolation and purification, has been developed to explore the general applicability of MCPs [80–82]. One unique advantage of MCTP is that the polymerization procedure could be designed to fine-tune the polymer structure as well as their sequence of functional units. For example, Tang, Hu, and coworkers have reported a room temperature metal-free four-component tandem polymerization of ester group-activated internal alkyne, two kinds of aromatic diamine, and formaldehyde (Scheme 6F) [76]. The two aromatic diamine monomers reacted with activated alkyne and formaldehyde to form enamine and Schiff base intermediates, respectively, which were then reacted together with excessive amount of formaldehyde to furnish the construction of tetrahydropyrimidines. Each of the diamine monomer was added at different stages of the polymerization, hence affording sequence-controlled poly(tetrahydropyrimidine) P**29** from a single polymerization. Although there was no conventional fluorophore existed in the polymer structure, P**29** possessed obvious emission in THF/*n*-hexane mixed solutions and its thin film emitted at 504nm with a Φ_F value of 4%, probably because of the formation of heteroatom clusters.

3 Structures of AIE polymers

With the rapid development of synthetic methodologies, AIE polymers with considerable structural diversity and multifunctionalities have been reported. Linear, dendritic, hyperbranched, star-shaped, and cross-linked AIE polymers have been developed with AIEgens installed in the polymer main chain, attached as polymer side chain, as terminal groups or central groups [83–88]. In this section, AIE polymers with unique structures such as polyelectrolytes, chiral polymers, hyperbranched polymers, porous polymers, and nonconventional fluorescent polymers are introduced.

3.1 Polyelectrolytes

Polyelectrolytes are charged polymers with positive or negative charges which are usually installed on the polymer main chain or side chains, and their solution properties are governed by electrostatic interactions. Polyelectrolytes with hydrophilic ionic side groups generally possess excellent water solubility, biocompatibility, low cytotoxicity, and electrostatic interaction. Hence, in AIE-active polyelectrolytes these unique features have been combined with luminescence to afford water-soluble luminescent materials with potential applications in bio-probes and bio-imaging [57,89–91].

Conjugated polyelectrolytes (CPEs) constituted with a π-conjugated backbone and hydrophilic ionic pendant groups are an important group of polyelectrolytes. Iyer's group synthesized a water-soluble and amphiphilic conjugated cationic polyelectrolyte P**30** containing TPE, fluorene, and benzothiadiazole units in the main chain (Chart 1A) [92]. Positive charges on the side chain were introduced through the postfunctionalization with 1-methyl imidazole to afford P**30** with Br^- as the counterion. Under UV irradiation, the solid state of P**30** emitted at 565 nm possesses a large Stokes shift of 105 nm. A straightforward, rapid, cost-effective fluorescence method for latent fingerprint detection on all sorts of surfaces such as glass, aluminum foil, steel, and adhesive tape, disregarding physical abrasion and chemical abrasion was established based on P**30** with high selectivity, high contrast, high resolution, no background interference, and no posttreatment. This detection was achieved through the hydrophobic interaction between the fatty components from the latent fingerprint and the polymer conjugated skeleton, and the electrostatic interaction between the sweat components from the fingerprint and the charged functional groups. Similarly, Zhao, Smith, Zhou, and coworkers have synthesized a TPE- and benzothiadiazole-containing conjugated polyelectrolyte P**31** with cationic quaternized ammonium groups on the side chains as a bio-probe for HeLa cells [93]. The polymer has obvious red emission at 606 nm in DMSO/THF mixtures with 90 vol% THF, and the fluorescence intensity is 32 times higher than that in pure DMSO. The Φ_F value of the polymer in the 90 vol% DMSO/THF mixtures reaches 26.7%.

Polyelectrolytes could also be designed as living cellular fluorescent probe with unique recognition and high sensitivity. For example, Bai, Zhao, and coworkers have reported a water-soluble polyelectrolyte P**32** with the central TPE unit serving as the fluorophore, and poly(acryloyl ethylene diamine) segments with ammonium salts on the side chains [94]. In H_2O/DMSO mixed solution of P**32** with 99 vol% DMSO, the fluorescence quantum efficiency reaches 38.38%; in dilute phosphate buffer solution with the polymer concentration lower than 20 μg/mL, no emission could be observed; however, when the concentration gradually increased, TPE units aggregated and the polymer started to emit fluorescence. The water-soluble fluorescent probe P**32** with low cytotoxicity and good signal-to-noise ratio could hence be applied in real-time bio-imaging for living HeLa cells.

AIEgen could also be installed as the terminal group of the polyelectrolyte to form AIE polyelectrolytes with excellent fluorescence property in the aggregated states. Tang, Wang, and coworkers have reported an AIE-active polyelectrolyte P**33** with rich carboxylate groups on the side chain and TPE unit at the end of polymer chain [95]. P**33** with adjustable M_n was synthesized through the controllable copolymerization of CO_2 and 1,2-epoxy-4-vinylcyclohexane in the presence of metalloporphyrin complex catalyst and using AIEgen-containing initiator. The hydrophobic polycarbonates were then quantitatively converted to water-soluble polyelectrolytes by the introduction of carboxyl groups in the side chains via thiol-ene click reaction, followed by deprotonation of carboxyl groups to afford P**33**. The carboxyl groups enabled the polyelectrolyte facile detection of Zn^{2+}.

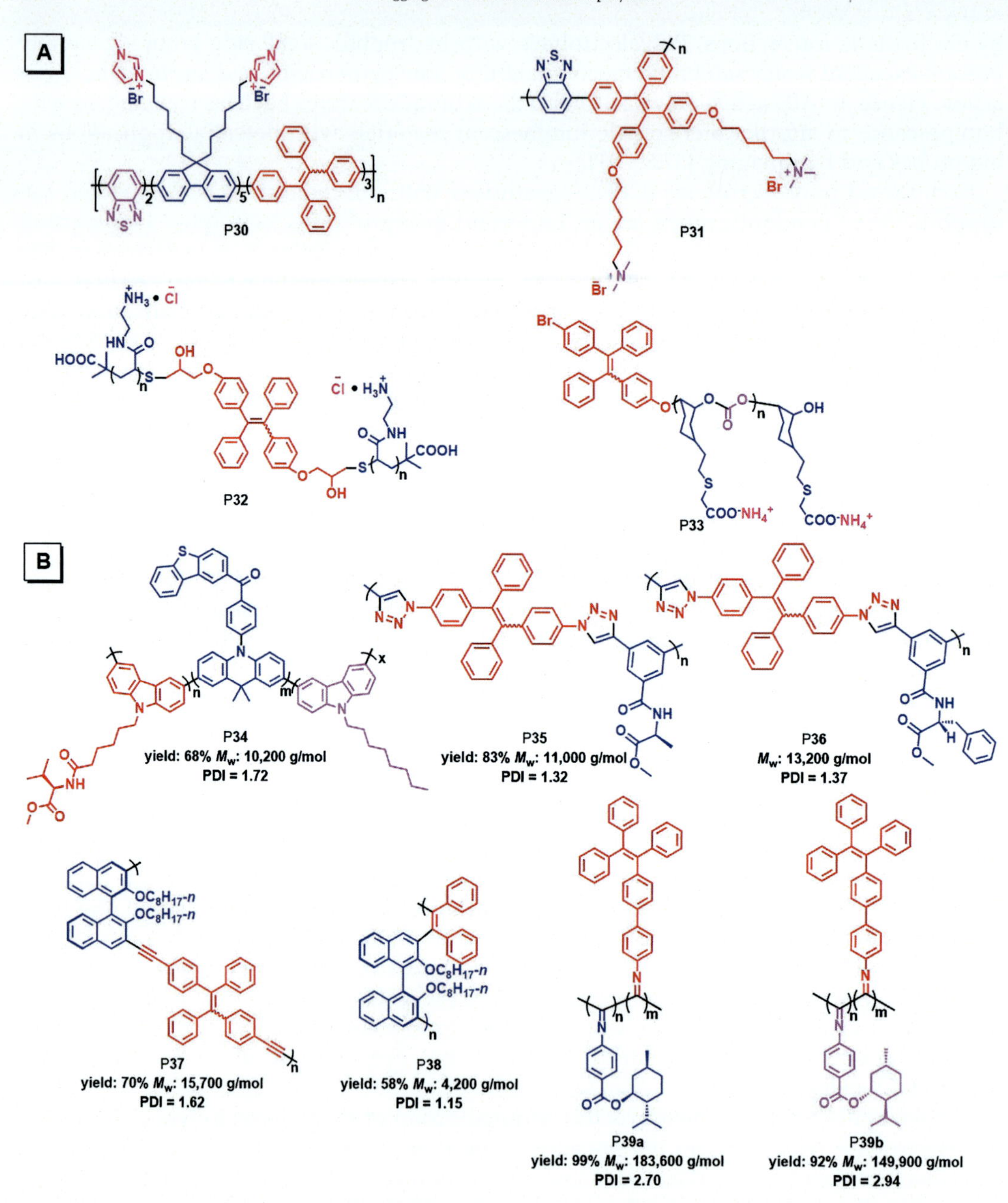

CHART 1 Chemical structures of (A) AIE polyelectrolytes [92–95] and (B) chiral AIE polymers [96–101].

3.2 Chiral polymers

AIE polymers with inherent chirality have been of great interest recently, because they could normally form helical structures and possess circularly polarized luminescence (CPL) property. Hence they may find potential applications in optoelectronic devices, chiral recognition, and self-assembly [102–104].

As a typical self-assembly directing motif, chiral amino acid moieties could induce the polymer chains to form helical conformation through the chirality transfer from pendant to polymer backbone. For example, Tang, Ma, Lam, and coworkers have designed and prepared a conjugated AIE-active polymer P**34** with electron-donating carbazole and acridine moieties in the polymer backbone, electron-withdrawing and aggregation-induced delayed fluorescence (AIDF)-active dibenzothiophen-2-yl(phenyl)methanone attached on the acridine units and chiral alanine pendants (Chart 1B) [96]. No obvious circular dichroism (CD) signal was observed from either the THF solution or 80vol% H_2O/THF mixture of P**34**, but strong Cotton effects at ~258 and 310nm were observed from the thin film of P**34**. Since the alanine pendants are CD-inactive beyond 300nm, the wide CD bands in the film states of P**34** at 350–450nm were attributed to the chirality of the conjugated polymer backbone, indicating that the chirality was successfully t*ran*sferred from alanine pendants to poly(carbazole-*ran*-acridine) backbone. Moreover, green emission at 552nm in 90vol% THF/H_2O solution was observed for P**34**, and its thin film exhibited enhanced emission at 542nm with a Φ_F value of 10.3% and a significantly prolonged fluorescence lifetime of 1.366μs.

Tang, Li, and coworkers have also constructed a chiral AIE-active polytriazole P**35** with the backbone of TPE and chiral amino ester pendant [97]. The intramolecular and intermolecular hydrogen bonds of amino acids not only drive the self-assembly of the polymer but also stabilize the helical conformation of the TPE-containing backbone. The polymer can self-assemble in THF/H_2O mixtures to form diverse structures such as spherical, pearl necklace, branched rods, helical nanofibers, and helical nanowires by tuning the water content and concentration. The Φ_F value of P**35** was increased from <1% to 16.8% from THF solution to the solid film state. Another chiral AIE-active polytriazole P**36** with similar structure and phenylalanine pendants was reported [98]. The deprotected carboxylic acid form of chiral polytriazole P**36** possessed higher PL intensity and fluorescence quantum efficiency. Moreover, *(1S,2S)*-cyclohexane-1,2-diamine could form host-guest polymer complexes through hydrogen bond formation between the diamine and phenylalanine carboxyl groups to promote the chirality transfer and hence increase the CPL intensity.

AIE-active conjugated polymers with main-chain chirality have also attracted wide attention as a promising CPL material, and the optically active 1,1′-binaphthyl building block has been widely adopted in the construction of chiral polymers with main-chain chirality. For example, Zhu, Cheng, and coworkers have designed and prepared a (*R*)-binaphthyl-containing AIE polymer P**37** [99]. The linking position with triple bond is crucial for the formation of the helical conformation of P**37**, and only when the triple bonds were linked through 3,3′-positions of the binaphthyl moiety, obvious CD signal and CPL properties could be observed owing to the formation of the helical conformation by self-assembly. With the increase of water content in the THF/H_2O mixed solution of P**37**, the Cotton peak redshifted, and the CPL signals were gradually enhanced, suggesting that the polymer may form certain chiral nanostructures upon aggregation. Cheng, Quan, and coworkers have also synthesized a similar binaphthol and AIEgen-containing polymer P**38** by Pd-catalyzed Suzuki coupling polymerization [100]. The left-handed helical polymer showed strong emission at 493 nm in the film state with the fluorescence quantum efficiency of 14.8%. A weak Cotton peak was observed at about 360 nm, suggesting that the chirality of binaphthalyl moiety has brought helical conformation of the polymer backbone. Most importantly, the doping-free device with P**38** as the emitting layer exhibited stable circularly polarized electroluminescence with the maximum brightness of 1669 cd m^{-2} and electroluminescence dissymmetry factor of 0.024.

Moreover, polyisocyanides with chiral pendants could also form helical structures and exhibit CD signals and AIE properties. For example, Li, Dong, and coworkers have prepared a pair of AIE-active single-handed helical random copolymers P**39a/b** via metal-free cationic copolymerization of (*1S,2R,5S*)/(*1R,2S,5R*)-2-isopropyl-5-methylcyclohexyl 4-isocyanobenzoate and TPE-containing achiral aryl isocyanide [101].

3.3 Hyperbranched polymers

Hyperbranched polymers with three-dimensional topology structures, a large number of branch points, abundant surface functional groups, excellent solubility, and interior cavities have been extensively studied. Through the introduction of AIE properties into hyperbranched polymers, potential applications such as fluorescence sensing, bio-imaging, and in situ monitoring of drug delivery can be found [105,106].

An AIE-active hyperbranched polytriazole P**40** was synthesized by metal-free phenylpropiolate-azide polycycloaddition [107], and the spatial effect of the hyperbranched structure results in a higher regioregularities of the polymer compared to its linear analogues (Chart 2). The fluorescence intensity of P**40** in THF/H_2O (1:9 v/v) at 484 nm was 115-fold higher than that in THF, suggesting its AIE characteristics, which was affected by the length of the flexible alkyl chains. The longer the alkyl chain spacer in the polymer, the larger the fluorescence enhancement that can be achieved. In another example, Tang, Qin, and coworkers have synthesized a hyperbranched polymer P**41** with AIE characteristics via spontaneous amino-yne click polymerization in an anti-Markovnikov addition manner [108]. The polymer showed no fluorescence in THF solution and could gradually emit blue-green light with the addition of poor solvent water, and reached a Φ_F value of 9.8% in 90 vol% aqueous solution. Owing to the existence of a large number of cavities in the three-dimensional network of the hyperbranched polymer, the detection sensitivity for explosive can be improved effectively. Furthermore, the hyperbranched polymer could be degraded upon the addition of trifluoroacetic acid, through the breakage from β-aminoacrylate moieties.

AIE-active hyperbranched polymers could also be constructed by multicomponent tandem polymerization to regulate the sequence structure of the functional units on the backbone. For example, Tang, Hu, and coworkers have designed and synthesized an AIE-active hyperbranched poly(tetrahydropyrimidine) P**42** via multicomponent tandem polymerization of a A_4-type aromatic amine, a A_2-type aromatic amine, and a diester group-activated internal alkyne [109]. The hydrophilic diethylene glycol monomethyl ether chains

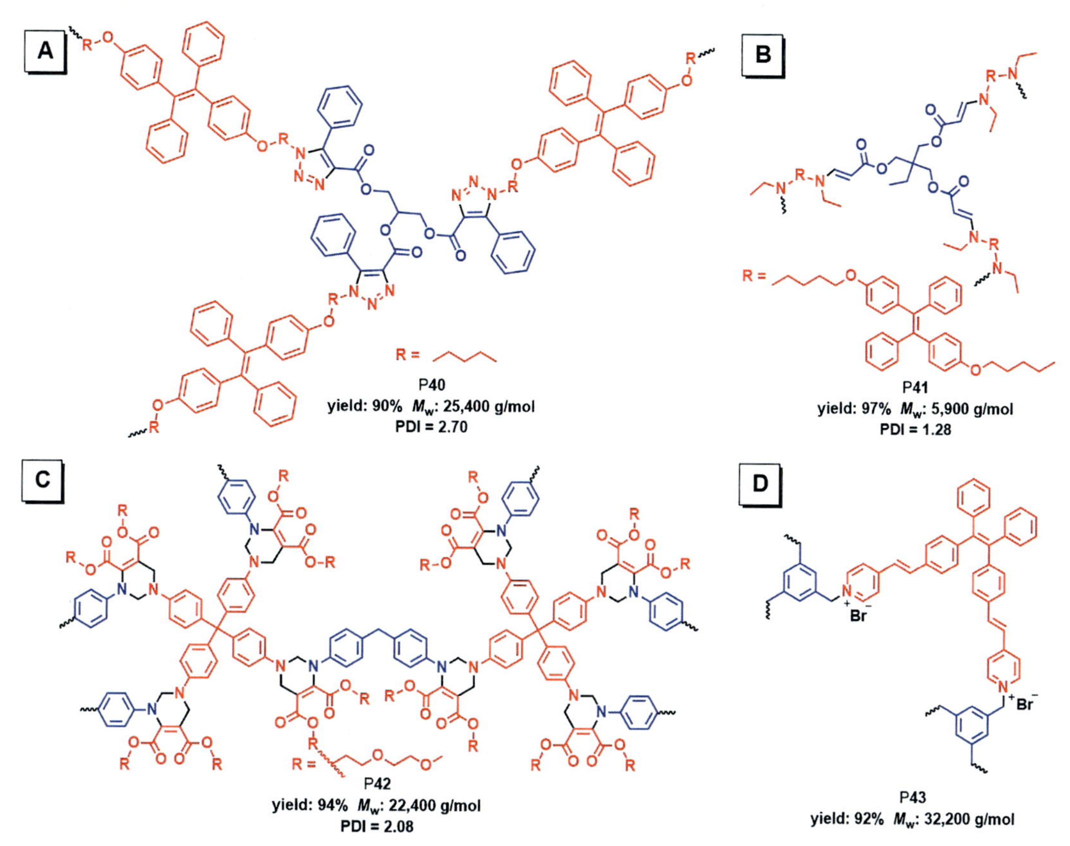

CHART 2 Chemical structure of hyperbranched AIE polymers [107–110].

ensure the polymer have good solubility in organic solvents and hence high molecular weight. Although there is no large π-conjugated fluorophore in the hyperbranched polymer, P**42** emits faintly in good solvent ($\Phi_F = 0.4\%$) and shows enhanced emission at 517 nm ($\Phi_F = 3\%$) in the thin film state. The AIE property could be attributed to the large number of carbonyl groups and heteroatoms which form "heteroatomic clusters" in the aggregated state.

AIE-active hyperbranched polyelectrolyte with good water solubility could detect substances with opposite charges by electrostatic complexation. For example, Tang, Sun, and coworkers have designed and synthesized a cationic hyperbranched polymer P**43** as a fluorescent probe for bio-imaging from pyridinium-modified TPE derivative via an "A_2 + B_3" type polymerization [110]. In DMF solution, P**43** emitted at 595 nm with a Φ_F value of 6.5%; in the solid state, the emission peak emerged at 609 nm with a pronounced shoulder peak at 630 nm owing to the existence of the *E*-isomers of TPE derivative. Moreover, a large Stokes shift of 202 nm was observed for P**43** in DMF solution, which is one of the largest values in the reported AIE polymers. The cationic pyridinium moieties could increase the viscosity of the solution, which is beneficial to the restriction of intramolecular motions. Ma's group has proven that the existence of a large number of cations in the positively charged hyperbranched polymer with tertiary ionized cores possessed good cell-membrane permeability, which may find broad application in biosensor and bio-imaging fields [111].

3.4 Porous polymers

Porous polymers with high surface area, inherent porosity, low density, adjustable chemical structure, and good stability have found potential applications in gas absorption and separation, heterogeneous catalysis, energy storage materials, photoelectric material, fluorescent sensing, and drug delivery [112,113]. Herein, AIE polymers with porous structures which combine luminescence and inherent porosity will be introduced.

For example, Zhang, Xu, and coworkers have designed and synthesized a TPE- and carbazole-containing photo-functional porous polymer film P**44** by electropolymerization of the 3,6-positions of the carbazole units (Chart 3) [114]. The electrochemical methods allow precise control of the shape, size, and thickness of polymer films with low cost and convenience without harsh reaction conditions. The fluorescence quantum efficiency Φ_F of the polymer film (25%) is higher than that of the corresponding monomer in the solid state (15%). During preparation, the boron trifluoride diethyl etherate can improve the electrochemical properties of the produced polymer films, and when prepared from the CH_2Cl_2-trifluoride diethyl etherate (v/v = 6:4) solution, the film emitted at 472 nm and the fluorescence lifetime of the polymer film was 3.9 ns.

Scherf's group synthesized a series of microporous polymer networks containing TPE core, electroactive carbazolyl or thienyl substituents through chemical oxidative coupling for microporous bulk polymer networks or electrochemical oxidative polymerization for microporous polymer film [115]. Bulk polymer P**45** with high surface area up to 2203 $m^2 g^{-1}$ was prepared using octafunctional monomers and additional 1,4-phenylene linkers. Carbazole-based microporous polymer networks showed higher surface area, Φ_F, and PL intensity than that prepared from thiophene building block. The polymer film possesses intense fluorescence at 520 nm with a high Φ_F value of 73%. Moreover, the polymer film produced by electrochemical coupling can be used for PL quenching-based sensing of nitroaromatic explosives.

Jiang's group synthesized an AIE-active porous organic polymer film P**46**, and the thickness of the film can be adjusted by varying the number of cyclic voltammetry [116]. The porous polymer film emitted green luminescence at 524 nm with a Φ_F value of 40%, and it also possessed high surface area up to 1020 $m^2 g^{-1}$, narrow pore size up to 1.5 nm, and high pore volume up to 0.61 $cm^3 g^{-1}$. Similarly, the porous polymer film is capable of detecting 2,4,6-trinitrophenol (TNP) in a highly sensitive and selective manner with a low detection limit of 10 ppm. Son, Ko, and coworkers have also synthesized a microporous organic polymer with water compatibility and hollow morphologies combining Sonogashira coupling polymerization and Suzuki coupling polymerization [118]. The polymer is highly dispersible and emissive in water with a Φ_F value of 78%, which shows excellent performance for sensing TNP in water with a low detection limit of 0.15 ppm.

Covalent organic frameworks are porous crystalline polymers with periodic two-dimensional or three-dimensional structures constructed from pure organic building blocks,

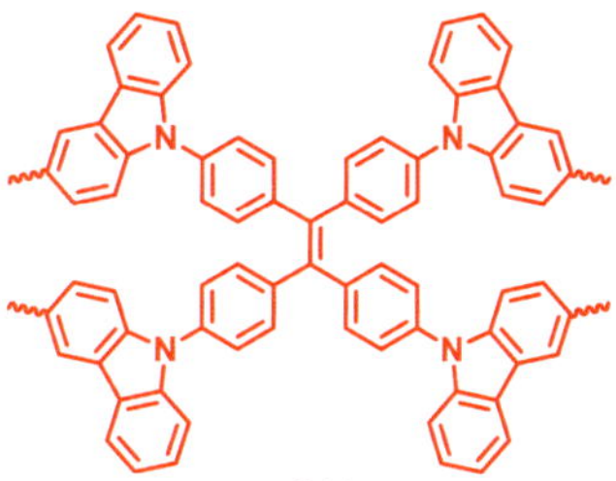

P44

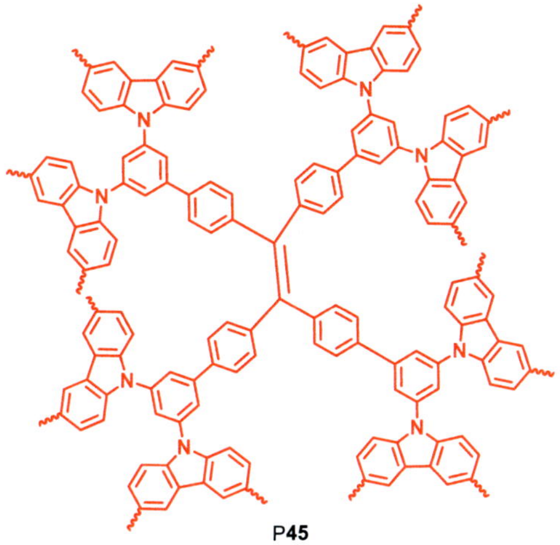

P45

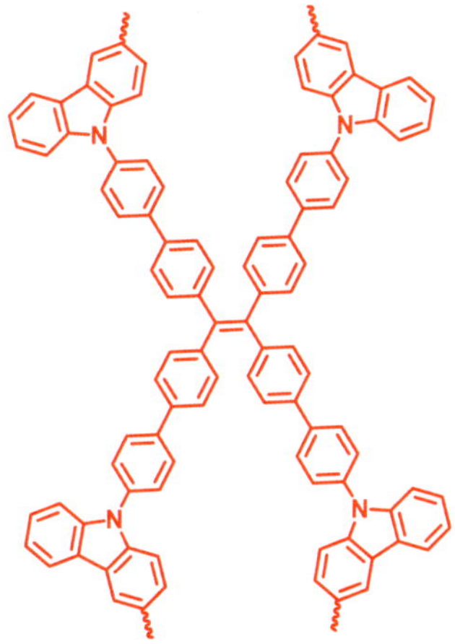

P46

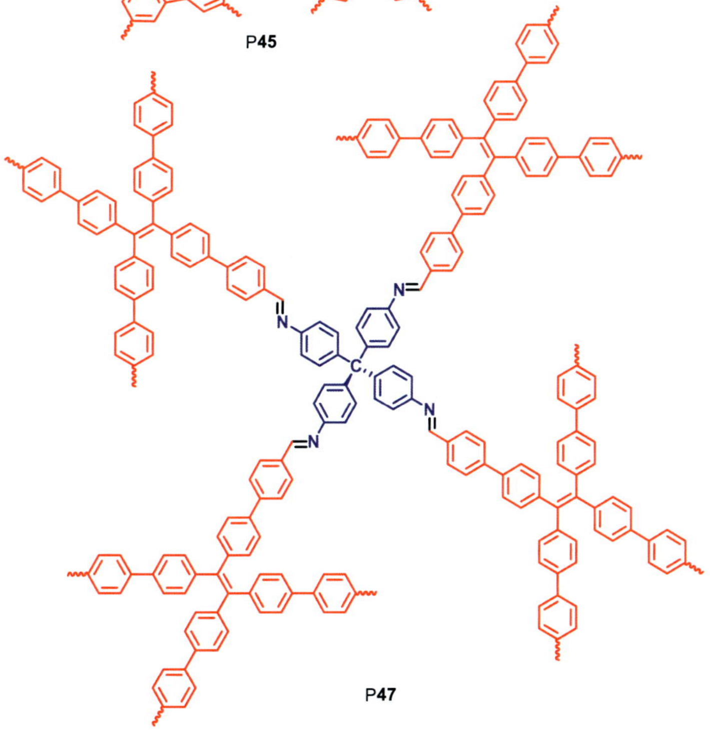

P47

CHART 3 Chemical structure of porous AIE polymers [114–117].

which usually enjoy high surface area, low density, and a large number of open sites [119]. Wang, Sun, and coworkers have designed and synthesized an AIEgen-based three-dimensional COF P**47** with a sevenfold interpenetrated pts topology (the tetrahedral and rectangle building blocks were conjugated) via [4+4] imine condensation reactions [117]. The TPE-based 3D COF possessed a high surface area of $1084\,m^2\,g^{-1}$ and total pore volume of $0.55\,cm^3\,g^{-1}$, which emitted yellow light with a λ_{em} at 543 nm and a Φ_F value of 20%.

Furthermore, when its film was coated on blue LED lamp, white light emission can be observed. Good stability was observed for the film with no degradation even after aging for 1200 h, demonstrating great potential in WLEDs.

3.5 Nonconventional fluorescent polymers

Besides conventional AIE polymers containing AIEgens as fluorophores in the polymer skeleton, a series of AIE polymers without conventional chromophore or conjugated structure have been developed based on clusteration-triggered emission [120,121]. These polymers are usually rich in heteroatoms, hydrogen bonds, or other complex structures that provide the potential to form heteroatom clusters. These intrachain and interchain interactions have facilitated the formation of clusters and conformation rigidification, which endows the polymer luminescence even without large conjugated structures.

For example, a nonconventional fluorescent polymer **P48** with C=C bonds and ester groups in the main chain, together with amide groups and methyl imidazole groups in the side chain was synthesized through multicomponent polymerization (Chart 4) [122]. Although without traditional AIEgens, **P48** emitted in both solid and solution states, and the PL intensity increased significantly in the THF/H_2O mixtures with an increase in water content from 60 to 90 vol%, demonstrating AIE property. This unique luminescence behavior was explained by the electron delocalization caused by interchain interaction between carbonyl groups and imidazole groups, which promotes the formation of clusters in the aggregated state.

wTang, Lam, and coworkers have also developed a multicomponent polymerization for the synthesis of a multisubstituted heteroatom-rich azetidine-containing nonconventional

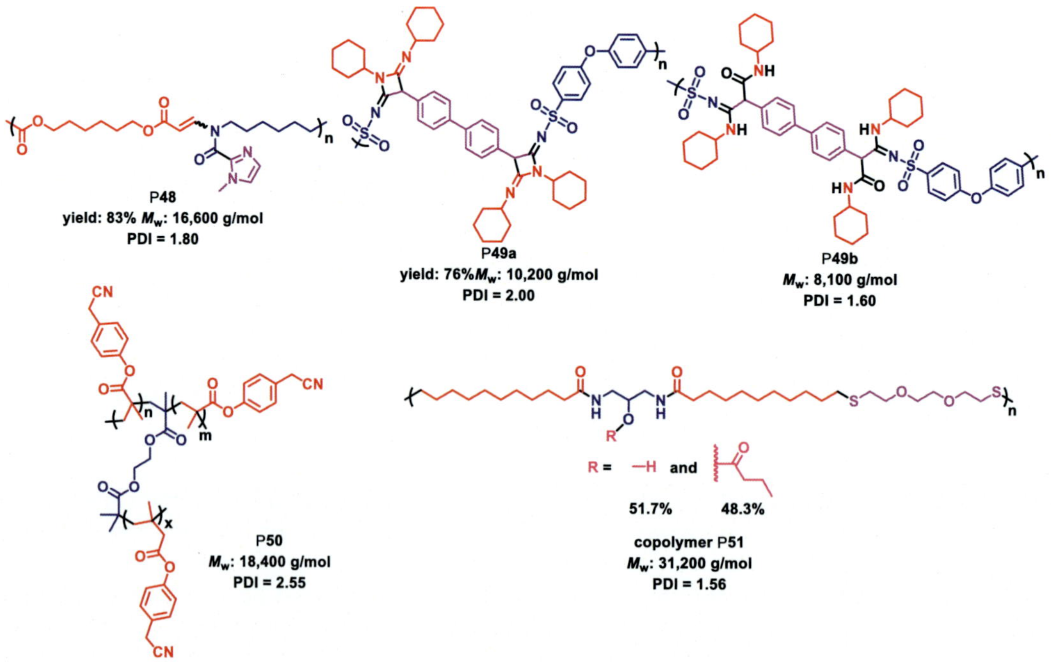

CHART 4 Chemical structure of nonconventional fluorescent polymers [122–125].

fluorescent polymer P**49a** without traditional conjugated structures and AIEgens [123]. The polymer was almost nonemissive in THF solution, but emitted greenish yellow light at 521 nm in the solid powder state, demonstrating AIE property. The strong lone-pair electronic interaction promoted the polymer to form through-space conjugated structure in the solid state, which may be destroyed in solution by active intramolecular motions. Interestingly, in the presence of HCl, the C—N bonds of azetidine in polymer backbone were broken to form amides and amidines to form the ring-opening polymer P**49b**. P**49b** showed brighter solid-state emission than P**49a**, which was attributed to rich intrachain and interchain hydrogen bonds. P**49b** possessed good cell staining ability, which could specifically locate and stain lysosomes in living HeLa cells.

Hyperbranched nontraditional fluorescent polymers were also developed. For example, Zhang's group designed and synthesized a hyperbranched polymer P**50** containing rich nitrile groups but without large-conjugated structure or chromophore [124]. The polymer solution emitted weak fluorescence, while the addition of poor solvent into the solution resulted in the formation of aggregated states and hence strong fluorescence, suggesting obvious AIE characteristics. The hyperbranched polymer possessed stronger AIE effect compared with its corresponding linear polymers, and the fluorescence intensity was higher when the polymer possessed higher degree of branching, exhibiting hyperbranching-enhanced-emission effect. Its AIE behavior is attributed to the collapse of the polymer chain caused by water, and the compact conformation of the polymer leads to the formation of nitrile clusters as the fluorophore.

Hydrogen bond interactions can also promote the formation of clusters which endow polymer without large conjugated structure/AIEgens with AIE property. For example, Tang, Wang, and coworkers have synthesized an AIE-active linear polyamide P**51** without chromophore [125]. The rich amide and hydroxyl groups in P**51** could form spatial conjugation by n-π^* or π-π^* transitions, and strong hydrogen bonding induces the formation of local amide clusters, which enable the polymer to emit strong blue light in the solid state, thus suggesting its AIE characteristics.

4 Applications of AIE polymers

After 20 years development of AIE field, the AIE research has brought a profound influence on photochemistry, photophysics, luminescent materials, biomedicine, and so on and has triggered a series of scientific and technological revolutions [22–29]. Compared with AIE small molecules, AIE polymers enjoy the advantages of good film-forming ability, good processability, functionality inherent from polymer backbones such as mechanical property and stimuli-responsive property, solution processing, signal amplification effect, and so on. In the following sections, the recent advances in the application of AIE polymers such as fluorescent sensors, stimuli-responsive materials, biomedical applications, and optoelectronic devices are introduced.

4.1 Fluorescent sensors

Environmental changes and external stimuli can be sensitively reflected on the luminescence behavior of AIE polymers through careful structure design. AIE polymers are ideal candidates for effective fluorescent sensors considering their high fluorescence efficiency, synergistic effect, and super-amplification effect in aggregate state, enabling their applications in explosive detection, metal ion detection, toxic pollutant detection, polymerization process monitoring, and so on [126–128].

The sensitive detection of explosive compounds has been widely concerned for national security, public safety, environmental protection, and so forth. There are extensive literatures on the utilization of AIE-active polymers for explosive detection, owing to their high sensitivity, convenience, and low cost, which usually use picric acid (PA) or 2,4,6-trinitrotoluene (TNT) as examples to demonstrate [129]. Recently, a hyperbranched poly(β-aminoacrylate)s P**41** with satisfactory regio- and stereo-specificity was synthesized by the metal-free amino-alkyne click polymerization of ester-activated triyne and TPE-containing diamine [108]. The emission intensity of P**41** with bluish-green light in THF was progressively enhanced upon the gradual addition of poor solvent water under the excitation of 322nm light (Fig. 1A), demonstrating its AIE property. In THF/H_2O mixtures (1:9v/v) of P**41**, upon gradual addition of PA, the fluorescence intensity decreased accordingly, enabling the sensitive detection of PA with a low detection limit of 1.39×10^{-7}M (Fig. 1B).

With the strong designability of polymer structures and functionalities, microenvironment sensitivity, high signal-to-noise ratio, and high photostability, AIE polymers can be designed as fluorescent sensors for various ions [131]. Lu's group has prepared an amphiphilic AIE polymer P**52** with AIE-active salicylaldehyde azine moieties via the aldehyde-hydrazine polycondensation as a fluorescence "turn-off" probe for the selective detection of Cu^{2+} in aqueous media. As shown in Fig. 1C, the fluorescence was selectively quenched by Cu^{2+} with a detection limit of 53nM, while changes in the emission intensity of other metal ions were negligible. Interestingly, when different anions were added into the resultant P**52**-Cu^{2+} solution, only S^{2-} could be detected from the "turn-on" fluorescence response, owing to the formation of CuS [130]. As shown in Fig. 1D, the previously quenched fluorescence was recovered only through the addition of S^{2-} with a detection limit of 0.24μM, while other anions brought negligible change on the emission.

The development of CN^- sensors with high sensitivity, high selectivity, and high efficiency have attracted extensive attention. Lin's group has synthesized an amphiphilic AIE copolymer P**53a** containing both TPE and spiropyran (SP) as bifluorophores (Fig. 2A). Upon UV and visible light irradiation, reversible photo-isomerization through the cleavage of the C—O bond in the SP ring between the close form of SP ring and the open form of merocyanine (MC) could take place causing emission change [132]. As shown in Fig. 2B, upon UV irradiation, the original green emission at 517nm derived from TPE moiety has gradually decreased, and a new red emission peak emerged at 627nm, attributed to the Förster resonance energy transfer (FRET) from TPE to MC chromophore. CN^- can react with MC moiety to cause reversible fluorescence change by diminishing the emission at 627nm and enhancing the emission at 517nm, P**53b** can hence be used as a fluorescent sensor for the detection of CN^- with a detection limit of 0.26μM (Fig. 2C). The selectivity test was conducted with CN^- and other 16 anions or nucleophiles in buffer solutions, revealing that only CN^- can show dramatic fluorescence change from red emission to green emission, indicating great selectivity of P**53b** as a CN^- sensor (Fig. 2D).

The real-time monitoring of polymerization process is of great importance for the understanding of polymerization mechanism. AIE polymers could be designed to be highly

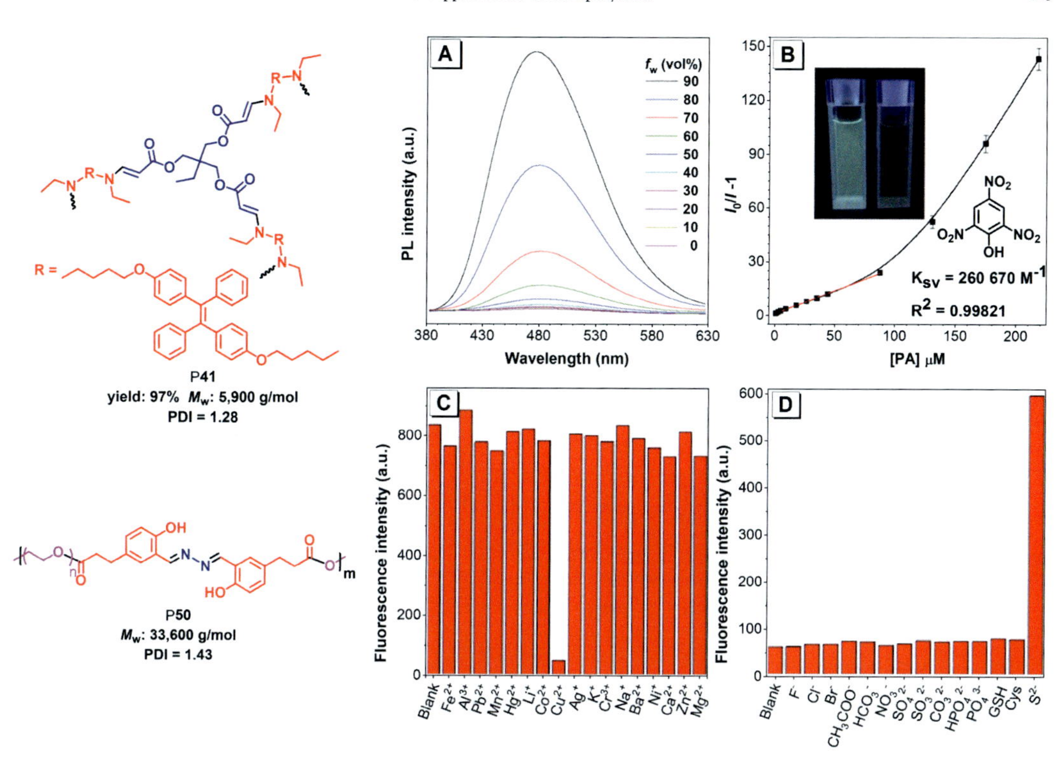

FIG. 1 (A) PL spectra of P**41** in THF/water mixtures with different water fractions (f_w). Concentration: 10 μM; λ_{ex}: 322 nm. (B) Stern-Volmer plots of $I_0/I-1$ of P**41** vs PA concentration, where I = peak intensity and I_0 = peak intensity without PA. Inset: solutions of P**41** in THF/water mixtures (f_w: 90%) without and with PA; photographs taken under illumination of a handheld UV lamp [108]. (C) Selectivity of the fluorescent sensor P**52** for Cu^{2+} over other representative metallic ions (counter anion for Fe^{2+}, Al^{3+}, Mn^{2+}, Hg^{2+}, Li^+, Co^{2+}, Cu^{2+}, K^+, Cr^{3+}, Na^+, Ba^{2+}, Ni^{2+}, Ca^{2+}, Zn^{2+}, and Mg^{2+} is Cl^-, for Pb^{2+} and Ag^+ is NO^{3-}) in aqueous solutions. (D) Selectivity of the fluorescent sensor P**52** for S^{2-} over various anions (counter cation: Na^+) and thiols in aqueous solution [130]. *Part (A, B) reproduced with permission from B. He, J. Zhang, J. Wang, Y. Wu, A. Qin, B.Z. Tang, Preparation of multifunctional hyperbranched poly(β-aminoacrylate)s by spontaneous amino-yne click polymerization, Macromolecules 53 (2020) 5248–5254. Copyright 2020, <American Chemical Society>; (C, D) reproduced with permission from J. Huang, H. Qin, H. Liang, J. Lu, An AIE polymer prepared via aldehyde-hydrazine step polymerization and the application in Cu^{2+} and S^{2-} detection, Polymer 202 (2020) 122663. Copyright 2020 <Elsevier>.*

sensitive to the change of microenvironment and hence can serve as an excellent candidate to probe the change of viscosity during the polymerization process. Tang, Lam, and coworkers have developed a facile, efficient, noninvasive approach to monitor RAFT polymerization in site through the combination of photochemistry and AIE technique. A TPE-containing initiator was designed for the blue light-induced RAFT polymerization of methyl methacrylate (MMA) monomers to synthesize PMMA P**54** [133]. With the gradual conversion of monomers, the viscosity of the polymerization solution increased, which led to the fluorescence enhancement of the solution (Fig. 3A). Moreover, M_n of the PMMA product increased exponentially with PL intensity (Fig. 3B). This real-time fluorescence visualization of RAFT polymerization is also applicable to different monomers such as butyl acrylate, methyl acrylate,

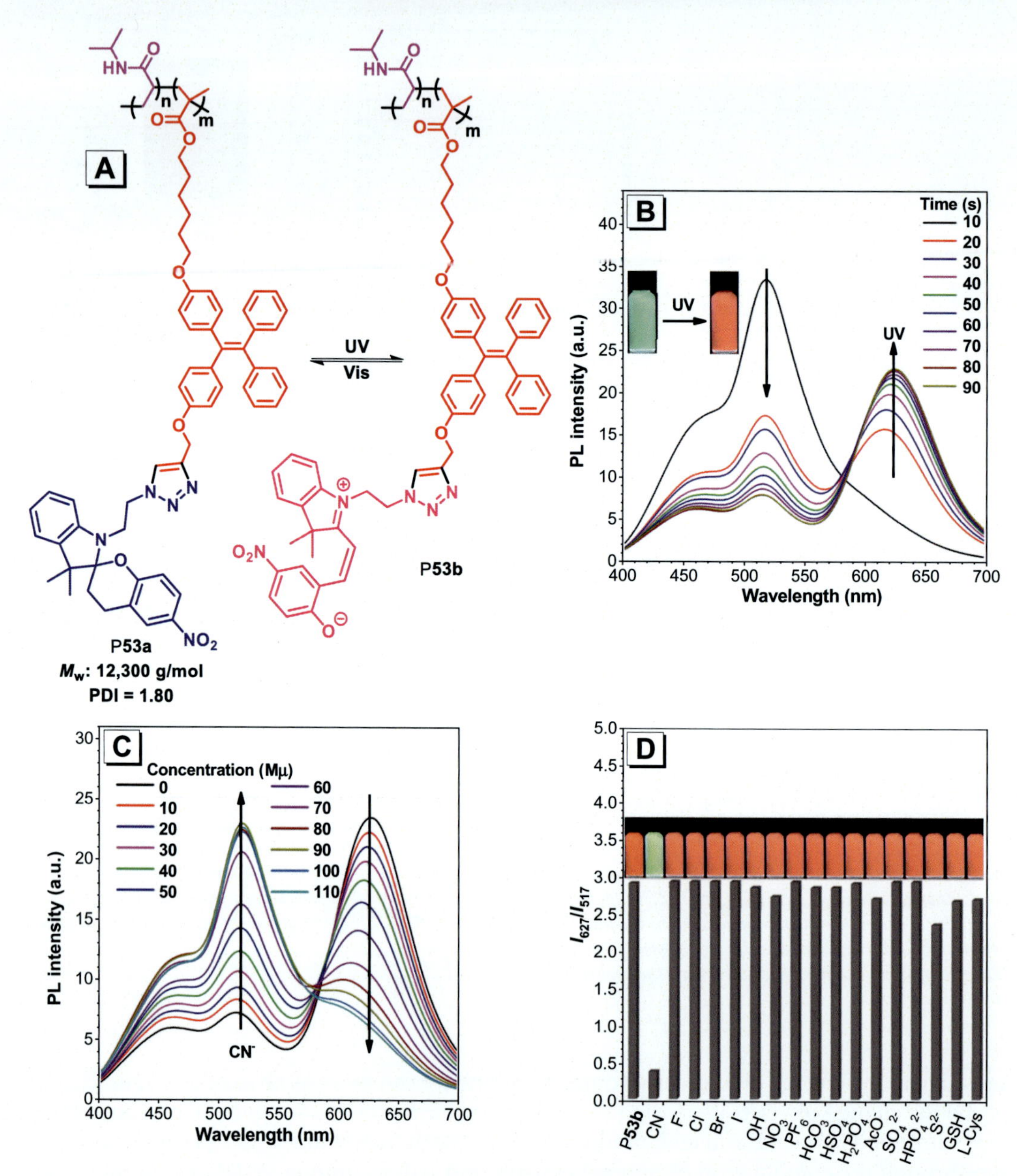

FIG. 2 (A) The reversible photo-isomerization between P**53a** and P**53b** upon UV and visible light irradiation. (B) PL spectra of P**53a** in water upon UV exposure for 90 s (1 g/L). Insets: Fluorescence change before and after UV irradiation (λ_{ex} = 365 nm). (C) Titration profile change and hypsochromic shift of P**53b** after addition of different concentrations of CN^-. (D) The ratio of fluorescence intensity at 627 and 517 nm in the presence of various analytes (200 μM). Inset: Fluorescence images of solutions of P**53b** in the presence of various analytes. Concentration: 1 g/L, λ_{ex}: 365 nm, 3-(*N*-morpholino)propanesulfonic acid buffer solution (10 mM, pH = 7.0) [132]. *Part (A, B, C, D) reproduced with permission from P.Q. Nhien, W.-L. Chou, T.T.K. Cuc, T.M. Khang, C.-H. Wu, N. Thirumalaivasan, B.T.B. Hue, J.I. Wu, S.-P. Wu, H.-C. Lin, Multi-stimuli responsive FRET processes of bifluorophoric AIEgens in an amphiphilic copolymer and its application to cyanide detection in aqueous media, ACS Appl. Mater. Interfaces 12 (2020) 10959–10972. Copyright 2020, <American Chemical Society>.*

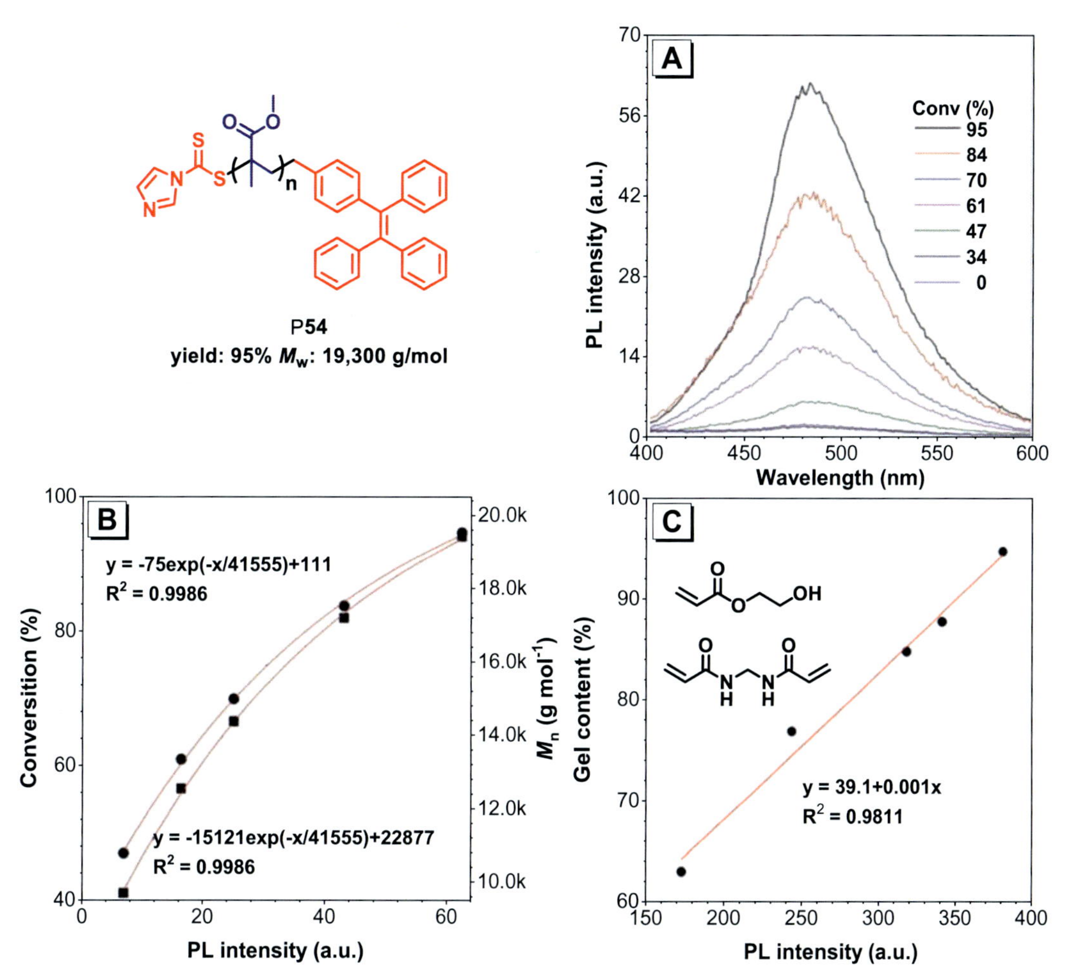

FIG. 3 (A) PL spectra of the polymerization mixtures with different monomer conversions. (B) The exponential relationship of monomer conversion and M_n with PL intensity. (C) The linear correlation between PL intensity and the gel content of the visible light-induced RAFT polymerization for 2-hydroxyethyl acrylate and *N,N′*-methylenebisacrylamide [133]. *Part (A, B, C) reproduced with permission from S. Liu, Y. Cheng, H. Zhang, Z. Qiu, R.T.K. Kwok, J.W.Y. Lam, B.Z. Tang, In situ monitoring of RAFT polymerization by tetraphenylethylene-containing agents with aggregation-induced emission characteristics, Angew. Chem. Int. Ed. 57 (2018) 6274–6278. Copyright 2018 <Wiley-VCH>.*

hexyl methacrylate, tert-butyl methacrylate, tert-butyl acrylate, 2-hydroxyethyl acrylate, and so on. In addition, the formation of highly cross-linked polymers could also be visualized through the gradual increase of the emission intensity, and a linear relationship between PL intensity and gel content (60%–95%) could be obtained as shown in Fig. 3C.

4.2 Stimuli-responsive materials

The combination of AIE characteristics with multifunctionalities from polymer backbone could generate a series of intelligent fluorescent materials with stimuli-responsive property, which show obvious response of fluorescence emission toward external stimuli such as temperature, pH, viscosity, pressure, and so on, indicating their prosperous future in imaging and sensing [134–137].

Firstly, pH-sensitive AIE polymers can be designed as promising smart materials, enabling the direct visualization of pH mapping in biological system. Recently, Lin, Yao, and coworkers have prepared an AIE-active azobenzene pendant-containing alternating amphiphilic copolymer **P55** via azide-yne click polymerization [138]. The amphiphilic copolymer can form large compound micelles with the diameter around 1 μm in aqueous solution; the self-assemble process is described in Fig. 4A. Upon the decrease in the pH value of the solution, the size of the large compound micelles increases, while the absorbance at 405 nm gradually decreases and a new absorption peak emerges at 550 nm (Fig. 4B). Accordingly, the color of the aqueous solution of **P55** changes from yellow to orange and eventually to purple, and reversible color change can be realized by tuning the pH value of the solution (Fig. 4C). The acid-triggered reversible color change of **P55** could be attributed to the protonation of the tertiary amine moieties in acidic solution, and the product was rearranged through intramolecular proton transfer. Moreover, acidity-enhanced fluorescence was observed for **P55**, owing to the protonation of the azo moiety, which prevented the π– conjugation, and the formation of large micelles, which further inhibited the intramolecular motion of the azo groups.

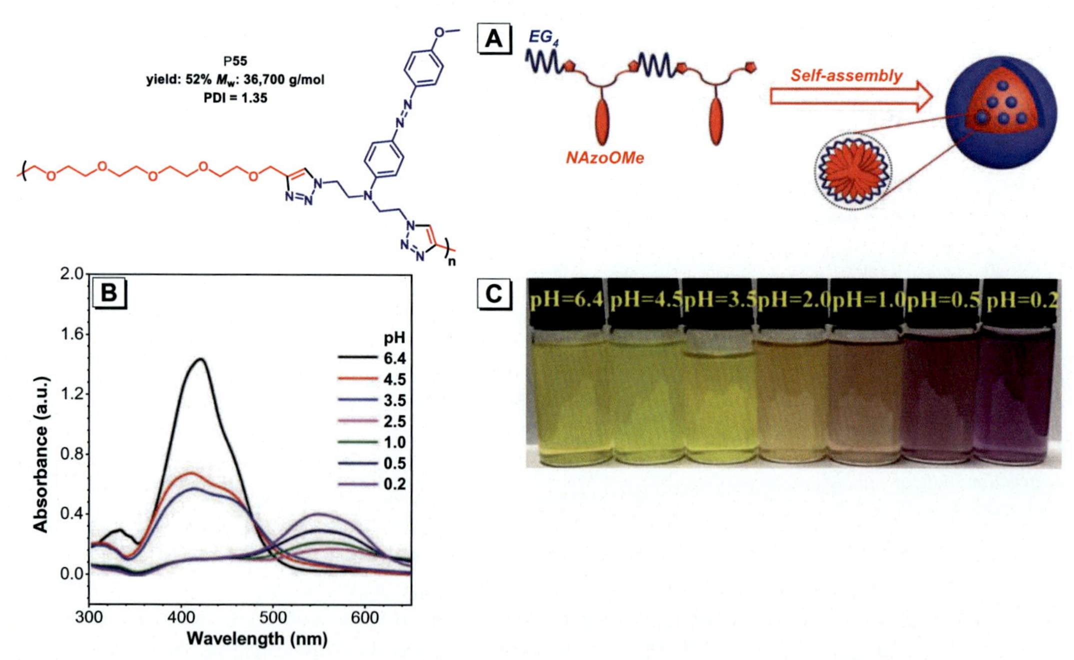

FIG. 4 (A) Self-assembly process of **P55**. (B) pH-dependent UV–Vis absorption spectra of **P55**. (C) pH-dependent color variation of aggregates of **P55** in aqueous solution [138]. *Part (A, B, C) reproduced with permission from J. Wu, B. Xu, Z. Liu, Y. Yao, Q. Zhuang, S. Lin, The synthesis, self-assembly and pH-responsive fluorescence enhancement of an alternating amphiphilic copolymer with azobenzene pendants, Polym. Chem. 10 (2019) 4025–4030. Copyright 2019, <Royal Society of Chemistry>.*

Temperature-sensitive AIE polymers can be used to detect various temperature transitions of polymers in situ such as lower critical solution temperature and glass-transition temperature owing to conformational transitions of the polymer. For example, Tang and coworkers have developed a thermo-responsive AIE polymer P**56** with fine-tuned lower critical solution temperature (LCST) and response temperature through the free radical copolymerization of thermosensitive *N*-isopropylacrylamide, oligo(ethylene glycol) methacrylate, and TPE-containing vinyl monomer [139]. By adjusting the loading ratio of hydrophilic oligo(ethylene glycol) methacrylate monomer, the hydrophilicity of the copolymer can be tuned, and hence the LCST and the responsive temperature range can be accurately controlled around the physiological temperature. When the solution temperature was increased from 32°C to 50°C, the emission at 480nm gradually increased (Fig. 5A), and reversible fluorescence intensity change can be observed when the heating and cooling cycles are repeated (Fig. 5B and C). Below LCST, the polymer dissolves in water and forms intermolecular hydrogen bonds with water molecules, showing weak emission; Above LCST, the intermolecular hydrogen bonds are replaced by the intrachain hydrogen bonds within the polymer chain, and the polymer gradually shrinks into nanoparticles to inhibit the

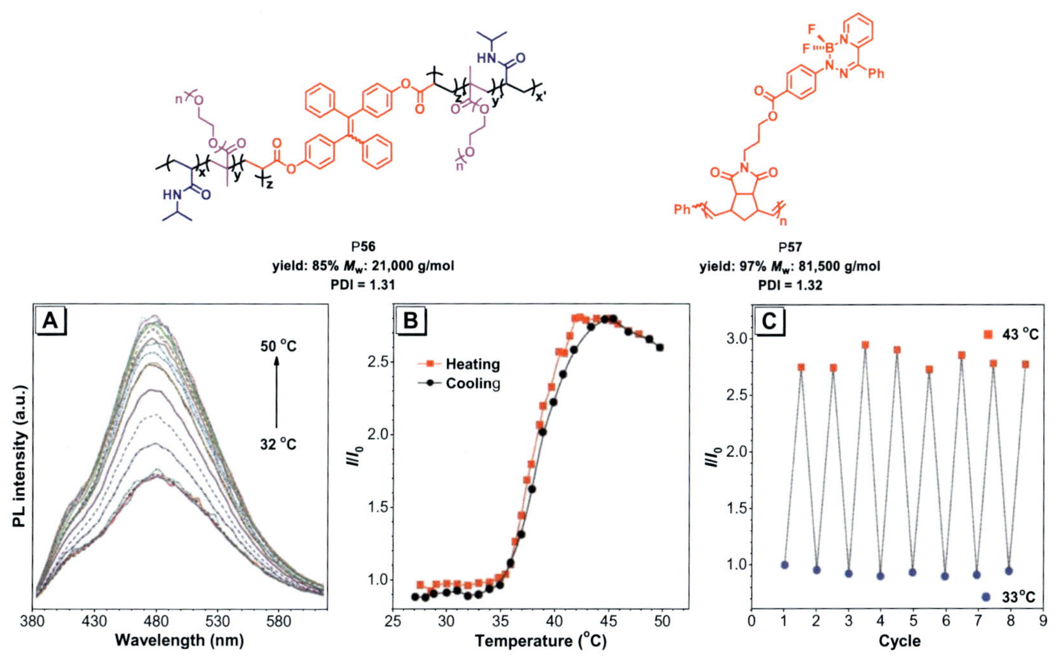

FIG. 5 (A) Temperature-dependent emission spectra of aqueous solution of P**56** from 32°C to 50°C. (B) Temperature-dependent fluorescence of aqueous solution of P**56** in a heating–cooling cycle. (C) Fluorescence intensities at 478nm recorded at 43°C and 33°C at each heating–cooling cycle. I is the emission intensity at different temperatures. I_0 is the emission intensity at 27°C. Concentration: 0.05mg/mL. Excitation wavelength: 310nm [139]. *Part (A, B, C) reproduced with permission from T. Li, S. He, J. Qu, H. Wu, S. Wu, Z. Zhao, A. Qin, R. Hu, B.Z. Tang, Thermoresponsive AIE polymers with fine-tuned response temperature, J. Mater. Chem. C 4 (2016) 2964–2970. Copyright 2016, <Royal Society of Chemistry>.*

intramolecular motion and induce strong emission. Combined with their good biocompatibility and cell permeability, such temperature-sensitive water-soluble AIE polymers may find potential applications in biological imaging.

Viscosity-sensitive AIE polymers have also been developed recently. For example, Gilroy's group synthesized a boron difluoride hydrazone (BODIHY)-containing AIE polymer **P57** via ring-opening metathesis polymerization [140]. The fluorescence quantum yield of the thin film of **P57** was sixfold higher than that of its solution, demonstrating typical AIE characteristics, attributed to the restriction of intramolecular motion of the BODIHY moieties in the aggregated state. Furthermore, the emission intensity of **P57** gradually increases with the increase in the volume ratio of DME in DMF/DME mixtures, demonstrating viscosity-dependent fluorescence of **P57**.

4.3 Biological applications

For the biological applications, compared with traditional fluorescent dyes, AIE materials enjoy the advantages of high brightness, low background signal, strong photobleaching resistance, low cytotoxicity, and so on. AIE polymers have also been widely used in biological imaging, photodynamic therapy, bacterial detection, and drug delivery in biomedical fields owing to their unique characteristics such as facile structural modification, polymer morphology and functionality, long in vivo tracking time, and efficient emission in the aggregated states [141–143].

For example, Zhao, Zhou, Smith, and coworkers have synthesized a series of conjugated polyelectrolytes with cyan to red emission, which can be used as a biological probe to stain HeLa cells with high fluorescence intensity and long fluorescence lifetime in the aggregated state [93]. Haddleton's group has synthesized an AIE-active amphiphilic copolymer **P58** for biological imaging through copper(0)-mediated reversible inactivation free radical polymerization in the presence of AIE-containing initiator [144]. The polymer product possessed low dispersity value, high end-group fidelity, and pH-responsive ability endowed by the PAA segments, together with its strong fluorescence, high photostability, excellent biocompatibility and good cell staining capability, its application for lysosome-specific probes was investigated. When the concentration of **P58** reached 20 μg/mL, good cell viability for 4T1 cells was still observed after cultured for 24h, indicating that the polymer had low cytotoxicity (Fig. 6A). Furthermore, compared with known lysosomal imaging reagents, the polymer was found to have specific localization and lysosomal staining abilities and their staining areas overlap nicely as shown in Fig. 6B. Furthermore, in contrast to the commercial lysosomal imaging probes, there was no obvious fluorescence quenching observed upon continuous irradiation, proving its excellent photostability (Fig. 6C).

Photodynamic therapy (PDT) refers to the activation of photosensitizers under the irradiation of laser with a specific wavelength, and the excited photosensitizers transfer energy to the surrounding oxygen to generate active oxygen species (ROS), which oxidizes with adjacent biomacromolecules to produce cytotoxic effects, leading to cell damage and even death. Photodynamic therapy has been widely studied for its advantages such as high spatiotemporal accuracy, noninvasive properties, low cytotoxicity, and low cost, and a series of AIE photosensitizers with electron-donating and electron-accepting structures have been designed

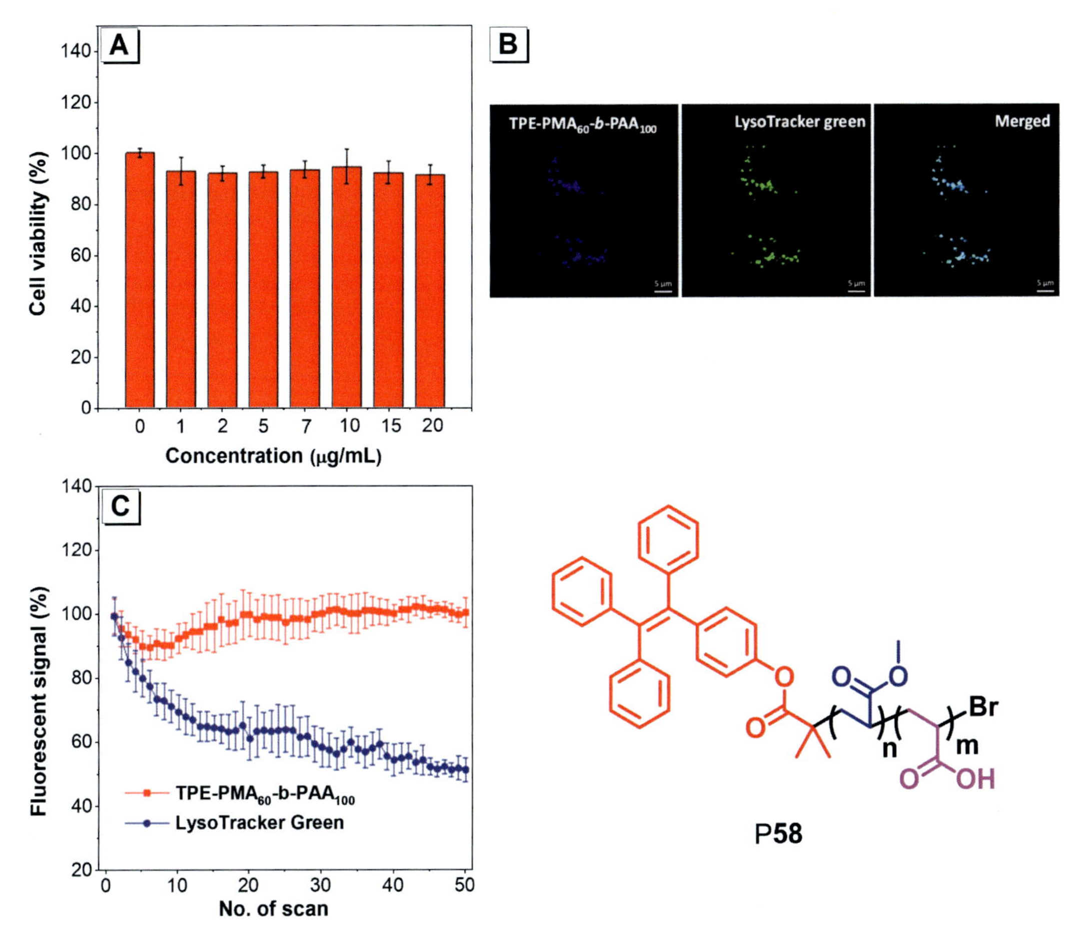

FIG. 6 (A) Cell viability of 4T1 cells in the presence of **P58** at different concentrations. (B) Colocalization images of 4T1 cells stained with **P58** (10 μg/mL, 3 h) and LysoTracker Green (1 μM, 30 min) and their merged image. (C) Change in the fluorescent signal from 4T1 cells stained with **P58** (10 μg/mL, 3 h) and LysoTracker Green (1 μM, 30 min) with the scan times. Excitation wavelength: 405 nm (**P58**) and 488 nm (LysoTracker Green); emission filter: 410–500 nm (**P58**) and 500–550 nm (LysoTracker Green). Scale bar = 5 μm [144]. *Part (A, B, C) reproduced with permission from C. Ma, T. Han, M. Kang, E. Liarou, A.M. Wemyss, S. Efstathiou, B.Z. Tang, D. Haddleton, Aggregation-induced emission active polyacrylates via cu-mediated reversible deactivation radical polymerization with bioimaging applications, ACS Macro Lett. 9 (2020) 769–775. Copyright 2020, <American Chemical Society>.*

and synthesized for PDT. For example, Wang, Liu, and coworkers have synthesized an AIE photosensitizer-bearing zwitterionic polymer **P59** [145]. As shown in Fig. 7A, the nanomicelles of **P59** possessed excellent ROS generation efficiency, indicated by the emission generated from the oxidation of dichlorofluorescein diacetate (DCFH-DA) by the in situ produced ROS. After they were incubated with Gram-positive *Staphylococcus aureus*, MRSA, and Gram-negative *Escherichia coli*, obvious emission was observed under acidic condition (Fig. 7B), while there was no emission observed for Gram-negative *E. coli* under normal

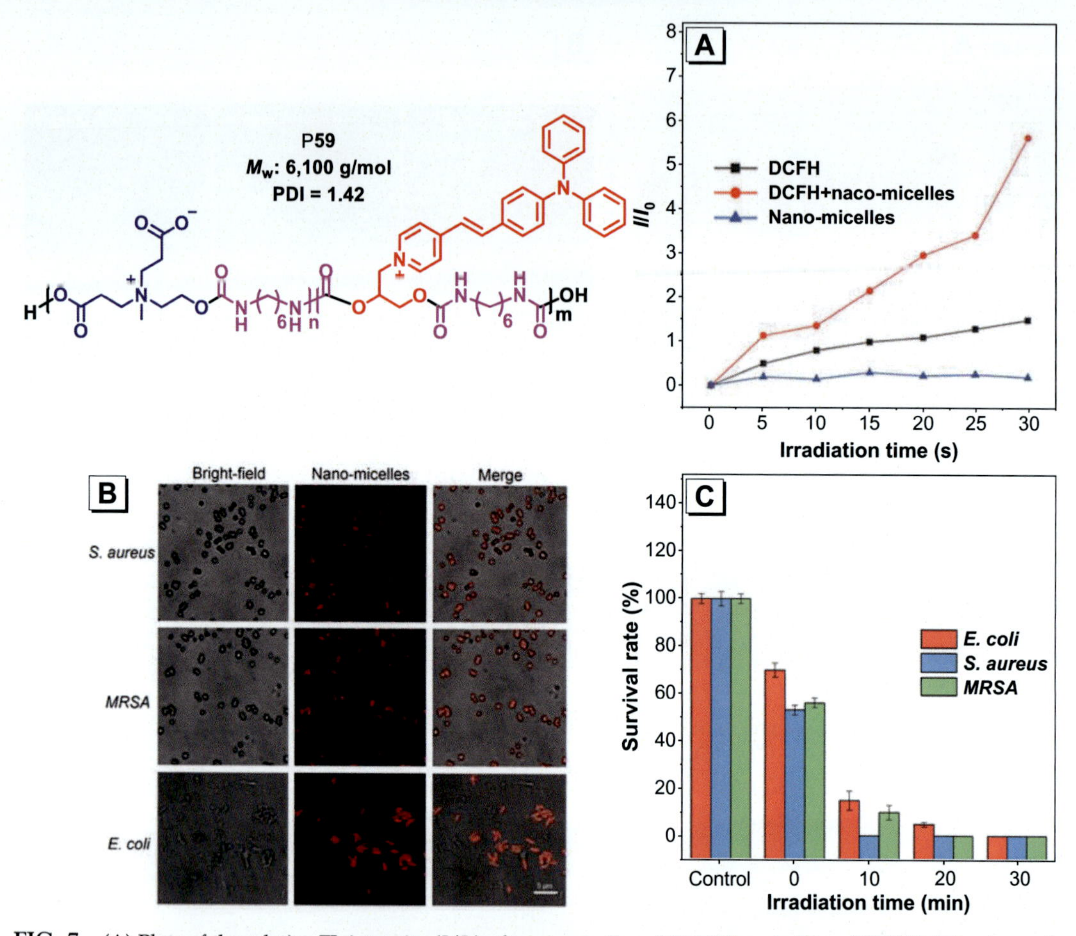

FIG. 7 (A) Plots of the relative FL intensity (I/I_0) of nano-micelles of P**59** (20mg/mL) and DCFH-DA mixture in water at 385nm vs the irradiation time. (B) Bright-field and fluorescence images of bacteria. S. aureus, MRSA, and E. coli incubated with 20mg/mL of nano-micelles of P**59** at pH5.4 for 30min before imaging. λ_{ex}: 488nm, λ_{em}: 600–700nm. (C) The antibacterial activities of nano-micelles (20mg/mL) toward E. coli, S. aureus, and MRSA under white light irradiation at pH5.4 for different time periods [145]. *Part (A, B, C) reprinted with permission from B. Ren, K. Li, Z. Liu, G. Liu, H. Wang, White light-triggered zwitterionic polymer nanoparticles based on an AIE-active photosensitizer for photodynamic antimicrobial therapy, J. Mater. Chem. B 8 (2020) 10754–10763. Copyright 2016, <Royal Society of Chemistry>.*

pH conditions. Most importantly, 99% of these pathogens were killed by the zwitterionic nano-micelles under the irradiation of white light for 30min in acidic conditions, showing excellent photodynamic antibacterial effect of the polymer (Fig. 7C).

In another work, Ma and coworkers have synthesized an AIE-active linear polyelectrolyte P**60** with positively charged pyridinium embedded on the polymer backbone [146], which emitted intense fluorescence at 550nm with a large Stokes shift of 164nm in water. When treating Gram-negative bacteria *E. coli* and Gram-positive bacteria *Staphylococcus epidermidis* with P**60**, the bacterials could be killed both in dark and upon light irradiation, and the efficiency is higher under light owing to the efficient ROS generation from the polymer (Fig. 8A

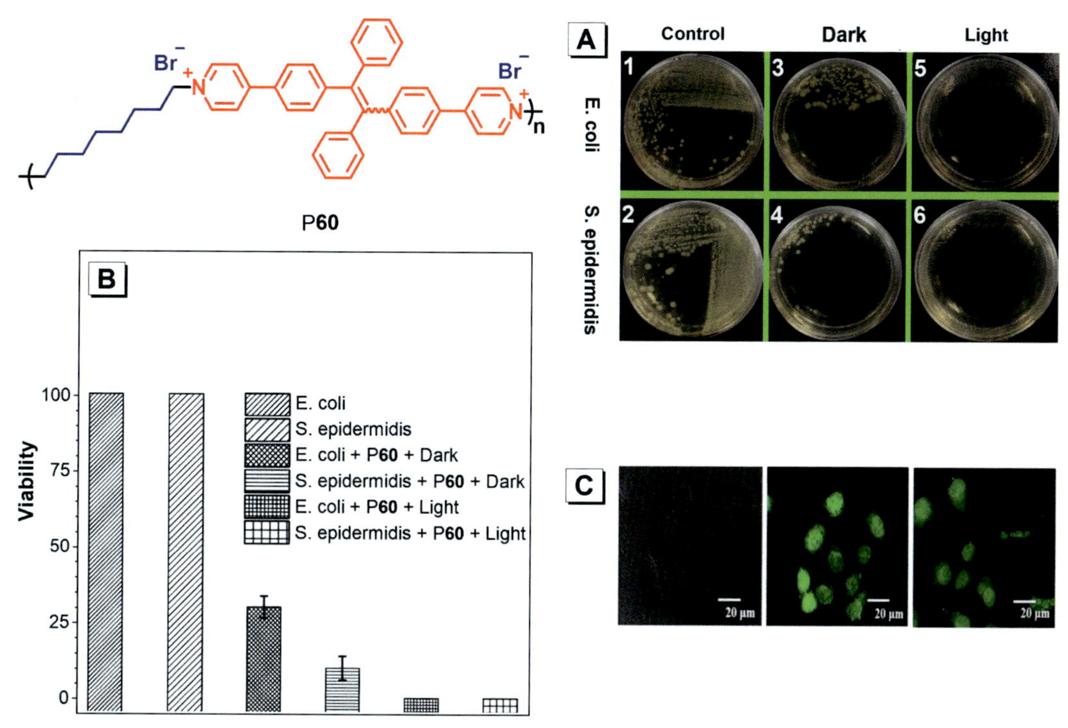

FIG. 8 (A) Plates of E. coli (top row) and S. epidermidis (bottom row): without **P60** (1 and 2), with **P60** and without light (3 and 4), and with **P60** and light irradiation for 1 h (5 and 6). For those treated with 150 μg/mL of **P60** for 3 h, followed by storage in the dark (3 and 4), or irradiation with room light for 3 h (5 and 6). (B) Killing efficiency of **P60** on E. coli and S. epidermidis in the absence and presence of room light irradiation. The bacteria were incubated with 150 μg/mL. (C) Fluorescence images of HepG-2 cells stained with **P60** (30 μg/mL) [146]. *Part (A, B, C) reproduced with permission from H. Ma, Y. Ma, L. Lei, W. Yin, Y. Yang, T. Wang, P. Yin, Z. Lei, M. Yang, Y. Qin, S. Zhang, Light-enhanced bacterial killing and less toxic cell imaging: multicationic aggregation-induced emission matters, ACS Sustain. Chem. Eng. 6 (2018) 15064–15071. Copyright 2018, <American Chemical Society>.*

and B). Moreover, **P60** showed strong cell-membrane permeability and could aggregate on cell nucleus with bright green fluorescence, which can also be used as cell nucleus imaging probes (Fig. 8C).

The superiority of AIE polymers in drug delivery and cancer therapy has also been demonstrated. For example, Cheng, Quan, Yuan, and coworkers have prepared a series of AIE-active conjugated polymers with color-tunable feature through adjusting their intramolecular FRET pairs [51]. The polymer nanoparticles could serve as drug carriers to load paclitaxel and could be further internalized by cell to release the drug efficiently, showing significant cytotoxicity against HeLa and A549 cancer cells.

4.4 Optoelectronic devices

With their excellent processability and efficient fluorescence emission in solid and thin film states, AIE polymers have been extensively explored for the applications of optoelectronic devices such as polymer light-emitting diode (PLED) and electrochromic devices, and as polymer materials with fluorescent liquid crystallinity and circularly polarized luminescence [147–149].

For the applications as emitters in OLED devices, compared with small molecules, polymers enjoy the advantages of convenient and economic solution-processed manufacturing, and introducing AIEgens into polymer materials may endow PLED devices with both excellent electroluminescence and photoluminescence performance. For example, Li and coworkers have designed a hyperbranched polymer **P61** with both TPE and tetraphenylcyclopentadiene chromophores, whose absolute fluorescence quantum yield in the thin film state is 50.1%. A PLED device was then designed using **P61** as the light-emitting layer and a bipolar thermally activated delayed fluorescence (TADF) material CzAcSF as the host [150]. The PLED device with host-doped film as emitting layer presented high luminescence of 2441 cd m^{-2}, good electroluminescence performance for the driving voltage of 4.4 V, the maximum power efficiency of 10.75 lm W^{-1}, the maximum current efficiency of 16.96 cd A^{-1}, and the maximum external quantum efficiency of 9.74% (Fig. 9A and B).

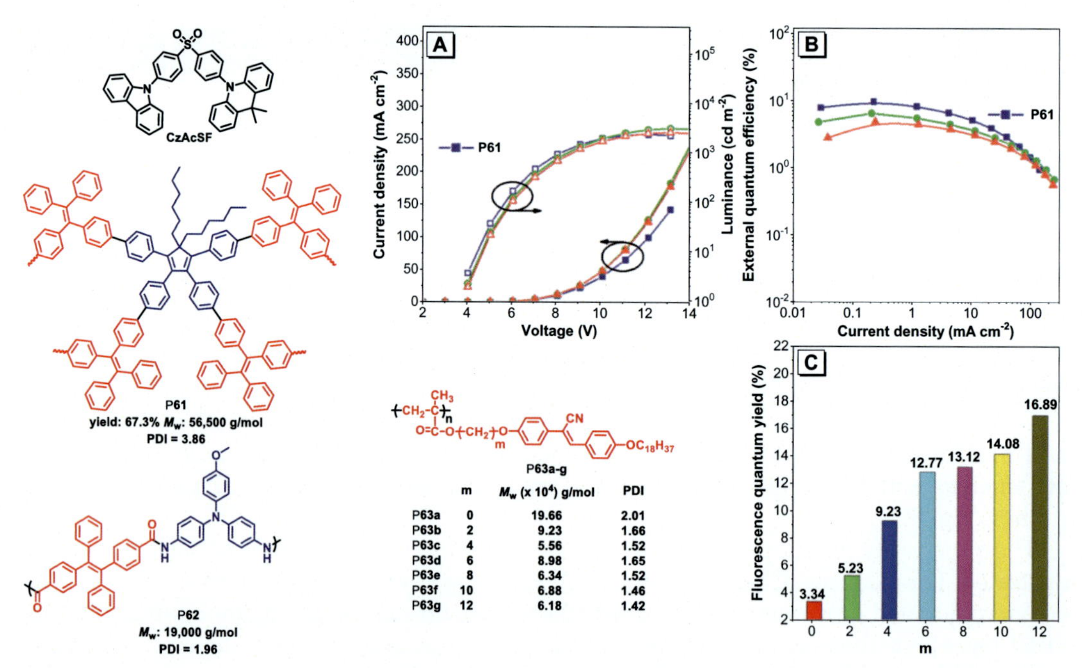

	m	M_w (x 10^4) g/mol	PDI
P63a	0	19.66	2.01
P63b	2	9.23	1.66
P63c	4	5.56	1.52
P63d	6	8.98	1.65
P63e	8	6.34	1.52
P63f	10	6.88	1.46
P63g	12	6.18	1.42

FIG. 9 (A) Current density-voltage-luminance (J-V-L) characteristics of the PLED devices fabricated from **P61**. (B) External EL quantum efficiency (EQE) vs current density of the PLED devices [150]. (C) Fluorescence quantum yield of **P63a-g** in the solid state; λ_{em} = 360 nm [151]. *Part (A, B) reproduced with modifications from Y. Xie, Y. Gong, M. Han, F. Zhang, Q. Peng, G. Xie, Z. Li, Tetraphenylcyclopentadiene-based hyperbranched polymers: convenient syntheses from one pot "A4 + B2" polymerization and high external quantum yields up to 9.74% in OLED devices, Macromolecules 52 (2019) 896–903. Copyright 2019, <American Chemical Society>. Part (C) reproduced with permission from Y. Yuan, J. Li, L. He, Y. Liu, H. Zhang, Preparation and properties of side chain liquid crystalline polymers with aggregation-induced emission enhancement characteristics, J. Mater. Chem. C 6 (2018) 7119–7127. Copyright 2018, <Royal Society of Chemistry>.*

Electrochromic materials with their stable and reversible color changes when electric field was applied have attracted much attention and have found applications in electrochromic intelligent glass, electrochromic display, and automatic antiglare rearview mirror. Through combining electrochromic (EC) active moieties and emission units into the material, high-performance electrofluorochromic (EFC) devices could be obtained, and AIE polymers are hence ideal candidates for such application. Liou, Tang, and coworkers have designed an aromatic polyamide P**62** with AIE-active TPE and electro-active triphenylamine units [152] and investigated its performance in EFC device. When P**62** was mixed with *n*-heptyl viologen (HV) as a counter EC layer for balancing charges, the device performance was improved because the polymer/HV displayed lower oxidation potential to enhance electrochemical switching ability. This EFC device showed electrochromic/electrofluorescence dual switching behavior with low trigger voltage, short response time of less than 4.9 s, and high PL contrast ratio (I_{off}/I_{on}) of 63.

The development of fluorescent liquid crystal materials has attracted much attention, which may find application in anisotropic light-emitting diodes, information storage, organic lasers, and so on. Through combining AIEgens with mesogenic units, liquid crystal materials with both fluorescence characteristics and liquid crystal properties can be constructed efficiently. For example, Zhang, Liu, and coworkers have developed a series of luminescent liquid crystalline polymers P**63a-g** with cyanostilbene moiety serving as both fluorophore and mesogenic unit [151]. The differential scanning calorimetry, polarized optical microscopy, and X-ray diffraction study suggested that all the polymers could form stable smectic textures, and the polymers were arranged in a double-layer smectic A phase. The fluorescence quantum yield of these polymers increased from 3.34% to 16.89% with the increase in the spacer length on the polymer side chains (Fig. 9C).

CPL materials with both fluorescence and chirality are promising advanced luminescent materials with the potential application in 3D displays, sensor, information processing, and so on. The introduction of AIEgens into chiral compounds is a common strategy to construct CPL materials, and helical chirality is also beneficial to achieve high luminescence asymmetry factor (g_{lum}) because of chiral amplification effect. Deng, Zhao, and coworkers have designed and synthesized a pair of AIE-active chiral helical polymers P64-R and P64-S by copolymerization of achiral TPE-containing acetylenic monomer and nonemissive chiral acetylenic monomer [153]. CD spectra and PL spectra suggested that the copolymer has optical activity and fluorescence properties (Fig. 10A and B), however, no CPL signal is observed in THF/H_2O mixtures with various ratios. When composite film was prepared with ordered arrangement or self-assembly, the CPL emission with a high g_{lum} values of 3.6×10^{-2} was obtained (Fig. 10C).

5 Conclusions and perspectives

In this chapter, the most recent progress in AIE polymers is introduced with an emphasis on the synthetic methodologies, unique structures, and applications. Living radical polymerizations such as ATRP and RAFT, and photo-induced ATRP or RAFT are utilized to synthesize polyethylenes with controllable molecular weight and low polydispersity, usually with

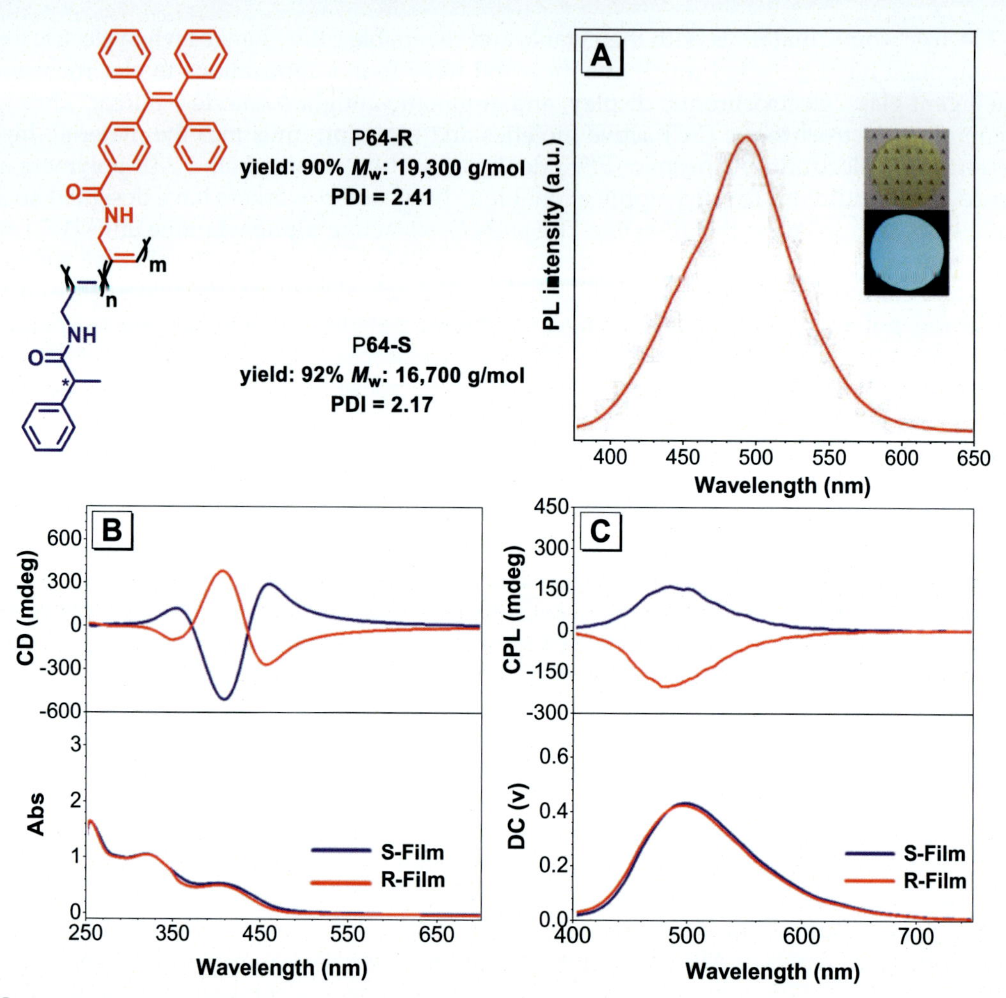

FIG. 10 (A) PL spectrum of the composite film of P64-S). Inset: Digital photo of the composite film under daylight and UV irradiation. (B) CD spectra and (C) CPL spectra of the composite films of P64-R and P64-S [153]. *Part (A, B, C) reproduced with permission from N. Lu, X. Gao, M. Pan, B. Zhao, J. Deng, Aggregation-induced emission-active chiral helical polymers show strong circularly polarized luminescence in thin films, Macromolecules 53 (2020) 8041–8049. Copyright 2020, <American Chemical Society>.*

AIEgens installed as terminal group or side chain; newly developed one-component, two-component, and multicomponent condensation polymerizations are used for the construction of AIE polymers with great structural novelty such as heteroatom or heterocycle-rich polymers, and the general strategy involves AIE-active monomers. Many of these approaches meet the requirement of green chemistry, demonstrating high efficiency and convenience.

With these various options of synthetic methods, special AIE polymer structures such as water-soluble polyelectrolytes with positive or negative charges on the side chain and AIEgen in the backbone, center, or terminal of the polymer chain, chiral polymers with chiral centers attached on the side chain or main chain helicity, hyperbranched polymers with three-dimensional topology structures and large number of terminal groups, porous polymers with high surface area and inherent porosity, as well as an unique type of AIE polymers without conventional fluorophore are afforded.

The efficient fluorescence of AIE polymers in the aggregated states, together with the mechanical property, stimuli-responsive property, water solubility, biocompatibility, or chirality, is endowed by the polymer backbone, these luminescent polymer materials have found potential applications in wide scope of fields. AIE polymer-based fluorescent sensor for explosive detection, metal ion detection, toxic and organic volatile detection, and polymerization process monitoring have been developed. Intelligent AIE polymer materials that show stimuli-responsive fluorescence change toward pH value, temperature, and viscosity have been explored; Biological applications of AIE polymers such as cell imaging, photodynamic therapy, bacterial detection, and drug delivery have been demonstrated. Optoelectronic devices fabricated from AIE polymers such as PLEDs and electrochromic devices, and AIE polymers with liquid crystallinity and circularly polarized luminescence are developed.

This prosperous field is full of opportunities with new synthetic approaches, new polymer structures, new functionalities, and new applications, which may play an important role for energy-, environment-, and health-related practical applications. It is anticipated that through the introduction of the recent trend of AIE polymers, this chapter could bring inspiration to scientists working in the area of chemistry, materials, biology, physics, and further accelerate the development of this interdisciplinary area.

Acknowledgment

This work was partially supported by the National Natural Science Foundation of China (21822102, 21788102, and 21774034), the Natural Science Foundation of Guangdong Province (2016A030306045), Guangdong Special Support Program (2017TQ04C901), and the Guangdong Provincial Key Laboratory of Luminescence from Molecular Aggregates (2019B030301003).

References

[1] J. Wang, L. Zhao, B. Yan, Indicator displacement assay inside dye-functionalized covalent organic frameworks for ultrasensitive monitoring of sialic acid, an ovarian cancer biomarker, ACS Appl. Mater. Interfaces 12 (2020) 12990–12997.

[2] Z.-Q. Yao, G.-Y. Li, J. Xu, T.-L. Hu, X.-H. Bu, A water-stable luminescent Zn^{II} metal-organic framework as chemosensor for high-efficiency detection of Cr^{VI}-anions ($Cr_2O_7^{2-}$ and CrO_4^{2-}) in aqueous solution, Chem. A Eur. J. 24 (2018) 3192–3198.

[3] S. Sadeghi, R. Melikov, H.B. Jalali, O. Karatum, S.B. Srivastava, D. Conkar, E.N. Firat-Karalar, S. Nizamoglu, Ecofriendly and efficient luminescent solar concentrators based on fluorescent proteins, ACS Appl. Mater. Interfaces 11 (2019) 8710–8716.

[4] W. Ren, G. Lin, C. Clarke, J. Zhou, D. Jin, Optical nanomaterials and enabling technologies for high-security-level anticounterfeiting, Adv. Mater. 32 (2020) 1901430.

[5] T. Yu, Z. Zhu, Y. Bao, Y. Zhao, X. Liu, H. Zhang, Investigation of novel carbazole-functionalized coumarin derivatives as organic luminescent materials, Dyes Pigm. 147 (2017) 260–269.

[6] T. Wang, X. Su, X. Zhang, W. Huang, L. Huang, X. Zhang, X. Sun, Y. Luo, G. Zhang, A combinatory approach towards the design of organic polymer luminescent materials, J. Mater. Chem. C 7 (2019) 9917–9925.

[7] H. Zhang, Q. Luo, Y. Mao, Y. Zhao, T. Yu, Synthesis and characterization of coumarin-biphenyl derivatives as organic luminescent materials, J. Photochem. Photobiol. A Chem. 346 (2017) 10–16.

[8] H. Shono, T. Ohkawa, H. Tomoda, T. Mutai, K. Araki, Fabrication of colorless organic materials exhibiting white luminescence using normal and excited-state intramolecular proton transfer processes, ACS Appl. Mater. Interfaces 3 (2011) 654–657.
[9] S. Mukherjee, P. Thilagar, Stimuli and shape responsive 'boron-containing' luminescent organic materials, J. Mater. Chem. C 4 (2016) 2647–2662.
[10] J. Luo, Z. Xie, J.W.Y. Lam, L. Cheng, H. Chen, C. Qiu, H.S. Kwok, X. Zhan, Y. Liu, D. Zhu, B.Z. Tang, Aggregation-induced emission of 1-methyl-1,2,3,4,5-pentaphenylsilole, Chem. Commun. 18 (2001) 1740–1741.
[11] Y. Hong, J.W.Y. Lam, B.Z. Tang, Aggregation-induced emission, Chem. Soc. Rev. 40 (2011) 5361–5388.
[12] Z. Zhao, J.W.Y. Lam, B.Z. Tang, Tetraphenylethene: a versatile AIE building block for the construction of efficient luminescent materials for organic light-emitting diodes, J. Mater. Chem. 22 (2012) 23726–23740.
[13] J. Li, J. Liu, J.W.Y. Lam, B.Z. Tang, Poly(arylene ynonylene) with an aggregation-enhanced emission characteristic: a fluorescent sensor for both hydrazine and explosive detection, RSC Adv. 3 (2013) 8193–8196.
[14] G. Yu, S. Yin, Y. Liu, J. Chen, X. Xu, X. Sun, D. Ma, X. Zhan, Q. Peng, Z. Shuai, B. Tang, D. Zhu, W. Fang, Y. Luo, Structures, electronic states, photoluminescence, and carrier transport properties of 1,1-disubstituted 2,3,4,5-tetraphenylsiloles, J. Am. Chem. Soc. 127 (2005) 6335–6346.
[15] Z. Zhao, Z. Wang, P. Lu, C.Y.K. Chan, D. Liu, J.W.Y. Lam, H.H.Y. Sung, I.D. Williams, Y. Ma, B.Z. Tang, Structural modulation of solid-state emission of 2,5-bis(trialkylsilylethynyl)-3,4-diphenylsiloles, Angew. Chem. Int. Ed. 48 (2009) 7608–7611.
[16] H. Lu, F. Su, Q. Mei, X. Zhou, Y. Tian, W. Tian, R.H. Johnson, D.R. Meldrum, A series of poly[*N*-(2-hydroxypropyl)methacrylamide] copolymers with anthracene-derived fluorophores showing aggregation-induced emission properties for bioimaging, J. Polym. Sci. A Polym. Chem. 50 (2012) 890–899.
[17] J. He, B. Xu, F. Chen, H. Xia, K. Li, L. Ye, W. Tian, Aggregation-induced emission in the crystals of 9,10-distyrylanthracene derivatives: the essential role of restricted intramolecular torsion, J. Phys. Chem. C 113 (2009) 9892–9899.
[18] M. Chen, L. Li, H. Nie, J. Tong, L. Yan, B. Xu, J.Z. Sun, W. Tian, Z. Zhao, A. Qin, B.Z. Tang, Tetraphenylpyrazine-based AIEgens: facile preparation and tunable light emission, Chem. Sci. 6 (2015) 1932–1937.
[19] Z. Xu, J. Gu, X. Qiao, A. Qin, B.Z. Tang, D. Ma, Highly efficient deep blue aggregation-induced emission organic molecule: a promising multifunctional electroluminescence material for blue/green/orange/red/white OLEDs with superior efficiency and low roll-off, ACS Photonics 6 (2019) 767–778.
[20] Y. Hong, J.W.Y. Lam, B.Z. Tang, Aggregation-induced emission: phenomenon, mechanism and applications, Chem. Commun. (2009) 4332–4353.
[21] J. Mei, Y. Hong, J.W.Y. Lam, A. Qin, Y. Tang, B.Z. Tang, Aggregation-induced emission: the whole is more brilliant than the parts, Adv. Mater. 26 (2014) 5429–5479.
[22] A. Qin, J.W.Y. Lam, B.Z. Tang, Luminogenic polymers with aggregation-induced emission characteristics, Prog. Polym. Sci. 37 (2012) 182–209.
[23] R. Hu, N.L.C. Leung, B.Z. Tang, AIE macromolecules: syntheses, structures and functionalities, Chem. Soc. Rev. 43 (2014) 4494–4562.
[24] Y.B. Hu, J.W.Y. Lam, B.Z. Tang, Recent progress in AIE-active polymers, Chin. J. Polym. Sci. 37 (2019) 289–301.
[25] S.-Y. Zhou, H.-B. Wan, F. Zhou, P.-Y. Gu, Q.-F. Xu, J.-M. Lu, AIEgens-lightened functional polymers: synthesis, properties and applications, Chin. J. Polym. Sci. 37 (2019) 302–326.
[26] R. Hu, A. Qin, B.Z. Tang, AIE polymers: synthesis and applications, Prog. Polym. Sci. 100 (2020), 101176.
[27] X. Zhang, K. Wang, M. Liu, X. Zhang, L. Tao, Y. Chen, Y. Wei, Polymeric AIE-based nanoprobes for biomedical applications: recent advances and perspectives, Nanoscale 7 (2015) 11486–11508.
[28] A.C.B. Rodrigues, J.S. Seixas de Melo, Aggregation-induced emission: from small molecules to polymers—historical background, mechanisms and photophysics, Top. Curr. Chem. 379 (2021) 15.
[29] R. Zhan, Y. Pan, P.N. Manghnani, B. Liu, AIE polymers: synthesis, properties, and biological applications, Macromol. Biosci. 17 (2017) 1600433.
[30] X. Guan, L. Wang, M. Liu, K. Wang, X. Yang, Y. Ding, J. Tong, Z. Lei, S. Lai, A versatile synthetic approach to tunable dual-emissive pdots with very small-size based on amphiphilic block copolymers for cell imaging, Mater. Chem. Front. 5 (2021) 355–367.
[31] R. Yang, Y. Wang, W. Luo, Y. Jin, Z. Zhang, C. Wu, N. Hadjichristidis, Carboxylic acid initiated organocatalytic ring-opening polymerization of *N*-sulfonyl aziridines: an easy access to well-controlled polyaziridine-based architectural and functionalized polymers, Macromolecules 52 (2019) 8793–8802.
[32] Y. Zhao, W. Zhu, L. Ren, K. Zhang, Aggregation-induced emission polymer nanoparticles with pH-responsive fluorescence, Polym. Chem. 7 (2016) 5386–5395.

[33] S.H. Ryu, D.H. Lee, Y.-J. Ko, S.M. Lee, H.J. Kim, K.C. Ko, S.U. Son, Aligned tubular conjugated microporous polymer films for the aggregation-induced emission-based sensing of explosives, Macromol. Chem. Phys. 220 (2019) 1900157.

[34] J. Gu, Z. Xu, D. Ma, A. Qin, B.Z. Tang, Aggregation-induced emission polymers for high performance PLEDs with low efficiency roll-off, Mater. Chem. Front. 4 (2020) 1206–1211.

[35] H. Wan, P. Gu, F. Zhou, H. Wang, J. Jiang, D. Chen, Q. Xu, J. Lu, Polyacrylic esters with a "one-is-enough" effect and investigation of their AIEE behaviours and cyanide detection in aqueous solution, Polym. Chem. 9 (2018) 3893–3899.

[36] Y. Jiang, N. Hadjichristidis, Tetraphenylethene-functionalized polyethylene-based polymers with aggregation-induced emission, Macromolecules 52 (2019) 1955–1964.

[37] J. Dong, R. Jiang, W. Wan, H. Ma, H. Huang, Y. Feng, Y. Dai, H. Ouyang, X. Zhang, Y. Wei, Two birds one stone: facile preparation of AIE-active fluorescent polymeric nanoparticles via self-catalyzed photo-mediated polymerization, Appl. Surf. Sci. 508 (2020), 144799.

[38] Y. Zhao, Y. Wu, S. Chen, H. Deng, X. Zhu, Building single-color AIE-active reversible micelles to interpret temperature and pH stimuli in both solutions and cells, Macromolecules 51 (2018) 5234–5244.

[39] R. Jiang, M. Liu, Q. Huang, H. Huang, Q. Wan, Y. Wen, J. Tian, Q.-y. Cao, X. Zhang, Y. Wei, Fabrication of multifunctional fluorescent organic nanoparticles with AIE feature through photo-initiated RAFT polymerization, Polym. Chem. 8 (2017) 7390–7399.

[40] T. Cheng, Y. Chen, A. Qin, B.Z. Tang, Single component polymerization of diisocyanoacetates toward polyimidazoles, Macromolecules 51 (2018) 5638–5645.

[41] H. Liu, S. Zhang, X. Yan, C. Song, J. Chen, Y. Dong, X. Li, Silylium cation initiated sergeants-and-soldiers type chiral amplification of helical aryl isocyanide copolymers, Polym. Chem. 11 (2020) 6017–6028.

[42] W. Yuan, W. Chi, R. Liu, H. Li, Y. Li, B.Z. Tang, Synthesis of poly(phenyltriazolylcarboxylate)s with aggregation-induced emission characteristics by metal-free 1,3-dipolar polycycloaddition of phenylpropiolate and azides, Macromol. Rapid Commun. 38 (2017) 1600745.

[43] K. Wang, H. Si, Q. Wan, Z. Wang, A. Qin, B.Z. Tang, Luminescent two-way reversible shape memory polymers prepared by hydroxyl–yne click polymerization, J. Mater. Chem. C 8 (2020) 16121–16128.

[44] R. Jiang, M. Liu, H. Huang, L. Mao, Q. Huang, Y. Wen, Q.-y. Cao, J. Tian, X. Zhang, Y. Wei, Fabrication of AIE-active fluorescent polymeric nanoparticles with red emission through a facile catalyst-free amino-yne click polymerization, Dyes Pigm. 151 (2018) 123–129.

[45] D. Huang, Y. Liu, S. Guo, B. Li, J. Wang, B. Yao, A. Qin, B.Z. Tang, Transition metal-free thiol–yne click polymerization toward Z-stereoregular poly(vinylene sulfide)s, Polym. Chem. 10 (2019) 3088–3096.

[46] H. Qin, J. Huang, H. Liang, J. Lu, Aggregation-induced emission-active fluorescent polymer: multi-targeted sensor and ROS scavenger, ACS Appl. Mater. Interfaces 13 (2021) 5668–5677.

[47] W.-W. Chi, R.-Y. Zhang, T. Han, J. Du, H.-K. Li, W.-J. Zhang, Y.-F. Li, B.Z. Tang, Facile synthesis of functional poly(methyltriazolylcarboxylate)s by solvent- and catalyst-free butynoate-azide polycycloaddition, Chin. J. Polym. Sci. 38 (2020) 17–23.

[48] H. Si, K. Wang, B. Song, A. Qin, B.Z. Tang, Organobase-catalysed hydroxyl–yne click polymerization, Polym. Chem. 11 (2020) 2568–2575.

[49] X. Chen, R. Hu, C. Qi, X. Fu, J. Wang, B. He, D. Huang, A. Qin, B.Z. Tang, Ethynylsulfone-based spontaneous amino-yne click polymerization: a facile tool toward regio- and stereoregular dynamic polymers, Macromolecules 52 (2019) 4526–4533.

[50] J. Du, D. Huang, H. Li, A. Qin, B.Z. Tang, Y. Li, Catalyst-free click polymerization of thiol and activated internal alkynes: a facile strategy toward functional poly(β-thioacrylate)s, Macromolecules 53 (2020) 4932–4941.

[51] Z. Wang, C. Wang, Y. Fang, H. Yuan, Y. Quan, Y. Cheng, Color-tunable AIE-active conjugated polymer nanoparticles as drug carriers for self-indicating cancer therapy via intramolecular FRET mechanism, Polym. Chem. 9 (2018) 3205–3214.

[52] X. Liu, X. Liang, Y. Hu, L. Han, Q. Qu, D. Liu, J. Guo, Z. Zeng, H. Bai, R.T.K. Kwok, A. Qin, J.W.Y. Lam, B.Z. Tang, Catalyst-free spontaneous polymerization with 100% atom economy: facile synthesis of photoresponsive polysulfonates with multifunctionalities, JACS Au 1 (2021) 344–353.

[53] H. Wan, S. Zhou, P. Gu, F. Zhou, D. Lyu, Q. Xu, A. Wang, H. Shi, Q. Xu, J. Lu, AIE-active polysulfates via a sulfur(VI) fluoride exchange (SuFEx) click reaction and investigation of their two-photon fluorescence and cyanide detection in water and in living cells, Polym. Chem. 11 (2020) 1033–1042.

[54] X. Liu, T. Chen, F. Yu, Y. Shang, X. Meng, Z.-R. Chen, AIE-active random conjugated copolymers synthesized by ADMET polymerization as a fluorescent probe specific for palladium detection, Macromolecules 53 (2020) 1224–1232.

[55] T. Han, Z. Zhao, J.W.Y. Lam, B.Z. Tang, Monomer stoichiometry imbalance-promoted formation of multisubstituted polynaphthalenes by palladium-catalyzed polycouplings of aryl iodides and internal diynes, Polym. Chem. 9 (2018) 885–893.

[56] Q. Gao, T. Han, Z. Qiu, R. Zhang, J. Zhang, R.T.K. Kwok, J.W.Y. Lam, B.Z. Tang, Palladium-catalyzed polyannulation of pyrazoles and diynes toward multifunctional poly(indazole)s under monomer nonstoichiometric conditions, Polym. Chem. 10 (2019) 5296–5303.

[57] X. Liu, M. Li, T. Han, B. Cao, Z. Qiu, Y. Li, Q. Li, Y. Hu, Z. Liu, J.W.Y. Lam, X. Hu, B.Z. Tang, *In Situ* generation of azonia-containing polyelectrolytes for luminescent photopatterning and superbug killing, J. Am. Chem. Soc. 141 (2019) 11259–11268.

[58] Y. Wu, L. Qu, J. Li, L. Huang, Z. Liu, A versatile method for preparing well-defined polymers with aggregation-induced emission property, Polymer 158 (2018) 297–307.

[59] Z. Zhang, Y. You, C. Hong, Multicomponent reactions and multicomponent cascade reactions for the synthesis of sequence-controlled polymers, Macromol. Rapid Commun. 39 (2018) 1800362.

[60] R. Kakuchi, Multicomponent reactions in polymer synthesis, Angew. Chem. Int. Ed. 53 (2014) 46–48.

[61] G. Liu, R. Pan, Y. Wei, L. Tao, The Hantzsch reaction in polymer chemistry: from synthetic methods to applications, Macromol. Rapid Commun. 42 (2021) 2000459.

[62] Y. Zhao, H. Wu, Z. Wang, Y. Wei, Z. Wang, L. Tao, Training the old dog new tricks: the applications of the Biginelli reaction in polymer chemistry, Sci. China Chem. 59 (2016) 1541–1547.

[63] T. Jia, N.-n. Zheng, W.-q. Cai, J. Zhang, L. Ying, F. Huang, Y. Cao, Microwave-assisted one-pot three-component polymerization of alkynes, aldehydes and amines toward amino-functionalized optoelectronic polymers, Chin. J. Polym. Sci. 35 (2017) 269–281.

[64] S. Saxer, C. Marestin, R. Mercier, J. Dupuy, The multicomponent Debus–Radziszewski reaction in macromolecular chemistry, Polym. Chem. 9 (2018) 1927–1933.

[65] X. Su, Q. Gao, D. Wang, T. Han, B.Z. Tang, One-step multicomponent polymerizations for the synthesis of multifunctional AIE polymers, Macromol. Rapid Commun. 42 (2021) 2000471.

[66] L. Xu, R. Hu, B.Z. Tang, Room temperature multicomponent polymerizations of alkynes, sulfonyl azides, and iminophosphorane toward heteroatom-rich multifunctional poly(phosphorus amidine)s, Macromolecules 50 (2017) 6043–6053.

[67] L. Xu, F. Zhou, M. Liao, R. Hu, B.Z. Tang, Room temperature multicomponent polymerizations of alkynes, sulfonyl azides, and *N*-protected isatins toward oxindole-containing poly(*N*-acylsulfonamide)s, Polym. Chem. 9 (2018) 1674–1683.

[68] L. Xu, T. Zhou, M. Liao, R. Hu, B.Z. Tang, Multicomponent polymerizations of alkynes, sulfonyl azides, and 2-hydroxybenzonitrile/2-aminobenzonitrile toward multifunctional iminocoumarin/quinoline-containing poly(*N*-sulfonylimine)s, ACS Macro Lett. 8 (2019) 101–106.

[69] Y. Huang, L. Xu, R. Hu, B.Z. Tang, Cu(I)-catalyzed heterogeneous multicomponent polymerizations of alkynes, sulfonyl azides, and NH_4Cl, Macromolecules 53 (2020) 10366–10374.

[70] Q. Gao, L.-H. Xiong, T. Han, Z. Qiu, X. He, H.H.Y. Sung, R.T.K. Kwok, I.D. Williams, J.W.Y. Lam, B.Z. Tang, Three-component regio- and stereoselective polymerizations toward functional chalcogen-rich polymers with AIE-activities, J. Am. Chem. Soc. 141 (2019) 14712–14719.

[71] Y. Hu, T. Han, N. Yan, J. Liu, X. Liu, W.-X. Wang, J.W.Y. Lam, B.Z. Tang, Visualization of biogenic amines and *in vivo* ratiometric mapping of intestinal pH by AIE-active polyheterocycles synthesized by metal-free multicomponent polymerizations, Adv. Funct. Mater. 29 (2019) 1902240.

[72] Y. Zhang, N.-W. Tseng, H. Deng, R.T.K. Kwok, J.W.Y. Lam, B.Z. Tang, BCl_3-mediated polycoupling of alkynes and aldehydes: a facile, metal-free multicomponent polymerization route to construct stereoregular functional polymers, Polym. Chem. 7 (2016) 4667–4674.

[73] T. Tian, R. Hu, B.Z. Tang, Room temperature one-step conversion from elemental sulfur to functional polythioureas through catalyst-free multicomponent polymerizations, J. Am. Chem. Soc. 140 (2018) 6156–6163.

[74] B. Song, B. He, A. Qin, B.Z. Tang, Direct polymerization of carbon dioxide, diynes, and alkyl dihalides under mild reaction conditions, Macromolecules 51 (2017) 42–48.

[75] H. Huang, R. Jiang, H. Ma, Y. Li, Y. Zeng, N. Zhou, L. Liu, X. Zhang, Y. Wei, Fabrication of claviform fluorescent polymeric nanomaterials containing disulfide bond through an efficient and facile four-component Ugi reaction, Mater. Sci. Eng. C 118 (2021), 111437.

[76] B. Wei, W. Li, Z. Zhao, A. Qin, R. Hu, B.Z. Tang, Metal-free multicomponent tandem polymerizations of alkynes, amines, and formaldehyde toward structure- and sequence-controlled luminescent polyheterocycles, J. Am. Chem. Soc. 139 (2017) 5075–5084.
[77] J. Zhang, W. Wang, Y. Liu, J. Sun, A. Qin, B.Z. Tang, Facile polymerization of water and triple-bond based monomers toward functional polyamides, Macromolecules 50 (2017) 8554–8561.
[78] B. Song, T. Bai, X. Xu, X. Chen, D. Liu, J. Guo, A. Qin, J. Ling, B.Z. Tang, Multifunctional linear and hyperbranched five-membered cyclic carbonate-based polymers directly generated from CO_2 and alkyne-based three-component polymerization, Macromolecules 52 (2019) 5546–5554.
[79] B. Song, K. Hu, A. Qin, B.Z. Tang, Oxygen as a crucial comonomer in alkyne-based polymerization toward functional poly(tetrasubstituted furan)s, Macromolecules 51 (2018) 7013–7018.
[80] X. Tang, L. Zhang, R. Hu, B.Z. Tang, Multicomponent tandem polymerization of aromatic alkynes, carbonyl chloride, and Fischer's base toward poly(diene merocyanine)s, Chin. J. Chem. 37 (2019) 1264–1270.
[81] C. Qi, C. Zheng, R. Hu, B.Z. Tang, Direct construction of acid-responsive poly(indolone)s through multicomponent tandem polymerizations, ACS Macro Lett. 8 (2019) 569–575.
[82] W. Tian, R. Hu, B.Z. Tang, One-pot multicomponent tandem reactions and polymerizations for step-economic synthesis of structure-controlled pyrimidine derivatives and poly(pyrimidine)s, Macromolecules 51 (2018) 9749–9757.
[83] J. Zhao, Z. Dong, H. Cui, H. Jin, C. Wang, Nanoengineered peptide-grafted hyperbranched polymers for killing of bacteria monitored in real time via intrinsic aggregation-induced emission, ACS Appl. Mater. Interfaces 10 (2018) 42058–42067.
[84] L. Mao, M. Liu, R. Jiang, Q. Huang, Y. Dai, J. Tian, Y. Shi, Y. Wen, X. Zhang, Y. Wei, The one-step acetalization reaction for construction of hyperbranched and biodegradable luminescent polymeric nanoparticles with aggregation-induced emission feature, Mater. Sci. Eng. C 80 (2017) 543–548.
[85] R. Jiang, M. Liu, T. Chen, H. Huang, Q. Huang, J. Tian, Y. Wen, Q.-y. Cao, X. Zhang, Y. Wei, Facile construction and biological imaging of cross-linked fluorescent organic nanoparticles with aggregation-induced emission feature through a catalyst-free azide-alkyne click reaction, Dyes Pigm. 148 (2018) 52–60.
[86] J. Liu, Y.-Q. Fan, S.-S. Song, G.-F. Gong, J. Wang, X.-W. Guan, H. Yao, Y.-M. Zhang, T.-B. Wei, Q. Lin, Aggregation-induced emission supramolecular organic framework (AIE SOF) gels constructed from supramolecular polymer networks based on tripodal pillar[5]arene for fluorescence detection and efficient removal of various analytes, ACS Sustain. Chem. Eng. 7 (2019) 11999–12007.
[87] K. Kokado, R. Taniguchi, K. Sada, Rigidity-induced emission enhancement of network polymers crosslinked by tetraphenylethene derivatives, J. Mater. Chem. C 3 (2015) 8504–8509.
[88] M. Chen, L. Li, H. Wu, L. Pan, S. Li, B. He, H. Zhang, J.Z. Sun, A. Qin, B.Z. Tang, Unveiling the different emission behavior of polytriazoles constructed from pyrazine-based AIE monomers by click polymerization, ACS Appl. Mater. Interfaces 10 (2018) 12181–12188.
[89] S. Lee, C.H. Jang, T.L. Nguyen, S.H. Kim, K.M. Lee, K. Chang, S.S. Choi, S.K. Kwak, H.Y. Woo, M.H. Song, Conjugated polyelectrolytes as multifunctional passivating and hole-transporting layers for efficient perovskite light-emitting diodes, Adv. Mater. 31 (2019) 1900067.
[90] B. Wang, B.N. Queenan, S. Wang, K.P.R. Nilsson, G.C. Bazan, Precisely defined conjugated oligoelectrolytes for biosensing and therapeutics, Adv. Mater. 31 (2019) 1806701.
[91] H. Yao, J. Dai, Z. Zhuang, J. Yao, Z. Wu, S. Wang, F. Xia, J. Zhou, X. Lou, Z. Zhao, Red AIE conjugated polyelectrolytes for long-term tracing and image-guided photodynamic therapy of tumors, Sci. China Chem. 63 (2020) 1815–1824.
[92] A.H. Malik, A. Kalita, P.K. Iyer, Development of well-preserved, substrate-versatile latent fingerprints by aggregation-induced enhanced emission-active conjugated polyelectrolyte, ACS Appl. Mater. Interfaces 9 (2017) 37501–37508.
[93] M. Gao, Y. Hong, B. Chen, Y. Wang, W. Zhou, W.W.H. Wong, J. Zhou, T.A. Smith, Z. Zhao, AIE conjugated polyelectrolytes based on tetraphenylethene for efficient fluorescence imaging and lifetime imaging of living cells, Polym. Chem. 8 (2017) 3862–3866.
[94] Y. Qian, H. Liu, H. Tan, Q. Yang, S. Zhang, L. Han, X. Yi, L. Huo, H. Zhao, Y. Wu, L. Bai, X. Ba, A novel water-soluble fluorescence probe with wash-free cellular imaging capacity based on AIE characteristics, Macromol. Rapid Commun. 38 (2017) 1600684.

[95] E. Wang, S. Liu, J.W.Y. Lam, B.Z. Tang, X. Wang, F. Wang, Deciphering structure–functionality relationship of polycarbonate-based polyelectrolytes by AIE technology, Macromolecules 53 (2020) 5839–5846.

[96] Y. Hu, F. Song, Z. Xu, Y. Tu, H. Zhang, Q. Cheng, J.W.Y. Lam, D. Ma, B.Z. Tang, Circularly polarized luminescence from chiral conjugated poly(carbazole-*ran*-acridine)s with aggregation-induced emission and delayed fluorescence, ACS Appl. Polym. Mater. 1 (2019) 221–229.

[97] Q. Liu, Q. Xia, S. Wang, B.S. Li, B.Z. Tang, *In situ* visualizable self-assembly, aggregation-induced emission and circularly polarized luminescence of tetraphenylethene and alanine-based chiral polytriazole, J. Mater. Chem. C 6 (2018) 4807–4816.

[98] Q. Liu, Q. Xia, Y. Xiong, B.S. Li, B.Z. Tang, Circularly polarized luminescence and tunable helical assemblies of aggregation-induced emission amphiphilic polytriazole carrying chiral L-phenylalanine pendants, Macromolecules 53 (2020) 6288–6298.

[99] S. Zhang, Y. Sheng, G. Wei, Y. Quan, Y. Cheng, C. Zhu, Aggregation-induced circularly polarized luminescence of an (R)-binaphthyl-based AIE-active chiral conjugated polymer with self-assembled helical nanofibers, Polym. Chem. 6 (2015) 2416–2422.

[100] L. Yang, Y. Zhang, X. Zhang, N. Li, Y. Quan, Y. Cheng, Doping-free circularly polarized electroluminescence of AIE-active chiral binaphthyl-based polymers, Chem. Commun. 54 (2018) 9663–9666.

[101] X. Yan, S. Zhang, P. Zhang, X. Wu, A. Liu, G. Guo, Y. Dong, X. Li, $[Ph_3C][B(C_6F_5)_4]$: a highly efficient metal-free single-component initiator for the helical-sense-selective cationic copolymerization of chiral aryl isocyanides and achiral aryl isocyanides, Angew. Chem. Int. Ed. 57 (2018) 8947–8952.

[102] K. Liu, Y. Shen, X. Li, Y. Zhang, Y. Quan, Y. Cheng, Strong CPL of achiral liquid crystal fluorescent polymer via the regulation of AIE-active chiral dopant, Chem. Commun. 56 (2020) 12829–12832.

[103] J. Liu, Z. Luo, L. Yu, P. Zhang, H. Wei, Y. Yu, A new soft-matter material with old chemistry: Passerini multicomponent polymerization-induced assembly of AIE-active double-helical polymers with rapid visible-light degradability, Chem. Sci. 11 (2020) 8224–8230.

[104] M. Hu, H.-T. Feng, Y.-X. Yuan, Y.-S. Zheng, B.Z. Tang, Chiral AIEgens-chiral recognition, CPL materials and other chiral applications, Coord. Chem. Rev. 416 (2020), 213329.

[105] N. Kalva, S. Uthaman, E.H. Jang, R. Augustine, S.H. Jeon, K.M. Huh, I.-K. Park, I. Kim, Aggregation-induced emission-active hyperbranched polymer-based nanoparticles and their biological imaging applications, Dyes Pigm. 186 (2021), 108975.

[106] W. Wu, S. Ye, G. Yu, Y. Liu, J. Qin, Z. Li, Novel functional conjugative hyperbranched polymers with aggregation-induced emission: synthesis through one-pot "A_2+B_4" polymerization and application as explosive chemsensors and PLEDs, Macromol. Rapid Commun. 33 (2012) 164–171.

[107] W. Chi, W. Yuan, J. Du, T. Han, H. Li, Y. Li, B.Z. Tang, Construction of functional hyperbranched poly(phenyltriazolylcarboxylate)s by metal-free phenylpropiolate-azide polycycloaddition, Macromol. Rapid Commun. 39 (2018) 1800604.

[108] B. He, J. Zhang, J. Wang, Y. Wu, A. Qin, B.Z. Tang, Preparation of multifunctional hyperbranched poly(β-aminoacrylate)s by spontaneous amino-yne click polymerization, Macromolecules 53 (2020) 5248–5254.

[109] Y. Huang, P. Chen, B. Wei, R. Hu, B.Z. Tang, Aggregation-induced emission-active hyperbranched poly(tetrahydropyrimidine) s synthesized from multicomponent tandem polymerization, Chin. J. Polym. Sci. 37 (2019) 428–436.

[110] R. Chen, X. Gao, X. Cheng, A. Qin, J.Z. Sun, B.Z. Tang, A red-emitting cationic hyperbranched polymer: facile synthesis, aggregation-enhanced emission, large stokes shift, polarity-insensitive fluorescence and application in cell imaging, Polym. Chem. 8 (2017) 6277–6282.

[111] H. Ma, Y. Qin, Z. Yang, M. Yang, Y. Ma, P. Yin, Y. Yang, T. Wang, Z. Lei, X. Yao, Positively charged hyperbranched polymers with tunable fluorescence and cell imaging application, ACS Appl. Mater. Interfaces 10 (2018) 20064–20072.

[112] K. Cousins, R. Zhang, Highly porous organic polymers for hydrogen fuel storage, Polymers 11 (2019) 690.

[113] L. Zhang, L. Yi, Z.-J. Sun, H. Deng, Covalent organic frameworks for optical applications, Aggregate 2 (2021), e24.

[114] X. Zhang, S. Zhen, L. Zhang, J. Chai, L. Zou, X. Xin, J. Xu, G. Zhang, The electrosynthesis of highly photofunctional porous polymer PTCPE and the effect of BFEE on its electrochemical polymerization and fluorescence property, Polymer 202 (2020), 122731.

[115] A. Palma-Cando, D. Woitassek, G. Brunklaus, U. Scherf, Luminescent tetraphenylethene-cored, carbazole- and thiophene-based microporous polymer films for the chemosensing of nitroaromatic analytes, Mater. Chem. Front. 1 (2017) 1118–1124.

[116] C. Gu, N. Huang, Y. Wu, H. Xu, D. Jiang, Design of highly photofunctional porous polymer films with controlled thickness and prominent microporosity, Angew. Chem. Int. Ed. 54 (2015) 11540–11544.

[117] H. Ding, J. Li, G. Xie, G. Lin, R. Chen, Z. Peng, C. Yang, B. Wang, J. Sun, C. Wang, An AIEgen-based 3D covalent organic framework for white light-emitting diodes, Nat. Commun. 9 (2018) 5234.

[118] S.H. Ryu, D.H. Lee, S.M. Lee, H.J. Kim, Y.-J. Ko, K.C. Ko, S.U. Son, Morphology engineering of a Suzuki coupling-based microporous organic polymer (MOP) using a Sonogashira coupling-based MOP for enhanced nitrophenol sensing in water, Chem. Commun. 55 (2019) 9515–9518.

[119] G. Lin, H. Ding, D. Yuan, B. Wang, C. Wang, A pyrene-based, fluorescent three-dimensional covalent organic framework, J. Am. Chem. Soc. 138 (2016) 3302–3305.

[120] Y. Chen, J.W.Y. Lam, R.T.K. Kwok, B. Liu, B.Z. Tang, Aggregation-induced emission: fundamental understanding and future developments, Mater. Horiz. 6 (2019) 428–433.

[121] L. Xu, X. Liang, S. Zhong, Y. Gao, X. Cui, Clustering-triggered emission from natural products: gelatin and its multifunctional applications, ACS Sustain. Chem. Eng. 8 (2020) 18816–18823.

[122] L. Dong, W. Fu, P. Liu, J. Shi, B. Tong, Z. Cai, J. Zhi, Y. Dong, Spontaneous multicomponent polymerization of imidazole, diacetylenic esters, and diisocyanates for the preparation of poly(β-aminoacrylate)s with cluster-induced emission characteristics, Macromolecules 53 (2020) 1054–1062.

[123] T. Han, H. Deng, Z. Qiu, Z. Zhao, H. Zhang, H. Zou, N.L.C. Leung, G. Shan, M.R.J. Elsegood, J.W.Y. Lam, B.Z. Tang, Facile multicomponent polymerizations toward unconventional luminescent polymers with readily openable small heterocycles, J. Am. Chem. Soc. 140 (2018) 5588–5598.

[124] L. Fang, C. Huang, G. Shabir, J. Liang, Z. Liu, H. Zhang, Hyperbranching-enhanced-emission effect discovered in hyperbranched poly(4-(cyanomethyl)phenyl methacrylate), ACS Macro Lett. 8 (2019) 1605–1610.

[125] L. Song, T. Zhu, L. Yuan, J. Zhou, Y. Zhang, Z. Wang, C. Tang, Ultra-strong long-chain polyamide elastomers with programmable supramolecular interactions and oriented crystalline microstructures, Nat. Commun. 10 (2019) 1315.

[126] B. Wang, C. Li, L. Yang, C. Zhang, L.-J. Liu, S. Zhu, Y. Chen, Y. Wang, Tetraphenylethene decorated with disulfide-functionalized hyperbranched poly(amido amine)s as metal/organic solvent-free turn-on AIE probes for biothiol determination, J. Mater. Chem. B 7 (2019) 3846–3855.

[127] Y. Li, K. Xu, Y. Si, C. Yang, Q. Peng, J. He, Q. Hu, K. Li, An aggregation-induced emission (AIE) fluorescent chemosensor for the detection of Al(III) in aqueous solution, Dyes Pigm. 171 (2019), 107682.

[128] Q. Li, X. Li, Z. Wu, Y. Sun, J. Fang, D. Chen, Highly efficient luminescent side-chain polymers with short-spacer attached tetraphenylethylene AIEgens via RAFT polymerization capable of naked eye explosive detection, Polym. Chem. 9 (2018) 4150–4160.

[129] H. Zhou, M.H. Chua, B.Z. Tang, J. Xu, Aggregation-induced emission (AIE)-active polymers for explosive detection, Polym. Chem. 10 (2019) 3822–3840.

[130] J. Huang, H. Qin, H. Liang, J. Lu, An AIE polymer prepared via aldehyde-hydrazine step polymerization and the application in Cu^{2+} and S^{2-} detection, Polymer 202 (2020), 122663.

[131] P. Alam, N.L.C. Leung, J. Zhang, R.T.K. Kwok, J.W.Y. Lam, B.Z. Tang, AIE-based luminescence probes for metal ion detection, Coord. Chem. Rev. 429 (2021), 213693.

[132] P.Q. Nhien, W.-L. Chou, T.T.K. Cuc, T.M. Khang, C.-H. Wu, N. Thirumalaivasan, B.T.B. Hue, J.I. Wu, S.-P. Wu, H.-C. Lin, Multi-stimuli responsive FRET processes of bifluorophoric AIEgens in an amphiphilic copolymer and its application to cyanide detection in aqueous media, ACS Appl. Mater. Interfaces 12 (2020) 10959–10972.

[133] S. Liu, Y. Cheng, H. Zhang, Z. Qiu, R.T.K. Kwok, J.W.Y. Lam, B.Z. Tang, *In situ* monitoring of RAFT polymerization by tetraphenylethylene-containing agents with aggregation-induced emission characteristics, Angew. Chem. Int. Ed. 57 (2018) 6274–6278.

[134] X. Guan, L. Meng, Q. Jin, B. Lu, Y. Chen, Z. Li, L. Wang, S. Lai, Z. Lei, A new thermo-, pH- and CO_2-responsive fluorescent four-arm star polymer with aggregation-induced emission for long-term cellular tracing, Macromol. Mater. Eng. 303 (2018) 1700553.

[135] J.-J. Hu, W. Jiang, L. Yuan, C. Duan, Q. Yuan, Z. Long, X. Lou, F. Xia, Recent advances in stimuli-responsive theranostic systems with aggregation-induced emission characteristics, Aggregate 2 (2021) 48–65.

[136] S. Lee, K.Y. Kim, S.H. Jung, J.H. Lee, M. Yamada, R. Sethy, T. Kawai, J.H. Jung, Finely controlled circularly polarized luminescence of a mechano-responsive supramolecular polymer, Angew. Chem. Int. Ed. 58 (2019) 18878–18882.

[137] Z. Tang, X. Lyu, L. Luo, Z. Shen, X.-H. Fan, White-light-emitting AIE/Eu^{3+}-doped ion gel with multistimuli-responsive properties, ACS Appl. Mater. Interfaces 12 (2020) 45420–45428.

[138] J. Wu, B. Xu, Z. Liu, Y. Yao, Q. Zhuang, S. Lin, The synthesis, self-assembly and pH-responsive fluorescence enhancement of an alternating amphiphilic copolymer with azobenzene pendants, Polym. Chem. 10 (2019) 4025–4030.

[139] T. Li, S. He, J. Qu, H. Wu, S. Wu, Z. Zhao, A. Qin, R. Hu, B.Z. Tang, Thermoresponsive AIE polymers with fine-tuned response temperature, J. Mater. Chem. C 4 (2016) 2964–2970.

[140] D. Cappello, A.E.R. Watson, J.B. Gilroy, A boron difluoride hydrazone (BODIHY) polymer exhibits aggregation-induced emission, Macromol. Rapid Commun. 42 (2020) 2000553.

[141] Z. Huang, Y. Chen, C. Zhou, K. Wang, X. Liu, L. Mao, J. Yuan, L. Tao, Y. Wei, Amphiphilic AIE-active copolymers with optical activity by chemoenzymatic transesterification and RAFT polymerization: synthesis, self-assembly and biological imaging, Dyes Pigm. 184 (2021), 108829.

[142] T. Zhou, R. Hu, L. Wang, Y. Qiu, G. Zhang, Q. Deng, H. Zhang, P. Yin, B. Situ, C. Zhan, A. Qin, B.Z. Tang, An AIE-active conjugated polymer with high ROS-generation ability and biocompatibility for efficient photodynamic therapy of bacterial infections, Angew. Chem. Int. Ed. 59 (2020) 9952–9956.

[143] Z. Dong, Y. Wang, C. Wang, H. Meng, Y. Li, C. Wang, Cationic peptidopolysaccharide with an intrinsic AIE effect for combating bacteria and multicolor imaging, Adv. Healthc. Mater. 9 (2020) 2000419.

[144] C. Ma, T. Han, M. Kang, E. Liarou, A.M. Wemyss, S. Efstathiou, B.Z. Tang, D. Haddleton, Aggregation-induced emission active polyacrylates via cu-mediated reversible deactivation radical polymerization with bioimaging applications, ACS Macro Lett. 9 (2020) 769–775.

[145] B. Ren, K. Li, Z. Liu, G. Liu, H. Wang, White light-triggered zwitterionic polymer nanoparticles based on an AIE-active photosensitizer for photodynamic antimicrobial therapy, J. Mater. Chem. B 8 (2020) 10754–10763.

[146] H. Ma, Y. Ma, L. Lei, W. Yin, Y. Yang, T. Wang, P. Yin, Z. Lei, M. Yang, Y. Qin, S. Zhang, Light-enhanced bacterial killing and less toxic cell imaging: multicationic aggregation-induced emission matters, ACS Sustain. Chem. Eng. 6 (2018) 15064–15071.

[147] Z. Liu, L. Zhang, X. Gao, L. Zhang, Q. Zhang, J. Chen, Highly efficient green PLED based on triphenlyaminesilole-carbazole-fluorene copolymers with TPBI as the hole blocking layer, Dyes Pigm. 127 (2016) 155–160.

[148] Y. Zhan, Z. Yang, J. Tan, Z. Qiu, Y. Mao, J. He, Q. Yang, S. Ji, N. Cai, Y. Huo, Synthesis, aggregation-induced emission (AIE) and electroluminescence of carbazole-benzoyl substituted tetraphenylethylene derivatives, Dyes Pigm. 173 (2020), 107898.

[149] C.-Y. Ke, M.-N. Chen, Y.-C. Chiu, G.-S. Liou, Luminescence behavior and acceptor effects of ambipolar polymeric electret on photorecoverable organic field-effect transistor memory, Adv. Electron. Mater. 7 (2021) 2001076.

[150] Y. Xie, Y. Gong, M. Han, F. Zhang, Q. Peng, G. Xie, Z. Li, Tetraphenylcyclopentadiene-based hyperbranched polymers: convenient syntheses from one pot "$A_4 + B_2$" polymerization and high external quantum yields up to 9.74% in OLED devices, Macromolecules 52 (2019) 896–903.

[151] Y. Yuan, J. Li, L. He, Y. Liu, H. Zhang, Preparation and properties of side chain liquid crystalline polymers with aggregation-induced emission enhancement characteristics, J. Mater. Chem. C 6 (2018) 7119–7127.

[152] S.-W. Cheng, T. Han, T.-Y. Huang, B.-Z. Tang, G.-S. Liou, High-performance electrofluorochromic devices based on aromatic polyamides with AIE-active tetraphenylethene and electro-active triphenylamine moieties, Polym. Chem. 9 (2018) 4364–4373.

[153] N. Lu, X. Gao, M. Pan, B. Zhao, J. Deng, Aggregation-induced emission-active chiral helical polymers show strong circularly polarized luminescence in thin films, Macromolecules 53 (2020) 8041–8049.

CHAPTER

4

Chiral aggregation-induced emission molecules: Design, circularly polarized luminescence, and helical self-assembly

Hongkun Li[a], *Bing Shi Li*[b], *and Ben Zhong Tang*[c]

[a]Laboratory of Advanced Optoelectronic Materials, College of Chemistry, Chemical Engineering and Materials Science, Soochow University, Suzhou, China [b]Key Laboratory of New Lithium-Ion Battery and Mesoporous Material, College of Chemistry and Environmental Engineering, Shenzhen University, Shenzhen, China [c]School of Science and Engineering, Shenzhen Key Laboratory of Functional Aggregate Materials, The Chinese University of Hong Kong, Shenzhen, Guangdong, China

1 Introduction

Fabrication of luminescent organic architectures with ordered structures has drawn considerable attention during the past decades because of their novel optical and photophysical properties and applications in optoelectronic devices [1–3]. Self-assembly of π-conjugated molecules, driven by noncovalent interactions such as π–π stacking, hydrogen bonding, electrostatic interaction, and dipole-dipole attraction, is an efficient approach for constructing supramolecular architectures [4–6]. Unfortunately, most "traditional" organic fluorophores emit efficiently in dilute solutions but exhibit weak luminescence or even no emission in the aggregate state, which is the notorious aggregation-caused quenching (ACQ) effect [7]. The ACQ effect limits the application of traditional luminescent molecules in the fabrication of efficient fluorescent micro-/nanostructures. The development of new luminophors with high efficiency in solid state is thus highly desirable.

https://doi.org/10.1016/B978-0-12-824335-0.00019-2

Aggregation-induced emission luminogens (AIEgens), as best exemplified by silole, tetraphenylethene (TPE), and their derivatives, are nonemissive in dilute solution but highly emissive in aggregated or solid state due to the restriction of intramolecular motions [8]. This unique luminescent property makes the AIEgens promising building blocks for the construction of fluorescent micro/nanomaterials with high efficiency. With the sustained efforts of researchers, a variety of luminescent supramolecular architectures based on AIEgens have been constructed [9–14]. Among them, helical micro-/nanostructures have attracted much interest in recent years because of their novel circularly polarized luminescence (CPL) properties [15–18], which make them an ideal candidate for 3D displays.

In this chapter, progress on the helical self-assembly of chiral AIEgens is summarized, which includes design rationales, chiroptical properties, and helical self-assembly behaviors of representative AIEgens [19–21].

2 Molecular design

Generally, there are two strategies for the fabrication of luminescent helical architectures with CPL properties by the self-assembly of AIEgens. The first strategy is to allow AIEgens to self-assemble according to the guidance of chiral templates or chiral host matrices [22–26]. For instance, the precursors of achiral AIEgens coassembled with chiral low-molecular-weight gelators and they formed single-handed helical polybissilsesquioxane nanotubes with endowed CPL activities and enhanced fluorescence efficiency [24]. The other strategy is to chemically attach chiral attachments to AIEgens to induce their helical self-assembly. According to this strategy, luminescent micro-/nanostructures were constructed using typical AIEgens such as silole and TPE as the fluorophors, and sugars, amino acids, and binaphthol as chiral pendants (Charts 1–5).

2.1 Silole derivatives

Since Tang's group reported that 1-methyl-1,2,3,4,5-pentaphenylsilole, a silole derivative, is a typical AIE-active organic fluorophor in 2001 [27], a number of silole-based AIEgens have

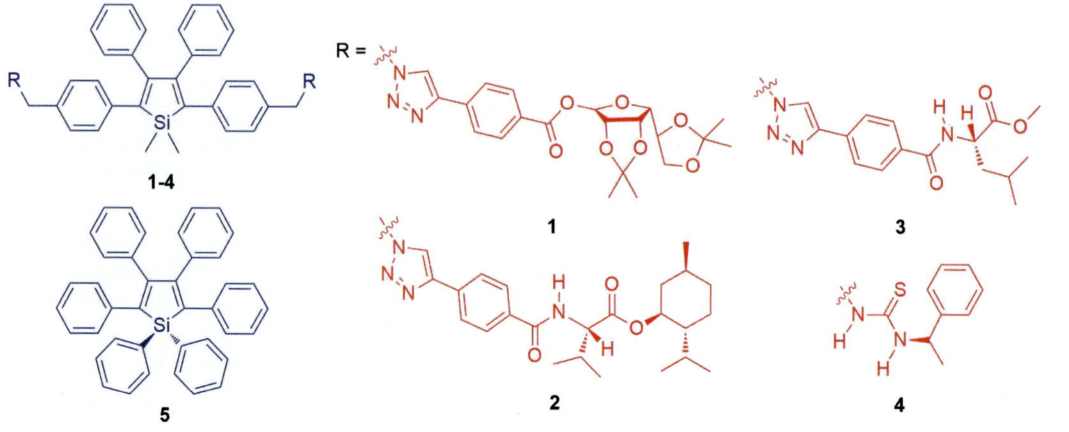

CHART 1 Molecular structures of silole derivatives **1–5**.

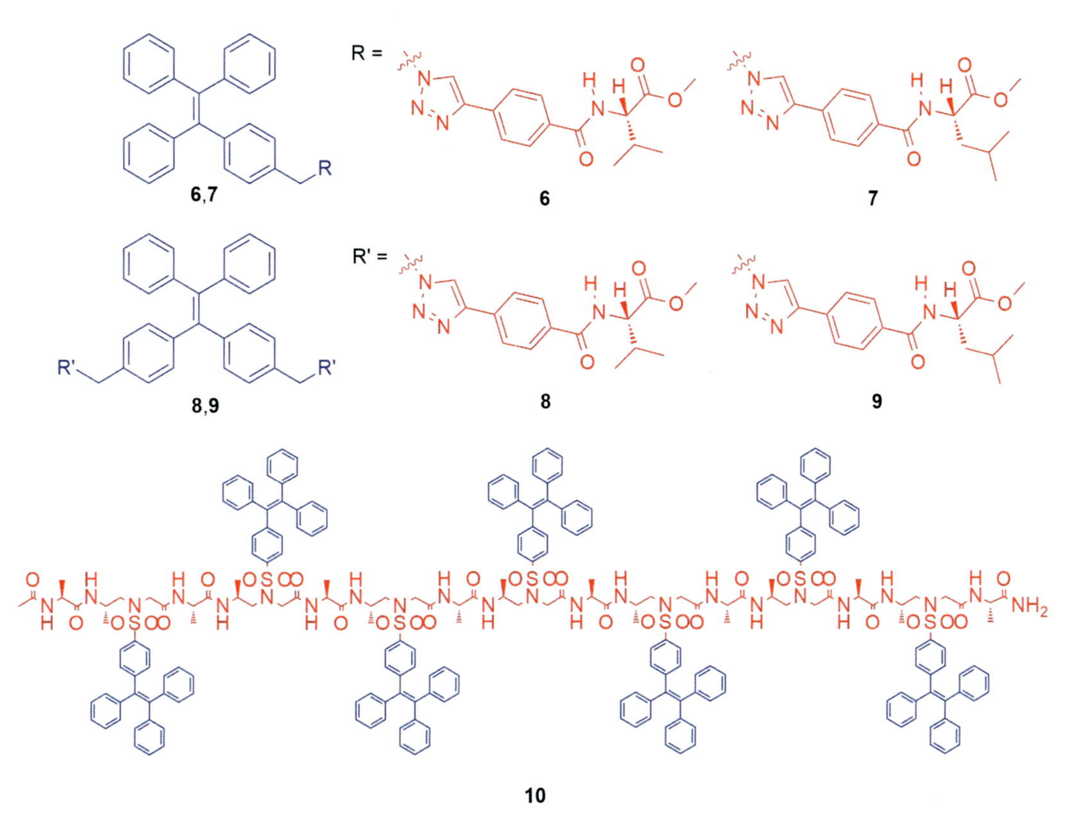

CHART 2 Molecular structures of TPE derivatives with point chirality, **6**–**10**.

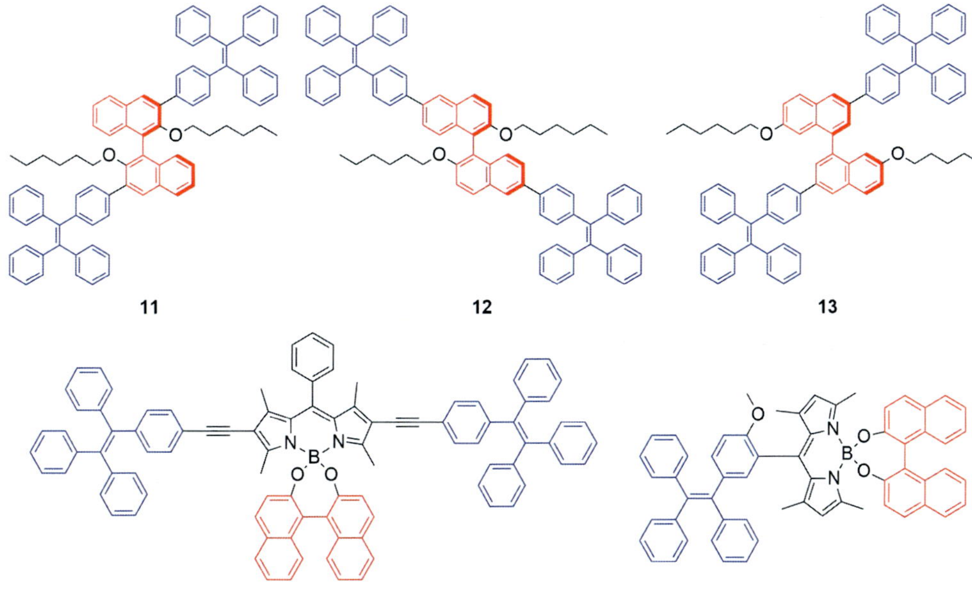

11 **12** **13**

R/*S*-**14** *R*/*S*-**15**

CHART 3 Molecular structures of chiral TPE derivatives with axial chirality, **11**–**15**.

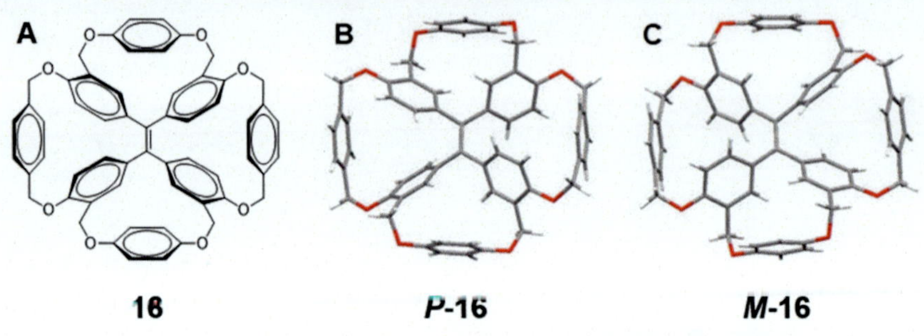

CHART 4 (A) Molecular structure of **16**. Crystal structures of *P*-**16** (B) and *M*-**16** (C).

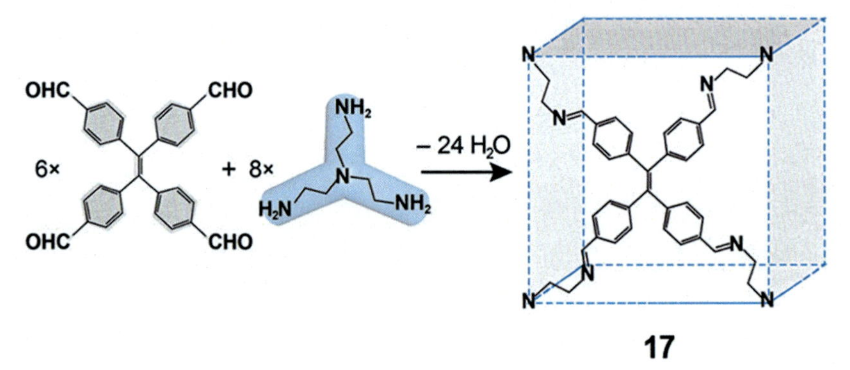

CHART 5 Schematic representation of synthesis of chiral organic cubes. *From H. Qu, Y. Wang, Z. Li, X. Wang, H. Fang, Z. Tian, X. Cao, Molecular face-rotating cube with emergent chiral and fluorescence properties, J. Am. Chem. Soc. 139 (2017) 18142–18145.*

been synthesized and widely used in optoelectronic and biomedical applications due to their high solid-state luminescence efficiency and photostability [28,29]. In 2012, a chiral silole-containing AIEgen **1** was first designed and synthesized by incorporating two mannose units into the tetraphenylsilole core by a Cu(I)-catalyzed azide-alkyne "click" reaction [30]. Using a similar strategy, chiral silole derivatives **2** and **3** with L-valine- and L-leucine-containing attachments were synthesized, respectively [31,32]. By introducing two chiral phenylethanamine moieties into a tetraphenylsilole unit, a thiourea-linked chiral silole derivative **4** was synthesized [33]. Considering that some achiral π-conjugated molecules can assemble into helical micro-/nanostructures [34,35], the achiral silole compound hexaphenylsilole bearing propeller-shaped structures (HPS, **5**) was synthesized to explore its self-assembly behaviors and CPL properties [36].

2.2 TPE derivatives

Although silole-based AIEgens generally possess high fluorescence quantum yields, their preparation process requires many reaction steps that involve the use of reactive organometallic species. Furthermore, the silole rings become unstable under basic conditions. These restrictions hinder the applications of silole-based AIEgens. In contrast, TPE, a star AIEgen, has the advantages of facile synthesis, easy functionalization, high luminescence efficiency, and good photo- and chemical stability [8]. TPE derivatives are ideal building blocks for constructing luminescent micro-/nanostructures and they have been more widely used in various applications in recent years.

In 2014, our groups designed and synthesized two chiral TPE derivatives, **6** and **7**, bearing L-valine- and L-leucine-containing attachments, respectively [37,38]. The effect of the number of chiral attachments on self-assembly behaviors and luminescence properties was also systematically explored by synthesizing two substituted chiral TPE derivatives, **8** and **9**, with bivaline and bileucine attachments, respectively [39,40]. By decorating TPE pendants to the right-handed peptidomimetic backbone, Cai et al. reported the synthesis of sulfono-γ-Aapeptides (**10**) [41]. Apart from these chiral TPE derivatives with point chirality, TPE derivatives with axial chirality, **11–15**, have also been synthesized by modifying TPE units with an axial chiral moiety, binaphthol [42–44].

Since the phenyl groups of TPE units are twisted around the molecular axes and adopt propeller-shaped conformations in crystal states, the enantiopure crystals of bare TPE, and some TPE core compounds, such as 1,2-bis(4-ethynylphenyl)-1,2-diphenylethene and tetrakis(4-ethynylphenyl)ethylene, can be obtained by spontaneous mirror-symmetry breaking in the absence of any chiral source [45,46]. However, they are usually racemes or mesomers and CD silent under conventional conditions. The helical conformation of TPE can be acquired by restricting phenyl ring rotation through covalent bonds. In 2016, Zheng and coworkers reported conformers with the propeller-like conformation of TPE fixed via intramolecular cyclization [47]. The fixed conformers (**16**) can be resolved into *M*- and *P*-enantiomer (Chart 4). In 2017, Cao et al. constructed chiral organic cages (**17**) by restricting the *P* or *M* rotational configuration of TPE faces through dynamic covalent chemistry (Chart 5) [48].

2.3 Other AIEgens

In addition to the silole and TPE derivatives, there is also exploration on other chiral AIEgens. We designed and synthesized two chiral phenanthro[9,10-*d*]imidazole (PIM) derivatives, **18** and **19**, by introducing L- and D-aniline pendants into PIM units [49], and two chiral AIEgens **20** and **21** by introducing chiral alanine to hydrazone derivatives [50]. We also reported chiral Au(I) complexes **22** and **23** possessing the BINOL core with AIE characteristics [51]. Compared with small molecular systems, the helical self-assembly of polymeric chiral AIEgens has been a less touched area [52–55]. We synthesized two TPE-based chiral conjugated polytriazoles, **24** and **25**, with L-alanine and L-phenylalanine, respectively, as pendants by the Cu(I)-catalyzed azide-alkyne click polymerization [54,55] (Chart 6).

3 Aggregation-induced emission

The compounds of **1–25** are all AIE-active. The photoluminescence (PL) spectra of **1** in solution and aggregated states are given and discussed as an example. It is almost nonemissive in the dichloromethane (DCM) solution and DCM-hexane mixtures with hexane fractions lower than 80%, but becomes highly emissive upon the addition of more poor

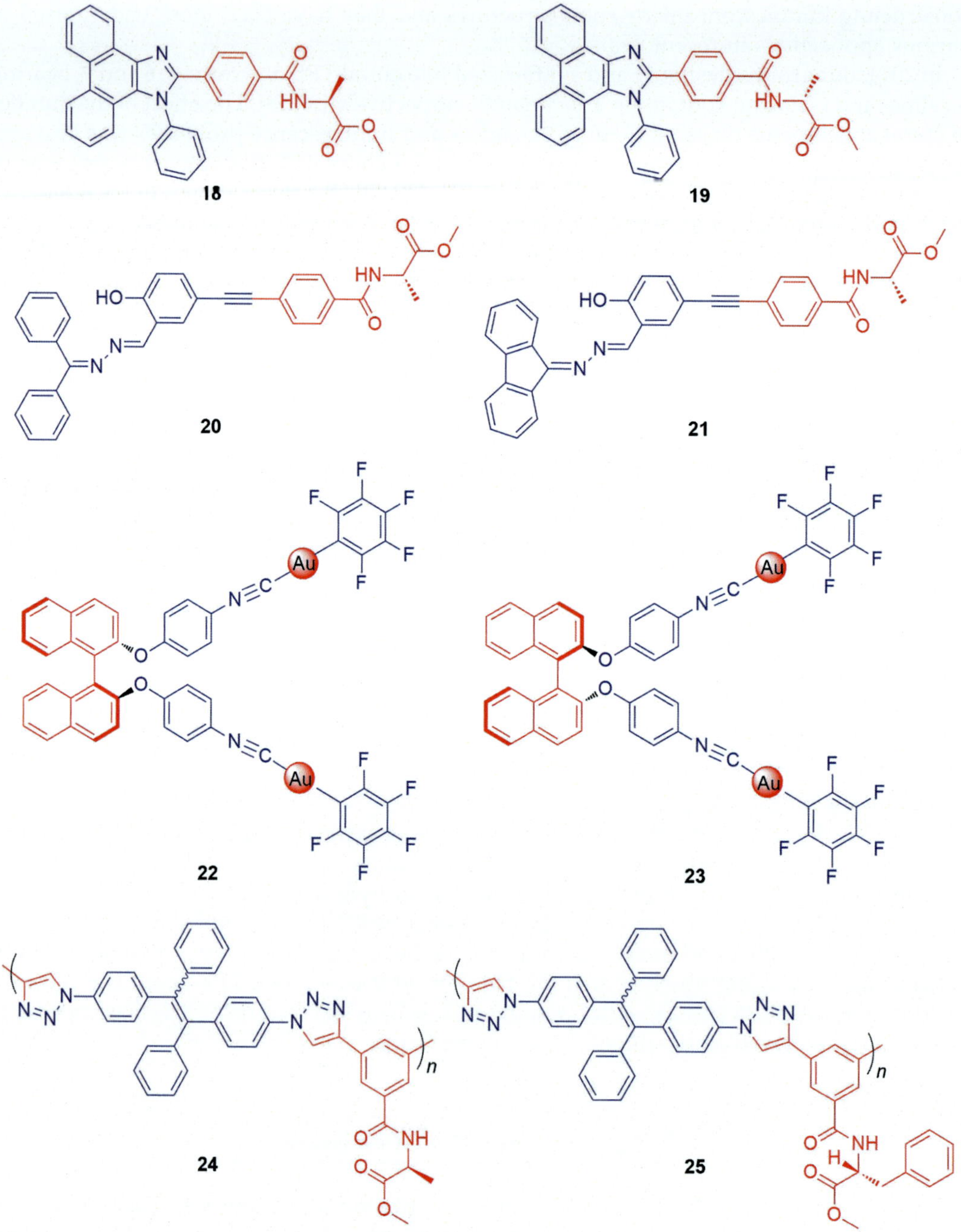

CHART 6 Molecular structures of other chiral AIEgens **18–25**.

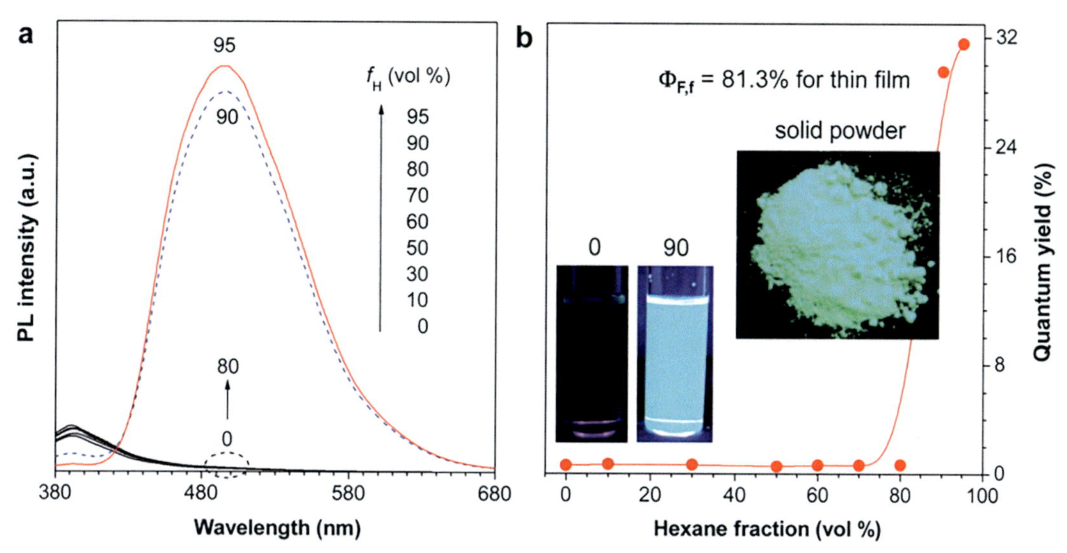

FIG. 1 (A) Photoluminescence (PL) spectra of **1** in dichloromethane (DCM) with different volume fractions of hexane (f_H) (Concentration: 10^{-5}M, λ_{ex}: 356 nm). (B) Variation of fluorescence quantum yields (Φ_F) of **1** versus f_H of the DCM-hexane mixture; the insets show the fluorescent photographs of **1** in DCM and DCM-hexane with f_H of 90% and its powder under irradiation with a handheld UV-lamp (365 nm). *Reproduced with permission from J. Liu, H. Su, L. Meng, Y. Zhao, C. Deng, J.C.Y. Ng, P. Lu, M. Faisal, J.W.Y. Lam, X. Huang, H. Wu, K.S. Wong, B.Z. Tang, Chem. Sci. 3 (2012) 2737–2747. Copyright 2012, the Royal Society of Chemistry.*

solvent (Fig. 1A). The fluorescence quantum yield (Φ_F) of **1** is about 0.6% in the DCM solution, but increases to 31.5% in the DCM-hexane mixture with a hexane fraction of 95% and reaches 81.3% at the film state (Fig. 1B), further confirming its AIE characteristic.

To reveal the underlying mechanism of AIE phenomenon, the dynamics emission of **1** was examined by measuring the fluorescence lifetime, and the influence of low-frequency motions on the fluorescence efficiency was investigated through theoretical calculations. Both experimental and theoretical results are in accordance with the proposed restricted intramolecular motions [8]. When **1** is dissolved in solution, its low-frequency intramolecular motions annihilate the excitons in a nonradiative way, making it weakly or even nonemissive, whereas in the aggregated state the intramolecular motions are greatly restricted, thus blocking the nonradiative energy decay and leading to intensive light emission.

4 Circular dichroism

Circular dichroism (CD) reflects the chiroptical information of molecules in the ground state. It is evaluated by using the absorptive dissymmetry factor, which is defined as $g_{abs} = 2(\varepsilon_L - \varepsilon_R)/(\varepsilon_L + \varepsilon_R)$, where ε_L and ε_R are the molar extinction coefficients of the left- and right-handed circularly polarized light, respectively.

The AIEgens, **1–3**, **5–10**, and **18–21**, are all CD-silent in the diluted solutions, but CD-active in the aggregate and solid state, exhibiting the aggregation-induced CD (AICD) effect. As exemplified with **1**, it shows two absorption peaks at the wavelengths of 279 and 360 nm in the DCM solution (Fig. 2A), which are ascribed to the absorption of the peripheral triazolylphenyl groups and the silole core, respectively. The DCM solutions of **1** with different concentrations give no CD signals (Fig. 2B). However, it shows Cotton effects at 249, 278, and 340 nm in the DCM-hexane mixture (1/9, v/v), and its intensities increase with an increase in the concentrations (Fig. 2C). The peaks at the wavelength of 278 and 249 nm are due to the absorption of the triazolylphenyl moiety, while the peaks at 340 nm are assigned to the absorption of the π–π conjugated system of the silole, suggesting that chirality has been transferred from the sugar-containing attachments to the silole core in the aggregated state. In addition, the g_{abs} values at 360 nm are increased from 1.59×10^{-3} to 2.23×10^{-3} with the concentration from 2×10^{-5} to 2×10^{-4} M. The CD spectra of **1** dispersed in the poly(methyl methacrylate) (PMMA) matrix with different weight fractions (wt%) and in neat film are shown in Fig. 2D. When the weight fraction of **1** in the PMMA matrix is 2.5 wt%, the cast film is CD-silent; at a higher weight fraction, CD signals occur and the intensities increase with an increase in the loading amounts, further confirming the AICD effect. When the loading ratio is small, molecule **1** remains in an isolated state in the PMMA matrix, whereas with an increase in the loading ratio the molecules tend to stack together induced by the intrinsic phase separation, affording the chiral aggregates with helical structures. Moreover, a significant red shift occurs in the film state than in the DCM/hexane mixture because of the formation of more helical structures, which enhances chirality transfer. This is further proved by the CD spectrum of the neat film of **1**.

Interestingly, the silole derivative **4** with chiral phenylethanamine pendants linked by thiourea displays a complexation-induced CD (CICD) effect [33]. When it is dissolved in good solvent or dispersed in poor solvents, no CD signals are observed, but when complexed with chiral acids, such as mandelic acids and phenyllactic acids, **4** becomes CD-active in thin films (Fig. 3), indicating that the complexation between the thiourea groups and the chiral acids efficiently induces the chirality transfer. It was found that such a CICD effect was the most effective in the complexation of **4** with mandelic acid due to their structural features.

Unlike the AIEgens with point chirality, the AIEgens with axial chirality, **11–13**, show an abnormal phenomenon of aggregation-annihilation CD (AACD). The CD signals of the AIEgens remained unchanged in THF/water mixtures with water fractions (f_w) lower than 40%, but decreased when f_w is higher than 40% (Fig. 4). A series of control experiments suggest that the AACD effect is mainly caused by the decrease in the twisted angle between two naphthalene rings in these AIEgens [42].

5 Circularly polarized luminescence

CPL, the emission analog of CD, refers to the differential emission of left- and right-handed circularly polarized light of chiral luminescent systems. During the past decades, functional materials with CPL property have attracted much attention for their potential applications in 3D displays, optical information storage and processing devices, asymmetric catalytic synthesis, and biosensing [56–61]. The performance of CPL-active materials is generally evaluated

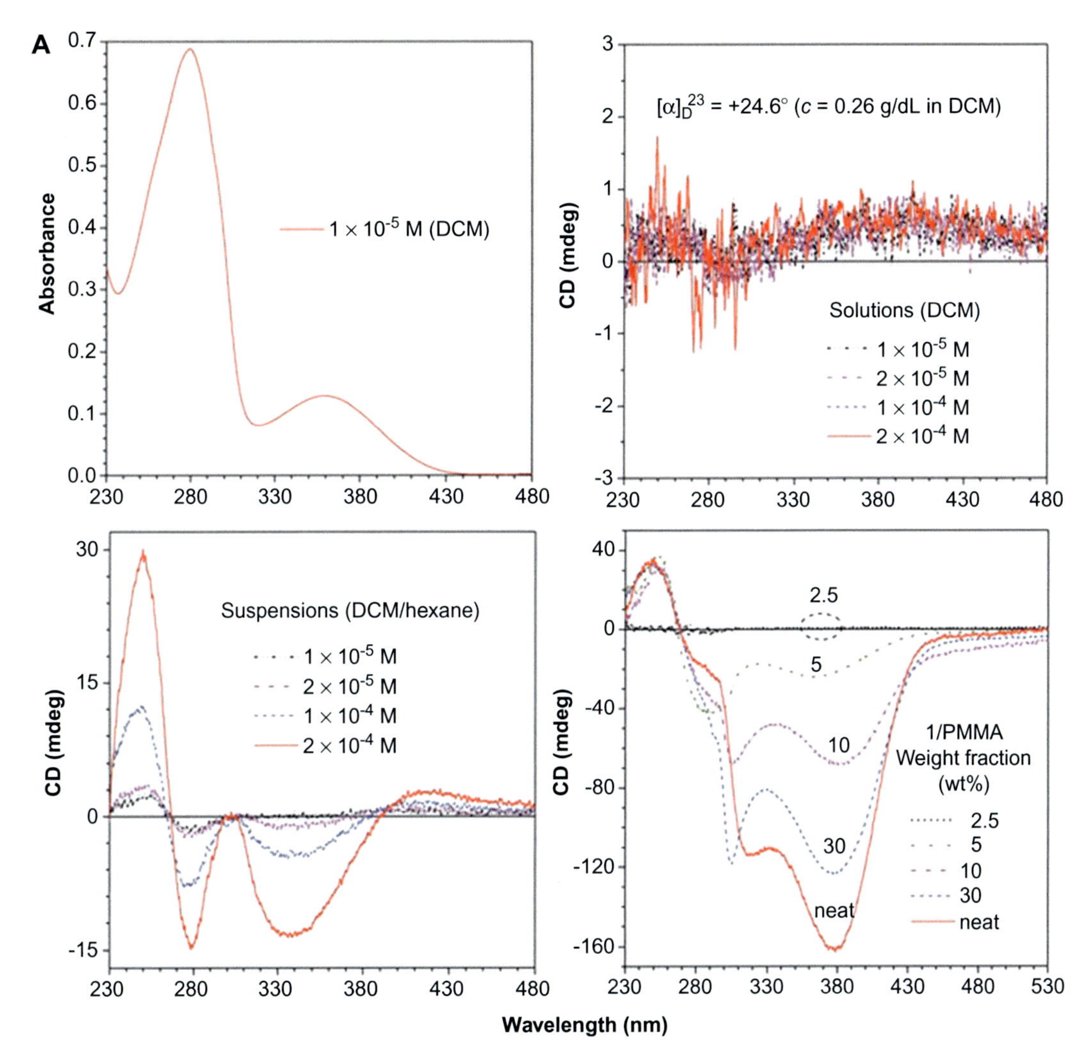

FIG. 2 (A) Absorption spectrum of **1** in DCM. (B and C) CD spectra of **1** with different concentrations in DCM solution and DCM/hexane (1/9, v/v) mixtures, respectively. (D) CD spectra of **1** with different weight fractions (wt%) dispersed in PMMA matrix prepared by drop casting of their mixed 1,2-dichloroethane (DCE) solutions. Concentration of PMMA: 10 mg mL^{-1}. CD spectrum of neat **1** prepared by natural evaporation of its DCE solution of 2 mg mL^{-1} on a quartz plate is also shown for comparison. *Reproduced with permission from J. Liu, H. Su, L. Meng, Y. Zhao, C. Deng, J.C.Y. Ng, P. Lu, M. Faisal, J.W.Y. Lam, X. Huang, H. Wu, K.S. Wong, B.Z. Tang, Chem. Sci. 3 (2012) 2737–2747. Copyright 2012, the Royal Society of Chemistry.*

by using two parameters. One is the emission dissymmetry factor (g_{em}), which is defined as $g_{em} = 2(I_L - I_R)/(I_L + I_R)$, where I_L and I_R are the emission intensities of the left- and right-handed circularly polarized light, respectively. The g_{em} values are in the range −2 to 2. The other is the luminescence efficiency, especially in solid state for practical use. Up to now, most of the reported CPL materials based on the conventional luminophors show

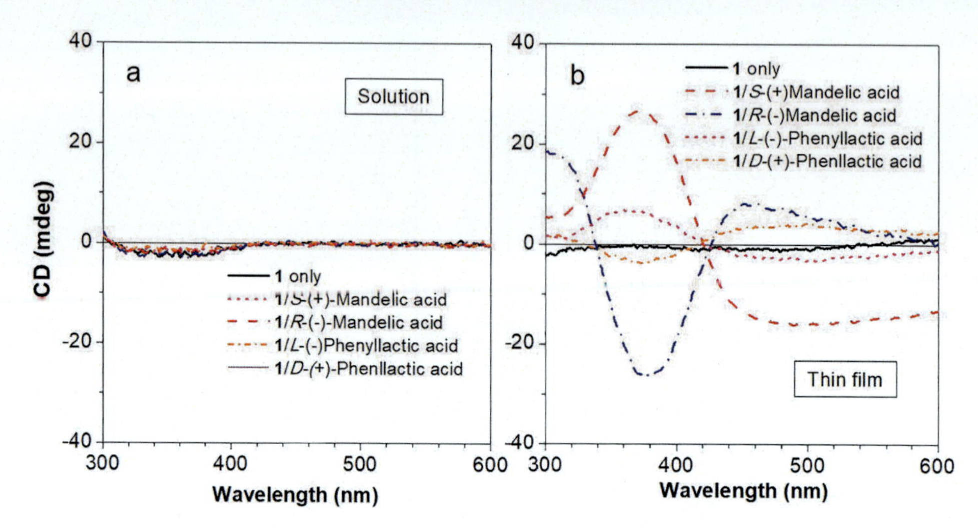

FIG. 3 CD spectra of **4** in the absence and presence of chiral hydroxyl acids in the (A) THF solution and (B) solid thin film states. Concentration of **4**: 1 mM; concentration of acid: 40 mM. *Reproduced with permission from J.C.Y. Ng, J. Liu, H. Su, Y. Hong, H. Li, J.W.Y. Lam, K.S. Wong, B.Z. Tang, J. Mater. Chem. C 2 (2014) 78–83. Copyright 2014, the Royal Society of Chemistry.*

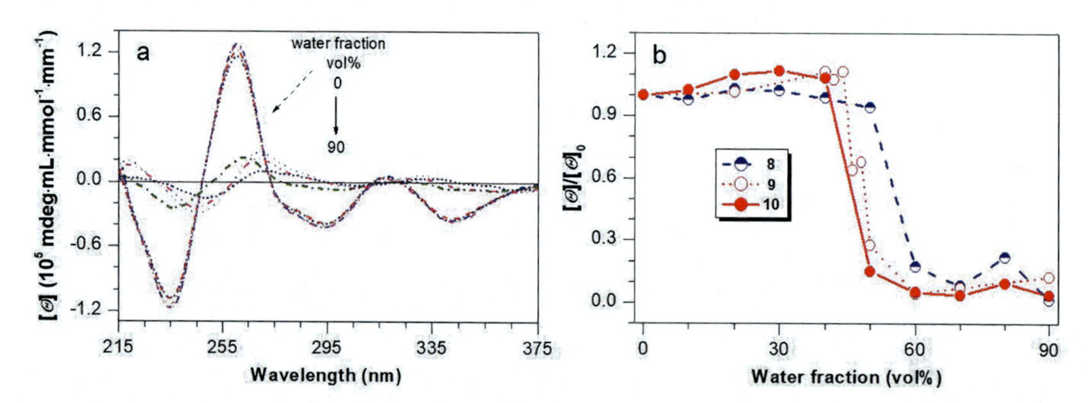

FIG. 4 (A) CD spectra of **11** in THF/water mixtures with different water fractions (f_w). Concentration: 10^{-4} M. (B) Plots of relative molar ellipticity of **11** (@260 nm), **12** (@280 nm), and **13** (@280 nm) versus f_w. $[\Theta]$ = molar ellipticity, $[\Theta]_0$ = molar ellipticity at $f_w = 0$. *Reproduced with permission from H. Zhang, H. Li, J. Wang, J. Sun, A. Qin, B.Z. Tang, J. Mater. Chem. C 3 (2015) 5162–5166. Copyright 2015, the Royal Society of Chemistry.*

the absolute g_{em} values in the range of 10^{-5}–10^{-2} in solutions, except for some rare earth compounds [62] and liquid crystalline conjugated polymers [63]. From solution to condensed state, the CPL performance usually becomes worse due to the ACQ effect. The development of efficient chiral AIEgens with both large g_{em} factors and high emission efficiency in solid state is thus highly desirable.

Thanks to their unique luminescent behaviors, AIEgens are ideal building blocks for the construction of efficient CPL-active materials in the condensed phase [19–21]. Given that the AIEgens **1–25** are both highly emissive and CD-active in the aggregate or solid state, they are anticipated to exhibit CPL activities. This is the case. In this part, we mainly discuss about the CPL performance of the sugar-containing silole derivative **1** as an example. The CPL activity of **1** was examined in various states besides in a good solvent [30]. Fig. 5A–F shows the fluorescence microscope images of its thin film obtained by natural evaporation of its 1,2-dichloroethane (DCE) solution, doped PMMA film, and well-resolved micropattern prepared by using a microfluidic technique. As it is nonluminescent and CD-silent in solution, **1** does not give any CPL signals in DCM, whereas in aggregate states it shows strong CPL signals (Fig. 5G), demonstrating an aggregation-induced CPL (AICPL) effect. The CPL spectra of **1** in the aggregated states show similar profiles but diverse g_{em} values (Fig. 5F). Its g_{em} values are in the range of -0.17 to -0.08 and show less dependence on the emission wavelength, in the DCM-hexane (v/v, 1/9) mixture, the cast film obtained from evaporating its DCE solution, and the **1** doped PMMA film. The g_{em} value of **1** can reach up to -0.32 in the micropatterned film prepared by using a Teflon-based microfluidic technique. This highest g_{em} may be caused by enhancement of the packing order of molecule **1** in the confined microchannel environment. The variation in the g_{em} values of **1** in different forms reveals that supramolecular structures play an important role in determining the g_{em} values. Moreover, the CPL activities of the samples can still remain even after storage for more than half a year under ambient conditions, showing good spectral stability. As far as we know, this CPL system represents the best result in the reported single component organic chiral conjugated materials in the light of luminescence efficiency, dissymmetry factor, and spectral stability.

Similar to **1**, the silole-based AIEgens **2** and **3** both show CPL activities in solid film state with the g_{em} values of about -0.05 and -0.016, respectively. The g_{em} values of TPE-based AIEgens with point chirality **6–10**, **24**, and **25** are about +0.03, +0.045, -0.003, +0.0032, +0.012, +0.0045, and +0.018, respectively, in the aggregated state. The TPE-based AIEgens with axial chirality, **14** and **15**, are CPL-active in both solution and aggregated states. The g_{em} values of *R/S*-**14** at 630 nm are about ±0.002. In the THF solution, the g_{em} values of *R/S*-**15** are -0.0027 and +0.0041, respectively, whereas in 90% aqueous suspension they show obvious CPL signals with g_{em} values of +0.0026 and -0.0028, respectively. Similarly, *M/P*-**16** displays the g_{em} values +0.0031 and -0.0033, respectively, in the THF solution, and +0.0062 and -0.0050, respectively, in 95:5 H_2O/THF suspension. The homodirectional cubes of (6*P*)-**17** and (6*M*)-**17** give positive and negative CPL signals in chloroform solutions, respectively, with the g_{em} values of ±0.0011. The CPL spectra of the heterodirectional cubes (4*P*2*M*)-**17** and (2*P*4*M*)-**17** are similar to those of (6*P*)-**17** and (6*M*)-**17**, respectively, with slightly smaller g_{em} values of ±0.00093. The cast films of chiral PIM-based AIEgens **18** and **19** exhibit their g_{em} values about +0.01 and -0.005, respectively. Interestingly, the chiral hydrazone derivative **20** shows a very weak CPL signal and low g_{em}, whereas **21** gives the g_{em} value of 0.013 in the film state due to the subtle difference in the chemical structure. The AIE-active chiral Au(I) complexes **22** and **23** can serve as chiral templates to coassemble with other achiral luminogens to generate CPL-active systems with the absolute g_{em} values in the range of $3–5 \times 10^{-3}$. It is worth noting that these g_{em} values are comparable to those ($|g_{em}| = 10^{-5}$–10^{-2}) of most reported CPL-active materials based on conventional organic fluorophores in the solution state [56–61]. These results demonstrate that combining AIEgens and chiral blocks is an efficient molecular design strategy for the construction of CPL-active materials in the condensed phase. Such high emission efficiency and g_{em} values in the solid state make these CPL materials promising candidates for high-tech applications [64–66].

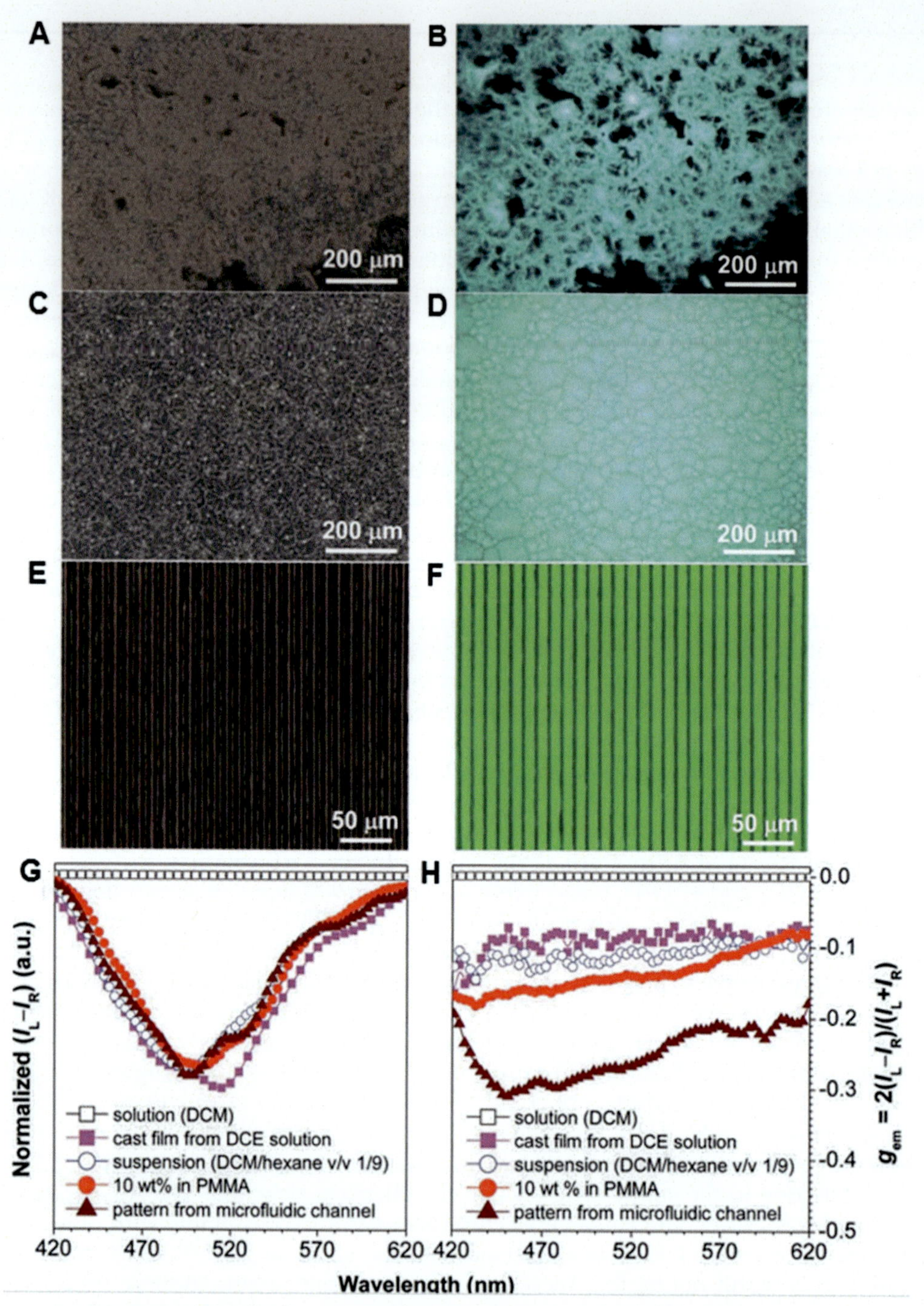

FIG. 5 (A–F) Fluorescence microscope images under normal laboratory lighting (left panels) and UV excitation (right panels) of **1**: (A and B) natural evaporation of the DCE solution, (C and D) dispersion in the PMMA matrix (10 wt%), and (E and F) evaporation of the DCM-toluene solution in microfluidic channels on quartz substrates. (G and H) Plots of (G) ($I_L - I_R$) and (H) CPL dissymmetry factor g_{em} versus wavelength for **1** existing in different formats: DCM solution, DCM-hexane (v/v, 1/9) mixture (suspension), neat cast film from the DCE solution of 2 mg mL^{-1}, dispersion in polymer matrix (10 wt% in PMMA), and fabricated micropattern by evaporation of the DCE solution in microfluidic channels. In DCM and the DCM-hexane mixture, concentration: 2×10^{-4} M, λ_{ex}: 356 nm. *Reproduced with permission from J. Liu, H. Su, L. Meng, Y. Zhao, C. Deng, J.C.Y. Ng, P. Lu, M. Faisal, J.W.Y. Lam, X. Huang, H. Wu, K.S. Wong, B.Z. Tang, Chem. Sci. 3 (2012) 2737–2747. Copyright 2012, the Royal Society of Chemistry.*

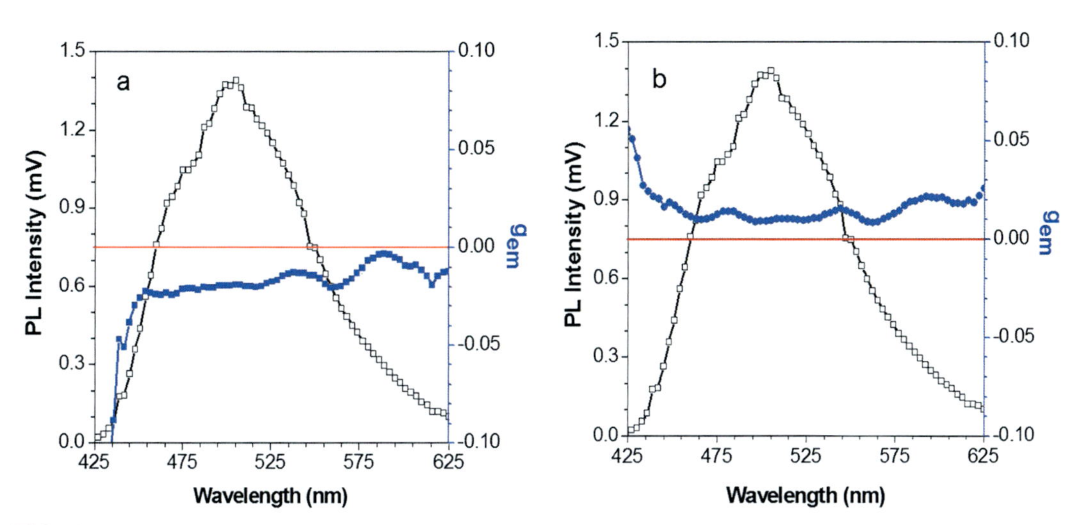

FIG. 6 Plots of PL intensity and g_{em} *versus* wavelength for **4** in the presence of (A) R-(−) or (B) S-(+)-mandelic acid ([**4**]/[acid] = 1: 40 by mole) in the film state, respectively. *Reproduced with permission from J.C.Y. Ng, J. Liu, H. Su, Y. Hong, H. Li, J.W.Y. Lam, K.S. Wong, B.Z. Tang, J. Mater. Chem. C 2 (2014) 78–83. Copyright 2014, the Royal Society of Chemistry.*

In contrast to the AIEgens described above, the neat film of silole derivative **4** is CPL-silent, implying that its conformation is randomly arranged. When complexed with R-(−)-mandelic acid and S-(+)-mandelic acid, the solid film of **4** shows the average g_{em} values of about −0.01 and +0.01, respectively, showing a complexation-induced CPL effect (Fig. 6). Furthermore, the predominant CPL of **4** can be tuned to emit left- or right-handed circularly polarized light through complexation with either enantiomer of mandelic acids. Owing to the AACD effect, the chiral AIEgens **11–13** show very weak CPL signals in the condensed phase.

Different from the AIEgens carrying chiral attachments, AIEgen **5** without chiral attachments is also CD- and CPL-active in the film state. The unexpected CD and CPL signals are only observed in the film state and they are likely related to the aggregation behavior of the molecules. The phenyl rings of **5** were arranged in a propeller shape, which is either clockwise or counterclockwise in solution and can cancel each other's CD absorption. In the aggregation state, molecules tend to adopt a consistent rotation and get it amplified with more molecules joining the aggregation, which thus leads to significantly improved CD and CPL performances. Imaging of their aggregates revealed that aggregation of the molecules indeed plays a critical role in the induced chirality and gave more clues to explain the generation of CD and CPL signals. The details of their aggregation behaviors are provided in the helical self-assembly section.

6 Helical self-assembly

Typical AIEgens generally possess propeller-shaped structures, which cause rather weak π–π stacking interactions. Therefore, precise self-assembling AIEgens into order architectures is a challenging task. The AIEgens described above are CPL-active in the aggregate state, indicating that predominantly one-handed helical structures are formed in the aggregates. Their self-assembly behaviors were thus explored by using microscopy techniques. Fig. 7 shows the transmission electron microscope (TEM) images of the aggregates of **1** generated in the DCM-hexane mixture (1/9, v/v), and by evaporation of its DCE solution [30]. The molecules were assembled into predominantly right-handed helical nanoribbons with an average width and helical pitch of about 30 and 120–150nm, respectively. The length of the nanoribbons can be up to micrometers. Elementary nanoribbons can further intertwine to form multistranded superhelical ropes with a right-handed twist. The driving force for the self-assembly process is the cooperative effect of the multiple unconventional weak hydrogen bonding interactions (C—H...O and C—H...N), π–π stacking, and stereo-shape matching between the sugar-containing pendants, through using the powder X-ray diffraction (XRD) technique and structural simulation.

The self-assembly properties of AIEgen **2** were investigated with an atomic force microscope (AFM). As shown in Fig. 8A, helical nanofibers with consistent left-handedness were readily obtained by evaporating the THF solution of **2** [31]. Helical nanostructures can also be formed upon the addition of the poor solvent, such as water or hexane, into its THF solution. By increasing the water content, the morphologies of the aggregates of **2** transformed from extended helical fibers to loops of helical fibers (Fig. 8B) and finally to the network of fibers (Fig. 8C), which all show left-handedness. When 10vol% of hexane was added, networked structures were generated (Fig. 9A). By increasing the hexane content to 50vol%, braids of fibers were arranged in a nebular-like morphology (Fig. 9B). Right-handed helical fibers were generated at the hexane fraction of 80% (Fig. 9C), indicating an inversion in the handedness of the helical fibers upon addition of more poor solvent.

The self-assembling behaviors of **3** were examined by using a scanning electron microscope (SEM), AFM, and a fluorescence microscope. The SEM image displays that the molecules assembled into left-handed helical fibers upon the evaporation of the DCE solution of **3** (Fig. 10A) [32]. The fibers show a broad distribution of the width, which is from 10 to 50nm due to their different association levels. The helical pitches of the thick fibers are longer than 100nm. The combined structures of braids of left-handed helical fibers and thin films with helical fibers at their edges are observed in the AFM images

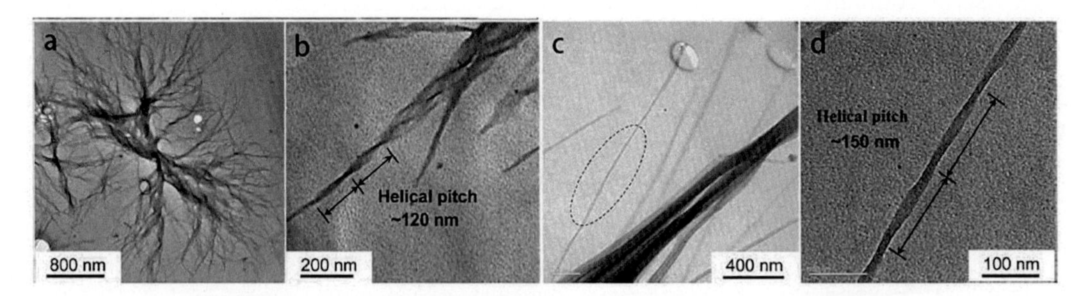

FIG. 7 (A–D) TEM images of the aggregates of **1** (A and B) in the DCM/hexane mixture (1/9, v/v), (C) from natural evaporation of the DCE solution of 2mgmL^{-1}, and (D) enlarged area circled in panel C. *Reproduced with permission from J. Liu, H. Su, L. Meng, Y. Zhao, C. Deng, J.C.Y. Ng, P. Lu, M. Faisal, J.W.Y. Lam, X. Huang, H. Wu, K.S. Wong, B.Z. Tang, Chem. Sci. 3 (2012) 2737–2747. Copyright 2012 the Royal Society of Chemistry.*

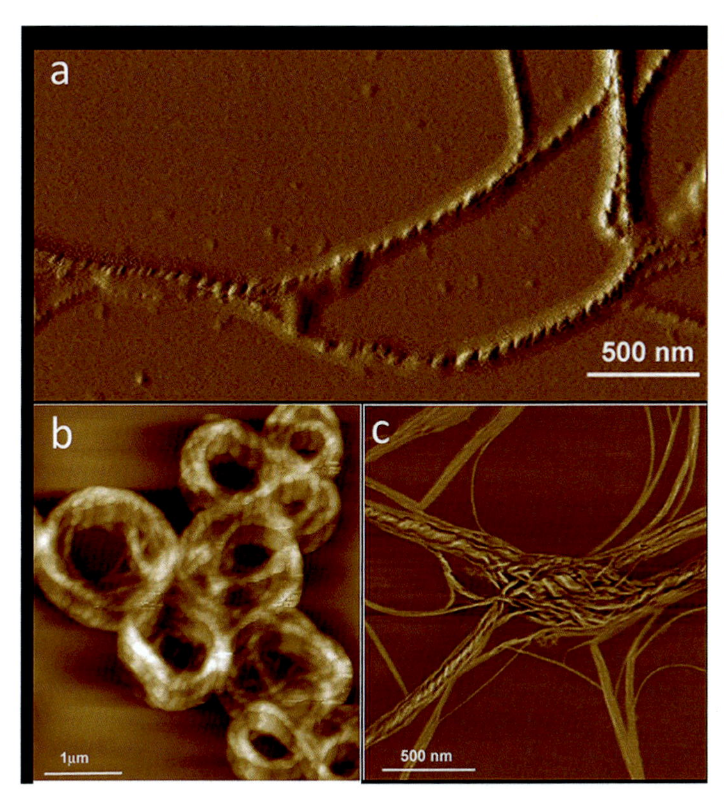

FIG. 8 AFM images of aggregates formed by **2** upon evaporation of its THF solution (**a**) and THF-water mixtures with water fractions of 20 vol% (**b**) and 90 vol % (**c**). *Reproduced with permission from J.C.Y. Ng, H. Li, Q. Yuan, J. Liu, C. Liu, X. Fan, B.S. Li, B.Z. Tang, J. Mater. Chem. C 2 (2014) 4615–4621. Copyright 2014, the Royal Society of Chemistry.*

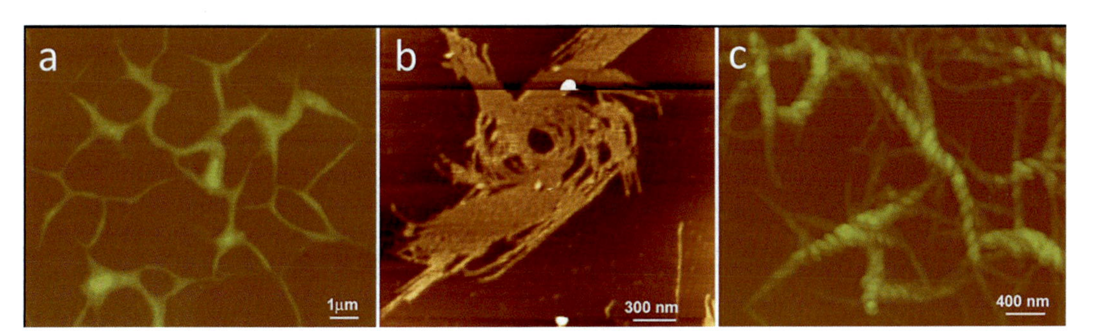

FIG. 9 AFM images of the manipulated assemblies formed by **2** upon evaporation of its different THF-hexane mixtures with hexane fractions of 10 vol% (A), 50 vol% (B), 80 vol% (C). *Reproduced with permission from J.C.Y. Ng, H. Li, Q. Yuan, J. Liu, C. Liu, X. Fan, B.S. Li, B.Z. Tang, J. Mater. Chem. C 2 (2014) 4615–4621. Copyright 2014, the Royal Society of Chemistry.*

(Fig. 10B and C), indicating that the helical fibers may be formed by the helical wrapping up of the thin films.

As shown in Fig. 11A, AIEgen **3** self-assembled into braids of helical fibers with left-handedness by evaporation of its DCE solution, which is consistent with the SEM image in Fig. 10 [32]. The helical fibers were arranged in circular frames in the DCM-hexane mixture

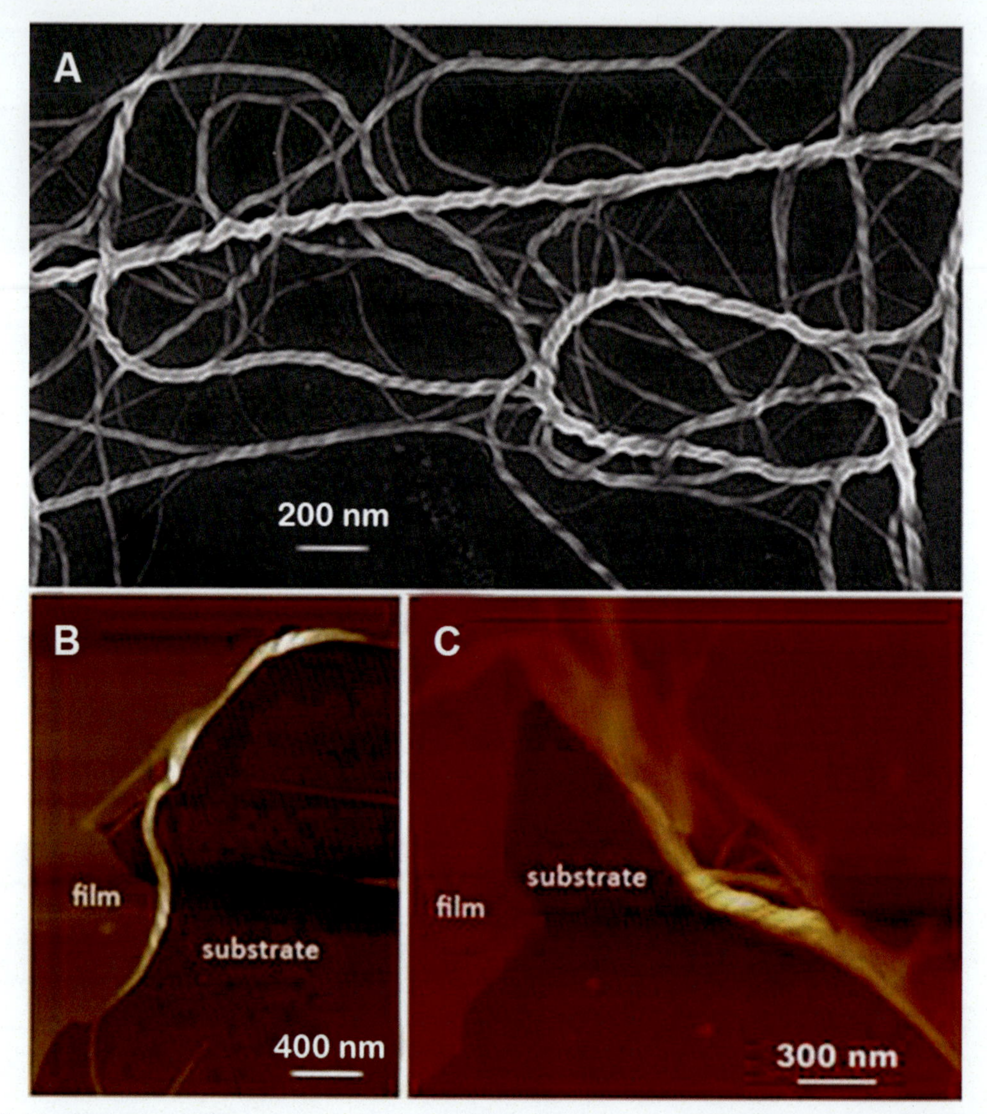

FIG. 10 SEM (A) and AFM (B and C) images of the aggregates formed by **3** upon the evaporation of its DCE solution. *Reproduced with permission from Ref. H. Li, S. Xue, H. Su, B. Shen, Z. Cheng, J.W.Y. Lam, K.S. Wong, H. Wu, B.S. Li, B.Z. Tang, Small 12 (2016) 6593–6601. Copyright 2016, Wiley-VCH.*

(1/1, v/v) (Fig. 11B). By increasing the content of poor solvent of hexane, the helical fibers formed networks (Fig. 11C) and became much thinner and more extended (Fig. 11D). Given that **3** possesses strong solid-state luminescence emission and good helical self-assembly property, its assemblies were explored by using a fluorescence microscope. Through the optimization of the condition of sample preparation and the ratio of good solvent and poor solvent, the helical arrangements of **3** formed by the evaporation of 40 μL of its DCE-hexane

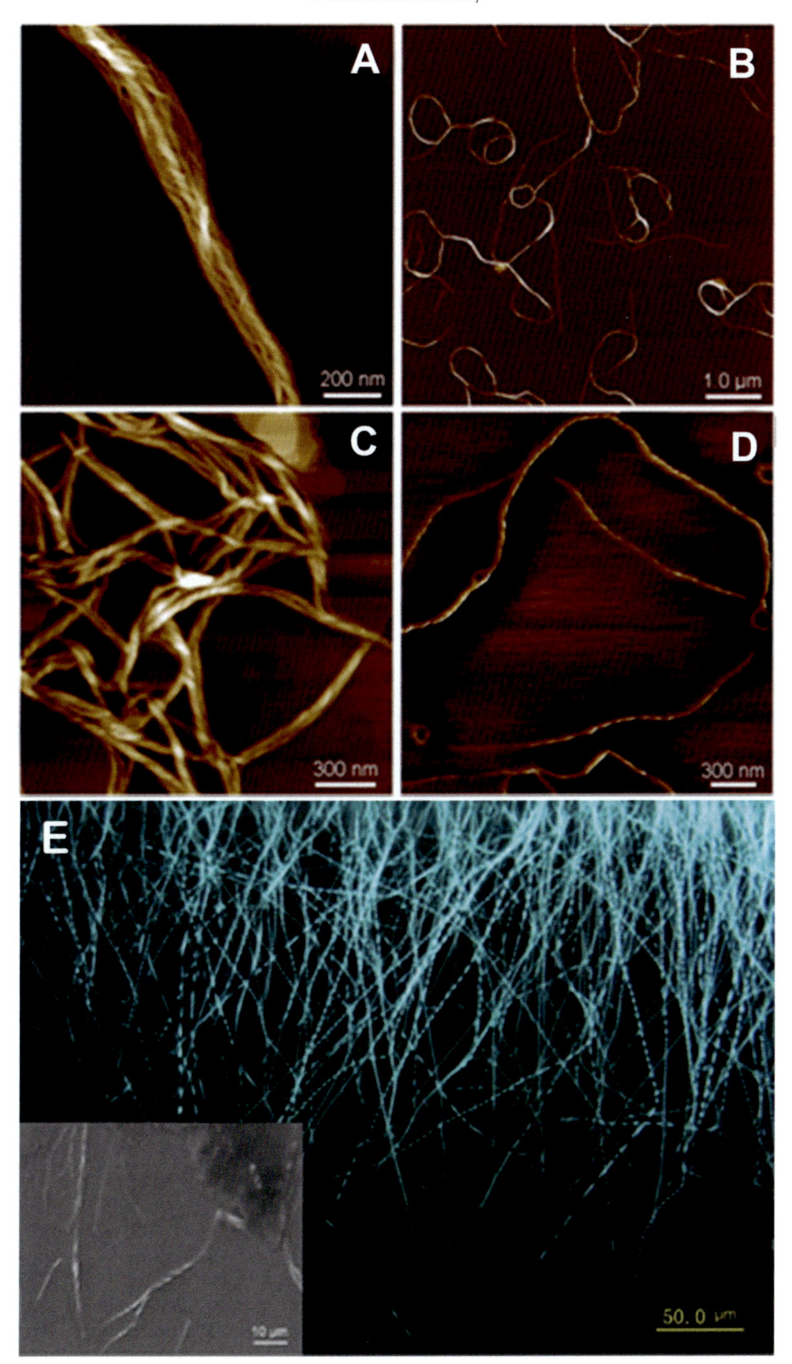

FIG. 11 AFM images of the aggregates formed by **3** upon the evaporation of its DCE solution (A), DCE/hexane (1:1, v/v) (B), DCE/hexane (1:4, v/v) (C), and DCE/hexane (1:9, v/v) (D) mixtures. Concentration: 10 μM. Fluorescence image (E) of the aggregates formed by **3** upon the evaporation of the mixture of DCE/hexane (1:1, v/v). The insert image is the SEM image of the sample for a fluorescence microscope. Concentration: 100 μM. *Reproduced with permission from H. Li, S. Xue, H. Su, B. Shen, Z. Cheng, J.W.Y. Lam, K.S. Wong, H. Wu, B.S. Li, B.Z. Tang, Small 12 (2016) 6593–6601. Copyright 2016, Wiley-VCH.*

mixture (1:1, v/v) can be visualized. As displayed in Fig. 11E, the blue dotted lines were observed. They are considered to be the helical fibers, which show the dotted contour due to the fluorescence contrast of the helical assembly. To prove that the dotted lines are caused by helicity rather than discontinuity, the morphology of the sample was further explored by SEM. As shown in the inset image in Fig. 11E, helical ribbons and fibers can be clearly observed, confirming that the bright dotted lines were reflected from the helicity of the fibers and ribbons.

AIEgen **5**, HPS, an achiral molecule, can also self-assemble into fluorescent helical nanostructures. As given in Fig. 12A, left-handed helical nanofibers were obtained by **5** upon

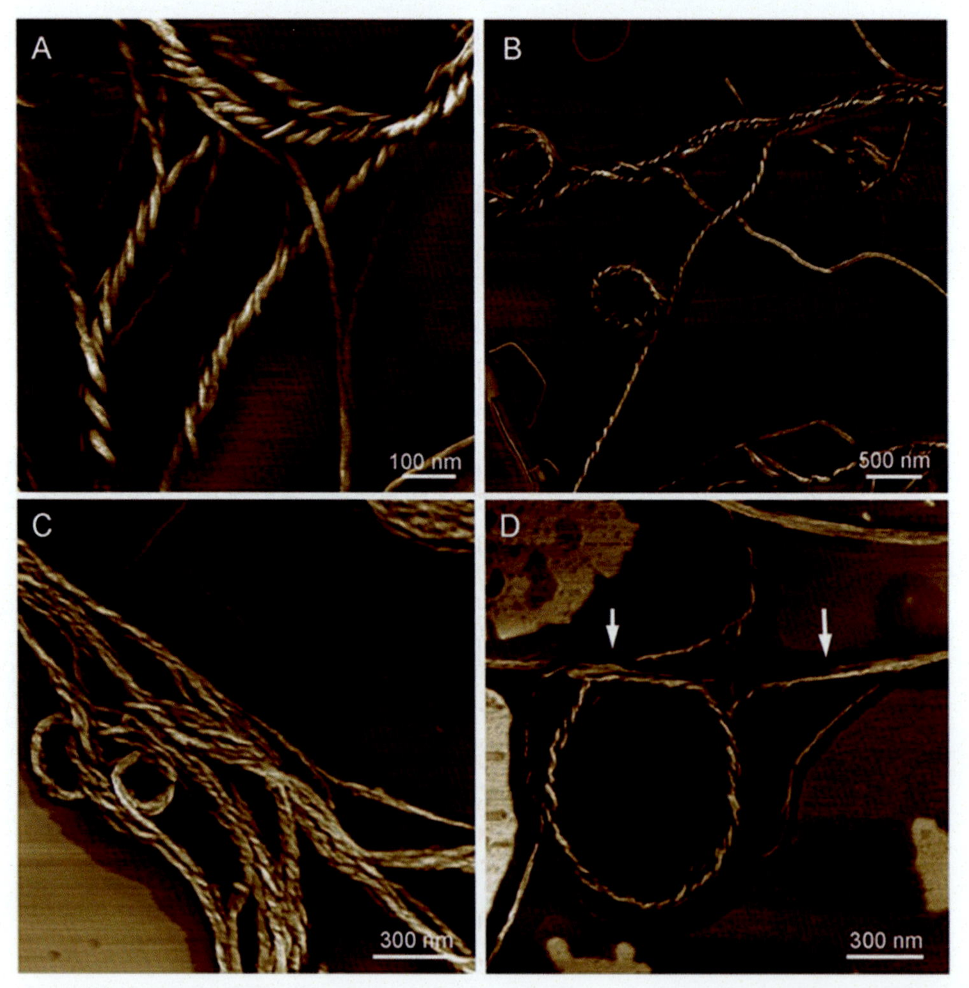

FIG. 12 AFM images of the helical fibers formed by **5** on the evaporation of its different THF-water solution. The water content is 0 (A), 50% (B), 80% (C), and 90% (D), concentration: 10 μM, ribbons curled into fibers most significant at the area labeled with white arrows (D). *Reproduced with permission from S. Xue, L. Meng, R. Wen, L. Shi, J.W. Lam, Z. Tang, B.S. Li, B.Z. Tang, RSC Adv. 7 (2017) 24841–24847. Copyright, 2017, the Royal Society of Chemistry.*

the evaporation of its THF solution, and they further intertwined to form suprahelical fibers with left-handedness [36]. Due to the different levels of helical rotation, the fibers show a broad distribution of width and helical pitches. When the poor solvent of water was added into its THF solution, **5** assembled into left-handed helical fibers (Fig. 12B–D). Additionally, the films jointed with the fibers, and the combined ribbons (marked with white arrows in Fig. 12D) were also observed, indicating that the helical fibers were generated by the curling of the thin films. The TEM images also show that **5** assembled into left-handed helical ribbons with the length of several micrometers. Along with them, tubular morphology was also observed. The hollow internalities of the fibers further suggested that they are probably formed by the curling of films.

In order to provide more information about the underlying principle of the helical assembly of **5**, the computational calculation was then performed. Through molecular dynamics simulations, it was proposed that HPS fibers are consisted of identical units that are packed together mostly with translocation with low energy. In this packing mode, two molecules of **5** are antiparallelly placed to form a basic unit (Fig. 13A) [36]. Multiple units are further packed together to form a periodic pattern. Molecular layers are generated by joining more molecules in the packing (Fig. 13B and C). The central symmetric arrangement of the dimmers in each unit cell leads to the tilting of the molecular layers, which will become wrapped up on a much larger scale. Fully wrapped layers show the morphology of helical tubes, while partially wrapped ones lead to helical fibers or ribbons (Fig. 13D). Through the theoretical calculation, the van der Waals interactions between the phenyl groups are found to contribute to the formation of the packing unit and packing mode.

Besides the silole derivatives, the TPE-based AIEgens can also self-assemble into helical micro-/nanostructures. As given in Fig. 14, the monofunctionalized TPE derivatives **6** and **7** with L-valine- and L-leucine-containing attachments, respectively, can readily form left-handed helical fibers and ribbons in their DCE-hexane mixtures (1/9, v/v) [37,38]. The individual helical fibers can further twine to form multistranded helices with left-handedness and the lengths of several micrometers (Fig. 14A). Fig. 14B and C show the coexistence of helical fiber, ribbons, and combined structures, suggesting that helical fibers are likely wrapped up

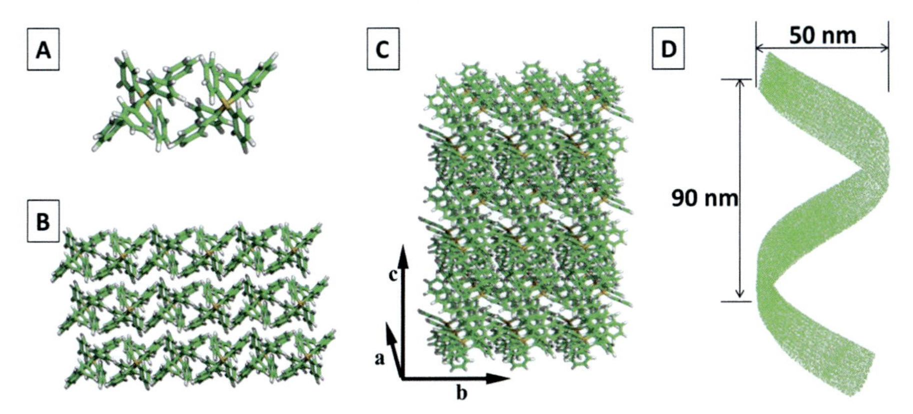

FIG. 13 Evolution of helical fibers from dimmers to molecular layers and helical fibers. The basic unit HPS aggregates with central symmetric property (A). Packing modes of molecular layers from different units from different views (B and C). Modeled helical fibers of HPS aggregates (D). *Reproduced with permission from S. Xue, L. Meng, R. Wen, L. Shi, J.W. Lam, Z. Tang, B.S. Li, B.Z. Tang, RSC Adv. 7 (2017) 24841–24847. Copyright 2014, the Royal Society of Chemistry.*

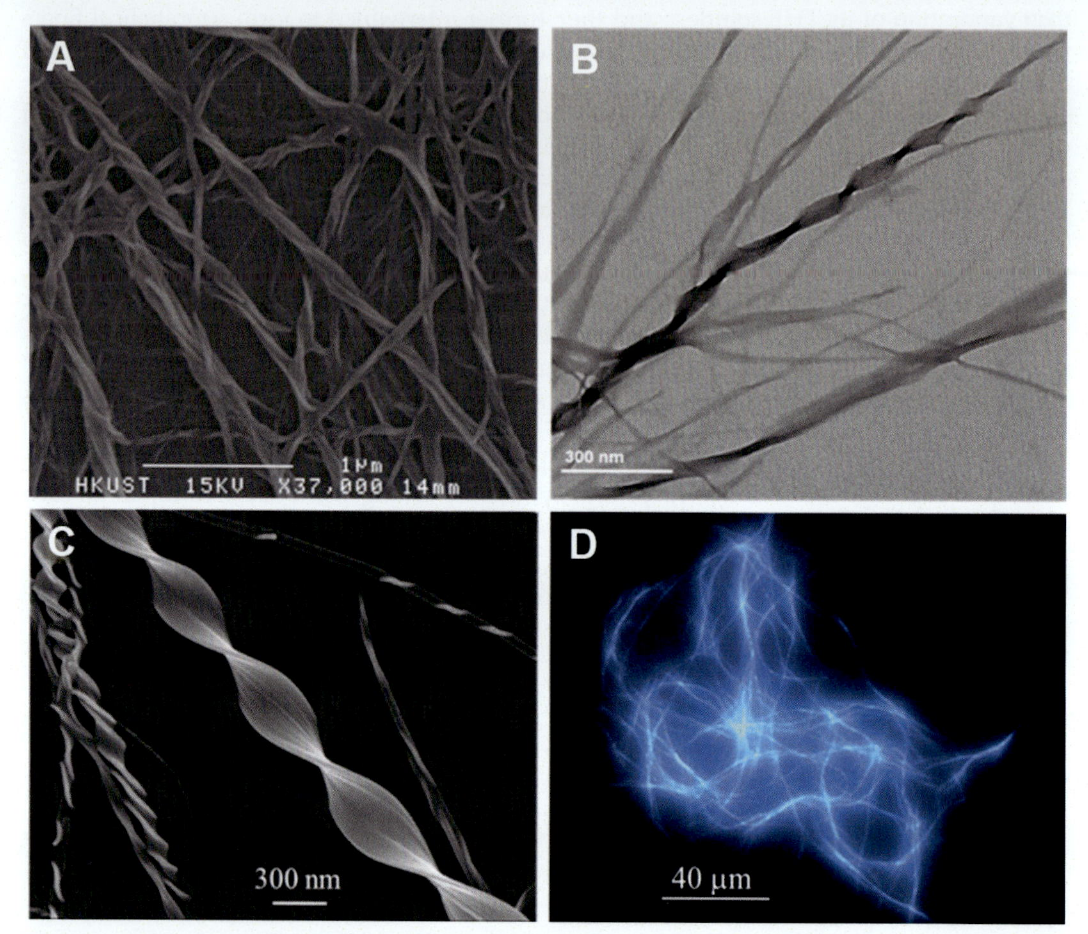

FIG. 14 SEM (A) and TEM (B) images of the aggregates of **6** formed upon the evaporation of the DCE-hexane solution (1/9, v/v), [**6**] = 10^{-4} M. SEM (C) and fluorescence microscopy (D) images of the aggregates of **6** formed upon the evaporation of the DCE-hexane solution (1/9, v/v), [**7**] = 10^{-4} M. *(A and B) Reproduced with permission from H. Li, J. Cheng, Y. Zhao, J. W.Y. Lam, K.S. Wong, H. Wu, B.S. Li, B.Z. Tang, Mater. Horiz. 1 (2014) 518–521. (C and D) Reproduced with permission from H. Li, J. Cheng, Y. Zhao, J. W.Y. Lam, K.S. Wong, H. Wu, B.S. Li, B.Z. Tang, Mater. Horiz. 1 (2014) 518–521; H. Li, J. Cheng, H. Deng, E. Zhao, B. Shen, J.W.Y. Lam, K.S. Wong, H. Wu, B.S. Li, B.Z. Tang, J. Mater. Chem. C 3 (2015) 2399–2404. Copyright 2014 and 2015, the Royal Society of Chemistry.*

by the ribbons. The fluorescence microscopy image of **7** exhibits that the fibers have the length up to millimeters and intense blue fluorescence (Fig. 14D), further confirming their strong self-assembly activities.

The disubstituted TPE derivatives, **8** and **9**, carrying two L-valine- and L-leucine-containing attachments, respectively, could readily self-assemble into helical fibers with consistent right-handedness in the DCE solution upon solvent evaporation [39,40]. Interestingly, molecule **9** could form left-handed helical nanofibers in its DCE-hexane mixtures (1/9, v/v), showing a solvent-controlled helical self-assembly behavior. Through XRD analysis and computational simulation, a synergistic effect from the multiple intermolecular hydrogen bonding interactions, π–π interactions, and the steric effect between the attachments is proposed to serve as the driving force for the assembling process.

The morphology of aggregates of *R*-**15** formed in THF/water with different water contents was varied. The molecules assemble into solid spheres with a mean diameter of 320 nm in 50% water (Fig. 15A) [44]. Moreover, there are some nanoribbons beside these spheres, indicating that nanospheres may be formed from nanoribbons by winding around each other. When the water content is higher than 60%, the resultant aggregates display bowl-like microspheres with a hole on their surface (Fig. 15B and C). The bowl-like spheres have a much larger size ranging from 1.5 to 3.3 mm and the holes on the surface are about 1.3 mm in diameter. All the holes are smooth, and the edges around the holes seemed to be bent toward the interior. Much bigger spherical structures with sunken surfaces were obtained in 90% aqueous solution (Fig. 15D). This indicates that these bowl-like structures may be produced by etching or dissolution of the solvent from the exterior, not by solvent evaporation from the interior.

To explore the influence of D- and L-amino acid-containing attachments on the self-assembly properties of the PIM-based AIEgens, the assembly behaviors of **18** and **19** were investigated. They can both form left-handed helical fibers upon the evaporation of their

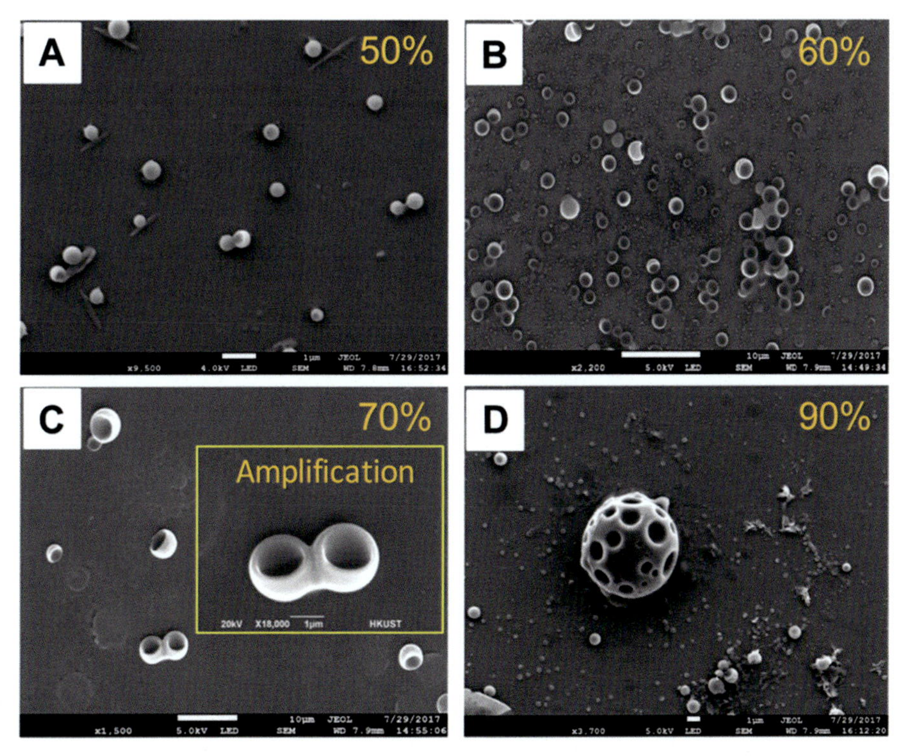

FIG. 15 SEM images of a suspension of *R*-**15** (0.1 mM) in a mixed solvent of THF/water, 50% (A), 60% (B), 70% (C), and 90% (D). *Reproduced with permission from H.-T. Feng, X. Gu, J.W.Y. Lam, Y.-S. Zheng, B.Z. Tang, J. Mater. Chem. C 6 (2018) 8934–8940. Copyright 2018, the Royal Society of Chemistry.*

THF solution (Fig. 16A and C), although they showed mirror CD and CPL spectra [49]. Compared with the uniform fibers formed by **18**, the fibers generated by **19** rotate more tightly, which are at multiple association levels. The helical pitches of the fibers vary in a wide range from several tens of nanometers to several hundreds of nanometers. When the poor solvent of water was added into their THF solution, **18** and **19** both took a more tight association to assemble into a loop-like morphology (Fig. 16B and D). Besides the helical fibers with left-handedness, the thin films were also obtained by the evaporation of their THF/water mixtures. The helical fibers occurred at the edge of the films, indicating that the elementary fibers may be wrapped up from the thin films.

The morphologies of their assemblies were then examined by a fluorescent microscope. As shown in Fig. 17A and C, the violet luminescent fibers formed by **18** and **19** upon the

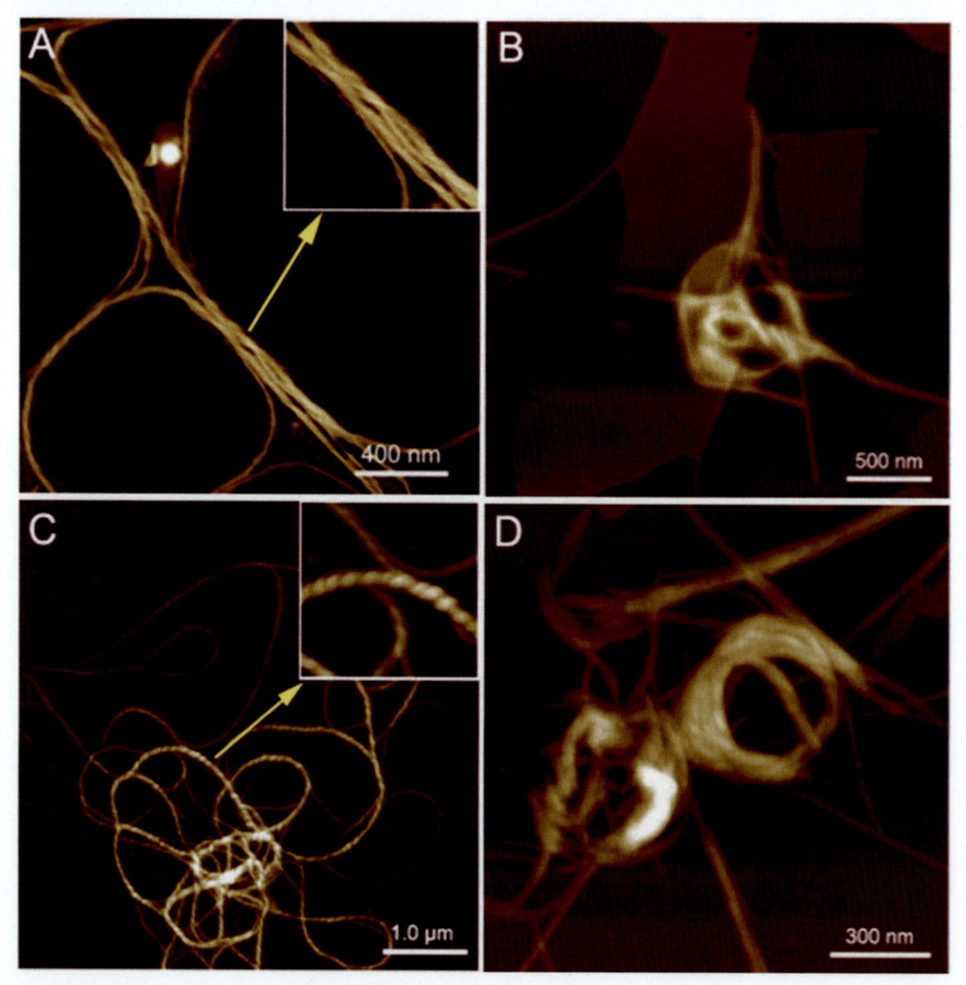

FIG. 16 AFM images of helical assemblies formed by **18** (A) and **19** (C) upon the evaporation of their THF solution, and aggregates formed by **18** (B) and **19** (D) in their THF-water mixture (1:9, v/v). The inserted images in A and C are the high-resolution images zoomed in from the corresponding labeled areas. Concentration: 10^{-5} M. *Reproduced with permission from B.S. Li, R. Wen, S. Xue, L. Shi, Z. Tang, Z. Wang, B.Z. Tang, Mater. Chem. Front. 1 (2017) 646–653. Copyright 2017, the Royal Society of Chemistry.*

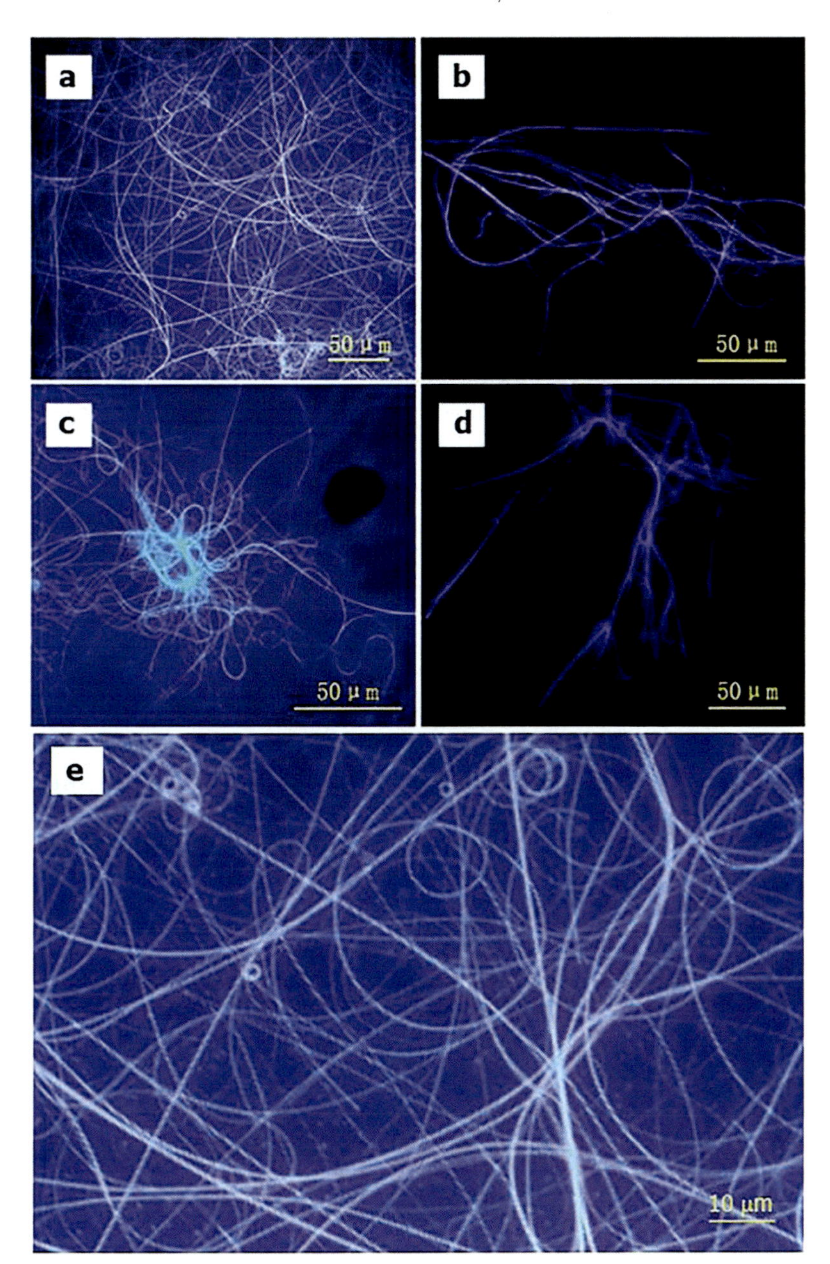

FIG. 17 Fluorescence images of fibers formed by **18** (A) and **19** (C) by the evaporation of their THF solution, and fibers formed by **18** (B) and **19** (D) by the evaporation of their THF-water (2:8, v/v) mixture. (E) Zoomed-in image of the labeled area in image A. Concentration: 10^{-3}M. *Reproduced with permission from B.S. Li, R. Wen, S. Xue, L. Shi, Z. Tang, Z. Wang, B.Z. Tang, Mater. Chem. Front. 1 (2017) 646–653. Copyright 2017, the Royal Society of Chemistry.*

evaporation of their THF solution are several hundreds of microns in length [49]. From the high-resolution image in Fig. 17E, the helical fibers exhibit a broad distribution of helical pitches in the range from 2 to 5 μm. They also display different handedness, which is complementary to the fibers with dominant left-handedness shown by AFM at the nanoscale. This suggests that each of **18** and **19** forms helical fibers with both handedness, but different dominant ones.

The chiral AIEgens **20** and **21** can both self-assemble into helical nanostructures. For instance, SEM images of **21** show that the molecules formed helical fibers with a width of about 50–70 nm and the length longer than several microns (Fig. 18) [50]. The helical fibers were left-handed and thinner fibers further interlaced with each other to form thicker ones, implying that the association of helical fibers was hierarchical. With the increase of water content, the helical fibers of **21** got much thicker and more twisted with each other.

The chiral Au(I) complexes **22** and **23** show impressive spontaneous hierarchical self-assembly transitions from vesicles to helical fibers [51]. As given in Fig. 19A, vesicles were formed by **23** upon aggregation in the freshly prepared THF/water mixtures, and their average diameters decreased with an increase in the water fractions. It was proposed that these two chiral enantiomers are more likely to self-assemble into spherical micelles first and then evolve into hollow vesicles upon entry of water molecules. According to their molecular structures, it was presumed that the hydrophilic pentafluorophenyl groups formed the interior and exterior coronas, while the more hydrophobic binaphthyl group probably became the membrane of the vesicles. Additionally, there exist multiple weak intra- and intermolecular interactions as well as weak hydrophilic interactions to assist the formation of membranes. After 1 h of incubation, the vesicles began to coalesce one by one and formed the "necklace"-like morphology. The area above the "necklaces" gradually grew into loosely twisted helical ribbons or fibers with axial elongation and fusion after 6 h. After 3 days, compactly twisted helical fibers with a helical pitch of ~300 nm (**22**) and ~400 nm

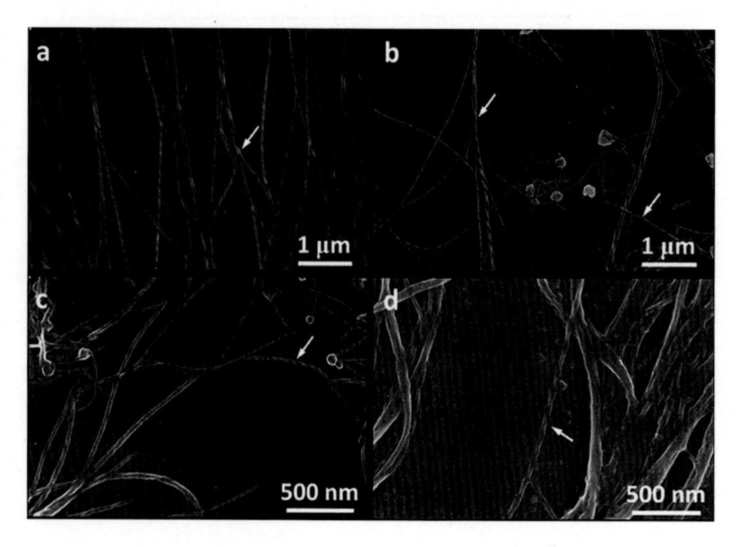

FIG. 18 SEM images of **21** at high magnification obtained from the evaporation of (A) THF, (B) THF/H_2O (5/5, v/v), (C) THF/H_2O (2/8, v/v), and (D) THF/H_2O (1/9, v/v). Solution concentration: 100 μM. *Arrows* in the images point to the areas where helices were most obviously discerned. *Reproduced with permission from G. Huang, R. Wen, Z. Wang, B.S. Li, B.Z. Tang, Mater. Chem. Front. 2 (2018) 1884–1892. Copyright 2018, the Royal Society of Chemistry.*

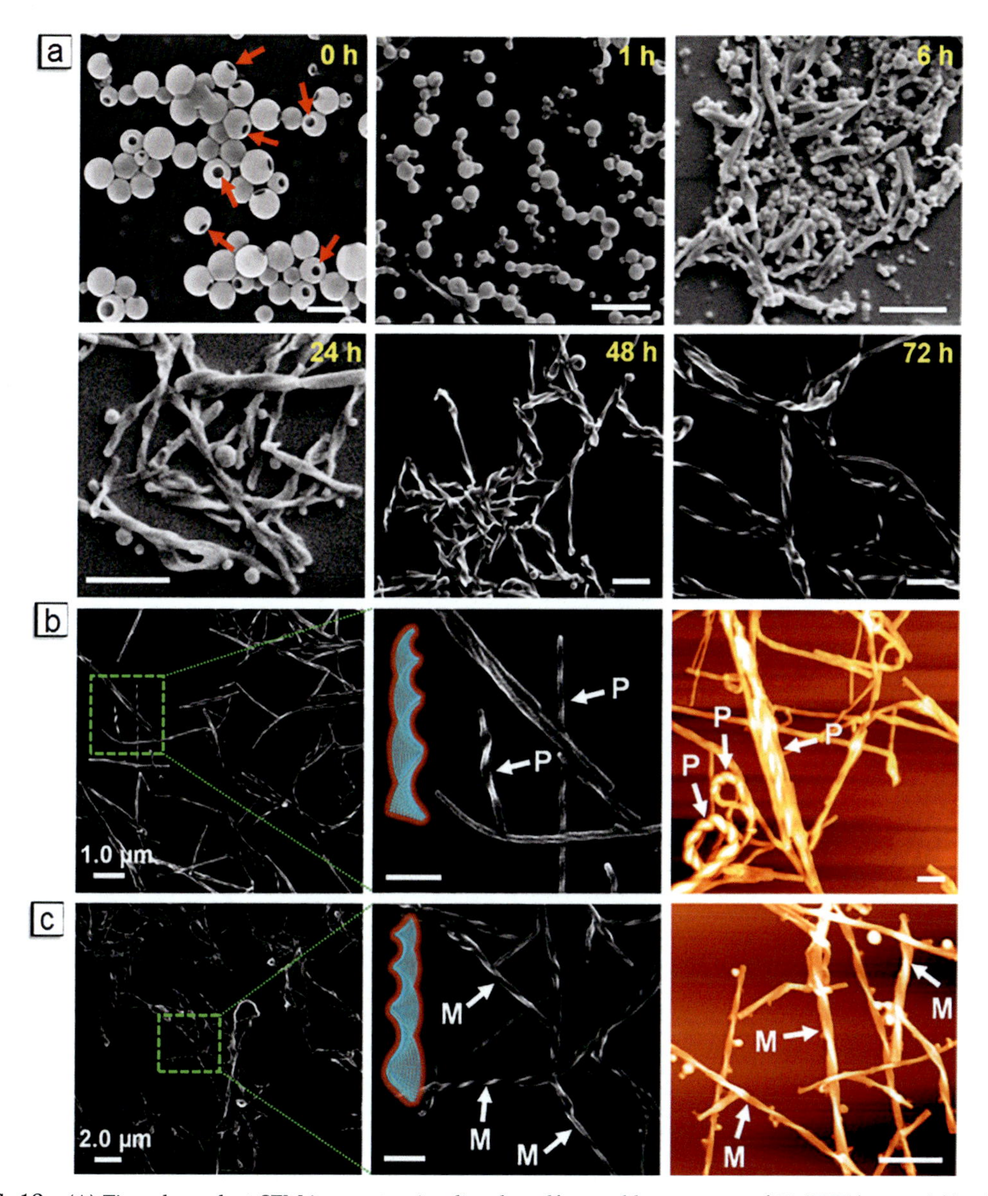

FIG. 19 (A) Time-dependent SEM images to visualize the self-assembly processes of **23** (THF/water: 1/4, v/v). (B, C) Helical fibers with inversed handedness formed by **22** (B) and **23** (C) after 72 h. Concentration: 1×10^{-4} M. Scale bar: 500 nm. *Reproduced with permission from J. Zhang, Q. Liu, W. Wu, J. Peng, H. Zhang, F. Song, B. He, X. Wang, H.H.Y. Sung, M. Chen, B.S. Li, S.H. Liu, J.W.Y. Lam, B.Z. Tang, ACS Nano 13 (2019) 3618–3628. Copyright 2019, American Chemical Society.*

(**23**) formed, where spirals could be clearly observed. The helical fibers formed by **22** are uniformly right-handed (*P*) (Fig. 19B), while the helical nanofibers formed by **23** are all left-handed (*M*) (Fig. 19C), i.e., the handedness of helices formed by the two enantiomers are perfectly opposite.

Compared with the research on the AIEgens based on small molecules, the helical self-assembly properties of polymeric AIEgens were seldom reported. The conjugated polymers with TPE units in the backbone and chiral amino acid-containing groups in the side chains, **24** and **25**, can both form helical assemblies [54,55]. The self-assembly property of the TPE- and alanine-containing polymer **24** in THF-water mixtures was discussed as an example [54]. As shown in Fig. 20a, the polymer assembled into spherical structures with diameters ranging from 80 to 220 nm at the water fraction of 50%. Exposed cavities and collapsed nanospheres were also formed (indicated by arrows in Fig. 20a2 and a3), indicating that the polymersomes are hollow. At the f_w of 60%, the polymersomes fused with one another and formed "pearl-necklace"-like structures (Fig. 20b). When the water fraction increased to 80%, the coexistence

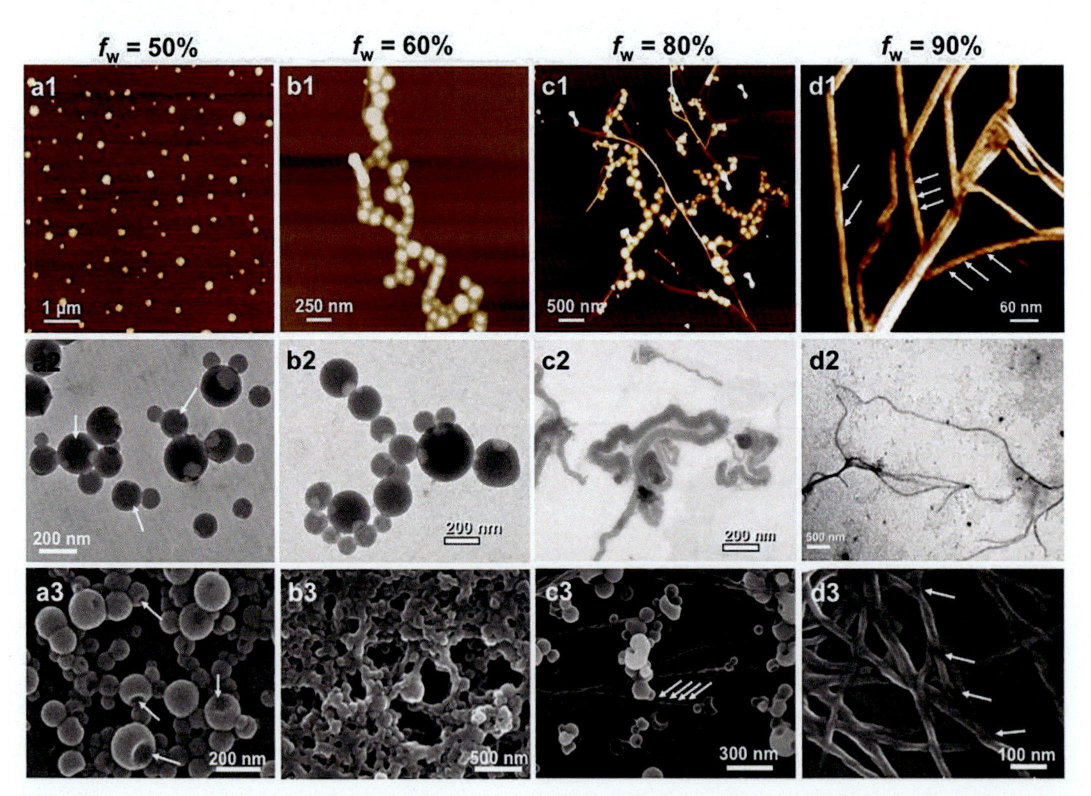

FIG. 20 AFM (a1–d1), TEM (a2–d2), and SEM images (a3–d3) of **24** obtained from THF-water mixtures at different volume ratios: (a1–a3) 5:5; (b1–b3) 4:6; (c1–c3) 2:8; (d1–d3) 1:9; concentration: 10^{-5}M. Collapsed nanospheres were also occasionally observed as indicated with the arrows in image a2 and a3. Helices of the helical fibers were most significantly exhibited in the area highlighted by arrows in image c3, d1, and d3. *Reproduced with permission from Q. Liu, Q. Xia, S. Wang, B.S. Li, B.Z. Tang, J. Mater. Chem. C 6 (2018) 4807–4816. Copyright 2018, the Royal Society of Chemistry.*

of "pearl-necklace"-like nanostructures with left- and right-handed helical nanowires was observed (Fig. 20c). At the f_w of 90%, nearly all the polymersomes have transformed to helical nanowires, and some of them intertwined into helical nanofibers (Fig. 20d). As indicated by the arrows in Fig. 20c3, d1, and d3, the knot-like screw pitches of the helical nanofibers are about 5–20nm, and the length of the fibers is up to several micrometers. The AFM images clearly reveal that the assembly process of **24** undergoes a morphological transition from pearl necklace to helical fibers with an increase in the water content.

7 Conclusions

We have summarized the progress of molecular design, CPL activities, and helical self-assembly of supramolecular molecular systems containing point chiral, axial chiral, and achiral AIEgens, respectively. They all possess high emission efficiency in solid state and show judicious optical activities such as AICD, CICD, and AACD effects.

Self-assembly of these deliberately designed AIEgens provides an efficient methodology for the construction of luminescent helical micro-/nanofibers with CPL properties. They could readily self-assemble into fluorescent helical micro-/nanostructures with predominantly one-handedness. This simple fabrication of luminescent helical architectures with AIEgens further paves the way for the future applications of these kinds of materials in optoelectronic and photonic devices and sensors. Thoroughly understanding the underlying principles of the helical self-assembly process is still a challenging task; however, these achievements are promising and will surely stimulate more future researches on exploiting new luminescent helical micro-/nanostructures formed with new AIEgens.

References

[1] L. Maggini, D. Bonifazi, Chem. Soc. Rev. 41 (2012) 211–241.
[2] C. Zhang, Y. Yan, Y.S. Zhao, J. Yao, Acc. Chem. Res. 47 (2014) 3448–3458.
[3] S. Chen, P. Slattum, C. Wang, L. Zang, Chem. Rev. 115 (2015) 11967–11998.
[4] M. Martínez-Abadía, R. Giménez, M.B. Ros, Adv. Mater. 30 (2018) 1704161.
[5] Y. Li, T. Liu, H. Liu, M.-Z. Tian, Y. Li, Acc. Chem. Res. 47 (2014) 1186–1198.
[6] A. Kaeser, A.P.H.J. Schenning, Adv. Mater. 22 (2010) 2985–2997.
[7] J.B. Birks (Ed.), Photophysics of Aromatic Molecules, John Wiley and Sons Ltd., 1970, pp. 403–489.
[8] J. Mei, N.L.C. Leung, R.T.K. Kwok, J.W.Y. Lam, B.Z. Tang, Chem. Rev. 115 (2015) 11718–11940.
[9] Z. Zhao, J.W.Y. Lam, B.Z. Tang, Soft Matter 9 (2013) 4564–4579.
[10] B.-K. An, J. Gierschner, S.Y. Park, Acc. Chem. Res. 45 (2012) 544–554.
[11] D. Yan, Chem. Eur. J. 21 (2015) 4880–4896.
[12] Y. Yan, J. Huang, B.Z. Tang, Chem. Commun. 52 (2016) 11870–11884.
[13] H.-T. Feng, C. Liu, Q. Li, H. Zhang, J.W.Y. Lam, B.Z. Tang, ACS Mater. Lett. 1 (2019) 192–202.
[14] H.-T. Feng, J.W.Y. Lam, B.Z. Tang, Coord. Chem. Rev. 406 (2020), 213142.
[15] D. Yang, P. Duan, L. Zhang, M. Liu, Nat. Commun. 8 (2017) 15727.
[16] Anuradha, D.D. La, M. Al Kobaisi, A. Gupta, S.V. Bhosale, Chem. Eur. J. 23 (2017) 3950–3956.
[17] F. Meng, Y. Sheng, F. Li, C. Zhu, Y. Quan, Y. Cheng, RSC Adv. 7 (2017) 15851–15856.
[18] L. Zhang, K. Liang, L. Dong, P. Yang, Y. Li, X. Feng, J. Zhi, J. Shi, B. Tong, Y. Dong, New J. Chem. 41 (2017) 8877–8884.
[19] J. Roose, B.Z. Tang, K.S. Wong, Small 12 (2016) 6495–6512.
[20] H. Li, B.S. Li, B.Z. Tang, Chem. Asian J. 14 (2019) 674–688.

[21] F. Song, Z. Zhao, Z. Liu, J.W.Y. Lam, B.Z. Tang, J. Mater. Chem. C 8 (2020) 3284–3301.
[22] D. Zhao, H. He, X. Gu, L. Guo, K.S. Wong, J.W.Y. Lam, B.Z. Tang, Adv. Optical Mater. 4 (2016) 534–539.
[23] J. Han, J. You, X. Li, P. Duan, M. Liu, Adv. Mater. 29 (2017) 1606503.
[24] X. Cai, J. Du, L. Zhang, Y. Li, B. Li, H. Li, Y. Yang, Chem. Commun. 55 (2019) 12176–12179.
[25] X. Li, W. Hu, Y. Wang, Y. Quan, Y. Cheng, Chem. Commun. 55 (2019) 5179–5182.
[26] S. Tsunega, R.-H. Jin, T. Nakashima, T. Kawai, ChemPlusChem 85 (2020) 619–626.
[27] J. Luo, Z. Xie, J.W.Y. Lam, L. Cheng, H. Chen, C. Qiu, H.S. Kwok, X. Zhan, Y. Liu, D. Zhu, B.Z. Tang, Chem. Commun. (2001) 1740–1741.
[28] J. Liu, J.W.Y. Lam, B.Z. Tang, J. Inorg. Organomet. Polym. 19 (2009) 249.
[29] Z. Zhao, B. He, B.Z. Tang, Chem. Sci. 6 (2015) 5347–5365.
[30] J. Liu, H. Su, L. Meng, Y. Zhao, C. Deng, J.C.Y. Ng, P. Lu, M. Faisal, J.W.Y. Lam, X. Huang, H. Wu, K.S. Wong, B.Z. Tang, Chem. Sci. 3 (2012) 2737–2747.
[31] J.C.Y. Ng, H. Li, Q. Yuan, J. Liu, C. Liu, X. Fan, B.S. Li, B.Z. Tang, J. Mater. Chem. C 2 (2014) 4615–4621.
[32] H. Li, S. Xue, H. Su, B. Shen, Z. Cheng, J.W.Y. Lam, K.S. Wong, H. Wu, B.S. Li, B.Z. Tang, Small 12 (2016) 6593–6601.
[33] J.C.Y. Ng, J. Liu, H. Su, Y. Hong, H. Li, J.W.Y. Lam, K.S. Wong, B.Z. Tang, J. Mater. Chem. C 2 (2014) 78–83.
[34] L. Dong, W. Wang, T. Lin, K. Diller, J.V. Barth, J. Liu, B.Z. Tang, F. Klappenberger, N. Lin, J. Phys. Chem. C 119 (2015) 3857–3863.
[35] Y. Zang, Y. Li, B. Li, H. Li, Y. Yang, RSC Adv. 5 (2015) 38690–38695.
[36] S. Xue, L. Meng, R. Wen, L. Shi, J.W. Lam, Z. Tang, B.S. Li, B.Z. Tang, RSC Adv. 7 (2017) 24841–24847.
[37] H. Li, J. Cheng, Y. Zhao, J.W.Y. Lam, K.S. Wong, H. Wu, B.S. Li, B.Z. Tang, Mater. Horiz. 1 (2014) 518–521.
[38] H. Li, J. Cheng, H. Deng, E. Zhao, B. Shen, J.W.Y. Lam, K.S. Wong, H. Wu, B.S. Li, B.Z. Tang, J. Mater. Chem. C 3 (2015) 2399–2404.
[39] H. Li, X. Zheng, H. Su, J.W.Y. Lam, K. Sing Wong, S. Xue, X. Huang, X. Huang, B.S. Li, B.Z. Tang, Sci. Rep. 6 (2016) 19277.
[40] H. Li, W. Yuan, H. He, Z. Cheng, C. Fan, Y. Yang, K.S. Wong, Y. Li, B.Z. Tang, Dyes Pigments 138 (2017) 129–134.
[41] Y. Shi, G. Yin, Z. Yan, P. Sang, M. Wang, R. Brzozowski, P. Eswara, L. Wojtas, Y. Zheng, X. Li, J. Cai, J. Am. Chem. Soc. 141 (2019) 12697–12706.
[42] H. Zhang, H. Li, J. Wang, J. Sun, A. Qin, B.Z. Tang, J. Mater. Chem. C 3 (2015) 5162–5166.
[43] S. Zhang, Y. Wang, F. Meng, C. Dai, Y. Cheng, C. Zhu, Chem. Commun. 51 (2015) 9014–9017.
[44] H.-T. Feng, X. Gu, J.W.Y. Lam, Y.-S. Zheng, B.Z. Tang, J. Mater. Chem. C 6 (2018) 8934–8940.
[45] L. Ding, L. Lin, C. Liu, H. Li, A. Qin, Y. Liu, L. Song, H. Zhang, B.Z. Tang, Y. Zhao, New J. Chem. 35 (2011) 1781–1786.
[46] D. Li, R. Hu, D. Guo, Q. Zang, J. Li, Y. Wang, Y.-S. Zheng, B.Z. Tang, H. Zhang, J. Phys. Chem. C 121 (2017) 20947–20954.
[47] J.-B. Xiong, H.-T. Feng, J.-P. Sun, W.-Z. Xie, D. Yang, M. Liu, Y.-S. Zheng, J. Am. Chem. Soc. 138 (2016) 11469–11472.
[48] H. Qu, Y. Wang, Z. Li, X. Wang, H. Fang, Z. Tian, X. Cao, J. Am. Chem. Soc. 139 (2017) 18142–18145.
[49] B.S. Li, R. Wen, S. Xue, L. Shi, Z. Tang, Z. Wang, B.Z. Tang, Mater. Chem. Front. 1 (2017) 646–653.
[50] G. Huang, R. Wen, Z. Wang, B.S. Li, B.Z. Tang, Mater. Chem. Front. 2 (2018) 1884–1892.
[51] J. Zhang, Q. Liu, W. Wu, J. Peng, H. Zhang, F. Song, B. He, X. Wang, H.H.Y. Sung, M. Chen, B.S. Li, S.H. Liu, J.W.Y. Lam, B.Z. Tang, ACS Nano 13 (2019) 3618–3628.
[52] S. Zhang, Y. Sheng, G. Wei, Y. Quan, Y. Cheng, C. Zhu, Polym. Chem. 6 (2015) 2416–2422.
[53] W. Zhang, X. Cai, H. Li, Y. Li, J. Funct. Polym. 32 (2019) 711.
[54] Q. Liu, Q. Xia, S. Wang, B.S. Li, B.Z. Tang, J. Mater. Chem. C 6 (2018) 4807–4816.
[55] Q. Liu, Q. Xia, Y. Xiong, B.S. Li, B.Z. Tang, Macromolecules 53 (2020) 6288–6298.
[56] R. Carr, N.H. Evans, D. Parker, Chem. Soc. Rev. 41 (2012) 7673–7686.
[57] E.M. Sánchez-Carnerero, A.R. Agarrabeitia, F. Moreno, B.L. Maroto, G. Muller, M.J. Ortiz, S. de la Moya, Chem. Eur. J. 21 (2015) 13488–13500.
[58] J. Kumar, T. Nakashima, T. Kawai, J. Phys. Chem. Lett. 6 (2015) 3445–3452.
[59] J. Han, S. Guo, H. Lu, S. Liu, Q. Zhao, W. Huang, Adv. Optical Mater. 6 (2018) 1800538.
[60] J.-L. Ma, Q. Peng, C.-H. Zhao, Chem. Eur. J. 25 (2019) 15441–15454.
[61] D.-W. Zhang, M. Li, C.-F. Chen, Chem. Soc. Rev. 49 (2020) 1331–1343.

[62] J. Yuasa, T. Ohno, K. Miyata, H. Tsumatori, Y. Hasegawa, T. Kawai, J. Am. Chem. Soc. 133 (2011) 9892–9902.
[63] B.A. San Jose, K. Akagi, Polym. Chem. 4 (2013) 5144–5161.
[64] F. Song, Z. Xu, Q. Zhang, Z. Zhao, H. Zhang, W. Zhao, Z. Qiu, C. Qi, H. Zhang, H.H.Y. Sung, I.D. Williams, J.W.Y. Lam, Z. Zhao, A. Qin, D. Ma, B.Z. Tang, Adv. Funct. Mater. 28 (2018) 1800051.
[65] M. Hu, H.-T. Feng, Y.-X. Yuan, Y.-S. Zheng, B.Z. Tang, Coord. Chem. Rev. 416 (2020), 213329.
[66] Z. Zhao, H. Zhang, J.W.Y. Lam, B.Z. Tang, Angew. Chem. Int. Ed. 59 (2020) 9888–9907.

CHAPTER

5

AIE-active supramolecular gel systems

Massimo Cametti[a] *and Zoran Džolić*[b]

[a]Department of Chemistry, Materials and Chemical Engineering "Giulio Natta", Politecnico di Milano, Milano, Italy [b]Ruđer Bošković Institute, Zagreb, Croatia

1 Introduction

Supramolecular gels are unique soft materials generated by the self-assembly of their molecular (or macromolecular) constituents into three-dimensional entangled networks that are capable of entrapping extremely large quantities of solvent molecules [1]. The underlying weak interactions that hold the whole architectures together can be of various nature. Hydrogen and halogen bonding, aromatic π–π stacking, and hydrophobic and electrostatic interactions have been all found to be suitable for the formation of fibrillar aggregates [2], which are the consequence of a hierarchical assembly operating at different length scales. These weak intermolecular forces which endow the gel state with characteristics somewhat intermediate between those of a solution and a solid are also responsible for their highly dynamic behavior, that is, a marked capability to react and adapt to different imposed stimuli. Materials that can respond to external chemical or physical stimuli are extremely important in several different technological applications. Nowadays, gel systems have been successfully applied to hybrid functional materials, drug delivery, enzyme immobilizations, biomedicine, and so on [3].

One of the most intriguing material properties is luminescence [4]. As far as the interaction of electromagnetic radiation with organic molecules is concerned, fluorescence corresponds to the radiative spin-allowed transition from the first excited singlet state to the ground state, and it is mainly characterized by three parameters: intensity (I_λ), efficiency (usually expressed as quantum yield, Φ_F), and lifetime (τ). Luminescence is at the basis of several technological applications and, consequently, it represents a highly desired property for organic compounds. Research efforts over several decades have discovered a large number of different chemical structures and functionalities that feature efficient emission in solution covering the entire visible spectrum and beyond (UV–vis and IR) [5]. However, the efforts toward their

Aggregation-Induced Emission (AIE)
https://doi.org/10.1016/B978-0-12-824335-0.00002-7

direct implementation into optoelectronic devices reveal a crucial limitation. Indeed, practical technological necessities call for the incorporation into the device of the active compounds in their aggregated form, i.e., as thin films or in the solid-state. Under these conditions, it is now well established that the majority of molecules that are emissive in solution suffer from severe emission quenching upon aggregation or crystallization. This phenomenon, called aggregation-caused quenching (ACQ) [6], is generally considered due to the formation of excimers and exciplexes, and it had critically limited the development of novel technologies based on fluorescent systems. This is especially true for light-emitting electrochemical cells, organic light-emitting diodes, and in general for solid-state devices. Likewise, ACQ also affects other fields of applications, e.g., luminescent soft materials such as gels.

In the wake of the century, the discovery of aggregation-induced emission (AIE) [7], the phenomenon by which a fluorophore is almost nonfluorescent in the molecular state (i.e., in dilute solution), but become highly emissive in the aggregate state, brings in new opportunities, as well as challenges, for the development of luminescent soft materials. Several mechanisms have been proposed in order to rationalize AIE phenomena, including aggregation-induced restriction of intramolecular motions, in particular, intramolecular rotation (RIR), J-aggregation, intramolecular planarization, and intramolecular charge transfer, among which RIR has been identified as the most common cause for AIE [8]. In the solution state, active intramolecular rotations may serve as a relaxation pathway for the excited state to decay nonradiatively, whereas in the aggregated state, these motions can be restricted, thus leaving emission as the preferential relaxation pathway. Accordingly, a number of AIE luminogens (AIEgens) have been designed and developed based on the RIR mechanism.

It is therefore clear that the introduction of AIE properties into soft material systems not only relieves the limitation posed by the ACQ phenomenon but also endows them with fascinating new properties. In line with this, the incorporation of AIE properties into supramolecular gelators/gels systems has given new vitality to the field providing supramolecular gels with bright emissions and high quantum yields. Gel morphology and fluorescence are mutually interdependent, and thus they can both be reversibly controlled via suitable external stimuli affecting the extent of aggregation. AIE-active supramolecular gels have thus greatly promoted the development of novel materials for applications in chemical and environmental sensing and in biomedicine [9]. Furthermore, as luminescence cannot be viewed only as an intrinsic property of a given molecule, for it has been shown to depend heavily on the way they may aggregate, studies which focus on the relation between gel-assembly architectures and emerging AIE properties are extremely useful. They can serve the dual purpose of better understanding the AIE properties of a gelator molecule but, at the same time, they provide an indirect mode of investigation on the gel self-assembly process itself. In other words, the incorporation of AIEgens into gelatable systems can not only afford excellent luminescent gel systems but also bring about built-in probes for gelation process monitoring.

Over the past two decades, several comprehensive reviews and books have been published on AIE-active supramolecular gels [10–14]. In this chapter, we have reviewed the most recent and representative research works on supramolecular gels with AIE properties, published within the 2014–20 period, highlighting the progress in preparation, properties characterization, and enhancement, and in proposing novel applications. Given the potential of AIE phenomena for developing new sensing systems, AIE-active supramolecular gels are considered a novel class of smart functional materials and have been widely studied as luminescent sensors and biomedical imaging agents. Also, we have illustrated the design principles for AIE-active supramolecular gels and how a modular approach, combining AIE luminogens with specific receptor units, can be a practical and successful tool for the construction of new AIE-based gel sensors. Finally, challenges and future perspectives on the development of AIE-based supramolecular gels have been presented

2 Strategies to build AIE-active supramolecular gel systems with examples

Rationally designed gelators with typical AIE luminogens such as tetraphenylethylene (TPE) [15], silole, and cyanostilbene derivatives [16,17], and nonclassical AIE luminogens such as carbazole-based dendrimers [18], phospholes [19], pillarenes [20], have shown an excellent self-assembling ability to generate highly emissive organogels or hydrogels. Although the number of papers related to AIE-active supramolecular gels is still limited, the great potential envisaged for this (sub)class of soft materials is finally starting to be recognized. The photophysical properties of supramolecular gels could be reversibly controlled by external stimuli, such as temperature, ultrasound, light, chemicals, etc. From this, it derives that incorporating an AIEgen into a supramolecular gel is a promising strategy for constructing responsive functional luminescent systems.

From a practical point of view, most commonly, the introduction of AIE effect into supramolecular gel systems can be achieved through one of these three options: (1) by linking a known AIE-active moiety to a self-complementary, or simply self-aggregating unit; (2) via multicomponent coassembly between a per se nongelling AIEgen unit and a known gelator; and (3) through AIEgen binding, exploiting specific host-guest interactions, to a host molecule itself capable of gel formation. A few examples for each of the abovementioned strategies are described.

The first approach can be easily exemplified by the works which focused on the outstanding performance of peptides with regard to self-assembly, thus combining an AIEgens moiety with a peptide unit to fabricate AIE-active supramolecular gels.

Tetraphenylethylene (TPE) derivatives have been widely employed as AIE-active gelators [21]. A very nice example of their properties, related to AIE-active gelation, is given by the work of Ma, Lei et al., where AIE properties of TPE derivatives **1** and **2** were compared to those of AIE-inactive **3**, devoid of two of the four phenyl group attached to the ethylene central core (Fig. 1A) [22].

AIE properties of **1** and **2** were initially tested by comparison of their fluorescence efficiency in solution and in the solid-state, as opposed to those of **3**. Results showed that both **1** and **2** do emit weakly in solution and strongly in the solid-state, with an up to 10 times increase of emission efficiency. On the contrary, this enhancement lacks with **3**. Compounds **1** and **2** also form gels in mixed solvents with good critical gelation concentrations (*cgc*), often below the 1 wt% (in $CH_3OH/CHCl_3$, 0.94 and 0.97 wt% for **1** and **2**, respectively), and with conspicuous emission, red-shifted in comparison to their sol phase (Fig. 1B and C). Notably, compound **3**, although being a gelator as well, does not show any AIE activity. In this case, aggregation was found to be due to attractive $C{=}O{\cdots}\pi$ interactions between the terminal ester carbonyls and the phenyls ring which imposed vicinity between double bonds of different molecules, as seen by NOESY NMR measurements.

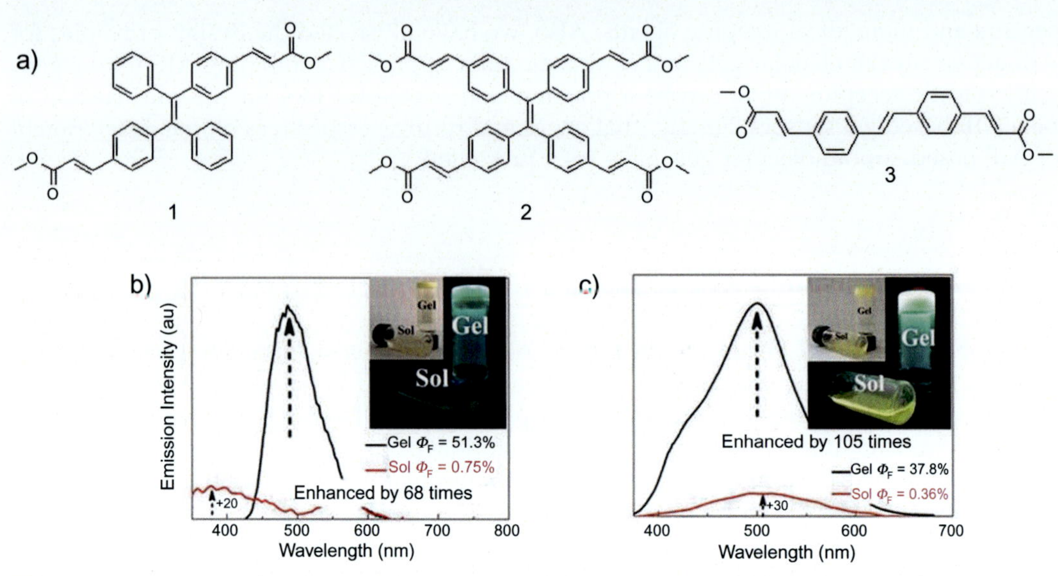

FIG. 1 (A) Chemical structures of gelators **1–3** and emission spectra of (B) **1** in $CH_3OH/CHCl_3$ (v/v = 3/1, 0.94 wt%), and (C) **2** in $CH_3OH/CHCl_3$ (v/v = 3/1, 0.97 wt%) in the sol and gel states. The inset photographs show the gel and sol states of **1** and **2** (under a 365 nm UV lamp); λ_{ex} = 280 nm for **1** and 295 nm for **2**. *Reproduced with permission from Y. Ma, H. Ma, Z. Yang, J. Ma, Y. Su, W. Li, Z. Lei, Methyl cinnamate-derived fluorescent rigid organogels based on cooperative $\pi-\pi$ stacking and C=O···π interactions instead of H-bonding and alkyl chains, Langmuir 31 (2015) 4916–4923. Copyright 2015, American Chemical Society.*

However, self-complementary interaction patterns between aggregating molecules can be obtained in several ways. The work by Galindo, Miravet et al. is a good example where amino acids are selected to boost molecular aggregation [23]. A series of 1,8-naphthalimide derivatives were used to create supramolecular hydrogels with the AIE effect (Fig. 2A). Compounds **4**, **5**, and **8** form gels in DMSO-water 8:2 *v*/v (*cgc* = 3–10 mol dm^{-3}) which can be characterized by fibrillar networks. Their aggregation behavior was first analyzed by UV–vis absorption by adding a different proportion of water to a dilute DMSO solution and correlating the observed shift to DFT computed models. In the tested solvents, compound **6** is soluble, compounds **7** and **9** do not form gels, while **10** generated nanoparticles (diameter 30 ± 5 nm). Fluorescence spectroscopic studies indicated that while all compounds (except **8**) present poor emission in DMSO solution, upon water addition, their fluorescence quantum yield (Φ_F) increases. Interestingly, there is a good correlation between gelation capability and emission enhancement, as nongelating molecules **6**, **9**, and **10** show decidedly lower enhancements. Compound **8** features a peculiar aptitude as it can act as hydrogelator, although showing poor AIE. This again shows how the arrangement of the fluorophores in the aggregated form can affect nonradiative decay, and this is further confirmed by the data on absolute Φ_F in dry solids which showed reduced emission ability. Finally, fluorescence microscopy was used to analyze the AIE of the fibrillar network of the wet samples, making it possible to distinguish the bluish emission of fibrillar materials from the greenish emissive

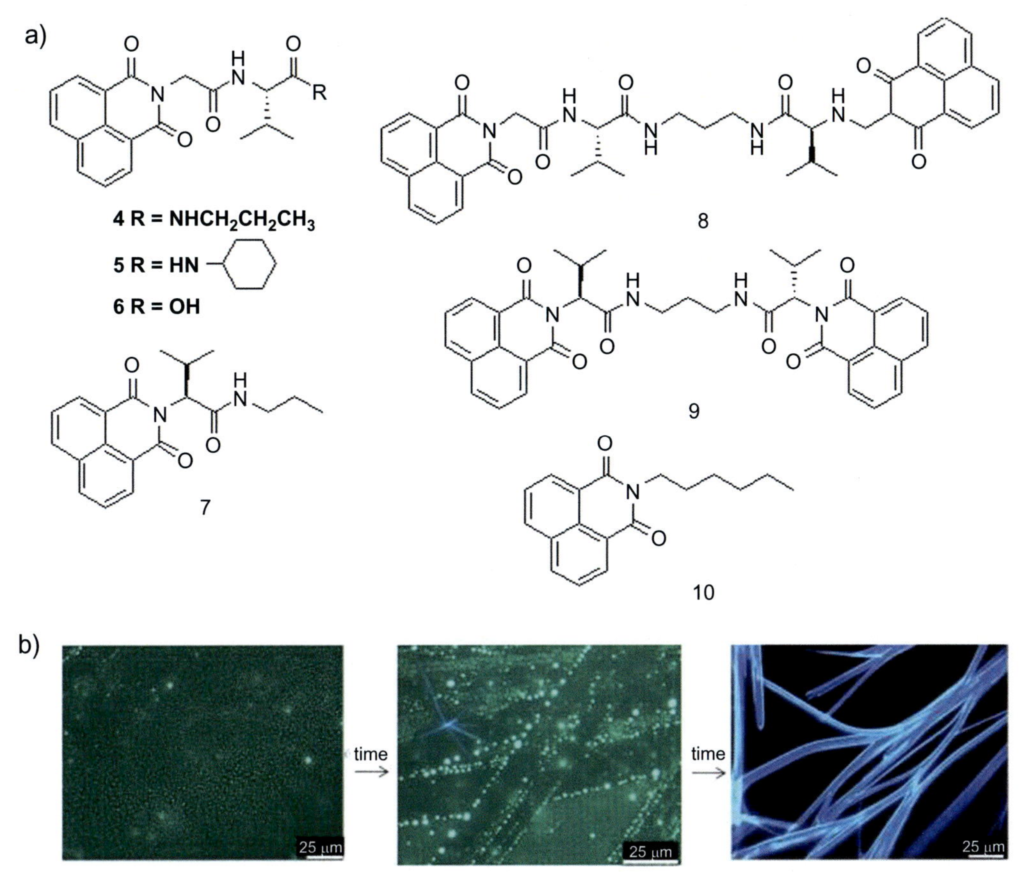

FIG. 2 (A) Chemical structures of 1,8-naphthalimide derivatives **4–10** and (B) time evolution (from left to right) from nanoparticles to fibrillar aggregates for compound **10**, showing the replacement of the greenish excimer nanoparticle emission with bluish emission by fibrils. Coexistence of the nanoparticles with a fibrillization nucleus can be observed (middle image). $\lambda_{ex} = 405\,nm$, $2\,mol\,dm^{-3}$ in 20/80 DMSO-water mixture. *Reproduced with permission from C. Felip-León, F. Galindo, J.F. Miravet, Insights into the aggregation-induced emission of 1,8-naphthalimide-based supramolecular hydrogels, Nanoscale 10 (2018) 17060–17069. Copyright 2018, Royal Society of Chemistry.*

nanoparticles made of **10** (Fig. 2B). These nanoparticles were found to represent an intermediate phase which then evolved into fibers with AIE properties similar to those obtained by **4**, **5**, and **8**.

A second approach to prepare AIE-active supramolecular gels is constituted by the coassembly of an AIEgen component with a gelator molecule. For example, compound **11** (Fig. 3A), composed of a TPE core functionalized with four phenyl glutamic acid dodecyl esters groups, is not fluorescent when dissolved in DMSO, but it tends to aggregate into strongly emitting thin and short fibers (Fig. 3B) [24]. In the presence of nonfluorescent *N,N′*-bis(octadecyl)-L-Boc-glutamic diamide cogelator (**12**), gels formed and they were characterized by fluorescence spectroscopy and rheology.

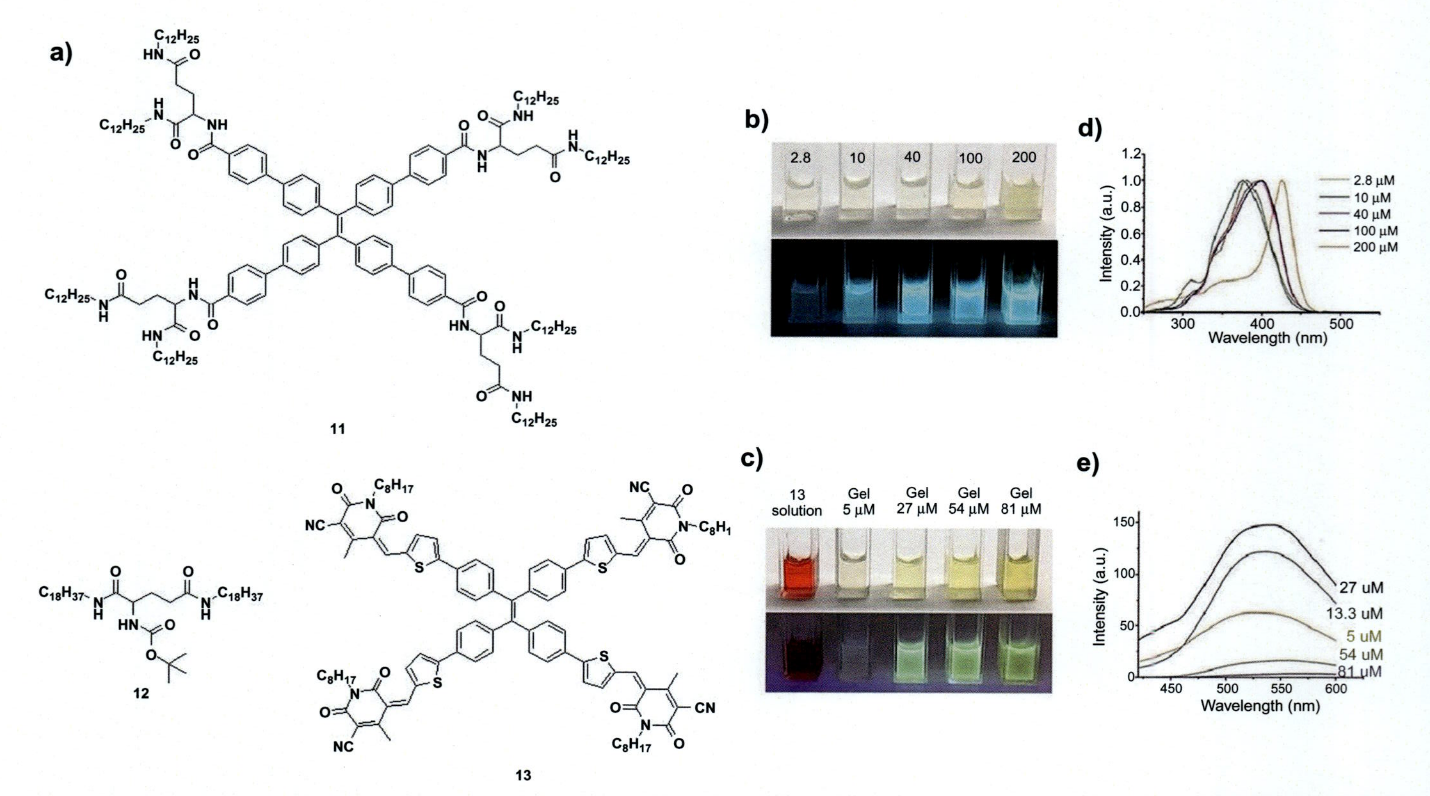

FIG. 3 (A) Chemical structures of TPE-base derivatives **11** and **13** and a cogelator **12** and images of the binary **11** + **12** (B) and **12** + **13** (C) gels taken under room light and UV lamp (365 nm) and fluorescence spectra of **11** + **12** gels (λ_{em} = 500 nm), (D) and excitation spectra of **12** – **13** gels (λ_{ex} = 400 nm, e). The concentration of **13** in the solution was 27 mol dm^{-3}. *Reproduced with permission from J.-Y. Chen, G. Kadam, A. Gupta, Anuradha, S.V. Bhosale, F. Zheng, C.-H. Zhou, B.-H. Jia, D.S. Dalal, J.-L. Li, A biomimetic supramolecular approach for charge transfer between donor and acceptor chromophores with aggregation-induced emission, Chem. A Eur. J. 24 (2018) 14668–14678. Copyright 2018, Wiley-VCH.*

The absorption and emission of **11** in the binary gel are highly dependent on its concentration, as seen in Fig. 3A for a concentration range between 2.8 and 200 mol dm^{-6}. This indicates marked concentration-dependent changes of its molecular packing within the two-component system, which probably hints at the fact that, upon concentration increase, **11** may aggregate onto the fibers surface of **12**. Similar studies were performed with a different TPE derivative, compound **13**, in which the TPE unit is terminally functionalized with four cyanopyridone moieties. The TPE derivative **13** is also not fluorescent in DMSO, but gels with **12** emit green fluorescence under UV light (Fig. 3E). It is also interesting to note that the peripheral functional groups on the TPE-core have a significant influence on the AIE emission maxima (λ_{max}). As expected, an increase in the concentration of **13** leads to the fluorescence enhancement of the gel, while however, upon further increase, a partial quenching occurs, up to reaching a complete lack of fluorescence at ca. 80 μM (Fig. 3C). This is an interesting observation which, in combination with the fact that powders of **12** have strong fluorescence whereas those of **13** do not, indicates a specific donor-acceptor structure with a self-quenching capability for the **12**–**13** gel system. Finally, ternary gels were produced and were investigated to study the effect of donor-acceptor complexes formation between **11** and **13**, which might mimic a protein-assisted charge-transfer system in nature.

Liu and coworkers used a C_3 symmetrical L-glutamic-acid derivative, gelator compound **14** to construct AIE-active hexagonal nanotubes in a wide range of solvent mixtures [25].

On account of the large inner space within the assembled nanotubes made of **14**, a wide range of AIEgens with different emission colors, such as TPE, hexaphenylsilole, and a series of styrylbenzene derivatives, was efficiently, noncovalently encapsulated and orderly arranged within them (Fig. 4). As the AIEgens assemble in a way that is strongly directed by the morphologies of the chiral nanotubes, the resultant gel system showed circularly polarized luminescent (CPL) signals.

Finally, the third strategy to the fabrication of AIE-active supramolecular gels is based on classical host-guest chemistry. Specific interactions between a host and a guest provide a controlled way to assemble the various active components into an integrated material. Its dynamic and reversible nature makes this approach quite promising and research toward AIE-active supramolecular gels based on host-guest interactions has been gradually reported.

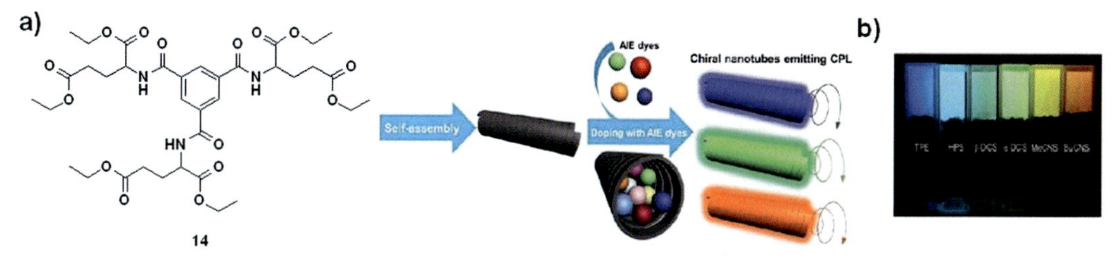

FIG. 4 (A) Chiral gelator **14** could form chiral nanotubes due to the intrinsic chirality of the substituted glutamate moieties. Various AIEgen dyes could be encapsulated into the nanotubes by simply cogel procedure; (B) The obtained **14**-AIE dyes cogels could emit circularly polarized light by exciting the AIEgen dyes. *Reproduced with permission from J. Han, J. You, X.G. Li,P. Duan, M. Liu, Full-color tunable circularly polarized luminescent nanoassemblies of achiral AIEgens in confined chiral nanotubes, Adv. Mater. 29 (2017) 1606503. Copyright 2017, Wiley-VCH.*

Yang et al. extensively studied inclusion complexes of TPE-bridged pillar[5]arene tetramers **15** and **16** with triazole-based neutral linkers **17–19**, including one of the first studies of pillarene-based supramolecular AIE gel system based on the assembly of TPE-bridged pillar[5]arene tetramers (**15**) with triazole-based neutral linker **17** in a 4:2 ratio (**17@15**) [26,27]. Gel formation, occurring in $CHCl_3$ and inducing strong emission ($\lambda_{ex}=405$; $\lambda_{em}=488$ nm), was the result of effective host-guest binding interactions between each pillarene ring belonging to the tetrameric host and the triazole-cyano moieties of the ditopic guest linkers (Fig. 5B and C). This gel could also undergo reversible sol–gel transition and presents switchable fluorescence intensity under temperature variation.

More recently, the design of the guests was also diversified, including tripodal C_3 symmetrical compounds **18** and **19**, with different carbon chain linker lengths. These new guests reacted with **15** and with a novel host **16** (with longer alkyl spacers) in a 4:3 ratio [27]. While all assemblies possessed intense fluorescence in the solid-state, only one host-guest combination (**18@16**) was successful in the formation of blue fluorescent supramolecular gel ($\lambda_{ex}=350$ nm, $\lambda_{em}=488$ nm, Fig. 5D–F). Although the present understanding of the gelation process is quite limited, the observed phenomenon was ascribed to a loose assembly structure formed by the host decorated with longer chains that could allow higher amounts of solvent molecules to be included in the supramolecular assembly.

The examples reported so far exemplified the main strategies by which an AIE-active supramolecular gel can be formed.

3 Stimuli-responsive AIE-active supramolecular gels and applications

The dynamic nature of noncovalent interactions and the unique structural features of supramolecular gels allow, in many circumstances, a rapid system response to external environmental stimuli. Distinct gel–sol phase transitions can be caused by numerous stimuli of different nature, such as temperature, pH, light, ultrasound, the addition of ions or neutral molecules, making supramolecular gels promising stimuli-responsive materials [28]. Consequently, the integration of AIE properties with the capability of supramolecular gel system to respond to a given stimulus, for instance via a phase transformation, provides a practical and effective route for the development of smart materials with broad biological and high-tech applications. In this section, recent developments in stimuli-responsive AIE-active supramolecular gels will be discussed through the description of selected examples that comprise UV-vis light, ions, and neutral molecules as the effective stimuli which elicit the material response.

3.1 Light

Photo-responsive supramolecular gels have attracted great attention due to their significance in smart optical and biological applications. They possess several advantages, for light is noninvasive, it can avoid the necessity for direct contact between different materials, and it can be accurately controlled and applied, also inducing a low thermal effect.

Cyanostilbene derivatives deserve to be considered one of the classic types of AIE-active materials constituents [29]. Furthermore, their ability to undergo Z/E photoisomerization can be exploited in the development of stimuli-responsive functional AIE gel systems. Indeed,

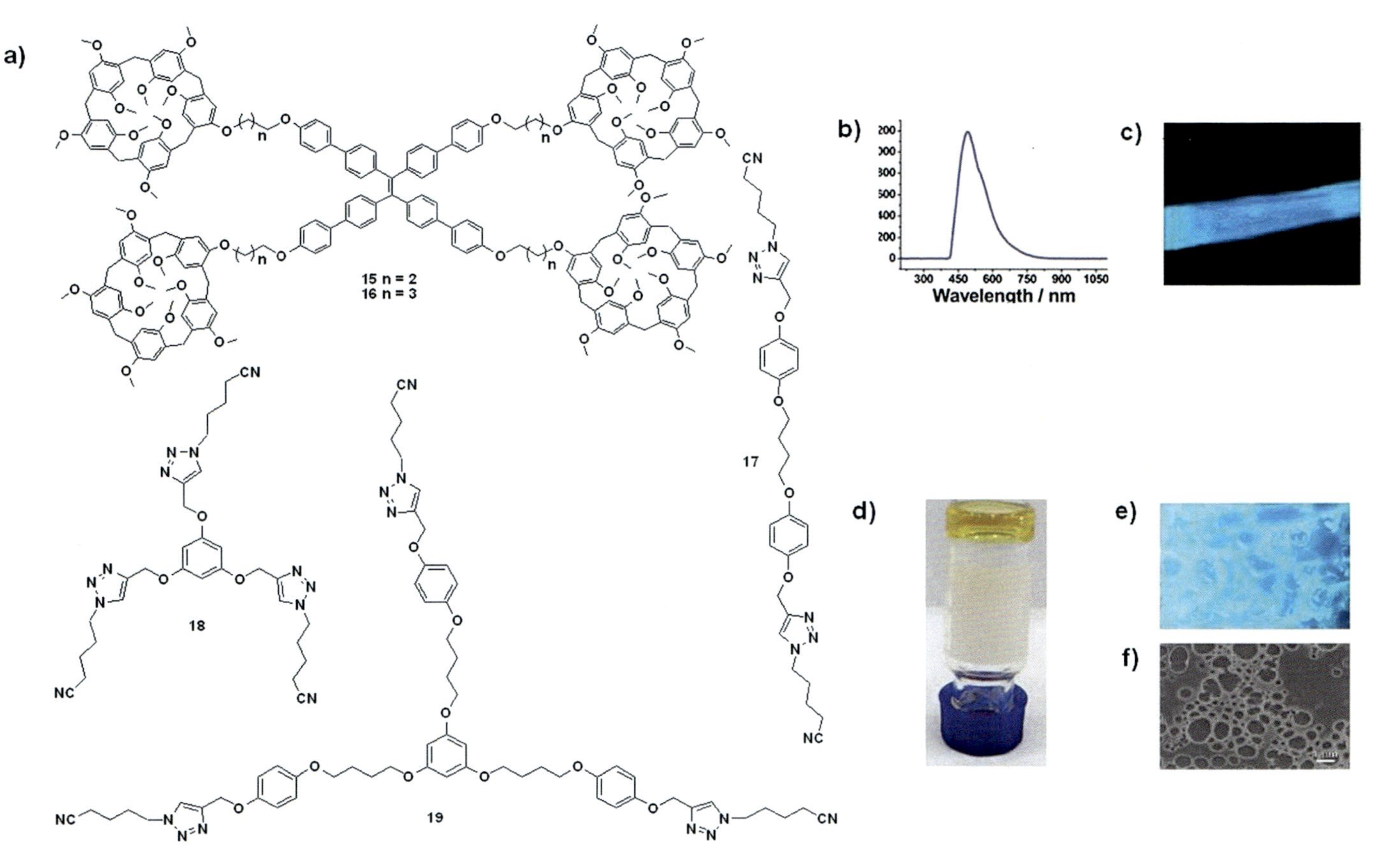

FIG. 5 (A) Chemical structures of TPE-bridged pillar[5]arene tetramers **15** and **16** and triazole-based linkers **17–19**; (B) the **17@15** gel in $CHCl_3$ exhibited a strong blue fluorescence emission with the maximum wavelength of 492nm as observed by solid-state fluorescence spectroscopy ($\lambda_{em}=488$nm); and (C) fluorescent image of the gel observed by the fluorescence microscopy ($\lambda_{ex}=365$nm, at magnification ×200). (D) Image of a supramolecular gel of **18@16** in $CHCl_3$ (D) and its fluorescence microscopy (E) and SEM (F) images (gel concentration 45 mmol dm^{-3}). *Panels (A–C): Reproduced with permission from N. Song, D.-X. Chen, Y.-C. Qiu, X.-Y. Yang, B. Xu, W. Tian, Y.-W. Yang, Stimuli-responsive blue fluorescent supramolecular polymers based on a pillar[5]arene tetramer, Chem. Commun. 50 (2014) 8231. Copyright 2014, Royal Society of Chemistry. Panels (D–F): Reproduced with permission from N. Song, X.-Y. Lou, W. Hou, C.-Y. Wang, Y. Wang, Y.-W. Yang, Pillararene-based fluorescent supramolecular systems: the key role of chain length in gelation, Macromol. Rapid Commun. 39 (2018) 1800593. Copyright 2018, Wiley-VCH.*

light-responsive solid-state materials based on α-cyanodiarylethenes have rarely been reported because of their tightly π–π stacked aggregates in which Z/E photoisomerization cannot proceed smoothly to any sufficient degree [30]. Gel state may provide a different, more flexible/adjustable environment. In this respect, Park et al. exploited this opportunity in a way to construct a multistate switchable organogel by using a two-component system comprising cyanostilbene gelator **20** and photochromic diarylethene dye **21** (Fig. 6) [31].

Cyanostilbene derivative **20**, already known as an AIE-active gelator in 1,2-dichloroethane and 1,1,2,2-tetrachloroethane [32], was studied in combination with diarylethene **21a**, a photochromic compound capable of a UV-triggered ring-closing reaction (to **21b**). The two components were selected based on the spectral overlap between the blue emission of **20** and the absorption of the closed-form **21b**, which was considered good enough to allow efficient energy transfer. Indeed, gel of **20** containing 1, 2, and 3 equivalents of **21a** exposed to UV-light irradiation showed a gradual decrease of the original blue gel emission (Fig. 6A–C) displaying only the green emission from **21b**. This green emissive gel could convert to a green fluorescent solution upon heating, due to thermal gel disruption. Notably, the fluorescence emission of this solution could be turned off upon visible-light irradiation via conversion of **21b** into **21a**, not emissive. This transformation cycle (Fig. 6D and E) can be repeated several times, indicating an underlying highly reversible process and suggesting that this mixed gel system might be useful as a combinational logic circuit (viz., two inputs: light and thermal stimuli; three on/off outputs: green and blue emission, and gel-to-solution transformation).

Park and coworkers also designed the light-responsive gelator **22**, based on α-cyanodiarylethene unit connected to a phenyl group decorated with three long alkoxy chains via an amidic bond (Fig. 7A) [33]. Self-assembly of **22** leads to a transparent yellow gel in cyclohexane (1 wt%) composed of an ultrafine nanometer-sized structure. Hydrogen bonding between the amide groups was proposed to drive the formation of monodimensional arrays where both the extremities, i.e., the rigid cyanostilbene moiety and the flexible alkyl chains, were disorderly arranged. This would thus give rise to a loose assembly. Such geometry facilitated the Z/E trans-cis photoisomerization of the α-cyanostilbene unit in the aggregated gel state. Although *Z*-**22** and *E*-**22** were virtually nonfluorescent in solution, the gel emitted strong bluish-green fluorescence due to its AIE behavior (Fig. 7B). The gel system collapsed into a viscous solution upon irradiation with a blue LED (460 nm) for 15 min. This gel-to-sol transition was induced by the Z/E photoisomerization of **22** and evidenced that the *E*-isomer perturbs the supramolecular organization of the gel (Fig. 7C and D).

Jiang and coworkers reported two examples of effective low molecular weight organic gelators based on C_2 symmetric V-shaped bis-α-cyanostilbene amide derivatives **23** and **24** (Fig. 8A) which exhibited excellent gelating ability in several solvents (*m*-xylene for **23** and acetonitrile and 1,2-dichloroethane for **24**) and marked response to light [34,35].

Indeed, the strongly emissive gels were transformed into a viscous nonfluorescent solution when exposed to UV light, through the Z/E photoisomerization of α-cyanostilbene; studied in detail in solution (Fig. 8B and C), this gel-to-sol transition could be reversed by heating. Although the behavior of **23** and **24** is qualitatively similar, **24** features a D-π-A structure which allowed a multistimuli responsive behavior and showed extremely different emission λ_{max} under aggregated and nonaggregated conditions.

Thereafter, Jiang et al. further reported another AIE gel system, **25**, involving the insertion of a third cyanostilbene unit into the gelator molecular structure [36]. The C_3 symmetric gelator **25**, consisting of a central benzene core and three α-cyanostilbenes, linked via an

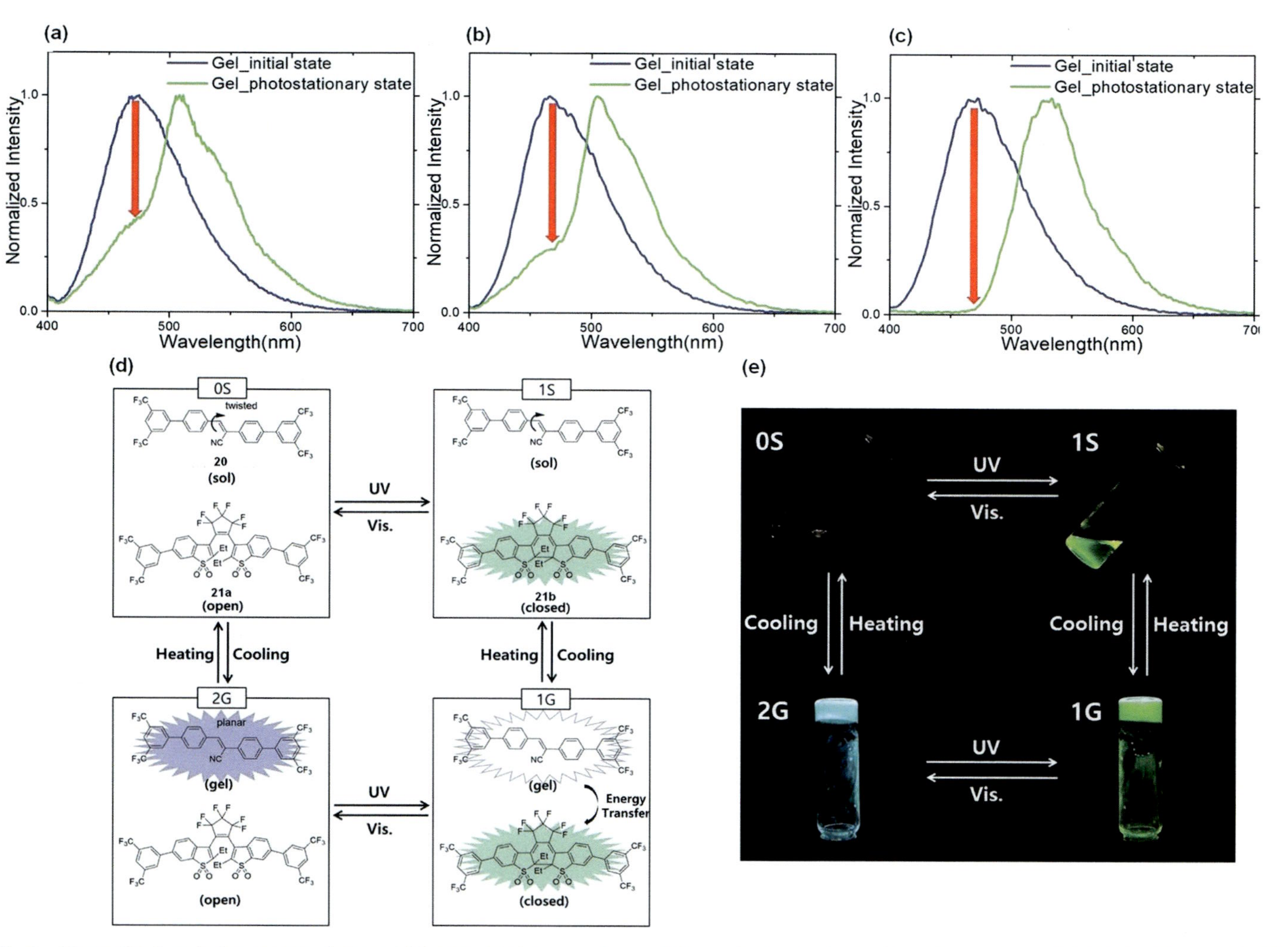

FIG. 6 Normalized emission spectra changes of the **20** + **21a** binary gel upon UV irradiation (weight ratio of the mixture = 1:1 (A), 1:2 (B), and 1:3 (C), respectively, concentration of **20** = 0.8 wt%; (D) schematic illustration and (E) photographs of the four states (0S, 1G, 1S, 2G) of the mixture. *Reproduced with permission from D. Kim, J.E. Kwon, S.Y. Park, Fully reversible multistate fluorescence switching: organogel system consisting of luminescent cyanostilbene and turn-on diarylethene, Adv. Funct. Mater. 28 (2018) 1706213. Copyright 2018, Wiley-VCH.*

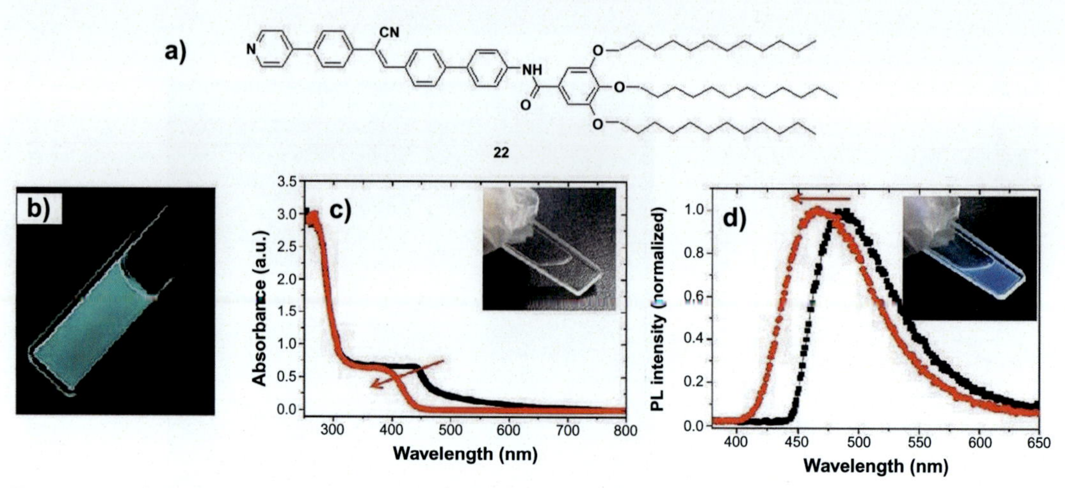

FIG. 7 (A) Chemical structure of gelator **22**; (B) Photograph of the gel of **22** in cyclohexane (1 wt%) under UV light; (C) UV–visible absorption spectra and (D) emission spectra of the gel of **22** in cyclohexane (0.25 wt%) before (black) and after (red) the blue light irradiation. Each inset shows the photo image of the sol state after the 465 nm light irradiation under room light (C) and UV light (D), respectively. *Reproduced with permission from J. Seo, J.W. Chung, J.E. Kwon, S.Y. Park, Photoisomerization-induced gel-to-sol transition and concomitant fluorescence switching in a transparent supramolecular gel of a cyanostilbene derivative, Chem. Sci. 5 (2014) 4845–4850. Copyright 2018, Royal Society of Chemistry.*

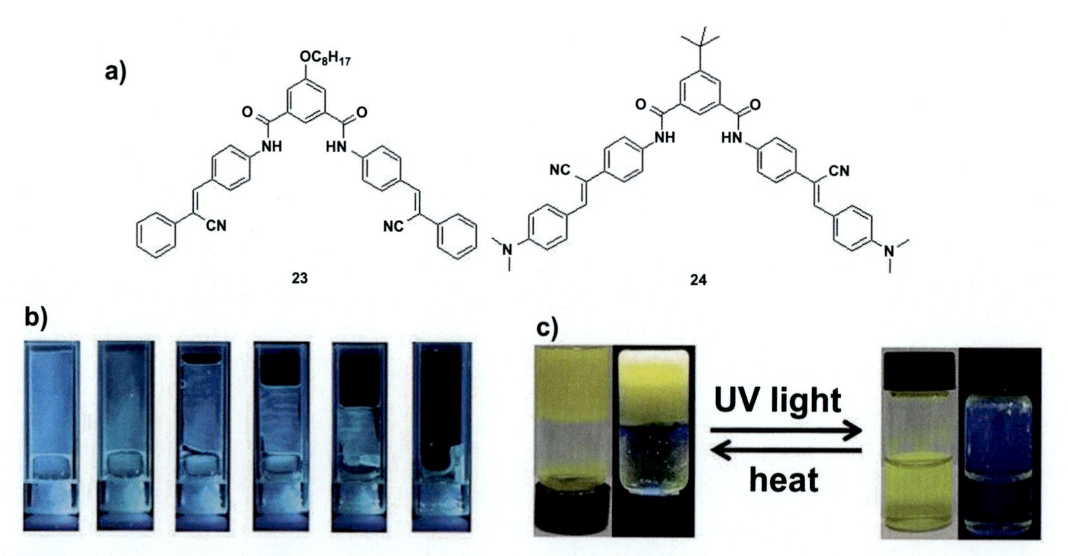

FIG. 8 (A) Chemical structures of cyanostilbene gelators **23** and **24** and (B) photographs recorded under UV light of toluene gel of **23** after photoirradiation with 365 nm light for (from left to right) 0, 5, 10, 20, 30, and 40 min. (C) The reversible responsive behavior of acetonitrile gel of **24** (8 mg mL^{-1}) upon UV irradiation at 365 nm and heating. *Panels (A, B): Reproduced with permission from Y. Ma, M. Cametti, Z. Džolić, S. Jiang, Responsive aggregation-induced emissive supramolecular gels based on bis-cyanostilbene derivatives, J. Mater. Chem. C 4 (2016) 10786–10790. Copyright 2016, Royal Society of Chemistry. Panel (C) Reproduced with permission from X. Wang, Z. Ding, Y. Ma, Y. Zhang, H. Shang, S. Jiang, Multi-stimuli responsive supramolecular gels based on a D–π–A structural cyanostilbene derivative with aggregation induced emission properties, Soft Matter 15 (2019) 1658–1665. Copyright 2019, Royal Society of Chemistry.*

amidic bond, formed two different gel systems, exhibiting different packing modes (as determined by P-XRD and FT-IR measurements), depending on the DMSO/water ratio in the solvent mixture employed.

Water content indeed affected both hydrogen-bonding and π–π interactions, thus at low water content, **25** assembled in tighter 1D hydrogen-bonded arrays (G-gel), although more loosely packed, whereas a gel material with a higher degree of interarrays aggregation (B-gel) was attained upon increasing water content, probably due to hydrophobic effect (Fig. 9). Different aggregation corresponded to different emission λ_{max}, with a redshift for the G-gel (green emission) compared to the B-gel (blue emission). Notably, upon irradiation with UV light, G-gel rapidly transformed into a viscous solution driven by the Z/E photoisomerization of its α-cyanostilbene moieties, while B-gel exhibited no visible change, confirming the need for α-cyanostilbene units for sufficient space for Z/E conversion to be allowed, a situation which in tighter packing might not materialize.

Cyanostilbene derivatives have been also employed as polymeric cross-linkers, or as supramolecular guests in gelating host-guest systems. As to the first case, recently, Park et al. reported a light-responsive polymer gel consisting of α-cyanostilbene derivative **26**, which can interact with poly(acrylic acid) (PAA) by N···HO-C(=O) interactions and induce gelation of the polymer component (Fig. 10) [37]. Blue emission, characteristic of monomeric species, was observed when PAA was added to a dilute solution of **26** in ethanol ($0.1 \times 10^{-3}\,mol\,dm^{-3}$). A 10-fold increase in the concentration of **26**, with the addition of PPA, resulted in largely red-shifted and enhanced fluorescence. This indicated that the molecular component of the system could interact with the polymeric one in a way that induced green excimer emission, although both samples did not gelate due to insufficient crosslinking

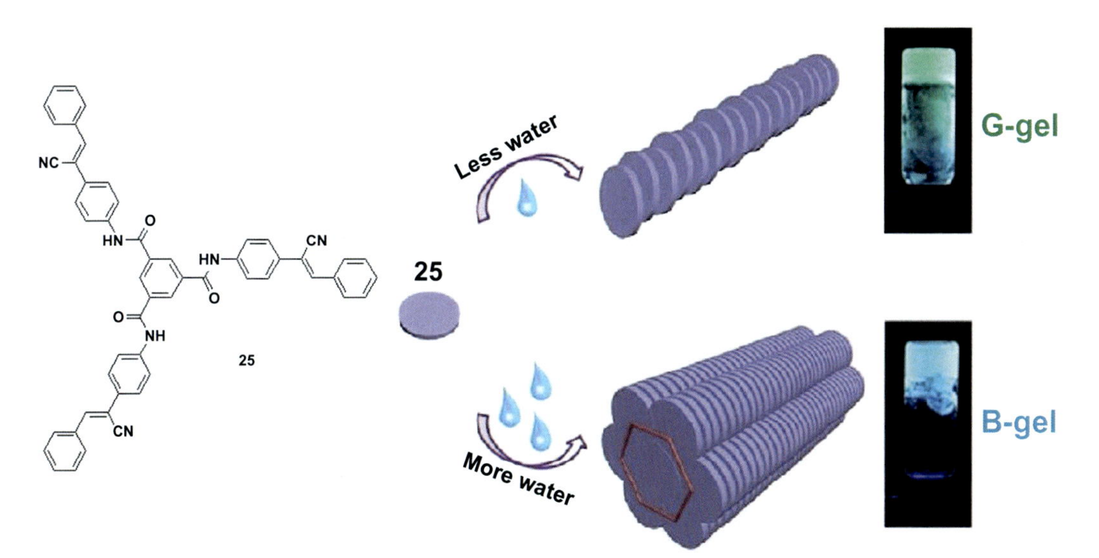

FIG. 9 Illustration of the proposed self-assembly behavior of gelator **25** in G-gel and B-gel. *Reproduced with permission from Z. Ding, Y. Ma, H. Shang, H. Zhang, S. Jiang, Fluorescence regulation and photoresponsivity in AIEE supramolecular gels based on a cyanostilbene modified benzene-1,3,5-tricarboxamide derivative, Chem. A Eur. J. 25 (2019) 315–322. Copyright 2019, Wiley-VCH.*

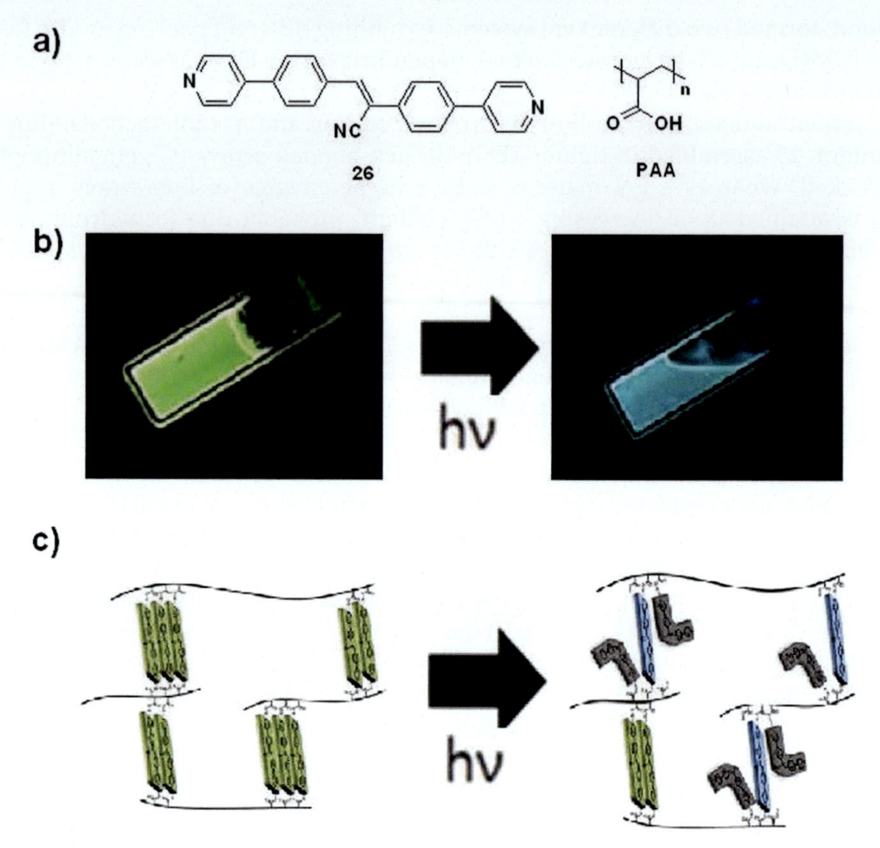

FIG. 10 (A) Chemical structures of cyanostilbene derivative **26** and poly(acrylic acid) (PAA). (B) Photographs of the **PAA**-templated **26** gel in ethanol and its photoinduced sol state under UV light and (C) cartoon of the mechanism by which photoisomerizaton-induced of **26** leads to sol and fluorescence switching. *Reproduced with permission from H.-J. Kim, H.J. Lee, J.W. Chung, D.R. Whang, S.Y. Park, A highly fluorescent and photoresponsive polymer gel consisting of poly(acrylic acid) and supramolecular cyanostilbene crosslinkers, Adv. Opt. Mater. 7 (2018) 1801348. Copyright 2018, Wiley-VCH.*

densities. Finally, by a further increase of **26** concentration to 10×10^{-3} mol dm^{-3}, a gel with strong green fluorescence was obtained. However, **26** can indeed act as a supramolecular cross-linker, whose degree of association with PPA influences its emission (Fig. 10B). Notably, also, in this case, the Z/E photoisomerization of **26** is accessible in the gel state and it leads to the disruption of supramolecular crosslinks, thus inducing a gel-to-sol transition with a concomitant switching in fluorescence color from green ($\lambda_{max}=541$ nm) to blue ($\lambda_{max}=506$ nm) and a decrease in emission efficiency (Φ_F from 0.68 to 0.12, respectively).

Another strategy to reach a balance between the ability to self-assemble and to photoisomerize was explored by Liu and coworkers [38]. They incorporated the α-cyanostilbene conjugated gelator **27** into cyclodextrin (CD) cavities through host-guest interactions (1:1 binding mode; K approximately 10^3 M^{-1}). Interestingly, although **27** could form gels with all CDs with different cavity sizes (α, β, and γ), it could undergo Z/E isomerization by exposure to UV-light only when included in γ-CD. This behavior was attributed to the difference in the inner cavity available space of CDs which, if too tight, would not allow for the required molecular rearrangement. Moreover, the morphology of the assembled nanostructure was observed to change from nanotubular to nanospherical due to Z/E transformation (Fig. 11).

Tang and coworkers also reported a light-responsive hydrogel formed through host-guest interactions between α-cyanostilbene-based molecule **28** and CDs (Fig. 12) [39]. However, **28** responded to UV light in a manner similar to what was reported previously [40]; that is, Z/E isomerization dominated the photoreaction initially, whereas photocyclization governed the photochemistry after 2h. Two kinds of hydrogel were obtained by self-assembly of **28** with CDs, depending on the CD cavity size (a 2:2 $\mathbf{28}_2$@γ-CD_2 complex or a 1:1 **28**@β-CD). While the formation of $\mathbf{28}_2$@γ-CD_2 greatly accelerated the photodimerization rate, the **28**@β-CD adduct deactivated the reaction. Interestingly, γ-CD-gel was employed as a solid-phase microreactor in which **28** within the γ-CD-gel were completely dimerized when irradiated with UV light. This process could be also visually detected thanks to a clear change of fluorescence from yellow to green. The photodimerization product (P) could be separated from the gel, hence the γ-CD-gel system is capable of manufacturing large amounts of dimer without any need for additional purification steps.

Hu and coworkers developed two photoresponsive AIE-active gelators **29** and **30** by connecting a cholesterol moiety to a TPE unit through a linker made of a central azobenzene group and alkyl spacers of different lengths (Fig. 13A) [41]. Both compounds form gels in dimethylformamide, acetone, ethyl acetate, and dichlorometane-ethyl acetate mixture, showing different morphologies and AIE properties. A photocontrolled sol–gel transition was observed for these gelators. Under UV-light irradiation (365nm), the DMF gel could be converted to a solution accompanied by a color change from yellow to orange (Fig. 13C). Such transition was investigated by UV–Vis spectroscopy and NMR which indicated the occurrence of a trans to cis-trans photoisomerization of the azobenzene moiety.

3.2 Metal ions and anion

Recognition and sensing of different analytes, being metal ions, anions, or neutral molecules, represent a core topic in supramolecular chemistry, and it is not surprising that gel systems have been reported for such applications [42]. In particular, AIE-based, supramolecular gel systems for fluorescence sensing have recently been investigated.

For example, Wei, Lin et al. designed a tripodal gelator **31** constructed by attaching terminal pillar[5]arene moieties to a central benzene ring through dihydrazide groups (Fig. 14) [43]. The compound was able to form a gel in cyclohexanol, although with a rather high *cgc* of 5 wt%. The assembly process was studied by ^{1}H-NMR, IR, P-XRD, and SEM revealing contributions by NH···O=C hydrogen bonds, CH–π, and π–π interactions toward the final assembly, which is constituted of a layered morphology. Despite the presence of π–π stacking interactions, the resulting gel is fluorescent, and thus AIE active, displaying a structured emission band centered at ca. 468nm. This feature was successfully used to selectively detect Hg^{2+} ions which induce emission quenching quite efficiently and selectively (no

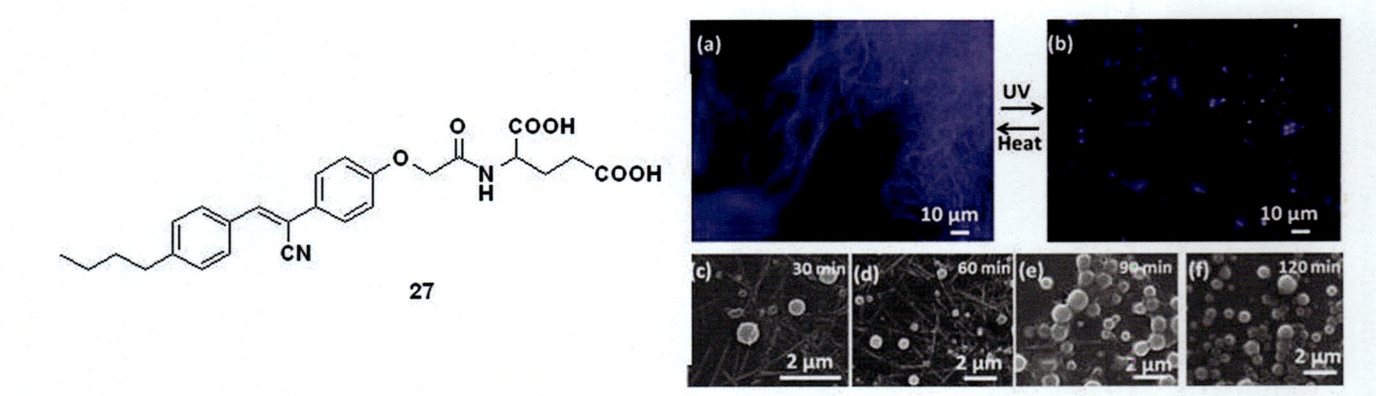

FIG. 11 Fluorescence microscope images of the (A) nanotubes and (B) the nanospheres. Morphological evolution (C)–(F) of **27**@γ-CD nanostructures with different UV light irradiation time. *Reproduced with permission from L. Ji, Q. He, D. Niu, J. Tan, G. Ouyang, M. Liu, Host-guest interaction enabled chiroptical photo-switching and enhanced circularly polarized luminescence, Chem. Commun. 55 (2019) 11747–11750. Copyright 2019, Royal Society of Chemistry.*

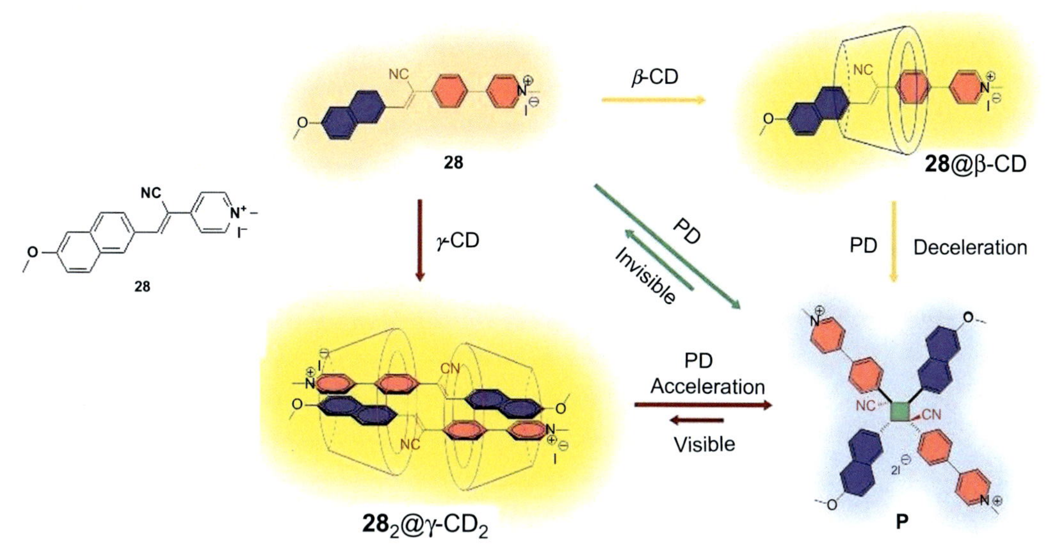

FIG. 12 Cartoon of the proposed mechanism for the visible and rate-controllable photodimerization. *Reproduced with permission from P. Wei, Z. Li, J.-X. Zhang, Z. Zhao, H. Xing, Y. Tu, J. Gong, T.S. Cheung, S. Hu, H.H.-Y Sung, I.D. Williams, R.T.K. Kwok, J.W.Y. Lam, B.Z. Tang, Molecular transmission: visible and rate-controllable photoreactivity and synergy of aggregation-induced emission and host-guest assembly, Chem. Mater. 31 (2019) 1092–1100. Copyright 2019, American Chemical Society.*

emission quenching observed in the presence of Zn^{2+}, Pb^{2+}, Cd^{2+}, Ca^{2+}, Mg^{2+}, Al^{3+}, Tb^{3+}, La^{3+}, Ba^{2+}, and Eu^{3+}). Notably, this gel system can be used to adsorb Hg^{2+} ions from a water solution if added in suspension (0.5 mg of the gel into a 0.8 mg cm^{-3} solutions of Hg^{2+} adsorb ca. 81% of the metal in 0.5 h, as detected by ICP), or as xerogel if drop-casted onto a glass surface.

In another report by the same group [44], the authors focused on the use of a different pillar [5]arene derivative combined with a naphtalimide unit, gelator **32** (Fig. 15A and B), which is capable of forming an AIE-active supramolecular gel in cyclohexanol with bright yellow emission ($\lambda_{ex}=297$ nm; $\lambda_{em}=530$ nm). In this case, emission can be quenched by Fe^{3+} (but not Hg^{2+}, Ag^{+}, Ca^{2+}, Cu^{2+}, Co^{2+}, Ni^{2+}, Cd^{2+}, Pb^{2+}, Zn^{2+}, Cr^{3+}, and Mg^{2+}), reportedly due to the fact that a transition from organo- to metallo-gels occurs; although no additional data were given. Interestingly, the emission can be recovered by addition of $H_2PO_4^-$ ions (but not with F^-, Cl^-, Br^-, I^-, AcO^-, HSO_4^-, ClO_4^-, SCN^-, and CN^-). The selective and reversible nature of the gel system was found to be applicable as a potential rewritable fluorescent display material and for ultrasensitive logic gates.

Schiff bases, traditionally studied as metal binders [45], have been discovered to be good candidates for AIE materials [46]. Thus, gelators decorated with Schiff base moiety have been considered as a valid approach for the formation of AIE-active gel systems.

Liu et al. developed gelator **33** (Fig. 16) with three pentafluorophenyl moieties attached to a central benzene ring via acylhydrazone groups [47]. The luminescence I_{473} ($\lambda_{ex}=380$ nm) of **33** in DMSO is very weak. This is possibly due to unrestricted conformational freedom which

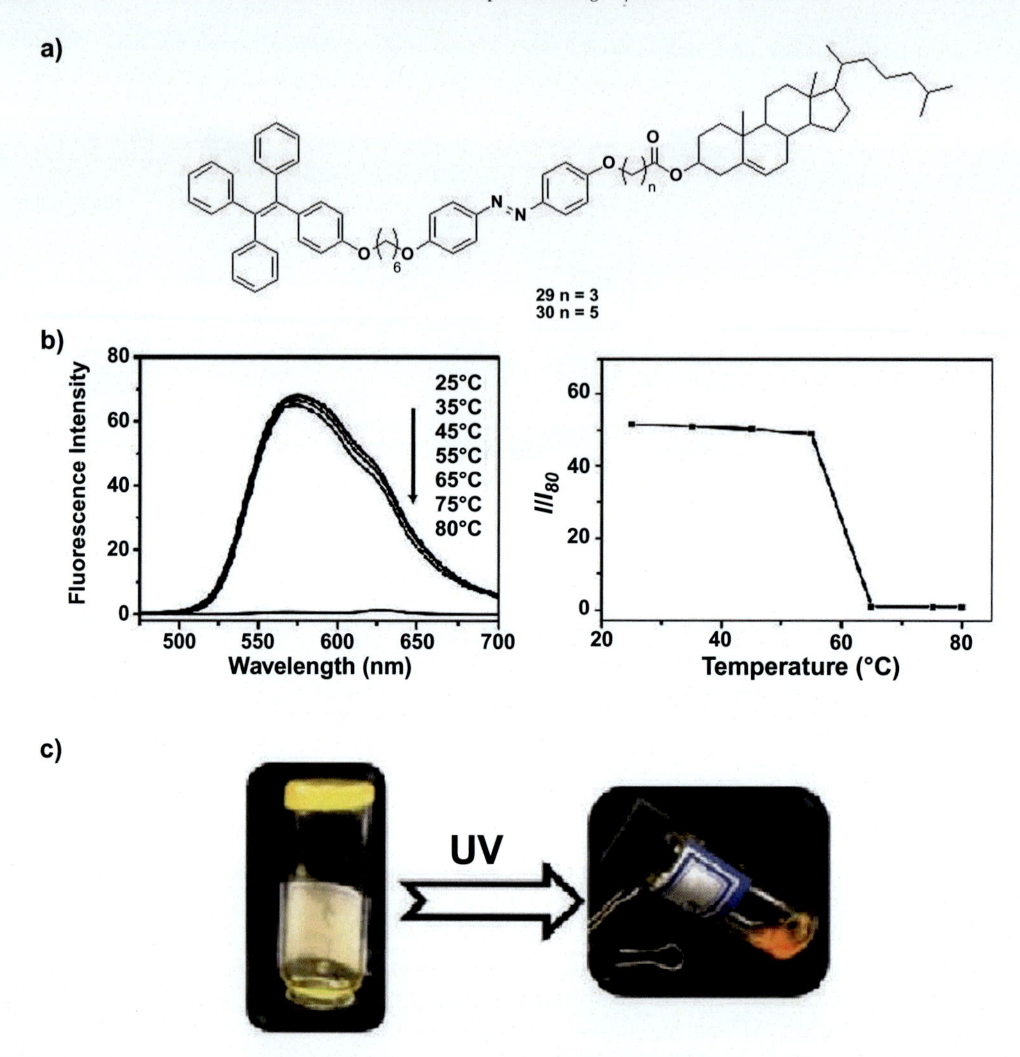

FIG. 13 (A) Chemical structures of AIE-active gelators **29** and **30**; (B) Temperature-dependent fluorescence spectra of **29** gel in DMF (conc. = 20 mmol dm^{-3} and plots of I/I$_{80}$ of **29** gel as a function of temperature from 25 to 80°C. I$_{80}$ = fluorescence peak intensity at 80°C, I = fluorescence peak intensity at other temperature (25°C, 35°C, 45°C, 55°C, 65°C, and 75°C), λ_{ex} = 310 nm. (C) Photographs showing gel–sol transition of **30** in DMF (c = 62 mmol dm^{-3}) after 15 min of UV light irradiation (90 mW cm^{-2}, 365 nm). *Reproduced with permission from X. Yu, H. Chen, X. Shi, P.-A. Albouy, J. Guo, J. Hu, M.-H. Li, Liquid crystal gelators with photo-responsive and AIE properties, Mater. Chem. Front. 2 (2018) 2245–2253. Copyright 2018, Royal Society of Chemistry.*

facilitates nonradiative decays, as well as a photoinduced electron transfer (PET) between a donor atom to the excited state of the luminophore, which is a well established process for aromatic Schiff base containing compounds [48]. In the gel state (DMSO-ethylene glycole, *cgc* 1.3 wt%), the luminescence of **33** is increased, by approximately 12-fold and the emission maxima are shifted from 473 to 489 nm. Upon addition of Al^{3+} ions, though, a fluorescence

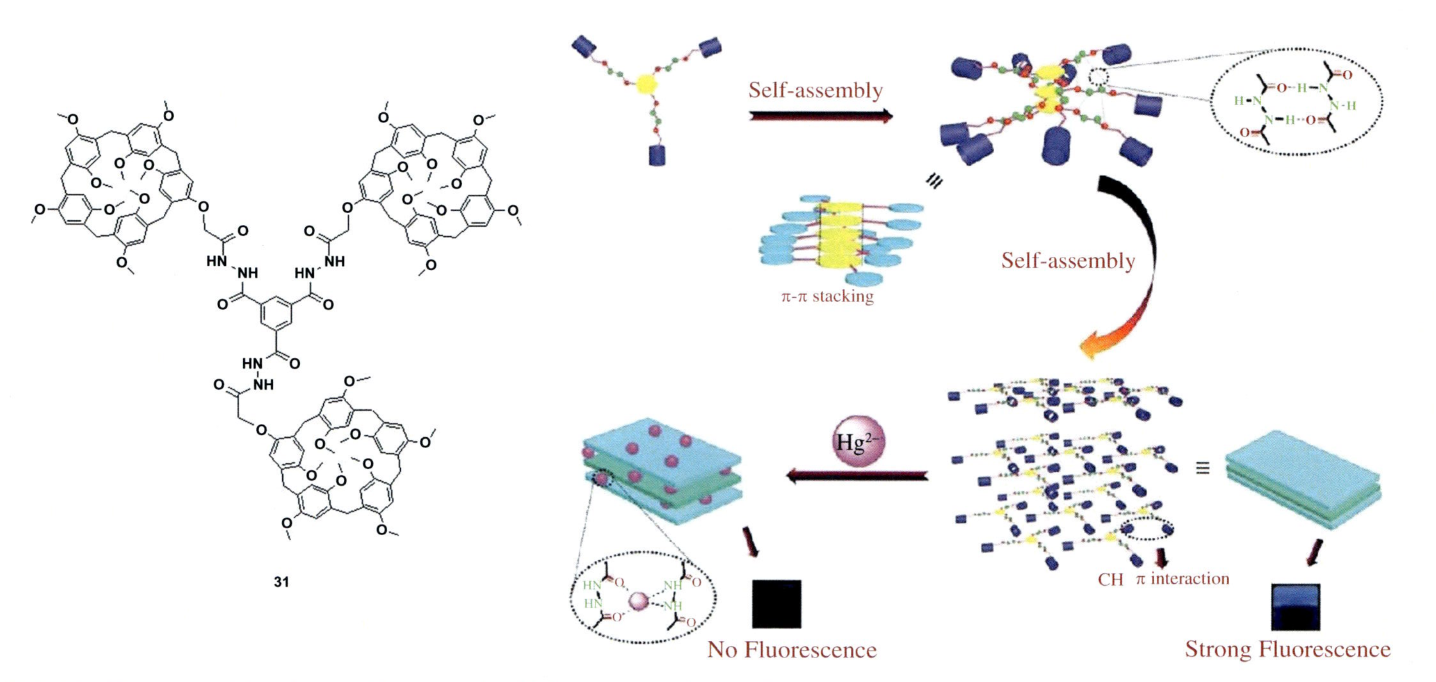

FIG. 14 Illustration of the self-assembly of **31** and its Hg^{2+} sensing mechanism. *Reproduced with permission from X.-M. Jiang, X.-J. Huang, S.-S. Song, X.-Q. Ma, Y.-M. Zhang, H. Yao, T.-B. Wei, Q. Lin, Tri-pillar[5]arene-based multi-stimuli responsive supramolecular polymer for fluorescent detection and separation of Hg2+, Polym. Chem. 9 (2018) 4625–4630. Copyright 2018, Royal Society of Chemistry.*

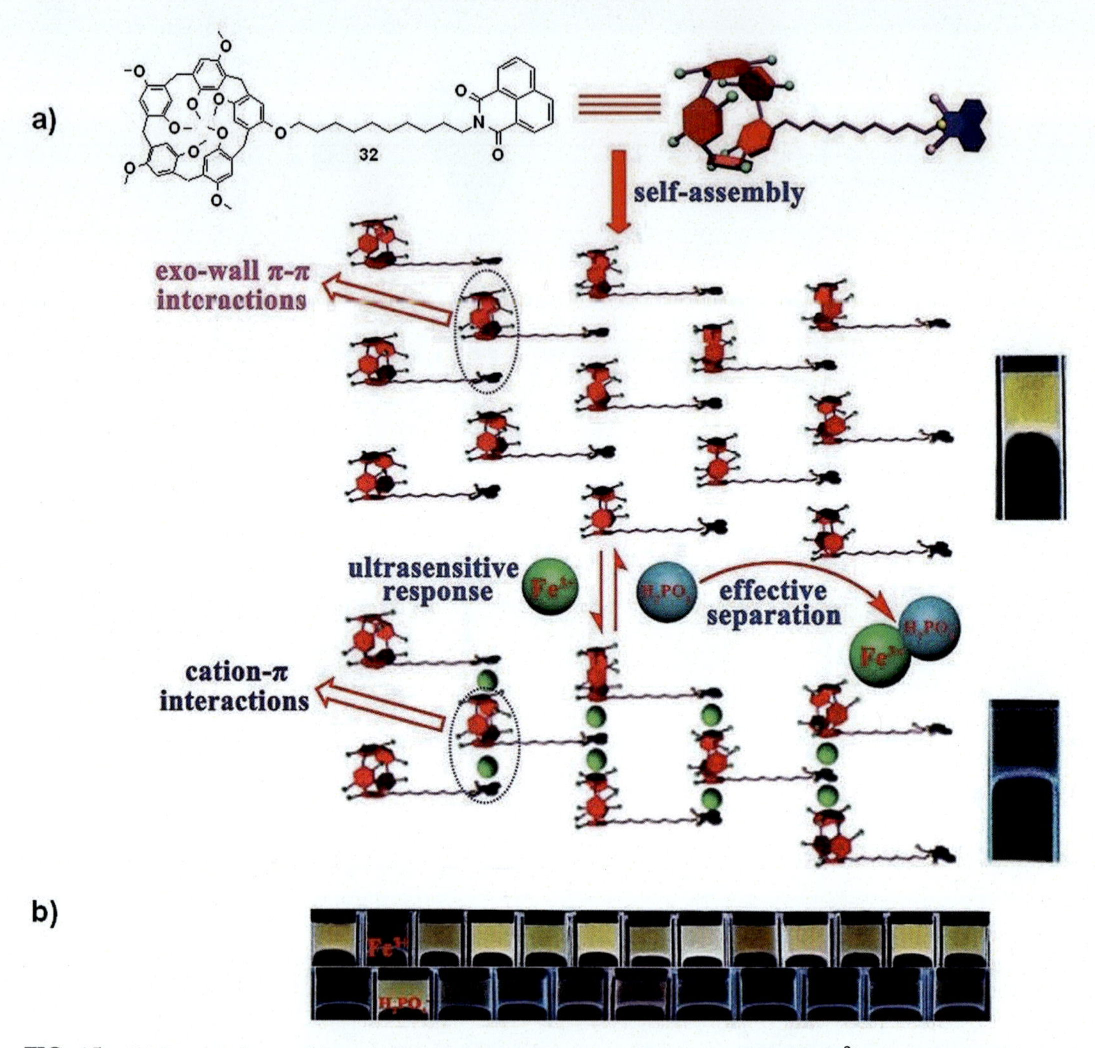

FIG. 15 (A) Proposed assembly and ultrasensitive-response mechanisms of **32** with Fe^{3+} and $H_2PO_4^-$; (B) **32** on adding various cations; and **32** upon adding various anions (under UV-lamp at 365nm, at room temperature). *Reproduced with permission from Y.-M. Zhang, W. Zhu, W.-J. Qu, K.-P. Zhong, X.-P. Chen, H. Yao, T.-B. Wei, Q. Lin, Competition of cation–π and exo-wall π–π interactions: a novel approach to achieve ultrasensitive response, Chem. Commun. 54 (2018) 4549–4552. Copyright 2018, Royal Society of Chemistry.*

OFF–ON change occurs (Fig. 16B), while other metal ions such as Cd^{2+}, Co^{2+}, Cr^{3+}, Cu^{2+}, Fe^{2+}, Fe^{3+}, Hg^{2+}, Mg^{2+}, Mn^{2+}, Na^+, Ni^{2+}, Pb^{2+}, and Zn^{2+} induced much weaker response. The selective turn-on of fluorescence by Al^{3+} selectivity was attributed to the interruption of the PET process upon metal coordination.

Similarly, Yao, Wei, Lin et al. reported a metallogel system based on gelator benzoimidazole/acylhydrazone/naphthol conjugate **34** (Fig. 17), which selectively detects Cr^{3+} ions with a limit of

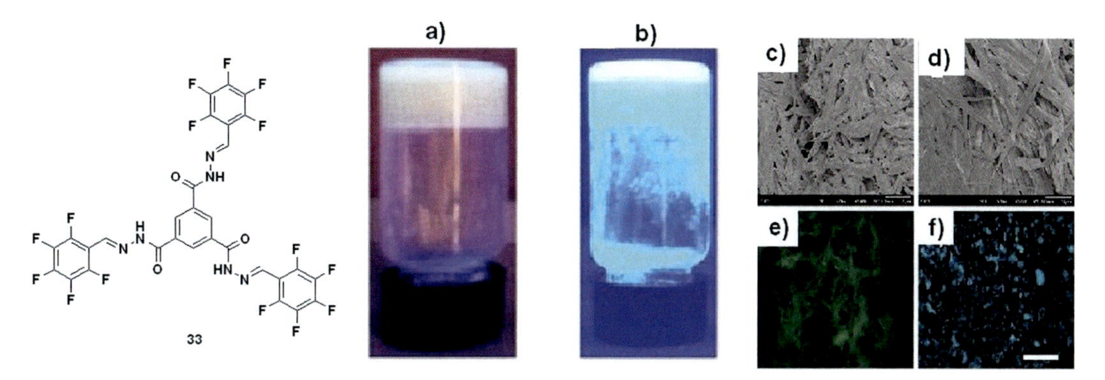

FIG. 16 Photographs of (A) organogel **33** in DMSO-ethylene glycol (1.3 wt%) and (B) **33**-Al metallogel (**33**:Al^{3+} = 1:1) under UV light at 365 nm. FE-SEM images of (C) **33** and (D) **33**-Al gel samples and fluorescent optical microscopy images of (E) **33** and (f) **33**-Al gel samples under UV at 365 nm (scale bar = 100 mm). *Reproduced with permission from X. Ma, Z. Zhang, H. Xie, Y. Ma, C. Liu, S. Liu, M. Liu, Emissive intelligent supramolecular gel for highly selective sensing of Al3+ and writable soft material, Chem. Commun. 54 (2018) 13674–13677. Copyright 2018, Royal Society of Chemistry.*

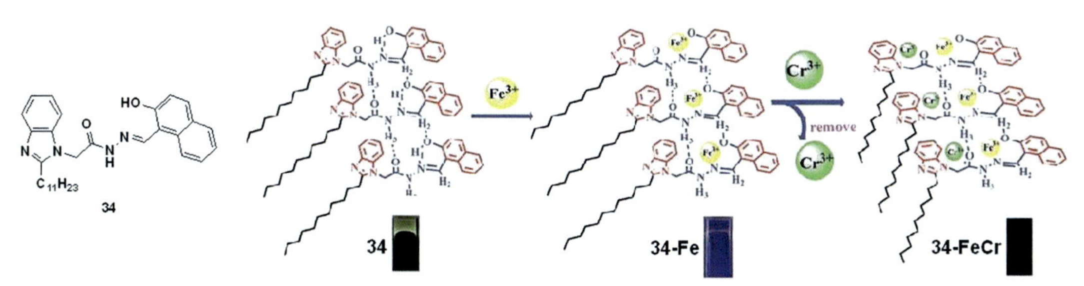

FIG. 17 Schematic representation of the proposed self-assembly and stimuli–response mechanisms of **34**-Fe (**34**:Fe^{3+} = 1:1) with Cr^{3+}. *Reproduced with permission from H. Yao, J. Wang, Q. Zhou, X.-W. Guan, Y.-Q. Fan, Y.-M. Zhang, T.-B. Wei, Q. Lin, Acylhydrazone functionalized benzimidazole-based metallogel for the efficient detection and separation of Cr3+, Soft Matter 14 (2018) 8390–8394. Copyright 2018, Royal Society of Chemistry.*

detection (LOD) of 2.62×10^{-8} mol dm^{-3}, also being capable to remove them from aqueous solution [49]. Compound **34** can form luminescent gels in glycerol (0.5 wt%) which can then interact with Fe^{3+} ions to afford a blue-emitting metallogel. In this case, the gelation, characterized by ^{1}H-NMR and fluorescence spectroscopies, induces AIE properties and the coordination to Fe^{3+} modulates the emission maxima (λ_{max}). Upon addition of the subsequent Cr^{3+} ions, instead, the emission of materials is quenched.

The authors proposed that a bimetallic Fe^{3+}/Cr^{3+} species, with a considerably different spatial organization of the AIEgens is formed, as suggested by FT-IR spectroscopy, P-XRD, and SEM. The system, in the form of xerogel, was then applied to detect and extract Cr^{3+} ions from water, by loading it onto a silica-gel plate.

Bimetallic gel systems are indeed very interesting. Another example showing that they can represent a way to modulate the AIE gel properties is that regarding the work on compound

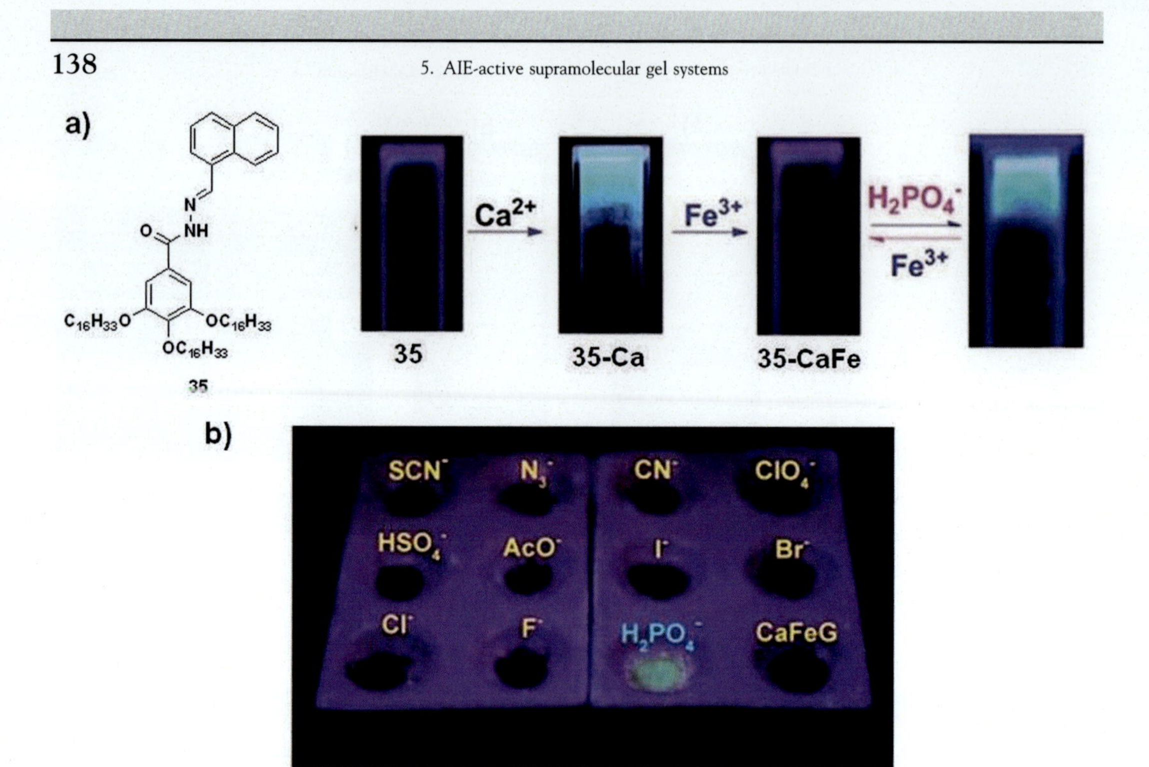

FIG. 18 (A) Photographs of organogel **35** (1 wt% in ethanol), metallogels **35-Ca**, **35-CaFe** (1 wt% in ethanol, for **35-Ca**, **35**: Ca^{2+} = 1:1; for **35-CaFe**, **35**: Fe^{3+}: Ca^{2+} = 1:2:1) and **35-CaFe** treated with $H_2PO_4^-$ and illuminated at 365 nm. (B) Photograph of metallogel **35-CaFe** (1%, in ethanol, **35**:Fe^{3+}:Ca^{2+} = 1:2:1) which selectively detects $H_2PO_4^-$ (5 eq., using 0.1 mol dm^{-3} NaH_2PO_4 water solution as the $H_2PO_4^-$ source) in water solution on a spot plate, under UV light at 365 nm. *Reproduced with permission from Q. Lin, B. Sun, Q.-P. Yang, Y.-P. Fu, X. Zhu, Y.-M. Zhang, T.-B. Wei, A novel strategy for the design of smart supramolecular gels: controlling stimuli-response properties through competitive coordination of two different metal ions, Chem. Commun. 50 (2014) 10669–10671. Copyright 2014, Royal Society of Chemistry.*

35 (Fig. 18) [50]. In EtOH, it forms a gel (*cgc* 1 wt%) which is only very weakly AIE-active, but its emission enhances (10 × increase, λ_{em} = 365 nm) upon addition of Ca^{2+} ions (other metals, such as Mg^{2+}, Cr^{3+}, Fe^{3+}, Co^{2+}, Ni^{2+}, Cu^{2+}, Zn^{2+}, Ag^+, Cd^{2+}, Hg^{2+}, and Pb^{2+} do not show changes).

Interestingly, the AIE effect can be disrupted by the addition of Fe^{3+} ions. Although the mechanism of this process is not clear, and the effective copresence of the two metal species needs further corroboration, this work shows evidently that the occurrence of different aggregation modes (which have been investigated with NMR, IR, and P-XRD) can be quite an effective way to reach AIE λ_{max} modulation. As the subsequent addition of $H_2PO_4^-$ restores the bluish fluorescence, the system can be also used for anion recognition, a topic which will be further described in the next section.

On the same note, the two tripodal amide-based cogelators **36** and **37** can be assembled in a DMSO-water mixture to afford slightly emissive gels [51]. Addition of Th^{4+} ions generates an emissive metallogel, which upon the subsequent addition of Hg^{2+} ions becomes again nonemissive (Fig. 19B). Thus, this supramolecular gel system can very efficiently detect

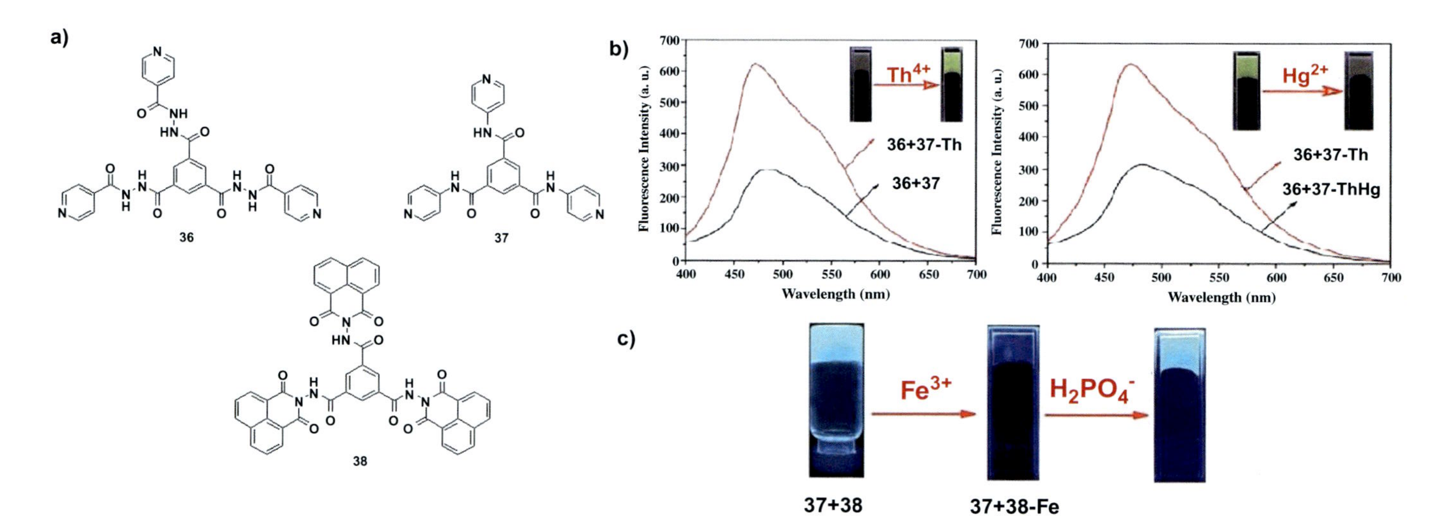

FIG. 19 (A) Chemical structures of tripodal amides **36–38**. (B) Fluorescence spectra of **36** + **37** and **36** + **37-Th** (left) and **36** + **37-Th** and **36** + **37-ThHg** (right). (C) Photographs of **37** + **38** gel in 26% water-DMSO solvent mixture (7 wt%) and fluorescence turn-off by the addition of Fe^{3+} and turn-on by the addition of $H_2PO_4^-$ under illumination at 365 nm. *Panels (A, B): Reproduced with permission from Y.-M. Zhang, W. Zhu, Q. Zhao, Q.-J. Qu, H. Yao, T.-B. Wei, Q. Lin, Th^{4+} tuned aggregation-induced emission: a novel strategy for sequential ultrasensitive detection and separation of Th^{4+} and Hg^{2+}, Spectrochim. Acta A 229 (2020) 117926. Copyright 2020, Elsevier Ltd. Panel (C): Reproduced with permission from G.-F. Gong, Y.-Y. Chen, Y.-M. Zhang, Y.-Q. Fan, Q. Zhou, H.-L. Yang, Q.-P. Zhang, H. Yao, T.-B. Wei, Q. Lin, A novel bis-component AIE smart gel with high selectivity and sensitivity to detect CN^-, Fe^{3+} and $H_2PO_4^-$, Soft Matter 15 (2019) 6348–6352. Copyright 2019, Royal Society of Chemistry.*

Th^{4+} and Hg^{2+} ions with LODs of 8.61×10^{-11} and 1.08×10^{-11} mol dm^{-3}, respectively. It can be also used in the form of xerogel for the separation of these two ions from an aqueous solution.

In another work on two-component gel system developed by the same group [52], compound **37** was mixed with C_3 symmetric tripodal **38** to form emissive gels in DMSO-water mixture ($\lambda_{ex}=365$ nm; $\lambda_{em}=470$ nm). The gel can selectively recognize Fe^{3+} ions over many other metal ions, such as Ca^{2+}, Mg^{2+}, Pb^{2+}, Ni^{2+}, Co^{2+}, Hg^{2+}, Zn^{2+}, Cd^{2+}, Ag^{+}, Cu^{2+}, Cr^{3+}, Al^{3+}, Tb^{3+}, Ba^{2+}, La^{3+}, and Eu^{3+}, in water ($c=0.1$ mol dm^{-3}). The fluorescence can be restored by the addition of $H_2PO_4^-$ which probably sequesters the ferric ions and reinstates the original gel architecture along with its fluorescence (Fig. 19C).

Nonemissive compound **39** forms fluorescent gels in several solvents and solvent mixtures [53]. Interestingly, the gel system obtained in DMSO-water solvent mixture can undergo an organogel-to-metallogel transformation upon addition of Cu^{2+} ions, also proven by SEM and rheology investigation (Fig. 20). In contrast, other common transition metal salts seem to have no effect on the AIE properties of the gel.

Wu and coworkers reported a two-stage assembly functional gelator **40** containing a Schiff base and a benzimidazole moieties (Fig. 21) [54]. In DMSO solution, compound **40** emits very weakly ($\lambda_{max}=409$ nm; $\lambda_{ex}=373$ nm), however, at concentrations above 1.3 wt% in a 1:3 DMSO-ethylene glycol mixture a highly emissive gel is formed. It displayed an emission enhancement of approximately 36-fold along with a λ_{max} redshift to 443 nm. Again, an RIR effect due to hydrogen-bonding-driven aggregation is probably responsible for the observed behavior. Notably, the addition of 2/3 equivalents of Cd^{2+} in DMSO solution leads to the formation of a bright blue-light-emitting metallogel, which was used to identify the formation of a metal complex. This amount of Cd^{2+} added was rationalized on the basis of titration data between **40** and Cd^{2+} (**40**:$CdCl_2$ ratio is 3:2).

As seen in the previous paragraphs, metal binding can be employed to turn on, turn off or modulate the emission of gel systems. Despite the need for a more careful evaluation of the whole set of possible mechanistic aspects which can be involved in these processes, a practical value to these observations must be recognized. One way to put them into use is, for example, in the recognition of anionic species.

Lin, Zhang et al. designed gelator **41** (Fig. 22) to contain an acylhydrazone moiety as a cation binder and a quinoline unit as fluorescence signaling unit [55]. The organogel of **41** in DMF showed aggregation-induced strong blue emission. In the presence of 0.5 equiv. of various metal ions such as Cu^{2+}, Fe^{3+}, Hg^{2+}, and Cr^{3+}, emission at 500 nm was almost quenched and the corresponding nonfluorescent metallogels were obtained. At variance with that, the diffusion of Zn^{2+} ions into the organogel of **41** resulted in an almost 40 nm redshift of the emission spectra, and the color of the gel emission changed to yellow. When treated with anions, **41**-Cu^{2+} and **41**-Fe^{3+} gels selectively responded to CN^- via fluorescence turn-on, while the **41**-Hg^{2+} and **41**-Cr^{3+} gels were nonresponsive (Fig. 22A). Notably, however, **41**-Hg^{2+} and **41**-Cr^{3+} gels showed fluorescence emission in the presence of SCN^- and S^{2-} ions (probably HS^-), respectively. On the other hand, **41**-Zn^{2+} gel was unresponsive to the abovementioned ions and showed quenching by I^- alone, while interaction with CN^- induces a shift in the emission to blue, restoring the original gel. Importantly, **41**-Cu^{2+} and **41**-Fe^{3+} gels could selectively sense CN^- ions from a mixture with other anions (multianalyte conditions) with LOD ca. 1.0×10^{-7}-mol dm^{-3} and 1.0×10^{-5} mol dm^{-3}, respectively. This metallogel system based on gelator **41**

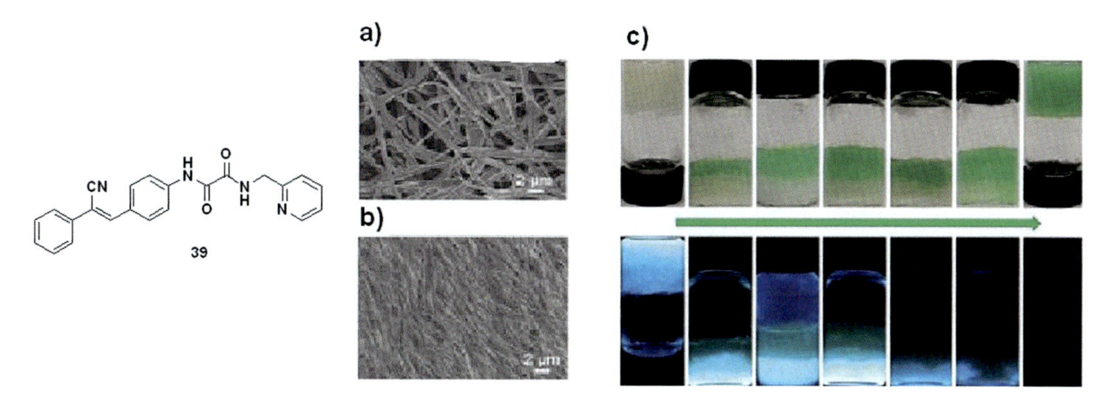

FIG. 20 (A) SEM images of the **39**-DMSO-H_2O gel and (B) **39**-$CuCl_2$ metallogel, respectively. (C) Photos of the gel of **39** in DMSO-H_2O (1:1 *v*/v) when layering a $CuCl_2$ solution above it (top: under ambient light, bottom: 365nm UV light). *Reproduced with permission from Y. Ma, M. Cametti, Z. Džolić, S. Jiang, Selective Cu(II) sensing by a versatile AIE cyanostilbene-based gel system, Soft Matter 15 (2019) 6145–6150. Copyright 2019, Royal Society of Chemistry.*

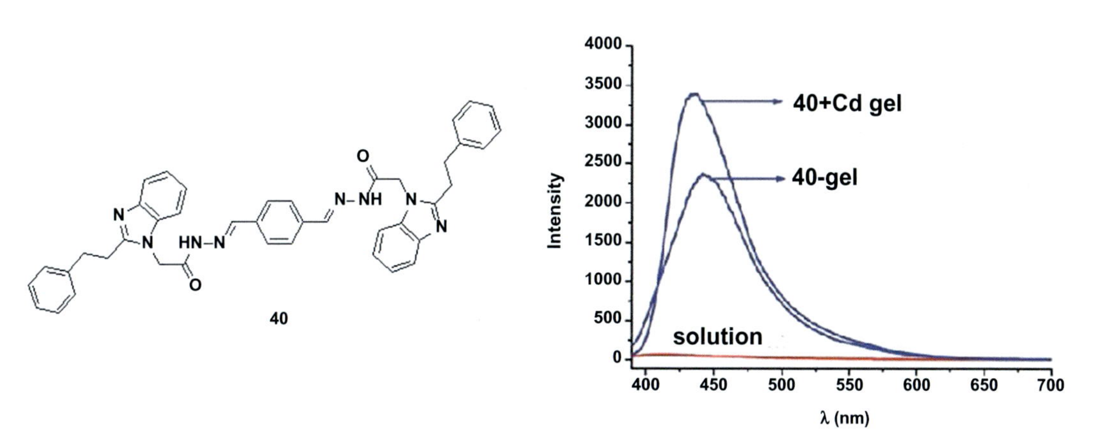

FIG. 21 Luminescence spectra of a **40** solution in DMSO (0.02g of **40** in 2mL DMSO), **40**-gel in DMSO-ethylene glycol mixture (1.3wt%), and **40**+**Cd**-gel (1.3wt% based on **40**, the molar ratio of **40**/$CdCl_2$ is 3:2); the samples are all excited at 373nm. *Reproduced with permission from X. Ma, J. Xie, N. Tang, J. Wu, AIE-caused luminescence of a thermally-responsive supramolecular organogel, New J. Chem. 40 (2016) 6584–6587. Copyright 2016, Royal Society of Chemistry.*

may be used as a sensor array composed of five different active species which could identify different anions in water with high accuracy and sensitivity. The construction of erasable security display materials was also proposed as a further potential application.

The coumarin -OH group in compound **42** is expected to be involved in intramolecular hydrogen bonding with the adjacent imine nitrogen and thus provide enhanced rigidity and an increase of the phenolic pK_a. Compound **42** showed considerable aggregation

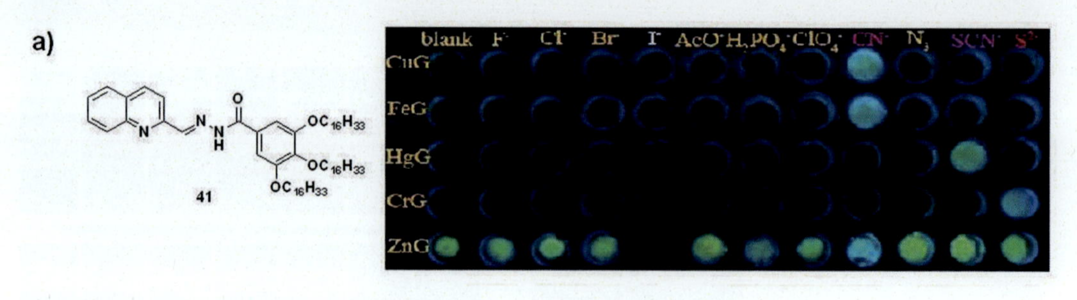

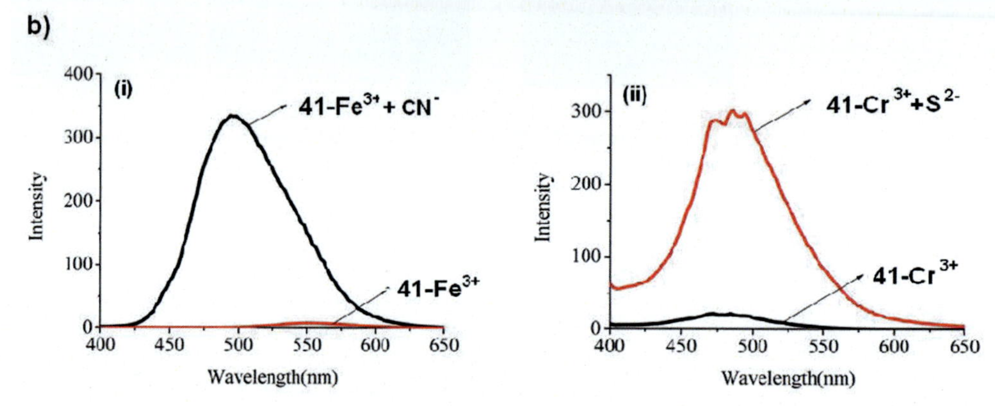

FIG. 22 (A) Fluorescence responses of the metallogel-based sensor array in the presence of 1 equiv. of various anions (using 0.1 $mol\,dm^{-3}$ anion sodium or potassium saltwater solution as anion source); (B) fluorescence spectra of metallogels (0.8 wt% in DMF) in the presence of various anions. (i) **41**-Fe^{3+} and **41**-Fe^{3+} gels treated with CN^- and (ii) **41**-Cr^{3+} and **41**-Cr^{3+} gels treated with S_2^-. *Reproduced with permission from Q. Lin, T.-T. Lu, X. Zhu, B. Sun, Q.-P. Yang, T.-B. Wei, Y.-M. Zhang, A novel supramolecular metallogel-based high-resolution anions sensor array, Chem. Commun. 51 (2015) 1635–1638. Copyright 2015, Royal Society of Chemistry.*

aptitude in solution (detected by concentration-dependent ^1H-NMR data) and forms a gel in n-BuOH-H_2O (9:1, v/v) [56]. The gel was treated with various anions including F^-, Cl^-, Br^-, I^-, $H_2PO_4^-$, AcO^-, ClO_4^-, HSO_4^-, SCN^-, N_3^-, and CN^- (c=0.1 M), and, surprisingly, it underwent a gel-to-gel transition in the presence of CN^- ions (and S^{2-}, probably HS^-), exhibiting blue-shifted emission (Fig. 23B). The detection was also confirmed visually, under UV excitation, through a color change of the gel from green-yellow to blue fluorescence. Spectroscopic studies suggested that during interaction with anions, the –OH group remained intact while the -NH group underwent deprotonation and produced the observed spectroscopic changes.

It is now interesting to compare the behavior observed for the **42**-gel with that obtained with gelator **34** in glicerol. In that case, in the gel state, the molecule underwent deprotonation of both the phenolic -OH and acylhydrazone -NH with CN^- and validates its visual sensing through a color change of the gel from yellow-green to blue (Fig. 23D) [57]. This color change was associated with a blue shift in the emission of the gel from 516 to 498 nm.

In light of all the knowledge obtained on the series of compounds **34, 41,** and **42** in terms of their interactions with metal ions and anions and the corresponding ways to modulate the fluorescence in those AIE systems, Lin, Zhang et al., prepared a new gelator **43** and proposed

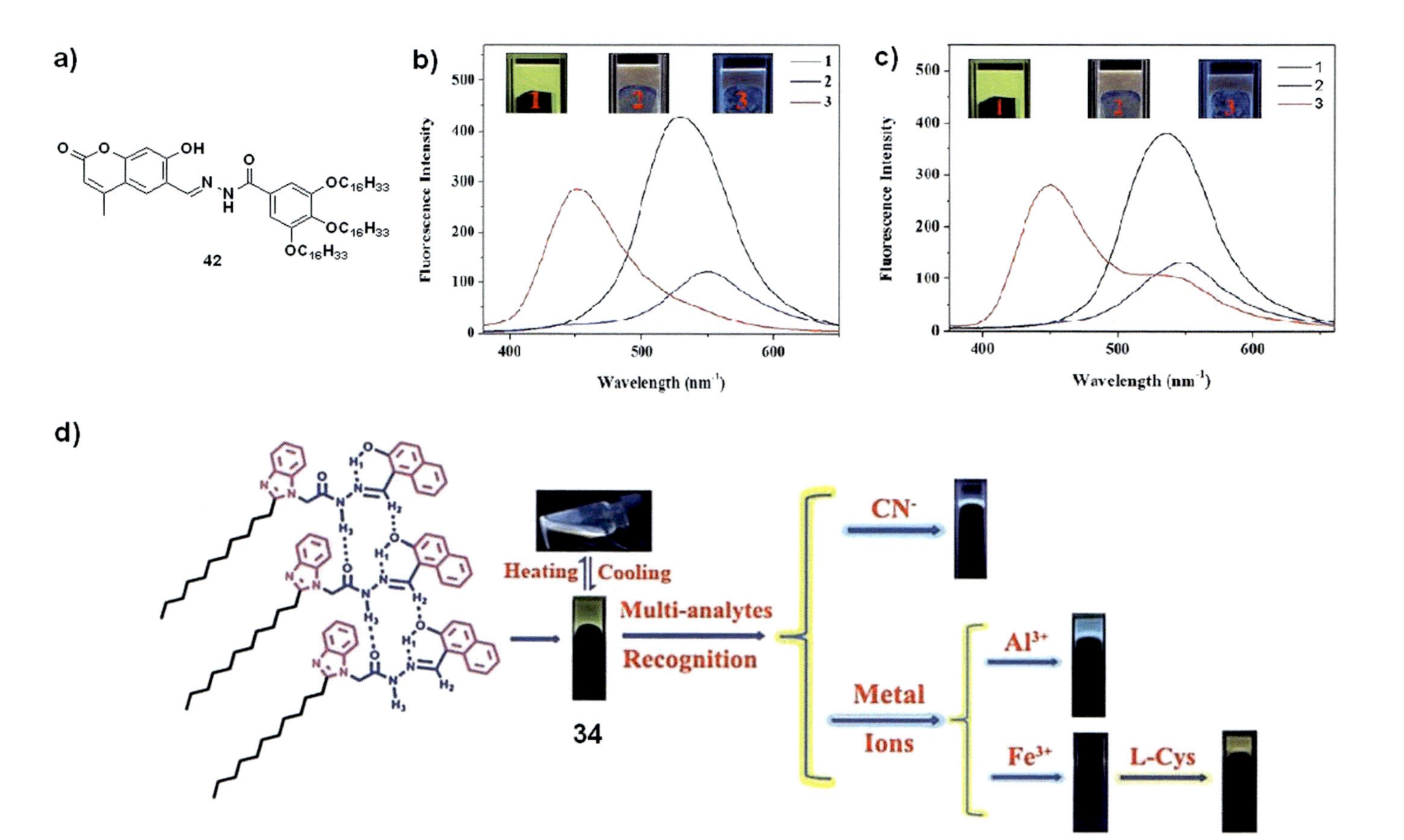

FIG. 23 (A) Chemical structure of gelator **42**, (B) fluorescence spectra of gel of **42** in *n*-BuOH-H_2O (9:1, v/v) with increasing quantities of CN^- and (C) with increasing quantities of S_2^- (probably HS^-). Inset: color change upon titration process (1: beginning; 2: middle; 3: end-point). (D) Proposed self-assembly mechanism of a supramolecular AIE gel of **34** and its multiple-stimuli responsive behavior. *Panels (A–C): Reproduced with permission from J.-H. Hu, Z.-Y. Yin, K. Gui, Q.-Q. Fu, Y. Yao, X.-M. Fu, H.-X. Liu, A novel supramolecular polymer gel based long-alkyl-chains functionalized coumarin acylhydrazone for sequential detection and separation of toxic ions, Soft Matter 16 (2020) 1029–1033. Copyright 2020, Royal Society of Chemistry. Panel (D): Reproduced with permission from H. Yao, J. Wang, S.-S. Song, Y.-Q. Fan, X.-W. Guan, Q. Zhou, T.-B. Wei, Q. Lin, Y.-M. Zhang, A novel supramolecular AIE gel acts as a multi-analyte sensor array, New J. Chem. 42 (2018) 18059–18065. Copyright 2018, Royal Society of Chemistry.*

a way to develop multianalytes sensor array [58]. The design is based on the schematic view in Fig. 24, where the initial organogel is subjected to a determined chemical stimulus (by the addition of one or two metal ions, or an anion). In this way, 22 different species were obtained (first column on the array in Fig. 24B).

These gel materials, either fluorescent or not, were then tested by the addition of anions or other metal cations. The results are shown in Fig. 24B. Together, they represent a quite remarkable sensor array that succeeds in several noteworthy sensory feats. For example, as far as anion recognition is concerned, **43** is capable of sensing fluoride anion by fluorescence turn-on seemingly without the interference of AcO^- or hydroxide [59]; HSO_4^- is the sole anion able to trigger fluorescence response of **43**-Fe and at the same time quench **43**-Al, **43**-BaCa, while **43**-Cu and **43**-Cr respond exclusively to SCN^- and S^{2-} (probably HS^-), respectively. Furthermore, **43**-AlHg senses Cl^-, with no interference by F^- or other halides; **43**-BaMg can recognize I^-, and finally, **43**-AlCu and **43**-AlFe recognize CN^-. As to metal sensing, Hg^{2+}, Pb^{2+}, Al^{3+}, Zn^{2+}, and Fe^{3+} are all specific analytes that can be effectively recognized in this array system. Again, although the mechanism of action of the different analytes contributing to the observed fluorescence changes is still partially unclear, the practical value of these systems cannot be overlooked.

Xu, Yang et al. presented a supramolecular cross-linked AIE-active supramolecular gel with multiple stimuli-responsive properties created by a hierarchical and very elegant self-assembly process [60]. The dipyridyl donor **44** was prepared by decorating the TPE core with two dipyridyl moieties and two pillar[5]arene units and reacted with bis-Pt(II) complexes **45** and **46** to form a rhomboidal metallacycle with four pillar[5]arene units and a hexagonal metallacycle with six pillar[5]arene units, respectively (Fig. 25A). Both metallamacrocycles could then undergo a supramolecular crosslinking process in the presence of neutral dinitrile **47**.

Owing to the restriction of the intramolecular motions of TPE within the three-dimensional polymeric networks, the supramolecular polymer gels showed AIE properties (Fig. 25B and C). Furthermore, the addition of TBABr led to gel-to-sol transition, quenching fluorescence. This effect was attributed to the competitive binding of the Br^- ions to the Pt^{2+} centers, leading to the disruption of the whole gel architecture. It must be added that this system is multiresponsive, as the addition of a bis-nitrile such as adiponitrile, which competes with **47** for binding with the pillarene moieties, also leads to gel disruption and fluorescence quenching.

A different metallamacrocycle, compound **48**, containing the TPE units was reported by Huang and Xu et al. to be also able to generate responsive AIEgels [61]. The AIE-active gels of **48** in acetone-water solvent mixture were found to be responsive to Br^- and F^- anions, via reversible gel-to-sol transition, as shown in Fig. 26.

3.3 Neutral molecules

Recognition and sensing of neutral molecules is also a field of great importance in terms of applications [62]. One clear example is given by CO_2 sensing, which has become a hot topic, given the necessity to develop sensitive methods for in situ detection and quantitative determination of CO_2 gas in real-world environments which could be carried out rapidly and with portable instrumentation. CO_2 sensors rely on the electrophilic nature of CO_2 and are based

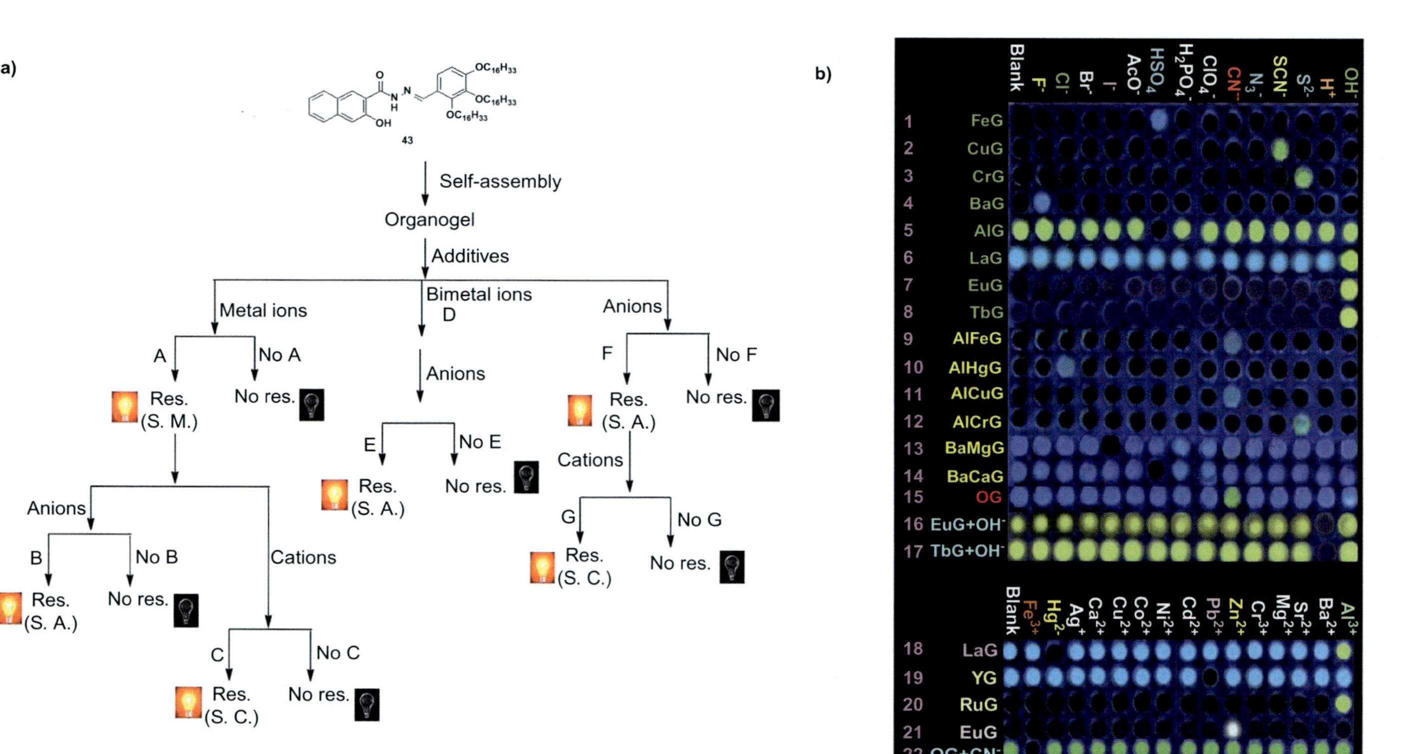

FIG. 24 (A) Competitive binding interactions: A, metal ions competitively coordinate with the gelator; B, anions competitively coordinate with the metal ions or gelator; C, competitive coordination between different metal ions and the gelator; D, bimetal ions competitively coordinate with the gelator; E, anions competitively coordinate with the bimetal ions; F, anions competitively binding with the gelator; G, cations competitively coordinate with the anions or gelators. Note for abbreviations: response (Res. or res.), sensing for anions (S. A.), sensing for Cations (S. C.) and sensing for metal ions (S. M.); (B) fluorescence responses of the supramolecular gel-based sensor array to the presence of various anions and cations. *Reproduced with permission from Q. Lin, T.-T. Lu, X. Zhu, T.-B. Wei, H. Li, Y.-M. Zhang, Rationally introduce multi-competitive binding interactions in supramolecular gels: a simple and efficient approach to develop multi-analyte sensor array, Chem. Sci. 7 (2016) 5341–5346. Copyright 2016, Royal Society of Chemistry.*

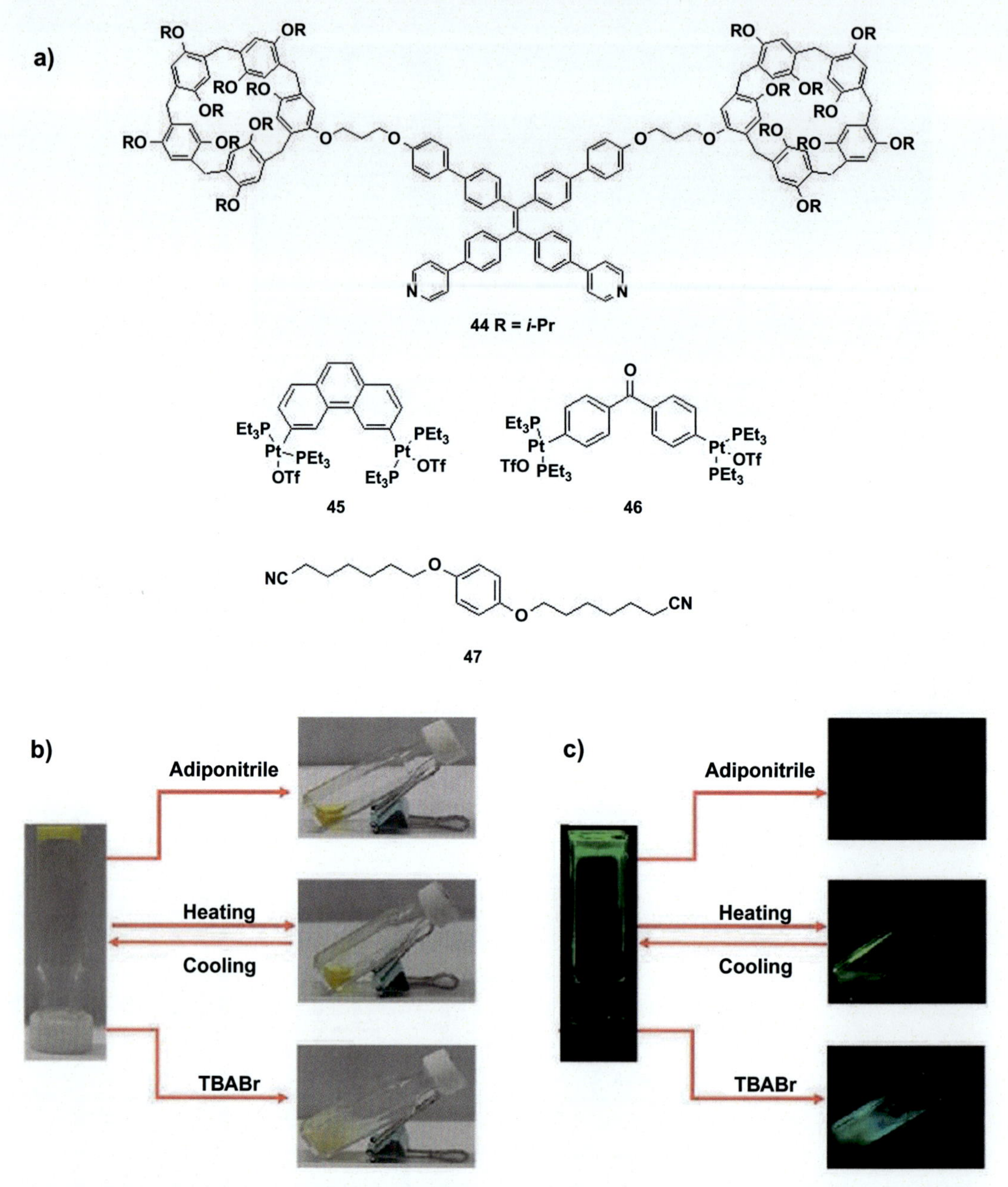

FIG. 25 (A) Chemical structures of TPE derivative **44**, bis-Pt(II) complexes **45** and **46** and dinitrile **47**; Different stimuli lead to gel–sol transitions of supramolecular polymer gel constructed by a hexagonal metallacycle with six pillar[5]arene units (**44**) and neutral dinitrile **47** under normal light (B) and under UV-light (C). *Reproduced with permission from C.-W. Zhang, B. Ou, S.-T. Jiang, G.-Q. Yin, L.-J. Chen, L. Xu, X. Li, H.-B. Yang, Cross-linked AIE supramolecular polymer gels with multiple stimuli-responsive behaviours constructed by hierarchical self-assembly, Polym. Chem. 9 (2018) 2021–2030. Copyright 2018, Royal Society of Chemistry.*

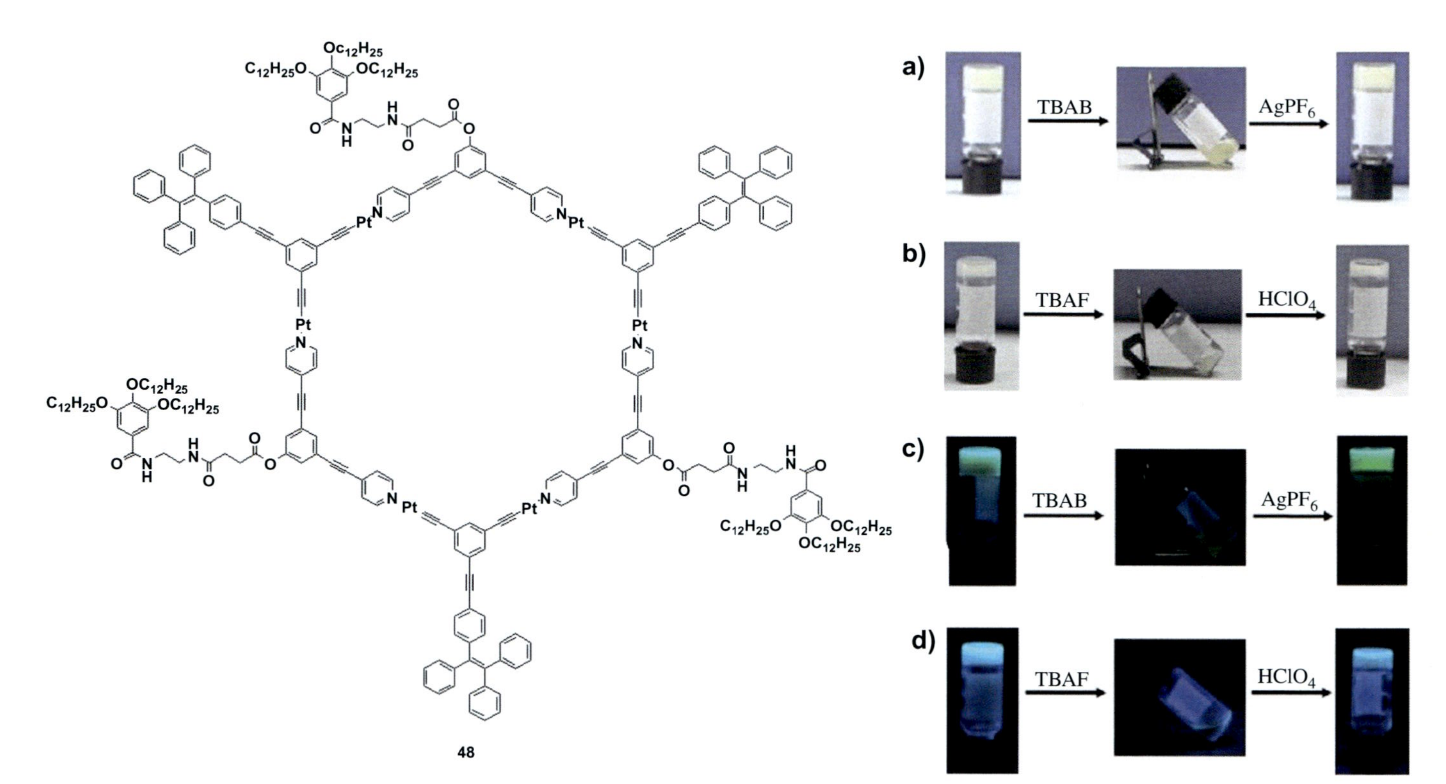

FIG. 26 Photographs demonstrating reversible stimuli-responsive gel-to-sol transitions of hexagonal metallacycle **48** in acetone-water (5:1) by the addition of (A) TBABr and $AgPF_6$ and (B) TBAF and $HClO_4$, under normal light; and under irradiation by a UV lamp at 365 nm (C and D). *Reproduced with permission from Y.-Y. Ren, Z. Xu, G. Li, J. Huang, X. Fand, L. Xu, Hierarchical self-assembly of a fluorescence emission-enhanced organogelator and its multiple stimuli-responsive behaviors, Dalton Trans. 46 (2017) 333–337. Copyright 2017, Royal Society of Chemistry.*

on either a direct modification of the sensor molecule upon reaction with CO_2 [63], or on physicochemical changes of the medium induced by CO_2 (polarity, viscosity, pH, etc.), which in turn influence the dissolved sensor molecules [64]. AIE effect can be quite useful in this respect.

Compounds **49–51** show the AIE effect in their DMSO gel state (13–69-fold fluorescence enhancements) as compared to the solution, with marked redshift of the emission peaks [65]. Above the T_{gel} of gelator **51** in DMSO, the mixture in solution state mainly gives weak emission at 439 nm attributed to pyrene monomer. The emission peak at 503 nm which is attributed to the pyrene excimer emission appears and gradually increases upon lowering the temperature, indicating that the AIE is due to the gradual self-assembly of gelator molecules into aggregates during the solution-to-gel process. As expected on the basis of previous works [66], fluoride ions can activate compound **49–51** to react with CO_2. Thus, bubbling of CO_2 into a DMSO solution of gelator **49** containing 3.5 equiv. of fluoride anions, induces gel formation (Fig. 27B). Notably, upon bubbling with N_2 at $T > T_{gel}$, the original transparent solution state could be restored.

Another example of an AIE-based CO_2-sensor gel system is given by Cametti and Jiang et al., based on the bis cyanostilbene derivative **52** (Fig. 27C) [67]. This compound can act as a good gelator in many aromatic solvents with *cgc* as low as 3.3 wt%. As depicted in Fig. 27D, the toluene gel of **52** is strongly fluorescent ($\lambda_{em} = 465$ nm; $\lambda_{ex} = 365$ nm), as compared to a practically nonemissive solution state. Interestingly, it was shown that **52** can interact in solution with relatively basic ions, such as acetate via hydrogen-bonding interactions. Notably, acetate ions also disrupt the gel state of **52**, provoking a gel-to-sol transformation. These findings represent the basis for a new concept of CO_2 sensor systems which relies upon a classic host-guest complex formation. Carbamate ions can be formed in situ by the reaction of CO_2 with an aliphatic amine (diethylamine in this case). The carbamate anion can bind to gelator **52**, similarly to acetate, affecting its aggregation state and consequently its emission. Thus, a gel-to-solution transformation along with emission quenching can be obtained by adding CO_2 into a dispersion of micrometer-sized gel aggregates of **52** in toluene in the presence of a controlled excess of DEA. In this setup, sensing of CO_2 can reach a LOD of 908 ppm. This sensitivity can be improved to reach 4.5 ppm by using xerogel films of **52** drop-casted onto a quartz template and exposed to diethylamine vapor (Fig. 27E and F). Furthermore, this second sensor setup was validated by comparison with a commercial NDIR detector and it can work in a dual mode, both by emission quenching and emission modulation. The changes in fluorescence due to CO_2_ sensing can be observed by the naked eye at high CO_2 concentration (4000 ppm).

Another class of neutral molecules that can be targeted by AIE gels is constituted by Brønsted acid and bases. There are several AIE-based gels which can respond to trifluoroacetic acid (TFA) [34], which being both volatile and a strong acid represents an ideal test for acid responsiveness.

Achalkumar et al., described the case of polyfunctional compound **53**, which can self-assemble into a columnar phase, in liquid crystalline and gel states, also showing marked AIE properties [68]. However, **53** is a symmetric molecule with a central pyridine ring, linked to 1,3,4-thiadiazoles at the 2 and 6 positions which are both further connected to trihexadecyloxy phenyl units. As far as its gelating properties are concerned, **53** forms gels in nonpolar aliphatic solvents (Fig. 28B). For example, in decane, its *cgc* was measured to be as low as ca. 0.5 wt%, quite a remarkable value for a gelator devoid of hydrogen-bonding

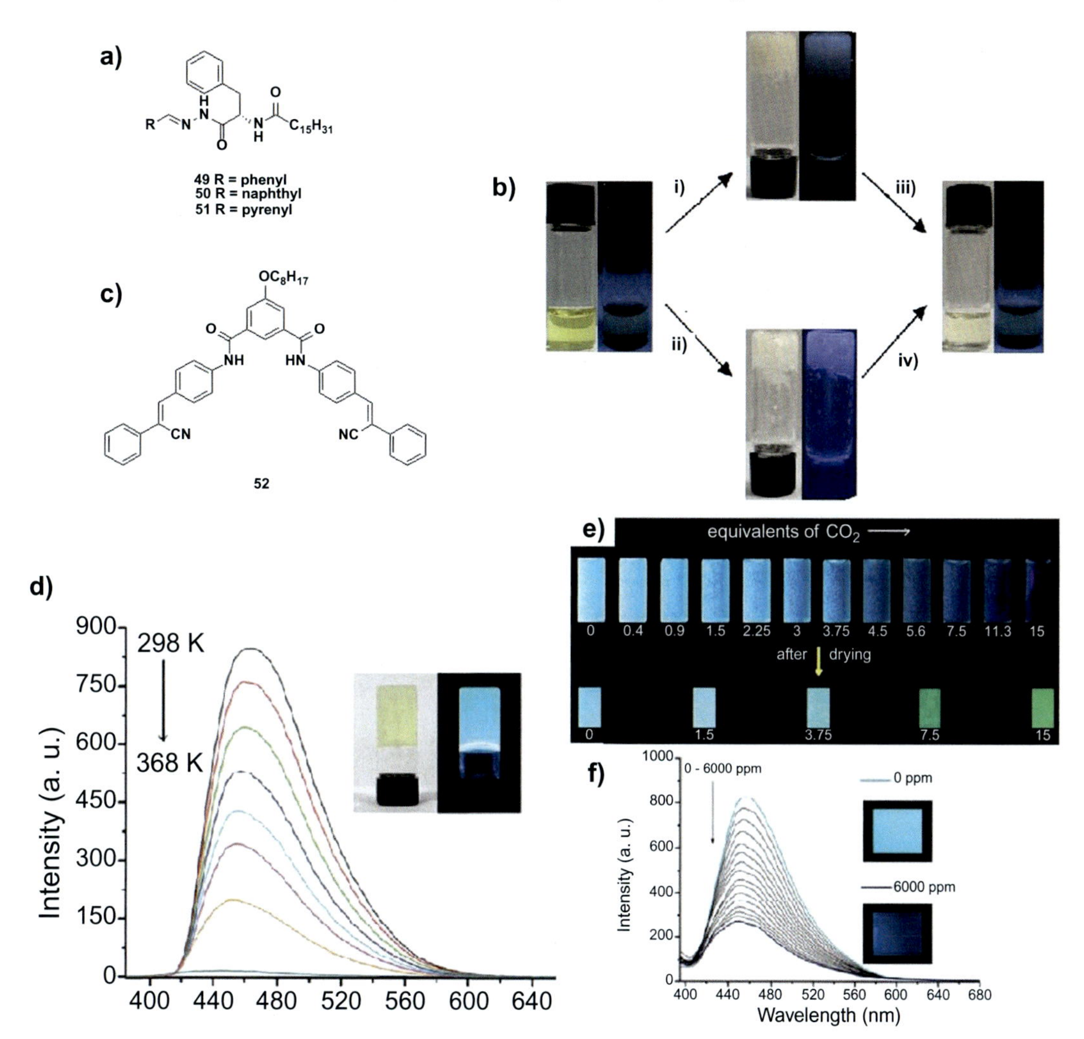

FIG. 27 (A) Chemical structures of gelators **49–51**; (B) photographs of the responsive processes of the transparent solution of gelator **49** (2.8 wt% in 0.5 mL of DMSO) in the presence of 3.5 equiv. of fluoride anions toward exposure to CO_2 (bubbling of CO_2 with the flow velocity of 6.0 mL min^{-1}) to the solution (i) or the exposure of the solution to air (ii), and the restoring process (iii) or (iv) of the solution state by heating and the simultaneously bubbling with N_2 the formed gels. (C) Chemical structure of cyanostilbene-based gelator **52**; (D) temperature-dependent fluorescent spectra of the toluene gel of **52** (λ_{ex} = 365 nm). Inset: Photographs of the gel under ambient and UV light; (E) fluorescent images of the gel aggregate samples exposed to different CO_2 concentrations and the films from the corresponding samples after drying; (F) fluorescent spectra of the xerogel films exposed to different CO_2 concentration, after drying. *Panels (A, B): Reproduced with permission from X. Zhang, H. Mu, H. Li, Y. Zhang, M. An, X. Zhang, J. Yoon, H. Yu, Dual-channel sensing of CO2: reversible solution-gel transition andgelation-induced fluorescence enhancement, Sens. Actuators B 255 (2018) 2764–2778. Copyright 2018, Elsevier Ltd. Panels (C–F) Reproduced with permission from Y. Ma, M. Cametti, Z. Džolić, S. Jiang, AIE-active bis-cyanostilbene-based organogels for quantitative fluorescence sensing of CO2 based on molecular recognition principles, J. Mater. Chem. C 6 (2018) 9232–9237. Copyright 2018, Royal Society of Chemistry.*

capability, also in light of its high mechanical strength. Initially, the effect of TFA on **53** was studied in solution. Upon addition of TFA to a solution of compound **53**, the color of the solution changed from colorless to light yellow, producing a new red-shifted shoulder in the absorption spectrum. This behavior was explained by the effect of protonation of the pyridine ring by TFA. Not surprisingly, the organogel, initially emitting in the blue region, collapsed into a yellow-colored solution quite rapidly as a drip of TFA was put on the top of the gel and let it penetrate into it, along with a complete emission quenching [69]. Interestingly, after 24h the gel state was recovered. Immediate reconversion to the gel state could be attained upon

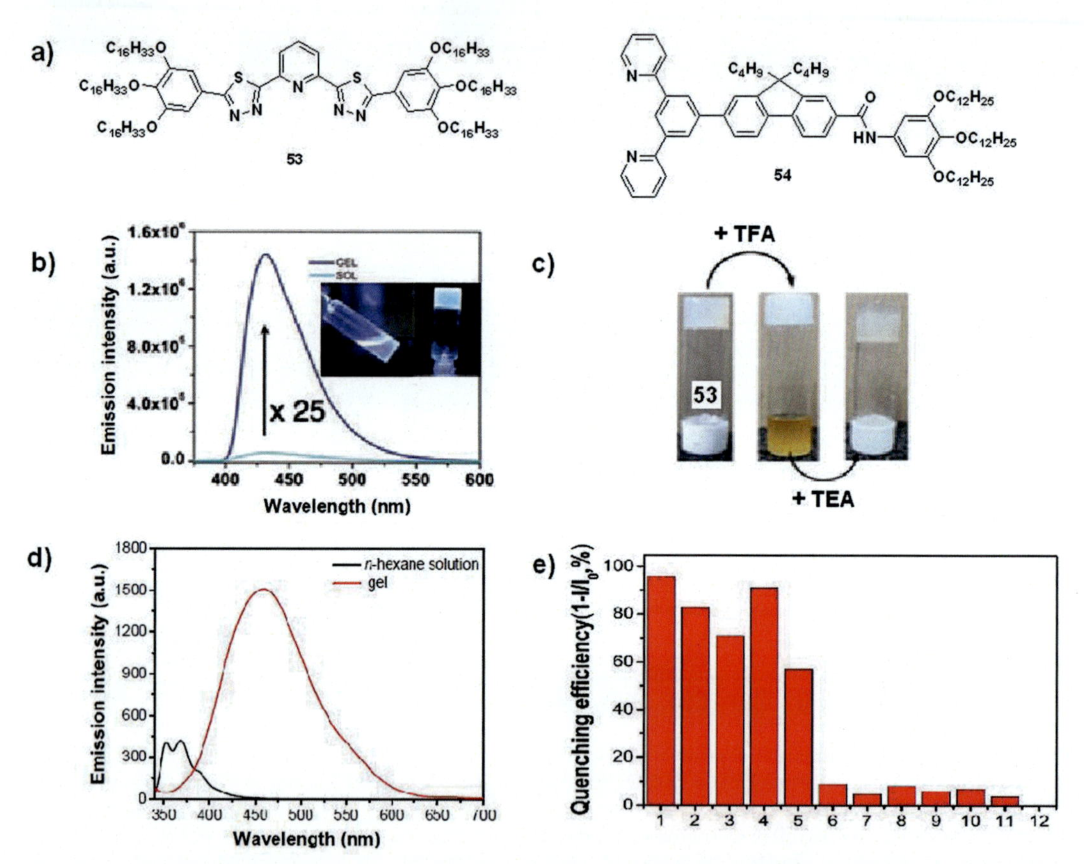

FIG. 28 (A) Chemical structure of gelators **53** and **54**; (B) emission spectra of **53** in *n*-decane (0.27 mmol dm^{-3}) showing AIE effect and (C) images of gel response toward TFA and TEA. (D) Emission spectra of **54** in *n*-hexane solution and in the gel state and (E) fluorescence response of the xerogel film of **54** (quenching efficiency $(1-I/I_0)$ recorded at 468 nm) upon exposure to different saturated acid and organic solvent vapors for 10 s: (1) trifluoroacetic acid; (2) formic acid; (3) acetic acid; (4) HCl; (5) HNO_3; (6) H_2SO_4; (7) tetrahydrofuran; (8) dichloromethane; (9) acetone; (10) ethyl acetate; (11) water. *Panels (A–C): Reproduced with permission from B. Pradhan, M. Gupta, S.K. Pal, A.S. Achalkumar, Multifunctional hexacatenar mesogen exhibiting supergelation, AIEE and its ability as a potential volatile acid sensor, J. Mater. Chem. C 4 (2016) 9669–9673. Copyright 2016, Royal Society of Chemistry. Panels (D, E): Reproduced with permission from X. Yang, Y. Liu, J. Li, Q. Wang, M. Yang, C. Li, A novel aggregation-induced-emission-active supramolecular organogel for the detection of volatile acid vapors, New J. Chem. 42 (2018) 17524–17532. Copyright 2018, Royal Society of Chemistry.*

the addition of triethylamine (TEA), showing the reversibility of the process, thus confirming the protonation of the central pyridine ring of **53** (Fig. 28C).

X. Yang, et al., designed and developed a novel organogelator, **54**, structurally composed of three different and functional sections: (i) an acid–base responsive group, the 3,5-di(pyridin-2-yl)phenyl unit; (ii) a fluorenyl group as an emissive linker unit; and (iii) hydrophobic tails made of 3,4,5-tris(dodecyloxy)phenyl groups [70]. **54** can self-assemble into one-dimensional nanofibers with an average diameter of ca. 300nm which, in turn, gives rise to stable gels in low polarity solvents (hexane, cyclohexane, toluene, and petroleum ether) with bright cyan emission. A film of such gels was then exposed to different acid vapors, including organic and inorganic acids and several common organic solvents. The fluorescence emission intensity of the films displayed marked changes after exposure to trifluoroacetic acid, formic acid, acetic acid, HCl, HNO_3, but not H_2SO_4 (Fig. 28E). Solvent vapors, instead, such as tetrahydrofuran, dichloromethane, acetone, ethyl acetate, and water caused no significant changes, indicating an acid–base reaction as the underlying process which is responsible for fluorescence quenching. TFA response was studied more quantitatively, and it was possible to determine the LOD for this acid to be equal to ca. 0.23ppm.

A very interesting responsive AIE-active gel system is that presented by Schmuck, Voskuhl et al., which can respond to both a volatile acid (TFA) and a volatile base (triethylamine) on the basis of the specific intergelator recognition pattern [71]. Bis-zwitterionic compound **55** is constituted of self-complementary guanidiniocarbonyl pyrrole carboxylate units which are known to form extremely strong hydrogen bond-assisted dimeric ion pairs. Due to this specific structural motif, **55** is capable of a marked self-assembly behavior which, depending on its concentration, may produce several types of adducts. Interestingly, as far as gels are concerned, **55** can form a stable gel in DMSO at 20mmoldm^{-3} concentration with AIE effect (Fig. 29B). The zwitterionic form, of the essence for strong interactions, is present however only within a narrow pH range, and thus any deviation from that leads to marked disruption of the self-assembly. Thus, the addition of TFA or triethylamine gives an immediate collapse of the gel which is based on the protonation or deprotonation of the guanidiniocarbonyl pyrrole carboxylate zwitterions leading to disassembly of the binding units (Fig. 29C). This change in the aggregation behavior goes along with a loss of the fluorescence emission since the self-association of the nonzwitterionic species is weaker and hence the restriction of the molecular rotation of the AIE rotors is reduced.

Another example of an AIE gel system capable of responding to both acids and bases is reported by Xue et al. [72], based on gelators **56** and **57** (Fig. 30A). Both compounds generate gel phases in aromatic hydrocarbon solutions, whereas only **56** (which features one CH_2 unit less than **57**) could also form gels in several alcohols. Gel morphology studies showed long nanoribbons with diameters in the 50–100nm range for **56**, and 20–50nm range for **57**, as shown in SEM and TEM images. Gelation induces marked AIE effects ($\lambda_{max}=508$nm). The gel of **56** in chlorobenzene can respond to TFA (Fig. 30B) as can be observed by the change in gel color, UV–Vis absorption spectrum, and by the severe loss of fluorescence after addition of 0.1 equivalent of TFA. TFA response can be also attained by using a xerogel based on **56**, reaching a response time of seconds and a LOD of ca. 0.5ppm. Likewise, the system can be also used for sensing aniline and other aromatic volatile amines via a PET mechanism, provided there existed compatible HOMO/LUMO energy levels between the donor amines and the acceptor gelator. In the case of aniline, the detection limit is as low as ca. 0.3ppb.

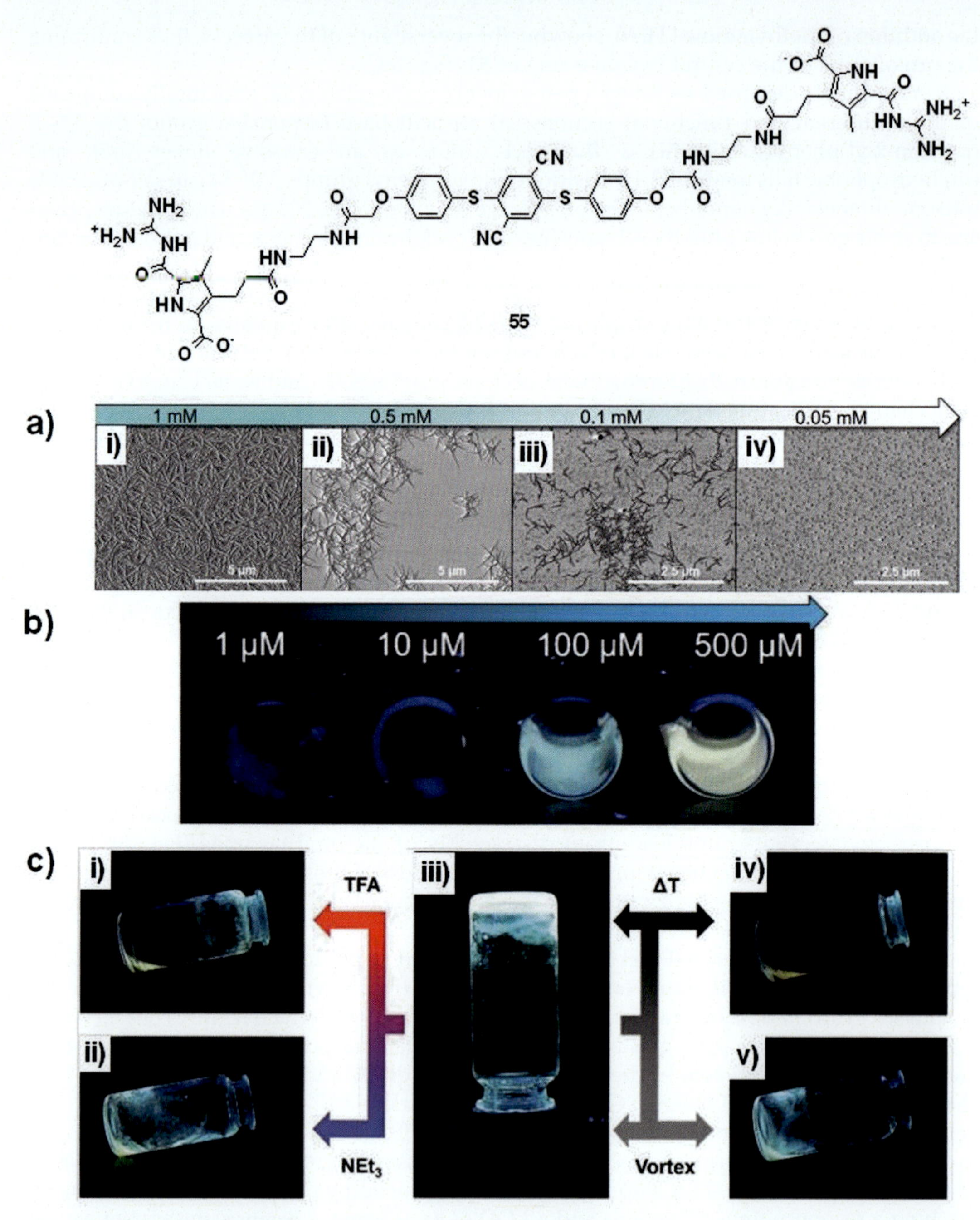

FIG. 29 (A) AFM images of the aggregates of **55** formed under decreasing concentration in DMSO (i–iv) and (B) photographs of different concentrations of **55** in DMSO under UV light irradiation ($\lambda = 365\,\text{nm}$); (C) photographs of the stimuli responsiveness of the gel obtained from **55** in DMSO ($c = 20\,\text{mmol}\,\text{dm}^{-3}$) under UV-light ($\lambda = 365\,\text{nm}$): (i) after addition of TFA, (ii) after addition of triethylamine, (iii) organogel of **55** in DMSO, (iv) after heating to 120°C, (v) after strong shaking. *Reproduced with permission from M. Externbrink, S. Riebe, C. Schmuck, J. Voskuhl, A dual pH-responsive supramolecular gelator with aggregation-induced emission properties, Soft Matter 14 (2018) 6166–6170. Copyright 2018, Royal Society of Chemistry.*

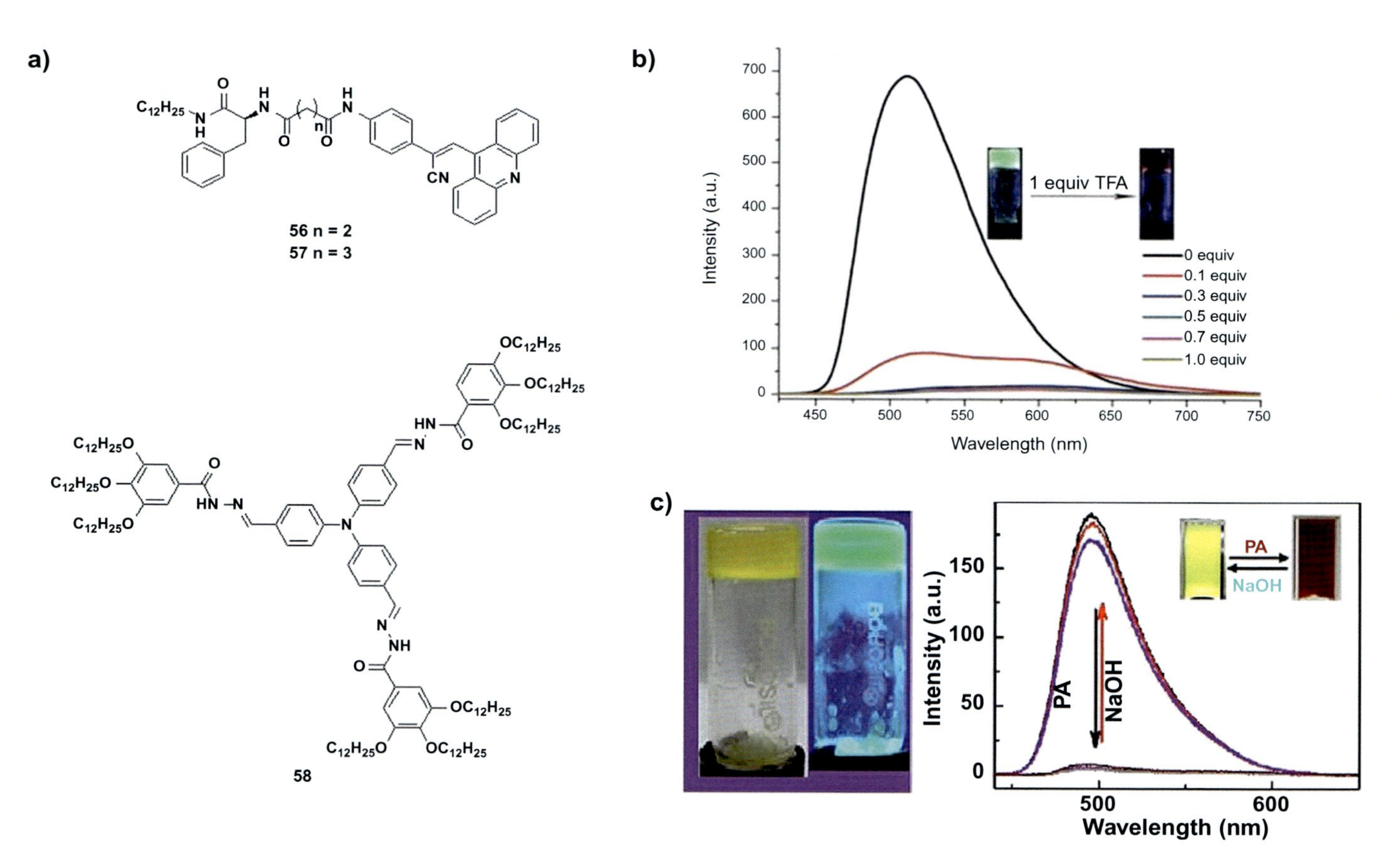

FIG. 30 (A) Chemical structures of gelators **56** and **57** and (B) fluorescence spectral changes for chlorobenzene gel of **56** (1.0 mg mL^{-1}) upon addition of TFA (λ_{ex} = 380 nm). Insets are photos under 365 nm light. (C) Photographs of **58** gel in 1-octanol under normal light and UV light (2.4 wt%) and (D) fluorescence titration of **58** gel in 1-octanol (8.5 mmol dm^{-3}) with picric acid and NaOH. Inset: visible color changes of **58**-gel on the action of picric acid and NaOH. *Panels (A, B): Reproduced with permission from P. Xue, J. Ding, Y. Shen, H. Gao, J. Zhao, J. Sun, R. Lu, Aggregation-induced emission nanofiber as dual sensor for aromatic amine and acid vapors, J. Mater. Chem. C 5 (2017) 11532–11541. Copyright 2017, Royal Society of Chemistry. Panels (C, D): Reproduced with permission from S. Mondal, P. Bairi, S. Das, A.-K. Nandi, Triarylamine-cored dendritic molecular gel for efficient colorometric, fluorometric, and impedometeric detection of picric acid, Chem. A Eur. J. 24 (2018) 5591–5600. Copyright 2017, Wiley-VCH.*

Another interesting category of target compounds is nitroaromatic. Organogelator **58**, having a triarylamine core and phenyl groups decorated with multiple long alkoxy chains at the extremities, produces AIE gels in 1-octanol which can selectively respond to picric acid [73]. Its gel state, constituted by a fibrillar network morphology, is endowed with a thixotropic behavior. Interestingly, gel fluorescence can be quenched in the presence of nitroaromatics, with the highest efficiency for picric acid, as demonstrated by the Stern-Volmer rate constant plots. Detailed studies indicated that the gelator molecules can act as an electron donor to the nitroaromatics in the electron transfer process. The gel system of **58** can be adapted for producing paper strips and detect picric acid by the naked eye upon UV light (365 nm) illumination within seconds. The corresponding colorimetric response is also quite obvious (Fig. 30C) and it is attributed to the formation of a charge-transfer complex.

4 Application of AIE-active supramolecular gel systems in BioSensing and bioimaging

The last set of applications described in this chapter involves the use of AIE gel systems in strongly biological-oriented fields.

Lin et al. demonstrated that hydrogelation can be attained with the TPE-peptide systems even by reducing the peptide fragment to a single amino acid [74]. In this case, two nonaromatic amino acids, serine and aspartic acid, were functionalized with TPE to develop AIE-based hydrogelators **59** and **60** for cell imaging purposes (Fig. 31A). The serine derivative **59** formed a stable hydrogel at 2 wt% at pH = 7.1, while **60** required slightly lower pH conditions. TEM characterization revealed nanosheet-like morphology with variable size and shapes, while CD spectra of the gel state showed bisignated Cotton effect signals in the 330 nm, matching the π–π^* absorption band, together with another strong CD signal in the 220 nm region. Hydrogels of both compounds exhibited strong AIE properties and they were found to be biocompatible with 3A6 and WS1 cells, at a concentration lower than 50 mmol dm^{-3} (Fig. 31B). Interestingly, distinct fluorescence can be observed in 3A6 cells after coincubation with the hydrogelators **59** and **60** for 24 h, which makes them good candidates for cell imaging applications.

More recently, the same group reported a tetraphenylethylene dipeptide hydrogelator containing tyrosine amino acid (**61**), which was able to form a self-supportive hydrogel in a broad pH range from 3.7 to 10.2 (Fig. 32A) [75]. The morphology of the obtained hydrogels is pH-dependent, with marked differences, as determined by TEM images (Fig. 32B). At a relatively high pH value (10.2), a weak transparent hydrogel with an entangled network of nanofibers (5 ± 2 nm diameter) was obtained. Upon a decrease in pH, however, flat and twisted nanobelts started to appear and progressively became dominant. At pH below 5.7, a turbid viscous gel and partial precipitation appeared, along with the thickening of the nanobelts. The pH-dependent gelation and aggregation processes were studied in detail by mono- and bi-dimensional ^{1}H-NMR, and by circular dichroism (CD) spectroscopy. The hydrogels exhibited excellent biocompatibility with 3A6 (human MSCs) and L929 (mouse fibroblast cells) cell lines (Fig. 32C). More notably, nanobelts made with **61** displayed a marked selective cell adhesion toward 3A6 cells in comparison with L929 and cancer cells (HeLa,

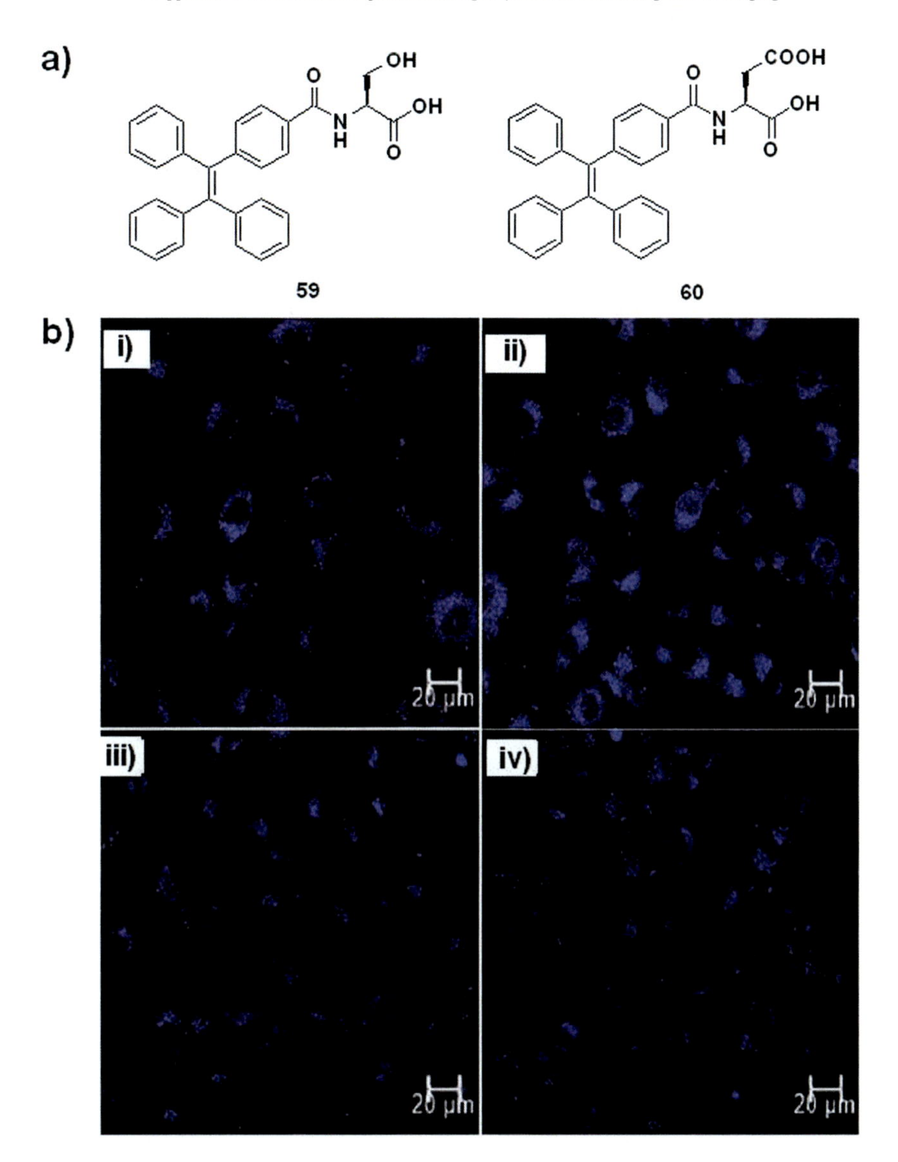

FIG. 31 (A) Chemical structures of hydrogelators **59** and **60** and (B) fluorescence microscopy images of 3A6 cells treated with different concentrations of **59** (i, ii = 50 μmol dm^{-3}, respectively) and **60** (iii, iv); concentrations are 10 (i, iii) and 50 (ii, iv) μmol dm^{-3}, respectively. *Reproduced with permission from N.-T. Chu, R.D. Chakravarthy, N.-C. Shih, Y.-H. Lin, Y.-C. Liu, J.-H. Lin, H.-C. Lin, Fluorescent supramolecular hydrogels self-assembled from tetraphenylethene (TPE)/single amino acid conjugates, RSC Adv. 8 (2018) 20922–20927. Copyright 2018, Royal Society of Chemistry.*

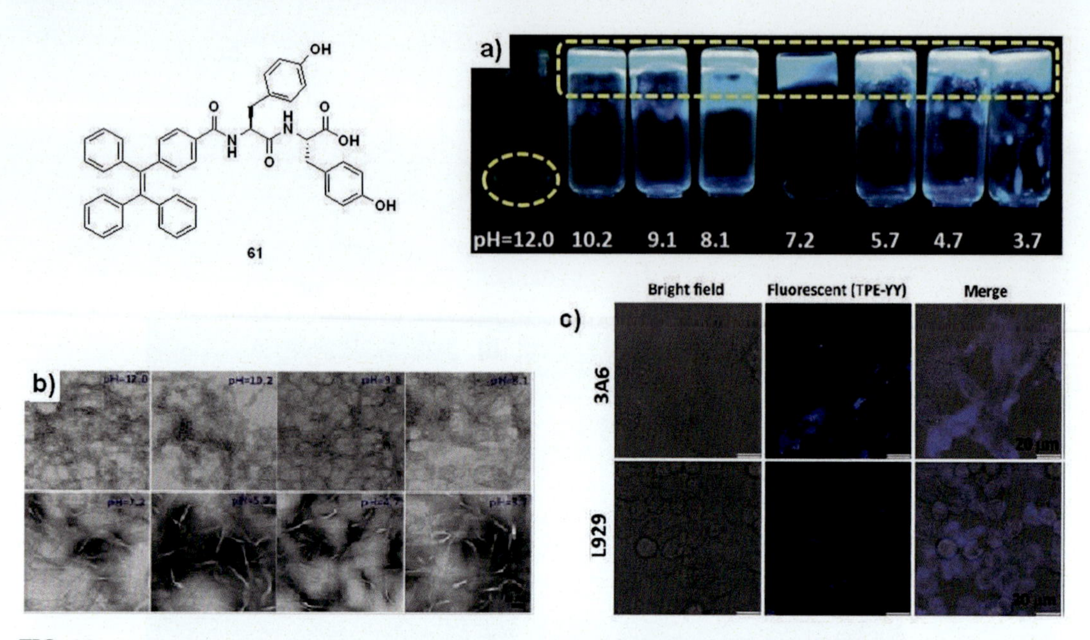

FIG. 32 (A) Optical fluorescent images and (B) negatively stained TEM micrographs of **61** hydrogels (2.0wt%) at various pH values ranging from 12.0 to 3.7: pH12.0–9.1 (entangled nanofibers), pH8.1 (nanofibers and flat nanobelts), pH7.2 (twisted nanobelts and nanofibers), pH5.7–3.7 (thickened twisted nanobelts). Scale bar: 50nm. (C) Confocal fluorescence microscopic images of 3A6 and L929 cells at 50 mmol dm^{-3} concentration for 90min of incubation time. Top: 3A6 cells; bottom: L929 cells. *Reproduced with permission from S.K. Talloj, M. Mohammed, H.-C. Lin, Construction of self-assembled nanostructures based tetraphenylethylene dipeptides: supramolecular nanobelts as biomimetic hydrogel for cell adhesion and proliferation, J. Mater. Chem. B 8 (2020) 7483–7493. Copyright 2020, Royal Society of Chemistry.*

MCF-7), and thus this system can be considered as a quite promising scaffold for stem cell-based therapies.

TPE can be also linked to specific peptidic sequences which are known to have well established functions. For example, Lin and Yeh et al. conjugated the TPE unit through a linker with the Asp-Gly-Glu-Ala motif which is a well-known binding sequence for the a2b1 integrins of cells, cellular structures which are involved in many cell functions also related to cancer progression. The bioprobes **62** and **63** were studied for their self-assembly and gelation properties and for the specific fluorescent labeling of the α2β1 integrins in PC-3 human prostate cancer cells (Fig. 33) [76]. Both compounds demonstrated to be AIE active gelators in water, with a slightly better performance given by **63** (up to 1 wt% under acidic conditions). TEM images showed nanobelt structures (107 ± 7nm) and highly entangled nanofibers (diameters of 11.8–2.0nm) for **62** and **63**, respectively. The aggregation aptitudes of **62** and **63** were also studied in more detail (DMSO-water mixture; CD and FT-IR) and it was demonstrated that **63** had a propensity to form β-sheet-like hydrogen-bonding structures. This finding prompted the testing of **63** as specific AIE active binders toward human prostate cancer cell lines PC-3. Indication of specific interactions of **63** with a2b1 integrins, also

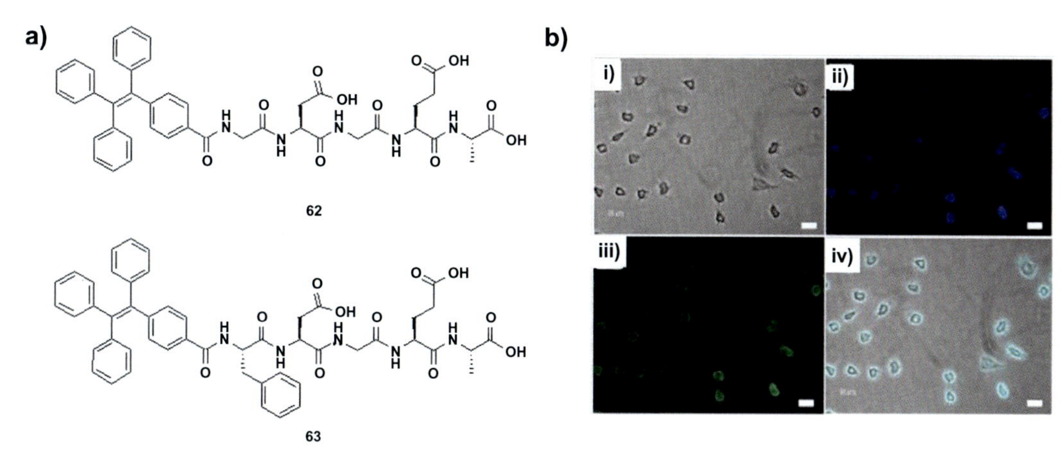

FIG. 33 (A) Chemical structures of AIE-active gelators **62** and **63** and (B) colocalization of **63** and integrin a2b1 antibodies in PC-3. (i) PC-3 were exposed to (ii) **63** (50 mmol dm^{-3}) for 4 h at 37°C followed by immunostaining with (iii) integrin a2b1 antibodies. (iv) The merge image of (i)–(iii) (scale bar: 20 mm). *Reproduced with permission from F.-K. Zhan, J.-C. Liu, B. Cheng, Y.-C. Liu, T.-S. Lai, H.-C. Lin, M.-Y. Yeh, Tumor targeting with DGEA peptide ligands: a new aromatic peptide amphiphile for imaging cancers, Chem. Commun. 55 (2019) 1060–1063. Copyright 2019, Royal Society of Chemistry.*

confirmed by additional immunostaining data which spatially correlated the AIE signals of **63** with the integrin a2b1 antibodies (Fig. 33B), gave hope for the use of **63** as a specific bioprobe.

Liang et al. have reported on **64** and **65** which are examples of TPE attached to the N-terminus of specific oligopeptide sequences. The two compounds feature a sequence inspired by the MAX1 (Ac-VKVKVKVKVDPPTVKVKVKVK-Am) and Q11 (Am-QQEFQFQFKQQ-Ac) peptides, both known to be capable to form β-sheet fibrillar structures under various conditions [77,78]. In both cases, a peptidic spacer was also introduced to increase water solubility and ensure no interference from the TPE group during the self-assembly process. Compound **64** self-assembled into hydrogels (0.5%wt) as the pH was raised from 6.0 to 10.0. This resulted in a concomitant AIE-based fluorescence turn-on, a process which could be reversibly switched off and on, over multiple cycles, by simply cycling between 6 and 10 pH range (Fig. 34B). This finding indicated that the system could be used as an intracellular pH sensor.

Compound **65** was found to self-assemble into a hydrogel network with bright fluorescence only in the presence of high salt concentrations [78]. The salt-induced aggregation process could be clearly visible, and thus easily monitored by the naked eye. Under UV illumination, the initially faintly emissive solution of **65** gradually evolved into a brightly fluorescent solution and finally to a fluorescent gel at a NaCl concentration of ca. 1.5 mol dm^{-3}, with an overall increase of emission λ_{466} of ca. 8 times (Fig. 35B). CD spectra recorded for **65** in water upon addition of increasing amounts of NaCl, indicated the formation of CD-active band at ca. 230 nm, attributed to π–π stacking interactions among TPE units, and suggesting a direct involvement of the AIE-active group in the gel formation. This

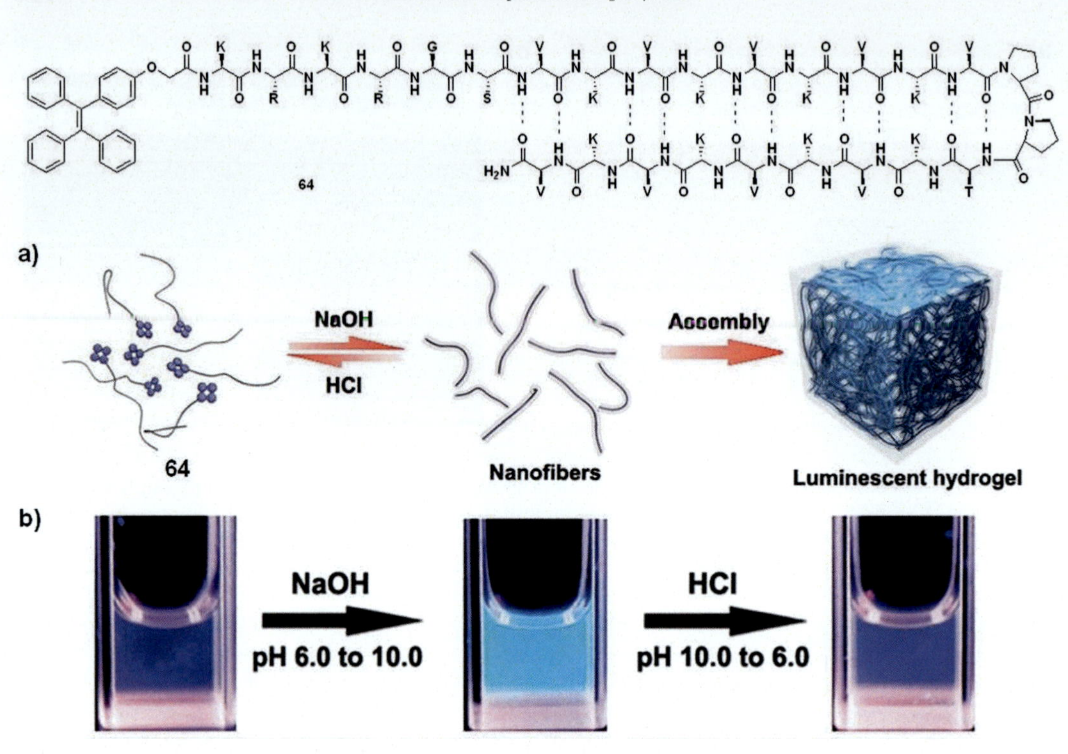

FIG. 34 (A) Self-assembly of tetraphenylethylene-capped MAX peptide **64** into luminescent hydrogels and (B) pH-dependent fluorescence of **64** showing switching of on and off fluorescent states upon changing of pH (bottom). *Reproduced with permission from C. Zhang, Y. Li, X. Xue, P. Chu, C. Liu, K. Yang, Y. Jiang, W.-Q. Chen, G. Zou, X.-J. Liang, A smart pH-switchable luminescent hydrogel, Chem. Commun. 51 (2015) 4168–4171. Copyright 2015, Royal Society of Chemistry.*

AIE-active gel system **65** was thus proposed as a potential biodegradable gel for cells encapsulation.

Lin and coworkers reported a naphthaleneimide protected phenylalanine-based supramolecular hydrogelator **66** which could form supramolecular microsized fibers at 1 wt% with a pH of 7.4 [79]. In comparison with relatively dilute solution, a strong AIE blue emission could be detected for the **66**-hydrogel under UV light (Fig. 36A). Human mesenchymal stem cells (hMSCs) viability in the presence of **66** was examined using colorimetric MTT [3-(4,5-dimethylthiazol-2-yl)-2,5-diphenyltetrazolium bromide] assay. After incubating the hMSCs with various concentrations of the hydrogelator, stem cells exhibited proliferation capacities with high survival ratios. The imaging of hMSCs within the fluorescent fibers network of **66** also verified the viability of cells providing a new dopant-free method for the development of high-contrast live cell-scaffold imaging (Fig. 36B).

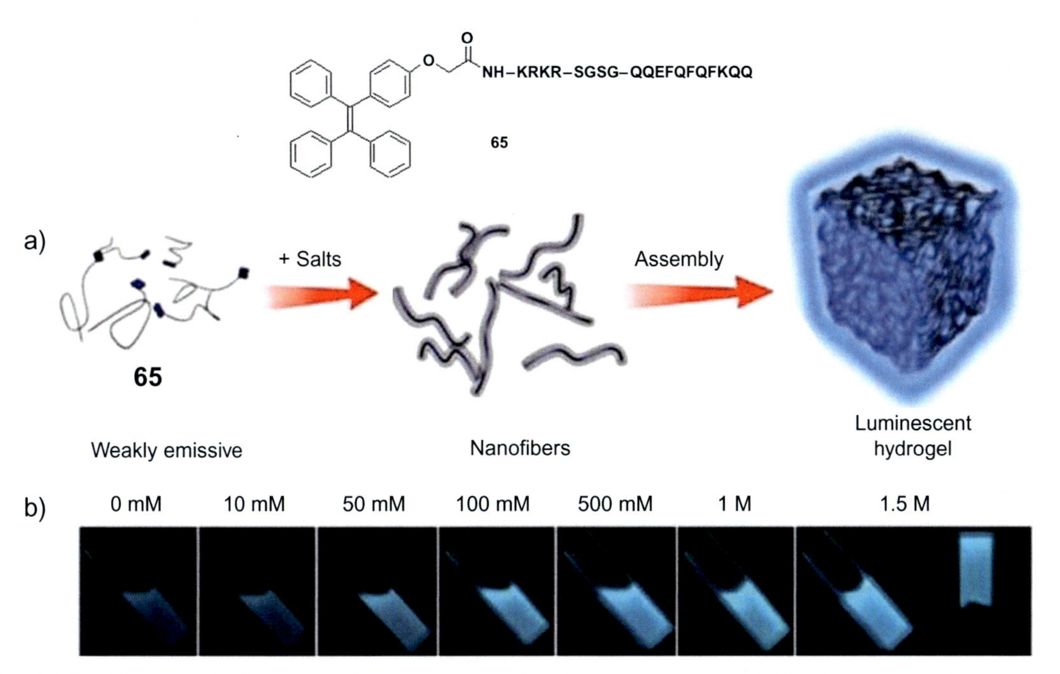

FIG. 35 (A) Schematic illustration of the formulation of luminescent hydrogel and (B) salt-responsive fluorescence and gelation of the TPE-caped Q19 peptide **65**. *Reproduced with permission from C. Zhang, C. Liu, X. Xue, X. Zhang, S. Huo, Y. Jiang, W.-Q. Chen, G. Zou, X.-J. Liang, Salt-responsive self-assembly of luminescent hydrogel with intrinsic gelation-enhanced emission, ACS Appl. Mater. Interfaces 6 (2014) 757–762. Copyright 2014, American Chemical Society.*

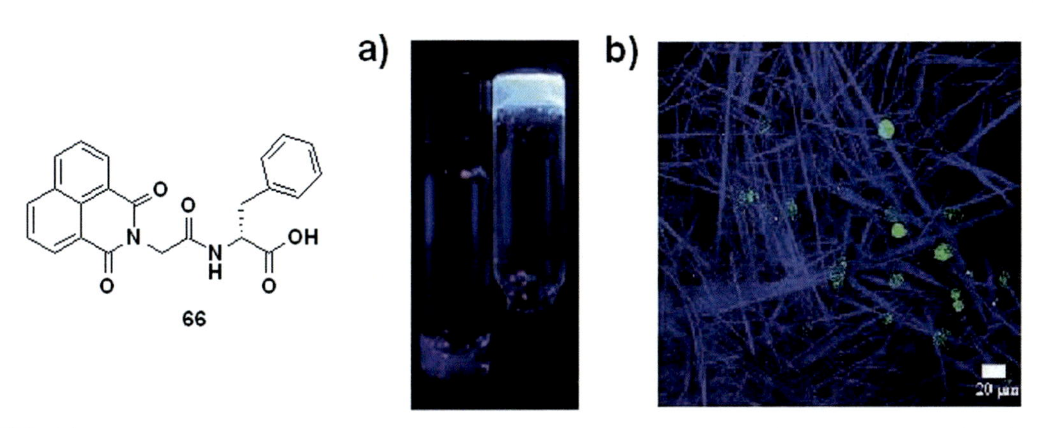

FIG. 36 Photographs of **66** solution (0.001 wt%, left vial) and hydrogel (1 wt%, right vial at pH7.4) and confocal microscopy image of the **66** gel; fluorescence images of hMSCs cultured in **66** gel for 72 h; luminescence emissions of fibrous structures (blue), live hMSCs stained with Calcein-AM (green). *Reproduced with permission from S.-M. Hsu, F.-Y. Wu, H. Cheng, Y.-T. Huang, Y.-R. Hsieh, D.-T.-H. Tseng, M.-Y. Yeh, S.-C. Hung, H.-C. Lin, Functional supramolecular polymers: a fluorescent microfibrous network in a supramolecular hydrogel for high-contrast live cell-material imaging in 3D environments, Adv. Healthc. Mater. 5 (2016) 2406–2412. Copyright 2018, Wiley-VCH.*

5 Conclusions

The combination of AIE features with the specific properties of supramolecular gels opens avenues for the development of novel fluorescent stimuli-responsive functional materials. In this chapter, we have highlighted a selection of important works in the field of AIE-active supramolecular gel materials. The discovery of AIE phenomena, characterized by systems with strong emission in the aggregated state as opposed to a weak one in solution, has significantly stimulated the development of novel luminescent materials. As far as gel materials are concerned, the supramolecular approach which relies on low molecular weight gelator systems surely has been demonstrated to bear substantial benefits to achieving efficient functional and responsive systems, with marked differences compared to more classical polymeric gels [11]. This view is well documented by extensive literature related to AIE-active supramolecular gels showing potential as functional materials useful for a variety of applications, in particular, for sensing.

There nonetheless remain many challenges yet to be resolved. They include the transfer of lab-manufactured system apparatuses to devices, and their improvements in terms of performances to meet real-world applications needs. Indeed, most of the reported AIE-active supramolecular gel-based sensors work well only qualitatively, while the quantitative detection of analytes is often still out of reach. Thus, more focus needs to be directed to improve detection performance in real samples and in real-world environments. This could be achieved by exploring new ways to combine AIE units with additional recognition moieties, stimuli-responsive, and/or functional groups into one multiresponsive supramolecular AIE gel system. Magnetism, electron and ionic conductivity, redox, and self-healing properties could be indeed envisaged as extra features which could widen the scope of AIE-active gel systems to specific technological uses [80]. Another aspect that should be taken into consideration is that related to the intrinsic fluorescent properties of the AIEgens, which mainly involve relatively short-wavelength emission (i.e., TPE, cyanostilbene, etc., and their derivatives usually emit in the green/blue region). This may limit their application in vivo imaging, for example, and developing AIE-active supramolecular hydrogels based on red or far-red/near-infrared emission is considered quite a desirable feature [81], which can considerably widen their scope in preclinical research and clinical applications. To this end, the design of AIE-active supramolecular hydrogels with enhanced biocompatibility and/or biodegradability, depending on the need, should also be pursued further.

AIE-active supramolecular gels constitute a class of novel functional and responsive materials that have received increasing attention in the last years and which are expected to contribute to the development of the next generation of smart soft materials. We hope that this chapter, which described the essential features of AIE-active supramolecular gels systems based on a selection of the existing works on the subject, had effectively presented the successes so far obtained, showed the future potential of this class of materials, and that it could inspire more strives toward the further development of the field.

Acknowledgment

We acknowledge the funding from the Croatian Science Foundation under the project IP-2016-06-5983. M. C. thanks MIUR for FFABR—Fondo finanziamento delle attività base di ricerca.

References

[1] R.G. Weiss, P. Terech (Eds.), Molecular Gels: Materials with Self-Assembled Fibrillar Networks, Springer, Netherlands, Dordrecht, 2006.

[2] B. Escuder, J.F. Miravet (Eds.), Functional Molecular Gels, Royal Society of Chemistry, Cambridge, 2014.

[3] D.B. Amabilino, D.K. Smith, J.W. Steed, Supramolecular materials, Chem. Soc. Rev. 46 (2017) 2404–2420.

[4] M. Gaft, R. Reisfeld, G. Panczer (Eds.), Modern Luminescence Spectroscopy of Minerals and Materials, Springer, 2015.

[5] L. Tao, M.-L. Li, K.-P. Yang, Y. Guan, P. Wang, Z. Shen, H.-L. Xie, Color-tunable and stimulus-responsive luminescent liquid crystalline polymers fabricated by hydrogen bonding, ACS Appl. Mater. Interfaces 11 (2019) 15051–15059.

[6] T. Förster, Excimers, Angew. *Chem. Int. Ed.* 8 (1969) 333–343.

[7] J. Luo, Z. Xie, J.W.Y. Lam, L. Cheng, H. Chen, C. Qiu, H.S. Kwok, X. Zhan, Y. Liu, D. Zhu, B.Z. Tang, Aggregation-induced emission of 1-methyl-1,2,3,4,5-pentaphenylsilole, Chem. Commun. (2001) 1740–1741.

[8] J. Mei, N.L.C. Leung, R.T.K. Kwok, J.W.Y. Lam, B.Z. Tang, Aggregation-induced emission: together we shine, united we soar! Chem. Rev. 115 (2015) 11718–11940.

[9] J. Li, J. Wang, H. Li, N. Song, D. Wang, B.Z. Tang, Supramolecular materials based on AIE luminogens (AIEgens): construction and applications, Chem. Soc. Rev. 49 (2020) 1144–1172.

[10] Z. Zhao, J.W.Y. Lamb, B.Z. Tang, Self-assembly of organic luminophores with gelation-enhanced emission characteristics, Soft Matter 9 (2013) 4564–4579.

[11] J. Tavakoli, A.J. Ghahfarokhi, Y. Tang, Aggregation-induced emission fluorescent gels: current trends and future perspectives, Top. Curr. Chem. 379 (2021) 9.

[12] X. Cao, A. Gao, J. Hou, T. Tao Yi, Fluorescent supramolecular self-assembly gels and their application as sensors: a review, Coord. Chem. Rev. 434 (2021), 213792.

[13] Z. Li, X. Ji, H. Xie, B.Z. Tang, Aggregation-induced emission-active gels: fabrications, functions, and applications, Adv. Mater. 33 (2021) 2100021.

[14] X. Yu, L. Geng, J. Guo, Preparation and sensing application of fluorescent organogels and hydrogels, in: T. Jiao (Ed.), Supramolecular Gels: Materials and Emerging Applications, WILEY-VCH GmbH, 2021, pp. 21–50.

[15] H. Chen, L. Zhou, X. Shi, J. Hu, J. Guo, P.-A. Albouy, M.-H. Li, AIE fluorescent gelators with thermo-, mechano-, and vapochromic properties, Chem. Asian J. 14 (2019) 781–788.

[16] M. Wang, D. Zhang, G. Zhang, D. Zhu, Fluorescence enhancement upon gelation and thermally-driven fluorescence switches based on tetraphenylsilole-based organic gelators, Chem. Phys. Lett. 475 (2009) 64–67.

[17] H. Shang, Z. Ding, Y. Shen, B. Yang, M. Liu, S. Jiang, Multi-color tunable circularly polarized luminescence in one single AIE system, Chem. Sci. 11 (2020) 2169–2174.

[18] I. Gracia, J.L. Serrano, J. Barberá, A. Omenat, Functional organogelators formed by liquid-crystal carbazole-containing bis-MPA dendrimers, RSC Adv. 6 (2016) 39734–39740.

[19] J. Rabah, A. Escola, O. Jeannin, P.-A. Bouit, M. Hissler, F. Camerel, Luminescent organogels formed by ionic self-assembly of AIE-active phospholes, ChemPlusChem 85 (2020) 79–83.

[20] Y.-Y. Chen, X.-M. Jiang, G.-F. Gong, H. Yao, Y.-M. Zhang, T.-B. Wei, Q. Lin, Pillararene-based AIEgens: research progress and appealing applications, Chem. Commun. 57 (2021) 284–301.

[21] D.D. La, Anuradha, A. Gupta, M. Al Kobaisi, A. Rananaware, S.V. Bhosale, Supramolecular chemistry of AIE-active tetraphenylethylene luminophores, in: P.J. Thomas, N. Revaprasadu (Eds.), Nanoscience: Volume 4, RSC Publishing, 2017, pp. 75–107.

[22] Y. Ma, H. Ma, Z. Yang, J. Ma, Y. Su, W. Li, Z. Lei, Methyl cinnamate-derived fluorescent rigid organogels based on cooperative $\pi-\pi$ stacking and $C{=}O\cdots\pi$ interactions instead of H-bonding and alkyl chains, Langmuir 31 (2015) 4916–4923.

[23] C. Felip-León, F. Galindo, J.F. Miravet, Insights into the aggregation-induced emission of 1,8-naphthalimide-based supramolecular hydrogels, Nanoscale 10 (2018) 17060–17069.

[24] J.-Y. Chen, G. Kadam, A. Gupta, Anuradha, S.V. Bhosale, F. Zheng, C.-H. Zhou, B.-H. Jia, D.S. Dalal, J.-L. Li, A biomimetic supramolecular approach for charge transfer between donor and acceptor chromophores with aggregation-induced emission, Chem. A Eur. J. 24 (2018) 14668–14678.

[25] J. Han, J. You, X.G. Li, P. Duan, M. Liu, Full-color tunable circularly polarized luminescent nanoassemblies of achiral AIEgens in confined chiral nanotubes, Adv. Mater. 29 (2017) 1606503.

[26] N. Song, D.-X. Chen, Y.-C. Qiu, X.-Y. Yang, B. Xu, W. Tian, Y.-W. Yang, Stimuli-responsive blue fluorescent supramolecular polymers based on a pillar[5]arene tetramer, Chem. Commun. 50 (2014) 8231.

[27] N. Song, X.-Y. Lou, W. Hou, C.-Y. Wang, Y. Wang, Y.-W. Yang, Pillararene-based fluorescent supramolecular systems: the key role of chain length in gelation, Macromol. Rapid Commun. 39 (2018) 1800593.

[28] C.D. Jones, J.W. Steed, Gels with sense: supramolecular materials that respond to heat, light and sound, Chem. Soc. Rev. 45 (2016) 6546–6596.

[29] M. Martínez-Abadía, R. Giménez, M.B. Ros, Self-assembled α-cyanostilbenes for advanced functional materials, Adv. Mater. 30 (2018) 1704161.

[30] J.W. Chung, S.-J. Yoon, B.-K. An, S.Y. Park, High-contrast on/off fluorescence switching via reversible E–Z isomerization of diphenylstilbene containing the α-cyanostilbenic moiety, J. Phys. Chem. C 117 (2013) 11285–11291.

[31] D. Kim, J.E. Kwon, S.Y. Park, Fully reversible multistate fluorescence switching: organogel system consisting of luminescent cyanostilbene and turn-on diarylethene, Adv. Funct. Mater. 28 (2018) 1706213.

[32] B.-K. An, D.-S. Lee, J.-S. Lee, Y.-S. Park, H.-S. Song, S.Y. Park, Strongly fluorescent organogel system comprising fibrillar self-assembly of a trifluoromethyl-based cyanostilbene derivative, J. Am. Chem. Soc. 126 (2004) 10232–10233.

[33] J. Seo, J.W. Chung, J.E. Kwon, S.Y. Park, Photoisomerization-induced gel-to-sol transition and concomitant fluorescence switching in a transparent supramolecular gel of a cyanostilbene derivative, Chem. Sci. 5 (2014) 4845–4850.

[34] Y. Ma, M. Cametti, Z. Džolić, S. Jiang, Responsive aggregation-induced emissive supramolecular gels based on bis-cyanostilbene derivatives, J. Mater. Chem. C 4 (2016) 10786–10790.

[35] X. Wang, Z. Ding, Y. Ma, Y. Zhang, H. Shang, S. Jiang, Multi-stimuli responsive supramolecular gels based on a D–π–A structural cyanostilbene derivative with aggregation induced emission properties, Soft Matter 15 (2019) 1658–1665.

[36] Z. Ding, Y. Ma, H. Shang, H. Zhang, S. Jiang, Fluorescence regulation and photoresponsivity in AIEE supramolecular gels based on a cyanostilbene modified benzene-1,3,5-tricarboxamide derivative, Chem. A Eur. J. 25 (2019) 315–322.

[37] H.-J. Kim, H.J. Lee, J.W. Chung, D.R. Whang, S.Y. Park, A highly fluorescent and photoresponsive polymer gel consisting of poly(acrylic acid) and supramolecular cyanostilbene crosslinkers, Adv. Opt. Mater. 7 (2018) 1801348.

[38] L. Ji, Q. He, D. Niu, J. Tan, G. Ouyang, M. Liu, Host-guest interaction enabled chiroptical photo-switching and enhanced circularly polarized luminescence, Chem. Commun. 55 (2019) 11747–11750.

[39] P. Wei, Z. Li, J.-X. Zhang, Z. Zhao, H. Xing, Y. Tu, J. Gong, T.S. Cheung, S. Hu, H.H.-Y. Sung, I.D. Williams, R.T.K. Kwok, J.W.Y. Lam, B.Z. Tang, Molecular transmission: visible and rate-controllable photoreactivity and synergy of aggregation-induced emission and host-guest assembly, Chem. Mater. 31 (2019) 1092–1100.

[40] P. Wei, J.-X. Zhang, Z. Zhao, Y. Chen, X. He, M. Chen, J. Gong, H.H.-Y. Sung, I.D. Williams, J.W.Y. Lam, B.Z. Tang, Multiple yet controllable photoswitching in a single AIEgen system, J. Am. Chem. Soc. 140 (2018) 1966–1975.

[41] X. Yu, H. Chen, X. Shi, P.-A. Albouy, J. Guo, J. Hu, M.-H. Li, Liquid crystal gelators with photo-responsive and AIE properties, Mater. Chem. Front. 2 (2018) 2245–2253.

[42] M.-O.M. Piepenbrock, G.O. Lloyd, N. Clarke, J.W. Steed, Metal- and anion-binding supramolecular gels, Chem. Rev. 110 (2010) 1960–2004.

[43] X.-M. Jiang, X.-J. Huang, S.-S. Song, X.-Q. Ma, Y.-M. Zhang, H. Yao, T.-B. Wei, Q. Lin, Tri-pillar[5]arene-based multi-stimuli responsive supramolecular polymer for fluorescent detection and separation of Hg^{2+}, Polym. Chem. 9 (2018) 4625–4630.

[44] Y.-M. Zhang, W. Zhu, W.-J. Qu, K.-P. Zhong, X.-P. Chen, H. Yao, T.-B. Wei, Q. Lin, Competition of cation–π and exo-wall π–π interactions: a novel approach to achieve ultrasensitive response, Chem. Commun. 54 (2018) 4549–4552.

[45] L. Fabbrizzi, Beauty in chemistry: making artistic molecules with schiff bases, J. Org. Chem. 85 (2020) 12212–12226.

[46] T. Han, Y. Hong, N. Xie, S. Chen, N. Zhao, E. Zhao, J.W.Y. Lam, H.H.Y. Sung, Y. Dong, B. Tong, B.Z. Tang, Defect-sensitive crystals based on diaminomaleonitrile-functionalized Schiff base with aggregation-enhanced emission, J. Mater. Chem. C 1 (2013) 7314–7320.

[47] X. Ma, Z. Zhang, H. Xie, Y. Ma, C. Liu, S. Liu, M. Liu, Emissive intelligent supramolecular gel for highly selective sensing of Al^{3+} and writable soft material, Chem. Commun. 54 (2018) 13674–13677.

[48] A. Ganguly, S. Ghosh, S. Kar, N. Guchhait, Selective fluorescence sensing of Cu(II) and Zn(II) using a simple Schiff base ligand: naked eye detection and elucidation of photoinduced electron transfer (PET) mechanism, Spectrochim. *Acta A Mol. Biomol. Spectrosc.* 143 (2015) 72–80.
[49] H. Yao, J. Wang, Q. Zhou, X.-W. Guan, Y.-Q. Fan, Y.-M. Zhang, T.-B. Wei, Q. Lin, Acylhydrazone functionalized benzimidazole-based metallogel for the efficient detection and separation of Cr^{3+}, Soft Matter 14 (2018) 8390–8394.
[50] Q. Lin, B. Sun, Q.-P. Yang, Y.-P. Fu, X. Zhu, Y.-M. Zhang, T.-B. Wei, A novel strategy for the design of smart supramolecular gels: controlling stimuli-response properties through competitive coordination of two different metal ions, Chem. Commun. 50 (2014) 10669–10671.
[51] Y.-M. Zhang, W. Zhu, Q. Zhao, W.-J. Qu, H. Yao, T.-B. Wei, Q. Lin, Th^{4+} tuned aggregation-induced emission: a novel strategy for sequential ultrasensitive detection and separation of Th^4+ and Hg^{2+}, Spectrochim. Acta A 229 (2020), 117926.
[52] G.-F. Gong, Y.-Y. Chen, Y.-M. Zhang, Y.-Q. Fan, Q. Zhou, H.-L. Yang, Q.-P. Zhang, H. Yao, T.-B. Wei, Q. Lin, A novel bis-component AIE smart gel with high selectivity and sensitivity to detect CN^-, Fe^{3+} and $H_2PO_4^-$, Soft Matter 15 (2019) 6348–6352.
[53] Y. Ma, M. Cametti, Z. Džolić, S. Jiang, Selective Cu(II) sensing by a versatile AIE cyanostilbene-based gel system, Soft Matter 15 (2019) 6145–6150.
[54] X. Ma, J. Xie, N. Tang, J. Wu, AIE-caused luminescence of a thermally-responsive supramolecular organogel, New J. Chem. 40 (2016) 6584–6587.
[55] Q. Lin, T.-T. Lu, X. Zhu, B. Sun, Q.-P. Yang, T.-B. Wei, Y.-M. Zhang, A novel supramolecular metallogel-based high-resolution anions sensor array, Chem. Commun. 51 (2015) 1635–1638.
[56] J.-H. Hu, Z.-Y. Yin, K. Gui, Q.-Q. Fu, Y. Yao, X.-M. Fu, H.-X. Liu, A novel supramolecular polymer gel based long-alkyl-chains functionalized coumarin acylhydrazone for sequential detection and separation of toxic ions, Soft Matter 16 (2020) 1029–1033.
[57] H. Yao, J. Wang, S.-S. Song, Y.-Q. Fan, X.-W. Guan, Q. Zhou, T.-B. Wei, Q. Lin, Y.-M. Zhang, A novel supramolecular AIE gel acts as a multi-analyte sensor array, New J. Chem. 42 (2018) 18059–18065.
[58] Q. Lin, T.-T. Lu, X. Zhu, T.-B. Wei, H. Li, Y.-M. Zhang, Rationally introduce multi-competitive binding interactions in supramolecular gels: a simple and efficient approach to develop multi-analyte sensor array, Chem. Sci. 7 (2016) 5341–5346.
[59] M. Cametti, K. Rissanen, Highlights on contemporary recognition and sensing of fluoride anion in solution and in the solid state, Chem. Soc. Rev. 42 (2013) 2016–2038.
[60] C.-W. Zhang, B. Ou, S.-T. Jiang, G.-Q. Yin, L.-J. Chen, L. Xu, X. Li, H.-B. Yang, Cross-linked AIE supramolecular polymer gels with multiple stimuli-responsive behaviours constructed by hierarchical self-assembly, Polym. Chem. 9 (2018) 2021–2030.
[61] Y.-Y. Ren, Z. Xu, G. Li, J. Huang, X. Fand, L. Xu, Hierarchical self-assembly of a fluorescence emission-enhanced organogelator and its multiple stimuli-responsive behaviors, Dalton Trans. 46 (2017) 333–337.
[62] B. Wang, P. Wang, L.-H. Xie, R.-B. Lin, J. Lv, J.-R. Li, B. Chen, A stable zirconium based metal-organic framework for specific recognition of representative polychlorinated dibenzo-*p*-dioxin molecules, Nat. Commun. 10 (2019) 3861.
[63] S. Kang, J. Kim, J.-H. Park, C.K. Ahn, C.-H. Rhee, M.S. Han, Intra-molecular hydrogen bonding stabilization based-fluorescent chemosensor for CO_2: application to screen relative activities of CO_2 absorbents, Dyes Pigm. 123 (2015) 125–131.
[64] Y. Liu, Y. Tang, N.N. Barashkov, I.S. Irgibaeva, J.W.Y. Lam, R. Hu, D. Birimzhanova, Y. Yu, B.Z. Tang, Fluorescent chemosensor for detection and quantitation of carbon dioxide gas, J. Am. Chem. Soc. 132 (2010) 13951–13953.
[65] X. Zhang, H. Mu, H. Li, Y. Zhang, M. An, X. Zhang, J. Yoon, H. Yu, Dual-channel sensing of CO_2: reversible solution-gel transition andgelation-induced fluorescence enhancement, Sens. Actuators B 255 (2018) 2764–2778.
[66] Y. Liu, D. Lee, X. Zhang, J. Yoon, Fluoride ion activated CO_2 sensing using sol-gel system, Dyes Pigm. 139 (2017) 658–663.
[67] Y. Ma, M. Cametti, Z. Džolić, S. Jiang, AIE-active bis-cyanostilbene-based organogels for quantitative fluorescence sensing of CO_2 based on molecular recognition principles, J. Mater. Chem. C 6 (2018) 9232–9237.
[68] B. Pradhan, M. Gupta, S.K. Pal, A.S. Achalkumar, Multifunctional hexacatenar mesogen exhibiting supergelation, AIEE and its ability as a potential volatile acid sensor, J. Mater. Chem. C 4 (2016) 9669–9673.
[69] Z. Džolić, M. Cametti, A. Dalla Cort, L. Mandolini, M. Žinić, Fluoride-responsive organogelator based on oxalamide-derived anthraquinone, Chem. Commun. (2007) 3535–3537.

[70] X. Yang, Y. Liu, J. Li, Q. Wang, M. Yang, C. Li, A novel aggregation-induced-emission-active supramolecular organogel for the detection of volatile acid vapors, New J. Chem. 42 (2018) 17524–17532.
[71] M. Externbrink, S. Riebe, C. Schmuck, J. Voskuhl, A dual pH-responsive supramolecular gelator with aggregation-induced emission properties, Soft Matter 14 (2018) 6166–6170.
[72] P. Xue, J. Ding, Y. Shen, H. Gao, J. Zhao, J. Sun, R. Lu, Aggregation-induced emission nanofiber as dual sensor for aromatic amine and acid vapors, J. Mater. Chem. C 5 (2017) 11532–11541.
[73] S. Mondal, P. Bairi, S. Das, A.K. Nandi, Triarylamine-cored dendritic molecular gel for efficient colorometric, fluorometric, and impedometeric detection of picric acid, Chem. A Eur. J. 24 (2018) 5591–5600.
[74] N.-T. Chu, R.D. Chakravarthy, N.-C. Shih, Y.-H. Lin, Y.-C. Liu, J.-H. Lin, H.-C. Lin, Fluorescent supramolecular hydrogels self-assembled from tetraphenylethene (TPE)/single amino acid conjugates, RSC Adv. 8 (2018) 20922–20927.
[75] S.K. Talloj, M. Mohammed, H.-C. Lin, Construction of self-assembled nanostructures based tetraphenylethylene dipeptides: supramolecular nanobelts as biomimetic hydrogel for cell adhesion and proliferation, J. Mater. Chem. B 8 (2020) 7483–7493.
[76] F.-K. Zhan, J.-C. Liu, B. Cheng, Y.-C. Liu, T.-S. Lai, H.-C. Lin, M.-Y. Yeh, Tumor targeting with DGEA peptide ligands: a new aromatic peptide amphiphile for imaging cancers, Chem. Commun. 55 (2019) 1060–1063.
[77] C. Zhang, Y. Li, X. Xue, P. Chu, C. Liu, K. Yang, Y. Jiang, W.-Q. Chen, G. Zou, X.-J. Liang, A smart pH-switchable luminescent hydrogel, Chem. Commun. 51 (2015) 4168–4171.
[78] C. Zhang, C. Liu, X. Xue, X. Zhang, S. Huo, Y. Jiang, W.-Q. Chen, G. Zou, X.-J. Liang, Salt-responsive self-assembly of luminescent hydrogel with intrinsic gelation-enhanced emission, ACS Appl. Mater. Interfaces 6 (2014) 757–762.
[79] S.-M. Hsu, F.-Y. Wu, H. Cheng, Y.-T. Huang, Y.-R. Hsieh, D.T.-H. Tseng, M.-Y. Yeh, S.-C. Hung, H.-C. Lin, Functional supramolecular polymers: a fluorescent microfibrous network in a supramolecular hydrogel for high-contrast live cell-material imaging in 3D environments, Adv. Healthc. Mater. 5 (2016) 2406–2412.
[80] B.P. Nowak, B.J. Ravoo, Magneto- and photo-responsive hydrogels from the co-assembly of peptides, cyclodextrins, and superparamagnetic nanoparticles, Faraday Discuss. 219 (2019) 220–228.
[81] T.L. Rapp, C.A. DeForest, Visible light-responsive dynamic biomaterials: going deeper and triggering more, Adv. Healthc. Mater. 9 (2020) 1901553.

CHAPTER

6

Mechanochromic luminescence in AIE luminogens

Yong Qiang Dong, Yue Si, and Yucong Zhang

Beijing Key Laboratory of Energy Conversion and Storage Materials, College of Chemistry, Beijing Normal University, Beijing, China

1 Introduction

Stimuli-responsive materials are a kind of smart materials that change their properties in response to single or multiple external stimuli. Mechanochromic (MC) luminogens refer to a kind of smart materials that change their emission color and/or intensity in response to mechanical stimuli such as pressing, grinding, crushing, stretching, or rubbing [1–5]. Luminescence is sensitive and visible to human eyes, whereas mechanical stimuli exist and play vital roles in our daily lives, from machines to the human body, from cells to organelles. Thus, MC luminogens have drawn much attention due to their promising application in security papers, optical storage, mechanical sensors of local environment, and the human body.

Emission of luminogens depends heavily on their molecular structure and aggregate morphologies, which may be altered by external mechanical stimuli, thus giving out signals of luminescence change including color, intensity, efficiency, lifetime, and so on. Limited examples of MC luminogens based on alteration of molecular structure have been reported due to the incomplete and irreversible chemical reactions in the solid state. The MC luminescence of most reported luminogens was achieved through modulation of their morphologies by mechanical stimuli, though each system may have its own characteristics.

To the best of our knowledge, the first organic MC luminogen was reported by Gawinecki and coworkers in 1993, where the yellow-green luminescence of 4-tert-butyl-1-(4′-dimethylaminobenzylideneamino)pyridinium perchlorate in solid state was quenched when the solid was squeezed with a spatula, affording the first example of a turn-off type MC luminogen [1].

Aggregation-Induced Emission (AIE)
https://doi.org/10.1016/B978-0-12-824335-0.00006-4

The MC luminescence of a 1,3,6,8-tetraphenylpyrene derivative was reported by Sagara and Araki et al. in 2007. The blue emissive solid of the luminogen transforms to a yellowish solid with a strong green luminescence upon exposure to pressure. The tight hydrogen bonding interactions between molecules was destroyed by pressure, which further caused the mismatched packings of pyrene, thus inhibiting the formation of H-aggregates [2]. Crystalline powders of two cyano-substituted oligo(*p*-phenylene vinylene) derivatives were also reported to exhibit emission change from blue to yellow color, which is attributed to the excimer formation caused by pressure [3].

Later, a series of heteropolycyclic donor-acceptor π-conjugated luminogens were found to display MC luminescence [4–6]. Since the difluoroboron avobenzone was first reported to show MC luminescence by Fraser's group [7], the relationship between molecular structures and the MC luminescence of many structurally similar compounds has been intensely investigated [8–10]. Some luminophore-doped polymer composites were also found to exhibit MC luminescence [11–13].

Although the mechanisms for MC luminescence are somewhat clear, few MC luminogens had been reported before 2008, probably due to two reasons. First, there is still no clear design strategy for MC luminogens. Second, the emission of many luminogens is totally or partly quenched upon aggregate formation due to the aggregation-caused quenching (ACQ) effect or traps or defects in the aggregates. Thus, the development of MC luminogens becomes a daunting task.

Exactly opposite to ACQ, Tang et al. found that some propeller-shaped molecules which are nonemissive in solution, but are highly emissive in aggregate state [14]. This unusual phenomenon is termed as aggregation-induced emission (AIE). The mechanism of the AIE process is attributed to the restriction of intramolecular motion (RIM), and many AIE active luminogens (AIEgens) have been developed based on the RIM mechanism. In addition to the AIE phenomenon, Tang and coworkers also found that crystals of many AIEgens exhibit higher emission intensity and bluer emission color compared with their amorphous solids [15–17]. Later, this crystallization-included emission enhancement (CIEE) effect or morphology-dependent emission was observed in many AIEgens [18,19].

AIEgens are emissive in the solid state and generally possess twisted and flexible conformations, and hence they can readily form polymorphs, and afford more loosely packing structures than traditional luminogens with planar structures. This facilitates the transformation of AIEgens between different morphologies by heating, solvent fuming, and mechanical stimuli and so on [16,20]. The crystals of many AIEgens can be readily transformed to amorphous solids upon exposure to mechanical stimuli such as grinding, rubbing, pressing, etc., with emission color and intensity changing in the process. These AIEgens therefore exhibit MC luminescence. Many of these AIEgens exhibiting MC luminescence have been developed by Chi [21–25], Tian [26–28], Tang [29–35] et al. and their coworkers. The relation between molecular structure and MC luminescence properties was also studied. However, examples exhibiting high-performance MC luminescence, such as a high contrast ratio, are rather rare [30,36,37].

In this chapter, we will briefly describe the discovery of MC luminescence of AIEgens, disclose the relation between molecular structure and MC luminescence, highlight recent work of organic AIEgens with high-performance MC luminescence, and point out possible future directions of MC luminescent materials.

2 From AIE to MCL

The AIE effect is caused by the RIM process and AIEgens are held together by weak intermolecular interactions. If static pressure is applied to the amorphous solid of AIEgens, the distance between molecules of AIEgens will decrease, and intermolecular interactions will be strengthened. The emission intensity may therefore be increased by the further reinforced RIM effect. Tang, Zou, and coworkers investigated the optical properties of amorphous film of a typical AIEgen, hexaphenylsilole (HPS) [29]. The emission intensity of an amorphous film of HPS increases swiftly by 9% with increasing pressure (up to 104 atm) but starts to slowly decrease when the film is exposed to further increased pressure (Fig. 1). It should be noted that, even at a pressure of 550 atm, the PL intensity is still 5% higher

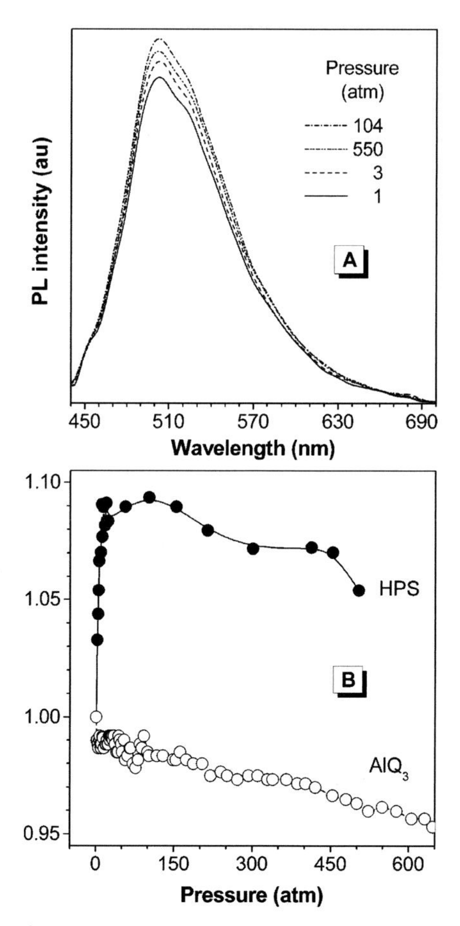

FIG. 1 (A) PL spectra of an HPS film under different pressures and (B) effects of pressure on the PL intensities of HPS and AlQ_3 films. *Reproduced with permission from X. Fan, J.L. Sun, F.Z. Wang, Z.Z. Chu, P. Wang, Y.Q. Dong, R.R. Hu, B.Z. Tang, D.C. Zou, Photoluminescence and electroluminescence of hexaphenylsilole are enhanced by pressurization in the solid state, Chem. Commun. (26) (2008) 2989–2991. Copyright 2008, The Royal Society of Chemistry.*

than that of the unpressurized film. The spectral profile of the film including its curve shape and peak position is practically unaffected by the application of external pressures, even at a pressure as high as 550 atm, indicating that the basic molecular structure of HPS is unaltered during the pressurization process. It is clear that the amorphous film of HPS is sensitive to the applied pressure, exhibiting a unique effect of MC luminescence. The result from a control experiment on a solid film of AlQ_3 shows that its PL intensity is slightly and monotonously weakened with increasing pressure. In addition to MC photoluminescence, the electroluminescence of the HPS device is also enhanced by pressurization.

Park and coworkers developed AIEgens of (2Z,2′Z)-2,2′-(1,4-phenylene)bis(3-(4-butoxyphenyl) acrylonitrile) (DBDCS), which is nearly nonemissive in solution with quantum yield (Φ_F) lowered to 0.26%, but exhibits a very high solid-state Φ_F of 62%. Solid DBDCS can form two different morphological phases, i.e., the metastable green-emitting G-phase and the thermodynamically stable blue-emitting B-phase. The B-phase crystal can be transformed to the G-phase by pressing, and the inverse process can be achieved by heating G-phase at 125°C for 5 min. The origin of the reversible luminescence switching is the two-directional shear-sliding capability of molecular sheets, which are formed via intermolecular multiple C—H⋯N and C—H⋯O hydrogen bonds. The two distinctive crystalline phases are promoted by different modes of local dipole coupling, which cause a substantial alternation of π–π overlap. A rewritable fluorescent optical recording film, which exhibited fast-responding and reversible multistimuli luminescence switching, was also successfully fabricated through the blending of luminogen with poly(methyl methacrylate) [38].

Almost within the same period, Chi [21–25], Tang, Dong, and their coworkers [27,29–34] found that crystals of many AIEgens can be amorphized upon exposure to mechanical stimuli, together with the morphology-dependent emission of AIEgens; many AIEgens exhibit MC luminescence, which is different from Park's mechanism. Later, based on the RIM mechanism of AIE, many AIEgens with MC luminogens have been developed.

3 MC luminogens with high contrast

3.1 Tetraphenylethylene derivatives

Tetraphenylethylene (TPE) is a star AIEgen as many excellent AIEgens have been developed from the TPE core as functional TPE derivatives can be readily synthesized. We select some TPE derivatives listed in Scheme 1.

Although many TPE cored CIEE luminogens are reported, the emission contrast between crystalline and amorphous states is still low. High CIEE contrast can be achieved through increasing the Φ_F of the crystalline state or decreasing that of the amorphous state. The Φ_F of the amorphous state can be decreased through construction of a looser molecular packing; for example, two luminogens may form two amorphous solids with different molecular packing patterns, and different degrees of disorder, which is affected by the molecular structure.

Dong and coworkers constructed TPE1 with low molecular symmetry [39]. The luminogen can form deep blue (420 nm Φ_F=50.4%, TPE1CA), sky blue (TPE1CB, 460 nm Φ_F=8.3%) emissive crystals, and a green emissive amorphous solid (TPE1Am, 496 nm Φ_F=7.6%) (Fig. 2). The Φ_F of TPE1CA is about six times higher than that of TPE1CB and TPE1Am, i.e., TPE1 exhibits CIEE effect with high contrast in both emission color and efficiency.

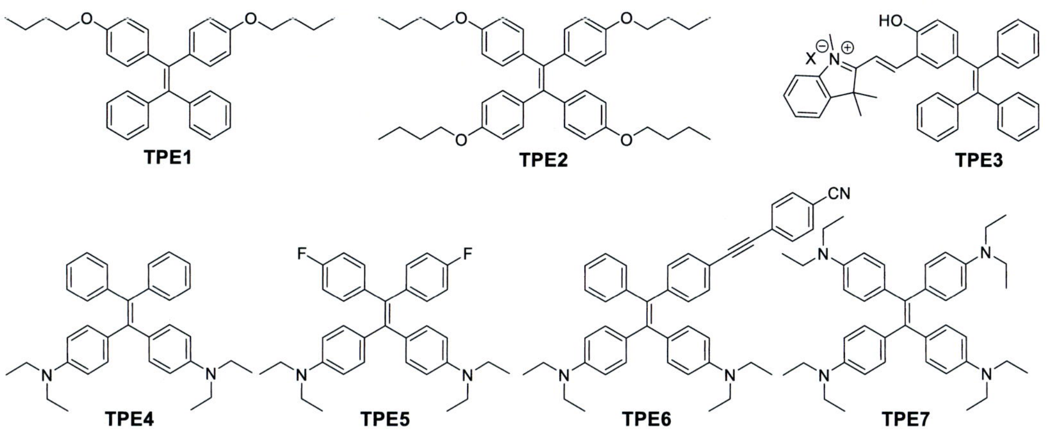

SCHEME 1 Chemical structure of TPE derivatives.

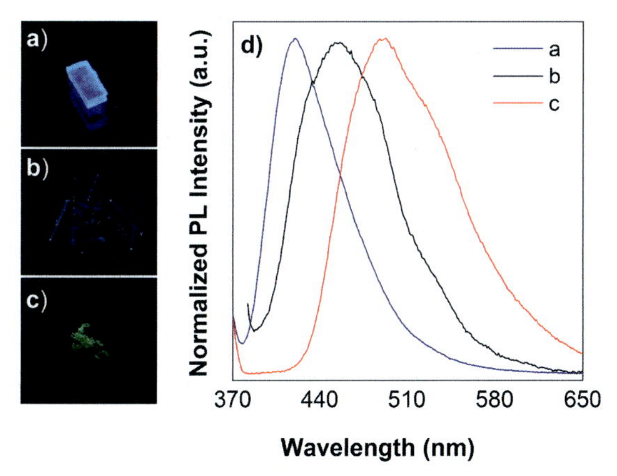

FIG. 2 Photos of (A) TPE1CA, (B) TPE1CB, (C) TPE1Am under 365nm UV illumination; (D) normalized PL spectra. Excitation wavelength: 350nm; exposure time: 1/25s. *Reproduced with permission from H. Tian, P. Wang, J. Liu, Y. Duan, Y.Q. Dong, Construction of a tetraphenylethene derivative exhibiting high contrast and multicolored emission switching, J. Mater. Chem. C 5 (48) (2017) 12785–12791. Copyright 2017, The Royal Society of Chemistry.*

The sharp contrast in both emission color and Φ_F of an AIEgen between two crystalline states may be the result of different molecular packing structures and conformations. This may be further implicated by CIEE mechanisms. Bond length alternation (BLA) was used to estimate the conjugation difference in TPE1CA and TPE1CB based on the exact molecular conformation in the crystals. BLA was calculated to be 0.14547 for TPE1CA, and 0.13924 for TPE1CB. The smaller BLA value for TPE1CB suggests better molecular coplanarity and conjugation, which coincide with its emission color, and thus, the red-shifted PL spectrum of amorphous solid may be induced by the more planar conformation.

In addition to the remarkable contrast in emission color among different morphologies of TPE1, the three morphologies of TPE1 also exhibit high contrast in Φ_F. The calculated density of TPE1CA (1.185 g/cm^3) is higher than that of TPE1CB (1.127 g/cm^3), suggesting tighter molecule packing in TPE1CA. The tighter packing pattern may further hinder the rotation and vibration of the phenyl rings, which will block the nonradiative transition and induce higher Φ_F. Similar to other TPE derivatives, there are no strong intermolecular interactions, such as π–π interaction or H/J-aggregation, in both TPE1CA and TPE1CB due to the twisted molecular conformation. In addition to the difference in density, there are more weak intermolecular interactions, such as C—H···O and C—H···π, in TPE1CA than in TPE1CB, which greatly hinder the rotation and vibration of all the phenyl rings in TPE1CA, blocking the nonradiative pathway, and resulting in high Φ_F. However, in the TPE1CB, all the phenyl rings of the molecule can rotate to some degree, which consumes more energy of the excited state, leading to low Φ_F.

The bright deep-blue emission of TPE1CA diminishes and the crystals transform to weakened green emissive powder during hard grinding due to the transformation of TPE1CA to amorphous powder; however, once the grinding stopped, green emission of the ground powder partially reverted back to deep-blue in 3 min at room temperature (about 30°C, Fig. 3). After 30 min, the emission of the ground powder completely turned to deep blue and intensified, due to further crystallization. Therefore, TPE1 exhibits high contrast and self-recovering MC luminescence.

TPE2 is also a CIEE-active crystal, which can emit either deep blue emission (446 nm with Φ_F of 54%) or sky-blue emission (460 nm with Φ_F of 48%). An amorphous solid of TPE2 obtained through quenching of its melt emits green fluorescence with Φ_F of 44%, which is slightly lower than its crystalline form, indicating that TPE2 is CIEE active with low contrast.

Through comparison of TPE1 and TPE2, it is clear that a luminogen with lower molecular symmetry exhibits high contrast CIEE effect due to the lower Φ_F of amorphous solid caused by the looser molecular packing in amorphous solid. In addition to TPE2 and TPE1, other examples have also been reported. TPE4-TPE6 are TPE derivatives with low molecular symmetry and all of them are weakly emissive in the amorphous state (on TLC plates, freshly prepared suspension, or ground powder). On the other hand, TPE4, TPE5, and TPE6 emit intense fluorescence with high Φ_F of 44.8%, 100%, and 17.5% in the crystalline state, respectively [40]. However, the similar derivative, TPE7, with high symmetry, is AIE active and exhibit low efficiency contrast between the amorphous and the crystalline state [41]. Thus, it is further clear that TPE derivatives with low molecular symmetry may afford higher contrast CIEE effect and high-performance MC luminescence. And the strategy may be further verified by more examples of TPE derivatives and other AIE active luminogens.

Liu and coworkers functionalized spiropyran with a TPE unit [42], to obtain luminogen TPE3 with low molecular symmetry, that exhibits a high-contrast CIEE effect. Microcrystals

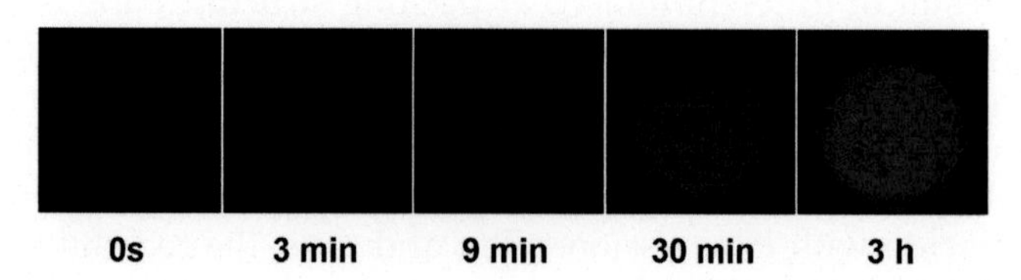

FIG. 3 Spontaneous recovery of TPE1CA at room temperature (30°C) after being ground in mortar. Photos are taken under 365 UV illumination at different times; exposed time: 1/8 s. *Reproduced with permission from H. Tian, P. Wang, J. Liu, Y. Duan, Y.Q. Dong, Construction of a tetraphenylethene derivative exhibiting high contrast and multicolored emission switching, J. Mater. Chem. C 5 (48) (2017) 12785–12791. Copyright 2017, The Royal Society of Chemistry.*

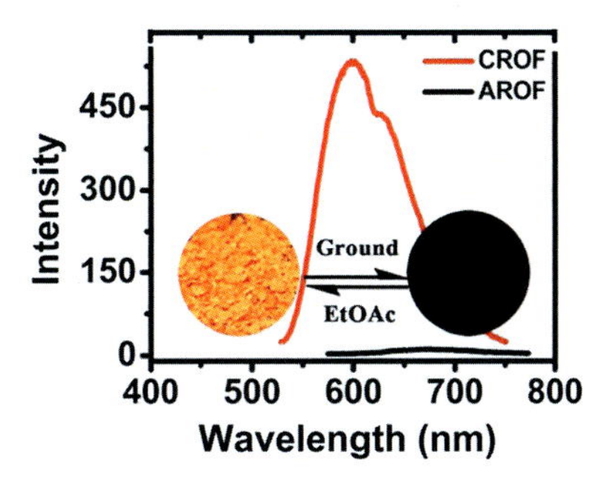

FIG. 4 Fluorescent images and PL spectra of TPE3 converting reversibly between crystal and amorphous in the solid states. *Modified with permission from X. Su, Y. Wang, X. Fang, Y.-M. Zhang, T. Zhang, M. Li, Y. Liu, T. Lin, S.X.-A. Zhang, A high contrast tri-state fluorescent switch: properties and applications, Chem. Asian J. 11 (22) (2016) 3205–3212. Copyright 2016, John Wiley and Sons.*

of TPE3 obtained by recrystallization exhibit orange emission (601 nm, $\Phi_F = 10.17\%$), while the amorphous solid of TPE3 is nearly nonemissive (678 nm, $\Phi_F < 0.01\%$) (Fig. 4). The HOMO of TPE3 distributes on the TPE unit, whereas the LUMO locates on the indolium moiety; thus, the orange emission of crystals of TPE3 arises from the intramolecular charge transfer (ICT) state. Although, the single crystal of TPE3 was not obtained, the authors found through the theoretical calculation that the optimized structure of TPE3 is a nearly planar conjugated system. They argue that molecules are interlocked by weak π–π interactions in the crystal structure of TPE3. The nonradiation transition is blocked and the π–π stacking is not strong enough to quench the fluorescence of ICT. If the crystal is crushed, the molecules will be more planar in conformation and the intermolecular distances will be smaller, which results in the enhancement of π–π interactions and dipole-dipole interactions; so, a bathochromic effect and quenching of the fluorescence appear. The red shift of the absorption spectra for the crystals (>100 nm) after grinding is consistent with the induction of a more planar conformation and enhancement of the π–π interactions in TPE3. The authors attribute the fluorescence quenching in the amorphous states of TPE3 to the change of π–π interaction and dipolar interaction. The high-contrast CIEE effect affords the high-contrast MC luminescence of TPE3; it also provides a possible design strategy for high-contrast MC luminogens: D-A structure with combination of twisted AIEgens as a donor and panel-like conjugation cation as an acceptor.

3.2 CIEE active luminogens with bulky conjugation core

A series of luminogens constructed with bulky conjugation core and peripheral aryl groups were found to exhibit high-contrast CIEE effect and MC luminescence. Herein, selected examples will be discussed to disclose the mechanism behind the phenomenon.

3.2.1 Dibenzofulvene (DBF) derivatives

A series of DBFs have been found to exhibit a distinct CIEE effect (Scheme 2). The freshly prepared suspension of DBF1 in a mixture of water and acetonitrile is nearly nonemissive due to its amorphous essence; however, the emission of the suspension is turned on 1 h later due to

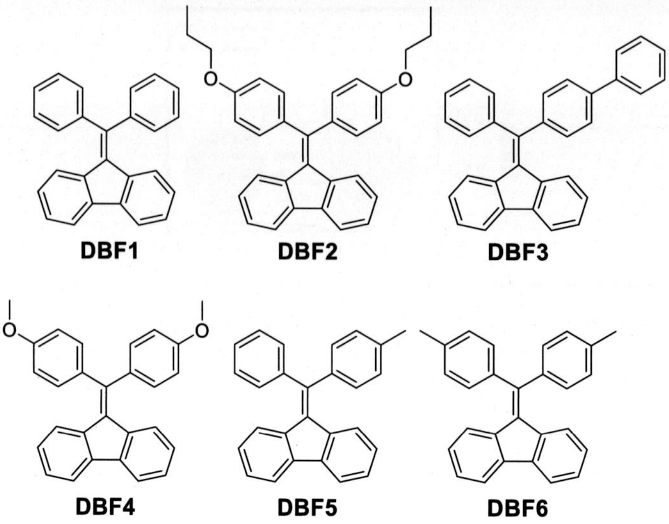

DBF1 DBF2 DBF3

DBF4 DBF5 DBF6

SCHEME 2 Structure of DBF derivatives.

the transformation of the suspension from amorphous to crystalline. Similar to TPE, DBF1 exhibits no change to shear stress as DBF1 can hardly be amorphized by mechanical stimuli.

To modulate the crystallization ability of DBF1, Tang and coworkers constructed DBF2 and DBF3. DBF2 can form two kinds of crystals of DBF2GC and DBF2YC, emitting at 500 and 545 nm with Φ_F of 82.1% and 56.2%, respectively. However, an amorphous solid of DBF2 (DBF2Am) exhibits a weak emission peaking at 580 nm with Φ_F lower than 1%, which is hardly detectable by integrating spheres. Thus, DBF2 is clearly CIEE active. DBF3 exhibits CIEE effect with highly emissive crystals (450 nm, $\Phi_F = 16\%$) and a nearly nonemissive amorphous solid (550 nm, $\Phi_F = 0.5\%$) [20]. Molecules in DBF2GC adopt a more twisted conformation than those in DBF2YC. Thus, a lower degree of conjugation of molecules in DBF2GC induces the bluer emission than that of DBF2YC. The higher emission efficiency of DBF2GC is attributed to more C—H···π and C—H···O intermolecular interactions in DBF2GC than those in DBF2YC [30].

The bright green emission of DBF2GC was found to be weakened and red-shifted upon grinding in a mortar (Fig. 5A, B) due to the transformation from crystalline to amorphous. However, after heating at 120°C for 1 min, the emission was recovered (Fig. 5C, E). The emission can be turned "off" again by grinding, and a simple pattern (Fig. 5D) can be obtained by grinding of selected areas. This reversible on-off process can be repeated many times, thus making the dye DBF2 a promising candidate for optical recording. DBF3 exhibits a similar MC fluorescence to DBF2. The emission of DBF3 turns "off" upon grinding and is recovered when heated, which is ascribed to the amorphization and crystallization of DBF3, respectively [30].

Emission of the ground DBF2 can be recovered spontaneously at room temperature. The ground solid of DBF3 is still weakly emissive after 1 day at room temperature, indicating that the amorphous solid of DBF3 is more stable than that of DBF2. Then, it is clear that stability or self-recovering ability of MC luminogens can be modulated by tuning the chemical structure [30].

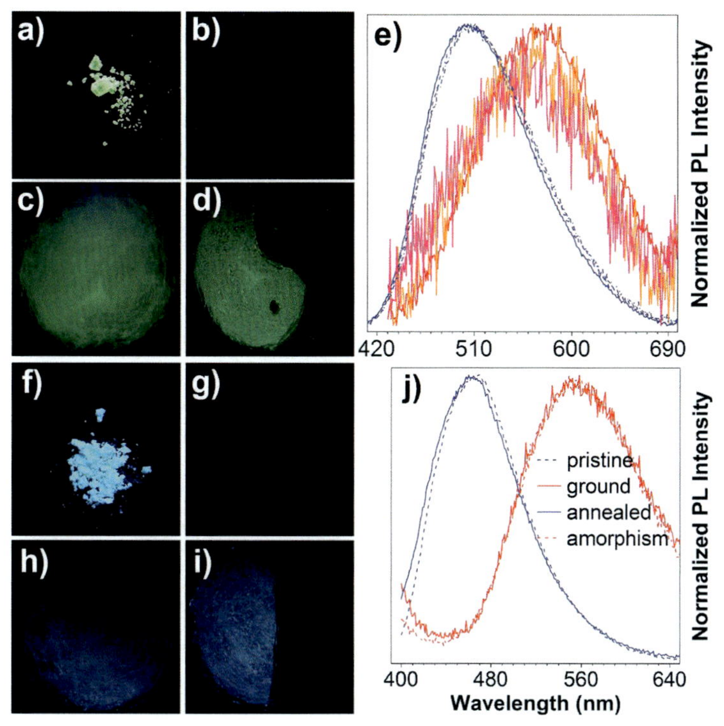

FIG. 5 Photographs of DBF2 (A–D) and DBF3 (F–I) before (A, F) and after grinding (B, G). Annealed DBF2 (C) and DBF3 (H) and regrinding of selected areas ((D) and (I), respectively). Photographs are taken under UV illumination. (E) Normalized PL spectra of DBF2 before grinding *(blue solid line)*, after grinding *(red lines)*, and annealing *(blue dashed and dash dot lines)* in the three repeating cycles. Normalized PL spectra of DBF3 are shown in panel (J). Annealing details: 120°C and 160°C for 1 min for DBF2 and DBF3, respectively; excitation wavelength: 370 nm. *Reproduced with permission From X.L. Luo, J.N. Li, C.H. Li, L.P. Heng, Y.Q. Dong, Z.P. Liu, Z.S. Bo, B.Z. Tang, Reversible switching of the emission of diphenyldibenzofulvenes by thermal and mechanical stimuli, Adv. Mater. 23 (29) (2011) 3261–3265. Copyright 2011, John Wiley and Sons.*

In addition to DBF2 and DBF3, many other DBF derivatives were also reported to exhibit high-contrast CIEE and MC luminescence, which seems to be a general property of DBFs [37,43–45].

3.2.2 9-([1,1′-Biphenyl]-4-ylphenylmethylene)-9H-xanthene

To develop more MC luminogens with high contrast, Dong and coworkers compared the CIEE effect of TPE8 and DBF3 (Scheme 3). They found that the cores of TPE8 and DBF3 are a double bond and dibenzofulvene, while the CIEE contrast (Φ_C/Φ_{Am}) of TPE8 and DBF3 are 2.7 and 32, respectively. Therefore, tiny differences in the chemical structure may greatly influence the CIEE effect of luminogens, and the combination of the bulky conjugation core and peripheral phenyl rings may afford luminogens with a high-contrast CIEE effect. The replacement of the dibenzofulvene core with 9-methylenexanthene, both of which are large conjugation cores, afford 9-([1,1′-biphenyl]-4-ylphenylmethylene)-9H-xanthene (BPPX). BPPX can form deep blue (BPPXBC, 432 nm) and green (BPPXGC 492 nm) emissive crystals with

	TPE8	DBF3	BPPX
Φ_{Am}(%)	12.7	0.5	0.4
Φ_{C}(%)	34.4	17.0	42.2 or 59.3
Φ_{C}/Φ_{Am}	2.7	34	105 or 148

SCHEME 3 Structure and emission efficiency of TPE8, DBF3, and BPPX.

Φ_F of 42.2% and 59.3%, respectively. Similar to DBF3, BPPX exhibit a high-contrast CIEE effect with a value of Φ_C/Φ_{Am} up to about 148. As aforementioned, different luminogens may form different amorphous solids with varied molecular packing density. For amorphous TPE8, the phenyl rings may adjust the torsion angle in response to the microenvironment, and form denser molecular packing. However, for DBF3 and BPPX, the bulky core cannot rotate as freely as the phenyl rings in TPE8, and may form more loosely packed amorphous solid, in which the phenyl rings can still rotate to some degree and quench the emission. Thus, the combination of a bulky core and peripheral phenyl rings may be a plausible design strategy for high-contrast CIEE luminogens [46].

The emission of both BPPXBC and BPPXGC are greatly weakened and red-shifted to dark orange upon grinding due to the amorphization process. Similar to BPPXAm, the ground powder can revert back to BPPXBC upon fuming with methanol vapor, and to BPPXGC upon annealing at 120°C or fuming with acetone vapor. Thus, the "dark" and "bright" process is reversible, and the emission of BPPX could be repeatedly switched between bright deep blue and dark orange, or between bright green and dark orange for many times through repeating grind and fuming or heating. Similar to BPPXBC obtained from fuming BPPXAm with methanol, BPPXBC obtained from fuming the ground solid with methanol, can also transform to BPPXGC upon heating. Then, it is clear that BPPX exhibit multicolored MC luminescence with high contrast in both emission color and efficiency (Fig. 6).

3.2.3 *Dicyanomethylenated acridones [47]*

A series of dicyanomethylenated acridones (DCNAC) with bulky conjugation cores exhibit AIE and high-contrast CIEE effect (Scheme 4). All the DCNAC derivatives are nearly nonemissive in amorphous state obtained by grinding the crystals. However, their crystals exhibit intense luminescence (Fig. 7) with different emission colors, i.e., all the four DCNACs exhibit high-contrast MC luminescence. The redder emission of DCNAC1C (647 nm, $\Phi_F = 5\%$) and DCNAC4C (707 nm, $\Phi_F = 16\%$) are caused by excimer and ICT processes, respectively. The π–π stacking and excimeric coupling are significantly weakened for luminogens with longer alkyl chains, resulting in the blue-shifted emission and enhanced Φ_F of DCNAC2C (572 nm, $\Phi_F = 30\%$) and DCNAC3C (562 nm, $\Phi_F = 41\%$) crystals.

To disclose the origin of the high-contrast CIEE, the author established a three-dimensional molecular arrangement and interaction image of amorphous solid of DCNAC1

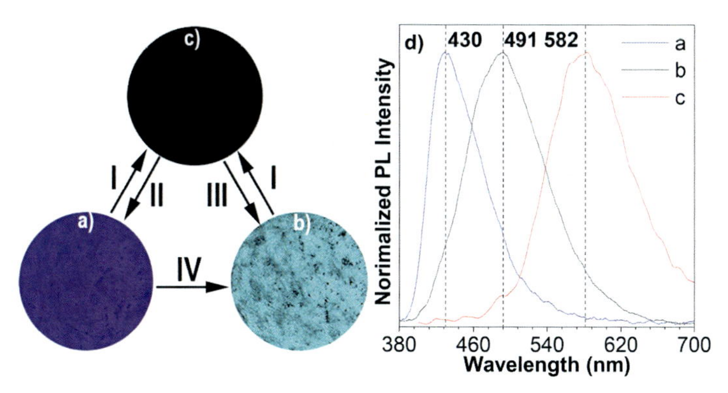

FIG. 6 Switching emission of BPPX among three different states through morphology tuning: (A, B) ground powder fumed by (A) methanol and (B) acetone vapor, and (C) ground powder. (D) Normalized PL spectra of samples a, b, and c. (Excitation wavelength: 360nm). Photos were taken under UV illumination. Conditions: (I) grinding; (II) fuming with methanol vapor, 3h; (III) fuming with acetone vapor, 5min; (IV) 160°C, annealing for 60min. *Reproduced with permission from Z. Zhao, T. Chen, S. Jiang, Z. Liu, D. Fang, Y.Q. Dong, The construction of a multicolored mechanochromic luminogen with high contrast through the combination of a large conjugation core and peripheral phenyl rings, J. Mater. Chem. C 4 (21) (2016) 4800–4804. Copyright 2016, The Royal Society of Chemistry.*

using the molecular mechanics force field by fitting quantum mechanical calculations. The simulation studies demonstrated that DCNAC1 molecules adopt a random packing without hydrogen bonding and π–π stacking in the amorphous phase. These intermolecular interactions play a key role in restricting the conformational transformation of the molecules. Therefore, the nonemissive properties of the amorphous solid are attributed to the excited-state quenching induced by torsional vibrations. However, in crystal lattices, these torsional vibrations can be restricted by multiple intermolecular interactions and thus the emission is turned on.

3.2.4 *Bis(diarylmethylene)dihydroanthracene*

Luminogens with bulky conjugation cores such as anthracene or derivatives of bis(diphenyl-methylene)-dihydroanthracene (R-DHA), are also found to be AIE and CIEE active, and exhibit high-contrast MC luminescence (Scheme 5) [48].

As aforementioned, in addition to the DBF derivatives, derivatives of DCNAC, BPPX, and DHA with large conjugation core do exhibit a high-contrast CIEE effect. In amorphous solid,

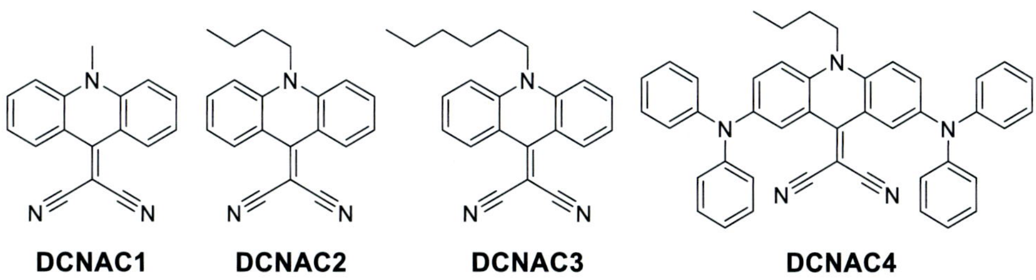

DCNAC1 **DCNAC2** **DCNAC3** **DCNAC4**

SCHEME 4 Molecular structures of the DCNAC derivatives.

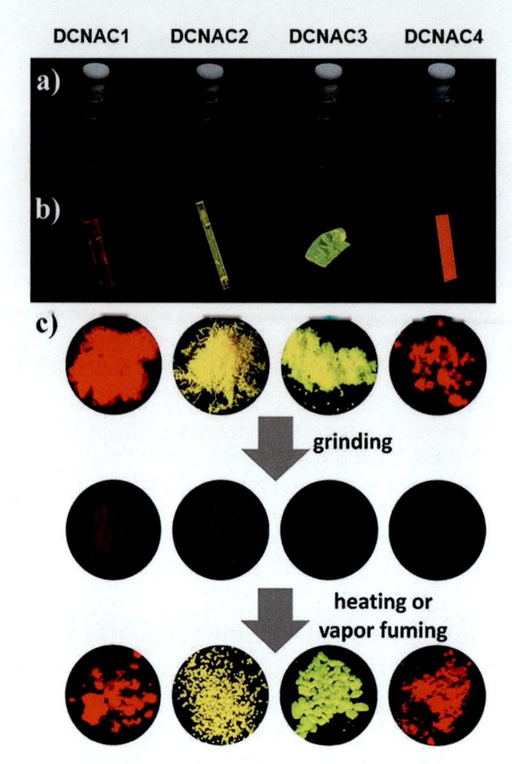

FIG. 7 Fluorescent photographs of the DCNAC derivatives (A) in solution (B) and in the crystalline state under 365nm light irradiation. (C) Fluorescent images of the pristine crystalline, ground, and ground/heated solids under 365nm light irradiation. *Modified with permission from W. Chen, S. Wang, G. Yang, S. Chen, K. Ye, Z. Hu, Z. Zhang, Y. Wang, Dicyanomethylenated acridone based crystals: torsional vibration confinement induced emission with supramolecular structure dependent and stimuli responsive characteristics, J. Phys. Chem. C 120 (1) (2016) 587–597. Copyright (2016) American Chemical Society.*

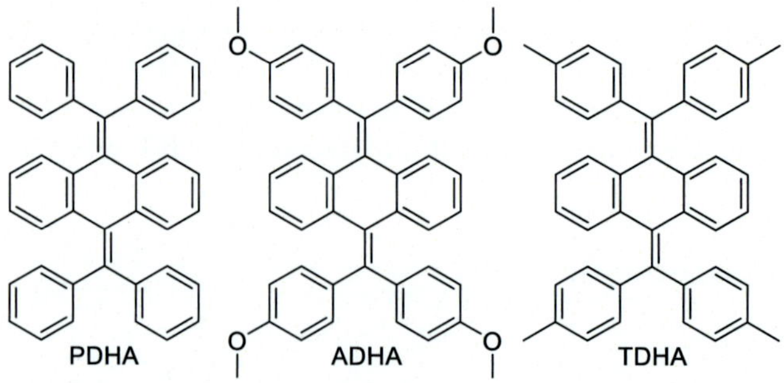

SCHEME 5 Molecular structures of the R-DHA derivatives.

the large conjugation core may facilitate looser packing patterns in which some groups such as phenyl rings may still have enough free space to rotate, hence quenching emission. In the crystalline phase, phenyl rings are locked by the weak intermolecular interactions, nonradiative pathways are blocked, and emission is recovered. Thus, the large conjugation core with peripheral phenyl rings may be a possible design strategy for high-contrast CIEE and MC luminogens.

3.3 Other CIEE luminogens

3.3.1 Diphenyl maleimide derivatives

A benzamide-based diphenyl maleimide (BADPMA) derivative exhibits a high-contrast CIEE effect (Fig. 8) [49]. Its amorphous films are nearly nonemissive with Φ_F lower than 0.1%, while its crystalline powder and single crystals are highly green emissive with Φ_F of 53% and 80%, respectively. The single crystal of BADPMA suggests that there are no specific strong intermolecular interactions (such as π–π stacking or H/J-aggregates) due to their twisted conformations. The twisted conformation molecules are arranged into molecular columns through weak N—H···O (2.08 Å) interactions and the molecular columns are held together by weak C—H···π (3.44 Å) and C—H···O (2.53 and 2.59 Å) interactions to form a 2D and 3D structure, respectively. These weak intermolecular interactions fix the molecular conformation of BADPMA in the crystals, and the rigidification of the twisted conformation of BADPMA inhibits the internal rotations and blocks the nonradioactive relaxation, thus contributing to the high Φ_F. For amorphous solid, through theoretical calculation, it is suggested that the benzamide group of the molecules in the excited states can freely rotate, accompanied by the formation of an intramolecular hydrogen bond between N—H and C=O, which quenches the emission.

The high-contrast CIEE effect affords the high-performance MC luminescence of BADPMA. The bright emissive crystalline powders transformed to nearly nonemissive amorphous solid after being ground for 10 min in the quartz plate. The emission of ground powder is turned on upon fuming with organic solvent or heating, and the off-on process can be repeated many times.

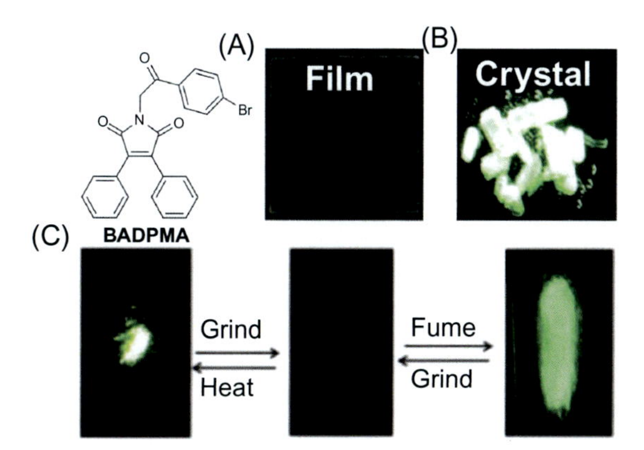

FIG. 8 Molecular structures and different aggregates of BADPMA, (A) film, (B) single crystal, and (C) powders with different treatment taken under UV light. *Modified with permission from R. Zheng, X. Mei, Z. Lin, Y. Zhao, H. Yao, W. Lv, Q. Ling, Strong CIE activity, multi-stimuli-responsive fluorescence and data storage application of new diphenyl maleimide derivatives, J. Mater. Chem. C 3 (39) (2015) 10242–10248. Copyright 2015, The Royal Society of Chemistry.*

3.3.2 2-Aminobenzophenone derivatives

Although many CIEE luminogens exhibiting high-contrast MC luminescence have been reported, most of them exhibit fluorescence turn-off response to mechanical stimuli, as their emissive crystals transform to weakly or nearly nonemissive amorphous solid. Comparatively, turn-on mode MC luminogens exhibit high sensitivity and low background noise and such MC materials are therefore highly desired. Tong and coworkers have developed a group of CIEE-active luminogens that exhibit a high-contrast turn-on mode response to mechanical stimuli [50].

Three 2-aminobenzophenone derivatives ABP1, ABP2 and ABP3 (Scheme 6) were found to exhibit a CIEE effect. Crystals of ABP1, ABP2, and ABP3 emit bright cyan, blue, and green light, with Φ_F of 11.74%, 2.30%, and 6.97%, respectively. However, their amorphous film is nearly nonemissive with efficiency of 0.07%, 0.02%, 0.03%, respectively, indicating that all the three ABPs are CIEE-active with high contrast. ABP2 have been reported to exhibit crystallization-induced phosphorescence at room temperature by Tang's group [51], while Tong and coworkers found that the lifetime of all the three ABPs falls in the range of 1–2 ns, indicating the emission of ABPs may be fluorescence arising from a crystalline phase rather than phosphorescence. The high luminescence efficiency of crystals of ABPs is attributed to the locking of conformation of ABP molecules in crystalline phases by the strong intramolecular N—H⋯O (d=2.05 Å) and intermolecular N—H⋯O (d=2.17 Å) interaction (Fig. 9).

The ethanol solution of ABPs was smeared on glass to form nonemissive amorphous films after evaporation of ethanol. Instead of crushing, grinding, shearing, and pressing to amorphize the crystals of luminogens, scratching is another way to transform amorphous to crystalline. As expected, the emissions of all the three ABPs are turned on after the transformation from amorphous to crystalline state. Interestingly, benzophenones without the amine group did not show the CIEE characteristics under scratching, indicating that amine groups may be essential for the intramolecular and intermolecular hydrogen bonds thus providing a possible way to achieve turn-on mode mechanical sensors with high-contrast CIEE luminogens.

3.4 Turn-on mode MC luminogens

MC AIEgens are generally brightly emissive in solid states, especially their crystalline state due to the CIEE effect, and thus most of them normally exhibit turn-off mode MC luminescence. However, researchers have developed some AIEgens that exhibit crystallization-caused quenching effect, which is useful for achieving turn-on mode MC luminescence.

SCHEME 6 Molecular structures of the ABP derivatives.

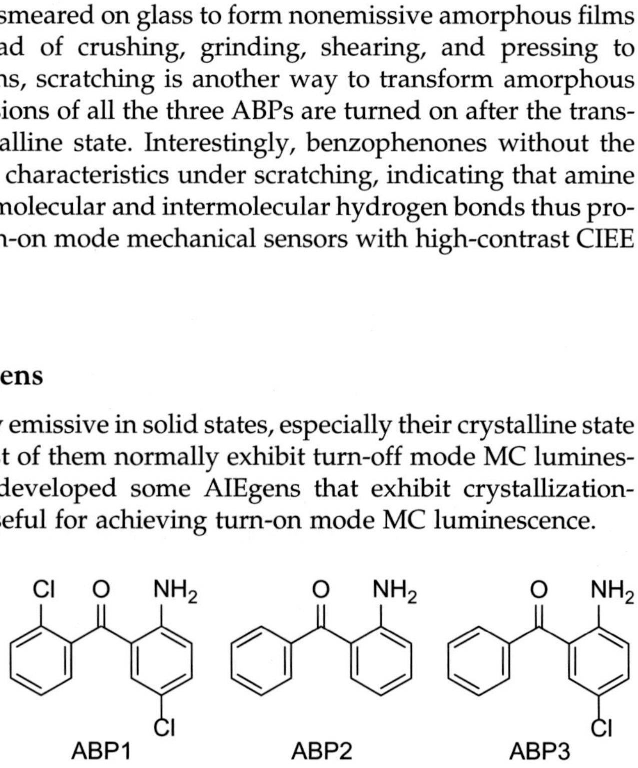

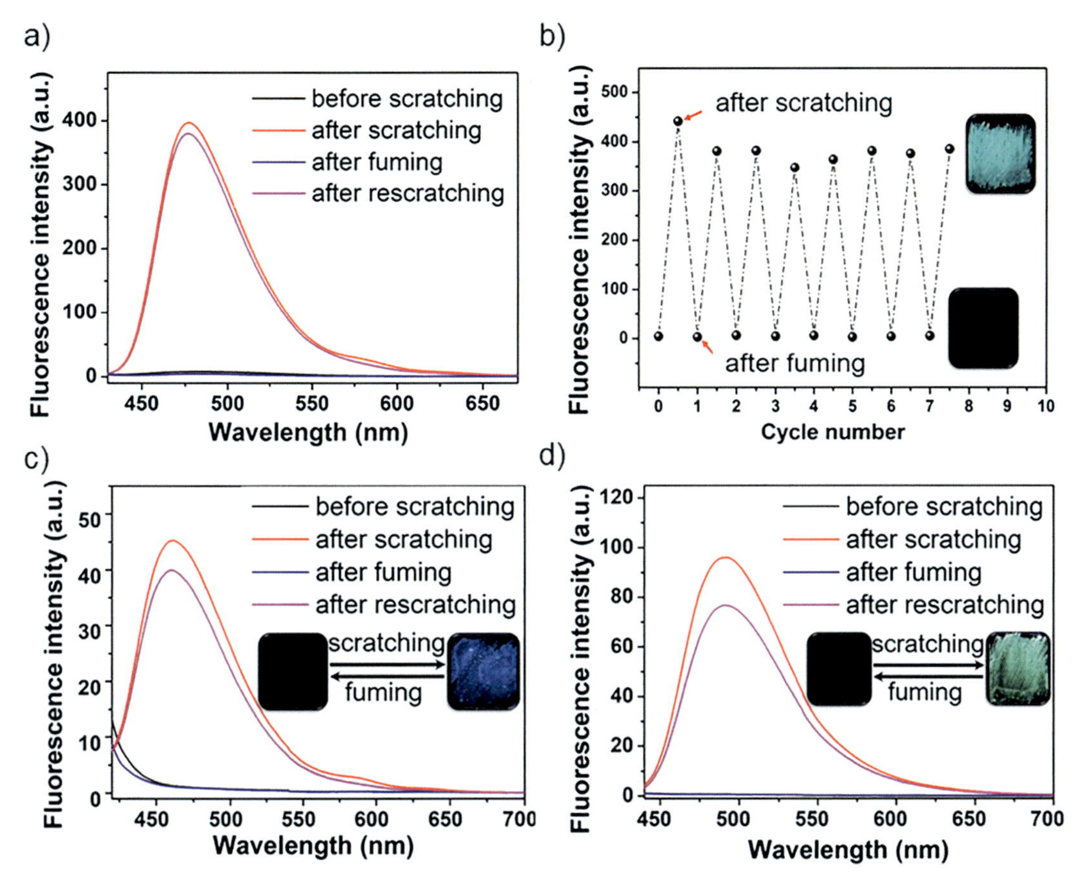

FIG. 9 (A) Fluorescence spectra and (B) fluorescence "off-on" cycles of ABP1 with scratching-fuming stimulation. (C and D) The corresponding fluorescence spectra of ABP2 and ABP3. Insets: images were taken under a 365 nm handheld UV lamp. *Modified with permission from X. Zheng, Y. Zheng, L. Peng, Y. Xiang, A. Tong, Mechanoresponsive fluorescence of 2-aminobenzophenone derivatives based on amorphous phase to crystalline transformation with high "off on" contrast ratio, J. Phys. Chem. C 121 (39) (2017) 21610–21615. Copyright 2017, American Chemical Society.*

3.4.1 TPE derivatives

Tian and coworkers have constructed AD-TPE that consists of a TPE unit as an electron donor and an acridonyl group as an electron acceptor (Fig. 10A) [52]. The crystalline powder of AD-TPE is weakly emissive with Φ_F of 1%, but generates extremely bright cyan emission with a Φ_F of 46% upon grinding, exhibiting turn-on MC luminescence (Fig. 10B). Moreover, the emission of the ground powders can return to the initial "dark-state" upon thermal annealing or fuming with organic solvents (Fig. 10C).

The molecular conformation and packing patterns of AD-TPE obtained from its single-crystal XRD is helpful in understanding the underlying mechanism of the emission quenching in the crystalline phase as well as the remarkable turn-on and color-tuned MC luminescence. The large distance between the adjacent molecules ruled out close π–π stacking in the crystals. Besides, the highly twisted TPE units afford the loose molecular packing, which

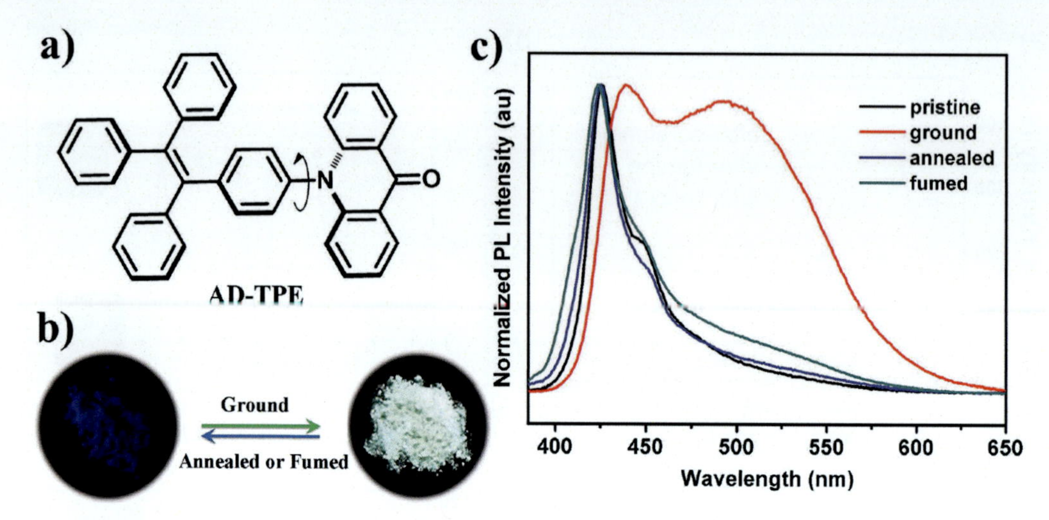

FIG. 10 (A) Structural formula of AD-TPE. (B) Fluorescence images and (C) corresponding fluorescence spectra of AD-TPE under different treatments. *Modified with permission from Q. Qi, J. Qian, X. Tan, J. Zhang, L. Wang, B. Xu, B. Zou, W. Tian, Remarkable turn-on and color-tuned piezochromic luminescence: mechanically switching intramolecular charge transfer in molecular crystals, Adv. Funct. Mater. 25 (26) (2015) 4005–4010. Copyright 2016, John Wiley and Sons.*

may be easily disrupted by external mechanical stimulus. The dipole moment (7.27 Debye) is pointed from the acridonyl group to the TPE group, and such strong polarization facilitates the coupling between the central TPE unit of the upper sheet and the acridonyl unit of the lower sheet, yielding antiparallel dimers with side-by-side alignment in the crystal lattice. Such an arrangement will promote the formation of H-type aggregation, which implies the forbidden exciton (0-0) transition from the second lowest excited state (S_2) to the ground state (S_0). In conclusion, the weak emission of the crystals can be attributed to the formation of H-aggregates via strong dipole-dipole interactions, which is further supported by density functional theory (DFT) calculations.

Bai et al. constructed CaDPE1 and CaDPE2 through embedding two of the phenyl rings of TPE in the calix[4]arenes skeleton [53]. The crystals of CaDPE1 emit blue light under UV light irradiation, but crystals of CaDPE2 are nearly nonemissive. By simply grinding the crystals of CaDPE2, a powder with blue emission was obtained (Fig. 11). It is clear that CaDPE2 exhibits turn-on mode MC luminescence. To study the origin of the turn-on mode MC luminescence of CaDPE2, the single crystals of both CaDPE1 and CaDPE2 were analyzed. In one unit cell, two CaDPE1 molecules adopt partial cone conformation with distinctive distortion, while two CaDPE2 molecules adopt cone conformation for the strong intramolecular hydrogen bonding in the bottom rim. The dihedral angles between the phenyl rings and the C=C bond in CaDPE2 are 88.556, 82.569, 84.298, and 25.601 degrees, which means three phenyl rings are almost perpendicular to the C=C bond and only one phenyl ring adopts a coplanar configuration relative to the C=C. TPE in such an abnormal condition is unlikely to emit light due to the destruction of cross-chromophore π conjugation. In compound CaDPE1, the dihedral angles of four phenyl rings range from 41.073 to 58.028 degrees, which are significantly smaller than those in CaDPE2. This result provides a direct evidence that the dihedral angles between

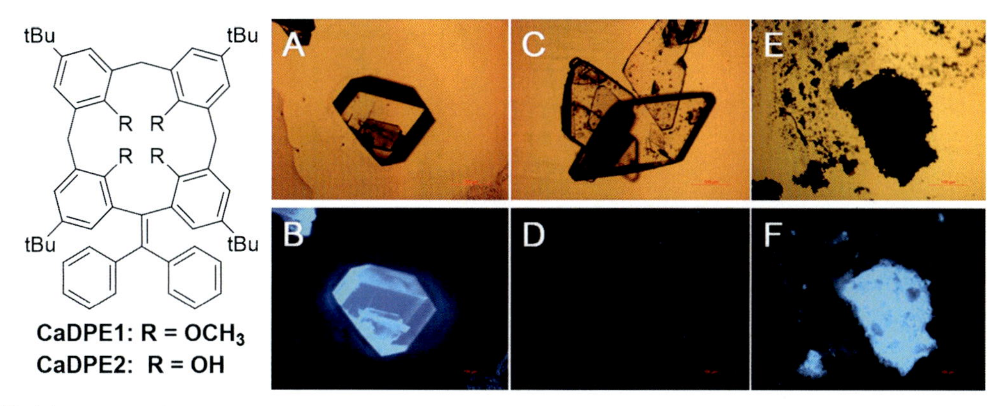

FIG. 11 Images of CaDPE1 (A and B), CaDPE2 (C and D) crystals and ground powder of CaDPE2 (E and F) (A, C, E under bright field; B, D, F under UV-light irradiation; wavelength: 330–380 nm). *Modified with permission from B. Han, X. Wang, Y. Gao, M. Bai, Constructing a nonfluorescent conformation of AIEgen: a tetraphenylethene embedded in the calix[4]arene's skeleton, Chem. Eur. J. 22 (45) (2016) 16037–16041. Copyright 2016, John Wiley and Sons.*

the phenyl rings and C=C could control the emission of TPE by altering the degree of conjugation. Perpendicular geometry may be destroyed and further be changed into a more stable conformation with a dihedral angle similar to pure TPE in response to mechanical force. Thus, CaDPE2 exhibits turn-on mode MC luminescence.

3.4.2 *Turn-on mode MC luminogens based on intersystem crossing*

A design strategy for constructing high-performance turn-on mode MC luminogens was proposed based on the control of intersystem crossing (ISC).

According to the perturbation theory, the decay rate of ISC, k_{ISC}, is influenced by the spin-orbit coupling constant (ξ_{ST}) and the energy gap (ΔE_{ST}) between involved singlet and triplet states, expressed as follows:

$$k_{ISC} \propto \frac{{\xi_{ST}}^2}{e^{\Delta {E_{ST}}^2}}$$

The nitrophenyl group with abundant lone pair electrons can boost the efficient ISC pathway with the aid of a great ξ_{ST} and a negligible ΔE_{ST}, where the spin-orbit interaction mixes the two states differing in both spin and electronic configurations (Fig. 12B) [54–57]. It is well known that ξ_{ST} and ΔE_{ST} are closely related to molecular conformations and electronic configurations and thus are highly sensitive to the surrounding environments such as the solid-state morphology. Therefore, it is promising to design new on-off MC luminogens by controlling the ISC process of nitrophenyl-substituted luminogens. Furthermore, the morphology of organic molecules with twisted conformations can be easily modulated by mechanical stimuli. Crystals of TPA-1N, TPA-2N, TPE-3N, and TPE-4N are nearly nonemissive with ultralow quantum yields, which indicates that nitrophenyl groups successfully open the nonradiative ISC channel and then quench the emission. ISC becomes the dominant decay pathway as nitrophenyl groups are efficient triplet state promoters (Fig. 12B). For comparison, crystals of TPE-1N and TPE-2N show intense emission with the Φ_F as high as 20.24% and 24.35%,

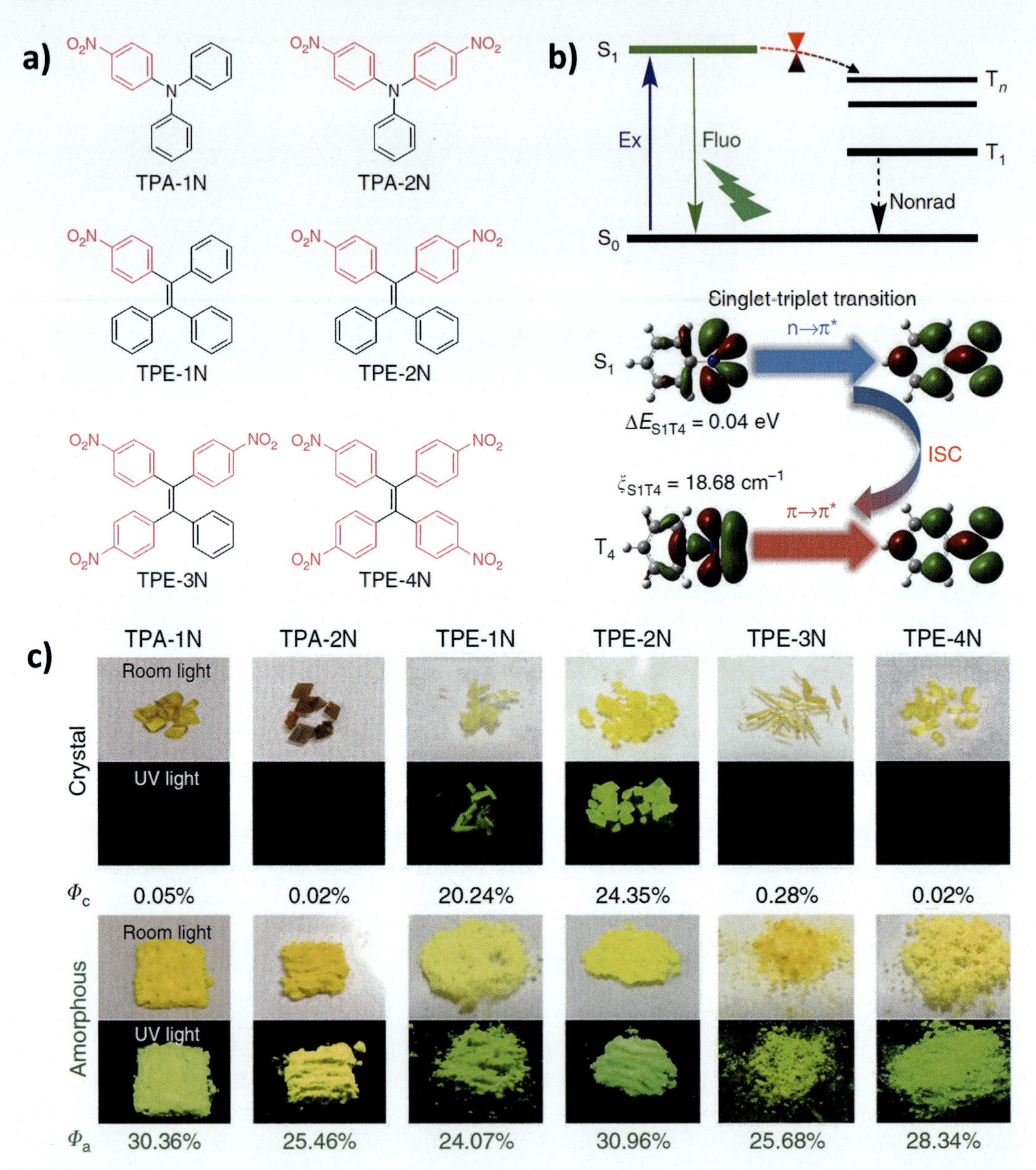

FIG. 12 (A)Molecular structures of nitro-TPAs and nitro-TPEs, (B) Nature transition orbitals of the lowest singlet excited state (S_1) and triplet excited state (T_4) of nitrobenzene. Since El-Sayed's rule states that the multiplicity change becomes highly efficient when the spin-orbit coupling mixes two states differing in both spin and electronic configuration, T_4 is the closet and a good "receiver state" in the intersystem crossing process which determines the ultrafast S_1 depletion through an favored $^1(n, \pi^*)$ to $^3(\pi, \pi^*)$ channel. (C) Photos of luminogens in crystalline and amorphous states taken under room and UV light irradiation with their quantum yields. Luminescence contrast ratio: $\alpha_\Phi = \Phi_a/\Phi_c$. *From W. Zhao, Z. He, Q. Peng, J.W.Y. Lam, H. Ma, Z. Qiu, Y. Chen, Z. Zhao, Z. Shuai, Y. Dong, B.Z. Tang, Highly sensitive switching of solid-state luminescence by controlling intersystem crossing, Nat. Commun. 9 (1) (2018) 3044. Copyright 2018, The Authors, published by Springer Nature.*

respectively, possibly because the number of nitrophenyl groups is not enough to overwhelm the AIE effect. Interestingly, in the amorphous state, all the six luminogens are found to be emissive exhibiting bright green or yellow fluorescence in thin films and in aggregates. The phenomena suggest that TPA-1N, TPA-2N, TPE-3N, and TPE-4N exhibit the proposed turn-on mode MC with a high luminescence contrast ratio. Meanwhile, the number of nitrophenyl groups introduced is also a critical factor in constructing the turn-on mode MC luminogens. The proposed design strategy thus provides a big step to expand the scope of the turn-on mode MC luminogens family.

3.4.3 *Schiff base derivatives*

A diaminomaleonitrile-functionalized Schiff base (A3MN) exhibiting turn-on MC luminescence was developed by Tang and coworkers [34]. The emission spectrum of the crystals of A3MN is nearly a flat line parallel to the abscissa with a low Φ_F of 3.77% (Fig. 13A).

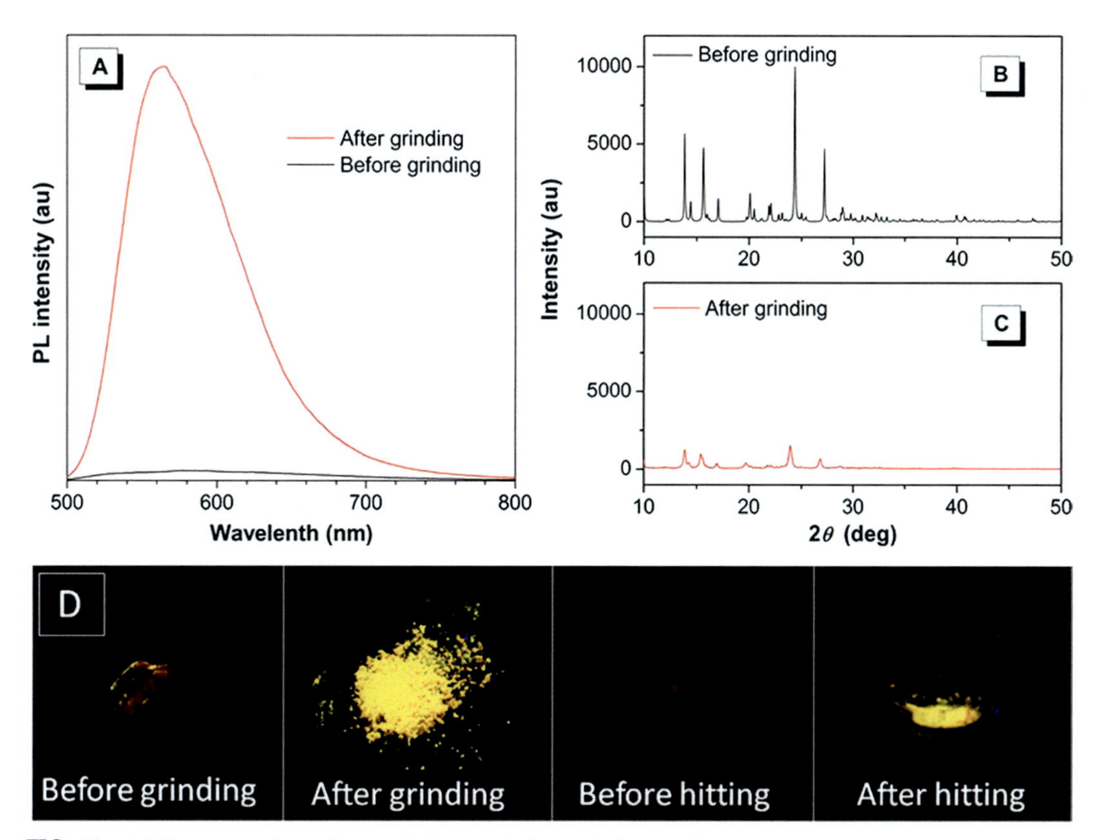

FIG. 13 (A) PL spectra of a single crystal of A3MN before and after grinding. Excitation wavelength: 445 nm. (B and C) XRD diffractograms of (B) the intact single crystal and (C) ground crystal. (D) Photograph of a single crystal of A3MN before and after grinding/hitting taken under a 365 nm UV light from a handheld UV lamp. *Modified with permission from T.Y. Han, Y.N. Hong, N. Xie, S.J. Chen, N. Zhao, E.G. Zhao, J.W.Y. Lam, H.H.Y. Sung, Y.P. Dong, B. Tong, B.-Z. Tang, Defect-sensitive crystals based on diaminomaleonitrile-functionalized Schiff base with aggregation-enhanced emission, J. Mater. Chem. C 1 (44) (2013) 7314–7320. Copyright 2013, The Royal Society of Chemistry.*

However, the crystals of A3MN turned to bright emissive with a large hypochromatic shift and a high Φ_F of 48.8% upon grinding. PXRD reflection peaks of A3MN still remain after grinding, indicating that it has partly maintained its crystalline state (Fig. 13B and C). The change of the fluorescence from dark red to bright yellow can be clearly detected by naked eyes under a 365 nm UV light (Fig. 13D), suggesting a high contrast ratio.

Molecules in the single crystal of A3MN adopt a planar conformation and form π-dimer in the crystal lattice (Fig. 14C), in which the adjacent molecules align in an antiparallel D-A coupling mode with the electron-rich diethylamine moiety (D unit) pointing to the electron-deficient maleonitrile moiety (A unit). The intermolecular distance between the coupled molecules of the π-dimer is 3.358 Å, which is within the range of intermolecular π–π interaction. Such D-A coupling further promotes a lower energy conformation as well as the decay of the excited molecules through nonradiative pathways. Thus, the turn-on MC luminescence of A3MN is ascribed to the formation of crystal defects, in which D-A coupling would be destructed to activate the radiative pathway. A3MN is sensitive to various kinds of

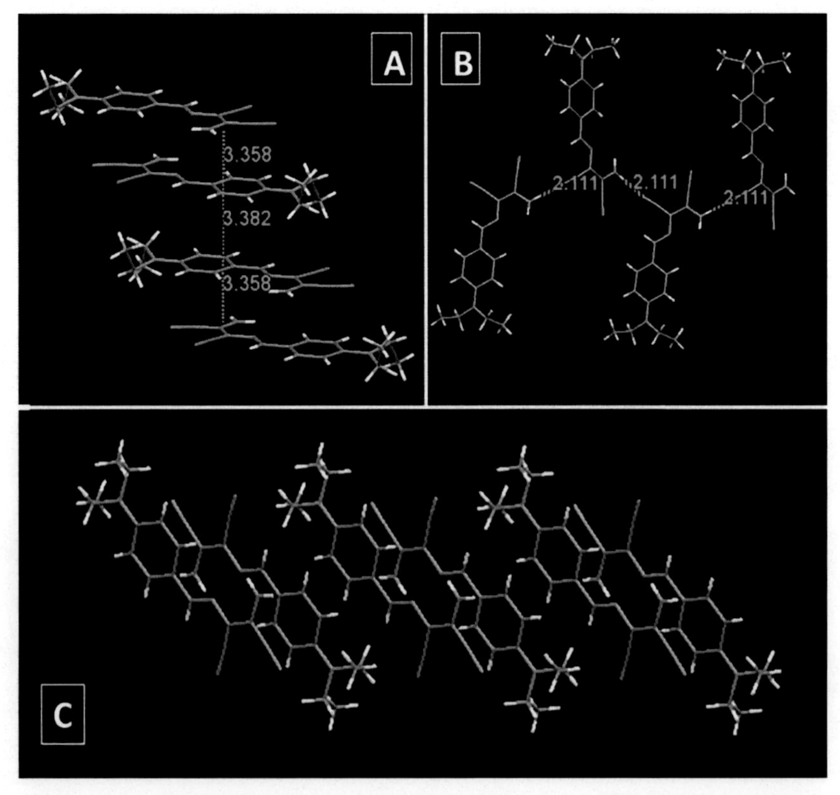

FIG. 14 Molecular packing showing (A) π–π interactions and (B) hydrogen bonding in the single crystal lattice of 26. (C) The top view of the molecular packing. *Modified with permission From T.Y. Han, Y.N. Hong, N. Xie, S.J. Chen, N. Zhao, E.G. Zhao, J.W.Y. Lam, H.H.Y. Sung, Y.P. Dong, B. Tong, B.Z. Tang, Defect-sensitive crystals based on diaminomaleonitrile-functionalized Schiff base with aggregation-enhanced emission, J. Mater. Chem. C 1 (44) (2013) 7314–7320. Copyright 2013, The Royal Society of Chemistry.*

mechanical actions, such as friction, hitting, sculpture, and ultrasonic vibration, which demonstrates the possibility of using A3MN for the analysis of mechanical force in various applications. For example, pressure monitoring experiments are performed by using a needle to apply pressure on the crystal surface of A3MN. The fluorescence intensity is proportional to the applied force, yielding the detection limit as low as 0.1 Newton owing to its turn on nature.

Murugesapandian and coworkers have developed two turn-on mode MC luminogens. The crystalline powders of both LH_3 [58] and SP [59] are weakly emissive with efficiency of 0.92% and 1.2%, while the emissions of LH_3 and SP are turned on upon grinding, with emission efficiencies increased to 13.54% and 18.5%, due to the amorphization of both luminogens (Fig. 15).

Xiang and coworkers reported a salicylaldehyde-based tri-Schiff bases (TSB) with a nonconjugated trimethylamine bridge. The pristine crystalline solids of TSB are weakly emissive, but after grinding, the resulting powders emit strong green emission under UV light. Fast evaporation of dichloromethane by reduced pressure distillation would bring about the same emissive solid. The powder wide-angle XRD curves of both the pristine crystal, ground, and solids obtained through fast evaporation of dichloromethane have intense and sharp reflection peaks, which are identical to the simulated pattern of a single crystal of TSB, indicating that all the solid samples are crystalline. Thus, the mechanism of MC luminescence of TSB is different from other AIEgens that exhibit the phase transition from crystalline to amorphous in response to mechanical stimuli. The amorphous TSB shows very weak fluorescence, and thus, the bright emission of the ground sample may be induced by

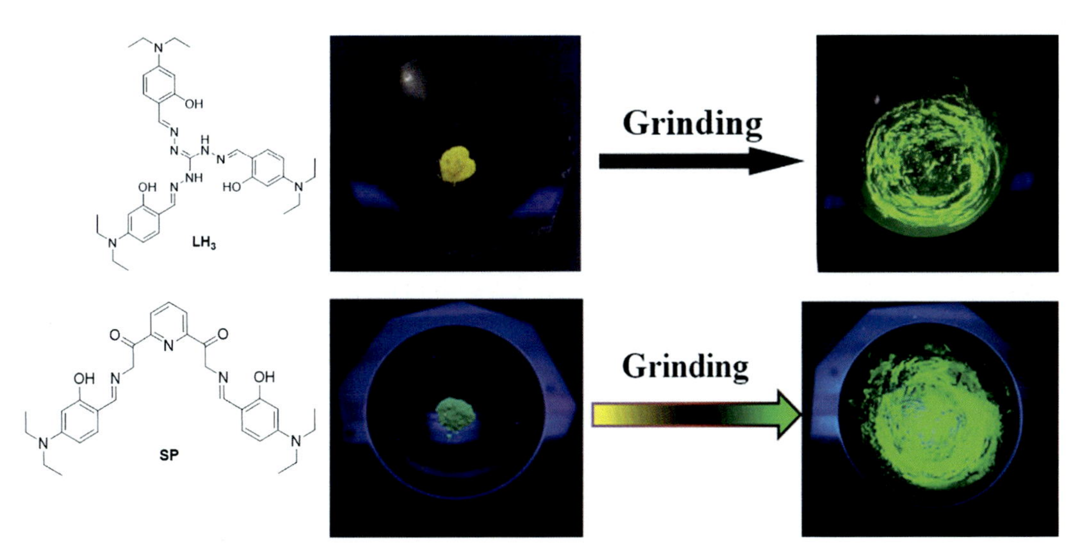

FIG. 15 Fluorescent photographs of pristine and ground sample of LH_3 and SP under a UV lamp. *LH_3 Modified with permission from B. Tharmalingam, M. Mathivanan, G. Dhamodiran, K.S. Mani, M. Paranjothy, B. Murugesapandian, Star-shaped ESIPT-active mechanoresponsive luminescent AIEgen and its on-off-on emissive response to Cu^{2+}/S^{2}, ACS Omega 4 (7) (2019) 12459–12469. https://doi.org/10.1021/acsomega.9b00845. Copyright 2019, American Chemical Society. SP Modified with permission from K. Santhiya, S.K. Sen, R. Natarajan, R. Shankar, B. Murugesapandian, D-A-D structured bis-acylhydrazone exhibiting aggregation-induced emission, mechanochromic luminescence, and Al(III) detection, J. Org. Chem. 83 (18) (2018) 10770–10775. Copyright 2018, American Chemical Society.*

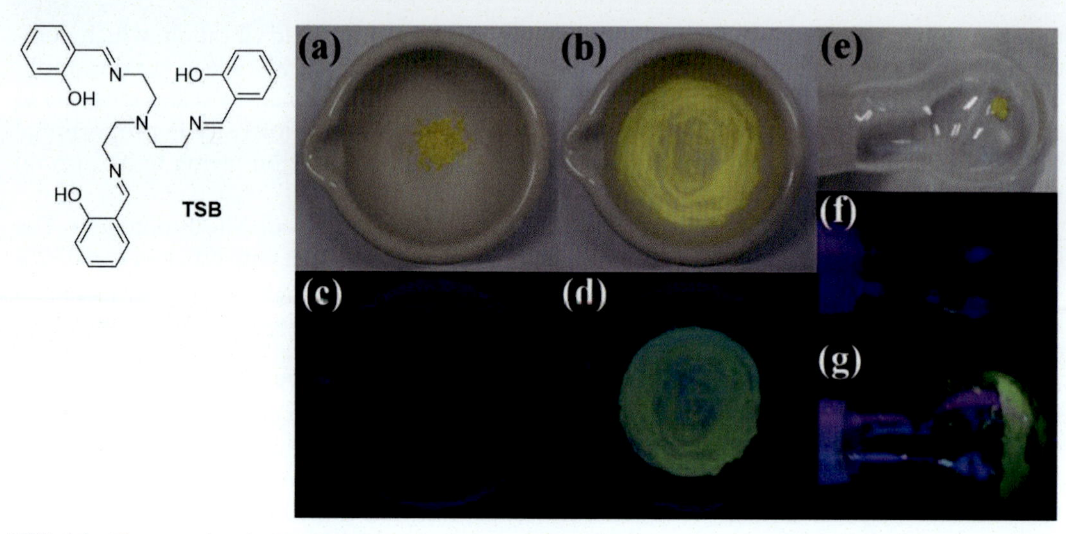

FIG. 16 Photographs of TSB solids ((A, C, E, and F) crystals; (B and D) ground solids; (G) solid obtained through fast evaporation of dichloromethane) under sunlight or a 360nm UV light. *Modified with permission from X. Zhang, J. Shi, G. Shen, F. Gou, J. Cheng, X. Zhou, H. Xiang, Non-conjugated fluorescent molecular cages of salicylaldehyde-based tri-Schiff bases: AIE, enantiomers, mechanochromism, anion hosts/probes, and cell imaging properties, Mater. Chem. Front. 1 (6) (2017) 1041–1050. Copyright 2017, The Royal Society of Chemistry.*

crystal defects in the ground solids, which is similar to the example of A3MN. Moreover, TSB in both poly(methyl methacrylate) casting film and frozen solvent also display strong fluorescence (Fig. 16) [60].

3.4.4 A-π-D-π-A fluorene derivatives

Li and coworkers developed a group of A-π-D-π-A fluorene derivatives with 1, 3-indandione acceptor but with different alkyl substituents in the 9, 9-position of the donor fluorenyl group (b-DIPF, o-DIPF, and d-DIPF). Through quick evaporation of dichloromethane under vacuum, o-DIPF grew into metastable nearly nonemissive 0D particles with emission efficiency of 0.12%. The emission of 0D particles is turned on upon mechanical grinding with efficiency of 19.6% due to the transformation of 0D particles from crystalline to amorphous. o-DIPF thus exhibit high contrast turn-on mode MC luminescence. The quenched emission of 0D particles is attributed to the π–π stacking of the aromatic rings, which is disrupted upon grinding, causing the emission to be turned on. Similar to o-DIPF, b-DIPF, and d-DIPF also display turn-on mode MC luminescence but with a lower contrast than o-DIPF (Fig. 17) [61].

3.5 MC luminescence in response to static pressure

In addition to the aforementioned mechanical stimuli including grinding, crushing, scratching, and so on, isotropic hydrostatic pressure from a diamond anvil cell (DAC) is also a useful tool to investigate structure-property relationships of AIE MC luminogens. By gradually applying relatively large pressure to molecular materials, gradual changes in emission color and intensity are observed, and the emission may be recovered to the original state after

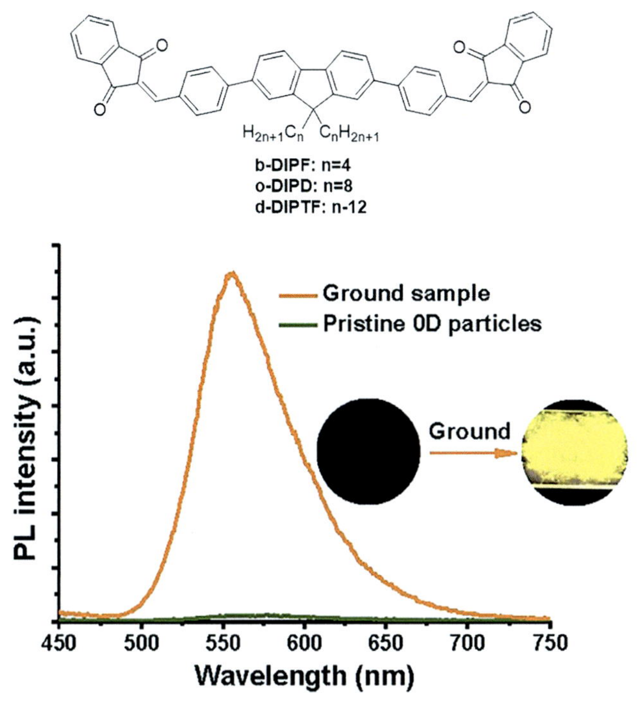

FIG. 17 PL spectra of the pristine 0D particles and the ground sample of o-DIPF. Insert: Fluorescence images of emission turn-on change of 0D particles of o-DIPF under mechanical force grinding. *Modified with permission from F. Zhang, R. Zhang, X. Liang, K. Guo, Z.; Han, X. Lu, J. Xie, J. Li, D. Li, X. Tian, 1,3-Indanedione functionalized fluorene luminophores: negative solvatochromism, nanostructure-morphology determined AIE and mechanoresponsive luminescence turn-on, Dyes Pigments 155 (2018), 225–232. Copyright 2018, Elsevier B.V.*

releasing the applied pressure, which is different from the aforementioned observations for the typical MC luminogens. Most luminescent materials show a gradually red-shifted and quenched emission as pressure increases. This is a consequence of the formation of a low-energy emission species and a nonradiative "dark" state (e.g., narrow bandgap excimer). Although the amorphous film of HPS have been reported to exhibit intensified emission under static pressure, many AIEgens crystals also exhibit red-shifted and weakened emission under increasing isotropic hydrostatic pressure. However, two TPE derivatives behave differently from the normal AIEgens.

As aforementioned, crystals of AD-TPE exhibit turn-on mode MC luminescence upon grinding. Tian and coworkers also found that the luminescence of AD-TPE single crystal can be switched on under hydrostatic pressure [52]. Hydrostatic pressure up to 3.00 GPa is applied by using a sapphire anvil cell technique, where silicone oil is used as a pressure-transmitting medium. With increase of pressure, the colorless and transparent single crystals gradually transferred to pale yellow and translucent, whereas the weak emission gradually intensified and turned bright cyan. The images of the single crystal under different hydrostatic pressures, as well as the absorption and emission spectra, are shown in Fig. 18. From Fig. 18B, C, it is clear that the change of emission follows an obvious [52] two-step process with the increase of the hydrostatic pressure. Initially, when pressure increased from 1 atm to 1.01 GPa, both the visible images and absorption spectra of the crystal had no obvious change (Fig. 18A, D), but the fluorescence images (Fig. 18B) became bright with no emission color change. From the emission spectra of the crystal (Fig. 18C), there was no new emission band, but the luminescence intensity gradually enhanced. With further increase of the

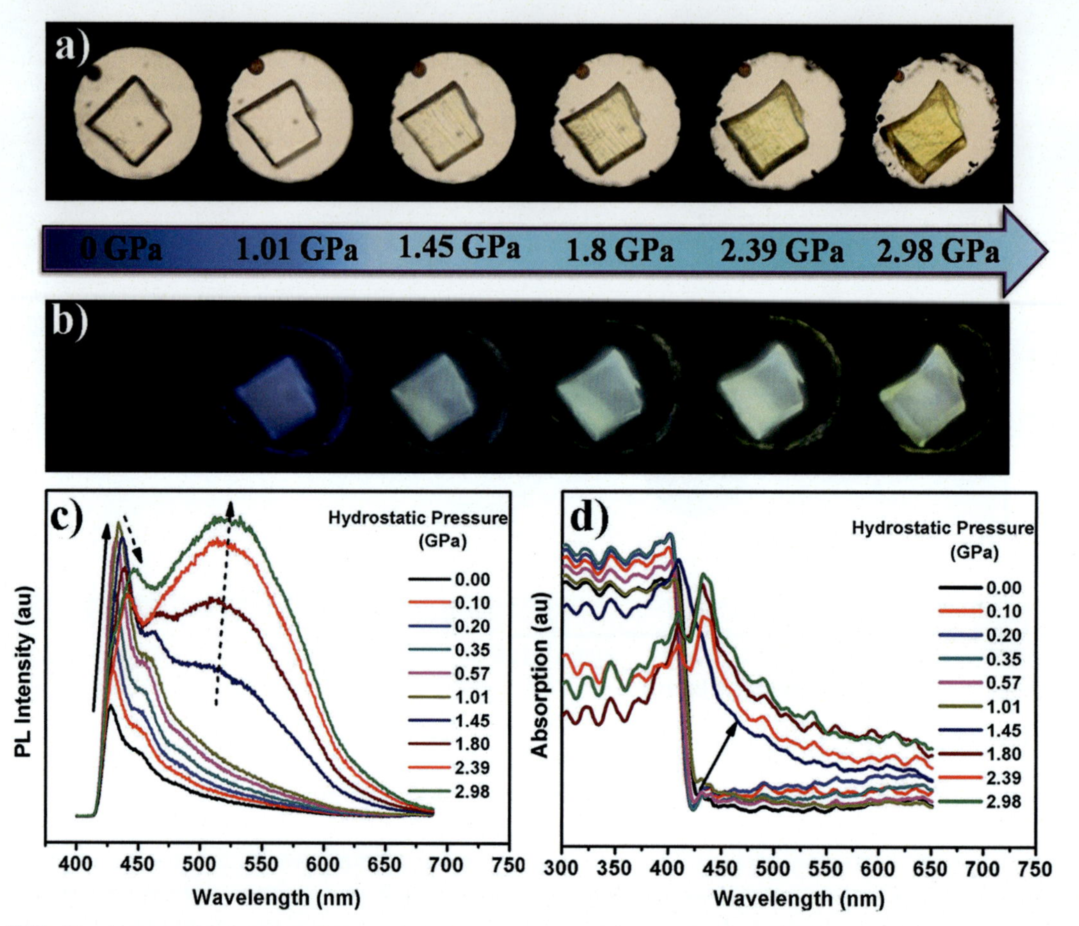

FIG. 18 (A, B) Visible and fluorescence images of a single crystal of AD-TPE under different hydrostatic pressures. (C, D) Corresponding fluorescence and absorption spectra. *Modified with permission from Q. Qi, J. Qian, X. Tan, J. Zhang, L. Wang, B. Xu, B. Zou, W. Tian, Remarkable turn-on and color-tuned piezochromic luminescence: mechanically switching intramolecular charge transfer in molecular crystals, Adv. Funct. Mater. 25 (26) (2015), 4005–4010. Copyright 2016, John Wiley and Sons.*

applied pressure, a red-shifted absorption and a new broad emission band at long wavelength regions concomitantly appeared. Eventually, the AD-TPE crystal showed a bright cyan emission at 2.98 GPa. The observed changes of both the emission intensity and wavelength of single crystal under hydrostatic pressure are similar to those of powders under grinding, implying that this unique MC luminescence of AD-TPE under different mechanical stimuli may have the same origin.

Wang and coworkers found that crystals of 9-(3-(1,2,2-triphenylvinyl)phenyl)-anthracene (mTPE-AN) exhibit pressure-induced blue-shifted and enhanced emission [62]. The crystal of mTPE-AN exhibit high tolerance to shearing stress, and no change was observed for its emission and XRD pattern even after grinding for 60 min, indicating strong intermolecular interactions. However, when mTPE-AN crystal was exposed to isotropic hydrostatic pressure directly exerted via a DAC, three-step variations were observed (Fig. 19). As pressure

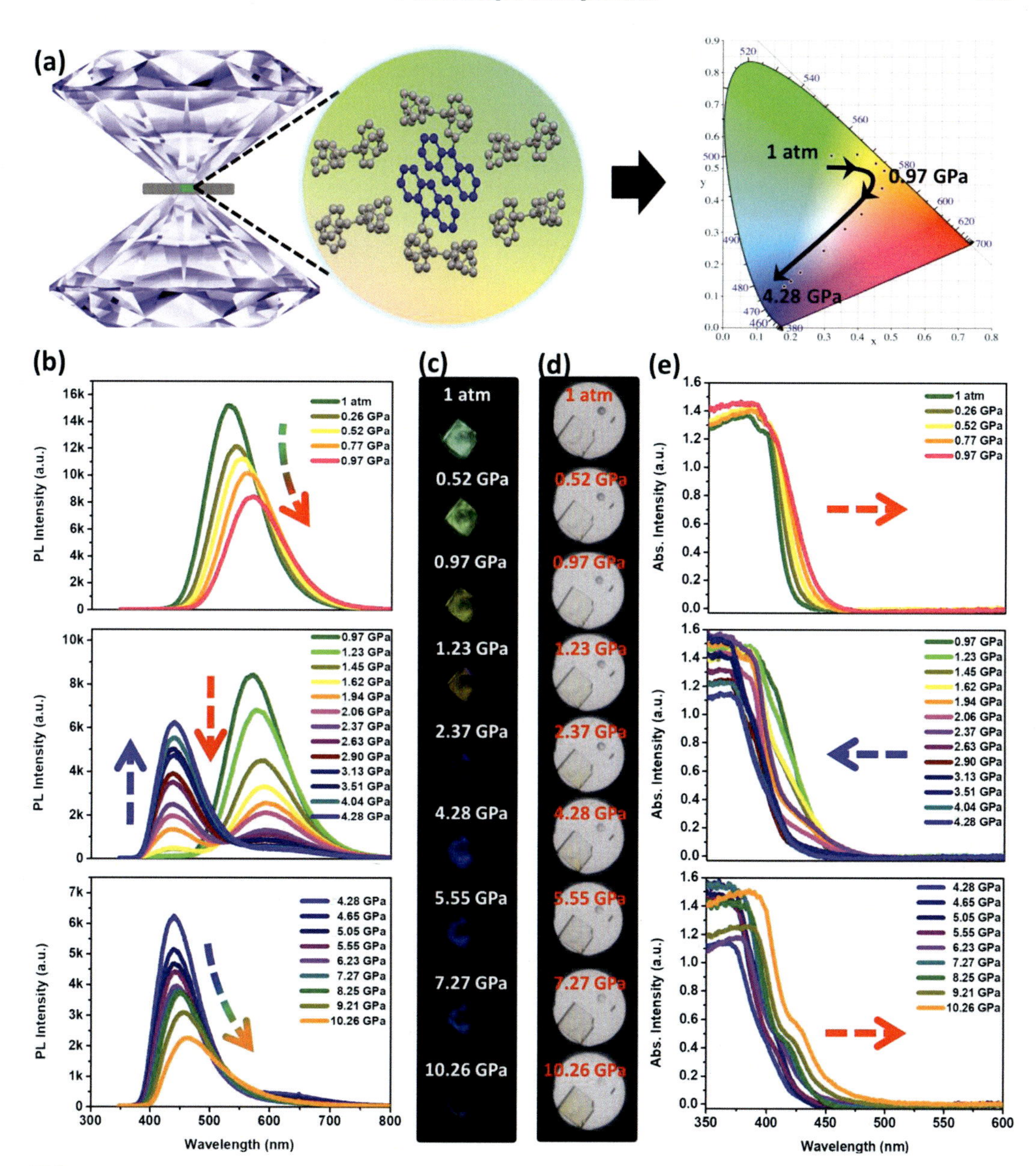

FIG. 19 (A) Schematic illustration of the DAC apparatus and molecular packing. (B) Emission spectra, (C) fluorescent photographs, (D) visible photographs, and (E) absorption spectra of mTPE-AN crystal under pressure from 1 atm to 10.26 GPa. *Modified with permission from H. Liu, Y. Gu, Y. Dai, K. Wang, S. Zhang, G. Chen, B. Zou, B. Yang, Pressure-induced blue-shifted and enhanced emission: a cooperative effect between aggregation-induced emission and energy-transfer suppression, J. Am. Chem. Soc. 142 (3) (2020) 1153–1158. Copyright 2020, American Chemical Society.*

increased from 1 atm to 1.23 GPa, the emission color of the mTPE-AN crystal changed from green to yellow, together with the decreased emission intensity. Slight red-shifting of absorption onsets was also recorded, which is similar to normal luminogens' response to isotropic hydrostatic pressure. Once pressure increased above 1.23 GPa, the emission at around 438 nm emerged, gradually increased, and reached its maximum intensity at 4.28 GPa, while a yellow emission band kept decreasing, giving rise to pure blue emission. The absorption onset also showed a blue-shifted trend until 4.28 GPa. With further compression of over 4.28 GPa, high- and low-energy emission bands demonstrated a red-shift and an intensity decline with red shift of absorption onset [62].

Theoretical and experimental investigations demonstrated that the blue-shifted emission band originates from the TPE units in the mTPE-AN crystal. The suppressed energy transfer from TPE to AN dimer induces the appearance of a blue-shifted absorption band as pressure increases from 1.23 to 4.28 GPa, and the AIE mechanism of TPE units contributes to the enhancement of the blue-shifted emission band. This work reports a novel principle for a new class of blue-shifted and enhanced MC luminescent materials formed by using the combination between energy transfer suppression and AIE activation. This study also presents an ideal model to improve the understanding of high-lying excited-state emission in fundamental photophysics.

3.6 Potential applications

3.6.1 Micro-embossing fluorescent patterns and haptic sensor

The application of TPE-4N was investigated due to its high contrast and sensitive turn-on mode MC luminescence. The rewriteable optical information storage system was described in Fig. 20A. A nonemissive film was prepared facilely through thermal annealing of the spin-coated film. After performing a finger pressure using a designed mold to the annealed film, a high-resolution micro-embossing fluorescent pattern can be easily replicated with a width of 10 μm and a spacing of 10 μm. It should be the first MC luminescent patterns with micrometer resolution. Also, the micro-embossing patterns are erased completely and film becomes rewriteable upon thermo-treatment.

TPE-4N films can also be fabricated on aluminum, ceramic, and wooden substrates using a simple brush coating process. As shown in Fig. 20B, bright fluorescence is observed on freshly brush-coated films. The fluorescence is switched off completely by thermal annealing. Upon pressing with thumb and fingers, the touched area turns on immediately and well-defined fingerprint patterns are observed by the naked eyes. In detail, the pressure is gradually enhanced from 0 to 0.98 MPa. In order to visualize monitoring, the fluorescent signal is digitized using image gray-scale processing. As shown in Fig. 20D, when the applied finger pressure reaches 0.15 MPa, fluorescent fingerprint signals can be detected with a 3.5-fold increase. 0.25 MPa results in a noticeable signal with 7.7-fold increase in fluorescence intensity. The signal intensity reaches its maximum value (15-fold) after 0.55 MPa. Also, thermal treatment can recover the film to the nonemissive state making the films reusable. The results suggest a promising fast responsive and reversible haptic sensor [63].

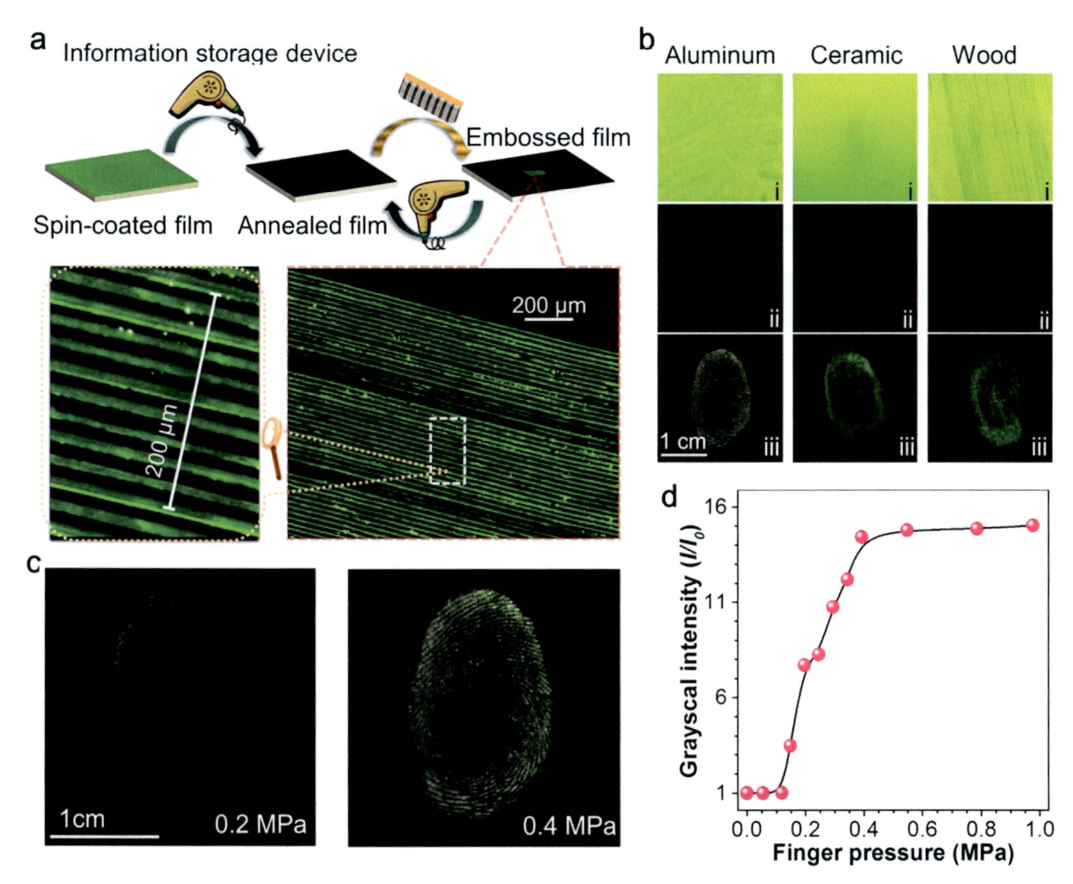

FIG. 20 Optical information storage and haptic sensor of TPE-4N thin film. (A) Procedures of the micro-embossing and recovery on the thin film of TPE-4N prepared by spin coating on a quartz plate and luminescent photos of micro-embossed patterns. Process I: heated by a handed heat gun at 150°C for 3 s; Process II: embossed with a mold with a width of 10 μm and a spacing of 10 μm. (B) Haptic photos of fingerprints on aluminum, ceramic, and wooden substrates coated with TPE-4N, (i) freshly brush-coated film; (ii) annealed film; (iii) pressed with finger. (C and D) Haptic sensor: luminescent photos C and gray-scale intensity changed with the increasing finger pressure. All luminescent photos were taken under UV irradiation at 365 nm. *From W. Zhao, Z. He, Q. Peng, J.W.Y. Lam, H. Ma, Z. Qiu, Y. Chen, Z. Zhao, Z. Shuai, Y. Dong, B.Z. Tang, Highly sensitive switching of solid-state luminescence by controlling intersystem crossing, Nat. Commun. 9 (1) (2018) 3044. Copyright 2018, The Authors, published by Springer Nature.*

3.6.2 Dynamic visualization of stress/strain distribution and fatigue crack

With the development of large-scale and complex structure components, full-field stress/strain measurements and defects monitoring are highly needed. TPE-4N can form a nonfluorescent, crystalline uniform film on the metal surface, which cracks into fluorescent amorphous fragments upon mechanical force. Therefore, the invisible information of the stress/strain distribution of the metal specimens is transformed to visible fluorescent signals, which generally matches well but provides more details than software simulation.

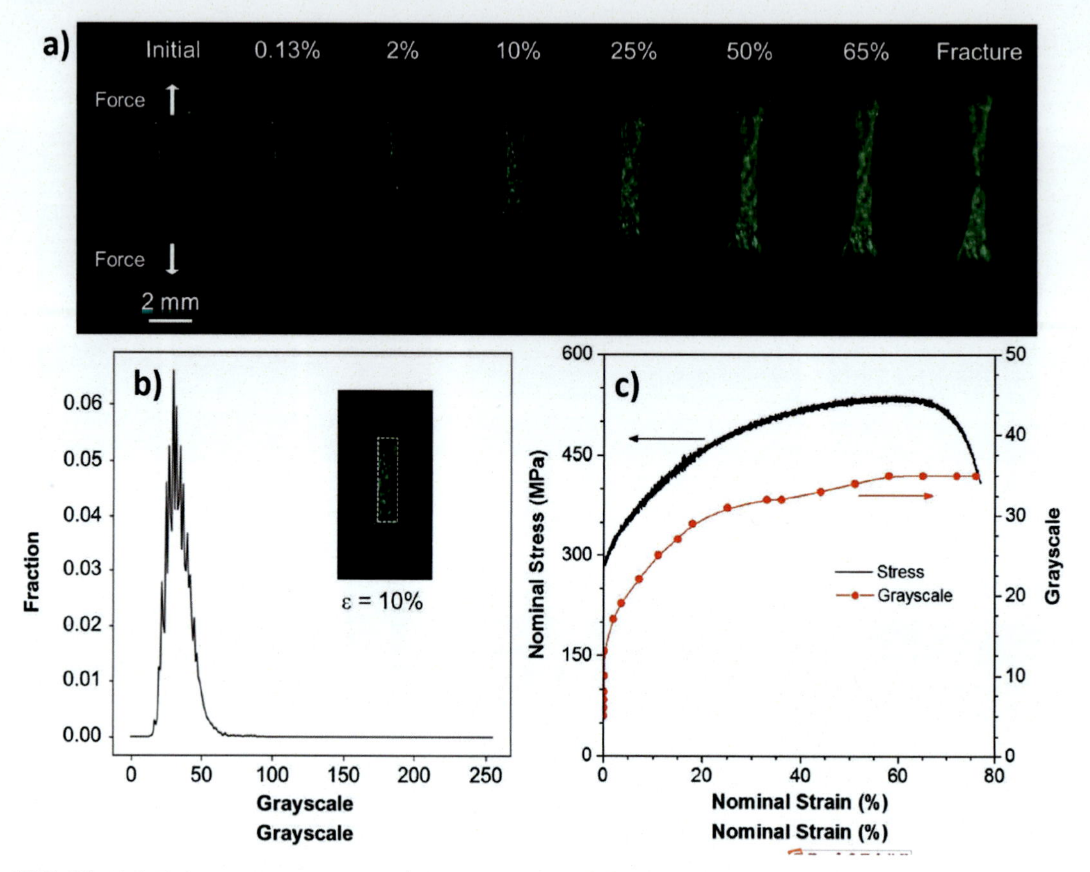

FIG. 21 (A) Fluorescence images of TPE-4N-coated steel tensile specimen at different strains (ε, %). Direction of stretching force: vertical. (B) Gray-scale distribution of the selected area at $\varepsilon = 10\%$. Inset: fluorescence image of TPE-4N-coated steel tensile specimen at $\varepsilon = 10\%$ and the selected area for gray-scale analysis. (C) Plots of strain against stress and gray scale of the TPE-4N-coated steel tensile specimen. *Modified with permission from Z. Qiu, W. Zhao, M. Cao, Y. Wang, J.W.Y. Lam, Z. Zhang, X. Chen, B.Z. Tang, Dynamic visualization of stress/strain distribution and fatigue crack propagation by an organic mechanoresponsive AIE luminogen, Adv. Mater. 30 (44) (2018) 1803924. Copyright 2018, John Wiley and Sons.*

Remarkably, fatigue crack propagation in stainless steel and aluminum alloy can be observed and predicted clearly, further demonstrating the ultrasensitivity and practicability of TPE-4N (Fig. 21) [64].

3.6.3 Optical recording

Tong and coworkers applied ABP1, ABP2, and ABP3 for mechanoresponsive fluorescence writing [50]. As shown in Fig. 22, ABP1, ABP2, and ABP3 were coated on glass substrates and showed no emission. After scratching, the fluorescent letters of "C," "I," and "F" in cyan, blue, and green were clearly written. The letters were completely erased by fuming in ethanol for 3 min. After rescratching, new letters of "T," "R," and "Y" were successfully written, and they

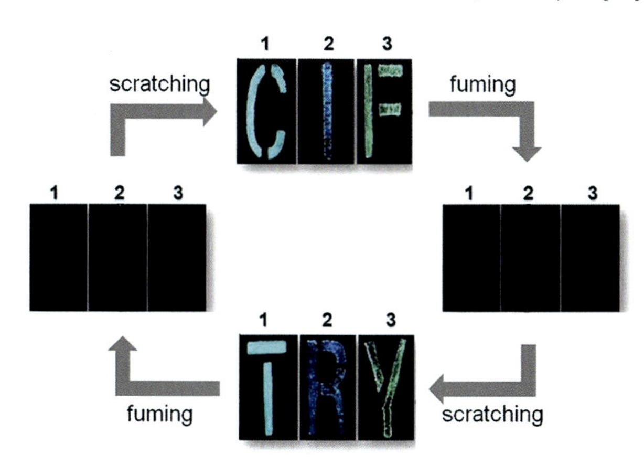

FIG. 22 Reversible mechanoresponsive fluorescence writing using ABP**1**, ABP**2**, and ABP**3**, respectively. Fluorescence images were taken under a 365 nm handheld UV lamp. *Modified with permission from X. Zheng, Y. Zheng, L. Peng, Y. Xiang, A. Tong, Mechanoresponsive fluorescence of 2-aminobenzophenone derivatives based on amorphous phase to crystalline transformation with high "off on" contrast ratio, J. Phys. Chem. C 121 (39) (2017) 21610–21615. Copyright 2017, American Chemical Society.*

could be erased again. These results demonstrated that the three luminogens could be applied for reversible mechanoresponsive fluorescence writing.

4 Summary and perspective

Many MC luminogens have been developed based on AIEgens that normally exhibit morphology-dependent emission and afford loose packing patterns to facilitate the transformation between different morphologies. In this chapter, we introduce the discovery of MC luminescence in AIEgens, discuss the relation between molecular structure and MC luminescence, and explore the plausible design strategy for such luminogens through the reports of the AIEgens with high-performance MC luminescence.

Some high-contrast CIEE luminogens can be constructed through lowering the molecular symmetry of AIEgens or a combination of a large conjugation core and peripheral phenyl rings. Most of the reported high-contrast CIEE luminogens show turn-off mode MC luminescence due to the transformation from crystals to amorphous solid upon exposure to mechanical stimuli. However, three 2-aminobenzophenone derivatives exhibit turn-on mode MC luminescence due to the transformation from amorphous to crystalline upon scratching, which provides a new route to develop turn-on mode MC luminogens with CIEE luminogens. Some AIEgens exhibiting crystallization-caused quenching have been developed through the formation of H-aggregates via strong dipole-dipole interaction, modulation of dihedral angles between the phenyl rings and C=C, control of the intersystem crossing with nitrophenyl, and so on. These AIEgens exhibit turn-on mode MC luminescence due to the transformation from crystals to amorphous solid. Although most luminogens show gradual red-shifted and quenched emission under increased static, some luminogens exhibiting enhanced or blue-shifted emission were also reported.

Although much work has been done in this area, many intriguing possibilities remain to be explored. (1) MC luminescence of most luminogens is related to the structure of aggregates before and after exposure to mechanical stimuli, which depend on both the molecular structure and the formation process of the aggregates. Thus, it is still difficult to establish general or specific design strategies for high-performance MC luminogens, as the origin of MC luminescence often varies significantly from one system to another. (2) It is necessary to develop MC sensors with the ability of quantitative and selective measurement of different kinds of mechanical stimuli. Much work needs to be done by interdisciplinary researchers to pave the way for real application of MC luminogens.

References

[1] R. Gawinecki, G. Viscardi, E. Barni, M.A. Hanna, 4-Tert-butyl-1-(4′-dimethylamino-benzylideneamino) pyridinium perchlorate (BDPP): a novel fluorescent dye, Dyes Pigments 23 (2) (1993) 73–78.
[2] Y. Sagara, T. Mutai, I. Yoshikawa, K. Araki, Material design for piezochromic luminescence: hydrogen-bond-directed assemblies of a pyrene derivative, J. Am. Chem. Soc. 129 (6) (2007) 1520–1521.
[3] J. Kunzelman, M. Kinami, B.R. Crenshaw, J.D. Protasiewicz, C. Weder, Oligo(p-phenylene vinylene)s as a "new" class of piezochromic fluorophores, Adv. Mater. 20 (1) (2008) 119–122.
[4] Y. Ooyama, G. Ito, H. Fukuoka, T. Nagano, Y. Kagawa, I. Imae, K. Komaguchi, Y. Harima, Mechanofluorochromism of heteropolycyclic donor–π-acceptor type fluorescent dyes, Tetrahedron 66 (36) (2010) 7268–7271.
[5] Y. Ooyama, Y. Kagawa, H. Fukuoka, G. Ito, Y. Harima, Mechanofluorochromism of a series of benzofuro[2,3-c] oxazolo[4,5-a]carbazole-type fluorescent dyes, Eur. J. Org. Chem. 2009 (31) (2009) 5321–5326.
[6] Y. Ooyama, Y. Kagawa, H. Fukuoka, G. Ito, Y. Harima, Mechanofluorochromism of a series of benzofuro 2,3-c oxazolo 4,5-a -carbazole-type fluorescent dyes, Eur. J. Org. Chem. 31 (2009) 5321–5326.
[7] G.Q. Zhang, J.W. Lu, M. Sabat, C.L. Fraser, Polymorphism and reversible mechanochromic luminescence for solid-state difluoroboron avobenzone, J. Am. Chem. Soc. 132 (7) (2010) 2160–2162.
[8] G.Q. Zhang, J.P. Singer, S.E. Kooi, R.E. Evans, E.L. Thomas, C.L. Fraser, Reversible solid-state mechanochromic fluorescence from a boron lipid dye, J. Mater. Chem. 21 (23) (2011) 8295–8299.
[9] N.D. Nguyen, G.Q. Zhang, J.W. Lu, A.E. Sherman, C.L. Fraser, Alkyl chain length effects on solid-state difluoroboron beta-diketonate mechanochromic luminescence, J. Mater. Chem. 21 (23) (2011) 8409–8415.
[10] T. Liu, A.D. Chien, J. Lu, G. Zhang, C.L. Fraser, Arene effects on difluoroboron β-diketonate mechanochromic luminescence, J. Mater. Chem. 21 (2011) 8401.
[11] C. Lowe, C. Weder, Oligo(p-phenylene vinylene) excimers as molecular probes: deformation-induced color changes in photoluminescent polymer blends, Adv. Mater. 14 (22) (2002) 1625–1629.
[12] K. Makyła, C. Müller, S. Lörcher, T. Winkler, M.G. Nussbaumer, M. Eder, N. Bruns, Fluorescent protein senses and reports mechanical damage in glass-fiber-reinforced polymer composites, Adv. Mater. 25 (19) (2013) 2701–2706.
[13] F. Ciardelli, G. Ruggeri, A. Pucci, Dye-containing polymers: methods for preparation of mechanochromic materials, Chem. Soc. Rev. 42 (3) (2013) 857–870.
[14] J.D. Luo, Z.L. Xie, J.W.Y. Lam, L. Cheng, H.Y. Chen, C.F. Qiu, H.S. Kwok, X.W. Zhan, Y.Q. Liu, D.B. Zhu, B.Z. Tang, Aggregation-induced emission of 1-methyl-1,2,3,4,5-pentaphenylsilole, Chem. Commun. 18 (2001) 1740–1741.
[15] Y. Dong, J.W.Y. Lam, A. Qin, Z. Li, J. Sun, B.Z. Tang, Vapochromism and crystallization-enhanced emission of 1,1-disubstituted 2,3,4,5-tetraphenylsiloles, J. Inorg. Organomet. Polym. Mater. 17 (4) (2007) 673–678.
[16] Y.Q. Dong, J.W.Y. Lam, Z. Li, A.J. Qin, H. Tong, Y.P. Dong, X.D. Feng, B.Z. Tang, Vapochromism of hexaphenylsilole, J. Inorg. Organomet. Polym. Mater. 15 (2) (2005) 287–291.
[17] Y.Q. Dong, J.W.Y. Lam, Z. Li, H. Tong, C.W. Law, X.D. Feng, B.Z. Tang, Vapochromism of hexaphenylsilole and its blends with poly(methyl methacrylate), Abstr. Pap. Am. Chem. Soc. 228 (2004) U472.
[18] J. Mei, N.L.C. Leung, R.T.K. Kwok, J.W.Y. Lam, B.Z. Tang, Aggregation-induced emission: together we shine, united we soar! Chem. Rev. 115 (21) (2015) 11718–11940.
[19] J. Mei, Y.N. Hong, J.W.Y. Lam, A.J. Qin, Y.H. Tang, B.Z. Tang, Aggregation-induced emission: the whole is more brilliant than the parts, Adv. Mater. 26 (31) (2014) 5429–5479.
[20] Y. Dong, J.W.Y. Lam, A. Qin, Z. Li, J. Sun, H.H.Y. Sung, I.D. Williams, B.Z. Tang, Switching the light emission of (4-biphenylyl)phenyldibenzofulvene by morphological modulation: crystallization-induced emission enhancement, Chem. Commun. 1 (2007) 40–42.
[21] H. Li, X. Zhang, Z. Chi, B. Xu, W. Zhou, S. Liu, Y. Zhang, J. Xu, New thermally stable piezofluorochromic aggregation-induced emission compounds, Org. Lett. 13 (4) (2011) 556–559.

[22] H.Y. Li, Z.G. Chi, B.J. Xu, X.Q. Zhang, X.F. Li, S.W. Liu, Y. Zhang, J.R. Xu, Aggregation-induced emission enhancement compounds containing triphenylamine-anthrylenevinylene and tetraphenylethene moieties, J. Mater. Chem. 21 (11) (2011) 3760–3767.
[23] B.J. Xu, Z.G. Chi, X.Q. Zhang, H.Y. Li, C.J. Chen, S.W. Liu, Y. Zhang, J.R. Xu, A new ligand and its complex with multi-stimuli-responsive and aggregation-induced emission effects, Chem. Commun. 47 (39) (2011) 11080–11082.
[24] X.Q. Zhang, Z.G. Chi, H.Y. Li, B.J. Xu, X.F. Li, W. Zhou, S.W. Liu, Y. Zhang, J.R. Xu, Piezofluorochromism of an aggregation-induced emission compound derived from tetraphenylethylene, Chem. Asian J. 6 (3) (2011) 808–811.
[25] Z. Chi, X. Zhang, B. Xu, X. Zhou, C. Ma, Y. Zhang, S. Liu, J. Xu, Recent advances in organic mechanofluorochromic materials, Chem. Soc. Rev. 41 (10) (2012) 3878–3896.
[26] Y. Dong, B. Xu, J. Zhang, X. Tan, L. Wang, J. Chen, H. Lv, S. Wen, B. Li, L. Ye, B. Zou, W. Tian, Piezochromic luminescence based on the molecular aggregation of 9,10-bis((E)-2-(pyrid-2-yl)vinyl)anthracene, Angew. Chem. Int. Ed. 51 (43) (2012) 10782–10785.
[27] Y.J. Dong, J.B. Zhang, X. Tan, L.J. Wang, J.L. Chen, B. Li, L. Ye, B. Xu, B. Zou, W.J. Tian, Multi-stimuli responsive fluorescence switching: the reversible piezochromism and protonation effect of a divinylanthracene derivative, J. Mater. Chem. C 1 (45) (2013) 7554–7559.
[28] Q. Qi, Y. Liu, X. Fang, Y. Zhang, P. Chen, Y. Wang, B. Yang, B. Xu, W. Tian, S.X.-A. Zhang, AIE (AIEE) and mechanofluorochromic performances of TPE-methoxylates: effects of single molecular conformations, RSC Adv. 3 (21) (2013) 7996–8002.
[29] X. Fan, J.L. Sun, F.Z. Wang, Z.Z. Chu, P. Wang, Y.Q. Dong, R.R. Hu, B.Z. Tang, D.C. Zou, Photoluminescence and electroluminescence of hexaphenylsilole are enhanced by pressurization in the solid state, Chem. Commun. 26 (2008) 2989–2991.
[30] X.L. Luo, J.N. Li, C.H. Li, L.P. Heng, Y.Q. Dong, Z.P. Liu, Z.S. Bo, B.Z. Tang, Reversible switching of the emission of diphenyldibenzofulvenes by thermal and mechanical stimuli, Adv. Mater. 23 (29) (2011) 3261–3265.
[31] X. Luo, W. Zhao, J. Shi, C. Li, Z. Liu, Z. Bo, Y.Q. Dong, B.Z. Tang, Switching emissions of tetraphenylethene derivatives among multiple colors with solvent vapor, mechanical, and thermal stimuli, J. Phys. Chem. C 116 (41) (2012) 21967–21972.
[32] J. Shi, N. Chang, C. Li, J. Mei, C. Deng, X. Luo, Z. Liu, Z. Bo, Y.Q. Dong, B.Z. Tang, Locking the phenyl rings of tetraphenylethene step by step: understanding the mechanism of aggregation-induced emission, Chem. Commun. 48 (86) (2012) 10675–10677.
[33] J. Wang, J. Mei, R. Hu, J.Z. Sun, A. Qin, B.Z. Tang, Click synthesis, aggregation-induced emission, E/Z isomerization, self-organization, and multiple chromisms of pure stereoisomers of a tetraphenylethene-cored luminogen, J. Am. Chem. Soc. 134 (24) (2012) 9956–9966.
[34] T.Y. Han, Y.N. Hong, N. Xie, S.J. Chen, N. Zhao, E.G. Zhao, J.W.Y. Lam, H.H.Y. Sung, Y.P. Dong, B. Tong, B.Z. Tang, Defect-sensitive crystals based on diaminomaleonitrile-functionalized Schiff base with aggregation-enhanced emission, J. Mater. Chem. C 1 (44) (2013) 7314–7320.
[35] W.Z. Yuan, Y. Tan, Y. Gong, P. Lu, J.W.Y. Lam, X.Y. Shen, C. Feng, H.H.Y. Sung, Y. Lu, I.D. Williams, J.Z. Sun, Y. Zhang, B.Z. Tang, Synergy between twisted conformation and effective intermolecular interactions: strategy for efficient mechanochromic luminogens with high contrast, Adv. Mater. 25 (20) (2013) 2837–2843.
[36] Y.Q. Dong, J.W.Y. Lam, B.Z. Tang, Mechanochromic luminescence of aggregation-induced emission luminogens, J. Phys. Chem. Lett. 6 (17) (2015) 3429–3436.
[37] C. Li, X. Luo, W. Zhao, Z. Huang, Z. Liu, B. Tong, Y. Dong, Switching the emission of di(4-ethoxyphenyl) dibenzofulvene among multiple colors in the solid state, Sci. China-Chem. 56 (9) (2013) 1173–1177.
[38] S.-J. Yoon, J.W. Chung, J. Gierschner, K.S. Kim, M.-G. Choi, D. Kim, S.Y. Park, Multistimuli two-color luminescence switching via different slip-stacking of highly fluorescent molecular sheets, J. Am. Chem. Soc. 132 (39) (2010) 13675–13683.
[39] H. Tian, P. Wang, J. Liu, Y. Duan, Y.Q. Dong, Construction of a tetraphenylethene derivative exhibiting high contrast and multicolored emission switching, J. Mater. Chem. C 5 (48) (2017) 12785–12791.
[40] Y. Lin, G. Chen, L. Zhao, W.Z. Yuan, Y. Zhang, B.Z. Tang, Diethylamino functionalized tetraphenylethenes: structural and electronic modulation of photophysical properties, implication for the CIE mechanism and application to cell imaging, J. Mater. Chem. C 3 (1) (2015) 112–120.
[41] Z. Wang, H. Nie, Z. Yu, A. Qin, Z. Zhao, B.Z. Tang, Multiple stimuli-responsive and reversible fluorescence switches based on a diethylamino-functionalized tetraphenylethene, J. Mater. Chem. C 3 (35) (2015) 9103–9111.

[42] X. Su, Y. Wang, X. Fang, Y.-M. Zhang, T. Zhang, M. Li, Y. Liu, T. Lin, S.X.-A. Zhang, A high contrast tri-state fluorescent switch: properties and applications, Chem. Asian J. 11 (22) (2016) 3205–3212.

[43] Y. Duan, H. Ma, H. Tian, J. Liu, X. Deng, Q. Peng, Y.Q. Dong, Construction of a luminogen exhibiting high contrast and multicolored emission switching through combination of a bulky conjugation core and tolyl groups, Chem. Asian J. 14 (6) (2019) 864–870.

[44] Y. Duan, X. Xiang, Y. Dong, Diphenyldibenzofulvene derivatives exhibiting reversible multicolored mechanochromic luminescence with high contrast, Acta Chim. Sin. 74 (11) (2016) 923–928.

[45] X. Gu, J. Yao, G. Zhang, Y. Yan, C. Zhang, Q. Peng, Q. Liao, Y. Wu, Z. Xu, Y. Zhao, H. Fu, D. Zhang, Polymorphism-dependent emission for di(p-methoxylphenyl)dibenzofulvene and analogues: optical waveguide/amplified spontaneous emission behaviors, Adv. Funct. Mater. 22 (23) (2012) 4862–4872.

[46] Z. Zhao, T. Chen, S. Jiang, Z. Liu, D. Fang, Y.Q. Dong, The construction of a multicolored mechanochromic luminogen with high contrast through the combination of a large conjugation core and peripheral phenyl rings, J. Mater. Chem. C 4 (21) (2016) 4800–4804.

[47] W. Chen, S. Wang, G. Yang, S. Chen, K. Ye, Z. Hu, Z. Zhang, Y. Wang, Dicyanomethylenated acridone based crystals: torsional vibration confinement induced emission with supramolecular structure dependent and stimuli responsive characteristics, J. Phys. Chem. C 120 (1) (2016) 587–597.

[48] Z. He, L. Zhang, J. Mei, T. Zhang, J.W.Y. Lam, Z. Shuai, Y.Q. Dong, B.Z. Tang, Polymorphism-dependent and switchable emission of butterfly-like bis(diarylmethylene)dihydroanthracenes, Chem. Mater. 27 (19) (2015) 6601–6607.

[49] R. Zheng, X. Mei, Z. Lin, Y. Zhao, H. Yao, W. Lv, Q. Ling, Strong CIE activity, multi-stimuli-responsive fluorescence and data storage application of new diphenyl maleimide derivatives, J. Mater. Chem. C 3 (39) (2015) 10242–10248.

[50] X. Zheng, Y. Zheng, L. Peng, Y. Xiang, A. Tong, Mechanoresponsive fluorescence of 2-aminobenzophenone derivatives based on amorphous phase to crystalline transformation with high "off on" contrast ratio, J. Phys. Chem. C 121 (39) (2017) 21610–21615.

[51] W.Z. Yuan, X.Y. Shen, H. Zhao, J.W.Y. Lam, L. Tang, P. Lu, C.L. Wang, Y. Liu, Z.M. Wang, Q. Zheng, J.Z. Sun, Y.-G. Ma, B.Z. Tang, Crystallization-induced phosphorescence of pure organic luminogens at room temperature, J. Phys. Chem. C 114 (13) (2010) 6090–6099.

[52] Q. Qi, J. Qian, X. Tan, J. Zhang, L. Wang, B. Xu, B. Zou, W. Tian, Remarkable turn-on and color-tuned piezochromic luminescence: mechanically switching intramolecular charge transfer in molecular crystals, Adv. Funct. Mater. 25 (26) (2015) 4005–4010.

[53] B. Han, X. Wang, Y. Gao, M. Bai, Constructing a nonfluorescent conformation of AIEgen: a tetraphenylethene embedded in the calix[4]arene's skeleton, Chem. Eur. J. 22 (45) (2016) 16037–16041.

[54] W. Zhao, Z. He, J.W. Lam, Q. Peng, H. Ma, Z. Shuai, G. Bai, J. Hao, B.Z. Tang, Rational molecular design for achieving persistent and efficient pure organic room-temperature phosphorescence, Chem 1 (4) (2016) 592–602.

[55] M. Takezaki, N. Hirota, M. Terazima, Nonradiative relaxation processes and electronically excited states of nitrobenzene studied by picosecond time-resolved transient grating method, J. Phys. Chem. A 101 (19) (1997) 3443–3448.

[56] M. Takezaki, N. Hirota, M. Terazima, Relaxation of nitrobenzene from the excited singlet state, J. Chem. Phys. 108 (11) (1998) 4685–4686.

[57] C. Xu, F.L. Gu, C.Y. Zhu, Ultrafast intersystem crossing for nitrophenols: ab initio nonadiabatic molecular dynamic simulation, Phys. Chem. Chem. Phys. (2018).

[58] B. Tharmalingam, M. Mathivanan, G. Dhamodiran, K.S. Mani, M. Paranjothy, B. Murugesapandian, Star-shaped ESIPT-active mechanoresponsive luminescent AIEgen and its on-off-on emissive response to Cu^{2+}/S^{2}, ACS Omega 4 (7) (2019) 12459–12469, https://doi.org/10.1021/acsomega.9b00845.

[59] K. Santhiya, S.K. Sen, R. Natarajan, R. Shankar, B. Murugesapandian, D-A-D structured bis-acylhydrazone exhibiting aggregation-induced emission, mechanochromic luminescence, and Al(III) detection, J. Org. Chem. 83 (18) (2018) 10770–10775.

[60] X. Zhang, J. Shi, G. Shen, F. Gou, J. Cheng, X. Zhou, H. Xiang, Non-conjugated fluorescent molecular cages of salicylaldehyde-based tri-Schiff bases: AIE, enantiomers, mechanochromism, anion hosts/probes, and cell imaging properties, Mater. Chem. Front. 1 (6) (2017) 1041–1050.

[61] F. Zhang, R. Zhang, X. Liang, K. Guo, Z. Han, X. Lu, J. Xie, J. Li, D. Li, X. Tian, 1,3-Indanedione functionalized fluorene luminophores: negative solvatochromism, nanostructure-morphology determined AIE and mechanoresponsive luminescence turn-on, Dyes Pigments 155 (2018) 225–232.

[62] H. Liu, Y. Gu, Y. Dai, K. Wang, S. Zhang, G. Chen, B. Zou, B. Yang, Pressure-induced blue-shifted and enhanced emission: a cooperative effect between aggregation-induced emission and energy-transfer suppression, J. Am. Chem. Soc. 142 (3) (2020) 1153–1158.

[63] W. Zhao, Z. He, Q. Peng, J.W.Y. Lam, H. Ma, Z. Qiu, Y. Chen, Z. Zhao, Z. Shuai, Y. Dong, B.Z. Tang, Highly sensitive switching of solid-state luminescence by controlling intersystem crossing, Nat. Commun. 9 (1) (2018) 3044.

[64] Z. Qiu, W. Zhao, M. Cao, Y. Wang, J.W.Y. Lam, Z. Zhang, X. Chen, B.Z. Tang, Dynamic visualization of stress/strain distribution and fatigue crack propagation by an organic mechanoresponsive AIE luminogen, Adv. Mater. 30 (44) (2018) 1803924.

CHAPTER

7

Photochromic and thermochromic luminescence in AIE luminogens

Ju Mei and He Tian

Key Laboratory for Advanced Materials, Feringa Nobel Prize Scientist Joint Research Center, Frontiers Science Center for Materiobiology and Dynamic Chemistry, Joint International Research Laboratory for Precision Chemistry and Molecular Engineering, Institute of Fine Chemicals, School of Chemistry & Molecular Engineering, East China University of Science & Technology, Shanghai, P. R. China

1 Fundamentals of photochromism and thermochromism

1.1 General introduction to photochromism

The term "photochromism" or "photochromic" is believed to originate from the Greek words φωτός (photos) and χρῶμα (chroma), which mean light and color, respectively. The photochromism can thus be simply defined as the feature to undergo a reversible change of color induced by light [1]. Photochromic materials can be easily found in things used in daily life such as in lenses of glasses that protect against sunshine damage, and in fashionable cosmetics and clothes. Photochromic materials are also widely applied in the transmission, gating, and storing of digital data [2,3]. For example, compact disc (CD) and digital versatile disc (DVD) are the media broadly used for data storage, in which information is written and erased by light, and the information is read by optical properties. Hence, the materials with photochromic characteristics satisfy the requirements of the rewritable recording media, by virtue of their reversible properties. More recently, photochromic materials made sufficient contributions to fluorescence microscopy imaging and hence have drawn considerable attention of the researchers from biology, chemistry, and materials disciplines. Since the resolution limit of conventional fluorescence microscopy is recognized to be 200 nm, "super-resolution imaging techniques" were developed. In the super-resolution imaging techniques, the resolution beyond the diffraction limit could be achieved via making use of controllable optical processes of fluorescent probes. Among them, photoswitchable fluorescent materials are very

Aggregation-Induced Emission (AIE)
https://doi.org/10.1016/B978-0-12-824335-0.00004-0

attractive and have been successfully applied in microscopy for subdiffractive imaging. In addition to the photoswitching of color and fluorescence, other properties including magnetic feature, electrical characteristics, redox ability, conductivity, solubility, and reactivity could also be switched by light.

1.1.1 Basic principles

The simple two-way reaction between A and B could be used as a common model to describe photochromism, where A refers to a photochromic molecule and B refers to a molecular species obtained after the light-switched chemical reaction of A. There is a potential barrier separating A and B according to which the photochromic systems could be divided into two subtypes. When the energy barrier is low, species B can be converted back to species A spontaneously and thus metastable. Such systems are termed as *T*-type, which refers to the thermal-induced conversion from B to A. On the other hand, the system possessing a high energy barrier is bistable. In this case, merely photons can induce the conversion reaction, and these systems are thus called *P*-type. That is to say, nothing would change without light.

As for a common photochromic system, A shows absorption in the UV or near-UV region, with a characteristic absorption band peaking at λ_A. As a photon at λ_A is absorbed, A is excited to the excited state from the ground state. The excited A has a possibility to generate B. On the other hand, when excited at λ_B, which is the absorption wavelength of B, B is converted back to A, with a pattern analogous to A. The spectral position of the absorption bands not only indicates the color of the light which is needed to induce the reaction but also suggests the color of the molecule itself.

1.1.2 Main mechanisms

From the very beginning of the 21st century, the studies of photochromism are focusing on no more than 10 families of compounds, such as diarylethenes, spiropyrans, spiroxazines, chromenes, salicylideneanilines, fulgides, hexaarylbiimidazoles, etc. Based on these photochromic systems, the main mechanisms of organic photochromism could be summarized as (1) proton transfer (e.g., salicylideneanilines, dinitrobenzylpyridines), (2) *trans-cis* photoisomerization (e.g., stilbenes, azobenzenes), (3) homolytic cleavage (e.g., hexaarylbiimidazoles), (4) photocyclization (e.g., spiropyrans, spirooxazines, chromenes, *cis*-stilbenes, diarylethenes, fulgides, fulgimides), and so on (Fig. 1).

For a photochromic system based on the intramolecular proton transfer, the intramolecular proton transfer is initiated upon the irradiation by UV light, rendering the change from enol form to the *cis*-keto form in the excited state, which is called the excited-state intramolecular proton transfer (ESIPT). Thereafter, an excited-state *cis-trans* isomerization occurs to give a *trans*-keto form. The whole process only takes place in a few picoseconds (ps) in solution and a few hundreds of ps even in the solid state. Moreover, salicylideneanilines are *T*-typed. The keto product turns back to the enol form quite fast, usually in a few milliseconds (ms) in solutions. While in the solid state, the reverting time may differ from a few seconds to several months. Dinitrobenzylpyridine derivatives are another family of most studied compounds controlled by the proton transfer process. The UV irradiation of the most stable colorless CH form gives rise to the OH and NH forms. Moreover, obvious color change between the phototautomers is usually observable.

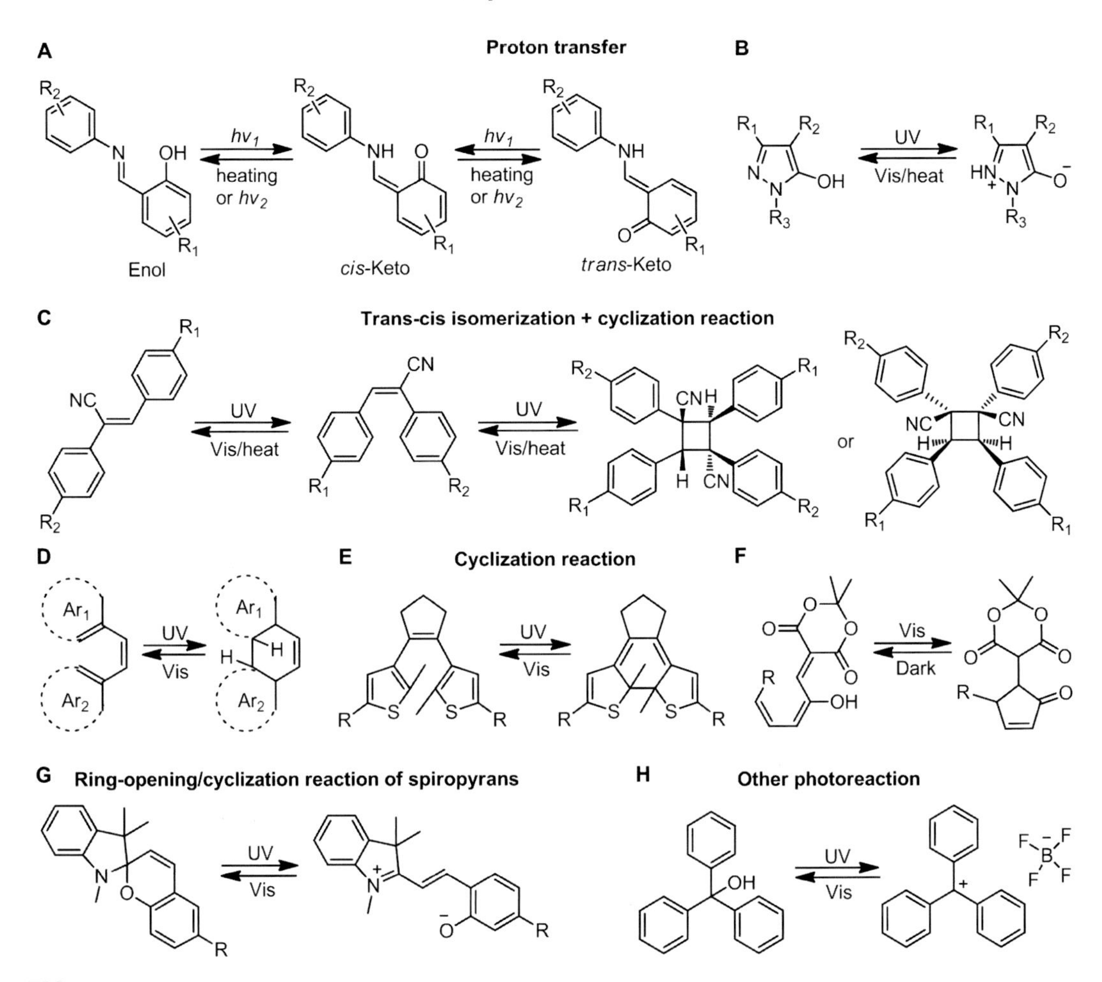

FIG. 1 The main photochromic mechanisms of AIE-active photochromic systems.

Stilbenes, azobenzenes, and their derivatives are the *trans-cis* reaction-modulated photochromic systems. Taking azobenzene, for instance, the color of azobenzene is usually yellow. The *cis*- and *trans*-isomers of azobenzenes both possess two characteristic bands, namely π-π^* and n-π^*. Because the n-π^* band is symmetry forbidden, it lies in a lower energy region and is less intense as compared to the π-π^* band. The difference in the absorption spectra between the *cis*- and *trans*-forms are usually not very evident as a result of a small difference in the conjugation between these two isomers, as such the color variation can hardly be witnessed by the naked eyes. On the contrary, the *cis-trans* isomerization causes a very dramatic change to the free volume of the molecules.

Hexaarylbiimidazoles are the representatives of the photochromic materials based on the homolytic cleavage of the C—N bond between the two imidazole rings. In these compounds, the homolytic cleavage can be triggered by light irradiation, heat, or pressure, yielding two triphenylimidazolyl radicals. Such radicals can be reverted to the original imidazole dimer, namely triphenylimidazolyl dimer, via thermal treatment and driven by the diffusion of radicals. The triphenylimidazolyl dimer merely absorbs in the UV light region and is therefore colorless, while the triphenylimidazolyl radical pair has an intense absorption in the visible light region. The photochromism of the hexaarylbiimidazole family is *T*-type. The cleavage of these compounds upon the irradiation of UV light occurs within no more than 100 femtoseconds (fs). Despite this, the recombination of radicals could take up to several minutes under ambient temperature.

Cyclization is the main mechanism of most organic photochromic materials reported. In most cases, the cyclization reaction is associated with six π electrons extended over six different atoms. Spiropyrans are among the most widespread cyclization-based photochromic systems in the past few decades. Upon UV irradiation, the colorless closed form of spiropyrans undergoes the C—O bond cleavage and the ring-opening reaction of the pyran ring and the subsequent *cis-trans* isomerization, and then the *spiro* form finally turns into the colored merocyanine (MC) form. The highly conjugated MC form has two mesomeric resonant forms, i.e., zwitterionic and quinonic. Such a feature is related to the strong absorption band in the visible light region and the color of the MC form. On the opposite, the conjugation is broken at the *spiro* carbon atom in the closed form, and as a result, the closed form is colorless and only shows absorption in the UV light region. Spiropyrans are also *T*-type photochromic materials. In most cases, the MC form is metastable while the closed form is more stable. Spirooxazines, where the CH group has been displaced by an N atom, are another type of cyclization-based photochromic compounds, and hence are also *T*-type. The photochromism of chromenes is similar to that of spiropyrans and spirooxazines.

The photochromic mechanisms of fulgides and fulgimides also involve the cyclization reaction via the conversion from π bond to σ bond resulting in the electrocyclization. Unlike the spiropyrans, the open form of fulgides and fulgimides is colorless species, whereas the cyclized form is colored. The fulgides and fulgimides can be *T*-type or *P*-type photochromic systems.

The diarylethenes in fulgides undergo a ring-closure reaction with the compound changing from a colorless open form to a colored closed form. Most of the diarylethenes involve heterocycles. Among these diarylethene derivatives, the dithienylethene subgroup is currently the most extensively studied. In the colorless open form, all the diarylethenes have a 1,3,5-hexatriene structure. The colored closed form, namely 1,3-cyclohexadiene, is generated under UV light irradiation. The reverse reaction is switched by the irradiation in the visible light region. The elongation of the π-conjugation accounts for the color of the closed form. Analogous to the dithienylethene systems, the *cis*-stilbene derivatives could also undergo cyclization reaction. It is worth mentioning that the diarylethene derivatives have shown some potential as *P*-type photochromic systems, although many diarylethenes are *T*-type [3].

Regardless of the mechanism, photochromic systems are usually supposed to feature the characteristics including (i) fatigue resistance, which is defined as the cycle number of a photochromic system able to complete; (ii) high quantum yield of the photoreaction; and (iii) a large absorption spectral shift between the A and B forms. Besides this, to facilitate real-world applications, photochromic systems should hold the capability to operate in the solid state.

1.2 General introduction to thermochromism

Similar to photochromism, the term "thermochromism" is simply defined as the phenomenon or characteristics of a substance to show a reversible change of color caused by temperature change [1]. However, thermochromism is not necessarily associated with a chemical reaction. Studies into the mechanisms of thermochromism began in the early 1970s. The mechanisms were found to greatly depend on the materials used. In general, there are mainly four types of thermochromic systems, namely, organic compounds, inorganic systems, polymers, and sol-gels. Particularly, the AIEgen-based thermochromic materials are generally organic compounds, aryl-substituted *o*-carborane/binary boranes, and metal complexes. As for the organic thermochromic AIE systems, the variation in the crystal structure, alteration in molecular conformation, and the change in the molecular arrangement is believed to be the most common mechanisms. Organic thermochromic systems often exhibit sharp color change and the factors to easily regulate the temperature are various. As to the aryl-substituted *o*-carborane/binary boranes, the change in the molecular packing is thought to be the most common mechanism. While for the metal complexes, the thermal transition most often results from an alteration in the crystalline phase, a change in ligand geometry, or the equilibria between different molecular structures. Obviously, materials possessing thermochromic properties can also be applied in the area like anticounterfeiting, and the AIE effect is believed to be conducive to their applications.

2 Photochromic and thermochromic AIE systems

2.1 Photochromic AIE systems

2.1.1 Structures and design principles

A system that not only shows photochromism but also exhibits aggregation-induced emission (AIE) properties is the so-called photochromic AIE system. As for how to integrate the photochromic feature with the AIE characteristics, there are generally three main strategies: (i) physically mixing the AIE motifs with the photochromic units; (ii) connecting AIE-active motifs to photochromic units via covalent bonds; (iii) according to the principle of restriction of the intramolecular motions (RIM), linking π-conjugated groups with photochromic motif(s) via single bonds or vibratory units. The photochromic mechanisms of these AIE systems are summarized in Fig. 1.

Photochromic AIE systems obtained by physically blending AIE motifs and photochromic units

The 3,3′-(perfluorocyclopent-1-ene-1,2-diyl)bis{5-[3,5-bis(trifluoromethyl)phenyl]-2-methylthiophene} (TFM-BTE), a trifluotomethyl (CF_3)-containing derivative of 1,2-bis (thienyl)ethane (BTE), was introduced as a photochromic unit, which displayed bistable and reversible photochromism. The 1-cyano-trans-1,2-bis[3′,5′-bis(trifluoromethyl)-biphenyl]ethylene (CN-TFMBE) was incorporated to play multiple roles as a gelator, an AIE motif, and a thermal-switching element. CN-TFMBE is almost nonemissive in the solution (sol) state, but highly fluoresces in the gel state, exhibiting a fluorescence enhancement of up to 100 folds. Physically mixing the thermal-responsive and AIE-active CN-TFMBE or

SS-TFMBE with the photoswitchable TFM-BTE, afforded the dual-mode switchable fluorescent organogel systems (Fig. 2A) [4]. Due to their profound AIE feature and outstanding self-assembly ability, both CN-TFMBE and SS-TFMBE show bright fluorescence in the gel state. When irradiated by a 300nm-UV light, a new absorption band emerged at around 580nm in the UV-Vis absorption spectrum. Accordingly, the solution of TFM-BTE turned from colorless to blue. The absorption spectrum of the original open form is restored by the subsequent irradiation of the closed form using visible light with a wavelength longer than 450nm.

The organogel system of TFM-BTE/CN-TFMBE displays reversible fluorescence on/off switching after being irradiated by UV or visible light. However, the fluorescence quenching in the closed form (photostationary state, PSS) is not very satisfactory. Obviously, the TFM-BTE/CN-TFMBE organogel shows intense fluorescence emission at the initial state because of the AIE feature. When irradiated by a 300-nm-UV light, the TFM-BTE is converted from the open form to the closed form. The intramolecular energy transfer, probably the Förster resonance energy transfer (FRET), from CN-TFMBE to closed-form TFM-BTE affected the fluorescence quenching degree. Irradiation of the closed form with visible light ($\lambda > 450$ nm) regenerated the original open form and recovered the initial emission.

Considering the unsatisfactory fluorescence quenching of the TFM-BTE/CN-TFMBE, (*E*)-2,3-bis{5-[3,5-bis(trifluoromethyl)phenyl]thiophen-2-yl}acrylonitrile (SS-TFMBE), which is a thiophene-containing derivative of CN-TFMBE, was synthesized to achieve a red-shifted fluorescence emission at 575nm in the gel state. The complete overlap between the emission spectrum of SS-TFMBE and the absorption spectrum of closed-form TFM-BTE results in the efficient FRET and complete fluorescence quenching when irradiating the open-form TFM-BTE with a 300nm-UV light, giving a fluorescence switching ratio greater than 166. As shown in Fig. 2B, the reproducibility and reversibility of the photoswitching of SS-TFMBE/TFM-BTE organogel are manifested by the repeated switchable fluorescence at 575nm with alternate UV and visible light irradiation.

Moreover, the on/off modulation of the strongly fluorescent SS-TFMBE/TFM-BTE organogel could be realized not only via light-driven photochromic isomerization but also via a thermal-driven gel-to-sol phase transition (Fig. 2C). The "OR" algorithm is displayed: both the sol and gel states of the closed form are nonfluorescent while the open form is nonemissive in the sol state but highly fluorescent in the gel state, the sol and gel states of both open and closed forms are switchable by thermal input. By virtue of the high-contrast and nondestructive fluorescence switching behavior, the SS-TFMBE/TFM-BTE organogel is applicable to high-density optical logic memory storage. The information recording and erasing procedures are depicted in Fig. 2D. Irradiating the nonfluorescent gel state of SS-TFMBE/TFM-BTE with visible light through an "AIEE"-containing a mask, the high-contrast fluorescent image of the "AIEE" letters could be generated. The "AIEE" letters could be erased either to the fluorescent form by the unmask irradiation with visible light or to the nonemissive form via thermal treatment. Exposure of the fluorescent gel to the 300nm-UV light with a "GEL"-coded mask, the fluorescent image with nonemissive "GEL" letters is obtained. As a result, the complete writing-erasing-rewriting cycle and nondestructive readout of a fluorescent optical recording on the basis of a binary "OR" gate manipulation are demonstrated. It is believed that such an AIE-active photochromic system is a promising material for erasable optical data recording and storage.

More recently, Park et al. reported another photochromic AIE system acquired by mixing a diarylethene derivative with CN-TFMBE [5]. In the fluorescence switching system shown in

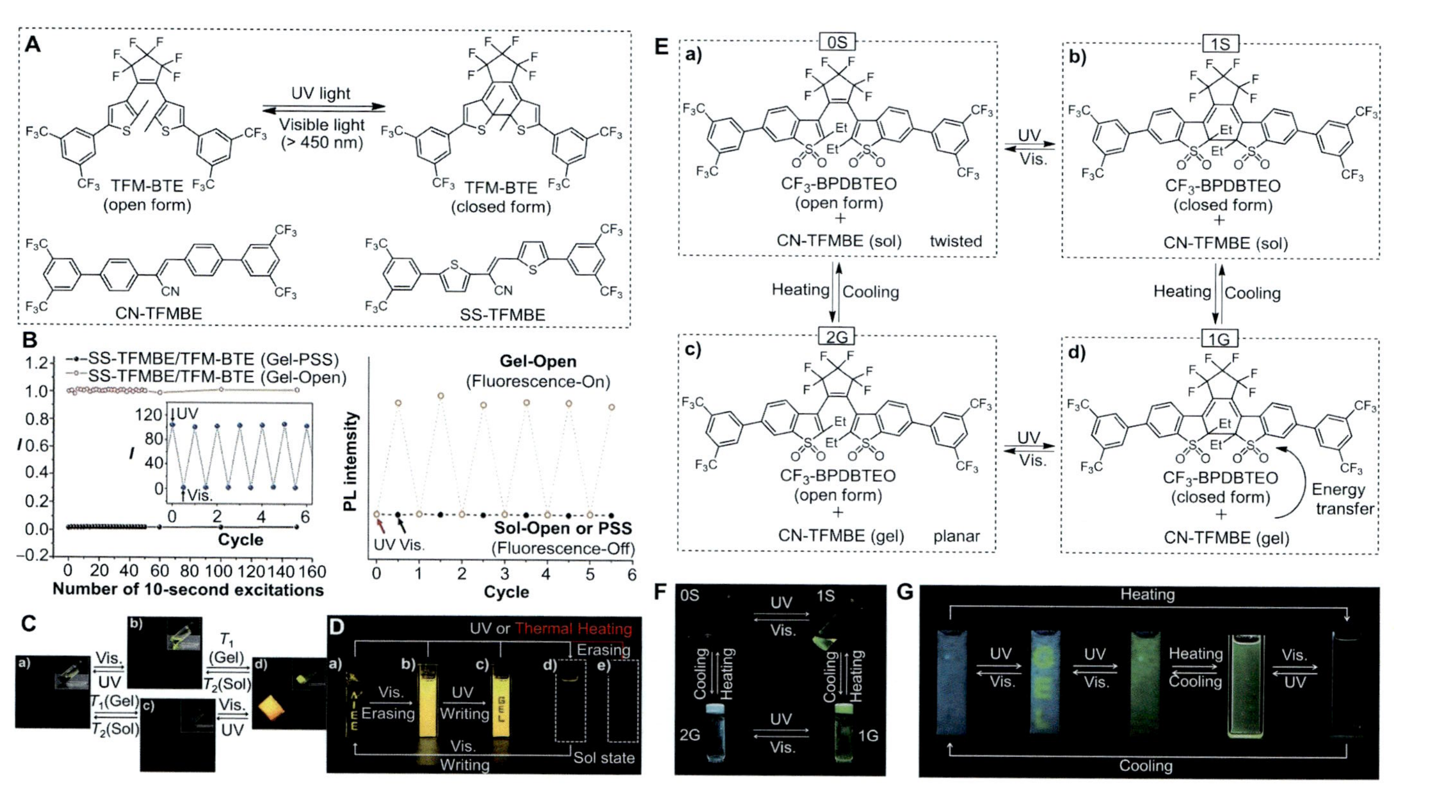

FIG. 2 The representative AIE-active photochromic organogel systems which were obtained by physically mixing AIE motifs with photochromic units and modulated by an intramolecular energy transfer mechanism. (A) The dual-mode switchable highly fluorescent organogels that were obtained by physically mixing TFM-BTE with CN-TFMBE or SS-TFMBE. (B) The graphs indicating the reproducible and reversible photoswitching capability. (C) The fluorescent photos of the TFM-BTE/SS-TFMBE organogel responding to photoirradiation or thermal treatment. (D) The reversible fluorescence switching of the TFM-BTE/SS-TFMBE organogel [4]. (E) The schematic illustration and (F) fluorescent photos of the states of 0S, 1S, 2G, and 1G. (G) Reversible optical writing and erasing processes of the gel system formed by the physical mixture of CF_3-BPDBTEO and CN-TFMBE [5]. *Panels (A–D) reproduced with permission from J.W. Chung, S.-J. Yoon, S.-J. Lim, B.-K. An, S.Y. Park, Dual-mode switching in highly fluorescent organogels: binary logic gates with optical/thermal inputs, Angew. Chem. Int. Ed. 48 (2009) 7030–7034. Copyright 2009 Wiley-VCH and (E–G) from K. Dojin, J.E. Kwon, S.Y. Park, Fully reversible multistate fluorescence switching: organogel system consisting of luminescent cyanostilbene and turn-on diarylethene, Adv. Funct. Mater. 28 (2018) 1706213. Copyright 2017 Wiley-VCH.*

Fig. 2E, similar to the TFM-BTE/CN-TFMBE organogel system, the CN-TFMBE in this system is also a thermal-responsive and AIE-active organogelator. The green-emissive diarylethene derivative (3,3′-(perfluorocyclopent-1-ene-1,2-diyl)bis(6-(3,5-bis(trifluoromethyl)phenyl)-2-ethylbenzo[*b*]thiophene-1,1-dioxide) (CF_3-BPDBTEO) acts as a turn-on photochromic unit. Multicolor and multistate fluorescence switching, namely interconversion among blue-fluorescent gel, nonfluorescent sol, green-fluorescent gel, and green-fluorescent sol, was achieved.

The open form of CF_3-BPDBTEO exhibits an absorption band in the UV light range (<380 nm) but practically has no fluorescence. When irradiated with UV light, a new absorption band peaked at about 443 nm gradually appeared by forming a closed isomer of CF_3-BPDBTEO with a bright green fluorescence peaked at 522 nm with a fluorescence quantum yield of 0.78. Irradiated by visible light with a wavelength longer than 420 nm, the reverse photoreaction of CF_3-BPDBTEO is initiated, completely recovering the original absorption of the open-form CF_3-BPDBTEO. Notably, these ring-closure/opening reactions and photoswitches can be conversably repeated for more than 30 cycles with no photodegradation. As mentioned above, CN-TFMBE displays no emission in the solution but exhibits brilliant blue emission at 472 nm with a quantum yield of 0.52 in the gel state as a result of the AIE characteristics.

Taking the gel system with a mixing weight ratio of CF_3-BPDBTEO/CN-TFMBE = 3/1 for example, when irradiated by a UV light, the gel displays a fairly significant color change to green (i.e., 1G state) from blue (i.e., 2G state). Upon heating, the 1G state of the green-fluorescent gel is reverted to a green-emissive solution (i.e., 1S state). Sharply contrasting the fluorescent photoswitching between blue and green in the gel state, with the visible-light irradiation, the fluorescence emission of this physically mixed organogel is switched off to afford the nonfluorescent 0S state (Fig. 2F). It is shown that the photo-induced color change in both gels (i.e., 1G, 2G) and solution (i.e., 0S, 2S) states could be repeated for many cycles. It means that by combining the UV/Vis light irradiation and heating-cooling operation, the interchange among the four states, namely, 0S, 1S, 1G, and 2G, can be fully addressed.

Making full use of the reversible and orthogonal switch among multiple states, a logic circuit composed of two distinct inputs can be established, where the inputs are light or thermal. Moreover, the potential of the CF_3-BPDBTEO/CN-TFMBE mixing system for information recording, color switching, and erasing was also demonstrated. As displayed in Fig. 2G, when starting with the blue fluorescent gel state (2G), after being irradiated with UV light via a "GEL"-containing mask, the high-contrast image of "GEL" is readily obtained with green fluorescence (1G). The overall fluorescence color could be reverted to green upon further irradiation with UV light without the photomask. On the other hand, the 2G could be switched to a green-emissive 1S state through heating. In order to erase the fluorescence, the 0S state is generated when exposed to visible light irradiation. As a consequence, the information with blue emission could also be recorded by a simple cooling down operation.

Apart from the physically mixing photochromic blocks and AIE-active units to afford photoswitchable organogel systems which are modulated via the energy transfer especially the FRET mechanism [4–8], physically blending AIE unit(s) with photochromic motif(s) via the host-guest interactions is also proved to be an efficient strategy to create AIE-active photochromic systems [9,10]. The systems shown in Fig. 3 are the representatives. Pillararene, featured with rigid structure, electron-donating cavities, and easy functionalization, has been

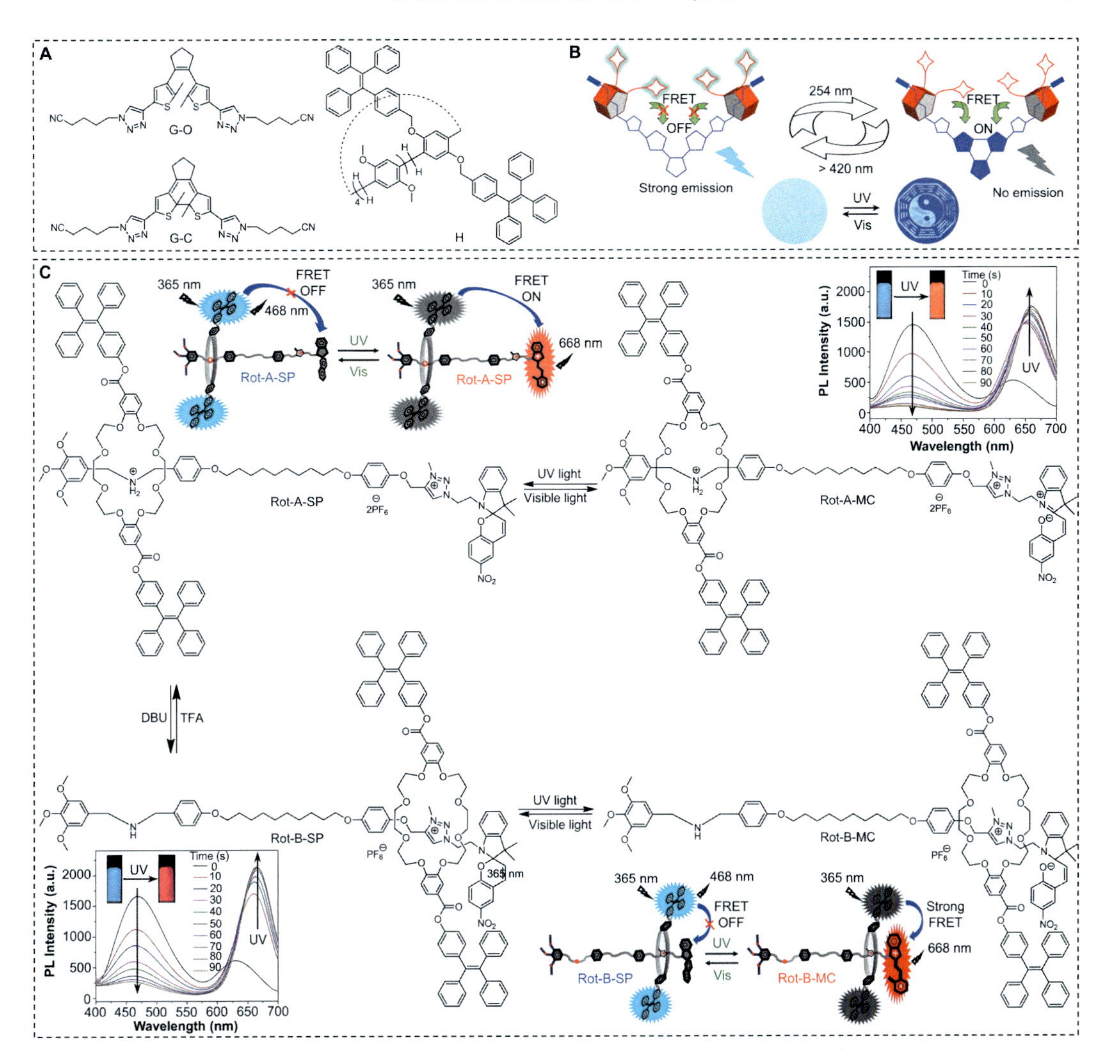

FIG. 3 The representative photochromic and AIE-active host-guest systems which were obtained by physically mixing AIE motifs with photochromic units. (A) The chemical structures of the dithienylethylene guest G and the AIE-active pillar[5]arene host H. (B) The cartoon illustration of the photo-induced fluorescence switching of the H/G complex [9]. (C) The photoisomerization of Rot-A-SP and Rot-B-SP to yield Rot-A-MC and Rot-B-MC, respectively, the highly reversible shuttling behavior between Rot-A-SP and Rot-B-SP under acid and base conditions. The insets are the schematic illustrations of the energy transfer processes from the AIE moiety to the merocyanine unit, and the time-dependent photoluminescence spectra of Rot-A-SP and Rot-B-SP in the THF/water (1/9, v/v) upon UV irradiation from 0 to 90s [10]. *Panels (A and B) reproduced with permission from L. Ma, S. Wang, C. Li, D. Cao, T. Li, X. Ma, Photo-controlled fluorescence on/off switching of a pseudo[3]rotaxane between an AIE-active pillar[5]arene host and a photochromic bithienylethene guest, Chem. Commun. 54 (2018) 2405–2408. Copyright 2018 Royal Society of Chemistry and (C) from P.Q. Nhien, T.T.K. Cuc, T.M. Khang, C.-H. Wu, B.T.B. Hue, J.I. Wu, B.W. Mansel, H.-L. Chen, H.-C. Lin, Highly efficient Förster resonance energy transfer modulations of dual-AIEgens between a tetraphenylethylene donor and a merocyanine acceptor in photo-switchable [2]rotaxanes and reversible photo-patterning applications, ACS Appl. Mater. Interfaces 12 (2020) 47921 – 47938. Copyright 2020 American Chemical Society.*

recognized as a novel attractive macrocyclic host composed of hydroquinone ether groups connected via methylene bridges. Ma and Cao, et al. integrated a pillar[5]arene unit with the prototype AIE motifs, namely tetraphenylethylene (TPE), via Williamson etherification reaction, generating a pillar[5]arene host with AIE attribute (i.e., H in Fig. 3A) [9]. When the pillar[5]arene-TPE adduct (H) interacts with the bithienylethene derivative containing two cyano-triazole branches (i.e., G-O or G-C in Fig. 3A), a pseudo[3]rotaxane is formed between H and G. The fluorescence properties of the resulting pseudo[3]rotaxane were investigated after mixing H ([H] = 10 μM) and G ([G] = 5 μM) in the THF/water mixture. The G⊂H complex showed profound AIE properties similar to the free H. The complex virtually did not display any fluorescence until the water fraction researched 50%, while afterward, the fluorescence intensity became higher and increased with the water content.

Irradiation of the G⊂H complex in the THF/water mixture (5/5, v/v) with 254 nm UV light for 10 min made the fluorescence intensity gradually quenched by 68%. Analogously, the emission of the G⊂H complex in the THF/water mixture (4/6, v/v) was reduced by 87% after being irradiated with 254 nm UV light for 18 min. On the contrary, irradiating these solutions with visible light, the quenched fluorescence gradually recovered to the original state. It is because that when irradiated by UV light, the guest was changed from the open form G-O to the closed-form G-C, yielding a new absorption band peaked at around 516 nm, which overlapped well with the emission of H. As a consequence, the FRET from the TPE groups in H to G-C took place and quenched the fluorescence. On the other hand, the fluorescence emission was gradually converted to the original level by the irradiation with visible light, because the closed form (G-C) reverted to the open form (G-O) which shut the FRET channel (Fig. 3B). The G⊂H complex-doped PMMA film (n_G:n_H=1:2, 1 wt%) displayed conversable fluorescence switching upon alternate UV and visible light irradiation. When irradiated with 254 nm-UV light for 10 min, the fluorescence intensity of the G⊂H complex was decreased by 84%. The irradiation with visible light gradually turned the emission intensity to the initial state. The PMMA film dopped with the G⊂H complex (1 wt%) fluoresces blue light when exposed to 300 nm-UV light (inset in Fig. 3B). The film covered with a Chinese "Eight Diagrams" pattern-containing photomask was subjected to the 300 nm-UV light irradiation for 5 min, the blue-fluorescent image was recorded on the film as the emission of the area exposed to UV light was quenched. The recorded image could be erased by the irradiation with visible light for 7 min. The film regenerated could be utilized for information storage again. The work not only offered a new approach for the preparation of fluorescence switch based on FRET and host-guest interaction but also provided fluorescence photoswitch applicable to rewritable information storage.

Most recently, Lin et al. reported another two photoswitchable rotaxanes as shown in Fig. 3C [10]. Rot-A-SP was obtained by grafting the two TPE moieties with a [2]rotaxane and complexing the TPE-functionalized [2]rotaxane with a guest (thread) containing spiropyran (SP) on the axle. The SP unit in the mechanically interlocked molecule serves as a photochromic stopper. After shuttling controlled by the base, the Rot-A-SP is reverted to Rot-B-SP. When triggered by the acid, Rot-B-SP is, in turn, converted back to Rot-A-SP again. The radiation with UV light transformed the rotaxanes Rot-A-SP and Rot-B-SP with mono AIEgen (i.e., TPE) and closed-formed photochromic moiety (namely nonfluorescent SP) into their respective isomers Rot-A-MC and Rot-B-MC with dual AIEgens and the open-formed photochromic unit (i.e., the red-fluorescent merocyanine). While the exposure of Rot-A-MC and Rot-B-MC to visible light would also revert them to Rot-A-SP and Rot-B-SP (Fig. 3C). Specifically, when using 1,8-diazabicyclo[5.4.0]undec-7-ene (DBU) and trifluoroacetic acid (TFA) as external stimuli, the highly reversible shuttling behaviors between Rot-A-SP and Rot-B-SP under respective acid and base conditions is confirmable by the ^{1}H NMR spectra.

It is reported that Rot-A-SP and Rot-B-SP are hardly emissive in pure THF and the THF/water mixtures with water fractions no more than 60% as a result of the intramolecular rotations of TPE groups, giving rise to the nonradiative decay of the excited state. When the water fraction reached 70%, along with the aggregation, the ring-opening reaction of the SP group in Rot-A-SP and Rot-B-SP spontaneously takes place owing to the high polarity of water. As a result, the SP unit was converted to the merocyanine (MC) group with the emergence of red fluorescence (~650 nm). When the water fraction of the THF/water mixtures is in the range of 70%–90%, a gradual increase in the fluorescence intensity at 468 nm is observed both in Rot-A-SP and Rot-B-SP. At 90% water content, intense blue emission is displayed, exhibiting the marked AIE effect of Rot-A-SP and Rot-B-SP.

The SP's ring-opened isomer, i.e., merocyanine (MC), was found to be AIE-active according to the RIM principle. Moreover, the absorption band of MC is partially overlapped with the TPE emission, leading to a similar FRET process from the TPE donor to the MC acceptor in Rot-A-MC and Rot-B-MC (insets in Fig. 3C). As the exposure time to UV light of Rot-A-SP and Rot-B-SP in the THF/water mixture (water fraction = 90%) increases from 0 to 90 s, the opening reaction of SP to MC unit proceeds. In the meantime, the initial emission of TPE donor at 468 nm is gradually reduced, and the emission of the newly generated MC acceptor at 668 nm is gradually intensified to different saturated intensities relying on the diverse FRET degrees induced by the distance-dependent dual-AIEgens. More specifically, the FRET efficiencies of Rot-A-MC and Rot-B-MC are 58.8% and 65.6%, respectively. It is worth mentioning that both Rot-A-MC and Rot-B-MC generally show AIE features. Moreover, the pH and temperature effects on the FRET processes were also systematically studied. Besides this, in light of its superb fluorescent and photochromic characteristics in both powder and solid film under UV/sunlight and visible light/heating processes, Rot-A-SP was successfully applied as fluorescent ink into a distinct and reversible fluorescent photopatterning [10]. The results imply that the photoswitchable [2]rotaxanes have the potential to be employed for real-world applications such as rewritable fluorescent inks for secret writings, photopatterning, and anticounterfeiting, etc. Apart from this, it is believed that this work may provide some guidance to the future design of smart photochromic fluorescent materials based on the FRET strategy.

Photochromic AIE systems achieved by chemically bonding AIE motifs and photochromic units

Besides the physical mixing or blending, the covalent bonding of AIE unit(s) to the photochromic group(s) is a more commonly used strategy to obtain systems possessing both AIE feature and photochromism characteristics. The representatives of such systems are displayed in Figs. 4 and 5.

BTEs are a group of classic P-typed photochromic molecules and therefore have been broadly utilized as a photochromic unit in the design of AIE-active photochromic systems. Meanwhile, as mentioned above, TPE is a "star" among AIE systems noted for its facile synthesis and structural modification, and hence have been widely used as an AIE-active

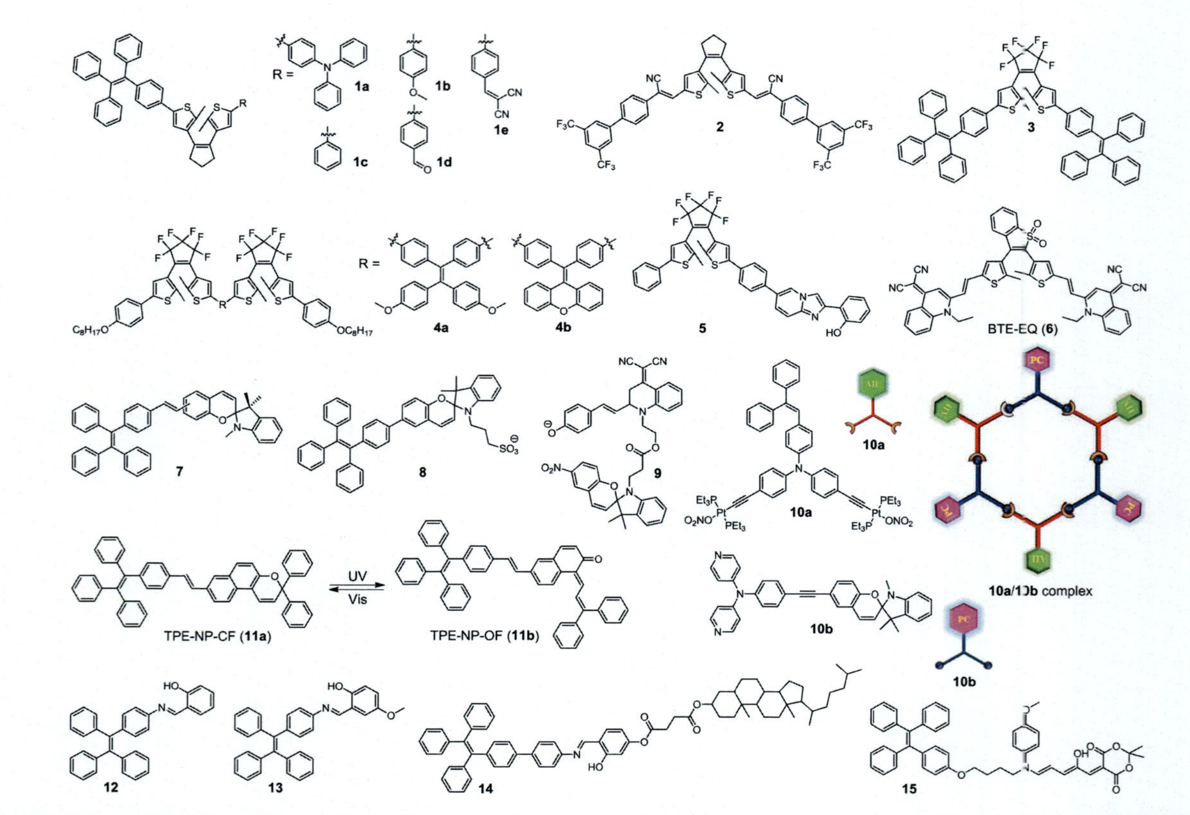

FIG. 4 Chemical structures of the representatives of the small molecular AIE-active photochromic systems which were obtained via chemically bonding AIE unit(s) to the photochromic group(s). Insets: Structures of **10a**, **10b**, and **10a/10b** complex. *Reproduced with permission from S. Bhattacharrya, A. Chowdhury, R. Saha, P.S. Mukherjee, Multifunctional self-assembled macrocycles with enhanced emission and reversible photochromic behavior, Inorg. Chem. 58 (2019) 3968–3981. Copyright 2019 American Chemical Society.*

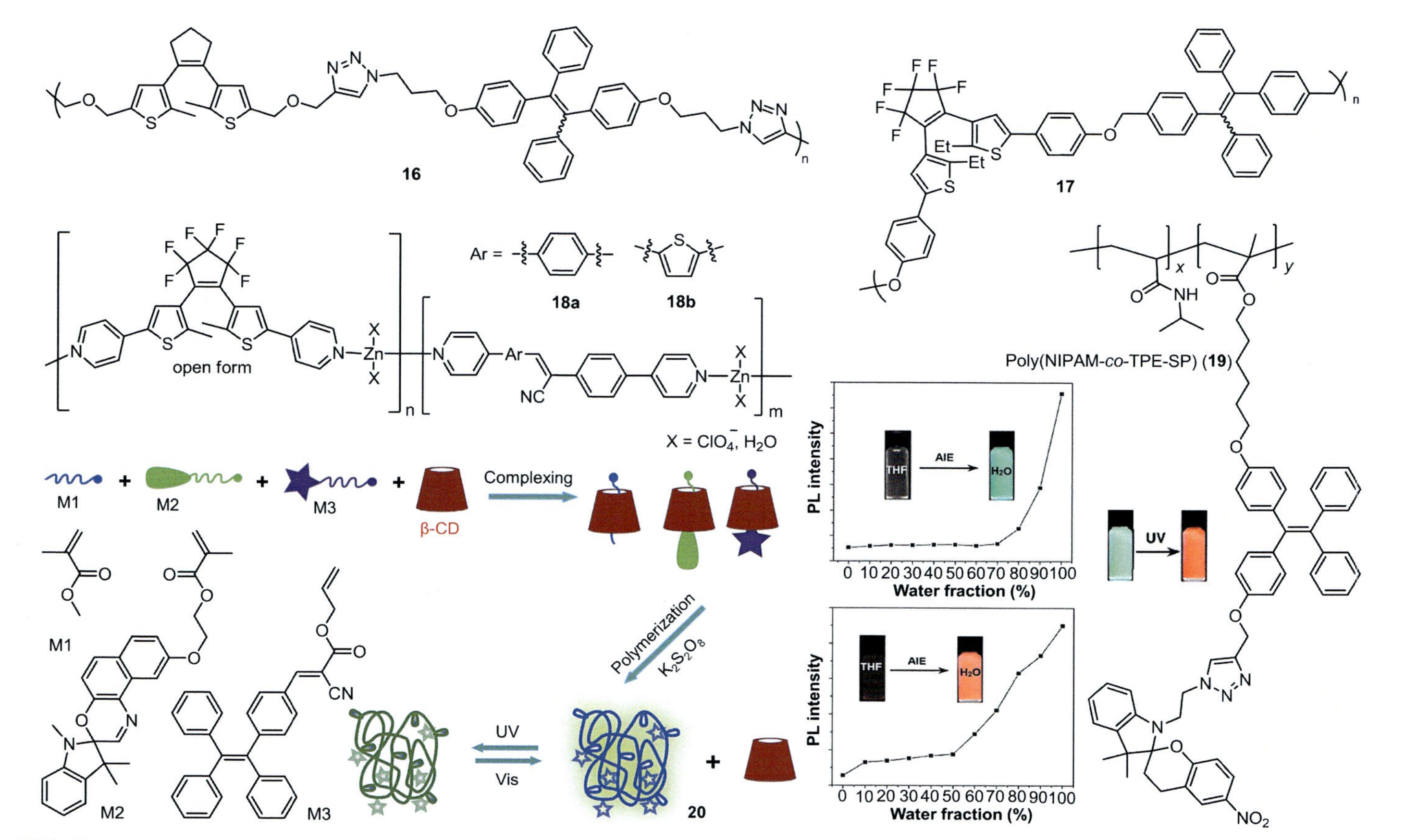

FIG. 5 Chemical structures of the representatives of the macromolecular AIE-active photochromic systems where the AIE unit(s) is/are chemically bound to the photochromic group(s). The insets indicate the typical AIE activity and photochromism of polymer **19** [11] and the construction scheme of **20** [12]. *Reproduced with permission from P.Q. Nhien, W.-L. Chou, T.T.K. Cuc, T.M. Khang, C.-H. Wu, N. Thirumalaivasan, B.T.B. Hue, J.I. Wu, S.-P. Wu, H.-C. Lin, Multi-stimuli responsive FRET processes of bifluorophoric AIEgens in an amphiphilic copolymer and its application to cyanide detection in aqueous media, ACS Appl. Mater. Interfaces 12 (2020) 10959 – 10972. Copyright 2020 American Chemical Society and P.Q. Nhiem, W.-L. Chou, T.T.K. Cuc, T.M. Khang, C.-H. Wu, N. Thirumalaivasan, B.T.B. Hue, J.I. Wu, S.-P. Wu, H.-C. Lin, Synthesis and properties of photochromic spirooxazine with aggregation-induced emission fluorophores polymeric nanoparticles, Dyes Pigments 142 (2017) 481–490. Copyright 2017 Elsevier.*

building block in the construction of AIE-active photochromic systems. For instance, compounds **1a–1e**, **3**, **4a**, and **4b** are the adducts of TPE and BTE derivatives, which show typical AIE and photochromism attributes. Benefiting from the AIE effect of TPE and the photochromism of the bithienylethene bridge, the unsymmetrical photochromic diarylethene dyads **1a–1e** are both AIE-active and photochromic [13]. As a result, the aggregates and solids of these compounds display remarkably fluorescence switches upon alternate UV and visible light irradiation. The fluorescence of these dyads in the THF/water mixture with a water content of 90% is gradually reduced upon exposure to the 365nm-UV light irradiation as a result of the FRET from the TPE donor to the closed-form BTE section (i.e., acceptor), which is recovered to the initial state by visible light irradiation. It is because the photo-induced cycloreversion reaction hindered the FRET process by changing the acceptor. Similar fluorescence "ON-OFF" switch in the solid state was also observed in **1a–1e** upon the manipulation by UV/Vis light. Moreover, the cyclization quantum yields are significantly dependent on the substituents and range from 0.02 to 0.44. Interestingly, **1a** could be utilized in the information recording and rewritable storage.

Among the developed AIEgens, 1-cyano-trans-1,2-bis-(4′-methylbiphenyl)-ethylene (CNMBE) is a very typical one owing to its unique J-type stacking and structural features, which reduce the nonradiative decay channels to a large degree upon aggregation or conformation rigidification. In view of this, photochromic AIEgen **2** was generated by symmetrically connecting one CNMBE block to each side of the BTE core [14]. Prior to the UV irradiation, the solution of open-formed **2** is light yellow, with absorption peaks at 276 and 390nm. On irradiation by 365nm light, a new absorption band peaking at a wavelength longer than 500nm ascribed to the close-formed **2** appears very slowly at the beginning, suggesting that the ring-closing reaction does not take place readily in the solution. Furthermore, the absorption at 390nm gradually decreases with a slow simultaneous increase in the absorption at 272nm and the isoabsorptive point at 319nm, implying that the *trans*-to-*cis* isomerization of the cyano ethylene part occurs before the ring-closing reaction of BTE. However, the isomerization from *cis* to *trans* renders the compound arrive at the PSS shortly, and the new absorption peak at 690nm starts to remarkably strengthen until arriving at the PSS of ring-closing reaction. Ultimately, the solution changes to green. The fluorescence spectra further proved the two-staged photochromic process, where the fluorescence hardly changes at the first stage, whereas the fluorescence at 433nm is weakened by around 50% due to the FRET between the CNMBE units and the closed-formed BTE along with the quenching effect of the *trans-cis* isomerization. Fluorescence "ON-OFF" switch with high contrast was successfully acquired in the 1D nanowire, the solid state of **2**, and the PMMA film loading **2**.

As a derivative of BTE, the perfluoro-substituted dithienylethene has also been widely used as a versatile building block for the construction of photochromic systems. For example, compounds **3**, **4a**, **4b**, and **5** are constructed with the perfluoro-substituted dithienylethene structure as the photochromic motif. Compound **3** is a conjugate of TPE and perfluoro-substituted dithienylethene fabricated by symmetrically coupling one TPE unit to each side of the perfluoro-substituted dithienylethene [15]. Compound **3** in THF solution, nanoparticle form, and solid state, undergoes a photo-induced cyclization reaction when irradiated by 365nm-UV light for 1min. The open form of **3** only shows absorption at a wavelength shorter than 400nm, while the closed form of **3** displays strong absorption in the visible region with the peak at 610nm. The reverse ring-opening process takes about 3min in THF, suggesting a process slightly slower than the ring-closing process. The highly reversible and bistable photochromism of **3** is demonstrated by the phenomenon that the absorption intensity of **3** is enhanced and weakened reversibly in solution, as nanoparticles are doped in PMMA films with alternate irradiation by UV and visible light (>440 nm). The gradual enhancement in the fluorescence intensity with increase in water fraction provides clear proof to the AIE feature of **3**. The open-formed **3** in the aggregated state is strongly green-fluorescent but nonemissive in solution, whereas the closed form of **3** is almost nonemissive in both solution and the aggregated state. The AIE and photochromism characteristics can be interpreted as follows: In solution, the extensive and active intramolecular motions of the open and closed forms of **3** open the nonradiative decay channels, leading to the quenching in fluorescence and a nonemissive state. The case of **3** in the aggregated state is more complicated. As a consequence of the RIM effect, the open form of **3** is strongly emissive. When transformed to the closed form, the new absorption peak at 600 nm overlaps well with the emission of TPE, giving rise to FRET process and the subsequent fluorescence quenching. Continuously being excited with 440 nm light, no obvious change in the fluorescence at 530 nm for either the open or the closed form of **3**, is indicative of the preservation of the bistability of **3**. Even after 1 h continuous excitation with a 440 nm light, merely a 7% reduction in the fluorescence intensity for the open form of **3** was caused, and a negligible increase in the fluorescence intensity for the nonemissive closed-form of **3** was observed. This result implies that both the open and closed forms of **3** are relatively photostable upon the excitation with 440 nm light, enabling a fluorescence readout with relatively low destruction. In addition to the application as a fluorogen for photoswitchable patterning, **3** could also be employed as an agent for localization-based super-resolution imaging. As suggested by the super-resolution fluorescence nanolocalization, the vicinal **3** exhibits a sub-100 nm resolution higher compared to the traditional fluorescent imaging.

Reducing the ratio of the AIE unit to photochromic moiety from 2:1 to 1:2, the resultant compounds **4a** and **4b** still preserve the AIE and photochromism properties [16]. The open form of **4a** is AIE-active and virtually nonfluorescent in solution while it exhibits strengthened bright cyan/yellow fluorescence in nanoparticles and the solid state. Changing the tetraarylethene unit from TPE to oxacycle TPE, namely 9-diphenylmethylene-9*H*-xanthene (OTPE), results in a red-shift in the fluorescence of more than 60 nm. For both **4a** and **4b**, merely the fully open forms are fluorescent. In contrast, the half-closed and fully closed forms of **4a** and **4b** are nonemissive. It means that only one closed perfluoro-substituted BTE is needed to quench the fluorescence. It is noteworthy that the increased number of photochrome units can lead to a dramatic improvement in the fluorescence quenching efficiency and the on/off ratio as compared to those constructed by connecting one TPE with one photochrome unit. FRET from excited tetraarylethene unit to closed-formed perfluoro-substituted BTE group would efficiently take place, owing to the good overlap between the emission of the tetraarylethene unit and the absorption of the closed-formed perfluoro-substituted BTE group. Therefore, the emission could be quenched in a closed form of **4a** and **4b**. In contrast, the open-formed perfluoro-substituted BTE group does not show any absorption at a wavelength longer than 450 nm, and hence its absorption would not overlap with the emission of the tetraarylethene unit. The FRET process is blocked and therefore the fluorescence of TPE or OTPE is permitted. Like the above-discussed AIE-active photochromic systems, photoswitchable patterns were achieved both with **4a** and **4b**. Moreover, super-resolution imaging with a resolution as high as sub 50nm was realized with **4a**, which would be discussed in detail in Section "Bioimaging and super-resolution imaging."

AIE systems exhibiting excited-state intramolecular proton transfer (ESIPT) feature have also been used to conjugate with the perfluoro-substituted BTE group. Compound **5** is such an example [17]. When irradiated by UV light, pink fluorescence is shown by **5** at the beginning. Further irradiation with UV light leads to reduced fluorescence along with the proceeding photo-induced cyclization reaction. As the water content in the THF/water mixtures increases to 10%–60%, the emission is gradually quenched, while as the water fraction exceeds 80%, strong and blue-shifted fluorescence can be observed, manifesting the AIE effect. The photoswitchable fluorescence was clearly shown by the crystal of **5**, which is green-fluorescent and can be suppressed by the formation of the closed-form isomer.

Even the cyclopentadiene unit in the BTE unit is changed to a more fused and complicated structure as that in compound **6**, the BTE core could still retain the photochromic property. Take the case of compound **6** which is an AIE-active photochromic BTE derivative achieved by covalently connecting two AIE-active quinolinemalononitrile (EQ) blocks to a derived form of BTE, for instance [18]. In solution, **6** is nonemissive and the photochromism could be observed via the change in the absorption spectra. **6** is red-emissive in the aggregated state with an emission maximum at 613nm. Unexpectedly, the aggregation-controlled photochromism was found in this compound, where the photochromic effect can be turned off and on reversibly via the control of the aggregation state during the AIE process. To be specific, in the THF/water mixture with a water fraction of 90%, there is no change in color and the spectra of the mixture when it was exposed to continuous irradiation with a 365nm-UV light, displaying a completely suppressed photochromic activity. In contrast, when the water fraction was decreased to 50% by adding THF into the above mixture, the photochromic activity was fully recovered. Normal photochromic conversion occurs upon UV irradiation in freshly prepared THF/water mixtures with a water fraction in the range of 10%–50%. It is believed that although a BTE analogue could have both antiparallel and parallel conformation, only the antiparallel one can display photoisomerization. Hence, the photochromic **6** might be confined to the parallel orientation in the aggregated state with high water fraction, and accordingly, the photochromic activity would be turned off. Such an aggregation-modulated photochromism offers a mild but efficient approach to control the photochromic activity of BTEs.

Apart from BTE and its derivatives, spiropyran (SP) and their derivatives are also well-known molecular photoswitches. They display quite different absorption characteristics during structural transformation between the ring-closed and ring-opened forms upon external stimuli such as UV/Vis light. Compounds **7–9** and **10a/10b** complex shown in Fig. 4 are representatives of small-molecular photochromic AIE systems generated via chemically attaching AIE group(s) with SP unit(s). **7** (i.e., the *meta*-**7** and *para*-**7**) are nearly nonemissive in THF or in the THF/water mixture with a water fraction lower than 70% [19]. The AIE properties of the TPE and SP conjugates *meta*-**7** and *para*-**7** were clearly demonstrated by the fluorescence enhancement as the increase in the water fraction of the THF/water mixtures. *meta*-**7** and *para*-**7** could undergo reversible photoisomerization between their unconjugated closed-form isomer and conjugated open-form isomer under the irradiation of a 254nm UV light. The absorption band peaking at 366nm for the closed form of *meta*-**7** is gradually weakened accompanied by a new peak assigned to the open form emerged at 528nm and enhanced with the exposure time. *para*-**7** shows a gradual decrease in the absorption band peaking at around 350 nm and a boosted absorption peak appeared at around 476 nm when it was continuously irradiated by UV light. Obviously, the open-form absorption of *meta*-**7** holds a red-shift of up to 52 nm as compared to that of *para*-**7**, suggesting that the open form of *meta*-**7** has a conjugation larger than that of *para*-**7**. In contrast to the photochromic absorption of *meta*-**7** and *para*-**7** in dichloromethane, the fluorescence change in this solution by UV light irradiation is much less obvious due to the AIE properties of TPE. On the contrary, in highly viscous phenylcarbinol, these two compounds can readily and partially transform into their open forms. Moreover, *meta*-**7** in phenylcarbinol shows superhigh sensitivity to light as the open form of *meta*-**7** can easily change to the closed form under the irradiation of visible light (>500 nm) and afterward the closed form converts to the open form when immediately left in the darkness. In the meantime, the out-appearance of *meta*-**7** and *para*-**7** in phenylcarbinol reversibly changes from colorless to red and the fluorescence becomes weakened in dark and recovers to intense blue emission under visible light.

8 is also an adduct of a SP derivative and TPE [20]. Similar to system **7**, the isomerization of **8** could also be switched by either visible light/darkness or acid/base in solution. Both the closed form and open form of **8** are AIE-active. The aggregates of the closed-formed **8** release a cyan fluorescence assigned to the local excited emission of the TPE unit. Whereas, the open-formed **8** possesses two different aggregated states that exhibit two red emissions due to the intramolecular charge transfer.

AIE-active and photochromic **9** is a conjugate of quinolinemalononitrile and SP derivative [21]. As **9** was generated by a H_2S-triggered cleavage, **9** and the compound yielding **9** will be introduced in detail in Section "Bioimaging and super-resolution imaging."

SP derivatives were even used as a photochromic block in the construction of self-assembled macrocycles with AIE and photochromic properties [22]. The macrocycle formed by the complexation of **10a** and **10b** is the first example of a self-assembled Pt(II) architecture which is featured with multiple functionalities including AIE, photochromism, and acidochromism. This macrocycle displayed enhanced emission driven by coordination and color change induced by light as compared to the original building blocks, i.e., **10a** and **10b**.

Naphthopyran is a photoswitchable group analogous to SP and its derivatives. By covalently linking a naphthopyran unit with a TPE group, **11a** was afforded [23]. **11a** is a photochromic AIEgen which hardly exhibits any fluorescence emission in the THF/water mixture with a water fraction less than 70%. In contrast, the emission band at 480 nm appears and is boosted when the water fraction is over 70%. Similar to systems **7** and **8**, the reversible photoisomerization between the unconjugated closed form (i.e., **11a**) and the conjugated open form (i.e., **11b**) can take place under the UV light irradiation in solution, resulting in the red-shift of the absorption maximum from 362 to 490 nm. The visible light irradiation can revert the absorption to the initial state. **11a** is highly blue-emissive and yellow-green in the aggregated state, while its open form (**11b**) is nonemissive and yellow-brown under parallel conditions. Therefore, repetitive photoswitchable patterns can be achieved in the solid state by alternate irradiation with UV and visible light.

As discussed in Section 1.1.2, salicylaldehyde hydrazone is a photochromic moiety that exhibits good reversibility and high resistance to fatigue [24]. When irradiated by UV light, the reversible tautomerism between the enol and keto forms would readily take place. Remarkably, the enol form of salicylaldehyde hydrazone is conducive to fluorescence emission owing to the ESIPT process. Moreover, the ESIPT systems easily possess the AIE effect. Li and Hou et al. smartly conjugated the salicylaldehyde hydrazone to the TPE unit and obtained an AIE-active system with reversible photochromism [25]. The resulting 4-(1,2,2-triphenylvinyl)aniline salicylaldehyde hydrazone **12** is yellow and displays intense yellowish-green fluorescence emission (545 nm) in the aggregated state prior to UV irradiation. **12** turned red accompanied by the significant decline in the fluorescence emission after UV irradiation. Thanks to its high resistance to fatigue, **12** can be utilized in reversible and erasable photopatterning and anticounterfeiting. Very interestingly, 2 years after the report of **12**, its analog (i.e., **13**) was reported [26]. TPE-based Schiff base 4-methoxy-2-(((4-(1,2,2-triphenylvinyl)phenyl)imino)methyl)phenol (i.e., **13**) has two distinct polymorphs which show completely different photochromic and fluorescence properties. The square block-like crystal of **13** (**13-a**) fluoresces very weakly and is reversibly photochromic. On the contrary, the needle-like crystal of **13** (**13-b**) shows relatively stronger orange fluorescence but no photochromism. It was found that such unique polymorph-dependent properties might be attributed to the distinct photo-inactivation ways originating from dramatically different crystal packing modes. In addition, **13-a** can be employed to fabricate rewritable and self-erasable paper responsive to visible light without the need for ink writing. As for the rewritable paper, the blue light (450 nm) can be used as the pen, whereas the white light and light with a wavelength ranging from 460 to 570 nm could be applied as the eraser. When combining **13-a** and **13-b**, reusable material for data encryption with both static and dynamic encryption-decryption could be obtained.

In Fig. 4, **14** exhibited is a conjugate of TPE, salicylideneaniline, and cholesterol, where the TPE serves as an AIE unit, salicylideneaniline functions as an ESIPT and photochromic moiety, and cholesterol acts as a group favoring gelation and self-assembly [27]. Unexpectedly, **14** could only gelate in cyclohexane. Its xerogel formed slowly by evaporating cyclohexane from the corresponding organogel, was made of a fibrous structure which is $\sim$100 nm wide and tens of micrometers long. Further crosslinking affords a 3D framework. The initial solution of **14** emits very weak fluorescence. The fluorescence is enhanced with the decrease in temperature from 70°C to 25°C, exhibiting a gelation-induced fluorescence enhancement. As anticipated, **14** inherits the typical AIE attribute of TPE. Under UV light irradiation, the absorption bands of **14** in the solutions of DMF and THF kept nearly unaltered at 280 and 358 nm. Prolonged irradiation of **14** in cyclohexane gel with UV light decreased the absorption bands at 280 and 358 nm. Prolonged irradiation of **14** in powder form made a new peak emerge at 425–575 nm and intensified with irradiation time.

Derivatives of donor-acceptor Stenhouse adduct (DASA) is a new group of photoswitches that show negative photochromism [28]. Irradiating the strongly colored elongated triene form using visible light initiates fast photobleaching which results in a colorless cyclopentenone. In comparison to traditional photochromic systems including azobenzene, BTEs, and SPs, which usually need UV irradiation to trigger their photoisomerization, the photochromic process of DASA is merely induced by visible light. He et al. introduced DASA moiety into the AIE system for the first time [28]. As a result, **15** is developed with TPE as the AIE-active part and DASA as the photochromic unit. By virtue of the FRET process, the excited energy of TPE can be transferred to the colored elongated triene form of DASA, simultaneously quenching the fluorescence of TPE. Once irradiated with visible light, the colored triene form which is thermodynamically stable converted to the colorless keto tautomer that is photostationary, recovering the intense emission of the TPE-based molecule. Apart from the photochromism in solution, the visible light-triggered fluorescence can be observed in solid films with a high on/off ratio. Moreover, heating treatment (at 55°C) could revert the absorption and emission spectra to the initial state. By virtue of the soft-lithographic contact printing method, visible-light-trigged fluorescent surface relief patterns have been prepared and supposed to have great potential in applications as distributed feedback lasers, optical molecular switches, and other devices driven by photo/light.

Besides the small-molecular photochromic AIE systems achieved by linking AIE motif with the photochromic unit via chemical bonds, there are also a number of macromolecular AIE-active photochromic systems obtained through chemical bonding. Amongst these macromolecules, systems **16**–**20** are chosen as representatives and are displayed in Fig. 5.

Polymer **16** is a photoswitch containing BTE and TPE groups and a triazole linker in the repeating unit and synthesized via click reaction [29]. The AIE behavior of TPE is with both the open and closed forms of BTE in **16** with high water fractions. The photoswitching of BTE from open to closed forms under UV irradiation in **16** takes a leading position in an organic solvent such as THF, THF/water mixture with water fraction as high as 90%, and acidic conditions. UV irradiation of **16** at 90% water fraction completely quenches the AIE effect through a FRET process from TPE to the cyclized BTE. Interestingly, the unusual monomeric fluorescence of TPE was unveiled by the photocyclization of BTE in **16** instead of the AIE emission of TPE with the open-formed BTE in **16** under conditions favoring aggregation, such as high water-content solutions (e.g., 90% water), and acidic conditions.

17 is a photochromic AIE-active polymer bearing perfluoro-substituted BTE groups and TPE units synthesized via a nucleophilic substitution reaction between 1,2-bis[5′-(4′-hydroxyphenyl)-2′-ethylthien-3′-yl]perfluorocyclopentene and 1,2-bis[4-(bromomethyl)phenyl]-1,2-diphenylethene [30]. Polymer **17** is soluble in diverse organic solvents and can be easily processed into a thin film. The perfluoro-substituted BTE chromophores in polymer **17** can be subjected to reversible photoisomerization between the open and closed forms upon alternate irradiation by UV and visible light. As a consequence, the emission of the AIE-active **17** can be switched by light in the aggregated state and solid film. To be specific, the THF solution of **17** shows an absorption maximum at around 283 nm. The exposure to the UV light of 300 nm makes the THF solution of **17** change to blue, corresponding to the new absorption band at 622 nm, indicating the occurrence of ring-closing reaction of perfluoro-substituted BTE group. The PSS is reached after being irradiated for 40 s. Upon irradiation with a 620 nm-visible light, the blue solution reverts to a colorless state, suggesting the conversion from ring-closed form to ring-opened form. The 622 nm absorption band completely vanishes after being irradiated by visible light for several minutes. A similar photochromic phenomenon is also observed in the solid state. In the aggregated or solid state, the fluorescence of **17** will be quenched by the irradiation with a 300 nm-UV light. Moreover, fluorescence will be recovered by irradiation with 620 nm light. The reversible photoisomerization of the perfluoro-substituted BTE unit enables the fluorescence of **17** to be switched repeatedly by alternate UV and visible light irradiation.

Perfluoro-substituted BTE has also been used as a ligand to afford coordination polymers with AIE properties and bistable solid-state fluorescence switching ability [31]. Polymers **18a** and **18b** are such examples. In these polymers, ligand 1,2-bis[2-methyl-5-(4-pyridyl)-3-thienyl]perfluorocyclopentene (perfluoro-substituted BTE) functions as a photochromic unit and the ligands (Z)-2,3-bis[4-(4-pyridyl)phenyl]acrylonitrile and (Z)-2-[4-(4-pyridyl)phenyl]-3-[5-(pyrid-4-yl)thiophen-2-yl] acrylonitrile acts as the AIE moiety, respectively, in polymers **18a** and **18b**. The coordination of ligands with zinc ions generates the coordinated polymers. The ring-closure yields for PSS upon being irradiated by a 365nm light were calculated to be 38% and 34% for **18a** and **18b**, respectively. Intense fluorescence was witnessed for the open forms of the infinite coordination polymers (**18a**: $\lambda_{em}=470$nm, fluorescence quantum yield=0.18; **18b**: $\lambda_{em}=577$nm, fluorescence quantum yield=0.19). The closed form generated by 365nm UV light irradiation barely displays any fluorescence as a result of the FRET process between the AIE-active ligand and the closed-form perfluoro-substituted BTE. The corresponding fluorescence on/off ratios is 21 and 17, for **18a** and **18b**, respectively. Well-ordered micropatterns of **18a** and **18b** fabricated via a soft-lithographic method show high-contrast photoswitchable fluorescence images.

Besides BTE and its derivatives, SP and its derivatives are also very attractive building blocks to generate AIE-active photochromic polymers such as polymers **19** and **20** [11,12]. **19** is an amphiphilic AIE copolymer that is composed of a hydrophilic unit (i.e., *N*-isopropylacrylamide (NIPAM)) and a fluorophoric group obtained by covalently linking TPE with SP via click reaction [11]. When exposed to UV light irradiation, the nonfluorescent closed-form SP in the copolymer can be transformed to the fluorescent opened-form MC in the copolymer in an aqueous solution, giving rise to the ratiometric fluorescence of AIEgens between cyan-emissive TPE and red-fluorescent MC at 517 and 627nm, respectively, through FRET. Multiple-stimuli responsiveness was observed for copolymer **19**. Different FRET processes of **19** are observable under alternating UV and visible light irradiations, acid-base treatment, heating and cooling, and CN^- interactions. As a consequence, the small variation in the environmental factors including light irradiation, pH value, temperature, and CN^- concentration could be probed in aqueous media with the aid of distinct ratiometric fluorescence changes of the FRET behavior in the amphiphilic copolymer **19**. Uniformly-spherical nanoparticles with a smooth surface of copolymer **20** possessing a poly (methylmethacrylate) backbone are obtained by semicontinuously polymerizing spirooxazine-pendent vinyl monomer, TPE-attached vinyl monomer, and methyl methacrylate monomer [12]. The nanoparticles exhibit remarkable AIE properties and superb photochromic properties as a result of the conversion from spirooxazine to the merocyanine form after UV irradiation. In consequence, the FRET process is activated which quenches the fluorescence of TPE in **20**. The fluorescence of the nanoparticles can thus be reversibly quenched or switched off upon the irradiation by UV and visible light.

Photochromic AIE systems created following the principle of restriction of the intramolecular motions (RIM)

In the past decade, a large number of photochromic AIE systems have been created by following the principle of restriction of the intramolecular motions (RIM). The representatives amongst them are displayed in Figs. 6 and 7. Simple esterification of carboxyl bithienylethene with 7-hydroxycoumarin produces compound **21** (*o*-B2C) [32]. **21** initially gives an absorption band at 315nm, whereas the solution of **21** changes from colorless to purple owing to the emergence of a new observable absorption band at 562nm originated from the formation of closed form. The purple solution is bleachable to colorless when irradiated with visible light (>550nm), indicative of a photochromism capability like other BTE-bearing systems.

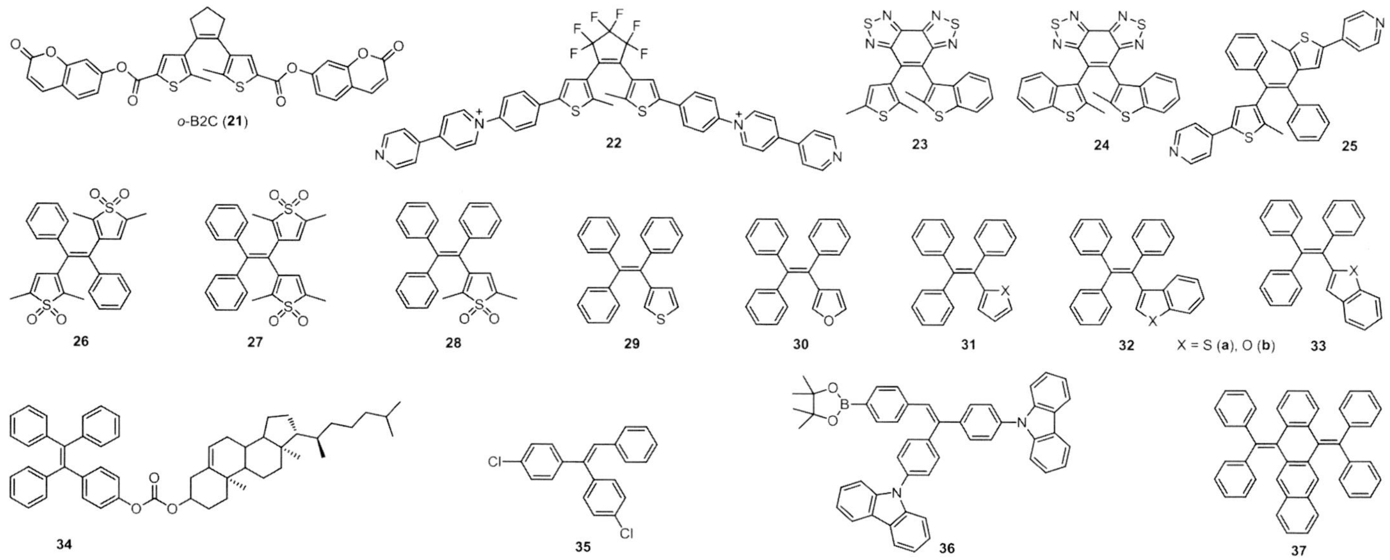

FIG. 6 Chemical structures of the partial representatives of the AIE-active photochromic systems achieved by RIM principle and ring-closing reaction.

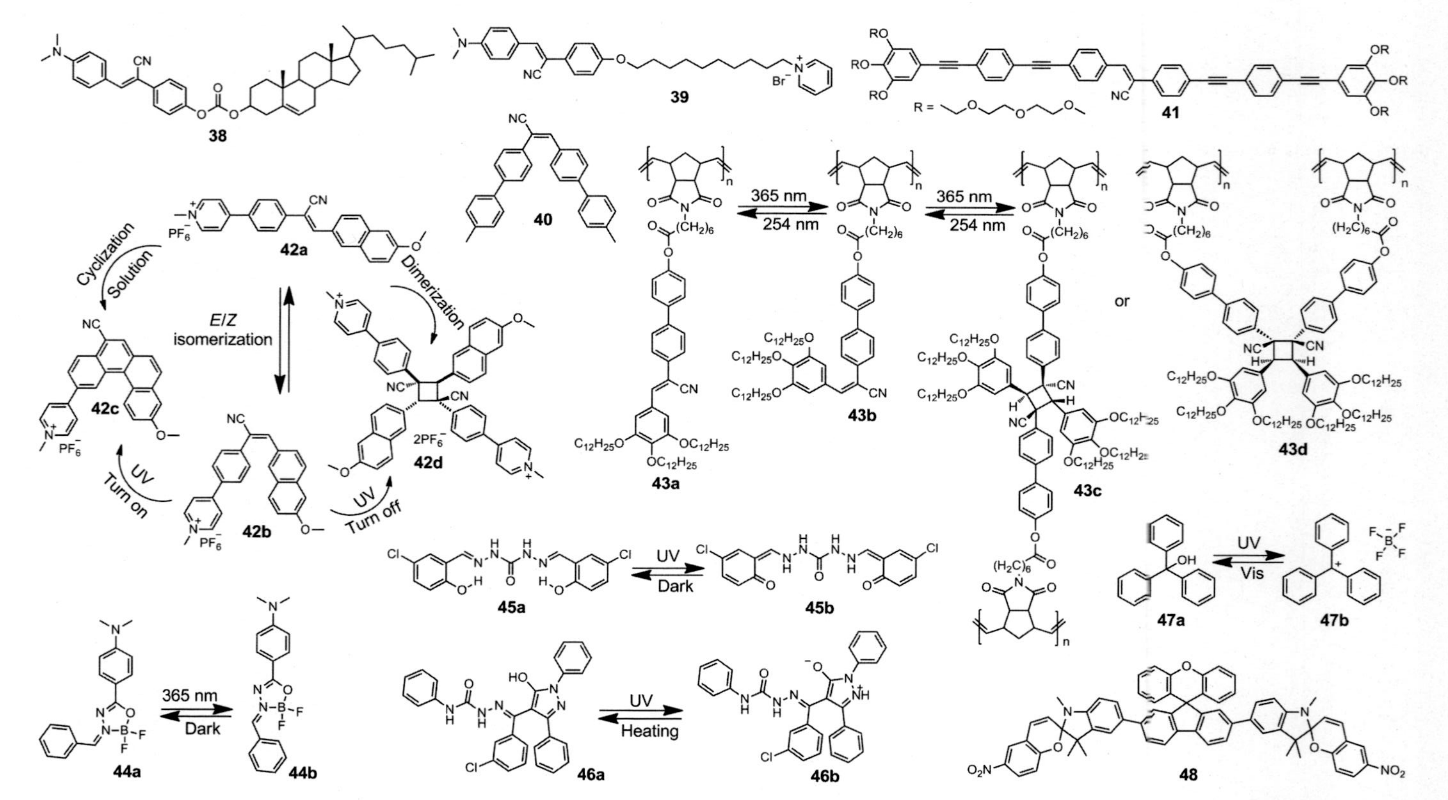

FIG. 7 Chemical structures of the other representatives of the AIE-active photochromic systems achieved by the RIM principle and other photochromic mechanisms.

Unlike most coumarin derivatives, **21** shows marked AIE properties, which are attributed to the RIM principle. **21** shows no emission with or without light irradiation. Whereas in the aggregated state, it is fluorescent and nonemissive before and after 254nm-UV irradiation, respectively. The visible light irradiation can also restore fluorescence. Perfluoro-substituted BTE was connected to bispyridinium moieties via single bonds to yield compound **22** [33]. It is AIE-active: Its methanol solution is very weakly fluorescent while the addition of glycerin to the methanol solution significantly enhances and blue-shifts the emission. By virtue of the photochromic perfluoro-substituted BTE, **22** is able to work as a fluorescent photoswitch in a highly viscous medium. The fluorescence of **22** will be significantly decreased by 93% in glycerin with 254nm light irradiation for 7min because the emissive open form changes to the nonfluorescent closed form via the photocyclization of the perfluoro-substituted BTE core. Furthermore, the fluorescence on/off ratio is estimated to be 14. On the other hand, the irradiation of **22** with visible light (>490nm) gives rise to the full restoration of the emission intensity and fluorescence color. The photocontrollable optical behaviors of **22** in the solid film state and its application will be further discussed in Section "Anticounterfeiting and information storage."

23 and **24** are photochromic systems with large steric hindrances [34]. These benzobis (thiadiazole)-bridged diarylethenes are both AIE-active and photochromic. In **23** and **24**, one or two bulky terminal units of benzothiophene are introduced into the photoswitchable system to tune the rotation or vibrational barrier of the single bonds linking the benzobis(thiadiazole) ethene core with side groups as well as the molecular packing to modulate the AIE activity. For the open forms of **23** and **24**, the restriction of intramolecular rotations (RIR) effect originated from large steric hindrance is responsible for the fluorescence properties, i.e., turned-on emission both in solution and aggregated state. As to the closed forms, the restriction of intramolecular vibrations (RIV) mechanism explains the AIE activity. Active intramolecular vibrations contributed by large steric hindrance effectively consume the excited-state energy and result in low emission of closed forms of **23** and **24**. In the meantime, the twisted conformations rendered by large steric hindrances remarkably hinder the intermolecular π-π stacking in the aggregated state. Along with the activated RIV process, the fluorescence of **23** and **24** are thus turned on to exhibit obvious AIE activity. Undoubtedly, **23** and **24** are photochromic and show significant reversible off/on fluorescence in the film state.

(*E*)-1,2-Bis(2-methyl-5-(pyridin-4-yl)thiophen-3-yl)-1,2-diphenylethene (**25**) is very analogous to the pyridinyl-substituted TPE. **25** was reported to be AIE-active [35]. It is because the aromatic moieties including phenyl rings, pyridinyls, and thiophenes are linked to the central ethene unit via single bonds, which could consume the excited-state energy via active intramolecular rotations and quench the emission in the solution state. Whereas, in the aggregated state, the intramolecular rotations are greatly restricted, which activates the radiative decay channels to release intense fluorescence. Meanwhile, the propeller-like conformation prevents π-π stacking and hinders the quenching of fluorescence. The reversible ring-closing and opening reaction between the phenyl and vicinal thiophene group endows **25** with the obvious photochromic property. After being irradiated with 365nm light for a few seconds, the THF solution turns from pale yellow to purple. While the irradiation by 500nm light, the solution changes from purple to pale yellow again. It is worth mentioning that when it coordinates to zinc ions to generate the coordination polymer, reversible photochromism in the solid state is observed.

It has been reported that tetraarylethenes similar to TPE and their oxides usually have the potential to be photochromic [36]. Analogous to TPE and its derivatives, the typical AIE effect has been observed in compounds **26–28**. Moreover, all these three AIEgens show reversible transformation between the open and closed forms when irradiated alternately with UV and visible light in the THF solution. Taking compound **27**, for example, the irradiation with 254 nm UV light leads to the gradual reduction of the absorption band at 351 nm while the band at 281 nm is boosted slowly with a new peak emerging at about 433 nm, making the solution turn from colorless to bright yellow. Upon illumination with visible light, the color faded completely with the absorption returning to the initial state. A similar photochromic phenomenon was also witnessed for the solid powder and PMMA films of **26–28**, suggestive of their possible application as fluorescent molecular photoswitches with simple structures. Very interestingly, the UV irradiation of the solutions of **26–28** turns on the fluorescence while the irradiation of **26–28** with UV light switches off the fluorescence in the solid state.

Like **28**, even only replacing one phenyl ring of TPE with aryl group such as 2,5-dimethylthiophene 1,1-dioxide (**28**), thienyl (**29**, **31a**), furan (**30**, **31b**), benzothienyl (**32a**, **33a**), and benzofuran group (**32b**, **33b**) would possibly afford AIE-active photochromic compounds with simple structures [37,38]. The TPE-analogous molecular structures endow compounds **29**, **30**, **31a**, **31b**, **32a**, **32b**, **33a**, and **33b** with typical AIE activity. The reversible photo-induced cyclization and cycloreversion reactions between the phenyl ring and the vicinal heterocyclic aryl groups give rise to the photochromism properties both in solution and in the solid state. Generally, their absorption maxima and colors are greatly red-shifted upon the irradiation by UV light, while the visible light illumination will completely revert the absorption and change the color to the initial state. These compounds are highly fluorescent before UV light irradiation but become almost nonemissive after UV light irradiation in the solid state. It is noteworthy that their photochromic properties in the solid form are more dramatic as compared to the ones in the solution state. Moreover, the systems with benzothienyl and benzofuran groups show more remarkable photochromism than the ones with thienyl and furan groups, respectively. In addition, the tetraarylethenes containing furan groups hold better photochromism compared to their thienyl counterparts. Such researches revealed that the position, conjugation, and types of substituents would all exert a significant impact on the photochromism in color and strength. With **32a** and **32b**, dual-mode patterning controlled by light irradiation on filter papers and anticounterfeiting were demonstrated.

Surprisingly, even without photoactive groups such as thienyl, furan, benzothienyl, and benzofuran group, TPE derivatives like **34** could also show photochromism [39], although TPE itself is reported to be not photochromic. **34** is AIE-active. With twisted and flexible 3D conformation, **34** also exhibits mechanofluorochromic property. The heated white powder, as well as the transparent crystal of **34**, could be turned to deep red after UV illumination for about 1 s. Once the UV light is removed, the color changes back to the initial state within 1 min. The photochromic behavior of **34** is closely associated with the solid-state photocyclization between the carbonic ester group-decorated phenyl and the adjacent phenyl ring at the same side of the vinyl group. The application of the multistimuli responsive **34** for anticounterfeiting will be discussed in Section "Anticounterfeiting and information storage."

Dichloro-substituted triphenylethylene derivative **35** is AIE-active [40]. It is nonfluorescent in solution but highly emissive in the aggregated state with an emission peak at 425 nm. Upon the UV light irradiation, the solids turn from white to bright red, and meanwhile, the blue emission gets quenched gradually. The red color reverts to white several seconds after switching off the UV light. Along with the photochromic bleaching process, the emission is recovered. Such a cycle could be repeatedly operated for many times without dramatic fatigue. It is believed that the photochromic response to UV light irradiation is attributed to the stilbene-type 6-π electron ring-closing reaction. The instability of the closed form suggests the low energy barrier during the photochromic process, which makes the closed form rapidly convert to the open form (**35**) even in dark at room temperature. After being irradiated for a number of cycles in air, the white solid of **35** changes into a pale yellow one, which is weakly fluorescent in solution. This was due to dehydrogenation of the closed form of **35**. Such a photooxidation reaction was also verified by the ^1H NMR characterization.

On the basis of **35**, the switchable photochromism of **36** is not hard to understand, as **36** is a carbazole-decorated triphenylethylene derivative [41]. With the AIE-active triphenylethylene framework and highly distorted 3D conformation, **36** possesses marked AIE activity and piezochromism. As a result of the stilbene-type intramolecular photocyclization similar to the case of **35**, good photochromic performance is shown by **36** in the crystalline state. The crystals of **36** exhibit strong fluorescence with a peak at 448 nm. Pressing the crystals leads to the crystal-to-amorphous transition, where the resulting amorphous state of **36** displays a distinct color and a bathochromically shifted emission band at 470 nm. Unexpectedly, the photochromism is found to be morphology-dependent: the photochromic properties of **36** are merely observable in the crystalline state but not in the amorphous state. As a result, the photochromic feature could be switched by the pressing/fuming and heating/fuming operations. Integrating the piezochromic properties with the photochromic properties into one molecule, the off/on switch could be achieved with different colors and emission properties.

The photochromism mechanism for AIE-active compound **37** is analogous to those for **35** and **36**. As revealed by the photochromism investigation, the unique and unexpected photochromic properties are associated with the photocyclization reaction of *cis*-stilene, the molecular configuration in the single crystal, and the tetracene skeleton as well. Moreover, the photochromism with rapid responsiveness, photo-reversibility, and thermo-irreversibility is promoted by aggregation. The crystals of **37** are colorless and needle-shaped with blue fluorescence under the illumination with UV light [42]. It was surprising to find that the crystals turned from colorless to red after taking away the UV light even if the irradiation merely lasted for as short as 1 s. In the meantime, the fluorescence was dramatically reduced. When the UV light-irradiated **37** is placed under room light, the crystals revert from red to colorless, along with the recovery of the blue emission of **37**. As such, the crystals of **37** exhibit a typical photochromism. The photochromic processes of crystals of **37** were then studied by UV-Vis reflectance spectroscopy. Before irradiation with UV light, an absorption maximum at 370 nm is shown by **37** in the crystalline state. Upon UV light irradiation, a strong absorption peak at 498 nm appears and increases with the photoirradiation. Such a photochromic process was repeated 10 times without any obvious degradation, reflecting the good fatigue resistance. Moreover, both the experiments and the theoretical calculations revealed that the dihydro intermediate is the most probable photochromic product generated via a 6-π electronic cyclization reaction.

38 is a cyanostilbene derivative tagged with a cholesterol unit that can self-assemble into vesicles in solution and then merge together to afford branched and capped nanotubes upon the irradiation by UV light [43]. By virtue of the AIE feature and the fluorescence enhancement induced by *trans*-to-*cis* photoisomerization, multiple color conversions are realized. **38** is weakly and blue-greenish fluorescent with a peak at 490 nm in THF, while in the aggregated state, **38** is strongly yellow-fluorescent with a peak at 550 nm. Upon UV light illumination, the fluorescence is greatly boosted for molecular species of **38**, whereas the one for the aggregates of **38** is gradually quenched and the emission color changes from yellow to blue-greenish as a result of the conformation transformation induced by UV light. It is noteworthy that the photoisomerization induces morphology transitions from vesicles to nanotubes. When the aggregates are destroyed by external stimuli, the fluorescence stem from the *cis*-isomer is able to be switched on, providing a novel means to probe the stimuli. As a result, H_2O_2 was selectively and quantitively detected under UV light illumination via this approach. **39** is also a cyanostilbene derivative. Moreover, it is amphiphilic and photoresponsive. Similar to **38**, **39** could also form vesicles via self-assembly [44]. Interestingly, when irradiated by light, the vesicles would reduce in size. When included by cucurbit[7]uril (CB[7]), the twisted intramolecular charge transfer (TICT) state of **39** is suppressed and the fluorescence of the supramolecular system is switched on. In the course of photoisomerization from *trans* to *cis* isomer, **39** exhibits hypochromicity and enhances emission as molecular species, similar to the case of inclusion by CB[7].

Symmetrically substituted cyanostilbene, (*Z*)-2,3-Bis(4′methylbiphenyl-4-ly)acrylonitrile (**40**), also exhibits typical photochromic effect [45]. The α-cyanostilbenic unit-containing diphenylstilbene **40** in powder state is almost nonluminescent under the illumination with 365 nm, while its *E*-isomer powders are brilliantly blue-emissive. Moreover, **40** and its *E*-isomer show absorption at 335 nm and 358 nm, respectively, while, both are also nonemissive in solution. The *E*-isomer of **40** is AIE-active. In dilute solution, **40** displays *Z* to *E* photoisomerization, and the *E*-isomer of **40** shows *E* to *Z* photoisomerization. The *E*-isomer of **40** in the solid state hardly exhibits apparent photochromism due to its dense packing and strong confinement in the intramolecular motions. However, the *E* to *Z* photoisomerization of the *E*-isomer of **40** could efficiently take place in the liquid crystalline mesophase state under UV light irradiation. In contrast, a fairly dramatic enhancement in the fluorescence is observed when **40** is illuminated by a 254 nm light under room temperature for 10 min, as a result of the loose packing originated from bent molecular conformation which facilitates the *Z* to *E* isomerization. Taking advantage of the photo/thermal reversible isomerization between *Z* and *E* forms of **40**, a fluorescent optical recording system with repetitive writing-erasing cycles was successfully demonstrated.

Making full use of the cyanostilbene derivatives, efficient and precise modulation over *E*/*Z* isomerization and [2+2] photocycloaddition with visible light in solution is achieved through a direct solvent-controlled means on the basis of supramolecular polymerization. For example, the linear cyanostilbene bolaamphiphilic **41**, in which the cyanostilbene framework is tethered with two oligophenyleneethynylene-based aromatic wedges featuring peripheral glycol chains, is both AIE-active and photochromic [46]. In the solution state, quantitative *E*/*Z* photoisomerization with high reversibility is displayed. In comparison, fluorescent J-type supramolecular polymers which are subjected to a highly efficient and selective [2+2] photocycloaddition when irradiated by visible light would be induced to form in an aqueous solution. It has been demonstrated that combining aromatic, hydrophobic, and CN···H interaction could render the accurate topochemical control for a particular cyclobutene (*anti* head-to-tail) to form efficiently. It means that the photoresponsiveness of cyanostilbene derivatives in the solution can be regulated in a precise manner through supramolecular polymerization controlled by solvents.

As suggested by compounds **38–41**, the reversible *cis*/*trans* or Z/E isomerization and/or [2+2] cycloaddition serve as the main mechanisms for the photochromism of at least the majority of cyanostilbene derivatives [43–47]. Even more, interestingly, Tang et al. found that in the unique AIEgen system **42**, a photocyclization process is also involved besides the Z/E isomerization and [2+2] cycloaddition-induced photodimerization [48]. **42a** has an absorption maximum at around 380nm. Under irradiation with room light, the Z to E isomerization of the cyanostilbene unit is promoted in solution. Such a photoisomerization is reversible and the E-isomer **42b** could be thoroughly reverted to **42a** via thermal treatment (heating at 75°C for 36 h). When irradiated with 365nm UV light, intensified strong yellow emission is displayed, suggesting that the emission process of **42a** in solution is activatable by light. Since **42a** is AIE-active, it is only weakly orange-fluorescent in solution, while the emission is dramatically enhanced and hypochromically shifts by 22 nm to 560nm in the presence of UV light irradiation. The fluorescence is enhanced with the irradiation time, and the absorption of **42a** at 380nm is progressively blue-shifted with reduced intensity as well. In the meantime, two new absorption bands emerge at 315 and 280nm and get enhanced with the irradiation progress. Such a process is assigned to the stilbene-type 6-π electron ring-closure photocyclization reaction which forms **42c**. **42a** forms microcrystals in an aqueous solution with a high water fraction, and the UV light irradiation could further transform **42a** to **42d**. Moreover, **42b** even has more ease to undergo topochemical [2+2] cycloaddition-induced photodimerization as compared to **42a**. **42a** weakly fluoresces dim orange light at 580nm with a quantum yield of 3.4% in its acetonitrile solution, and its fluorescence could be intensified to a quantum yield of 6.5% in the acetonitrile/water mixture with a water fraction of 99%, as a result of the AIE effect. The facile photocyclization in the solution generates the yellow-fluorescent **42c** with a quantum yield of up to 42.0%. The microcrystals of **42a** affording a faintly blue-fluorescent **42d** with a quantum yield of 1.3% via photodimerization. The difference in the quantum yields between solution and aggregates after photoreaction is 32 times. A well-resolved 2D fluorescence pattern with a high signal-to-background ratio and clear edges was achieved with **42a**.

Cyanostilbene could also retain photoactivity even when incorporated into polymers. System **43** is such an example [49]. Polynorbornene **43a** contains α-cyanostilbene-decorated side chains and could display reversible fluorescence switching induced by light in the liquid crystalline state. With three alkoxy tails, **43a** has a hexagonal columnar structure. **43a** also preserves the AIE feature of α-cyanostilbene. The fluorescence is intensified gradually with the continuous addition of water to the THF/water mixtures accompanied by the red-shift of emission maximum. Upon irradiation with 365nm UV light, **43a** can undergo reversible Z/E isomerization in solution. In noncrystalline thin films, photochromic fluorescence switching is induced through different UV irradiation, because of the E/Z isomerization and [2+2] cyclodimerization, where the fluorescence gradually is blue-shifted from 476 to 462nm. The fluorescence could be recovered via a thermal annealing as well as the irradiation by a 254nm light.

As a green fluorescent protein chromophore analog, **44a** is a difluoroboronate-furnished acylhydrozone [50]. It is faintly emissive in the solution state but strongly fluoresces in the powder and film state, with a solid-state fluorescence quantum yield of 32.0%, suggestive of its marked AIE characteristics, owing to the RIM process. **44a** is also photochromic. When irradiated by intense UV light at 365nm, photoisomerization from *Z*-isomer to *E*-isomer (**44b**) is induced. As a result, the visible absorption is decreased, and the fluorescence maximum red-shifts from 555 to 574nm with the concomitant enhancement on fluorescence intensity, when excited at 450nm. Such a process could be reversed by placing the chromophore in a totally dark environment, away from irradiation source and even devoid of ambient light.

ESIPT systems generally have the potential to be photochromic and ESIPT fluorogens are often found to be AIE-active. As such, some ESIPT compounds such as **45a** themselves are both AIE-active and photochromic [51]. This Schiff base displayed reversible color changes controlled by light in solid states owing to the dramatic ESIPT effect. After being irradiated by UV light, the crystal changes from colorless to yellow and then returns to the colorless state again in the dark. The photochromism is attributed to be the photon-induced intramolecular proton transfer. Pyrazolone phenylsemicarbazone derivative **46a** is also an AIE-active ESIPT fluorogen which shows good photochromism performance with high fatigue resistance in the solid state when irradiated with UV light or heated [52].

Triphenylmethanol **47a** is initially nonemissive, but when irradiated by UV light at 254nm, intense blue emission is observed on a solid surface or in the aggregated state [53]. In contrast, irradiation with a UV light at 365nm quenches the fluorescence, enabling the photoactivatable emission reversible. The spot of **47a** on silica gel turned from colorless to slightly yellow after being irradiated by 254nm light, which is the characteristic color of the tritylium cation, indicating the generation of tritylium (**47b**) induced by photoirradiation might be the cause for the photoactivatable fluorescence. Such a reversible activation-deactivation of fluorescence at the solid surface is repetitive for no less than 5 times. Moreover, the photochromism of **47a** is aggregation-facilitated and promoted, because for one thing the tritylium cation is much more stable in the aggregated state as compared to the solution state, and for another, the fluorescence is more intense in the aggregated state as a result of RIM effect in comparison to that of the solution state. **47a** was also employed as a writable and erasable fluorophore for solid-state photopatterning.

48 is an AIE-active fluorescent molecular switch obtained through the covalent conjugation of the aggregation-caused quenching (ACQ) luminophore (i.e., spiro[fluorene-9,9′-xanthene]) with the photochromic unit (i.e., SP) [54]. The spiropyran-functionalized spiro[fluorene-9,90-xanthene] derivative **48** not only retains the isomerization ability under visible light/dark and acid/base condition in the solution state but also exhibits high-contrast fluorescence between the closed and open form of SP group in **48** in the solid states.

2.1.2 *Applications*

The most common applications of photochromic materials are in the areas of anticounterfeiting and information storage, bioimaging, and super-resolution imaging. The AIE-active photochromic materials also perform very well and sometimes even better as compared with the ACQ-active ones in these high-tech application areas.

Anticounterfeiting and information storage

The applications of photochromic materials in the area of anticounterfeiting and information storage have been partially demonstrated by the above-discussed systems such as the those shown in Fig. 2 and compounds **1a, 3, 7, 11a, 12, 13, 22, 29–34, 43, 47**, and **48**. In this section, we further demonstrate the high performance of photochromic AIE systems in the field of anticounterfeiting and information storage.

The system shown in Fig. 8A is a reversible photoswitchable luminescent liquid crystalline nanoparticle system (LCNPs) fabricated via a mini-emulsion polymerization approach [6]. The LCNPs are composed of an AIE-active monomer based on dicyanodistyrylbenzene, a liquid crystalline cross-linker, and a photochromic switch based on a perfluoro-substituted BTE derivative. As a result, the luminescence of the LCNPs can undergo reversible and high-contrast ON-OFF switch with superb repeatability when irradiated alternately with the 365nm UV and visible light ($\lambda > 450$nm). Such a photochromic feature is originated from the ring-closure and ring-opening photoisomerization of the perfluoro-substituted BTE unit based on the FRET process between the AIE-active monomer and the BTE unit inside the particles. The aqueous dispersion of the photoswitchable fluorescent LCNPs was employed as a security ink and used for information encryption and anticounterfeiting as well.

The LNCPs depicted in Fig. 8A hold good water-dispersibility, which enables the photoswitchable fluorescent LNCPs to be dispersible in aqueous media and hence makes their aqueous solution applicable as green and environmental-friendly inks for information encryption. The ink was fabricated by mixing the precursor solution with the suspension of LCNPs, and then it was printed on a paper-based substrate via an HP Deskjet 3800 printer. The patterning, information encryption, and decryption are hence realized utilizing the LCNPs as ink (Fig. 8B and C). At the initial stage, invisible ink was prepared by using the fluorescence "ON" state of the LCNPs, where the ink is colorless under ambient light as BTE is in the open form. The printing patterns were facilely printed onto the position as desired utilizing a nozzle. When dried thoroughly, the pattern is invisible under ambient light. In this way, the information is encrypted. The information could be read in two ways (Fig. 8B). Firstly, the pattern printed is observable under the excitation by a 365nm light, as bright green fluorescence is emitted by the pattern. As a consequence, the printed information is decrypted. Secondly, when the pattern is irradiated by the UV light for 30s, the perfluoro-substituted BTE is transformed from the open form to a colored closed form. In this manner, the pattern is changed from invisible to visible under natural light, and accordingly, the information is decrypted as well. Such a mechanism also applies to the dynamic fluorescent anticounterfeiting. As shown in Fig. 8C, the QR code printed on a ticket with the LCNPs-ink is invisible under natural light but could be clearly visualized by cell phone under UV light. When irradiated by UV light for 30s, the QR code becomes invisible under UV light but clearly visible under ambient light. Moreover, the fluorescence "OFF" state can be recovered to the "ON" state after being irradiated with visible light for 90s. Both the color and fluorescence of the QR codes are very stable even after many cycles of photoswitch.

Inspired by the excellent photocontrollable photophysical behaviors in the solid state, **22** was further applied as dual-functional luminescent ink possesses both photo and vapor erasable capability (Fig. 8D) [33]. When written on the filter paper employing the solution containing **22** as ink, the English letters exhibit quite faint red fluorescence at the initial stage. However, the letters gradually become very strongly yellow-fluorescent after a few seconds, as the solvent volatilizes in the process of air-drying. Without a doubt, the intense yellow emission could be converted to the fairly weak red-emissive state after being fumed by the vapor of acetonitrile. Clearly, the red-shifted and quenched emission results from the de-aggregation of **22**. Moreover, the letters could be erased by UV light illumination, and

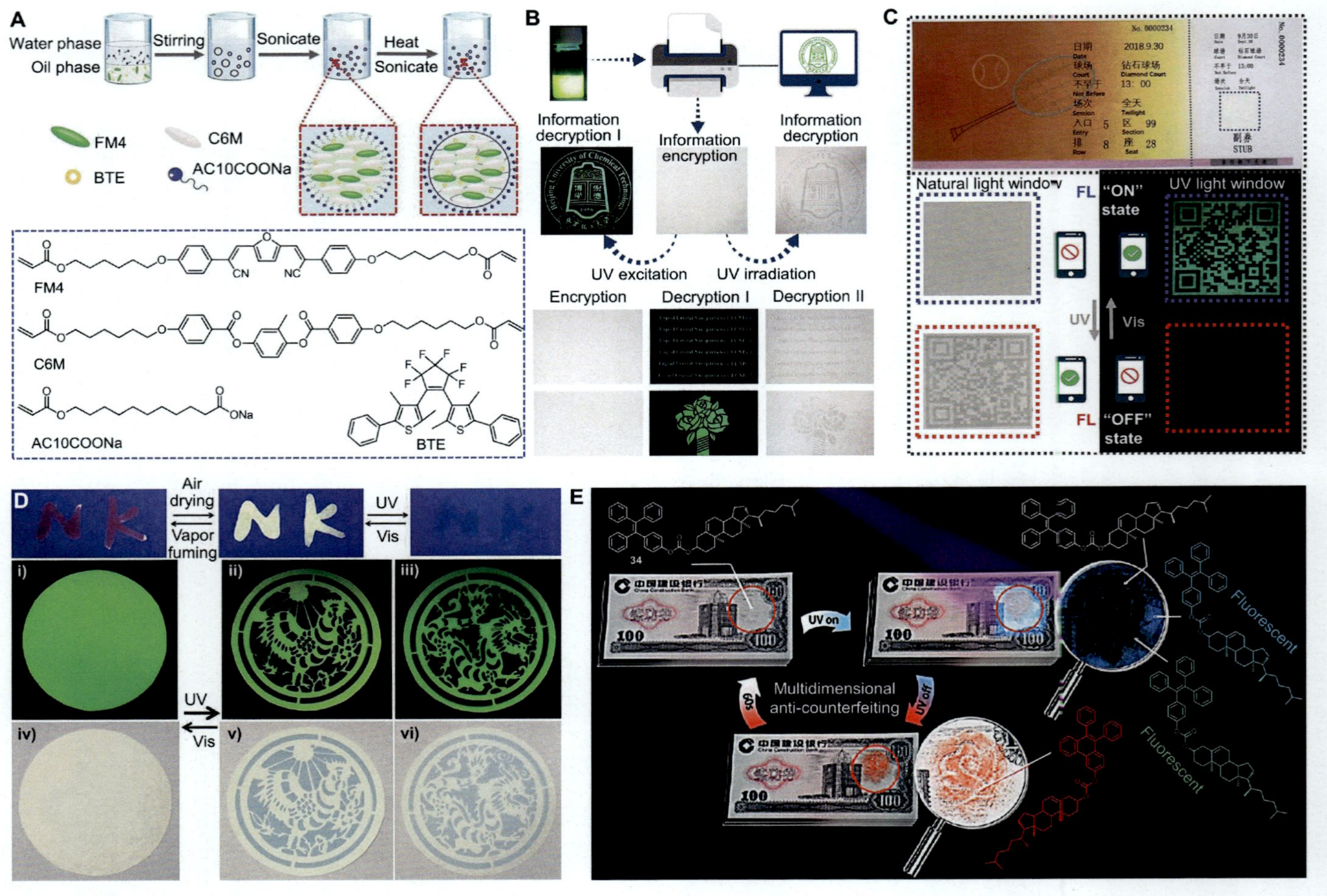

FIG. 8 (A) The synthesis of LCNPs in an aqueous suspension via the mini-emulsion process and thermal polymerization. (B)The application of the LCNPs-ink for information encryption and decryption, and (C) dynamic fluorescent anticounterfeiting [6]. (D) The writing and erasing processes of **22** on the filter paper or **22**-doped polymeric films [33]. (E) The application of **34** for multidimensional anticounterfeiting with a model banknote as a prototype [39]. *Panels (A–C) reproduced with permission from J. Li, M. Tian, H. Xu, X. Ding, J. Guo, Photoswitchable fluorescent liquid crystal nanoparticles and their inkjet-printed patterns for information encrypting and anti-counterfeiting, Part. Part. Syst. Charact. 36 (2019) 1900316. Copyright 2019 Wiley-VCH, (D) from G. Liu, Y.-M. Zhang, L. Zhang, C. Wang, Y. Liu, Controlled photoerasable fluorescent behaviors with dithienylethene-based molecular turnstile, ACS Appl. Mater. Interfaces 10 (2018) 12135–12140. Copyright 2017 American Chemical Society, (E) from G. Huang, Q. Xia, W. Huang, J. Tian, Z. He, B.S. Li, B.Z. Tang, Multiple anti-counterfeiting guarantees from a simple tetraphenylethylene derivative-high-contrasted and multi-state mechanochromism and photochromism, Angew. Chem. Int. Ed. 58 (2019) 17814–17819. Copyright 2019 Wiley-VCH.*

be revisualized by the irradiation with visible light, for many times. Apart from this, the reversible photoswitchable solid-state luminescence renders the achievement of **22**-doped PVDF and PMMA films which could work as modulated fluorescence photowritable and erasable materials (i–vi in Fig. 8D). Diverse patterns could be easily written on the polymeric films by light with a hollowed-out mold. When covering the mold on the film, merely the parts exposed can be illuminated by UV light and then transformed to the closed form of **22** with the fluorescence quenched, while the parts protected by mold preserve the original emission in the ring-open form. In this manner, pretty and vivid patterns can be printed on the films doped with **22** in high on/off contrast. Then, upon exposure to the visible light with a wavelength longer than 490 nm, the pattern printed on the film can be thoroughly erased owing to the recovery of the quenched fluorescence state. Very importantly, the photocontrollable writing-erasing process can be repeated many times, suggestive of the recyclability of these luminescent films.

34 possesses outstanding mechanochromic and photochromic properties and therefore holds great potential in anticounterfeiting by virtue of the following merits: (i) **34** has two fluorescence "ON" and one "OFF" states when exposed to external stimuli and appears in distinct colors modulated by UV illumination. The high contrast and reversibility between multiple states facilitate the complicated anticounterfeiting; (ii) The rapid light-responsiveness and recovery of **34** make the repeated verification possible in a short time; (iii) The commonly used anticounterfeit inks generally have one functionality [39]. In comparison, **34** has multifunctionality and can respond to multiple stimuli with changes in both fluorescence and color tunable by facile UV irradiation. Such a multistimuli responsiveness guarantees and enhances the reliability of anticounterfeiting methods, therefore increasing the difficulty of counterfeiting. The application ability of **34** in practical anticounterfeiting was demonstrated with a model banknote as a prototype. As exhibited in Fig. 8E, fuming the ground powder of **34** afforded the blue-fluorescent background. Heating at the middle area generates a rose pattern. The rose petals become red with the enhanced UV light irradiation. With the removal of UV light, the rose blooms with bright red color and gradually becomes lighter and lighter until it completely vanishes. Such a dynamic picture can be uncovered repetitively. Similar to this case, diverse patterns or paintings can be devised according to specific need for personalized anticounterfeiting.

Bioimaging and super-resolution imaging

As switchable signals could help to conquer the intrinsic limitations of the imaging modality based on the photon, like the interferences brought about by light absorption, scattering, and autofluorescence from biological substrates, photochromic materials have shown their unique charm in the area of bioimaging. Moreover, the switching by light is an effective way to improve the detectability, spatio-resolution, and the capacity to signal identification of fluorescence bioimaging. In light of this, a few AIE-active photochromic systems have been established to fulfill the task as high-performance bioimaging probes [7,8,21], such as the ones shown in Fig. 9A–C.

The photoswitchable binary composite nanococktails (NCs) shown in Fig. 9A and B are achieved by integrating the conjugated polymers based on a cyanovinylene skeleton (cvCP) with a perfluoro-substituted BTE, where the cvCPs function as a nanoemitter and the BTE moiety acts as a photoswitching regulator [7]. Such a physically blending nanosystem

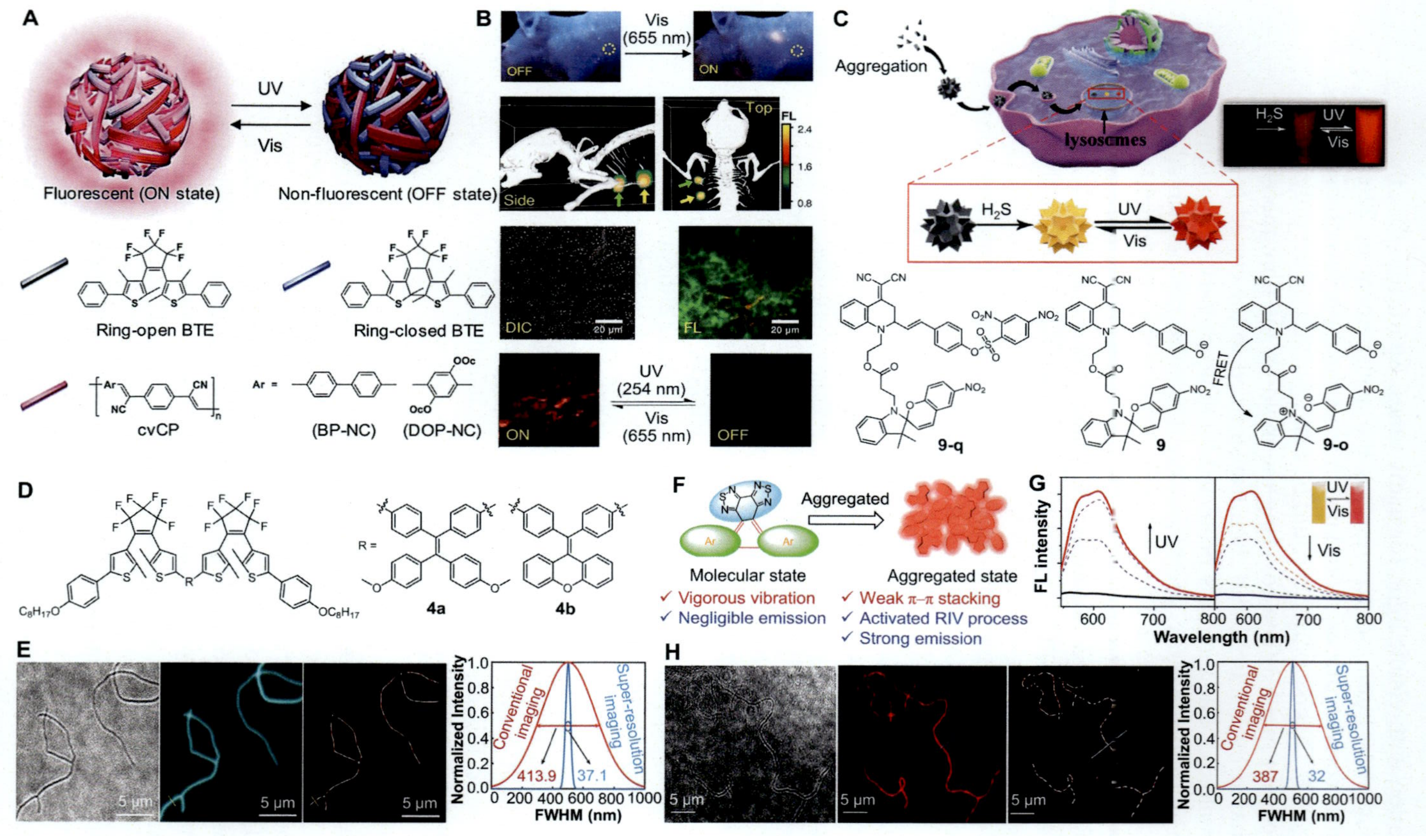

FIG. 9 (A) Schematic illustration of the binary NCs formed by integrating the conjugated polymer with a cyanovinylene skeleton (cvCP) and a perfluoro-substituted BTE, and their photoswitching via UV/Vis light irradiation. (B) In vivo and ex vivo fluorescence imaging of SLNs with the physically integrated NCs composed of cvCP and a perfluoro-substituted BTE [7]. (C). The design and working principle for the sensing of the lysosomal endeogenous H_2S and reversible dual-color imaging on the basis of **9-q**, **9**, and **9-o** [21]. (D) The molecular structures and the (E) super-resolution imaging performance of **4a** and **4b** [16]. (F) The schematic illustration of the AIE mechanism of **24**. (G) Changes in the emission spectra of the PDLLA film doped with 1.0 wt% of **24** upon alternate UV and visible light irradiation. (H) The super-resolution imaging using **24** as the probe [34]. *Panels (A and B) reproduced with permission from K. Jeong, S. Park, Y.-D. Lee, C.-K. Lim, J. Kim, B.H. Chung, I.C. Kwon, C.R. Park, S. Kim, Conjugated polymer/photochromophore binary nanocooktails: bistable photoswitching of near-infrared fluorescence for in vivo imaging, Adv. Mater. 25 (2013) 5574–5580. Copyright 2013 Wiley-VCH, (C) from Y. Hong, P. Zhang, H. Wang, M. Yu, Y. Gao, J. Chen, Photoswitchable AIE nanoprobe for lysosomal hydrogen sulfide detection and reversible dual-color imaging, Sensors Actuators B Chem. 272 (2018) 340–347. Copyright 2018 Elsevier, (D and E) from C. Li, K. Xiong, Y. Chen, C. Fan, Y.-L. Wang, H. Ye, M.-Q. Zhu, Visible-light-driven photoswitching of aggregated-induced emission-active diarylethenes for super-resolution imaging, ACS Appl. Mater. Interfaces 12 (2020) 27651 – 27662. Copyright 2020 American Chemical Society, (F, G, H) from H. Yang, M. Li, C. Li, Q. Luo, M.-Q. Zhu, H. Tian, W.-H. Zhu, Unraveling dual aggregation-induced emission behavior in steric-hindrance photochromic system for super resolution Imaging, Angew. Chem. Int. Ed. 59 (2020) 8560–8570. Copyright 2019 Wiley-VCH.*

features the following characteristics: (i) high luminosity in the near-infrared (NIR) region with deep-tissue penetration ability and small interference with autofluorescence; (ii) high on/off contrasted photoswitching with good reversibility and repeatability, (iii) outstanding on/off memory bistability for nondestructive fluorescence readout under physiological conditions, and (iv) small particle size with satisfactory water-dispersibility for in vivo delivery. The luminescence can be switched on and off via alternate photochromic switching between the colorless and the colored nonemissive states, in which the latter can result in fluorescence quenching via FRET. Such a densely assembled tiny nanoparticle system enables easy integration and optimization between dual functions, namely the NIR luminescence ($\lambda_{em}=728\,nm$) and photochromism in the solid state, to successfully achieve in vivo and ex vivo photoswitching and biomedical imaging of sentinel lymph node (SLN) in a live mouse model.

Specifically, when the forepaw pad of the mouse is administrated with the turned-off DOP-NC, no luminescence signal was witnessed by the naked eye or even with a highly sensitive imaging system (Fig. 9B) [7]. A strong 655 nm laser was applied to switch on the fluorescence of DOP-NC, with the laser beam shined on a spot in a diameter of 2 mm around the axilla at 5 min after probe injection. Through such a photoswitching manipulation, a luminescent spot emerges near the position illuminated by the laser, clearly visualizing a subcutaneous SLN with brilliant fluorescence which can be recognized in a noninvasive fashion even by the naked eye. Post the photoswitching-on, the 3D-reconstructed representation clearly displays that the injected NCs are accumulated at two regions like a node in as short as 5 min after injection. It is noteworthy that merely part of the administrated dose arrives at the SLNs with the majority undrained in typical cases of SLN mapping, unavoidably leading to significant background noise at the injection site when utilizing unswitchable luminescent probes. Sharply contrasted to the general situation, the utilization of the switchable NCs generates signals with high contrast merely at the SLNs irradiated by laser without any undesirably produced background from the remnant probes in the administration site. Consequently, contrasts with high accuracy can be unambiguously exhibited at the targeted tissue under the in vivo conditions which are rich in autofluorescence. It is manifested by the histological analysis of the resected nodes that the NCs are up-taken by cells in the sinusoidal area, in which the NIR fluorescence spectrum of the turned-on DOP-NC is clearly differentiable from the autofluorescent background. It is important that the NCs transported to the node site preserve their reversible photoswitching ability, as demonstrated by the alternate irradiation with 254 and 655 nm lights in the ex vivo microscopic setup. The resulted ON-OFF contrast as high as $\sim$15 indicates that the physically formed NCs retain the original structural integrity nondestructively in the in vivo and ex vivo experiments.

Compound **9-q** specifically reacts to H_2S, giving a sensitive light-up fluorescent response at around 592 nm, affording the compound **9** [21]. The AIE nanoprobe shows a high sensitivity of about 5 nM and good specificity. As discussed in Section "Photochromic AIE systems achieved by chemically bonding AIE motifs and photochromic units," compound **9** in Fig. 9C generated by the cleavage of O-S of **9-q** induced by H_2S is both AIE-active and photochromic. The luminescence of the AIE nanoprobe activated by H_2S can be reversibly switched by the FRET process from the AIE unit to the open form of SP through alternative illumination by UV/Vis light. Furthermore, as demonstrated by the mapping of the endogenous H_2S generation in live cells and the reversible dual-color fluorescence imaging, the developed photoswitchable AIE nanoprobe holds good feasibility to be used in biomedical imaging with high spatiotemporal resolution.

Although fluorescent molecular switches match the principle of super-resolution imaging on the basis of the localization of a single molecule, the majority of the diarylethenes-containing photoswitchable molecules depend on the illumination by at least one UV light and do not have a fluorescence on-off ratio high enough to acquire a dark state that is rare in super-resolution imaging. As **4a** and **4b** display marked photoswitchable AIE properties, they can be employed for super-resolution imaging (Fig. 9D) [16]. **4a** and **4b** were used as super-resolution probes to image the nanostructure of cylindrical micelles self-assembled by an amphiphilic block copolymer, via incorporating **4a** and **4b** into the hydrophobic phase. As can be seen from the reconstructed super-resolution images depicted in Fig. 9E, the images are dim in a bright field. The traditional fluorescence imaging can merely reveal the rough morphology of the cylindrical micelles with the diameter determined to be in the range of 300–500 nm. As a result of the restriction on optical diffraction, it is hard to achieve the real nanostructure with a size smaller than 100 nm with conventional fluorescence microscopes. In sharp contrast, the staining of the cylindrical micelles with **4a** and **4b** enables the clear visualization of the morphology and more details of these micelles. The average photo numbers are detected to be 668 and 408 for **4a** and **4b**, respectively. Through the analysis of the profile of the micelles in the super-resolution images, the full-width at half-maximum (fwhm) is estimated to be about 47 nm, revealing that the resolution of the images is below 50 nm. As compared to the conventional imaging technique, the resolution is improved by about 10 folds. The overall resolutions of **4a** and **4b** are determined to be 50.2 and 56.4 nm, respectively, indicating the outstanding resolution of imaging using **4a** and **4b** as probes.

24 could undergo an excellent photochromism and reversible off-on fluorescence switching process in the film state (Fig. 9F) [34]. The yellow film fabricated via doping **24** into poly(D,L-lactic acid) is almost nonfluorescent when excited at 522 nm due to the negligible absorption. Whereas, the film color gradually turns red with an enhancing fluorescence peak at about 607 nm when irradiated with UV light at about 365 nm, as a result of the formation of closed form of **24** with high luminosity (Fig. 9G). The resulting PSS film is returned back to the original state with reduced fluorescence by visible light irradiation with a wavelength longer than 510 nm. The superb light-up fluorescence switching feature of **24** driven by light is demonstrated in the super-resolution imaging. Similar to the **4a** and **4b** systems, **24** was doped into the polystyene-*block*-poly(ethylene oxide) to stain the cylindrical micelles formed by this polymer. The reconstructed images shown in Fig. 9H exhibits significantly improved resolution far beyond the diffraction limit as compared to conventional fluorescence imaging. It is impressive that the fwhm in super-resolution imaging is around 32 nm, which shows a 12-fold enhancement in comparison to that in conventional imaging of around 387 nm, with the overall resolution calculated to be 29.6 nm. Such results indicate the application potential in the area of nanoimaging.

2.2 Thermochromic AIE systems

AIE systems that exhibit thermochromism can be classified into three groups: (1) organic compounds, (2) aryl-substituted *o*-carborane/binary boranes, and (3) metal complexes. Accordingly, in this Section, the introduction and discussions are also divided into three subsections.

2.2.1 Thermochromic AIE systems based on organic AIEgens

Liquid crystalline (LC) systems have been found to hold great potential to be thermochromic. AIEgen **49** is such an example [55]. (2*Z*,2′*Z*)-2,2′-(1,4-Phenylene)bis(3-(3,4,5-tris(dodecyloxy)phenyl)acrylonitrile), i.e., **49**, is a phasmidic molecule based on dicyanodistyrylbenzene. The molecules of **49** can self-assemble into supramolecular disks composed of a pair of molecules in a side-by-side packing aided by secondary bonding interactions of the cyano groups on the lateral positions, forming a hexagonal columnar LC phase under room temperature (Fig. 10A). AIE-active **49** in liquid or solid crystalline states emits considerably strong green or yellow light with a quantum yield of 0.25 or 0.45, respectively. On the other hand, **49** in the solution or isotropic melt state is totally nonfluorescent with a quantum yield as low as 0.011 (THF) or 0.032 (a liquid state at 120°C), respectively (Fig. 10A). This means that **49** is both AIE-active and thermochromic, and the fluorescence intensity and emission color could be easily tuned by the variation in temperature. The thermochromism was illustrated to be attributed to the change in the interdisk stacking in a given column which is induced by the specific local dipole coupling between molecular disks. In turn, the alteration in stacking leads to the difference in the excited-state dimeric coupling degree to release different fluorescence. The fluorescent images of **49** depicted in Fig. 10A manifest the continuous changes in the fluorescence intensity and color as the temperature decreases from 200°C to room temperature. These fluorescence variations with temperature in the whole range could be divided into four regions: (1) Firstly, in the liquid state at high temperature (200–150°C), there is no obvious change in both emission intensity and color (λ_{em} = 492–495 nm) with temperature decrease, which is the so-called nonemissive/N-phase probably resulted from the dominance of nonradiative decay activated by thermal. (2) Secondly, in the liquid state at a lower temperature (150–60°C), the fluorescence is gradually intensified and red-shifted from 495 to 537 nm, which is denoted as planarization/P-phase. In this case, the intramolecular planarization is most likely caused by the restriction of intramolecular torsions by reducing the thermal energy and increasing the medium viscosity. (3) Thirdly, in the LC state (60–25°C), namely the green/G-phase, **49** shows a gradual increase in fluorescence intensity but constant emission color as the rise of temperature. It is because the intramolecular planarization and dimeric coupling to molecular disk have already finished in this region to give an unchanged emission color (λ_{em} = 537–538 nm), and despite this, the intramolecular rotations of the molecular disks in a given column are still temperature-reliable in this phase. (4) Finally, the liquid-crystalline G-phase was turned into a crystalline state, i.e., the yellow/Y-phase at a temperature low enough or over several hours at room temperature, with a dramatic enhancement in the fluorescence intensity and a further red-shift from 538 to 558 nm. The further emission increase is due to the complete restriction of the rotations of molecular disks together with the preferable intra-disk slanted stacking interactions. Temperature-dependent steady-state and time-resolved fluorescence measurements were systematically implemented to study the complicated photophysical properties of **49**.

Although the thermochromism mechanism of **50** is also linked to phase transition, the case of **50** is still rather different from that of **49**. Diphenyldibenzofulvene derivative **50** exhibits both crystallization-induced emission enhancement and AIE effect [56]. Moreover, it also displays thermochromism and mechanochromism. The two kinds of crystals formed by **50** are green- (**50**GSC) and yellow-fluorescent (**50**YSC), respectively, with quantum yields of 82.1% and 56.2%. In contrast, the amorphous state of **50** (**50**Am) very weakly fluoresces with the

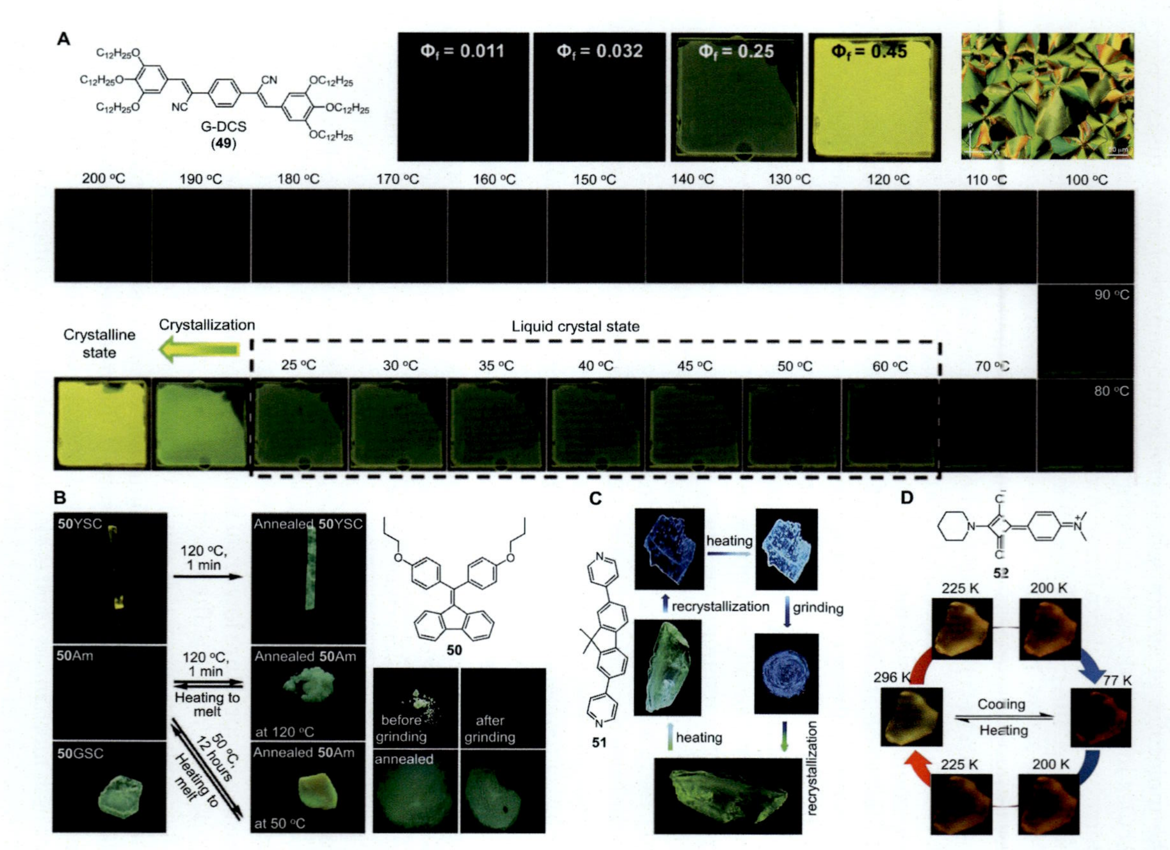

FIG. 10 The representatives of the thermochromic AIE systems based on organic AIEgens. (A) The molecular structure, the distinct fluorescence properties of the solution, LC phase, and crystal states, and the thermochromic fluorescence switching behaviors of the dicyanodistyrylbenzene derivative **49** [55]. (B) The molecular structure, the irreversible transition between the crystals, the reversible change between the amorphous and the green-emissive crystal state, and the mechanochromic behaviors of **50** [56]. (C) The molecular structure and the tricolor luminescence switching behaviors of **51** [57]. (D) The molecular structure and thermochromic behaviors of **52** [58]. *Panel (A) reproduced with permission from S.-J. Yoon, J.H. Kim, K.S. Kim, J.W. Chung, B. Heinrich, F. Mathevet, P. Kim, B. Donnio, A.J. Attias, D. Kim, S.Y. Park, Mesomorphic organization and thermochromic luminescence of dicyanodistyrylbenzene-based phasmidic molecular disks: uniaxially aligned hexagonal columnar liquid crystals at room temperature with enhanced fluorescence emission and semiconductivity, Adv. Funct. Mater. 22 (2012) 61–69. Copyright 2012 Wiley-VCH, (B) from X. Luo, J. Li, C. Li, L. Heng, Y.Q. Dong, Z. Liu, Z. Bo, B.Z. Tang, Reversible switching of the emission of diphenyldibenzofulvenes by thermal and mechanical stimuli, Adv. Mater. 23 (2011) 3261–3265. Copyright 2011 Wiley-VCH, (C) from J. Guan, F. Xu, C. Tian, L. Pu, M.-S. Yuan, J. Wang, Tricolor luminescence switching by thermal and mechanical stimuli in the crystal polymorphs of pyridyl-substituted fuorene, Chem. Asian J. 14 (2019) 216–222. Copyright 2019 Wiley-VCH, and (D) from S. Yang, P.-A. Yin, L. Li, Q. Peng, X. Gu, G. Gao, J. You, B.Z. Tang, Crystallization-induced reversal from dark to bright excited states for construction of solid-emission-tunable squaraines, Angew. Chem. Int. Ed. 59 (2020) 10136–10142. Copyright 2019 Wiley-VCH.*

quantum yield below 1%. When heated but not to a melted state, the transformation from **50**YSC to **50**GSC takes place (Fig. 10B). However, such a process is irreversible. The reversible "dark-bright" switch in the fluorescence of **50** could be repeatedly operated via modulating the transition between **50**Am and **50**GSC employing heating-cooling cycles. The emission of the crystals of **50** can also be switched "dark" and "bright" through grinding and annealing, as a result of the amorphization and crystallization induced by these processes. Meanwhile, the fluorescence of the ground **50** even can spontaneously revert to green light under room temperature. There are some other similar systems reported after this work [59–62]; however, due to the page limitation, we cannot discuss here.

Most of the reported thermochromic systems are based on reversible two-color emission switching, and the switching between multicolors is still very rare. **51** is such a valuable example [57]. It is a very simple derivative of fluorene and has three distinct single-crystal polymorphs, namely, V-**51**, B-**51**, and G-**51**, which have different morphologies and emission colors (Fig. 10C). More specifically, the initial as-prepared powders of **51** show intense blue fluorescence at 432 nm with a quantum yield of 0.47. The V-**51** is crystallized by slow diffusion of the dry dichloromethane solution of the as-prepared powders. V-**51** is a colorless and rectangular sheet in shape which emits violet light at 398 nm with a quantum yield of 0.36. Meanwhile, the slow diffusion of the THF solution affords the irregular sheet crystals B-**51** which displays blue fluorescence at 430 nm with a quantum yield of 0.52. The yellowish block crystals G-**51** which are green-emissive at 515 nm with a quantum yield of 0.39 are generated by slowly diffusing *n*-hexane into the chloroform solution of as-prepared samples. The stimuli-responsive fluorescence switching behaviors of these three crystal polymorphs are shown in Fig. 10C. When heated to about 147°C, V-**51** exhibits a fluorescence color change from violet to blue with the spectra changing from 398 to 432 nm. While the heated V-**51** or B-**51** is ground, a violet emission similar to that of V-**51** would appear. The green-fluorescent form G-**51** can be converted into B-**51** that is thermodynamically more stable via heating at 190°C. In this way, a tricolor luminescence switching from violet to blue to green is realized simply via physical stimuli. It has been identified that such fluorescence properties are closely associated with their intermolecular interactions and molecular packing modes in the aggregated state. The phase transition along with the luminescence switching among the three crystal polymorphs are linked to the structural characteristics of **51**, namely the appropriate π-conjugated plane of the fluorene unit, weak hydrogen bonding, and a twisted molecular conformation, which facilitate the competition between weak π-π stacking and weak hydrogen bonding.

Squaraine (SQ) derivatives have a special central electron-deficient structure and are widely used as organic functional materials in the area of bioimaging, nonlinear optics, photovoltaics, photodynamic therapy, chemical/biological sensing, photocatalysis, and so on. However, the ACQ effect greatly hampered the application of SQ in solid states. Recently, a new AIE mechanism of crystallization-induced dark-bright reversal of the excited state was proposed, through which, the squaric acid-based AIE molecule, CIEE-SQ, with high-efficiency crystalline emission is achieved [58]. CIEE-SQ (**52**) is a squaric acid derivative containing a flexible and bendable piperidine group (Fig. 10D). It hardly emits light in THF solution and has a quantum efficiency of only 0.5%; in sharp contrast, it emits remarkably in the crystal state with the quantum efficiency increased to 2.84%. According to single crystal analysis, there are three different conformations in CIEE-SQ crystals, with different torsion angles, intermolecular C-H⋯π interaction distances, and interaction sites. The theoretical calculation of QM/MM shows that the S_1 excited states of these three conformations are very different, including the near-planar conformation similar to the excited state in the free state, and two twisted conformations. According to this, the crystallization process induces the electronic configuration inversion from the dark excited state to the bright excited state, resulting in the enhanced luminescence of the CIEE-SQ aggregated state.

The CIEE-SQ crystal has the flexible conformation and thus exhibits a temperature-responsive single crystal to single crystal (SCSC) reversible transition which is accompanied by a reversible change in luminescence. During the cooling process from 296 to 77 K, the emission is red-shifted and weakened. While in the heating process, the luminescence properties were reversibly restored, and a reversible phase transition was observed by DSC. According to the theoretical calculation results, the three initial conformations are all reversibly transformed into a new conformation during the SCSC process (Fig. 10D). The reversion of the electronic configuration still takes place in the newly generated conformation with the bright S_1 having the property of $^1(\pi, \pi^*)$. Because this conformation has more regular molecular packing and tighter intermolecular interactions, it causes emission red-shift and attenuation. In addition, when using chloroform vapor to fumigate the CIEE-SQ crystal, the luminescence is blue-shifted and enhanced, with the quantum yield significantly increasing to 28%. The fumigated crystals are dried in the air, and the luminescence presents a reversible change of red-shift and weakening. Interestingly, this property can be applied to information encryption and decryption.

Most of the thermochromic AIE systems discussed above are merely changeable between two colors. Even though those change among three colors, the color change is not continuous. Moreover, the thermochromic mechanisms of these systems are generally linked to phase transformation. Therefore, such systems can hardly be used for temperature sensing. For the purpose of sensing temperature, it is better to build AIE systems with conformation that can continuously change in response to the temperature variation. Some fantastic thermometers built on the basis of AIE and another novel luminescence mechanism, namely vibration-induced emission (VIE), can display dynamic and continuous color change over a very broad spectral range from blue to orange-red passing the white-light region to temperature.

The term of VIE was put forward by Tian et al. in 2015 [63,64], which defines the abnormally dual-emissive properties of a special kind of fluorogens, such as the *N,N′*-disubstituted-dihydrophenazine derivatives. When in solution or other unconstrained states, the VIE-active luminogens predominately show weak but observable orange-red emission which originates from the excited-state planarization promoted by intramolecular vibrations. When the intramolecular motions including vibrations and rotations are restricted, the fluorescence spectra would be dominated by the intense intrinsic blue emission. The fluorescence properties of VIE-active luminogens are sensitive to external conditions including temperature and viscosity, because of their flexible conformations. In other words, all the VIE systems are potentially thermochromic materials. At low temperature, when the intramolecular motions are constrained, the molecules take on a saddle-shaped conformation which limits the conjugation length and hence mainly gives out blue fluorescence. As the temperature increases, the molecules get unconstrained and are prone to undergo intramolecular motions such as bent-to-planar vibrations which elongate the conjugation length, and the intramolecular rotations that consume the excited-state energy via nonradiative decay channels, resulting in the emergence of orange-red emission and the decrease of blue emission.

It means that the VIE-active compounds have the potential to show dynamic and continuous thermochromic effects in a suitable temperature range. In view of this, the typical VIE system based on 9,14-diphenyl-9,14-dihydrodibenzo[*a*,*c*]phenazine (DPAC) framework is tailored as excellent thermochromic materials to meet the requirements of different applications in the daily-concerned temperature region (Fig. 11) [65–67].

Firstly, flexible side chains like long alkyl or siloxane moieties were attached to the DPAC core in order to fluidize the DPAC derivatives and thereby enhance their applicability in the bulk state [65]. The long alkyl groups made **53** (DPC) fluidic at 40°C, and the plasticizing effect of siloxane chains made **54** (DPSi-1) and **55** (DPSi-2) nonvolatile liquids at room temperature (Fig. 11A). The emission properties of **53–55** are all temperature-dependent. The thermochromic properties of **54** are presented in Fig. 11B–D as an example. **54** shows marked dual-emissions: one is the blue fluorescence peaked at 469–490 nm and the other one is the orange-red fluorescence situated at 567–587 nm. The fluorescence behaviors altered significantly with the temperature change. More specifically, as the temperature raised from 5°C to 20°C, merely very subtle change in the fluorescence was observed. The fluorescence decrease became more obvious as the temperature further increased. The orange-red emission band emerged at around 50°C, which was much weaker than the blue-fluorescence band. As the temperature further increases from 50°C to 80°C, the orange-red fluorescence became stronger. The blue and orange-red fluorescence bands are comparable at about 80°C. The orange-red fluorescence remained almost constant but predominated as the temperature rose from 80°C to 105°C, meanwhile the blue fluorescence continuously reduced. The blue emission further decreased with the temperature reducing from 105°C to 135°C, and faded at 130° C, while the orange-red emission only slightly reduced and displayed moderate intensity. Such a unique thermochromic feature endowed these compounds with superb temperature-sensing ability. As a result, the temperature information which is invisible over a wide range from 5°C to 135°C was able to be visualized by these VIE compounds, with a sensitivity up to 3.40% per °C, a linear sensing region as broad as 85°C (50–135°C), and satisfactory reversibility in a solvent-free fashion which is observable by naked eyes. Accordingly, the fluorescence color varied from blue to greenish-blue to near white and ultimately to orange-red. It is very important that these thermochromic compounds were able to be coated on the surface of any objects, despite the material, shape, size, or state of the objects, and used for facile and rapid large-area temperature sensing (Fig. 11E), exhibiting the practicability which is superior to other fluorescent thermometers.

On the basis of the above work, a series of temperature-responsive molecular liquids were developed with the DPAC backbone and applied for dynamic multicolor-fluorescent anticounterfeiting and information encryption [66]. Fluorescent molecular liquids are a class of organic soft materials that can be used without the need for solvent or dopant. The VIE attribute endows these alkyls chain-decorated DPAC derivatives (DPAC-*n*, $n=5$, 10, 15, and 20, Fig. 11F) with dynamic multicolor fluorescence in response to temperature change. The complex viscosities, the intramolecular motions, and the luminescence behaviors are closely linked to the length of alkyl chains. By virtue of the VIE effect, these luminogens mainly fluoresce blue emission at low temperature with high complex viscosity while predominantly orange-red emission at a high temperature (80°C) with low complex viscosity (Fig. 11G). Compound **56** (DPAC-20) shown in Fig. 11F is an example of them. As a result of the lowest complex viscosity among the DPAC-*n* ($n=5$, 10, 15, and 20) derivatives, **56** with

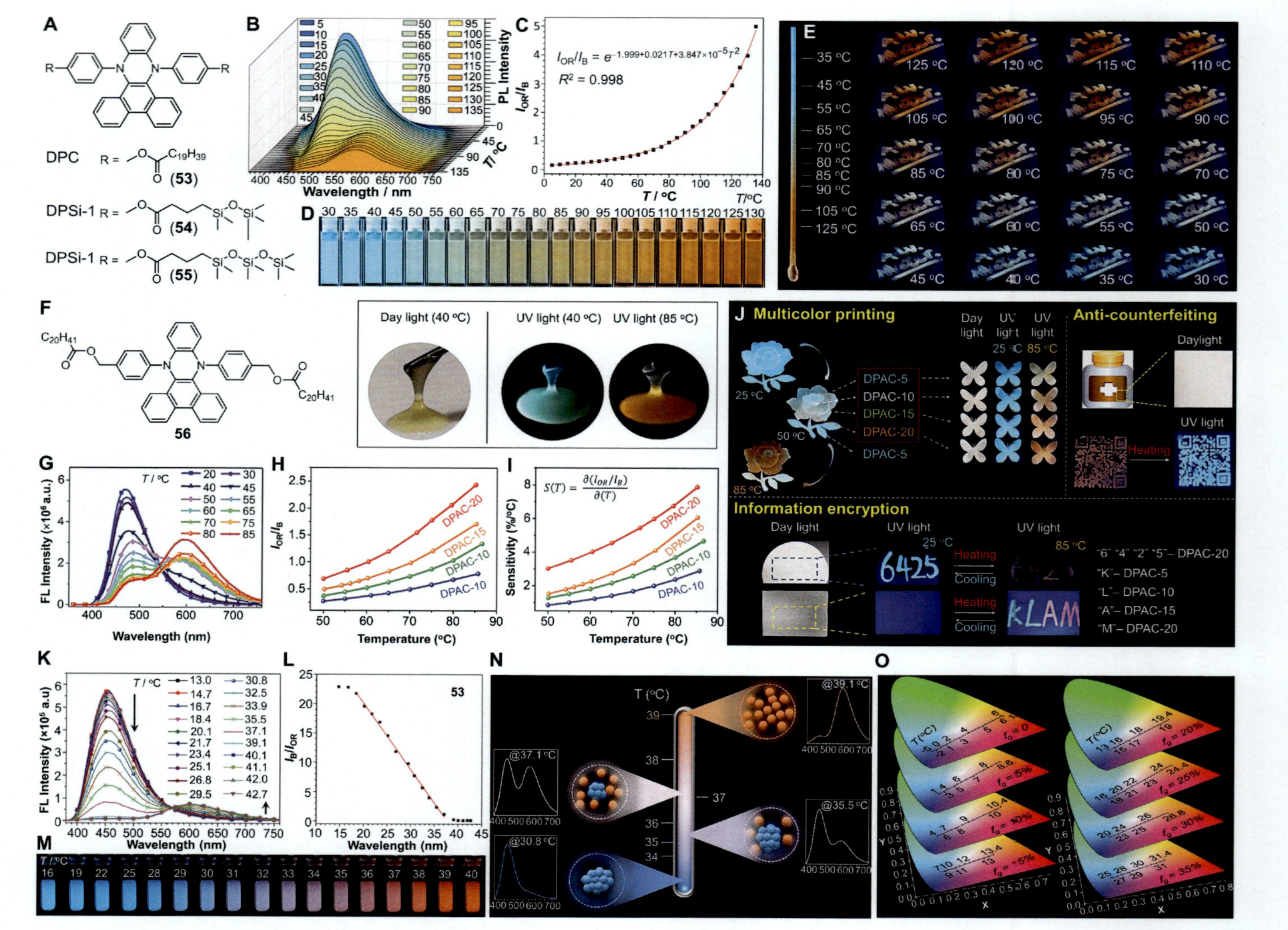

FIG. 11 The thermochromic systems which are based on VIE-active DPAC derivatives. (A) and (F) The molecular structures of **53**–**56**. (B) The fluorescence spectra and (C) the plot of I_{OR}/I_B versus temperature as the temperature rose from 5°C to 135°C. (D) The fluorescent images of noumenon **54** in cuvettes were taken under a temperature varying from 30°C to 130°C. (E) Fluorescent photographs of **54** in a melting-point tube when it was subjected to thermal treatment and the ones of **54** scribbled on a ceramic heater taken at a varied temperature [65]. (G) The fluorescence spectra, (H) the curve of I_{OR}/I_B versus temperature, (I) the sensitivities of **56** in bulk as the temperature increased from 50°C to 80°C, (J) the application of **56** in multicolor printing, information encryption and anticounterfeiting areas [66]. (K) The fluorescence spectra,

Continued

the longest alkyl chain has the most remarkable thermochromis and temperature probing ability, showing the highest I_{OR}/I_B value at the same temperature as well as the highest sensitivity within the range of temperature from 50°C to 85°C (Fig. 11H and I). These compounds were first employed as solvent/dopant-free fluorescent inks for advanced anticounterfeiting and information encryption (Fig. 11J). Images printed on filter paper with DPAC-*n* derivatives exhibit distinct fluorescence at a temperature below 50°C. In this way, with these thermochromic VIE-active molecular liquids, temperature-sensitive multicolor printing is realized. Moreover, DPAC-*n* derivatives were also utilized as a type of security ink for number encryption. Under daylight, the number code of "6425" handwritten with the solvent-free liquid of DPAC-20 on filter paper cannot be recognized, but can be clearly and correctly decrypted under the UV lamp. Moreover, the emission of the numbers dramatically altered from blue to orange-red as the temperature elevated. In the meantime, an advanced code with the letters "KLAM" were written employing DPAC-*n*-based ($n=5$, 10, 15, and 20) inks. The information is hardly visible under ambient light, because of the intense blue emission from the commercial A4 paper. Very interestingly, the encrypted letters gradually emerged on the paper with the temperature increase. The decrypted letters could be encrypted again through the cooling operation. The encryption-decryption course is reversible and repeatable, suggesting the high potential of these compounds to achieve advanced multilevel encryption with high reliability. Furthermore, advanced dynamic anticounterfeiting is accomplished with the ability to guarantee higher-level security against counterfeiting. The 2D code fabricated with DPAC-20 on the medicine label is not visible under daylight, whereas, the pattern could be witnessed by the intense blue fluorescence under UV light. It is worth noting that the blue-fluorescent 2D code was turned into an orange-red-fluorescent one when the temperature increased. Undoubtedly, such a process is reversible and can be repeated for many cycles under the heating-cooling treatment. In this way, various patterns are designable in view of the specific needs for personalized advanced anticounterfeiting. By taking advantage of the VIE feature and thermochromism, multicolor fluorescent printing, dynamic information encryption, and advanced anticounterfeiting can be achieved.

It is very fantastic that besides the utilization in bulk state, the dibenzo[*a*,*c*]phenazine-9,14-diylbis-(4,1-phenylene))bis(methylene) bis(icosanoate) (**53**) was also employable as thermometer in the other aggregated states [67]. The excited-state conformation transformation can be modulated via facile manipulation of the disaggregation and aggregation of **53**, as such ratiometric temperature sensing is realized. Take **53** in the mixture of ethanol and glycerol with a f_g of 40% for example. There was no obvious change observed in the fluorescence when

FIG.11, CONT'D (L) the curve of I_B/I_{OR} versus temperature, and (M) the fluorescent images of **53** in the mixture of ethanol/glycerol with the glycerol fraction (f_g) of 40% as the temperature increased from 16°C to 40°C. (N) The schematic illustration of the working mechanism of the ratiometric thermometers of **53** on the basis of aggregation and disaggregation of **53**. (O) Fluorescence color-temperature scales made with the fluorescence color of **53** in the ethanol/glycerol mixtures with $f_g=0\%$, 5%, 10%, 15%, 20%, 25%, 30%, and 35% under different temperatures [67]. *Panels (A–F) reproduced with permission from L. Shi, W. Song, C. Lian, W. Chen, J. Mei, J. Su, H. Liu, H. Tian, Dual-emitting dihydrophenazines for highly sensitive and ratiometric thermometry over a wide temperature range, Adv. Optical Mater. 6 (2018) 1800190. Copyright 2018 Wiley-VCH, (G–J) from H. Liu, W. Song, X. Chen, J. Mei, Z. Zhang, J. Su, Temperature-responsive molecular liquids based on dihydrophenazines for dynamic multicolor fluorescent anti-counterfeiting and encryption, Mater. Chem. Front. 5 (2021) 2294–2302. Copyright 2021 Royal Society of Chemistry, and (K–O) from W. Song, W. Ye, L. Shi, J. Huang, Z. Zhang, J. Mei, J. Su, H. Tian, Smart molecular butterfly: an ultra-sensitive and range-tunable ratiometric thermometer based on dihydrophenazines, Mater. Horiz. 7 (2020) 615–623. Copyright 2020 Royal Society of Chemistry.*

the temperature rose from 13.0°C to 25.1°C. The orange-red fluorescence continuously increased with the decrease of blue fluorescence when the temperature elevated from 25.1°C to 37.1°C. The orange-red emission became constant and dominating with the disappearance of blue fluorescence when the temperature arrived at 39.1°C (Fig. 11K–M). In converse, the reduction in the orange-red fluorescence and the emergence and enhancement of blue fluorescence take place in the cooling course from 42.2°C to 16.7°C. Very interestingly, the temperature-response range and sensitivity could be fine-tuned via the change in the composition of the ethanol/glycerol mixtures (Fig. 11N), with an overall linear range as wide as 49.1°C (−11.4 to 37.3°C). Unexpectedly, besides the superb repeatability, and the relative sensitivity of up to 2000% per °C that is the highest among all reported fluorescent thermometers, the thermometer can even enable the temperature to be read out precisely from the fluorescence color as a result of the establishment of the accurate functional relationship between the CIE coordinates of the fluorescence color and the temperature (Fig. 11O). It means that such a luminescent thermometer does not need to depend on professional spectrometers, which accordingly enlarges the application scope of luminescent thermometers.

2.2.2 Thermochromic AIE systems based on aryl-substituted o-carborane/binary borane

o-Carborane, as a polyhedral boron cluster bearing two adjacent carbon atoms in the cluster cage, exhibits unique features including neutron-capturing capability and thermal and chemical stability. So far, a number of *o*-carborane derivatives are reported to be AIE-active and *o*-carborane has been regarded as an "element-block" for the construction of AIE systems. The fluorescence bands result from the ICT state where the *o*-carborane moiety often functions as an electron-acceptor due to the electron-deficient attribute of boron clusters constituted of 3-center 2-electron bonds. Furthermore, the intramolecular vibrations at the C—C bond in *o*-carborane significantly cause fluorescence annihilation through the σ^*-π^*-involved electronic conjugation. Thus, the strong fluorescence would be released not only via the restriction of nonradiative decay induced by intramolecular vibrations but also via preventing ACQ due to the steric hindrance of the *o*-carborane group to intermolecular interactions in the condensed phase. Up till now, a few AIE-active *o*-carborane derivatives have been reported to show thermochromism (Fig. 12A) [68–74].

A series of anthracene-decorated *o*-carborane dyads (Fig. 12A) were developed with a substituent at the adjacent carbon atom varying from hydrogen (ANT-H), methyl (ANT-Me), phenyl (ANT-Ph) to trimethylsilyl (ANT-TMS), which exhibited bright yellow or orange fluorescence in the range of 563–624 nm in the aggregated state, exhibiting typical AIE or aggregation-enhanced emission (AEE) effect [68]. These anthracene-containing *o*-carborane derivatives also show thermochromism properties. Generally, their fluorescence can be red-shifted and decreased by heating and recovered by cooling, and such a process can be repeated many times. The reason for their thermochromism can be interpreted as follows: Higher temperature could facilitate the thermal motions at the substituents, releasing the crystal packing, enhancing the structural relaxation, consequently increasing the chance of forming the π-conjugation-favorable conformation, and finally leading to the bathochromic shifts. In the meantime, the intramolecular motions enhanced by the higher temperature cause the annihilation of emission. Particularly, ANT-Ph exhibits the most dramatic thermochromic effect among all these dyads, which might be because the phenyl unit is more suitable to extend the conjugation.

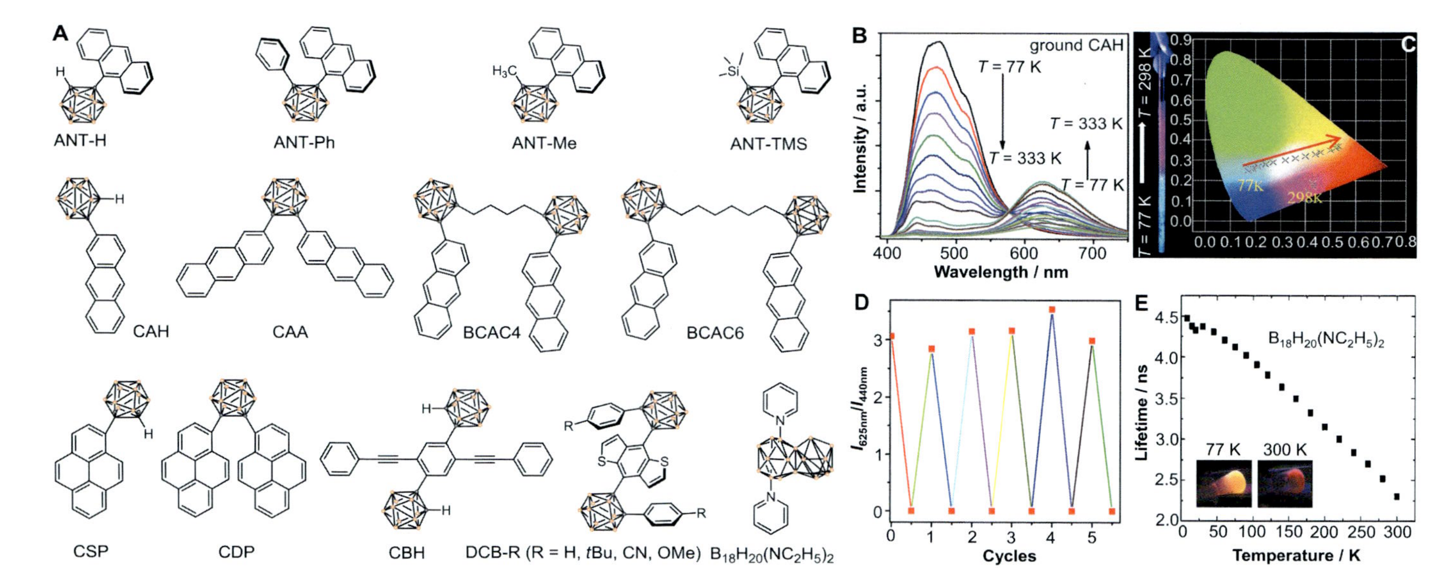

FIG. 12 (A) The molecular structures of the thermochromic AIE systems based on *o*-carboranes. (B) The photoluminescence spectra of ground CAH in the solid state when heated from 77 to 333 K. (C) A photo and CIE diagram exhibiting the color and CIE coordinates change of ground CAH with temperature. (D) The reversible and reproducible thermochromic behaviors of ground CAH over 5 cycles. (E) The fluorescence decay of $B_{18}H_{20}(NC_5H_5)_2$ at a temperature ranging from 8 to 300 K. Inset: the fluorescent photographs of solid-state $B_{18}H_{20}(NC_5H_5)_2$ at 77 K *(left)* and 300 K *(right)*, excitation: 365 nm. *Panels (A–D) reproduced with permission from X. Wu, J. Guo, Y. Cao, J. Zhao, W. Jia, Y. Chen, D. Jia, Mechanically triggered reversible stepwise tricolor switching and thermochromism of anthracene-o-carborane dyad, Chem. Sci. 9 (2018) 5270–5277. Copyright 2018 Royal Society of Chemistry and (D) from M.G.S. Londesborough, J. Dolanský, L. Cerdán, K. Lang, T. Jelínek, J.M. Oliva, D. Hnyk, D. Roca-Sanjuán, A. Francés-Monerris, J. Martinčík, M. Nikl, J. Kennedy, Thermochromic fluorescence from B18H20(NC5H5)2: an inorganic-organic composite luminescent compound with an unusual molecular Geometry, Adv. Optical Mater. 5 (2017) 1600694. Copyright 2017 Wiley-VCH.*

CAH is an isomer to ANT-H (Fig. 12A) and also AIE-active [69]. Although CAH is also thermochromic, the detailed behaviors are quite different from those of ANT-H. CAH firstly displayed a tricolored mechanochromism. The thermochromism would be triggered by mechanical stimuli. To be specific, the brightly blue-emissive (440nm) pristine CAH powder can be changed to be very bright yellow-fluorescent (530nm) via gentle grinding. With the increase of the grinding force, the fluorescence further changes to pink (625nm). Very interestingly, the CAH which has been heavily ground exhibits typical thermochromism (Fig. 12B–D). The fluorescence of the pristine-stated CAH peaks at 440nm and becomes weak with the increase in temperature, which is assigned to the locally excited (LE) emission. Whereas the heavily ground CAH exhibits the emission originated from the TICT state, and the LE emission and another wide fluorescence band attributed to the anthracene excimer will be observed as the temperature decreases from 333 to 77K (Fig. 12B and C). An evident color change from pink to blue was observed in the cooling process. In this process, the TICT fluorescence band gets gradually reduced with the CIE coordinates that vary nearly linearly from yellow to sky blue. Such a thermochromic response can be fully reversible after being repeated for five cycles (Fig. 12D).

Unlike CAH, although compounds CAA, BCAC4, and BCAC6 are also anthracene-*o*-carborane dyads, the ratios of anthracene to *o*-carborane in one molecule and the linking modes are different (Fig. 12A) [70]. It was reported that the linking of two *o*-carborane blocks together via tailored alkyl chains can greatly enhance the solid-state emission efficiency of anthracene-*o*-carborane dyads. The introduction of alkyl spacers not only hinders the energy-dissipating intramolecular motions of *o*-carborane including rotations and vibrations but also provides good solution processability. The solid-state fluorescence quantum yield increases from CAH (4.3%) to CAA (36.0%) to BCAC4 (40.8%) and to BCAC6 (51.4%), which are much higher than that of the corresponding solution state, suggesting their typical AIE properties. Furthermore, the fluorescence of the crystal powder samples of CAH, CAA, BCAC4, and BCAC6 showed different degrees of responses to temperature change in the range of 77–363K. Their fluorescence would be blue-shifted by the decreasing temperature and recovered by increasing temperature, and the change in the fluorescence of BCAC6 is the largest. The thermochromic mechanisms are believed to be similar to those of ANT-H, ANT-Me, ANT-Ph, and ANT-TMS.

Besides anthracene, other aryl groups such as pyrene have also been used to conjugate with *o*-carborane to afford functional optoelectronic materials (Fig. 12A) [72]. CSP and CDP are pyrenyl-decorated *o*-carborane derivatives, which display multiluminescence properties. They are AIE-active or AEE-active. In the solid state, CSP shows dual emissions and the thermochromism between orange and blue, while CDP displays ternary fluorescence and less obvious thermochromism than CSP as merely its fluorescence intensity could be changed by temperature variation. More specifically, as the temperature decreases from 333 to 77K, the emission at 597nm of CSP increases at first and then weakens with the appearance and gradual enhancing of a broad and structureless emission band at about 470nm. The fluorescence color clearly changed from orange to azure during the course. The fluorescence peaked at 597nm is attributed to the ICT emission, while the one at 470nm stems from the LE and the excimer structure in the excited state. As for the CDP solid, mere enhancement in the deep-red fluorescence without change in the CIE coordinates was observed as temperature decreases from 333 to 77K. It was speculated that there might be no intermolecular interactions in the aggregated state of CDP and the fluorescence is only contributed by the ICT state, i.e., the intramolecular through-bond interaction between the pyrenyl moieties and the *o*-carborane cage.

CBH is a conjugate of 1,4-bis(phenylethynyl)benzene and *o*-carborane, where the two *o*-carborane groups are symmetrically substituted on the 1,4-bis(phenylethynyl)benzene framework [73]. In the pristine state, CBH displayed dual fluorescence which is associated with the LE and TICT states in solution. The intensity of the TICT emission is enhanced by heating and only LE fluorescence can be observed at 77 K. As a result, obvious solvatochromism and thermochromism are observed. It is worth mentioning that the TICT fluorescence could even be observed in the solid state, making thermochromism observable in the solid state as well. Moreover, CBH shows AIE and crystallization-enhanced emission (CEE) effects.

The DCB-R derivatives with R varying from H to *t*Bu, CN, and OMe are also aryl-modified *o*-carborane derivatives (Fig. 12A), which can undergo thermochromism without change in molecular structure and phase transformation [71]. These bis(*o*-carborane)-substituted benzobithiophenes all show solid-state fluorescence with the ICT features, and the fluorescence color greatly depends not only on the substituents at the *para*-position of the benzyl ring but also on the molecular distribution in the solid state. As anticipated, all these four compounds are AIEgens with efficient crystal-state emissions. Fast and reversible thermochromic emission behaviors are exhibited in the crystalline phase. It is believed that the distorted aromatic center generated by the bis(*o*-carborane)-substitution might be responsible for the thermochromism. The intramolecular motions in the crystalline packing can be activated by heating, which generates the space to yield thermally stable conformations. As a consequence, the elongation of conjugation could take place, leading to the red-shifts in the fluorescence. Immediate and conversable responses can be attained due to the subtle change in the structure. Despite this, owing to the robustness of their crystals, the thermochromic changes are not very dramatic.

Apart from *o*-carborane, binary borane *anti*-$B_{18}H_{22}$, the centrosymmetric isomer of octadecaborane(22), has also been utilized as a building block to construct thermochromic AIE systems [74]. *anti*-$B_{18}H_{22}$ is an intensely blue-fluorescent compound with a quantum yield of up to 0.97. $B_{18}H_{20}(NC_5H_5)_2$ which is generated via the reaction of *anti*-$B_{18}H_{22}$ with pyridine as an example of the two conjoined boron hydride subclusters of *nido* and *arachno* geometry. The solutions of $B_{18}H_{20}(NC_5H_5)_2$ show emission at 690 nm at room temperature. While in the solid state, the fluorescence is blue-shifted to 620 nm and enhanced as a result of the RIM process. Obvious thermochromism in the luminescence is shown by $B_{18}H_{20}(NC_5H_5)_2$ in the solid state (Fig. 12E). Further blue-shift in the fluorescence to 585 nm and a twofold enhancement in the intensity is caused by cooling to 8 K. The molecular slip and restriction on the intramolecular motions are believed to be responsible for the solid-state thermochromic fluorescence behaviors of the inorganic-organic composite compound $B_{18}H_{20}(NC_5H_5)_2$.

2.2.3 Thermochromic AIE systems based on metal complexes

In addition to the thermochromic AIE systems obtained with organic compounds and aryl-decorated *o*-carborane/binary borane derivatives, the AIE properties and thermochromism can also be simultaneously exhibited by some metal complexes. To date, there have been several thermochromic and AIE-active metal complexes reported, as shown in Fig. 13 [75–79].

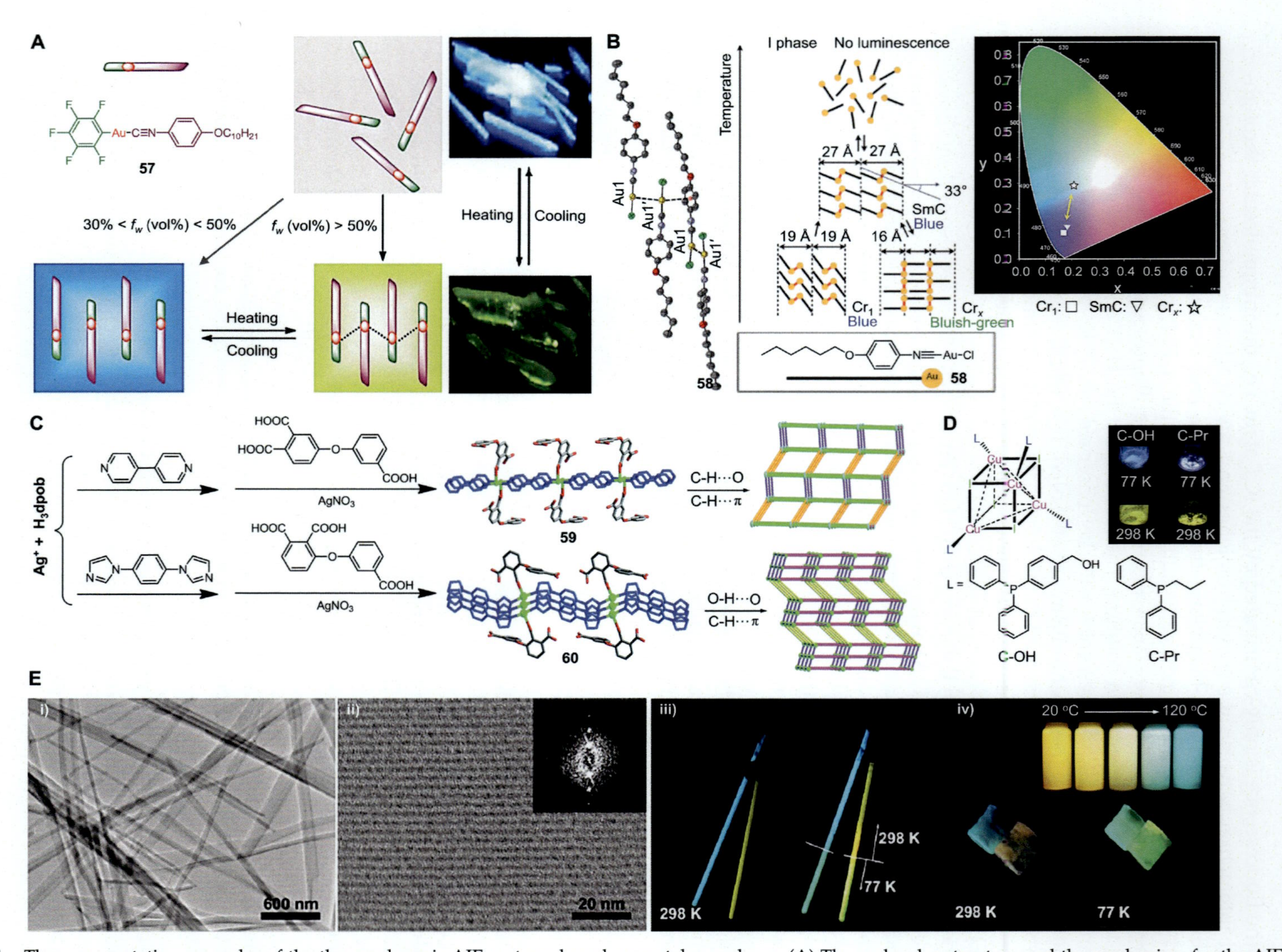

FIG. 13 The representative examples of the thermochromic AIE systems based on metal complexes. (A) The molecular structure and the mechanism for the AIE and thermochromic effects of **57** [75]. (B) The molecular structure, crystal packing, thermochromic mechanism, and CIE coordinates of **58** [76]. (C) The fabrication scheme and the structures of **59** and **60** [77]. (D) The molecular structures and thermochromic luminescence of C-OH and C-Pr [78]. (E) The morphologies and the thermochromism effect

Continued

57, which is both AIE-active and thermochromic, is a gold(I) compound with a 1-(decyloxy)-4-isocyanobenzene tail (Fig. 13A). The emission is reversibly switchable between blue and yellowish-green fluorescence as temperature alters [75]. Crystals of **57** exhibit two emission peaks at 407 and 428 nm, which originate from the intra-ligand localized π—π^* excited state. An emission band at 530 nm became observable after the sample was gently heated (>55°C), which might be attributed to the new ligand-to-metal-metal charge-transfer excited state caused by aurophilic interactions. When the temperature decreases to 25°C, the initial blue emission is recovered. Such a blue-to-yellowish green fluorescence cycle is reversible. Thermochromism was attributed to changes in molecular packing and the intermolecular Au(I)-Au(I) interactions upon heating and the rearrangement of the amorphous phase into the more stable crystalline state on cooling. Moreover, the multiple weak π—π interactions and intermolecular C—H⋯F and C⋯F interactions facilitate molecular packing. When the good solvent is added with a poor solvent such as water or when the blue-emissive powder is heated, nanoaggregates or an amorphous form are generated with aurophilic interactions, which result in the lower energy yellowish-green fluorescence.

Similar to **57**, **58** is also a gold(I) complex-based AIEgen that shows thermochromic luminescence properties (Fig. 13B) [76]. It was found to take a rod-like conformation in a dimeric packing form. The emission shape and maxima of **58** in the crystalline state (Cr_1 and Cr_x) are identical to those in the smectic C (SmC) phase upon heating. Nevertheless, the emission of Cr_x obtained via recrystallization from its melt state after cooling shows a 35 nm red-shift compared to those of the Cr_1 and SmC phases. The luminescence is deep-blue in the Cr_1 and SmC states, and pale bluish-green in the Cr_x phase: the CIE coordinates are (0.17, 0.10) in the Cr_1 state at 40°C on first heating, (0.18, 0.13) in the SmC state at 130°C, and (0.21, 0.29) in the Cr_x phase at 40°C. When the temperature is above 130°C, the luminescence is switched OFF as the sample melts. In other words, thermochromism stems from the transformation between the Cr_x and SmC phases.

Some Ag(I) complexes have been reported to be AIEgens showing thermochromic luminescent behaviors [77]. Ag(I) supramolecular architectures **59** and **60** displayed in Fig. 13C were synthesized through hydrothermal approaches with mixed ligands including H_3dpob (i.e., 3-(3′,4′-dicarboxylphenoxy)benzonic acid (in **59**) or 3-(2′,3′-dicarboxylphenoxy)benzonic acid (in **60**), and 4,4′-bipyridine (in **59**), and 1,4-bis(1-imidazoly)benzene (in **60**). As suggested by the crystal structures, the hydrogen bonding including O—H⋯O, C—H⋯O, and C—H⋯π, and π-π interactions play crucial roles in the formation of the extended

FIG.13, CONT'D of the luminescence of the Cu nanoclusters [79]. *Panel (A) reproduced with permission from J. Liang, Z. Chen, J. Yin, G.-A. Yu, S.H. Liu, Aggregation-induced emission (AIE) behavior and thermochromic luminescence properties of a new gold(I) complex, Chem. Commun. 49 (2013) 3567–3569. Copyright 2013 Royal Society of Chemistry, (B) from K. Fujisawa, Y. Okuda, Y. Izumi, A. Nagamatsu, Y. Rokusha, Y. Sadaike, O. Tsutsumi, Reversible thermal-mode control of luminescence from liquid-crystalline gold(I) complexes, J. Mater. Chem. C 2 (2014) 3549–3555. Copyright 2014 Royal Society of Chemistry, (C) from Y. Song, R. Fan, X. Du, K. Xing, P. Wang, Y. Dong, Y. Yang, Effect of noncovalent interactions on Ag(I)/Cu(I) supramolecular architecture for dual-functional luminescence and semiconductive properties, CrystEngComm, 18 (2016) 6411–6424. Copyright 2016 Royal Society of Chemistry, (D) from R. Utrera-Melero, J.-Y. Mevellec, N. Gautier, N. Stephant, F. Massuyeau, S. Perruchas, Aggregation-induced emission properties of copper iodide clusters, Chem. Asian J. 14 (2019) 3166–3172. Copyright 2019 Wiley-VCH, and (E) from Z. Wu, J. Liu, Y. Gao, H. Liu, T. Li, H. Zou, Z. Wang, K. Zhang, Y. Wang, H. Zhang, B. Yang. Assembly-induced enhancement of Cu nanoclusters luminescence with mechanochromic property, J. Am. Chem. Soc. 137 (2015) 12906 – 12913. Copyright 2015 American Chemical Society.*

supramolecular architecture. **59** and **60** show very faint luminescence in good solvent and intensified emission as the addition of poor solvent, indicating their typical AIE effect. Moreover, the narrowing fwhm and the red-shift of the spectra of **59** (from 480 to 516nm) and **60** (from 479 to 511nm) as the temperature is lowered from 298 to 77K collectively suggest the remarkable thermochromism.

The copper iodide ([$Cu_4I_4L_4$]) clusters C-OH and C-Pr are phosphorescent and obtained as colorless crystalline solids [78]. The fluorescence spectra of C-OH and C-Pr in the solid state at 293K are featured with a wide band emission in the greenish-yellow region peaked at 540 and 568nm, with a quantum yield of 54% and 72%, respectively (Fig. 13D). At 77K, a new emission band emerges at 472 and 425nm, for C-OH and C-Pr, respectively. The presence of these two fluorescence bands for each complex, i.e., the low-energy and high-energy ones, whose intensities alter as a function of temperature, gives rise to the high-contrasted color change in the fluorescence from greenish-yellow to purple-blue. The low-energy band results from halide-to-metal charge transfer and metal-metal transitions, while the high-energy one is associated with the (X,M) ligand charge transfer. Such thermochromic luminescence is caused by the thermal equilibrium between the two corresponding excited states. Significant AIE properties of the clusters are proven by the intensified emission of both C-OH and C-Pr as the aggregation occurs, which are attributed to the blocking of nonradiative decay channels in the excited states.

Apart from C-OH and C-Pr, the Cu nanoclusters (Cu NCs) (Fig. 13E) are also a class of phosphors featured with AIE and thermochromism [79]. The Cu NCs capped with 1-dodecanethiol show the greatly intensified luminescence via self-assembly strategy. Compact and ordered structures are consequently formed, resulting in the originally nonemissive Cu NCs emitting intense phosphorescence. Moreover, polymorphism of Cu NCs assemblies is enabled by the flexible self-assembly. The ribbons of self-assembled Cu NCs exhibit a strong blue-green emission at 490nm with a fwhm of ca. 86nm and quantum yield of 6.5%. Sheets of Cu NCs with loose aggregation structures are fabricated via self-assembly, which has a red-shifted yellow phosphorescence peaked at 547nm with a fwhm about 84nm and a lower quantum yield of 3.6%. A clear relationship between the compactness of assemblies and the phosphorescent emission is established. Firstly, high compactness enhances the cuprophilic Cu(I)···Cu(I) interaction of inter- and intrananoclusters, and in the meantime, the intramolecular motions of the capping 1-dodecanethiol ligand are hindered, and the phosphorescence intensity of Cu NCs is thus reinforced. Secondly, since the emission energy relies on the Cu(I)···Cu(I) distance, the average distance of Cu(I)···Cu(I) is increased by the improved compactness via inducing additional cuprophilic interaction and thereby results in the blue-shift of the phosphorescence of Cu NCs. Due to the compactness-dependent emission properties, the phosphorescence of Cu NCs shows a strong dependence on the temperature as the compactness is closely related to temperature. The emission changes from yellow to bluish green as the annealing temperature increases from 20°C to 120°C, along with the increase in the emission intensity, while the absorption remains unchanged. Although the emission of ribbons and sheets is fairly distinct at room temperature, the difference is reduced when the samples are dipped in liquid nitrogen (77K). The initially bluish-green emissive ribbons display a peak at 517nm, with an emission color of bright green, while the originally yellow-phosphorescent sheets also show an emission band at 517nm. It is worth noting that a new phosphorescent band shows up at 690nm both for ribbons and sheets as the temperature decreases. Apart from this, the emission enhancement at 77 K is observed both in the ribbons and sheets, with the quantum yield of ribbons being as high as 20%. This thermochromic phosphorescence is conversable as cycling the temperature from 77 to 298 K. The emission shift along with temperature change is very complicated. Generally, the phosphorescence of ribbons shows red-shift at low temperature, whereas the phosphorescence of the sheets exhibits a blue-shift. Despite this, the phosphorescence of the sheets first red-shifts until 90 K and blue-shifts afterward. Moreover, the mutable coprophilic interactions enable them to show nonlinear thermosensitivity in the range of 77–298 K.

3 Challenges and outlook

In summary, a large number of photochromic AIE systems varying from the ones obtained by the physical blending of AIEgens with photochromic compounds to those achieved by chemical bonding of AIE units with photochromic motif(s) to the ones generated following the RIM principle are discussed in detail. The discussions focus on their working principle, AIE properties, photochromic features, and the potential applications in the fields of anticounterfeiting, information storage, bioimaging, and super-resolution imaging. Likewise, a series of AIE-active thermochromic luminophores are also systematically introduced and classified into three categories as the thermochromic AIE systems based on organic AIEgens, the ones on the basis of aryl-substituted *o*-carborane/binary boranes, and those based on metal complexes. As can be seen from these systems, the AIE effect could add extra benefits to the photochromic materials as the AIE attribute can afford high off/on or on/off contrasted photoswitching and enable the photoswitch of fluorescence in the aggregated state. Similarly, the AIE feature is also conducive to the aggregated-state thermochromism of luminescence. In turn, photochromism and thermochromism also benefit AIE research as they can endow the AIE materials with the possibility to be used in the areas where smart or stimuli-responsive materials are usually applied.

In spite of the successes in the photochromic or thermochromic AIE systems, there remain several challenges that need to be overcome. For example, (i) the variety and diversity of the photochromic or thermochromic AIE systems need to be enhanced. So far, the AIE units used to construct photochromic AIE systems are mainly restricted to TPE, cyanostilbene derivatives, and a few simple ESIPT compounds. Moreover, no more than 60 examples of the photochromic AIE systems exist to date, and those of AIE-active thermochromic systems are even much fewer. Therefore, the scope of photochromic or thermochromic AIE systems should be greatly expanded in the future. (ii) The clear relationship between the structure and the photochromic or thermochromic properties of the photochromic or thermochromic AIE systems should be expounded. The limited photochromic or thermochromic AIE systems might result from the lack of versatile and universal design principles obtained on the basis of the structure-property relationship. (iii) Promising photochromic or thermochromic AIE systems need to be translated into real-world technology and products in the near future. Furthermore, the commercialization or even industrialization of photochromic or thermochromic AIE systems that hold great potential in practical areas should also be a key point in future investigation. Thus, it can be envisioned that in pursuit of more and various high-performance practically usable photochromic or thermochromic AIE systems with rational design principles, the research related to AIE and photochromic/thermochromic areas will be very active and attract increasing the attention of researchers from the areas of chemistry, functional materials, biology, life science, medical science, and so on.

References

[1] H. Bouas-Laurent, H. Durr, Organic photochromism, Pure Appl. Chem. 73 (2001) 639–665.
[2] K. Szacilowski, Digital information processing in molecular systems, Chem. Rev. 108 (2008) 3481–3548.
[3] H. Tian, J. Zhang (Eds.), Photochromic Materials: Preparation, Properties, and Applications, Wiley, Weinheim, 2016.
[4] J.W. Chung, S.-J. Yoon, S.-J. Lim, B.-K. An, S.Y. Park, Dual-mode switching in highly fluorescent organogels: binary logic gates with optical/thermal inputs, Angew. Chem. Int. Ed. 48 (2009) 7030–7034.
[5] K. Dojin, J.E. Kwon, S.Y. Park, Fully reversible multistate fluorescence switching: organogel system consisting of luminescent cyanostilbene and turn-on diarylethene, Adv. Funct. Mater. 28 (2018) 1706213.
[6] J. Li, M. Tian, H. Xu, X. Ding, J. Guo, Photoswitchable fluorescent liquid crystal nanoparticles and their inkjet-printed patterns for information encrypting and anti-counterfeiting, Part. Part. Syst. Charact. 36 (2019) 1900346.
[7] K. Jeong, S. Park, Y.-D. Lee, C.-K. Lim, J. Kim, B.H. Chung, I.C. Kwon, C.R. Park, S. Kim, Conjugated polymer/photochromophore binary nanococktails: bistable photoswitching of near-infrared fluorescence for in vivo imaging, Adv. Mater. 25 (2013) 5574–5580.
[8] W. Zhong, X. Zeng, J. Chen, Y. Hong, L. Xiao, P. Zhang, Photoswitchable fluorescent polymeric nanoparticles for rewritable fluorescence patterning and intracellular dual-color imaging with AIE-based fluorogens as FRET donors, Polym. Chem. 8 (2017) 4849–4855.
[9] L. Ma, S. Wang, C. Li, D. Cao, T. Li, X. Ma, Photo-controlled fluorescence on/off switching of a pseudo[3]rotaxane between an AIE-active pillar[5]arene host and a photochromic bithienylethene guest, Chem. Commun. 54 (2018) 2405–2408.
[10] P.Q. Nhien, T.T.K. Cuc, T.M. Khang, C.-H. Wu, B.T.B. Hue, J.I. Wu, B.W. Mansel, H.-L. Chen, H.-C. Lin, Highly efficient Förster resonance energy transfer modulations of dual-AIEgens between a tetraphenylethylene donor and a merocyanine acceptor in photo-switchable [2]rotaxanes and reversible photo-patterning applications, ACS Appl. Mater. Interfaces 12 (2020) 47921–47938.
[11] P.Q. Nhien, W.-L. Chou, T.T.K. Cuc, T.M. Khang, C.-H. Wu, N. Thirumalaivasan, B.T.B. Hue, J.I. Wu, S.-P. Wu, H.-C. Lin, Multi-stimuli responsive FRET processes of bifluorophoric AIEgens in an amphiphilic copolymer and its application to cyanide detection in aqueous media, ACS Appl. Mater. Interfaces 12 (2020) 10959–10972.
[12] P.Q. Nhiem, W.-L. Chou, T.T.K. Cuc, T.M. Khang, C.-H. Wu, N. Thirumalaivasan, B.T.B. Hue, J.I. Wu, S.-P. Wu, H.-C. Lin, Synthesis and properties of photochromic spirooxazine with aggregation-induced emission fluorophores polymeric nanoparticles, Dyes Pigments 142 (2017) 481–490.
[13] L. Ma, C. Li, Q. Yan, S. Wang, W. Miao, D. Cao, Unsymmetrical photochromic bithienylethene-bridge tetraphenylethene molecular switches: synthesis, aggregation-induced emission and information storage, Chin. Chem. Lett. 31 (2020) 361–364.
[14] S. Wang, F. Wang, C. Li, T. Li, D. Cao, X. Ma, Photo-induced morphology transition of a multifunctional photochromic bisthienylethene molecule with switchable aggregation-induced emission, Sci. China Chem. 61 (2018) 1301–1306.
[15] C. Li, W.-L. Gong, Z. Hu, M.P. Aldred, G.-F. Zhang, T. Chen, Z.-L. Huang, M.-Q. Zhu, Photoswitchable aggregation-induced emission of a dithienylethene-tetraphenylethene conjugate for optical memory and super-resolution imaging, RSC Adv. 3 (2013) 8967–8972.
[16] C. Li, K. Xiong, Y. Chen, C. Fan, Y.-L. Wang, H. Ye, M.-Q. Zhu, Visible-light-driven photoswitching of aggregated-induced emission-active diarylethenes for super-resolution imaging, ACS Appl. Mater. Interfaces 12 (2020) 27651–27662.
[17] L. Kono, Y. Nakagawa, A. Fujimoto, R. Nishimura, Y. Hattori, T. Mutai, N. Yasuda, K. Koizumi, S. Yokojima, S. Nakamura, K. Uchida, Aggregation-induced emission effect on turn-off fluorescent switching of a photochromic diarylethene, Beilstein J. Org. Chem. 15 (2019) 2204–2212.

[18] S. Chen, W. Li, X. Li, W.-H. Zhu, Aggregation-controlled photochromism based on a dithienylethene derivative with aggregation-induced emission, J. Mater. Chem. C 5 (2017) 2717–2722.
[19] Q. Qi, J. Qian, S. Ma, B. Xu, S.X.-A. Zhang, W. Tian, Reversible multistimuli-response fluorescent switch based on tetraphenylethene-spiropyran molecules, Chem. Eur. J. 21 (2015) 1149–1155.
[20] Q. Yu, X. Su, T. Zhang, Y.-M. Zhang, M. Li, Y. Liu, S.X.-A. Zhang, Non-invasive fluorescence switch in polymer films based on spiropyran-photoacid modified TPE, J. Mater. Chem. C 6 (2018) 2113–2122.
[21] Y. Hong, P. Zhang, H. Wang, M. Yu, Y. Gao, J. Chen, Photoswitchable AIE nanoprobe for lysosomal hydrogen sulfide detection and reversible dual-color imaging, Sensors Actuators B Chem. 272 (2018) 340–347.
[22] S. Bhattacharrya, A. Chowdhury, R. Saha, P.S. Mukherjee, Multifunctional self-assembled macrocycles with enhanced emission and reversible photochromic behavior, Inorg. Chem. 58 (2019) 3968–3981.
[23] S. Peng, J. Wen, M. Hai, Z. Yang, X. Yuan, D. Wang, H. Cao, W. He, Synthesis and application of reversible fluorescent photochromic molecules based on tetraphenylethylene and photochromic groups, New J. Chem. 43 (2019) 617–621.
[24] S. Biswas, R. Mengji, S. Barman, V. Venugopal, A. Jana, N.D.P. Singh, 'AIE + ESIPT' assisted photorelease: fluorescent organic nanoparticles for dual anticancer drug delivery with real-time monitoring ability, Chem. Commun. 54 (2018) 168–171.
[25] L. Wang, Y. Li, X. You, K. Xu, Q. Feng, J. Wang, Y. Liu, K. Li, H. Hou, An erasable photo-patterning material based on a specially designed 4-(1,2,2-triphenylvinyl)aniline salicylaldehyde hydrazone aggregation-induced emission (AIE) molecule, J. Mater. Chem. C 5 (2017) 65–72.
[26] H. Sun, S.-S. Sun, F.-F. Han, Z.-H. Ni, R. Zhang, M.-D. Li, A new tetraphenylethene-based Schiff base: two crystalline polymorphs exhibiting totally different photochromic and fluorescence properties, J. Mater. Chem. C 7 (2019) 7053–7060.
[27] M. Luo, S. Wang, M. Wang, S. Huang, C. Li, L. Chen, X. Ma, Novel organogel harnessing Excited-State Intramolecular Proton Transfer process with aggregation induced emission and photochromism, Dyes Pigments 132 (2016) 48–57.
[28] B. Wu, T. Xue, W. Wang, S. Li, J. Shen, Y. He, Visible light triggered aggregation-induced emission switching with a donor-acceptor Stenhouse adduct, J. Mater. Chem. C 6 (2018) 8538–8545.
[29] R. Singh, H.-Y. Wu, A.K. Dwivedi, A. Singh, C.-M. Lin, P. Raghunath, M.-C. Lin, T.-K. Wu, K.-H. Wei, H.-C. Lin, Monomeric and aggregation emissions of tetraphenylethene in a photo-switchable polymer controlled by cyclization of diarylethene and solvent conditions, J. Mater. Chem. C 5 (2017) 9952–9962.
[30] G. Sinawang, J. Wang, B. Wu, X. Wang, Y. He, Photoswitchable aggregation-induced emission polymer containing dithienylethene and tetraphenylethene moieties, RSC Adv. 6 (2016) 12647–12651.
[31] J. Kim, Y. You, S.-J. Yoon, J.H. Kim, B. Kang, S.K. Park, D.R. Whang, J. Seo, K. Cho, S.Y. Park, Bistable solid-state fluorescence switching in photoluminescent, infinite coordination polymers, Chem. Eur. J. 23 (2017) 10017–10022.
[32] L. Chen, J. Zhang, Q. Wang, L. Zou, Photo-controllable and aggregation-induced emission based on photochromic bithienylethene, Dyes Pigments 123 (2015) 112–115.
[33] G. Liu, Y.-M. Zhang, L. Zhang, C. Wang, Y. Liu, Controlled photoerasable fluorescent behaviors with dithienylethene-based molecular turnstile, ACS Appl. Mater. Interfaces 10 (2018) 12135–12140.
[34] H. Yang, M. Li, C. Li, Q. Luo, M.-Q. Zhu, H. Tian, W.-H. Zhu, Unraveling dual aggregation-induced emission behavior in steric-hindrance photochromic system for super resolution Imaging, Angew. Chem. Int. Ed. 59 (2020) 8560–8570.
[35] C. Xiong, L. Wang, W. Li, Q. Luo, Photochromic monomer and coordination polymer based hybrid tetraarylethenes, Chem. Lett. 47 (2018) 1127–1130.
[36] C. Wang, L. Yan, W. Ding, Q. Luo, Preparation and properties of multifunctional tetraarylethenes and their oxides, Dyes Pigments 177 (2020), 108264.
[37] S. Zhou, S. Guo, W. Liu, Q. Yang, H. Sun, R. Ding, Z. Qian, H. Feng, Rational design of reversibly photochromic molecules with aggregation-induced emission by introducing photoactive thienyl and benzothienyl groups, J. Mater. Chem. C 8 (2020) 13197–13204.
[38] S. Guo, S. Zhou, J. Chen, P. Guo, R. Ding, H. Sun, H. Feng, Z. Qian, Photochromism and fluorescence switch of furan-containing tetraarylethene luminogens with aggregation-induced emission for photocontrolled interface-involved applications, ACS Appl. Mater. Interfaces 12 (2020) 42410–42419.

[39] G. Huang, Q. Xia, W. Huang, J. Tian, Z. He, B.S. Li, B.Z. Tang, Multiple anti-counterfeiting guarantees from a simple tetraphenylethylene derivative-high-contrasted and multi-state mechanochromism and photochromism, Angew. Chem. Int. Ed. 58 (2019) 17814–17819.

[40] D. Ou, T. Yu, Z. Yang, T. Luan, Z. Mao, Y. Zhang, S. Liu, J. Xu, Z. Chi, M.R. Bryce, Combined aggregation induced emission (AIE), photochromism and photoresponsive wettability in simple dichloro-substituted triphenylethylene derivatives, Chem. Sci. 7 (2016) 5302–5306.

[41] T. Yu, D. Ou, L. Wang, S. Zheng, Z. Yang, Y. Zhang, Z. Chi, S. Liu, J. Xu, M.P. Aldred, A new approach to switchable photochromic materials by combining photochromism and piezochromism together in an AIE-active molecule, Mater. Chem. Front. 1 (2017) 1900–1904.

[42] Z. He, L. Shan, J. Mei, H. Wang, J.W.Y. Lam, H.H.Y. Sung, I.D. Williams, X. Gu, Q. Miao, B.Z. Tang, Aggregation-induced emission and aggregation promoted photochromism of bis(diphenylmethylene)dihydroacenes, Chem. Sci. 6 (2015) 3538–3543.

[43] P. Xing, H. Chen, L. Bai, Y. Zhao, Photo-triggered transformation from vesicles to branched nanotubes fabricated by a cholesterol-appended cyanostilbene, Chem. Commun. 51 (2015) 9309–9312.

[44] P. Xing, H. Chen, M. Ma, X. Xu, A. Hao, Y. Zhao, Light and cucurbit[7]uril complexation dual-responsiveness of a cyanostilbene-based self-assembled system, Nanoscale 8 (2016) 1892–1896.

[45] J.W. Chung, S.-J. Yoon, B.-K. An, S.Y. Park, High-contrast on/off fluorescence switching via reversible E – Z isomerization of diphenylstilbene containing the α-cyanostilbenic moiety, J. Phys. Chem. C 117 (2013) 11285–11291.

[46] T. Dünnebacke, K.K. Kartha, J.M. Wiest, R.Q. Albuquerque, G. Fernandez, Solvent-controlled E/Z isomerization vs. [2 + 2] photocycloaddition mediated by supramolecular polymerization, Chem. Sci. 11 (2020) 10405–10413.

[47] J.W. Chung, Y. You, H.S. Huh, B.-K. An, S.-J. Yoon, S.H. Kim, S.W. Lee, S.Y. Park, Shear- and UV-induced fluorescence switching in Stilbenic π-dimer Crystals Powered by Reversible [2+2] Cycloaddition, J. Am. Chem. Soc. 131 (2009) 8163–8172.

[48] P. Wei, J.-X. Zhang, Z. Zhao, Y. Chen, X. He, M. Chen, J. Gong, H.H.Y. Sung, I.D. Williams, J.W.Y. Lam, B.Z. Tang, Multiple yet controllable photoswitching in a single AIEgen system, J. Am. Chem. Soc. 140 (2018) 1966–1975.

[49] Y. Wu, S. Zhang, J. Pei, X.-F. Chen, Photochromic fluorescence switching in liquid crystalline polynorbornenes with α-cyanostilbene side-chains, J. Mater. Chem. C 8 (2020) 6461–6469.

[50] C. Yu, E. Hao, X. Fang, Q. Wu, L. Wang, J. Li, L. Xu, L. Jiao, W.-Y. Wong, AIE-active difluoroboronated acylhydrozone dyes (BOAHY) emitting across the entire visible region and their photo-switching properties, J. Mater. Chem. C 7 (2019) 3269–3277.

[51] J. Chai, Y. Wu, B. Yang, B. Liu, The photochromism, light harvesting and self-assembly activity of a multi-function Schiff-base compound based on the AIE effect, J. Mater. Chem. C 6 (2018) 4057–4064.

[52] T. Ning, L. Liu, D. Jia, X. Xie, D. Wu, Aggregation-induced emission, photochromism and self-assembly of pyrazolone phenlysemicarbazones, J. Photochem. Photobiol. A Chem. 291 (2014) 48–53.

[53] Y. Zheng, X. Zheng, Y. Xiang, A. Tong, Photoactivatable aggregation-induced emission of triphenylmethanol, Chem. Commun. 53 (2017) 11130–11133.

[54] L. Wang, W. Xiong, H. Tang, D. Gao, A multistimuli-responsive fluorescent switch in the solution and solid states based on spiro[fluorene-9,9′-xanthene]-spiropyran, J. Mater. Chem. C 7 (2019) 9102–9111.

[55] S.-J. Yoon, J.H. Kim, K.S. Kim, J.W. Chung, B. Heinrich, F. Mathevet, P. Kim, B. Donnio, A.J. Attias, D. Kim, S.Y. Park, Mesomorphic organization and thermochromic luminescence of dicyanodistyrylbenzene-based phasmidic molecular disks: uniaxially aligned hexagonal columnar liquid crystals at room temperature with enhanced fluorescence emission and semiconductivity, Adv. Funct. Mater. 22 (2012) 61–69.

[56] X. Luo, J. Li, C. Li, L. Heng, Y.Q. Dong, Z. Liu, Z. Bo, B.Z. Tang, Reversible switching of the emission of diphenyl-dibenzofulvenes by thermal and mechanical stimuli, Adv. Mater. 23 (2011) 3261–3265.

[57] J. Guan, F. Xu, C. Tian, L. Pu, M.-S. Yuan, J. Wang, Tricolor luminescence switching by thermal and mechanical stimuli in the crystal polymorphs of pyridyl-substituted fuorene, Chem. Asian J. 14 (2019) 216–222.

[58] S. Yang, P.-.A. Yin, L. Li, Q. Peng, X. Gu, G. Gao, J. You, B.Z. Tang, Crystallization-induced reversal from dark to bright excited states for construction of solid-emission-tunable squaraines, Angew. Chem. Int. Ed. 59 (2020) 10136–10142.

[59] Y. Zhang, G. Zhuang, M. Ouyang, B. Hu, Q. Song, J. Sun, C. Zhang, C. Gu, Y. Xu, Y. Ma, Mechanochromic and thermochromic fluorescent properties of cyanostilbene derivatives, Dyes Pigments 98 (2013) 486–492.

[60] Y. Zhang, Q. Song, K. Wang, W. Mao, F. Cao, J. Sun, L. Zhan, Y. Lv, Y. Ma, B. Zhu, C. Zhang, Polymorphic crystals and their luminescence switching of triphenylacrylonitrile derivatives upon solvent vapour, mechanical, and thermal stimuli, J. Mater. Chem. C 3 (2015) 3049–3054.
[61] Y. Zhou, L. Qian, M. Liu, X. Huang, Y.X. Wang, G. Wu, Y. Cheng, W. Gao, H. Wu, 5-(2,6-Bis((*E*)-4-(dimethylamino)styryl)-1-ethylpyridin-4(1*H*)-ylidene)-2,2-dimethyl-1,3-dioxane-4,6-dione: aggregation-induced emission, polymorphism, mechanochromism, and thermochromism, J. Mater. Chem. C 5 (2017) 9264–9272.
[62] Z. Wu, S. Mo, L. Tan, B. Fang, Z. Su, Y. Zhang, M. Yin, Crystallization-induced emission enhancement of a deep-blue luminescence material with tunable mechano- and thermochromism, Small 14 (2018) 1802524.
[63] Z. Zhang, Y.S. Wu, K.C. Tang, C.L. Chen, J.W. Ho, J. Su, H. Tian, P.T. Chou, Excited-state conformational/electronic responses of saddle-shaped *N,N′*-disubstituted-dihydrodibenzo[*a*,c]phenazines: wide-tuning emission from red to deep blue and white light combination, J. Am. Chem. Soc. 137 (2015) 8509–8520.
[64] W. Huang, L. Sun, Z. Zheng, J. Su, H. Tian, Colour-tunable fluorescence of single molecules based on the vibration induced emission of phenazine, Chem. Commun. 51 (2015) 4462–4464.
[65] L. Shi, W. Song, C. Lian, W. Chen, J. Mei, J. Su, H. Liu, H. Tian, Dual-emitting dihydrophenazines for highly sensitive and ratiometric thermometry over a wide temperature range, Adv. Optical Mater. 6 (2018) 1800190.
[66] H. Liu, W. Song, X. Chen, J. Mei, Z. Zhang, J. Su, Temperature-responsive molecular liquids based on dihydrophenazines for dynamic multicolor fluorescent anti-counterfeiting and encryption, Mater. Chem. Front. 5 (2021) 2294–2302.
[67] W. Song, W. Ye, L. Shi, J. Huang, Z. Zhang, J. Mei, J. Su, H. Tian, Smart molecular butterfly: an ultra-sensitive and range-tunable ratiometric thermometer based on dihydrophenazines, Mater. Horiz. 7 (2020) 615–623.
[68] H. Naito, K. Nishino, Y. Morisaki, K. Tanaka, Y. Chujo, Highly-efficient solid-state emissions of anthracene-*o*-carborane dyads with various substituents and their thermochromic luminescence properties, J. Mater. Chem. C 5 (2017) 10047–10054.
[69] X. Wu, J. Guo, Y. Cao, J. Zhao, W. Jia, Y. Chen, D. Jia, Mechanically triggered reversible stepwise tricolor switching and thermochromism of anthracene-*o*-carborane dyad, Chem. Sci. 9 (2018) 5270–5277.
[70] X. Wu, J. Guo, W. Jia, J. Zhao, D. Jia, H. Shan, Highly-efficient solid-state emission of tethered anthracene-*o*-carborane dyads and their visco- and thermo-chromic luminescence properties, Dyes Pigments 162 (2019) 855–862.
[71] K. Nishino, Y. Morisaki, K. Tanaka, Y. Chujo, Design of thermochromic luminescent dyes based on the bis(ortho-carborane)-substituted benzobithiophene structure, Chem. Asian J. 14 (2019) 789–795.
[72] X. Wu, J. Guo, J. Zhao, Y. Che, D. Jia, Multifunctional luminescent molecules of *o*-carborane-pyrene dyad/triad: flexible synthesis and study of the photophysical properties, Dyes Pigments 154 (2018) 44–51.
[73] H. Mori, K. Nishino, K. Wada, Y. Morisaki, K. Tanaka, Y. Chujo, Modulation of luminescence chromic behaviors and environment-responsive intensity changes by substituents in bis-*o*-carborane-substituted conjugated molecules, Mater. Chem. Front. 2 (2018) 573–579.
[74] M.G.S. Londesborough, J. Dolanský, L. Cerdán, K. Lang, T. Jelínek, J.M. Oliva, D. Hnyk, D. Roca-Sanjuán, A. Francés-Monerris, J. Martinčík, M. Nikl, J. Kennedy, Thermochromic fluorescence from $B_{18}H_{20}(NC_5H_5)_2$: an inorganic-organic composite luminescent compound with an unusual molecular Geometry, Adv. Optical Mater. 5 (2017) 1600694.
[75] J. Liang, Z. Chen, J. Yin, G.-A. Yu, S.H. Liu, Aggregation-induced emission (AIE) behavior and thermochromic luminescence properties of a new gold(I) complex, Chem. Commun. 49 (2013) 3567–3569.
[76] K. Fujisawa, Y. Okuda, Y. Izumi, A. Nagamatsu, Y. Rokusha, Y. Sadaike, O. Tsutsumi, Reversible thermal-mode control of luminescence from liquid-crystalline gold(I) complexes, J. Mater. Chem. C 2 (2014) 3549–3555.
[77] Y. Song, R. Fan, X. Du, K. Xing, P. Wang, Y. Dong, Y. Yang, Effect of noncovalent interactions on Ag(I)/Cu(I) supramolecular architecture for dual-functional luminescence and semiconductive properties, CrystEngComm 18 (2016) 6411–6424.
[78] R. Utrera-Melero, J.-Y. Mevellec, N. Gautier, N. Stephant, F. Massuyeau, S. Perruchas, Aggregation-induced emission properties of copper iodide clusters, Chem. Asian J. 14 (2019) 3166–3172.
[79] Z. Wu, J. Liu, Y. Gao, H. Liu, T. Li, H. Zou, Z. Wang, K. Zhang, Y. Wang, H. Zhang, B. Yang, Assembly-induced enhancement of Cu nanoclusters luminescence with mechanochromic property, J. Am. Chem. Soc. 137 (2015) 12906–12913.

CHAPTER

8

AIE-active rare-metal-free phosphorescent materials

Masaki Shimizu

Faculty of Molecular Chemistry and Engineering, Kyoto Institute of Technology, Kyoto, Japan

1 Introduction

Phosphorescence is a radiation that occurs when a molecule in the lowest excited triplet state (T_1) returns to the ground state (S_0) (Fig. 1). Due to the nature of spin-flipping, phosphorescence is a forbidden transition and the T_1 state has a long lifetime, which is much longer than that of fluorescence, enough to undergo nonradiative decays that originate from molecular motions occurring at room temperature. Since molecular oxygen in the ground state adopts the triplet state, deactivation via energy transfer from the T_1 state to molecular oxygen also occurs when the luminophore is in close proximity to molecular oxygen. Thus, the intensity and lifetime of phosphorescence are very sensitive to the temperature and concentration of molecular oxygen. Therefore, organic molecules that phosphoresce under ambient conditions have high potentials for applications in the fields of bioimaging, chemical sensing including oxygen and temperature monitoring, and data encryption/anticounterfeiting technology.

Due to the progress of organic light-emitting diodes (OLEDs), organometallic compounds containing a rare metal such as iridium and platinum have attracted extensive attention as materials that exhibit room temperature phosphorescence (RTP) [1–4]. This is because rare metals efficiently accelerate both intersystem crossing (ISC) and radiation of phosphorescence due to the significant heavy atom effect, resulting in excellent phosphorescence efficiency. However, the resources of rare metals are limited and expensive. Owing to the significant promotion of spin-flipping transitions, the phosphorescence lifetimes of such organometallic compounds are generally on the order of a few microseconds, which is suitable for OLED applications but not for afterglow applications. Furthermore, the scope of molecular design for rare-metal-containing molecules is limited compared with that for rare-metal-free luminophores because the variation of ligands for phosphorescent metal complexes is restricted.

https://doi.org/10.1016/B978-0-12-824335-0.00005-2

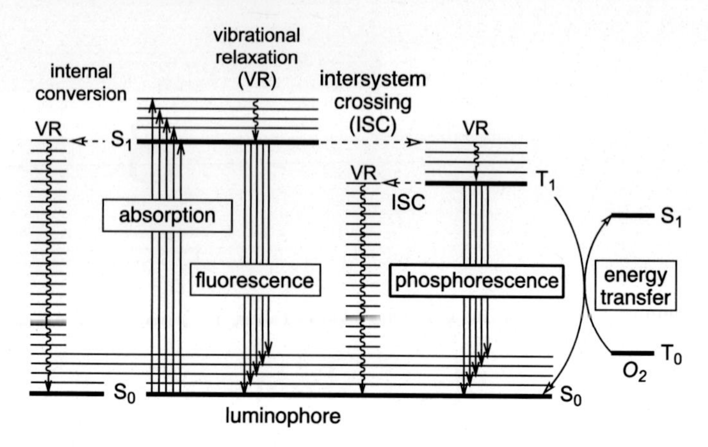

FIG. 1 Photophysical processes of organic molecules.

From these backgrounds, extensive studies on the development of rare-metal-free organic small molecules and polymers that exhibit RTP are currently in progress [5–29]. As described above, the T_1 state is easily deactivated by molecular motions of phosphors. In addition, collisions with surrounding molecules such as solvents also cause phosphorescence quenching. As a result, it is very difficult for rare-metal-free luminophores to exhibit RTP in solution, although very few exceptions are reported [30–33]. Therefore, when rare-metal-free small molecules and polymers exhibit RTP in solid states such as crystal, powder, film, and nanoparticles, their RTP-emissive behaviors can be regarded as aggregation-induced emission (AIE) [34]. As many reviews on rare-metal-free phosphors are now available, this chapter focuses on recent examples of rare-metal-free small phosphors reported in 2018–2020 [35] and phosphorescent polymers reported in 2007–2020. Due to the limited space, the following classes of RTP materials are excluded from this chapter: carbon dots [36], cocrystals including doped crystals [37–39], inclusion complexes [40–42], nonconventional luminophores [43,44], small phosphors dispersed in polymers [45,46]. The RTP-emitters reviewed in this chapter are categorized into the following compound classes: arylcarbonyl compounds, diaryl sulfones, *N*-heterocyclic compounds, boron compounds, phosphorus compounds, diaryl sulfides, poly(lactic acid)s, polyacrylates/poly(acrylic acid)s, polystyrenes, polyurethanes, and polyimides. The last section includes a summary of this chapter.

2 RTP-emissive small molecules

2.1 Arylcarbonyl compounds

Benzophenone undergoes ISC smoothly with a quantum yield of near unity from an S_1 (n, π^*) state to an energetically close T_2 (π, π^*) state that rapidly relaxes to the T_1 (n, π^*) state [47,48]. Hence, molecular modification of a benzophenone skeleton is one of the most reliable molecular design strategies of RTP-active luminophores. Aggregates of 4,4′-bis(carbazole) moiety-substituted benzophenone **1** were demonstrated to exhibit dual phosphorescence under vacuum at room temperature with phosphorescence maxima (λ_P) of 505 and 563 nm and

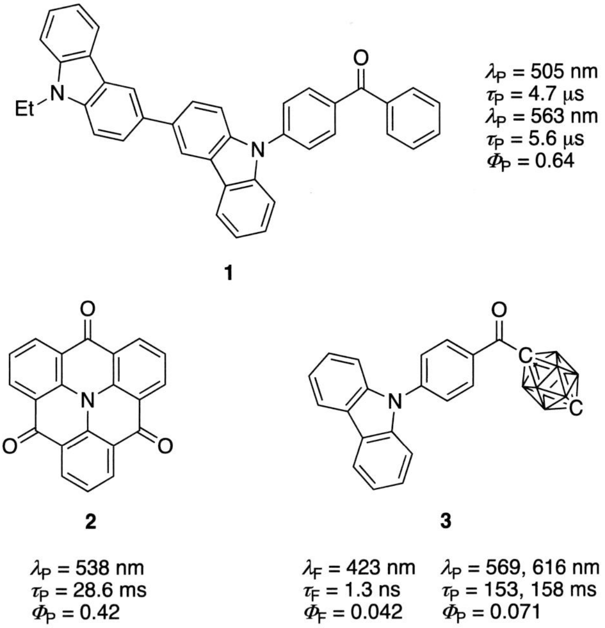

λ_P = 505 nm
τ_P = 4.7 μs
λ_P = 563 nm
τ_P = 5.6 μs
Φ_P = 0.64

1

2

λ_P = 538 nm
τ_P = 28.6 ms
Φ_P = 0.42

3

λ_F = 423 nm
τ_F = 1.3 ns
Φ_F = 0.042

λ_P = 569, 616 nm
τ_P = 153, 158 ms
Φ_P = 0.071

FIG. 2 Molecular structures of aryl ketones **1**–**3**.

lifetimes (τ_P) of 4.7 and 5.6 μs, respectively (Fig. 2) [49]. The RTP quantum yield (Φ_P) was remarkably high (0.64). Time-dependent (TD)-density functional theory (DFT) calculations indicated that the dominant configurations of the S_1 and T_2 states are n–π* transitions and those of the T_1 and T_3 states are π–π* transitions, while those of the T_2 and T_3 states are degenerated. Thus, according to El-Sayed's rule [50,51], both ISCs from S_1 to T_1 and S_1 to T_3 degenerated with T_2 are allowed. The relatively large energy gap between T_2 and T_1 hampers the IC and thus results in dual phosphorescence. Aggregation-induced phosphorescence of **1** is attractive for application to nondoped OLEDs. Indeed, an OLED device fabricated with **1** as a light-emitting layer showed electroluminescence with an external quantum efficiency (EQE) of 5.8% at maximum, which exceeds the theoretical limit value for fluorescent OLEDs.

Triply carbonyl-bridged triphenylamine **2** in single crystal phosphoresced at 538 nm in air with an excellent Φ_P of 0.42 and τ_P of 28.6 ms (Fig. 2) [52]. Molecules in the crystal are almost planar and form π-stacked one-dimensional columns with an interplanar distance of 3.412 Å. The adjacent molecules in each column are virtually eclipsed, implying that the dihedral angle of the transition dipoles ($N \cdots C{=}O \leftrightarrow N^+ \cdots C–O^-$) is almost zero. TD-DFT calculations of the π-stacked dimer of **2** suggested that the value of oscillator strength for $S_1 \rightarrow S_0$ transition (fluorescence) increases from 0 to 0.06 as the dihedral angle between neighboring N···C=O dipoles is changed from 0° to 60°. Thus, the eclipsed stacking in the crystal is a key for suppressing fluorescence and thus accelerating ISC involving transitions between n–π* and π–π* configurations. H-aggregation is proposed to stabilize the triplet excited states and result in the emergence of persistent RTP [53]. However, the phosphorescence quantum yields of such systems are generally low. Hence, it is noteworthy that highly efficient RTP (Φ_P=0.42) is attained with densely packed crystals.

Dual emission of fluorescence and RTP was observed for 4-carbazolyl-1-(*p*-carboranylcarbonyl)benzene (**3**) with quantum yields of 0.042 and 0.071, respectively (Fig. 2) [54]. The RTP quantum yield of **3** was higher than those of *meta*- (0.009) and *ortho*-carboranyl analogues (0.058), and the ISC rate of **3** was the fastest compared with the analogues. X-ray diffraction analysis of a single crystal of **3** disclosed that there were intermolecular interactions such as C–H···O (2.537 Å) and B–H···π interactions (2.718 and 2.848 Å) in the crystal, which effectively suppressed molecular motions resulting in nonradiative decays of the T_1 state. TD-DFT calculations revealed that the optical excitation of **3** involves intramolecular CT from the carbazolyl group to the *p*-carboranylcarbonyl moiety, and T_n states ($n = 1$ to 4) generated through ISC have a mixed character of $^3(\pi,\pi^*)$ and $^3(n,\pi^*)$ configurations.

The lifetime-prolonging effect of chlorine is demonstrated with 2-carbonyl-9,9-dimethylxanthenes **4** and **5** (Fig. 3) [55]. Their time-gated photoluminescence spectra showed phosphorescence emission maxima at 560–572 nm with lifetimes of 52 (**4a**), 139 (**4b**), 484 (**5a**), and 601 (**5b**) milliseconds, respectively. Thus, the phosphorescence lifetimes were prolonged by the presence of a chlorine atom. X-ray single crystal analysis revealed that the dimethylxanthene core adopted planar conformation and formed parallel dimers in all cases. The intermolecular distances of the dimers of chlorine-containing **4b** and **5b** were shorter than those of nonchlorinated **4a** and **5a** due to the intermolecular interactions occurring at the chlorine atom. The resulting dense packing with multiple intermolecular interactions suppressed molecular motions and contributed to the prolongation of the lifetimes. In vivo bioimaging of HeLa cervical cancer cells was demonstrated with a high signal-to-noise ratio of 376 using nanoparticles prepared from **5b** and water-soluble polymer F127.

Bis(amide)s **6a**–**6c** in crystal emitted greenish-white, white, and pale greenish-yellow light, respectively, which consisted of fluorescence and RTP (Fig. 4) [56]. The phosphorescence quantum yields were 0.102 (**6a**), 0.057 (**6b**), and 0.098 (**6c**), respectively, and their lifetimes were over 300 ms. As a long lifetime and efficient quantum yield are generally incompatible for phosphorescence, it is noteworthy that those are simultaneously realized in the example of **6a**–**6c**. X-ray crystallographic analyses of **6a** and **6c** were successful and revealed that there are several C–H···O intermolecular interactions between benzene hydrogen and carbonyl oxygen atoms and intermolecular π–π stacking between carbazole moieties, which effectively restrain intramolecular motions. Ground solids of **6b** and **6c** also exhibited dual emission under ambient conditions. Based on the robustness of RTP, anticounterfeiting and encryption were demonstrated with **6b**. In addition, in vivo afterglow imaging with a high signal-to-

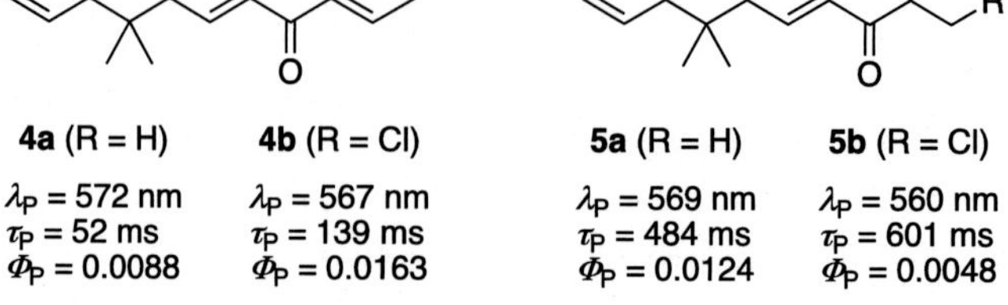

4a (R = H): λ_P = 572 nm, τ_P = 52 ms, Φ_P = 0.0088
4b (R = Cl): λ_P = 567 nm, τ_P = 139 ms, Φ_P = 0.0163
5a (R = H): λ_P = 569 nm, τ_P = 484 ms, Φ_P = 0.0124
5b (R = Cl): λ_P = 560 nm, τ_P = 601 ms, Φ_P = 0.0048

FIG. 3 Molecular structures of 2-carbonylxanthenes **4** and **5**.

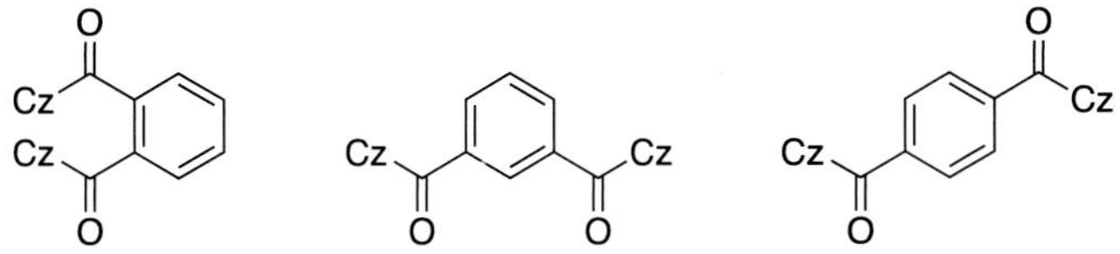

6a
Φ_F = 0.134
λ_P = 529, 572, 623 nm
τ_P = 344 ms
Φ_P = 0.102

6b
Φ_F = 0.078
λ_P = 529, 572, 623 nm
τ_P = 711 ms
Φ_P = 0.057

6c
Φ_F = 0.112
λ_P = 529, 572, 623 nm
τ_P = 312 ms
Φ_P = 0.098

FIG. 4 Molecular structures of aromatic amides **6** (Cz: carbazolyl).

background ratio was realized using nanoparticles of **6b** or **6c** encapsulated by amphiphilic copolymer lipid-PEG_{2000}.

Crystalline powder of phthalimide **7** concurrently exhibited blue fluorescence and yellow phosphorescence at 457 and 548 nm, respectively, resulting in white emission whose values of the Commission Internationale de l'Eclairage (CIE) coordinates were (0.29, 0.35) (Fig. 5) [57]. Interestingly, the yellow phosphorescence was intensified upon UV irradiation and reached a maximum after exposure to UV irradiation for 0.8 s. Thus, the emission color changed from blue to white and then to yellow when a UV lamp was turned on and off. The photoluminescence spectrum of **7** in 2-methyltetrahydrofuran at 77 K had no emission maximum in the yellow region, suggesting that the yellow RTP was originated from the aggregation. Scratching powder of **7** by a spatula induced intense blue emission followed by yellow afterglow observable by naked eyes, indicating that **7** has a mechanoluminescent character. X-ray diffraction analysis revealed that the crystal packing structure was noncentrosymmetric and each asymmetric unit consisted of two kinds of dimers whose dipole moments were almost double those of the monomer. Hence, the emergence of mechanoluminescence is closely related to breaking the crystal structure with large dipole moments. TD-DFT calculations confirmed that the ISC channels of the dimers were much larger than those of a monomer, implying that the dimer formation is also advantageous for efficient generation of the T_1 state.

Swallow-tailed bromonaphthalimide **8** with a glass transition temperature (T_g) of −38.7°C was developed as an organic phosphor that exhibits RTP at 594 nm in liquid (Fig. 5) [58]. Based on photophysical properties of **8** dispersed in a thin film of poly(methyl methacrylate), intermolecular Br···O interactions are proposed to be essential for the emergence of RTP in liquid. The viscous nature may also play a key role in suppressing nonradiative decays of

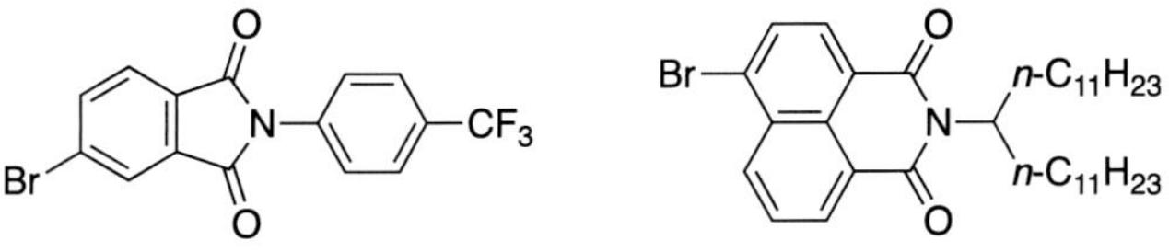

7
λ_F = 457 nm λ_P = 548 nm
Φ_F = 0.018 Φ_P = 0.041
τ_F = 2.6 ns τ_P = 102 ms
(x,y) = (0.29, 0.35)

8
T_g = −38.7 °C
λ_P = 594 nm
Φ_P = 0.001
τ_P = 5.7 ms

FIG. 5 Molecular structures of aromatic amides **7** and **8**.

the T_1 state. Although the phosphorescence quantum yield of **8** in neat liquid was quite low (0.001), the yield was enhanced to 0.02 by mixing **8** with terephthalaldehyde in a 1:1 ratio. The mixture was also liquid at room temperature with a T_g of −33.4°C, and an RTP-emissive large area film (10 cm × 10 cm) of the mixture was fabricated.

2.2 Diaryl sulfones

Diaryl sulfone moieties are also typical platforms for rare-metal-free RTP-emitters [59,60]. Crystals of phenothiazine-*S*,*S*-dioxide **9** exhibited dual emission of blue fluorescence at 411 nm and green phosphorescence at 486 and 514 nm with a good Φ_P (0.082) and a very long τ_P (876 ms) under ambient conditions (Fig. 6) [61]. Several intermolecular interactions such as C–H···π and C–H···O were observed in the crystal, which effectively suppressed molecular motions of each luminophore, resulting in nonradiative decay processes of the triplet states. Crystals of *N*-acetylphenothiazine, the nonoxidized counterpart of **9**, phosphoresced with a τ_P of 238 ms that was 3.7 times shorter than that of **9**. TD-DFT calculations indicated that the presence of d-pπ bonds of the S=O moieties contributes to imparting nearly pure π–π* configurations to the T_1 state. Therefore, it is proposed that the long RTP lifetime of **9** is attributed to the d-pπ bond of the S=O moiety. Based on the large difference of afterglow between **9** and the nonoxidized counterpart, application to data encryption was demonstrated.

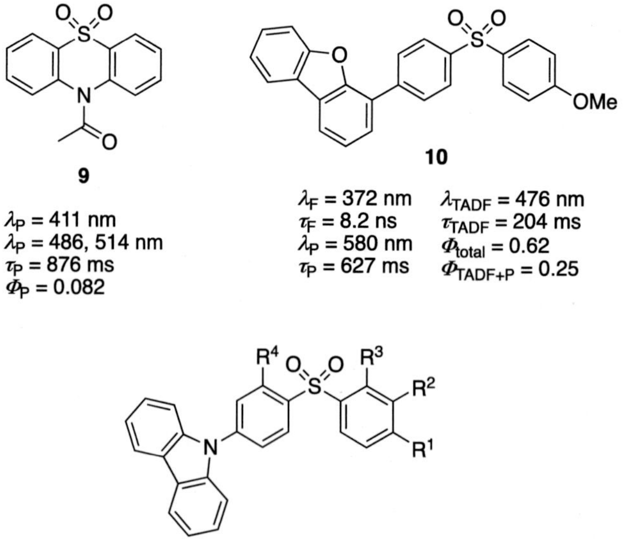

9

λ_P = 411 nm
λ_P = 486, 514 nm
τ_P = 876 ms
Φ_P = 0.082

10

λ_F = 372 nm
τ_F = 8.2 ns
λ_P = 580 nm
τ_P = 627 ms
λ_{TADF} = 476 nm
τ_{TADF} = 204 ms
Φ_{total} = 0.62
Φ_{TADF+P} = 0.25

	R¹	R²	R³	R⁴	Φ_F	τ_F (ns)	Φ_P	τ_P (ms)
11	Br	H	H	H	0.44	2.7	0.06	120
12	H	Br	H	H	0.51	4.9	0.03	530
13	H	H	Br	H	0.10	5.3	0.52	180
14	H	H	H	Br	0.006	1.9	0.27	265

FIG. 6 Molecular structures of diaryl sulfones **9**–**14**.

Benzofuran and methoxy-substituted diphenyl sulfone **10** in crystalline form concurrently exhibited prompt fluorescence at 372 nm, delayed fluorescence at 476 m, and phosphorescence at 580 nm with a total quantum yield of 0.62 (Fig. 6) [62]. The delayed fluorescence was confirmed to be thermally activated (TADF) due to the dependency of luminescence intensity on temperature. X-ray structural analysis of a single crystal revealed that intermolecular hydrogen bonds were formed between the benzene hydrogen of the benzofuran moiety and the methoxy oxygen atom in two adjacent molecules, which contributed to the suppression of molecular motions resulting in the deactivation of the T_1 state. TD-DFT calculations of a monomer and a dimer suggested that intermolecular charge transfer is operative with a dimer. Hence, it is concluded that the blue delayed fluorescence emitted from the S_1 state of dimer, which was generated through intermolecular charge transfer. The lifetimes of the blue delayed fluorescence and orange phosphorescence were largely distinct ($\tau_{TADF}=204$ and $\tau_P=627$ ms) so the color of the afterglow changed from cold-white to orange over time. This color change of long-lived luminescence with a single component is unique and may find applications in the fields of information encryption and data anticounterfeiting. When the crystalline state of **10** was converted into an amorphous state by grinding, the afterglow became very weak ($\Phi_{TADF+P}=0.01$), probably due to the breaking of the intermolecular hydrogen bonding. This corresponds to the deduction that the TADF was originated from dimers constructed by intermolecular hydrogen bonds, and the hydrogen bonds also played a key role in suppressing nonradiative decays of the excited states.

Bromine and carbazolyl-substituted diphenyl sulfones **11**–**14** in the crystalline state exhibited dual emission of blue fluorescence and yellow phosphorescence under ambient conditions (Fig. 6) [63]. Compounds **11** and **12** in which a bromine atom was attached at the *para*- or *meta*-position of the sulfonyl group emitted intense fluorescence along with weak phosphorescence (**11**: $\Phi_F=0.44$, $\Phi_P=0.06$; **12**: $\Phi_F=0.51$, $\Phi_P=0.03$), while phosphorescence was a major component of dual emission for *ortho*-brominated derivatives **13** ($\Phi_P=0.52$, $\Phi_F=0.10$) and **14** ($\Phi_P=0.27$, $\Phi_F=0.006$). The ISC rates of **13** and **14** were much higher than those of **11** and **12**. X-ray analysis of single crystals of **13** and **14** disclosed that intramolecular hydrogen bonding between the bromine and oxygen atoms existed. TD-DFT calculations suggested that the spin-orbit coupling (SOC) matrix elements between S_1 and lower-lying triplet states (T_n) of **13** and **14** were significantly larger than those of **11** and **12**, implying that effective ISC channels were available for **13** and **14**. Therefore, construction of intramolecular halogen bonding in diaryl sulfones is effective for promotion of ISC, immobilization of molecular conformation of the excited states, and thus realization of highly efficient RTP.

2.3 N-heterocyclic compounds

To develop RTP luminophores, acceleration of the ISC process is essential. One approach to accelerate ISC is to enhance SOC via the incorporation of arylcarbonyl groups, halogen atoms such as bromine and iodine, and *N*-heterocyclic compounds like carbazole and triazine. Enhanced SOC also causes acceleration of radiative decay of the T_1 state, which should shorten the lifetime of phosphorescence. Therefore, the creation of persistent RTP luminophores requires an approach which is different from SOC enhancement. It is to reduce the energy gap between S_1 and T_1, which is the same as the design principle for TADF luminophores. This is

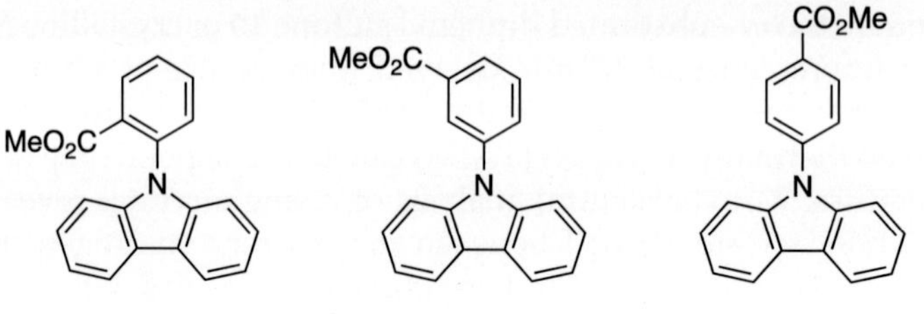

15a	15b	15c
$\Phi_F = 0.115$	$\Phi_F = 0.250$	$\Phi_F = 0.194$
$\lambda_P = 550, 600$ nm	$\lambda_P = 545, 595$ nm	$\lambda_P = 560, 590$ nm
$\tau_P = 619$ ms	$\tau_P = 795$ ms	$\tau_P = 129$ ms
$\Phi_P = 0.020$	$\Phi_P = 0.021$	$\Phi_P = 0.012$

FIG. 7 Molecular structures of *N*-(methoxycarbonylphenyl)carbazoles **15**.

the case of *N*-(methoxycarbonylphenyl)carbazoles **15** that consisted of a twisted donor-acceptor electronic structure (Fig. 7) [64]. The τ_P values of **15a**–**15c** were 619, 795, and 129 ms, respectively. Dihedral angles between the carbazolyl and *N*-phenyl groups, determined by X-ray crystal analysis, were 81.09° for **15a**, 48.37° for **15b**, and 50.30° for **15c**. The twisted conformation was beneficial not only for the spatial separation of the HOMOs and LUMOs, but also for reducing concentration quenching of luminescence. TD-DFT calculations revealed that the S_1 and T_n ($n=1$–3) states of **15a** and **15b** adopt pure π–π^* configurations, and the calculated SOC constants and energy gaps between the S_1 and T_3 states are small. This suggests that the ISC of **15a** and **15b** proceeds due to the small energy gap and the radiative decay of the T_1 state should be very slow. Meanwhile, the energy gap between the S_1 and T_3 states of **15c** was relatively large and the electronic configuration of the T_1 state was a hybrid of π–π^* and n–π^* ones, suggesting that the ISC of **15c** was primarily promoted by SOC and the phosphorescence decay became faster than those of **15a** and **15b**. This analysis is well matched with the experimental results.

The intramolecular interaction between bromine and a nitrogen atom of a carbazole ring is also effective for attaining highly efficient RTP. The crystal of 2,5-dibromo-1,4-dicarbazolylbenzene (**16b**) exhibited RTP at 546 nm with a Φ_P of 0.38, which was larger than that of the nonbrominated counterpart **16a** by 15.2 times (Fig. 8) [65]. It is shown that C–Br···π and C–Br···N interactions and H-aggregation exist in the crystal of **16b** by X-ray diffraction analysis. TD-DFT calculations confirmed that the presence of the bromine atoms greatly enhanced SOC constants between S_1 and T_n ($n=7$, 5, and 3), and the SOC constant between the T_1 and S_0 states remained small irrespective of the presence or absence of bromine atoms.

N-Dibenzofurylcarbazoles **17a** and **17b** in crystalline powder exhibited dual emission of fluorescence at 400–500 nm and phosphorescence at >520 nm (Fig. 8) [66]. In the case of **17b**, the color of prompt emission was white with CIE coordinates of (0.34, 0.30) due to the balanced fluorescence and phosphorescence. The emission spectra of **17a** and **17b** are quite similar to that of carbazole, indicating that the carbazole functioned as a luminescent part. To understand why both fluorescence and phosphorescence were originated from only the carbazole moiety and the phosphorescence was persistent, the following mechanism is proposed. Photoexcitation of **17** occurs via a π-π^* transition of the carbazole portion. The generated S_1 state smoothly undergoes ISC from the carbazole portion to the dibenzofuran

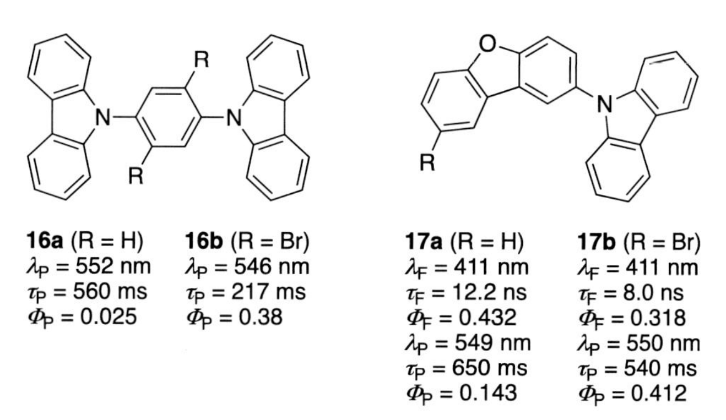

FIG. 8 Molecular structures of *N*-arylcarbazoles **16** and **17**.

moiety via a spin-vibronic coupling mechanism [67,68], in which the dibenzofuran moiety serves as an ISC facilitator. The generated T_n state is efficiently converted into the T_1 state through intramolecular triplet-triplet energy transfer from the dibenzofuran moiety to the carbazole portion possessing a π-π* configuration that leads to persistent phosphorescence.

Dibenzo[*a*,*c*]phenazine (**18**) in powder form exhibited fluorescence at 440 nm and dual phosphorescence simultaneously at 489 and 569 nm, emitting white light with CIE coordinates of (0.28, 0.33) under ambient conditions (Fig. 9) [69]. TD-DFT calculations revealed that the configurations of the S_1 and T_2 states are n-π* transitions and those of the S_2 and T_1 states are π-π* transitions. Thus, both ISC processes involving $S_1 \rightarrow T_1$ and $S_2 \rightarrow T_2$ are favored according to El-Sayed's rule, and the radiation from the T_2 state was possible and preferred to IC

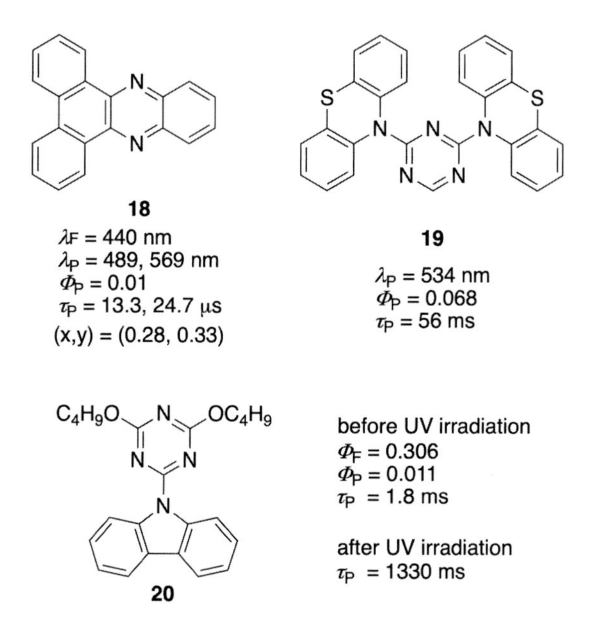

FIG. 9 Molecular structures of phenazine **18** and triazines **19** and **20**.

from T_2 to T_1 owing to the n-π* nature of the T_2 state. The SOC constants of benzo[*b*]triphenylene (a nitrogen-free counterpart of **16**) for ISC are significantly small, indicating that the nitrogen atoms in **18** are crucial for the multiemission.

Crystals of phenothiazine-substituted triazine **19** obtained by recrystallization from ethyl acetate solution exhibited green RTP at 534 nm with a Φ_P of 0.068 and τ_P of 56 ms, whose afterglow was observed by naked eyes (Fig. 9) [70]. In contrast, recrystallization from chloroform solution produced chloroform-embedded crystals whose phosphorescence was less efficient ($\Phi_P = 0.012$) and shorter-lived ($\tau_P = 1.5$ ms). The incorporation of chloroform molecules in the crystal resulted in the acceleration of k_r and k_{nr} by 6.7 and 39.7 times, respectively. Based on these observations, **19** in the solid state was demonstrated to serve as a chloroform probe. Thus, when chloroform-free crystalline powder exhibiting an afterglow was fumed with chloroform, the afterglow disappeared. Heating the fumed sample at 100°C for 1 h led to the revival of the afterglow. The sensitivity limitation for chloroform detection was as low as 5 ppm. Good repeatability and high selectivity were also demonstrated.

Dynamic ultralong RTP was observed with triazine **20** (Fig. 9) [71]. Crystalline form of **20** exhibited fluorescence-phosphorescence dual emission with quantum yields of 0.306 and 0.011, respectively, under ambient conditions. When the crystals were irradiated by a UV lamp (365 nm) for 8 min, the phosphorescence intensity increased by 30 times. The phosphorescence lifetime was prolonged from 1.8 to 1330 ms upon UV irradiation for 10 min. The photoactivated **20** was reverted to the original state by standing under ambient conditions for 160 min or heating at 338 K for 5 min. X-ray diffraction analysis of a single crystal at 100 K revealed that intermolecular distances between adjacent molecules in the photo-irradiated crystal were shorter than those of the crystal before irradiation. Thus, UV irradiation resulted in the restriction of molecular motions by making each molecule closer, which led to the prolongation of the phosphorescence lifetime. The dynamic RTP properties can be tuned simply by changing the butoxy groups to other alkoxy groups, and the application of multilevel anticounterfeiting was demonstrated with a device containing **20** with its methoxy, propoxy, and pentoxy derivatives as luminescent components.

2.4 Boron compounds

Crystals of boryl-substituted triphenylamine **21** exhibited dual emission of fluorescence and phosphorescence with quantum yields of 0.76 and 0.13, respectively, in vacuum (Fig. 10) [72]. In air, the phosphorescence quantum yield was decreased to 0.014 and phosphorescence lifetimes were greatly shortened from 348 to 12.8 ms at 525 nm and from 354 to 13.3 ms at 565 nm, indicating that **21** in crystal form was sensitive to molecular oxygen. This was because there were enough vacancies in the crystals for molecular oxygen to penetrate. Each molecule in the crystals formed dimers, which favored narrowing of the energy gap between the S_1 and T_1 states and increase of the ISC channels compared with the monomer, suggesting that dimer formation induced the RTP. RTP properties of a thin film of poly(ethylene oxide)/poly(propylene oxide) triblock copolymer doped with **21** were also sensitive to molecular oxygen. The rapid and quantitative sensing system for molecular oxygen was established using the polymer film.

Tris(2,6-dimethylphenyl)borane (**22**) in crystalline form exhibited violet fluorescence and greenish-yellow RTP with a long phosphorescence lifetime of 478 ms (Fig. 10) [73]. It is worth

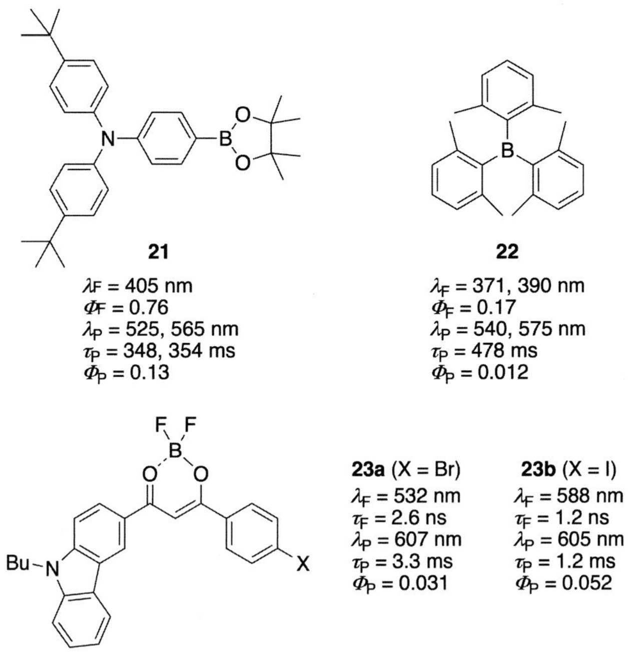

21
λ_F = 405 nm
Φ_F = 0.76
λ_P = 525, 565 nm
τ_P = 348, 354 ms
Φ_P = 0.13

22
λ_F = 371, 390 nm
Φ_F = 0.17
λ_P = 540, 575 nm
τ_P = 478 ms
Φ_P = 0.012

23a (X = Br)
λ_F = 532 nm
τ_F = 2.6 ns
λ_P = 607 nm
τ_P = 3.3 ms
Φ_P = 0.031

23b (X = I)
λ_F = 588 nm
τ_F = 1.2 ns
λ_P = 605 nm
τ_P = 1.2 ms
Φ_P = 0.052

FIG. 10 Molecular structures of boron-containing compounds **21–23**.

noting that **22** is free of the lone pair of electrons because most organic phosphors have lone pair-containing heteroatoms such as nitrogen and oxygen for smooth ISC. TD-DFT calculations showed that the optical excitation involved π-B transition, which is a transition from π-orbitals of the aromatic rings to the empty *p*-orbital on the boron atom, and ISC was accelerated by $(\sigma, Bp) \rightarrow (\pi, Bp)$ and $(\pi, Bp) \rightarrow (\sigma, Bp)$ transitions.

Nanocrystals of carbazole-containing difluoroboron β-diketonates **23a** and **23b**, which were prepared by the rapid injection of a THF solution of **23** into an aqueous solution of cetyltrimethyl ammonium bromide micelle, exhibited fluorescence-phosphorescence dual emission in water under ambient conditions (Fig. 10) [74]. Surprisingly, no RTP was observed with the single crystals of **23a** and **23b**, which were obtained by recrystallization from an ethyl acetate/acetone solution. Instead, both crystals emitted TADF that was completely overlapped with prompt fluorescence. X-ray diffraction patterns of single crystals and nanocrystals were different from one another, suggesting that the emergence of RTP or TADF is closely related to the packing mode of the crystals. Long-lived RTP with nanocrystals in water is attractive for bio-related applications.

2.5 Phosphorus compounds

Dicarbazolyl(phenyl)phosphine oxide **24** and sulfide **25** exhibited dual emission of fluorescence and RTP with long phosphorescence lifetimes of several hundred milliseconds (Fig. 11) [75]. The ISC rates of **24** and **25** in powders at room temperature were 2.6×10^6 and $9.7 \times 10^6 s^{-1}$, respectively, which were much faster than that ($1.2 \times 10^5 s^{-1}$) of di(9*H*-carbazol-9-yl)(phenyl)phosphine (**26**) by 22 and 81 times. Resonance of N–P=O/N–P=S

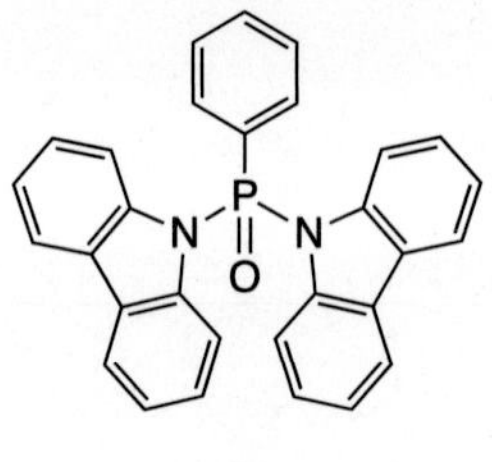

24
λ_F = 430 nm, Φ_F = 0.299
λ_P = 537, 582 nm
τ_P = 670, 660 ms
Φ_P = 0.028

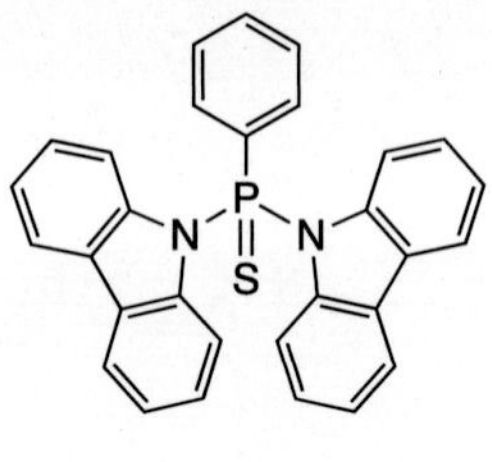

25
λ_F = 430 nm, Φ_F = 0.105
λ_P = 530, 577, 627 nm
τ_P = 510, 470, 430 ms
Φ_P = 0.040

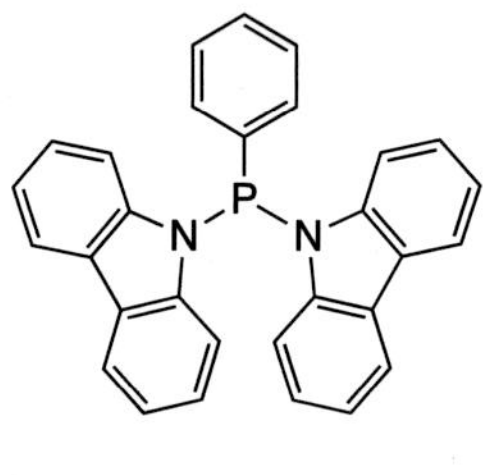

26
λ_F = 411, 430 nm, Φ_F = 0.0489
λ_P = 587, 644 nm
τ_P = 230, 210 ms
Φ_P = 0.0008

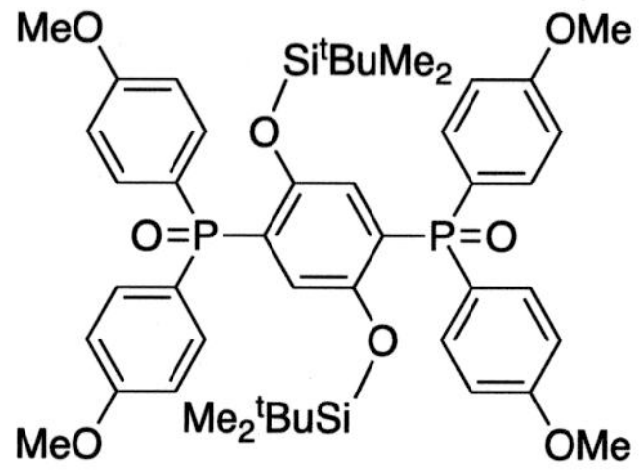

27
λ_F = 366 nm, τ_F = 1.1 ns
λ_P = 457 nm, τ_P = 59.1 ms
Φ_{F+P} = 0.37

FIG. 11 Molecular structures of phosphorus-containing compounds **24**–**27**.

moieties is proposed to account for the ISC acceleration. The long lifetimes were attributed to H-aggregation which was found in the single-crystal structures. The time-resolved and color-coded QR code that can deal with multiple information is demonstrated with **24** with the aid of its long-lived RTP and large color difference between the prompt dual emission and afterglow.

As described above, most examples of fluorescence-phosphorescence dual-emissive luminophores are the ones that emit intense fluorescence and faint RTP, in which emission peaks of RTP are only detectable with time-gated measurement. On the other hand, organic light-emitting compounds that exhibit dual emission of fluorescence and RTP with a bimodal spectrum in which both emission peaks are easily detected without time-gated measurement are very limited [76]. Such balanced dual emission was achieved with 2,5-disiloxy-1,4-bis(diarylphosphinyl)benzene **27** (Fig. 11) [77]. The steady-state photoluminescence spectrum of **27** in powder under vacuum consisted of two emission maxima at 366 and 457 nm with an intensity ratio (I_P/I_F) of 0.62. As 1,4-dibenzoyl-2,5-disiloxybenzenes in crystalline form exhibited only RTP [78], the choice of a diarylphosphinyl group as acceptors in the electronic structure of 1,4-di(acceptor)-2,5-disiloxybenzene is a key for the realization of the balanced dual emission. An arylsulfonyl group was also found to serve as an inducer of balanced dual emission. The fluorescence-phosphorescence intensity ratios of **27** in powder and a thin film of PMMA were sensitive to temperature and molecular oxygen, respectively, with excellent fittings to the calculated calibration curves, demonstrating that 2,5-disiloxy-1,4-bis (diarylphosphinyl)benzenes have high potential as probes for ratiometric luminescence sensing of temperature and molecular oxygen.

2.6 Diaryl sulfides

Cyanophenyl methoxyphenyl sulfide **28a** in crystalline form under ambient conditions exhibited triple luminescence of prompt and delayed fluorescence and RTP (Fig. 12) [79]. On the other hand, cyanophenyl pyridyl sulfide **28b** did not phosphoresce under the same conditions. However, when **28b** was exposed to vapor of HCl or AcOH, the resulting **28c** showed bright RTP with a lifetime of 58 μs, which was much shorter than that of **28a** ($\tau_P = 82.5$ ms), suggesting the possibility of **28b** as a turn-on-type luminescent probe for acid. Indeed, the phosphorescence intensity was linearly intensified as the HCl concentration increased. The detection limit was 8.3 mg m^{-3}, which is below the allowed concentration (14.9 mg m^{-3}) for humans, indicating that **28b** is applicable to practical use. Given that RTP-based sensing is generally based on quenching, this turn-on sensing system is intriguing and widens the possibility of RTP-based functional materials. Theoretical calculations suggested that the protonation induced a charge transfer (CT) character of the T_1 state, which resulted in the bright RTP.

Hexakis(arylthio)benzenes in the solid state are known to exhibit highly efficient RTP [80,81]. Powder of 1,4-dicyano-2,3,5,6-tetrakis(4-hydroxyphenylthio)benzene (**29a**) [82] and 1,4-dicyano-2,3,5,6-tetrakis(3-hydroxyphenylthio)benzene (**29b**) phosphoresced under ambient conditions with quantum yields of 0.014 and 0.052 and lifetimes of 0.4 and 0.7 μs,

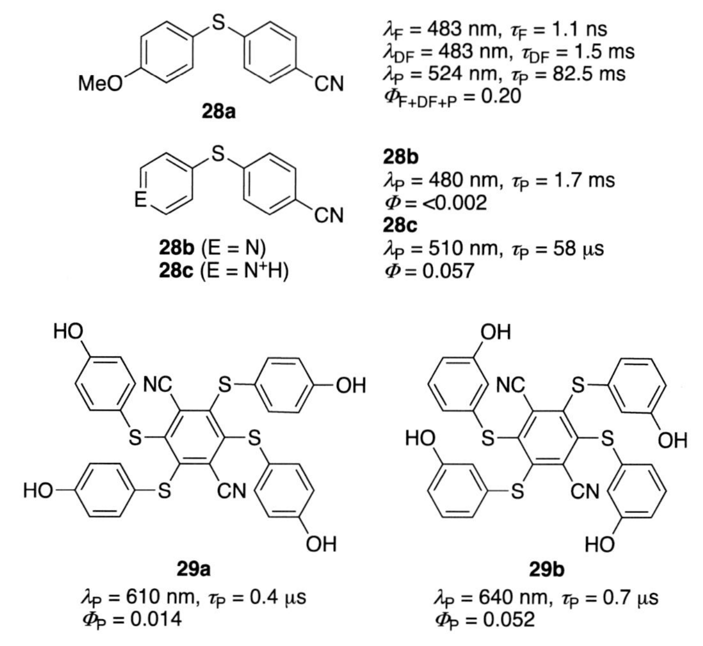

FIG. 12 Molecular structures of diaryl sulfides **28** and **29**.

respectively (Fig. 12) [83]. As THF solutions of **29a** and **29b** showed no emission and weak phosphorescence ($\Phi_P = 0.011$), respectively, the solid-state luminescence of **29b** is classified as aggregation-induced enhanced phosphorescence. DFT calculations indicated that the HOMOs are located on two hydroxyphenyl groups ortho to the cyano groups, while the LUMOs are developed over the NC-phenylene-CN moiety, suggesting that the intramolecular charge transfer (ICT) promoted ISC causing the emergence of RTP.

3 RTP-emissive polymers

3.1 Poly(lactic acid)s

Since Fraser and coworkers demonstrated that poly(lactic acid)s (PLAs) containing a difluoroboron dibenzoylmethane moiety at a molecular terminal position exhibited dual emission of prompt and delayed fluorescence along with RTP under vacuum [84], PLAs have attracted attention as phosphorescent polymers. PLAs **30b** and **30c** were reported to exhibit fluorescence-phosphorescence dual emission whose luminescence spectra were bimodal (Fig. 13) [85]. In contrast, the low-molecular-weight PLA **30a** in powder and film showed

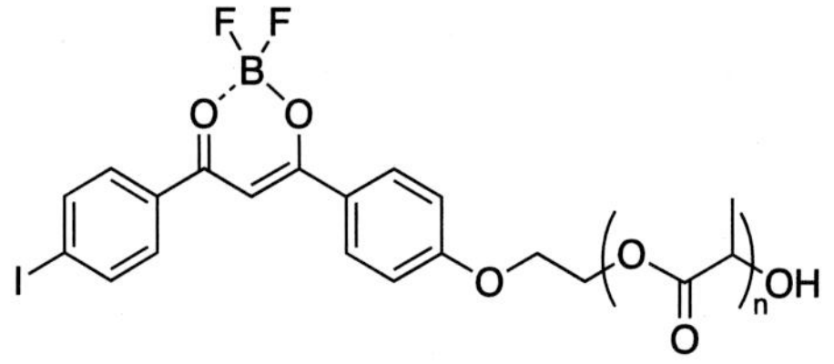

		powder				film			
30	n	λ_F (nm)	τ_F (ns)	λ_P (nm)	τ_P (ms)	λ_F (nm)	τ_F (ns)	λ_P (nm)	τ_P (ms)
30a	27	485	0.37	535	4.06	458	0.48	532	4.25
30b	95	470	0.42	527	4.39	445	0.54	526	4.37
30c	234	456	0.43	525	4.50	438	0.64	523	4.41

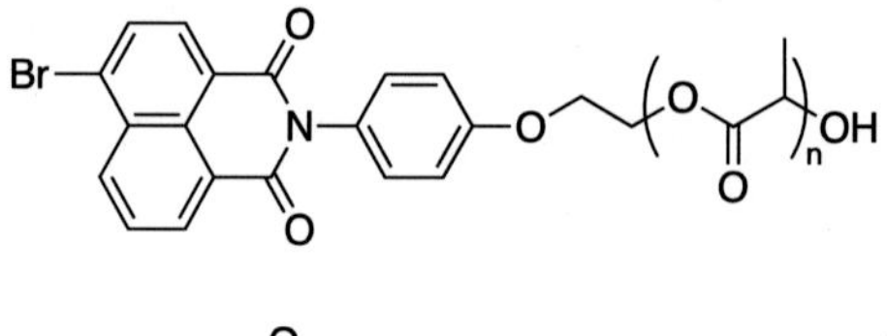

31
λ_P = 611 nm
τ_P = 3.6 ms
Φ_P = 0.046

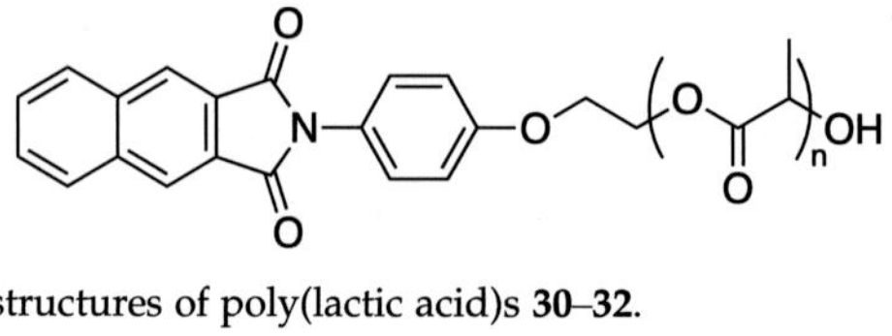

32
λ_P = 510 nm
τ_P = 1200 ms
Φ_P = 0.039

FIG. 13 Molecular structures of poly(lactic acid)s **30–32**.

monomodal spectra consisting of intense RTP at 535 and 532 nm with small fluorescence shoulders at 485 and 458 nm, respectively. The luminescence data shown in Fig. 12 are the ones collected under nitrogen atmosphere. Both fluorescence and phosphorescence emission maxima of **30** were increasingly blue-shifted as the molecular weight increased. Fluorophore-fluorophore interactions in the polymer were proposed as the cause of the observed emission color change, depending on the polymer molecular weight [86]. The bimodal spectra are advantageous for ratiometric luminescence sensing of oxygen with high accuracy, and indeed quantitative imaging of tumor hypoxia was demonstrated with the nanoparticles of **30b**.

N-phenylnaphthalimide-terminated PLAs **31** and **32** in film under vacuum phosphoresced at 611 and 510 nm, respectively, without any detectable fluorescence (Fig. 13) [87]. ICT from the alkoxyphenylene part connected to the nitrogen atom to the naphthalimide moiety reduced ΔE_{ST} and thus enhanced ISC leading to the triplet state with π–π^* configuration. Natural biomacromolecules such as chitosan and bovine serum albumin were converted into RTP-emissive polymers by incorporating the bromonaphthalimide moiety.

3.2 Polyacrylates and poly(acrylic acid)s

Poly(methacrylate) **33**, in which 4-bromo-4-formyl-2,5-dioxybenzene moieties were incorporated as phosphorescent cross-linkers (1.2 wt%) with the aid of the Diels–Alder reaction, exhibited RTP at 513 nm in film (Fig. 14) [88]. The quantum yield of the cross-linked film was 0.26, which was larger than that (0.16) of 2,5-bis(hexyloxy)-4-

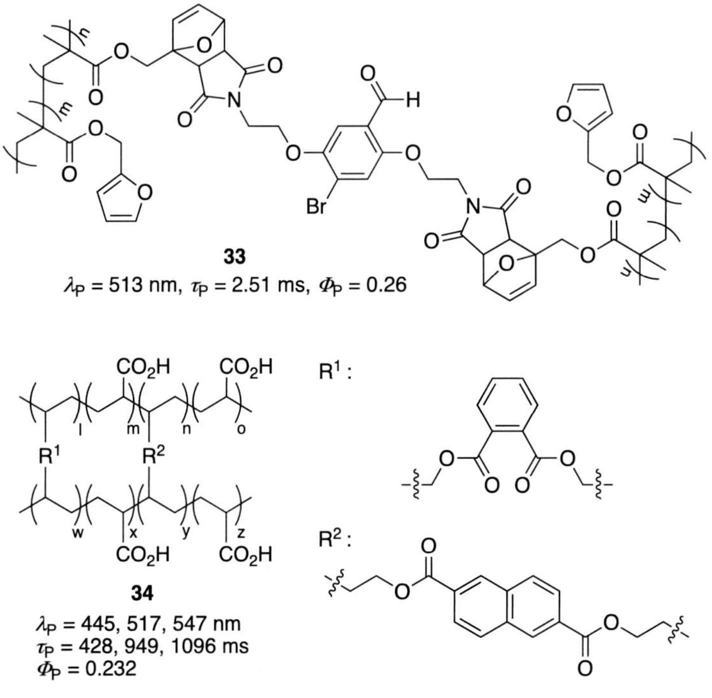

33
λ_P = 513 nm, τ_P = 2.51 ms, Φ_P = 0.26

34
λ_P = 445, 517, 547 nm
τ_P = 428, 949, 1096 ms
Φ_P = 0.232

FIG. 14 Molecular structures of poly(methacrylate) **33** and poly(acrylic acid) **34**.

bromo-4-formylbenzene-doped poly(furfuryl methacrylate) film, indicating that cross-linking of the polymer chains is advantageous to the suppression of molecular motions and thus enhancing phosphorescence quantum yields. Cross-linked structures are also beneficial for reducing Dexter-type triplet energy transfer. The present cross-linking approach was applicable to various kinds of copolymers such as poly(furfuryl methacrylate-*r*-styrene), poly(furfuryl methacrylate-*r*-acrylamide), poly(furfuryl methacrylate-*r*-acrylonitrile), and poly(furfuryl methacrylate-*r*-4-vinylbenzyl chloride).

Radical copolymerization of acrylic acid, dipropenyl phthalate, and dibutenyl naphthalene dicarboxylate in the presence of AIBN produced cross-linked multicomponent copolymers **34** that efficiently phosphoresced in film under ambient conditions with quantum yields ranging from 0.13 to 0.38 (Fig. 14) [89]. The phthalate and naphthalene dicarboxylate moieties in the cross-linkers R^1 and R^2 served as blue- and green-emissive components, respectively, and the emission color could be altered from blue to yellow by changing an excitation wavelength from 254 to 365 nm and the feeding molar ratio of the monomers. It is noteworthy that UV irradiation at 365 nm induced yellow RTP with a very long lifetime over 1 s. Both the cross-linking structure and hydrogen bonds formed with the carboxyl groups effectively restricted the molecular motions and suppressed the quenching of triplet states by an external stimulus such as oxygen and moisture, resulting in the persistent emission. Using the excitation wavelength dependency of phosphorescence colors, multilevel data encryption was demonstrated.

Amorphous powder of copolymer **35**, which was prepared from acrylamide and its derivative that contained 2-bromo-5-butoxybenzaldehyde moiety as a phosphor, exhibited yellow-green phosphorescence at 510 nm with a Φ_P of 0.114 under ambient conditions (Fig. 15) [90]. When **35** was dissolved in a mixture of DMF and water, the suspension formed, which also exhibited RTP. As the water content increased, the RTP intensity gradually decreased, suggesting that the hydrogen bonding between polymer chains played an important role in achieving efficient RTP. Copolymer **36** containing a 4-bromonaphthalene-1,8-dicarboximide moiety as a phosphor exhibited an orange RTP at 580 nm with a Φ_P of 0.074. Information encryption was performed using the aqueous solution of **36** to demonstrate the usefulness of amide-based phosphorescent polymers as functional materials.

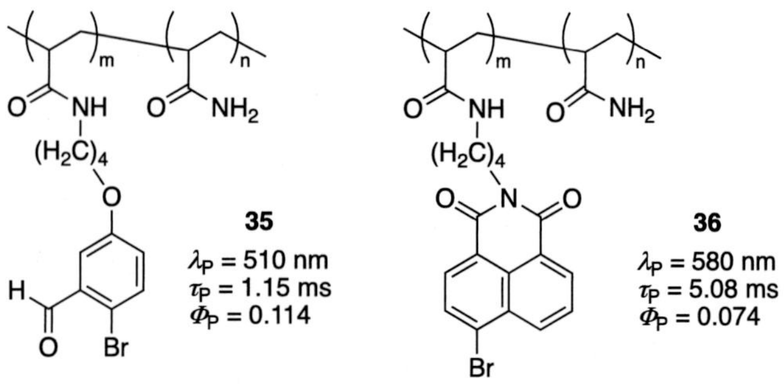

FIG. 15 Molecular structures of poly(acrylamide)s **35** and **36**.

37

λ_F = 398 nm
τ_F = 1.05 ns
λ_P = 496 nm
τ_P = 1010 ms
Φ_{F+P} = 0.077

SO_3H

38	38a	38b
E	N	$N^+(CH_2)_4SO_3^-$
λ_F (nm)	420	418
τ_F (ns)	4.1	2.5
λ_P (nm)	510	524
τ_P (ms)	0.57	102
Φ_P	0.036	0.064

FIG. 16 Molecular structures of polystyrene **37** and poly(vinylpyridine)s **38**.

3.3 Polystyrenes

Very long-lived RTP was attained with a dried amorphous film of sulfonated polystyrene **37** (Fig. 16) [91]. Upon UV irradiation at 365 nm, the film exhibited white-blue emission and the color turned to green after irradiation was stopped. The lifetime of the green RTP reached 1.01 s. As commercially available polystyrene, which was used as the starting material for **37**, showed only weak fluorescence at 356 nm, the sulfonic acid groups were essential for the ultralong lifetime RTP. The higher the introduction ratio of sulfonic acid groups, the longer the RTP lifetime and the higher the photoluminescence quantum yield. On the other hand, the RTP was quenched when the dried sample was subjected to moisture which can cleave hydrogen bonds. Therefore, the emergence of the RTP is attributed to both the intra- and intermolecular hydrogen bonds between sulfonic acid groups. Metal and ammonium salts of **37** were also demonstrated to exhibit long-lived RTP that was induced by ionic bonding [92].

Poly(4-vinylpyridine) **38a** in the solid state under ambient conditions exhibited dual emission of blue fluorescence and green phosphorescence with a very short τ_P of 0.57 ms (Fig. 16) [93]. No afterglow was observed with the naked eye. In contrast, **38b**, which was the ionized counterpart of **38a**, exhibited long-lived RTP with a τ_P of 102 ms, which was observed with the naked eye as green emission after the excitation light turned off. The quantum yield of **38b** was also higher than that of **38a**. The prolonging lifetime and enhanced quantum yield are ascribed to the restriction of molecular motions by the ionic interactions between sulfonic acid and the poly(4-vinylpyridine) unit. Furthermore, the emission color of phosphorescence was found to be dependent on the excitation wavelength. The excitation-dependent multicolor RTP suggests that **38b** in the solid state includes isolated phosphors and multiple types of aggregated phosphors, which have distinct excited triplet states.

3.4 Polyurethanes

Polyurethane **39**, which contained dibenzophenone moieties as phosphors (1 wt%) and ammonium carboxylate salts as internal emulsifiers in an aqueous solution, was developed as a fluorescence-phosphorescence dual-emissive polymer (Fig. 17) [94]. The dried film of **39** fluoresced at 448 nm with a quantum yield of 0.80 in air, which was much larger than that (0.12) of the emulsion of **39**. The increase of a fluorescence quantum yield was attributed to the rigidity of the polyurethane matrix [95]. Under vacuum or nitrogen, the film exhibited dual emission of blue fluorescence and green phosphorescence. When the phosphor content

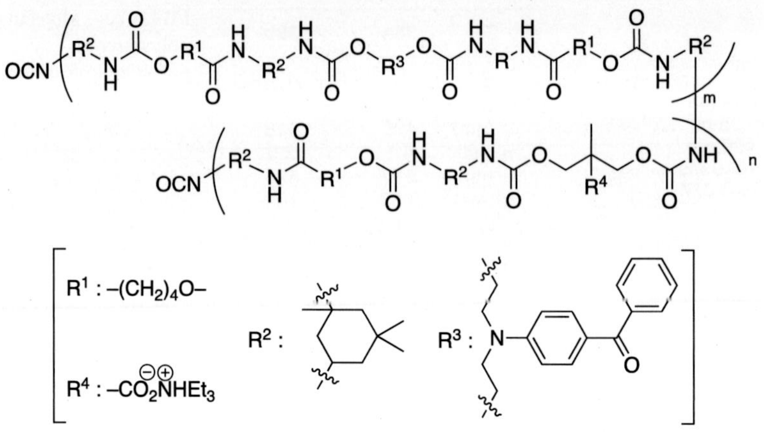

R^1 : $-(CH_2)_4O-$

R^2 :

R^3 :

R^4 : $-CO_2^{\ominus}\overset{\oplus}{N}HEt_3$

39

λ_F = 445 nm, τ_F = 2.2 ns, Φ_F = 0.80, λ_P = 505 nm, τ_P = 53.5 ms

FIG. 17 Molecular structure of polyurethane **39**.

increased up to 20 wt%, the fluorescence and phosphorescence maxima became closer and merged, indicating that the aggregation of the luminophores decreased the energy gap between the S_1 and T_1 states. In view of the water solubility, applications as dual-emissive probes applicable in an aqueous environment is expected.

3.5 Polyimides

Halogenated polyimides **40a** (X = Br) and **40b** (X = I) in film phosphoresced under ambient conditions at 592 and 586 nm with quantum yields of 0.03 and 0.01, respectively (Fig. 18) [96]. As the corresponding nonhalogenated polyimide showed only very weak fluorescence ($\Phi_F \leq 0.001$), the observed RTP was attributed to the heavy atom effect of bromine and iodine

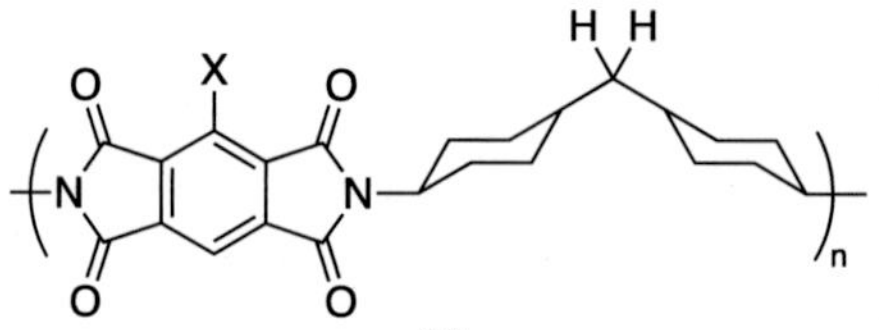

FIG. 18 Molecular structures of polyimides **40** and **41**.

40	40a	40b
X	Br	I
λ_P (nm)	592	586
τ_P (ms)	2.3	1.0
Φ_P	0.03	0.01

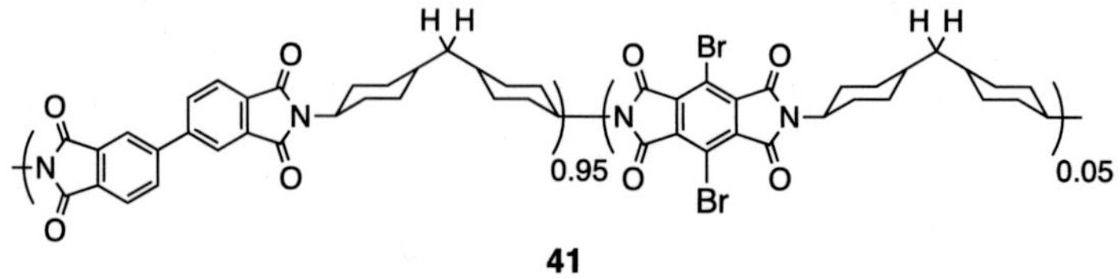

40

41

λ_F = 418 nm, τ_F = 1.05 ns, Φ_F = 0.01, λ_P = 549 nm, τ_P = 3.5 ms, Φ_P = 0.038

atoms. The substitution of the halogen atom also resulted in the enhancement of refractive indices and reduction of birefringence. A thin film of polyimide **41** consisted of 4,4′-biphthalimide- and dibromopyromellitimide-based polyimides in a ratio of 95:5 emitted white light whose CIE coordinates were (0.28, 0.30) (Fig. 18) [97]. The white emission was the result of the dual emission of blue fluorescence from the 4,4′-biphthalimide moieties and orange phosphorescence from the dibromopyromellitimide moieties. As the Förster resonance energy transfer from the blue fluorophore to the orange phosphor was involved in the photoluminescent processes, the luminescence color could be altered from blue to orange by changing the ratio between the luminophores.

4 Summary

This chapter reviewed rare-metal-free small molecules reported in 2018–20 and phosphorescent polymers reported in 2007–20, which were classified by molecular structures: arylcarbonyl compounds, diaryl sulfones, *N*-heterocyclic compounds, boron compounds, phosphorus compounds, diaryl sulfides, poly(lactic acid)s, polyacrylates/poly(acrylic acid) s, polystyrenes, polyurethanes, and polyimides. The key issues for attaining RTP with rare-metal-free molecular structures are how to promote ISC and to suppress nonradiative decays of the T_1 state. A conventional approach for the ISC acceleration is the choice of luminophores whose excitation involves n–π transition and the introduction of heavy atoms such as bromine and iodine. Resonance- and ICT-promoted ISC have emerged as novel approaches. Recently, molecular aggregation is proposed to induce ISC in some benzophenones [98]. The suppression of nonradiative decays of the T_1 state was realized by restricting molecular motions with the designed intra- and intermolecular interactions such as hydrogen bonding, halogen bonding [99], and ionic bonding. In the case of polymers, cross-linking is also effective for the rigidification of the polymer chains. The creation of rare-metal-free phosphorescent materials is challenging, while the materials have high potentials for applications that cannot achieve with fluorophores. It is hoped that this article would stimulate many materials chemists and promote the development and applications of rare-metal-free phosphorescent materials.

References

[1] L. Ravotto, P. Ceroni, Coord. Chem. Rev. 346 (2017) 62–76.
[2] V. Sathish, A. Ramdass, P. Thanasekaran, K.-L. Lu, S. Rajagopal, J. Photochem. Photobiol. C Photchem. Rev. 23 (2015) 25–44.
[3] Y. Chi, P.-T. Chou, Chem. Soc. Rev. 39 (2010) 638–655.
[4] R.C. Evans, P. Douglas, C.J. Winscom, Coord. Chem. Rev. 250 (2006) 2093–2126.
[5] S. Mukherjee, P. Thilagar, Chem. Commun. 51 (2015) 10988–11003.
[6] S. Wang, W.Z. Yuan, Y. Zhang, in: M. Fujiki, B. Liu, B.Z. Tang (Eds.), ACS Symposium Series, American Chemical Society, Washington, DC, 2016, pp. 1–26.
[7] J. Yuan, Y. Tang, S. Xu, R. Chen, W. Huang, Sci. Bull. 60 (2015) 1631–1637.
[8] Y. Liu, G. Zhan, Z.-W. Liu, Z.-Q. Bian, C.-H. Huang, Chin. Chem. Lett. 27 (2016) 1231–1240.
[9] C.-R. Wang, Y.-Y. Gong, W.-Z. Yuan, Y.-M. Zhang, Chin. Chem. Lett. 27 (2016) 1184–1192.
[10] S. Hirata, Adv. Opt. Mater. 5 (2017) 1700116.
[11] M. Baroncini, G. Bergamini, P. Ceroni, Chem. Commun. 53 (2017) 2081–2093.

[12] H. Yuasa, S. Kuno, Bull. Chem. Soc. Jpn. 91 (2018) 223–229.
[13] A. Forni, E. Lucenti, C. Botta, E. Cariati, J. Mater. Chem. C 6 (2018) 4603–4626.
[14] J. Zhao, K. Chen, Y. Hou, Y. Che, L. Liu, D. Jia, Org. Biomol. Chem. 16 (2018) 3692–3701.
[15] M. Hayduk, S. Riebe, J. Voskuhl, Chem. A Eur. J. 24 (2018) 12221–12230.
[16] N. Gan, H. Shi, Z. An, W. Huang, Adv. Funct. Mater. 28 (2018) 1802657.
[17] L. Xiao, H. Fu, Chem. A Eur. J. 25 (2019) 714–723.
[18] P. Data, Y. Takeda, Chem. Asian J. 14 (2019) 1613–1636.
[19] H. Ma, A. Lv, L. Fu, S. Wang, Z. An, H. Shi, W. Huang, Ann. Phys. 531 (2019) 1800482.
[20] C. Kenry, B.L. Chen, Nat. Commun. 10 (2019) 2111.
[21] W. Jia, Q. Wang, H. Shi, Z. An, W. Huang, Chem. A Eur. J. 26 (2020) 4437–4448.
[22] J. Yang, M. Fang, Z. Li, InfoMat 2 (2020) 791–806.
[23] J. Zhi, Q. Zhou, H. Shi, Z. An, W. Huang, Chem. Asian J. 15 (2020) 947–957.
[24] T. Zhang, X. Ma, H. Wu, L. Zhu, Y. Zhao, H. Tian, Angew. Chem. Int. Ed. 59 (2020) 11206–11216.
[25] L. Gu, X. Wang, M. Singh, H. Shi, H. Ma, Z. An, W. Huang, J. Phys. Chem. Lett. 11 (2020) 6191–6200.
[26] L. Huang, C. Qian, Z. Ma, Chem. A Eur. J. 26 (2020) 11914–11930.
[27] W. Zhao, Z. He, B.Z. Tang, Nat. Rev. Mater. 5 (2020) 869–885.
[28] J. Wang, X. Lou, Y. Wang, J. Tang, Y. Yang, Macromol. Rapid Commun. 42 (2021) 2100021.
[29] A.D. Nidhankar, Goudappagouda, V.C. Wakchaure, S.S. Babu, Chem. Sci. 12 (2021) 4216–4236.
[30] J. Xu, A. Takai, Y. Kobayashi, M. Takeuchi, Chem. Commun. 49 (2013) 8447–8449.
[31] S. Kuila, K.V. Rao, S. Garain, P.K. Samanta, S. Das, S.K. Pati, M. Eswaramoorthy, S.J. George, Angew. Chem. Int. Ed. 57 (2018) 17115–17119.
[32] A. Lv, W. Ye, X. Jiang, N. Gan, H. Shi, W. Yao, H. Ma, Z. An, W. Huang, J. Phys. Chem. Lett. 10 (2019) 1037–1042.
[33] H. Shu, H. Li, J. Rao, L. Chen, X. Wang, X. Wu, H. Tian, H. Tong, L. Wang, J. Mater. Chem. C 8 (2020) 14360–14364.
[34] Review on AIE: J. Mei, N.L.C. Leung, R.T.K. Kwok, J.W.Y. Lam, B.Z. Tang, Chem. Rev. 115 (2015) 11718–11940.
[35] For rare-metal-free organic small molecules exhibiting room temperature phosphorescence reported by 2017, see M. Shimizu, in: Y. Tang, B.Z. Tang (Eds.), Principles and Applications of Aggregation-Induced Emission, Springer International Publishing, Cham, 2019, pp. 43–76.
[36] J. Jia, W. Lu, Y. Gao, L. Li, C. Dong, S. Shuang, Talanta 231 (2021), 122350.
[37] M. Singh, K. Liu, S. Qu, H. Ma, H. Shi, Z. An, W. Huang, Adv. Opt. Mater. 9 (2021) 2002197.
[38] O. Bolton, K. Lee, H.-J. Kim, K.Y. Lin, J. Kim, Nat. Chem. 3 (2011) 205–210.
[39] T. Ono, A. Taema, A. Goto, Y. Hisaeda, Chem. A Eur. J. 24 (2018) 17487–17496.
[40] S. Guo, W. Dai, X. Chen, Y. Lei, J. Shi, B. Tong, Z. Cai, Y. Dong, ACS Mater. Lett. 3 (2021) 379–397.
[41] G. Qu, Y. Zhang, X. Ma, Chin. Chem. Lett. 30 (2019) 1809–1814.
[42] W. Jun Jin, in: A. Douhal (Ed.), Cyclodextrin Materials Photochemistry, Photophysics and Photobiology, Elsevier, Amsterdam, 2006, pp. 137–153.
[43] Q. Li, Y. Tang, W. Hu, Z. Li, Small 14 (2018) 1801560.
[44] X. Dou, T. Zhu, Z. Wang, W. Sun, Y. Lai, K. Sui, Y. Tan, Y. Zhang, W.Z. Yuan, Adv. Mater. 32 (2020) 2004768.
[45] H. Thomas, D.L. Pastoetter, M. Gmelch, T. Achenbach, A. Schlögl, M. Louis, X. Feng, S. Reineke, Adv. Mater. 32 (2020) 2000880.
[46] M. Louis, H. Thomas, M. Gmelch, A. Haft, F. Fries, S. Reineke, Adv. Mater. 31 (2019) 1807887.
[47] D.R. Kearns, W.A. Case, J. Am. Chem. Soc. 88 (1966) 5087–5097.
[48] G. Dormán, H. Nakamura, A. Pulsipher, G.D. Prestwich, Chem. Rev. 116 (2016) 15284–15398.
[49] T. Wang, X. Su, X. Zhang, X. Nie, L. Huang, X. Zhang, X. Sun, Y. Luo, G. Zhang, Adv. Mater. 31 (2019) 1904273.
[50] M.A. El-Sayed, Acc. Chem. Res. 1 (1968) 8–16.
[51] S.K. Lower, M.A. El-Sayed, Chem. Rev. 66 (1966) 199–241.
[52] E. Hamzehpoor, D.F. Perepichka, Angew. Chem. Int. Ed. 59 (2020) 9977–9981.
[53] Z. An, C. Zheng, Y. Tao, R. Chen, H. Shi, T. Chen, Z. Wang, H. Li, R. Deng, X. Liu, W. Huang, Nat. Mater. 14 (2015) 685–690.
[54] D. Tu, S. Cai, C. Fernandez, H. Ma, X. Wang, H. Wang, C. Ma, H. Yan, C. Lu, Z. An, Angew. Chem. Int. Ed. 58 (2019) 9129–9133.
[55] Q. Liao, Q. Gao, J. Wang, Y. Gong, Q. Peng, Y. Tian, Y. Fan, H. Guo, D. Ding, Q. Li, Z. Li, Angew. Chem. Int. Ed. 59 (2020) 9946–9951.
[56] Z. He, H. Gao, S. Zhang, S. Zheng, Y. Wang, Z. Zhao, D. Ding, B. Yang, Y. Zhang, W.Z. Yuan, Adv. Mater. 31 (2019) 1807222.

[57] J. Li, J. Zhou, Z. Mao, Z. Xie, Z. Yang, B. Xu, C. Liu, X. Chen, D. Ren, H. Pan, G. Shi, Y. Zhang, Z. Chi, Angew. Chem. Int. Ed. 57 (2018) 6449–6453.
[58] Goudappagouda, A. Manthanath, V.C. Wakchaure, K.C. Ranjeesh, T. Das, K. Vanka, T. Nakanishi, S.S. Babu, Angew. Chem. Int. Ed. 58 (2019) 2284–2288.
[59] Z. Mao, Z. Yang, Y. Mu, Y. Zhang, Y.-F. Wang, Z. Chi, C.-C. Lo, S. Liu, A. Lien, J. Xu, Angew. Chem. Int. Ed. 54 (2015) 6270–6273.
[60] J. Yang, X. Zhen, B. Wang, X. Gao, Z. Ren, J. Wang, Y. Xie, J. Li, Q. Peng, K. Pu, Z. Li, Nat. Commun. 9 (2018) 840.
[61] S. Tian, H. Ma, X. Wang, A. Lv, H. Shi, Y. Geng, J. Li, F. Liang, Z. Su, Z. An, W. Huang, Angew. Chem. Int. Ed. 58 (2019) 6645–6649.
[62] J. Chen, T. Yu, E. Ubba, Z. Xie, Z. Yang, Y. Zhang, S. Liu, J. Xu, M.P. Aldred, Z. Chi, Adv. Opt. Mater. 7 (2019) 1801593.
[63] Z. Yang, C. Xu, W. Li, Z. Mao, X. Ge, Q. Huang, H. Deng, J. Zhao, F.L. Gu, Y. Zhang, Z. Chi, Angew. Chem. Int. Ed. 59 (2020) 17451–17455.
[64] Y. Xiong, Z. Zhao, W. Zhao, H. Ma, Q. Peng, Z. He, X. Zhang, Y. Chen, X. He, J.W.Y. Lam, B.Z. Tang, Angew. Chem. Int. Ed. 57 (2018) 7997–8001.
[65] H. Shi, L. Song, H. Ma, C. Sun, K. Huang, A. Lv, W. Ye, H. Wang, S. Cai, W. Yao, Y. Zhang, R. Zheng, Z. An, W. Huang, J. Phys. Chem. Lett. 10 (2019) 595–600.
[66] W. Zhao, T.S. Cheung, N. Jiang, W. Huang, J.W.Y. Lam, X. Zhang, Z. He, B.Z. Tang, Nat. Commun. 10 (2019) 1595.
[67] T.J. Penfold, E. Gindensperger, C. Daniel, C.M. Marian, Chem. Rev. 118 (2018) 6975–7025.
[68] M.K. Etherington, J. Gibson, H.F. Higginbotham, T.J. Penfold, A.P. Monkman, Nat. Commun. 7 (2016) 13680.
[69] C. Zhou, S. Zhang, Y. Gao, H. Liu, T. Shan, X. Liang, B. Yang, Y. Ma, Adv. Funct. Mater. 28 (2018) 1802407.
[70] Q. Wu, H. Ma, K. Ling, N. Gan, Z. Cheng, L. Gu, S. Cai, Z. An, H. Shi, W. Huang, ACS Appl. Mater. Interfaces 10 (2018) 33730–33736.
[71] L. Gu, H. Shi, M. Gu, K. Ling, H. Ma, S. Cai, L. Song, C. Ma, H. Li, G. Xing, X. Hang, J. Li, Y. Gao, W. Yao, Z. Shuai, Z. An, X. Liu, W. Huang, Angew. Chem. Int. Ed. 57 (2018) 8425–8431.
[72] Y. Zhou, W. Qin, C. Du, H. Gao, F. Zhu, G. Liang, Angew. Chem. Int. Ed. 58 (2019) 12102–12106.
[73] Z. Wu, J. Nitsch, J. Schuster, A. Friedrich, K. Edkins, M. Loebnitz, F. Dinkelbach, V. Stepanenko, F. Würthner, C.M. Marian, L. Ji, T.B. Marder, Angew. Chem. Int. Ed. 59 (2020) 17137–17144.
[74] X. Wang, W. Guo, H. Xiao, Q. Yang, B. Chen, Y. Chen, C. Tung, L. Wu, Adv. Funct. Mater. 30 (2020) 1907282.
[75] Y. Tao, R. Chen, H. Li, J. Yuan, Y. Wan, H. Jiang, C. Chen, Y. Si, C. Zheng, B. Yang, G. Xing, W. Huang, Adv. Mater. 30 (2018) 1803856.
[76] M. Shimizu, T. Sakurai, ChemPlusChem 86 (2021) 446–459.
[77] M. Shimizu, S. Nagano, T. Kinoshita, Chem. A Eur. J. 26 (2020) 5162–5167.
[78] M. Shimizu, R. Shigitani, M. Nakatani, K. Kuwabara, Y. Miyake, K. Tajima, H. Sakai, T. Hasobe, J. Phys. Chem. C 120 (2016) 11631–11639.
[79] L. Huang, B. Chen, X. Zhang, C.O. Trindle, F. Liao, Y. Wang, H. Miao, Y. Luo, G. Zhang, Angew. Chem. Int. Ed. 57 (2018) 16046–16050.
[80] G. Bergamini, A. Fermi, C. Botta, U. Giovanella, S. Di Motta, F. Negri, R. Peresutti, M. Gingras, P. Ceroni, J. Mater. Chem. C 1 (2013) 2717–2724.
[81] A. Fermi, G. Bergamini, M. Roy, M. Gingras, P. Ceroni, J. Am. Chem. Soc. 136 (2014) 6395–6400.
[82] S. Riebe, C. Vallet, F. van der Vight, D. Gonzalez-Abradelo, C. Wölper, C.A. Strassert, G. Jansen, S. Knauer, J. Voskuhl, Chem. A Eur. J. 23 (2017) 13660–13668.
[83] W. Xi, J. Yu, M. Wei, Q. Qiu, P. Xu, Z. Qian, H. Feng, Chem. A Eur. J. 26 (2020) 3733–3737.
[84] G. Zhang, J. Chen, S.J. Payne, S.E. Kooi, J.N. Demas, C.L. Fraser, J. Am. Chem. Soc. 129 (2007) 8942–8943.
[85] G. Zhang, G.M. Palmer, M.W. Dewhirst, C.L. Fraser, Nat. Mater. 8 (2009) 747–751.
[86] G. Zhang, S.E. Kooi, J.N. Demas, C.L. Fraser, Adv. Mater. 20 (2008) 2099–2104.
[87] X. Chen, C. Xu, T. Wang, C. Zhou, J. Du, Z. Wang, H. Xu, T. Xie, G. Bi, J. Jiang, X. Zhang, J.N. Demas, C.O. Trindle, Y. Luo, G. Zhang, Angew. Chem. Int. Ed. 55 (2016) 9872–9876.
[88] M.S. Kwon, Y. Yu, C. Coburn, A.W. Phillips, K. Chung, A. Shanker, J. Jung, G. Kim, K. Pipe, S.R. Forrest, J.H. Youk, J. Gierschner, J. Kim, Nat. Commun. 6 (2015) 8947.
[89] L. Gu, H. Wu, H. Ma, W. Ye, W. Jia, H. Wang, H. Chen, N. Zhang, D. Wang, C. Qian, Z. An, W. Huang, Y. Zhao, Nat. Commun. 11 (2020) 944.
[90] H. Chen, X. Yao, X. Ma, H. Tian, Adv. Opt. Mater. 4 (2016) 1397–1401.

[91] T. Ogoshi, H. Tsuchida, T. Kakuta, T. Yamagishi, A. Taema, T. Ono, M. Sugimoto, M. Mizuno, Adv. Funct. Mater. 28 (2018) 1707369.
[92] S. Cai, H. Ma, H. Shi, H. Wang, X. Wang, L. Xiao, W. Ye, K. Huang, X. Cao, N. Gan, C. Ma, M. Gu, L. Song, H. Xu, Y. Tao, C. Zhang, W. Yao, Z. An, W. Huang, Nat. Commun. 10 (2019) 4247.
[93] H. Wang, H. Shi, W. Ye, X. Yao, Q. Wang, C. Dong, W. Jia, H. Ma, S. Cai, K. Huang, L. Fu, Y. Zhang, J. Zhi, L. Gu, Y. Zhao, Z. An, W. Huang, Angew. Chem. Int. Ed. 58 (2019) 18776–18782.
[94] C. Zhou, T. Xie, R. Zhou, C.O. Trindle, Y. Tikman, X. Zhang, G. Zhang, ACS Appl. Mater. Interfaces 7 (2015) 17209–17216.
[95] R.O. Loutfy, Macromolecules 14 (1981) 270–275.
[96] K. Kanosue, S. Ando, ACS Macro Lett. 5 (2016) 1301–1305.
[97] M. Nara, R. Orita, R. Ishige, S. Ando, ACS Omega 5 (2020) 14831–14841.
[98] J. Zhang, S. Mukamel, J. Jiang, J. Phys. Chem. B 124 (2020) 2238–2244.
[99] W. Wang, Y. Zhang, W.J. Jin, Coord. Chem. Rev. 404 (2020), 213107.

CHAPTER

9

AIE luminogens exhibiting thermally activated delayed fluorescence

Xiaojie Chen[a], Xiangyu Ge[a], Zhan Yang[a], Juan Zhao[b], Zhiyong Yang[a], Yi Zhang[a], and Zhenguo Chi[a,b]

[a]PCFM Lab, GDHPPC Lab, Guangdong Engineering Technology Research Center for High-performance Organic and Polymer Photoelectric Functional Films, State Key Laboratory of OEMT, School of Chemistry, Sun Yat-sen University, Guangzhou, China [b]School of Materials Science and Engineering, Sun Yat-sen University, Guangzhou, China

1 Introduction

In organic light-emitting diodes (OLEDs), the electrogenerated excitons have a 1:3 ratio of singlet (S_1) and triplet (T_1) excitons, according to spin statistics. For conventional fluorescent OLEDs, only singlet excitons can be utilized for light emission to render a limited internal quantum efficiency (IQE) of 25%, and a theoretical external quantum efficiency (EQE) of up to 5%. Although phosphorescence-based OLEDs can achieve an IQE of nearly 100%, the high price and potential environmental contamination of rare-metal complexes are the main limiting factors for its mass production. Meanwhile, it remains challenging for the development of efficient pure blue and deep blue phosphorescent complexes. In recent years, pure organic thermally activated delayed fluorescence (TADF) materials have been considered as the most potential candidate to overcome the above problems. Similar to phosphorescent materials, TADF emitters can also harvest both singlet and triplet excitons simultaneously through a reverse intersystem crossing (RISC) process, leading to an IQE of 100%. In 2012, Adachi et al. [1] first reported some TADF-based OLEDs achieving high luminous efficiencies, which are comparable with the state-of-the-art phosphorescence-based OLEDs. Thereafter, TADF materials have attracted great attention and interest in both academics and industry, representing the third-generation luminescent materials for OLEDs. However, traditional TADF molecules tend to easily aggregate through π-π interactions, resulting in aggregation-caused emission quenching (ACQ) phenomenon, and meanwhile,

Aggregation-Induced Emission (AIE)
https://doi.org/10.1016/B978-0-12-824335-0.00018-0

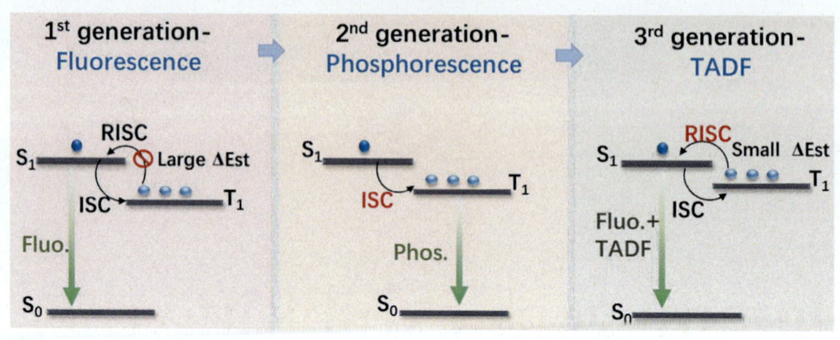

FIG. 1 Jablonski diagrams for fluorescence, phosphorescence, and TADF emitters.

exciton annihilation processes are likely to occur owing to the long lifetime of excited states. Although host-guest doping technology has been widely used to address the above issues, precise control of the doping concentration and guest-host phase separation are very troublesome. In particular, the severe efficiency roll-off at high luminance still remains a great challenge for TADF-based doped OLEDs. Aggregation-induced emission (AIE) refers to an interesting photophysical phenomenon in which luminescent molecules show weak emissions in dilute solutions but give out strong emissions in the aggregated state, which could help to suppress concentration quenching and exciton annihilation, thus making AIE materials ideal for nondoped OLEDs application. In light of this, the integration of AIE and TADF, named as AIE-TADF or aggregation-induced delayed fluorescence (AIDF), is anticipated as an effective solution to overcome the shortcomings of TADF. The AIE-TADF concept was first reported in 2015 by Chi and coworkers [2]. Afterwards, numerous researchers began to explore the application of AIE luminescent materials with TADF behavior in electroluminescent devices, especially in nondoped OLEDs. In this chapter, we systematically summarize recent developments of AIE-TADF materials, including molecular design strategies, photophysical properties, and device performances, which will provide valuable insights into molecular designs and applications of AIE-TADF materials (Fig. 1).

2 AIE-TADF materials

2.1 Aryl sulfoxide derivatives

In 2015, Chi et al. *[2]* reported the first AIE-TADF compounds **1** and **2** with inspiration from a normal TADF molecule **3**. They proposed an asymmetric molecular structure strategy to design high-efficiency TADF molecules with phenothiazine moiety. Compound **1** exhibited artful packing and highly ordered alignment without any π-π interactions, and showed a high photoluminescence quantum yield (PLQY) of 93.3% in solid state. Simultaneously, **1** exhibited strong mechanoluminescence (ML, which means the fluorescent materials can emit light spontaneously upon mechanical forces) without any pretreatment. This molecular design strategy has an important guiding role in designing this kind of compounds to integrate TADF, AIE, and ML features into one compound. Thereafter, they presented highly efficient OLEDs based on **1** [3], which demonstrated universal applications in both doped and nondoped OLEDs with high performance, ascribing to its high PLQY and excellent AIE and TADF properties. The doped and nondoped OLEDs by employing **1** achieved maximum EQEs of 26.2% and 20.7%, respectively, which were rarely reported among TADF-based OLEDs. These results highlight bright prospects for material development by using the easily realizable AIE-TADF strategy, which drives intrinsic performance improvement toward simple and efficient OLEDs. Furthermore, Chi et al. [4] presented a novel and effective structural design strategy for an asymmetric TADF emitter (**4**), wherein donor (D) and acceptor (A) groups were linked at ortho-position to afford a spatially close D-A interaction, leading to reduced vibrations and suppressed nonradiative pathways. Remarkably, **4** possessed dual-charge transfer pathways (through-bond charge transfer and through-space charge transfer), resulting in rapid prompt and delayed decays. **4** exhibited a highly twisted conformation and a high PLQY (91.9%) in neat film (AIE property), as well as excellent thermal stability. Most importantly, **4** afforded a record-high EQE of 28.7% among previously reported nondoped OLEDs. The presented design strategy opens up new possibilities to develop high-efficiency TADF emitters and devices.

Based on asymmetric structure, Lee et al. [5] reported a compound **5** which behaved as both AIE and TADF emitters, contributing to nondoped OLEDs with a high EQE of 17.0%. This result further confirms the effectiveness of the asymmetric molecular design for the development of high-efficiency nondoped OLEDs.

By using electron-deficient diphenyl sulphone and dibenzothiophene-S,S-dioxide as acceptors, and electron-rich phenoxazine and phenothiazine as donors, Tang et al. [6] designed a series of luminogens **2**, **6–8**. A large twisted angle between donor and acceptor led to efficient separation of the highest occupied molecular orbitals (HOMOs) and lowest unoccupied molecular orbitals (LUMOs), giving rise to a small energy gap between singlet and triplet states (ΔE_{ST}) (from 0.06 to 0.26 eV) and thus delayed fluorescence. The nonplanar conformations of these molecules were anticipated to suppress strong intermolecular π-π interactions and also decreased emission quenching, which was responsible for the AIE characteristics. **2**, **6**, **7**, and **8** showed low PLQYs in dilute tetrahydrofuran (THF) solutions (3.1%, 1.0%, 1.6%, and 1.2%), which were obviously increased (16.3%, 6.7%, 20.3%, and 10.7%) in solid state, confirming AIE-active luminogens (Scheme 1).

In 2015, Adachi et al. [7] reported a sky-blue TADF emitter **9**, bis[4-(9,9-dimethyl-9,10-dihydroacridine)phenyl]sulfone, which enabled a nondoped OLED with an EQE up to 19.5%. The relatively large Stokes shift and weak π-π stacking interactions were responsible for the concentration insensitive property of **9**. As inferred by its nondoped PLQY (54%) and EQE, **9** displayed the AIE feature although it was not mentioned. Similar to compound **9**, Guo et al. [8] reported a diphenyl sulfone derivative **10** by combing a diphenyl sulfone core with bi-9,9-dimethyl-9,10-dihydroacridine substituted at 3,3′-position. The highly twisted zig-zag configuration endowed the compound with desired frontier orbital distribution to favor small ΔE_{ST} (0.02 eV) of TADF molecule and AIE characteristic. The amorphous **10** displayed enhanced PLQY and excellent TADF property. As for **10**-based nondoped OLEDs, the blue device with Commission Internationale del'Eclairage (CIE) coordinates of (0.18, 0.32) achieved a maximum current efficiency (CE), power efficiency (PE), and EQE of 31.7 cd A^{-1}, 28.4 lm W^{-1}, and 14.0%, respectively, which were higher than those of its linear isomer **9**. In addition, Yang et al. [9] also investigated three isomers **9–11**. It was found that the meta-linking compound **10** obtained the highest PLQY of 76% in neat film. In comparison

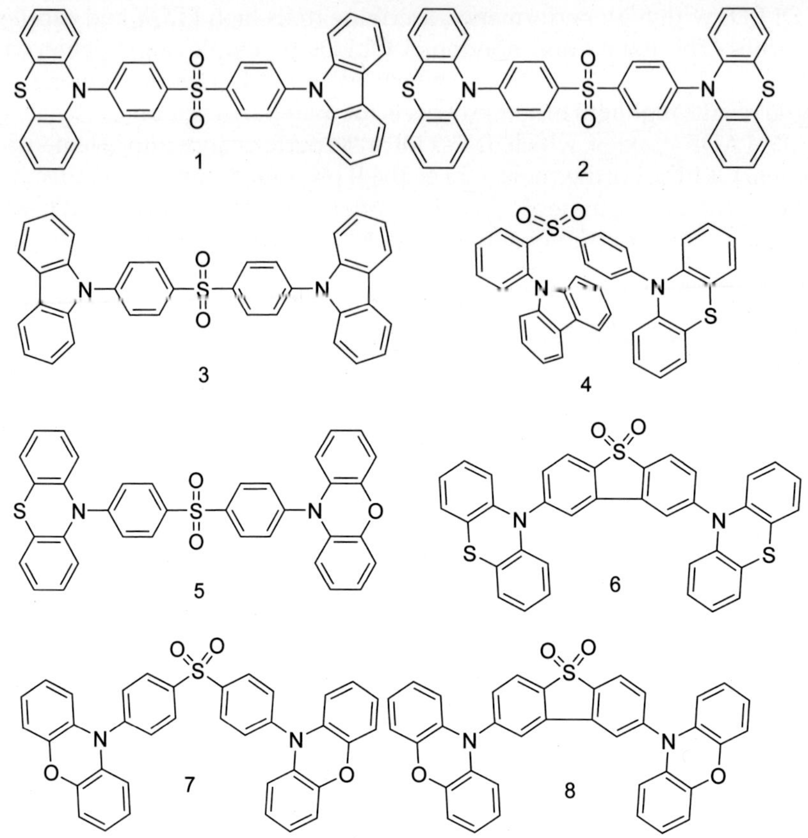

SCHEME 1 Chemical structures of compounds **1–8**.

to **9**-/**11**-based OLEDs, **10**-based nondoped OLEDs by solution processing achieved higher performance such as an EQE of 17.2%, CE of 37.9 cd A^{-1}, and PE of 23.8 lm W^{-1}, which were among the highest values of nondoped sky-blue TADF OLEDs by solution processing, and even superior to the vacuum-deposited ones.

The isomeric strategy was of great significance in molecular design owing to its profound influence on physical properties. Yang et al. [10] design three isomeric green and sky-blue AIE-TADF molecules **12–14**. All three emitters had small ΔE_{ST} of 0.02, 0.03, and 0.09 eV and delayed fluorescence lifetimes of 3.1, 3.3, and 4.4 ms in doped films, respectively. As for the meta-para linking isomer **14**, it showed a proper torsion angle between D and A, accounting for the highest PLQY of 77% with sky-blue emission. Accordingly, the best electroluminescence (EL) performance was achieved in **14**-based doped OLED, such as an EQE of 20.5%, which was among the best efficiencies recorded among sky-blue TADF-based OLEDs. This work further verified that this simple isomerization strategy could provide important guidance to explore the relationship between molecular structure and functions for developing more efficient TADF emitters. Moreover, they also reported two blue emitters **15** and **16** [11]. Both the compounds possessed typical AIE, TADF, and room temperature phosphorescence (RTP) properties. **15** exhibited distinct mechanoluminescence properties. However, when a methyl unit was introduced, which resulted in ML-inactive **16** due to the regulation of intermolecular interactions and packing mode. The simply structured **16** was the first example showing TADF, RTP, AIE, and ML simultaneously at that time.

Wang et al. [12] modified the AIE-TADF chromophore with four host-substituents including 1,4-di(9H-carbazol-9-yl)benzene (DCB), 1,3-di(9H-carbazol-9-yl)benzene (mCP), 3-(4-(9H-carbazol-9-yl)phenyl)-9H-carbazole (pPhDCz), and 3-(3-(9H-carbazol-9-yl)-phenyl)-9H-carbazole (mPhDCz), resulting in four self-host TADF emitters **17–20**. By changing the bonding positions of host-substituents, the twist angles in the molecules were adjusted accordingly, thus impacting the AIE and PL properties. In comparison to DCB and mCP substituted AIE-TADF emitters (**17** and **18**), pPhDCz and mPhDCz substituted ones **19** and **20** presented stronger AIE effect, due to relatively large twist angles between host-substituents and diphenyl sulfone-phenoxazine. As a consequence, neat films of **19** and **20** exhibited higher PLQYs of 56% and 55%, respectively. As for **19**- and **20**-based nondoped OLEDs, their maximum EQEs were 17.1% and 18.1%, respectively. These results imply that it is a valid approach to develop high-efficiency AIE-TADF molecules by modifying with suitable host-substituents (Scheme 2).

Later, Wang et al. [13] reported two AIE-TADF compounds **21** and **22** with a D-A-D′ structure. Diphenyl sulfone was used as the acceptor, 9-phenylcarbazole was the donor, while phenoxazine and 9,9-dimethylacridine were separately employed as the other donor D′. Both luminogens exhibited obvious AIE property, good thermal stability, and high PLQY in neat films. Due to the stronger electron-donating capability of phenoxazine than 9,9-dimethylacridine, **21** manifested red-shifted emission enhanced AIE effect and increased RISC rate constant (k_{RISC}) when compared with **22**. For the EL performance, **21**-based nondoped OLEDs achieved a maximum EQE of 17.9%, which remained 14.5% at a luminance of 1000 cd m^{-2}, showing low efficiency roll-off, while the EQE of **22**-based nondoped OLED was 9.1%.

By combining the advantages of both asymmetric diphenyl sulfone and alkyne linker as an electron acceptor, Zeng et al. [14] reported an asymmetric molecule **23**, which not only displayed multiple functions including AIE, TADF, and mechanochromic luminescence (MCL, which is defined as the phenomenon that shows a major change in emission colors or strengths of fluorescent materials when responding to mechanical stimuli) but also exhibited three emission colors from white to orange, in which TADF properties could be maintained. The spectrum analysis and theoretical calculations revealed that the twist structure helped to reduce orbital overlap for realizing a relatively small ΔE_{ST} (0.17 eV), while the crystal structure accounted for the rotation nature of diphenylethyne, which was related to the MCL behavior. **23** could give out reasonable white-light emission by a combination of dual emission bands peaked at 410 and 550 nm.

Wang et al. [15] reported a compound **24** with AIE and TADF properties in which an electron acceptor based on 10-phenyl-10H-phenothiazine-5,5-dioxide was linked with the electron donor phenoxazine. The compound had a highly stereoscopic structure due to multiple nearly vertical dihedral angles between adjacent groups and a small ΔE_{ST} (0.02 eV). Importantly, **24** enabled both doped and nondoped OLEDs with high efficiencies. A green-doped OLED realized a maximum EQE of 16.3%, CE of 43.8 cd A^{-1}, and PE of 35.2 lm W^{-1}, whereas the corresponding nondoped OLED were 16.4%, 44.9 cd A^{-1} and 32.0 lm W^{-1}, respectively. The results indicated that 10-phenyl-10H-phenothiazine-5,5-dioxide group could be an excellent acceptor for AIE-TADF materials. Sun et al. [16] performed a theoretical study on the

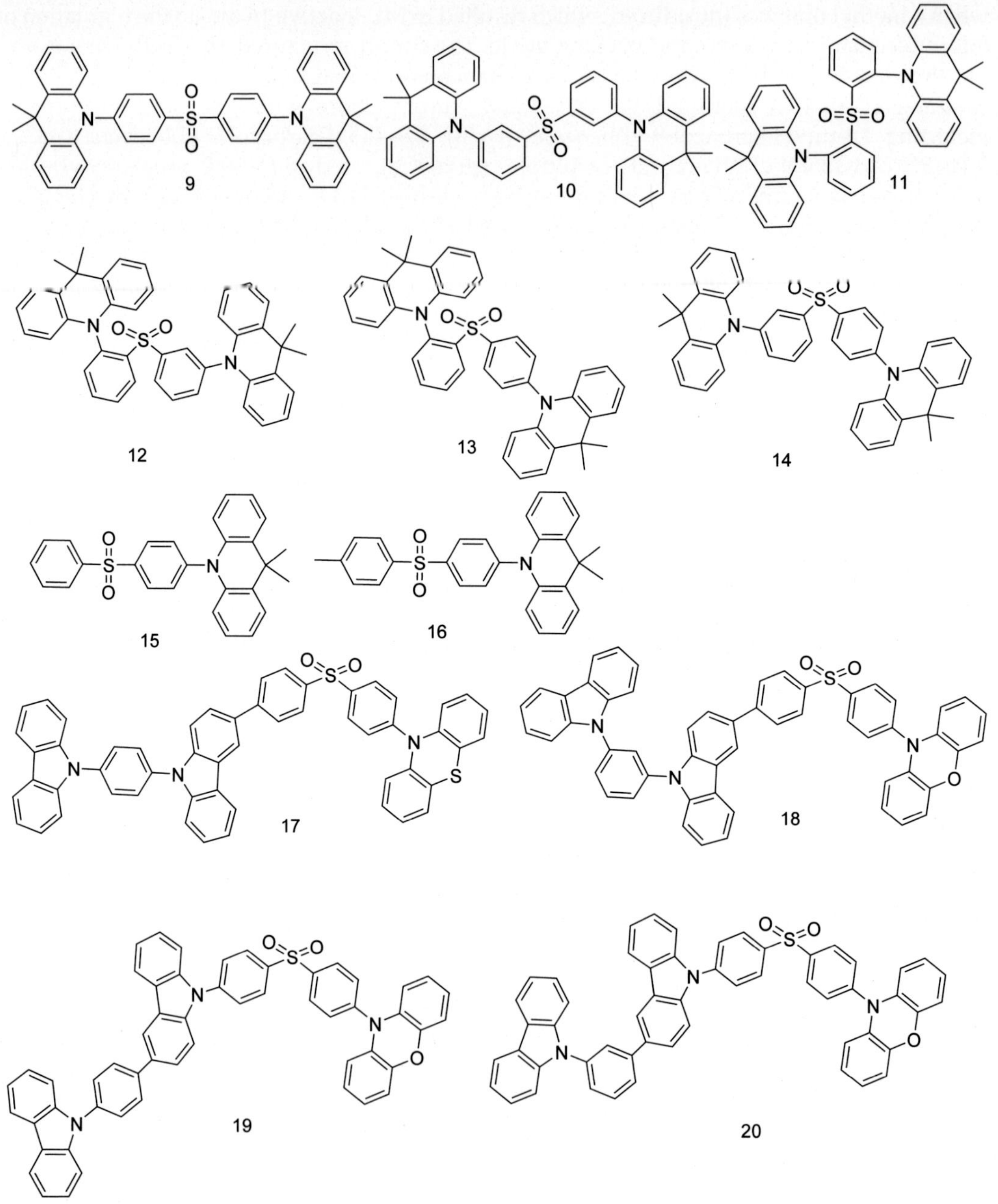

SCHEME 2 Chemical structures of compound **9–20**.

luminescent mechanism of **24** based on first-principle calculations. For simulation of the surrounding environments for molecules in toluene solution and in solid phase, polarizable continuum model (PCM) and quantum mechanics and molecular mechanics method (QM/MM) were used, respectively. It was found that, in the solid phase, the rigid surrounding environment led to the restriction of the intramolecular rotation (RIR) effect, which then suppressed molecular vibrations. Meanwhile, the geometrical changes during the excitation were also restricted due to enhanced intermolecular interactions, as confirmed by an analysis of weak intermolecular interactions in a dimer. The excited-state kinetics of **24** showed that, in the toluene solution, the nonradiative decay process was dominant, as a result of the relatively loose and free surrounding environment. However, in solid phase, the consumption energies of nonradiative decay decreased significantly, accompanied by enhanced radiative rate, revealing the AIE mechanism of **24**. Meanwhile, the intersystem crossing and reverse intersystem crossing rates were comparable with the radiative rate from S_1 to S_0, showing the potential of **24** as an efficient TADF emitter. The theoretical study gave a reasonable explanation of experimental results and provided valuable information for designing AIE-TADF emitters.

Sun et al. [17] reported two thianthrene-9,9′,10,10′-tetraoxide derivatives **25** and **26**. At room temperature, AIE-active **25** emitted intensely with a high PLQY of 92% in solid state. By increasing the number of carbazole units, **26** exhibited both AIE and TADF in solid state with a PLQY of 19% in air and 41% in N_2 atmosphere. The time-dependent density functional theory (DFT) calculations showed that the ΔE_{ST} values of **25** and **26** were 0.29 and 0.06 eV, respectively, and the minimized ΔE_{ST} of **26** was stemmed from separated HOMO and LUMO that was induced by the increased proportion of carbazole units.

In general, AIE-TADF molecules could not be guaranteed by simply and directly connecting the AIE core group with the TADF core group in a molecule, while sometimes such compounds can be obtained. Tang et al. [18] reported two luminescent materials **27** and **28** by integrating tetraphenyl/triphenyl-ethene (TPE/TrPE), carbazole group, and thianthrene-9,9,10,10-tetraoxide unit. These molecules were nonemissive in pure dimethylsulfoxide (DMSO) but highly emissive in the aggregated state. As learned from their phosphorescence and fluorescence spectra, the ΔE_{ST} values of **27** and **28** were determined to be 0.12 and 0.08 eV, respectively, suggesting a high potential for the occurrence of RISC. In addition, the decay lifetimes of **27** and **28** were 2.12 and 3.81 ms, respectively, within the lifetime range of delayed fluorescence, confirming TADF properties. Therefore, **27** and **28** were AIE-TADF molecules. Based on **27** and **28**, blue-green and green nondoped OLEDs were constructed with EL peaks at 517 and 494 nm, respectively, as well as maximum current efficiency of 2.47 and 3.43 cd A^{-1}, respectively. Later, they reported a luminescent molecule **29 [19]**, which was an isomer of **28**. The AIE and TADF properties of **29** were evidenced from its high solid-state PLQY and small ΔE_{ST} (0.04 eV). **29**-based nondoped OLED obtained a maximum CE of 3.96 cd A^{-1} with bluish green light emission.

2.2 Aromatic ketone derivatives

Aromatic ketone (including diphenyl ketone, xanthone, anthraquinone, thioxanthone, etc.) derivatives, which contain a highly electron-withdrawing carbonyl group with a twist angle in the center, are known as typical organic phosphorescent molecules with efficient intersystem crossing [20]. The efficient intersystem crossing of aromatic ketone probably makes them easy to trigger efficient RISC, with the assistance of rational molecular design for tuning ΔE_{ST} low enough to realize TADF. Moreover, diphenyl ketone is a typical crystallization-induced phosphorescence material [20].

At the early time of TADF, Adachi et al. [7] reported an AIDF emitter **30** by adopting 9,9-dimethyl-9,10-dihydroacridine as donor and benzophenone as acceptor. This compound endowed a green nondoped OLED with a maximum EQE of 18.9% and luminance of about 50,000 cd m^{-2}. The author claimed the main reasons for the concentration insensitive property of **30** were the relatively large Stokes shift and weak π-π stacking interactions. In fact, this compound displayed typical AIE property although it was not mentioned in the work (Scheme 3).

By coupling a phenothiazine donor unit with a xanthone or benzophenone acceptor unit, Yasuda et al. [21] reported highly luminescent D-A molecules (**31** and **32**). The two molecules were almost nonluminescent in pure THF solution, but emitted strong yellow delayed fluorescence upon their aggregation in THF/water mixtures or neat film, showing AIE behavior. The result revealed that the TADF characteristics of these molecules were generated from molecular aggregation. Based on AIE-TADF **31** and **32**, nondoped OLEDs achieved high EQEs of 11% and 7.6%, respectively.

Tao et al. [22] reported two soluble AIE-TADF emitters (**33** and **34**), using triphenylamine as an end-capping electron-donating group and isophthaloyl or terephthaloyl as central electron-withdrawing moiety. The optoelectronic properties could be rationally tuned by altering the linkage modes (from para- to meta-linkage) on the central phenyl ring of phthaloyl. **34** showed PLQYs of 2%/39% in solution/solid states, compared to 46%/75% for **33**. Based on the sky blue **33** and green **34**, solution-processed nondoped OLEDs obtained EQEs of 2.4% and 3.7%, respectively. Furthermore, solution-processed green TADF OLEDs with a double host/dopant TADF system were fabricated and exhibited EQEs of 13.0% and 9.0% by using **33** and **34** as the host, respectively. These high performances were associated with their high singlet and triplet energy levels for efficient energy transfer as well as efficient suppression of reverse energy transfer from TADF dopant to TADF host.

Tang et al. [23] reported an asymmetrical emitter **35** containing 9,9-dimethyl-9,10-dihydroacridine which intrigued AIE, TADF, and ML properties. The AIE nature allowed **35** to exhibit strong solid-state PL and prominent mechanoluminescence. The neat film of **35** was stable and showed a higher PL efficiency than the doped films of **35** in 4,4′-bis(9H-carbazol-9-yl)biphenyl (CBP). In the case of the doped OLEDs, a low doping concentration afforded high device efficiencies but obvious efficiency roll-off, like most doped OLEDs with TADF emitters, while increasing the doping concentration led to reduced device efficiencies but lowered efficiency roll-off. In contrast, the nondoped OLED achieved a high EQE of 14.2% with extremely low efficiency roll-off, indicating the nondoped OLED outperformed the doped OLED at high luminance. This result further certified that the combination of AIE and TADF merits was a promising molecular design principle for exploring ideal emitters to enable efficient and stable OLEDs with simplified structures. Wang et al. [24] performed a detailed study on the photoelectric properties of **35** based on first-principles calculations, helping to deeply understand TADF and AIE mechanisms of **35**.

When 9,9-dimethyl-9,10-dihydroacridine in **35** was replaced by phenoxazine and phenothiazine, Tang et al. [25] obtained two compounds **36** and **37**. As expected, like **35**, compounds **36** and **37** exhibited AIE and TADF characteristics, allowing not only high solid-state PL efficiencies but also low concentration quenching and exciton annihilation. **36**-based OLED

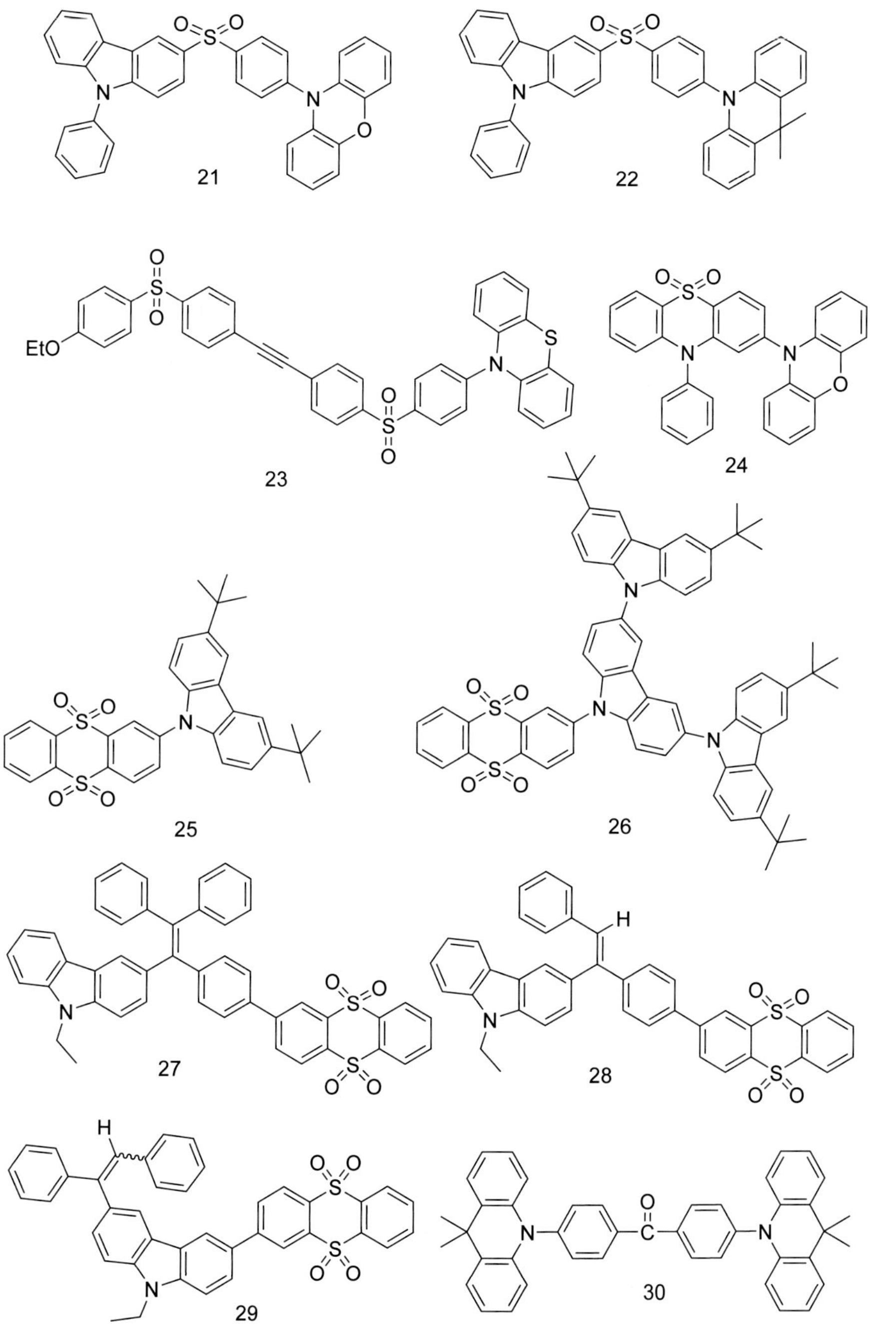

SCHEME 3 Chemical structures of compounds **21**–**30**.

achieved outstanding EQE, CE, and PE of 19.2%, 60.6 cd A^{-1}, and 59.2 lm W^{-1}, respectively, and notably very low-efficiency roll-off. The **37**-based nondoped device showed an EQE of 9.7% with a much lower efficiency roll-off. These results further support that the design strategy of combing AIE and TADF for developing robust emissive materials can help to solve severe efficiency roll-off in common doped TADF OLEDs.

Moreover, Tang et al. [26] reported AIE-TADF materials **38–40** which were similar to **35–37**. Their nondoped OLEDs provided excellent CE, PE, and EQE of 59.1 cd A^{-1}, 65.7 lm W^{-1}, and 18.4%, respectively, along with negligible efficiency roll-off of 1.2% at 1000 cd m^{-2}. They continuously reported a series of luminogens **41–43** containing carbonyl core (A) and chlorine-substituted groups (D). [27] The luminogens exhibited intriguing AIE-TADF property, with higher PLQYs (70%–73%) and shorter delay lifetimes ($\tau_{delayed}$) (0.42–0.76 s) than those of chlorine-free compound **40** (58% and 2.1 s). The small ΔE_{ST} and greatly enhanced spin-orbit coupling caused by the strengthened heavy atom effect significantly accelerated RISC and then shortened $\tau_{delayed}$. The nondoped OLEDs of the luminogens exhibited high EL performance, such as CE, PE, and EQE of 76.6 cd A^{-1}, 75.2 lm W^{-1}, and 21.7%, respectively, with very small efficiency roll-off. The luminogens could also function robustly in doped OLEDs in a wide range of doping concentrations (5–90 wt%) with small efficiency variations. Impressively, remarkable CE, PE, and EQE of 100.1 cd A^{-1}, 104.8 lm W^{-1}, and 29.1%, respectively, were attained in doped OLEDs of **41**. These findings demonstrated the AIE-TADF luminogens with fast RISC were promising candidates to fulfill various demands for OLED application (Scheme 4).

TADF emitters applicable to both doped and nondoped OLEDs with high device performance were rare. Chi et al. [28] investigated an AIE-TADF yellow emitter 4-((10H-phenothiazin-10-yl)phenyl)(4-(diphenylphosphoryl)phenyl)methanone (**44**) using a benzophenone core to link phenothiazine (D) and diphenylphosphoryl (A) groups. Interestingly, **44**-based doped and nondoped OLEDs achieved maximum EQEs of 26.7% and 16.6%, respectively, along with very low efficiency roll-off, representing a rare TADF emitter that afforded both doped and nondoped OLEDs with excellent device performances. Furthermore, they proposed a TADF and AIE hybrid device design strategy by combing **44** with a blue AIE emitter, and simple and efficient white OLEDs were successfully realized.

Based on the asymmetrical D-A-D′ structure, Tang et al. [29] reported two luminogens **45** and **46**. The D-A structure was formed by twisted electron-donating 9,9-dimethyl-9,10-dihydro-acridine donor and electron-accepting carbonyl group, resulting in the separation of HOMO and LUMO to reduce ΔE_{ST} for delayed fluorescence. To further investigate intermolecular π-π stacking, two donors of fluorene derivatives (9,9-dimethylfluorene and 9,9-diphenylfluorene) with different degrees of steric hindrance, were introduced. **45** and **46** showed AIE-TADF property that accounted for good EL performance of nondoped OLEDs. **45**- and **46**-based OLEDs showed low turn-on voltages of 2.8–3.9 V with strong green light and maximum EQEs of 5.7% and 13.2%, respectively, along with small efficiency roll-off. Similarly, they have also reported three AIE-TADF compounds **47–49** [30], wherein phenoxazine and fluorene derivatives donors were coupled with a benzoyl core acceptor. It was found that **47–49** displayed greatly enhanced fluorescence with increased delayed component upon aggregate formation, attributing to suppressed internal conversion channels and promoted intersystem crossing processes in the aggregated state. The unusual anti-Kasha behavior was also observed, revealing that the emissions were originated from a higher energy electronic excited state instead of the lowest ones.

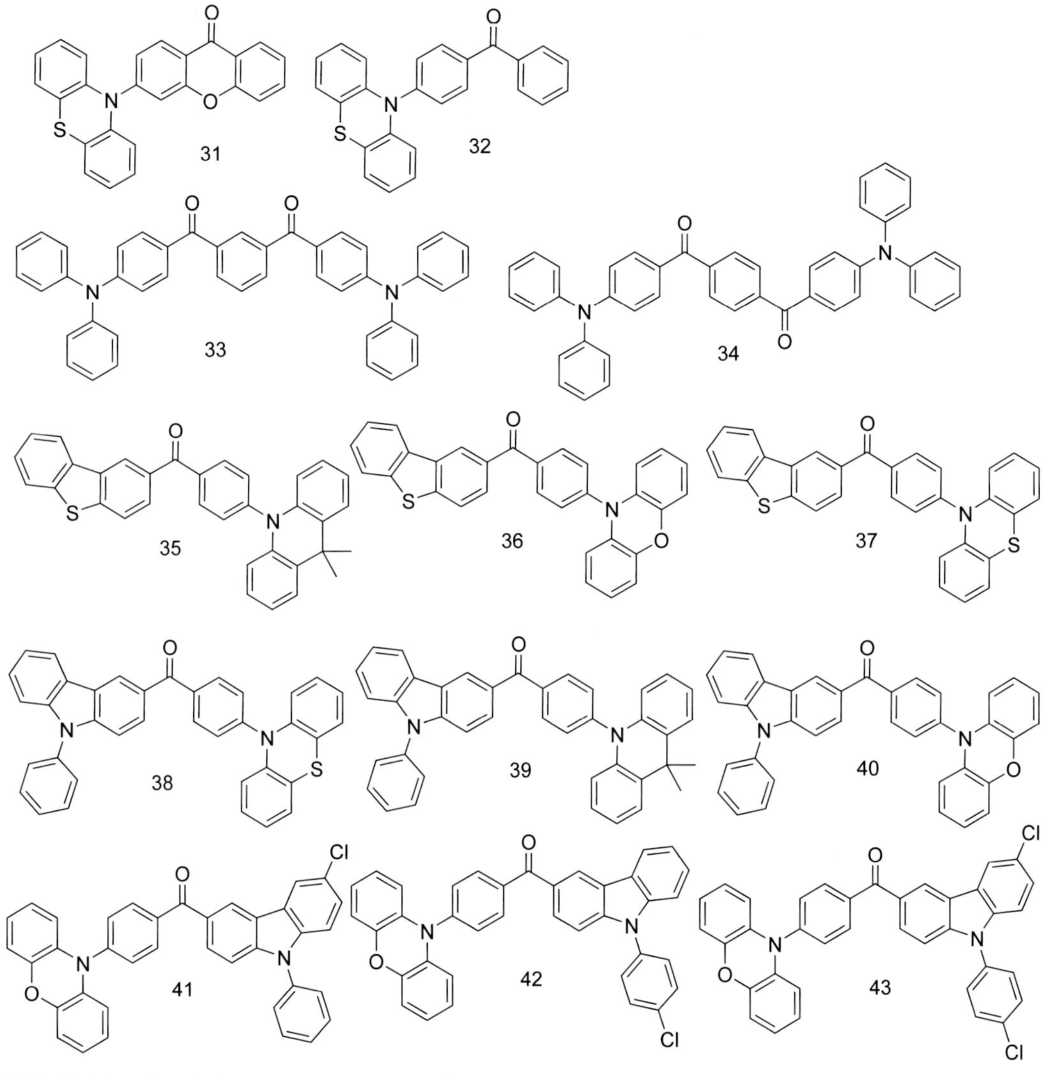

SCHEME 4 Chemical structures of compounds **31**–**43**.

In solution-processed TADF-OLEDs, efficiency roll-off tends to be a troublesome problem. Therefore, solution-processable TADF emitters enabling devices with low-efficiency roll-off are in high demand. To improve the film-forming ability for solution processing, Tang et al. [31] reported two AIDF molecules **50** and **51** by incorporating with long alkyl chains, while benzoyl worked as electron acceptor and 9-hexylcarbazole and phenoxazine (or 9,9-dimethyl-9,10-dihydroacridine) as electron donors. As a result, **51**-based solution-processed nondoped OLEDs had an EQE of 9.02%, while **50**-based doped devices achieved an EQE of 12.1%, and importantly, very low efficiency roll-off was observed in these solution-processed OLEDs. It was believed that the AIE-TADF character played an essential role in achieving high performance, inferring a great potential of AIE-TADF for developing solution-processed OLEDs with high performance (Scheme 5).

To explore efficient luminogens for nondoped OLEDs, Tang et al. [32] reported four molecules (**52–55**) by grafting AIDF moiety, 4-(phenoxazin-10-yl)benzoyl, within common host materials including DCB, CBP, mCP, and mCBP. The highly twisted molecular geometry, on one hand, helped to reduce ACQ; on the other hand, prevented excitons accumulation at the central benzoyl acceptor from short-range Dexter energy transfer. As a result, high-concentration excitons under high currents could be efficiently harvested for light emission without long-distance diffusion and severe annihilation in neat films. In addition, excitonic

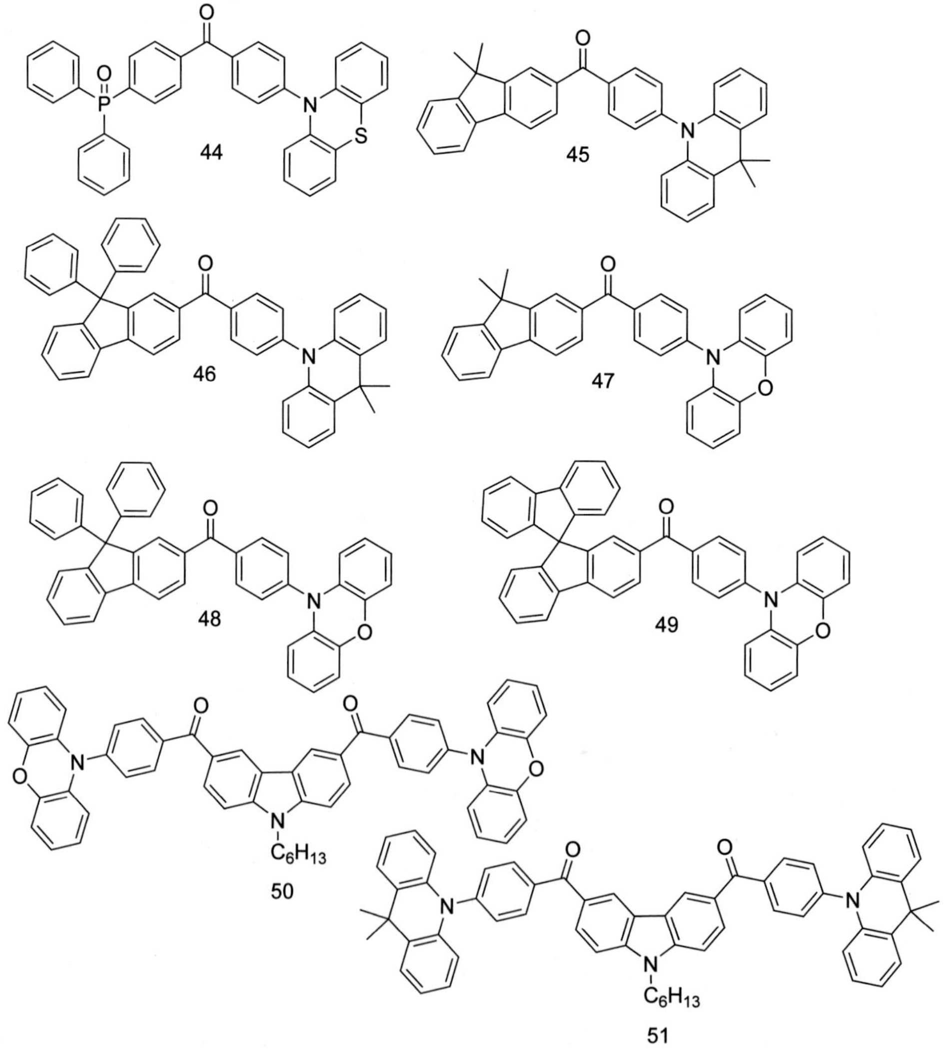

SCHEME 5 Chemical structures of compounds **44–51**.

energy transfer from peripheral carbazole moieties to benzoyl part might occur and contribute to EL efficiencies. Additionally, these luminogens were likely to enable relatively good ambipolar charge carrier transport, thus favoring electron and hole balance. Nondoped OLEDs of **52–55** demonstrated remarkable performances, such as a maximum luminance of ca. 100,000 cd m^{-2}, an EQE of 22.6% with negligible efficiency roll-off at 1000 cd m^{-2}. These results certified the feasibility and versatility of the presented molecular design to explore new AIE-TADF luminogens for highly efficient and stable OLEDs.

Moreover, Tang et al. [33] reported an AIE-TADF compound **56**, bearing carbazole as the skeleton, 9,9-dimethyl-9,10-dihydroacridine group (D) and benzophenone group (A). The **56**-based nondoped and doped OLEDs were explored. The nondoped device emitting green light exhibited a maximum luminance of 123,371 cd m^{-2} and a maximum EQE of 8.15%. Among the doped OLEDs, the best performances were achieved to be a maximum luminance of 116,100 cd m^{-2} and a maximum EQE of 19.67%.

Also, Wang et al. [34] reported two solution-processable small molecules (**57** and **58**) containing triazatruxene as donor and benzophenone as acceptor. These triazatruxene-based molecules are featured with AIE, TADF, and MCL. **57** and **58** displayed similar TADF emissions, considering similar twisted D-A-D conformation and energy level distribution. Despite that, different substituents on triazatruxene unit led to different AIE and MCL behaviors. The hexyl substituted **57** showed a color-changing AIE behavior in THF/water mixtures and a large red-shift of mechanochromic emission (59 nm) from 483 to 542 nm upon mechanical grinding. The phenyl substituted **58** displayed an AIE behavior without color change, and a smaller red-shift of mechanochromic emission (36 nm) because of the steric hindrance effect. As for solution-processed nondoped OLEDs, **58** achieved a higher EQE of 6.0% with a smaller efficiency roll-off (Scheme 6).

To develop solution-processable AIE-TADF emitters, Qi et al. [35] demonstrated a class of AIE-TADF materials **59–64** by adopting phenyl(pyridyl)methanone as acceptor and di(tert-butyl)carbazole and 9,9-dimethyl-9,10-dihydroacridine (or phenoxazine) as donors. As the positions of a nitrogen atom in the acceptor changed, intramolecular hydrogen-bonding interactions varied, transforming to eventual rigidity of the molecular structures. The crystal structure and theoretical simulation certified the existence of intramolecular hydrogen-bonding interactions, accounting for lowered ΔE_{ST}, suppressed nonradiative decay, and increased fluorescent efficiency in solid state. These TADF compound-based OLEDs gave out EL emission from sky blue to orange and a nondoped solution-processed device of **60** presented an EQE of 11.4%. This work suggested incorporating intramolecular hydrogen-bonding interactions into the AIE-TADF emitters is a promising design strategy to develop efficient emitters for solution-processed OLEDs.

Qi et al. [36] also reported two emitters **65** and **66** with triphenylamine donor and phenyl ketone acceptor groups [36]. The emitters were designed to present the following features. First, the distance between adjacent ortho-linked D and A units was proximal, thus inducing through-space electron interaction that could minimize ΔE_{ST} to enable TADF. Second, the electron clouds of HOMO and LUMO overlapped around the linking bridge, which was different from that of conventional TADF emitters. Meanwhile, the electron clouds of LUMO were partially extended to the ortho-linked donor unit through spatial π-π interactions, thus benefitting radiative transition to enhance efficiency. Finally, the highly twisted molecular configurations with weak intermolecular interactions facilitated the AIE effect, contributing

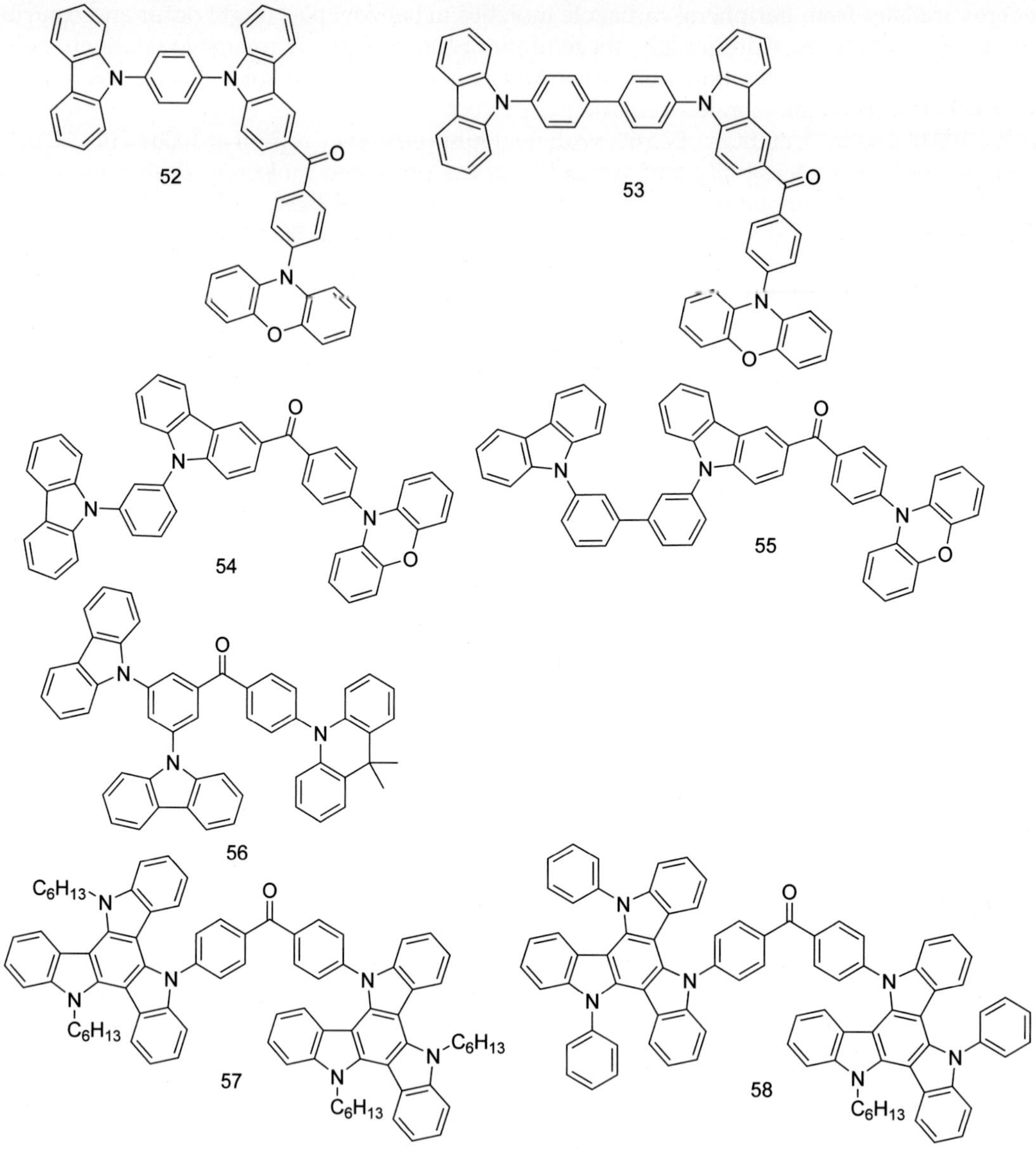

SCHEME 6 Chemical structures of compounds **52–58**.

to high fluorescent efficiency in solid state. Accordingly, **66**-based nondoped OLED achieved a higher EQE of 8.2% than the **65**-based device (4.4%), due to its higher PLQY (60.5%) and reduced concentration quenching in solid state.

Yang et al. [37] reported two D-A molecules **67** and **68** containing electron donors 9,9-diphenyl-9,10-dihydroacridine or 9,9-dimethyl-9,10-dihydroacridine, which were para-connected to phenyl ring of 4-benzoylpyridine acceptor. The large torsional angles between acridine units and 4-benzoylpyridine units were responsible for their TADF and AIE characters. Due to relatively high PLQYs in neat films, maximum EQEs of 8.4% and 9.7% were achieved in vacuum-deposited nondoped devices of **67** and **68**, respectively. Moreover, the EQE of **68**-based OLED remained almost 9.0% at 1000 cd m^{-2}. Furthermore, a higher EQE of 11.0% was achieved in the solution-processed nondoped device of **68** (Scheme 7).

With the combination of triazatruxene donor and benzophenone acceptor, Wang et al. [38] reported two AIE-TADF molecules **69** and **70**. Due to the rigid structure and high steric hindrance of triazatruxene, a large D-A twist angle was formed, which facilitated HOMO-LUMO separation and ΔE_{ST} reduction. Meanwhile, benzophenone was a classical phosphor with high intersystem crossing efficiency, which might induce efficient RISC with a small ΔE_{ST}. For solution-processable nondoped OLEDs, the **70**-based device presented better EL performance such as a maximum EQE of 9.8% (which was 6.4% for **69**-based device), possibly owning to improved charge carrier balance by introducing one more benzophenone unit, as well as lower efficiency roll-off due to a shorter delay fluorescence lifetime (0.54 ms) of **70** than that of **69** (0.79 ms).

Qi et al. [39] reported a series of AIE-active TADF emitters **71–76** by adopting phenyl ketone as acceptor and 9,9-diphenyl-9,10-dihydroacridine (or 9,9-dimethyl-9,10-dihydroacridine and phenoxazine) as the donor. The TADF and AIE properties of the dendrimers and the reference nondendritic luminogens were comparatively studied. In the dendrimers, a highly twisted molecular skeleton could be formed by incorporating a propeller-structured triphenylamine unit into the central core. In addition, steric repulsion between D and A units occurred, rendering the circumambient arms with a nearly orthogonal configuration. In the aggregated state of these dendrimers, improved PLQYs and promoted RISC processes were simultaneously realized, stemming from suppressed intermolecular π-π interaction. Moreover, the introduced branched alkyl chains helped to increase free volumes, thus playing a role in preparing pinhole-free uniform films during solution processing. The PLQYs of dendrimers **74**, **75**, and **76** were 46.1%, 64.8%, and 66.5%, respectively, while that of nondendritic **71–73** were 39.8%, 55.7%, and 58.6%, respectively. **76** and **73** were utilized for fabricating solution-processed nondoped and doped OLEDs, resulting in EQEs of 12.1% and 17.6%, respectively. This work provided an approach to design dendritic AIE-TADF luminogens for solution-processed OLEDs with high performance (Scheme 8).

Zhao et al. [40] proposed a design strategy for AIE-TADF luminogens by introducing functional group of 4-(phenoxazin10-yl)benzoyl to common chromophores such as 2,7-di(9H-carbazol-9-yl)-9,9-dimethyl-9H-fluorene, 2,7-di(9H-carbazol-9-yl)-9,9-diphenyl-9H-fluorene, and 2,7-di(9H-carbazol-9-yl)-9,9′-spirobi (fluorine). The influence of subtle modulations on bulkiness and stiffness of target molecules (**77**, **78**, and **79**), as well as PL and EL properties, were studied. **77** and **78** neat films showed higher PLQYs of 88.5% and 89.0%, respectively, than that of **79** (39.6%) due to its relatively poor molecular π – conjugation and strong intermolecular π-π interaction. In nondoped OLEDs, **77** and **78** enabled high EQEs of 19.0% and 18.5%, respectively, while **79** afforded an EQE of 3.3%, resulting from the low PLQY and unbalanced charge carrier transport ability. Interestingly, **77**, **78**, and **79** endowed doped OLEDs with high EQEs of 21.7%, 24.4%, and 22.3%, respectively, given improved PLQYs and balanced charge carrier transport. All the devices displayed very low efficiency roll-off. Therefore, the proposed design strategy could be applicable for designing efficient AIE-TADF emitters.

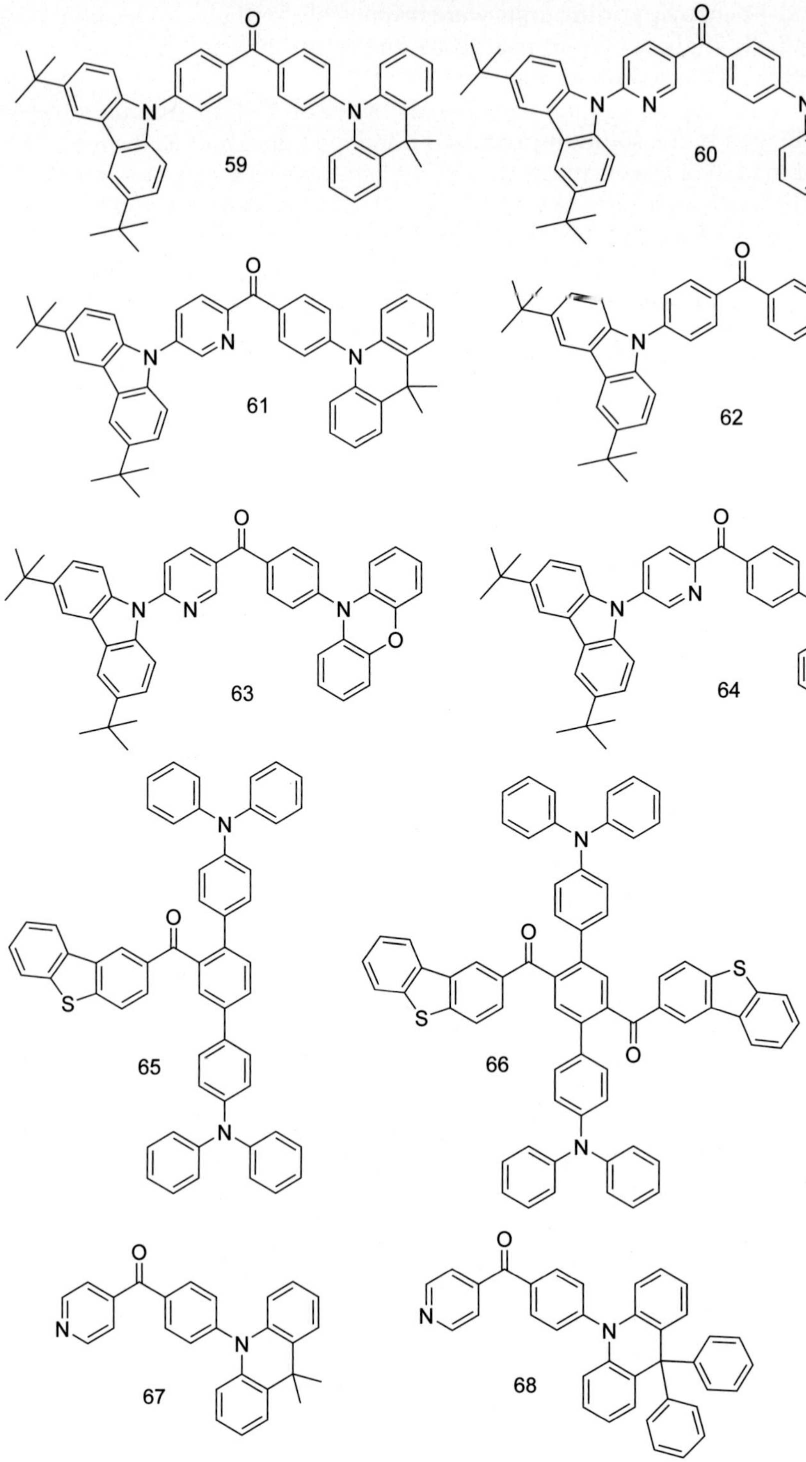

59 60 61 62 63 64 65 66 67 68

SCHEME 7 Chemical structures of compounds **59–68**.

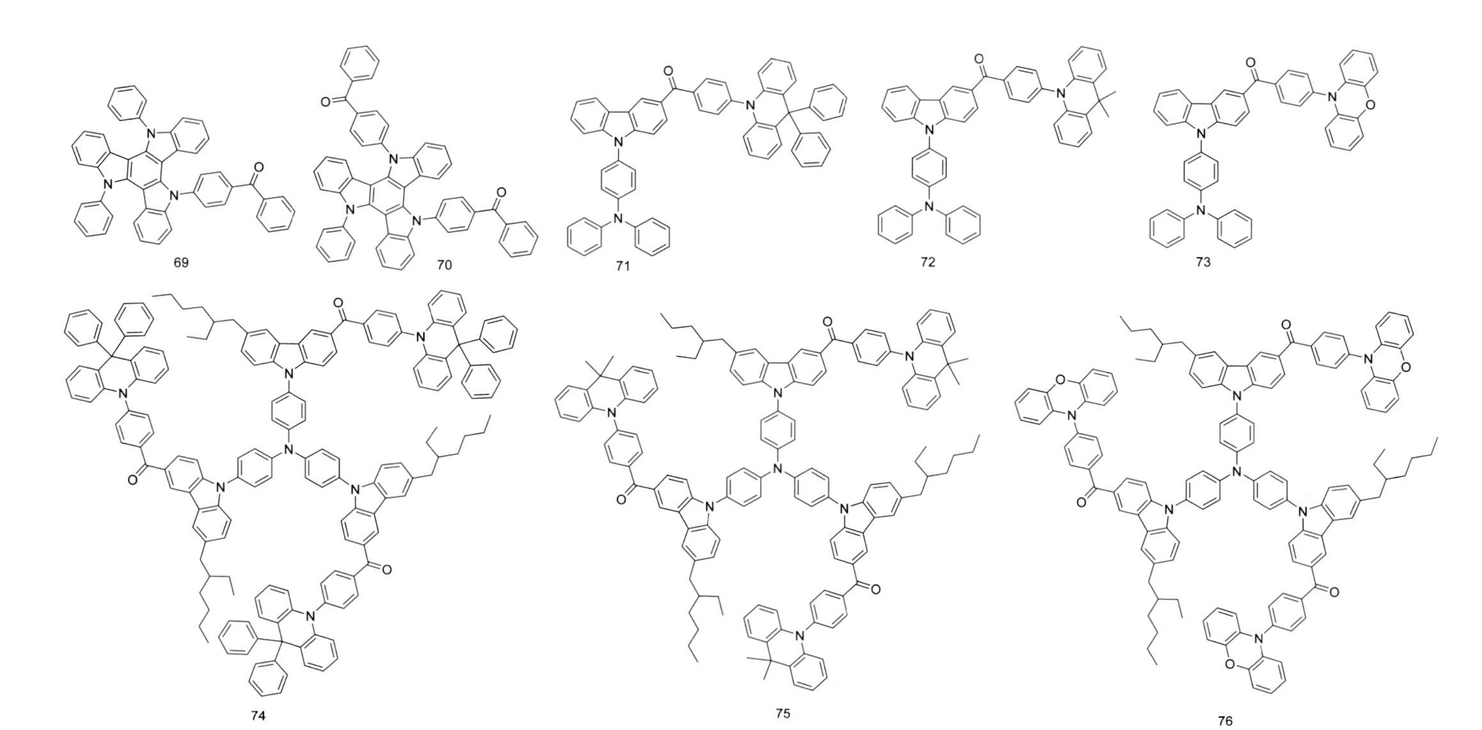

SCHEME 8 Chemical structures of compounds **69–76**.

It was conceived as a feasible strategy to improve charge carrier transport ability of AIE-TADF materials with the incorporation of bipolar charge carrier transport materials. Zhao et al. [41] designed two compounds **80** and **81** by integrating an AIE-TADF moiety, 4-(phenoxazin-10-yl)benzoyl, with bipolar carrier transport materials, 3,5-bis((9H-carbazol-9-yl)-3,1-phenylene)pyridine and 2,6-bis(3-(9H-carbazol-9-yl)phenyl)pyridine. **80** and **81** exhibited AIE-TADF properties with high PLQYs in the aggregated state. The nondoped OLEDs of **80** and **81** showed maximum EQEs of 17.3% and 16.1%, respectively, which remained as high as 17.2% and 15.9% at 1000 cd m^{-2}, indicating negligible efficiency roll-off (Scheme 9).

SCHEME 9 Chemical structures of compounds **77**–**81**.

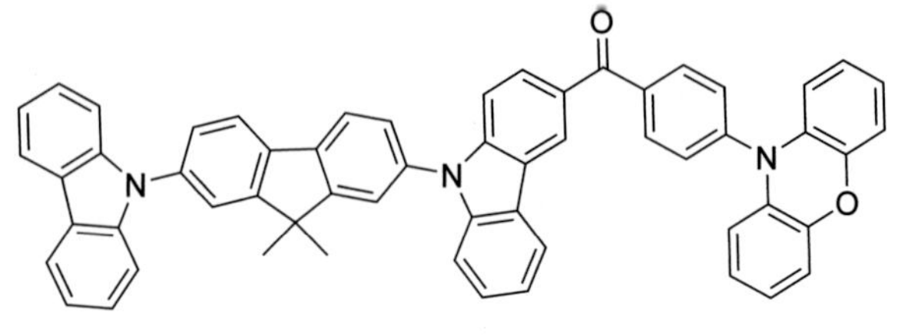

77

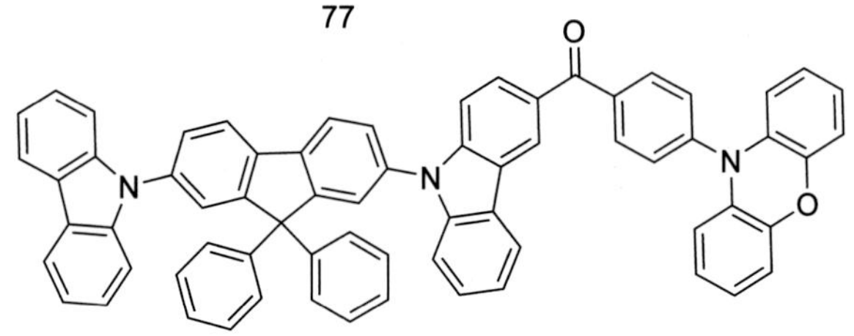

78

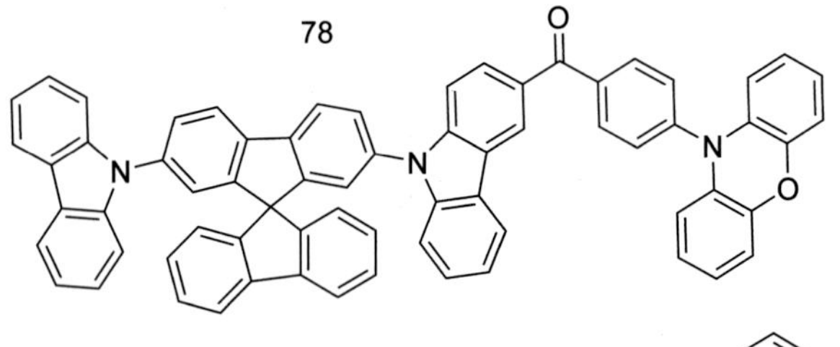

79

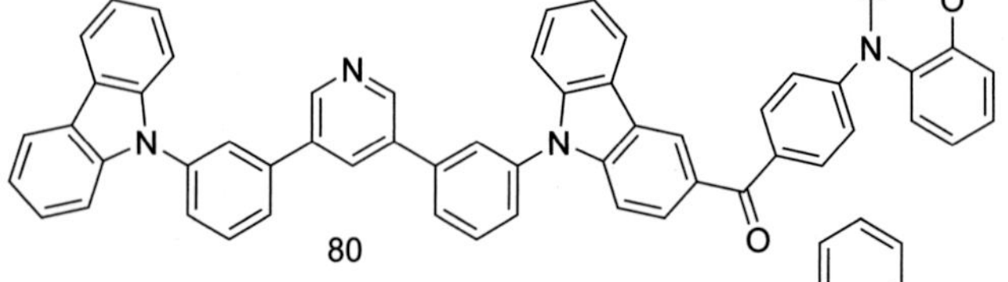

80

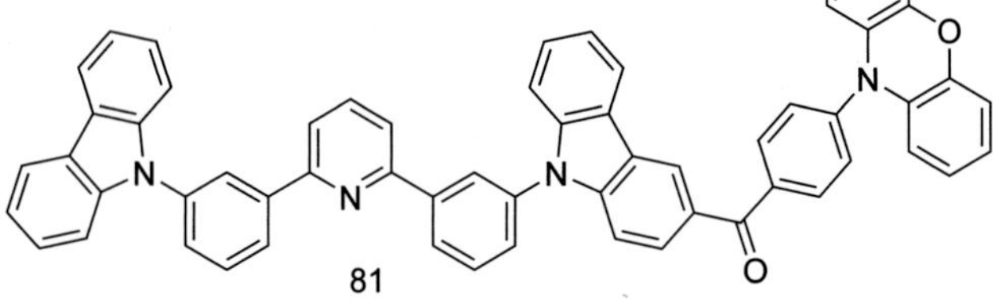

81

Theoretical studies on photophysical properties of AIE-TADF molecules in solid phase are of importance for revealing their emission mechanisms. Fan et al. [42] performed theoretical studies on AIE-TADF molecule **82** in solution and solid states. Solvent environment effect in THF was simulated by polarizable continuum model, and solid-state effect was considered by combining the QM/MM method. The excited-state energy consumption process was investigated, based on a combination of thermal vibration correlation function (TVCF) theory and first-principles calculation. As calculated, the prompt fluorescence efficiency, delayed fluorescence efficiency and total fluorescence efficiency for molecule in solid phase (14.4%, 31.5%, and 45.9% respectively) were obviously enhanced than that in THF (3.0%, 0.4%, and 3.0% respectively), suggesting the AIE feature. To detect the inner mechanisms, the geometrical structures, Huang-Rhys (HR) factors and reorganization energies as well as excited-state transition properties were analyzed. In comparison to the solution state, HR factors and reorganization energies were found to decrease in the solid phase, due to the out-plane vibration of 9,9-dimethyl-9,10-dihydroacridine and the rotation of the molecule in a rigid environment, accounting for the AIE property. Meanwhile, a smaller ΔE_{ST} (0.07 eV) in the solid state helped to promote RISC process for triggering TADF. Moreover, T_1 state showed a locally excited state feature in the solid phase but hybridized local and charge-transfer features in THF. This work provided information for deeply understanding the AIE-TADF mechanism of **82** from a theoretical perspective, which was helpful for molecular design.

Tang et al. [43] reported two emitters **83** and **84** consisting of electron-withdrawing benzoyl and electron-donating phenoxazine and 9,9-dihexylfluorene. The emitters were weakly emissive in dilute solution, but strongly emissive in the aggregated state with prominent delayed fluorescence, verifying the AIE-TADF character. As for **83-** and **84**-based OLEDs, EQEs of 12.49% and 14.69% were achieved in solution-processed devices, respectively, while EQEs of 14.86% and 14.12% were realized in vacuum-deposited ones, respectively. It is noted that solution-processed OLED of **83** displayed a very low efficiency roll-off, demonstrating excellent efficiency stability.

In addition, Su et al. [44] reported three stimuli-responsive asymmetric AIE-TADF luminogens, consisting of a fused N-heterocycle diarylketone acceptor, and phenoxazine, phenothiazine, and 9,9-dimethyl-9,10-dihydroacridine function as electron donors for **85**, **86**, and **87**, respectively. The donor groups were attached to a rotatable phenyl ring to generate twisted conformations, which was a crucial prerequisite to initiate both AIE and TADF. The acceptor embedded with two nitrogen atoms possessed a strong electron-withdrawing ability, thus favoring the formation of a strong intramolecular charge transfer (ICT) effect and the realization of long-wavelength emission. Moreover, the rigid planar conformation of pyridylbenzimidazole helped to enhance PLQY in the aggregated state by reducing nonradiative energy dissipation. The luminogens showed an emission peak over 570 nm in the neat film with a short lifetime of delayed fluorescence within 1 μs. **86** demonstrated two kinds of crystals (green and orange crystals) with distinct intermolecular interactions between pyridylbenzimidazole moieties. The green crystal with a looser packing mode presented significant morphology-dependent stimuli-responsive behavior with a 56-nm wavelength shift. Nondoped devices presented an EQE of 7.04% with negligible efficiency roll-off, while nondoped **85** devices gave out orange emission. The doped **85** devices showed a high EQE of 15.77% but a large efficiency roll-off.

In contrast to AIE perspective, Yasuda et al. [45] reported a series of 9H-xanthen-9-one derivatives **88–91** and explained the high luminous efficiency in solid states from another view of point. They pointed out that a small modulation in molecular geometric structures had a great influence on the concentration quenching of TADF, thus enhancing solid-state PL and EL efficiencies. The exciton-quenching rates of TADF emitters show an exponential dependence on their intermolecular distance in thin films. As described by the Dexter energy-transfer model, electron-exchange interactions of triplet excitons dominated the concentration quenching of TADF. This result was different from conventional fluorescent and phosphorescent molecules involving Förster energy-transfer mechanism. The proposed concept and strategy for inhibiting concentration quenching of TADF provided useful guidelines for developing more high-efficiency TADF emitters and devices (Scheme 10).

Based on anthraquinone derivatives, Sun et al. [46] synthesized two red AIE-TADF compounds, 2-carbazol-9-yl-anthraquinone (**92**) and 2-(4-diphenylamino-phenyl)-anthraquinone (**93**). As for solution-processed OLEDs of **92** and **93**, orange and red emission lights with EL peaks at 572 and 612 nm, and maximum EQEs of 5.8% and 7.8%, respectively, were achieved. Moreover, through controlling film formation in different solvents, the solution-processed nondoped devices with **92** displayed dual emission peaks at 600 and 680 nm [47]. Also, **92** possessed mechanochromic luminescence. In addition, Lee et al. [48] reported an anthraquinone derivative, 2-(phenothiazine-10-yl)-anthraquinone (**94**), which existed five different aggregation states, including three different crystal structures. The compound exhibited AIE, TADF, and ML properties, and the five aggregation states showed different emission colors from green to deep red. According to single-crystal structural study, the emission shifted from yellow to red was caused by enhanced π-π interaction. The aggregation state was of importance to the TADF properties. Through controlling the molecular stacking modes, a red crystal of **94** exhibited efficient TADF with a high PLQY (84.8%) and a small ΔE_{ST} (0.01 eV). This was the first example showing the TADF property could be systematically adjusted by aggregation state.

Moreover, Shi et al. [49] also reported AIE-TADF emitters with ML property. Two compounds **95** and **96** were derived from (9H-carbazol-9-yl)(phenyl)methanone. The one bromine substituted **95** presented AIE, TADF, and ML properties simultaneously, while two bromine substituted **96** only showed TADF and AIE properties. The crystal structures analysis and theoretical calculation stated that the ML property of **95** was mainly related to compact molecular packing and intense intermolecular interactions of the crystal. This work provided a more fundamental understanding of the ML mechanism and offered possible directions to develop prominent ML luminophores in the future.

2.3 Nitrogen-containing six-membered aromatic heterocycles derivatives

Nitrogen-containing six-membered heterocycles (including triazine, pyrimidine, pyrazine, quinoxaline, etc.) derivatives are the suitable building blocks for the construction of organic semiconductor molecules because they are highly electron-deficient groups. Many of reported D-A-type nitrogen-containing six-membered aromatic heterocycle derivatives display TADF emission, resulting from low ΔE_{ST} induced by an effective separation between HOMO and LUMO.

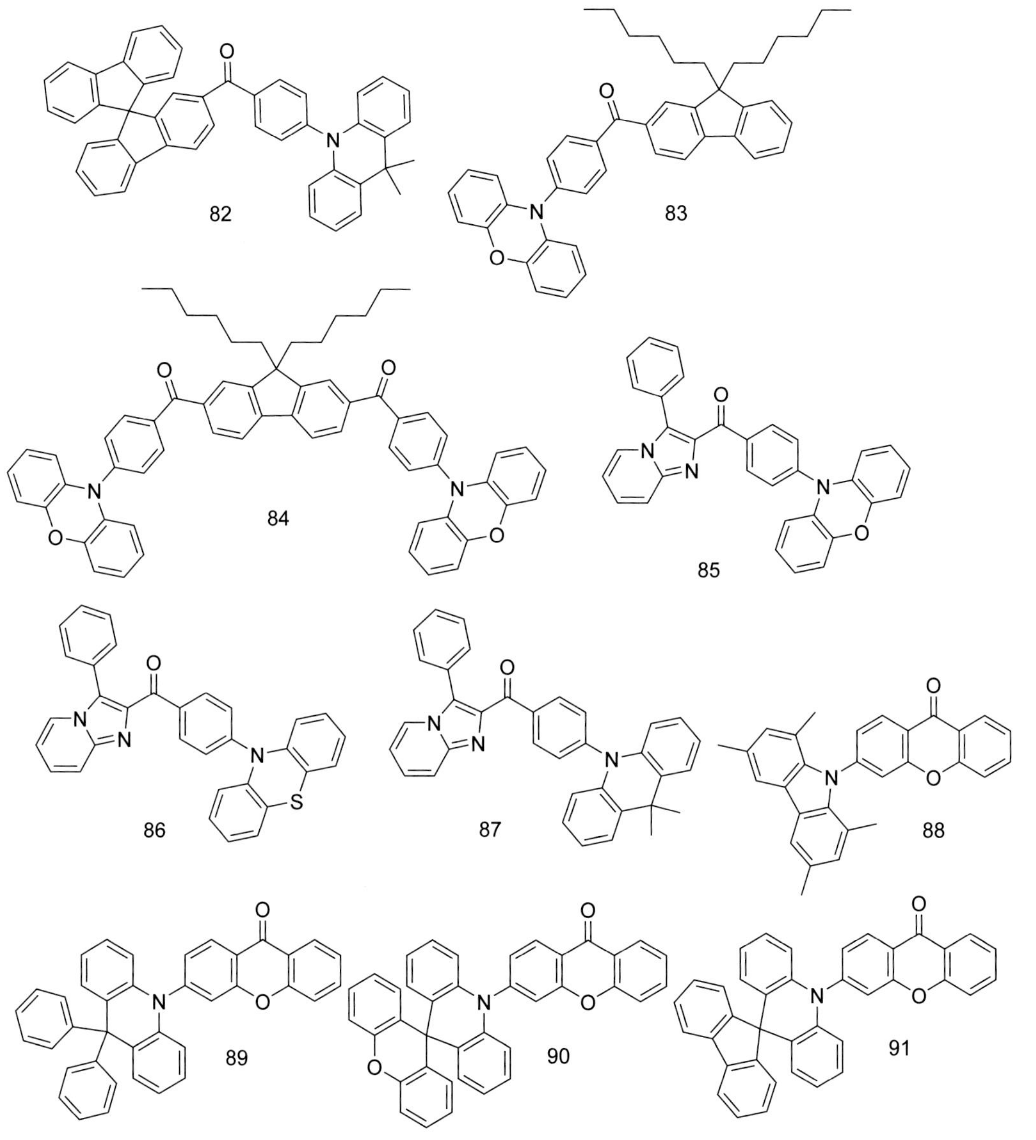

82 83 84 85 86 87 88 89 90 91

SCHEME 10 Chemical structures of compounds **82–91**.

Wu et al. [50] reported an efficient TADF emitter **97** that enabled high PLQYs in both doped film (90%) and neat film (83%), indicating low concentration quenching. Consequently, when **97** was used for OLEDs fabrication, the doped and nondoped devices achieved high EQEs of 26.5% and 20%, respectively, and these device efficiencies were comparable to that of the best-performing phosphorescent OLEDs at that time.

The through-space charge transfer is mediated by spatial π-π interactions because D and A groups are placed in close proximity. In a through-space conjugation system, the charge is transferred through aromatic π bonds linking between a donor and an acceptor, which is beneficial for achieving high-efficiency TADF emitters. To obtain intramolecular through-space D-A π-π interactions, Swager et al. [51] synthesized three 9,9-dimethylxanthene bridged D-A U-shaped molecules (**98–100**), comprising of phenothiazine, carbazole or 3,6-di-tert-butylcarbazole as donor groups. The rigid placement of a donor and an acceptor into a cofacial arrangement at a distance of 3.3–3.5 Å led to the quantitative formation of a charge transfer excimer structure. These luminogens exhibited delayed fluorescence in the absence of triplet-quenching oxygen in both solution and solid states, while the AIE property was evidenced by enhanced PLQYs in the solid state. The crystal structure analysis suggested that C-H---π interactions promoted a rigid environment that decreased nonradiative deactivation. **98**-based OLEDs obtained an EQE of 10%, indicating both triplet and singlet excitons contributed to light emissions (Scheme 11).

Two luminogens **101** and **102** were reported by Tang et al. [52] by connecting acridine and phenoxazine donors and triazine acceptor with through-space conjugated hexaphenylbenzene (HPB). Due to peripheral phenyls in HPB being closely aligned in a propeller-like fashion, efficient through-space charge transfer between D and A occurred. In solution state, **101** and **102** showed weak fluorescence with a negligible delayed component. Upon aggregation, strong fluorescence with highly enhanced delayed components was observed. These results

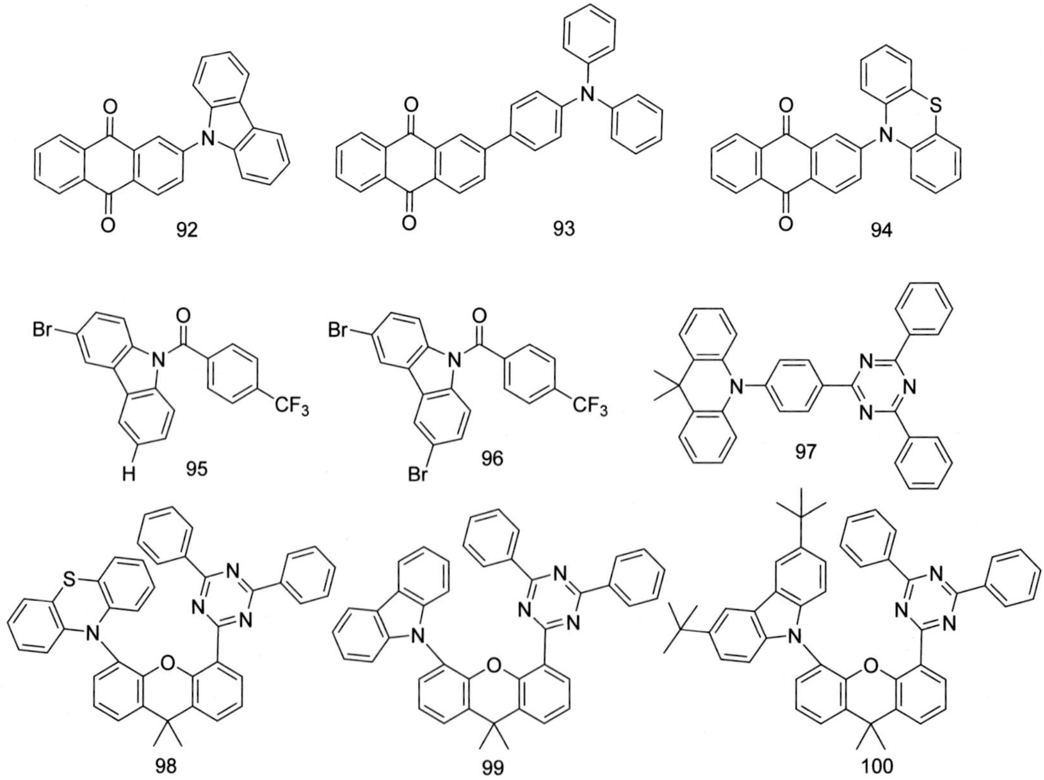

SCHEME 11 Chemical structures of compounds **92–100**.

confirmed their AIE-TADF property. The PLQYs of **101** and **102** neat films were 61.5% and 51.8%, respectively. Accordingly, nondoped OLEDs based on **101** and **102** achieved considerable high EQE of 12.7% with low efficiency roll-off. This work suggests that the through-space charge transfer strategy has the potential to develop AIE-TADF emitters for nondoped OLEDs with high performance.

In addition, Wang et al. [53] proposed a strategy to design AIE-TADF luminogens with through-space charge transfer, consisting hexaarylbenzenes with circularly arrayed donors (acridan/dendritic triacridan) and acceptors (triazine) in the periphery of hexaphenylbenzene core. Spatial π-π interactions between D and A accounted for efficient through-space charge transfer emission. In comparison to **103** with acridan donor, **104** with stronger dendritic teracridan donor had higher PLQYs of 36% and 63% in neat and doped films, respectively. Consequently, the doped solution-processed OLEDs of **104** presented higher EL performance such as a maximum EQE of 14.2%, which was among the best efficiencies regarding solution-processed TADF OLEDs (Scheme 12).

By using phenyl spacer (π-bridge) to connect triazine (acceptor) and carbazole dendrites (donors), Wang et al. [54] reported two D-π-A-type emitters **105** and **106**, while phosphorus oxygen groups were added to increase electron transport ability. As expected, a highly twisted conformation was formed between the carbazole dendrites and phenyl spacer, leading to AIE and TADF properties. Theoretical calculations and experimental results revealed that, as the number of carbazole dendrites increased, the fluorescence oscillator strength was enhanced and meanwhile ΔE_{ST} was reduced. Due to its reasonable molecular design, **105** showed high oscillator strength and a small ΔE_{ST}, which promoted the RISC process and fluorescence efficiency. The solution-processed nondoped OLEDs of **105** and **106** achieved EQEs of 12.8% and 9.1%, respectively, along with very low efficiency roll-off, owning to well-balanced charge carrier transfer, and shortened exciton lifetimes and reduced exciton concentration.

Wang et al. [55] reported two luminogens **107** and **108** with a highly twisted structure containing an acridine-carbazole hybrid donor. The two luminogens showed both TADF and AIE features. Compared to that of **108**, the neat film of **107** had a higher PLQY of 67% due to a more prominent AIE character. Consequently, the OLEDs of **107** outperformed that of **108**. A nondoped OLED of **107** achieved a maximum EQE of 14.1%, while its doped OLED exhibited a higher EQE of 22.6%, which remained as high as 20.1% at 5000 cd m^{-2}.

A series of AIE-TADF emitters **109–112** were reported by Yang et al. [56] by the employment of 9,9-dimethyl-9,10-dihydroacridine (DMAC) or 10H-phenoxazine (PXZ) as donor units into a quinoxaline framework. DMAC and PXZ had large spatial conformations to favor HOMO-LUMO separation, leading to small ΔE_{ST}. As the electron-donating capability of donor and the number of donor were adjusted, photophysical properties of the compounds were systematically regulated, with emissions ranging from green to red. When adopted to construct OLEDs, the **110**-based doped green device achieved a maximum EQE of 22.4%, which was 14.1% in **112**-based doped orange devices with low efficiency roll-off. For **110**-based nondoped devices, a maximum EQE of 12.0% was obtained, which was associated with the TADF and AIE features. They reported luminogens **113–115** by introducing a fluorine atom into the quinoxaline system to integrate AIE and TADF features [57]. As for mono-fluorinated **113**, it enabled a doped OLED with a maximum EQE of 23.5%, and a nondoped orange device with an EQE over 10%. In comparison, the nonfluorinated **115**

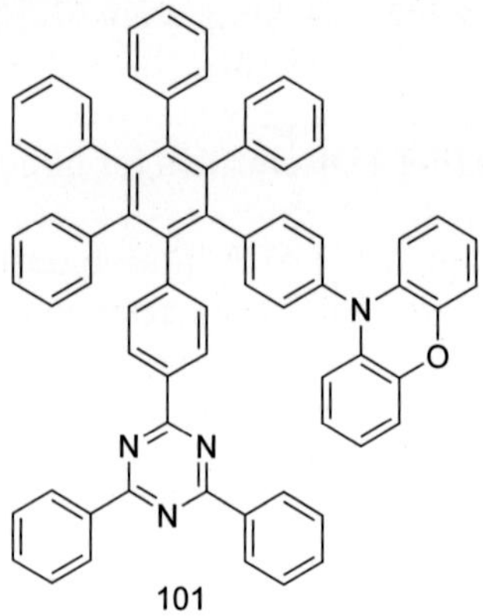

101

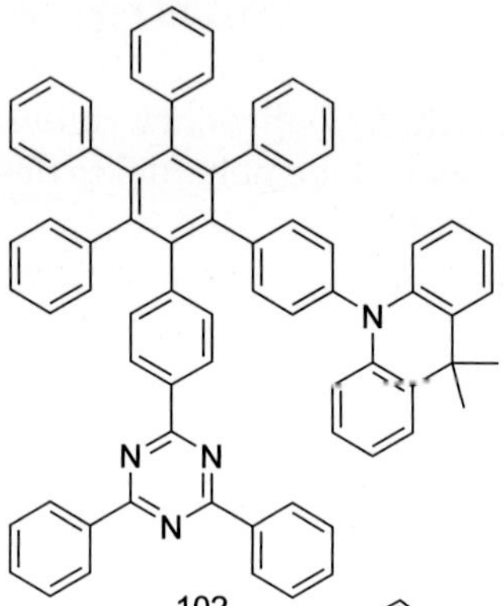

102

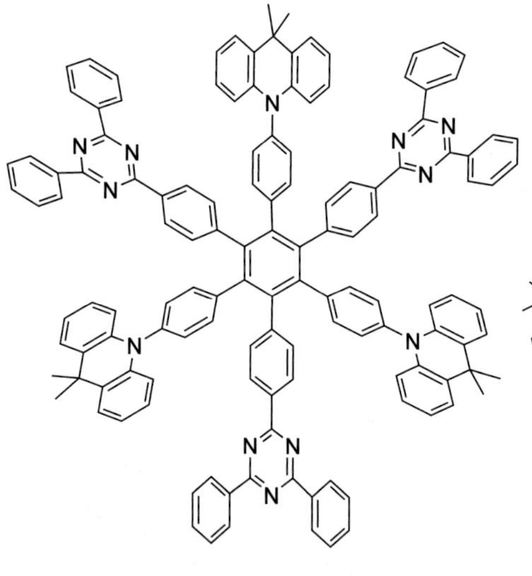

103

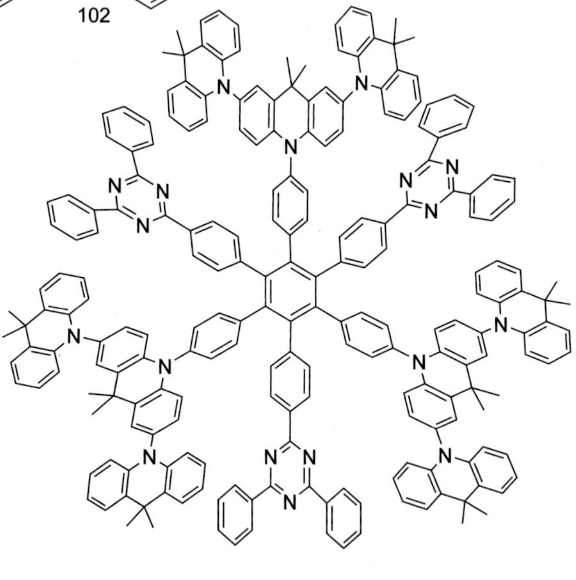

104

SCHEME 12 Chemical structures of compounds **101–104**.

enabled a nondoped yellow device with an EQE of 8.8%, suggesting higher efficiency of fluorine-substituted luminogens (Scheme 13).

Furthermore, Yang et al. [58] reported two orange-red emitters **116** and **117** with TADF and AIE characteristics, bearing 9,9-dimethyl-9,10dihydroacridine donor and rigid quinoxaline together with cyanobenzene acceptor. The fluorine atom had a similar atomic radius but stronger electron-withdrawing ability to that of hydrogen, which could bring about a similar packing mode but different photophysical property. Both emitters displayed polymorph-dependent TADF emissions from yellow to red and multicolor MCL in powder state. Four and five aggregation states were observed in **116** and **117**, respectively. The molecular level to aggregation state studies revealed that the molecular structure determined the TADF property of the molecules, while the aggregation state played a role in emission colors, lifetimes, and PLQYs.

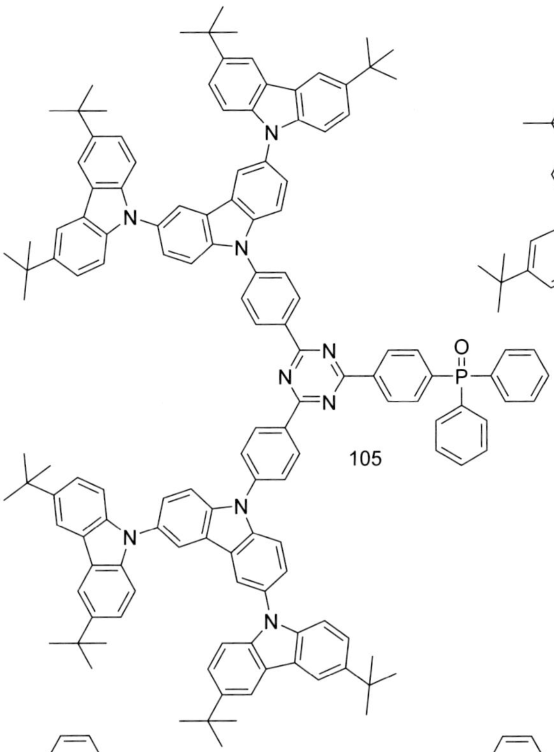

105

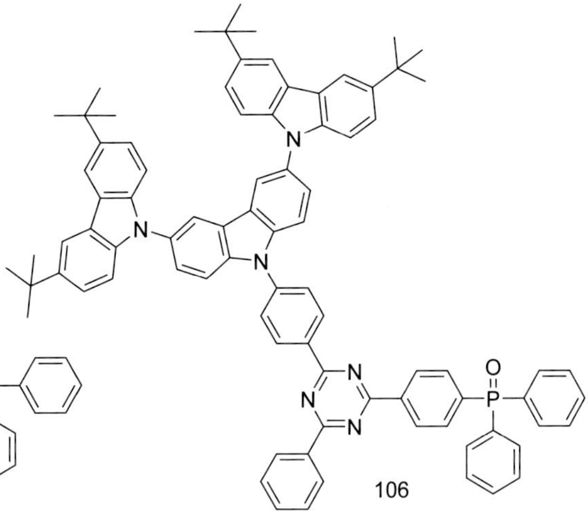

106

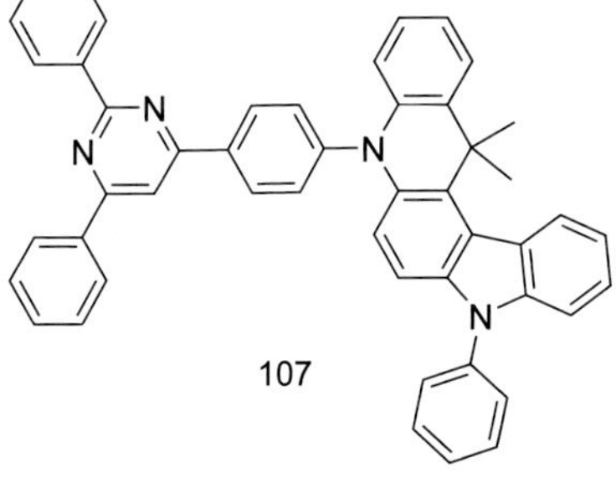

107

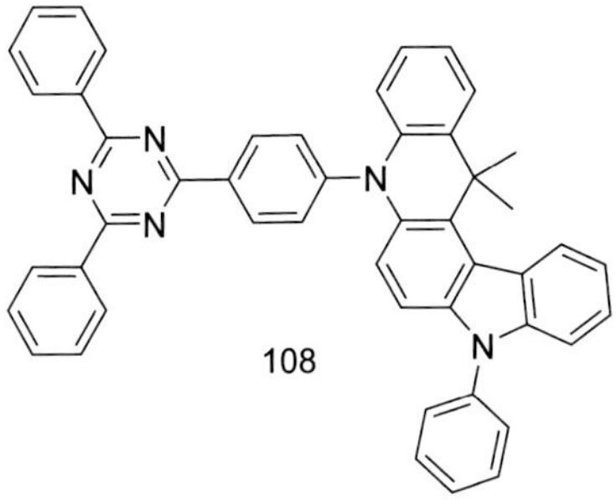

108

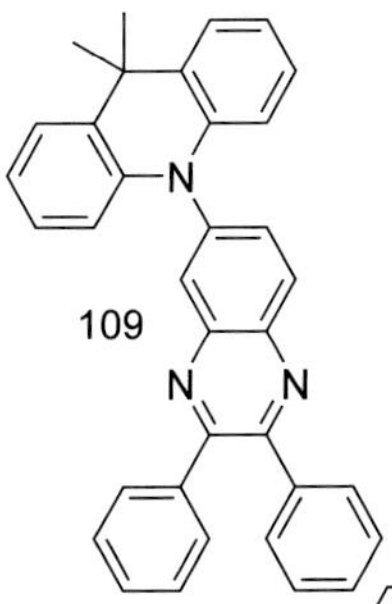

109

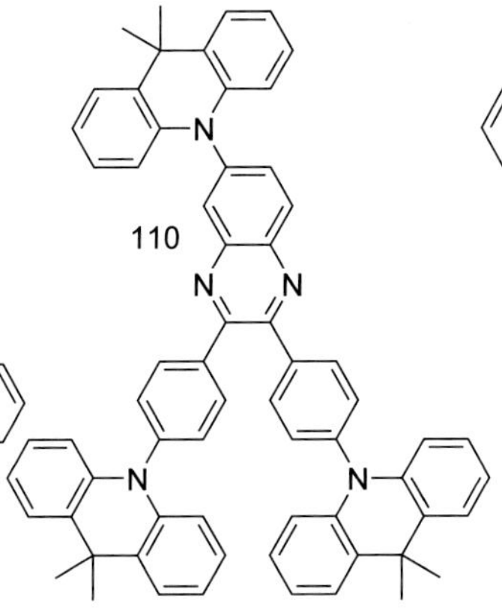

110

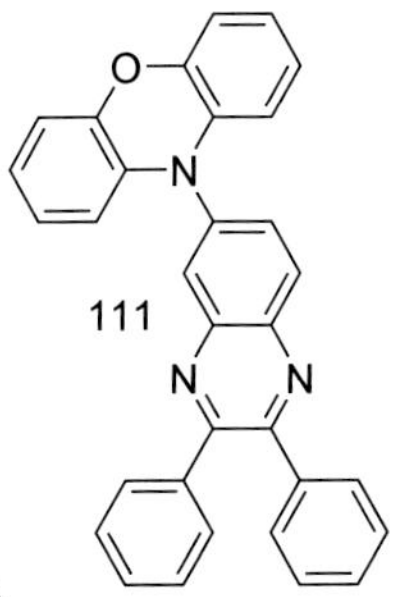

111

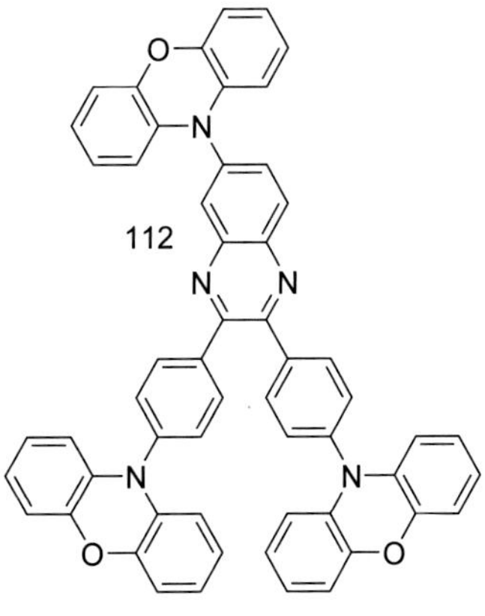

112

SCHEME 13 Chemical structures of compounds **105–112**.

Tian et al. [59] reported an AIE-TADF luminogen **118**, which was used to form fluorescent nanoparticles (NPs) by reprecipitation. The fluorescent NPs exhibited excellent fluorescence brightness and photostability, as well as a long fluorescence lifetime. In detail, an amphiphilic polymer with the oxygen blocking property was utilized as the host matrix, and small-molecular-weight antifade agents (singlet oxygen quencher and antioxidant) and a triplet energy barrier were incorporated in the hybrid nanoparticles. The triplet energy barrier helped to confine triplet excitons within the molecules, and the AIE property allowed **118** with high fluorescing ability. The reasons for the excellent photostability of the hybrid nanoparticles were firstly, the oxygen blocking property of the polymer matrix could separate **118** from molecular oxygen; secondly, the incorporated antifade agents could suppress oxygen-mediated photodegradation of **118** by scavenging reactive oxygen species (ROS). Moreover, the TADF nature of **118** together with the protective role of the matrix polymer and antifade agents bestowed the target nanoparticles with long fluorescence lifetimes. The preliminary cell fluorescence imaging results obtained in intracellular physiological milieu showed high-contrast images with the target fluorescent nanoparticles highlighted in the regions of interest, enabling high-resolution visualization of fine subcellular architecture. These results confirmed the practicability of the nanoparticles, which could be used for biological fluorescence imaging applications rather than OLED.

Yang et al. [60] reported four heterocyclic compounds **119–122** with TADF, AIE, mechanochromism, polymorphism, and on/off acidochromic properties. **120** exhibited reversible tricolor switching mechanochromic behaviors. The doped OLED of **119** obtained a moderate EQE of 11.3%. **119** was applied to ink-free rewritable paper in acid/base-responsive on/off chemosensors. **120** had two crystalline polymorphs with bright yellow and orange-yellow fluorescence, and diverse molecular packing patterns unveiled the tricolor-changing mechanochromism mechanism and different TADF features in aggregation states (Scheme 14).

With 1,8-naphthyridine as an electron acceptor, Chen et al. [61] reported three AIE-TADF emitters **123–125**. Doped OLEDs of **123**, **124,** and **125** achieved maximum EQEs of 14.1%, 13.4%, and 13.0%, respectively, as well as low efficiency roll-off. Notably, a nondoped device of **123** realized a luminance of 20,000 cd m^{-2} at a voltage of 6.5 V, and an EQE of 33.6% at 10,000 cd m^{-2} with low efficiency roll-off. The results certified that these emitters are promising for constructing high-performance OLEDs.

Yang et al. [62] reported two emitters **126** and **127**, bearing naphthyridine or cyano-naphthyridine as acceptor and acridine as the donor. Due to a nearly orthogonal molecular configuration, both emitters exhibited TADF and AIE features, with **126** possessing higher PLQYs in doped film (66%) and neat film (55%). As a result, the yellow doped and nondoped OLEDs of **126** achieved EQEs of 16.8% and 12.0%, respectively, while the orange doped device of **127** had a moderate EQE of 8.4%.

Three quinoline-based AIE-TADF molecules were reported by Chen et al. [63], including 9,9-dimethyl-10-(quinolin-2-yl)-9,10-dihydroacridine (**128**), 10-(quinolin-2-yl)-10*H*-phenoxazine (**129**), and 10-(quinolin-2-yl)-10*H*-phenothiazine (**130**). **128**, **129**, and **130** exhibited ΔE_{ST} of 0.06, 0.10, and 0.04 eV, and delay lifetimes of 2.15, 1.86, and 15.76 μs, respectively, supporting their TADF properties. The nondoped OLEDs of **128**, **129**, and **130** showed maximal EQEs of 7.7%, 17.3%, and 14.8%, respectively.

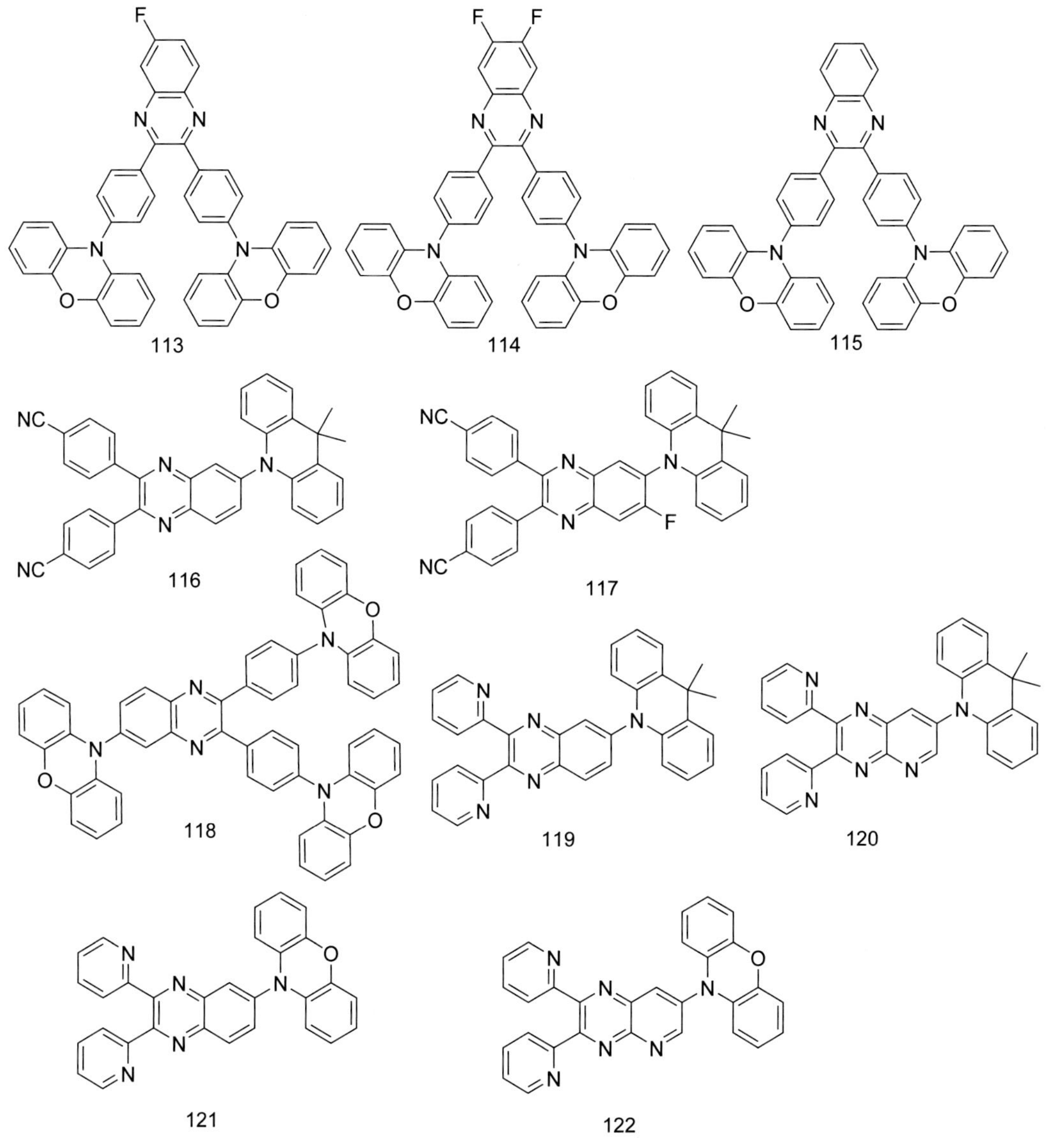

SCHEME 14 Chemical structures of compounds **113–122**.

2.4 Triarylboron derivatives

Triarylboron derivatives have a vacant p-orbital on the central boron atom, leading to attractive electron-accepting properties. D-A systems with a triarylboron acceptor and amine-based donor groups have received considerable interest because of their strong ICT properties.

Lu et al. [64] reported two blue TADF compounds, **131** and **132**, with triarylboron as acceptor and carbazole or 3,6-di-*tert*-butylcarbazole as the donor. The donor groups were ortho-substituted on one of the aryl rings and parallel to another aryl ring of triarylboron in close proximity. With the assistance of significant intramolecular D-A interactions, a combined charge transfer pathway was formed and contributed to small ΔE_{ST} and high efficiencies. Besides, the strong intramolecular interactions could suppress intramolecular vibration relaxation to boost radiationless decay. The highly twisted structures and alkyl substituents of these molecules helped to inhibit serious quenching that was usually caused by intermolecular π-π stacking, and thereby restricting concentration quenching in films. A high EQE reaching 19.1% was achieved in a nondoped blue OLED of **132** by solution processing. Therefore, the combined charge-transfer pathway is an effective strategy to develop efficient TADF materials.

In the research on blue TADF emitters, boron-fused acceptors are of particular significance, because the coordination of an external aryl group to the boron center could readily disrupt extended π-conjugation, restricting unsolicited red-shifted emission. Hatakeyama et al. [65] reported nitrogen-bridged cyclized boron-based TADF emitters, proposing a concept of multiresonance TADF. Given the opposite resonance effects of boron and nitrogen atoms, clear separation of HOMO and LUMO distributions were observed. The multiresonance effect enhanced oscillator strength between S_1 and singlet ground (S_0) states, resulting in a small full-width at half-maximum. Moreover, Choi et al. [66] investigated three blue AIE-TADF emitters **133–135** by tethering various electron-rich aromatic entities containing carbazole derivatives as electron donors to the boron-fused electron-acceptor core (TB). **133–135** exhibited excellent thermal robustness ($T_g > 270°C$). With structural modulation of donor groups, **133–135** emitted fluorescence ranging from deep blue to sky blue, with high PLQYs (90%–99%) in neat films. For solution-processed nondoped OLEDs of **133**, **134**, and **135**, maximum EQE and CIE coordinates of 9.90% and (0.17, 0.07), 6.13% and (0.15, 0.08), and 6.04% and (0.18, 0.40), respectively, were obtained. The EQE of **133** was the best one among the reported solution-processable nondoped OLEDs using deep-blue TADF emitters. Notably, **133** and **134** displayed deep-blue emission with CIE coordinates approaching the National Television Standards Committee's (NTSC) standard of blue, exhibiting a CIE y value close to 0.1, indicating the great potential of **133** and **134** as deep-blue luminogens in OLEDs (Scheme 15).

With help of through-space charge transfer, Wang et al. [67] reported a series of blue polymers **133–138** comprising of acridan donor and oxygen-bridged triphenylboron acceptors in the side chain of nonconjugated polystyrene backbone. The polymers had small ΔE_{ST} (<0.1 eV) due to physically separated HOMO and LUMO, resulting in a typical TADF effect. Meanwhile, the polymers exhibited AIE effect with ~27-fold enhancement of emission intensity from solution to aggregation states. The blue emissive polymers showed PLQYs up to 70% in solid-state film. Different substituents including tert-butyl, hydrogen, and fluorine to the acceptors were introduced for tuning charge transfer strength, while emission colors were regulated from deep blue (414 nm) to sky blue (480 nm). Solution-processed OLEDs based on these polymers presented high performance such as a maximum EQE of 15.0% with CIE coordinates of (0.16, 0.27), which was the highest efficiency among blue TADF polymers reported so far.

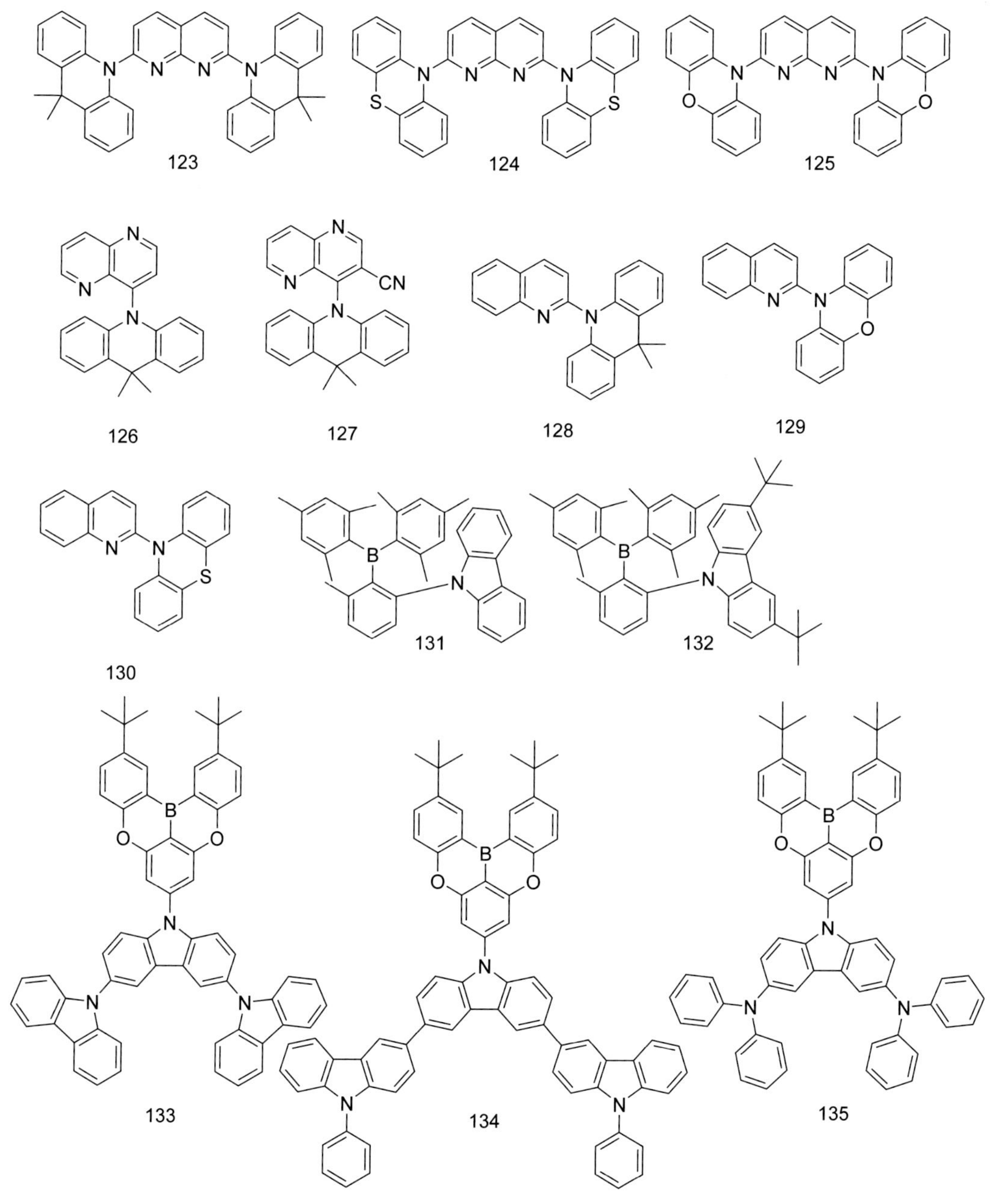

SCHEME 15 Chemical structures of compounds **123–135**.

2.5 Perfluorinated benzene derivatives

The first example by using perfluorobiphenyl as an acceptor for TADF molecule design was reported by Grazulevicius et al. [68], while nonsubstituted and tert-butyl substituted 9,9-dimethyl-9,10-dihydro-acridine functioned as the donor. Accordingly, they presented a series of multifunctional luminophores **142–145** with TADF, AIEE, and sky blue to blue color-changing features. The D-A compound without tert-butyl groups emitted deep blue light with very low efficiency of TADF, while the D-A compound with tert-butyl showed bi-color solid-state emissions (blue and sky blue) with efficient TADF. As for D-A-D type ones, the compounds with and without tert-butyl substitute exhibited efficient TADF with the sky blue and blue emissions, respectively. **143** had two crystalline polymorphs showing different emission colors, while the color could be changed under external stimuli including solvents, temperature, and mechanical treatments. According to single-crystal X-ray structural analysis, the formation of crystalline polymorphs was related to tert-butyl groups, π-π and unique C-F…π interactions and C-H…F hydrogen bonds. When used to fabricate doped and nondoped OLEDs, a doped device of **145** achieved the highest EQE of 16.3%, which was 6.6% in its nondoped one. It is noted that Song et al. [69] also made studies on the above **142–145** from a theoretical perspective. In comparison to the D-A-type molecules, it was found that the tert-butyl substituted D-A-D molecules presented higher luminous performance because of more efficient ISC and RISC processes, leading to better TADF features (Scheme 16).

2.6 Cyano-substituted benzene derivatives

Owing to strong electron-withdrawing ability, cyano (CN) group has been widely employed as a strong acceptor to design organic TADF molecules with intramolecular D-A structure, and generally is in the form of cyano-substituted benzene derivatives. Moreover, CN and derivatives groups are also utilized to achieve AIE molecules.

Choi et al. [70] reported a triad TADF molecule **146**, wherein 9-phenyl-9H-carbazole units as host were tethered to a green-emitting core 4,5-bis(3-phenyl-9H-carbazol-9-yl) phthalonitrile through a nonconjugated cyclohexane unit. **146** possessed AIE property due to the pendant phenyl-carbazole moiety at both ends, and exhibited a high PLQY of 89%, leading to a maximum EQE of 13.4% in solution-processed OLEDs.

Lin et al. [71] reported two binaphthalene-containing circularly polarized luminescence (CPL) molecules **147** and **148** with AIE and TADF features. The light-emitting properties of enantiomers were hardly influenced except for the electronic circular dichroism (ECD). The theoretical calculations considering the decay rates and adiabatic excitation energy of excited states showed that **147** and **148** had different TADF mechanisms. For **147** the upconversion process mainly happened between T_1 and S_1, while **148** involved a two-step upconversion process. **148** also demonstrated more efficient RISC than **147**, despite of a larger ΔE_{ST}. These theoretical calculations suggested a higher luminous efficiency of **148** than **147**, which matched well with the experimental results (Scheme 17).

Tang et al. [72] reported three AIE compounds **149–151**, which were transformed from an ACQ molecule 2,3,4,5,6-penta(9H-carbazol-9-yl)benzonitrile (5CzBN) by simply modifying 5CzBN core with alkyl chain-linked spirobifluorene dendrons. As the number of flexible

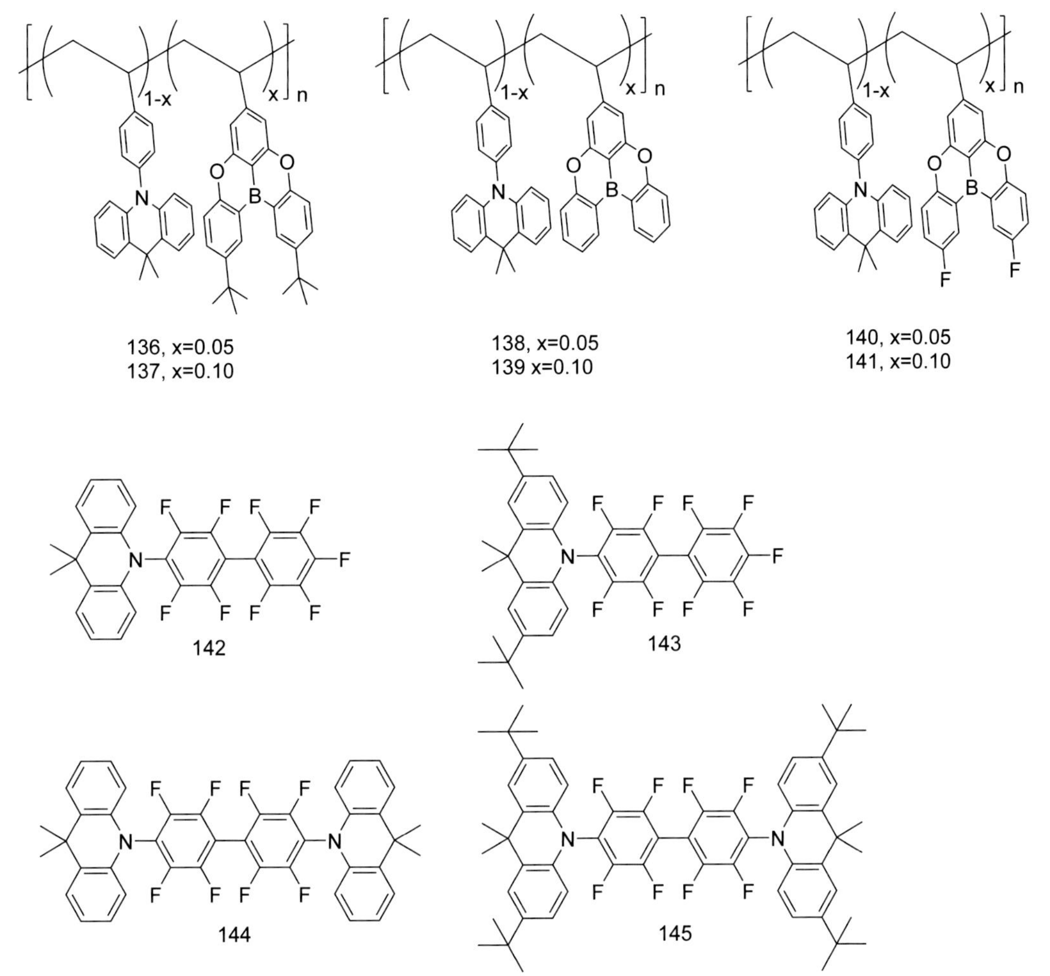

136, x=0.05
137, x=0.10

138, x=0.05
139 x=0.10

140, x=0.05
141, x=0.10

142

143

144

145

SCHEME 16 Chemical structures of compounds **136–145**.

dendrons increased, the compounds not only showed improved AIE-TADF property and uniform film morphology but also enhanced resistance to isopropyl alcohol, which was advantageous for constructing fully solution-processed OLEDs. **151**-based OLED exhibited a high EQE of 20.1%, which outperformed that of fully solution-processed OLEDs with traditional TADF materials. This result indicated that the common fluorophores even ACQ molecules could be transformed into new AIE molecules by adjusting the number of alkyl-chain linked spirobifluorene dendrons that were attached to the TADF cores, thus opening up a new route to develop efficient AIE-TADF emitters (Scheme 18).

Jiang et al. [73] reported two TADF luminogens **152** and **153**, derived from 2,4,5,6-tetra (carbazol-9-yl)-1,3-dicyanobenzene (4CzIPN) framework. Methoxy groups were employed to decorate the carbazole fragment of 4CzIPN, phenyl-bridges were inserted between the carbazole and central benzene ring, and AIE behavior was observed in **153**. The phenyl-bridge

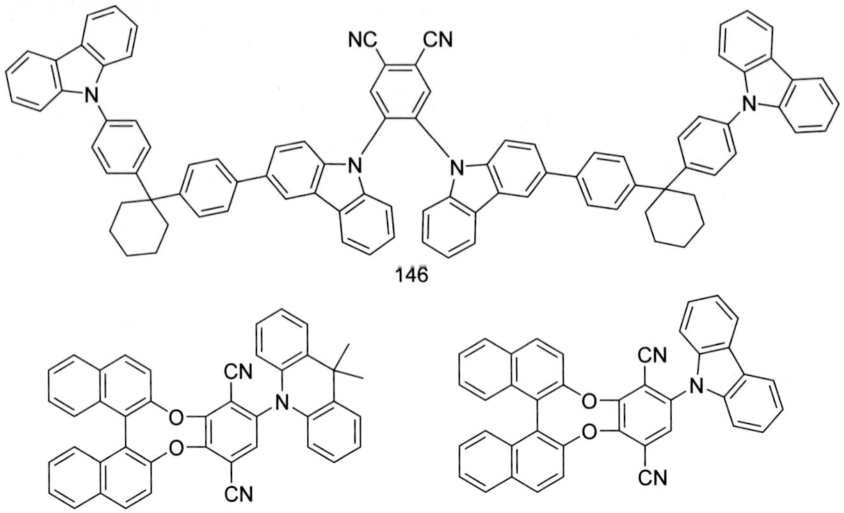

SCHEME 17 Chemical structures of compounds **146–148**.

acting like an axle helped to alleviate the sterically hindering effect of donor units, resulting in the AIE property. Accordingly, the nondoped solution-processed OLED of **153** provided a high EQE of 14.5% with low efficiency roll-off. In addition, the doped solution-processed OLED of **152** presented a high EQE of 16.2% but with serious efficiency roll-off at high currents.

2.7 Imide derivatives

Imide groups with strong electron-withdrawing ability had been widely adopted to design organic semiconductors with excellent ICT features, such as polyimides. As one of the imide groups, 1,8-naphthalimide had been used as an acceptor for D-A-type TADF emitters. Nevertheless, the reported 1,8-naphthalimide derivatives were orange-red emissive (emission peaks $\leq$600 nm) without AIE property.

In 2019, Chi et al. [74] synthesized two D-π-A-type 1,8-naphthalimide derivatives **154** and **155**. Two nonplanar strong electron-rich groups including 10-H-phenothiazine and 9,9-dimethyl-9,10-dihydroacridine acted as donors, and 2,6-dimethylphenyl (DM) group was used as a π bridge. It was revealed that the separated 1,8-naphthalimide and donor units were almost parallel to each other but they were both nearly vertical to the π bridge, and such a twist structure facilitated HOMO and LUMO separation to induce a small ΔE_{ST} triggering TADF emission. Additionally, the π bridge possessed strong steric hindrance regarding both 1,8-naphthalimide and donor groups, contributing to more rigid molecules, thus restraining molecular vibration and suppressing nonradiative processes in solid state. Therefore, **154** and **155** exhibited AIE properties, giving out red (624 nm) and orange-red (595 nm) emissions with PLQYs of 55% and 39% in solid state, respectively. By using **154**-doped films for OLED fabrication, an EQE of 7.13% with an EL peak around 635 nm was achieved. The easily available

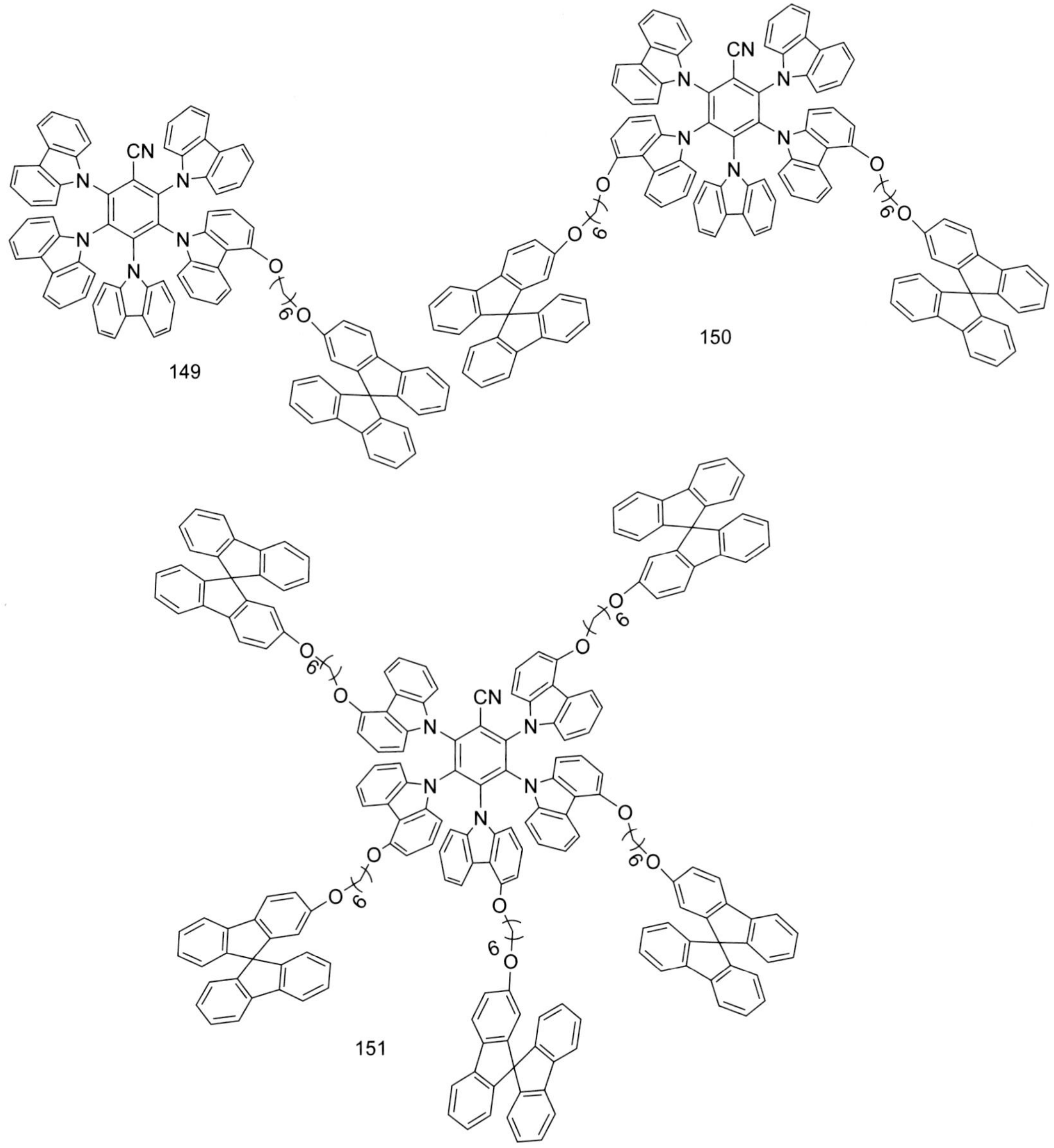

SCHEME 18 Chemical structures of compounds **149–151**.

1,8-naphthalimide derivatives provided a simple and effective approach to develop AIE-TADF red emitters.

Peng et al. [75] developed two orange to red emitters **156** and **157** with a central naphthalimide acceptor core and arylamine donor units. In comparison to the D-A-type **156**, a phenyl bridge (π-bridge) was introduced in D-π-A-type **157**, while the π-bridge played a positive role in PLQY, ΔE_{ST} and k_{RSIC}, due to large steric hindrance and dihedral angles. The **157**-doped film had a higher PLQY (87%) than that of **156** (62%). The nondoped OLEDs of **156** and **157** exhibited maximum EQEs of 1.53% and 1.39%, respectively, with CIE coordinates of (0.64, 0.35) and (0.53, 0.46), respectively, while another nondoped OLED of **156** showed CIE coordinates of (0.65, 0.34), close to the NTSC standard of red (0.67, 0.33). The doped OLEDs of **156** and **157** with low doping concentrations offered higher EQEs of 4.81% and 7.59%, respectively (Scheme 19).

By using a novel heptagonal diimide acceptor, (*N*-(4-(tert-butyl)phenyl)-1,1′-biphenyl-2,2′-dicarboximide, You et al. [76] reported a D-A-D triad **158**. The heptagonal structure was endowed with well-balanced rigidity and rotatability, and the rigidity contributed to restriction of excessive intramolecular rotation to promote radiative transition rate, while moderate rotatability impeded close intermolecular π-π stacking to suppress exciton quenching in the aggregated state. The theoretical calculation revealed a highly twisted geometry of **158** with well-separated HOMO and LUMO, leading to a calculated ΔE_{ST} of 0.02 eV that was low enough to trigger TADF. **158** neat film showed an AIDF feature with a high PLQY of 95.8%, which enabled its nondoped OLEDs with a high EQE of 24.7% and very low efficiency roll-off.

2.8 *o*-Carborane derivatives

o-Carborane is known as an electron-deficient icosahedral boron cluster with three-center two-electron bonds and exhibits a highly polarizable *s*-aromatic character. This unique feature allows the carborane cage to interact electronically with π-conjugated systems attached through carbon atoms. Luminescent materials consisting of boron clusters, such as carboranes, have attracted immense interest in recent years [77–79].

Yasuda et al. [77] provided a novel design strategy for AIE-TADF materials and reported an organic-inorganic conjugated system (**159–161**) by directly bonding *o*-carboranes to π-conjugated donor and acceptor units. These *o*-carborane derivatives had both AIE and TADF properties, emitting strong yellow-to-red lights in aggregated states or neat films. As for the OLEDs based on **159**, **160**, and **161**, the maximum EQEs were obtained to be 11.0%, 10.1%, and 9.2%, respectively. Su et al. [78] also made investigations on the two AIE-TADF *o*-carborane derivatives **159** and **160**. Theoretical studies suggested that the separated HOMO and LUMO led to small ΔE_{ST} and favorable TADF features. With regard to the AIE phenomenon, in comparison to the isolated molecule, the molecule in aggregated state demonstrated reduced energy dissipation and enhanced luminescence efficiency. Moreover, they also replaced the acceptor units of **159** and **160** with two electron-withdrawing groups and obtained **162–165**, which were supposed with TADF and AIE properties and red-emissive materials as expected (Scheme 20).

Although carboranes are well-known boron clusters with unique properties to enable optoelectronic applications. The molecular design of carborane-containing luminophores remains a great challenge. Lee et al. [79] reported two series of nido-carborane-appended triarylboranes, with the nido-carborane being meta- or para-substituted on the phenyl ring of dimesitylphenylborane acceptor, affording in D-A compounds with various 8-R groups (R=H, Me, *i*-Pr, pH) (**166–169** and **170–173**). According to crystal structures and optimized geometries studies, the meta-substituted **167–169** showed highly twisted connectivity between D and A units, whereas the para-substituted **170–173**, except for nido-p3 with a bulky *i*-Pr group, exhibited relatively small torsion angles between

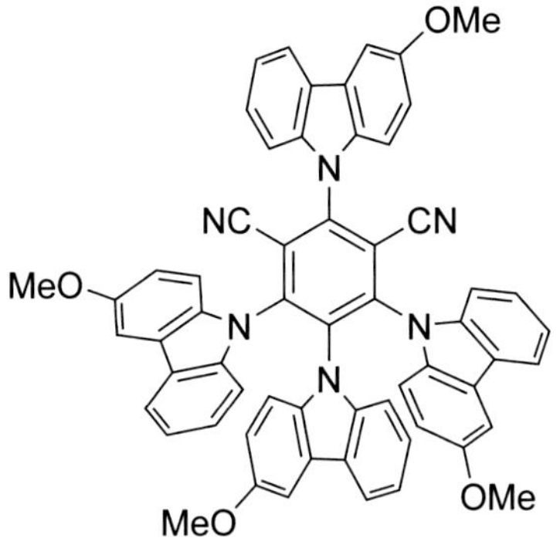

OMe
NC CN
MeO N N N
MeO OMe
152

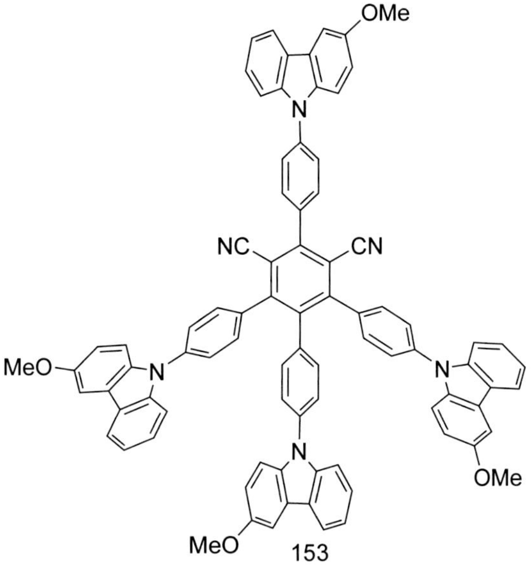

OMe
N
NC CN
MeO N N
N
MeO OMe
153

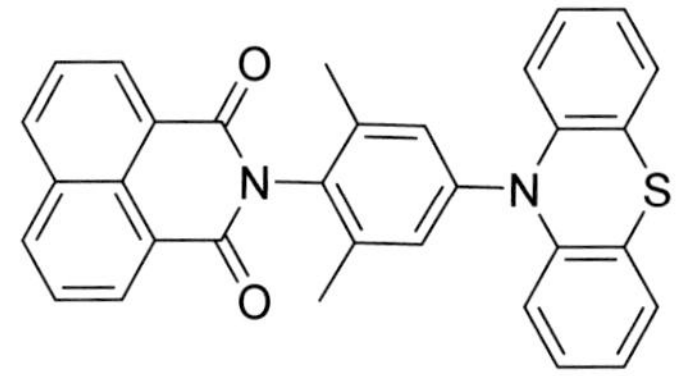

O N S O
154

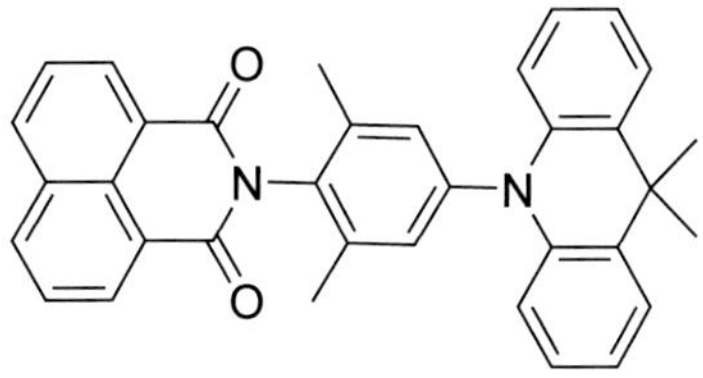

O N N O
155

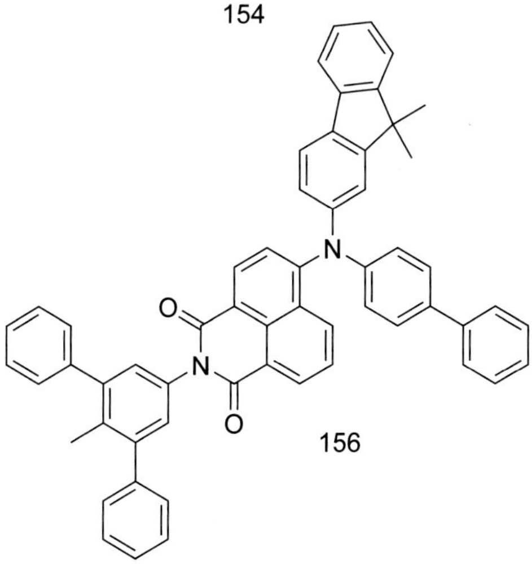

N O N O
156

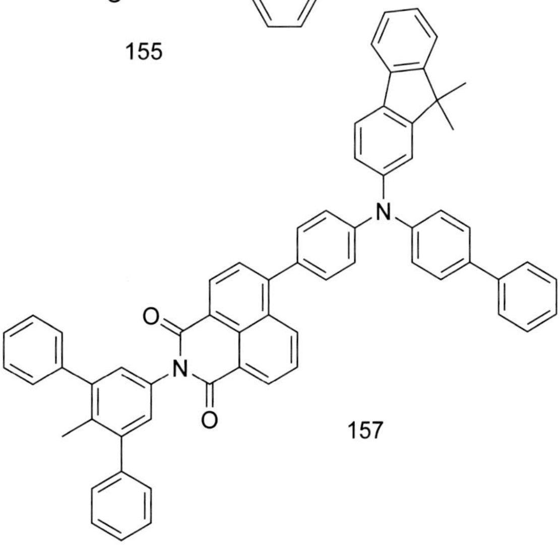

N O N O
157

SCHEME 19 Chemical structures of compounds **152–157**.

SCHEME 20 Chemical structures of compounds **158–165**.

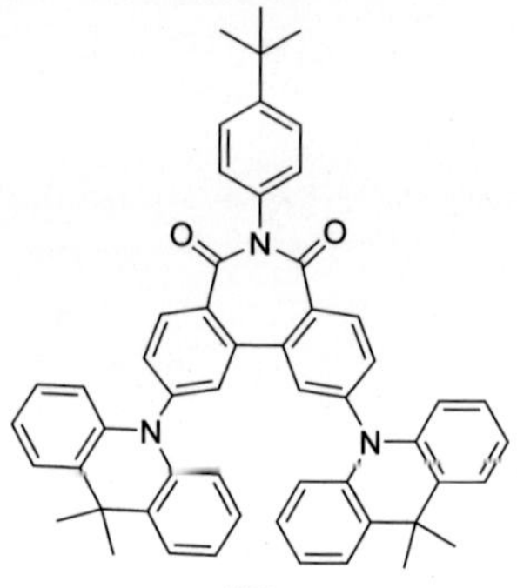

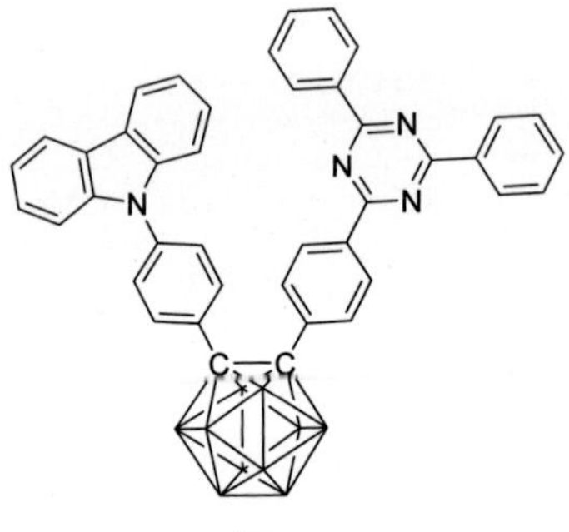

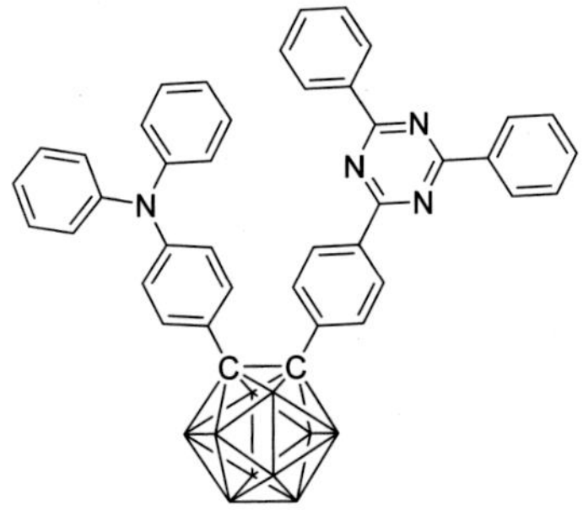

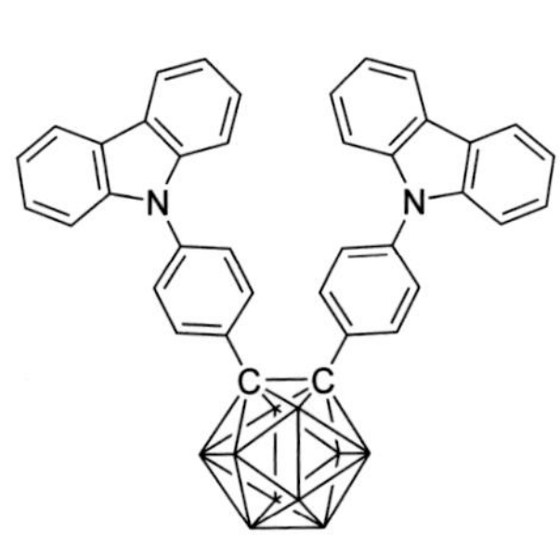

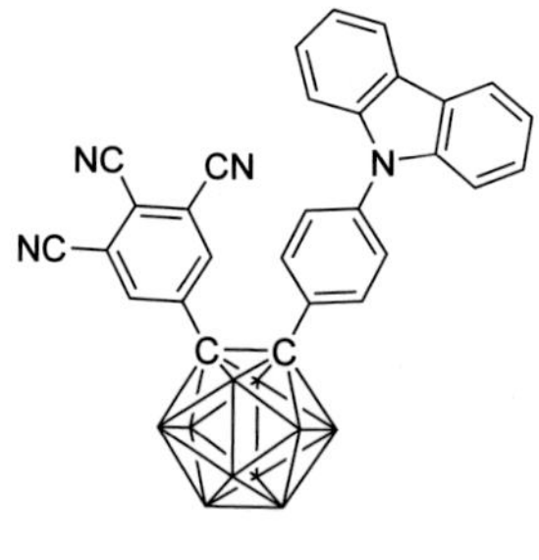

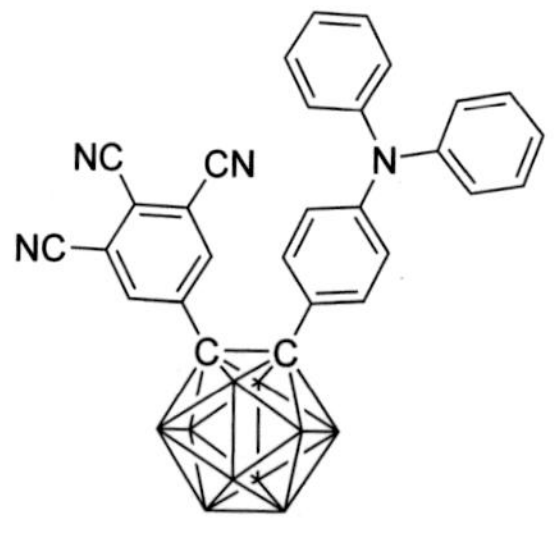

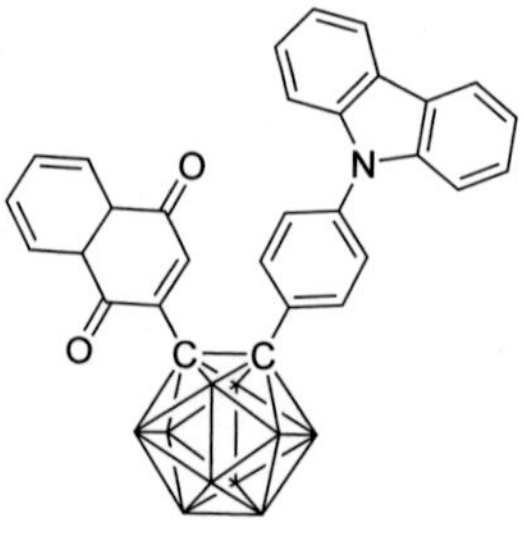

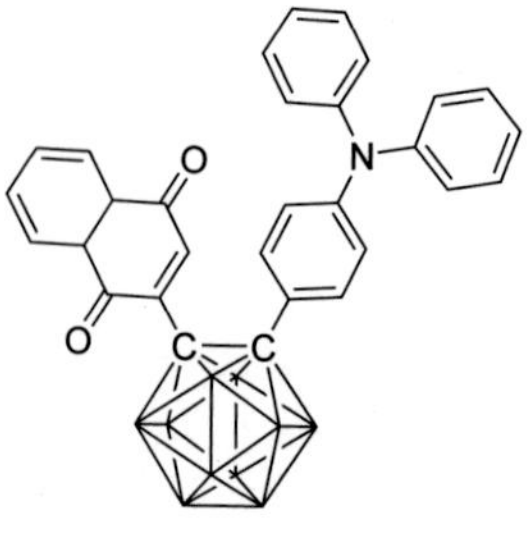

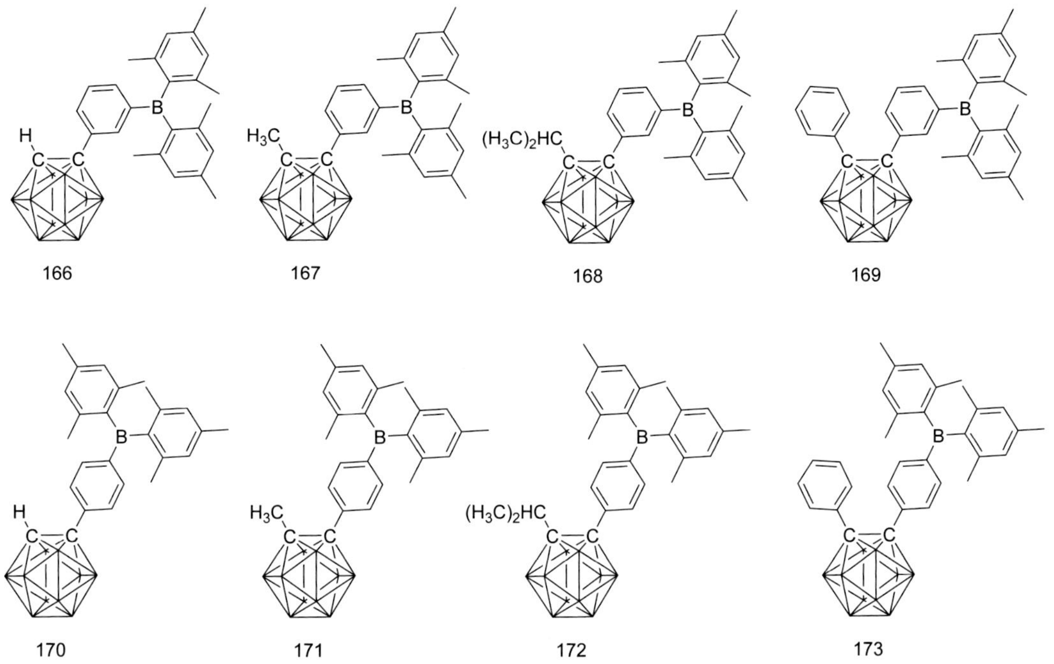

166 167 168 169

170 171 172 173

SCHEME 21 Chemical structures of compounds **166–173**.

nido-carborane and phenylene ring. Small ΔE_{ST} values (<0.05 eV) were found in **167**, **168**, and **172**, producing efficient TADF, while AIE behaviors were also observed in **167** and **168**. These findings highlight that both the substituted position and the 8-R group of nido-carborane have significant effects on the TADF properties of compounds (Scheme 21).

2.9 Organometallic complexes

Wang et al. [80] reported two isomeric Cu complexes (**174** and **175**) with AIE and TADF features. Theoretical studies indicated that the geometrical changes between S_0 and S_1 were hindered in the solid phase by the RIR effect, leading to smaller Huang-Rhys factors. Moreover, the dihedral angle was reduced, decreasing the reorganization energy from solution to solid phase. Consequently, nonradiative energy consumption process of S_1 was hindered and AIE property was realized in **174** and **175**. In comparison to **175**, more efficient TADF was observed in **174** with stronger intramolecular and intermolecular interactions, which contributed to reduced nonradiative decay rate and increased IRSC rate in **174**. The calculations reasonably elucidated the experimental results and helped to understand the AIE and TADF mechanisms of these Cu complexes (Scheme 22).

SCHEME 22 Chemical structures of compounds **174–175**.

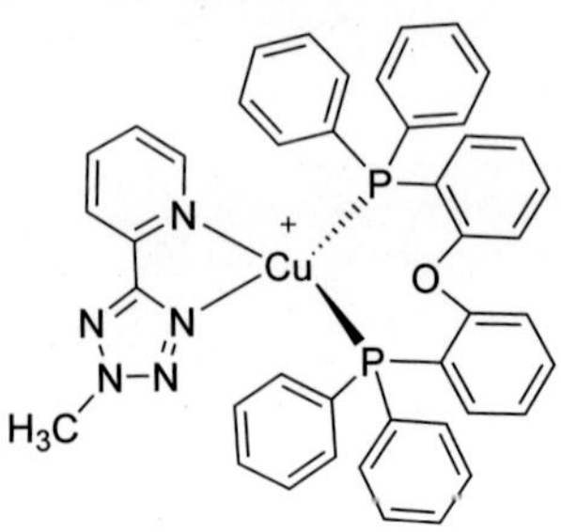

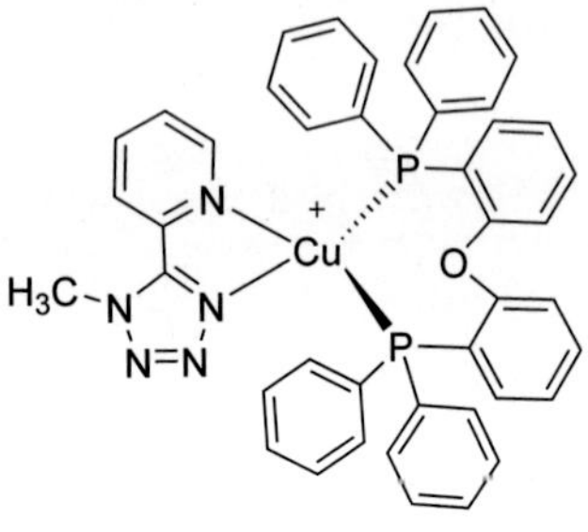

3 Conclusions and outlook

The AIE-TADF research is developing fastly in recent years, resulting in the accumulation of a wealth of information on the structural design of AIE-TADF luminogens and mechanism understanding of the TADF processes. Some reviews are very helpful for interested readers to retrieve [81–84]. Although much work has been done, many possibilities still remain to be explored. For example, new molecular design strategies are in urgent need, and novel TADF units that are different from traditional ones are worth exploring. Systematic theoretical studies could be of great help for revealing explicit mechanisms of TADF emitters. Also, the EL performance of TADF-based OLEDs, particularly solution-processed ones should be further enhanced to enable its practical applications. Moreover, attention can be paid to the potential applications of TADF materials in other fields, such as organic photodetectors, bioimaging, and fluorescence probes. We enthusiastically look forward to new advancements in this exciting area of research. In this chapter, we provide a comprehensive summary of the development of AIE-TADF materials, and we hope it may provide a clear panorama of these novel functional materials to researchers in different areas and attract more researchers to devote to this interesting field.

Acknowledgment

This work was financially supported by the National Natural Science Foundation of China (NSFC: 51733010, 52073316, 51973239) and the Fundamental Research Funds for the Central Universities.

References

[1] H. Uoyama, K. Goushi, K. Shizu, H. Nomura, C. Adachi, Highly efficient organic light-emitting diodes from delayed fluorescence, Nature 492 (2012) 234–238.
[2] S. Xu, T. Liu, Y. Mu, Y.F. Wang, Z. Chi, C.C. Lo, S. Liu, Y. Zhang, A. Lien, J. Xu, Angew. Chem. Int. Ed. 54 (2015) 874–878.
[3] J. Zhao, X. Chen, Z. Yang, T. Liu, Z. Yang, Y. Zhang, J. Xu, Z. Chi, J. Phys. Chem. C 123 (2019) 1015–1020.
[4] Z. Yang, Z. Mao, C. Xu, X. Chen, J. Zhao, Z. Yang, Y. Zhang, W. Wu, S. Jiao, Y. Liu, M.P. Aldred, Z. Chi, Chem. Sci. 10 (2019) 8129–8134.
[5] I.H. Lee, W. Song, J.Y. Lee, Org. Electron. 29 (2016) 22–26.
[6] S. Gan, W. Luo, B. He, L. Chen, H. Nie, R. Hu, A. Qin, Z. Zhao, B.Z. Tang, J. Mater. Chem. C 4 (2016) 3705–3708.

[7] Q. Zhang, D. Tsang, H. Kuwabara, Y. Hatae, B. Li, T. Takahashi, S.Y. Lee, T. Yasuda, C. Adachi, Adv. Mater. 27 (2015) 2096–2100.
[8] J. Li, R. Zhang, Z. Wang, B. Zhao, J. Xie, F. Zhang, H. Wang, K. Guo, Adv. Opt. Mater. 6 (2018) 1701256.
[9] K. Wu, Z. Wang, L. Zhang, C. Zhong, S. Gong, G. Xie, C. Yang, J. Phys. Chem. Lett. 9 (2018) 1547–1553.
[10] L. Zhan, Y. Xiang, Z. Chen, K. Wu, S. Gong, G. Xie, C. Yang, J. Mater. Chem. C 7 (2019) 13953–13959.
[11] L. Zhan, Z. Chen, S. Gong, Y. Xiang, F. Ni, X. Zeng, G. Xie, C. Yang, Angew. Chem. Int. Ed. 58 (2019) 17651–17655.
[12] P. Leng, S. Sun, R. Guo, Q. Zhang, W. Liu, X. Lv, S. Ye, L. Wang, Org. Electron. 78 (2020), 105602.
[13] R. Guo, P. Leng, Q. Zhang, Y. Wang, X. Lv, S. Sun, S. Ye, Y. Duan, L. Wang, Dyes Pigments 184 (2021), 108781.
[14] S. Zheng, T. Liu, Z. Song, Z. He, Z. Yang, H. Wang, Z. Zeng, Dyes Pigments 176 (2020), 108204.
[15] S. Xiang, Z. Huang, S. Sun, X. Lv, L. Fan, S. Ye, H. Chen, R. Guo, L. Wang, J. Mater. Chem. C 6 (2018) 11436–11443.
[16] H. Sun, X. Yin, Z. Liu, S. Wei, J. Fan, L. Lin, Y. Sun, Int. J. Quantum Chem. 121 (2020), e26490.
[17] K. Sun, W. Jiang, X. Ban, B. Huang, Z. Zhang, M. Ye, Y. Sun, RSC Adv. 6 (2016) 22137–22143.
[18] X. Dong, S. Wang, C. Gui, H. Shi, F. Cheng, B.Z. Tang, Tetrahedron 74 (2018) 497–505.
[19] Z. Shi, X. Zhang, H. Wang, J. Huo, H. Zhao, H. Shi, B.Z. Tang, Org. Electron. 70 (2019) 7–13.
[20] W.Z. Yuan, X.Y. Shen, H. Zhao, J.W.Y. Lam, L. Tang, P. Lu, C. Wang, Y. Liu, Z. Wang, Q. Zheng, J.Z. Sun, Y. Ma, B.Z. Tang, J. Phys. Chem. C 114 (2010) 6090–6099.
[21] N. Aizawa, C.-J. Tsou, I.S. Park, T. Yasudas, Polym. J. 49 (2016) 197–202.
[22] J. Hu, X. Zhang, D. Zhang, X. Cao, T. Jiang, X. Zhang, Y. Tao, Dyes Pigments 137 (2017) 480–489.
[23] J. Guo, X.-L. Li, H. Nie, W. Luo, S. Gan, S. Hu, R. Hu, A. Qin, Z. Zhao, S.-J. Su, B.Z. Tang, Adv. Funct. Mater. 27 (2017) 1606458.
[24] J. Fan, L. Lin, C. Wang, J. Mater. Chem. C 5 (2017) 8390–8399.
[25] J. Guo, X. Li, H. Nie, W. Luo, R. Hu, A. Qin, Z. Zhao, S. Su, B.Z. Tang, Chem. Mater. 29 (2017) 3623–3631.
[26] J. Huang, H. Nie, J. Zeng, Z. Zhuang, S. Gan, Y. Cai, J. Guo, S. Su, Z. Zhao, B.Z. Tang, Angew. Chem. Int. Ed. 56 (2017) 12971–12976.
[27] J. Xu, X. Zhu, J. Guo, J. Fan, J. Zeng, S. Chen, Z. Zhao, B.Z. Tang, ACS Mater. Lett. 1 (2019) 613–619.
[28] X. Chen, Z. Yang, Z. Xie, J. Zhao, Z. Yang, Y. Zhang, M.P. Aldred, Z. Chi, Mater. Chem. Front. 2 (2018) 1017–1023.
[29] J. Zeng, J. Guo, H. Liu, J.W.Y. Lam, Z. Zhao, S. Chen, B.Z. Tang, Chem. Asian J. 14 (2019) 828–835.
[30] J. Guo, J. Fan, L. Lin, J. Zeng, H. Liu, C. Wang, Z. Zhao, B.Z. Tang, Adv. Sci. 6 (2019) 1801629.
[31] J. Huang, Z. Xu, Z. Cai, J. Guo, J. Guo, P. Shen, Z. Wang, Z. Zhao, D. Ma, B.Z. Tang, J. Mater. Chem. C 7 (2019) 330–339.
[32] H. Liu, J. Zeng, J. Guo, H. Nie, Z. Zhao, B.Z. Tang, Angew. Chem. Int. Ed. 57 (2018) 9290–9294.
[33] Y. Zhao, W. Wang, C. Gui, L. Fang, X. Zhang, S. Wang, S. Chen, H. Shi, B.Z. Tang, J. Mater. Chem. C 6 (2018) 2873–2881.
[34] Y. Chen, S. Wang, X. Wu, Y. Xu, H. Li, Y. Liu, H. Tong, L. Wang, J. Mater. Chem. C 6 (2018) 12503–12508.
[35] F. Ma, G. Zhao, Y. Zheng, F. He, K. Hasrat, Z. Qi, ACS Appl. Mater. Interfaces 12 (2020) 1179–1189.
[36] F. Ma, Y. Cheng, Y. Zheng, H. Ji, K. Hasrat, Z. Qi, J. Mater. Chem. C 7 (2019) 9413–9422.
[37] X. Zhou, Y. Xiang, F. Ni, Y. Zou, Z. Chen, X. Yin, G. Xie, S. Gong, C. Yang, Dyes Pigments 176 (2019), 108179.
[38] Y. Liu, X. Wu, Y. Chen, L. Chen, H. Li, W. Wang, S. Wang, H. Tian, H. Tong, L. Wang, J. Mater. Chem. C 7 (2019) 9719–9725.
[39] F. Ma, X. Zhao, H. Ji, D. Zhang, K. Hasrat, Z. Qi, J. Mater. Chem. C 8 (2020) 12272–12283.
[40] H. Liu, H. Liu, J. Fan, J. Guo, J. Zeng, F. Qiu, Z. Zhao, B.Z. Tang, Adv. Opt. Mater. 8 (2020) 2001027.
[41] Y. Fu, H. Liu, X. Zhu, J. Zeng, Z. Zhao, B.Z. Tang, J. Mater. Chem. C 8 (2020) 9549–95573.
[42] Y. Zhang, Y. Ma, K. Zhang, Y. Song, L. Lin, C. Wang, J. Fan, Spectrochim. Acta A Mol. Biomol. Spectrosc. 241 (2020), 118634.
[43] Z. Cai, H. Chen, J. Guo, Z. Zhao, B.Z. Tang, Front. Chem. 8 (2020) 193.
[44] Z. Yang, Y. Zhan, Z. Qiu, J. Zeng, J. Guo, S. Hu, Z. Zhao, X. Li, S. Ji, Y. Huo, S. Su, ACS Appl. Mater. Interfaces 12 (2020) 29528–29539.
[45] J. Lee, N. Aizawa, M. Numata, C. Adachi, T. Yasuda, Adv. Mater. 29 (2017) 1604856.
[46] B. Huang, Y. Ji, Z. Li, N. Zhou, W. Jiang, Y. Feng, B. Lin, Y. Sun, J. Lumin. 187 (2017) 414–420.
[47] B. Huang, Z. Li, H. Yang, D. Hu, W. Wu, Y. Feng, Y. Sun, B. Lin, W. Jiang, J. Mater. Chem. C 5 (2017) 12031–12034.
[48] B. Huang, W. Chen, Z. Li, J. Zhang, W. Zhao, Y. Feng, B.Z. Tang, C. Lee, Angew. Chem. Int. Ed. 57 (2018) 12473–12477.
[49] C. Liu, J. Chen, C. Xu, H. Hao, B. Xu, D. Hu, G. Shi, Z. Chi, Dyes Pigments 174 (2020), 108093.
[50] W.L. Tsai, M.H. Huang, W.K. Lee, Y.J. Hsu, K.C. Pan, Y.H. Huang, H.C. Ting, M. Sarma, Y.Y. Ho, H.C. Hu, C.C. Chen, M.T. Lee, K.T. Wong, C.C. Wu, Chem. Commun. 51 (2015) 13662–13665.

[51] H. Tsujimoto, D.G. Ha, G. Markopoulos, H.S. Chae, M.A. Baldo, T.M. Swager, J. Am. Chem. Soc. 139 (2017) 4894–4900.
[52] P. Zhang, J. Zeng, J. Guo, S. Zhen, B. Xiao, Z. Wang, Z. Zhao, B.Z. Tang, Front. Chem. 7 (2019) 199.
[53] X. Wang, S. Wang, J. Lv, S. Shao, L. Wang, X. Jing, F. Wang, Chem. Sci. 10 (2019) 2915–2923.
[54] J. Wang, C. Liu, C. Jiang, C. Yao, M. Gu, W. Wang, Org. Electron. 65 (2019) 170–178.
[55] Q. Zhang, S. Sun, W. Liu, P. Leng, X. Lv, Y. Wang, H. Chen, S. Ye, S. Zhuang, L. Wang, J. Mater. Chem. C 7 (2019) 9487–9495.
[56] L. Yu, Z. Wu, G. Xie, W. Zeng, D. Ma, C. Yang, Chem. Sci. 9 (2018) 1385–1391.
[57] L. Yu, Z. Wu, G. Xie, C. Zhong, Z. Zhu, D. Ma, C. Yang, Chem. Commun. 54 (2018) 1379–1382.
[58] K. Zheng, F. Ni, Z. Chen, C. Zhong, C. Yang, Angew. Chem. Int. Ed. 59 (2020) 9972–9976.
[59] X. Luo, J. Meng, B. Li, A. Peng, Z. Tian, New J. Chem. 43 (2019) 10735–10743.
[60] W. Yang, Y. Yang, L. Zhan, K. Zheng, Z. Chen, X. Zeng, S. Gong, C. Yang, Chem. Eng. J. 390 (2020), 124626.
[61] C. Chen, H. Lu, Y. Wang, M. Li, Y. Shen, C. Chen, J. Mater. Chem. C 7 (2019) 4673–4680.
[62] X. Zhou, H. Yang, Z. Chen, S. Gong, Z. Lu, C. Yang, J. Mater. Chem. C 7 (2019) 6607–6615.
[63] Y. Shen, M. Li, W. Zhao, Y. Wang, H. Lu, C. Chen, Mater. Chem. Front. 5 (2021) 834–842.
[64] X.L. Chen, J.H. Jia, R. Yu, J.Z. Liao, M.X. Yang, C.Z. Lu, Angew. Chem. Int. Ed. 56 (2017) 15006–15009.
[65] T. Hatakeyama, K. Shiren, K. Nakajima, S. Nomura, S. Nakatsuka, K. Kinoshita, J. Ni, Y. Ono, T. Ikuta, Adv. Mater. 28 (2016) 2777–2781.
[66] H.J. Kim, M. Godumala, S.K. Kim, J. Yoon, C.Y. Kim, H. Park, J.H. Kwon, M.J. Cho, D.H. Choi, Adv. Opt. Mater. 8 (2020) 1902175.
[67] F. Chen, J. Hu, X. Wang, S. Shao, L. Wang, X. Jing, F. Wang, Sci. China Chem. 63 (2020) 1112–1120.
[68] I. Hladka, D. Volyniuk, V. Kinzhybalo, O. Bezvikonnyi, T.J. Bednarchuk, Y. Danyliv, A. Lazauskasd, J.V. Grazulevicius, J. Mater. Chem. C 6 (2018) 13179–13189.
[69] Q. Lu, G. Jiang, F. Li, L. Lin, C. Wang, J. Fan, Y. Song, Spectrochim. Acta A Mol. Biomol. Spectrosc. 229 (2020), 117964.
[70] H.J. Kim, S.K. Kim, M. Godumala, J. Yoon, C.Y. Kim, J.E. Jeong, H.Y. Woo, J.H. Kwon, M.J. Cho, D.H. Choi, Chem. Commun. 55 (2019) 9475–9478.
[71] X. Yin, J. Fan, J. Liu, L. Cai, H. Sun, Y. Sun, C. Wang, L. Lin, Phys. Chem. Chem. Phys. 21 (2019) 7288–7297.
[72] D. Liu, J.Y. Wei, W.W. Tian, W. Jiang, Y.M. Sun, Z. Zhao, B.Z. Tang, Chem. Sci. 11 (2020) 7194–7203.
[73] K. Sun, D. Liu, W. Tian, F. Gu, W. Wang, Z. Cai, W. Jiang, Y. Sun, J. Mater. Chem. C 8 (2020) 11850–11859.
[74] Y. Wu, X. Chen, Y. Mu, Z. Yang, Z. Mao, J. Zhao, Z. Yang, Y. Zhang, Z. Chi, Dyes Pigments 169 (2019) 81–88.
[75] S. Chen, P. Zeng, W. Wang, X. Wang, Y. Wu, P. Lin, Z. Peng, J. Mater. Chem. C 7 (2019) 2886–2897.
[76] Z. Huang, Z. Bin, R. Su, F. Yang, J. Lan, J. You, Angew. Chem. Int. Ed. 59 (2020) 9992–9996.
[77] R. Furue, T. Nishimoto, I.S. Park, J. Lee, T. Yasuda, Angew. Chem. Int. Ed. 55 (2016) 7171–7175.
[78] Y. Duan, Y. Gao, Y. Geng, Y. Wu, G. Shan, L. Zhao, M. Zhang, Z. Su, J. Mater. Chem. C 7 (2019) 2699–2709.
[79] N.V. Nghia, S. Jana, S. Sujith, J.Y. Ryu, J. Lee, S.U. Lee, M.H. Lee, Angew. Chem. Int. Ed. 57 (2018) 12483–12488.
[80] J. Fan, Y. Zhang, K. Zhang, J. Liu, G. Jiang, L. Lin, C. Wang, Org. Electron. 71 (2019) 113–122.
[81] Z. Yang, Z. Mao, Z. Xie, Y. Zhang, S. Liu, J. Zhao, J. Xu, Z. Chi, M.P. Aldred, Chem. Soc. Rev. 46 (2017) 915–1016.
[82] J. Guo, Z. Zhao, B.Z. Tang, Adv. Opt. Mater. 6 (2018) 1800264.
[83] F. Rizzo, F. Cucinotta, ISR J. Chem. 58 (2018) 874–888.
[84] D. Barman, R. Gogoi, K. Narang, P.K. Iyer, Front. Chem. 8 (2020) 483.

CHAPTER

10

Aggregation-induced emission luminogens for organic light-emitting diodes

Suraj Kumar Pathak and Chuluo Yang

College of Materials Science and Engineering, Shenzhen University, Shenzhen, China

1 Introduction

Organic light-emitting diodes (OLEDs) are panel display and solid-state lighting devices that are based on the principle of energy released in the form of light from mutual annihilation of electrically generated holes and electrons. OLEDs are considered as a radically innovative and relatively fast-growing technology for flexible, high-resolution displays and solid-state lighting, appealing deeply to the scientific and industrial community [1]. Fluorescent OLEDs are known as the first-generation OLEDs where traditional fluorophores are used as emitters. Fluorescent OLEDs have been extensively studied thanks to their outstanding stability and somewhat prolonged lifetime [2]. The external quantum efficiency of an OLED (η_{ext}) can be established using the mathematical expression $\eta_{ext} = \eta_{int} \times \eta_{out}$, where η_{int} represents internal quantum efficiency and η_{out} stands for light out-coupling factor. Usually, for OLEDs without the optical out-coupling layer, the η_{out} is approximately 20%–30%. The equation, $\eta_{int} = \gamma \times \beta \times \Phi_{PL}$, can be employed to obtain η_{int} value, where the carrier balance ratio of holes and electrons is designated as γ, the fraction of excitons that are capable of radiative decay is designated as β, and the intrinsic photoluminescence quantum yield (PLQY) of the emitting layer is Φ_{PL} [3]. According to the spin statistics theorem, out of the total excitons generated by electrical excitation, 25% constitute singlet excitons and the remaining 75% constitute triplet excitons. The singlet excitons can radiatively decay to ground state as fluorescence. For that reason, β for conventional fluorescent OLEDs harvesting only the singlet excitons is restricted to 25% and the maximum internal quantum efficiency (η_{int}) is limited to 25% and the theoretical maximum η_{ext} of the device is about 5%–7.5%, even though γ and Φ_{PL} of a fluorophore are

https://doi.org/10.1016/B978-0-12-824335-0.00016-7

in full potential [3]. Therefore, it is highly desirable to develop efficient and stable fluorescent materials with high Φ_{PL} values. Notably, conventional fluorophores typically emit strongly in dilute solution or in isolated states; nevertheless, partial or complete quenching effect is observed in the aggregated state, which is coined as aggregation-caused quenching (ACQ) [4]. The ACQ effect is thought to be regulated by the creation of delocalized excitons via strong intermolecular π-π stacking interactions, resulting in red-shifted emissions and low Φ_{PL} values [4]. The ACQ effect plays a major role in lowering the Φ_{PL} and hence impedes the application of conventional fluorophores toward highly efficient OLEDs.

Aggregation-induced emission (AIE) is a novel photophysical phenomenon where the aggregation of chromophores induces strong fluorescence instead of emission quenching, which is basically the reverse of ACQ [5]. The typical AIE phenomenon can be vividly demonstrated by the classical AIE luminogen (AIEgen), 1,1,2,3,4,5-hexaphenylsilole (HPS) [6], which becomes highly fluorescent upon aggregation as shown in Fig. 1.

Over the last two decades, much effort has been devoted to the thorough understanding of the AIE phenomenon via methodical experiments and theoretical interpretation. These investigations point out that the AIE phenomenon is mostly attributed to the restriction of intramolecular motions (RIM), including the restriction of intramolecular rotations (RIR) as well as the restriction of intramolecular vibrations (RIV) [6]. In general, molecules undergo active intramolecular motions in the solution state, resulting in the energetic decay from the excited state to the ground state via nonradiative channels, and quench the emission. Conversely, these types of molecular motions are significantly restricted in the aggregated state, which prevents the energy dissipation via nonradiative decay channels and yields the efficient emission.

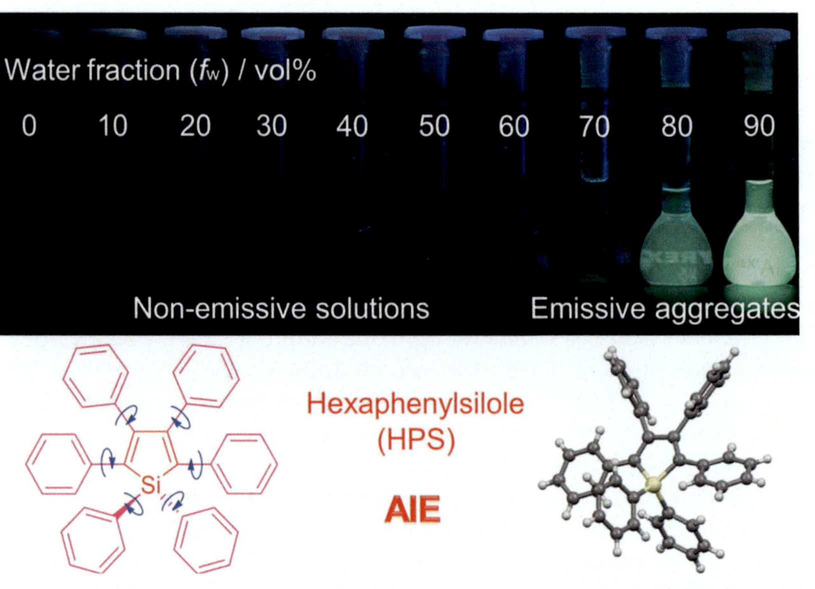

FIG. 1 Fluorescence photographs of solutions and suspensions of hexaphenylsilole (HPS; 20 mM) in THF/water mixtures with different fractions of water. *Reproduced with permission from J. Mei, Y. Hong, J.W.Y. Lam, A. Qin, Y. Tang, B.Z. Tang, Aggregation-induced emission: the whole is more brilliant than the parts, Adv. Mater. 26 (2014) 5429–5479. Copyright 2014 John Wiley & Sons.*

The AIE phenomenon exhibits practical significance in the development of highly efficient conventional fluorophores (the first-generation OLED luminescent materials) toward commercially viable OLEDs. However, according to spin statistics [7], the conventional fluorescent OLEDs can harvest only singlet excitons (25%) from electrically generated excitons and the remaining 75% of the generated excitons (triplet excitons) can be employed through other strategies such as TADF [8], phosphorescence [9], in order to enhance the efficiencies of OLEDs (Fig. 2). In 1998, OLEDs based on phosphorescent materials (second-generation luminescent materials) such as Os (II) and Pt (II) complexes can obtain nearly 100% η_{int} via harvesting both the electrically generated singlet and triplet excitons for light emission, which is facilitated by the intersystem crossing due to the strong spin-orbit coupling of heavy metals [10]. Nonetheless, there are certain limitations to the large-scale utilization of phosphorescent materials, such as high costs and potential toxicity to the environment. Efficient AIE-active phosphorescent materials are less reported and still in high demand for OLEDs. In order to explore suitable alternatives for rare metal-based emitters, using pure organic thermally activated delayed fluorescence (TADF) emitters (third-generation luminescent materials) has been a promising strategy toward the fabrication of high-performance OLEDs. TADF materials can harvest both the singlet and triplet excitons to emit light, thus providing a large β to achieve a nearly 100% η_{int}. In terms of nondoped OLEDs, emitters featuring AIE properties are highly desired to address the ACQ issue. In this context, introduction of AIE properties to TADF molecules is one attractive approach for developing robust AIEgens for high-performance OLEDs. The classical schematic energy diagrams of different generations of luminescent materials are shown in Fig. 2. This chapter presents a detailed review on the latest development of high-performance OLEDs based on varieties of organic luminescent materials and an in-depth analysis of conventional AIE fluorescence emitters, aggregation-induced delayed fluorescence (AIDF) and aggregation-induced phosphorescence (AIP) luminogens in terms of molecular design strategies, photophysical properties, and device performances. It also provides a comprehensive study on the advancement of emissive materials for high-performance electroluminescence (EL) devices.

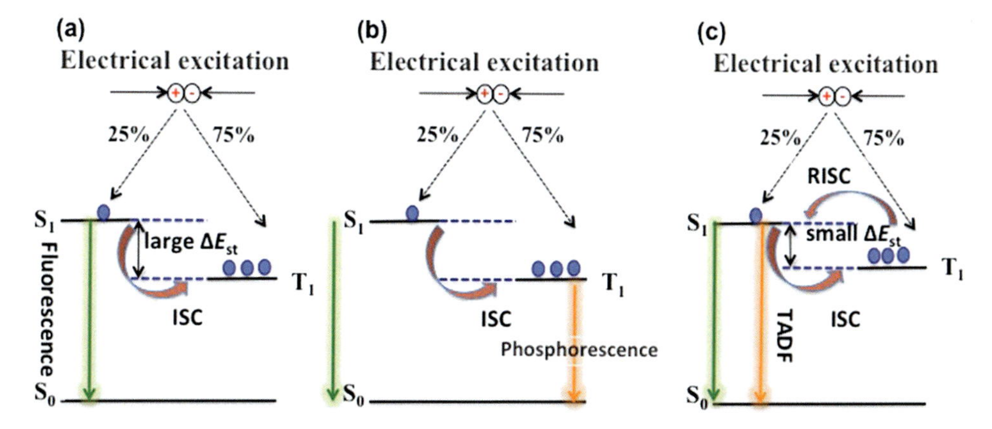

FIG. 2 Schematic energy diagrams of fluorescent (first-generation) (A), phosphorescent (second-generation) (B), and TADF molecules (third-generation) (C) and their corresponding decay pathways, when used in OLEDs.

2 Conventional fluorescent AIEgens

Many researchers have devoted much effort to harnessing the full potential of the AIE phenomenon after getting the deeper understanding of the AIE mechanism, and a wide variety of AIEgens have been developed, aiming to provide alternative approaches to solve the problems caused by fluorescence quenching [11]. Nondoped fluorescent OLEDs with a simple and stable structure have been fabricated employing fluorescent AIEgens with high fluorescence quantum yield (Φ_F) values as emitters in solid films, specifically silole [12] and tetraphenylethene (TPE) [13] derivatives. However, few of these AIEgen-based OLEDs are able to reach the theoretical limit of electroluminescence (EL) performance.

2.1 Silole-based AIEgens

The first reported propeller-shaped silole with AIE characteristics drew much research attention in the field of OLEDs. Most siloles having inherent AIE features present high solid-state PLQY values. Siloles display exclusive σ^*-π^* conjugation which happens from the interaction between the σ^* orbital of two exocyclic single carbon-silicon bonds and the π^* orbital of the butadiene moiety, which endows them with superior electron affinity and high electron mobility due to the low-lying LUMO (lowest unoccupied molecular orbital) levels, granting them the ability to transport electrons in OLEDs [12]. Furthermore, the high thermal stability, high morphological stability, and good solubility in common solvents of siloles can facilitate smooth film preparation by vapor deposition or solution processing techniques. The fascinating amalgamated AIE performance of siloles displays their potential in nondoped OLEDs, and hence many efficient silole-based fluorescent materials have recently been developed for OLEDs.

Fluorene-based substituents are known for their intense emission and superior thermal stability, which are found to be widely used as efficient emitters for OLEDs as a result of their promising properties. By skillfully incorporating dimethylfluorene as substituents at the 2,5-positions of the silole ring helps in achieving excellent PL and EL properties in the resultant materials [14]. For example, the AIE nature of **MFMPS** (shown in Scheme 1) ensured that its

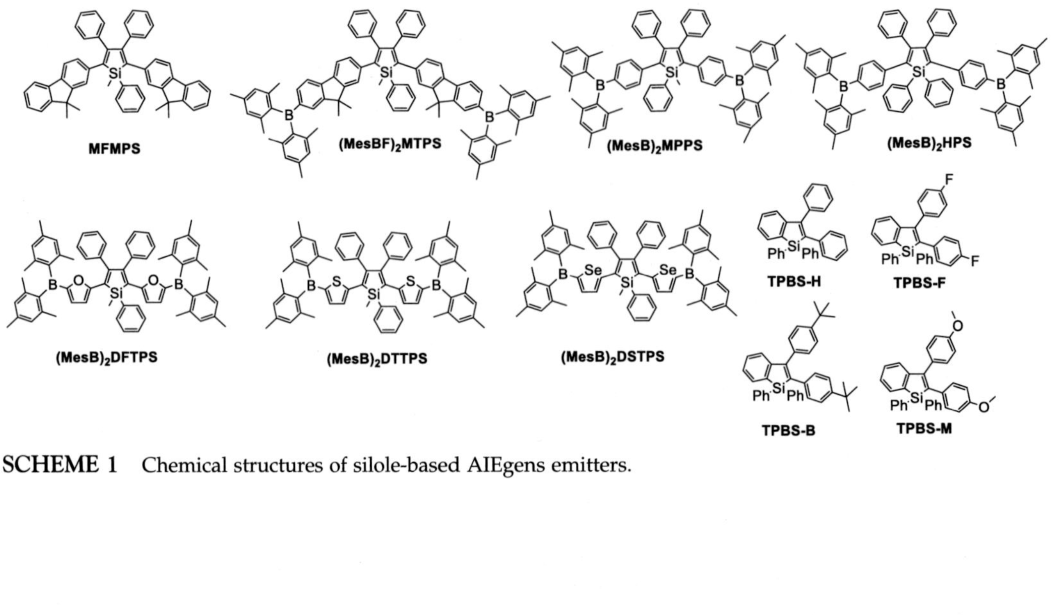

SCHEME 1 Chemical structures of silole-based AIEgens emitters.

neat film possessed a high solid-state Φ_F value (88%) with a strong fluorescence peak at 534 nm. Inspired by the solid-state luminescence properties, the corresponding nondoped OLEDs based on **MFMPS** with the structure of ITO/NPB (60 nm)/**MFMPS** (20 nm)/TPBi (40 nm)/LiF (1 nm)/Al (100 nm) showed outstanding EL performance, achieving a maximum current efficiency ($\eta_{C,max}$) of 16.0 cd A^{-1}, a maximum power efficiency ($\eta_{P,max}$) of 13.5 lm W^{-1}, and a maximum external quantum efficiency ($\eta_{ext,max}$) of 4.8%, with a yellow emission at 544 nm and CIE coordinates of (0.37, 0.57). Furthermore, the nondoped OLED based on the device configuration of ITO/MoO_3 (5 nm)/NPB (60 nm)/**MFMPS** (20 nm)/TPBi (60 nm)/LiF (1 nm)/Al (100 nm) displayed a turn-on voltage (V_{on}) of 3.3 V with a yellow EL peak at 540 nm and CIE coordinates of (0.36, 0.57), and provided a $\eta_{C,max}$ and $\eta_{P,max}$ of 18.3 cd A^{-1} and 15.7 lm W^{-1}, respectively. Especially, the optimized device achieved a $\eta_{ext,max}$ of 5.5%, which is comparable with the theoretical limit (5.0%–7.5%) of OLEDs based on the first-generation fluorescent materials.

As described above, siloles with excellent AIE features and the exclusive electronic configuration enable them to serve as efficient emitters and electron-transporting materials in nondoped OLEDs. Interestingly, silole-based efficient bifunctional materials of electron transporters and emitters can also be prepared, which can simplify the OLED structures and reduce the fabrication costs. The silole and dimesityboryl hybrids of **(MesB)$_2$MPPS** and **(MesB)$_2$HPS** shown in Scheme 1 are both good electron transporters [15]. The tetraphenylsilole moiety present in both the materials furnishes AIE characteristics and excellent solid-state emissions, while the dimesitylboryl moiety can lower the LUMO energy level and enhance the electron-transporting ability of the material on account of the empty low-lying π orbital on the boron center. As a result, **(MesB)$_2$MPPS** and **(MesB)$_2$HPS** exhibited deep LUMO energy levels of 3.06 and 3.10 eV, respectively, demonstrating their promising application as electron-transporting materials for OLEDs. The thin films prepared from **(MesB)$_2$MPPS** and **(MesB)$_2$HPS** exhibited strong fluorescence at around 524 and 526 nm, with quantum yields of 58% and 62%, respectively. Hence, the inherent bifunctional properties of **(MesB)$_2$MPPS** and **(MesB)$_2$HPS** justified their application in the double-layer OLEDs with the simple structure of ITO/NPB (60 nm)/**(MesB)$_2$MPPS** or **(MesB)$_2$HPS** (60 nm)/LiF (1 nm)/Al (100 nm), where **(MesB)$_2$MPPS** or **(MesB)$_2$HPS** concurrently acted as both the light-emitting layer (emissive layer, EML) and electron-transporting layer (ETL). These two devices presented much higher performances with $\eta_{ext,max}$ (4.35%), $\eta_{C,max}$ (13.9 cd A^{-1}), and $\eta_{P,max}$ (11.6 lm W^{-1}) than those accomplished using the more complicated triple-layer devices with an extra TPBi as the electron-transporting layer. It was proposed that the efficient electron transport and suitable LUMO levels of **(MesB)$_2$MPPS** or **(MesB)$_2$HPS** close to the cathode work function are the main factors responsible for the outstanding performance of these OLEDs. Therefore, from the above studies, it can be concluded that dimesitylboryl substituted siloles are capable bifunctional materials for fabricating high-performance OLEDs with simple device structures.

Similarly, **(MesBF)$_2$MTPS** is also a silole derivative formed by incorporating dimesitylboryl substituents into the framework of **MFMPS** [16]. With an excellent Φ_F of 88% and the potential for electron transport, **(MesBF)$_2$MTPS** can also act as a bifunctional material of both the emitter and electron transporter, providing good EL performances (Table 1). In order to explore silole-based red AIEgens and study the impact of electron-donating bridges such as furan, thiophene, and selenophene on the photophysical property of the AIE luminogens, a series of new red fluorescent silole derivatives, namely

TABLE 1 Electroluminescent key data of AIE emitters for 1st generation OLEDs.

Emitter	Active layer in device	λ_{EL} (nm)	V_{on}	L_{max} (cd m^{-1})	$\eta_{c,max}/\eta_{p,max}/\eta_{ext,max}$ (cd A^{-1}/lm W^{-1}/%)	Ref.
MFMPS	NPB (60nm)/MFMPS (20nm)/TPBi (40nm)	544	3.2	31,900	16/13.5/4.8	[14]
MFMPS	MoO_3 (5nm)/NPB (60nm)/MFMPS (20nm)/TPBi (60nm)	544	3.3	37,800	18.3/15.7/5.5	[14]
$(MesBF)_2MTPS$	NPB (60nm)/$(MesBF)_2MTPS$ (20nm)/TPBi (40nm)	554	3.8	48,348	12.3/8.8/4.1	[16]
$(MesB)_2HPS$	NPB (60nm)/$(MesB)_2HPS$ (60nm)	524	4.3	12,200	13.9/11.6/4.35	[15]
$(MesB)_2DFTPS$	ITO/TPD (60nm)/$(MesB)_2DFTPS$ (20nm)/TPBi (40nm)/LiF (1nm)/Al (100nm)	591	4.2	9600	2.8/2.0/1.3	[17]
$(MesB)_2DTTPS$	ITO/TPD (60nm)/$(MesB)_2DTTPS$ (20nm)/TPBi (40nm)/LiF (1nm)/Al (100nm)	589	4.7	13,300	4.3/2.9/1.8	[17]
$(MesB)_2DSTPS$	ITO/TPD (60nm)/$(MesB)_2DSTPS$ (20nm)/TPBi (40nm)/LiF (1nm)/Al (100nm)	615	4.7	3126	0.9/0.6/0.5	[17]
TPBS-H	ITO/(PEDOT: PSS)/(TFB) (40nm)/TPBS-H (about 15nm)/(TmPyPB) (30nm)/LiF (1.5nm)/Al (120nm)	438	2.8	2258	3.15/3.3/3.5	[18]
TPBS-F	ITO/(PEDOT: PSS)/(TFB) (40nm)/TPBS-F (about 15nm)/(TmPyPB) (30nm)/LiF (1.5nm)/Al (120nm)	438	2.8	2139	3.28/3.4/3.6/	[18]
TPBS-B	ITO/(PEDOT: PSS)/(TFB) (40nm)/TPBS-B (about 15nm)/(TmPyPB) (30nm)/LiF (1.5nm)/Al (120nm)	438	2.8	2281	2.89/3.0/3.1	[18]
TPBS-M	ITO/(PEDOT: PSS)/(TFB) (40nm)/TPBS-M (about 15nm)/(TmPyPB) (30nm)/LiF (1.5nm)/Al (120nm)	438	2.8	2390	3.13/3.3/3.4	[18]
TPTPE	ITO/(NPB) (60nm)/TPTPE (20nm)/(TPBi) (40nm)/LiF (1nm)/Al (100nm)	488	4.2	10,800	5.8/3.5/2.7	[19]
BTPTPE	ITO/(NPB) (60nm)/BTPTPE (20nm)/(TPBi) (40nm)/LiF (1nm)/Al (100nm)	448	5.2	3530	2.8/1.4/1.6	[19]
BTPE	ITO/NPB (60nm)/BTPE (20nm)/TPBi (10nm)/Alq_3 (30nm)/LiF(1nm)/Al (100nm)	488	4		7.26/−/−3.17	[20]
TTPEPy	NPB (60nm)/TTPEPy (40nm)/TPBI (20nm)	492	4.7	18,000	10.6/5/4.04	[21]
TTPEPy	NPB (60nm)/TTPEPy (26nm)/TPBi (20nm)/	488	3.6	36,300	12.3/7/4.95	[21]
TPEPY	ITO/NPB (60nm)/TPEPY (20nm)/TPBi (10nm)/Alq_3 (30nm)/LiF (1nm)/Al (100nm)	484	3.6	13,400	7.3/5.6/3.0	[22]

TABLE 1 Electroluminescent key data of AIE emitters for 1st generation OLEDs—cont'd

Emitter	Active layer in device	λ_{EL} (nm)	V_{on}	L_{max} (cd m^{-1})	$\eta_{c,max}/\eta_{p,max}/\eta_{ext,max}$ (cd A^{-1}/lm W^{-1}/%)	Ref.
TPEBPy	ITO/NPB (60nm)/TPEBPy (20nm)/TPBi (10nm)/Alq_3 (30nm)/LiF (1nm)/Al (100nm)	516	4.8	13,370	5.8/2.0/2.0	[22]
TPEBPy	ITO/NPB (60nm)/TPEBPy (20nm)/TPBi (40nm)/LiF (1nm)/Al (100nm)	516	4.6	25,500	6.0/2.7/2.1	[22]
TPE-2-Np	ITO/NPB (60nm)/TPE-2-Np (20nm)/TPBi (10nm)/Alq_3 (30nm)/LiF (1nm)/Al (100nm)	494	4.4	13,500	6.1/3.7/2.5	[22]
TPPyE	ITO/NPB (60nm)/TPPyE (20nm)/TPBi (30nm)/LiF (1nm)|Al (100nm)	508	4.6	15,450	3.7/2.1/1.4	[23]
TPPyE	ITO/NPB (60nm)|TPPyE (20nm/|TPBi (10nm)/Alq_3 (30nm)/LiF (1nm)/Al (100nm)	504	4.6	25,470	4.0/2.7/2.0	[23]
DPDPyE	ITO/NPB (60nm)/DPDPyE (20nm)/TPBi (30nm)/LiF (1nm)/Al (100nm)	520	5.3	45,550	9.1/4.1/2.9	[23]
DPDPyE	ITO/NPB (60nm)/DPDPyE (20nm)/TPBi (10nm)/Alq_3 (30nm)/LiF (1nm)/Al (100nm)	516	3.2	49,830	10.2/9.2/3.3	[23]
BPBAPE	ITO/2-TNATA (60nm)/NPB (15nm)/BPBAPE (30nm)/Alq_3 (30nm)/LiF (1nm)/Al (200nm).	475	8.1	–	10.33/4.00/–	[24]
TPVAn	ITO/PEDOT /TFTPA (30nm)/(TPVAn) (40nm)/(TPBI) (40nm)/Mg:Ag (100nm)/Ag (100nm)	456	4.9	–	5.3/2.8/5.3	[25]
MethylTPA-3pTPE	MoO_3 (10nm)/NPB (60nm)/MethylTPA-3pTPE (15nm)/TPBi (35nm)	480	3.1	13,639	8.03/7.04/3.99	[26]
MethylTPA-3pTPE	MoO_3 (10nm)/MethylTPA-3pTPE (75nm)/TPBi (35nm)	469	2.9	15,089	6.51/6.88/3.39	[26]
2TPATPE	NPB (40nm)/2TPATPE (20nm)/TPBi (10nm)/Alq_3 (30nm)	514	3.4	32,230	12.3/10.1/4.0	[27]
2TPATPE	2TPATPE (60nm)/TPBi (10nm)/Alq_3 (30nm)	512	3.2	33,770	13/11/4.4	[27]
TPA-TPE	TPA-TPE (65nm)/Bphen (35nm)	510	2.6	48,300	8.3/8.7/3.6	[28]
TPA-TPE	NPB (40nm)/TPA-TPE (25nm)/Bphen (35nm)	515	2.6	58,300	14.3/15/4.5	[28]
PDA-TPE	PDA-TPE (65nm)/Bphen (35nm)	523	2.4	54,200	14.4/14.1/4.5	[28]
PDA-TPE	NPB (40nm)/PDA-TPE (25nm)/Bphen (35nm)	523	2.4	53,600	15.9/16.2/5.9	[28]

Continued

TABLE 1 Electroluminescent key data of AIE emitters for 1st generation OLEDs—cont'd

Emitter	Active layer in device	λ_{EL} (nm)	V_{on}	L_{max} (cd m^{-1})	$\eta_{c,max}/\eta_{p,max}/\eta_{ext,max}$ (cd A^{-1}/lm W^{-1}/%)	Ref.
DTDAE	ITO/HTL/DTDAE/TPBi/LiF/Al	524	3.6	40,940	11.2/−/−	[29]
DTDAE	ITO/DTDAE/TPBi/LiF/Al	524	6.0	18,400	8.7/−/−	[29]
4TPEDTPA	ITO/4TPEDTPA/TPBi (10nm)/Alq_3 (30nm)/LiF (1nm)/Al (100nm)	488	4.1	10,723	8.0/5.2/3.7	[30]
2,5-BTPEMTPS	ITO/NPB (60nm)/2,5-BTPEMTPS (40nm)/TPBi (20nm)/LiF (1nm)/Al (100nm	552	5.2	12,560	6.40/2.45/1.98	[31]
3,4-BTPEMTPS	ITO/NPB (60nm)/3,4-BTPEMTPS (40nm)/TPBi (20nm)/LiF (1nm)/Al (100nm	520	6.2	3980	4.96/2.05/1.66	[31]
TPECaP	ITO/TPECaP (80nm)/TPBi (40nm)/LiF (1nm)/Al (100nm)	496	4.4	12,930	5.5/3.8/2.2	[32]
TTPECaP	ITO/TTPECaP (20nm)/TPBi (40nm)/LiF (1nm)/Al (100nm)	490	3.6	9048	6.3/4.1/2.3	[32]
BTPEBCF	ITO/NPB (60nm)/BTPEBCF (20nm)/TPBi (10nm)/Alq_3 (30nm)/LiF (1nm)/Al (100nm)	502	4.5	6400	7.9/3.7/2.9	[33]
TPEDMesB	ITO/NPB (60nm)/TPEDMesB (60nm)/LiF (1nm)/Al (100nm)	496, 512	6.3	5170	7.13/3.2/2.7	[34]
TPE-NPA-BTD	NPB (80nm)/TPE-NPA-BTD (20nm)/TPBi (40nm)	604	3.2	16,396	7.3/7.5/3.9	[35]
TPE-TPA-BTD	NPB (80nm)/TPE-TPA-BTD (20nm)/TPBi (40nm)	604	3.2	15,584	6.3/6.4/3.5	[35]
TTPEBTTD	NPB (60nm)/TTPEBTTD (20nm)/TPBi (40nm)	650	4.2	3750	−/2.4/3.7	[36]
BTPETD	NPB (60nm)/BTPETD (20nm)/TPBi (10nm)/Alq_3 (30nm)	540	3.9	13,540	5.2/3.0/1.5	[37]
BTPETTD	NPB (60nm)/BTPETTD (20nm)/TPBi (10nm)/Alq_3 (30nm)	592	5.4	8330	2.9/6.4/3.1	[37]
BTPEBTTD	NPB (60nm)/BTPEBTTD (20nm)/TPBi (10nm)/Alq_3 (30nm)	668	4.4	1640	0.4/0.5/1.0	[37]
T2BT2	ITO/NPB(60nm)/T2BT2 (20nm)/TPBi (40nm)/LiF (1.0nm)/Al (100nm)	590	4.3	13,535	6.81/4.96/2.88	[38]
BTPE-PI	NPB (60nm)/BTPE-PI (20nm)/TPBi (40nm)	463	3.2	20,300	5.3/5.9/4.4	[39]
BTPE-PI	NPB (40nm)/BTPE-PI (20nm)/TPBi (40nm)	450	3.2	16,400	4.4/4.9/4.0	[39]
TPE-PNPB	NPB (60nm)/TPE-PNPB (20nm)/TPBi (40nm)	516	3.2	49,993	12.9/15.7/5.12	[40]
TPE-PNPB	TPE-PNPB (80nm)/TPBi (40nm	516	3.2	13.678	14.4/16.8/5.35	[40]

TABLE 1 Electroluminescent key data of AIE emitters for 1st generation OLEDs—cont'd

Emitter	Active layer in device	λ_{EL} (nm)	V_{on}	L_{max} (cd m^{-1})	$\eta_{c,max}/\eta_{p,max}/\eta_{ext,max}$ (cd A^{-1}/lm W^{-1}/%)	Ref.
TPE-PBN	ITO/PEDOT:PSS (40nm)/mCP: TPE-PBN (99:1, 20nm)/TmPyPB (50nm)/Liq (1nm)/Al (100nm)	469	9.0	1308	6.2/2.0/4.1	[41]
TPE-2PBN	ITO/PEDOT:PSS (40nm)/mCP: TPE-2PBN (99:1, 20nm)/TmPyPB (50nm)/Liq (1nm)/Al (100nm)	478	9.0	1612	2.7/0.6/1.5	[41]
Hpz-3C12	ITO/PEDOT:PSS/CBP:Emitter(97:3)/TPBi/LiF/Al	–	5.1	251	0.4/0.3/1.6	[42]
OPVT1	ITO (125nm)/PEDOT:PSS (35nm)/CBP doped with 0.5wt% of the emitters OPVT1 (20nm)/TPBi (40nm)/LiF (1nm)/Al (100nm)	392	4.7	–	0.9/0.6/4.0	[43]

Abbreviations: *λ_{EL} = the electroluminescence peak; V_{on} = turn-on voltage at 1 cd m^{-2}; L_{max} = maximum luminance; $\eta_{C,max}$ = maximum current efficiency; $\eta_{P,max}$ = maximum power efficiency; $\eta_{ext,max}$ = maximum external quantum efficiency; PEDOT:PSS = Poly(3,4-ethylenedioxythiophene)-poly(styrenesulfonate; TFB = Poly(9,9-dioctylfluorene-*alt*-N-(4-s-butylphenyl)-diphenylamine); TmPyPB = 1,3,5-Tris (3-pyridyl-3-phenyl)benzene, NPB = N,N′-di(1-naphthyl)-N,N′-diphenylbenzidine; TPBi = 1,3,5-tris(N-phenylbenzimidazol- 2-yl)benzene; Bphen = 4,7-diphenyl-1,10-phenanthroline; Alq_3=Tris-(8-hydroxyquinoline)aluminum. PEDOT:PSS, NPB and TFB functions as a hole-transporting layer (HTL); TPBi and Bphen serve as an electron-transporting layer (ETL) and a hole-blocking layer (HBL), respectively; Alq_3 functions as ETL; and MoO_3 serves as a hole-injection layer (HIL).*

(MesB)$_2$DFTPS, **(MesB)$_2$DTTPS**, and **(MesB)$_2$DSTPS** (Scheme 1), were developed via introducing different electron-donating bridges between the silole core and dimesitylboranyl substituent. Both thiophene-based **(MesB)$_2$DTTPS** and selenophene-based **(MesB)$_2$DSTPS** show typical AIEE features, whereas the furan-based **(MesB)$_2$DFTPS** exhibited opposite behaviors. AIE features endowed **(MesB)$_2$DTTPS** with a high PLQY of 27% compared to other red silole derivatives. Nondoped OLED based on **(MesB)$_2$DTTPS** as the emitting layer exhibited a low turn-on voltage of 4.7V and an EL emission peak at 589nm, and the L_{max}, $\eta_{C,max}$, $\eta_{P,max}$, and $\eta_{ext,max}$ were 13,300 cd m^{-2}, 4.3 cd A^{-1}, 2.9 lm W^{-1}, and 1.8%, respectively [17].

Over the past decade, HPS has been known as a classical building block to construct AIEgens for high-performance OLEDs. However, HPS suffers from certain drawbacks such as synthetic complication and low rigidity induced by several rotatable phenyl groups, which makes it less convenient for developing deep-blue AIEgens. To address this issue, four new deep-blue AIEgens based on the TPBS building blocks, namely **TPBS-H**, **TPBS-F**, **TPBS-B**, and **TPBS-M** (Scheme 1), were designed and synthesized by fusing one phenyl group with the silole unit and replacing the two phenyl groups of HPS at the 2,3-position as shown in Scheme 1. All the four TPBS-based emitters exhibited typical AIE characteristics similar to HPS and displayed deep-blue emission in solid state, which resulted in highly efficient solution-processed, nondoped OLEDs with EL emission at 438nm and EQEs of 3.1%–3.6% [18].

2.2 TPE-based AIEgens

The tetraphenyleneethene (TPE) building block is one of the most classic AIEgen moieties. The factors behind its uniqueness include (i) the propeller-shaped conformation, which can adequately restrict the π-π stacking interactions that are likely to increase the nonradiative transitions and (ii) the multiple C-H•••π interactions between the hydrogen atoms of phenyl rings of the TPE unit and the π electrons of the phenyl rings of the adjacent molecule that can lock and rigidify the molecular conformation. Introducing TPE units into ACQ fluorophores produces a wide variety of new fluorescent AIEgens with high efficiencies of solid-state emission for the fabrication of efficient OLEDs. Several examples of TPE-based AIEgens have been reported (Scheme 2–9), where slight structural modifications lead to a wide range of emission colors from blue to red. Based on these TPE-based AIEgens, a number of high-performance OLEDs were fabricated [11,13].

For full-color displays and solid-state lighting toward commercial applications, efficient blue OLEDs, especially deep-blue ones, are highly desirable. However, robust blue solid-state emitters that function efficiently in OLEDs are still rare, mainly due to the intrinsic large band

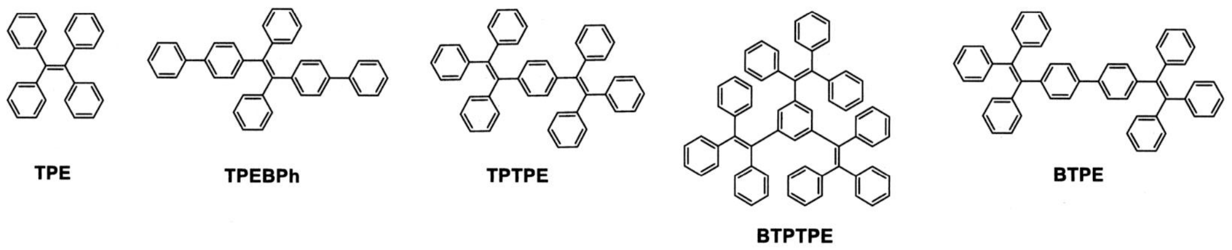

TPE TPEBPh TPTPE BTPTPE BTPE

SCHEME 2 Molecular structures of TPE-hydrocarbon based AIE emitters.

TPEPy TPVAn TPE-2-Np TPPyE TTPEPy BPBAPE DPDPyE

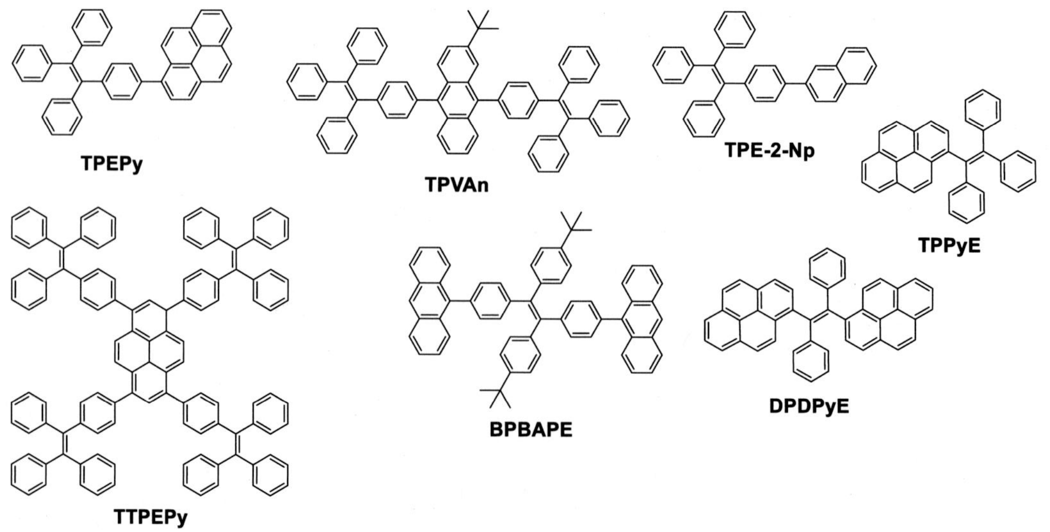

SCHEME 3 Molecular structures of TPE-aromatic hydrocarbon based AIE emitters.

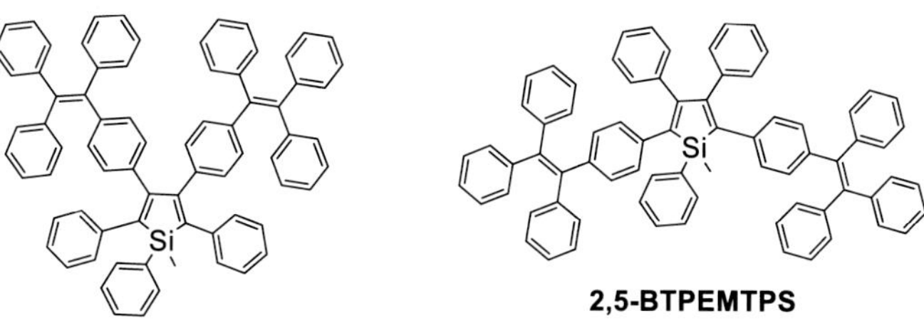

SCHEME 4 Molecular structures of TPE-silane based AIE emitters.

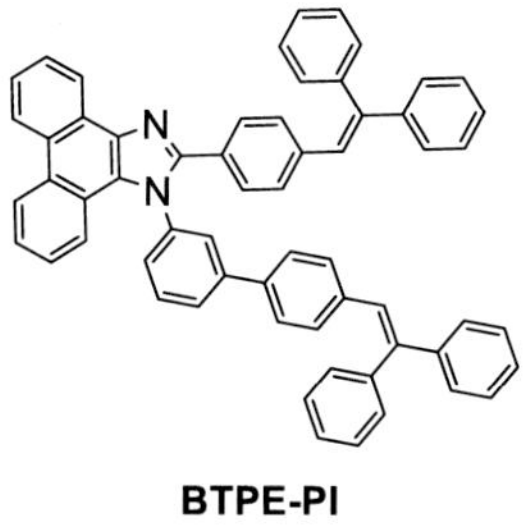

BTPE-PI

SCHEME 5 Molecular structures of triphenylethene-based AIE emitters.

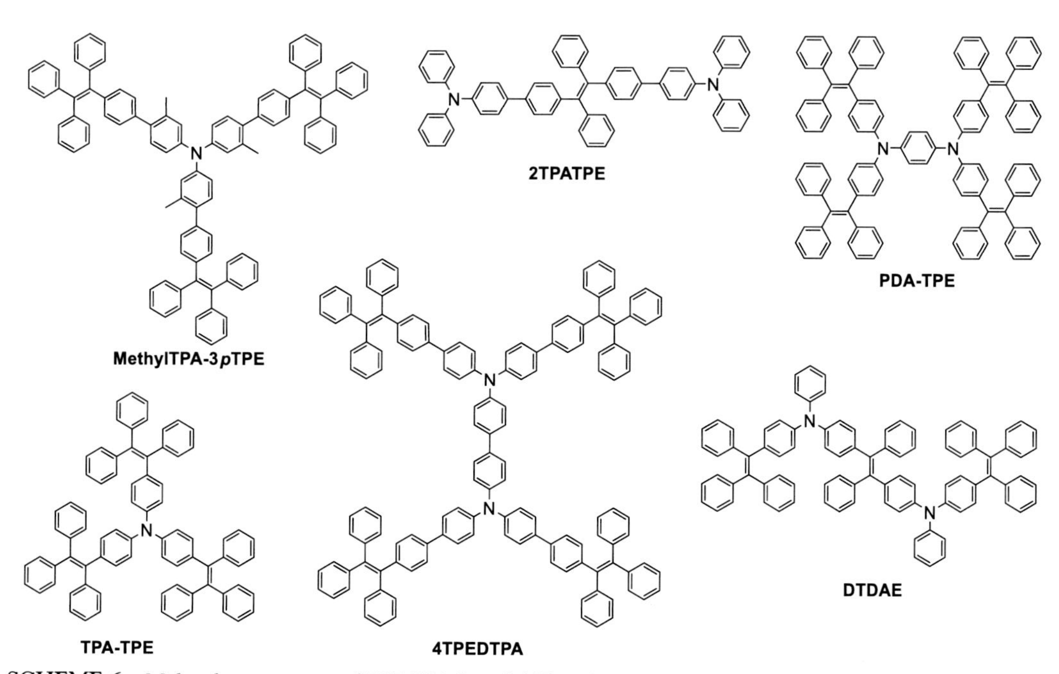

SCHEME 6 Molecular structures of TPE-TPA based AIE emitters.

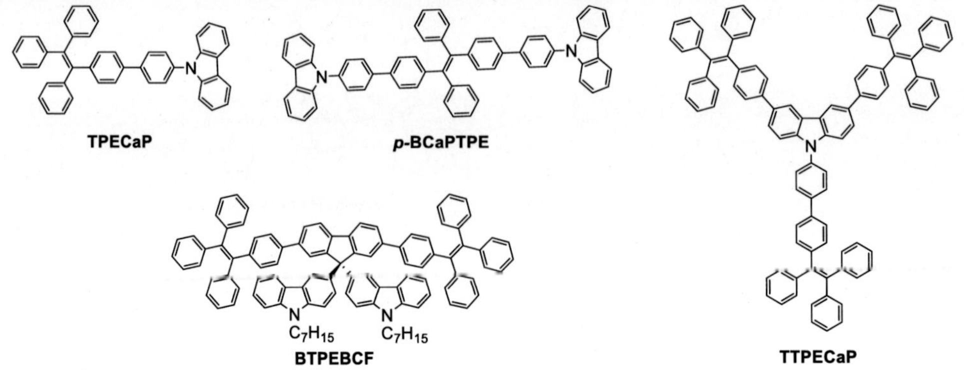

SCHEME 7 Molecular structures of TPE-carbazole based AIE emitters.

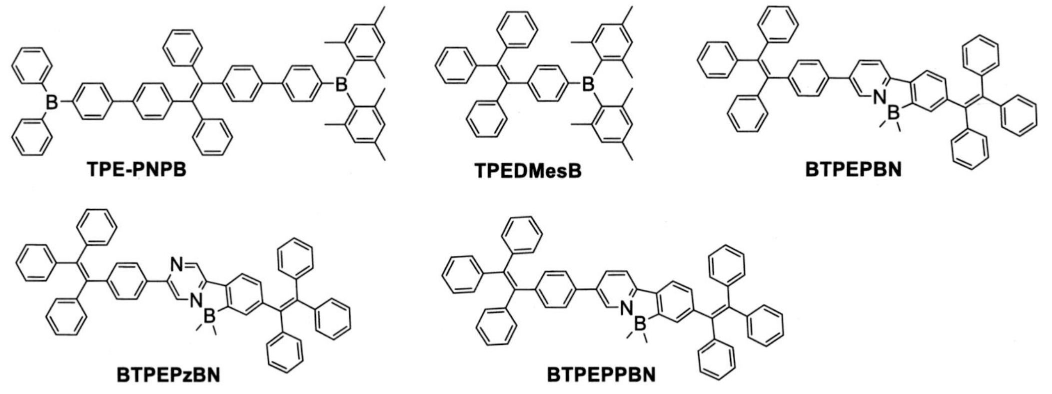

SCHEME 8 Molecular structures of TPE-boron based AIE emitters.

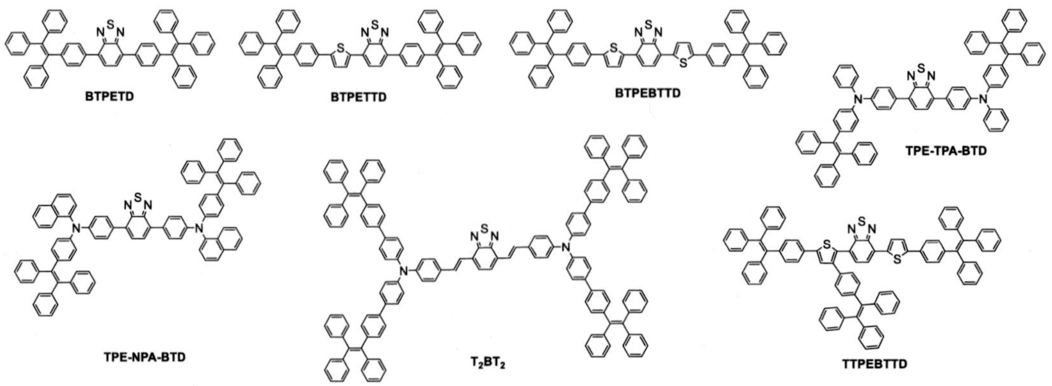

SCHEME 9 Molecular structures of TPE-heterocycle-based AIE emitters.

gaps. Substantial work has been devoted to the development of purely organic blue fluorophores for improving the stability and durability of the resulting OLEDs. The TPE unit possesses a simple chemical structure and can act as a promising scaffold for designing luminescent materials. Via the McMurry coupling reaction, pristine TPE and its derivatives can be easily synthesized (Scheme 2). Through the facile organic reactions such as Suzuki coupling, integration of TPE with pristine hydrocarbons and heterocyclic rings can be easily realized. Pure TPE exhibits a deep-blue EL emission peak at 445 nm but with poor device performances, exhibiting a $\eta_{c,\max}$, $L_{\max}$, and $\eta_{ext,\max}$ of 0.45 cd A^{-1}, 1800 cd m^{-2}, and 0.4%, respectively. Adding two more phenyl rings to the TPE unit gave rise to **TPEBPh** with better device performance. The OLED with a device configuration of ITO/NPB (50 nm)/**TPEBPh** (30 nm)/BCP (20 nm)/Alq_3 (10 nm)/LiF (1 nm)/Al (150 nm) exhibited a blue EL peak at 476 nm with a $\eta_{c,\max}$, $L_{\max}$, and $\eta_{ext,\max}$ of 5.15 cd A^{-1}, 10,680 cd m^{-2}, and 2.56%, respectively. **TPTPE**-based OLED displayed a slightly improved device performance with sky-blue EL at 488 nm, an $L_{\max}$ of 10,800 cdm^{-2}, $\eta_{c,\max}$ of 5.8 cd A^{-1}, and $\eta_{ext,\max}$ of 2.7% [19]. The starburst **BTPTPE** showed a bluer emission than **TPTPE** with an EL peak at 448 nm but lower efficiencies (2.8 cd A^{-1} and 1.6%) [19]. Out of these luminogens, a **BTPE**-based device with a configuration of ITO/NPB (60 nm)/**BTPE** (20 nm)/TPBi (10 nm)/Alq_3 (30 nm)/LiF (1 nm)/Al (100 nm) displayed a sky-blue emission peak at 488 nm and achieved the best device performances, with an $L_{\max}$,$\eta_{c,\max}$, and $\eta_{ext,\max}$ of 11,180 cd m^{-2}, 7.26 cd A^{-1}, and 3.17%, respectively [20].

Pyrene is a typical ACQ fluorophore that shows excellent PL efficiency in solution but weak emission in solid state, arising from the strong intermolecular π-π interactions. This ACQ problem can be removed by introducing one TPE unit at the periphery of the pyrene moiety, which offers a new AIE luminogen (**TPEPy**) with a high $\Phi_F = 100\%$ in the film state (Scheme 3). As a result, using **TPEPy** as the emitting layer, the OLED with a device configuration of ITO/NPB (60 nm)/**TPEPy** (20 nm)/TPBi (10 nm)/Alq_3 (30 nm)/LiF (1 nm)/Al (100 nm) exhibited sky-blue EL emission with device performances of 7.3 cd A^{-1}, 13,400 cd m^{-2}, 5.6 lm W^{-1}, and 3.0% [22]. Again, introducing four TPE units to the periphery of pyrene afforded a new fluorophore of **TTPEPy** with AIE features and a high solid-state fluorescence yield ($\Phi_F = 70\%$) (Scheme 3) [21]. **TTPEPy**-based nondoped OLED emitted sky-blue light peaking at ~490 nm, with outstanding performances ($\eta_{C,\max}$ up to 12.3 cd A^{-1}- and $\eta_{ext,\max}$ of 4.95%). Another example of AIE luminogen is **TPE-2-Np**, which is a combination of TPE and conventional aromatic hydrocarbons (Scheme 5). The OLED using **TPE-2-Np** as the emitter demonstrated good EL performances of 19,800 cd m^{-2}, 7.2 cd A^{-1}, 4.2 lm W^{-1}, and 2.7% [22]. A series of blue AIEgens comprising TPE and an anthracene building block (Scheme 3) were introduced by Park and coworkers [24]. In order to achieve blue emission by restraining the bathochromic shift arising from strong intermolecular interactions, two *tert*-butyl groups were introduced to the TPE core, giving rise to the AIE-active **BPBAPE**. The nondoped OLED based on **BPBAPE** presented a blue EL emission at 475 nm with a maximum current efficiency of 10.33 cd A^{-1}. Similarly, Shu and coworkers [25] also designed and synthesized an AIEgen named **TPVAn** (Scheme 3) by introducing a *tert*-butyl group in the anthracene ring. The emitter displayed a high $\Phi_F = 89\%$ in solid-state film in comparison with that in solution (6%). As a result, nondoped OLED based on **TPVAn** displayed an intense blue emission with $\eta_{C,\max}$ and $\eta_{EQE,\max}$ of 5.3 cd A^{-1} and 5.3%, respectively.

Replacing the phenyl group of the TPE unit by polyfused aromatic hydrocarbon units can also result in various AIE-active AIEgens. The pyrene-substituted ethenes named **TPPyE** and **DPDPyE** (Scheme 3) obtained through substitution of the phenyl ring(s) in TPE with pyrene ring(s) displayed intense emission in the aggregated state. The Φ_F values for **TPPyE** and **DPDPyE** (2.8% and 9.8%) in dilute THF solution are much lower than their amorphous films (61% and 100%) [23]. Both emitters are AIE active because of various intermolecular interactions (π-π interaction, C-H•••π hydrogen bonds) that adequately limit the intramolecular rotation (IMR) process [44], ultimately blocking the nonradiative channels and, as a result, both **TPPyE** and **DPDPyE** performed brilliantly as emitters in OLED devices [23]. For illustration, the **DPDPyE**-based device with a configuration of ITO/NPB (60nm)/**DPDPyE** (20nm)/TPBi (10nm)/Alq_3 (30nm)/LiF (1nm)/Al (100nm) exhibited a green emission with a maximum at 516nm. The device displayed a low turn-on voltage of 3.2V, an L_{max} of 49,830 cd m^{-2}, $\eta_{C,max}$ of 10.2 cd A^{-1}, and $\eta_{ext,max}$ of 3.3%. The substantial π-π stacking between the pyrene rings, which can enhance the carrier mobility of **DPDPyE**, and the excellent PL efficiency of **DPDPyE** in solid states are the contributing factors for the superior EL performance.

TPE-substituted silanes such as methylpentaphenylsilole (**MPPS**) offer AIE features with high PLQYs in the solid state. The extended conjugation decreases the LUMO level, which in turn endows them with high electron mobility and high electron affinity [45]. Melding TPE units with MPPS units gives rise to various new AIEgens (Scheme 4). Among them, **2,5-BTPEMTPS** exhibited superior thermal and morphological stability, which demonstrated intense and red-shifted emission in solution and an aggregated state because of high conjugation compared to its regioisomer of **3,4-BTPEMTPS**. As a result, OLED based on the structure of ITO/NPB (60nm)/**2,5-BTPEMTPS** (40nm)/TPBi (20nm)/LiF (1nm)/Al (100nm)] emitted yellow EL emission peaking at 552nm with an L_{max} of 12,560 cd A^{-1} and a $\eta_{C,max}$ of 6.4 cd A^{-1}, much higher than those obtained for the **3,4-BTPEMTPS**-based device (3980 cd m^{-2}, and 4.96 cd A^{-1}) [31].

Triphenylethene is another valuable AIE unit. Triphenylethene is a promising building block as it holds a shorter conjugation length compared to TPE and emits blue solid-state emission, suitable for the construction of efficient solid-state blue emitters. The combination of triphenylethene and a phenanthro[9,10-*d*]imidazole (PI) group resulted in an efficient deep-blue AIEgen, **BTPE-PI** (Scheme 5) [39]. A nondoped multilayer OLED based on **BTPE-PI** [ITO/NPB (40nm)/**BTPE-PI** (20nm)/TPBi (40nm)/LiF (1nm)/Al (100nm)] demonstrated a deep-blue EL emission peaking at 450nm, exhibiting an excellent $\eta_{ext,max}$ of 4.4% and a small efficiency roll-off with CIE coordinates of (0.15, 0.12) [39].

Lately, Li's group reported a series of blue AIE luminogens using steric hindrance to control the balance between molecular rotation and conjugation, achieved by different connection patterns and intramolecular twisting of the molecular structure [26]. Among them, **MethylTPA-3*p*TPE**, comprising a methyl-substituted triphenylamine (TPA) core and three TPE units in peripheries, was a promising AIEgen which displayed interesting EL properties (Scheme 6). A multilayer OLED device based on **MethylTPA-3*p*TPE** with a configuration of ITO/MoO_3 (10nm)/NPB (60nm)/**MethylTPA-3*p*TPE** (15nm)/TPBi (35nm)/LiF(1nm)/Al exhibited a blue emission peaking at 480nm with CIE coordinates of (0.17, 0.28). The L_{max}, $\eta_{C,max}$, $\eta_{P,max}$, and $\eta_{ext,max}$ of the device were 13,639 cd m^{-2}, 8.03 cd A^{-1}, 7.04 lm W^{-1}, and 3.99%, respectively. The good hole-transporting properties of the methyl-substituted TPA in the molecular structure of **MethylTPA-3*p*TPE** rendered a simplified device in the absence of HTL with a configuration of MoO_3 (10nm)/**MethylTPA-3*p*TPE** (75nm)/TPBi (35nm)/LiF (1nm)/Al, which achieved equivalent EL performances (6.51 cd A^{-1}, 6.88 lm W^{-1}, and 3.39%) to those of the device with an HTL and exhibited a blue emission at 469nm with CIE coordinates of (0.18, 0.25) [26]. Both efficient solid-state emission and high charge carrier mobilities of an emitter are highly desirable. Using multifunctional materials in OLEDs (e.g., concurrently serve as emitting layer and charge-transporting layer) can simplify the device configuration, curtain fabrication process and lower the manufacturing costs [13]. Due to the intrinsic hole-injection/transporting ability, triphenylamine (TPA) has been widely adopted as an electron donor to construct various optoelectronic functional materials. Nonetheless, pristine TPA suffers from the notorious ACQ effects in the aggregated state. An original and extremely versatile semiconductor **2TPATPE** featuring AIE characteristics was constructed by fusing the TPA groups with the TPE unit (Scheme 6) [27]. Based on its excellent hole mobility up to 5.2×10^{-4} $cm^2 V^{-1} S^{-1}$ determined by the time-of-flight technique and the excellent Φ_F value (~100%) in solid state, **2TPATPE** endowed its OLED of a simplified configuration without HTL [ITO/**2TPATPE** (60nm)/TPBi (10nm)/Alq_3 (30nm)/LiF/Al (200nm)] with outstanding EL performances (4.4%, 13.0cd A^{-1}, and 11.0lm W^{-1}). These performances were better than those of the device with HTL [ITO/NPB (40nm)/**2TPATPE** (20nm)/TPBi (10nm)/Alq_3 (30nm)/LiF/Al (200nm)] (4.0%, 12.3cd A^{-1}, and 10.1lm W^{-1}) [27]. Two starburst AIE luminogens, **PDA-TPE** and **TPA-TPE** (Scheme 3), were designed and developed by Adachi's group via incorporating a hole-transporting *N,N,N′,N′*-tetraphenyl*p*-phenylenediamine (PDA) or a TPA core with the AIE group [28]. Both the AIEgens showed strong fluorescence with higher Φ_F values of 56%–73% and enhanced hole mobility than the archetypal classical hole transporter, *N,N′*-di(1-naphthyl)-*N,N′*-diphenylbenzidine. Owing to the existence of the PDA or TPA unit, plus the starburst molecular structure ensuring the spontaneous molecular orientation, the simplified OLEDs in which **PDA-TPE** or **TPA-TPE** acted as both the emitter and hole-transporting layer achieved excellent EL performances (Table 1). On account of the improved η_{out} attributed to the spontaneous molecular orientation and favorable charge balance, the three-layer OLED [ITO/NPB (40nm)/**PDA-TPE** or **TPA-TPE** (25nm)/BPhen (35nm)/LiF (0.8nm)/Al (70nm)] demonstrated remarkably high $\eta_{ext,max}$ values approaching 5.9% with EL emissions at 510–530nm. Similarly, **4TPEDTPA** (Scheme 6) can also act as both the emitter and hole-transporting layer in simplified OLED, with an L_{max} of 10,723 cd m^{-2}, $\eta_{C,max}$ of 8.0 cd A^{-1}, $\eta_{P,max}$ of 5.2 lm W^{-1}, and $\eta_{ext,max}$ of 3.7% [30]. On the other hand, the combination of TPE and TPA hybrids to develop various oligomers and polymers can also attain AIE-active emitters with high emission efficiency in the aggregated state and good hole-transporting property. As an illustration, the OLED based on **DTDAE** exhibited good EL performances with an L_{max} and $\eta_{C,max}$ of 40,940 cd m^{-2} and 11.2 cd A^{-1}, respectively [29].

Carbazole and its derivatives are usually used as highly efficient light-emitting and hole-transporting materials with high thermal stability [46–48]. However, the ACQ issue of the carbazole chromophore limits its wide applications in the EL devices. One effective strategy is to functionalize the carbazole unit with TPE to endow the materials with AIE characteristics in the aggregated film state with high PLQYs. The target luminogens enjoyed high thermal and morphological stabilities. For example, the T_g and T_d of **TTPECaP** (Scheme 7) are 179°C and 554°C, respectively [32]. The hole-transporting property of the carbazole moiety endowed the EL devices of the TPE-carbazole emitters with high performances in the absence of HTL. For instance, the simplified EL device using **TPECaP** concurrently as an emitter layer and hole-transporting layer [ITO/**TPECaP** (80nm)/TPBi (40nm)/LiF (1nm)/Al(100nm)] demonstrated an L_{max}, $\eta_{C,max}$, and $\eta_{ext,max}$ of 12,930cdm^{-2}, 5.5cd A^{-1}, and 2.2%, respectively [32]. The EL efficiencies can also be observed for ***p*-BCaPTPE** and **TTPECaP** without HTL, offering a decent $\eta_{C,max}$ of 6.6 and 6.3cd A^{-1}, respectively [49]. Luminophores with three-dimensional structures (3D) represent a novel strategy of delivering high emission efficiency and good thermal stability, achieved by suppressing the close packing of the molecules. For example, the AIE-active 3D luminogen of **BTPEBCF**, composed of carbazole and TPE units bridged by fluorene, exhibited a high $[Fcy]_F$ value of 100% in the aggregated film state [33]. The corresponding OLED using **BTPEBCF** as both the emitting layer and hole-transporting layer with a device architecture of ITO/NPB (60nm)/**BTPEBCF** (20nm)/TPBi (10nm)/Alq_3 (30nm)/LiF (1nm)/Al (100nm) achieved EL performances of 7.9cd A^{-1} and 2.9%.

Bipolar luminescent materials based on electron donors and acceptors (D-A) represent an interesting type of design, which can not only contribute to balancing the injection and transport of carriers in OLEDs, but also simplify the device structure. Tang's group designed a bipolar AIEgen of **TPE-PNPB** by linking an electron donor (diphenylamino) and an electron acceptor (dimesitylboryl) through a TPE unit, leading to a D-A skeleton with an AIE unit [40]. The AIE-active **TPE-PNPB** exhibited weak D-A interactions with a high $[Fcy]_F$ value of 94% in the solid state (Scheme 8). OLED utilizing the new bipolar AIEgen of **TPE-PNPB** with a configuration of ITO/NPB (60nm)/**TPE-PNPB** (20nm)/TPBi (40nm)/LiF (1nm)/Al (100nm) exhibited a green emission at 516nm, with a low turn-on voltage of 3.2V, and a high L_{max} of 49,993cdm^{-2}. This device presented excellent EL performances, achieving an $\eta_{C,max}$, $\eta_{P,max}$, and $\eta_{ext,max}$ of 15.7cd A^{-1}, 12.9lmW^{-1}, and 5.12%, respectively. Furthermore, in a bilayer OLED in which **TPE-PNPB** acted as both the EML and HTL, a high $\eta_{ext,max}$ of 5.35% was recorded.

The boron atom is characterized by a vacant p orbital, which made it fascinating as an electron acceptor. By combining the boron-based acceptor and the AIE-active TPE unit, unique bifunctional materials as both the emitters and electron transporter can be developed. TPE substituted boranes (Scheme 8) are highly emissive in the solid state and possess good thermal stability. For example, **TPEDMesB** with a simple structure emits intensely at 478nm because of its AIE features and a high $[Fcy]_F$ value of unity [34]. Inspired by the AIE features and electron-transporting ability of the boron center, OLED based on **TPEDMesB** with a simplified device of ITO/NPB (60nm)/**TPEDMesB** (60nm)/LiF (1nm)/Al (100nm) exhibited high EL performance with a $\eta_{C,max}$ and $\eta_{ext,max}$ of 7.13cd A^{-1} and 2.7%, respectively. However, **TPEDMesB** showed lower device performances when it imply acted as the emitter with the device configuration of ITO/NPB (60nm)/**TPEDMesB** (20nm)/TPBi(40nm)/LiF (1nm)/Al (100nm), showing a $\eta_{ext,max}$ of 5.78% and 2.3%. It has been observed that AIEgens with four coordinate organo-boron usually exhibit good EL properties. For example, the **BTPEPBN**-based device exhibited a $\eta_{C,max}$, $\eta_{P,max}$, and $\eta_{ext,max}$ of 4.43cd A^{-1}, 1.64lmW^{-1}, and 1.52%, respectively. Similar to **BTPEPBN**, other luminogens such as **BTPEPPBN** and **BTPEPzBN** displayed superior device efficiencies, demonstrating the importance of B-N coordination in the EL efficiencies.

The development of efficient blue, green, and red fluorescent emitters is highly desirable for the manufacture of full-color displays. Similar to blue OLEDs, the performance of the current red fluorescent OLEDs is also not satisfactory. By and large, it has been seen that many established red fluorophores are prepared from the planar polycyclic aromatic hydrocarbon (PAH) group which exhibits strong ACQ effects and results in poor emission in the aggregated state due to the inherent π-π stacking [35]. These demerits can be addressed by utilizing AIE-active groups to create efficient red fluorophores. For instance, **BTPETTD** as a new orange-red AIEgen (Scheme 9), consisting of two TPE units connected to a famous red-emitting core of benzo-2,1,3-thiadiazole and thiophene, was prepared [37], which exhibited AIE features and emitted strong fluorescence in the neat film state with a Φ_F value of 55%. Utilizing **BTPETTD** as the emitter, the OLEDs exhibited an orange-red emission at 592 nm and showed a high $\eta_{c,\ max}$ and $\eta_{ext,\ max}$ of 6.1 cd A^{-1} and 3.1%, respectively [37].

Again, the introduction of a thiophene unit in the **BTPETD** luminogen can tune the emission color with the aid of push-pull interactions and extended molecular conjugation. All the luminogens named **BTPETD** and **BTPEBTTD** emit green and red photoluminescence with AIE features and practically exhibit high Φ_F in the solid state (Scheme 9). The EL devices of these luminogens with a configuration of the ITO/NPB (60 nm)/emitter (20 nm)/TPBi (10 nm)/Alq_3 (30 nm)/LiF (1 nm)/Al (100 nm) exhibited EL emission peaks at 540 nm for **BTPETD** and 668 nm for **BTPEBTTD**. The **BTPETD**-based nondoped OLED displayed an L_{max}, $\eta_{C,max}$, and $\eta_{ext,max}$ of 13,540 cd m^{-2}, 5.2 cd A^{-1}, and 1.5%, respectively. In comparison, the **BTPEBTTD**-based nondoped OLED exhibited poorer performance, with an L_{max} and $\eta_{ext,\ max}$ of 1640 cd m^{-2} and 1.0%, respectively [37]. **TPE-TPA-BTD** and **TPE-NPA-BTD** were two red AIEgens constructed based on the TPE, benzo-2,1,3-thiadiazole, and arylamino units (Scheme 9) [35] with high Φ_F values of 48.8% and 63.0%, respectively. The nondoped OLEDs using **TPE-NPA-BTD** as the emitter emitted at 604 nm with a high $\eta_{ext,max}$ value approaching 3.9%. Interestingly, the presence of arylamino units in the molecular structures endows these new AIEgens with good hole-transporting characteristics, and using these materials can achieve efficient double-layer OLEDs in which those materials function as both the emitters and hole transport layer (Table 1). The introduction of more TPE moieties in the conjugated backbone gave rise to **TTPEBTTD** with a highly twisted conformation, which can effectively suppress the intermolecular interactions (Scheme 10). The neat film of **TTPEBTTD** showed red emission with a PL maximum at 646 nm. Utilizing the **TTPEBTTD** as EML in nondoped OLED achieved an EL emission at 650 nm with high L_{max} and $\eta_{ext,max}$ values of 3750 cd m^{-2} and 3.7%, respectively [50]. Similar results were also reported by other research groups. For instance, Xu's group [36] prepared a red luminophore (**T_2BT_2**) with a star-shaped structure composed of benzo-2,1,3-thiadiazole core and TPE peripheries linked by TPA bridges (Scheme 9). Because of the presence of multiple TPE moieties, **T_2BT_2** displays AIEE characteristics and high thermal stability. The nondoped EL device based on **T_2BT_2** as the emitting layer demonstrated a red EL emission with an L_{max} and $\eta_{C,max}$ of 13,535 cd m^{-2} and 6.81 cd A^{-1}, respectively.

2.3 AIEgens based on liquid crystals

Hitherto, much effort has been devoted to developing luminescent mesogens [38,51], with the most prevalent methodology being of introducing terminal flexible chains in the emissive core [52]. Kato and coworkers have developed various luminescent mesogens based on a rigid π-conjugated core (such as pyrene and anthracene) functionalized with peripheral

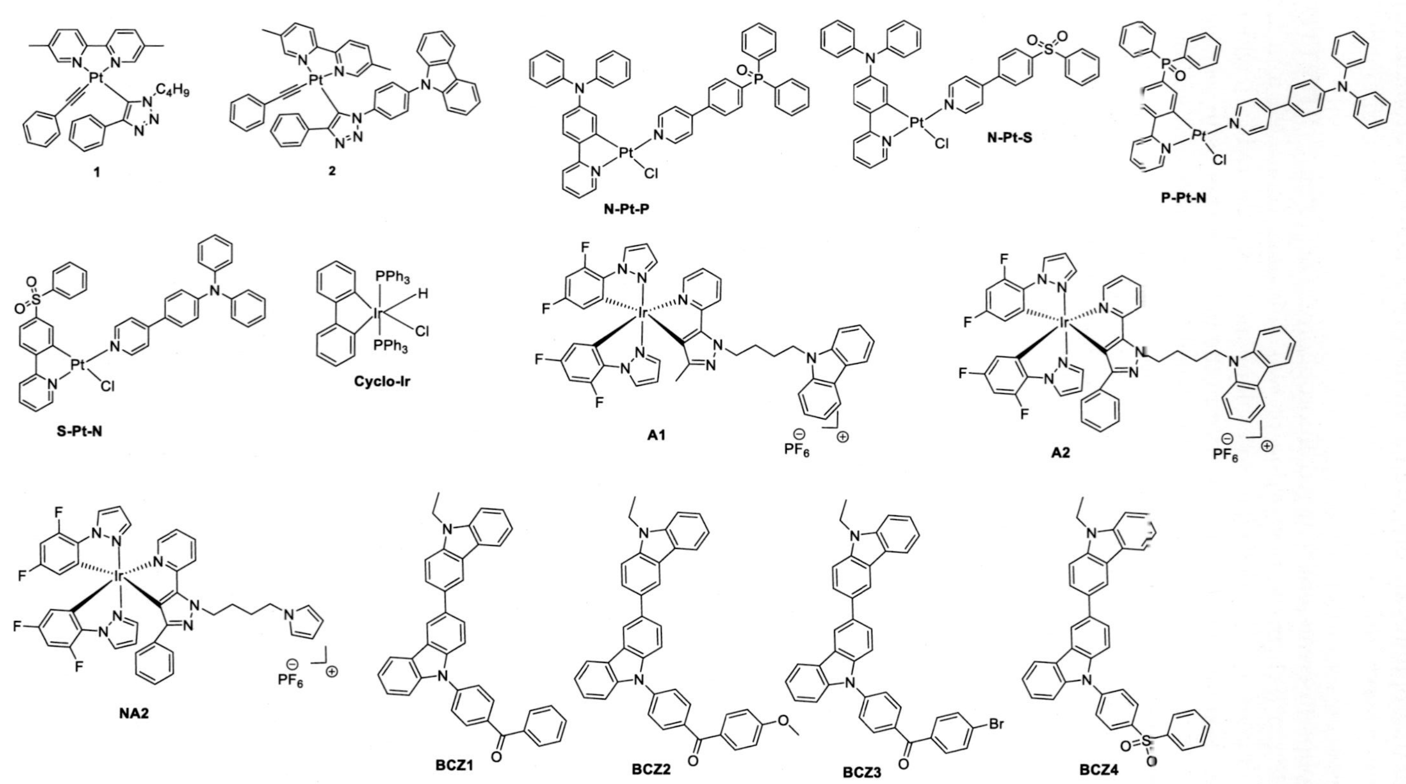

1

2

N-Pt-P

N-Pt-S

P-Pt-N

S-Pt-N

Cyclo-Ir

A1

A2

NA2

BCZ1

BCZ2

BCZ3

BCZ4

SCHEME 10 Molecular structures of metal complexes based Aggregation-induced phosphorescence and AIE-active RTP molecules for OLEDs.

alkoxy chains [53]. Despite the fact that this methodology can adequately produce liquid crystallinity, the emission quantum yield (Φ_{PL}) is usually low, presumably due to the nonradiative transitions by means of intramolecular vibration brought about by the periphery flexible chains and the aggregation-induced quenching of the rigid π-conjugated core. Above all, incorporation of the phenyl pyridine substituted TPE unit in the mesogenic block of 4-cyanobiphenyl with flexible tails leads to intense blue/bluish-green light with high PLQYs of 71% and 83% in the solid state for **TPE-PBN** and **TPE-2PBN**, respectively (Fig. 3). Employing **TPE-PBN** and **TPE-2PBN** as the emitting materials, both nondoped and doped devices were fabricated with the configuration of ITO/PEDOT:PSS (40nm)/compounds (40nm)/TmPyPB (50nm)/Liq (1nm)/Al (100nm) and ITO/PEDOT:PSS (40nm)/mCP:compound (99:1, 20nm)/TmPyPB (50nm)/Liq (1nm)/Al (100nm), respectively. It was found that **TPE-PBN**-based doped OLED showed better device performances with a maximum external quantum efficiency (EQE) of 4.1%, which was among the highest of the blue AIE-active fluorescent OLEDs. [41]

Generally, it was observed that the discotics typically show ACQ behaviors in the aggregated state. But in the following case, ***s*-Heptazine** [42]-based electron-deficient columnar LC materials exhibited AIE characteristics, which made them suitable emitters for OLEDs (Fig. 3). The newly developed AIE-active **Hpz-3C12** showed strong sky-blue emission in a solid state while being weakly emissive in the solution state. Deep-blue solution-processed OLEDs based on the **Hpz-3C12** emitter doped in different hosts were fabricated. The devices showed the best performances with the CBP host and 3wt% dopant concentration. In another report,

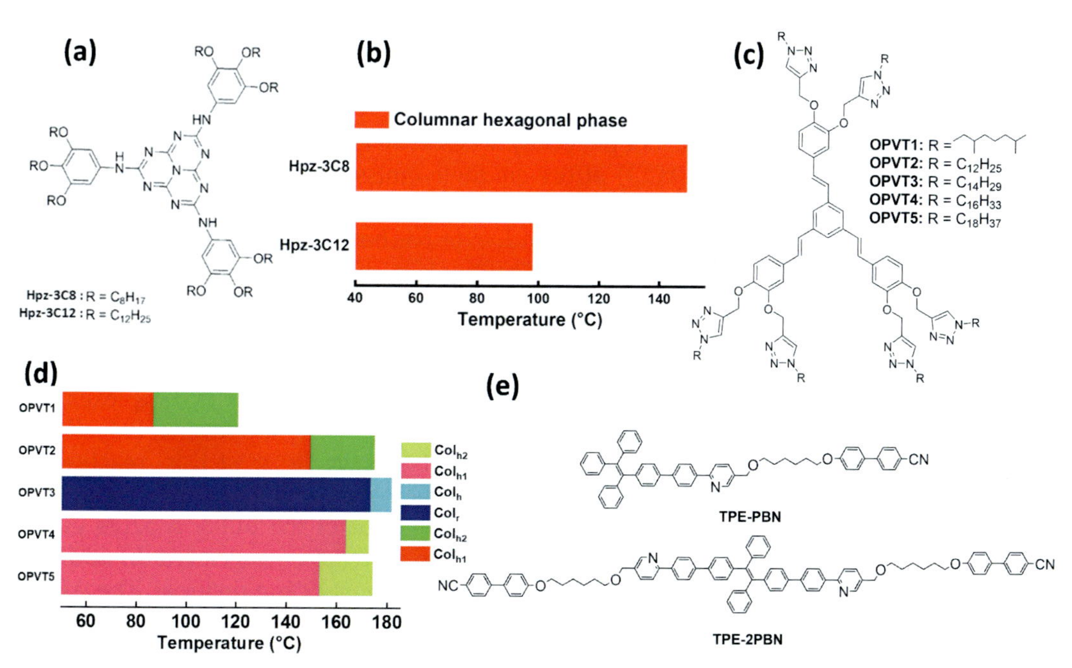

FIG. 3 Molecular structures of AIE-mesogens (A), (C) and (E); Bar graph representing the thermal behavior of mesogen derivatives (B) and (D).

the notorious problem of ACQ in the columnar luminomesogen arising from π-π stacking between the columns was addressed via a new design strategy, in which all the derivatives (**OPVT1-OPVT5**) [43] were integrated by the triazole moiety using click chemistry. In addition to this, these materials exhibited the Col liquid crystalline (LC) phase at room temperature over a wide temperature range with desirable alignment properties, which is suitable for organic electronic devices. It was found that **OPVT1** (Table 1) was AIE active and can act as highly efficient emitter. The corresponding OLEDs achieved a maximum external quantum efficiency of 4.0% for the first time in Col LCs with CIE coordinates of (0.17, 0.07), which closely matched the National Television System Committee (NTSC) standard, corresponding to pure deep-blue color.

3 Phosphorescence AIEgens for OLEDs

Phosphorescent materials and devices are conventionally based on metal–organic complexes. The presence of heavy metal atom(s) in the complexes facilitates the harvesting of both electrically generated singlet and triplet excitons via efficient intersystem crossing arising from the strong spin-orbit coupling (SOC) effect of heavy metal atoms, leading to a considerable upgrading of their OLED performances with a maximal internal quantum efficiency of 100% [54–57]. In recent years, a few room temperature phosphorescent materials have been reported on purely organic compounds in the absence of rare, toxic, and expensive heavy metal atoms [58]. Instead, triplet-state enhancing functional groups such as carbonyl and/or halide substitution are introduced to enhance the SOC effect in the sustainable cost effective metal free phosphorescent material, aiming to populate the electrically generated triplet excitons for high efficiency electroluminescence.

Like fluorescence materials, most of the phosphorescence emitters usually suffer from the aggregation-caused quenching (ACQ) effect on their emissions in the aggregated state. Hence, to circumvent the hindrance of ACQ in the solid state in order to improve the EL performance of phosphorescent materials, much endeavor has been committed. The AIE discovery by Tang in 2001 provides a stimulating motivation to the design and synthesis of phosphorescence luminogens with improved EL performance via altering the emitter behavior from the ACQ to AIE characteristics in the aggregate state. To date, very few reports are available for phosphorescence with the AIE characteristics.

Yam's group explored efficient PHOLEDs established on transition metal complexes with the AIE nature. These two metal complexes built on unsymmetric bipyridine–Pt^{II}–alkynyl complexes (**1** and **2**) (Scheme 10) were synthesized via a postclick reaction of bis(arylalkynyl)platinum(II) complexes through the alkyne-azide click reaction. Both the compounds show interesting AIE behaviors in the presence of the bulky triazole group; the phenyl unit prevents the stacking interaction due to its twist arrangement. In addition, **compound 2** with a carbazole substituent acts as the hole-transporting material. The AIE effect of **compound 1** has been further supported by the huge difference in PLQY of 0.03 and 0.72 in DCM and mCP doped film, respectively. Solution processed OLEDs with a device architecture of ITO/PEDOT:PSS (70 nm)/5% **1** or **2**:mCP (60 nm)/BmPyPhB (30 nm)/LiF (0.8 nm)/Al (100 nm) have been fabricated with broad structureless EL emission at 511 nm showing a $\eta_{C,max}$ of 18.4 cd A^{-1} and a $\eta_{ext,max}$ of 5.8% for device **1** (Table 2). The better

TABLE 2 Electroluminescent key data are for AIE phosphorescence emitters.

Emitter	Active layer in device	λ_{EL} (nm)	V_{on}	L_{max} ($cd\,m^{-1}$)	$\eta_{c,max}/\eta_{p,max}/\eta_{ext,max}$ ($cd\,A^{-1}/lm\,W^{-1}/\%$)	Ref.
1	ITO/PEDOT:PSS (70nm)/5% 1: mCP (60nm)/BmPyPhB (30nm)/LiF(0.8nm)/Al (100nm)	511	–	–	18.4/8.0/5.8	[54]
N–Pt–P (8.0wt %)	ITO/PEDOT:PSS (45nm)/N–Pt–P 8%:CBP (40nm)/TPBi (45nm)/LiF (1nm)/Al (100nm)	535	3.9	18,575	64.9/46.5/23.2	[55]
N–Pt–S (8.0wt %)	ITO/PEDOT:PSS (45nm)/N–Pt–S 8%:CBP (40nm)/TPBi (45nm)/LiF (1nm)/Al (100nm)	535	3.4	13,874	54.0/43.0/19.4	[55]
P–Pt–N (8.0wt %)	ITO/PEDOT:PSS (45nm)/P–Pt–N 8%:CBP (40nm)/TPBi (45nm)/LiF (1nm)/Al (100nm)	506, 540	3.7	15,163	57.4/44.7/21.7	[55]
S–Pt–N (8.0wt %)	ITO/PEDOT:PSS (45nm)/S–Pt–N 8%:CBP (40nm)/TPBi (45nm)/LiF (1nm)/Al (100nm)	512, 544	3.6	13,926	75.9/62.7/28.4	[55]
A1	(ITO)/PEDOT:PSS/A1:BMIMPF$_6$ (5: 1)/Al	496	4.1	2655	25.7/–/7.6	[56]
NA2	(ITO)/PEDOT:PSS/NA2:BMIMPF$_6$ (5: 1)/Al	480	5.5	33	1.7/–/0.7	[56]
Cyclo-Ir (LB)	ITO/NPD/Cyclo-Ir/Bphen/Al	–	3.0	260	–/–0.35	[57]
BCZ1	ITO/m-MTDATA /TAPC (20nm)/BCZ1 (40nm)/TmPyPB (30nm)/Al	–	3.3	3295	13.4/10.5/5.8	[58]
BCZ2	ITO/m-MTDATA /TAPC (20nm)/BCZ2 (30nm)/TmPyPB (30nm)/Al	–	3.2	4019	4.8/3.5/2.0	[58]
BCZ4	ITO/m-MTDATA /TAPC (20nm)/BCZ4 (40nm)/TmPyPB (30nm)/Al	–	4.5	2123	2.8/1.2/1.6	[58]

Abbreviations: *λ_{EL} = the electroluminescence peak; V_{on} = turn-on voltage at 1 cd m^{-2}; L_{max} = maximum luminance; $\eta_{C,max}$ = maximum current efficiency; $\eta_{P,max}$ = maximum power efficiency; $\eta_{ext,max}$ = maximum external quantum efficiency; CBP = 4,4′-Bis(N-carbazolyl)-1,1′-biphenyl; m-MTDATA = (4,4′,4″-Tris[(3-methylphenyl)phenylamino]triphenylamine); PEDOT:PSS = Poly(3,4-ethylenedioxythiophene)-poly (styrenesulfonate; TAPC = 1,1-Bis[(di-4-tolylamino)phenyl]cyclohexane; TPBi = 1,3,5-tris(N-phenylbenzimidazol-2-yl)benzene; Bphen = 4,7-diphenyl-1,10-phenanthroline;BMIMPF$_6$ = 1-butyl-3-methyl-imidazolium hexafluorophosphate (Ionic liquid); BmPyPhB = 1,3-Bis[3,5-di (pyridin-3-yl)phenyl]benzene; PEDOT:PSS andTAPC functions as a hole-transporting layer (HTL); TPBi, BmPyPhB and Bphen serve as an electron-transporting layer (ETL) and a hole-blocking layer (HBL), respectively; m-MTDATA serves as a hole-injection layer (HIL).*

solubility and comparatively higher PLQY of **1** explained the higher EQE of the device doped with **1** than with **2** [54].

Wu and coworker made an effort to design and synthesize four AIE-active heteroleptic Pt^{II}(C∘N)(N-donor ligand)Cl type complexes **N-Pt-P**, **N-Pt-S**, **P-Pt-N**, and **S-Pt-N** (Scheme 10). All the Pt^{II} Complexes exhibit strong phosphorescence in PMMA film with lifetimes in the range of 0.3–2.3 μs and high PLQYs of up to 72%. In contrast, the PLQYs in solution are only lower than 7.8%. The AIE behavior ensures that all the Pt^{II} complexes attain improved device performance. In addition, the balanced charge carrier injection/transporting capabilities of Pt^{II} complexes bearing $-SO_2Ph$ and $-POPh_2$ units with efficient electron injection/transporting potentiality and the $-NPh_2$ moiety with hole injection/transporting property helps to improve the EL performance. Among them, an **S-Pt-N** (8wt% in CBP) based solution processed OLED with the configuration of ITO/PEDOT:PSS (45nm)/Pt x%:CBP (40nm)/TPBi (45nm)/LiF (1nm)/Al (100nm) displays high EL efficiencies with a maximal external η_{ext} of 28.4%, maximal η_L of 75.9cd A^{-1}, and maximal η_P of 62.7lmW^{-1} (Table 2) and (Fig. 4) [55].

Su's group successfully demonstrated the first highly efficient solution processed nondoped OLEDs employing AIE-active iridium(III) complexes as the emitting layers. In this study, three iridium(III) complexes, namely **A1**, **A2**, and **NA2**, have been synthesized (Scheme 10). Complexes **A1** and **A2** containing the carbazole moiety were found to be AIE active while **NA2** was AIE inactive, showing the key role of the carbazolyl group in the

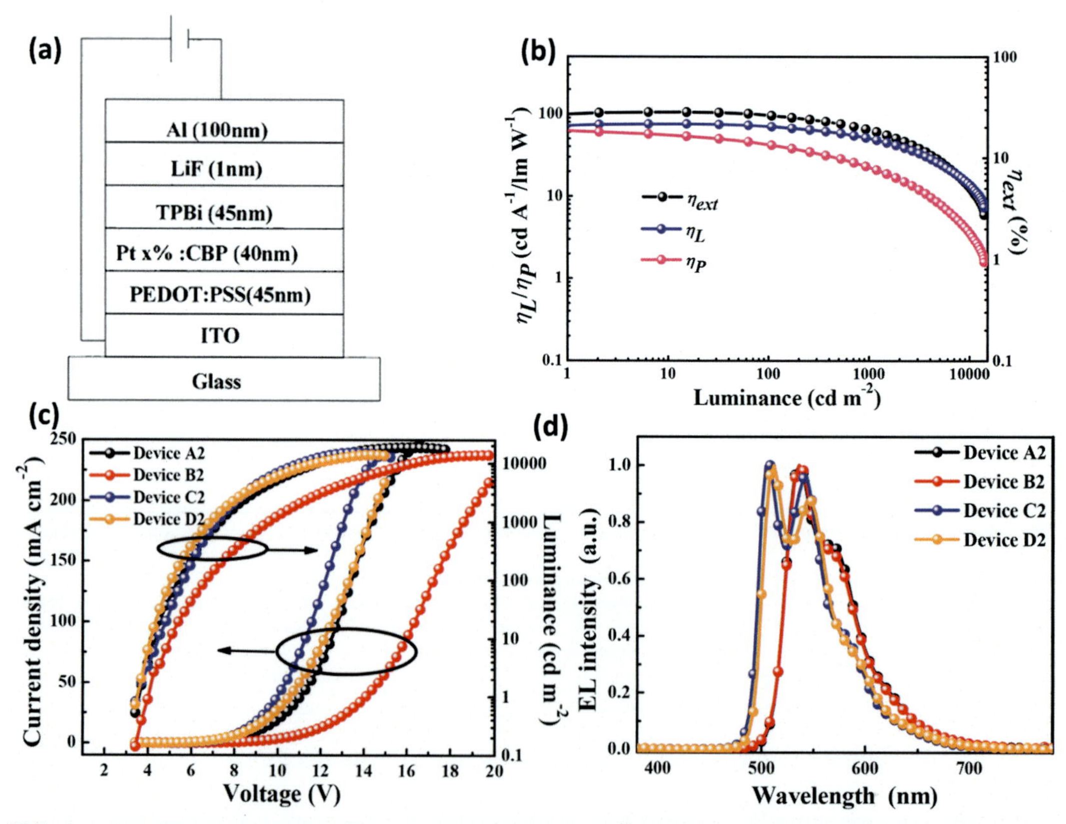

FIG. 4 (A) Configuration of the OLEDs made from AIE-active Pt^{II} complexes; (B) EL efficiency-luminance curves for the optimized device D2 [**S-Pt-N**(8.0wt%)]; (C) Current density-voltage-luminance (*J-V-L*) relationships of the optimized devices [A2: **N-Pt-P** (8.0wt%); B2: **N-Pt-S** (8.0wt%); C2: **N-Pt-N** (8.0wt%)]; (D) EL spectra for the optimized devices at 10V. *Reproduced with permission from J. Zhao, Z. Feng, D. Zhong, X. Yang, Y. Wu, G. Zhou, Z. Wu, Cyclometalated platinum complexes with aggregation-induced phosphorescence emission behavior and highly efficient electroluminescent ability, Chem. Mater. 30 (2018) 929–946. Copyright 2018 American Chemical Society.*

realization of AIE activity for these types of Ir(III) complexes. Moreover, the improved charge carrier injection and transport capability associated with the carbazolyl group in turn enhanced the EL performance of the devices based on the AIE Ir(III) emitters. The phosphorescence emissions of complexes **A1**, **A2**, and **NA2** were confirmed by their emission lifetimes of 0.71, 0.80, and 0.62 μs, respectively. Employing a neat film of **A1** or **NA2** as the emitting layer in a single nondoped OLED with a simplified configuration of glass/indium tin oxide (ITO)/PEDOT:PSS/emitter: BMIMPF6 (5: 1)/Al demonstrates that the device based on **A1** has a maximum luminance of 2655 cd m^{-2} and a peak EQE of 7.6% which are higher than the device based on **NA2** with corresponding values of 33 cd m^{-2} and 0.7% (Table 2) [56].

The formation of branched molecular wires with the support of exclusive aggregation properties of an AIE-active complex has been reported for the first time by Acharya's group [57]. The cyclometalated complex of Ir(III), namely, [Ir(ppy)$(PPh_3)_2$(H)(Cl)), (**Cyclo-Ir**) (Scheme 10), is revealed in plane J-aggregation at the air-water interface and finally takes the form of branched supramolecular wires. Its assembly stretches over 50 μm^2 while maintaining the close packing. As a result, the supramolecular wire demonstrates large amplification of photoluminescence (PL) because of RIR and RIV in the ordered aggregates. Thus, the branched supramolecular wires obtained from the Langmuir–Blodgett technique used as the light-emitting layer in the fabrication of light-emitting diodes exhibited lower threshold voltage with 260 cd m^{-2} and improved EQE of 0.35% than spin cast OLEDs [57]. Overall, this report shows the significance of ordered assembly of AIE complexes to accomplish the most favorable luminescence properties.

The pure organic aggregation-induced phosphorescence (AIP) luminogens, namely, **BCZ1**, **BCZ2**, **BCZ3**, and **BCZ4** (Scheme 10), were designed and developed by the Zhang group. The presence of an identical biscarbazole subunit and varying functional groups such as benzophenone, bromo benzophenone, methoxy benzophenone, and sulfone strengthens the ISC and thus RTP processes. All of the molecules in dilute solutions are either nonemissive (**BCZ2** and **BCZ3**) or only weakly emissive (**BCZ1** and **BCZ4**) but show increased PL intensity upon aggregation on addition of more than 50% (*v*/v) water. The reasons behind the AIE phenomena include restricted intermolecular motion (RIM), twisted intramolecular charge transfer (TICT) process, and reduced π-π stacking interaction. These AIE-active **BCZ** molecules deliver strong dual RTP with PLQYs in the range of 40%–64%. Using these unique RTP AIEgens as the emitting layer in the fabrication of nondoped OLED devices (configuration: ITO/m-MTDATA/TAPC (20 nm)/EML (30–40 nm)/TmPyPB (30 nm)/Al) achieved an external quantum efficiency of 5.8% and a low turn-on voltage of 3.3 V for the acetophenone modified **BCZ**. Therefore, this design strategy of metal-free AIE-based RTP organic molecules (commonly known as AIP organic molecules) was employed to realize nondoped OLEDs with high efficiency [58].

4 Aggregation-induced delayed fluorescence (AIDF)

The introduction section discussed conventional OLEDs in which the use of electrogenerated excitons is restricted to singlet excitons and hence restrains the η_{int} to 25%. On the basis of the assumption that in the absence of an optical out-coupling layer, η_{out} is approximately 20%–30%. As a result, the theoretically highest EQE is restricted to 5%–7.5%

regardless of the γ and photoluminescence quantum yield of a fluorophore approaching unity [59]. Therefore, great efforts have been dedicated to exploiting the nonemissive triplet excitons. A monumental breakthrough had occurred in 1998 where OLED using Ir(III), Pt(II), and Os(II) complexes were successfully confirmed the utilization of triplet excitons and broke the limits of the theoretical highest efficiency barrier [9]. In fact, on account of the strong spin-orbit coupling of heavy metals (e.g., iridium, platinum, and osmium) in the complexes, phosphorescent OLEDs (PhOLEDs) can realize nearly 100% IQE [10]. Conversely, PhOLEDs suffer from certain limitations for long-term mass production because of the lofty price and implicit environmental hazard due to the use of expensive metal complexes. To surmount such limitations, thermally activated delayed fluorescence (TADF) materials have aroused interest as the most potential candidates. In 1960, the "E-type" delayed fluorescence phenomenon came into the light for the first time, which is currently known as TADF [60]. It had been observed that eosin [60], fullerene [61], and porphyrin [62] derivatives with pure organic skeletons demonstrated TADF characteristics. Adachi and coworkers, particularly in 2012, had shown the scientific and industrial community their various contributions to improving the OLED performances via utilizing TADF molecules [63]. Organic TADF luminogens like the phosphorescent materials can also simultaneously utilize singlet and triplet excitons through the activated reverse intersystem crossing (RISC) process with the absorption of thermal energy from the environment, achieving theoretically 100% IQE.

5 AIDF (aggregation-induced delayed fluorescence) based on the conventional donor-acceptor (D-A) structural design

According to Boltzmann statistics, the prerequisite condition for an effective RISC process at a certain temperature is the sufficiently small singlet–triplet energy gap (ΔE_{ST}) [64], which can lead to small exchange energy (J) of the two unpaired electrons at the excited states [65]. By designing a specifically twisted donor-acceptor molecular structure, one can realize a small ΔE_{ST} via spatial wave function separation of the highest occupied molecular orbital (HOMO) and the lowest unoccupied molecular orbital (LUMO).

Over the past years, a number of efficient TADF emitters have been synthesized following the established theory; for instance, cores like triazine [66], sulfone [67], benzophenone [68], spirofluorene [69], xanthone [70,71], quinoxaline [72], etc. have been used to construct TADF materials. Based on the extensive design feasibility and ideal device performance, TADF luminogens have been considered as the third-generation light-emitting materials for OLEDs. However, the long lifetime of the excited states and strong intermolecular interaction in designed TADF emitters usually leads to concentration-caused emission quenching and serious exciton annihilation processes, which severely hampers the performance of most TADF-OLEDs. To thwart these problems, a host-guest doped light-emitting layer is usually adopted. Highly efficient doped TADF-OLEDs have been developed with the EQEs approaching 37.5% for blue [73], 31.3% for green [74], 21.5% for yellow [8e], 29.2% for orange [75], and 26.0% for orange-red TADF emitters [76]. But the accurate management of the doping concentration is intricate, and phase separation and device reproducibility are some of the tricky problems that stay with the host-guest doping OLEDs. What is more, particularly doped OLEDs based on TADF emitters suffer a severe efficiency roll-off at high luminance [77]. The combination of TADF and AIE is expected to create robust solid luminescent emitters with better exciton utilization, impressive EL efficiencies, and a reduced efficiency roll-off. Certainly, the aggregation-induced delayed fluorescence (AIDF) luminogens may play an essential part for nondoped OLEDs [78]. In this section, we will discuss in detail molecular design strategy, optical properties, and OLED performance using novel AIDF emitters. This section will present a better understanding of the progress of delayed fluorescence AIEgens with the improved performance of doped and nondoped OLEDs.

To validate the idea of AIDF, the Tang group reported a series of four D-A type small molecules, **DPS-PXZ**, **DBTO-PXZ**, **DPS-PTZ**, and **DBTO-PTZ** (Scheme 11), with diphenylsulfone (DPS) or dibenzothiophene-S,S-dioxide (DBTO) as electron-acceptor moieties and phenoxazine (PXZ) or phenothiazine (PTZ) as electron-donating moieties [79].

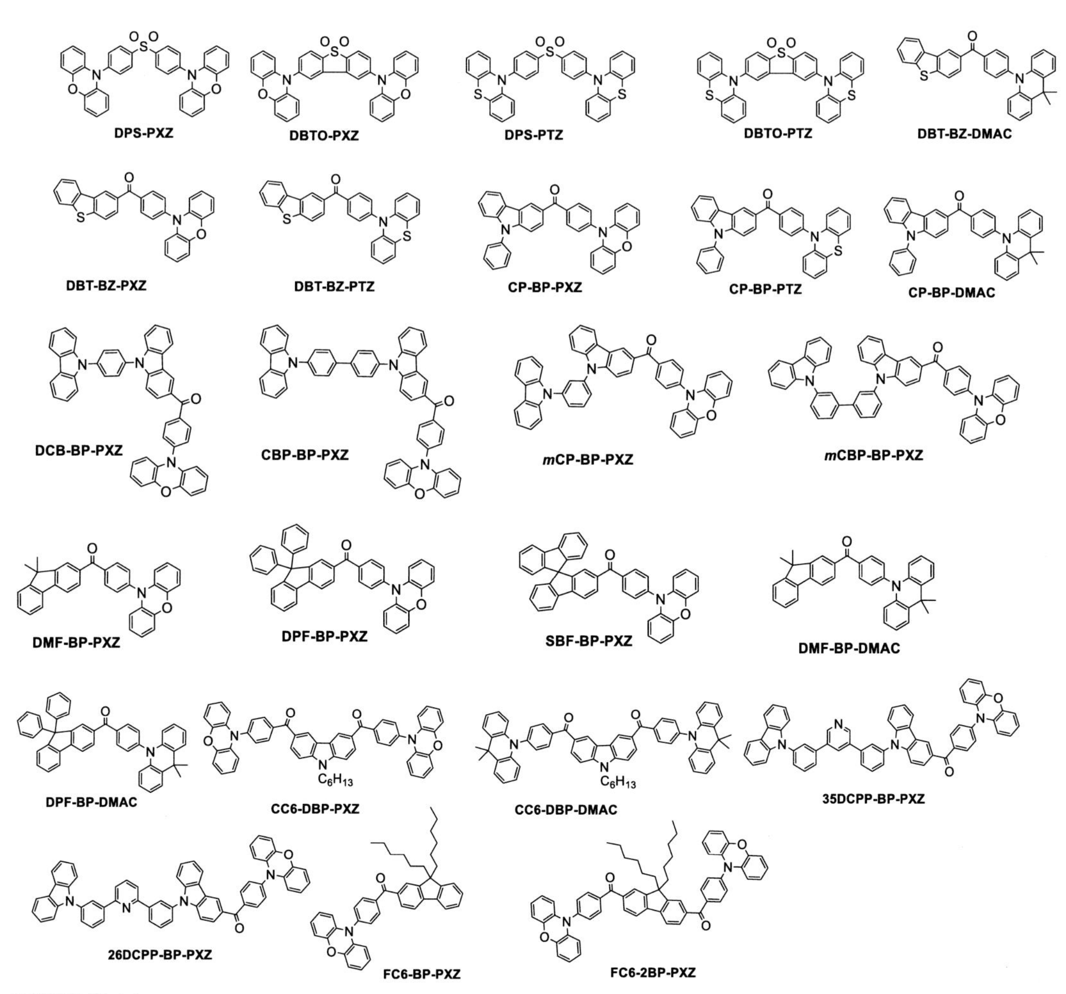

SCHEME 11 Chemical structures of small AIDF molecules reported by B.Z. Tang's group for OLEDs.

Noticeably, the twisted configurations of the D-A molecular design help in actualizing the efficient separation of HOMO–LUMO along with inhibiting the strong intermolecular π-π interactions. These fluorescent materials possessing both AIE and TADF properties can act as suitable emitters in nondoped OLEDs.

The unsymmetrical D-A-D′ luminogen, **DBT-BZ-DMAC**, was designed and developed by the Tang group by incorporating DMAC and dibenzothiophene (DBT) donors into the benzoyl (BZ) core (Scheme 11). The AIEgens displayed excellent PLQY of 80.2% in neat film and high exciton utilization in the device because of its AIDF feature. The **DBT-BZ-DMAC** based nondoped OLED with a configuration of ITO/TAPC/**DBT-BZ-DMAC**/TmPyPB/LiF/Al was fabricated with a considerable EL performance of up to 43.3 cd A^{-1} for $\eta_{C,max}$, 35.7 lm W^{-1} for $\eta_{P,max}$, and 14.2% for $\eta_{ext,max}$. Negligible efficiency roll-off and outstanding efficiency stability were recorded, which are comparable with most doped TADF-OLEDs [80]. To explain the difference in performance between doped and nondoped OLEDs, the same device configuration was maintained during fabrication apart from using doped films of **DBT-BZ-DMAC** in CBP with different doping concentrations. It was observed that the maxima EL efficiencies are decreased with the increase in doping concentration but extremely negligible efficiency roll-offs. From these outcomes, it can be concluded that the AIE character has a decisive part in inhibiting exciton annihilation and attaining good operational performance in OLEDs.

In the same year, the same group synthesized two new luminogens, **DBT-BZ-PXZ** and **DBT-BZ-PTZ**, by replacing DMAC with PXZ or PTZ [81]. Due to the efficient charge carrier transport and injection from the two emitters, **DBT-BZ-PXZ** and **DBT-BZ-PTZ** in simplified nondoped OLEDs with a similar configuration radiated light at a low voltage of 2.9 and 2.7 V, respectively. The device of **DBT-BZ-PTZ** achieved good performance with the efficiencies of approaching 26.5 cd A^{-1}, 29.1 lm W^{-1}, and 9.7% (Table 3). Notably, the nondoped OLEDs

TABLE 3 Electroluminescent key data of representative AIDF emitter-based OLEDs.

Emitter	Active layer in device	λ_{EL} (nm)	V_{on}	L_{max} (cd m^{-1})	$\eta_{c,max}/\eta_{p,max}/\eta_{ext,max}$ (cd A^{-1}/lm W^{-1}/%)	Ref.
PTZ-XT	ITO (100 nm)/α-NPD (40 nm)/mCBP (10 nm)/PTZ-XT (15 nm)/B3PyPB (55 nm)/Liq (1 nm)/Al (80 nm).	553	–	–	–/–/11.1	[71b]
PTZ-BP	ITO (100 nm)/α-NPD (40 nm)/mCBP (10 nm)/PTZ-BP (15 nm)/B3PyPB (55 nm)/Liq (1 nm)/Al (80 nm).	577	–	–	–/–/7.6	[71b]
DBT-BZ-PXZ	ITO/TAPC (25 nm)/DBT-BZ-PXZ (35 nm)/TmPyPB (55 nm)/LiF (1 nm)/Al	557	2.9	–	26.6/27.9/9.2	[81]
DBT-BZ-PTZ	ITO/TAPC (25 nm)/DBT-BZ-PTZ (35 nm)/TmPyPB (55 nm)/LiF (1 nm)/Al.	563	2.7	–	26.5/29.1/9.7	[81]
DBT-BZ-DMAC	ITO/TAPC (25 nm)/DBT-BZ-DMAC (35 nm)/TmPyPB (55 nm)/LiF (1 nm)/Al	516	2.7	27,270	43.3/35.7/14.2	[80]
CP-BP-PXZ	ITO/TAPC (25 nm)/CP-BP-PXZ (35 nm)/TmPyPB (55 nm)/LiF (1 nm)/Al	548	2.5	100,290	59.1/65.7/18.4	[82]

TABLE 3 Electroluminescent key data of representative AIDF emitter-based OLEDs—cont'd

Emitter	Active layer in device	λ_{EL} (nm)	V_{on}	L_{max} (cd m^{-1})	$\eta_{c,max}/\eta_{p,max}/\eta_{ext,max}$ (cd A^{-1}/lm W^{-1}/%)	Ref.
CP-BP-PTZ	ITO/TAPC (25nm)/CP-BP-PTZ (35nm)/TmPyPB (55nm)/LiF (1nm)/Al	554	2.5	46,820	46.1/55.7/15.3	[82]
CP-BP-DMAC	ITO/TAPC (25nm)/CP-BP-DMAC (35nm)/TmPyPB (55nm)/LiF (1nm)/Al	502	2.7	37,680	41.6/37.9/15.0	[82]
SBDBQ-DMAC	ITO/MoO_3 (10nm)/TAPC (50nm)/mCP (10nm)/SBDBQ-DMAC (20nm)/Bphen (45nm)/LiF/Al	544	2.8	14,578	35.4/32.7/10.1	[72a]
DBQ-3DMAC	ITO/MoO_3 (10nm)/TAPC (50nm)/mCP (10nm)/DBQ-3DMAC (20nm)/Bphen (45nm)/LiF/Al	548	2.6	29,843	41.2/45.4/12.0	[72a]
SBDBQ-PXZ	ITO/MoO_3 (10nm)/TAPC (50nm)/mCP (10nm)/SBDBQ-PXZ (20nm)/Bphen (45nm)/LiF/Al	608	2.4	21,050	10.5/12.0/5.6	[72a]
DBQ-3PXZ	ITO/MoO_3 (10nm)/TAPC (50nm)/mCP (10nm)/DBQ-3PXZ (20nm)/Bphen (45nm)/LiF/Al	616	2.8	13,167	7.5/6.2/5.3	[72a]
PCZ-CB-TRZ	ITO/α-NPD (35nm)/mCP (10nm)/PCZ-CB-TRZ (20nm)/PPT (40nm)/LiF (0.8nm)/Al (80nm)	586	6.3	4530	–/–/11.0	[71a]
TPA-CB-TRZ	ITO/α-NPD (35nm)/mCP (10nm)/TPA-CB-TRZ (20nm)/PPT (40nm)/LiF (0.8nm)/Al (80nm)	590	4.4	–	–/–/10.1	[71a]
2PCZ-CB	ITO/α-NPD (35nm)/mCP (10nm)/2PCZ-CB (20nm)/PPT (40nm)/LiF (0.8nm)/Al (80nm)	631	4.4	–	–/–/9.2	[71a]
SFDBQPXZ	ITO/MoO_3 (10nm)/TAPC (50nm)/mCP (10nm)/SFDBQPXZ (30nm)/Bphen (45nm)/LiF (1nm)/Al	584	–	–	24.3/22.5/10.1	[72b]
SFDBQPXZ	ITO/MoO_3 (10nm)/TAPC: MoO_3 (20%, 50nm)/TAPC (20nm)/CBP: SFDBQPXZ (10%, 20nm)/Bphen (45nm)/LiF (1nm)/Al	548	2.7	31,790	78.3/91.1/23.5	[72b]
DFDBQPXZ	ITO/MoO_3 (10nm)/TAPC (50nm)/mCP (10nm)/DFDBQPXZ (30nm)/Bphen (45nm)/LiF (1nm)/Al	588	–	–	21.0/20.6/9.8	[72b]
DFDBQPXZ	ITO/MoO_3 (10nm)/TAPC: MoO_3 (20%, 50nm)/TAPC (20nm)/CBP: DFDBQPXZ (10%, 20nm)/Bphen (45nm)/LiF (1nm)/Al	548	2.8	31,099	55.9/57.8/16.8	[72b]
CCDC	ITO/NPB/TCTA/CCDC /TPBi/LiF/Al	480	3.3	3000	4.0/4.0/2.0	[83]

Continued

TABLE 3 Electroluminescent key data of representative AIDF emitter-based OLEDs—cont'd

Emitter	Active layer in device	λ_{EL} (nm)	V_{on}	L_{max} (cd m^{-1})	$\eta_{c,max}/\eta_{p,max}/\eta_{ext,max}$ (cd A^{-1}/ lm W^{-1}/%)	Ref.
CCDD	–	543	2.4	14,600	39.8/41.7/12.7	[83]
o-ACSO2	ITO/PEDOT:PSS/*o*-ACSO2/DPEPO/TmPyPB/Liq/Al	490	–	≈1000	14.1/7.8/5.9	[72c]
m-ACSO2	ITO/PEDOT:PSS/*m*-ACSO2/DPEPO/TmPyPB/Liq/Al	486	–	≈2000	37.9/23.8/17.2	[72c]
DCPDAPM	ITO/HATCN/TAPC/DCPDAPM / TmPyPB/LiF/Al	522	3.2	123,371	26.9/15.6/8.2	[84]
Fene	ITO/TAPC (30nm)/TCTA (5nm)/Fene (15nm)/(TmPyPB) (65nm)/LiF (1nm)/Al (100nm)	570	3.2	–	42.2/31.57/14.9	[85]
Fens	ITO/TAPC (30nm)/TCTA (5nm)/Fens (15nm)/(TmPyPB) (65nm)/LiF (1nm)/Al (100nm)	568	3.2	–	36.8/30.40/13.1	[85]
Yad	ITO/TAPC (30nm)/TCTA (5nm)/Yad (15nm)/(TmPyPB) (65nm)/LiF (1nm)/Al (100nm)	534	3	–	58.14/57.01/17.4	[85]
CzTAZPO	ITO/PEDOT:PSS (40nm)/CzTAZPO (50nm)/TmPyPB (20nm)/Ca (10nm)/Al (100nm)	537	4.5	9776	29.1/−12.8	[86]
sCzTAZPO	ITO/PEDOT:PSS (40nm)/sCzTAZPO (50nm)/TmPyPB (20nm)/Ca (10nm)/Al (100nm)	531	4.1	8283	20.6/−9.6	[86]
pipd-BZ-PXZ	ITO/HATCN (5nm)/TAPC (25nm)/TCTA (5nm)/*pipd*-BZ-PXZ (35nm)/TmPyPB (55nm)/LiF (1nm)/Al	570	3.2	24,474	19.86/17.35/7.04	[87]
pipd-BZ-PTZ	ITO/HATCN (5nm)/TAPC (25nm)/TCTA (5nm)/*pipd*-BZ-PTZ (35nm)/TmPyPB (55nm)/LiF (1nm)/Al	576	2.6	18,006	17.35/18.64/6.90	[87]
pipd-BZ-DMAC	ITO/HATCN (5nm)/TAPC (25nm)/TCTA (5nm)/*pipd*-BZ-DMAC (35nm)/TmPyPB (55nm)/LiF (1nm)/Al	528	3.0	4436	7.16/6.25/2.58	[87]
3-CCP-BP-PXZ	ITO/HATCN (5nm) /TAPC (30nm)/TCTA (5nm)/3-CCP-BP-PXZ (20nm)/TmPyPB (40nm)/LiF (1nm)Al	540	2.8	88,750	76.6/75.2/21.7	[88]
9-CCP-BP-PXZ	ITO/HATCN (5nm) /TAPC (30nm)/TCTA (5nm)/9-CCP-BP-PXZ (20nm)/TmPyPB (40nm)/LiF (1nm)Al	537	2.6	29,510	72.5/53.5/20.4	[88]
3,9-CCP-BP-PXZ	ITO/HATCN (5nm) /TAPC (30nm)/TCTA (5nm)/3,9-CCP-BP-PXZ (20nm)/TmPyPB (40nm)/LiF (1nm)Al	541	2.8	35,580	72.1/65.4/20.6	[88]
		560	2.7	27,331	39.9/38.0/13.3	[89a]

TABLE 3 Electroluminescent key data of representative AIDF emitter-based OLEDs—cont'd

Emitter	Active layer in device	λ_{EL} (nm)	V_{on}	L_{max} (cd m^{-1})	$\eta_{c,max}/\eta_{p,max}/\eta_{ext,max}$ (cd A^{-1}/lm W^{-1}/%)	Ref.
DMF-BP-PXZ	(ITO)/TAPC (25nm)/DMF-BP-PXZ (35nm)/TmPyPB (55nm)/LiF (1nm)/Al					
DPF-BP-PXZ	(ITO)/TAPC (25nm)/DMF-BP-PXZ (35nm)/TmPyPB (55nm)/LiF (1nm)/Al	560	2.6	31,422	41.6/45.0/14.3	[89a]
SBF-BP-PXZ	(ITO)/TAPC (25nm)/DMF-BP-PXZ (35nm)/TmPyPB (55nm)/LiF (1nm)/Al	560	2.5	33,990	36.8/37.9/12.3	[89a]
DMF-BP-DMAC	ITO/HATCN (5nm)/TAPC (20nm)/TCTA (5nm)/DMF-BP-DMAC (35nm)/TmPyPB (55nm)/LiF (1nm)/Al	528	3.2	23,270	19.9/16.4/5.7	[89b]
DMF-BP-DMAC	ITO/HATCN (5nm)/NPB (30nm)/mCP (10nm)/DMF-BP-DMAC (30nm)/TPBi (50nm)/LiF (1nm)/Al	526	3.9	32,460	21.6/14.6/6.4	[89b]
DPF-BP-DMAC	ITO/HATCN (5nm)/TAPC (20nm)/TCTA(5nm)/DPF-BP-DMAC (35nm)/TmPyPB (55nm)/LiF (1nm)/Al	534	2.8	50,170	43.8/40.2/13.2	[89b]
DPF-BP-DMAC	ITO/HATCN (5nm)/NPB (30nm)/mCP (10nm)/DPF-BP-DMAC (30nm)/TPBi (50nm)/LiF (1nm)/Al	524	3.1	52,560	42.3/30.2/14.4	[89b]
DCB-BP-PXZ	(ITO)/TAPC (25nm)/emitter (35nm)/TmPyPB (55nm)/LiF (1nm)/Al	548	2.5	95,577	72.9/81.8/22.6	[90]
CBP-BP-PXZ	(ITO)/TAPC (25nm)/emitter (35nm)/TmPyPB (55nm)/LiF (1nm)/Al	546	2.5	98,089	69.0/75.0/21.4	[90]
mCP-BP-PXZ	(ITO)/TAPC (25nm)/emitter (35nm)/TmPyPB (55nm)/LiF (1nm)/Al	542	2.5	100,126	72.3/79.0/22.1	[90]
mCBP-BP-PXZ	(ITO)/TAPC (25nm)/emitter (35nm)/TmPyPB (55nm)/LiF (1nm)/Al	542	2.5	96,815	70.4/76.5/21.8	[90]
PXZ2PTO	ITO/MoO_3 (10nm)/TAPC (40nm)/mCP (20nm)/PXZ2PTO (30nm)/DPEPO (20nm)/TPBi (15nm)/LiF (1nm)/Al (100nm)	504	4.3	–	44.9/32.0/16.4	[91]
PXZ2PTO	ITO/MoO_3 (10nm)/TAPC (30nm)/mCP (25nm)/80wt% PXZ2PTO doped DPEPO (30nm)/DPEPO (5nm)/TPBi (30nm)/LiF (1nm)/Al (100nm)	500	3.8	–	43.8/35.2/16.3	[91]
TATC-BP		549	2.6	–	17.8/20.0/5.9	[92]

Continued

TABLE 3 Electroluminescent key data of representative AIDF emitter-based OLEDs—cont'd

Emitter	Active layer in device	λ_{EL} (nm)	V_{on}	L_{max} (cd m^{-1})	$\eta_{c,max}/\eta_{p,max}/\eta_{ext,max}$ (cd A^{-1}/lm W^{-1}/%)	Ref.
	ITO/PEDOT:PSS (25nm)/TATC-BP (25nm)/TmPyPB (55nm)/LiF (1nm)/Al (150nm)					
TATP-BP	ITO/PEDOT:PSS (25nm)/TATC-BP (25nm)/TmPyPB (55nm)/LiF (1nm)/Al (150nm).	541	2.7	–	18.9/19.2/6.0	[92]
CC6-DBP-DMAC	ITO/PEDOT:PSS (50nm)/PVK (30nm)/ CC6-DBP-DMAC (60nm) /TmPyPB (40nm)/LiF (1nm)/Al	505	4.2	14,366	25.08/11.25/9.02	[93]
CC6-DBP-PXZ	ITO/PEDOT:PSS (50nm)/PVK (30nm)/ CC6-DBP-PXZ (40nm) /TmPyPB (40nm)/ LiF (1nm)/Al	568	2.9	30,644	22.23/16.11/7.73	[93]
PFBP-1b	ITO/MoO_3/NPB/PFBP-1b:mCP/TPBi/ Ca/Al	480	3.8	1850	6.6/4.4/3.3	[94a]
PFBP-2a	ITO/MoO_3/NPB/PFBPb-2a:TCz1/TPBi/ Ca/Al	487	3.9	550	3.7/2.5/3.9	[94a]
PFBP-2b	ITO/MoO_3/NPB/PFBP-2b/TPBi/Ca/Al	499	7.0	3000	16.3/5.8/6.6	[94a]
PFBP-2b	ITO/MoO_3/NPB/PFBP-2b:TCz1/TPBi/ Ca/A	499	3.0	3900	30.8/22.7/16.3	[94a]
NAI–BiFA	ITO/HAT-CN (15nm)/TAPC (40nm)/ TCTA (5nm)/NAI–BiFA (20nm)/TmPyPB (40nm)/LiF (1nm)/Al (100nm)	628	2.9	10,200	2.44/1.85/1.53	[95]
NAI–BiFA	ITO)/HAT-CN (15nm)/TAPC (40nm)/ TCTA (5nm)30wt%- NAI–BiFA:CBP (20nm)/TmPyPB (40nm)|LiF (1nm)Al (100nm)	602	3.1	15,210	15.25/13.30/7.21	[95]
NAI–PhBiFA	ITO)/HAT-CN (15nm)/TAPC (40nm)/ TCTA (5nm)/NAI–PhBiFA (20nm)/ TmPyPB (40nm)/LiF (1nm)/Al (100nm)	583	2.9	14,310	3.71/2.39/1.39	[95]
NAI–PhBiFA	ITO)/HAT-CN (15nm)/TAPC (40nm)/ TCTA (5nm)/5wt%- NAI–PhBiFA:CBP (20nm)/TmPyPB (40nm)/LiF (1nm)/Al (100nm)	538	3.4	51,220	27.95/20.97/7.59	[95]
ND-AC	ITO/TAPC (30nm)/TCTA (5nm)/ND-AC (15nm) (EML)/TmPyPB (65nm)/LiF (1nm)/Al (100nm).	558	–	–	38.5/30.2/12.0	[96]
ND-AC	ITO/TAPC (30nm)/TCTA (5nm)/ND-AC (9wt%):CBP (15nm) (EML)/TmPyPB (65nm)/LiF (1nm)/Al (100nm).	542	–	–	38.9/16.1/11.3	[96]

TABLE 3 Electroluminescent key data of representative AIDF emitter-based OLEDs—cont'd

Emitter	Active layer in device	λ_{EL} (nm)	V_{on}	L_{max} (cd m^{-1})	$\eta_{c,max}/\eta_{p,max}/\eta_{ext,max}$ (cd A^{-1}/lm W^{-1}/%)	Ref.
CND-AC	ITO/TAPC (30nm)/TCTA (5nm)/CND-AC (1.5wt%):CBP (15nm) (EML)/TmPyPB (65nm)/LiF (1nm)/Al (100nm)	588	–	–	21.3/14.6/8.4	[96]
DMAC-ND	ITO/TAPC (30nm)/TCTA (5nm)/ emitting layer (20nm)/1,3,5- (Tm3PyPB) (60nm)/LiF (1nm)/Al (100nm)	514	2.9	–	34.7/32.0/11.0	[97]
PTZ-ND	ITO/TAPC (30nm)/TCTA (5nm)/ emitting layer (20nm)/1,3,5- (Tm3PyPB) (60nm)/LiF (1nm)/Al (100nm)	534	3.3	–	30.7/26.2/9.7	[97]
PXZ-ND	ITO/TAPC (30nm)/TCTA (5nm)/ emitting layer (20nm)/1,3,5- (Tm3PyPB) (60nm)/LiF (1nm)/Al (100nm)	568	3.0	–	10.2/8.8/3.7	[97]
34AcCz-PM	ITO/MoO_3 (10nm)/TAPC/TCTA/ 34AcCz-PM (15) /TmPyPB/LiF (1nm)/Al	548	3.10	–	45.2/40.0/14.1	[98]
34AcCz-Trz	ITO/MoO_3 (10nm)/TAPC/TCTA/ 34AcCz-Trz (10) /TmPyPB/LiF (1nm)/Al	576	3.35	–	18.0/15.5/7.3	[98]
TPA-DCPP	ITO/NPB (80nm)/TCTA (5nm)/TPA-DCPP (20nm)/TPBi (30nm)/LiF (0.5nm)/ Al	710	4.0	591	0.24/0.19/2.1	[99]
TPA-DCPP	ITO/NPB (80nm)/TCTA (5nm)/ TPBi:20wt% TPA-DCPP (20nm)/TPBi (30nm)/LiF (0.5nm)/Al	668	3.1	3805	4.0/4.0/9.8	[99]
TAT-BP	ITO/PEDOT:PSS (25nm)/TAT-BP (25nm)/TmPyPB (55nm)/LiF (1nm)/Al (150nm)	530	2.5	–	20.9/21.8/6.4	[100]
TAT-2BP	ITO/PEDOT:PSS (25nm)/TAT-2BP (25nm)/TmPyPB (55nm)/LiF (1nm)/Al (150nm)	530	2.5	–	32.3/33.0/9.8	[100]
35DCPP-BP-PXZ	ITO/HATCN (5nm)/TAPC(30nm)/TCTA (5nm)/35DCPP-BP-PXZ (20nm)/TmPyPB (50nm)/LiF (1nm)/Al	538	3.0	38,237	57.6/49.7/17.3	[101]
26DCPP-BP-PXZ	ITO/HATCN (5nm)/TAPC(30nm)/TCTA (5nm)/26DCPP-BP-PXZ (20nm)/TmPyPB (50nm)/LiF(1nm)/Al	542	3.0	25,106	53.2/37.0/16.1	[101]
4CzIPN-MO	ITO/PEDOT:PSS (60nm)/4CzIPN-MO(40nm)/TPBi (30nm)/Cs_2CO_3 (2nm)/ Al (100nm)	574	3.1	7615	15.6/12.8/5.6	[102]
4CzPhIPN-MO	ITO/PEDOT:PSS (60nm)/4CzPhIPN-MO (40nm)/TPBi (30nm)/Cs_2CO_3 (2nm)/Al (100nm)	536	3.4	16,682	45.1/36.0/14.5	[102]

Continued

TABLE 3 Electroluminescent key data of representative AIDF emitter-based OLEDs—cont'd

Emitter	Active layer in device	λ_{EL} (nm)	V_{on}	L_{max} ($cd\,m^{-1}$)	$\eta_{c,max}/\eta_{p,max}/\eta_{ext,max}$ ($cd\,A^{-1}$/ $lm\,W^{-1}$/%)	Ref.
FC6-BP-PXZ	ITO/TAPC(25nm)/CBP:30wt% FC6-BP-PXZ (35nm) /TmPyPB (55nm)/LiF (1nm)/Al	544	3.2	80,507	48.02/38.91/14.86	[103]
FC6-2BP-PXZ	ITO/PEDOT:PSS (50nm)/PVK (30nm)/ CBP:10wt% FC6-2BP-PXZ (50nm) / TmPyPB (40nm)/LiF (1nm)/Al	568	3.9	22,530	44.83/32.03/14.69	[103]
OTQx	ITO/MoO_3(0.5nm)/NPB (35nm)/OTQx or OCQx (10wt%):TCTA (20nm)/TSPO1 (7nm)/TPBi (30nm)/LiF (0.5nm)/Al	544	4.08	48,809	33.64/21.08/10.53	[94b]

Abbreviations: *λ_{EL}=the electroluminescence peak; V_{on}=turn-on voltage at 1 $cd\,m^{-2}$; L_{max}=maximum luminance; $\eta_{C,max}$=maximum current efficiency; $\eta_{P,max}$=maximum power efficiency; $\eta_{ext,max}$=maximum external quantum efficiency; mCP=1,3-bis(Ncarbazolyl)Benzene; TCz1=3,6-Bis(carbazol-9-yl)-9-(2-ethyl-hexyl)-9H–carbazole; CBP=4,4′-Bis(N-carbazolyl)-1,1′-biphenyl; PEDOT:PSS=Poly(3,4-ethylenedioxythiophene)-poly(styrenesulfonate; TAPC=1,1-Bis[(di-4-tolylamino)phenyl]cyclohexane; HAT-CN=1,4,5,8,9,11-Hexaazatriphenylenehexacarbonitrile; TCTA=tris(4-carbazoyl-9-ylphenyl)amine; α-NPD=naphtyl-substituted benzidine derivative; PVK =poly(9-vinylcarbazole); TmPyPB=1,3,5-Tris(3-pyridyl-3-phenyl)benzene, NPB=N,N′-di(1-naphthyl)-N,N′-diphenylbenzidine; TPBi=1,3,5-tris(N-phenylbenzimidazol- 2-yl)benzene; Bphen=4,7-diphenyl-1,10-phenanthroline; B3PyPB=1,3-Bis(3,5-dipyrid-3-ylphenyl) benzene; DPEPO=Bis[2-(diphenylphosphino)phenyl] ether oxide; PPT=2,8-bis(diphenylphosphoryl)dibenzo[b,d]thiophene); Alq_3=Tris-(8-hydroxyquinoline)aluminum. PEDOT:PSS, NPB, TAPC, TCTA, α-NPD and PVK functions as a hole-transporting layer (HTL); B3PyPB, TmPyPB,TPBi, PPT, DPEPO, and Bphen serve as an electron-transporting layer (ETL) and a hole-blocking layer (HBL), respectively; Alq_3 functions as ETL; and MoO_3, HATCN serves as a hole-injection layer (HIL).*

secure very small efficiency roll-offs, by using materials having AIE and TADF features which support the suppression of strong intermolecular interactions and exciton annihilation. Therefore, to design novel organic luminogens for high-performance and stable nondoped OLEDs, the alliance of TADF and AIE is highly desirable.

To understand the underlying relationship between AIE and TADF, the same group put forward a new idea of AIDF. A series of new luminogens, namely **CP-BP-PXZ, CP-BP-PTZ,** and **CP-BP-DMAC**, were designed and synthesized through this concept (Scheme 11) [82]. All luminogens illustrated weak emission and negligible delayed fluorescence in dilute THF solution but strong emission with aggregation-induced delayed fluorescence upon the formation of aggregates on adding a large volume of water into the THF solution (Fig. 5). The reason for high emission efficiencies and effective exciton utilization in neat films was related to the suppression of concentration-caused emission quenching and exciton annihilation. Hence, nondoped OLEDs with a simplified architecture of ITO/TAPC/emitter/ TmPyPB/LiF/Al were fabricated, where AIDF luminogens in the neat film state acted as the emitting layer. The **CP-BP-PXZ**-based OLEDs presented outstanding electroluminescence efficiencies of 59.1 $cd\,A^{-1}$, 65.7 $lm\,W^{-1}$, and 18.4% and a negligible efficiency roll-off of 1.2% at 1000 $cd\,m^{-2}$ and excellent device stability. The device performance for **CP-BP-PTZ** and **CP-BP-DMAC** emitters with the same device configuration illuminated yellow and green emissions at 554 and 502 nm and EL performances of 46.1 $cd\,A^{-1}$, 55.7 $lm\,W^{-1}$, 15.3% and 41.6 $cd\,A^{-1}$, 37.9 $lm\,W^{-1}$, 15.0%, respectively. Remarkably, the EL performances of nondoped devices, showing the advantage of a smaller efficiency roll-off, are comparable

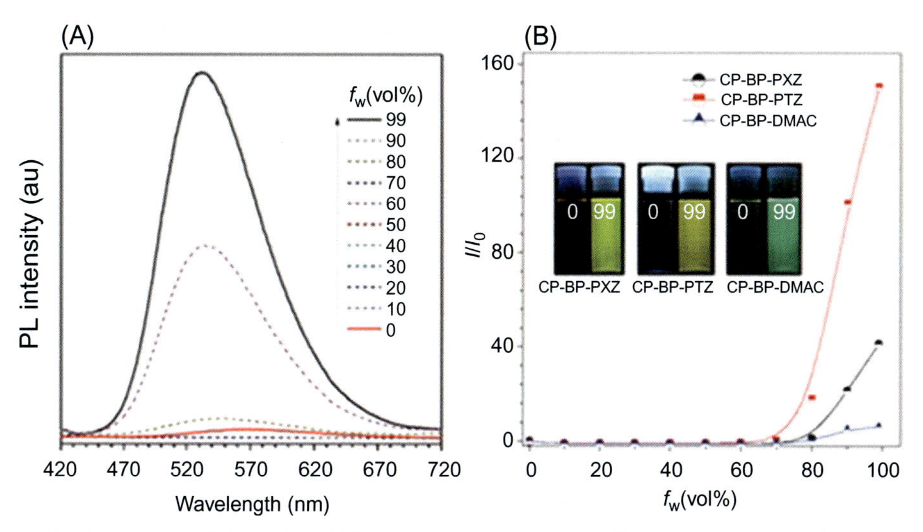

FIG. 5 (A) PL spectra of **CP-BP-PXZ** in THF/water mixtures with a different water fraction (f_w). (B) Plot of I/I_o versus f_w. I_o is the PL intensity in pure THF. Inset: Photographs of different luminogens in the THF/water mixtures ($f_w = 0$ and 90%), taken under 365 nm excitation. *Reproduced with permission from J. Huang, H. Nie, J. Zeng, Z. Zhuang, S. Gan, Y. Cai, J. Guo, S.-J. Su, Z. Zhao, B.Z. Tang, Highly efficient nondoped OLEDs with negligible efficiency roll-off fabricated from aggregation-induced delayed fluorescence luminogens, Angew. Chem. 129 (2017) 13151–13156; Angew. Chem. Int. Ed. 56 (2017), 12971–12976. Copyright 2017 John Wiley and Sons.*

to those of doped devices. In the nondoped OLEDs, balanced electrons and holes transporting ability leads to efficiently recombination of both the charges that afforded for the almost full exciton utilization and high performance.

The same group then reported another four AIDF luminogens, namely **DCB-BP-PXZ**, **CBP-BP-PXZ**, ***m*CP-BP-PXZ**, and ***m*CBP-BP-PXZ**, which were acquired via tagging of the AIDF (4-(phenoxazin-10-yl)benzoyl unit to ordinary host materials comprising DCB, CBP, *m*CP, and *m*CBP (Scheme 11) [90]. Utilizing the above unique AIDF luminogens as the emitting layer in nondoped OLEDs helps in realizing nearly 100% exciton utilization and providing outstanding EL performances with an L_{max} of ≈100,000 cd m^{-2}, excellent EQE of up to 22.6%, and negligible efficiency roll-off at 1000 cd m^{-2} (Table 3; Fig. 6). These outcomes further confirmed the potentiality of AIDF materials to be used as light-emitting layers for fabricating the high-performance and highly stable nondoped OLEDs.

Interestingly, Tang and coworkers put forward comprehensive insights into the basic mechanism of AIDF materials following the Anti-kasha behavior. To explain the underlying mechanism, a series of new AIDF luminogens **DMF-BP-PXZ**, **DPF-BP-PXZ**, and **SBF-BP-PXZ** (Scheme 11) were designed and synthesized by using benzoyl as the electron withdrawing group and phenoxazine and fluorine derivatives as the electron-donating group [89a]. These luminogens demonstrated weak emission without delayed fluorescence in solutions but strong emission with delayed components in the aggregate and thin-film states, representing the classical characteristics of AIDF features. The theoretical calculations of **DMF-BP-PXZ** elucidated that its effective emission arises from the higher energy electronic excited states (S_2) rather than the S_1 state, substantiating the anti-kasha behavior. On account of the excellent

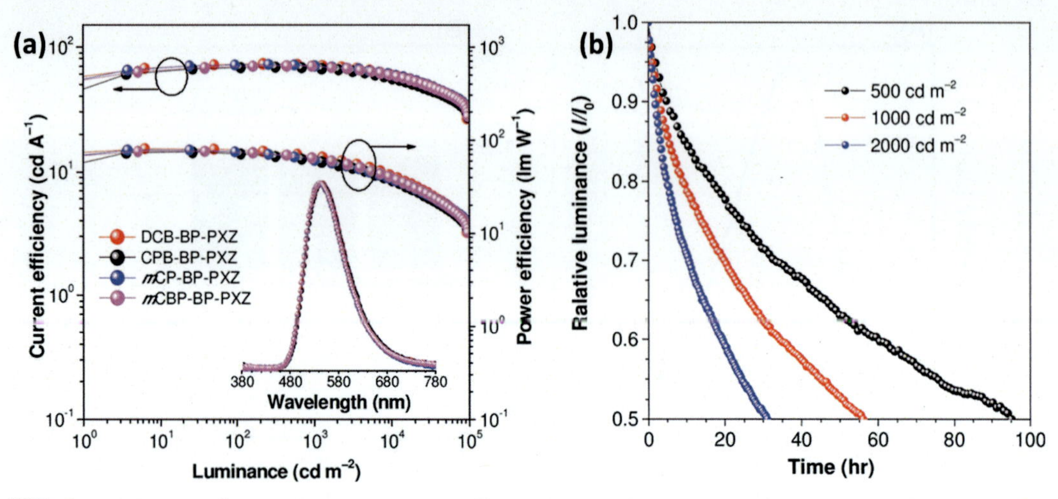

FIG. 6 (A) Current efficiency-luminance-power efficiency curves of nondoped OLEDs. The inset of figure (A) is the EL spectra obtained at a luminance of $5000\,cd\,m^{-2}$. (B) Relative luminance-time curves of **CBP-BP-PXZ**-based nondoped OLEDs at different initial luminances. *Reproduced with permission from H. Liu, J. Zeng, J. Guo, H. Nie, Z. Zhao, B.Z. Tang, High-performance non-doped OLEDs with nearly 100% exciton use and negligible efficiency roll-off, Angew. Chem. Int. Ed. 57 (2018) 9290–9294. Copyright 2018 John Wiley & Sons.*

AIDF property, these luminogens achieved high exciton utilization, thus rendering superb EL performance and low efficiency roll-off for both nondoped and doped devices with the same configuration of indium tin oxide (ITO)/TAPC (25nm)/emitter (35nm)/TmPyPB (55nm)/LiF (1nm)/Al, where the neat film of AIDF materials was adopted as the light-emitting layer for nondoped devices and 30wt% AIDF emitter with CBP host as the light-emitting layer for doped devices. The best EL performance of nondoped devices based on **DPF-BP-PXZ** presented $41.6\,cd\,A^{-1}$, $45.0\,lm\,W^{-1}$, 14.3% accompanied by low efficiency roll-off (1.4%) at $1000\,cd\,m^{-2}$ (Table 3).

Similar behaviors were observed for the other two AIDF luminogens, **DMF-BP-DMAC** and **DPF-BP-DMAC**, which were designed with an asymmetric D-A-D′ molecular skeleton [89b]. The AIDF nature supported the simplified nondoped OLEDs fabrication, demonstrating green EL emission with a low turn-on voltage of 3.1V together with a device performance of $42.3\,cd\,A^{-1}$, $30.2\,lmw^{-1}$, and 14.4% for **DPF-BP-DMAC** (Table 3). In the same year, the same group reported two new fluorescent small molecules **CC6-DBP-PXZ** and **CC6-DBP-DMAC** (Scheme 11), comprised of the benzoyl group as the electron acceptor and 9-hexylcarbazole and the PXZ or DMAC unit as the electron donor [93]. Both the molecules emitted a weakly nondelayed component in dilute solution but strong delayed fluorescence after aggregation. Accordingly, **CC6-DBP-DMAC** can perform admirably as light-emitting layers in solution-processed nondoped OLEDs with a configuration of ITO/PEDOT:PSS (50nm)/PVK (30nm)/emitter/TmPyPB (40nm)/LiF (1nm)/Al, which revealed a high $\eta_{ext,max}$ of 9.02% and negligible efficiency roll-off at the luminance of $1000\,cd\,m^{-2}$. Conversely, the two materials both performed potently as dopants in a CBP host at diverse doping concentrations, allowing the solution-processed doped OLEDs reaching as high of 12.1% ($\eta_{ext,max}$) (Table 3). Recently, Tang et al. developed two new compounds, namely **35DCPP-**

BP-PXZ and **26DCPP-BP-PXZ** (Scheme 11), by inserting an AIDF moiety [4-(phenoxazin-10-yl)benzoyl] into bipolar carrier transport materials, namely 3,5-bis((9H-carbazol-9yl)-3,1-phenylene)pyridine (35DCPP) and 2,6-bis(3-(9H-carbazol-9-yl)phenyl)pyridine (26DCPP) [101]. Notably, both compounds possessed the AIDF property with high PLQY in neat film that fulfilled the request of acting as the light-emitting layer in nondoped devices. Moreover, the excellent bipolar charge transport abilities of both compounds helped to improve the EL performances of up to 57.6 cd A^{-1}, 49.7 lm W^{-1}, 17.3%, and efficient device stability with roll-off of 0.6% at the luminance of 1000 cd m^{-2} (Table 3).

Furthermore, Tang and coworkers reported two new emitters **FC6-BP-PXZ** and **FC6-2BP-PXZ** (Scheme 11) with the electron acceptor of benzoyl and electron donor of phenoxazine and 9,9-dihexylfluorene [103]. Both of the compounds showed weak emission in dilute solution but emitted delayed fluorescence efficiently in the aggregated state. Therefore, both compounds executed outstandingly as emitting layers in solution-processed OLEDs, achieving a high EQE of up to 14.69% and negligible efficiency roll-off at the high luminance of 1000 cd m^{-2}, reflecting the excellent efficiency stability (Table 3). Alternatively, vacuum-deposited OLEDs using the two emitters also exhibited good $\eta_{ext,max}$ approaching 14.86% and insignificant efficiency RO at 1000 cd m^{-2}. The above outcome points out the importance of utilizing small AIDF molecules in the solution-processed and vacuum-deposited OLEDs with high EL performance.

During the same period, Yasuda and coworkers [71a] also recognized the significance of luminogens possessing both AIE and TADF properties. In their design, D-A-A′ type molecules related to organic–inorganic conjugated systems were developed with a electron-deficient icosahedral boron cluster, *o*-carborane, as the common acceptor. The three luminogens (**PCZ-CB-TRZ**, **TPA-CB-TRZ**, and **2PCZ-CB**) were found to demonstrate distinct AIE characteristics, exhibiting a high PLQY of 97% in neat films (Scheme 12). These *o*-carborane derivatives possessed small ΔE_{ST} values (0.003–0.146 eV) that are sufficient to favor an efficient RISC process and thus emit delay fluorescence. Utilizing **PCZ-CB-TRZ** as the emitter, the nondoped OLEDs displayed a maximum EQE of 11.0% and maximum luminance of 4530 cd m^{-2}.

In 2017, the same group revealed two new AIDF luminogens, **PTZ-XT** and **PTZ-BP**, which are composed of the PTZ donor and xanthone or benzophenone electron acceptor moieties (Scheme 13) [71b]. The modest PLQY values of 53 and 31% for **PTZ-XT** and **PTZ-BP** with emission centered on 545 and 565 nm were recorded in neat films, respectively. The nondoped OLEDs using **PTZ-XT** and **PTZ-BP** revealed maximum EQE values of 11.1% and 7.6% (Table 3), respectively, indicating that luminogens with unique AIDF characteristics can act as prospective emitters for nondoped OLEDs.

Our group has also demonstrated several AIDF luminogens-based nondoped OLEDs, displaying exceptional EL performance with a low efficiency roll-off along with excellent EL stabilities. For instance, our group competently designed and synthesized four AIDF compounds, namely **SBDBQ-DMAC**, **DBQ-3DMAC**, **SBDBQ-PXZ**, and **DBQ-3PXZ** (Scheme 14), by linking DMAC or PXZ units with a quinoxaline unit [72a]. By regularly tuning the electron-donating ability and amount of donor units, these luminogens unveiled diverse emissions ranging from green to red (541, 551, 594, and 618 nm). Notably, a **DBQ-3DMAC**-based nondoped device furnished the best EL efficiencies in the region of orange emission with 41.2 cd A^{-1}, 45.4 lm W^{-1}, and 12.0% (Table 3).

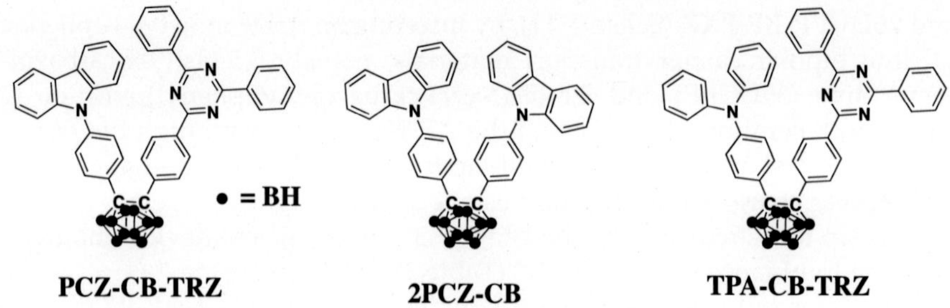

SCHEME 12 Molecular structures of *o*-carborane based AIDF emitters.

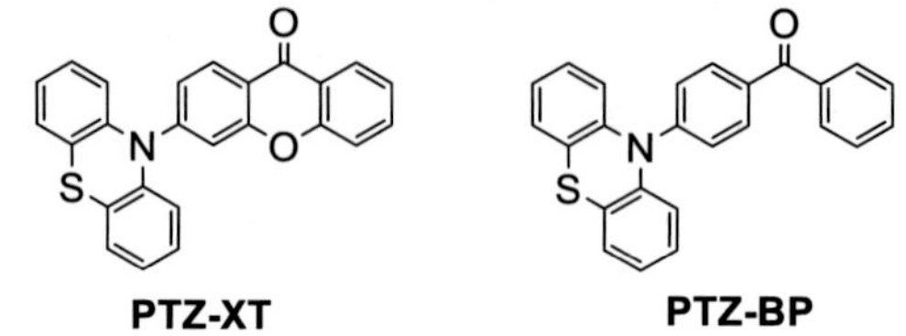

SCHEME 13 Chemical structures of xanthone or benzophenone-based AIDF emitters.

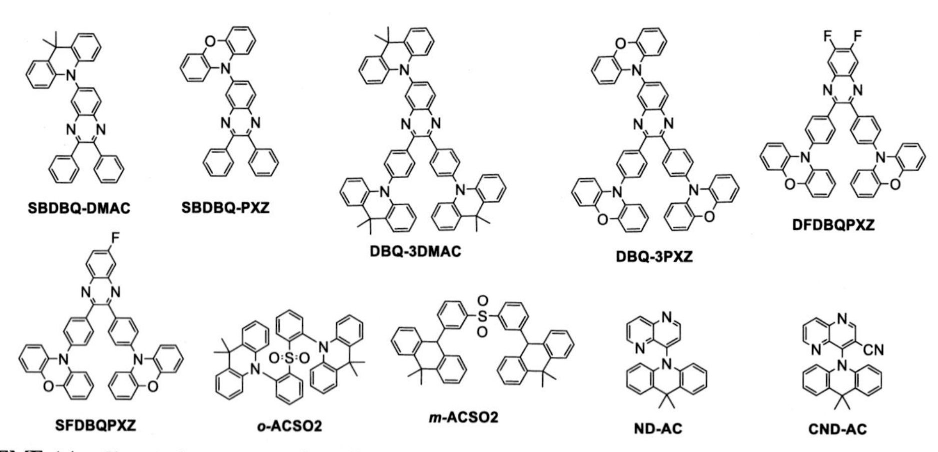

SCHEME 14 Chemical structures of small AIDF molecules reported by Yang's group.

Instantly after the aforesaid report, another quinoxaline system was utilized by our group to design new efficient AIDF emitters (**SFDBQPXZ** and **DFDBQPXZ**) via inducting fluorine atoms (Scheme 14) [72b]. Interestingly, the corresponding nondoped OLEDs involving **SFDBQPXZ** and **DFDBQPXZ** realized an efficient orange EL emission with peaks at 584 and 588 nm, maximum EQE values close to 10.1% and 9.8%, respectively, and maintaining small efficiency roll-offs.

In the same year, our group designed two new blue emitters based on an isomeric platform, namely ***o*-ACSO2** and ***m*-ACSO2** (Scheme 14). The two emitters revealed unique AIDF properties, outstanding solubility, and excellent morphology in neat films [72c]. The meta-linking isomer ***m*-ACSO2** showcased a small ΔE_{ST} of 0.07 eV with a high PLQY of 76%. As expected, the excellent morphology in neat film supported the solution-processed sky-blue OLEDs and gave out an optimized EQE of 17.2% (Table 3).

Besides, our group also investigated two new compounds, namely **ND-AC** and **CND-AC** (Scheme 14), which consisted of an electron withdrawing naphthyridine or cyano-naphthyridine unit and an electron-donating acridine unit arranged in an orthogonal D-A molecular configuration [96]. The nearly orthogonal molecular configuration adopted in the two luminogens not only favored small $\triangle E_{st}$s for securing the TADF property but also conferred an incredible AIE feature. Thus, the multilayer OLEDs based on the two emitters with a device architecture of ITO/TAPC (30 nm)/TCTA (5 nm)/**ND-AC**: CBP (15 nm) or **CND-AC**:CBP (15 nm) (EML)/(TmPyPB) (65 nm)/LiF (1 nm)/Al (100 nm) were fabricated with an ideal doping concentration of 9 wt% for **ND-AC** and 1.5 wt% for **CND-AC**. The best EL performance for the doped and nondoped devices with similar architecture exhibit an excellent maximum EQE of 16.8% and 12.0%, respectively.

Several research groups also reported some interesting work in the development and application of AIDF materials for OLEDs. Two D-A type TADF molecules (**CCDC** and **CCDD**) (Scheme 15) were designed and synthesized by Su and coworkers [83] via a combination of the stable benzophenone acceptor and branch-shaped donors, in which the branch-shaped donor can be obtained by linking two carbazoles or DPA units with a carbazole group. Regardless of the structural similarities of both molecules, their PL behaviors in dilute THF solution showed significant differences. The PLQY of **CCDC** is quite high in contrast to that of **CCDD**. Notably, the PL intensity of **CCDC** decreases immensely upon adding plenty of water into the THF solution, supporting a classic ACQ effect. However, **CCDD** demonstrated the AIE phenomenon. Unsurprisingly, the AIDF luminogen **CCDD**-based nondoped device manifested the EL performance with 39.8 cd A^{-1}, 41.7 lm W^{-1}, and 12.7%, with comparatively excellent efficiency stability (Table 3). What is more, the **CCDC**-based nondoped device only revealed the maximum EQE of simply 2.0% because of the severe ACQ effect. Undoubtedly, AIE plays a key role in improving the performance of nondoped OLEDs.

The same group further developed three stimuli-responsive AIDF luminogens, ***pipd*-BZ-PXZ**, ***pipd*-BZ-PTZ**, and ***pipd*-BZ-DMAC** (Scheme 15), adopting a D-A type asymmetric design with imidazo[1,2-*a*]pyridin-2-yl(phenyl)methanone (pipd) performing as the electron acceptor (A) and phenoxazine (PXZ), phenothiazine (PTZ), or 9,9-dimethyl-9,10-dihydroacridine (DMAC) unit as the electron donor [87]. The highly twisted conformation in the three luminophores caused a loose crystals packing mode and generated interesting mechanochromic luminescence (MCL) behavior confirmed by XRD analysis. The emissions of three luminophores are weak and mostly generated from prompt decay in the dilute solution, but became strong upon aggregation with a distinct delay component. Because of the well-separated HOMO and LUMO of the donor and acceptor, small $\triangle E_{st}$ presented in neat films and significantly sped up the RISC process and shortened the lifetime of delayed fluorescence. Thus, nondoped OLED based on ***pipd*-BZ-PXZ** demonstrated emission at peak 570 nm and reinforced a small current efficiency roll-off of 2.3% at 1000 cd m^{-2} (Table 3) [87].

Wang and coworkers described a unique AIDF luminogen, **PXZ2PTO** (Scheme 15), established on phenothiazine oxide that acted as an electron acceptor and phenoxazine as the strong electron donor [91]. This molecular design allowed a highly stereoscopic structure

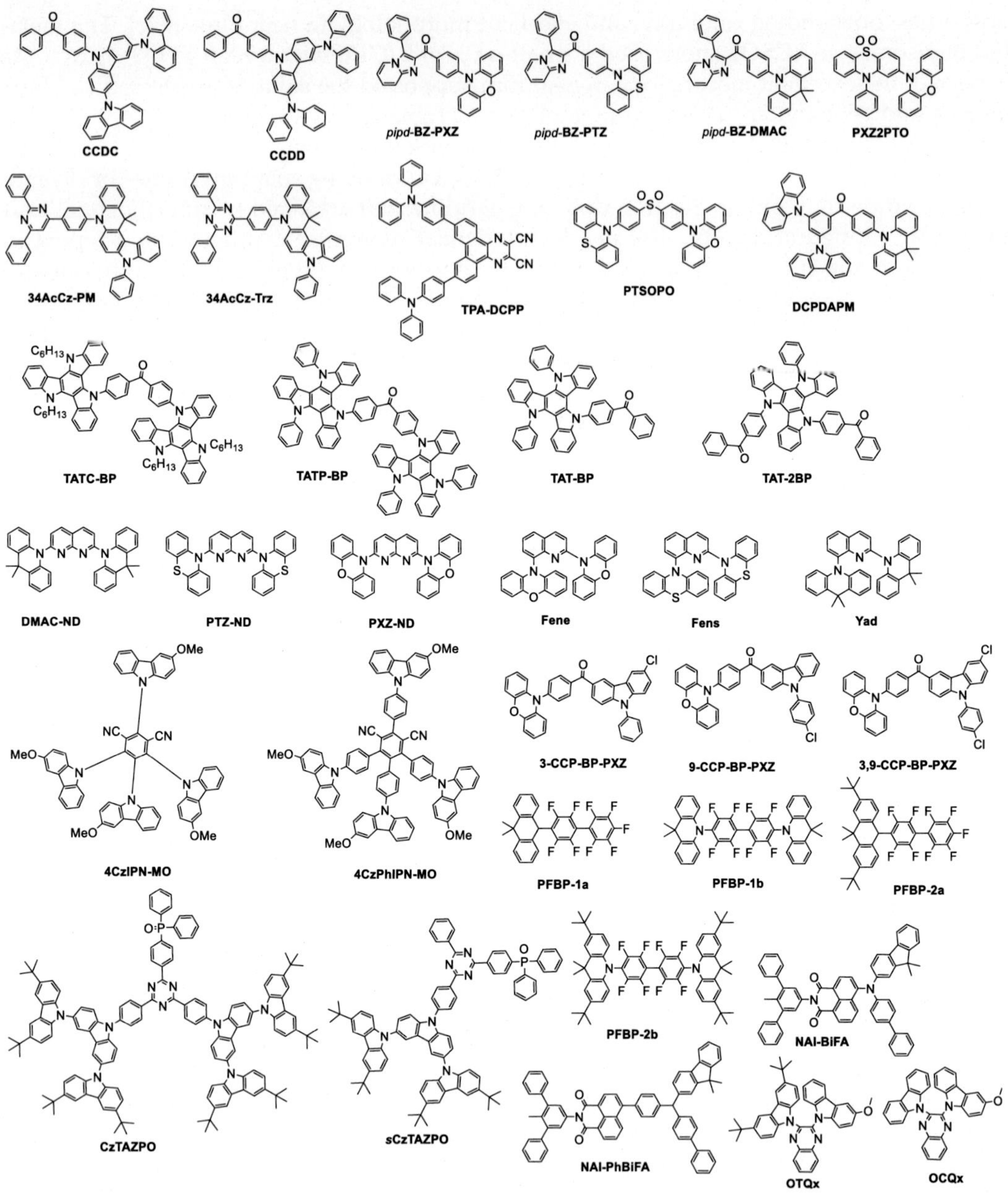

SCHEME 15 Chemical structures of small AIDF molecules designed and synthesized by different groups.

because of its multiple nearly vertical dihedral angles. **PXZ2PTO** also displayed both AIDF properties with a small ΔE_{st} of 0.02 eV and a high PLQY. As a result, the OLED devices utilizing this emitter with a device configuration of ITO/MoO_3 (10 nm)/TAPC (30 nm)/mCP (25 nm)/80 wt% **PXZ2PTO**-doped DPEPO (30 nm)/DPEPO (5 nm)/TPBi (30 nm)/LiF (1 nm)/Al (100 nm) presented green EL emission with a performance of 16.3%, 43.8 cd A^{-1}, and 35.2 lm W^{-1}. Likewise, the fabricated nondoped devices were optimized with a structure of ITO/MoO_3 (10 nm)/TAPC (40 nm)/mCP (20 nm)/**PXZ2PTO** (30 nm)/DPEPO (20 nm)/TPBi (15 nm)/LiF (1 nm)/Al (100 nm), which exhibited a device performance of 16.4%, 44.9 cd A^{-1}, 32.0 lm W^{-1} (Table 3).

The same group also developed a new acridine-carbazole hybrid donor, namely, 34AcCz: 13,13-dimethyl-8-phenyl-8,13-dihydro-5*H*-indolo[3,2-*a*]acridine, which consisted of two donor segments of DMAc and Cz. By introducing two acceptors pyrimidine and triazine on this novel donor, two new AIDF emitters, namely **34AcCz-PM** and **34AcCz-Trz** (Scheme 15), were obtained [98]. Because of the noteworthy AIE characteristics, **34AcCz-PM** procured a high PLQY. As a result, the device performance approached a maximum current efficiency of 73.3 cd A^{-1} and an EQE of 22.6% for the doped OLED of **34AcCz-PM** with a device architecture of ITO/MoO_3 (10 nm)/TAPC/TCTA/EML/TmPyPB/LiF (1 nm)/Al. The well-organized device maintained a high performance of 65.5 cd A^{-1} and 20.1% even at a high luminance of 5000 cd m^{-2}. Remarkably, nondoped OLED of **34AcCz-PM** disclosed outstanding EL performance of 45.2 cd A^{-1}/14.1% (Table 3).

The first NIR TADF emitter **TPA-DCPP** (Scheme 15) was reported by Wang et al. [99] **TPA-DCPP** was composed of 2,3-dicyanopyrazino phenanthrene (DCPP) as the electron-deficient core and DPA as the electron-rich core, featuring the AIE characteristic. The acceptor and donor were linked through a phenyl. Notably, there was a small twisting between DCPP and the peripheral phenyl ring, and hence the dihedral angle comes to around 35°, anticipating a near-planar configuration and reasonable orbital overlaps. A small ΔE_{ST} of 0.13 eV was realized in **TPA-DCPP**. The neat film of **TPA-DCPP** revealed an NIR emission peak at 710 nm, achieving an astonishing PLQY of 14%. **TPA-DCPP**-based nondoped OLED with a configuration of ITO/NPB/TCTA/**TPA-DCPP**/TPBi/LiF/Al confirmed NIR emission with an L_{max} of 591 cd m^{-2} and a $\eta_{ext,max}$ of 2.1%. Further, the doped device achieved a high EQE of 9.8% with maximum peak at 668 nm (Table 3).

Lee and coworker [104] demonstrated that the asymmetric molecule, 10-(4-((4-(10*H*-phenothiazin-10-yl) phenyl)sulfonyl)phenyl)-10*H*-phenoxazine (**PTSOPO**) (Scheme 15), with excellent AIE and TADF properties was designed as an effective emitter in a nondoped device with a $\eta_{ext,max}$ of 17%. In addition, the device performance based on **DPS-PTZ** (Scheme 11) (symmetric molecular structure) was far inferior to that of the **PTSOPO** device. The group reasonably explained the advantage of the asymmetric backbone in the molecular design of **PTSOPO** in controlling the efficiency roll-off in a nondoped device.

Shi et al. [84] communicated an unsymmetrical D-A-D′ type molecule **DCPDAPM** (Scheme 15) via linking the benzophenone core to the DMAC and carbazole moieties. A classic AIDF molecule (**DCPDAPM**) exhibited a high solid-state PLQY and a small ΔE_{ST} of 0.10 eV. The **DCPDAPM**-based nondoped OLED provided good device performance with 123,371 cd m^{-2}, 26.88 cd A^{-1}, 15.63 lm W^{-1}, and EQE of 8.15% (Table 3).

Wang et al. published two small molecules, **TATC-BP** and **TATP-BP**, consisting of two electron donor triazatruxene segments connected by an electron acceptor benzophenone unit (Scheme 15) [92]. Both of the luminogens featured TADF, AIE, and mechanochromic luminescence properties. Owing to steric hindrance, **TATP-BP** with a phenyl substituent showed an MCL emission with a smaller red-shift (36 nm), whereas **TATC-BP** with hexyl substituents ascribed to less steric hindrance enjoyed a clear MCL with a broad emission color ranging from 483 to 542 nm against mechanical grinding. Moreover, the solution-processed nondoped OLEDs of **TATP-BP** and **TATC-BP** with a device configuration of ITO/PEDOT:PSS (25 nm)/

emitters (25 nm)/TmPyPB (55 nm)/LiF (1 nm)/Al (150 nm) exhibited a better EL performance of 18.9 cd A^{-1}, 6.0%, a negligible tiny roll-off of 3.3% at a high luminance of 1000 cd m^{-2} and 17.8 cd A^{-1}, 5.9%, and an efficiency roll-off (18.6%) at a high luminance of 1000 cd m^{-2}, respectively (Table 3). **TATP-BP** displayed a much lower efficiency roll-off than **TATC-BP**, indicating the effect of rigidity and bulky phenyl groups. Concurrently, the doped devices based on **TATC-BP** and **TATP-BP** with a similar device configuration presented a tiny efficiency roll-off of 3.8% and 7.8%, respectively.

The same group also designed and synthesized two new green TADF emitters, namely, **TAT-BP** and **TAT-2BP** (Scheme 15) by linking one or two electron acceptor benzophenone into electron donor phenyl substituted triazatruxene [100]. Both of the luminogens exhibited distinct AIE and TADF behaviors. However, **TAT-2BP** revealed a shorter delay fluorescence lifetime of 0.54 μs than did **TAT-BP** (0.79 μs) in neat film. Therefore, **TAT-2BP**-based solution-processed nondoped OLEDs with a device architecture of ITO/PEDOT:PSS (25 nm)/**TAT-BP** or **TAT-2BP**(25 nm)/TmPyPB (55 nm)/LiF (1 nm)/Al (150 nm) presented a low turn-on voltage of 2.5 V with EL efficiencies up to 32.3 cd A^{-1}, 33.0 lm W^{-1}, and 9.8% (Table 3). Astonishingly, **TAT-2BP**-based devices showed a very small efficiency roll-off of 1.0% at 1000 cd m^{-2}, comparatively much lower than did **TAT-BP**, signifying the importance of **TAT-2BP** with a short delay fluorescence lifetime to allow a negligible efficiency roll-off.

Chen and coworker reported three 1,8-naphthyridine-based TADF emitters, namely **DMAC-ND**, **PTZ-ND**, and **PXZ-ND** (Scheme 15), with 1,8-naphthyridine as the electron acceptor and PXZ, PTZ, or DMAC as the electron donor [97]. The three novel emitters with twisted structures and small $\triangle E_{st}$ demonstrated admirable AIDF properties (Fig. 7). The

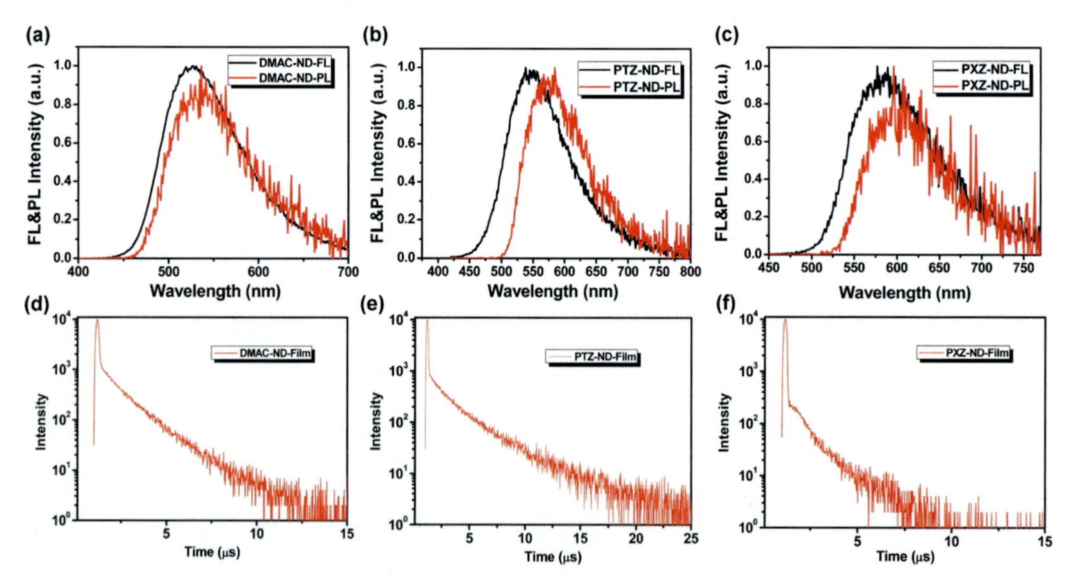

FIG. 7 Normalized photoluminescence and phosphorescence spectra of (A) **DMAC-ND**, (B) **PTZ-ND**, and (C) **PXZ-ND** in neat films at 77 K; transient PL spectra of (D) **DMAC-ND**, (E) **PTZ-ND**, and (F) **PXZ-ND** in THF and film states at 300 K. *Reproduced with permission from C. Chen, H.-Y. Lu, Y.-F. Wang, M. Li, Y.-F. Shen, C.-F. Chen, Naphthyridine-based thermally activated delayed fluorescence emitters for multi-color organic light-emitting diodes with low efficiency roll-off, J. Mater. Chem. C 7 (2019) 4673—4680. Copyright 2018 John Wiley & Sons.*

maximum EQE values of 10.6%, 8.0%, and 3.5% were exhibited by multicolor nondoped OLEDs using **DMAC-ND, PTZ-ND,** or **PXZ-ND** with a device configuration of ITO/TAPC (30nm)/(TCTA) (5nm)/emitting layer (20nm)/Tm3PyPB (60nm)/LiF (1nm)/Al (100nm), whereas doped devices achieved a maximum EQE of 14.1%, 13.4%, and 13.0%, respectively (Table 3). Because of their twisted conformation and the AIDF properties, the fabricated devices exhibited relatively low turn-on voltages and low efficiency roll-off ranging from 4.5% to 20% at the luminance of 1000 $cd\,m^{-2}$. Interestingly, the emitter **DMAC-ND**-based device approached the luminance of 20,000 $cd\,m^{-2}$ at the voltage of 6.5V and favored a very negligible efficiency roll-off of 4.5% at 1000 $cd\,m^{-2}$ (Fig. 8).

The same group then revealed three new luminogens, namely **Fene**, **Fens**, and **Yad** (Scheme 15), featuring with quinoline as the electron acceptor and PXZ, PTZ, or DMAC as the electron donor [85]. As expected, all the emitters with their twisted structure facilitated small $\triangle E_{st}$, providing TADF properties as well as AIE feature. Furthermore, the AIDF feature and decent PLQY enabled these emitters to be used in nondoped OLEDs with a configuration of ITO/TAPC (30nm)/TCTA (5nm)/emitter (15nm)/(TmPyPB) (65nm)/LiF (1nm)/Al (100nm), demonstrating high EL performance with emission peaks at 570, 568, and 534nm and maximum EQE of 14.9%, 13.1%, and 17.4%, respectively (Table 3).

Two novel orange to red emitters, **NAI-BiFA** and **NAI-PhBiFA** (Scheme 15), were designed and synthesized by Peng et al. by integrating a naphthalimide as the electron acceptor core and arylamine as the electron donor segment [95]. The insertion of a phenyl linker between the D and A segments in **NAI–PhBiFA** played an important role in creating a spatially twisted structure and ensuring the blue-shift emission as compared with **NAI–BiFA**. **NAI–PhBiFA**-based nondoped OLED displayed orange EL emission and demonstrated an external quantum efficiency of only 1.39%, whereas a doped OLED using **NAI–PhBiFA** as the light-emitting layer displayed a notable enhancement of device efficiency of 7.59% (Table 3). Furthermore, an **NAI–BiFA**-based host-free OLED device acquired a standard red CIE (0.65, 0.34). These outcomes showed that the donor-linker-acceptor molecular design strategy is highly encouraging to realize high-performance OLEDs.

Sun and coworker [102] demonstrated the importance of methoxy groups and phenyl bridges in the twisted molecular conformation in terms of TADF and the AIE property. Two TADF luminophores, **4CzIPN-MO** and **4CzPhIPN-MO** (Scheme 15), consisting of the methoxy substituted carbazole fragment and insertion of phenyl bridges between carbazole and the central benzene ring, were designed and synthesized. Interestingly, an unanticipated AIE behavior was witnessed for **4CzPhIPN-MO** because the phenyl bridge alleviated the steric hindrance, whereas the methoxy group assisted in shallowing the HOMO level, which in turn decreased the injection barrier in the device. Therefore, **4CzPhIPN-MO** demonstrated a high PLQY of 86% and excellent TADF property as compared to **4CzIPN-MO**. As a result, solution-processed OLED with a simplified configuration of the ITO/PEDOT:PSS (60nm)/emitter (40nm)/TPBi (30nm)/Cs_2CO_3 (2nm)/Al (100nm) exhibited an EQE_{max} of 14.5% with an EL emission peak at 536nm for **4CzPhIPN-MO** as compared to **4CzIPN-MO** (5.6%) (Table 3). In addition, the **4CzPhIPN-MO**-based nondoped device displayed extremely small driving voltage and efficiency roll-offs, while the doped device based on **4CzPhIPN-MO** in 10wt% CBP with a similar configuration exhibited the best EL performance of 44.2 $cd\,A^{-1}$, 16.9 $lm\,W^{-1}$, and 16.2%, respectively.

Zhao et al. [88] put forward a comprehensive insight into understanding the importance of fast RISC in the performance and stability of OLEDs. To explain this, a series of luminogens,

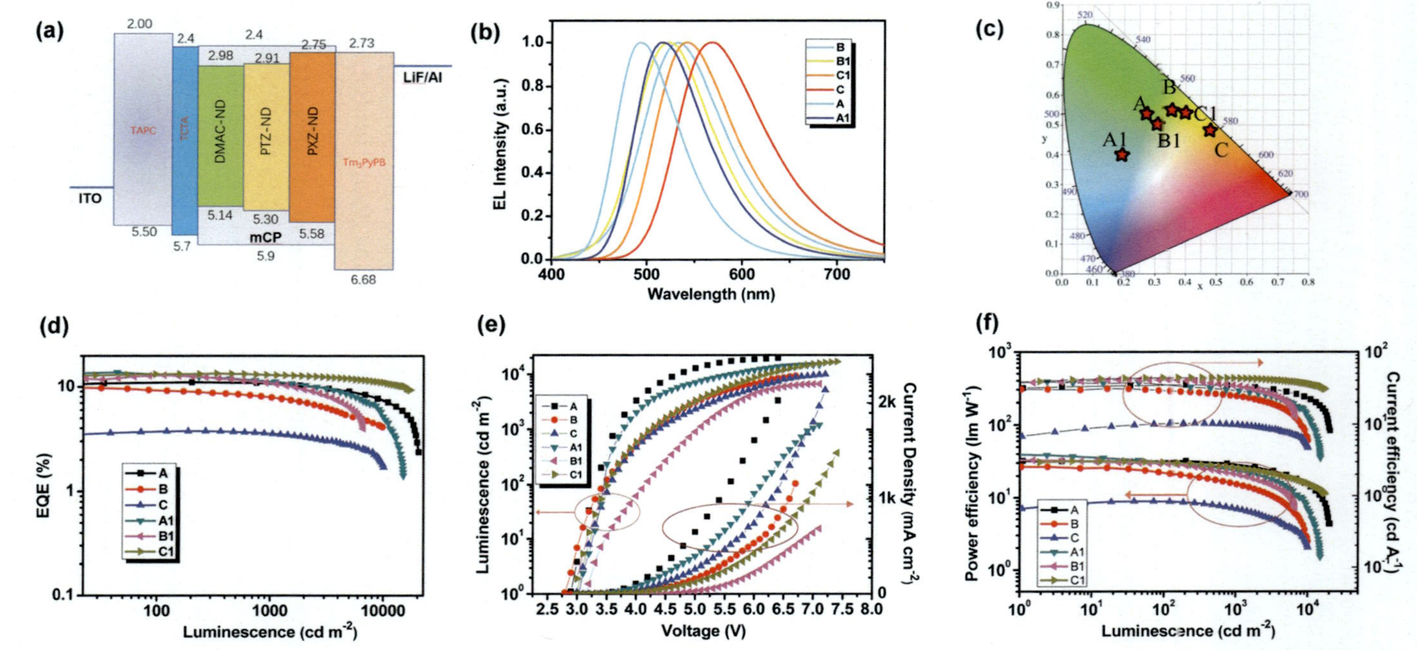

FIG. 8 (A) Energy level diagram of the materials employed in the devices; (B) electroluminescence spectra of devices A: **DMAC-ND**, B: **PTZ-ND**, C: **PXZ-ND**, A1: [**DMAC-ND** (10 wt%): mCP], B1: [**PTZ-ND** (7.5 wt%): mCP] and C1: [**PXZ-ND** (3.5 wt%): mCP]; (C) the CIE coordinate of the EL; (D) the EQE versus luminance curves; (E) luminance-voltage and current density-voltage curves; and (F) power efficiency and current efficiency for devices A–C and A1–C1. *Reproduced with permission from C. Chen, H.-Y. Lu, Y.-F. Wang, M. Li, Y.-F. Shen, C.-F. Chen, Naphthyridine-based thermally activated delayed fluorescence emitters for multi-color organic light-emitting diodes with low efficiency roll-off, J. Mater. Chem. C 7 (2019) 4673—4680. Copyright 2018 John Wiley & Sons.*

3-CCP-BP-PXZ, **9-CCP-BP-PXZ**, and **3,9-CCP-BP-PXZ** (Scheme 15), consisting of an electron-withdrawing carbonyl core and chlorine substituted electron-donating segments, were designed and synthesized. The new luminophores demonstrated an interesting AIDF property with a high PLQY and short delayed fluorescence lifetime. Because of the heavy atom effect, small $\triangle E_{st}$ and greatly improved SOC values in neat film considerably sped up the RISC and shortened the delayed fluorescence lifetime as a consequence. The best EL performance of **3-CCP-BP-PXZ**-based host-free OLEDs with a device configuration of ITO/HATCN (5nm)/TAPC (30nm)/TCTA (5nm)/emitter (20nm)/TmPyPB (40nm)/LiF (1nm) Al came out to be $\eta_{C,max}$ of 76.6cd A^{-1}, $\eta_{P,max}$ of 75.2lm W^{-1}, and $\eta_{ext,max}$ of 21.7% with a negligible efficiency roll-off, whereas the **3-CCP-BP-PXZ**-based doped devices achieved an impressive EL performance of 100.1cd A^{-1}, 104.8lmW^{-1}, and 29.1% (Table 3). Hence, the notable AIDF character, fast RISC features in AIDF luminogens can help to improve the performance and stability of OLEDs.

During the same period, Wang and coworkers developed a molecular design strategy based on highly twisted D-π-A carbazole dendritic structures, namely **CzTAZPO** and **sCzTAZPO** (Scheme 15) [86]. This molecular design endowed AIE, TADF, and bipolar charge transport properties that allowed **CzTAZPO** to furnish high PLQY and small $\triangle E_{st}$. Solution-processed host-free OLEDs utilizing both **CzTAZPO** and **sCzTAZPO** as the emitting layer present excellent EL efficiencies with a $\eta_{C,max}$ of 29.1cd A^{-1} and a $\eta_{ext,max}$ of 12.8% for **CzTAZPO**, and a $\eta_{C,max}$ of 20.4cd A^{-1} and a $\eta_{ext,max}$ of 9.1% for **sCzTAZPO**. The efficiency roll-offs were found to be 1.6% and 1.1% at a high luminance of 1000cdm^{-2} (Table 3). From this outcome, it can be concluded that the molecular design strategy combined with AIE, TADF, and bipolar charge transport properties can be an ideal method for developing materials for high-performance stable OLEDs with a simplified structure and low efficiency roll-off.

Grazulevicius and coworker [94a] developed four luminogens, namely **PFBP-1a**, **PFBP-1b**, **PFBP-2a**, and **PFBP-2b** (Scheme 15), which were designed with perfluorobiphenyl as the acceptor and nonsubstituted and tert-butyl substituted 9,9-dimethyl-9,10-dihydro-acridine as the appropriate donor. These D-A- or D-A-D-type emitters displayed multifunctional features such as delay fluorescence, AIEE, and blue to sky-blue reversible color changing. Interestingly, luminogens **PFBP-1a** and **PFBP-1b** exhibited deep-blue emission in solid state, whereas luminogens **PFBP-2a** and **PFBP-2b** with tert-butyl radiated sky-blue emission in neat film. Single crystal analysis suggested the formation of two polymorphs of **PFBP-2a** and explained that tert-butyl groups, π–π and unique C–F——π interactions, and C–H——F hydrogen bonds were responsible for the sky-blue and blue emission, respectively. Both the doped and nondoped TADF OLEDs were fabricated using the **PFBP-2b** emitter, demonstrating sky-blue emission with maximum external quantum efficiencies of 6.6 and 16.3%, respectively (Table 3).

Recently, the same group reported new multifunctional AIDF emitters, namely **OTQx** and **OCQx** (Scheme 15), which were synthesized with quinoxaline as the acceptor and different substituted carbazole units as the donor. **OTQx** produced more proficient TADF, improved thermal stability, stronger mechanoluminescence. Both of the luminogens showed AIEE behavior with decent PLQYs. The EL device based on **OTQx** with the configuration of ITO/MoO_3(0.5nm)/NPB (35nm)/**OTQx** or **OCQx** (10wt%):TCTA (20nm)/TSPO1 (7nm)/TPBi (30nm)/LiF (0.5nm)/Al exhibited an EL emission peak at 544nm and an L_{max} of 48,809cdm^{-2}, $\eta_{C,max}$ of 33.6cd A^{-1}, $\eta_{P,max}$ of 21.1lmW^{-1}, and $\eta_{ext,max}$ of 10.5% (Table 3) [94b].

6 AIDF based on through-space charge transfer (TSCT) for OLEDs

As we have already discussed in the previous section, TADF materials with a restricted overlap between HOMO and LUMO result in small exchange energies (J), leading to small ΔE_{ST} and an activated RISC process. The majority of the reported TADF designs adopt the twisted donor-acceptor link mode to minimize the HOMO-LUMO exchange energy. In addition, there is another approach of through-space charge transfer (TSCT), which introduces spatial π-π interaction between the electron donor and the electron acceptor and thus sufficiently controls the HOMO-LUMO separation to attain small singlet−triplet ΔE_{ST} [105]. In this section, we will discuss the new molecular design strategy that provides a different viewpoint to construct novel organic AIDF materials for the fabrication of high-performance OLEDs.

To demonstrate the concept of through-space interaction, the Swager group designed and synthesized two luminogens of **TPA-QNX-(CN)2** and **TPA-PRZ(CN)2** (Scheme 16), derived from the donor-acceptor triptycenes [105]. Importantly, homoconjugation between the physically separated donor and acceptor groups linked to different parts of the three-dimensional triptycene scaffold favored the small ΔE_{ST} and thus the TADF property. Multilayer OLED devices based on **TPA-QNX-(CN)2** and **TPA-PRZ(CN)2** as the EML with the following configuration of ITO (132nm)/MoO_3 (5nm)/TCTA (30nm)/10wt% EML: mCP (30nm)/ TmPyPb (40nm)/LiF (0.8nm)/Al (100nm) presented a decent EL performance with the emission peak at 573 and 572nm and an EQE of 9.4% and 4.0%, respectively (Table 4) [105].

The same group then put forward three chromophores based on 9,9-dimethylxanthene, namely **XPT**, **XCT**, and **XtBuCT**, via the bridged donor-acceptor (D-b-A) design (Scheme 16) [106]. Both in solution and thin film states, the U-shaped AIEgens unveiled TADF characteristics in the absence of oxygen. The single crystal analysis revealed that C-H—π interactions help in upgrading a firm environment that ultimately enhanced the quantum yields in the thin film state. **XPT** and **XtBuCT**-based thermal evaporation OLED devices with the configuration of ITO (100nm)/TAPC (70nm)/DPEPO: Dopant (10%) (30nm)/DPEPO (2nm)/ TmPyPb (45nm)/LiF (1nm)/Al (100nm) demonstrated an EL performance with emission peaks at 584 and 488nm and EQE of 10% and 4%, respectively (Table 4). An overall study suggests that the U-shaped molecular design can endow with a promising route in the development of various AIDF materials [106].

Wang et al. employed a novel concept based on a nonconjugated polyethylene backbone with TSCT effect between the physically separated and spatially proximate donor and acceptor units to achieve blue TADF emission. To study the influence of the distance between the D and A units on the CT interactions of the polymers, four TSCT-based polymers (**P-Ac95-TRZ05**, **P-Ac50-TRZ50**, **P-TBAc95-TRZ05**, and **P-TBAc50-TRZ50**) (Scheme 16) consisting of nonconjugated polyethylene as the backbone, 9,9-dimethyl-10-phenyl-acridan or 9,9-bis (1,3-ditert-butylphenyl)-10-phenyl-acridan (TBAc) as the pendant electron donor, and 2,4,6-triphenyl-1,3,5-triazine (TRZ) as the pendant electron acceptor were developed [107]. Among them, **P-Ac50-TRZ50** and **P-Ac95-TRZ05** with 50 and 5mol% acceptor segments, respectively, possess sufficient small ΔE_{ST}s of 0.019eV and 0.021eV, respectively, to facilitate delayed emission under thermal activation, displaying a high PLQY of 60% and 51% in polymer film under nitrogen, respectively. The logically TSCT type nonconjugated polymer

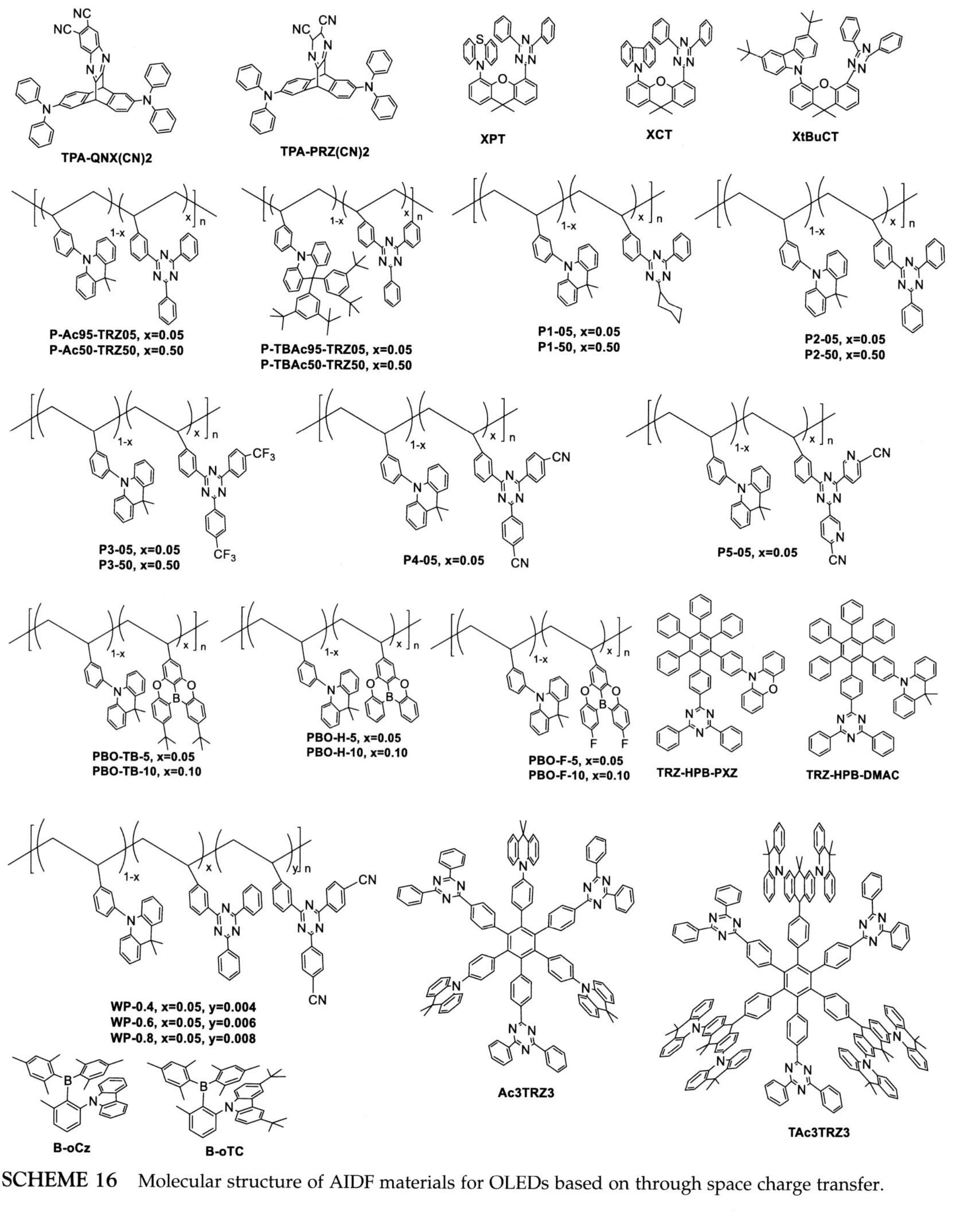

TPA-QNX(CN)2
TPA-PRZ(CN)2
XPT
XCT
XtBuCT

P-Ac95-TRZ05, x=0.05
P-Ac50-TRZ50, x=0.50

P-TBAc95-TRZ05, x=0.05
P-TBAc50-TRZ50, x=0.50

P1-05, x=0.05
P1-50, x=0.50

P2-05, x=0.05
P2-50, x=0.50

P3-05, x=0.05
P3-50, x=0.50

P4-05, x=0.05

P5-05, x=0.05

PBO-TB-5, x=0.05
PBO-TB-10, x=0.10

PBO-H-5, x=0.05
PBO-H-10, x=0.10

PBO-F-5, x=0.05
PBO-F-10, x=0.10

TRZ-HPB-PXZ
TRZ-HPB-DMAC

WP-0.4, x=0.05, y=0.004
WP-0.6, x=0.05, y=0.006
WP-0.8, x=0.05, y=0.008

Ac3TRZ3
TAc3TRZ3
B-oCz
B-oTC

SCHEME 16 Molecular structure of AIDF materials for OLEDs based on through space charge transfer.

TABLE 4 Electroluminescent key data of OLEDs based on TSCT TADF AIEgens.

Emitter	Active layer in device	λ_{EL} (nm)	V_{on}	L_{max} (cd m^{-1})	$\eta_{c,max}/\eta_{p,max}/\eta_{ext,max}$ (cd A^{-1}/lm W^{-1}/%)	Ref.
TPA-QNX-CN)2	ITO (132nm)/MoO_3 (5nm)/TCTA (30nm)/ 10wt% EML: mCP (30nm)/TmPyPb (40nm)/ LiF (0.8nm)/Al (100nm)	573	–	–	–/–/9.4	[105]
TPA-PRZ-CN)2	ITO (132nm)/MoO_3 (5nm)/TCTA (30nm)/ 10wt% EML: mCP (30nm)/TmPyPb (40nm)/ LiF (0.8nm)/Al (100nm)	572	–	–	–/–/4.0	[105]
XPT	ITO (100nm)/(TAPC) (70nm)/ (DPEPO): XPT (10%) (30nm)/DPEPO (2nm)/ (TmPyPb) (45nm)/LiF (1nm)/Al (100nm).	584	–	–	–/–/10	[106]
XtBuCT	ITO (100nm)/(TAPC) (70nm)/ (DPEPO): XtBuCT (10%) (30nm)/DPEPO (2nm)/(TmPyPb) (45nm)/LiF (1nm)/Al (100nm).	488	–	–	–/–/4	[106]
P-Ac95-TRZ05	ITO/PEDOT:PSS (40nm)/P-Ac95-TRZ05 (40nm)/TSPO1 (8nm)/TmPyPB(42nm)/LiF (1nm)/Al (100nm)	472	3.2	6150	24.8/–/12.1	[107]
P1–05	ITO/PEDOT:PSS (40nm)/P1–05 (40nm)/ TSPO1 (8nm)/TmPyPB(42nm)/LiF (1nm)/Al (100nm)	455	3.2	1902	10.6/–/7.1	[108]
P2–05	ITO/PEDOT:PSS (40nm)/P2–05 (40nm)/ TSPO1 (8nm)/TmPyPB(42nm)/LiF (1nm)/Al (100nm)	472	3.2	6150	24.8/–/12.1	[108]
P3–05	ITO/PEDOT:PSS (40nm)/P3–05 (40nm)/ SPPO13 (50nm)/LiF (1nm)/Al (100nm)	525	3.0	10,273	50.3/–/16.2	[108]
P4–05	ITO/PEDOT:PSS (40nm)/P4–05 (40nm)/ SPPO13 (50nm)/LiF (1nm)/Al (100nm)	568	3.2	5339	21.1/–/7.8	[108]
P5–05	ITO/PEDOT:PSS (40nm)/P5–05 (40nm)/ SPPO13 (50nm)/LiF (1nm)/Al (100nm)	616	4.6	1283	1.7/–/1.0	[108]
Ac3TRZ3	ITO/PEDOT:PSS (40nm)/Ac3TRZ3 (40nm)/ TSPO1 (8nm)/TmPyPB (42nm)/LiF (1nm)/Al (100nm)	520	3.4	6910	11.4/–/3.5	[109]
Ac6: Ac3TRZ3 (10wt%)	ITO/PEDOT:PSS (40nm)/Ac6: Ac3TRZ3 (10wt%) (40nm)/TSPO1 (8nm)/TmPyPB (42nm)/LiF (1nm)/Al (100nm)	492	2.9	6175	30.3/–/11.0	[109]
TAc3TRZ3	ITO/PEDOT:PSS (40nm)/TAc3TRZ3 (40nm)/ TSPO1 (8nm)/TmPyPB (42nm)/LiF (1nm)/Al (100nm)	538	2.9	7860	10.2/–3.1	[109]
Ac6: TAc3TRZ3 (10wt%)	ITO/PEDOT:PSS (40nm)/Ac6: TAc3TRZ3 (10wt%) (40nm)/TSPO1 (8nm)/TmPyPB (42nm)/LiF (1nm)/Al (100nm)	503	2.9	9689	40.6/–/14.2	[109]

TABLE 4 Electroluminescent key data of OLEDs based on TSCT TADF AIEgens—cont'd

Emitter	Active layer in device	λ_{EL} (nm)	V_{on}	L_{max} ($cd\,m^{-1}$)	$\eta_{c,max}/\eta_{p,max}/\eta_{ext,max}$ ($cd\,A^{-1}/lm\,W^{-1}/\%$)	Ref.
PBO-TB-5	ITO/PEDOT: PSS (40nm)/PBO-TB-5 (40nm)/ TSPO1 (8nm)/TmPyPB (42nm)/LiF (1nm)/Al (100nm)	449	3.8	405	4.6/3.6/3.5	[110]
PBO-TB-10	ITO/PEDOT: PSS (40nm)/PBO-TB-10 (40nm)/TSPO1 (8nm)/TmPyPB (42nm)/LiF (1nm)/Al (100nm)	451	3.8	578	5.8/5.1/3.8	[110]
PBO-H-5	ITO/PEDOT: PSS (40nm)/PBO-H-5 (40nm)/ TSPO1 (8nm)/TmPyPB (42nm)/LiF (1nm)/Al (100nm)	453	3.4	1190	8.1/7.5/5.2	[110]
PBO-H-10	ITO/PEDOT: PSS (40nm)/PBO-H-10 (40nm)/ TSPO1 (8nm)/TmPyPB (42nm)/LiF (1nm)/Al (100nm)	455	3.4	1409	10.3/9.5/6.1	[110]
PBO-F-5	ITO/PEDOT: PSS (40nm)/PBO-F-5 (40nm)/ TSPO1 (8nm)/TmPyPB (42nm)/LiF (1nm)/Al (100nm)	471	3.2	1532	25.3/24.9/14.4	[110]
PBO-F-10	ITO/PEDOT: PSS (40nm)/PBO-F-10 (40nm)/ TSPO1 (8nm)/TmPyPB (42nm)/LiF (1nm)/Al (100nm)	474	3.2	1650	30.7/30.2/15.0	[110]
B-oCz	ITO/PEDOT:PSS (40nm)/TFB (10nm)/B-oCz (40nm)/DPEPO (10nm)/TyPMPB (30nm)/ LiF (1nm)/Al (100nm)	463	4.4	–	–/–/8	[111]
B-oTC	ITO/PEDOT:PSS (40nm)/TFB (10nm)/B-oTC (40nm)/DPEPO (10nm)/TyPMPB (30nm)/ LiF (1nm)/Al (100nm)	474	3.9	–	–/–/19.1	[111]
TRZ-HPB-PXZ	ITO/HATCN (5nm)/TAPC (20nm)/TCTA (5nm)/TRZHPB-PXZ (35nm)/TmPyPB (55nm)/LiF (1nm)/Al.	544	2.5	40,382	41.2/44.9/12.7	[112]
TRZ-HPB-DMAC	ITO/HATCN (5nm)/TAPC (20nm)/TCTA (5nm)/TRZ-HPB-DMAC (20nm)/TmPyPB (55nm)/LiF (1nm)/Al.	521	3.1	15,460	21.4/17.6/6.5	[112]

Abbreviations: *λ_{EL} = the electroluminescence peak; V_{on} = turn-on voltage at 1 $cd\,m^{-2}$; L_{max} = maximum luminance; $\eta_{C,max}$ = maximum current efficiency; $\eta_{P,max}$ = maximum power efficiency; $\eta_{ext,max}$ = maximum external quantum efficiency; CBP = 4,4′-Bis(N-carbazolyl)-1,1′-biphenyl; mCP = 1,3-bis(Ncarbazolyl)Benzene; NPB = N,N′-di(1-naphthyl)-N,N′-diphenylbenzidine; PEDOT:PSS = Poly(3,4-ethylenedioxythiophene)-poly(styrenesulfonate; TAPC = 1,1-Bis[(di-4-tolylamino)phenyl]cyclohexane; HATCN=1,4,5,8,9,11-Hexaazatriphenylenehexacarbonitrile; TFB = Poly(9,9-dioctylfluorene-alt-N-(4-s-butylphenyl)-diphenylamine); TCTA = tris(4-carbazoyl-9-ylphenyl)amine; DPEPO = Bis[2-(diphenylphosphino)phenyl] ether oxide; TmPyPB = 1,3,5-Tris(3-pyridyl-3-phenyl)benzene, TSPO1 = Diphenyl[4-(triphenylsilyl)phenyl] phosphine oxide; SPPO13 = 2,7-bis(diphenylphosphoryl)-9,9′-spirobifluorene; Bphen = 4,7-diphenyl-1,10-phenanthroline; PEDOT:PSS, TAPC, NPB, TCTA, and TFB functions as a hole-transporting layer (HTL); TPBi, TmPyPB, TSPO1, SPPO13 and Bphen serve as an electron-transporting layer (ETL) and a hole-blocking layer (HBL), respectively; and MoO_3, HATCN serves as a hole-injection layer (HIL).*

delivered a superb backbone to manage the degree of π-π overlap between D and A to accomplish the delicate balance between small ΔE_{ST} and high PLQY. The resulting polymer **P-Ac95-TRZ05** with a 5mol% acceptor unit has been used as the emitting layer in polymer OLEDs with a configuration of ITO/PEDOT:PSS (40nm)/polymer (40nm)/TSPO1 (8nm)/TmPyPB (42nm)/LiF (1nm)/Al (100nm), ensuring outstanding blue EL with a high η_{ext} of 12.1% and a negligible efficiency roll-off of 4.9% at 1000cd m^{-2} [107].

The same group then applied the TSCT polymers, namely **P1-05, P1-50, P2-05, P2-50, P3-05, P3-50, P4-05** and **P5-05** (Scheme 16), consisting of a nonconjugated polystyrene motif and spatially proximate acridan as the donor and different triazine acceptor units in the side chain for solution-processed OLEDs to bestow full color and white emission [108]. The weak electron acceptor, 2,4, diphenyl 6 cyclohexyl-1,3,5-triazine (TRZ-Cy), bearing cyclohexane as the substituent was designed to decrease the electron-accepting ability of acceptors. The strong electron-accepting triazine core bearing substituents of trifluoromethylphenyl (TRZ-CF3), 4-cyanophenyl (TRZ-CN), and 4-cyanopyridyl (TRZ-PyCN) were also chosen to modulate the CT strength between donors and adjacent acceptors, allowing the polymers to exhibit emission from deep blue (455nm) to red (616nm), along with a PLQY of 6%–74% in solid states.Interestingly, by incorporating two kinds of donor/acceptor pairs in a single polymer would generated duplex TSCT channel that help in emitting blue and yellow emission at the same time to attain white EL from a single polymer. The TSCT polymers revealed TADF emission with delayed component lifetimes in the range of 0.36–1.98μs and unanticipated aggregation-induced emission. The emission intensity can be enhanced up to 117-fold from solution to the aggregation state. Utilizing the TSCT polymer as an EML, the solution-processed OLEDs revealed encouraging EL performance with maximum η_{ext}s of 7.1%, 16.2%, 1.0%, and 14.1% for deep-blue, green, red, and white emission, respectively (Table 4) [108].

The same group was also inspired to develop a simple, synthetically feasible small organic molecule in which through-space charge transfer hexaarylbenzenes (TSCT-HABs) revealed both TADF and AIE effects. Hence, two types of TSCT-HABs, namely, **Ac3TRZ3** and **TAc3TRZ3** (Scheme 16) were designed by connecting acridan or dendritic acridan donor and triazine acceptors in the arms of the hexaphenylbenzene center, respectively [109]. The CT emission of TSCT-HABs depended exclusively on spatial π-π interaction by controlling the overlap between the electron cloud of the donor and acceptor through space charge transfer. Consequently, both **Ac3TRZ3** and **TAc3TRZ3** revealed small ΔE_{ST}s of 0.08 and 0.04eV with delayed fluorescence lifetimes of 3.45 and 3.16μs, respectively. In addition, **Ac3TRZ3** and **TAc3TRZ3** demonstrated a PLQY of 20% and 36% in neat films, respectively, indicating that dendritic acridan of **TAc3TRZ3** help in improving efficient charge transfer transition by π-π interaction than **Ac3TRZ3**. Using the **TAc3TRZ3** doped film as the emissive layer, the solution-processed OLEDs with a device architecture of ITO/PEDOT:PSS (40nm)/EML (40nm)/TSPO1 (8nm)/TmPyPB (42nm)/LiF (1nm)/Al (100nm) showed the best device performance with a $\eta_{ext,max}$ of 14.2% and a $\eta_{C,max}$ of 40.6cd A^{-1} [109].

The Wang group continued their research on TSCT-based TADF materials by attempting to develop three types of different blue polymers based on TSCT, namely **PBO-TB, PBO-H,** and **PBO-F** (Scheme 16), which derived from the nonconjugated polystyrene backbone along with the spatially separated acridan donor and oxygen-bridged triphenylboron acceptors with different substituents of *tert*-butyl, hydrogen, or fluorine [110]. The backbone of the polymer endowed weaker electron coupling between the spatially proximate donor and acceptor, aiming to attain the blue emission as well as minimizing the HOMO and LUMO overlap to furnish small ΔE_{ST}. TADF features were found in these polymers. In addition, AIE phenomena with emission intensity increased up to 27-fold were recorded from dilute solution to aggregated state. The different substituent on the acceptors in the polymers helped to tune the CT emission from deep blue to sky blue, providing promising PLQYs of up to 70% in thin film. Accordingly, the TSCT-based blue TADF polymers can be used as a light-emitting layer in solution-processed OLEDs, which can exhibit an outstanding EL performance with a $\eta_{C,max}$ of 30.7 cd A^{-1} and a $\eta_{ext,max}$ of 15.0% [110].

In addition, Lu and coworker reported two new blue TADF luminogens, **B-oCz** and **B-oTC** (Scheme 16), composed of ortho-donor (D)-acceptor (A) positioning to attain a combination of two charge transfer pathways via an aryl linker and through space at the same time (Fig. 9) [111]. As a result of having two charge transfer pathways, both the compounds **B-oCz** and **B-oTC** exhibited small ΔE_{ST}s of 0.06 and 0.05 eV, and a decent PLQY of 61% and 94%, respectively. Notably, **B-oCz** and **B-oTC** with nonplanar arrangement inhibited the concentration quenching and the presence of alkyl substituents in **B-oTC** further restricted belligerent quenching arising from intermolecular π-π stacking in the film state. Remarkably, utilizing **B-oCz** and **B-oTC** as a blue TADF emitting layer, the nondoped OLEDs with a simplified configuration of ITO/PEDOT:PSS (40 nm)/TFB (10 nm)/EML (40 nm)/DPEPO (10 nm)/TyPMPB (30 nm)/LiF (1 nm)/Al (100 nm) exhibited an efficient EL performance with peaks at 463 and 474 nm and maximum η_{ext}s of 8% and 19.1%, respectively (Fig. 9) [111].

Tang et al. also reported two customized HPB-based luminogens **TRZ-HPB-PXZ** and **TRZ-HPB-DMAC** (Scheme 16) with phenoxazine or acridine as the electron donor and triazine as the acceptor [112]. The designed luminogens **TRZ-HPB-PXZ** and **TRZ-HPB-DMAC** displayed TSCT interaction between the donor and acceptor that brings about sufficient separation between HOMO and LUMO, achieving small ΔE_{ST}s of 0.02 and 0.09 eV. In addition, both compounds showed an AIE effect upon aggregation from dilute solution to aggregated state. **TRZ-HPB-PXZ** and **TRZ-HPB-DMAC** showed delay fluorescence lifetimes of 2.1 and 4.7 μs, respectively, and decent PLQYs of 61.5% and 51.8%. **TRZ-HPB-PXZ** or **TRZ-HPB-DMAC**-based solution-processed multilayer nondoped OLEDs with a simplified architecture of ITO/HATCN (5 nm)/TAPC (20 nm)/TCTA (5 nm)/emitter/TmPyPB (55 nm)/LiF (1 nm)/Al were fabricated, presenting EL emission peaks at 544 and 521 nm with maximum η_{ext}s of 12.7% and 6.5% as well as low efficiency roll-offs of 2.7% and 7.0%, respectively, at 1000 cd m^{-2} [112].

7 Conclusions and perspective

Since the first report of the AIE phenomenon, the AIEgens have attracted much interest of the OLED community and made great progress, with the EL spectra covering the whole range of visible light and the efficiencies reaching the internal theoretical limit of 100%. In this chapter, we have comprehensively summarized the AIE, AIP, and AIDF luminogens in terms of molecular design, photophysical properties, and EL device performances. To develop extremely robust AIDF by the combination of pure organic TADF materials with AIE effect has been demonstrated as an effective strategy for high-performance OLEDs. Comparatively,

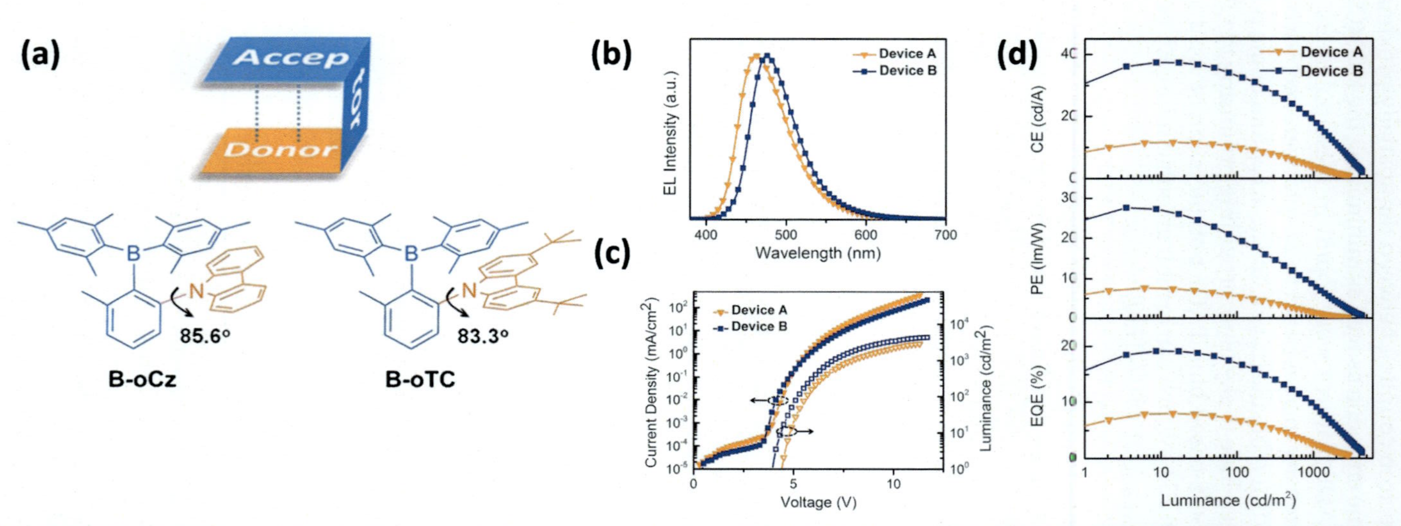

FIG. 9 (A) Configuration and chemical structures of the compounds **B-oCz** and **B-oTC**; (B) EL spectra (at 8 V), (C) Voltage (V) versus current density (*J*) and luminance (*L*) characteristics, and (D) efficiency versus luminance relationships of the nondoped devices. *Reproduced with permission from X.-L. Chen, J.-H. Jia, R. Yu, J.-Z. Liao, M.-X. Yang, C.-Z. Lu, Combining charge-transfer pathways to achieve unique thermally activated delayed fluorescence emitters for high-performance solution-processed, non-doped blue OLEDs, Angew. Chem. Int. Ed. 56 (2017) 15006–15009. Copyright 2018 John Wiley & Sons.*

AIE-active phosphorescent materials may not be desirable emitters for OLEDs due to the expensive precious metal and potential environment contaminants. Looking to the future, we still need to develop highly efficient AIEgens for OLEDs. Besides, the deliberate device structures suitable for the AIEgens should be taken into account, and the corresponding device mechanism should be deeply studied. The intrinsic characters of AIEgens could provide OLEDs with high efficiency and low efficiency roll-offs at a high doping level, even in a nondoped emitting layer. To reach such a goal, we are still on the way.

References

[1] M. Zhu, C. Yang, Blue fluorescent emitters: design tactics and applications in organic light-emitting diodes, Chem. Soc. Rev. 42 (2013) 4963–4976.

[2] M. Liu, X.-L. Li, D.-C. Chen, Z. Xie, X. Cai, G. Xie, K. Liu, J. Tang, S.-J. Su, Y. Cao, Study of configuration differentia and highly efficient, deep-blue, organic light-emitting diodes based on novel naphtho[1,2-*d*]imidazole derivatives, Adv.Funct.Mater. 25 (2015) 5190–5198.

[3] H. Kaji, H. Suzuki, T. Fukushima, K. Shizu, K. Suzuki, S. Kubo, T. Komino, H. Oiwa, F. Suzuki, A. Wakamiya, Y. Murata, C. Adachi, Purely organic electroluminescent material realizing 100% conversion from electricity to light, Nat. Commun. 6 (2015) 8476.

[4] Y. Hong, J.W.Y. Lam, B. Tang, Z. Aggregation-induced emission, Chem. Soc. Rev. 40 (2011) 5361–5388.

[5] J. Luo, Z. Xie, J.W.Y. Lam, L. Cheng, H. Chen, C. Qiu, H.S. Kwok, X. Zhan, Y. Liu, D. Zhu, B.Z. Tang, Aggregation-induced emission of 1-methyl-1,2,3,4,5-pentaphenylsilole, Chem. Commun. (2001) 1740–1741.

[6] J. Mei, Y. Hong, J.W.Y. Lam, A. Qin, Y. Tang, B.Z. Tang, Aggregation-induced emission: the whole is more brilliant than the parts, Adv. Mater. 26 (2014) 5429–5479.

[7] (a) M.A. Baldo, D.F. O'Brien, M.E. Thompson, S.R. Forrest, Excitonic singlet-triplet ratio in a semiconducting organic thin film, Phys. Rev. B 60 (1999) 14422. (b) L.J. Rothberg, A.J. Lovinger, Status of and prospects for organic electroluminescence, J. Mater. Res. 11 (1996) 3174.

[8] (a) A. Endo, K. Sato, K. Yoshimura, T. Kai, A. Kawada, H. Miyazaki, C. Adachi, Efficient up-conversion of triplet excitons into a singlet state and its application for organic light emitting diodes, Appl. Phys. Lett. 98 (2011), 083302. (b) S. Hirata, Sakai, K. Masui, H. Tanaka, S.Y. Lee, H. Nomura, N. Nakamura, M. Yasumatsu, H. Nakanotani, Q. Zhang, K. Shizu, H. Miyazaki, C. Adachi, Highly efficient blue electroluminescence based on thermally activated delayed fluorescence, Nat. Mater. 14 (2015) 330–336. (c) J.-J. Liang, Y. Li, Y. Yuan, S.-H. Li, X.-D. Zhu, S. Barlow, M.-K. Fung, Z.-Q. Jiang, S.R. Marder, L.-S. Liao, A blue thermally activated delayed fluorescence emitter developed by appending a fluorene moiety to a carbazole donor with *meta*-linkage for high-efficiency OLEDs, Mater. Chem. Front. 2 (2018) 917–922. (d) S. Gong, Y. Chen, J. Luo, C. Yang, C. Zhong, J. Qin, D. Ma, Bipolar tetraarylsilanes as universal hosts for blue, green, orange, and white electrophosphorescence with high efficiency and low efficiency roll-off, Adv. Funct. Mater. 21 (2011) 1168. (e) H. Wang, L. Xie, Q. Peng, L. Meng, Y. Wang, Y. Yi, P. Wang, Novel thermally activated delayed fluorescence materials–thioxanthone derivatives and their applications for highly efficient OLEDs, Adv. Mater. 26 (2014) 5198. (f) Y. Seino, S. Inomata, H. Sasabe, Y.-J. Pu, J. Kido, High-performance green OLEDs using thermally activated delayed fluorescence with a power efficiency of over 100 lm W^{-1}, Adv. Mater. 28 (2016) 2638–2643. (g) Q. Zhang, B. Li, S. Huang, H. Nomura, H. Tanaka, C. Adachi, Efficient blue organic light-emitting diodes employing thermally activated delayed fluorescence, Nat. Photonics 8 (2014) 326–332.

[9] (a) Y. Ma, H. Zhang, J. Shen, C. Che, Electroluminescence from triplet metal—ligand charge-transfer excited state of transition metal complexes, Synth. Met. 94 (1998) 245–248. (b) M.A. Baldo, D.F. O'Brien, Y. You, A. Shoustikov, S. Sibley, M.E. Thompson, S.R. Forrest, Highly Efficient Phosphorescent Emission from Organic Electroluminescent Devices, vol. 395, Nature, 1998, pp. 151–154.

[10] (a) C. Adachi, M.A. Baldo, M.E. Thompson, S.R. Forrest, Nearly 100% internal phosphorescence efficiency in an organic light-emitting device, J. Appl. Phys. 90 (2001) 5048. (b) H. Sasabe, J. Kido, Recent progress in phosphorescent organic light-emitting devices, Eur. J. Org. Chem. 2013 (2013) 7653–7663. (c) B. Minaev, G. Baryshnikov, H. Agren, Principles of phosphorescent organic light emitting devices, Phys. Chem. Chem. Phys. 16 (2014) 1719–1758.

[11] J. Mei, N.L.C. Leung, R.T.K. Kwok, J.W.Y. Lam, B.Z. Tang, Aggregation-induced emission: together we shine, united we soar, Chem. Rev. 115 (2015) 11718–11940.

[12] Z. Zhao, B. He, B.Z. Tang, Aggregation-induced emission of siloles, Chem. Sci. 6 (2015) 5347–5365.
[13] Z. Zhao, J.W.Y. Lam, B.Z. Tang, Tetraphenylethene: a versatile AIE building block for the construction of efficient luminescent materials for organic light-emitting diodes, J. Mater. Chem. 22 (2012) 23726–23740.
[14] B. Chen, Y. Jiang, L. Chen, H. Nie, B. He, P. Lu, H.H.Y. Sung, I.D. Williams, H.S. Kwok, A. Qin, Z. Zhao, B.Z. Tang, 2,5-Difluorenyl-substituted Siloles for the fabrication of high-performance yellow organic light-emitting diodes, Chem. Eur. J. 20 (2014) 1931–1939.
[15] L. Chen, Y. Jiang, H. Nie, P. Lu, H.H.Y. Sung, I.D. Williams, H.S. Kwok, F. Huang, A. Qin, Z. Zhao, B.Z. Tang, Creation of bifunctional materials: improve Electron-transporting ability of light emitters based on AIE-active 2,3,4,5-tetraphenylsiloles, Adv. Funct. Mater. 24 (2014) 3621–3630.
[16] C. Quan, H. Nie, R. Hu, A. Qin, Z. Zhao, B.Z. Tang, A silole-based efficient electroluminescent material with good electron-transporting potential, Chin. J. Chem. 33 (2015) 842–846.
[17] B. Chen, H. Nie, R. Hu, A. Qin, Z. Zhao, B.-Z. Tang, Red fluorescent siloles with aggregation-enhanced emission characteristics, Sci. China Chem. 59 (2016) 699–706.
[18] W. Feng, Q. Su, Y. Ma, Z. Džolić, F. Huang, Z. Wang, S. Chen, B.Z. Tang, Tetraphenylbenzosilole: an AIE building block for deep-blue emitters with high performance in nondoped spin-coating OLEDs, J. Org. Chem. 85 (2020) 158–167.
[19] C.Y.K. Chan, Z. Zhao, J.W.Y. Lam, J. Liu, S. Chen, P. Lu, M. Faisal, X. Chen, H.H.Y. Sung, H.S. Kwok, Y. Ma, I.D. Williams, K.S. Wong, B.Z. Tang, Efficient light emitters in the solid state: synthesis, aggregation-induced emission, electroluminescence, and sensory properties of luminogens with benzene cores and multiple triarylvinyl peripherals, Adv. Funct. Mater. 22 (2012) 378–389.
[20] Z. Zhao, S. Chen, X. Shen, F. Mahtab, Y. Yu, P. Lu, J.W.Y. Lam, H.S. Kwok, B.Z. Tang, Aggregation-induced emission, self-assembly, and electroluminescence of 4,40-bis(1,2,2-triphenylvinyl)biphenyl, Chem. Commun. 46 (2010) 686–688.
[21] Z. Zhao, S. Chen, J.W.Y. Lam, P. Lu, Y. Zhong, K.S. Wong, H.S. Kwok, B.Z. Tang, Creation of highly efficient solid emitter by decorating pyrene core with AIE-active tetraphenylethene peripheries, Chem. Commun. 46 (2010) 2221–2223.
[22] Z. Zhao, S. Chen, C.Y.K. Chan, J.W.Y. Lam, C.K.W. Jim, P. Lu, Z. Chang, H.S. Kwok, H. Qiu, B.Z. Tang, A facile and versatile approach to efficient luminescent materials for applications in organic light-emitting diodes, Chem. Asian J. 7 (2012) 484–488.
[23] Z. Zhao, S. Chen, J.W.Y. Lam, Z. Wang, P. Lu, M. Faisal, H.H.Y. Sung, I.D. Williams, Y. Ma, H.S. Kwok, B.Z. Tang, Pyrene-substituted ethenes: aggregation-enhanced excimer emission and highly efficient electroluminescence, J. Mater. Chem. 21 (2011) 7210–7216.
[24] S.-K. Kim, Y.-I. Park, I.-N. Kang, J.-W. Park, New deep-blue emitting materials based on fully substituted ethylene derivatives, J. Mater. Chem. 17 (2007) 4670–4678.
[25] P.-I. Shih, C.-Y. Chuang, C.-H. Chien, E.W. Diau, C.-F. Shu, Highly efficient non-doped blue-light-emitting diodes based on an anthrancene derivative end-capped with tetraphenylethylene groups, Adv. Funct. Mater. 17 (2007) 3141–3146.
[26] J. Huang, N. Sun, J. Yang, R. Tang, Q. Li, D. Ma, Z. Li, Blue aggregation-induced emission luminogens: high external quantum efficiencies up to 3.99% in LED device, and restriction of the conjugation length through rational molecular design, Adv. Funct. Mater. 24 (2014) 7645–7654.
[27] Y. Liu, S. Chen, J.W.Y. Lam, P. Lu, R.T.K. Kwok, F. Mahtab, H.S. Kwok, B.Z. Tang, Tuning the electronic nature of aggregation-induced emission luminogens with enhanced hole-transporting property, Chem. Mater. 23 (2011) 2536–2544.
[28] J.Y. Kim, T. Yasuda, Y.S. Yang, C. Adachi, Bifunctional star-burst amorphous molecular materials for OLEDs: achieving highly efficient solid-state luminescence and carrier transport induced by spontaneous molecular orientation, Adv. Mater. 25 (2013) 2666–2671.
[29] Y. Liu, X. Chen, Y. Lv, S. Chen, J.W.Y. Lam, F. Mahtab, H.S. Kwok, X. Tao, B.Z. Tang, Systemic studies of tetraphenylethene–triphenylamine oligomers and a polymer: achieving both efficient solid-state emissions and hole-transporting capability, Chem.–Eur. J. 18 (2012) 9929–9938.
[30] W. Yuan, P. Lu, S. Chen, J.W.Y. Lam, Z. Wang, Y. Liu, H.S. Kwok, Y. Ma, B.Z. Tang, Changing the behavior of chromophores from aggregation-caused quenching to aggregation-induced emission: development of highly efficient light emitters in the solid state, Adv. Mater. 22 (2010) 2159–2163.
[31] Z. Zhao, S. Chen, J.W.Y. Lam, C.K.W. Jim, C.Y.K. Chan, Z. Wang, P. Lu, C. Deng, H.S. Kwok, Y. Ma, B.Z. Tang, Steric hindrance, electronic communication, and energy transfer in the photo- and electroluminescence processes of aggregation-induced emission luminogens, J. Phys. Chem. C 114 (2010) 7963–7972.

[32] Z. Zhao, C.Y.K. Chan, S. Chen, C. Deng, J.W.Y. Lam, C.K.W. Jim, Y. Hong, P. Lu, Z. Chang, X. Chen, P. Lu, H.S. Kwok, H. Qiu, B.Z. Tang, Using tetraphenylethene and carbazole to create efficient luminophores with aggregation-induced emission, high thermal stability, and good hole-transporting property, J. Mater. Chem. 22 (2012) 4527–4534.

[33] Z. Zhao, S. Chen, C. Deng, J.W.Y. Lam, C.Y.K. Chan, P. Lu, Z. Wang, B. Hu, X. Chen, P. Lu, H.S. Kwok, Y. Ma, H. Qiu, B.Z. Tang, Construction of efficient solid emitters with conventional and AIE luminogens for blue organic light-emitting diodes, J. Mater. Chem. 21 (2011) 10949–10956.

[34] W.Z. Yuan, S. Chen, J.W.Y. Lam, C. Deng, P. Lu, H.H.Y. Sung, I.D. Williams, H.S. Kwok, Y. Zhang, B.Z. Tang, Towards high efficiency solid emitters with aggregation-induced emission and electron-transport characteristics, Chem. Commun. 47 (2011) 11216–11218.

[35] W. Qin, J.W.Y. Lam, Z. Yang, S. Chen, G. Liang, W. Zhao, H.S. Kwok, B.Z. Tang, Red emissive AIE luminogens with high hole-transporting properties for efficient non-doped OLEDs, Chem. Commun. 51 (2015) 7321–7324.

[36] H. Li, Z. Chi, X. Zhang, B. Xu, S. Liu, Y. Zhang, J. Xu, New thermally stable aggregation-induced emission enhancement compounds for non-doped red organic light-emitting diodes, Chem. Commun. 47 (2011) 11273–11275.

[37] Z. Zhao, C. Deng, S. Chen, J.W.Y. Lam, W. Qin, P. Lu, Z. Wang, H.S. Kwok, Y. Ma, H. Qiu, B.Z. Tang, Full emission color tuning in luminogens constructed from tetraphenylethene, benzo-2,1,3-thiadiazole and thiophene building blocks, Chem. Commun. 47 (2011) 8847–8849.

[38] (a) M. O'Neill, S.M. Kelly, Liquid crystals for charge transport, luminescence, and photonics, Adv. Mater. 15 (2003) 1135–1146. (b) Y. Wang, J. Shi, J. Chen, W. Zhu, E. Baranoff, Recent progress in luminescent liquid crystal materials: design, properties and application for linearly polarised emission, J. Mater. Chem. C 3 (2015) 7993–8005.

[39] W. Qin, Z. Yang, Y. Jiang, J.W.Y. Lam, G. Liang, H.S. Kwok, B.Z. Tang, Construction of efficient deep blue aggregation-induced emission luminogen from triphenylethene for nondoped organic light-emitting diodes, Chem. Mater. 27 (2015) 3892–3901.

[40] L. Chen, Y. Jiang, H. Nie, R. Hu, H.S. Kwok, F. Huang, A. Qin, Z. Zhao, B.Z. Tang, Rational design of aggregation-induced emission luminogen with weak electron donor–acceptor interaction to achieve highly efficient undoped bilayer OLEDs, ACS Appl. Mater. Interfaces 6 (2014) 17215–17225.

[41] Y. Wang, Y. Liao, C.P. Cabry, D. Zhou, G. Xie, Z. Qu, D.W. Bruce, W. Zhu, Highly efficient blueish-green fluorescent OLEDs based on AIE liquid crystal molecules: from ingenious molecular design to multifunction materials, J. Mater. Chem. C 5 (2017) 3999–4008.

[42] I. Bala, L. Ming, R.A.K. Yadav, J. De, D.K. Dubey, S. Kumar, H. Singh, J.-H. Jou, K. Kailasam, S.K. Pal, Deep-blue OLED fabrication from heptazine columnar liquid crystal based AIE-active sky-blue emitter, Chemistry Select 3 (2018) 7771–7777.

[43] J. De, W.-Y. Yang, I. Bala, S.P. Gupta, R.A.K. Yadav, D.K. Dubey, A. Chowdhury, J.-H. Jou, S.K. Pal, Room-temperature columnar liquid crystals as efficient pure deep-blue emitters in organic light-emitting diodes with an external quantum efficiency of 4.0%, ACS Appl. Mater. Interfaces 11 (2019) 8291–8300.

[44] (a) R. Nandy, M. Subramoni, B. Varghese, S. Sankararaman, Intramolecular π-stacking interaction in a rigid molecular hinge substituted with 1-(Pyrenylethynyl) units, J. Org. Chem. 72 (2007) 938–944. (b) S. Sankararaman, G. Venkataramana, B. Varghese, Conformational isomers from rotation of diacetylenic bond in an ethynylpyrene-substituted molecular hinge, J. Org. Chem. 73 (2008) 2404–2407.

[45] (a) M. Uchida, T. Izumizawa, T. Nakano, S. Yamaguchi, K. Tamao, K. Furukawa, Structural optimization of 2,5-Diarylsiloles as excellent electron-transporting materials for organic electroluminescent devices, Chem. Mater. 13 (2001) 2680–2683. (b) Z. Zhao, D. Liu, M. Faisal, L. Xin, Z. Shen, Y. Yu, C.Y.K. Chan, P. Lu, J.W.Y. Lam, H.H.Y. Sung, I.D. Williams, B. Yang, Y. Ma, B.Z. Tang, Synthesis, structure, aggregation-induced emission, self-assembly, and electron mobility of 2,5-bis(triphenylsilylethynyl)-3,4-diphenylsiloles, Chem.–Eur. J. 17 (2011) 5998–6008.

[46] (a) J.-Y. Shen, X.-L. Yang, T.-H. Huang, J.T. Lin, T.-H. Ke, L.-Y. Chen, C.-C. Wu, M.-C.P. Yeh, Ambipolar conductive 2,7-carbazole derivatives for electroluminescent devices, Adv. Funct. Mater. 17 (2007) 983–995. (b) K.R. J. Thomas, M.J. Velusamy, T. Lin, Y.-T. Tao, C.-H. Chuen, Cyanocarbazole derivatives for high-performance electroluminescent devices, Adv. Funct. Mater. 14 (2004) 387–392.

[47] (a) V. Promarak, M. Ichikawa, T. Sudyoadsuk, S. Saengsuwan, S. Jungsuttiwong, T. Keawin, Synthesis of electrochemically and thermally stable amorphous hole-transporting carbazole dendronized fluorine, Synth. Met. 157 (2007) 17–22. (b) J. Li, C. Ma, J. Tang, C.-S. Lee, S.T. Lee, Novel starburst molecule as a hole injecting and transporting material for organic light-emitting devices, Chem. Mater. 17 (2005) 615–619. (c) Q. Zhang, J.

Chen, Y. Cheng, L. Wang, D. Ma, X. Jingand, F. Wang, Novel hole-transporting materials based on 1,4-bis (carbazolyl)benzene for organic light-emitting devices, J. Mater. Chem. 14 (2004) 895.

[48] (a) S. Lamansky, P. Djurovich, D. Murphy, A.-R. Feras, H.-E. Lee, C. Adachi, P.E. Burrows, S.R. Forrest, M.E. Thompson, Highly phosphorescent bis-cyclometalated iridium complexes: synthesis, photophysical characterization, and use in organic light emitting diodes, J. Am. Chem. Soc. 123 (2001) 4304–4312. (b) K.R.J. Thomas, M. Velusamy, J.T. Lin, Y.-T. Tao, C.-H. Chuen, Cyanocarbazole derivatives for high-performance electroluminescent devices, Adv. Funct. Mater. 14 (2004) 387–392. (c) Z. Yang, Z. Chi, T. Yu, X. Zhang, M. Chen, B. Xu, S. Liu, Y. Zhang, J. Xu, Triphenylethylene carbazole derivatives as a new class of AIE materials with strong blue light emission and high glass transition temperature, J. Mater. Chem. 19 (2009) 5541–5546.

[49] Z. Zhao, J.W.Y. Lam, C.Y.K. Chan, S. Chen, J. Liu, P. Lu, M. Rodriguez, J.-L. Maldonado, G. Ramos-Ortiz, H.H.Y. Sung, I.D. Williams, H. Su, K.S. Wong, Y. Ma, H.S. Kwok, H. Qiu, B.Z. Tang, Stereoselective synthesis, efficient light emission, and high bipolar charge mobility of chiasmatic luminogens, Adv. Mater. 23 (2011) 5430–5435.

[50] Z. Zhao, J. Geng, Z. Chang, S. Chen, C. Deng, T. Jiang, W. Qin, J.W.Y. Lam, H.S. Kwok, H. Qiu, B. Liu, B.Z. Tang, A tetraphenylethene-based red luminophor for an efficient non-doped electroluminescence device and cellular imaging, J. Mater. Chem. 22 (2012) 11018–11021.

[51] S. Setia, S. Sidiq, J. De, I. Pani, S.K. Pal, Applications of liquid crystals in biosensing and organic light-emitting devices: future aspects, Liq. Cryst. 43 (2016) 2009–2050.

[52] (a) H. Singh, A. Balamurugan, M. Jayakannan, Solid state assemblies and photophysical characteristics of linear and bent-core π-conjugated oligophenylenevinylenes, ACS Appl. Mater. 5 (2013) 5578–5591. (b) X.M. He, J.B. Lin, W.H. Kan, P.C. Dong, S. Trudel, T. Baumgartner, Molecular engineering of "click"-phospholes towards self-assembled luminescent soft materials, Adv. Funct. Mater. 24 (2014) 897–906. (c) S. Thiery, B. Heinrich, B. Donnio, C. Poriel, F. Camerel, Luminescence modulation in liquid crystalline phases containing a dispiro[fluorene-9,11′-indeno[1,2-*b*]fluorene-12′,9″-fluorene] core, J. Mater. Chem. C 2 (2014) 4265–4275. (d) B. Zhang, C.H. Hsu, Z.Q. Yu, S. Yang, E.Q. Chen, A reproducible mechano-responsive luminescent system based on a discotic crown ether derivative doped with fluorophores: taking advantage of the phase transition of a matrix, Chem. Commun. 49 (2013) 8872–8874. (e) R. Gimenez, M. Pinol, J.L. Serrano, Luminescent liquid crystals derived from 9,10-Bis(Phenylethynyl)anthracene, Chem. Mater. 16 (2004) 1377–1383. (f) S. Moyano, J. Barbera, B.E. Diosdado, J.L. Serrano, A. Elduque, R. Gimenez, Self-assembly of 4-aryl-1*H*-pyrazoles as a novel platform for luminescent supramolecular columnar liquid crystals, J. Mater. Chem. C 1 (2013) 3119–3128. (g) E. Beltran, J.L. Serrano, T. Sierra, R. Gimenez, Functional star-shaped tris(triazolyl)triazines: columnar liquid crystal, fluorescent, solvatofluorochromic and electrochemical properties, J. Mater. Chem. 22 (2012) 7797–7805. (h) R. Cristiano, J. Eccher, I.H. Bechtold, C.N. Tironi, A.A. Vieira, F. Molin, H. Gallardo, Luminescent columnar liquid crystals based on tristriazolotriazine, Langmuir 28 (2012) 11590–11598.

[53] (a) Y. Sagara, T. Kato, Stimuli-responsive luminescent liquid crystals: change of photoluminescent colors triggered by a shear-induced phase transition, Angew. Chem. Int. Ed. 47 (2008) 5175–5178. (b) S. Yamane, Y. Saqara, T. Mutai, K. Araki, T. Kato, Mechanochromic luminescent liquid crystals based on a bianthryl moiety, J. Mater. Chem. C 1 (2013) 2648–2656. (c) S. Yamane, Y. Saqara, T. Kato, Steric effects on excimer formation for photoluminescent smectic liquid-crystalline materials, Chem. Commun. 49 (2013) 3839–3841. (d) M. Mitani, S. Ogata, S. Yamane, M. Yoshio, M. Haseqawa, T. Kato, Mechanoresponsive liquid crystals exhibiting reversible luminescent color changes at ambient temperature, J. Mater. Chem. C 4 (2016) 2754–2760.

[54] Y. Li, D.P.-K. Tsang, C.K.-M. Chan, K.M.-C. Wong, M.-Y. Chan, V.W.-W. Yam, Synthesis of unsymmetric bipyridine–PtII–alkynyl complexes through post-click reaction with emission enhancement characteristics and their applications as phosphorescent organic light-emitting diodes, Chem. Eur. J. 20 (2014) 13710–13715.

[55] J. Zhao, Z. Feng, D. Zhong, X. Yang, Y. Wu, G. Zhou, Z. Wu, Cyclometalated platinum complexes with aggregation-induced phosphorescence emission behavior and highly efficient electroluminescent ability, Chem. Mater. 30 (2018) 929–946.

[56] P. Li, Q.-Y. Zeng, H.-Z. Sun, M. Akhtar, G.-G. Shan, X.-G. Hou, F.-S. Li, Z.-M. Su, Aggregation-induced emission (AIE) active iridium complexes toward highly efficient single-layer non-doped electroluminescent devices, J. Mater. Chem. C 4 (2016) 10464–10470.

[57] S. Maji, P. Alam, G.S. Kumar, S. Biswas, P.K. Sarkar, B. Das, I. Rehman, B.B. Das, N.R. Jana, I.R. Laskar, S. Acharya, Induced aggregation of AIE-active mono-cyclometalated Ir(III) complex into supramolecular branched wires for light-emitting diodes, Small 13 (2017) 1603780.

[58] T. Wang, X. Su, X. Zhang, X. Nie, L. Huang, X. Zhang, X. Sun, Y. Luo, G. Zhang, Aggregation-induced dual-phosphorescence from organic molecules for nondoped light-emitting diodes, Adv. Mater. 31 (2019) 1904273.

[59] D.Y. Kondakov, Role of triplet–triplet annihilation in highly efficient fluorescent devices, J. Soc. Inf. Disp. 17 (2009) 137.

[60] C.A. Parker, C.G. Hatchard, Triplet-singlet emission in fluid solutions. Phosphorescence of eosin, Trans. Faraday Soc. 57 (1961) 1894–1904.

[61] M.N. Berberan-Santos, J.M.M. Garcia, Unusually strong delayed fluorescence of C_{70}, J. Am. Chem. Soc. 118 (1996) 9391–9394.

[62] A. Endo, M. Ogasawara, A. Takahashi, D. Yokoyama, Y. Kato, C. Adachi, Thermally activated delayed fluorescence from Sn^{4+}–porphyrin complexes and their application to organic light emitting diodes—anovel mechanism for electroluminescence, Adv. Mater. 21 (2009) 4802–4806.

[63] (a) H. Uoyama, K. Goushi, K. Shizu, H. Nomura, C. Adachi, Highly efficient organic light-emitting diodes from delayed fluorescence, Nature 492 (2012) 234. (b) Q. Zhang, J. Li, K. Shizu, S. Huang, S. Hirata, H. Miyazaki, C. Adachi, Design of efficient thermally activated delayed fluorescence materials for pure blue organic light emitting diodes, J. Am. Chem. Soc. 134 (2012) 14706–14709.

[64] Y. Tao, K. Yuan, T. Chen, P. Xu, H. Li, R. Chen, C. Zheng, L. Zhang, W. Huang, Thermally activated delayed fluorescence materials towards the breakthrough of organoelectronics, Adv. Mater. 26 (2014) 7931.

[65] K. Shizu, Y. Sakai, H. Tanaka, S. Hirata, C. Adachi, H. Kaji, Meta-linking strategy for thermally activated delayed fluorescence emitters with a small singlet-triplet energy gap, ITE Trans. Media Technol. Appl. 3 (2015) 108–113.

[66] H. Tanaka, K. Shizu, H. Miyazaki, C. Adachi, Efficient green thermally activated delayed fluorescence (TADF) from a phenoxazine–triphenyltriazine (PXZ–TRZ) derivative, Chem. Commun. 48 (2012) 11392–11394.

[67] (a) I. Lee, J.Y. Lee, Molecular design of deep blue fluorescent emitters with 20% external quantum efficiency and narrow emission spectrum, Org. Electron. 29 (2016) 160–164. (b) F.B. Dias, J. Santos, D.R. Graves, P. Data, R.S. Nobuyasu, M.A. Fox, A.S. Batsanov, T. Palmeira, M.N. Berberan-Santos, M.R. Bryce, A.P. Monkman, The role of local triplet excited states and D-A relative orientation in thermally activated delayed fluorescence: photophysics and devices, Adv. Sci. 3 (2016) 1600080.

[68] Lee, S.Y.; Yasuda, T.; Yang, Y.S.; Zhang, Q.; Adachi, C. Luminous butterflies: efficient exciton harvesting by benzophenone derivatives for full-color delayed fluorescence OLEDs. Angew. Chem. 2014, 126, 6520; Angew. Chem. Int. Ed. 2014, 53, 6402–6406.

[69] (a) Y.-K. Wang, Q. Sun, S.-F. Wu, Y. Yuan, Q. Li, Z.-Q. Jiang, M.-K. Fung, L.-S. Liao, Thermally activated delayed fluorescence material as host with novel spiro-based skeleton for high power efficiency and low roll-off blue and white phosphorescent devices, Adv. Funct. Mater. 26 (2016) 7929–7936. (b) K. Nasu, T. Nakagawa, H. Nomura, C.-J. Lin, C.-H. Cheng, M.-R. Tseng, T. Yasuda, C. Adachi, A highly luminescent spiro-anthracenone-based organic light-emitting diode exhibiting thermally activated delayed fluorescence, Chem. Commun. 49 (2013) 10385–10387.

[70] J. Lee, N. Aizawa, M. Numata, C. Adachi, T. Yasuda, Versatile molecular functionalization for inhibiting concentration quenching of thermally activated delayed fluorescence, Adv. Mater. 29 (2017) 1604856.

[71] (a) R. Furue, T. Nishimoto, I.S. Park, J. Lee, T. Yasuda, Aggregation-induced delayed fluorescence based on donor/acceptor-tethered janus carborane triads: unique photophysical properties of nondoped OLEDs, Angew. Chem. Int. Ed. 55 (2016) 7171–7175. (b) N. Aizawa, C.-J. Tsou, I.S. Park, T. Yasuda, Aggregation-induced delayed fluorescence from phenothiazine-containing donor–acceptor molecules for high-efficiency non-doped organiclight-emitting diodes, Polym. J. 49 (2017) 197–202.

[72] (a) L. Yu, Z. Wu, G. Xie, W. Zeng, D. Ma, C. Yang, Molecular design to regulate the photophysical properties of multifunctional TADF emitters towards high-performance TADF-based OLEDs with EQEs up to 22.4% and small efficiency roll-offs, Chem. Sci. 9 (2018) 1385–1391. (b) L. Yu, Z. Wu, G. Xie, C. Zhong, Z. Zhu, D. Ma, C. Yang, An efficient exciton harvest route for high-performance OLEDs based on aggregation-induced delayed fluorescence, Chem. Commun. 54 (2018) 1379–1382. (c) K. Wu, Z. Wang, L. Zhan, C. Zhong, S. Gong, G. Xie, C. Yang, Realizing highly efficient solution-processed homojunction-like sky-blue OLEDs by using thermally activated delayed fluorescent emitters featuring an aggregation-induced emission property, J. Phys. Chem. Lett. 9 (2018) 1547–1553.

[73] T.A. Lin, T. Chatterjee, W.L. Tsai, W.K. Lee, M.J. Wu, M. Jiao, K.C. Pan, C.L. Yi, C.L. Chung, K.T. Wong, C.C. Wu, Sky-blue organic light emitting diode with 37% external quantum efficiency using thermally activated delayed fluorescence from spiroacridine-triazine hybrid, Adv. Mater. 28 (2016) 6976–6983.

[74] K.C. Pan, S.W. Li, Y.Y. Ho, Y.J. Shiu, W.L. Tsai, M. Jiao, W.K. Lee, C.C. Wu, C.L. Chung, T. Chatterjee, Y.S. Li, K.-T. Wong, H.C. Hu, C.C. Chen, M.T. Lee, Efficient and tunable thermally activated delayed fluorescence emitters

having orientation-adjustable CN-substituted pyridine and pyrimidine acceptor units, Adv. Funct. Mater. 26 (2016) 7560–7571.

[75] W. Zeng, H.-Y. Lai, W.-K. Lee, M. Jiao, Y.-J. Shiu, C. Zhong, S. Gong, T. Zhou, G. Xie, M. Sarma, K.-T. Wong, C.-C. Wu, C. Yang, Achieving nearly 30% external quantum efficiency for orange–red organic light emitting diodes by employing thermally activated delayed fluorescence emitters composed of 1,8-naphthalimide-acridine hybrids, Adv. Mater. 30 (2018) 1704961.

[76] J. Liang, C. Li, Y. Cui, Z. Li, J. Wang, Y. Wang, Rational design of efficient orange-red to red thermally activated delayed fluorescence emitters for OLEDs with external quantum efficiency of up to 26.0% and reduced efficiency roll-off, J. Mater. Chem. C 8 (2020) 1614–1622.

[77] X. Cao, D. Zhang, S. Zhang, Y. Tao, W. Huang, CN-containing donor–acceptor-type small-molecule materials for thermally activated delayed fluorescence OLEDs, J. Mater. Chem. C 5 (2017) 7699–7714.

[78] M.R. Bryce, Sci., Aggregation-induced delayed fluorescence (AIDF) materials: a new break-through for nondoped OLEDs, China: Chem. 60 (2017) 1561–1562.

[79] S. Gan, W. Luo, B. He, L. Chen, H. Nie, R. Hu, A. Qin, Z. Zhao, B.Z. Tang, Integration of aggregation-induced emission and delayed fluorescence into electronic donor–acceptor conjugates, J. Mater. Chem. C 4 (2016) 3705–3708.

[80] J. Guo, X.-L. Li, H. Nie, W. Luo, S. Gan, S. Hu, R. Hu, A. Qin, Z. Zhao, S.-J. Su, B.Z. Tang, Achieving high-performance nondoped OLEDs with extremely small efficiency roll-off by combining aggregation-induced emission and thermally activated delayed fluorescence, Adv. Funct. Mater. 27 (2017) 1606458.

[81] J. Guo, X.-L. Li, H. Nie, W. Luo, R. Hu, A. Qin, Z. Zhao, S.-J. Su, B.Z. Tang, Robust luminescent materials with prominent aggregation-induced emission and thermally activated delayed fluorescence for high-performance organic light-emitting diodes, Chem. Mater. 29 (2017) 3623–3631.

[82] J. Huang, H. Nie, J. Zeng, Z. Zhuang, S. Gan, Y. Cai, J. Guo, S.-J. Su, Z. Zhao, B.Z. Tang, Highly efficient nondoped OLEDs with negligible efficiency roll-off fabricated from aggregation-induced delayed fluorescence luminogens, Angew. Chem. 129 (2017) 13151–13156. Angew. Chem. Int. Ed. 56 (2017), 12971–12976.

[83] H. Zhao, Z. Wang, X. Cai, K. Liu, Z. He, X. Liu, Y. Cao, S.-J. Su, Highly efficient thermally activated delayed fluorescence materials with reduced efficiency roll-off and low on-set voltages, Mater. Chem. Front. 1 (2017) 2039–2046.

[84] Y. Zhao, W. Wang, C. Gui, L. Fang, X. Zhang, S. Wang, S. Chen, H. Shi, B.Z. Tang, Thermally activated delayed fluorescence material with aggregation-induced emission properties for highly efficient organic light-emitting diodes, J. Mater. Chem. C 6 (2018) 2873–2881.

[85] L. Zhang, Y.-F. Wang, M. Lia, Q.-Y. Gao, C.-F. Chen, Quinoline-based aggregation-induced delayed fluorescence materials for highly efficient non-doped organic light-emitting diodes, Chin. Chem. Lett. (2020), https://doi.org/10.1016/j.cclet.2020.07.041.

[86] J. Wang, C. Liu, C. Jiang, C. Yao, M. Gu, W. Wang, Solution-processed aggregation-induced delayed fluorescence (AIDF) emitters based on strong π-accepting triazine cores for highly efficient nondoped OLEDs with low efficiency roll-off, Org. Elect. 65 (2019) 170–178.

[87] Z. Yang, Y. Zhan, Z. Qiu, J. Zeng, J. Guo, S. Hu, Z. Zhao, X. Li, S. Ji, Y. Huo, S.-J. Su, Stimuli-responsive aggregation-induced delayed fluorescence emitters featuring the asymmetric D–A structure with a novel diarylketone acceptor toward efficient OLEDs with negligible efficiency roll-off, ACS Appl. Mater. Interfaces 12 (2020) 29528–29539.

[88] J. Xu, X. Zhu, J. Guo, J. Fan, J. Zeng, S. Chen, Z. Zhao, B.Z. Tang, Aggregation-induced delayed fluorescence luminogens with accelerated reverse intersystem crossing for high-performance OLEDs, ACS Materials Lett. 1 (2019) 613–619.

[89] (a) J. Guo, J. Fan, L. Lin, J. Zeng, H. Liu, C.-K. Wang, Z. Zhao, B.Z. Tang, Mechanical insights into aggregation-induced delayed fluorescence materials with anti-kasha behavior, Adv. Sci. 6 (2019) 1801629. (b) J. Zeng, J. Guo, H. Liu, J.W.Y. Lam, Z. Zhao, S. Chen, B.Z. Tang, Aggregation-induced delayed fluorescence luminogens for efficient organic light-emitting diodes, Chem. Asian J. 14 (2019) 828–835.

[90] H. Liu, J. Zeng, J. Guo, H. Nie, Z. Zhao, B.Z. Tang, High-performance non-doped OLEDs with nearly 100% exciton use and negligible efficiency roll-off, Angew. Chem. Int. Ed. 57 (2018) 9290–9294.

[91] S. Xiang, Z. Huang, S. Sun, X. Lv, L. Fan, S. Ye, H. Chen, R. Guo, L. Wang, Highly efficient non-doped OLEDs using aggregation-induced delayed fluorescence materials based on 10-phenyl-10*H*-phenothiazine 5,5-dioxide derivatives, J. Mater. Chem. C 6 (2018) 11436–11443.

[92] Y. Chen, S. Wang, X. Wu, Y. Xu, H. Li, Y. Liu, H. Tong, L. Wang, Triazatruxene-based small molecules with thermally activated delayed fluorescence, aggregation-induced emission and mechanochromic luminescence properties for solution-processable nondoped OLEDs, J. Mater. Chem. C 6 (2018) 12503–12508.
[93] J. Huang, Z. Xu, Z. Cai, J. Guo, J. Guo, P. Shen, Z. Wang, Z. Zhao, D. Ma, B.Z. Tang, Robust luminescent small molecules with aggregation-induced delayed fluorescence for efficient solution-processed OLEDs, J. Mater. Chem. C 7 (2019) 330–339.
[94] (a) I. Hladka, D. Volyniuk, O. Bezvikonnyi, V. Kinzhybalo, T.J. Bednarchuk, Y. Danyliv, R. Lytvyn, A. Lazauskas, J.V. Grazulevicius, Polymorphism of derivatives of tert-butylsubstituted acridan and perfluorobiphenyl as sky-blue OLED emitters exhibiting aggregation induced thermally activated delayed fluorescence, J. Mater. Chem. C 6 (2018) 13179–13189. (b) R. Pashazadeh, G. Sych, S. Nasiri, K. Leitonas, A. Lazauskas, D. Volyniuk, P.J. Skabara, J.V. Grazulevicius, Multifunctional asymmetric D-A-D′ compounds: Mechanochromic luminescence, thermally activated delayed fluorescence and aggregation enhanced emission, Chem. Eng. J. 401 (2020), 125962.
[95] S. Chen, P. Zeng, W. Wang, X. Wang, Y. Wu, P. Lina, Z. Peng, Naphthalimide–arylamine derivatives with aggregation induced delayed fluorescence for realizing efficient green to red electroluminescence, J. Mater. Chem. C 7 (2019) 2886–2897.
[96] X. Zhou, H. Yang, Z. Chen, S. Gong, Z.-H. Lu, C. Yang, Naphthyridine-based emitters simultaneously exhibiting thermally activated delayed fluorescence and aggregation-induced emission for highly efficient non-doped fluorescent OLEDs, J. Mater. Chem. C 7 (2019) 6607–6615.
[97] C. Chen, H.-Y. Lu, Y.-F. Wang, M. Li, Y.-F. Shen, C.-F. Chen, Naphthyridine-based thermally activated delayed fluorescence emitters for multi-color organic light-emitting diodes with low efficiency roll-off, J. Mater. Chem. C 7 (2019) 4673–4680.
[98] Q. Zhang, S. Sun, W. Liu, P. Leng, X. Lv, Y. Wang, H. Chen, S. Ye, S. Zhuang, L. Wang, Integrating TADF luminogens with AIE characteristics using a novel acridine–carbazole hybrid as donor for high-performance and low efficiency roll-off OLEDs, J. Mater. Chem. C 7 (2019) 9487.
[99] S. Wang, X. Yan, Z. Cheng, H. Zhang, Y. Liu, Y. Wang, Highly efficient near-infrared delayed fluorescence organic light emitting diodes using a phenanthrene-based charge-transfer compound, Angew. Chem. Int. Ed. 54 (2015) 13068–13072.
[100] Y. Liu, X. Wu, Y. Chen, L. Chen, H. Li, W. Wang, S. Wang, H. Tian, H. Tong, L. Wang, Triazatruxene-based thermally activated delayed fluorescence small molecules with aggregation-induced emission properties for solution-processable nondoped OLEDs with low efficiency roll-off, J. Mater. Chem. C 7 (2019) 9719–9725.
[101] Y. Fu, H. Liu, X. Zhu, J. Zeng, Z. Zhao, B.Z. Tang, Efficient aggregation-induced delayed fluorescent materials based on bipolar carrier transport materials for the fabrication of high-performance nondoped OLEDs with very small efficiency roll-off, J. Mater. Chem. C 8 (2020) 9549–9557.
[102] K. Sun, D. Liu, W. Tian, F. Gu, W. Wang, Z. Cai, W. Jiang, Y. Sun, Manipulation of the sterically hindering effect to realize AIE and TADF for high-performing nondoped solution-processed OLEDs with extremely low efficiency roll-off, J. Mater. Chem. C 8 (2020) 11850–11859.
[103] Cai, Z.; Chen, H.; Guo, J.; Zhao, Z.; Tang, B.Z. Efficient aggregation-induced delayed fluorescence luminogens for solution-processed OLEDs with small efficiency roll-off. Front. Chem. 8: 193. doi:https://doi.org/10.3389/fchem.2020.00193.
[104] I.H. Lee, W. Song, J.Y. Lee, Aggregation-induced emission type thermally activated delayed fluorescent materials for high efficiency in non-doped organic light-emitting diodes, Org. Electron. 29 (2016) 22–26.
[105] K. Kawasumi, T. Wu, T. Zhu, H.S. Chae, T.V. Voorhis, M.A. Baldo, T.M. Swager, Thermally activated delayed fluorescence materials based on homoconjugation effect of donor–acceptor triptycenes, J. Am. Chem. Soc. 137 (2015) 11908–11911.
[106] H. Tsujimoto, D.-G. Ha, G. Markopoulos, H.S. Chae, M.A. Baldo, T.M. Swager, Thermally activated delayed fluorescence and aggregation induced emission with through-space charge transfer, J. Am. Chem. Soc. 139 (2017) 4894–4900.
[107] S. Shao, J. Hu, X. Wang, L. Wang, X. Jing, F. Wang, Blue thermally activated delayed fluorescence polymers with nonconjugated backbone and through-space charge transfer effect, J. Am. Chem. Soc. 139 (2017) 17739–17742.
[108] J. Hu, Q. Li, X. Wang, S. Shao, L. Wang, X. Jing, F. Wang, Developing through-space charge transfer polymers as a general approach to realize full-color and white emission with thermally activated delayed fluorescence, Angew. Chem. Int. Ed. 58 (2019) 8405–8409.

[109] X. Wang, S. Wang, J. Lv, S. Shao, L. Wang, X. Jing, F. Wang, Through-space charge transfer hexaarylbenzene-dendrimers with thermally activated delayedfluorescence and aggregation-induced emission for efficient solution-processed OLEDs, Chem. Sci. 10 (2019) 2915–2923.

[110] F. Chen, J. Hu, X. Wang, S. Shao, L. Wang, X. Jing, F. Wang, Through-space charge transfer blue polymers containing acridan donor and oxygen-bridged triphenylboron acceptor for highly efficient solution-processed organic light-emitting diodes, Sci. China Chem. 63 (2020), https://doi.org/10.1007/s11426-020-9750-9.

[111] X.-L. Chen, J.-H. Jia, R. Yu, J.-Z. Liao, M.-X. Yang, C.-Z. Lu, Combining charge-transfer pathways to achieve unique thermally activated delayed fluorescence emitters for high-performance solution-processed, non-doped blue OLEDs, Angew. Chem. Int. Ed. 56 (2017) 15006–15009.

[112] Zhang, P.; Zeng, J.; Guo, J.; Zhen, S.; Xiao, B.; Wang, Z.; Zhao, Z.; Tang, B.Z. New Aggregation-induced delayed fluorescence luminogens with through-space charge transfer for efficient non-doped OLEDs. Front. Chem. 7: 199. doi:https://doi.org/10.3389/fchem.2019.00199.

CHAPTER

11

Liquid crystalline aggregation-induced emission luminogens for optical displays

Kyohei Hisano, Osamu Tsutsumi, and Supattra Panthai

Department of Applied Chemistry, Ritsumeikan University, Kusatsu, Japan

1 Introduction

Optical displays such as liquid crystal (LC) and organic light-emitting diode (OLED) displays are ubiquitous in modern technology and have attracted significant attention for visualizing and communicating information [1–3]. LCs, which have been described as the fourth state of matter, exhibit both liquid-like fluidity and crystal-like orientational order [4]. LC materials are at the forefront of not only displays, but also various optical applications, owing to the molecular anisotropy of their polarized luminescence, polarization conversion, carrier transportation, etc. Interestingly, long-range communication can occur between LC molecules (molecular cooperative effect) and the molecular orientational order can be changed by external stimuli such as temperature, shear, topographic effects, and electromagnetic fields. Thus, the optical properties based on molecular anisotropy can be easily manipulated by controlling the long-range molecular orientation [5–8]. Owing to intensive investigations of these control methods, various LC-based optical applications with controlled molecular orientations have been achieved [9–12]. In recent years, fundamental advances in the development of optical displays have been based not only on the molecular synthetic approaches for LCs and OLEDs but also on the device structure designs using quantum dots, microLEDs, and holographic technologies [1–3]. These advanced materials have enabled the realization of new conceptual applications, such as smart windows, smart glasses, see-through displays, rollable/bendable/foldable displays, three-dimensional displays, virtual reality, and augmented reality [13–15]. Thus, the most attractive avenue for developing next-generation optical displays is to fabricate state-of-the-art materials.

Aggregation-Induced Emission (AIE)
https://doi.org/10.1016/B978-0-12-824335-0.00008-8

As an innovative material, light-emitting LCs combine the advantages of both OLEDs and LCs. Depending on their molecular anisotropy and long-range orientation, these materials may also exhibit other unique properties such as linearly/circularly polarized luminescence and bright luminescence owing to their high carrier mobility [16–19]. To control the luminescence properties, such as polarization states and/or brightness of LC materials, both the molecular orientation and the molecular structure play crucial roles. Typical luminescent molecules are highly emissive in solution but are quenched in aggregates, which is termed the aggregation-caused quenching (ACQ) effect. To overcome this issue, Tang et al. first proposed a novel aggregation-induced emission (AIE) phenomenon for a series of propeller-shaped molecules with hexaphenylsilole (HPS) or tetraphenylethene (TPE) units [20]. When an AIE-active luminogen (AIEgen) aggregates, internal molecular motion, which results in the nonradiative decay of the excited state, is restricted by the twisted molecular configuration. Therefore, in contrast to those of ACQ luminogens, the luminescence quantum yields of AIEgens are dramatically enhanced in aggregates. This pioneering research has inspired the emerging field of AIEgens, and more recently, AIE combined with thermally activated delayed phosphorescence has been applied practically in OLEDs [21]. However, various challenges remain for the design of AIE-active LCs. In particular, LC phase formation is often hindered by the bulky propeller shape of typical AIEgens.

The molecular structure of an organic LC typically consists of a mesogenic core, which is a rigid anisotropic subunit with a rod-like (calamitic LC) or disk-like shape (discotic LC), and terminal flexible chain(s) with alkoxy and alkyl subunits. A promising route for synthesizing AIE-active LCs is the incorporation of flexible alkoxy or alkyl chains into rigid AIE-active units [22–26]. Because of the propeller-shaped molecular geometry of typical AIE units, AIE-active LCs generally show columnar discotic or smectic phases. A method for amalgamating AIEgens with mesogenic subunits has been developed [27]. In such molecules, the AIE subunit and the mesogenic subunit are designed separately; thus, AIE-active LCs showing nematic or chiral nematic phases can also be obtained. This is the favorable feature for generating linearly/circularly polarized luminescence property. To generate a variety of mesophases, the exploration of new concepts for designing AIE-active molecules with calamitic LC subunits is also highly desirable. The characteristics of calamitic LC compounds include d^{10} transition-metal complexes such as Au(I) complexes. Metal complexes with AIE behavior are of considerable interest because they exhibit inherently high luminescence intensities in condensed phases and unique luminescence behavior depending on the aggregate structure, which has been attributed to the interactions between metals, termed metallophilic interactions [28–30]. Such metal complexes also act as mesogen units, called metallomesogens, and exhibit the same mesophases as typical organic LCs, thus enabling the realization of AIE-active LCs. Broadening the variety of AIE-active LCs would facilitate the production of next-generation optical displays, especially those with polarized emission behavior.

In this chapter, we review and highlight some recent examples of molecular design strategies for AIE-active LCs based on propeller-shaped units, metallomesogen units, and amalgamated AIEgens with LCs and provide significant applications with potential for use as next-generation optical displays. Some important reviews on AIEgens have been published in the past few years [17,18,19], which we recommend for a more detailed understanding.

We hope that this chapter will contribute toward the development of strategies for designing AIE-active LCs and the realization of a wide variety of high-performance applications, such as linearly/circularly polarized luminescent LCs and high carrier mobility in OLEDs. In particular, AIE-active LCs are promising materials for producing cost-effective optical displays with simpler device designs, lower power consumption, higher brightness and color contrast, and wider viewing angles. Herein, we first explore the general design concept by outlining some recent advances in AIE-active LCs and propose some key applications of these materials before finally summarizing the remaining challenges and future perspectives.

2 Molecular design of AIE-active LC materials

2.1 Discotic and calamitic LCs with π-conjugated cores

In a pioneering study, AIE-active LCs were developed by synthesizing a series of low-molecular-weight organogelators, which underwent self-assembly to form supramolecular networks through intermolecular interactions, e.g., π-π stacking, hydrogen bonding, and van der Waals interactions [28–31]. In 2007, Li et al. reported a system for synthesizing highly luminescent organogelators with bisurea-functionalized naphthalene units. Hydrogen bonding and π-π stacking interactions between the gelators and the luminophore resulted in the formation of a columnar LC mesophase. Interestingly, strong fluorescence emission was observed, depending on the aggregation of the luminophore, and the fluorescence behavior could be reversibly tuned by altering the aggregates using temperature and chemical stimuli [32]. Soon after, Lai et al. successfully synthesized two novel organogelators based on 2,3,4,5-tetraphenylsilole, a well-known AIE unit, by functionalization with terminal long alkoxydiacylamido chains (compounds **1** and **2**) (Fig. 1) [26]. Conventionally, such tetraphenylsilole derivatives (**1**,**2**) have noncoplanar geometry, which prevents aggregation and generally hinders the formation of LC phases and/or gels. However, these drawbacks were overcome by incorporating amide groups into the tetraphenylsilole units, which induced intermolecular hydrogen bonding. The compounds prepared using this approach exhibit both AIE behavior and a columnar mesophase over a wide temperature range (Fig. 1) [26]. Following these innovative studies, including an early report on metallogens [29], many other AIE-active organogels have been developed, as reviewed in 2014 [33,34].

For the development of luminescent LCs, impressive success has been demonstrated using the simple and powerful strategy of functionalizing a luminescent unit with terminal flexible chains. For example, Kato et al. synthesized luminescent LCs by incorporating peripheral alkoxy chains into rigid, π-conjugated mesogens of pyrene and anthracene [35]. A wide variety of AIE-active LCs (e.g., compounds **3**–**10**) have been designed using AIEgens as mesogenic units with various flexible alkyl/alkoxy chains, as summarized in Fig. 2. In 2016, Cho et al. investigated the effect of peripheral nonpolar and polar chains on the AIE properties of bulk TPE derivatives (compounds **11**–**13**, Fig. 3) [36]. The bulk TPE derivatives with nonpolar dodecyl (**11**,**12**) and polar di(ethylene oxide) chains (**13**) were found to form columnar LC and polycrystalline phases, respectively, at room temperature. Interestingly, differences in structural mobility arising from the number of peripheral flexible chains and their intrinsic polarity

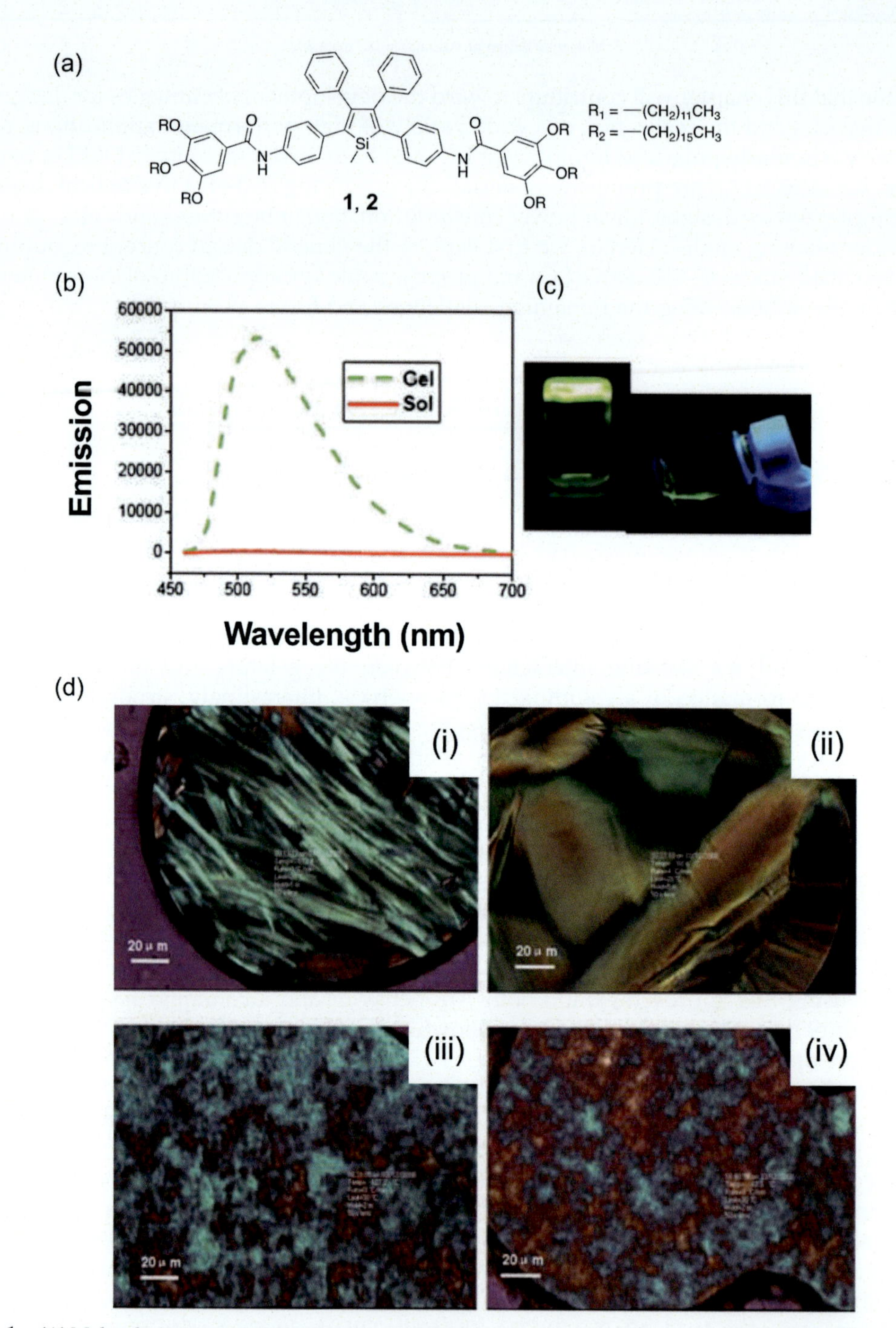

FIG. 1 (A) Molecular structures of the AIE-active organogelators of compounds **1** and **2**. (B) Fluorescence spectra of compound **1** in gel state and CH_2Cl_2 solution with the same concentration (14 mg/mL). (C) Photographic images of the gel and solution observed upon UV irradiation at 365 nm. (D) Polarized optical microscope (POM) images of compounds **1** (i and ii) and **2** (iii and iv) at the liquid crystalline temperature of (i) 112.4°C, (ii) 60.6°C, (iii) 127.2°C, and (iv) 67.6°C. *Reproduced from J.H. Wan, L.Y. Mao, Y.B. Li, Z.F. Li, H.Y. Qiu, C. Wang, G.Q. Lai, Self-assembly of novel fluorescent silole derivatives into different supramolecular aggregates: fibre, liquid crystal and monolayer, Soft Matter 6 (14) (2010) 3195–3201. https://doi.org/10.1039/b925746b with permission from the Royal Society of Chemistry.*

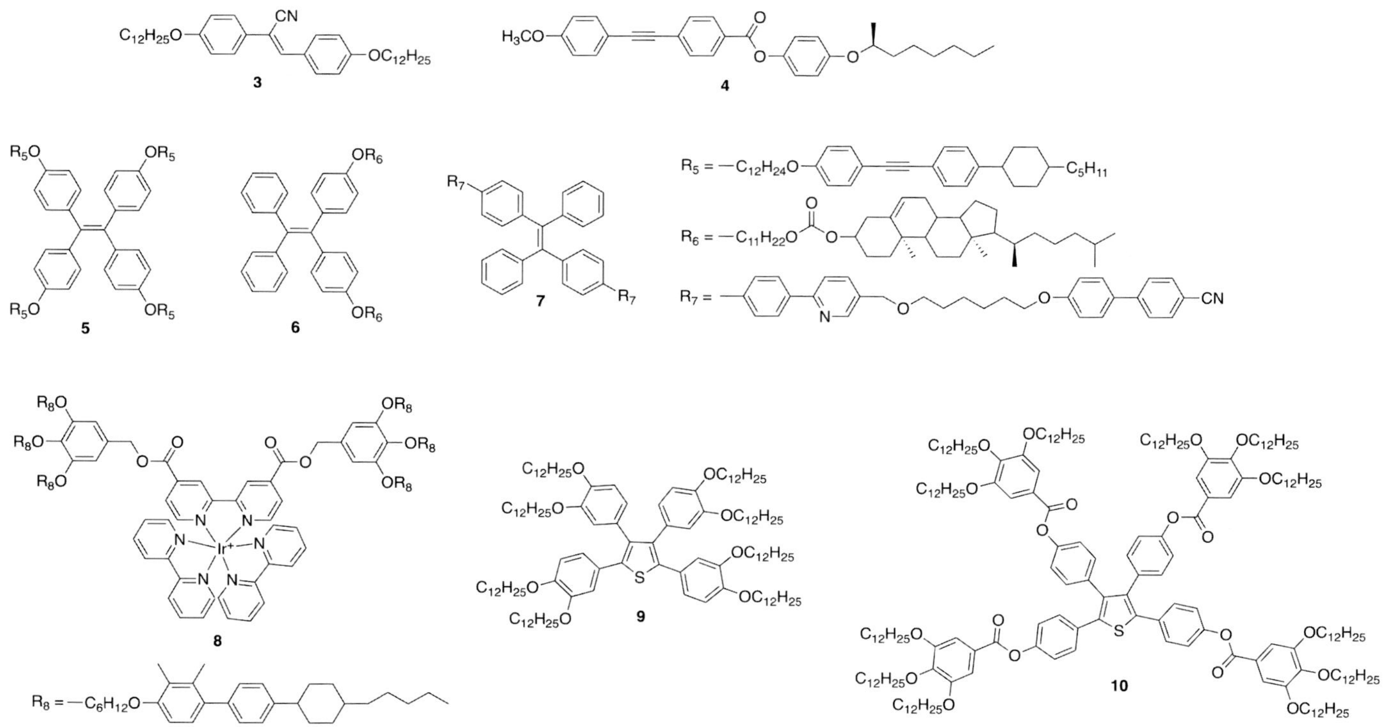

FIG. 2 Some representative molecular structures of AIE-active LCs have been reported in the literature.

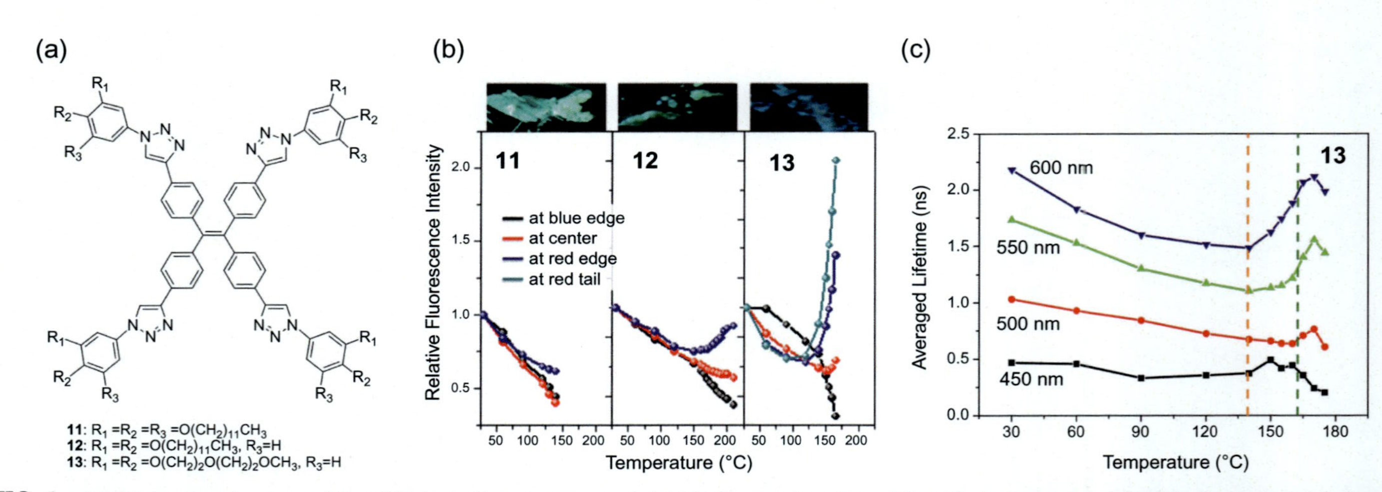

FIG. 3 (A) Molecular structure of the TPE derivatives (compounds **11–13**) for investigation of the effect of peripheral flexible chains on LC behavior. (B) Temperature-dependent relative emission intensities at various detection wavelengths from the blue-to-red-edge region of the emission bands of the TPE derivatives. (C) The averaged emission lifetimes of the balk phase of compound **13** at various wavelengths. The orange and green dashed lines represent 140°C and the isotropic transition temperature, respectively. *Adapted with permission from H.T. Bui, J. Kim, H.J. Kim, B.K. Cho, S. Cho, Advantages of mobile liquid-crystal phase of AIE luminogens for effective solid-state emission, J. Phys. Chem. C 120 (47) (2016) 26695–26702. https://doi.org/10.1021/acs.jpcc.6b10026. Copyright (2016) American Chemical Society.*

might lead to numerous crystallographic defects, resulting in an undesirable nonemissive electronic state. Thus, for high AIE efficiency, the structural mobility of flexible chains is crucial to minimize the number of undesirable nonemissive local sites in the aggregates of TPE derivatives. In addition to the effects of peripheral flexible chains on an existing AIE core, mesogenic molecules with intermolecular interactions, such as dipole-dipole interactions, hydrogen bonding, halogen bonding, and metallophilic interactions, have been revealed to act as organogelators and, in some cases, lead to the formation of smectic or discotic LC phases.

To date, most AIE-active LCs have had conventional propeller-shaped AIEgens such as TPE and HPS (forming smectic and/or columnar mesophases). In addition, to form other LC phases, calamitic LCs with rod-like molecular shapes, such as cyanostilbene, tolane, and metallomesogens, have attracted much attention. As calamitic LCs typically exhibit nematic phases, such compounds have great potential for achieving linearly/circularly polarized luminescence, which is advantageous for AIE-based next-generation optical displays such as light-emitting LC displays. Unfortunately, considering the geometrical restrictions of conventional AIEgens such as TPE and HPS, the synthesis of AIE-active calamitic LCs is challenging and has been rarely reported. In 2013, Tang et al. proposed a new synthetic strategy for applying the AIE phenomenon to the molecular design of LCs by employing tolane, a typical LC molecule, as a mesogenic unit. The obtained tolane-based AIE-active LCs, e.g., compounds **14** and **15** in Fig. 4, showed nematic and/or smectic phases [37,38]. In the tolane-based luminogen, which can be viewed as a propeller-like molecule with two blades (phenyl groups), aggregation restricts the intramolecular rotation of the blades around the ethynyl unit and thus leads to AIE properties.

Another promising method for achieving AIE-active nematic LCs is the simple amalgamation of an exceedingly small portion of an AIEgen (a few wt%) with a host nematic LC. Recently, Tang et al. proposed an efficient route to fabricate AIE-active LCs by decorating a TPE core as an AIE unit with a terminal LC chain (such as a tolane derivative) to obtain a material that has enhanced miscibility with host LCs. For example, TPE-PPE (**16**) exhibited thermotropic smectic phase behavior and AIE properties (Fig. 5) [39]. Doping TPE-PPE (**16**) into a host nematic LC resulted in a nematic phase. More importantly, this mixture overcame the ACQ effect of conventional luminescent LCs. This AIE-active nematic LC (**16**) also exhibited linearly/circularly polarized luminescent behavior, which is favorable for next-generation optical displays, as described in Section 3.2. Interestingly, in 2017, Wang et al. reported that such π-conjugated systems with terminal LC chains can also act as single-component AIE-active LCs (Fig. 6) [40]. By employing a TPE derivative as an AIE core and a

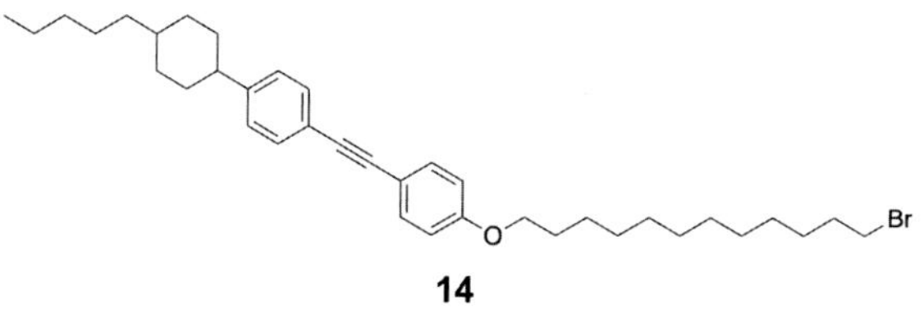

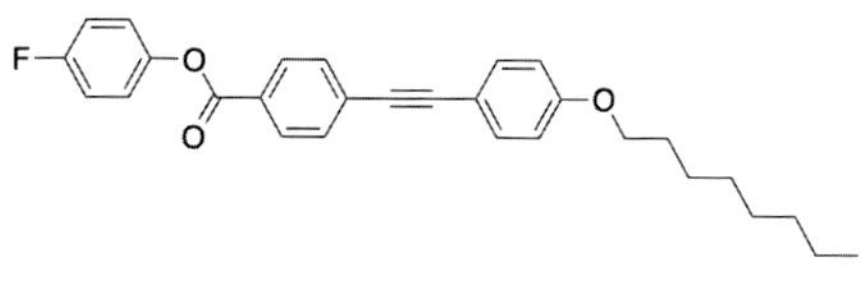

FIG. 4 Representative molecular structures of tolane-based AIE-active LCs (**14** and **15**) reported in the literature [37,38].

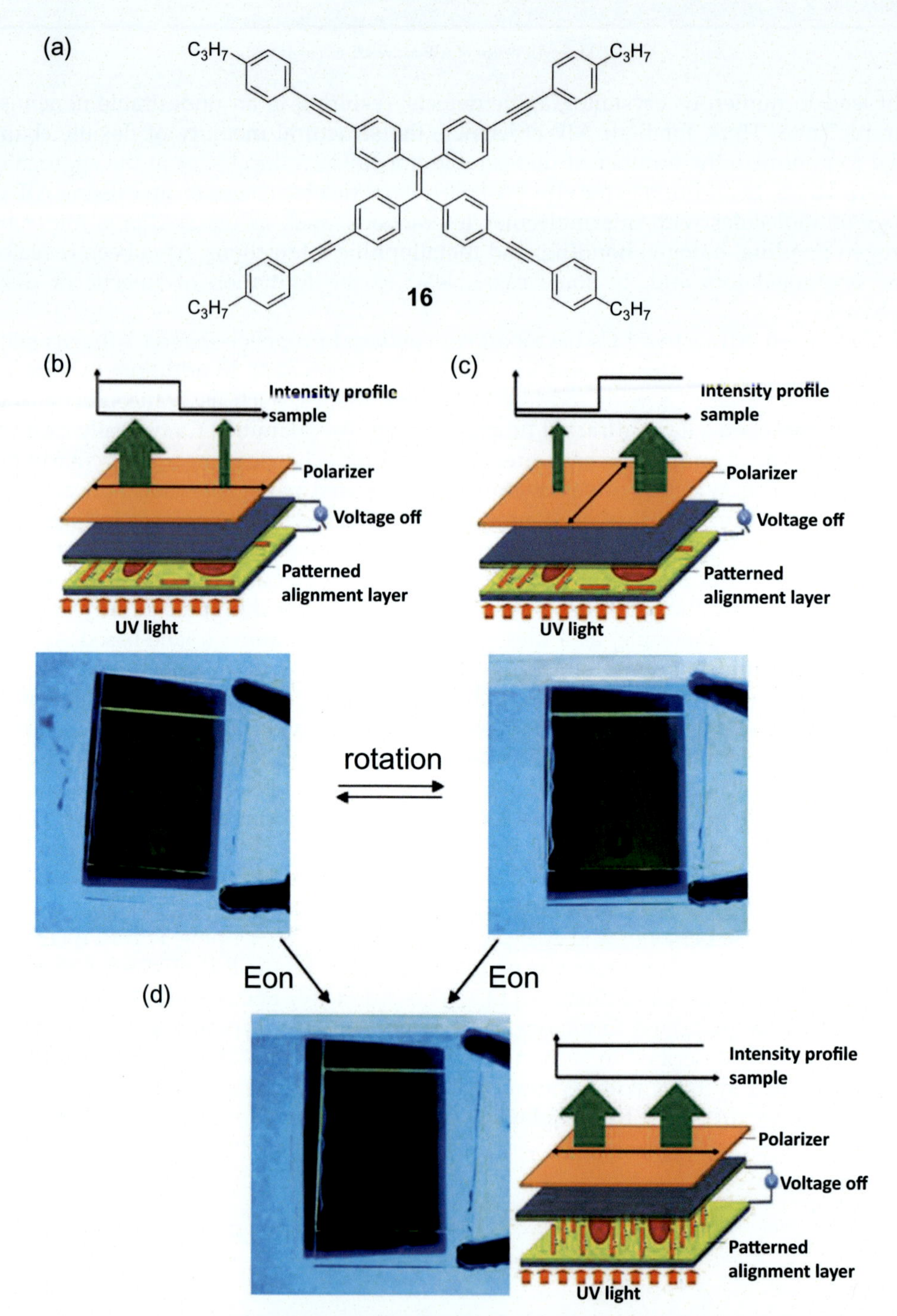

FIG. 5 (A) Molecular structure of TPE-PPE (**16**). (B–D) The device structure and the photographs of light-emitting LC display with LC orientation patterned by photoalignment technology. (B and C) The device in the electric field "off" state under UV irradiation, and rotated by 90° from each other. The bright region has the LC orientation along the transmission direction of a polarizer. (D) The device in the electric field "on" state along the thickness direction that generates out-of-plane orientation of LC and in-plane linearly polarized luminescence behavior is disappeared. *Adapted from D. Zhao, F. Fan, J. Cheng, Y. Zhang, K.S. Wong, V.G. Chigrinov, H.S. Kwok, L. Guo, B.Z. Tang, Light-emitting liquid crystal displays based on an aggregation-induced emission luminogen, Adv Opt Mater 3 (2) (2015) 199–202. https://doi.org/10.1002/adom.201400428 with permission from Wiley.*

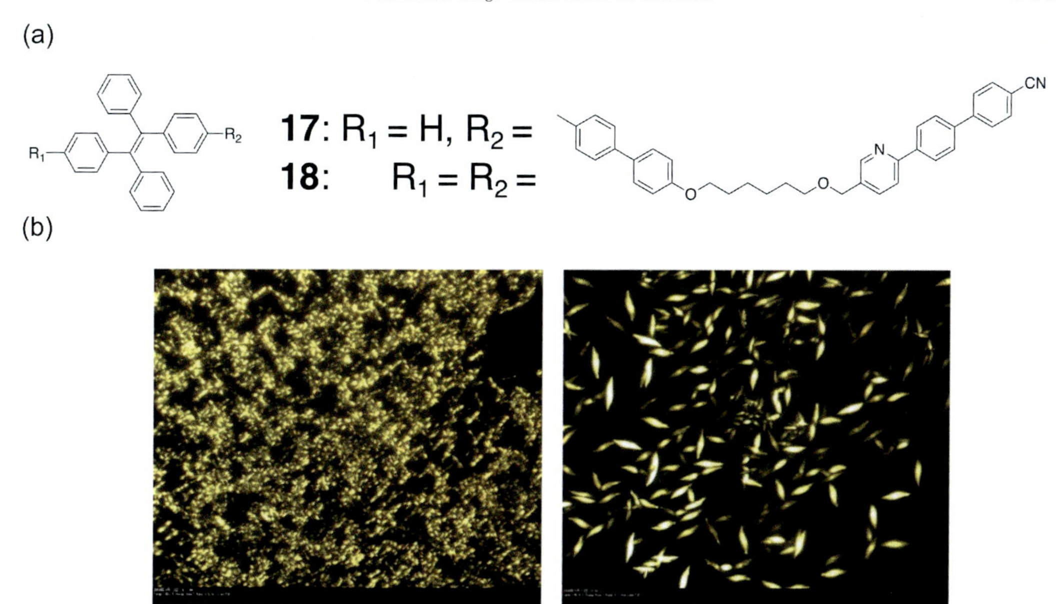

FIG. 6 (A) Molecular structure of TPE-PBN (**17**) and TPE-2PBN (**18**). (B) POM images of TPE-PBN (**17**) and TPE-2PBN (**18**) at the liquid crystalline temperatures of 138°C and 148°C, respectively. *Y. Wang, Y. Liao, C.P. Cabry, D. Zhou, G. Xie, Z. Qu, D.W. Bruce, W. Zhu, Highly efficient blueish-green fluorescent OLEDs based on AIE liquid crystal molecules: from ingenious molecular design to multifunction materials, J. Mater. Chem. C 5 (16) (2017) 3999–4008. https://doi.org/10.1039/c7tc00034k with permission from the Royal Society of Chemistry.*

4-cyanobiphenyl moiety as a mesogenic unit, two AIE-active LCs (TPE-PBN, compound **17**, and TPE-2PBN, compound **18**) were synthesized. These molecules (**17** and **18**) showed LC phases and exhibited clear AIE with high quantum efficiencies of 71% and 83%, respectively, in the solid state. Owing to these properties, neat TPE-PBN (**17**) and PTE-2PBN (**18**) films showed bluish-green emission with high hole mobilities in the range of 10^{-4} cm^2 V^{-1} s^{-1}. Notably, the OLED based on TPE-PBN (**17**) exhibited the highest external quantum efficiency (4.1%) among blue AIE-based OLEDs.

2.2 AIE-active LCs of metallomesogens

Another fascinating approach for the development of AIE-active LCs is the introduction of a metallomesogen unit as an AIE core [30]. Various metallomesogens with designed coordination geometries have been explored to modify physical properties such as conductivity, color, luminescence, and magnetism. Among advanced metallomesogens, Au(I) has a strong affinity for linear coordination and metal-metal interactions, thus allowing the generation of supramolecular structures and the formation of both calamitic and discotic mesogenic structures, resulting in various LC mesophases. Several AIE-active LCs containing Au(I) ions show strong photoluminescence in condensed phases.

Recently, our group designed and synthesized a series of Au(I) complexes with rod-like and disk-like geometries based on a simple concept to improve the intermolecular aurophilic

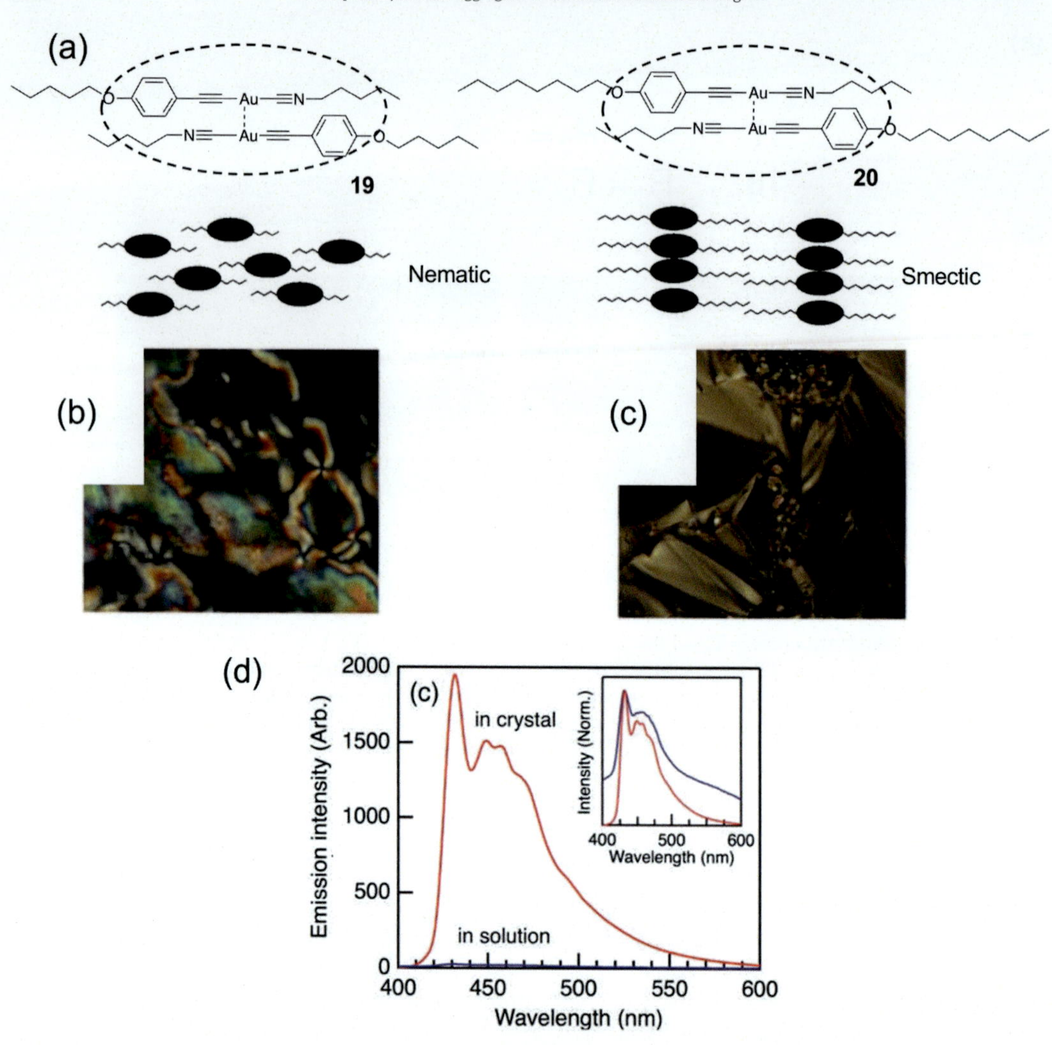

FIG. 7 (A) (Top) Structural model of a mesogen unit formed by aurophilic interaction of Au(I) complexes (**19**) (left) and (**20**) (right) in the condensed phase. (Bottom) Schematic illustration of molecular packing in LC phases. (B and C) POM images of Au(I) complexes (**19**) and (**20**), respectively. (F) Photoluminescence spectra of the Au(I) complex (**19**) (*red*; excitation, 340 nm) and CH_2Cl_2 solution (*blue*, 2.7×10^{-4} mol L^{-1}; excitation, 330 nm). The normalized photoluminescence spectra both in the crystal and in the solution are shown in the inset. *Reproduced from K. Fujisawa, N. Kawakami, Y. Onishi, Y. Izumi, S. Tamai, N. Sugimoto, O. Tsutsumi, Photoluminescent properties of liquid crystalline gold(I) isocyanide complexes with a rod-like molecular structure, J. Mater. Chem. C 1 (34) (2013) 5359–5366. https://doi.org/10.1039/c3tc31105h with permission from The Royal Society of Chemistry.*

interaction, which is a crucial factor for realizing AIE activity. These complexes possessed simple coordinating ligands to minimize the steric hindrance around the Au(I) ions. For rod-like mononuclear complexes, an ethylbenzene group was conjugated to alkoxy and *n*-alkylisocyanide chains of various lengths (**19** and **20**, Fig. 7) [23,41], whereas for disk-like trinuclear complexes, long *n*-alkyl side chains were introduced at the 4-position of pyrazole

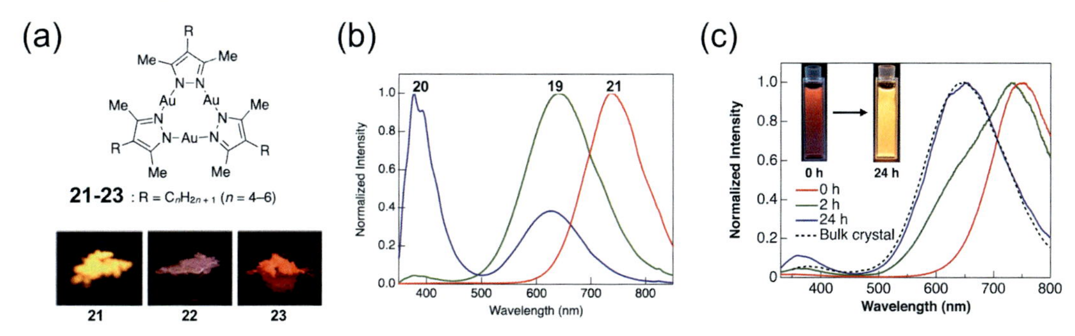

FIG. 8 (A) Molecular structure (top) and photographs under UV irradiation (bottom; $\lambda = 254$ nm) of the synthesized trinuclear Au(I) complexes with the different length of a terminal alkyl chain (**21**–**23**). (B) Photoluminescence spectra (excitation at 280 nm) of (**21**–**23**) bulk crystals at room temperature and in the presence of air. (C) Time evolution of the photoluminescence spectra (excitation at 280 nm) of (**21**) crystals in a methanol suspension prepared by precipitation from $CHCl_3$/methanol (50 µL/10 mL). The insets show photographs of (**21**) crystal suspensions taken under UV irradiation. *Reproduced from Y. Kuroda, M. Tamaru, H. Nakasato, K. Nakamura, M. Nakata, K. Hisano, K. Fujisawa, O. Tsutsumi, Observation of crystallisation dynamics by crystal-structure-sensitive room-temperature phosphorescence from Au(I) complexes, Commun. Chem. 3 (1) (2020). https://doi.org/10.1038/s42004-020-00382-1 and licensed under CC BY 4.0.*

derivatives (**21**–**23**) (Fig. 8) [42,43]. Detailed investigations revealed that the rod-like complexes (**19** and **20**) formed dimer structures through intermolecular aurophilic interactions, even in the isotropic phase; thus, the dimeric structure acted as a mesogen unit to induce nematic and smectic phases. These LC phases were observed at relatively low temperatures (~100 °C), which is preferable for designing and constructing light-emitting LCs because high-temperature processing can degrade devices and high-temperature conditions can reduce luminescence quantum yields by enhancing the nonradiative transitions of excited states. Interestingly, a series of Au(I) complexes with rod-like molecular shapes (**19** and **20**) showed bright room-temperature phosphorescence at the peak wavelength of ~450 nm with high quantum yields of >50%. In addition, the trinuclear Au(I) complexes with disk-like molecular shapes (**21**–**23**) exhibited highly efficient room-temperature phosphorescence (quantum yield >75%) owing to intermolecular aurophilic interactions at multiple sites, which increased the crystal packing density [42]. In addition, we recently found that the crystal size of our trinuclear Au(I) complexes (**21**) drastically affected the phosphorescence wavelength, as the crystal packing structure changed as the crystal size increased (Fig. 8) [43]. This phenomenon provided insights into the behavior of nano-to-micro-sized solids.

In 2018, Gimeńez et al. reported a series of trinuclear Au(I)-pyrazole complexes [44]. When dispersed in a poly(methyl methacrylate) film, these complexes exhibited bright red/deep red room-temperature phosphorescence with exceptionally high quantum yields of 90% and large Stokes shifts of ~400 nm. The 4-hexyl-3,5-dimethylpyrazolate ligand bears only one short hexyl chain at the 4-position of the pyrazole ring. This results in an appreciable solubility for mixing with polymer matrices and film processing and also induces columnar LC phases. Furthermore, the Au(I) complexes showed red-to-blue thermochromism undercooling and color changes in response to silver ions, indicating their applicability as sensing materials.

Complexes based on other metals (Ag and Cu) have also been synthesized. Detailed analyses of crystallographic data and thermal and luminescence properties revealed that the room-temperature phosphorescence arises from metallophilic interactions, with stronger metallophilic interactions (Ag > Au > Cu) typically resulting in the formation of stable mesophases (the phase transition temperatures to the columnar phase are 145°C and 193°C for Au and Ag, respectively, whereas there is no mesophase for Cu). Color tunability and "on/off" switching by external stimuli are particularly useful in optical sensing applications. Therefore, these metallomesogen-based AIE-active LCs have the potential not only to function as platforms for generating various calamitic and discotic LCs for optical displays but also for advancing sensing applications.

2.3 Supramolecular self-assembly for AIE-active LCs

As most design approaches to afford AIE-active LCs involve step-wise synthesis with relatively low overall yield, ionic self-assembly (ISA) has been proposed as a unique alternative synthetic strategy for developing new AIE systems that form LC phases. However, there are few reports on this approach. In ISA, oppositely charged tectonic discotic units are bound together via electrostatic interactions, enabling the facile and efficient formation of supramolecular columnar LCs. In 2016, Ren et al. first applied the ISA process to form AIE-active LCs, in which a TPE unit acted as an AIE unit and dimethyldioctadecylammonium bromide acted as an ionic element [45]. The resultant supramolecule successfully formed a hierarchical supramolecular structure with a columnar mesophase and excellent luminescent properties. The fluorescence quantum yields of the supramolecules in DMF solution and the solid state were found to be 0.68% and 46%, respectively, which resulted in an AIE factor (quantum yield in the solid state divided by that in solution) of 67.6, rendering this the most AIE-active LC ever reported. In 2020, a series of novel AIE-active ionic LCs (**24** and **25**) was proposed by Yang et al. based on diphenylacrylonitrile-imidazole salts linked by flexible spacers with different counter anions (Fig. 9) [46]. These ionic LCs formed supramolecular lamellar (**24**) and columnar LC phases (**25**) depending on the number of alkyl chains on the phenyl ring. Increasing the number of alkyl chains and the size of the counter anion allowed the mesophases to be observed at lower temperatures over a wider range of mesomorphic temperatures.

Yang et al. proposed another design concept for synthesizing AIE-active supramolecular LCs, in which the functionalities were separated by using LC-forming units (porphyrin derivative) and an AIE unit (cyanostilbene) [24]. Porphyrins have a planar π-conjugated aromatic core, and the incorporation of alkyl/alkoxy chains on the porphyrin unit results in a columnar mesophase through supramolecular self-assembly. Generally, porphyrin forms a discotic LC phase that unfortunately exhibits ACQ. To avoid ACQ, a cyanostilbene derivative of a polyglycol-diphenylacrylonitrile unit was introduced into the porphyrin unit. The obtained supramolecule of compound **26** showed a hexagonal columnar LC phase between 70°C and 120°C with strong AIE (Fig. 10). Owing to its AIE properties in the aggregate state, including an emission wavelength of ~425 nm, the stilbene unit acts as a photo-antenna for fluorescence resonance energy transfer (FRET) to porphyrin, which

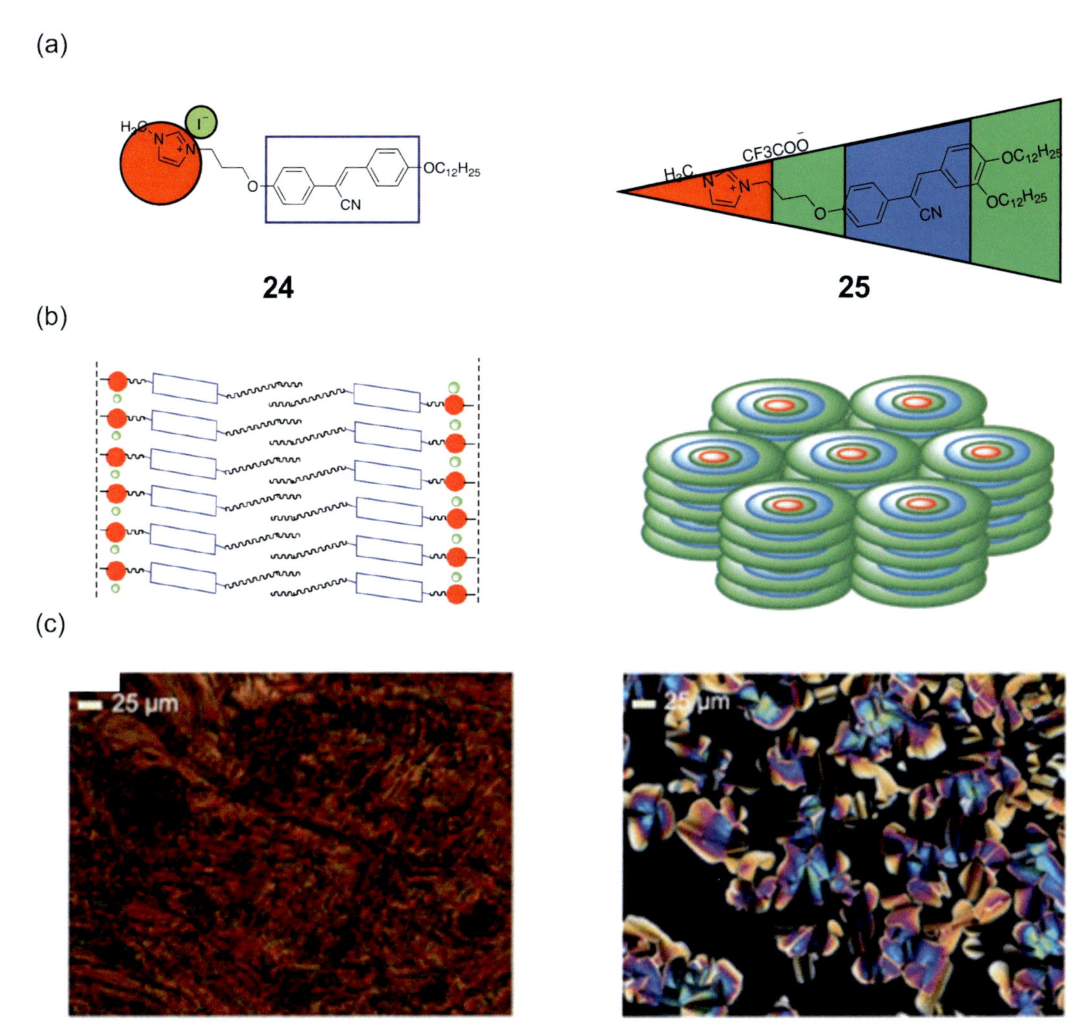

FIG. 9 (A) Molecular structure and schematic diagrams of AIE-active ionic LC derivatives (**24** and **25**). (B and C) The schematic image of stacking structure and POM images of smectic lamellar of (**24**) (left) and hexagonal columnar stacking of (**25**) (right), respectively. *Reproduced from reference M. Schadt, Milestone in the history of field-effect liquid crystal displays and materials, Jpn. J. Appl. Phys. 48 (3) (2009). https://doi.org/10.1143/JJAP.48.03B001 by permission of The Royal Society of Chemistry.*

has an absorption wavelength in the same region. This FRET-based AIE system (**26**) exhibited a large pseudo-Stokes shift of ~210 nm and a high quantum yield of 12% in the solid state. Such supramolecular LC systems could provide a pathway for synthesizing various types of AIE-active LCs, even with common luminogens with desirable luminescence properties.

FIG. 10 A representative molecular structure of porphyrin-based AIE-active supramolecular LCs (**26**) reported in the literature [24].

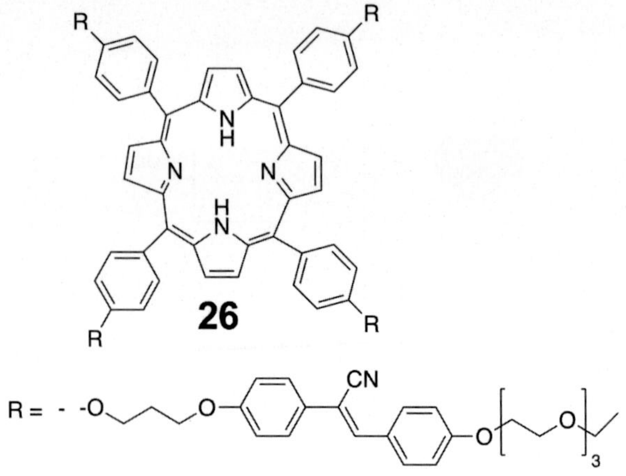

3 AIE-active LCs toward optical applications

3.1 Polymeric films of AIE-active LCs

Although low-molecular-weight AIE-active LCs have enjoyed impressive success, polymeric AIE-active LCs, in which AIEgens are incorporated into polymer networks, have been less explored [47,48]. In contrast to low-molecular-weight AIE-active LCs, polymeric AIE-active LCs are attractive for realizing thin and flexible optical displays because they are expected to exhibit enhanced performance for low-cost film processing, hierarchical molecular orientations, and fascinating functionalities.

Recently, two main research trends have emerged for polymeric AIE-active LCs: (1) polymeric OLEDs and (2) mechano-optical applications [49,50]. For optical display applications, extensive research on polymeric OLEDs, mostly with ACQ luminogens, has been conducted since the first report in 1990 [51]. However, polymeric OLEDs based on AIE-active polymeric LCs have potential advantages owing to their polarized luminescence behavior, high luminescence efficiency resulting from restricted molecular motion in the film state, facile film processability, and high thermal stability [49]. Nevertheless, there are still limited studies on AIE-active polymeric LCs. In contrast, for mechano-optical applications, AIE-active LC elastomers, which are emerging polymeric LC systems, have been explored. LC elastomers have a polymer network strongly coupled to the oriented LCs and thus exhibit reversible shape changes coincident with orientational order changes under external stimuli. For example, a polydomain LC elastomer with macroscopically randomly oriented LCs transformed into a monodomain LC phase with unidirectional orientation under external stimuli such as uniaxial mechanical strain. After the strain was removed, the elastomer gradually returned to its initial shape and molecular orientation. This unique reversibility of both the macroscopic shape and the microscopic LC orientation will enable the achievement of next-generation optical applications with stimuli-responsiveness, e.g., sensors and actuators. Considering that the photophysical properties of AIEgens are very sensitive to the external environment as well as their aggregate structure, AIE-active LC elastomers are expected to exhibit unique mechano-optical properties.

The first example of an AIE-active LC elastomer was reported by Yang et al. in 2018. This system consisted of a TPE derivative (**27**) as an AIE unit cross-linked with an LC polymer network of mesogenic phenylbenzoate units using an in situ two-step acrylic diene polymerization/cross-linking method (Fig. 11) [52]. In the first stage, stretching gave a monodomain AIE-active LC film that was then cross-linked to form an LC elastomer. The elastomer with an AIE-active cross-linker (**27**) exhibited the following phase transition behavior: smectic C—89°C—nematic—110°C—isotropic (on heating) and isotropic—103°C—nematic—82°C—smectic C (on cooling). Moreover, this elastomer had almost perfect in-plane uniaxial LC alignment with an order parameter of >0.9 (perfect orientation = 1) in the nematic phase. Owing to this near-perfect alignment, the AIE-active LC elastomer showed bright fluorescence in the solid state. Interestingly, when immersed in THF, the fluorescence of this elastomer was quickly quenched and the fluorescence intensity drastically decreased to ~0.03 times that of the original dried state. This behavior was due to the swelling of the elastomer by the solvent, leading to a random orientation of AIE-active LCs. Similarly, when heated over the isotropic transition temperature, the fluorescence intensity drastically decreased owing to the relaxation of aggregated TPE derivatives (**27**) as well as the polymer network, with macroscopic deformation. These results suggest that such AIE-active LC elastomers can provide a new design concept for thermomechanically controllable fluorescent soft actuators. In particular, AIE-active polymeric LCs are promising candidates because the relaxation/orientation of the polymer network directly affects AIEgen aggregation, which can potentially be exploited for stimuli-responsive optical applications.

3.2 Linearly polarized luminescence

As described in the previous section, extensive work on AIE-active LCs has resulted in the development of various systems that have potential as platforms for next-generation optical applications such as linearly/circularly polarized luminescence. LC displays, as the most common optical display application, are now in widespread use in daily life. More recently, light-emitting LC and/or OLED displays have attracted research interest because they can overcome various drawbacks of LC displays, including low brightness and low energy efficiency resulting from the output light passing through a commercial polarizer from a white unpolarized backlight, which can lead to further issues such as a low contrast ratio and a viewing angle dependence [40,53,54,55]. As an alternative to conventional backlit devices, light-emitting LC displays based on AIE-active LCs only emit linearly polarized luminescence from the nematic phase in the localized region of voltage application. As a result, next-generation optical displays could have a simplified design, be more cost effective, consume less energy, and exhibit increased brightness, a higher contrast ratio, and a wider viewing angle.

As a linearly polarized luminescent system, a light-emitting LC display based on the TPE-PPE (**16**) was first proposed by the group of Tang and coworkers in 2015 [39,53]. To realize a uniform LC orientation, they employed photoalignment technology, by which the desired LC orientation was inscribed by exposing a photosensitive layer of azobenzene derivatives

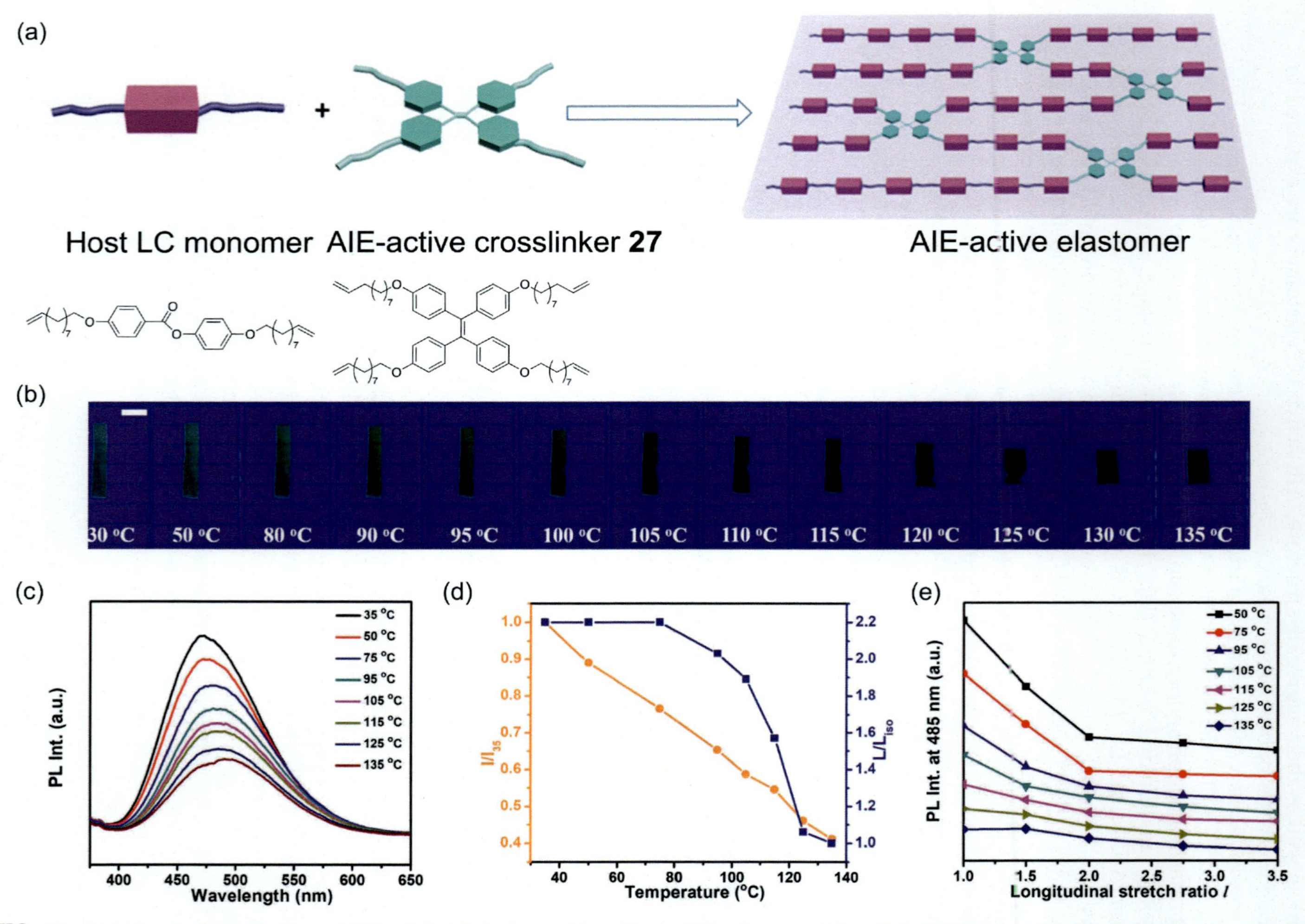

FIG. 11 (A) Chemical composition of AIE-active LC elastomer film with an AIE-active crosslinker (**27**). (B) Photographs of the thermally induced actuation behaviors of the elastomer film with UV irradiation ($\lambda = 365$ nm). The scale bar is of 5 mm. (C and D) Fluorescence spectra (*orange line*) and the longitudinal shape deformation (*blue line*) of the elastomer film as a function of temperature (excitation, 350 nm), respectively. (E) Correlation of longitudinal stretch ratio and fluorescence intensity of the film measured at different temperatures. *Reproduced with permission from L. Liu, M. Wang, L.X. Guo, Y. Sun, X.Q. Zhang, B.P. Lin, H. Yang, Aggregation-induced emission luminogen-functionalized liquid crystal elastomer soft actuators, Macromolecules 51 (12) (2018) 4516–4524. https://doi.org/10.1021/acs.macromol.8b00677. Copyright (2018) American Chemical Society.*

coated on a substrate in a sample cell to linearly polarized light. Using a photomask, an LC orientation pattern was introduced on a mixture of host nematic LC and TPE-PPE (**16**), whereas the other regions were initially aligned perpendicular to the polarized light. As a result, a number pattern was generated under the polarizer, and the linearly polarized light-emitting LC display concept was confirmed by rotating the polarizer to reverse the light-emitting and dark regions. Furthermore, the application of an electric field resulted in vertical LC orientation, causing the optical image to vanish (Fig. 5). Although this system had a low quantum yield (~21%), this does not fundamentally limit the practical applications of light-emitting LC displays.

3.3 Circularly polarized luminescence

Circularly polarized luminescence (CPL) is another fascinating phenomenon that has attracted considerable interest for advanced optical applications such as three-dimensional displays and polarization sensors [56,57,58]. The purity of right- or left-CPL can be theoretically determined from the luminescence dissymmetry factor, g_{lum}, which represents the ratio of the intensity difference between the right- and left-CPL divided by the average total luminescence intensity. Furthermore, the dissymmetry factor is related to the electric dipole transition moment μ^{gn} and the imaginary magnetic dipole transition moment m^{gn}, as follows [56]:

$$g_{\text{lum}} \propto \frac{\mu^{\text{gn}} \cdot m^{\text{gn}}}{|\mu^{\text{gn}}|^2 + |m^{\text{gn}}|^2}$$

A promising route for increasing the g_{lum} value is the synthesis of molecules with forbidden electric dipole transitions but allowing magnetic transitions. However, typical luminescent materials have much smaller magnetic dipole transition moments than electric dipole transition. Thus, most CPL organic molecules with chiral molecular structures show high luminescence intensities but very low g_{lum} values in the range of 10^{-5} to 10^{-2}. In contrast, recent work has revealed that controlling hierarchical structures on the nano- to micro-level can enable strong CPL signals with high g_{lum} values, even with an achiral molecular skeleton.

CPL organic systems with controlled hierarchical structures were first reported in 2012 by Akagi et al. In these systems, chiral or racemic polymers with a lyotropic LC phase formed highly ordered chiral molecules via self-assembly with a chiral dopant. Interestingly, the lyotropic chiral nematic phase and cast film exhibited CPL with a high g_{lum} value on the order of 10^{-1}, regardless of the presence of chiral or racemic molecules [59]. However, the lyotropic LC system (including some solvent) and the cast film had relatively low quantum yields of ~30% and ~10%, respectively. In the same period, the supramolecular concept for generating strong CPL with a high g_{lum} value was applied to the AIE system by Liu et al. They synthesized an AIE-active tetraphenylsilole derivative with chiral units (mannose-containing peripheral chains) that formed a chiral supramolecular structure. This supramolecule was found to exhibit strong and bright CPL with a g_{lum} value on the order of 10^{-1}. An investigation of the AIE factor of this system revealed that the fluorescence quantum yield increased by 136 fold (quantum yield of 0.6% in the solution versus that of 81.3% in the solid state) [60]. Such supramolecular system for generating CPL was extensively researched in the last decade [61].

In addition to such helical supramolecular structures, cholesteric LC (chiral nematic) phases have been explored for diverse applications. In 1999, Chen et al. first reported an interesting phenomenon in which the g_{lum} value of a cholesteric LC medium was drastically enhanced by merely doping with a typical achiral luminogen of a fluorene derivative, named as Exalite 428 [62]. As the helically twisted molecular orientation in a cholesteric LC phase reflects circularly polarized light at a certain peak wavelength with the same handedness and periodic structure based on Bragg's law, this CPL enhancement might be partially explained by the selective reflection of circularly polarized light, which is a unique optical phenomenon for cholesteric LC phases. Recently, Cheng et al. investigated the effect of AIE-active achiral LCs mixed with a cholesteric LC medium. They achieved a very high quantum yield of 20.4% with g_{lum} values as high as +1.42/−1.39 at the reflection peak wavelength of the cholesteric LC medium. Interestingly, the luminescence peak wavelength could be tuned from 403 to 601 nm by merely changing the doped AIEgen [63]. Thus, the bright CPL with a very high g_{lum} value is not related only to the selective reflection by the periodic structure of the cholesteric LC phase. The drastic enhancement of g_{lum} values in the cholesteric LC medium is affected not only by the simple reflection of circularly polarized light but also by an effective chirality transfer phenomenon. Based on this fundamental knowledge, cholesteric LC phases have great potential for creating CPL systems with high quantum yields of over 50%. Tang et al., who proposed a concept for developing an AIE-active LC system with high CPL efficiency, synthesized a single-component system of AIE-active cholesteric LCs (**28–31**) without any host AIE-inactive LCs (Fig. 12) [64,65]. The AIE-active cholesteric LCs incorporated a chiral LC unit covalently bonded to the AIE core. This single-component system has the advantage of no phase separation to degrade the optical performance. The resulting CPL materials exhibited high g_{lum} values ($>10^{-1}$) and large quantum yields (up to ~70%) without phase separation and/or CPL switching behavior depending on their self-assembled aggregate structures.

Until now, high-performance optical displays with linearly/circular polarized luminescence properties and high quantum yields in the solid state (e.g., films) have mainly employed the limited number of available AIE cores based on TPE and HPS. We believe that the recent development of AIE-active LCs with various alternative AIE cores and systems, such as metallomesogens, supramolecules, and cholesteric LC phases, will pave the way for the further development of next-generation optical displays.

4 Conclusions

The development of next-generation optical displays such as smart glasses, see-through displays, and three-dimensional displays is an emerging topic for communicating and visualizing information in the modern era. AIE-active LCs are considered promising for significantly upgrading conventional LC displays and/or OLEDs owing to their low-energy consumption, simplified device structures, and linearly/circularly polarized luminescence

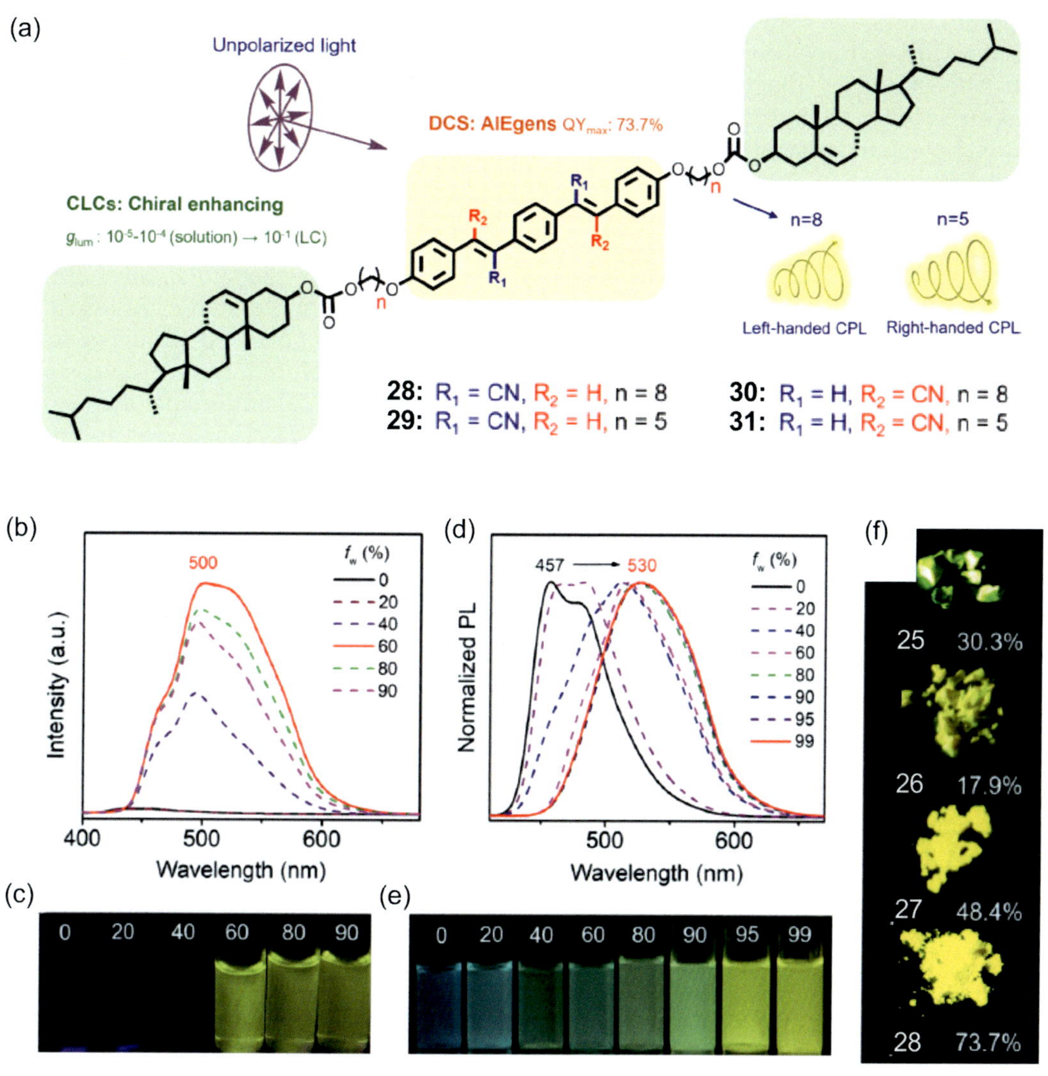

FIG. 12 (A) Rational design of AIE-active LCs with CPL behavior showing both high values of glum and quantum yield. (B) Emission spectra of 25 in H_2O/THF mixtures (10 μM) with different water fractions (fw). (C) Photographs of compound **28** in H_2O/THF under UV irradiation. (D) Normalized emission spectra of compound **30** in H_2O/THF with different fw. (E) Photographs of compound **30** under UV irradiation. (F) Photographs and fluorescent quantum yield of compound **28** (30.3%), 26 (17.9%), 27 (48.4%), and 28 (73.7%) in solid powder states under UV irradiation. *Y. Wu, L.H. You, Z.Q. Yu, J.H. Wang, Z. Meng, Y. Liu, X.S. Li, K. Fu, X.K. Ren, B.Z. Tang, Rational design of circularly polarized luminescent aggregation-induced emission luminogens (AIEgens): promoting the dissymmetry factor and emission efficiency synchronously, ACS Mater. Lett. 2 (5) (2020) 505–510. https://doi.org/10.1021/acsmaterialslett.0c00063. Reproduced with permission from reference D. Dunmur, K. Toriyama, Handbook of Liquid Crystals, Wiley-VCH, 2014, pp. 231–280. Copyright (2020) American Chemical Society.*

properties. Although various types of low-molecular-weight AIEgens, for example, TPE and HPS derivatives, have been widely explored, comparable progress has not been achieved with AIE-active LCs, likely because general design concepts for AIE-active LCs have not yet been established. In this chapter, we highlighted key examples of AIE-active LCs to shed light on potential design strategies. Generally, the molecular skeletons of LCs are rigid with anisotropic mesogenic core units and terminal flexible chains (e.g., alkoxy and alkyl chains). Thus, the design of AIE-active LCs exhibiting arbitrary LC phases requires the careful investigation of both the mesogenic core, which should be an AIE unit, and the terminal chain. Many studies have investigated TPE and HPS derivatives as conventional AIE units directly bonded to alkyl/alkoxy chains, which have been, in some cases, successfully applied for realizing light-emitting LC display systems with linearly/circularly polarized luminescence. In addition, a new series of AIE-active LC cores have also been explored. For example, supramolecular and metal complex systems are formed by introducing intermolecular interactions such as hydrogen bonding, halogen bonding, metallophilic interactions, and electrostatic interactions. This enrichment of the number of synthesized AIE-active LC cores will allow the design of suitable material systems for optical displays. We believe that continued efforts toward the development of AIE-active LCs can pave the way for next-generation and high-performance optical displays.

References

[1] C. Hai-Wei, L. Jiun-Haw, L. Bo-Yen, C. Stanley, W. Shin-Tson, Liquid crystal display and organic light-emitting diode display: present status and future perspectives, Light Sci. Appl. (2018) 17168, https://doi.org/10.1038/lsa.2017.168.

[2] Y. Huang, E.L. Hsiang, M.Y. Deng, S.T. Wu, Mini-LED, micro-LED and OLED displays: present status and future perspectives, Light Sci. Appl. 9 (1) (2020), https://doi.org/10.1038/s41377-020-0341-9.

[3] M. Schadt, Milestone in the history of field-effect liquid crystal displays and materials, Jpn. J. Appl. Phys. 48 (3) (2009), https://doi.org/10.1143/JJAP.48.03B001.

[4] J.W. Goodby, R.J. Mandle, E.J. Davis, T. Zhong, S.J. Cowling, What makes a liquid crystal? The effect of free volume on soft matter, Liq. Cryst. 42 (5–6) (2015) 593–622, https://doi.org/10.1080/02678292.2015.1030348.

[5] V.G. Chigrinov, V.M. Kozenkov, H.S. Kwok, Photoalignment of Liquid Crystalline Materials: Physics and Applications, John Wiley & Sons, Ltd., 2008.

[6] T. Ikeda, O. Tsutsumi, Optical switching and image storage by means of azobenzene liquid-crystal films, Science 268 (5219) (1995) 1873–1875, https://doi.org/10.1126/science.268.5219.1873.

[7] C. Tschierske, Liquid crystal engineering—new complex mesophase structures and their relations to polymer morphologies, nanoscale patterning and crystal engineering, Chem. Soc. Rev. 36 (12) (2007) 1930–1970, https://doi.org/10.1039/b615517k.

[8] J.W. Goodby, P.J. Collings, T. Kato, C. Tschierske, H.F. Gleeson, P. Raynes, Handbook of Liquid Crystals, Wiley-VCH, 2014.

[9] H. Finkelmann, S.T. Kim, A. Muñoz, P. Palffy-Muhoray, B. Taheri, Tunable mirrorless lasing in cholesteric liquid crystalline elastomers, Adv. Mater. 13 (14) (2001) 1069–1072, https://doi.org/10.1002/1521-4095(200107)13:14<1069::AID-ADMA1069>3.0.CO;2-6.

[10] M. O'Neill, S.M. Kelly, Liquid crystals for charge transport, luminescence, and photonics, Adv. Mater. 15 (14) (2003) 1135–1146, https://doi.org/10.1002/adma.200300009.

[11] T.J. White, D.J. Broer, Programmable and adaptive mechanics with liquid crystal polymer networks and elastomers, Nat. Mater. 14 (11) (2015) 1087–1098, https://doi.org/10.1038/nmat4433.

[12] M. Xie, K. Hisano, M. Zhu, T. Toyoshi, M. Pan, S. Okada, O. Tsutsumi, S. Kawamura, C. Bowen, Flexible multifunctional sensors for wearable and robotic applications, Adv. Mater. Technol. 4 (3) (2019), https://doi.org/10.1002/admt.201800626.

[13] R. Baetens, B.P. Jelle, A. Gustavsen, Properties, requirements and possibilities of smart windows for dynamic daylight and solar energy control in buildings: a state-of-the-art review, Sol. Energy Mater. Sol. Cells 94 (2) (2010) 87–105, https://doi.org/10.1016/j.solmat.2009.08.021.

[14] Z. He, X. Sui, G. Jin, L. Cao, Progress in virtual reality and augmented reality based on holographic display, Appl. Optics 58 (5) (2019) A74–A81, https://doi.org/10.1364/AO.58.000A74.

[15] P. Jongchan, L. KyeoReh, P. YongKeun, Ultrathin wide-angle large-area digital 3D holographic display using a non-periodic photon sieve, Nat. Commun. (2019), https://doi.org/10.1038/s41467-019-09126-9.

[16] A.E.A. Contoret, S.R. Farrar, P.O. Jackson, S.M. Khan, L. May, M. O'Neill, J.E. Nicholls, S.M. Kelly, G.J. Richards, Polarized electroluminescence from an anisotropic nematic network on a non-contact photoalignment layer, Adv. Mater. 12 (13) (2000) 971–974, https://doi.org/10.1002/1521-4095(200006)12:13<971::AID-ADMA971->3.0.CO;2-J.

[17] H. Zhang, Z. Zhao, A.T. Turley, L. Wang, P.R. McGonigal, Y. Tu, Y. Li, Z. Wang, R.T.K. Kwok, J.W.Y. Lam, B.Z. Tang, Aggregate science: from structures to properties, Adv. Mater. 32 (36) (2020), https://doi.org/10.1002/adma.202001457.

[18] Z. Zhao, H. Zhang, J.W.Y. Lam, B.Z. Tang, Aggregation-induced emission: new vistas at the aggregate level, Angew. Chem. Int. Ed. 59 (25) (2020) 9888–9907, https://doi.org/10.1002/anie.201916729.

[19] S. Liu, Y. Li, R.T. Kwok, J.W. Lam, B.Z. Tang, Structural and process controls of AIEgens for NIR-II theranostics, Chem. Sci. (2021), https://doi.org/10.1039/d0sc02911d.

[20] J. Luo, Z. Xie, Z. Xie, J.W.Y. Lam, L. Cheng, H. Chen, C. Qiu, H.S. Kwok, X. Zhan, Y. Liu, D. Zhu, B.Z. Tang, Aggregation-induced emission of 1-methyl-1,2,3,4,5-pentaphenylsilole, Chem. Commun. 18 (2001) 1740–1741, https://doi.org/10.1039/b105159h.

[21] D. Liu, J.Y. Wei, W.W. Tian, W. Jiang, Y.M. Sun, Z. Zhao, B.Z. Tang, Endowing TADF luminophors with AIE properties through adjusting flexible dendrons for highly efficient solution-processed nondoped OLEDs, Chem. Sci. 11 (27) (2020) 7194–7203, https://doi.org/10.1039/d0sc02194f.

[22] M. Castillo-Vallés, A. Martínez-Bueno, R. Giménez, T. Sierra, M.B. Ros, Beyond liquid crystals: new research trends for mesogenic molecules in liquids, J. Mater. Chem. C 7 (46) (2019) 14454–14470, https://doi.org/10.1039/c9tc04179f.

[23] K. Fujisawa, N. Kawakami, Y. Onishi, Y. Izumi, S. Tamai, N. Sugimoto, O. Tsutsumi, Photoluminescent properties of liquid crystalline gold(i) isocyanide complexes with a rod-like molecular structure, J. Mater. Chem. C 1 (34) (2013) 5359–5366, https://doi.org/10.1039/c3tc31105h.

[24] H. Guo, S. Zheng, S. Chen, C. Han, F. Yang, A first porphyrin liquid crystal with strong fluorescence in both solution and aggregated states based on the AIE-FRET effect, Soft Matter 15 (41) (2019) 8329–8337, https://doi.org/10.1039/c9sm01174a.

[25] L.X. Guo, Y.B. Xing, M. Wang, Y. Sun, X.Q. Zhang, B.P. Lin, H. Yang, Luminescent liquid crystals bearing an aggregation-induced emission active tetraphenylthiophene fluorophore, J. Mater. Chem. C 7 (16) (2019) 4828–4837, https://doi.org/10.1039/c9tc00448c.

[26] J.H. Wan, L.Y. Mao, Y.B. Li, Z.F. Li, H.Y. Qiu, C. Wang, G.Q. Lai, Self-assembly of novel fluorescent silole derivatives into different supramolecular aggregates: fibre, liquid crystal and monolayer, Soft Matter 6 (14) (2010) 3195–3201, https://doi.org/10.1039/b925746b.

[27] W.Z. Yuan, Z.Q. Yu, P. Lu, C. Deng, J.W.Y. Lam, Z. Wang, E.Q. Chen, Y. Ma, B.Z. Tang, High efficiency luminescent liquid crystal: aggregation-induced emission strategy and biaxially oriented mesomorphic structure, J. Mater. Chem. 22 (8) (2012) 3323–3326, https://doi.org/10.1039/c2jm15712h.

[28] M. Bardají, Gold liquid crystals in the XXI century, Inorganics 2 (3) (2014) 433–454, https://doi.org/10.3390/inorganics2030433.

[29] K. Binnemans, Luminescence of metallomesogens in the liquid crystal state, J. Mater. Chem. 19 (4) (2009) 448–453, https://doi.org/10.1039/b811373d.

[30] A. Kishimura, T. Yamashita, T. Aida, Phosphorescent organogels via\metallophilic\interactions for reversible RGB-color switching, J. Am. Chem. Soc. 127 (1) (2005) 179–183, https://doi.org/10.1021/ja0441007.

[31] Y. Sagara, T. Mutai, I. Yoshikawa, K. Araki, Material design for piezochromic luminescence: hydrogen-bond-directed assemblies of a pyrene derivative, J. Am. Chem. Soc. 129 (6) (2007) 1520–1521, https://doi.org/10.1021/ja0677362.

[32] H. Yang, T. Yi, Z. Zhou, Y. Zhou, J. Wu, M. Xu, F. Li, C. Huang, Switchable fluorescent organogels and mesomorphic superstructure based on naphthalene derivatives, Langmuir 23 (15) (2007) 8224–8230, https://doi.org/10.1021/la7005919.

[33] A.R. Hirst, B. Escuder, J.F. Miravet, D.K. Smith, High-tech applications of self-assembling supramolecular nanostructured gel-phase materials: from regenerative medicine to electronic devices, Angew. Chem. Int. Ed. 47 (42) (2008) 8002–8018, https://doi.org/10.1002/anie.200800022.

[34] S.S. Babu, V.K. Praveen, A. Ajayaghosh, Functional π-gelators and their applications, Chem. Rev. 114 (4) (2014) 1973–2129, https://doi.org/10.1021/cr400195e.

[35] Y. Sagara, T. Kato, Mechanically induced luminescence changes in molecular assemblies, Nat. Chem. 1 (8) (2009) 605–610, https://doi.org/10.1038/nchem.411.

[36] H.T. Bui, J. Kim, H.J. Kim, B.K. Cho, S. Cho, Advantages of mobile liquid-crystal phase of AIE luminogens for effective solid-state emission, J. Phys. Chem. C 120 (47) (2016) 26695–26702, https://doi.org/10.1021/acs.jpcc.6b10026.

[37] J. Tong, Y.J. Wang, Z. Wang, J.Z. Sun, B.Z. Tang, Crystallization-induced emission enhancement of a simple tolane-based mesogenic luminogen, J. Phys. Chem. C 119 (38) (2015) 21875–21881, https://doi.org/10.1021/acs.jpcc.5b06088.

[38] C. YaFei, L. JieSheng, Y. Wang Zhang, Y. Zhen Qiang, L.J. Wing Yip, T.B. Zhong, 1-((12-Bromododecyl)oxy)-4-((4-(4-pentylcyclohexyl)phenyl)ethynyl) benzene: liquid crystal with aggregation-induced emission characteristics, Sci. China Chem. (2013) 1191–1196, https://doi.org/10.1007/s11426-013-4950-5.

[39] D. Zhao, F. Fan, J. Cheng, Y. Zhang, K.S. Wong, V.G. Chigrinov, H.S. Kwok, L. Guo, B.Z. Tang, Light-emitting liquid crystal displays based on an aggregation-induced emission luminogen, Adv. Opt. Mater. 3 (2) (2015) 199–202, https://doi.org/10.1002/adom.201400428.

[40] Y. Wang, Y. Liao, C.P. Cabry, D. Zhou, G. Xie, Z. Qu, D.W. Bruce, W. Zhu, Highly efficient blueish-green fluorescent OLEDs based on AIE liquid crystal molecules: from ingenious molecular design to multifunction materials, J. Mater. Chem. C 5 (16) (2017) 3999–4008, https://doi.org/10.1039/c7tc00034k.

[41] S. Yamada, S. Yamaguchi, O. Tsutsumi, Electron-density distribution tuning for enhanced thermal stability of luminescent gold complexes, J. Mater. Chem. C 5 (31) (2017) 7977–7984, https://doi.org/10.1039/c7tc00728k.

[42] O. Tsutsumi, M. Tamaru, H. Nakasato, S. Shimai, S. Panthai, Y. Kuroda, K. Yamaguchi, K. Fujisawa, K. Hisano, Highly efficient aggregation-induced room-temperature phosphorescence with extremely large stokes shift emitted from trinuclear gold(I) complex crystals, Molecules 24 (24) (2019), https://doi.org/10.3390/molecules24244606.

[43] Y. Kuroda, M. Tamaru, H. Nakasato, K. Nakamura, M. Nakata, K. Hisano, K. Fujisawa, O. Tsutsumi, Observation of crystallisation dynamics by crystal-structure-sensitive room-temperature phosphorescence from Au(I) complexes, Commun. Chem. 3 (1) (2020), https://doi.org/10.1038/s42004-020-00382-1.

[44] J. Cored, O. Crespo, J.L. Serrano, A. Elduque, R. Giménez, Decisive influence of the metal in multifunctional gold, silver, and copper metallacycles: high quantum yield phosphorescence, color switching, and liquid crystalline behavior, Inorg. Chem. 57 (20) (2018) 12632–12640, https://doi.org/10.1021/acs.inorgchem.8b01778.

[45] H. Jing, L. Lu, Y. Feng, J.F. Zheng, L. Deng, E.Q. Chen, X.K. Ren, Synthesis, aggregation-induced emission, and liquid crystalline structure of tetraphenylethylene-surfactant complex via ionic self-assembly, J. Phys. Chem. C 120 (48) (2016) 27577–27586, https://doi.org/10.1021/acs.jpcc.6b09901.

[46] H. Guo, Q. Yu, Y. Xiong, F. Yang, Room-temperature AIE ionic liquid crystals based on diphenylacrylonitrile-imidazole salts, Soft Matter 16 (45) (2020) 10368–10376, https://doi.org/10.1039/d0sm01474e.

[47] R. Hu, N.L.C. Leung, B.Z. Tang, AIE macromolecules: syntheses, structures and functionalities, Chem. Soc. Rev. 43 (13) (2014) 4494–4562, https://doi.org/10.1039/c4cs00044g.

[48] Y.B. Hu, J.W.Y. Lam, B.Z. Tang, Recent progress in AIE-active polymers, Chin. J. Polym. Sci. 37 (4) (2019) 289–301, https://doi.org/10.1007/s10118-019-2221-4.

[49] J. Gu, Z. Xu, D. Ma, A. Qin, B.Z. Tang, Aggregation-induced emission polymers for high performance PLEDs with low efficiency roll-off, Mater. Chem. Front. 4 (4) (2020) 1206–1211, https://doi.org/10.1039/d0qm00012d.

[50] T. Han, X. Wang, D. Wang, B.Z. Tang, Functional polymer systems with aggregation-induced emission and stimuli responses, Top. Curr. Chem. 379 (1) (2021), https://doi.org/10.1007/s41061-020-00321-7.

[51] J.H. Burroughes, D.D.C. Bradley, A.R. Brown, R.N. Marks, K. Mackay, R.H. Friend, P.L. Burns, A.B. Holmes, Light-emitting diodes based on conjugated polymers, Nature 347 (6293) (1990) 539–541, https://doi.org/10.1038/347539a0.

[52] L. Liu, M. Wang, L.X. Guo, Y. Sun, X.Q. Zhang, B.P. Lin, H. Yang, Aggregation-induced emission luminogen-functionalized liquid crystal elastomer soft actuators, Macromolecules 51 (12) (2018) 4516–4524, https://doi.org/10.1021/acs.macromol.8b00677.

[53] D. Zhao, F. Fan, V.G. Chigrinov, H.S. Kwok, B.Z. Tang, Aggregate-induced emission in light-emitting liquid crystal display technology, J. Soc. Inf. Disp. 23 (5) (2015) 218–222, https://doi.org/10.1002/jsid.382.

[54] J. Koo, S.I. Lim, S.H. Lee, J.S. Kim, Y.T. Yu, C.R. Lee, D.Y. Kim, K.U. Jeong, Polarized light emission from uniaxially oriented and polymer-stabilized AIE luminogen thin films, Macromolecules 52 (4) (2019) 1739–1745, https://doi.org/10.1021/acs.macromol.8b02513.

[55] N. Mochizuki, R. Morita, Development of transparent emissive LCD using novel polarized-light-emitting film, J. Soc. Inf. Disp. 26 (11) (2018) 670–674, https://doi.org/10.1002/jsid.738.

[56] E.M. Sánchez-Carnerero, A.R. Agarrabeitia, F. Moreno, B.L. Maroto, G. Muller, M.J. Ortiz, S. De La Moya, Circularly polarized luminescence from simple organic molecules, Chem. A Eur. J. 21 (39) (2015) 13488–13500, https://doi.org/10.1002/chem.201501178.

[57] J. Han, S. Guo, H. Lu, S. Liu, Q. Zhao, W. Huang, Recent progress on circularly polarized luminescent materials for organic optoelectronic devices, Adv. Opt. Mater. 6 (17) (2018), https://doi.org/10.1002/adom.201800538.

[58] J. Roose, B.Z. Tang, K.S. Wong, Circularly-polarized luminescence (CPL) from chiral AIE molecules and macrostructures, Small 12 (47) (2016) 6495–6512, https://doi.org/10.1002/smll.201601455.

[59] B.A. San Jose, S. Matsushita, K. Akagi, Lyotropic chiral nematic liquid crystalline aliphatic conjugated polymers based on disubstituted polyacetylene derivatives that exhibit high dissymmetry factors in circularly polarized luminescence, J. Am. Chem. Soc. 134 (48) (2012) 19795–19807, https://doi.org/10.1021/ja3086565.

[60] J. Liu, H. Su, L. Meng, Y. Zhao, C. Deng, J.C.Y. Ng, P. Lu, M. Faisal, J.W.Y. Lam, X. Huang, H. Wu, K.S. Wong, B.Z. Tang, What makes efficient circularly polarised luminescence in the condensed phase: aggregation-induced circular dichroism and light emission, Chem. Sci. 3 (9) (2012) 2737–2747, https://doi.org/10.1039/c2sc20382k.

[61] Y. Sang, J. Han, T. Zhao, P. Duan, M. Liu, Circularly polarized luminescence in nanoassemblies: generation, amplification, and application, Adv. Mater. 32 (41) (2020), https://doi.org/10.1002/adma.201900110.

[62] S.H. Chen, D. Katsis, A.W. Schmid, J.C. Mastrangelo, T. Tsutsui, T.N. Blanton, Circularly polarized light generated by photoexcitation of luminophores in glassy liquid-crystal films, Nature 397 (6719) (1999) 506–508, https://doi.org/10.1038/17343.

[63] X. Li, W. Hu, Y. Wang, Y. Quan, Y. Cheng, Strong CPL of achiral AIE-active dyes induced by supramolecular self-assembly in chiral nematic liquid crystals (AIE-N*-LCs), Chem. Commun. 55 (35) (2019) 5179–5182, https://doi.org/10.1039/c9cc01678c.

[64] F. Song, Y. Cheng, Q. Liu, Z. Qiu, J.W.Y. Lam, L. Lin, F. Yang, B.Z. Tang, Tunable circularly polarized luminescence from molecular assemblies of chiral AIEgens, Mater. Chem. Front. 3 (9) (2019) 1768–1778, https://doi.org/10.1039/c9qm00332k.

[65] Y. Wu, L.H. You, Z.Q. Yu, J.H. Wang, Z. Meng, Y. Liu, X.S. Li, K. Fu, X.K. Ren, B.Z. Tang, Rational design of circularly polarized luminescent aggregation-induced emission luminogens (AIEgens): promoting the dissymmetry factor and emission efficiency synchronously, ACS Mater. Lett. 2 (5) (2020) 505–510, https://doi.org/10.1021/acsmaterialslett.0c00063.

CHAPTER

12

Electrofluorochromism in AIE luminogens

Guey-Sheng Liou[a] and Hung-Ju Yen[b]

[a]Institute of Polymer Science and Engineering, National Taiwan University, Taipei, Taiwan
[b]Institute of Chemistry, Academia Sinica, Taipei, Taiwan

1 Principle/conception of electrofluorochromism

Fluorescent molecules whose emission intensity can be switched by external stimuli have attracted attentions for several applications in analytical chemistry, display technology, molecular devices, and single-molecule detection. Although the fluorescent molecules have been developed and investigated for decades, the term "electrofluorochromism (EFC)," which deals with the electrically manipulated changes in the fluorescence, has been rarely studied [1–8]. Compared with electrochromism (EC) [9] or electrochemiluminescence [10], research in EFC is at an initial stage. The EFC materials summarized in this chapter were grouped into small molecules and polymers, as well as non-AIE or AIE materials, with a focus on their molecular structures. Small EFC molecules can serve as fluorophores or dyads with separated electroactive and fluorescent units. The first case of EFC molecules was reported where a quinone redox state was used to switch on and off photoluminescence of a ruthenium-bipyridine complex [11]. Since then, several EFC devices derived from small organic molecules [2,12–14], inorganic materials [15–17], conjugated polymers [18,19], high-performance polymers [7,20], and composites [21] have been developed and studied.

The EFC characteristics could be acquired from three mechanisms: (1) by enabling or forbidding proton- or electron transfer between the redox-active moiety and luminophore, (2) through the energy transfer resulted from the overlap between absorption of the redox unit and fluorescence of the luminophore, and (3) by directly tuning the electroactive luminophore with its redox states (Fig. 1) [1].

Dyads or triads with redox and luminophore moieties, directly conjugated or nonconjugated linked or through a bridge, were used as models for Mechanisms 1 and 2 in Fig. 1 that can exhibit emission changes during the electrochemical reactions. The

Aggregation-Induced Emission (AIE)
https://doi.org/10.1016/B978-0-12-824335-0.00012-X

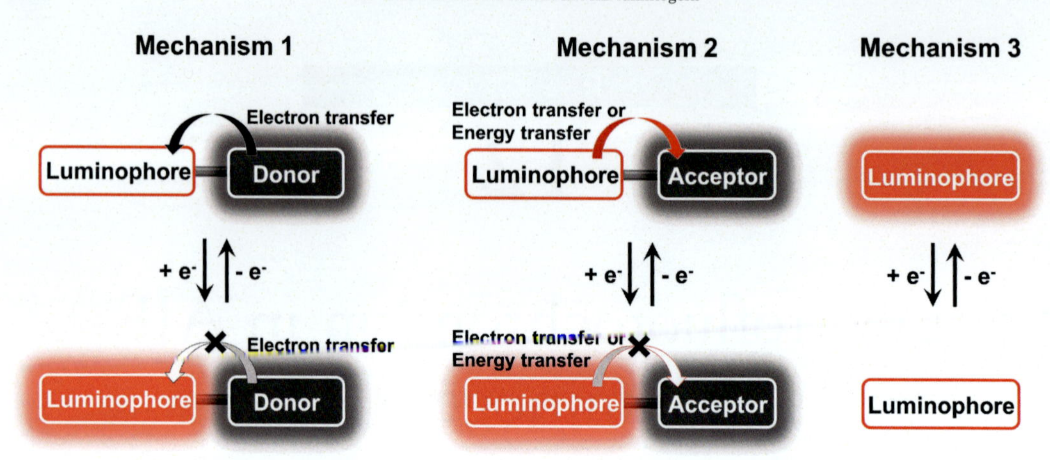

FIG. 1 Mechanism of switchable dyad systems through (1) electron transfer, (2) energy transfer, and (3) direct electrochemical switch of the luminophore.

appropriate match between the redox units with luminophores is a crucial parameter for developing new EFC systems. Although ionic radicals and dications (or dianions) are mostly neither stable nor emissive, only a few stable ones containing the counterparts, such as triarylamines [20] and tetrazines [22] have been explored and studied as promising EFC materials.

2 Electrofluorochromic luminogens and devices

The first devices based on a tetrazine/polymer electrolyte blend were reported by Audebert and Kim et al., which clearly proved the potential of in situ operated electrochemical device with fluorescent switching simultaneously (Fig. 2) [13]. The configuration of EFC devices is conceptually the same as typical EC devices, including EFC-active materials, transparent electrodes (i.e., working and counter electrodes), and electrolyte. Two-electrode and three-electrode devices are the two commonly applied multilayer structures for EFC devices, which only differ from whether the insertion of a silver reference electrode is included. A three-electrode device provides a tunable electrical conversion than two-electrode one and is beneficial for multicolor EFC devices [1]. Moreover, the complementary layer is usually a second EFC component which is used as the counter electrode. When voltage is applied between the electrodes, the emission of the EFC films can be tuned as the electrode ions are extracted from or doped into the EFC films (Fig. 3) [23]. Therefore, the transporting ability between both electron/ion and the completion of a circuit with high ion conductivity is a necessity for EFC layers [24]. Another charge balance layer is beneficial for EFC performance regarding optical contrasts, operational voltages, response time, emission efficiency, and cyclability [25–27]. For instance, Prussian blue or heptyl viologen is one of the most frequently utilized complementary materials in EFC devices. In the following part of this chapter, the reported EFC materials will be categorized and discussed extensively.

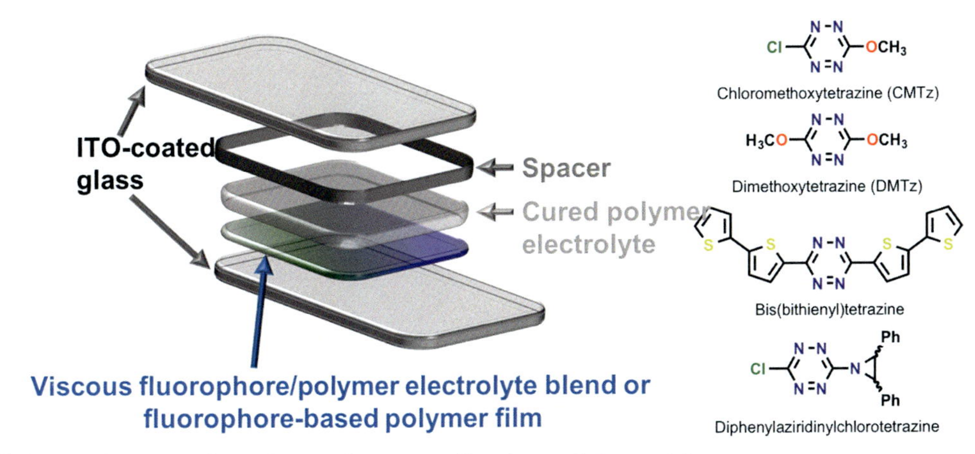

FIG. 2 The first electrofluorochromic device as well as the applied materials.

2.1 Non-AIE small molecules

2.1.1 Quinones

Quinones are one of the most recognized organic electroactive dyads, and the representative benzoquinone can be two-electron reduced to hydroquinone with two protons involved [28]. The first EFC molecule M1 was based on an emissive $Ru(bpy)_2^{2+}$ with a quinone molecule (Scheme 1) [11]. Since quinone molecule quenches the triplet state of $Ru(bpy)_2^{2+}$, the emission can be therefore restored by reducing quinone to hydroquinone.

Later on, a combination of fluorescence measurement and electrochemical microscopies was performed to explore the possibility of applying a water-soluble M2, resorufin, as redox mediator and fluorophore. M2 was selected as it is a highly emissive electroactive molecule whose fluorescence can be easily quenched by reduction. In the feedback mode of scanning electrochemical microscopy (SEM), it was shown that the fluorescence modulation amplitude is sensitive to the substrate and tip-substrate distance and therefore can be used to record optical approach curves. Further changing the polarization of the ITO substrate enables the move from positive to negative feedback. The fluorescence modulation amplitude is also sensitive to this change. Furthermore, substrate generation-tip collection mode was also conducted for detecting species produced at the substrate by fluorescence intensity with a higher accuracy than the conventional electrochemical current [29].

2.1.2 Tetrathiafulvalenes

Tetrathiafulvalenes are electrochemically stable and reversible molecules, which are another well-known electroactive group and often utilized in molecular and supramolecular switching. The molecule M3 is nonaromatic and electrochemically active, which is also stable at its oxidative state with reversible behaviors due to the increasing aromaticity (Scheme 1) [12]. In addition, a π-extended tetrathiafulvalene-boradiazaindacene M4 was reported to undergo sequential one-electron oxidations, accompanied with EC and EFC behaviors in near-IR region (emission ranged from 950 to 1450 nm with the peak at 1185 nm) of the electromagnetic spectrum [30].

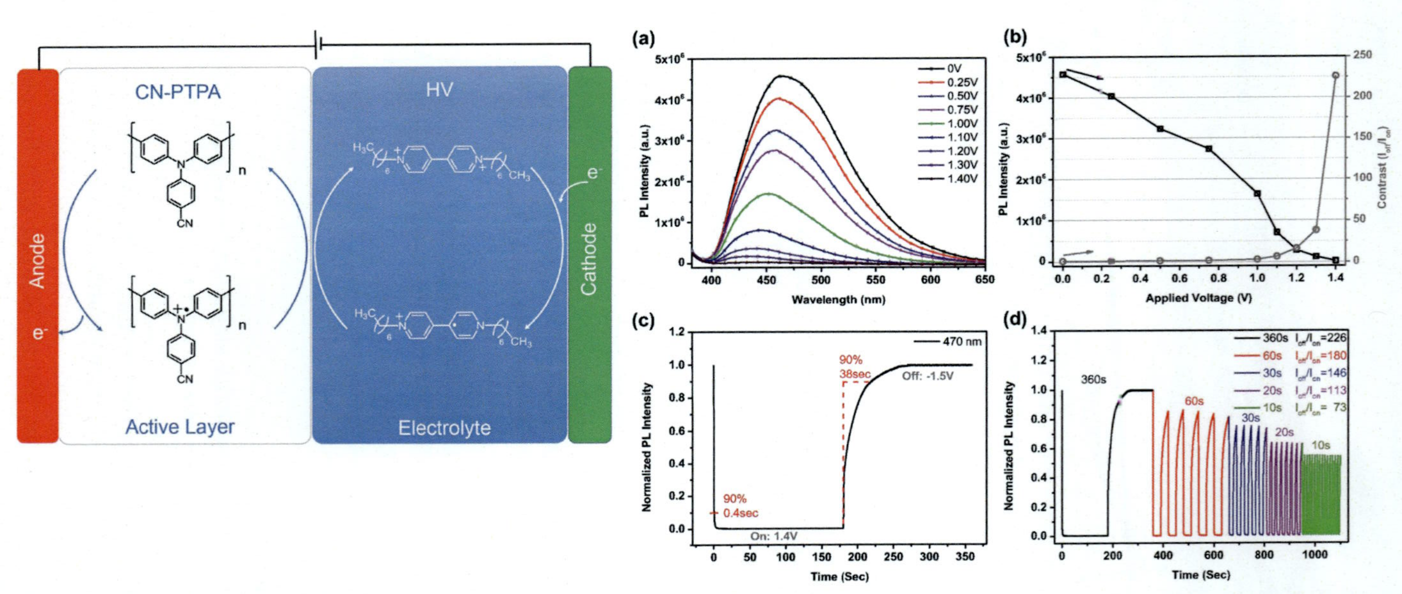

FIG. 3 Working principle of CN-PTPA/HVEFC device as well as its EFC performance. *Reproduced with permission from R.J. Mortimer, Electrochromic materials, Ann. Rev. Mater. Res. 41 (1) (2011) 241–268. Copyright 2014, John Wiley & Sons Ltd.*

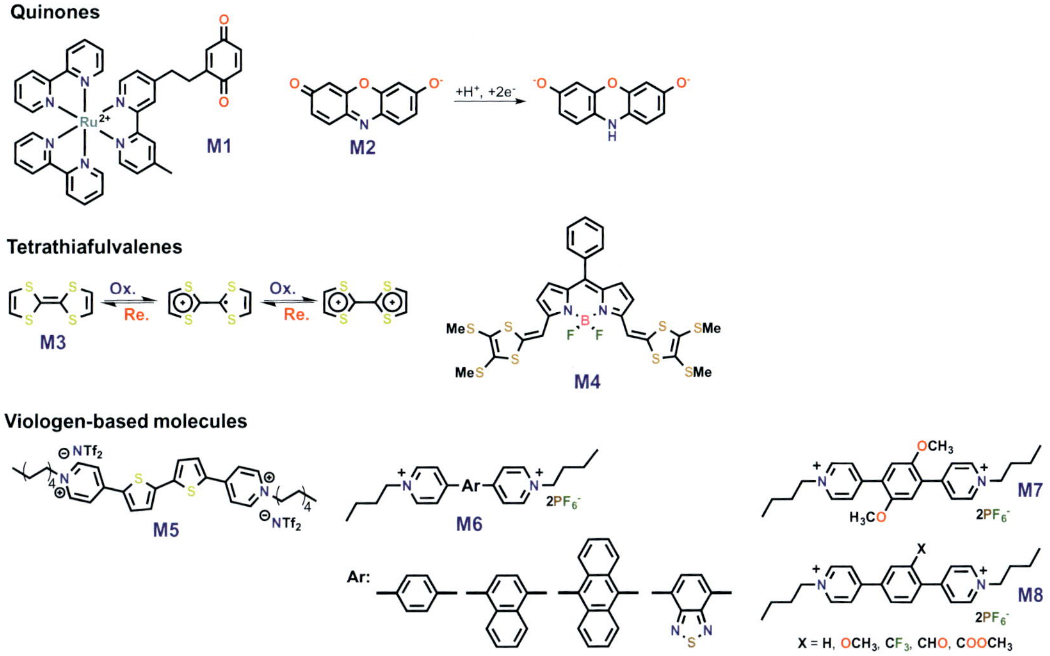

SCHEME 1 Structure of non-AIE small molecules M1–M8.

2.1.3 *Viologen-based molecules*

As bipyridinium salts, thienoviologens possess strong electron-withdrawing capability and exhibit two-electron reductive reactions. An EFC ionic liquid crystal thienoviologen M5, derived from the typical EC viologen, was reported by Beneduci et al. and exhibited an unordered isotropic phase with two ordered phases (columnar and calamitic) [31]. Upon reduction, the fluorescence of M5 was detected as the radical anion is formed. The isotropic phase was found to convert into an ordered phase at a reduced state, therefore the gradual increase in fluorescence intensity was accompanied by a redshift in emission. It has also been molecularly hypothesized [32] that the acceptor-donor-acceptor structure not only features an extended π-conjugation with relatively high fluorescence [33], but also rendered thienoviologens two reduction couples and EFC characteristics.

Later on, Xu et al. reported a series of aryl-substituted viologen derivates M6 by introducing benzene, naphthalene, anthracene, and benzothiadiazole rings between two pyridine rings. Most of the corresponding EFC devices revealed exceptional EC properties, high optical contrast, and fast response time under applied bias. Owing to the increased degree of conjugated π-framework, the fabricated devices could achieve intense fluorescence emission compared to conventional viologens. The good stability of EC devices over thousands of cycles made M6 promising candidates for EFC applications [34].

Another series of extended skeletal pyridine π-system, M7, and M8, with various aromatic substituents were prepared and exhibited unprecedented EC and EFC behaviors among previous reported viologen-based systems [35].

2.1.4 Tetrazines

Small molecule tetrazines were also used as EFC fluorophores [13,36], even though they generally only reveal weak absorption coefficients of around 500–1000 $mol^{-1} L cm^{-1}$, thus restricting their EFC performance in terms of switching contrast and fluorescence intensity [37].

Audebert et al., therefore, incorporated electroactive triphenylamines (TPAs) to tetrazines for synthesizing bichromophores M9 (Scheme 2) [38,39], whereas their emission was restored when oxidized but with low quantum yield due to the spectral overlap [40]. To prevent the spectral overlap, electron-rich groups were incorporated into TPA moieties, which can efficiently shift the absorption of the cationic TPA radical to a longer wavelength [41].

In another example, naphthalimide was attached to tetrazine and resulted in efficient light absorption, therefore exhibiting a longer lifetime and stronger fluorescence than tetrazine itself because of the quasi-complete energy transfer [42]. In addition, electropolymerizable monomers, M11 and M12, were synthesized based on bis(thienyl)benzene groups with pendant tetrazine [43]. While the carbazole moiety usually gives a poorly stable polymer film, on the other hand, 1,4-bis(thiophen-2-yl)benzene moiety allows well-defined films with controlled thicknesses. Dual EFC at ~400 and 570 nm could be observed for the electropolymerized films arising from both the polymer's and the pending tetrazine's emission, respectively.

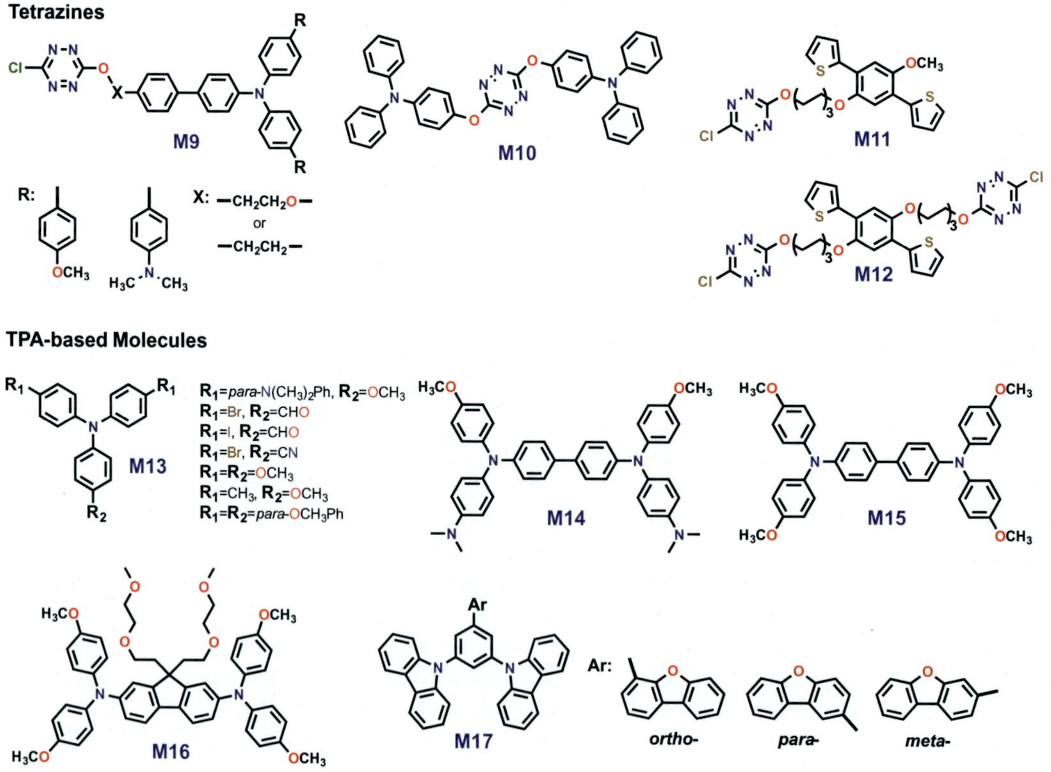

SCHEME 2 Structure of non-AIE small molecules M9–M17.

2.1.5 TPA-based molecules

Alain-Rizzo et al. [44,45] studied the chemical and EFC performances of systematically designed a series of TPA derivatives M13 (Scheme 2). The TPA-based molecules M13 were characterized by cyclic voltammetry, UV–vis, and fluorescence spectroscopy. Theoretical calculations were also performed in order to corroborate the experimental results. This EFC switching was achieved for both electro-donating and electro-withdrawing substituted TPA, but the use of donor group was found to be more favorable since the radical-cation is more stable.

Liou et al. reported EFC molecules M14 and M15 with strong electro-donating dimethylamine and methoxy groups. The redox-active M15 was found to exhibit the crucial optical characteristics of high PL quantum yields in both solution (34.5%) and solid (32.4%) states. This unique PL characteristic was the key for fabricating liquid-type and gel-type electrochemical devices with EC and EFC behaviors. The devices based on M14 and M15 possessed excellent EFC performance, including fast switching response times, high PL intensity contrast ratio values of up to 374, from strong emission to truly dark, and remarkable levels of electrochemical stability. In addition, devices based on M15 can be transformed between being highly transparent and colorless devices to truly black ones through switching between their original and oxidative states [46,47].

The electro-donating methoxy-arylamine groups were also incorporated to fluorene for preparing an efficient mixed-valence EFC molecule M16. This new fluorene-based mixed-valence EFC system allows a fast EFC switching property, high fluorescence on/off contrast, and long-term cycling stability with device cyclability of more than 10,000 cycles, which are all distinctive properties for a real-life application of EFC devices [48]. Recently, Meng et al. reported a series of cathodically coloring molecules M17 based on carbazole and dibenzofuran derivatives, which exhibit transmissive-to-multicolored EC and fluorescent-to-nonemissive EFC behavior. The reversible emission quenching on their electrochemical reduction to radical anions has been demonstrated with the feasibility of designing functional EC/EFC devices by exploiting electrochemical properties of traditional fluorescent small molecules [49].

2.1.6 Metal coordination

Lanthanide complexes are another classic luminophores with favorable properties, such as a long luminescent lifetime, large Stokes shift, and a narrow emission band. Therefore, this series of complexes are also designed to combine with electroactive units for generating EFC molecules.

Miomandre et al. reported the synthesis and characterization of gold and silver nanoparticles, M18-M20 (Scheme 3), functionalized by luminescent and electroactive iridium complexes. Cyclometalated iridium complexes coupled with modified phenanthroline ligands bearing a pyridine end group are able to cap gold nanoparticles without changing their size and shape, while the size of silver slightly increases and aggregation starts to occur but without any flocculation. Colloidal suspensions of these nanocomposites in 1,2-dichlorobenzene are stable and can be kept for long times with retained plasmonic properties. The luminescence of iridium is partially quenched by gold nanoparticles even though the interactions with the complex do not involve surface functionalization. The EFC contrast is much lower in the case of silver, and

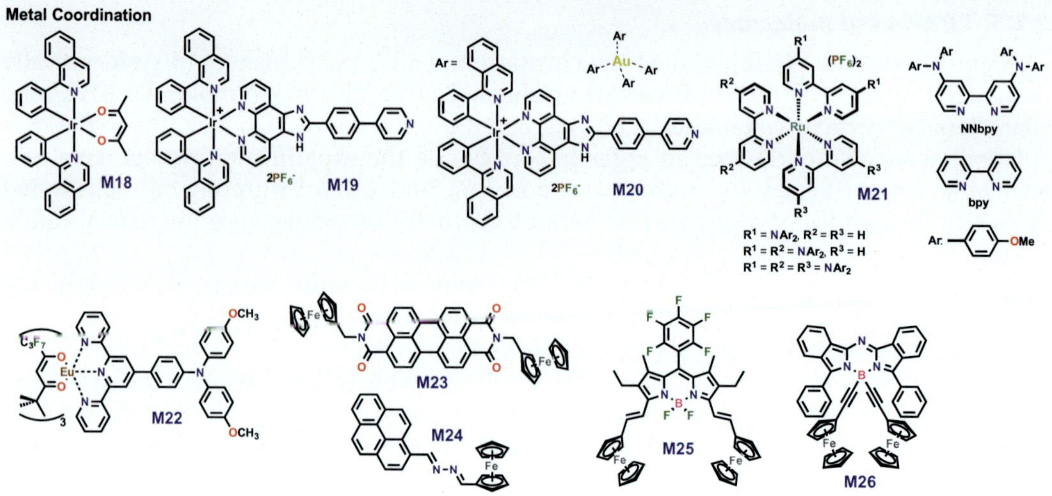

SCHEME 3 Structure of non-AIE small molecules M18–M26.

capped nanoparticles retain the same luminescence as the free complex. All iridium complexes display EFC behaviors, that is, a reversible electrochemically driven luminescence switch when changing the redox state of the metal center [50].

Zhong et al. also reported that the attachment of multiple electron-donating and redox-active amine units can significantly decrease the redox potential of ruthenium-tris(-bipyridine) complexes M21 and therefore modulate the spin density distribution of the complex in the one-electron-oxidized state. These amine-containing complexes display deep-red to NIR EC and EFC in response to various electrochemical potentials. Considering the low operation potential and high stability in the one-electron-oxidized state, they are considered attractive materials for NIR EC and EFC devices and information storage [51].

Europium (Eu) complex M22 reported by Yano et al. was prepared by direct coordination of TPA-bonded terpyridine ligand with Eu tris(β-diketonate) [52]. The results showed that the luminescence from Eu^{3+} can be switched reversibly, which could be attributed to the intramolecular energy transfer from complexes to TPA radical cation.

By coupling another redox-active Fc with chromophore through conjugated bridges, the triads could demonstrate photoinduced electron transfer (PET) or intramolecular CT in the neutral form. At their redox states, however, PET or CT will be restricted thus switching the emission wavelength [53]. In this regard, Zhang et al. prepared a triad M23 containing redox-active Fc and fluorescent perylene diimide [54]. During the redox reactions, the fluorescence of M23 was turned on and off reversibly due to the release and suppression of the PET process. Similarly, the EFC-active triad azine M24 reported by Martínez et al. was prepared by bridging Fc and pyrene [4]. By increasing the applied potential to oxidized Fc, the fluorescence of the triad was turned on, and the fluorescence intensity could be further turned off by lowing the applied potential back to its neutral state, which is due to the recovered electron transfer from Fc to pyrene.

Differ from the abovementioned Fc-based molecules M23 and M24, whose EFC behaviors were depending on photoinduced electron transfer, boron dipyrromethene dyes were

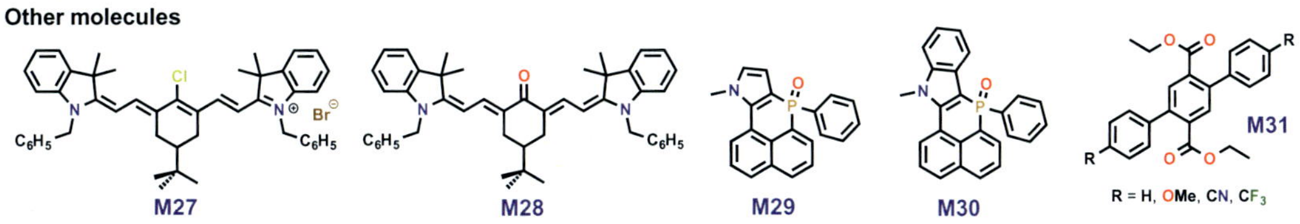

SCHEME 4 Structure of non-AIE small molecules M27–M31.

prepared by conjugating Fc either to its two side rings (M25) [55,56] or the boron atom (M26) [56]. It was found that M26 with two Fc groups attached to the boron atom promoted the intersystem crossing from singlet to the triplet state, which is a phenomenon often associated with the presence of heavy atoms. This crossing allowed an extended (excited state) lifetime of the dyad by three orders of magnitude. These Fc-type triads M25 and M26 are intriguing for their uses in molecular probes and sensors of redox reactions.

2.1.7 Other molecules

Two polymethine dyes, M27 and M28, were used as a fluorophore and an electroactive modulator in order to achieve reversible EFC switching in the near-infrared (NIR) region (Scheme 4). M27 displayed high absorption and emission in the NIR region with a reversible electrochemical reaction. In contrast, a keto group bridged M28 showed irreversible electrochemistry, which is possibly due to the keto group disrupting the full extent of conjugation of the entire molecule. The reversible redox reaction of M27 allowed the EFC switching in the NIR region, which was visually observable through a visible light cut-off filter with a cyclability of over 100 times [57].

Another family of electroluminescent organophosphorus materials, M29 and M30, for solution-processed organic light-emitting diodes and EFC devices, was synthesized and reported. The investigated systems present a six-membered phosphorus heterocycle fused with a pyrrole or benzopyrrole moiety. When fabricated as EFC devices, the color emission coordinates of M29 and M30 are tuned upon applying an electric potential. These findings set the bedrock for the development of new generations of organophosphorus light-emitting devices [58]. In addition, a new series of phthalate-based EC materials M31 were prepared and investigated by Xu et al. The introduced functional substituents on phthalate pi-framework have a significant influence on the optical and electrochemical behaviors. To the authors' surprise, the phthalate with methoxyl substituent reveals an unanticipated PL nature, which renders the ECD of M31-OMe exhibits EFC nature [59].

2.2 Non-AIE polymers

2.2.1 Conjugated polymers

Poly(phenylene-vinylene)s are a classical conjugated polymers which were recently reported as potential materials for EFC devices [60]. For example, the fluorescence deactivation of P1 (Scheme 5) was studied by in situ electrochemical fluorescence spectroscopy during p-doped scanning [61]. The critical points, such as electrolyte diffusion, processes of electrochemical doping, and degradation processes, were extensively investigated to understand the EFC processes. This application of poly(phenylene-vinylene)s derivatives into EFC devices has been studied extensively. A multistate fluorescence switching induced by iodide ions was reported by using triazine bridged poly(phenylene-vinylene)s P2 [3]. P2-doped gel layer and polymer electrolyte layer were then applied to fabricate the solid-state two-electrode EFC device with an iodide and iodine couple.

Kim et al. examined the multicolor EFC switching from P3 and P4 thin films. P3 and P4 showed vivid blue and yellow fluorescence, respectively, at neutral state but their emission was quenched upon oxidized to radical cation states. The fluorescence of polymer films could be reversibly recovered to vivid color when the films were reduced back to their neutral states. The EFC switching contrast ratio for P3 was about four times higher than that of

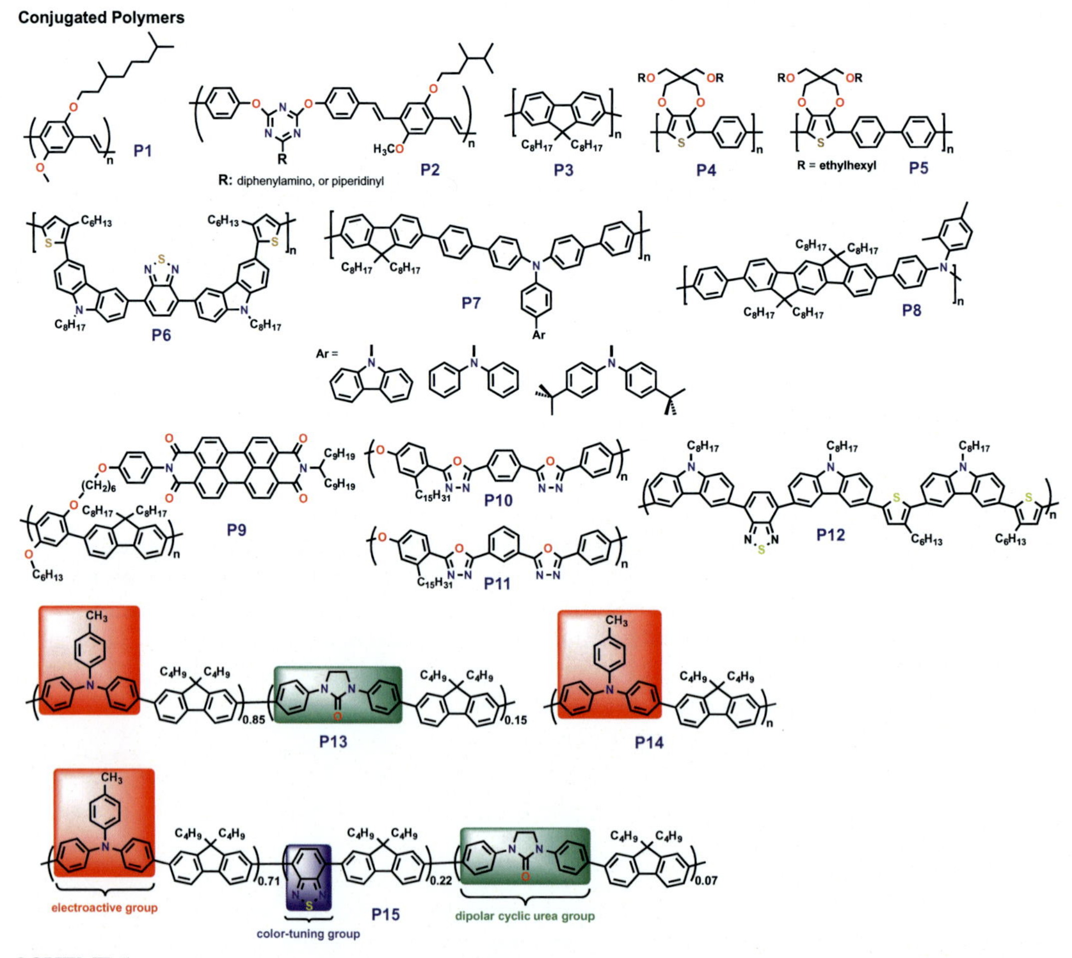

SCHEME 5 Structure of non-AIE polymers P1–P15.

the P4 (13.5 vs. 3.3). Because both polymer films possess different colors and working potentials, a multicolor EFC device was fabricated by coating P3 and P4 films onto working and counter electrodes, respectively. The fabricated EFC device showed fluorescence switching from blue to white then to yellow and vice versa, depending on the applied potential [62].

Xu et al. have reported the synthesis of polythiophenes P5 (Scheme 5) by introducing biphenyl into the thiophene backbone to form a twisting conformation, alleviating common aggregation-caused quenching and achieving synchronous EC and EFC switching. The polymer can emit intense fluorescence in both solution and solid states while continue to exhibit excellent EC properties. When an external potential was applied, the sprayed polymer film can switch to a nonfluorescent state from the yellow fluorescent state with comparable large transmittance change (35%) and fast response time [63].

Moreover, an EFC conjugated copolymer P6 containing carbazole and benzothiadiazole moieties was synthesized by electrochemical polymerization of monomers, resulting in a nanoporous EFC polymer electrode. The EFC electrode exhibits high sensitivity and selectivity when detecting cyanide anions (CN^-) in largely aqueous electrolyte (67 vol% water) because electrochemical oxidation of P6 leads to a significant fluorescence quenching, and the presence of various concentrations (1–100 μM) of CN^- in the electrolyte can weaken the oxidative quenching to different extents [64].

Wang et al. have synthesized a series of TPA-containing aromatic dibromides and further prepared three kinds of conjugated oligomers P7 by Suzuki coupling reaction of newly synthesized dibromo monomers with fluorene-based boronic acid dipinacol ester. In EC behavior, P7 showed high optical contrast from a yellowish neutral state to blue oxidized state and reversible EC properties that keep good stability after 500 s. The EFC properties can be also tuned by electrochemistry redox with high contrast [65].

Furthermore, an inkjet-printed dual-mode display was fabricated to demonstrate the use of P8 as a promising material for display applications. The optical properties of P8 are suitable to build multilayer EFC devices in combination with well-known materials for increasing either the optical contrast in the visible region or the color gamut. This work demonstrated the first inkjet-printed EFC device, which can be operated in a reflective or passive-emissive mode. These results also highlight the applicability of digitally printed dual-mode EFC devices for signage or advertisement [66].

Notably, the emission of π-conjugated polymers can be tuned over a wide range by introducing electron-donating or accepting moieties into the polymer backbone [67]. The rapid transport of excitons along the π-conjugation chain and high degree of π-delocalization result in sensitive fluorescent responses [19]. For example, Montilla et al. reported a fluorene-*alt*-phenylene copolymer P9 with electron-accepting perylene diimide pendant unit, and its PL was largely suppressed upon undergoing oxidation due to the photoinduced electron transfer from conjugated chain to neutral perylene diimide aggregates [14]. The fluorescence was then recovered after reduction because of the inhibited photoinduced electron transfer and formation of the radical anion. When applying more negative potentials, the Förster resonance energy transfer to perylene diimide radical dianion occurs thus quenching the emission.

Basically, highly soluble polymers are capable of facile solution processing. Accordingly, fluorescent and solution-processable poly(1,3,4-oxadiazole)s P10 and P11 with flexible, long alkyl groups were reported by E. Kim et al., and the derived three-electrode EFC devices were fabricated via spin coating [7]. The patterned EFC films through photo-cleavage of P10 displayed reversible EFC behaviors under applied voltage between +1.8 and −1.8 V. During the reduction, P10 with π-conjugation through C1–C20 exhibited excellent electrochemical reversibility and stability over 1000 cyclic switching. However, P11-based EFC device was not really responsive due to the interrupted π-electron delocalization.

EFC-active polymers can also be utilized to detect ions. For instance, Lu et al. reported the use of benzothiadiazole-based EFC conjugated polymer P12 for detecting cyanide (CN^-) [18]. Since nucleophilic CN^- interacted with electron-deficient benzothiadiazole, the oxidative fluorescence quenching effect was greatly weakened. Therefore, this kind of EFC polymeric sensor can be also applied to quantify ion concentration.

Leung et al. reported conjugated copolymers P13–P15 containing fluorene and TPA in the polymer backbone [68]. Fluorene-based P14 exhibited high fluorescence with EFC behavior which was generated by TPA radical cations. During the electrochemical process, P13 with bipolar cyclic urea revealed ion transport and showed blue fluorescence with high quantum yields of ~65% and 40% in solutions and thin-film states, respectively, which could be quenched reversibly during oxidation steps. Interestingly, the EFC devices possess a faster response time because of the stabilized radical cations and electrolyte diffusing. In addition, the electron acceptor benzothiadiazole segment was introduced into P15 for fine-tuning the emission and bandgap via intramolecular charge transfer [19]. The synthesized P15 was thermally stable, organo-soluble and exhibit yellow PL. It was further applied to obtain a white-light EFC device by blending with P13. Thus, the EFC devices derived from P15 and P13 have their fluorescence simultaneously quenched upon a proper applied voltage, resulting in a white/dark EFC switching.

2.2.2 TPA-based polymers

Wang, Chao, and Zhao reported a series of electroactive polyamides (P16–P24) [69–76] and polyimides (P25 and P26) [77,78] from diphenylamine-fluorene and diphenylamine-pyrene skeletal diamino monomers with dicarboxylic acids and dianhydrides, respectively (Scheme 6). These polymers were amorphous and readily soluble in many organic solvents. They displayed outstanding thermal stability and high glass transition temperatures. The polymer films showed reversible electrochemical redox and satisfactory EC performance with long-term stability, high coloration efficiency, and acceptable switching times. Furthermore, the fluorescence can be effectively electro-switched with a high contrast ratio of up to 83.

Chao et al. reported the design and synthesis of an EC and EFC dual-switching polyamic acid P27 and polyurea P28-bearing oligoaniline and fluorescent TPA units. P27 and P28 revealed a good and reversible electroactivity in a wide potential range of 0.0–1.6 V. The coexistence of two electroactive units resulted in a unique EC performance with good optical contrast, reasonable switching times, and moderated coloration efficiency in the NIR region. Furthermore, their EFC behaviors have also been studied, demonstrating that the oligoaniline groups could be served as electroactive units in the EFC system [79,80].

Another case reported by Chen et al. was a series of nonconjugated poly(aryl amino ketone) P29 with carbazolyl TPA units. These polymers were first used to study EC and EFC behaviors and were found to exhibit some exciting EC and EFC characteristics, such as good switching time, high optical contrast up to 87%, and fast fluorescence response [81].

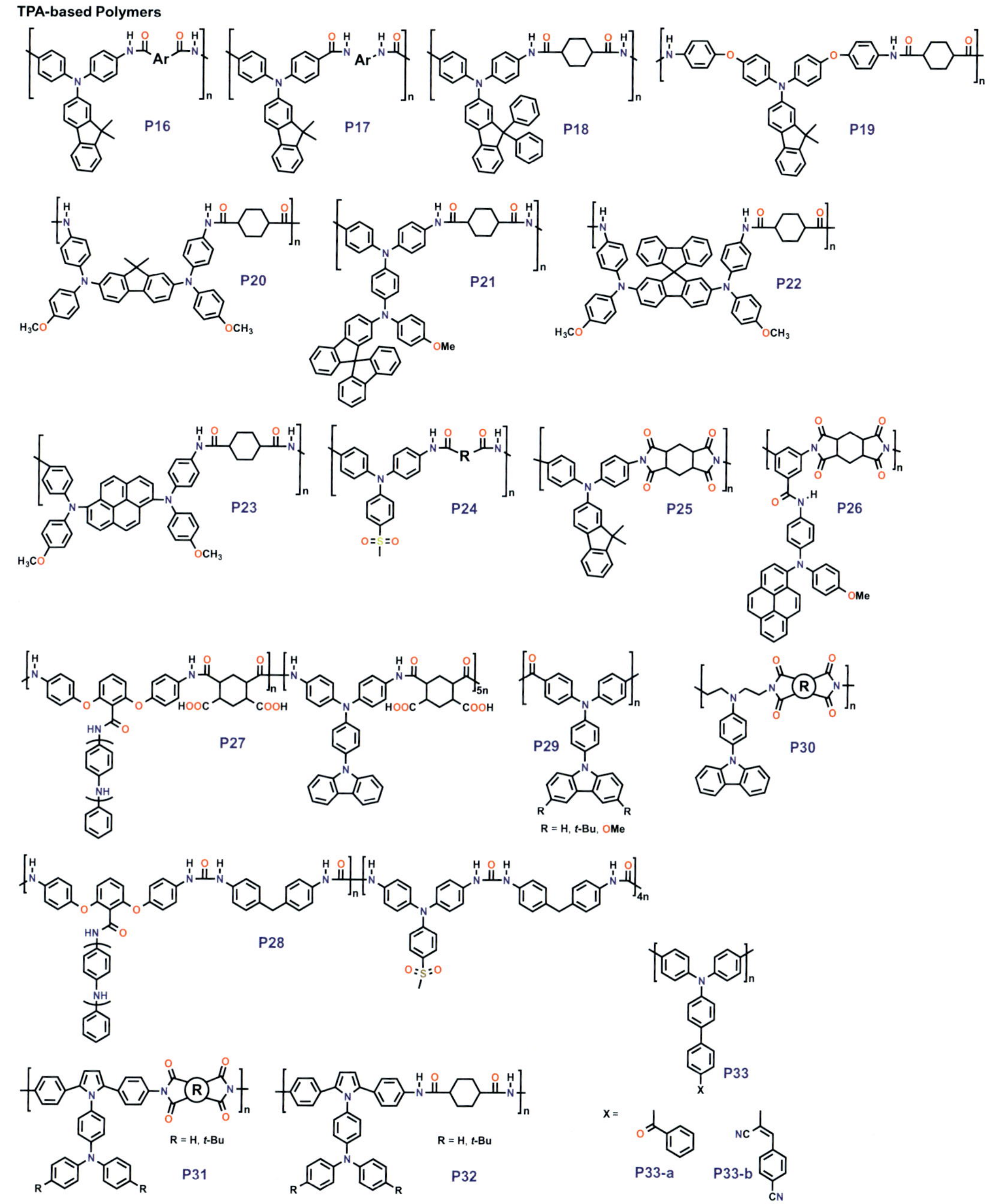

TPA-based Polymers

P16 P17 P18 P19 P20 P21 P22 P23 P24 P25 P26 P27 P28 P29 P30 P31 P32 P33

R = H, t-Bu, OMe

R = H, t-Bu

X =

P33-a P33-b

SCHEME 6 Structure of non-AIE polymers P16–P33.

Niu and Wang reported the preparation of polyimides P30 containing flexible carbazole blocks by polycondensation of a diamine monomer with five anhydrides. These polyimides not only showed high solubility and thermal stability but also demonstrate stable electrochemical oxidation behavior and anodic EC properties, as well as good EFC properties [82]. Pyrrole core with TPA groups were then designed and incorporated into polyamide P31 and polyimide P32. The polymers exhibited obvious EC behavior with good thermal stability, film-forming ability, and reversible electrochemical stability. Additional to the reversible EFC behaviors, P31 showed a rise/decay of the photovoltaics when the light was switched on and off, revealing its potential as photoelectric conversion materials [83,84].

Liang et al. designed two donor-acceptor polymers P33 that contain the same electron-rich and anodic polyTPA as π-backbone, yet with different electron-deficient ketone and cyano units as pendant groups, respectively. They both exhibit solvatochromic effects due to intrinsic characteristics of intramolecular charge transfer. Compared to P33-a, P33-b shows a stronger intramolecular charge transfer, which leads to a higher electrochemical oxidation potential and lower ion diffusion coefficient. Both polymers present simultaneous EC and EFC behaviors with multistate coloring stages and remarkably rapid fluorescence responses [85].

2.2.3 *Polysilsesquioxane and others*

Two dendritic macromolecules P34 containing electroactive TPA unit as end-capped have been rationally designed and synthesized with good yield (Scheme 7). Polymers P34 were further formed through oxidative electrochemical polymerization and found to be capable of showing the simultaneous operation in EC and EFC properties. Both polymers not only produced reversible multicolor EC changes with high coloration efficiency and optical contrast ($\Delta T\% = 65\%–71\%$) but also exhibited high EFC contrast ratio up to 179 at low working voltage with excellent electro-switching stability even after hundredth cycles [86].

Chao et al. reported the synthesis of hydrolyzable cross-linked siloxane monomers via nucleophilic reaction, followed by drop-casted onto the hydrophilic ITO substrate. After hydrolysis during cyclic voltammetry, the aforementioned functional siloxane monomers are transformed into a network of polymers P35 and P36. The resulting polymer film exhibits good EC performance with outstanding stability, due to its inherent robust cross-linking structure and strong adhesion to ITO substrate. Moreover, due to the interplay between electroactive oligoaniline and fluorescent fluorene groups, the EFC behavior was also been reported [87,88].

Wang et al. synthesized materials P37 with EC and EFC properties through the combination of TPA and ladder-like polysilsesquioxanes. The ladder-like structure was confirmed by Fourier transform infrared spectroscopy, X-ray diffraction, and ^{29}Si NMR techniques. These polymers exhibit up to 37.3% fluorescence quantum efficiency and show strong solid-state fluorescence because TPA groups are effectively isolated in ladder-like backbones. The P37 films exhibit a reversible change from colorless of bleached state to blue or green of oxidized state. Among them, P37-a film revealed a high optical contrast of 83.0% at 787 nm with good cycle stability over 1000 s. The unique structure and good electrochemical activity of P37 make them promising materials for the dual-mode device [89].

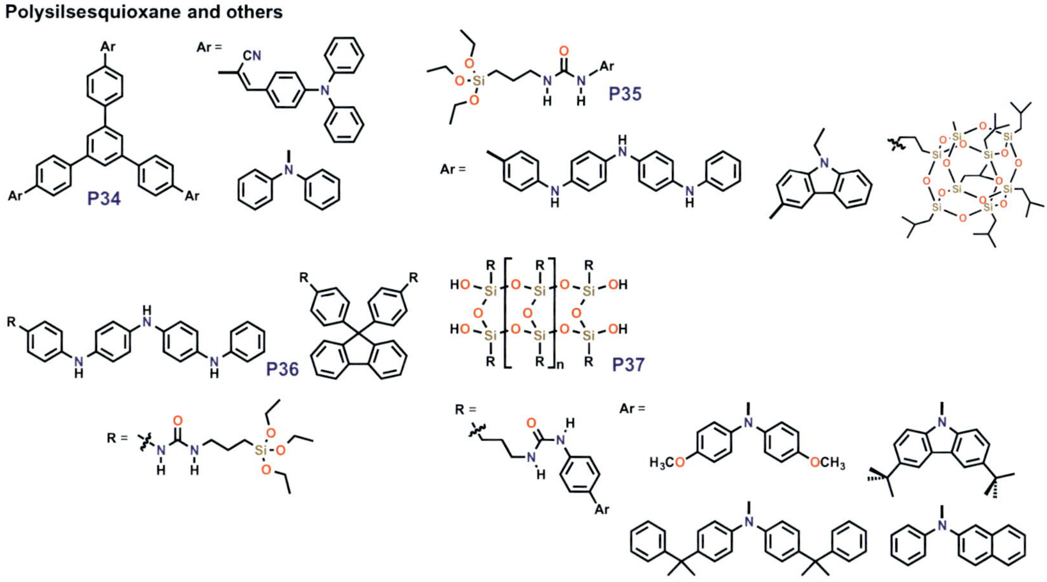

SCHEME 7 Structure of non-AIE polymers P34–P37.

3 Electrofluorochromic AIE molecules

3.1 TPA-based molecules

Liou et al. reported new high-performance electroactive AIEgens based on TPA with TPE (M32, M33) and benzo[*b*]thiophene-1,1-dioxide (M34, M35) (Scheme 8). All the obtained materials exhibit high fluorescence quantum yield in solid state (M32: 98.3%, M33: 91.1%, M34: 25.3%, and M35: 47.9%) and nearly nonemissive in the solution state. The structure–property relationship of EC and EFC behaviors in cross-linking gel-type devices derived from the prepared materials were systematically investigated [90]. Thereafter, Liou et al. further synthesized several AIE- and electro-active α-cyanostilbene-based TPA-containing derivatives M36 with different substituents and investigate the effects of substituents on the PL properties, EC, and EFC behavior of gel-type ECD devices. EFC devices based on M36 were fabricated by combining these AIE- and electro-active materials with cathodic EC heptyl viologen into the gel-type electrolyte system to enhance the emission intensity, on/off contrast ratio, and response capability. Furthermore, different alkyl chain lengths and anions were incorporated into M37 through a simple alkylation reaction and ion exchange process. The optical absorption characteristics in solution and solid states for M37 changed very slightly despite replacing Br- with BF_4- but exhibited a bathochromic effect when a longer alkyl chain length was incorporated. The intrinsically ambipolar system in M37 was generated by incorporating the pyridinium moiety, which serves as a charge storage unit to effectively balance the charge during the electrochemical oxidation redox process [91,92].

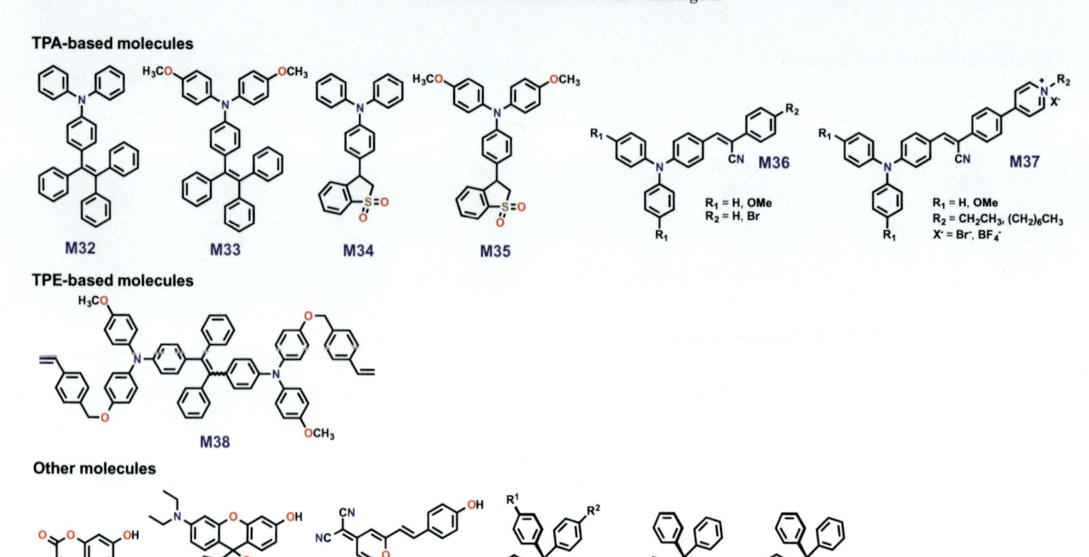

SCHEME 8 Structure of AIE molecules M32–M40.

3.2 TPE-based molecules

With the development of the EFC system based on TPA derivatives, Joseph et al. synthesized a cross-linkable, symmetric diphenylamine-TPE molecule M38 that exhibits transmissive-to-black EC and fluorescent-to-dark EFC dual behaviors (Scheme 9). The thermal cross-linking of small molecule M38 via styryl end groups ensures the formation of flawless films on electrode substrates with long-term stability. EC devices of the cross-linked M38 showed an EC switching between a highly transmissive, light yellow colored neutral state, and a black colored oxidized state with optical contrasts up to 50% and reasonably switching times (<10 s). Under the potential sweep, the fluorescence of the device based on cross-linked M38 also switched between a highly fluorescent yellowish green neutral state with an emission maximum around 524 nm and a strongly quenched dark oxidized state [93].

3.3 Other molecules

Zhang et al. reported the fabrication of RGB color-tunable "turn-on" EFC devices based on M39-B, M39-G, and M39-R. The EFC devices exhibited high color purity (M39-B: 457 nm; M39-G: 539 nm; M39-R: 641 nm), relatively quick response/fading speeds, and remarkable fluorescence contrast ratios. This approach exhibits great potential for increasingly important multistage encrypted information storage and displays [94].

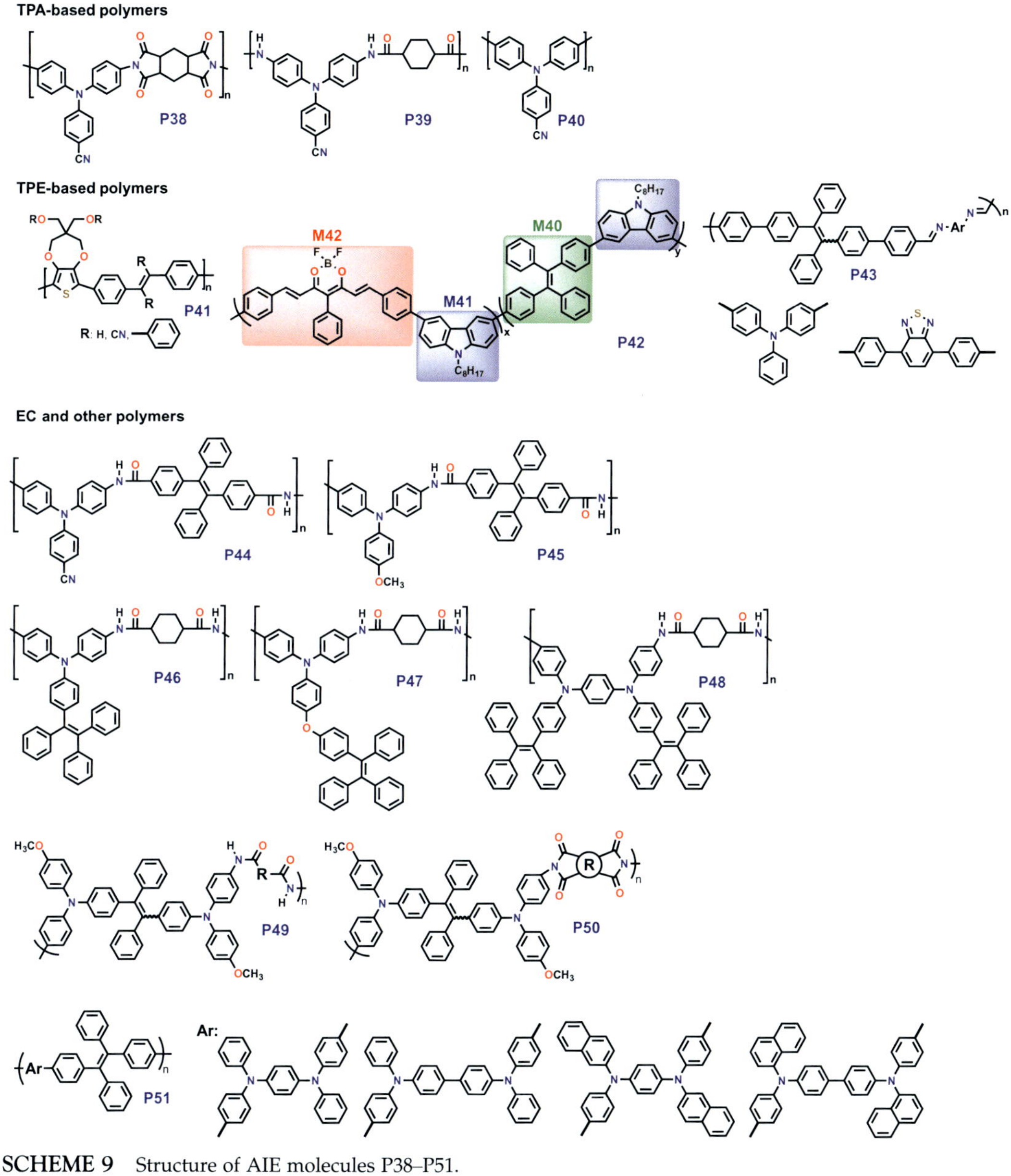

TPA-based polymers

TPE-based polymers

EC and other polymers

SCHEME 9 Structure of AIE molecules P38–P51.

Highly substituted triazoline-based M40 was prepared by reacting triarylvinyl Grignard reagents with functionalized organic azides. The heterocycles M40 are fluorescent in the solid state, and few of them can display AIE behavior. Upon oxidation, M40 formed stable radical cations with altered photophysical properties. Therefore, M40 represented rare structural examples of solid-state emitters with intrinsic EFC behavior [95].

4 Electrofluorochromic AIE polymers

4.1 TPA-based polymers

Two cyanoTPA-based polyimide P38 and polyamide P39 were prepared, which were further fabricated as flexible EFC devices. The two polymers revealed AIE properties and a remarkable EFC contrast ratio of up to 152 [20]. The results demonstrated that the incorporation of AIE-active moieties is a feasible approach for developing high-performance EFC materials. Nevertheless, the response times were relatively long even with high contrast ratios. Therefore, a fully conjugated P40 with TPA as building blocks was synthesized and further fabricated as a high-performance EFC device [26]. Thin films of P40 revealed a strong PL at 470 nm and were rapidly quenched in the oxidized state. The fabricated EFC device based on P40 had an enhanced charge transport capacity possessing a high contrast ratio of 242 and a rapid response time of less than 0.4 s.

4.2 TPE-based polymers

New polythiophene derivatives P41 containing trans-stilbene and fumaronitrile groups were designed and synthesized. P41 emitted intense fluorescence in both solution and film states, attributing to the restriction of intramolecular rotation caused by multiple C-H⋯π bonds existed in the neighboring polymer backbones. Driven by different applied potentials,the polymer films show reversible redox reactions with colors changing from yellow-green to sky-blue and red to rufous, respectively, with EFC behavior occurring synchronously [96].

Three-component conjugated polymers P42 with strong donor-acceptor type were synthesized by Pd-catalyzed Suzuki coupling polymerization reaction of TPE-based M40 with carbazole-based M41 and dioxaborinine-based M42. Among them, P42-1 and P42-2 with high TPE ratios of 0.95 and 0.9 showed obvious AIE behavior; but P42-3 with a low TPE ratio at 0.8 showed an ACQ phenomenon. In particular, the P42 polymer dots (Pdots) exhibited a lower electrochemiluminescence potential by 200 mV than M42 and M41 due to their strong donor-acceptor electronic structure. Furthermore, P42-1 Pdots with the strongest electrochemiluminescence signal were successfully used as electrochemiluminescence biosensors for the detection of catechol, epinephrine, and dopamine with detection limits of 1, 7, and 3 nM, respectively. This work develops sensitive electrochemiluminescence biosensors by the structure design of the AIE-active Pdots [97].

Not all TPE-based materials are AIE-active. Two TPE-based polyazomethines P43 were prepared from an AIE-active dialdehyde monomer. Fluorescent testing shows that dialdehyde monomer owns intense AIE characteristics, but unfortunately, P43 does not have AIE property. It may be attributed that the low energy of molecules being consumed at the excited state in the form of electron transfer, thus reducing the proportion of energy outputted from light and causing the disappearance of macroscopic fluorescence. Another guess might be that the TPE unit takes part of the conjugated backbone of the polymer, therefore eliminating its AIE behavior [98].

4.3 EC and other polymers

EC and AEE-active TPA-based polyamides, P44 and P45, were prepared from 4-cyanoTPA, 4-methoxyTPA with TPE moieties via condensation polymerization. The emission from the polyamides could be quenched from the neutral to oxidized states effectively due to the structural planarization and optical absorption shift of TPA units during electrochemical switching. With the introduction of heptyl viologen into the device system as a counter EC layer for balancing charges, the resulting high-performance EFC devices based on P44 as a photoluminescent (with a fluorescence quantum yield of up to 46% in the film state) and redox-active layer showed a high EFC contrast ratio of 105 [99].

Later on, Chen, Wang, and Zhao also tried to incorporate TPE units into the TPA-based polymer backbone to prepare AIE-active polyamides P46–P48. These polyamides showed high fluorescence quantum yield up to 69% at solid film state and stable electrochemical stability with reversible color and emission switching. The as-prepared porous P46 polymer film enhanced the response speed of switching, and thus also resulted in high fluorescence contrast, faster response speed, high stability, and facile processing properties [100–102].

Guan et al. reported TPE-centered polyamide P49 and polyimide P50 by replacing the core-phenyl within tetraphenyl-*p*-phenylenediamine by TPE moieties. Both polymers exhibited good solubility, thermal stability, and AIE behavior. Among these, polyamide P50 showed colorless-to-black EC properties with low voltage, high optical contrasts (up to 88%), fast switching times, and high fluorescence properties. In addition, polyamide P50 also showed good continuous switching stability for EC property and acceptable stability for EFC property [103].

Four functional EC polymers P51 based on TPE and TPA units synthesized via Stille coupling reaction were reported by Niu and Wang. Electrochemical impedance spectroscopy (EIS) measurements found that P51 had low resistance values which improved their electrochemical and EC properties. In addition, the maximum coloration efficiency value is 126 cm^2/C, and bleaching and coloring response times are 2.6 and 1.6 s, respectively. There is no significant change in performance after 100 cyclic scans. Photoelectric responsive results show that the four polymers are potential candidates in the field of photodetectors. In addition to their excellent EC properties, P51 also exhibits a strong AIE effect. Furthermore, the fluorescence color changed from bright green to black when voltage was applied positively. Four polymers also responded rapidly to light and maintained their stability over 1000 s [104].

5 Potential applications

AIEgens have been extensively studied and widely used in optoelectronic devices, chemosensors, and fluorescent bioprobes, and researchers have also striven to explore the potentials in other areas. The AIEgens have demonstrated superior advantages such as highly intensive emissions in the aggregated/solid state and thus exhibited much higher performances as compared with traditional luminescent materials. As applied in EFC devices, AIEgens are required to improve the contrast ratio, reduce the response time, and increase the stability of long-term operation [105].

5.1 Optoelectronics

AIEgens have been realized progressive achievements in developing blue light-emitting materials, and this section will focus on the representative advancements of optoelectronic applications and their performance based on AIEgens [106].

Leung et al. reported the development of multicolor emissive devices composed of complementary conjugated copolymers P13–P15 [68], in which the emission of each polymeric layer could be individually controlled. The concept of applying complementary layers for EC devices has been widely utilized to provide a wide emissive region, high contrast ratio, and fast switching performance. To form a complementary pair, polymers P13–P15 have to fulfill several conditions: (1) The polymers should possess stable and reversible electrochemistry under redox reactions. (2) The emission intensity should be retained upon applying a negative bias. (3) Barriers for ion diffusion should be low to achieve a fast EFC switching.

In this regard, a greenish-yellow emitting P15 and a blue-emitting P13 as complementary partners were fabricated for multicolor EFC devices. Meanwhile, a dipolar cyclic urea moiety was also incorporated to promote ion transport during the electrochemical process and facilitate the formation of TPA radical cation. As a result, a white-light EFC device in which the emission spectral range covers two merged complementary blue and yellow colors was achieved [25].

One ingenious method to obtain desirable EFC materials is to separate the redox skeleton from the fluorophore and utilize proton transfer to fine-tune the emission. Generally, photoacid was added to switch the intermolecular fluorescence, while an electro-base with alkalinity can be controlled by redox reactions and is beneficial for developing EFC devices. Accordingly, *p*-benzoquinone as an electro-base was chosen to "turn-on" RGB (M39-B, M39-G, M39-R) color-tunable EFC devices by Zhang et al., which not only exhibited high color purity but also revealed fast responsive speed and high contrast ratios (Fig. 4) [94].

5.2 Energy storage devices

EC supercapacitors that are able to reveal their light absorption by changing color have been considered promising applications in the future. However, a reflective mode is rarely applied in conditions with low ambient light, which encourages the development of light-emitting indicators. Chao et al. prepared a network polysiloxane-containing fluorescent cyanophenethylene groups and electroactive oligoaniline groups via an electrochemistry-assisted hydrolytic cross-linking reaction. The resulting polysiloxane revealed EC and EFC behaviors with a high optical contrast ratio, a good fluorescence contrast ratio of 85%, and good cycling durability. The prepared polysiloxane was then fabricated as a symmetrical supercapacitor device, which integrates the functions of energy storage with EC and EFC properties. Therefore, the energy levels can be simultaneously obtained by tunable colors from transmissive green-yellow to dark green, accompanied with EFC switching, which represents a new approach toward smart energy storage systems (Fig. 5) [107].

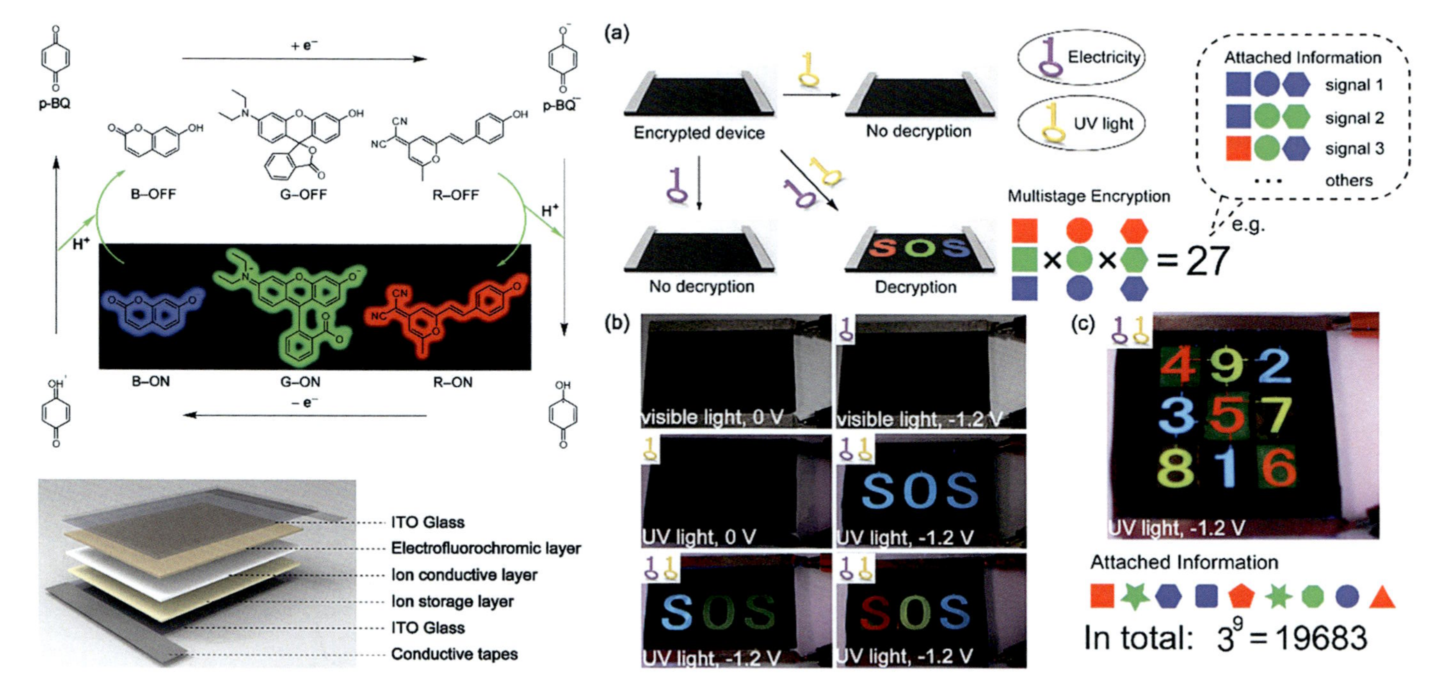

FIG. 4 The schematic diagram, mechanism, and device configuration for an encrypted information storage and display device derived from M39. *Reproduced with permission from X. Wang, W. Li, W. Li, C. Gu, H. Zheng, Y. Wang, et al., An RGB color-tunable turn-on electrofluorochromic device and its potential for information encryption, Chem. Commun. 53 (81) (2017), 11209–11212. Copyright 2017, Royal Society of Chemistry.*

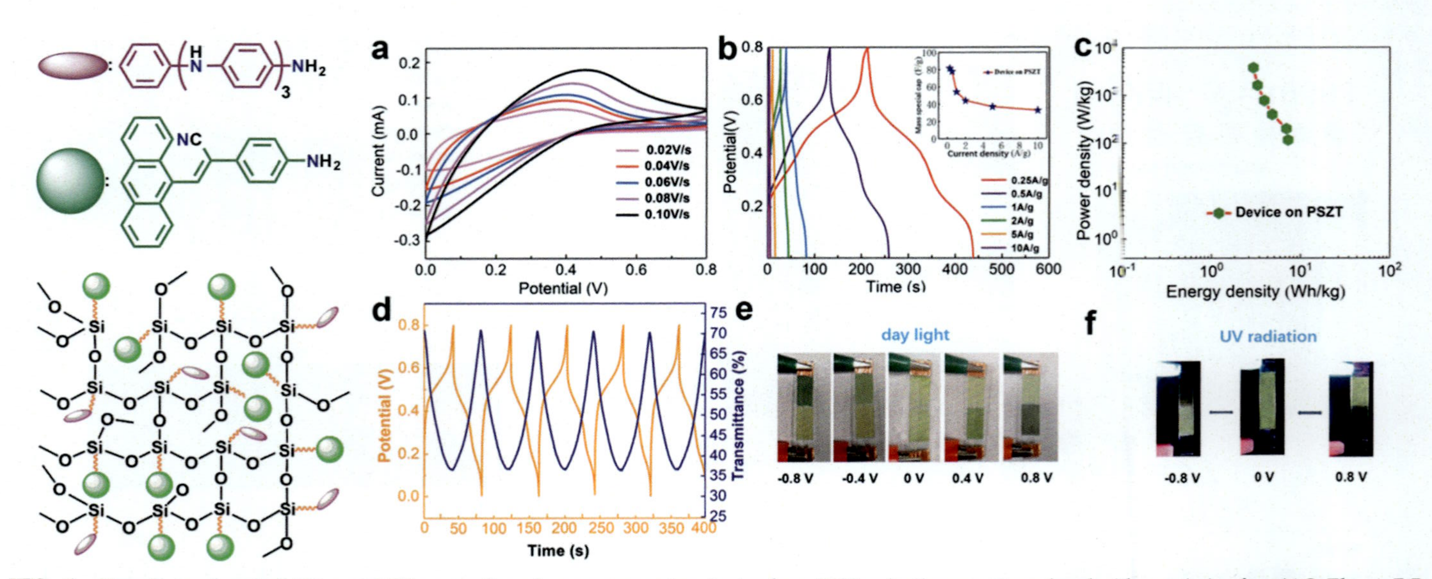

FIG. 5 The electrochemical, EC, and EFC properties of a supercapacitor device from EFC polysiloxane. *Reproduced with permission from Y.C. Zhang, E.B. Berda, X.T. Jia, Z. Lu, M.H. Zhu, D.M. Chao, Electrochromic/electrofluorochromic supercapacitor based on a network polysiloxane bearing oligoaniline and cyanophenethylene groups, ACS Appl. Polym. Mater. 2 (7) (2020) 3024–3033. Copyright 2020, American Chemical Society.*

5.3 Sensors

Electrochemiluminescence has been considered as a luminescence emission process on electrode surfaces via the excited states formed by an exergonic electron transfer reaction. Various bio-probes have been explored due to their high sensitivity and low detection. Compared with traditional electrochemiluminescence materials, Pdots have been recognized as one of the most promising electrochemiluminescence luminogens due to their low toxicity and versatile structural modification at a well-defined molecular level. P42 Pdots with TPE, carbazole, and dioxaborinine moieties have demonstrated that its electrochemiluminescence response signals can be greatly improved for the detection of catechol, with a detection limit down to 1 nM [97].

Other than this, EFC-active polymers can also be used to selectively detect ions, especially the toxic cyanide. Conjugated copolymers P6 [64] and P12 [18] with carbazole and benzothiadiazole moieties possessed interactions with nucleophilic cyanide, therefore demonstrating high sensitivity and selectivity in the EFC detection.

5.4 Biomedical applications

Mucin-1 is a glycoprotein that participates in the metastasis and invasion of multiple kinds of tumors. In common cells, the expression of mucin-1 is really low, whereas the upregulation is associated with invasion, proliferation, and survival of tumor cells. Therefore, specific and sensitive detection of mucin-1 on the cancer cell surface is critical. However, currently reported methods cannot simultaneously achieve signal visualization, prevent cell damage issues to the cells themselves, obtain high spatial and temporal resolution imaging. Therefore, the combination of electrical and fluorescent signals accompanied with both electrochemical sensitivity and fluorescent visualization, enabling optical information to be switchable by the electric field in situ, will be of great importance for human health and life science development.

Therefore, an EFC strategy was applied to visualize cancer cell surface glycoprotein mucin 1 [28]. The device includes two separate cells, anodic sensing cells and cathodic reporting cells, which were connected by a screen-printing electrode on poly(ethylene terephthalate) membrane. Under an applied voltage, the ferrocene was oxidized and 1,4-benzoquinone in the cathodic reporting cell was reduced, which further turned on the fluorescence of a pH-responsive fluorescent molecule 2-(2-(4-hydroxystyryl)-6-methyl-4Hpyran-4-ylidene) malononitrile coexisting in the cathode for both spectrophotometric detection and imaging (Fig. 6). This strategy allows sensitive detection of mucin 1 at a relatively low concentration of 10 fM. This approach also possesses high temporal and spatial resolution, fast response time, and high contrast ratio.

6 Conclusions and perspectives

In this chapter, we mainly focused on the EFC properties of AIE luminogens and non-AIE materials, which can be further categorized into small molecules and polymers. EFC that deals with the electrically driven reversible optical changes of fluorescence has only recently

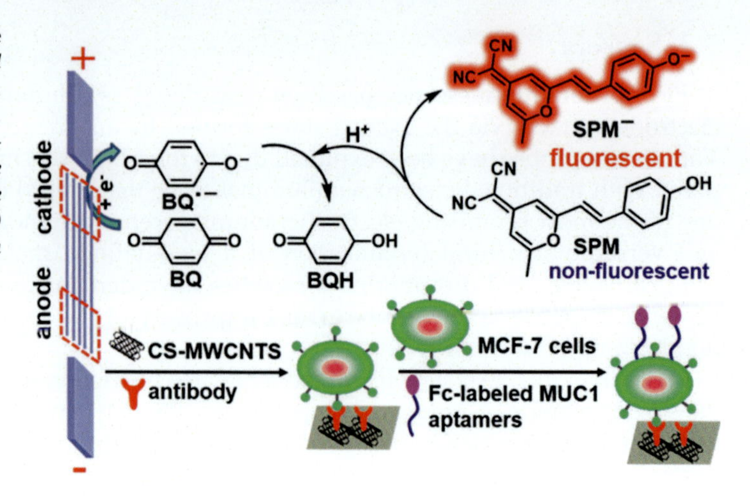

FIG. 6 Sensing principle for the detection of cancer cells. *Reproduced with permission from Z. Tian, L. Mi, Y. Wu, F. Shao, M. Zou, Z. Zhou, S. Liu, Visual electrofluorochromic detection of cancer cell surface glycoprotein on a closed bipolar electrode chip, Anal. Chem. 91 (12) (2019) 7902–7910. Copyright 2019, American Chemical Society.*

been coined, and the related researches are at a relatively early state compared to the electrochemical manipulation of optical properties such as EC (redox-switchable color) or electrochemical light generation by electrochemiluminescence. EFC materials whose fluorescence can be switched by external stimuli have garnered attention for implications in optoelectronic, energy storage devices, sensors, biomedical applications, which were reviewed and discussed in this chapter. Although the reports on the applications are still scarce, we believe that new challenges in optoelectronic or sensors will follow these pioneering investigations.

References

[1] P. Audebert, F. Miomandre, Electrofluorochromism: from molecular systems to set-up and display, Chem. Sci. 4 (2) (2013) 575–584.

[2] S. Seo, Y. Kim, Q. Zhou, G. Clavier, P. Audebert, E. Kim, White electrofluorescence switching from electrochemically convertible yellow fluorescent dyad, Adv. Funct. Mater. 22 (17) (2012) 3556–3561.

[3] J. Yoo, T. Kwon, B.D. Sarwade, Y. Kim, E. Kim, Multistate fluorescence switching of s-triazine-bridged p-phenylene vinylene polymers, Appl. Phys. Lett. 91 (24) (2007) 241107.

[4] R. Martinez, I. Ratera, A. Tarraga, P. Molina, J. Veciana, A simple and robust reversible redox-fluorescence molecular switch based on a 1,4-disubstituted azine with ferrocene and pyrene units, Chem. Commun. 36 (2006) 3809–3811.

[5] R. Zhang, Y. Wu, Z. Wang, W. Xue, H. Fu, J. Yao, Effects of photoinduced electron transfer on the rational design of molecular fluorescence switch, J. Phys. Chem. C 113 (6) (2009) 2594–2602.

[6] G. Hennrich, H. Sonnenschein, U. Resch-Genger, Redox switchable fluorescent probe selective for either Hg(II) or Cd(II) and Zn(II), J. Am. Chem. Soc. 121 (21) (1999) 5073–5074.

[7] S. Seo, Y. Kim, J. You, B.D. Sarwade, P.P. Wadgaonkar, S.K. Menon, A.S. More, E. Kim, Electrochemical fluorescence switching from a patternable poly(1,3,4-oxadiazole) thin film, Macromol. Rapid Commun. 32 (8) (2011) 637–643.

[8] G. Zhang, D. Zhang, X. Guo, D. Zhu, A new redox-fluorescence switch based on a triad with tetrathiafulvalene and anthracene units, Org. Lett. 6 (8) (2004) 1209–1212.

[9] R.J. Mortimer, D.R. Rosseinsky, P.M.S. Monk, Electrochromic Materials and Devices, Wiley, 2015.

[10] L. Hu, G. Xu, Applications and trends in electrochemiluminescence, Chem. Soc. Rev. 39 (8) (2010) 3275–3304.

[11] V. Goulle, A. Harriman, J.-M. Lehn, An electro-photoswitch: redox switching of the luminescence of a bipyridine metal complex, J. Chem. Soc. Chem. Commun. 12 (1993) 1034–1036.

[12] D. Canevet, M. Salle, G. Zhang, D. Zhang, D. Zhu, Tetrathiafulvalene (TTF) derivatives: key building-blocks for switchable processes, Chem. Commun. 17 (2009) 2245–2269.
[13] Y. Kim, E. Kim, G. Clavier, P. Audebert, New tetrazine-based fluoroelectrochromic window; modulation of the fluorescence through applied potential, Chem. Commun. 34 (2006) 3612–3614.
[14] F. Montilla, R. Esquembre, R. Gómez, R. Blanco, J.L. Segura, Spectroelectrochemical study of electron and energy transfer in poly(fluorene-alt-phenylene) with perylenediimide pendant groups, J. Phys. Chem. C 112 (42) (2008) 16668–16674.
[15] H. Gu, L. Bi, Y. Fu, N. Wang, S. Liu, Z. Tang, Multistate electrically controlled photoluminescence switching, Chem. Sci. 4 (12) (2013) 4371–4377.
[16] F. Miomandre, R.B. Pansu, J.F. Audibert, A. Guerlin, C.R. Mayer, Electrofluorochromism of a ruthenium complex investigated by time resolved TIRF microscopy coupled to an electrochemical cell, Electrochem. Commun. 20 (Supplement C) (2012) 83–87.
[17] Y.-X. Yuan, Y. Chen, Y.-C. Wang, C.-Y. Su, S.-M. Liang, H. Chao, L.-N. Ji, Redox responsive luminescent switch based on a ruthenium(II) complex [Ru(bpy)2(PAIDH)]2+, Inorg. Chem. Commun. 11 (9) (2008) 1048–1050.
[18] G. Ding, H. Zhou, J. Xu, X. Lu, Electrofluorochromic detection of cyanide anions using a benzothiadiazole-containing conjugated copolymer, Chem. Commun. 50 (6) (2014) 655–657.
[19] C.-P. Kuo, C.-N. Chuang, C.-L. Chang, M.-k. Leung, H.-Y. Lian, K.C.-W. Wu, White-light electrofluorescence switching from electrochemically convertible yellow and blue fluorescent conjugated polymers, J. Mater. Chem. C 1 (11) (2013) 2121–2130.
[20] H.J. Yen, G.S. Liou, Flexible electrofluorochromic devices with the highest contrast ratio based on aggregation-enhanced emission (AEE)-active cyanotriphenylamine-based polymers, Chem. Commun. 49 (84) (2013) 9797–9799.
[21] L. Jin, Y. Fang, D. Wen, L. Wang, E. Wang, S. Dong, Reversibly electroswitched quantum dot luminescence in aqueous solution, ACS Nano 5 (6) (2011) 5249–5253.
[22] G. Clavier, P. Audebert, S-tetrazines as building blocks for new functional molecules and molecular materials, Chem. Rev. 110 (6) (2010) 3299–3314.
[23] R.J. Mortimer, Electrochromic materials, Annu. Rev. Mater. Res. 41 (1) (2011) 241–268.
[24] D.R. Rosseinsky, R.J. Mortimer, Electrochromic systems and the prospects for devices, Adv. Mater. 13 (11) (2001) 783–793.
[25] C.-P. Kuo, C.-L. Chang, C.-W. Hu, C.-N. Chuang, K.-C. Ho, M.-K. Leung, Tunable electrofluorochromic device from electrochemically controlled complementary fluorescent conjugated polymer films, ACS Appl. Mater. Interfaces 6 (20) (2014) 17402–17409.
[26] J.-H. Wu, G.-S. Liou, High-performance electrofluorochromic devices based on electrochromism and photoluminescence-active novel poly(4-Cyanotriphenylamine), Adv. Funct. Mater. 24 (41) (2014) 6422–6429.
[27] K. Kanazawa, K. Nakamura, N. Kobayashi, Electroswitching of emission and coloration with quick response and high reversibility in an electrochemical cell, Chem. Asian J. 7 (11) (2012) 2551–2554.
[28] Z. Tian, L. Mi, Y. Wu, F. Shao, M. Zou, Z. Zhou, S. Liu, Visual electrofluorochromic detection of cancer cell surface glycoprotein on a closed bipolar electrode chip, Anal. Chem. 91 (12) (2019) 7902–7910.
[29] L. Guerret-Legras, J.F. Audibert, I.M.G. Ojeda, G.V. Dubacheva, F. Miomandre, Combined SECM-fluorescence microscopy using a water-soluble electrofluorochromic dye as the redox mediator, Electrochim. Acta 305 (2019) 370–377.
[30] N.L. Bill, J.M. Lim, C.M. Davis, S. Bähring, J.O. Jeppesen, D. Kim, J.L. Sessler, π-Extended tetrathiafulvalene BODIPY (ex-TTF-BODIPY): a redox switched "on–off–on" electrochromic system with two near-infrared fluorescent outputs, Chem. Commun. 50 (51) (2014) 6758–6761.
[31] A. Beneduci, S. Cospito, M. La Deda, L. Veltri, G. Chidichimo, Electrofluorochromism in π-conjugated ionic liquid crystals, Nat. Commun. 5 (2014) 3105.
[32] A. Beneduci, S. Cospito, M.L. Deda, G. Chidichimo, Highly fluorescent thienoviologen-based polymer gels for single layer electrofluorochromic devices, Adv. Funct. Mater. 25 (8) (2015) 1240–1247.
[33] S. Cospito, A. Beneduci, L. Veltri, M. Salamonczyk, G. Chidichimo, Mesomorphism and electrochemistry of thienoviologen liquid crystals, Phys. Chem. Chem. Phys. 17 (27) (2015) 17670–17678.
[34] Y. Shi, J. Liu, M. Li, J. Zheng, C. Xu, Novel electrochromic-fluorescent bi-functional devices based on aromatic viologen derivates, Electrochim. Acta 285 (2018) 415–423.
[35] Y. Shi, G. Wang, Q. Chen, J. Zheng, C. Xu, Electrochromism and electrochromic devices of new extended viologen derivatives with various substituent benzene, Sol. Energy Mater. Sol. Cells 208 (2020), 110413.

[36] Y. Kim, J. Do, E. Kim, G. Clavier, L. Galmiche, P. Audebert, Tetrazine-based electrofluorochromic windows: modulation of the fluorescence through applied potential, J. Electroanal. Chem. 632 (1) (2009) 201–205.

[37] Q. Zhou, P. Audebert, G. Clavier, R. Méallet-Renault, F. Miomandre, Z. Shaukat, T.-T. Vu, J. Tang, New tetrazines functionalized with electrochemically and optically active groups: electrochemical and photoluminescence properties, J. Phys. Chem. C 115 (44) (2011) 21899–21906.

[38] C. Quinton, V. Alain-Rizzo, C. Dumas-Verdes, G. Clavier, P. Audebert, Original electroactive and fluorescent bichromophores based on non-conjugated tetrazine and triphenylamine derivatives: towards more efficient fluorescent switches, RSC Adv. 5 (61) (2015) 49728–49738.

[39] S. Pluczyk, P. Zassowski, C. Quinton, P. Audebert, V. Alain-Rizzo, M. Lapkowski, Unusual electrochemical properties of the electropolymerized thin layer based on a s-tetrazine-triphenylamine monomer, J. Phys. Chem. C 120 (8) (2016) 4382–4391.

[40] C. Quinton, V. Alain-Rizzo, C. Dumas-Verdes, G. Clavier, F. Miomandre, P. Audebert, Design of new tetrazine–triphenylamine bichromophores – fluorescent switching by chemical oxidation, Eur. J. Org. Chem. 2012 (7) (2012) 1394–1403.

[41] C. Quinton, V. Alain-Rizzo, C. Dumas-Verdes, F. Miomandre, P. Audebert, Tetrazine–triphenylamine dyads: influence of the nature of the linker on their properties, Electrochim. Acta 110 (Supplement C) (2013) 693–701.

[42] Z. Qing, P. Audebert, G. Clavier, R. Meallet-Renault, F. Miomandre, J. Tang, Bright fluorescence through activation of a low absorption fluorophore: the case of a unique naphthalimide-tetrazine dyad, New J. Chem. 35 (8) (2011) 1678–1682.

[43] L. Guerret-Legras, P. Audebert, J.-F. Audibert, C. Niebel, T. Jarrosson, F. Serein-Spirau, J.-P. Lère-Porte, New TBT based conducting polymers functionalized with redox-active tetrazines, J. Electroanal. Chem. 840 (2019) 60–66.

[44] C. Quinton, V. Alain-Rizzo, C. Dumas-Verdes, F. Miomandre, G. Clavier, P. Audebert, Redox- and protonation-induced fluorescence switch in a new triphenylamine with six stable active or non-active forms, Chemistry 21 (5) (2015) 2230–2240.

[45] C. Quinton, V. Alain-Rizzo, C. Dumas-Verdes, F. Miomandre, G. Clavier, P. Audebert, Redox-controlled fluorescence modulation (electrofluorochromism) in triphenylamine derivatives, RSC Adv. 4 (65) (2014) 34332–34342.

[46] J.-T. Wu, H.-T. Lin, G.-S. Liou, Synthesis and characterization of novel triarylamine derivatives with dimethylamino substituents for application in optoelectronic devices, ACS Appl. Mater. Interfaces 11 (16) (2019) 14902–14908.

[47] H.-T. Lin, J.-T. Wu, M.-H. Chen, G.-S. Liou, Novel electrochemical devices with high contrast ratios and simultaneous electrochromic and electrofluorochromic response capability behaviours, J. Mater. Chem. C 8 (36) (2020) 12656–12661.

[48] G.A. Corrente, E. Fabiano, M. La Deda, F. Manni, G. Gigli, G. Chidichimo, A.-L. Capodilupo, A. Beneduci, High-performance electrofluorochromic switching devices using a novel arylamine-fluorene redox-active Fluorophore, ACS Appl. Mater. Interfaces 11 (13) (2019) 12202–12208.

[49] Y. Sun, M. Shi, Y. Zhu, I.F. Perepichka, X. Xing, Y. Liu, C. Yan, H. Meng, Multicolored cathodically coloring electrochromism and electrofluorochromism in regioisomeric star-shaped carbazole dibenzofurans, ACS Appl. Mater. Interfaces 12 (21) (2020) 24156–24164.

[50] L. Guerret, J.F. Audibert, A. Debarre, M. Lepeltier, P. Haghi-Ashtiani, G.V. Dubacheva, F. Miomandre, Investigation of photophysical and electrofluorochromic properties of gold nanoparticles functionalized by a luminescent electroactive complex, J. Phys. Chem. C 120 (4) (2016) 2411–2418.

[51] H.-J. Nie, W.-W. Yang, J.-Y. Shao, Y.-W. Zhong, Ruthenium-tris(bipyridine) complexes with multiple redox-active amine substituents: tuning of spin density distribution and deep-red to NIR electrochromism and electrofluorochromism, Dalton Trans. 45 (25) (2016) 10136–10140.

[52] M. Yano, K. Matsuhira, M. Tatsumi, Y. Kashiwagi, M. Nakamoto, M. Oyama, K. Ohkubo, S. Fukuzumi, H. Misaki, H. Tsukube, "ON-OFF" switching of europium complex luminescence coupled with a ligand redox process, Chem. Commun. 48 (34) (2012) 4082–4084.

[53] D.R. van Staveren, N. Metzler-Nolte, Bioorganometallic chemistry of ferrocene, Chem. Rev. 104 (12) (2004) 5931–5986.

[54] R. Zhang, Z. Wang, Y. Wu, H. Fu, J. Yao, A novel redox-fluorescence switch based on a triad containing ferrocene and perylene diimide units, Org. Lett. 10 (14) (2008) 3065–3068.

[55] O. Galangau, I. Fabre-Francke, S. Munteanu, C. Dumas-Verdes, G. Clavier, R. Méallet-Renault, R.B. Pansu, F. Hartl, F. Miomandre, Electrochromic and electrofluorochromic properties of a new boron dipyrromethene–ferrocene conjugate, Electrochim. Acta 87 (Supplement C) (2013) 809–815.

[56] E. Maligaspe, T.J. Pundsack, L.M. Albert, Y.V. Zatsikha, P.V. Solntsev, D.A. Blank, V.N. Nemykin, Synthesis and charge-transfer dynamics in a ferrocene-containing organoboryl aza-BODIPY donor–acceptor triad with boron as the hub, Inorg. Chem. 54 (8) (2015) 4167–4174.

[57] S. Seo, S. Pascal, C. Park, K. Shin, X. Yang, O. Maury, B.D. Sarwade, C. Andraud, E. Kim, NIR electrochemical fluorescence switching from polymethine dyes, Chem. Sci. 5 (4) (2014) 1538–1544.

[58] P. Hindenberg, J. Zimmermann, G. Hernandez-Sosa, C. Romero-Nieto, Lighting with organophosphorus materials: solution-processed blue/cyan light-emitting devices based on phosphaphenalenes, Dalton Trans. 48 (22) (2019) 7503–7508.

[59] Y. Shi, Q. Chen, J. Zheng, C. Xu, Electrochromism of substituted phthalate derivatives and outstanding performance of corresponding multicolor electrochromic devices, Electrochim. Acta 341 (2020), 136023.

[60] L. Akcelrud, Electroluminescent polymers, Prog. Polym. Sci. 28 (6) (2003) 875–962.

[61] F. Montilla, R. Mallavia, In situ electrochemical fluorescence studies of PPV, J. Phys. Chem. B 110 (51) (2006) 25791–25796.

[62] H. Lim, S. Seo, C. Park, H. Shin, X. Yang, K. Kanazawa, E. Kim, Multi-color fluorescence switching with electrofluorochromic polymers, Opt. Mater. Express 6 (6) (2016) 1808–1816.

[63] J. Liu, M. Li, J. Wu, Y. Shi, J. Zheng, C. Xu, Electrochromic polymer achieving synchronous electrofluorochromic switching for optoelectronic application, Org. Electron. 51 (2017) 295–303.

[64] G. Ding, T. Lin, R. Zhou, Y. Dong, J. Xu, X. Lu, Electrofluorochromic detection of cyanide anions using a nanoporous polymer electrode and the detection mechanism, Chem. Eur. J. 20 (41) (2014) 13226–13233.

[65] C. Yang, W. Cai, X. Zhang, L. Gao, Q. Lu, Y. Chen, Z. Zhang, P. Zhao, H. Niu, W. Wang, Multifunctional conjugated oligomers containing novel triarylamine and fluorene units with electrochromic, electrofluorochromic, photoelectron conversion, explosive detection and memory properties, Dyes Pigments 160 (2019) 99–108.

[66] M. Pietsch, T. Rödlmeier, S. Schlisske, J. Zimmermann, C. Romero-Nieto, G. Hernandez-Sosa, Inkjet-printed polymer-based electrochromic and electrofluorochromic dual-mode displays, J. Mater. Chem. C 7 (23) (2019) 7121–7127.

[67] Y.-J. Cheng, S.-H. Yang, C.-S. Hsu, Synthesis of conjugated polymers for organic solar cell applications, Chem. Rev. 109 (11) (2009) 5868–5923.

[68] C.-P. Kuo, Y.-S. Lin, M.-k. Leung, Electrochemical fluorescence switching properties of conjugated polymers composed of triphenylamine, fluorene, and cyclic urea moieties, J. Polym. Sci. A Polym. Chem. 50 (24) (2012) 5068–5078.

[69] N. Sun, F. Feng, D. Wang, Z. Zhou, Y. Guan, G. Dang, H. Zhou, C. Chen, X. Zhao, Novel polyamides with fluorene-based triphenylamine: electrofluorescence and electrochromic properties, RSC Adv. 5 (107) (2015) 88181–88190.

[70] N. Sun, S. Meng, Z. Zhou, J. Yao, Y. Du, D. Wang, X. Zhao, H. Zhou, C. Chen, High-contrast electrochromic and electrofluorescent dual-switching materials based on 2-diphenylamine-(9,9-diphenylfluorene)-functionalized semi-aromatic polymers, RSC Adv. 6 (70) (2016) 66288–66296.

[71] N. Sun, Z. Zhou, D. Chao, X. Chu, Y. Du, X. Zhao, D. Wang, C. Chen, Novel aromatic polyamides containing 2-diphenylamino-(9,9-dimethylamine) units as multicolored electrochromic and high-contrast electrofluorescent materials, J. Polym. Sci. A Polym. Chem. 55 (2) (2017) 213–222.

[72] N. Sun, Z. Zhou, S. Meng, D. Chao, X. Chu, X. Zhao, D. Wang, H. Zhou, C. Chen, Aggregation-enhanced emission (AEE)-active polyamides with methylsulfonyltriphenylamine units for electrofluorochromic applications, Dyes Pigments 141 (2017) 356–362.

[73] K. Su, N. Sun, Z. Yan, S. Jin, X. Li, D. Wang, H. Zhou, J. Yao, C. Chen, Dual-switching electrochromism and electrofluorochromism derived from diphenylamine-based polyamides with spirobifluorene/pyrene as bridged fluorescence units, ACS Appl. Mater. Interfaces 12 (19) (2020) 22099–22107.

[74] K. Su, N. Sun, X. Tian, S. Guo, Z. Yan, D. Wang, H. Zhou, X. Zhao, C. Chen, High-performance blue fluorescent/electroactive polyamide bearing p-phenylenediamine and asymmetrical SBF/TPA-based units for electrochromic and electrofluorochromic multifunctional applications, J. Mater. Chem. C 7 (16) (2019) 4644–4652.

[75] S. Meng, N. Sun, K. Su, F. Feng, S. Wang, D. Wang, X. Zhao, H. Zhou, C. Chen, Optically transparent polyamides bearing phenoxyl, diphenylamine and fluorene units with high-contrast of electrochromic and electrofluorescent behaviors, Polymer 116 (2017) 89–98.

[76] N. Sun, S. Meng, D. Chao, Z. Zhou, Y. Du, D. Wang, X. Zhao, H. Zhou, C. Chen, Highly stable electrochromic and electrofluorescent dual-switching polyamide containing bis(diphenylamino)-fluorene moieties, Polym. Chem. 7 (39) (2016) 6055–6063.

[77] K. Su, N. Sun, X. Tian, S. Guo, Z. Yan, D. Wang, H. Zhou, X. Zhao, C. Chen, Highly soluble polyimide bearing bulky pendant diphenylamine-pyrene for fast-response electrochromic and electrofluorochromic applications, Dyes Pigments 171 (2019), 107668.

[78] N. Sun, S. Meng, F. Feng, Z. Zhou, T. Han, D. Wang, X. Zhao, C. Chen, Electrochromic and electrofluorochromic polyimides with fluorene-based triphenylamine, High Perform. Polym. 29 (10) (2017) 1130–1138.

[79] Y. Yan, X. Jia, M. Feng, C. Wang, D. Chao, Synthesis and electrochemical characterization of polyamic acid containing oligoaniline and triphenylamine, J. Polym. Sci. A Polym. Chem. 55 (10) (2017) 1669–1673.

[80] Y. Yan, N. Sun, X. Jia, X. Liu, C. Wang, D. Chao, Electrochromic and electrofluorochromic behavior of novel polyurea bearing oligoaniline and triphenylamine units, Polymer 134 (2018) 1–7.

[81] Y. Han, Y. Lin, D. Sun, Z. Xing, Z. Jiang, Z. Chen, Poly(aryl amino ketone)-based materials with excellent electrochromic and electrofluorochromic behaviors, Dyes Pigments 163 (2019) 40–47.

[82] R. Zheng, T. Huang, Z. Zhang, Z. Sun, H. Niu, C. Wang, W. Wang, Novel polyimides containing flexible carbazole blocks with electrochromic and electrofluorescencechromic properties, RSC Adv. 10 (12) (2020) 6992–7003.

[83] S. Cai, H. Niu, Synthesis and characterization of triarylamine-based polyimides for electrochromic and optoelectronic conversion behaviors, J. Appl. Polym. Sci. 137 (24) (2020) 48808.

[84] S. Cai, S. Wang, D. Wei, H. Niu, W. Wang, X. Bai, Multifunctional polyamides containing pyrrole unit with different triarylamine units owning electrochromic, electrofluorochromic and photoelectron conversion properties, J. Electroanal. Chem. 812 (2018) 132–142.

[85] J. Sun, Z. Liang, Swift electrofluorochromism of donor–acceptor conjugated polytriphenylamines, ACS Appl. Mater. Interfaces 8 (28) (2016) 18301–18308.

[86] D.C. Santra, S. Nad, S. Malik, Electrochemical polymerization of triphenylamine end-capped dendron: Electrochromic and electrofluorochromic switching behaviors, J. Electroanal. Chem. 823 (2018) 203–212.

[87] Y. Li, Y. Zhou, X. Jia, D. Chao, Synthesis and characterization of a dual electrochromic and electrofluorochromic crosslinked polymer, Eur. Polym. J. 106 (2018) 169–174.

[88] Y. Li, Y. Zhou, X. Jia, D. Chao, Dual functional electrochromic and electrofluorochromic network polymer film prepared from two hydrolysable crosslinked siloxane monomers, J. Electroanal. Chem. 823 (2018) 672–677.

[89] W. Zhang, H. Niu, C. Yang, Y. Wang, Q. Lu, X. Zhang, H. Niu, P. Zhao, W. Wang, Electrochromic and electrofluorochromic bifunctional materials for dual-mode devices based on ladder-like polysilsesquioxanes containing triarylamine, Dyes Pigments 175 (2020), 108160.

[90] H.-T. Lin, C.-L. Huang, G.-S. Liou, Design, synthesis, and electrofluorochromism of new triphenylamine derivatives with AIE-active pendent groups, ACS Appl. Mater. Interfaces 11 (12) (2019) 11684–11690.

[91] S.-Y. Chen, M.-H. Pai, G.-S. Liou, Effects of alkyl chain length and anion on the optical and electrochemical properties of AIE-active α-cyanostilbene-containing triphenylamine derivatives, J. Mater. Chem. C 8 (22) (2020) 7454–7462.

[92] S.-Y. Chen, Y.-W. Chiu, G.-S. Liou, Substituent effects of AIE-active α-cyanostilbene-containing triphenylamine derivatives on electrofluorochromic behavior, Nanoscale 11 (17) (2019) 8597–8603.

[93] S. Abraham, S. Mangalath, D. Sasikumar, J. Joseph, Transmissive-to-black electrochromic devices based on cross-linkable tetraphenylethene-diphenylamine derivatives, Chem. Mater. 29 (23) (2017) 9877–9881.

[94] X. Wang, W. Li, W. Li, C. Gu, H. Zheng, Y. Wang, Y.-M. Zhang, M. Li, S. Xiao-An Zhang, An RGB color-tunable turn-on electrofluorochromic device and its potential for information encryption, Chem. Commun. 53 (81) (2017) 11209–11212.

[95] A.A. Suleymanov, A. Ruggi, O.M. Planes, A.-S. Chauvin, R. Scopelliti, F. Fadaei Tirani, A. Sienkiewicz, A. Fabrizio, C. Corminboeuf, K. Severin, Highly substituted Δ3-1,2,3-Triazolines: solid-state emitters with electrofluorochromic behavior, Chem. Eur. J. 25 (27) (2019) 6718–6721.

[96] J. Wu, Y. Han, J. Liu, Y. Shi, J. Zheng, C. Xu, Electrofluorochromic and electrochromic bifunctional polymers with dual-state emission via introducing multiple C—H π bonds, Org. Electron. 62 (2018) 481–490.

[97] Z. Wang, N. Wang, H. Gao, Y. Quan, H. Ju, Y. Cheng, Amplified electrochemiluminescence signals promoted by the AIE-active moiety of D–A type polymer dots for biosensing, Analyst 145 (1) (2020) 233–239.

[98] F. Liu, Z. Cong, G. Yu, H. Niu, Y. Hou, C. Wang, S. Wang, Novel D-A-D conjugated polymers based on tetraphenylethylene monomer for electrochromism, Opt. Mater. 100 (2020), 109658.

[99] S.-W. Cheng, T. Han, T.-Y. Huang, B.-Z. Tang, G.-S. Liou, High-performance electrofluorochromic devices based on aromatic polyamides with AIE-active tetraphenylethene and electro-active triphenylamine moieties, Polym. Chem. 9 (33) (2018) 4364–4373.
[100] N. Sun, K. Su, Z. Zhou, Y. Yu, X. Tian, D. Wang, X. Zhao, H. Zhou, C. Chen, AIE-active polyamide containing diphenylamine-TPE moiety with superior electrofluorochromic performance, ACS Appl. Mater. Interfaces 10 (18) (2018) 16105–16112.
[101] N. Sun, X. Tian, L. Hong, K. Su, Z. Zhou, S. Jin, D. Wang, X. Zhao, H. Zhou, C. Chen, Highly stable and fast blue color/fluorescence dual-switching polymer realized through the introduction of ether linkage between tetraphenylethylene and triphenylamine units, Electrochim. Acta 284 (2018) 655–661.
[102] N. Sun, K. Su, Z. Zhou, X. Tian, D. Wang, N. Vilbrandt, A. Fery, F. Lissel, X. Zhao, C. Chen, Synergistic effect between electroactive tetraphenyl-p-phenylenediamine and AIE-active tetraphenylethylene for highly integrated electrochromic/electrofluorochromic performances, J. Mater. Chem. C 7 (30) (2019) 9308–9315.
[103] T. Yu, Y. Han, H. Yao, Z. Chen, S. Guan, Polymeric optoelectronic materials with low-voltage colorless-to-black electrochromic and AIE-activity electrofluorochromic dual-switching properties, Dyes Pigments 181 (2020), 108499.
[104] Q. Lu, C. Yang, X. Qiao, X. Zhang, W. Cai, Y. Chen, Y. Wang, W. Zhang, X. Lin, H. Niu, W. Wang, Multifunctional AIE-active polymers containing TPA-TPE moiety for electrochromic, electrofluorochromic and photodetector, Dyes Pigments 166 (2019) 340–349.
[105] M. Yu, R. Huang, J. Guo, Z. Zhao, B.Z. Tang, Promising applications of aggregation-induced emission luminogens in organic optoelectronic devices, PhotoniX 1 (1) (2020) 11.
[106] H.-J. Yen, G.-S. Liou, Design and preparation of triphenylamine-based polymeric materials towards emergent optoelectronic applications, Prog. Polym. Sci. 89 (2019) 250–287.
[107] Y.C. Zhang, E.B. Berda, X.T. Jia, Z. Lu, M.H. Zhu, D.M. Chao, Electrochromic/electrofluorochromic supercapacitor based on a network polysiloxane bearing oligoaniline and cyanophenethylene groups, ACS Appl. Polym. Mater. 2 (7) (2020) 3024–3033.

CHAPTER

13

AIE-active materials for photovoltaics

Andrea Pucci

Department of Chemistry and Industrial Chemistry of the University of Pisa, Pisa, Italy

1 Introduction

The sun is the only real energy source we have on the Earth: its irradiation creates atmospheric temperature differences which give birth to the wind, the wind generates waves, and the evaporation of water forms clouds and rain due to sunlight [1]. Moreover, solar energy can be directly converted into electricity, heat, and chemical energy. Solar energy arrives on the Earth as radiation, which can be captured as excited electron-hole pairs in a semiconductor, a dye, or a chromophore, or as heat in a thermal storage medium, usually water. Excited electrons and holes can be immediately converted to electrical power or transferred to biological or chemical molecules for conversion to fuel, as happens during photosynthesis. Covering the 0.16% of the land on the Earth, the 10% of solar-electricity conversion would provide 20 TW of power. Considering that in 2015 the global energy consumption was 17.4 TW, it seems more than reasonable to focus on finding efficient ways to capture, convert, and store this enormous amount of energy coming from the Sun [2]. Solar cells are classified into first-, second-, and third-generation photovoltaic cells (PV) [3]. The first-generation solar cells are the conventional ones, made of silicon, which has a bandgap of 1.11 eV, i.e., very close to 1.4 eV that corresponds to the maximum efficiency attainable. The silicon used can be monocrystalline, polycrystalline, or amorphous. Monocrystalline Si solar PV cells are based on single-crystal Si with high production costs. Polycrystalline silicon PV cells made out of recrystallized silicon are more economical but are characterized by a higher number of defects in the form of grain boundaries, thus causing a drop of the devices' efficiency compared to that of monocrystalline Si cells. A third cost-effective option is that of amorphous silicon. Nevertheless, because of the high defect density, amorphous silicon solar cells' record efficiency is only 14.0%, which is still far from 26.1% achieved by crystalline silicon solar cells [4].

Second-generation cells are thin-film cells, with thickness ranging from few nanometers to tens of micrometers. In these devices, active semiconductor layers are sandwiched between a transparent conductive oxide layer and the electric back contact. The materials used

Aggregation-Induced Emission (AIE)
https://doi.org/10.1016/B978-0-12-824335-0.00014-3

have a high solar absorption coefficient, requiring a layer thickness of ~2.5 μm, compared to ~170–250 μm for silicon solar cells. In order to have mechanical stability, the thin films are deposited onto a substrate made of glass, metal, or polymer foil. Examples of these devices are GaAs, CIGS (copper indium gallium diselenide), and CdTe solar cells [5]. Although these features are very interesting for solar cell research because of their easy manufacturing methods and the possibility of integration in an urban environment, thin-film cells generally have a lower efficiency than c-Si solar cells, with GaAs being an exception [6]. The reduced spectral response of such PV cells in the sub-500 nm region often requires luminescent down-shifting layers made of highly efficient fluorophores to effectively match the spectral response of the solar cells. Fluorophores with large Stokes shift and reduced tendency to π-stack are, however, required to avoid dissipative phenomena that adversely affect radiative de-excitation [7].

Finally, third-generation solar cells comprehend all those novel technologies to achieve high efficiencies at low cost, and hence are compatible with large-scale implementation [8]. For example, organic photovoltaics (OPV) use conductive organic polymers or small organic molecules as absorber layers with maximum efficiency of 18.2%. The organic molecules used are usually cheap, flexible, and have high absorption coefficient. Dye-sensitized solar cells (DSSC) and perovskite solar cells (PSC) also belong to this category. DSSC are characterized by high throughput, low cost, easy fabrication, semitransparency, and multicolor that suggest their potential integration in the urban architecture [9]. DSSC are photoelectrochemical systems composed of a photosensitized anode and an electrolyte solution and were firstly proposed by O'Regan and Grätzel in 1991 [10]. Since then, intense work has been addressed to increase the cell efficiency that has nowadays reached the maximum value of 13.0% [4].

A perovskite solar cell (PSC) is a PV cell based on a light-harvesting active layer composed by a perovskite-structured material, consistently a hybrid organic-inorganic Pb or Sn halide-based material. Perovskite compounds like methylammonium Pb halides and all-inorganic Cs-Pb halide, are simple to produce and cost-effective. Perovskite solar cell technology has made outstanding progress since the publication of the first paper on the topic in 2009 [11]. Since then, the power conversion efficiencies (PCEs) of these devices evolved from 3.8% to the current 25.2% record, approaching the top values achieved with single-crystalline silicon solar cells [12]. The growing interest in perovskite materials comes from the significant number of advantageous properties they have, like tunable bandgap, high absorption coefficient, high power conversion energy, low exciton binding energy, easy crystallization at low temperature, and simple fabrication.

Lastly, it should be reminded that one of the most significant limitations of solar cells is that they can only absorb a section of the solar spectrum corresponding to the bandgap of the absorber material used. To overcome this issue, multijunction solar cells can be used, where several cell materials with different bandgaps are combined to maximize the amount of sunlight that they can harvest. On this account, two or more cells are stacked onto each other: the top cell has the highest bandgap and will absorb and convert the short-wavelength light, while the bottom cell has the lowest bandgap and absorbs the long-wavelength light. Geometrical or luminescent solar concentrators (LSCs) have been also proposed to maximize solar harvesting by collecting photons over a large area and conveying them to a smaller solar cell [13] (Fig. 1).

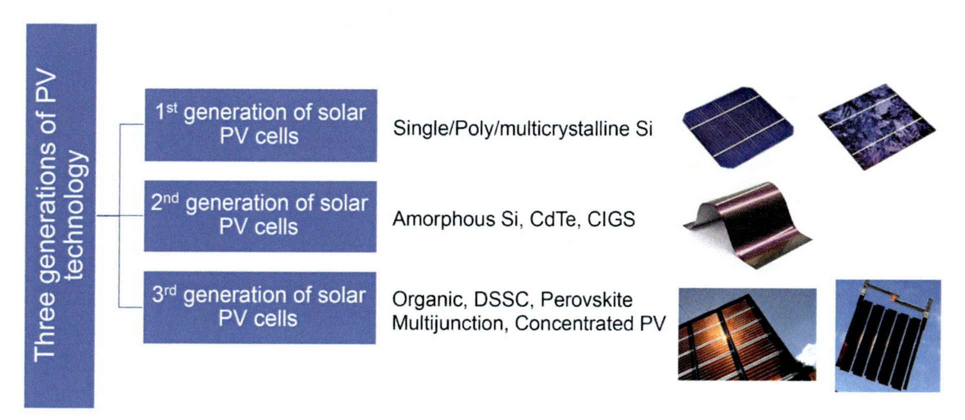

FIG. 1 Scheme of the three generations of PV solar cells.

In most of the recently developed third generation of solar PV cells, the aggregation of the chromophore moieties contained in the device strongly affects the overall efficiencies of the photovoltaic process. For example, in DSSC solar cells, dye-dye interactions on the metal oxide surface possibly lead to a hypsochromic shift of the absorption that in turn invariably decreases the light-harvesting capability of the system, thus reducing the photocurrent generated [14]. Such issues are often accompanied by phase separation of the main components that causes efficiency losses as those due to donor-acceptor interface recombination in organic PV cells. Also, fluorophore aggregation leads often to quenching of the emission (i.e., aggregation-caused quenching effect, ACQ), which is a severe drawback in LSC devices [13].

Therefore, it is clear how the aggregation of the organic chromophoric components in modern PV cells plays a fundamental role in the photovoltaic performances. The development of molecules whose structure could prevent the detrimental generation of supramolecular aggregates while boosting the required optical features is highly desirable. In 2001, Ben Zhong Tang proposed for the first time a class of molecules with the characteristics mentioned above and called them aggregation-induced emission (AIE) fluorophores [15]. AIE luminogens (AIEgen) generally consist of phenyl rings in a twisted propeller-shaped conformation that renders intermolecular π-π interactions difficult in the aggregate state. Prototypical AIEgens were three-dimensional propeller-shaped molecules like hexaphenylsilole (HPS) and tetraphenylethene (TPE) [16], whose central silole or olefin stators are surrounded by phenyl rotors (Fig. 2).

The phenyl rings can dynamically rotate against each other in dilute solution, thus rendering AIEgens faintly emissive. Conversely, upon aggregate formation, the restriction of intramolecular motions (RIM) suppresses the nonradiative pathway and enabling the emission of bright fluorescence. Molecular engineering has recently allowed the production of AIEgens with rotor structures, modulable absorption and emission features, and large Stokes shift with striking impacts on energy, optoelectronics, life science, and the environment. Most of the published documents are in the area of chemistry (33%), materials science (22%), and chemical engineering (15%) [17,18].

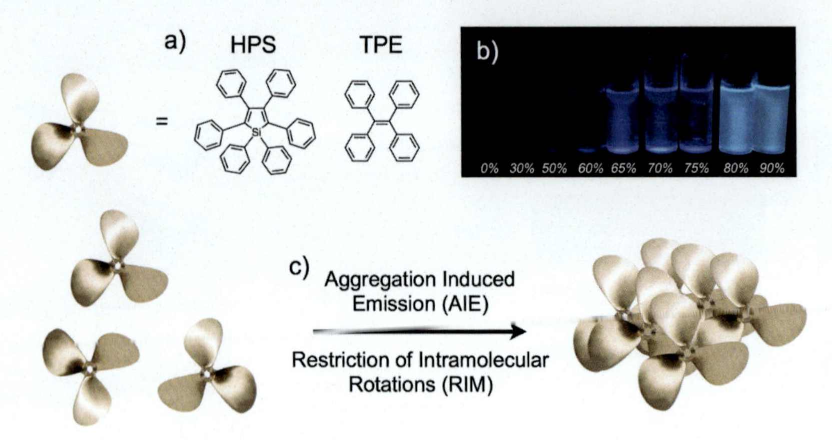

FIG. 2 (A) Chemical structures of hexaphenylsilole (HPS) and tetraphenylethene (TPE), (B) photo taken by exciting TPE dilute solutions in dioxane/water mixtures with different water contents, and (C) AIE working mechanism. *Reprinted with permission from reference G. Iasilli, A. Battisti, F. Tantussi, F. Fuso, M. Allegrini, G. Ruggeri, et al., Aggregation-induced emission of tetraphenylethylene in styrene-based polymers, Macromol. Chem. Phys. 215 (6) (2014) 499–506 with permission from John Wiley and Sons.*

In this contribution, the use of AIEgens is described as excellent dopants for the second and the third generations of PV solar cells. AIEgens are introduced as luminescent down-shifting components for CdTe PV cells and then as rotor structures in DSSC and OPV solar cells. Illustrative examples of their use in perovskite PV cells and as luminescent dopants in LSC will be eventually described. A final perspective will be provided on the role of AIEgen in the modern photovoltaic technologies.

2 AIEgens in CdTe PV cells

CdTe PV cells consist in a thin-film technology since the active layers are just a few microns thick due to their prominent absorption coefficient and solar harvesting features, which are to the benefit of material supply and fabrication costs. CdTe solar cells are structured in a multilayer configuration as reported in Fig. 3 [19].

A transparent conducting oxide (TCO) layer based on SnO_2:F (FTO) or In_2O_3:Sn (ITO) transports current efficiently and transmits visible light to the intermediate layers such as CdS, which is the *n*-part of the junction. CdTe is the absorbing core of the cell (bandgap of 1.45 eV) and the *p*-part of the junction that is in contact with the metal back electrical contact. The CdTe film acts as the primary photoconversion layer and absorbs most visible light within the first micron of the material. The sandwich layers composed by CdTe, CdS, and TCO generate an electric field that converts the light absorbed by CdTe into electrical current with a maximum efficiency of 22.1% [20]. Nevertheless, the absorption of the *n*-type layer of CdS cadmium sulfide (bandgap of 2.42 eV) causes a poor spectral response at wavelength less than 500 nm that adversely affects the overall PV cell efficiency. Among the different strategies proposed to limit the detrimental effect of CdS, the use of a luminescent down-shifting (LDS) layer seems the most promising. An LDS layer generally consists of a sheet of polymer

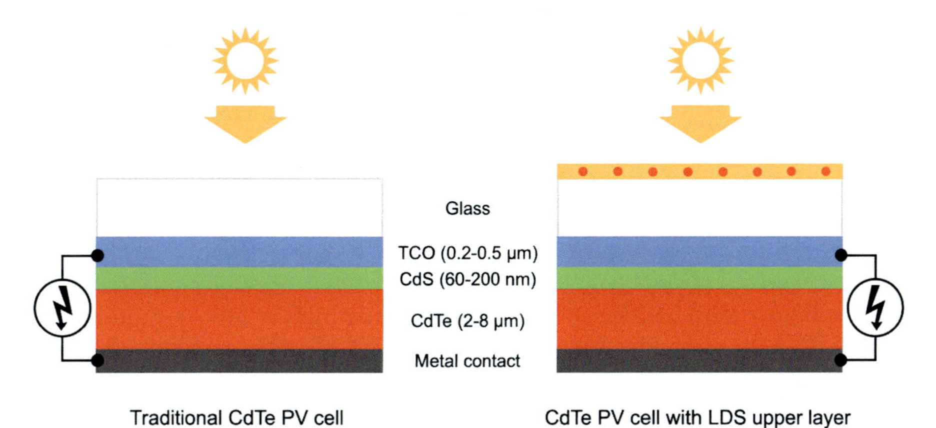

FIG. 3 Schematic structure of a typical CdTe PV cell (left) and that equipped with the LDS layer.

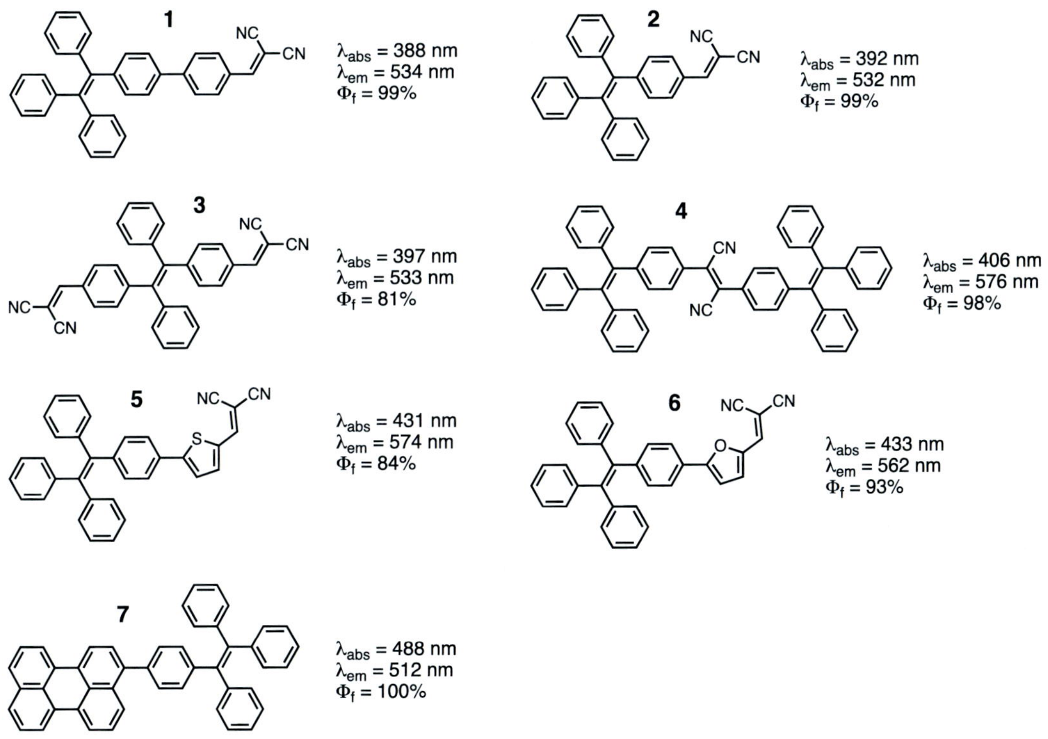

1 NC CN λ_{abs} = 388 nm λ_{em} = 534 nm Φ_f = 99%

2 NC CN λ_{abs} = 392 nm λ_{em} = 532 nm Φ_f = 99%

3 NC CN NC CN λ_{abs} = 397 nm λ_{em} = 533 nm Φ_f = 81%

4 CN NC λ_{abs} = 406 nm λ_{em} = 576 nm Φ_f = 98%

5 NC CN S λ_{abs} = 431 nm λ_{em} = 574 nm Φ_f = 84%

6 NC CN O λ_{abs} = 433 nm λ_{em} = 562 nm Φ_f = 93%

7 λ_{abs} = 488 nm λ_{em} = 512 nm Φ_f = 100%

FIG. 4 Chemical structures and optical properties of **1–7** AIEgens for LDS layers in CdTe PV cells.

doped with a luminescent molecule able to absorb photons in the short-wavelength window of light and emit light at a longer wavelength to match the spectral response of the CdTe layer. Currently, perylene-based fluorophores like Lumogen F Yellow 083 (F083) have been employed in LDS layers but the limited Stokes shift (64 nm) and the inherent characteristics to generate π-stacking aggregates strongly affect the down-shifting efficiency. The attention has been therefore directed to fluorescent dopants able to show a shorter wavelength absorption and a more red-shifted emission to maximize the matching with the spectral response of CdTe. This strategy has been also flanked by the selection of AIEgens to prevent the typical ACQ behavior of perylene-based fluorophores. On this account, Dong et al. proposed the use of AIEgens made of the TPE nucleus bearing electron-withdrawing malonitrile groups as dopants in LDS layers based on PMMA (Fig. 4) [21,22].

The characterization in solution confirmed the AIE features when the water content (nonsolvent) reached 60 vol%. The molecular architecture of **1–6** enables fluorescence peaking at wavelengths higher than 530 nm and with significantly high relative quantum yields (QY) with respect to F083, thanks to the large Stokes shift of 129–170 nm. Once embedded in PMMA LDS layers with thickness of 85 μm, the AIEgens improved the performances of the CdTe PV cells with short-circuit current density enhancements (ΔJ_{sc}) comprised between 6% and 10%, i.e., higher than that of about 4% provided by F083 (Fig. 5).

The authors addressed such efficiency improvements to a combination of beneficial effects including the larger absorption of short-wavelength photons and the wider Stokes shift that enabled a better spectral matching of the fluorescence with the spectral response of the CdTe cell also at high AIEgen loading. Such features were also flanked by the high Φ_F originated by the AIE characteristics of the designed molecules that prevent fluorescence quenching due to aggregation. More recently, the same authors addressed the environmental issue caused by the malononitrile moieties attached to the conjugated TPE core of the AIEgens [23]. Such groups are considered toxic and could exert their activity during synthesis and might be released into the environment caused by AIEgen photodegradation [24]. To solve this issue, a perylene-based AIEgen was proposed (Figs. 4 and 7) and characterized by intramolecular energy transfer (IET), restricted internal rotation (RIR) enhanced fluorescence lifetime, and Φ_F when embedded in PMMA thin films (50 μm of thickness). Notably, the comparable optical properties of **7** with respect those of Y083 in terms of Stokes shift and QY, provided to the CdTe PV cell similar J_{sc} enhancement, i.e., 3.30% against 3.28%, respectively.

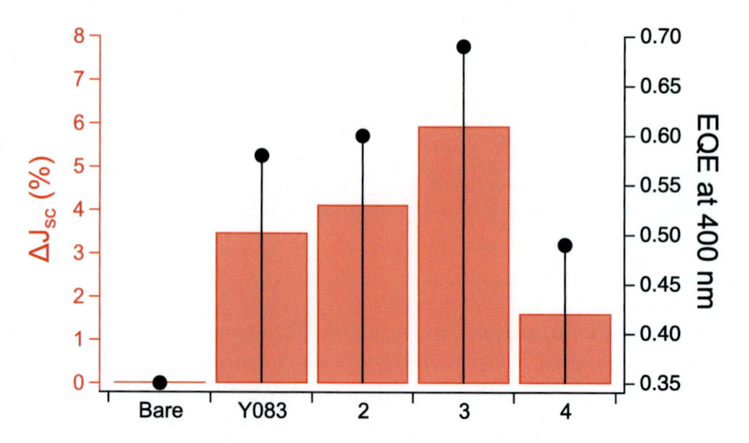

FIG. 5 Short-circuit current density enhancements (*red bars*, ΔJ_{sc}) and external quantum efficiency (*black circles*, EQE) of a CdTe PV cell with an LDS film containing 2 wt% of **2–4** and Y083 fluorophores.

3 AIEgens in dye-sensitized solar cells (DSSC)

In a DSSC, a dye sensitizer is used as an absorber bound to a wide-bandgap n-type semiconductor, usually TiO_2, deposited onto a fluorine-doped tin oxide-coated glass substrate (transparent conducting oxide, TCO). Upon excitation of the dye by light exposure, an electron from the highest occupied molecular orbital (HOMO, VB) reaches the lowest unoccupied molecular orbital (LUMO, CB). The excited-state dye then injects an electron in the metal oxide semiconductor's conduction band, leaving a hole behind in the oxidized dye. The electron from the semiconductor reaches the TCO by diffusion and can then travel through an outer circuit to reach the counter electrode (CE). The latter is in contact with a redox electrolyte solution, usually iodide/triiodide (I^-/I_3^-), which accepts the electron from the counter electrode and consequentially closes the circuit by reducing the dye back to its initial condition (Fig. 6).

The DSSC was conceptualized starting from sensitizers based on metal complexes anchored to TiO_2 particles due to their high molar absorptivity and possibility of redshifts of visible transitions and electron transfer to the semiconductor. Recently, numerous organic dyes for efficient DSSCs have been also reported and promoted by cost reduction and easier tunability of their optical features [25]. Nevertheless, in both cases, the aggregation of the chromophoric units of the sensitizers is reported as an important drawback since it adversely affect the electron injection capacity and the binding ability of these chromophores over the TiO_2 surface, thus lowering J_{sc} [26]. In connection with these findings, the use of AIEgens as sensitizers would be therefore beneficial for the photovoltaic efficiency of DSSC PV cells. The first report on the use of AIEgens was proposed by C. Chen et al. in 2012 [27], even if examples of dyes with triphenylethylene group as rotor architecture were reported earlier, as recently reviewed by K.-H. Ong and B. Liu [7]. Notably, Chen et al. reported the synthesis of **8–11** twisted structures of AIEgens for DSSC PV cells with a D-D-π-A configuration that was

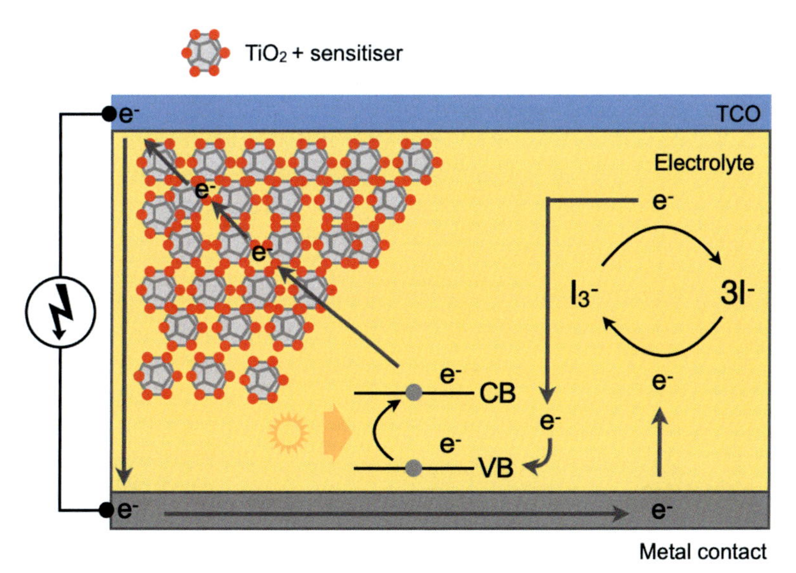

FIG. 6 Working principle of a typical DSSC PV cell.

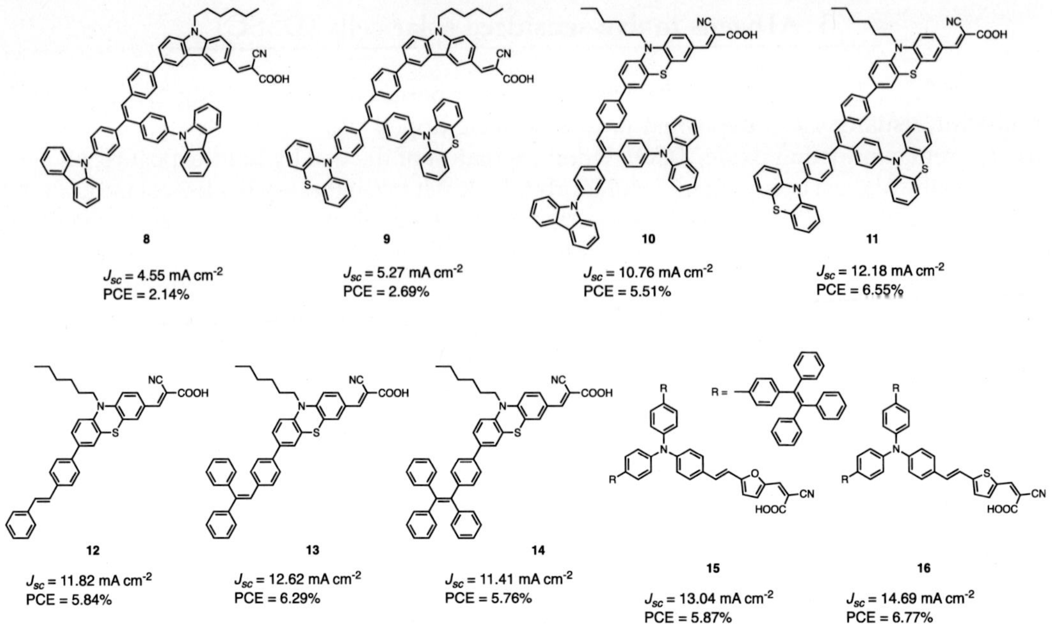

8
J_{sc} = 4.55 mA cm^{-2}
PCE = 2.14%

9
J_{sc} = 5.27 mA cm^{-2}
PCE = 2.69%

10
J_{sc} = 10.76 mA cm^{-2}
PCE = 5.51%

11
J_{sc} = 12.18 mA cm^{-2}
PCE = 6.55%

12
J_{sc} = 11.82 mA cm^{-2}
PCE = 5.84%

13
J_{sc} = 12.62 mA cm^{-2}
PCE = 6.29%

14
J_{sc} = 11.41 mA cm^{-2}
PCE = 5.76%

15
J_{sc} = 13.04 mA cm^{-2}
PCE = 5.87%

16
J_{sc} = 14.69 mA cm^{-2}
PCE = 6.77%

FIG. 7 Chemical structures of **8–16** AIEgens and photovoltaic performances of the derived DSSC PV cells.

designed to extend the absorption region and to modulate the energy of the VB and CB frontiers (Fig. 7). The presence of the electron donors triphenylethylene phenothiazine or the triphenylethylene carbazole moieties was reported to enhance the hole transport features of the AIEgens. It was observed that the triphenylethylene phenothiazine moiety has larger solar harvesting characteristics than the triphenylethylene carbazole one and the more twisted architecture of the former confers the AIEgen superior performances. Notably, the DSSC PV cell based on the **11** AIEgen with the three phenothiazine moieties reached maximum PCE of 6.55% flanked by a J_{sc} of 12.18 mA cm^{-2} that appeared remarkable for PV cells based on metal-free organic dyes. The authors addressed these performances on the twisted nonplanar structure of the designed AIEgens that appeared very beneficial for the high PCE of the DSSC PV cells thanks to the suppression of charge recombination in the nonplanar structures.

The same authors extended the use of the phenothiazine anchoring group in the preparation of dyes based on diphenylethylene (**12**), triphenylethylene (**13**), and tetraphenylethylene (TPE) (**14**) end groups [28]. **13** and **14** were found AIE active, and the derived DSSC PV cells showed PCE comprised between 5.79% and 6.29%. Notably, the authors found that the maximum voltage available from the solar cell (V_{oc}) increased from 0.759 to 0.804 V with increasing numbers of phenyl rings on the end-capping group due to the reduction of the injected electrons' recombinations.

In the same year, Shi et al. proposed two propeller-shaped D-π-A organic sensitizers based on TPE moieties in the donor part of the triphenylamine group (**15** and **16**) [29]. The presence of the two TPE rotors extended the conjugation of the chromophoric core thus red-shifting the absorption peak toward 480–490 nm when the dye is anchored to TiO_2 and providing molar extinction coefficients higher than those of Ru complexes utilized in DSSC. The authors proposed the use of chenodeoxycholic acid (CDCA) as an antiaggregation coadsorbent in combination of the designed AIEgens and maximum PV performances was reached by **16** with J_{sc} of 14.69 mA cm^{-2} and PCE of 6.77%.

PCE higher than 6 with $J_{sc} = 13.20$ mA cm^{-2} was also more recently gathered by Zhang et al. that reported the use of two *N,N*-diethylaniline (DEA) moieties and a twisted 1,1,2,2-tetraphenylethene (TPE) structure with two cyanoacrylic acid-anchoring groups [30]. The AIEgen showed a broad absorption spectrum from 300 to 550 nm thanks to the intramolecular charge-transfer (ICT) between the DEA donor and the cyanoacrylic acid acceptor that generates an efficient charge-separation excited state. The authors again confirmed the need for propeller-like structures to prevent the tendency of the sensitizers to aggregate at the surface of the TiO_2 semiconductor. Nevertheless, they suggest future effective rational design for the preparation of more powerful D-A AIEgens for high-performance DSSC PV cells.

Very recently, the rotor molecular architecture suggested by C. Chen et al. has been elaborated with the introduction of thienyl groups as π-spacers to form a hetero-propeller-like triarylethylene (TAE) or benzene-rich triphenylethene (TPE) cores (Fig. 8) [31].

The authors reported that the nature of the TAE-type π-spacer considerably affected the ICT transition from the donor TPA to the cyanoacrylic acid acceptor that changed from 430 to 490 nm. Notably, the trithienylethene (TTE)-based dye **17** displayed the longest absorption but surprisingly did not provide the highest short-circuit current in iodine as electrolyte (8.75 mA cm^{-2}, PCE = 3.67% against 9.74 mA cm^{-2}, PCE = 5.28% for **19**). Specifically, all the benzene-based TAE π-bridges worked better than the TTE-based π-spacer in providing higher DSSC PV cell performance in the I^-/I_3^- redox couple and the DSSC performances were not totally addressed to the solar harvesting features of the selected propeller-shaped dyes. The authors addressed such behavior to the effective sulfur-iodine interactions in TTE-based dyes like **17** that cause an insufficient hindrance to the electrolyte penetration through the dye layer that

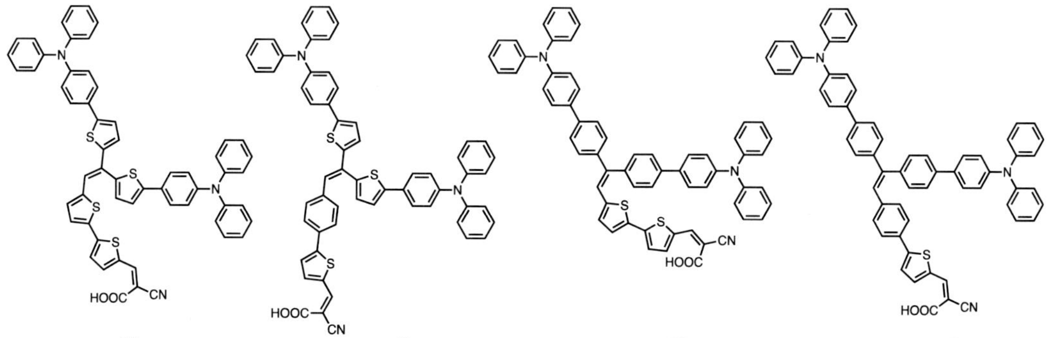

17	18	19	20
J_{sc} = 11.41 mA cm^{-2}	J_{sc} = 9.32 mA cm^{-2}	J_{sc} = 9.74 mA cm^{-2}	J_{sc} = 8.13 mA cm^{-2}
PCE = 5.84%	PCE = 4.38%	PCE = 5.28%	PCE = 4.01%

FIG. 8 Chemical structures of **17–20** AIEgens and photovoltaic performances of the derived DSSC PV cells. The photovoltaic performances of **17** and **19** were reported by using the Co(phen) $_3^{2+/3+}$ electrolyte.

and, in turn, adversely affected the overall PV cell performances. Accordingly, replacing iodine electrolyte with the larger volume $Co(phen)_3^{2+/3+}$ redox shuttle, the less effective sulfur-electrolyte interaction demonstrated to be more suitable for blocking the penetration of the Co complex than iodine electrolyte, thus maximizing cell performances up to 11.41 mA cm^{-2} and PCE = 5.84% for **17**.

4 AIEgens in organic photovoltaics (OPV)

Organic photovoltaic (OPV) solar cells was designed to use earth-abundant materials to obtain electrical current at a lower cost than the first two generation of PV technology [32]. Moreover, soluble organic molecules and polymers as well as the possibility of using flexible substrates enabled roll-to-roll preparation procedures, thus broadening their final use compared to rigid PV modules [33]. Since OPV devices are based on organic-colored absorbers, this technology is particularly suitable for the building-integrated PV market. OPV cells are composed of electron-donor and electron-acceptor materials (Fig. 9). Conjugated molecules and polymers possessing highly delocalized π-electrons composed the electron-donor region, where exciton electron-hole pairs are formed thanks to the UV-Vis light excitation. In bulk heterojunction OPV cells, exciton charge separation occurs at the interface between an electron donor and acceptor blend. This process activates the charge transport through the interfacial layers to the appropriate electrodes where they are eventually collected to produce electrical current.

OPV has achieved maximum efficiencies of 18.2% [4] but long-term reliability caused by material photodegradation is the major limitation.

The first paper concerning the application of AIEgens on OPV technology was presented in 2005 by Tang and coworkers who developed morphologically stable mono- and bis-carbazolylsiloles as donors in organic solar cells with a maximum PCE of 2.19% [34]. Notwithstanding the low PV efficiency, the results gathered by authors were useful to understand the role of the prepared AIEgens on the solar cell performance. Specifically, the

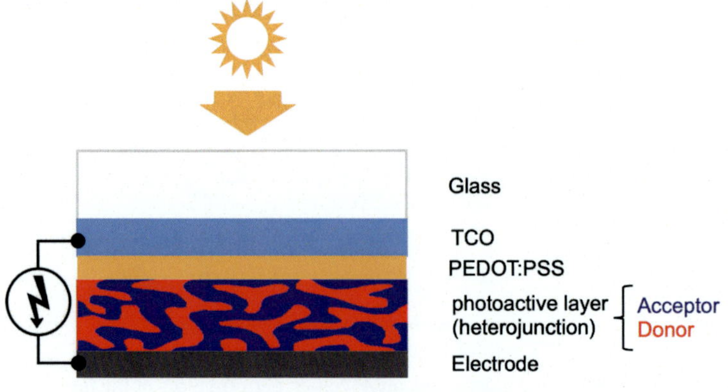

FIG. 9 Schematic structure of a typical bulk heterojunction OPV cell.

bis-carbazolylsiloles performed better than the mono-substituted silole due to the higher photo-responsive character of the former that in turn caused a more effective charge dissociation at the D-A heterojunction.

AIEgens with rotor structure have been then developed to substitute fullerene acceptor derivatives in OPV. [6,6]-Phenyl-C_{61}-butyric acid methyl ester (PCBM) represented the most utilized acceptor in bulk heterojunction solar cells thanks to the high electron mobility and small size that renders charge separation more efficient. Nevertheless, the high tendency to aggregate in large domains and the expensive production limit the large-scale diffusion of OPV technology and encourage the design of nonfullerene small organic acceptors [35]. On this account, AIEgens based on TPE moieties or tetraphenylpyrazine (TPPz) rotor structure endowing perylene bisimide [36–38] or diketopyrrolopyrrole [36,39] chromophores were the most representative, as recently reviewed by Ong and Liu [7]. In detail, the use of **21** (Fig. 10) composed by the TPE core functionalized by four diketopyrrolopyrrole units in ITO/PEDOT:PSS (38 nm)/poly(3-hexylthiophene):35/Ca (20 nm)/Al (100 nm) OPV cell provided a maximum J_{sc} of 5.17 mA cm^{-2} and PCE of 3.86% against the 2.85% of efficiency reached by the same cell structure using PCBM as acceptor [36]. This result was attributed to the excellent solubility of the AIEgen in the organic solvent utilized for the cell preparation and promoted by the rotor structure that in turn favored a more intimate blend with the donor poly(3-hexylthiophene).

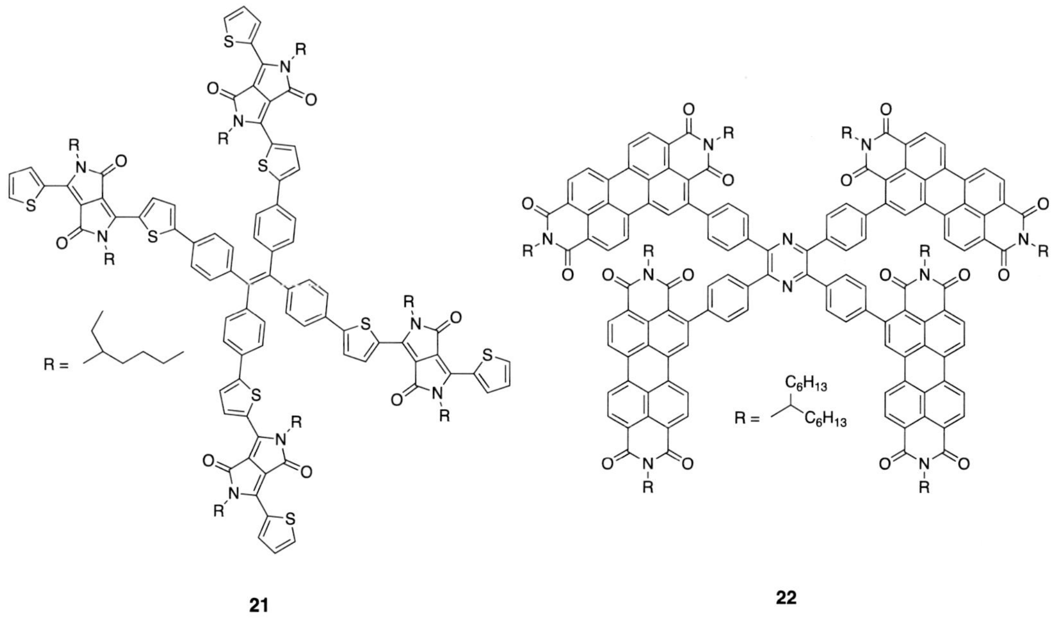

21

J_{sc} = 5.17 mA cm^{-2}
PCE = 3.86%

22

J_{sc} = 12.5 mA cm^{-2}
PCE = 7.1%

FIG. 10 Chemical structures of **21–22** AIEgens and photovoltaic performances of the derived OPV cells.

AIEgen **22** based on the tetraphenylpyrazine (TPPz) rotor structure and decorated with four perylene bisimide nuclei (Fig. 10) applied to the ITO/ZnO/difluorobenzothiadizole-oligothiophene-polymer:**22**/MoO_3/Al cell noteworthy PCE of 7.1% with J_{sc} of 12.5 mA cm^{-2}, notwithstanding the less pronounced twisted structure with respect to the TPE core [37]. Such excellent PV performances were attributed to the relatively small donor/acceptor domain size in the bulk heterojunction as revealed by AFM microscopy, coupled with the higher charge transfer process than those exhibited by analogous structures based on the TPE central rotor.

Very recently, Adil and coworkers proposed to optimize the bulk heterojunction interface by utilizing a ternary blend based on the incorporation of TPE as AIEgen [40]. The optimization of the interface with the introduction of TPE zwitterionic derivatives in the active blend was previously proposed with success by C. Wang and coworkers (PCE of 8.94%) [41] and here optimized to further boost cell efficiency and charge mobility. Specifically, 5–20 wt% of TPE was blended with the binary blend composed of poly[[5,6-difluoro-2-(2-hexyldecyl)-2*H*-benzotriazole-4,7-diyl]-2,5-thiophenediyl[4,8-bis[5-(tripropylsilyl)-2-thienyl] benzo[1,2-*b*:4,5-*b*′] dithiophene-2,6-diyl]-2,5-thiophenediyl] (J71) and 3,9-bis(2-methylene-(3-(1,1-dicyanomethylene)-indanone))-5,5,11,11-tetrakis(4-hexylphenyl)-dithieno [2,3-*d*:20,30-*d*′]-s-indaceno[1,2-*b*,5,6-*b*′]dithiophene (ITIC) as acceptor. Specifically, the ternary blend

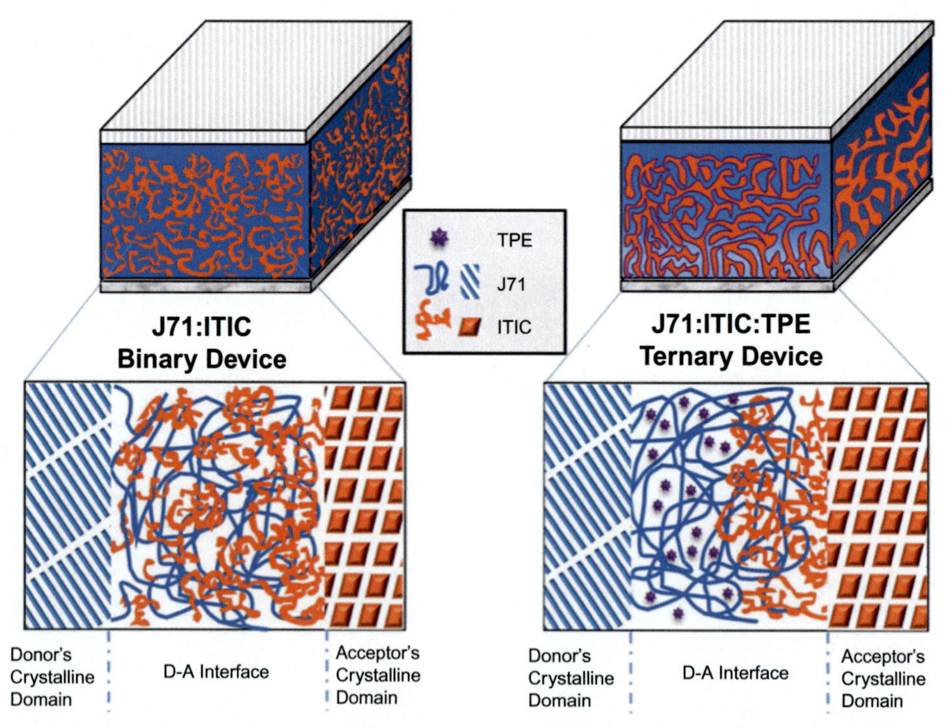

FIG. 11 Proposed effect of TPE on the domain architecture and average purity at the bulk heterojunction interface. *Reprinted with permission from reference M.A. Adil, J. Zhang, Y. Wang, J. Yu, C. Yang, G. Lu, et al., Regulating the phase separation of ternary organic solar cells via 3D architectured AIE molecules, Nano Energy 68 (2020) 104271 with permission from the Royal Society of Chemistry.*

composed of the 15 wt% of TPE showed the highest PCE of 12.16% with J_{sc} of 18.13 mA cm^{-2}, which is more than 20% of enhancement with respect to the pristine binary blend. The authors addressed these performances on the ability of the TPE AIEgen to promote better charge transportation abilities among the components of the bulk heterojunction. This feature is enabled by the rearrangement of the donor-acceptor interface caused by the presence of TPE, which pushed ITIC away from the blend being well interacting with J71. This phenomenon caused the formation of a more crystalline content of J71 at the interface and a relatively higher average acceptor domain purity (Fig. 11).

5 AIEgens in perovskite solar cells

Among all the possible perovskite compositions, lead halide perovskites display the semiconducting properties that are desired for photovoltaic applications and have therefore been the most studied (Fig. 12) [42]. Halide perovskites are ionic crystals that exhibit semiconducting properties. This double nature gives these materials a very advantageous characteristic, that is, their optical absorption wavelength can vary widely depending on the molar ratio of different halides (I, Br, Cl). In particular, it was found that the methylammonium lead iodide ($MAPbI_3$) bandgap of 1.57 eV could be increased up to 2.29 eV for pure methylammonium lead bromide ($MAPbBr_3$) by controlling the I/Br ratio in the precursor solution. Being characterized as an intrinsic semiconductor, MAPbI3 exhibits excellent mobility of photogenerated electrons and holes, which is one of the main factors that makes it so popular as a photovoltaic material. Defects formed by either I^- or Pb^{2+} vacancies generate trap states that reside either within or near the bands. When carriers are trapped in these shallow defects, they can be de-trapped easily and still contribute to the generation of electric current: this characteristic is responsible for the large carrier diffusion lengths of perovskites, which range from 1 μm in polycrystalline films to over 100 μm for single crystals [43]. Perovskite solar cells work as n-i-p (regular planar) and p-i-n (inverted planar) solar cells, where perovskite works as an intrinsic absorber sandwiched between two selective contacts (p and n). A TCO layer,

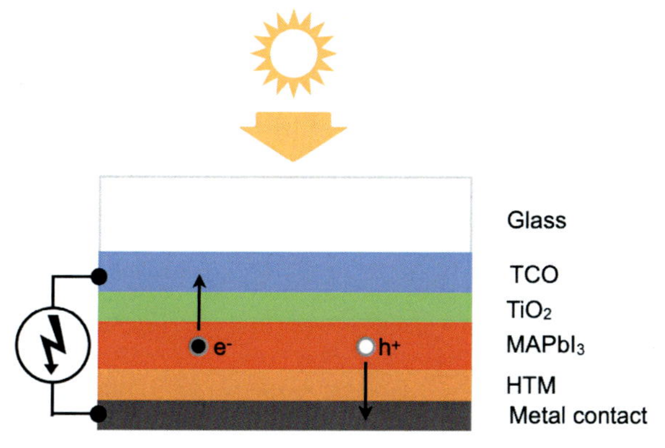

FIG. 12 Schematic structure of a perovskite solar cell.

usually FTO (fluorine-doped tin oxide, SnO_2:F) or ITO, is used as the front contact and metallic gold or silver as the back contact. In the n-i-p structure, a layer of TiO_2 or SnO_2 is deposited, which works as an n-type electron transport material (ETM), while perovskite functions as the intrinsic (i) absorber and an hole transport material (HTM) such as spiroOMeTAD (an organic molecule), poly(3-hexylthiophene) (P3HT) and poly(triarylamine) (PTAA) work as a p-type contact [44].

Despite the popularity and fast growth of perovskite solar cells, there are several issues that still need to be addressed. One of these is the stability with respect to external conditions, such as moisture, oxygen, heat, and light, which will be crucial for commercialization of the devices. Z. Chen and coworkers for the first time introduced the typical AIEgen tetra(4-methoxyphenyl) ethylene (TMOE, **23**) into $MAPbI_3$ thin films to investigate the possible energy transfer between them in the solid state [45]. The authors demonstrated that **23** increased the radiative recombination ratio of free carriers (FCs) versus excitons in perovskite films. Since such recombination processes play a fundamental role in photovoltaics, ITO/SnO_2/$MAPbI_3$:**23**/Spiro-OMeTAD/Au cells were prepared and tested. Notably, the addition of $1.6\,mg\,mL^{-1}$ of **23** to $MAPbI_3$ improved the incident photon-to-current conversion efficiency (IPCE) in the UV-light region due to a decreased radiative recombination of the perovskite after doping. As a consequence, J_{sc} increased from 20.36 to $21.19\,mA\,cm^{-2}$, thus enhancing PCE from 14.96% to 15.21%. However, further increasing the doping concentration of **23** had a detrimental effect caused by the reduction of $MAPbI_3$ grain size and crystallinity (Fig. 13).

More recently, Chen and coworkers reported two **24** and **25** AIEgens based on a triphenylethylene core and peripheral diphenylamine/triphenylamine moieties as cost-effective substitutes of Spiro-OMeTAD in hole-transporting materials (HTMs) [46]. The triphenylethylene AIE stator is surrounded by three phenyl rings that are highly twisted due to steric hindrance, thus forming a 3D charge-transporting network, useful in facilitating high solubility in organic solvents, low staking tendency, and high hole transport features. PV cells with a n-i-p configuration of FTO/TiO_2/$MAPbI_3$/HTM/Au were prepared by using **24**, **25**, and

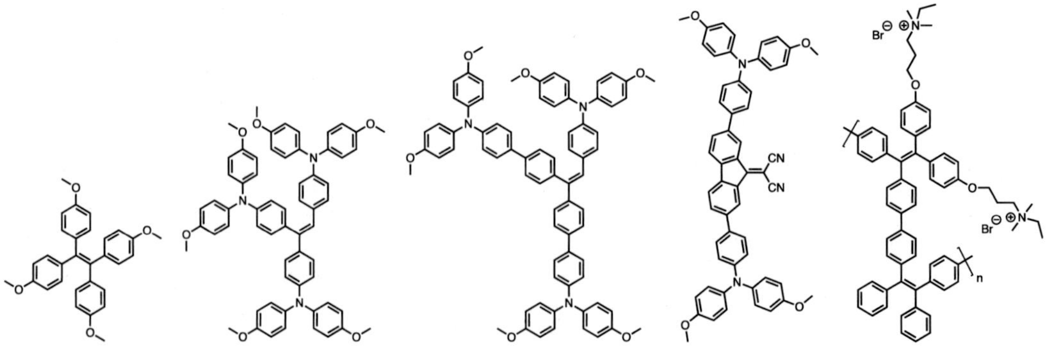

23 J_{sc} = 21.19 mA cm^{-2} PCE = 15.21%

24 J_{sc} = 22.32 mA cm^{-2} PCE = 18.56%

25 J_{sc} = 18.57 mA cm^{-2} PCE = 12.57%

26 J_{sc} = 22.68 mA cm^{-2} PCE = 16.03%

27 J_{sc} = 21.89 mA cm^{-2} PCE = 17.44%

FIG. 13 Chemical structures of **23**–**27** AIEgens and photovoltaic performances of the derived perovskite solar cells.

Spiro-OMeTAD as reference HTM. Hole mobility was found similar for all the HTM materials, i.e., 5.76×10^{-5}, 6.27×10^{-5}, and 9.71×10^{-5} $cm^2 V^{-1}$ for **24**, **25**, and Spiro-OMeTAD, respectively. **24** showed higher PV performances than **25** possibly due to a more favorable hole extraction being a HOMO level higher than that of $MAPbI_3$. Specifically, **24** delivered a J_{sc} of 22.32 $mA cm^{-2}$ and a PCE of 18.56%, close to those of 22.72 $mA cm^{-2}$ and 18.69% provided by using the reference Spiro-OMeTAD. Notwithstanding the similar PV performances, the low production costs of **24** and **25** render the AIE technology attractive in the formulation of modern HTM.

This strategy was recently confirmed by the work of Y. Cao and coworkers that utilized the 2-(2,7-bis(4-(bis(4-methoxyphenyl)amino)phenyl)-9H-fluoren-9-ylidene)malononitrile (TFM, **26**) as HTM in ITO/HTM/CsFAMA perovskite/C_{60}/BCP/Ag and PEDOT:PSS as reference material [47]. Optimal J_{sc} of 22.68 $mA cm^{-2}$ and PCE of 16.03% was found higher than that of 14.95% obtained by the device using the reference HTM.

With the aim to reduce the work function of the Ag electrode to enhance cathode interfacial contact and accelerate electron transport mobility, J. Tu and coworkers proposed to use a polymeric AIEgen (PTN-Br, **27**) based on tetraphenylethylene (TPE) decorated with quaternary ammonium salt side chains as a cathode interfacial layer between Ag and the perovskite layers in a p-i-n solar cell [48]. **27** exhibited an absorption band at 341–385 nm in the solid film state, thus also preventing the adverse direct illumination of the perovskite layer by near-UV radiation. When **27** was deposited as a thin layer beneath the Ag electrode, its working function decreased to 4.10 eV, i.e., matching with the E_{LUMO} (−4.0 eV) of the HTM layer based on [6,6]-phenyl-C_{61}-butyric acid methyl ester (PCBM). This feature was beneficial for the charge extraction and ohmic contact at the electrode interlayer. Atomic force microscopy investigations evidenced that **27** helped the reorganization of the HTM layer into a more compact structure that hampers the direct contact of the Ag electrode with $MAPbI_3$. Overall, the addition of the layer of **27** enhanced the J_{sc} from 19.13 to 21.89 $mA cm^{-2}$ and PCE from 15.14% to 17.44%.

6 AIEgens in luminescent solar concentrators (LSC)

A promising path to cost-effective photovoltaic (PV) systems is sunlight concentration. Over classical nonimaging geometrical concentrators, luminescent solar concentrators (LSCs) possess several advantages: low cost and lightweight, elevated theoretical concentration factors, and aptitude to efficiently work with diffuse light without tracking or cooling equipment [49,50]. LSCs are slabs of transparent host material such as poly(methyl methacrylate) (PMMA) or polycarbonate (PC) doped with 2–3 wt% max of high quantum yield (QY) fluorophores. The slab can be totally made of plastic or of high-purity glass coated with a thin film of doped polymer [51]. The host refractive index higher than that of air allows the trapping of a fraction of the emitted photons through total internal reflection. Photons are then collected at the edges of the LSC to produce electric power by PV cells (Fig. 14), thus allowing their effective integration in the building architecture.

Nevertheless, LSCs are afflicted by severe drawbacks that hinder their ability to efficiently deliver light to PV cells, including fluorescence quenching. Most fluorophores display a

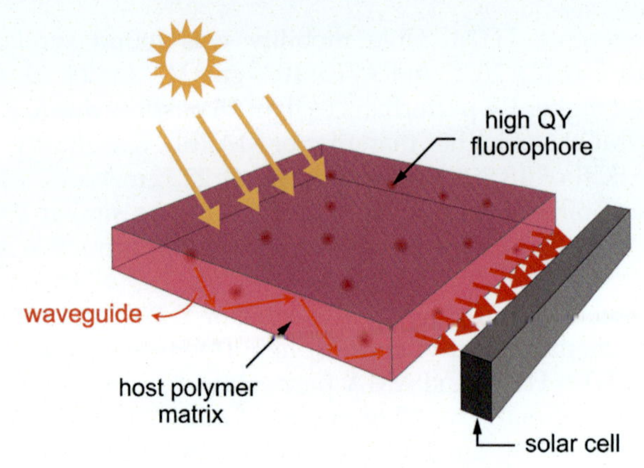

FIG. 14 Working principle of a typical luminescent solar concentrator.

tendency to π-stack when concentrated, which affects their QY due to the typical aggregation-caused quenching (ACQ) phenomenon. A low fluorophore concentration in the host polymer is considered useless as its capability to harvest sunlight would be modest. Although considerable progress has been made, current LSC-PV systems attain low power conversion efficiencies (PCE), with profitless maximum values around 7% [52,53]. To overcome this issue, AIEgens have been presented as excellent alternatives for photostable dopants in LSCs as recently reviewed by A. Pucci [54]. Indeed, in the aggregate solid form, the RIM of AIEgens allow them to emit bright fluorescence and enhance the optical efficiency of LSCs. Moreover, the easy chemical modification of the AIE core opens to economic and scalable synthetic routes that enable the functionalization of the AIE unit with donor-acceptor functional groups. This strategy favors the downshifting of the fluorescence band to the wavelength window characterized by the highest light/electric current conversion of the PV cell.

The first examples of AIEgens designed for LSC was proposed by the works of J. L. Banal and coworkers, who proposed the dispersion of 10wt% of tetraphenylethylene (TPE) and its conjugated forms in PMMA thin films reaching QY of about 40% and very low auto-absorption losses even at geometric factor (G) >100 [55]. However, TPE microcrystal formation strongly contributed to the formation of scattering points that affected the LSC optical efficiencies to deliver light to the solar cells (Fig. 15).

To increase the LSC performances, the authors blended two different fluorophores within the waveguide to combine their absorption spectra and preserve QY through efficient energy cascade pathways such as fluorescence resonance energy transfer (FRET). The strong emitting DPATPAN (**28**) AIEgen was able to transfer its excitation to the PITBT-TPE (**29**) AIEgen acceptor, whose red-shifted emission greatly limited auto-absorption, thus boosting solar harvesting and LSC efficiency (Fig. 16) [56]. Notably, **29** displayed a fluorescence band that peaked at about 650nm and QY in PMMA of 45%. The authors identified the best doping concentration to be 250 and 22.5mM for **28** and **29**, respectively, which allowed an 8% improvement of the LSC performances at a geometrical gain of 25 with respect to the device containing **28** alone, attributed to the substantial reduction of dissipative auto-absorption phenomena.

De Nisi proposed a red-emitting TPE derivative (TPE-AC, **30**) as a potential AIEgen for high-performance LSCs [57]. The TPE core was decorated with dimethylamine as donor

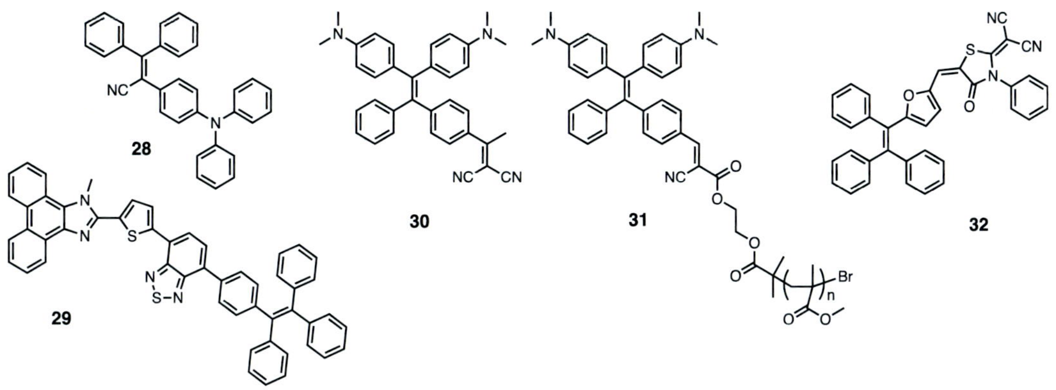

FIG. 15 Chemical structures of **28**–**32** AIEgens utilized as high QY fluorophores in luminescent solar concentrators.

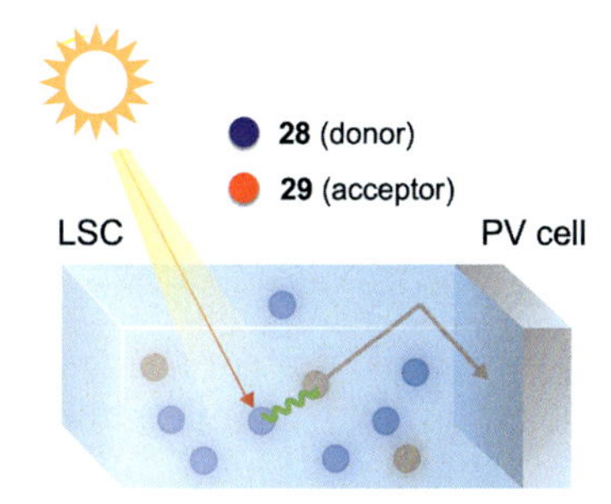

FIG. 16 Working principle of the LSC with the typical cascade mechanism via FRET by donor (**28**) and acceptor (**29**) AIEgens.

and malononitrile as acceptor groups and displayed intense emission in the aggregated state close to the NIR window and good photostability. Different PMMA and PC films were prepared and tested as a function of the AIEgen content (0.1–1.5 wt%) and showed fluorescence at 600–620 nm, Stokes shift larger than 100 nm and QY max of 50% that declined to 30% at the highest AIEgen content due to the autoabsortion phenomena. With TPE-AC being highly compatible with the PC matrix, well homogeneous films with superior light-harvesting features were obtained and a maximum optical efficiency (η_{opt}) of 6.7% was recorded. This data is promising but still lower than those recorded in the literature for the traditional fluorophore-based LSCs (around 10%).

R. Mori and coworkers considered the compatibility issues that adversely affected LSC performances based on **30**. Specifically, red-emitting PMMA polymer (PMMA_TPE_RED, **31**) was designed to maximize the fluorophore-polymer interactions and prepared via atom transfer radical polymerization (ATRP) [58]. ATRP was exploited to obtain the desired polymeric AIEgen architecture to provide phase stability between the fluorophore and the polymer matrix, an essential characteristic of high-performance LSC. In particular, thin films based on the **31** polymer exhibited a QY of 26.5% and their blend with 50 wt% of PMMA provided an LSC with max η_{opt} of 10%. This result was reported as the highest ever registered with the same G factor (16.6) and therefore consistent with the utilization of AIEgen in LSC technology.

More recently, Ma and coworkers designed and prepared AIEgen nanoparticles (TPFE-Rho, **32**) with a size of about 110nm, good photostability, large Stokes shift, and low toxicity and based on the triphenylfuranethylene core decorated with a Rhodanic-CN derivative [59]. TPFE-Rho/PMMA LSC, with 0.25cm of thickness and containing different AIEgen concentrations (0.02–0.10wt%), showed fluorescence peaking at around 570nm with a Stokes' shift of about 100nm. Such features allowed the LSC with the 0.8wt% of **32** reaching a maximum J_{SC} of 9.97mA/cm^2 and η_{opt} of 4.09.

7 Conclusions and outlook

Since the discovery of the first AIEgen in 2001 [15], tremendous advancements in many areas of research have been achieved by exploiting AIE, and these include optoelectronics, environmental monitoring, and biomedical research. Tang and collaborators' inspiring reviews [17,60–62] and talks at several international conferences illuminated the global scientific community in providing AIEgen with even more intriguing characteristics and applications. In this regard, this contribution aimed to convince the readership about the potential offered by the AIE technology in the design and formulation of new AIEgens whose rotor structures and optical features help in the advancement of photovoltaic solar cell research. For second and third generation of solar cells, AIE propeller structures decorated with conjugated donor and acceptor groups have been properly designed to prevent the formation of π-stacked structures and to favor the more intimate interactions with the active layers of solar devices and match their optical bandgap. Specifically, AIEgen have been effectively utilized in luminescent down-shifting (LDS) layers in CdTe PV cells thanks to beneficial effects provided by the large absorption of short-wavelength photons and the wide Stokes shift that enable a better spectral matching with the cell spectral response. Also, AIEgens with rotor structure have been developed to substitute expensive fullerene acceptor derivatives in OPV and as metal-free sensitizers in DSSC to extend the absorption region and to modulate the energy of the VB and CB frontiers. Effective solutions have been proposed in modern perovskite solar cells and in fluorescent solar collectors. HTM layers in perovskite cells have been improved by highly twisted AIEgen that form a 3D charge-transporting network, useful to provide high solubility in organic solvents, low staking tendency, and high hole transport features. In LSC, AIEgens are proposed to substitute the traditional ACQ fluorophores with propeller-like high QY fluorogens able to increase solar harvesting and to maintain unaltered the emission features also at high doping.

Further optimization of AIEgens in terms of broad absorption range, Stokes shift, and QY may inspire additional PV cells' exploitation to boost their efficiencies and cost-effectiveness. The versatile chemistry of AIEgens will enable the preparation of even more efficient propeller-like structures whose dispersibility within the organic/inorganic active layers of solar cells is one of the critical factors of PV cells' performance improvements. In luminescent down-shifting layers and solar collectors, attention will be paid to the design of AIEgens with maximum quantum efficiency for a broad region of the visible spectrum of light.

Acknowledgments

Miss Rima Charaf is kindly acknowledged for the helpful discussion on perovskite solar cells. Financial support from PRA_2020_21—SUNRISE is acknowledged.

References

[1] A. Smets, K. Jäger, O. Isabella, R. van Swaaij, M. Zeman, Solar Energy the Physics and Engineering of Photovoltaic Conversion, Technologies and Systems, UIT Cambridge Ltd., Cambridge, UK, 2016.
[2] N. Armaroli, V. Balzani, Solar electricity and solar fuels: status and perspectives in the context of the energy transition, Chem. A Eur. J. 22 (1) (2016) 32–57.
[3] C.J. Cleveland, Encyclopedia of Energy, Elsevier Academic Press, Amsterdam; Boston, 2004.
[4] NREL, Best Research-Cell Efficiency Chart, NREL, 2021.
[5] J. Ramanujam, U.P. Singh, Copper indium gallium selenide based solar cells—a review, Energ. Environ. Sci. 10 (6) (2017) 1306–1319.
[6] M.A. Green, K. Emery, Y. Hishikawa, W. Warta, E.D. Dunlop, Solar cell efficiency tables (version 44), Prog. Photovolt. Res. Appl. 22 (7) (2014) 701–710.
[7] K.-H. Ong, B. Liu, Applications of fluorogens with rotor structures in solar cells, Molecules 22 (6) (2017). 897/1-/19.
[8] G. Conibeer, Third-generation photovoltaics, Mater. Today 10 (11) (2007) 42–50.
[9] A. Hagfeldt, G. Boschloo, L. Sun, L. Kloo, H. Pettersson, Dye-sensitized solar cells, Chem. Rev. 110 (11) (2010) 6595–6663.
[10] B. O'Regan, M. Grätzel, A low-cost, high-efficiency solar cell based on dye-sensitized colloidal TiO_2 films, Nature 353 (6346) (1991) 737–740.
[11] A. Kojima, K. Teshima, Y. Shirai, T. Miyasaka, Organometal halide perovskites as visible-light sensitizers for photovoltaic cells, J. Am. Chem. Soc. 131 (17) (2009) 6050–6051.
[12] A.K. Jena, A. Kulkarni, T. Miyasaka, Halide perovskite photovoltaics: background, status, and future prospects, Chem. Rev. 119 (5) (2019) 3036–3103.
[13] A. Pucci, Luminescent solar concentrators based on aggregation induced emission, Isr. J. Chem. 58 (8) (2018) 837–844.
[14] L. Zhang, J.M. Cole, Dye aggregation in dye-sensitized solar cells, J. Mater. Chem. A 5 (37) (2017) 19541–19559.
[15] J. Luo, Z. Xie, J.W.Y. Lam, L. Cheng, H. Chen, C. Qiu, et al., Aggregation-induced emission of 1-methyl-1,2,3,4,5-pentaphenylsilole, Chem. Commun. 18 (2001) 1740–1741.
[16] G. Iasilli, A. Battisti, F. Tantussi, F. Fuso, M. Allegrini, G. Ruggeri, et al., Aggregation-induced emission of tetraphenylethylene in styrene-based polymers, Macromol. Chem. Phys. 215 (6) (2014) 499–506.
[17] J. Mei, Y. Hong, J.W.Y. Lam, A. Qin, Y. Tang, B.Z. Tang, Aggregation-induced emission: the whole is more brilliant than the parts, Adv. Mater. 26 (31) (2014) 5429–5479.
[18] J. Mei, N.L.C. Leung, R.T.K. Kwok, J.W.Y. Lam, B.Z. Tang, Aggregation-induced emission: together we shine, united we soar! Chem. Rev. 115 (21) (2015) 11718–11940.
[19] B.E. McCandless, J.R. Sites, Cadmium telluride solar cells, in: Handbook of Photovoltaic Science and Engineering, John Wiley & Sons, Ltd, 2010, pp. 600–641.
[20] M. Green, E. Dunlop, J. Hohl-Ebinger, M. Yoshita, N. Kopidakis, X. Hao, Solar cell efficiency tables (version 57), Prog. Photovolt. Res. Appl. 29 (1) (2021) 3–15.
[21] Y. Li, Z. Li, T. Ablekim, T. Ren, W.-J. Dong, Rational design of tetraphenylethylene-based luminescent down-shifting molecules: photophysical studies and photovoltaic applications in a CdTe solar cell from small to large units, Phys. Chem. Chem. Phys. 16 (47) (2014) 26193–26202.
[22] Y. Li, Z. Li, Y. Wang, A. Compaan, T. Ren, W.-J. Dong, Increasing the power output of a CdTe solar cell via luminescent down shifting molecules with intramolecular charge transfer and aggregation-induced emission characteristics, Energ. Environ. Sci. 6 (10) (2013) 2907–2911.
[23] Y. Li, J. Olsen, W.-J. Dong, Enhancing the output current of a CdTe solar cell via a CN-free hydrocarbon luminescent down-shifting fluorophore with intramolecular energy transfer and restricted internal rotation characteristics, Photochem. Photobiol. Sci. 14 (4) (2015) 833–841.

[24] E. Jaszczak, Ż. Polkowska, S. Narkowicz, J. Namieśnik, Cyanides in the environment—analysis—problems and challenges, Environ. Sci. Pollut. Res. 24 (19) (2017) 15929–15948.

[25] C.-P. Lee, C.-T. Li, K.-C. Ho, Use of organic materials in dye-sensitized solar cells, Mater. Today 20 (5) (2017) 267–283.

[26] A. Mishra, M.K.R. Fischer, P. Bäuerle, Metal-free organic dyes for dye-sensitized solar cells: from structure: property relationships to design rules, Angew. Chem. Int. Ed. 48 (14) (2009) 2474–2499.

[27] C. Chen, J.-Y. Liao, Z. Chi, B. Xu, X. Zhang, D.-B. Kuang, et al., Metal-free organic dyes derived from triphenylethylene for dye-sensitized solar cells: tuning of the performance by phenothiazine and carbazole, J. Mater. Chem. 22 (18) (2012) 8994–9005.

[28] C. Chen, J.Y. Liao, Z. Chi, B. Xu, X. Zhang, D.B. Kuang, et al., Effect of polyphenyl-substituted ethylene end-capped groups in metal-free organic dyes on performance of dye-sensitized solar cells, RSC Adv. 2 (20) (2012) 7788–7797.

[29] J. Shi, J. Huang, R. Tang, Z. Chai, J. Hua, J. Qin, et al., Efficient metal-free organic sensitizers containing tetraphenylethylene moieties in the donor part for dye-sensitized solar cells, Eur. J. Org. Chem. 2012 (27) (2012) 5248–5255.

[30] F. Zhang, J. Fan, H. Yu, Z. Ke, C. Nie, D. Kuang, et al., Nonplanar organic sensitizers featuring a tetraphenylethene structure and double electron-withdrawing anchoring groups, J. Org. Chem. 80 (18) (2015) 9034–9040.

[31] Y.-Q. Yan, Y.-Z. Zhu, J. Han, P.-P. Dai, M. Yan, J.-Y. Zheng, Fine tuning of the photovoltaic properties of triarylethylene-bridged dyes by altering the position and proportion of phenyl/thienyl groups, Dyes Pigm. 183 (2020), 108630.

[32] O. Inganäs, Organic photovoltaics over three decades, Adv. Mater. 30 (35) (2018) 1800388.

[33] D. Angmo, M. Hösel, F.C. Krebs, Roll-to-roll processing of polymer solar cells, in: Organic Photovoltaics, Wiley-VCH Verlag GmbH & Co. KGaA, 2014, pp. 561–586.

[34] B. Mi, Y. Dong, Z. Li, J.W.Y. Lam, M. Häußler, H.H.Y. Sung, et al., Making silole photovoltaically active by attaching carbazolyl donor groups to the silolyl acceptor core, Chem. Commun. 28 (2005) 3583–3585.

[35] X. Zhan, S.R. Marder, Non-fullerene acceptors inaugurating a new era of organic photovoltaic research and technology, Mater. Chem. Front. 3 (2) (2019) 180.

[36] A. Rananaware, A. Gupta, J. Li, A. Bilic, L. Jones, S. Bhargava, et al., A four-directional non-fullerene acceptor based on tetraphenylethylene and diketopyrrolopyrrole functionalities for efficient photovoltaic devices with a high open-circuit voltage of 1.18 V, Chem. Commun. 52 (55) (2016) 8522–8525.

[37] H. Lin, S. Chen, H. Hu, L. Zhang, T. Ma, J.Y.L. Lai, et al., Reduced intramolecular twisting improves the performance of 3D molecular acceptors in non-fullerene organic solar cells, Adv. Mater. 28 (38) (2016) 8546–8551.

[38] Y. Liu, C. Mu, K. Jiang, J. Zhao, Y. Li, L. Zhang, et al., A tetraphenylethylene core-based 3D structure small molecular acceptor enabling efficient non-fullerene organic solar cells, Adv. Mater. 27 (6) (2015) 1015–1020.

[39] S.-Y. Liu, W.-Q. Liu, C.-X. Yuan, A.-G. Zhong, D. Han, B. Wang, et al., Diketopyrrolopyrrole-based oligomers accessed via sequential CH activated coupling for fullerene-free organic photovoltaics, Dyes Pigm. 134 (2016) 139–147.

[40] M.A. Adil, J. Zhang, Y. Wang, J. Yu, C. Yang, G. Lu, et al., Regulating the phase separation of ternary organic solar cells via 3D architectured AIE molecules, Nano Energy 68 (2020), 104271.

[41] C. Wang, Z. Liu, M. Li, Y. Xie, B. Li, S. Wang, et al., The marriage of AIE and interface engineering: convenient synthesis and enhanced photovoltaic performance, Chem. Sci. 8 (5) (2017) 3750–3758.

[42] C.C. Stoumpos, M.G. Kanatzidis, The renaissance of halide perovskites and their evolution as emerging semiconductors, Acc. Chem. Res. 48 (10) (2015) 2791–2802.

[43] Q. Dong, Y. Fang, Y. Shao, P. Mulligan, J. Qiu, L. Cao, et al., Electron-hole diffusion lengths > 175 μm in solution-grown $CH_3NH_3PbI_3$ single crystals, Science 347 (6225) (2015) 967.

[44] C.H. Teh, R. Daik, E.L. Lim, C.C. Yap, M.A. Ibrahim, N.A. Ludin, et al., A review of organic small molecule-based hole-transporting materials for meso-structured organic–inorganic perovskite solar cells, J. Mater. Chem. A 4 (41) (2016) 15788–15822.

[45] Z. Chen, Z. Chen, H. Li, X. Zhao, M. Zhu, M. Wang, Investigation on charge carrier recombination of hybrid organic-inorganic perovskites doped with aggregation-induced emission luminogen under high photon flux excitation, Adv. Opt. Mater. 6 (15) (2018) 1800221.

[46] J. Chen, J. Xia, H.-J. Yu, J.-X. Zhong, X.-K. Wu, Y.-S. Qin, et al., Asymmetric 3D hole-transporting materials based on triphenylethylene for perovskite solar cells, Chem. Mater. 31 (15) (2019) 5431–5441.

[47] Y. Cao, W. Chen, H. Sun, D. Wang, P. Chen, A.B. Djurisic, et al., Efficient perovskite solar cells with a novel aggregation-induced emission molecule as hole-transport material, Solar RRL. 4 (2) (2020) 1900189.

[48] J. Tu, C. Liu, Y. Fan, F. Liu, K. Chang, Z. Xu, et al., Enhanced performance and stability of p-i-n perovskite solar cells by utilizing an AIE-active cathode interlayer, J. Mater. Chem. A 7 (26) (2019) 15662–15672.
[49] J. Roncali, Luminescent solar collectors: Quo Vadis? Adv. Energy Mater. 10 (36) (2020) 2001907.
[50] I. Papakonstantinou, M. Portnoi, M.G. Debije, The hidden potential of luminescent solar concentrators, Adv. Energy Mater. 11 (2020) 2002883.
[51] Y. Li, X. Zhang, Y. Zhang, R. Dong, C.K. Luscombe, Review on the role of polymers in luminescent solar concentrators, J. Polym. Sci. A Polym. Chem. 57 (3) (2019) 201–215.
[52] L.H. Slooff, E.E. Bende, A.R. Burgers, T. Budel, M. Pravettoni, R.P. Kenny, et al., A luminescent solar concentrator with 7.1% power conversion efficiency, Phys. Status Solidi RRL 2 (6) (2008) 257–259.
[53] M. Debije, Renewable energy better luminescent solar panels in prospect, Nature 519 (7543) (2015) 298–299.
[54] A. Pucci, Luminescent solar concentrators based on aggregation induced emission, Isr. J. Chem. 58 (8) (2018) 837–844.
[55] J.L. Banal, B. Zhang, D.J. Jones, K.P. Ghiggino, W.W.H. Wong, Emissive molecular aggregates and energy migration in luminescent solar concentrators, Acc. Chem. Res. 50 (1) (2017) 49–57.
[56] B. Zhang, J.L. Banal, D.J. Jones, B.Z. Tang, K.P. Ghiggino, W.W.H. Wong, Aggregation-induced emission-mediated spectral downconversion in luminescent solar concentrators, Mater. Chem. Front. 2 (3) (2018) 615–619.
[57] F. De Nisi, R. Francischello, A. Battisti, A. Panniello, E. Fanizza, M. Striccoli, et al., Red-emitting AIEgen for luminescent solar concentrators, Mater. Chem. Front. 1 (2017) 1406–1412.
[58] R. Mori, G. Iasilli, M. Lessi, A.B. Muñoz-García, M. Pavone, F. Bellina, et al., Luminescent solar concentrators based on PMMA films obtained from a red-emitting ATRP initiator, Polym. Chem. 9 (10) (2018) 1168–1177.
[59] W. Ma, W. Li, M. Cao, R. Liu, X. Zhao, X. Gong, Large stokes-shift AIE fluorescent materials for high-performance luminescent solar concentrators, Org. Electron. 73 (2019) 226–230.
[60] Y. Hong, J.W.Y. Lam, B.Z. Tang, Aggregation-induced emission: phenomenon, mechanism and applications, Chem. Commun. 29 (2009) 4332–4353.
[61] Y. Hong, J.W.Y. Lam, B.Z. Tang, Aggregation-induced emission, Chem. Soc. Rev. 40 (11) (2011) 5361–5388.
[62] J. Mei, N.L.C. Leung, R.T.K. Kwok, J.W.Y. Lam, B.Z. Tang, J. Mei, et al., Aggregation-induced emission: together we shine, united we soar! Chem. Rev. 115 (21) (2015) 11718–11940.

CHAPTER

14

AIE molecular probes for biomedical applications

Alex Y.H. Wong, Fei Wang, Chuen Kam, and Sijie Chen

Ming Wai Lau Centre for Reparative Medicine, Karolinska Institutet, Hong Kong, China

1 Introduction

Fluorescence techniques have contributed to important biological discoveries and studies for over a century. From the beginning of the late 16th century, many scientists have made the developmental jump from white light microscopy to fluorescence microscopy, leading to the modern advanced fluorescence microscopes nowadays that have spatial resolution beyond the optical diffraction limit [1]. Fluorescent probes are powerful tools with diverse applications which can be engineered to detect and measure diverse biological processes and structures. The fluorescence image, which is usually taken by a digital camera, can be analyzed in terms of its pixel intensity and spatial information. The use of chemical fluorescent probes in biomedical applications mainly focuses on biosensing and bioimaging.

Nowadays, fluorescence imaging continues to gain greater traction in the research field because of the diverse variety of fluorescent reagents available, high temporal and spatial resolution, strong signal and quantifiability, economical and wide applicability, etc. Popular fluorescent reagents such as organic dyes, fluorescent proteins (e.g., green fluorescent protein), and inorganic quantum dots (QDs) have also been developed in recent years [2]. However, inorganic QDs are not particularly biocompatible and exhibit a high degree of cytotoxicity [3]. Fluorescent proteins have a low molar absorptivity and require genetic modification of the biological targets. For small organic fluorophores, many conventional fluorophores suffer from aggregation-caused quenching (ACQ) which greatly impedes their biological applications [4]. AIE is a phenomenon that is the opposite of ACQ. Since the discovery of the AIE phenomenon in 2001, an increasing number of fluorophores with AIE properties have been synthesized. AIEgens are usually nonemissive or weakly emissive in the solution state. When the molecular motions of these molecules are restricted (e.g., in the aggregated state), they become highly emissive. They can be used for the labeling or

Aggregation-Induced Emission (AIE)
https://doi.org/10.1016/B978-0-12-824335-0.00011-8

sensing of different biological targets specifically by decorating the AIE fluorophore core with different targeting or functional groups. Many AIEgens are also highly biocompatible. They are therefore desirable tools for bioimaging and a variety of biomedical applications.

2 Protein detection

Proteins are essential biomacromolecules that carry out most of the processes in all living organisms and are also components of organelles in eukaryotes. Their conformational changes as physical three-dimensional structures or chemical properties may take place under physiological or pathological conditions. Proteins have served as an important diagnostic marker of diseases for over 190 years since the development of protein tests for urine albumin, which is a marker for kidney disease by Bright in 1827 [5]. Because of the diverse yet significant roles played by the protein, quantitative and qualitative detection of proteins is central to biochemical and biomedical research. The first work utilizing water-soluble AIEgens for serum albumin detection was published in 2006. [6] Ever since then, various AIEgen-based probes for sensing serum albumin proteins in water solutions have been developed. AIEgens can also be modified to selectively label an entire class of biomacromolecules, a specific analyte, organelles, or pathological biomarkers such as amyloid plaques. There are countless possible intuitive designs engineered to label the desired target. For example, selective labeling of a certain protein in a system containing a plethora of different protein types may warrant an AIEgen with a peptide moiety that senses the protein of interest. The use of AIE probes in protein detection and characterization has been summarized in detail in our recent review [7]. Here we will introduce a few examples of the applications of small-molecule AIE probes in protein detection.

2.1 Protein gel imaging

Polyacrylamide gel electrophoresis (PAGE) is an important and widely used method for protein separation according to the isoelectric points or molecular weight of proteins prior to downstream detection or analysis in biochemical research. Protein gel staining, which helps to visualize the proteins in the gel, is critical for protein analysis. Back in 2012 and 2013, amine-reactive and thiol-reactive tetraphenylethylene (TPE) derivatives were applied in protein gel staining [8,9].

Silver staining is a sensitive protein gel staining method with a detection limit of the nanogram scale [10]. The classic silver staining utilizes the special bioaffinity of silver ions to proteins and the chromogenic reduction of silver ions to visualize the protein bands in gels. Although this method is sensitive, it is rather capricious and often lacks reproducibility; it suffers from high levels of a nonspecific background and a low signal-to-noise ratio (SNR). In 2018, a fluorescent silver staining method based on an AIE silver ion (Ag^+) sensitive probe named **TPE-4TA** was developed. **TPE-4TA** is a TPE derivative conjugated with four tetrazolate anions which act as Ag^+-specific targeting groups to trigger aggregation, while the core TPE endows its AIE properties. **TPE-4TA** is water soluble and has almost no fluorescent background in water when it exists in the salt form. The addition of Ag^+ will trigger

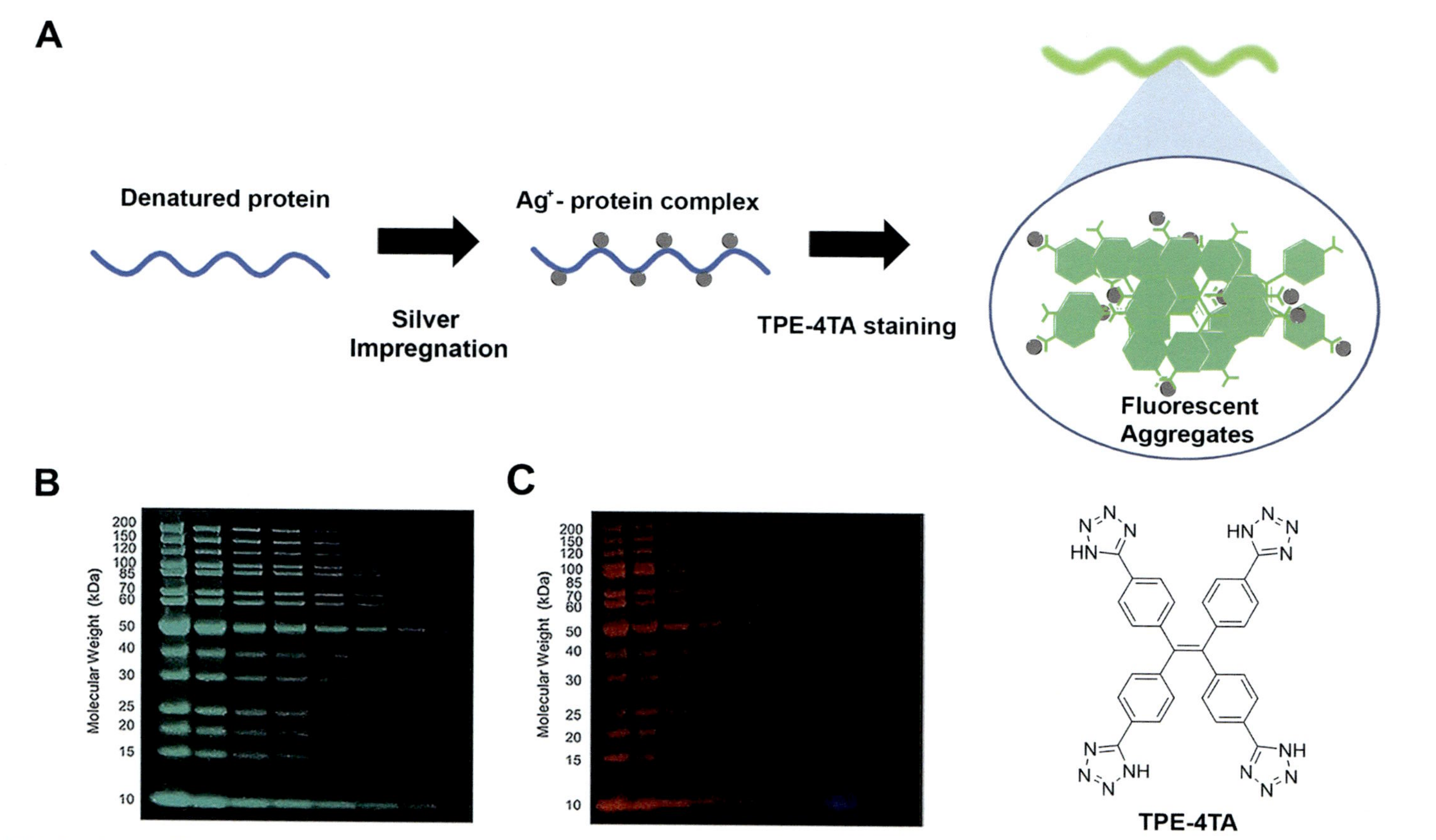

FIG. 1 Silver-AIE staining strategy in PAGE. (A) The protein gel is impregnated with Ag^+ which binds to the denatured protein to form an Ag^+-protein complex. After washing, the gel is then stained with the Ag^+-sensitive **TPE-4TA** probe which produces a fluorescence signal when bound and aggregated to the Ag^+-protein complex. (B) The silver-AIE stain and the (C) SYPRO Ruby stain imaged in parallel under a handheld UV lamp with 365 nm irradiation. *Reprinted (adapted) with permission from S. Xie, A.Y.H. Wong, R.T.K. Kwok, Y. Li, H. Su, J.W.Y. Lam, S. Chen, B.Z. Tang, Fluorogenic Ag+-tetrazolate aggregation enables efficient fluorescent biological silver staining. Angew. Chem. Int. Ed. 2018. Copyright (2018) from John Wiley and Sons.*

the aggregation of the probe and result in emission of strong greenish-blue fluorescence with maximum intensity at 504 nm. The fluorescent silver staining method also exploits the affinity of Ag^+ to proteins. Instead of reducing the protein-bound silver ions to dark-colored metallic silver nanoparticles in the classical silver staining method, a turn-on of the Ag^+ probe, **TPE-4TA**, is used to visualize the silver-bound proteins in the fluorescent silver staining method (Fig. 1A) [7,11,12]. This fluorescent silver staining method offers sensitive in-gel protein detection with a broad linear dynamic range that outperforms the conventional silver nitrate stain and the sensitive fluorescent SYPRO Ruby stain (Fig. 1B and C). The high contrast and SNR of the AIE-based fluorescent silver stain is more suitable for quantification and analysis than its conventional chromogenic counterparts. Thanks to the high sensitivity for silver ions, **TPE-4TA** has a 1–16-fold lower limit of detection for proteins compared with some silver nitrate protocols [13] and is 1–18-fold more sensitive than the SYPRO Ruby stain [11]. It is an attractive alternative to both the conventional silver stain and commercially available fluorescent protein stains.

2.2 Protein fibril detection

AIE probes have also been used for protein fibril detection. In Alzheimer's disease (AD), insoluble amyloid fibrils arise under pathological conditions. One of the reasons accounting for the formation of insoluble aggregates is protein misfolding which can occur due to the mutations in Presenilin1/2 genes. Presenilin1/2 are the catalytic subunits of the γ-secretase complex which cleave the amyloid precursor protein (APP) to β-amyloid (Aβ) in varied length. The production of $A\beta_{42}$, which is hydrophobic, leads to fibrillization and ultimately aggregates into plaques in familial AD. Another example of protein associated with fibrillization is insulin, a hormone responsible for regulating blood glucose homeostasis. Insulin can undergo misfolding and amyloidosis due to certain destabilizing conditions in vitro such as long-distance shipping and long-term storage. Therefore, insulin fibrillogenesis is a good model not only for the study of amyloidogenesis but also for the development of therapeutic delivery in diabetes treatments. 1,2-bis[4-(3-sulfonatopropoxyl)phenyl]-1,2-diphenylethene salt (**BSPOTPE**) (Fig. 2A) is a negatively charged water-soluble AIEgen. It is nonemissive when dissolved in native insulin in buffer and becomes fluorescent in the presence of preformed insulin fibrils. It is highly distinguishable in native and fibrillar insulin and can produce high-contrast images of the protein fibril (Fig. 2D). The ex situ monitoring of amyloidogenesis kinetics is demonstrated in the PL of **BSPOTPE** with increments of fibrillar insulin versus native insulin at increasing concentrations (Fig. 2B and C) [14]. This will be a useful model for studying pathological protein aggregation in neurodegenerative diseases and nonneuropathic diseases.

To study amyloid plaques, extrinsic fluorescent probes such as thioflavin T (ThT) and thioflavin S (ThS) have been used as a gold standard for over half a century [15]. These fluorophores, including other extrinsic probes such as rhodamine analogues and inorganic QDs, can be aggregated in situ when staining dense-core structures like amyloid plaques. The π–π interaction between the stacked aromatic rings will promote the formation of detrimental species such as excimers and exciplexes which can eventually lead to severe self-quenching, making the fluorophores unsuitable for quantitative analysis [16]. In contrast, the twistable conformational structure of AIEgens can avoid the stacked intramolecular π–π interactions in the aggregated hydrophobic regions of the amyloid microenvironment [17]. The "turn-on"

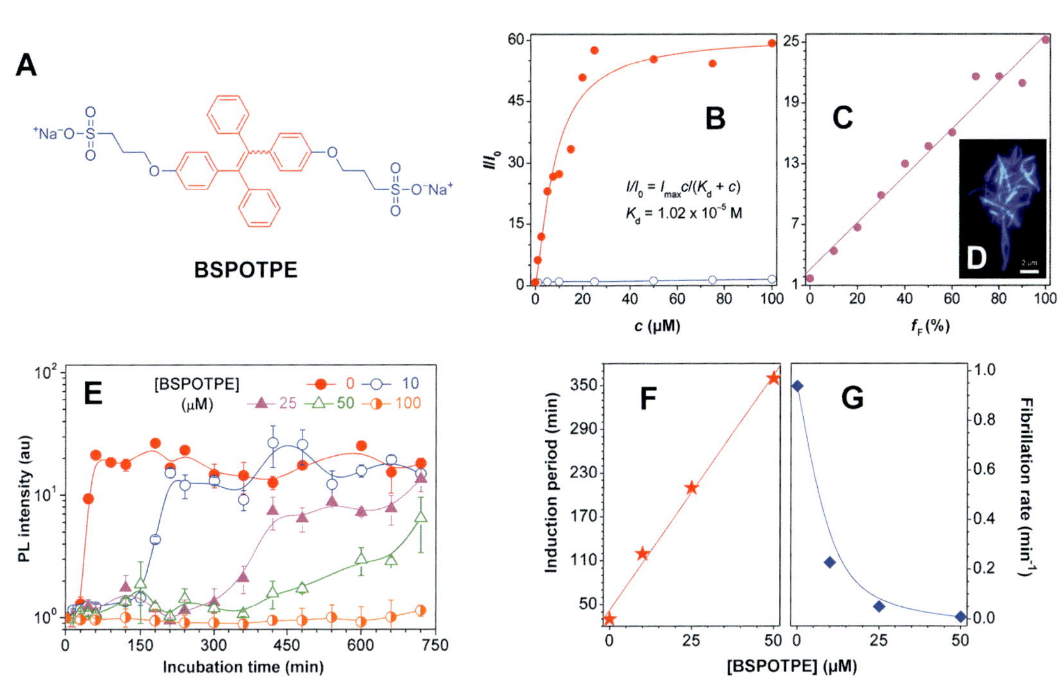

FIG. 2 Visualization of insulin fibrils by **BSPOTPE**. (A) Molecular structure of 1,2-Bis[4-(3-sulfonatopropoxyl)phenyl]-1,2-diphenylethene salt (**BSPOTPE**). (B) Photoluminescence (PL) intensity change of **BSPOTPE** at 470 nm at different concentrations of fibrillar (solid circles) or native (open circle) insulin. (C) PL intensity of **BSPOTPE** in different molar fraction of fibrillar insulin (f_F). (D) The fluorescence image of insulin fibrils stained by **BSPOTPE**. Scale bar: 2 μm. (E) Fibrillogenesis of insulin in the presence of 0–100 μM of BSPOTPE. (F) Induction period and (G) fibrillation rate of insulin in the presence of **BSPOTPE** at various concentrations. *Reprinted (adapted) with permission from Y. Hong, L. Meng, S. Chen, C.W.T. Leung, L.T. Da, M. Faisal, D.A. Silva, J. Liu, J.W.Y. Lam, X. Huang, B.Z. Tang, Monitoring and inhibition of insulin fibrillation by a small organic fluorogen with aggregation-induced emission characteristics, J. Am. Chem. Soc. 134 (3) (2012) 1680–1689. Copyright (2018) American Chemical Society.*

mechanism of **BSPOTPE** is triggered by the docking of the probe to the hydrophobic surface of the insulin aggregate facilitated by its aromatic core [14]. The dose response of insulin solutions (500 μM) to **BSPOTPE** (0–100 μM) (Fig. 2E) shows a delay in the nucleation phase (Fig. 2F) and decreased growth rate in the elongation phase (Fig. 2G), suggesting that the probe can hinder insulin amyloidogenesis. The proposed mechanism for the nonemission of **BSPOTPE** is due to the electrostatic repulsion of the probe and native insulin as they are both negatively charged. Additionally, **BSPOTPE** is unable to bind to the hydrophobic residues buried in the inner core of the native insulin structure through hydrophobic interactions. In the insulin fibrils, extended β-strand structures are assembled which allow the phenyl rings of **BSPOTPE** to stack on the surfaces through hydrophobic interaction. This causes the probe to bind noncovalently to the partially folded insulin fibrils, which in turn trigger fluorescence emission through the restricted intramolecular rotation (RIR) in the AIE phenomenon. Through the docking simulation, the partially folded B-chain helix (B11–B19) of insulin exposes hydrophobic residues under fibril-forming conditions which favor the binding of **BSPOTPE**. This suggests how the probe mechanistically impedes amyloidogenesis as the B-chain helix participates in the nucleation phase of fibrillogenesis [14,16].

Although thioflavin and its derivatives are commonly used for histological staining of amyloid fibrils, it is not suitable for in vivo studies. Ideal designs of the probe for AD are having near-infrared (NIR) emission for good tissue penetrability, a high SNR, discriminative labeling of Aβ plaques, and most importantly high permeability to the blood–brain barrier (BBB). Recently, a BBB permeable AIE probe has been developed for in vivo Aβ plaques imaging. **QM-FN-SO$_3$** is a water-soluble AIE-active molecule for high fidelity "off–on" (Fig. 3A) NIR responses (720nm) to Aβ_{42} complexes with a higher binding affinity than ThS [18]. The selective binding mechanism of **QM-FN-SO$_3$** to Aβ plaques is due to the interactions of the binding unit *N,N′*-dimethylamine and the hydrophobic pockets of the aggregated amyloid fibrils. This then activates RIR in **QM-FN-SO$_3$** which produces high contrast fluorescence. **QM-FN-SO$_3$** in the histological staining of the AD mice model (APP/PS1 transgenic mice) demonstrates specific staining of plaques and identifies more plaques than ThS (Fig. 3B–H). The high biocompatibility and BBB permeability of **QM-FN-SO$_3$** allow for in vivo mapping of the Aβ plaques (Fig. 3I). The in vivo imaging of 22-month-old male APP/PS1 transgenic mice confirms the feasibility and selectivity of in situ Aβ plaques staining by the intravenous tail injection of **QM-FN-SO$_3$**. Almost all the fluorescence signals captured are localized at the brain compartments (Fig. 3J–M) which is further validated in ex vivo staining using the anti-Aβ antibody #2454 (Fig. 3N–P). This probe could potentially allow studies to assess the effectiveness of candidate drugs in clearing or reducing Aβ plaques in live animals.

2.3 Detecting and evaluating enzymatic activity

Based on the AIE mechanism, bioprobes for enzyme detection and enzymatic activity evaluation can also be easily designed. Many of these probes are based on AIE bioconjugates [7]. Here we introduce an example of the enzyme detection system based on AIE small molecules.

Diseases in the brain may also arise from imbalances of neurotransmitters which lead to illnesses ranging from depression to dementia [19,20]. Low levels of the neurotransmitter acetylcholine (ACh) in the hippocampus and cortex (which affects memory and cognition) are a characteristic of AD and a leading cause of dementia in the elderly [21]. Acetylcholinesterase (AChE) is an enzyme responsible for the hydrolysis of acetylcholine into acetic acid and choline which normally prevents the accumulation of ACh in the synaptic cleft to terminate synaptic transmission. Since the 1990s, AChE inhibitors have shown therapeutic benefits and relief for memory deficits in patients with AD by slowing down the hydrolysis of ACh, which leads to its excessive accumulation in the synaptic cleft. This increases the stimulation of muscarinic and nicotinic receptors [22]. Examples of the AChE inhibitor are pesticides of organophosphorus, which is one of the common causes of poisoning worldwide. Drugs such as donepezil, rivastigmine, and tacrine are AChE inhibitors for the clinical treatment of AD and have good BBB permeability [23]. It is thus important to develop reliable assays to quantify and monitor AChE activity as well as screen inhibitors and evaluate their efficacy. Conventionally, AChE activity and inhibition are monitored with Ellman's reagent using absorption spectroscopy or enzyme-based cascade reactions where AChE activity is coupled with the detection of hydrogen peroxide. However, these methods are time-consuming, citing demands for a dynamic, convenient, and reliable method.

In developing a novel AChE and its inhibitor screening assay, a fluorometric AIE probe is applied. The platform consists of three compounds (Scheme 1): compound 1(**BSPOTPE**) is a

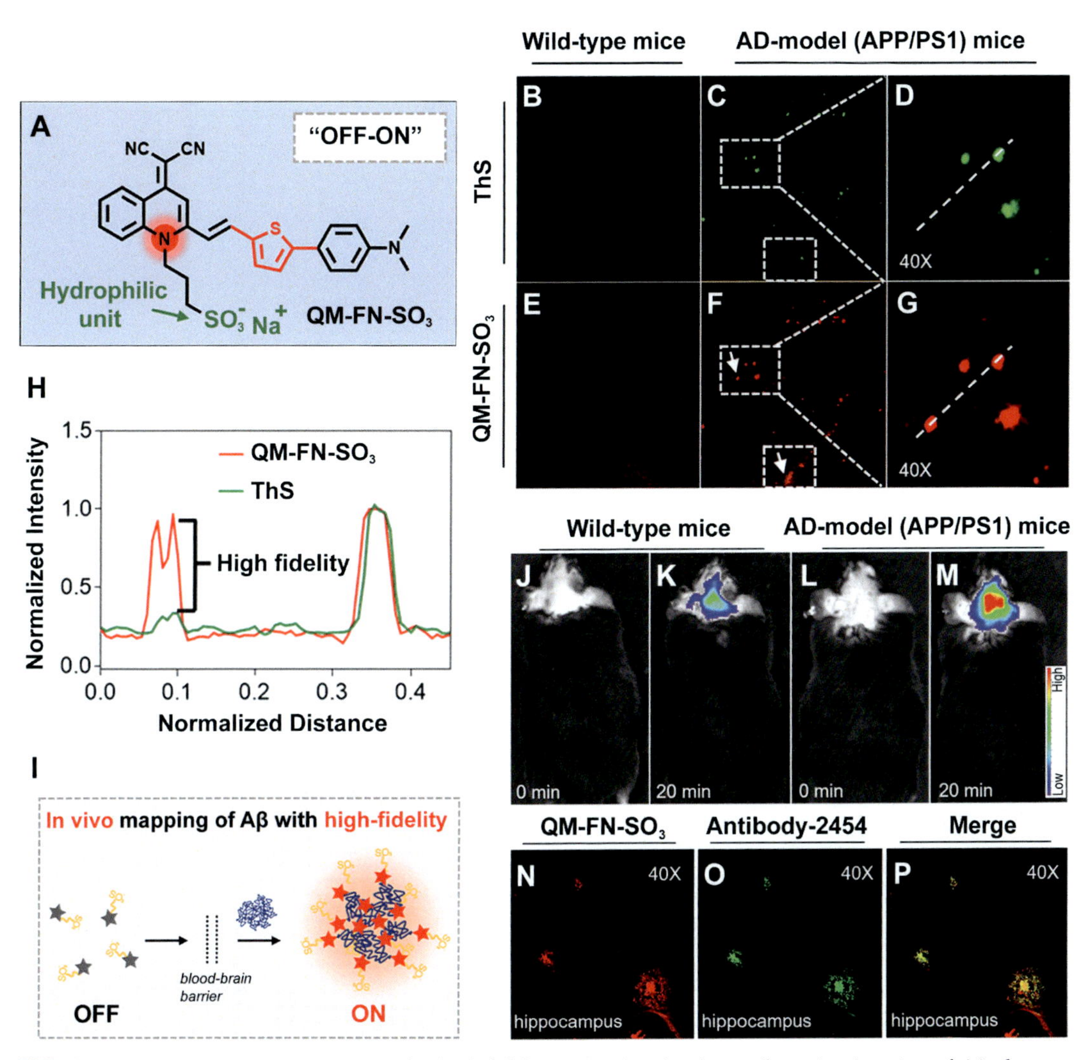

FIG. 3 AIE-based NIR emissive probe for high fidelity in in vivo, in situ, and ex vivo imaging of Aβ plaques. (A) Sulfonate substituted position (in red) for fluorescence-off state in the "off–on" binding mechanism. (B–G) Histological staining of Aβ plaques in paraffin-embedded 8-μm brain tissue sections from the hippocampal region of wild-type mice and APP/PS1 mice using ThS (B–D) or **QM-FN-SO$_3$** (E–G). (H) The linear fluorescence intensity profiles of **ThS** from the dotted line in (D) and **QM-FN-SO$_3$** in (G). (I) **QM-FN-SO$_3$** is in the "OFF" state outside of the BBB. After crossing the BBB, fluorescence "turn-on" is triggered in the presence of Aβ plaques. (J–M) Heatmap of the fluorescence intensity in the wild-type and APP/PS1 mice injected with **QM-FN-SO$_3$** shows that the fluorescence signal mainly comes from the brain region. (N–P) Ex vivo fluorescence imaging of brain slices from APP/PS1 mice after injection of **QM-FN-SO$_3$** (N) and labeling with anti-Aβ antibody #2454 in the adjacent section (O). (P) Merged image of (N) and (O). *Reprinted (adapted) with permission from W. Fu, C. Yan, Z. Guo, J. Zhang, H. Zhang, H. Tian, W.H. Zhu, Rational design of near-infrared aggregation-induced-emission-active probes: in situ mapping of amyloid-β plaques with ultrasensitivity and high-fidelity, J. Am. Chem. Soc. 141 (7) (2019) 3171–3177. Copyright (2019) American Chemical Society.*

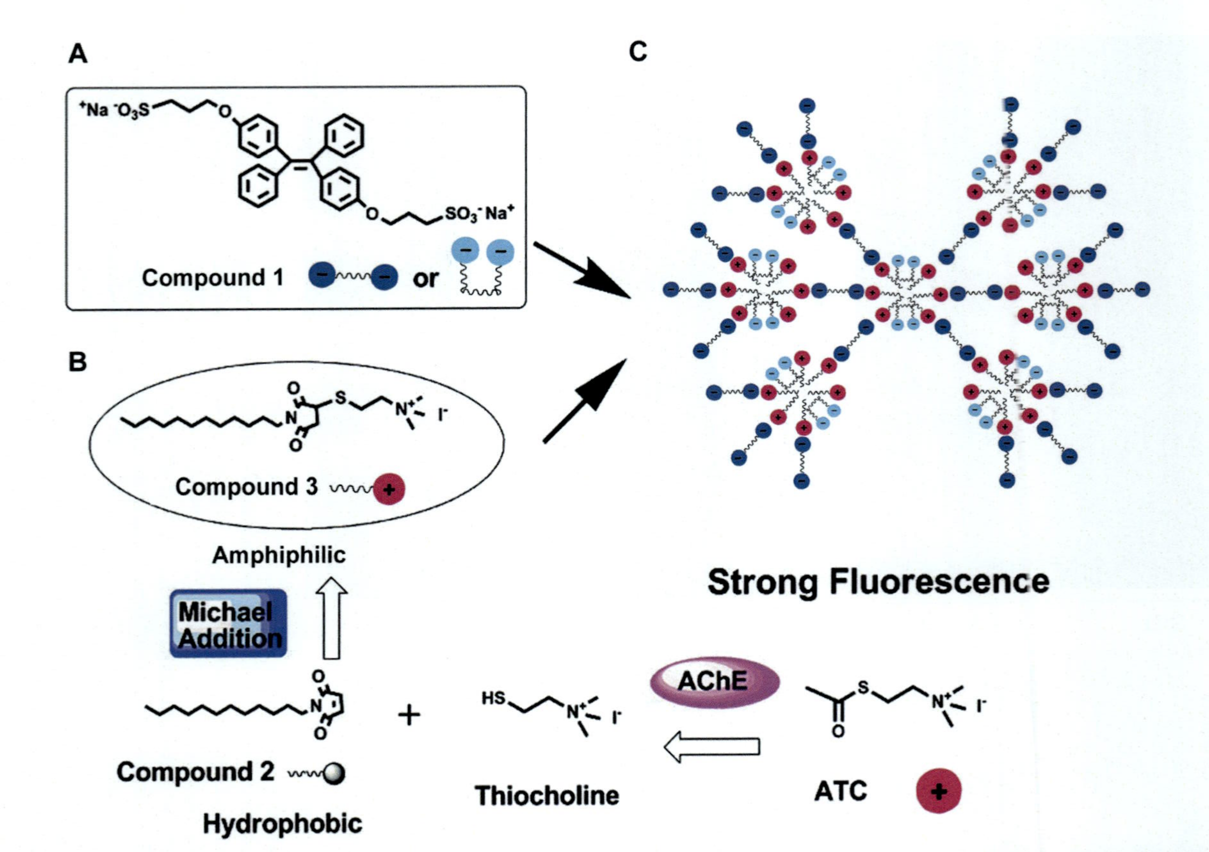

SCHEME 1 (A) Chemical structure of compound 1 and representative schematic diagram of the charged molecule under two different states. (B) Series of reactions starting from enzymatic reaction of the positively charged ATC (and representative schematic) by AChE, producing Thiocholine, which react the hydrophobic compound 2 (with a representative schematic) to make the amphiphilic compound 3 via Michael Addition. (C) Illustration of the aggregation of compound 1 in the presence of compound 3. *Reprinted (adapted) with permission from L. Peng, G. Zhang, D. Zhang, J. Xiang, R. Zhao, Y. Wang, D. Zhu, A fluorescence "Turn-on" ensemble for acetylcholinesterase activity assay and inhibitor screening, Org. Lett. 11 (17) (2009) 4014–4017. Copyright (2009) American Chemical Society.*

TPE probe with two sulfonate (-SO_3^-) units, while compound 2 is a hydrophobic compound with a maleimide group which leads to the formation of compound 3 myristoylcholine, an amphiphilic compound with one ammonium group that induces aggregation of the compound [24]. This mechanism is dynamic in the sense that the fluorescence intensity of the AIE probe compound 1 is based on the activity of AChE in real time. The working mechanism is initiated by AChE hydrolysis of the acetylthiocholine iodide (ATC) substrate into thiocholine. The thiocholine then reacts with the maleimide in compound 2. This produces the amphiphilic compound 3 which then creates an ensemble causing aggregation of the AIE compound 1 and increases its fluorescence significantly. In the presence of an AChE inhibitor, the activity of ACh hydrolysis is reduced which produces less amphiphilic compound 3 (myristoylcholine). This in turn decreases the level of fluorescence enhancement.

3 Nucleic acid detection

Deoxyribonucleic acid (DNA) is a polymeric double-stranded molecule encoding genetic information that can be inherited. It consists of the nucleotides guanine, cytosine, adenine, and thymine in unidirectional sequences. Ribonucleic acid (RNA) is a polymer transcribed from DNAs, located in the nucleus and the cytoplasm. In essence, they are translated into polypeptide chains consisting of a sequence of amino acids which are fundamental for protein synthesis [25]. Determination of DNA/RNA sequences is important for biomedical studies as we can study and associate genetic variants in certain genes with diseases, which allows us to understand a person's risk of developing an illness. One of the most practical methods to detect nucleic acids such as DNA and RNA in a mixture is to conjugate a fluorescent reporter with a nucleic acid sequence complementary to the target for hybridization. Conventional methods utilizing radiolabeled nucleotides are sensitive but long-term exposure is potentially hazardous to researchers. Fluorescence methods have several advantages: not only are they less labor intensive, but they are also compatible with existing DNA microarray technologies. The simplest nucleic acid detection system utilizes the electrostatic attraction-aided complexation or aggregation of negatively charged biomacromolecules such as DNAs and RNAs and positively charged water-soluble AIE molecules [6]. To achieve better selectivity and sensitivities, AIE-based nucleic acid detection systems with a more complicated design are developed. In this section, we introduce some examples of the detection of DNAs with a specific sequence or specific secondary structures using AIE probes.

3.1 DNA detection

Back in 2006, researchers first tried to apply water-soluble AIE probes for the detection of biomacromolecules [6]. One of the first two analytes they tested was DNA. The structure of the AIE they used is shown in Fig. 4A. It is a water-soluble cationic **TPE** derivative. Thanks to its good water solubility and typical AIE properties, this probe is nonemissive in the water buffer. In the presence of DNA, a negatively charged biomacromolecule, this positively charged probe binds to the DNA through electrostatic interactions and gives out strong blue emission.

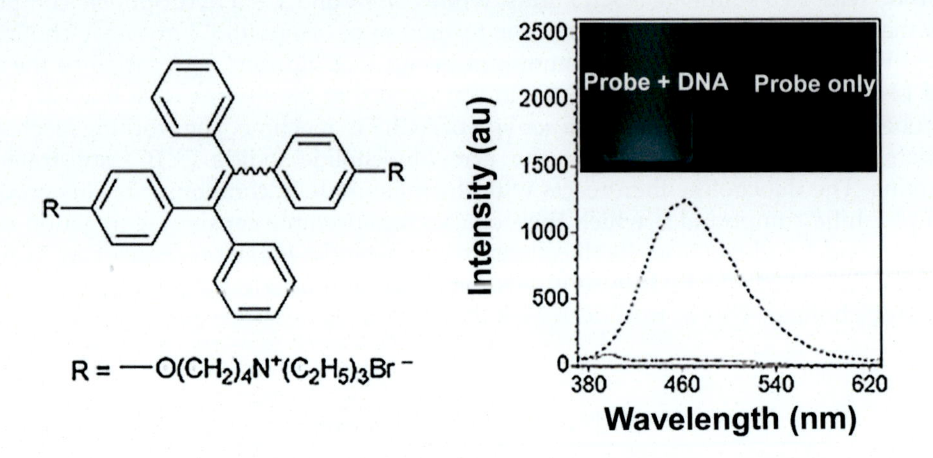

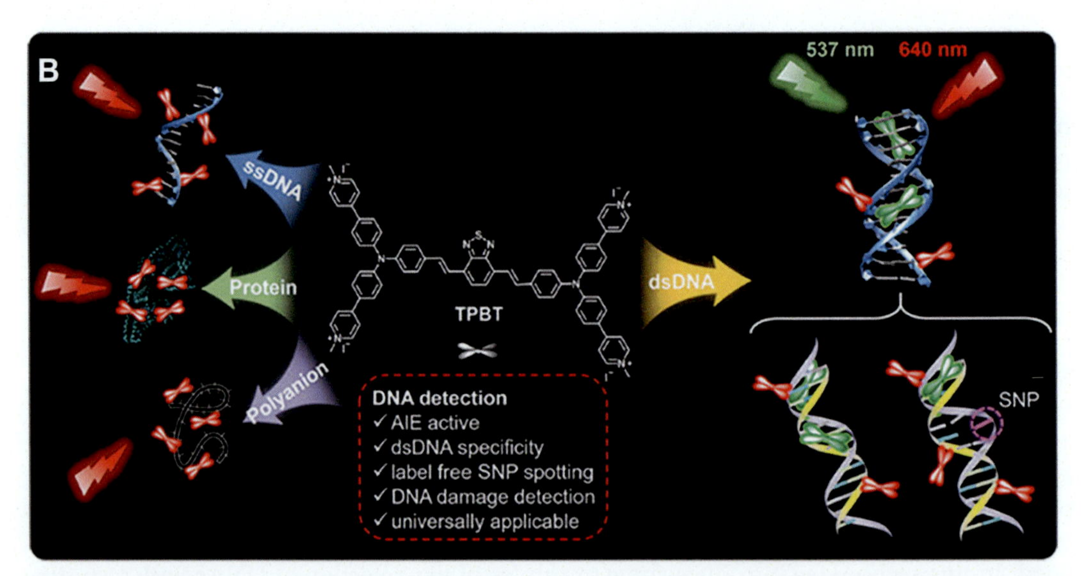

FIG. 4 DNA Detection with AIE probes. (A) water-soluble **TPE** derivative and its application in DNA sensing. (B) Schematic diagram showing the chemical structure of TPBT with red fluorescence emission in ssDNAs, proteins, and polyanions. Ratiometric emission in green and red in the presence of dsDNAs can be used as an indicator of SNPs. *Panel (A): Reprinted (adapted) with permission from H. Tong, Y. Hong, Y. Dong, M. Häußler, J.W.Y. Lam, Z. Li, Z. Guo, Z. Guo, B.Z. Tang, Fluorescent "Light-up" bioprobes based on tetraphenylethylene derivatives with aggregation-induced emission characteristics, Chem. Commun. 35 (2006) 3705–3707. Panel (B): Reproduced with permission from Y. Gao, Z. He, X. He, H. Zhang, J. Weng, X. Yang, F. Meng, L. Luo, B.Z. Tang, Dual-color emissive AIEgen for specific and label-free double-stranded DNA recognition and single-nucleotide polymorphisms detection, J. Am. Chem. Soc. 141 (51) (2019) 20097–20106. Copyright (2019) American Chemical Society.*

There is another AIE probe, namely, **TPBT** (Fig. 4B), which can detect double-stranded DNA (dsDNA) by binding through electrostatic and/or hydrophobic interactions in addition to polyanions and negatively charged albumin. **TPBT** can spectrally distinguish single-stranded DNA (ssDNA) and dsDNA simultaneously. This cationic probe is unique as it has a dual-color fluorescence emission of red (~640 nm) and green (~537 nm) wavelengths, and it can accurately detect UV damage to DNAs and SNPs [26]. **TPBT**'s green emission is exclusively dependent on binding of dsDNAs; this response is elicited because of conformational changes in groove binding. The red-color emission is observed when TPBT binds to ssDNAs, proteins, and other polyanionic molecules. To demonstrate **TPBT**'s capability for detecting SNPs or DNA damage, the wild-type (WT) HIV-1 DNA is compared with a single-base mismatch DNA and a completely mismatched HIV-1 DNA in a **TPBT**-heparin system. While the WT HIV-1 DNA emits the highest fluorescence intensity at 537 nm, the single-base mismatch DNA and the completely mismatched HIV-1 DNA, respectively, exhibit modest and the weakest fluorescence at 537 nm. This suggests that defects in the DNA-duplex affect the groove binding of **TPBT.** This is further validated in a similar experiment with the WT tau DNA, tau R406W mutation (a single-base mismatch), and a completely mismatched DNA. The fluorescence intensity of **TPBT** decreases upon higher tau R406W concentration. With a different duration of UV irradiation to DNAs, longer UV irradiation results in weaker fluorescence signals at 537 nm. Therefore, the unique characteristics of **TPBT** can potentially be scaled up for high-content screening of genetic mutations or variations.

3.2 G-quadruplex sensing

G-quadruplexes are guanine-rich DNAs or RNAs which can fold into noncanonical secondary structures [27]. They are associated with important genomic functions such as transcription, replication, and recombination [27–30]. The presence of G-quadruplex structures in telomeres, which cap at the end of chromosomes, suggests their roles in chromosome stability. However, their isolation with small-molecule ligands often leads to cell death [31]. Not only can they influence gene expression, they can also inhibit telomerase activity in cancer cells [32]. The diverse roles of G-quadruplexes in genomic function and stability make them an interesting topic in genetics.

1,1,2,2-Tetrakis[4-(2-triethylammonioethoxy)phenyl]ethene tetrabromide (**TTAPE-Et**, Scheme 2) is a simple AIE dye used in a label-free DNA assay system [32]. It is applied as a sensitive stain to visualize guanine-rich DNAs resolved by PAGE. **TTAPE-Et** binds to

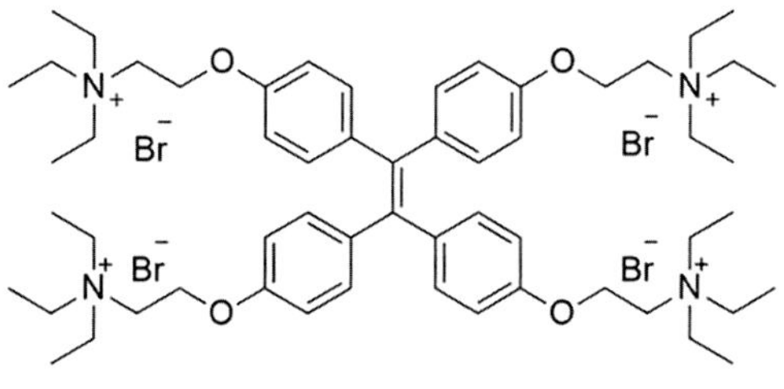

SCHEME 2 Structure of **TTAPE-Et.**

the human telomeric G-quadruplex (G1: 5′-GGGTTAGGGTTAGGGTTAGGG-3′) by electrostatic attraction. When DNA molecules are added into a solution of **TTAPE-Et** in the tris(hydroxymethyl)aminomethane (Tris)-HCl buffer system, **TTAPE-Et** begins to luminesce at 470nm as it forms a **TTAPE-Et**-G1 complex by electrostatic attraction which restricts the intramolecular rotations. In the presence of potassium ions, G1 will fold into a G-quadruplex structure. The initial addition of potassium ions causes the fluorescence intensity of **TTAPE-Et**-G1 complexes to drop drastically as the potassium ions have better binding to DNAs in comparison with **TTAPE-Et** due to their small size and high concentration. As the **TTAPE-Et**-G1 complex begins to form, there is recovery in the fluorescence intensity; the emission of the **TTAPE-Et**-G1 complex shows an approximately 20-nm red-shifted emission in the excitation and emission which peaks at 492nm. It specifically allows for hybridization of complementary strands where the **TTAPE-Et** probes are released back into the solution as the potassium ions compete for the ssDNAs. Mismatched sequences, on the other hand, do not disassemble the **TTAPE-Et**-G1 complex structure. This work provides us with a label-free method for G-quadruplex sensing, and this probe can potentially be used in high-throughput quadruplex-targeting anticancer drug screening.

4 Cellular organelles and structures

The eukaryotic cell is made up of different organelles which include mitochondria, lysosomes, the endoplasmic reticulum, the Golgi apparatus, the plasma membrane, etc. They are vital for cellular functions and their phenotypic features serve as indicators of diseases which allow monitoring of disease progression. Many conventional fluorophores used to visualize organelles are vulnerable to self-quenching at high concentrations. AIE dyes used for imaging of subcellular organelles such as mitochondria, lysosomes, and lipid droplets (LDs) benefit from their AIE properties, thus enjoying high SNR and good photostability [33,34]. There are review papers summarizing the AIE probes for imaging the subcellular organelles [35]. In this section, we will introduce some AIE probes for mitochondria imaging, lysosome imaging, and their applications in biomedical sciences.

4.1 Mitochondria imaging

Mitochondria are known as the "powerhouse of the cell" which provides energy to eukaryotic cells by ATP synthesis through oxidative phosphorylation [36]. Mitochondrial dysfunction occurs in neurodegenerative diseases, cancers, and cardiovascular diseases. Mitochondrial membrane potential ($\Delta\psi_m$) reflects the mitochondrial functional status and is an indicator in cell health. In sperms, sperm motility depends on mitochondrial activity, [37] which is evidenced by low $\Delta\psi_m$ in individuals with oligozoospermia (low sperm count and abnormalities) [38]. Ultimately, asthenozoospermia caused by abnormal $\Delta\psi_m$ may lead to male infertility. Therefore, the abilities to track the mitochondrial dynamics and function are beneficial for medical diagnosis and biomedical research.

TPE-Ph-In is a red-emitting cationic AIE probe with an indolium salt incorporated in TPE as an electron-accepting moiety which is crucial for mitochondrial specificity [39]. Classic cationic dyes developed to target mitochondria such as Rhodamine 123 often suffer from self-quenching problems. This means low concentrations of the dye must be used to reduce quenching problems. However, dyes with lower concentrations are often vulnerable to photobleaching. **TPE-Ph-In** and **TPE-In** are initially designed and synthesized together, but it is later found that the addition of an extra phenyl ring to **TPE-In** can significantly enhance its biocompatibility and the AIE effect (Fig. 5A and B). Costaining experiments of **TPE-Ph-In** and CellLight Mitochondria-GFP (Mito-GFP), a commercial mitochondrial targeting green fluorescent protein, confirm its selectivity and uniform labeling to mitochondria. Imaging without genetic modification is one of the advantages of chemical probes. On the other hand, Mito-GFP expression gives varied labeling because of the inhomogeneous transfection rates and different expression levels among different cells (Fig. 5C). Superior photostability of **TPE-Ph-In** is observed when staining mitochondria in HeLa cells in comparison to the commercial fluorescent mitochondrial probe MitoTracker Red FM (MT). The dye also displays low levels of cytotoxicity as shown in the MTT assay even when the concentration is as high as 8 μM. **TPE-Ph-In** stained sperm cells have high fluorescence intensity in healthy sperms compared to little or no fluorescence in the sperms with poor motility (Fig. 5D–F).

It is found that metabolic activities of mitochondria in cancer cells are different and distinguishable from those in normal cells. The mitochondria-specific AIE bioprobe **TPN** (Fig. 6A) was applied in identifying circulating tumor cells (CTCs) in the blood from cancer patients [40]. An interesting finding is that **TPN** has greater fluorescence intensity in cancer cell lines when compared to leukocytes, which are the most abundantly nucleated cells in blood (Fig. 6B and C). This suggests that the increased $\Delta\psi_m$ and number of mitochondria in cancer cells account for the stronger fluorescence signal. **TPN** has an absorption maximum at 405 nm and an emission peak at 636 nm. It is a cell-permeable "light-up" probe which simply "turns on" upon binding to mitochondria in live cells. Compared with immunofluorescence labeling using the cytokeratin marker, **TPN** staining in A549 cells is compatible with downstream

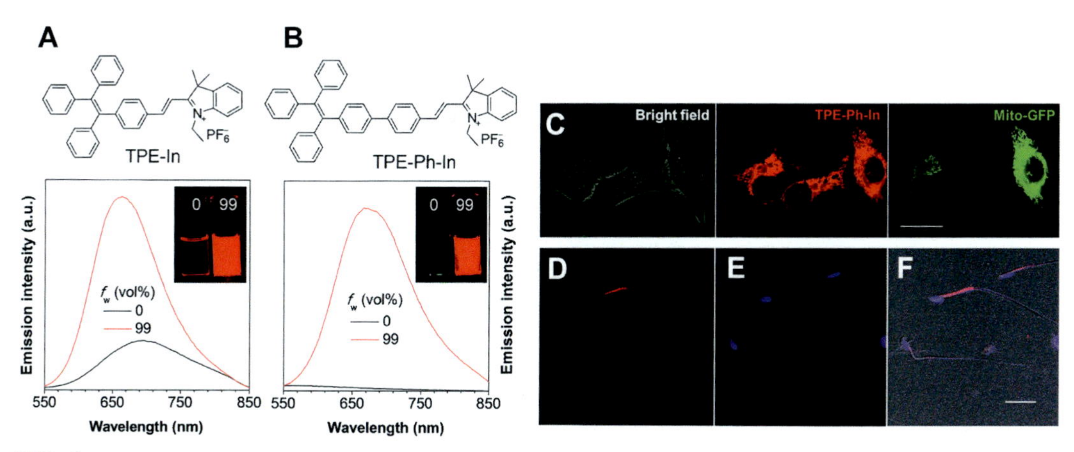

FIG. 5 Advantages of labeling mitochondria with the AIE probe **TPE-Ph-In**. Chemical structure and emission spectra of (A) **TPE-In** and (B) **TPE-Ph-In** in DMSO or 99% water fraction with DMSO (red; f_w). (C) Co-stained mitochondria in HeLa cells by **TPE-Ph-In** (red) and Mito-GFP (green). Scale bar: 20 μm. (D–F) Confocal images of mouse sperms stained with **TPE-Ph-In** (D), Hoechst 33342 (E), and the merged image (F). Scale bar: 20 μm. *Reprinted (adapted) with permission from N. Zhao, S. Chen, Y. Hong, B.Z. Tang, A red emitting mitochondria-targeted AIE probe as an indicator for membrane potential and mouse sperm activity, Chem. Commun. 51 (71) (2015) 13599–13602. Copyright (2018) American Chemical Society.*

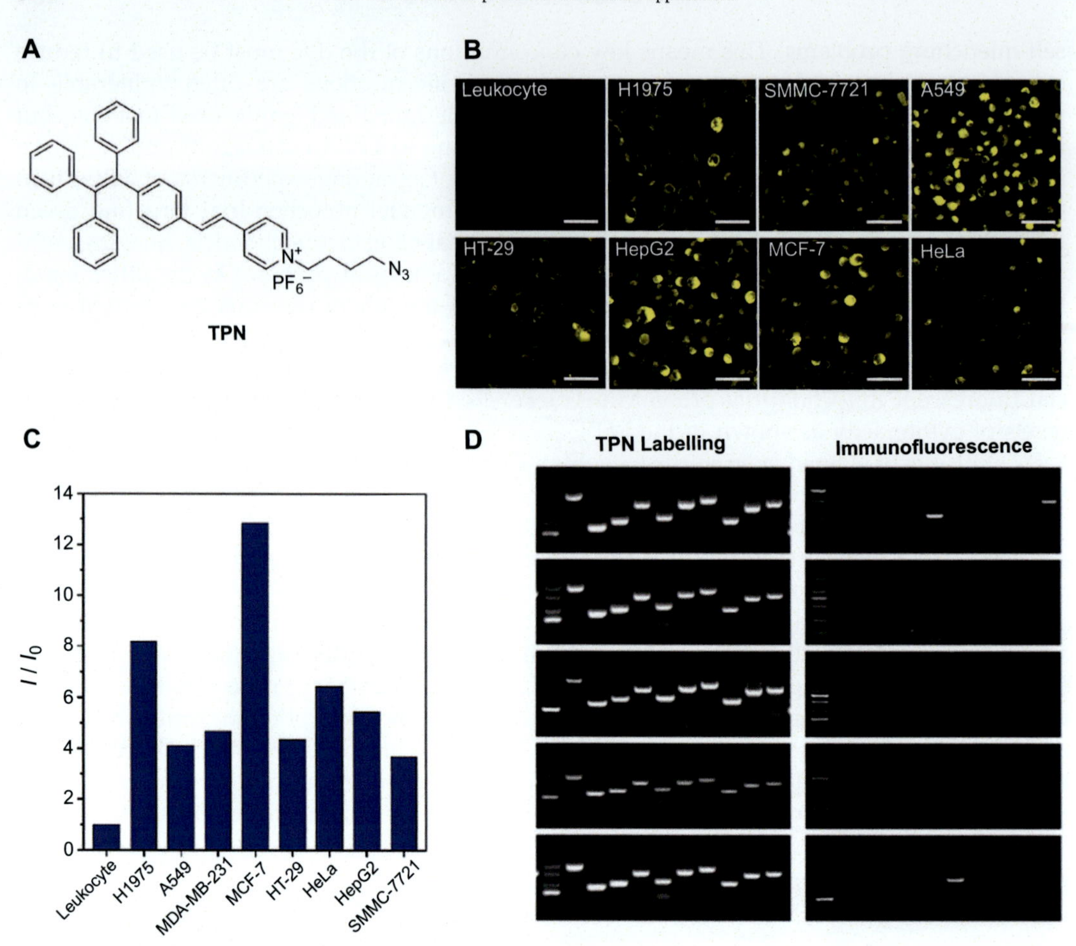

FIG. 6 CTC detection by the mitochondria-specific AIE probe **TPN**. (A) Chemical structure of **TPN**. (B) Fluorescence images of different cell lines and leukocytes with **TPN** staining (2 μM) for 10 min. (C) Relative fluorescence intensity analyzed by flow cytometry of TPN-stained leukocytes and different cancer cell lines. (D) Electropherogram of 10 genomic loci amplified by multiple displacement amplifications from TPN-labeled or cytokeratin-labeled A549 cells. *Reprinted with permission from B. Situ, X. Ye, Q. Zhao, L. Mai, Y. Huang, S. Wang, J. Chen, B. Li, B. He, Y. Zhang, J. Zou, B.Z. Tang, X. Pan, L. Zheng, Identification and single-cell analysis of viable circulating tumor cells by a mitochondrion-specific AIE bioprobe, Adv. Sci. 7 (4) (2020) 1902760. (copyright 2020) John Wiley and Sons.*

single-cell genomic analysis as all ten genomic loci can be covered and amplified. This contrasts with just one to two expected bands shown in the conventional immunofluorescence approach (Fig. 6D).

AIE-based mitochondria probes have also been used in selective staining and killing of cancer cells [41]. The use of light as treatment (phototherapy) was applied by ancient civilizations over three thousand years ago to treat diseases such as psoriasis and skin cancer [42–44]. Abnormal cancer cells have a higher demand for energy for its high metabolism, which is reflected in its abnormally higher $\Delta\psi_m$. **TPE-IQ-2O** is also a $\Delta\psi_m$-sensitive AIE-active probe. It has a strong absorption peak at 430 nm with a strong emission peak at

620 nm in the aggregated state. Since cancer cells have a more hyperpolarized $\Delta\psi_m$ than normal healthy cells, **TPE-IQ-2O** stained cancer cell lines, such as MDA-MB-231, MCF-7, PC-9, A549, HCC-827, and HepG2, show greenish yellow emissions. In comparison, **TPE-IQ-2O** stained noncancer cell lines, such as LX-2, COS-7, HEK-293, and MDCK-II, show a significantly weaker fluorescence. More importantly, **TPE-IQ-2O** is not only a probe that can differentiate cancer cells from normal cells but also a promising photosensitizer (PS) for PDT. When costained with the ROS indicator **H_2DCF-DA** and **TPE-IQ-2O**, HeLa cells show increased **H_2DCF-DA** signal upon white light illumination, suggesting a high level of ROS production. Since **TPE-IQ-2O** is more selective to mitochondria in cancer cells, it shows selectivity in killing cancer cells and exhibits almost no toxic effect on COS-7 cells under light illumination. This probe therefore is a promising theranostic agent with great potential in biomedicine and therapeutic research [41].

4.2 Lysosome imaging

Lysosomes are membrane-bound organelles with digestive capabilities, containing over 60 kinds of proteases, lipases, nucleases, and hydrolytic enzymes which break down unwanted polymeric biomacromolecules and organelles in eukaryotic cells [45]. In return, metabolites and ions can be recycled to maintain cellular homeostasis. The lysosome also protects cells by killing invading pathogens [46]. Since all lysosomal enzymes are acid hydrolases, the optimal luminal pH in lysosomes is 4–5. As a result, alkalization of lysosomes can impede autophagosome degradation [47]. Limited investigation has been done on tracking lysosomal pH and tissue regeneration in vivo. The cyanostilbene derivative (Z)-3-(4-(4-methylpiperazin-1-yl)phenyl)-2-(4-(pyridin-4-yl)phenyl)acrylonitrile, abbreviated as **CSMPP**, is an AIE molecule that has a crystal structure with a slightly twisted conformation with dihedral angles of 10.95° and 11.08° and adjacent phenyl rings to the acetonitrile group that are freely rotatable in a solvent. It shows weak emission in acetonitrile (Fig. 7A) but a sevenfold higher intensity in acetonitril/water mixed solution with 90% water fraction (f_w) (Fig. 7B) [48]. As the pH decreases from 6.8 to 2.6, the peak at 503 nm gradually shifts from green to red at 615 nm as it is protonated. This ratiometric probe can be used for pH monitoring as the $\log[(R-R_{min})/(R_{max}-R)]$ has a good linear relationship with the pH which is important for studying the pH response of lysosomes in biological events. For example, during tissue regeneration in the caudal fin of the medaka larvae 24–48 hours Postamputation (hpa), lysosomal pH becomes more acidic. This decrease in pH can be reported by **CSMPP**, especially at the peak of the regeneration during 24–48 hpa (Fig. 7C). **CSMPP**'s use in biomedical research can therefore be focused on monitoring tissue regeneration or lysosome physiology during pathological progression found in living systems [49].

5 Lipids

Lipids are important constituents of the membranes in eukaryotic cells. The organization and distribution of lipids vary in the plasma membrane or organelles depending on the cell type. They play roles in cell signaling, vesicle trafficking, and energy storage. Eukaryotic cells can produce over 1000 lipid species and approximately 5% of their genes are responsible for

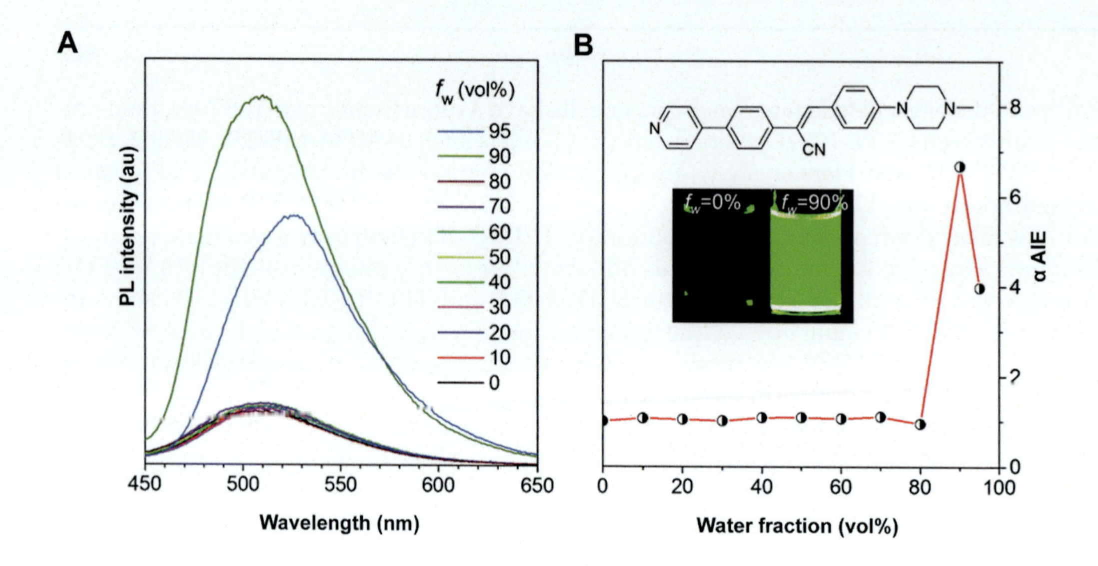

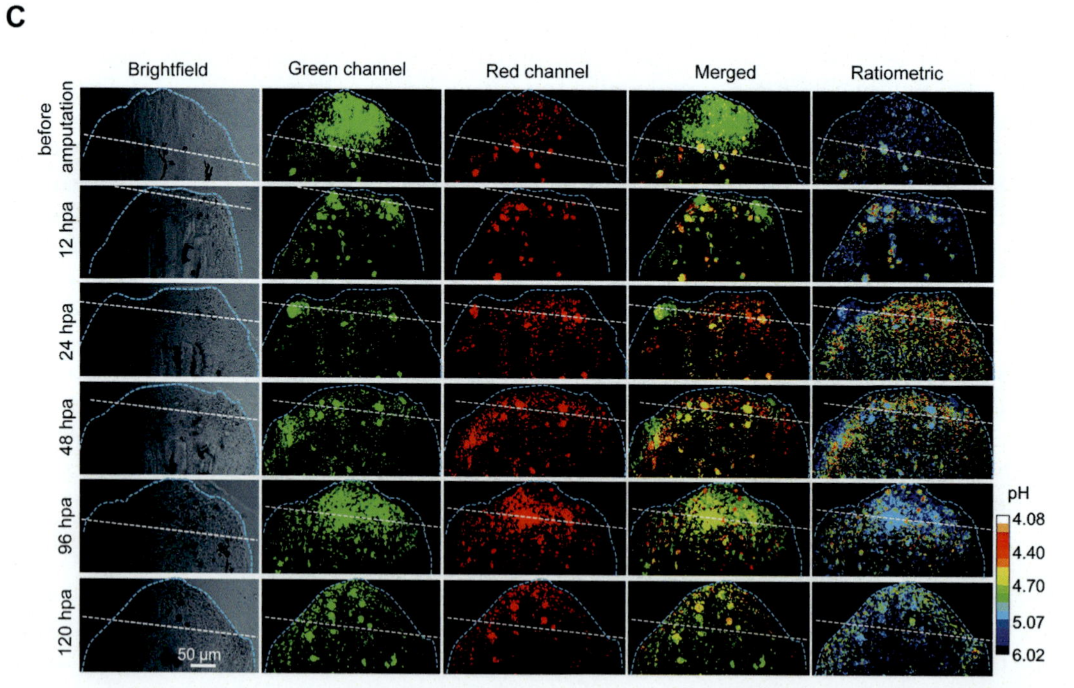

FIG. 7 Lysosomal labeling by **CSMPP** for tracking of tissue regeneration. (A) **CSMPP** PL spectra in acetonitrile/water mixtures with different water fractions (f_w). (B) Chemical structure of **CSMPP** and ratio of PL intensities in different acetonitrile/water mixtures (I) or in pure acetonitrile (I_0) at 509 nm. α_{AIE}: I/I_0. Fluorescence images are **CSMPP** solutions with $f_w = 0\%$ (left) and $f_w = 90\%$ (right) under a handheld UV lamp irradiation with an excitation wavelength at 365 nm. (C) Confocal images of the caudal fin in medaka larvae before amputation and after amputation at different time points (12, 24, 48, 96, and 120 hpa) on the white dotted line show the amputation plane. The ratiometric image shows the lysosomal pH distribution with a pH color scale. The ratiometric images are calculated by red emission/green emission. Cyan dashed lines indicate the outline of the caudal fin. Hpa denotes hours postamputation. Scale bar: 50 μm. *Reprinted (adapted) from X. Shi, N. Yan, G. Niu, S.H.P. Sung, Z. Liu, J. Liu, R.T.K. Kwok, J.W.Y. Lam, W.X. Wang, H.H.Y. Sung, I.D. Williams, B.Z. Tang, In vivo monitoring of tissue regeneration using a ratiometric lysosomal AIE probe, Chem. Sci. 11 (12) (2020) 3152–3163.*

synthesizing lipids. The full utilization of the complete repertoire of eukaryotic lipids is still a mystery and to be elucidated [50,51]. Lipids used for energy storage are primarily stored in LDs which are fundamentally made up of triacylglycerols and steryl esters [51]. This reservoir of energy is tapped into the body when deprived. However, in metabolic diseases such as obesity or metabolic syndrome, LDs can be accumulated to abnormal levels in the liver known as hepatic steatosis. There are different disorders that are attributed to hepatic steatosis (also known as fatty liver disease, FLD); for instance, obesity-related steatosis and nonalcoholic FLD (NAFLD) which have a global incidence rate of 25%. Around 30%–40% of those patients manifest other complications such as cirrhosis or hepatocellular carcinoma [52].

Fluorescent probes are useful for colocalization studies as well as monitoring subcellular localization and dynamics of certain lipids in live cells [50]. It is rather difficult to label lipids of interest with antibodies since procedures such as fixation and permeabilization can dissolve or alter the distribution of lipids present in the cell membrane and intracellular compartments [53]. LDs are dynamic organelles involved in many physiological processes, and changes in their numbers and activities are related to cancers and other diseases.

5.1 Lipid imaging for hepatic steatosis study

Examples of AIE probes that can visualize LDs are **FAS** and **DPAS** (Fig. 8A). **FAS** and **DPAS** (Fig. 8A) can be synthesized from readily available commercial raw materials. The differences in their conformation result in orange (595 nm) and yellow (565 nm) fluorescence for **FAS** and **DPAS**, respectively. Their good biocompatibility is demonstrated in the MTT assay. [54] They are small molecules with a lack of specific functional groups to avoid localization preference to other subcellular organelles. Both the dyes show good selectivity to LDs in HepG2 cells which were used as an in vitro model of hepatic steatosis (Fig. 8A). Their low background and high specificity also allow for qualitative and quantitative studies of the lipids in the biological samples [55,56]. The staining mechanism behind these two AIE probes is their hydrophobic nature, as they can be internalized into the cells via endocytosis followed by labeling the hydrophobic LDs [54]. They outperform commercial LD probes like Nile red and some of the BODIPY-based probes in terms of photostability (Fig. 8B).

Besides cultured cell lines, LDs can also be visualized in liver tissue sections; this is of interest as FLD is a common medical condition. Nonalcoholic FLD is reversible but severe and advanced stages can manifest into more life-threatening illnesses such as hepatic carcinoma or cirrhosis. There is high demand for a sensitive and rapid histological staining of LDs. Oil Red O is widely used for clinical diagnosis of FLD, but its performance is questionable and there is a lack of selectivity toward LDs as well. In addition, preparing Oil Red O is time consuming and requires a large amount of organic solvents such as ethanol for the staining protocol because of its low solubility which can cause artificial fusion of LDs [57].

The AIEgen **ABCXF** is a donor-acceptor incorporated fluorescent two-photon compound for LD labeling. It has a nonaromatic rotor (CF_3) on the conjugated core which contributes to the probe's AIE properties. It shows a planar intramolecular charge transfer (PICT) effect; having an intramolecular charge transfer is advantageous for red shift in the emission, increasing the intensity of fluorescence and electron mobility. **ABCXF**'s performance in visualizing LDs in FLD tissues is investigated using guinea pigs fed with high-fat diets and normal

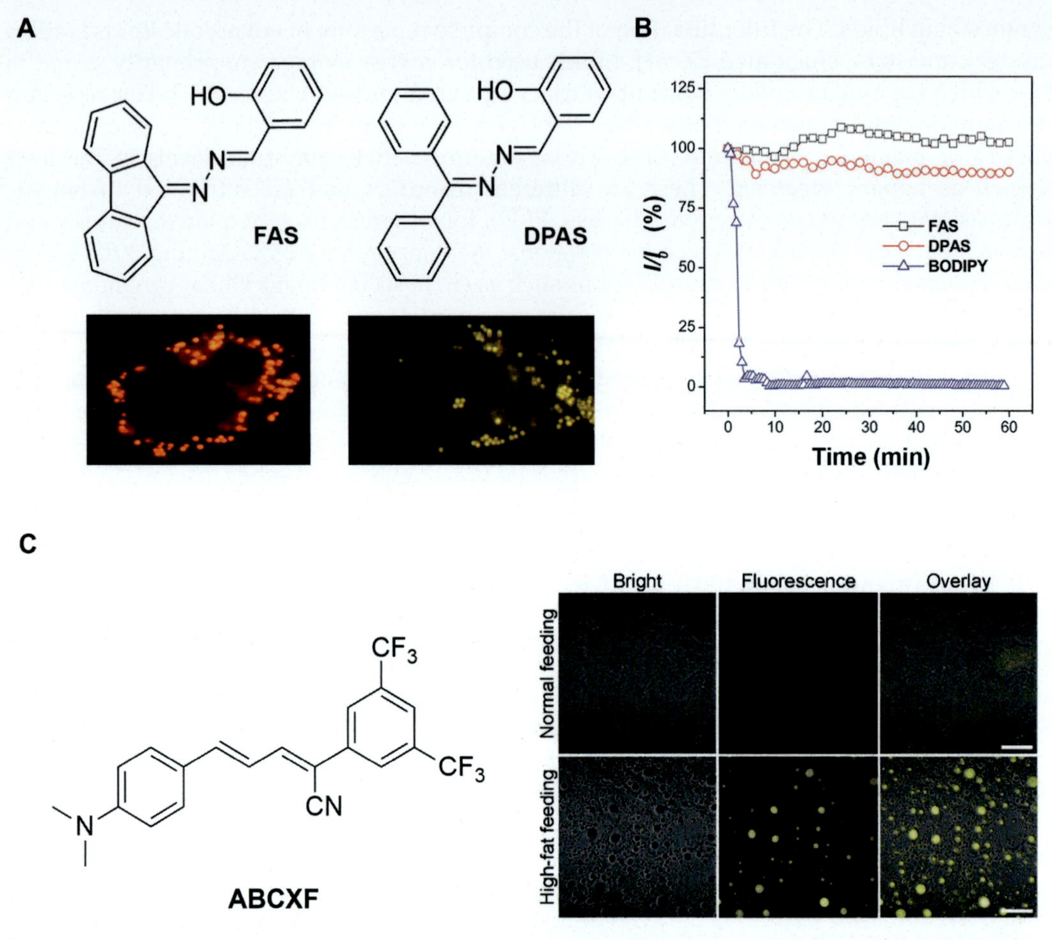

FIG. 8 AIE probes **FAS**, **DPAS**, and **ABCXF** for LD staining. (A) Chemical structures of the AIE probes **FAS** and **DPAS** and their corresponding fluorescence live-cell images. (B) Fluorescence intensity of **FAS**, **DPAS**, and BODIPY under illumination over time to show photostability. (C) Left: chemical structure of **ABCXF**. Right: brightfield and one-photon fluorescence images of liver tissues stained with **ABCXF** from normal or high-fat fed guinea pigs. Scale bars: 20 μm. *Panels (A, B): Reprinted (adapted) with permission from Z. Wang, C. Gui, E. Zhao, J. Wang, X. Li, A. Qin, Z. Zhao, Z. Yu, B.Z. Tang, Specific fluorescence probes for lipid droplets based on simple AIEgens, ACS Appl. Mater. Interfaces 8 (16) (2016) 10193–10200. Copyright (2016) American Chemical Society. Panel (C): Reprinted (adapted) with permission from H. Park, S. Li, G. Niu, H. Zhang, Z. Song, Q. Lu, J. Zhang, C. Ma, R.T.K. Kwok, J.W.Y. Lam, K.S. Wong, X. Yu, Q. Xiong, B.Z. Tang, Diagnosis of fatty liver disease by a multiphoton-active and lipid-droplet-specific AIEgen with nonaromatic rotors, Mater. Chem. Front. 5 (2021) 1853–1862. Copyright (2021) Royal Society of Chemistry.*

diets as a control (Fig. 8C). Sectioned tissues from the liver are incubated in **ABCXF** and imaged with single-photon (λ_{ex} = 488 nm) or two-photon (λ_{ex} = 850 nm) imaging with an emission of 646 nm in the solid state; two-photon imaging displays a higher SNR and a deeper tissue penetration than single-photon imaging. Compared to Oil Red O which starts to fade dramatically after 48 h, **ABCXF** retains 90% of its fluorescence intensity. Most commercial fluorescent probes such as BODIPY dyes are susceptible to photobleaching due to their low working concentration under constant irradiation by high intensity two-photon lasers. The low concentrations are used instead because a high concentration of the commercial probes can cause the undesirable ACQ effect. **ABCXF** is a convenient, reliable, and robust alternative for histological LD staining in tissue sections. Its nonaromatic rotor design and PICT-based AIE properties could also guide the developments for an NIR-emissive probe for in vivo biomedical imaging [58].

5.2 Lipid imaging for atherosclerotic plaques study

Cardiovascular disease (CVD) is one of the leading causes of premature mortality worldwide. There were 422.7 million cases of CVD (95% uncertainty interval: 415.53 to 427.87 million cases) and 17.92 million CVD deaths (95% uncertainty interval: 17.59–18.28 million CVD deaths) in 2015 [59]. According to the World Health Organization (WHO), around 23.6 million people are predicted to die from CVD annually by 2030 [60]. Therefore, treatments for atherosclerosis are of great concern to reduce premature mortality. Histological methods are commonly used to visualize atherosclerotic plaques or vascular lipids by chemical staining of tissue sections which is tedious and time-consuming. **IND** is a smart fluorescent switchable AIE probe that allows for high-resolution imaging of atherosclerotic plaques in situ. It can be imaged with multiphoton microscopy with an absorption cross section of 33.4 GM (Göeppert-Mayer) at 900 nm [61].

The lipid-targeting mechanism of the dye is due to the two alkyl tails in **IND** which are designed for lipid targeting. The experiment demonstrating the lipid-sensing capabilities is performed by adding **IND** to a mixture of sunflower oil and water (Fig. 9A). As a dual-color probe, **IND** emits green fluorescence at 522 nm in the presence of lipids and red fluorescence with a peak at 640 nm in water (Fig. 9B). This can be explained by the "packaging modes" of **IND** which exist as monomers in the lipid environment and pairwise dimers in aqueous solution (Fig. 9C). The compatibility of **IND** with two-photon microscopy and a high SNR allows the lipids within foam cells (lipid-laden macrophages) and plaques to be visualized at high resolution and reconstructed in three-dimensional images for quantification (Fig. 9D–F).

This probe will allow for a more accurate presentation of plaques in three-dimensional studies of atherosclerosis compared to conventional analysis in two-dimensional sections. Because of the compatibility with multiphoton microscopy and good performance with a high SNR, **IND** can also be used for lipid/water analysis in other organs due to its excellent tissue penetrability, biocompatibility, and dual-color emissive properties.

6 Inflammation and cancer research

6.1 HClO sensing for inflammation and cancer study

Chronic inflammation can increase the risk of cancers as they are casually related [62]. It is of concern as it involves long-term dysregulated tissue homeostasis and loss of protective response to tissue damage. Hypochlorous acid (HClO) is an ROS that is naturally generated in physiological processes such as the reaction of hydrogen peroxide (H_2O_2) and chloride ions (Cl^-). Abnormal increases in HClO levels, however, are related to a variety of inflammatory diseases. Although it is believed that inflammatory signals are the main driving force of

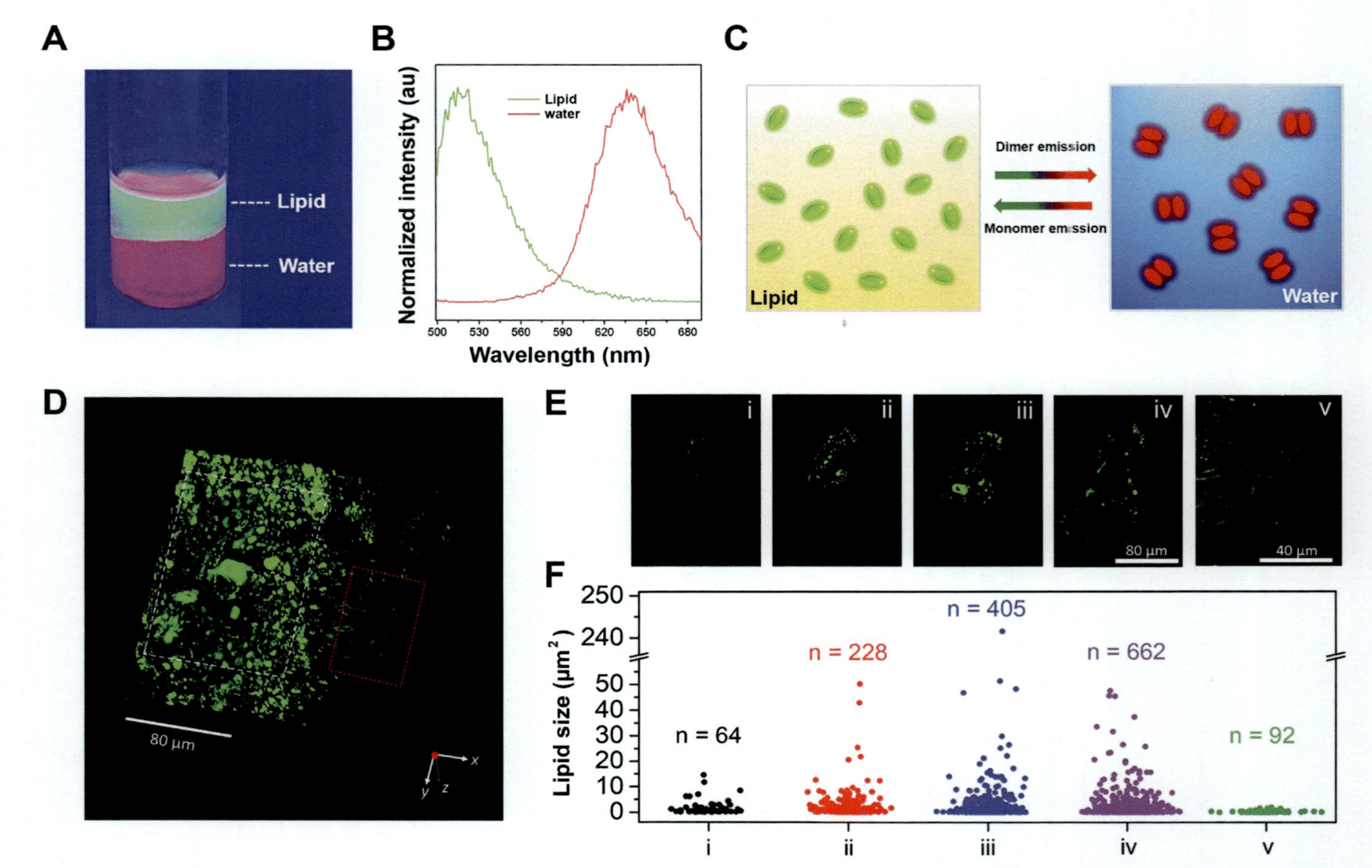

FIG. 9 Visualization of atherosclerotic plaque structure with the AIE-based probe **IND**. (A) Photograph of **IND** (10 μM) taken under UV irradiation at 365 nm in a lipid/water mixture. (B) Normalized intensity (PL spectra) of **IND** (10 μM) in lipid (green) or water (red). (C) Illustration of **IND** dimers in water exhibiting red emission and monomers in the lipid environment showing green fluorescence. (D) Volumetric reconstruction of an atherosclerotic plaque from a two-photon z-stack. (E) Z-stack slices at the imaging depths of 5, 10, 15, and 25 mm (i–iv) from the red-dotted ROI in (D). The zoomed image of the adjacent normal arterial intima corresponding to the red boxed area in panel D (v). (F) Quantitative measurement of fluorescence signals in panel E. *Reprinted (adapted) with permission from B. Situ, M. Gao, X. He, S. Li, B. He, F. Guo, C. Kang, S. Liu, L. Yang, M. Jiang, Y. Hu, B.Z. Tang, L. Zheng, A two-photon AIEgen for simultaneous dual-color imaging of atherosclerotic plaques, Mater. Horiz. 6 (3) (2019) 546–553. Copyright (2019) Royal Society of Chemistry.*

cancers, several reports also suggest that increasing HClO levels and ROS using an anticancer drug candidate such as elesclomol can treat cancers by inducing cancer cell apoptosis [63]. This calls for a more efficient and specific ROS detection method in certain concentration ranges to study the ROS impact in cancers and drug discovery.

Two aqueous-soluble AIE probes CH_3O-TPE-Py^+-N^+ (**COTN**) and OH-TPE-Py^+-N^+ (**HOTN**) are designed for HClO-sensing [64]. The presence of HClO oxidizes the double bond of **COTN** and **HOTN** with the cleavage of the Py^+-N^+ group. The resulting products **COT** and **HOT** thus have an increased hydrophobicity and enhanced emission, respectively (Fig. 10A). In comparison, the fluorescence intensity of **COT** is only a quarter of that of **HOT**. This is because only **HOT** can form an intermolecular hydrogen bond between the pair of molecules, leading to molecular bending and complex stability in creating a more significant fluorescence response (Fig. 10B). The AIE response can also be observed for in situ staining of the liver, where a cancerous liver can be distinguished from healthy liver by the fluorescence intensity (Fig. 10C).

6.2 Intraoperative pathological diagnosis of hepatocellular carcinoma

The working mechanism of **TPE-IQ-2O** used for cancer cell staining is based on the hyperpolarized mitochondrial membrane potential in cancer cells. This is, however, not applicable to intraoperative histological examination because of the need for fixing the specimens immediately during the surgical operation. In this regard, Chen et al. developed an AIE probe of dicyanoacrylate-containing hexaphenyl-1,3-butadiene (**ZZ-HPB-NC**, Fig. 10D) having two absorption peaks at 318 and 397 nm. In pure THF, the energy of the excited state can be dissipated by freely rotatory benzine rings in **ZZ-HPB-NC** through nonradiative transitions. In polar solvents, **ZZ-HPB-NC** exhibits a 16-fold higher fluorescence emission peaking at 560 nm. Despite the working mechanism not being fully understood, **ZZ-HPB-NC** stains paraformaldehyde-fixed hepatocellular carcinoma (HCC) but not noncancer cells. The fluorescence turn-on phenomenon of **ZZ-HPB-NC** is further verified in the tumor foci of the intraoperative frozen sections from patients with liver neoplasms, suggesting that **ZZ-HPB-NC** is practically functional in the tumor microenvironment. Importantly, this detection method is sensitive enough to distinguish malignant HCC foci from benign focal nodular hyperplasia (FNH) and paratumor cirrhosis/normal liver tissues with an increased fluorescence intensity of 4.3-fold and 12.2-fold, respectively (Fig. 10D). Instead of performing the time-consuming hematoxylin and eosin (H&E) stain in the intraoperative frozen-section diagnosis, **ZZ-HPB-NC** staining shortens the workflow to approximately 10 min and reduces the workload of clinical pathologists. This offers great potential clinical applications to intraoperative HCC diagnosis.

7 Cytogenetic studies

Chromosomes, which are composed of DNAs and proteins, store genetic information in cells. Cytogenetics focuses on the study of the chromosome number, structure, gene locus in the chromosome, and related abnormalities. Researchers have also explored the applications of AIE probes in cytogenetic studies. In 2013, Hong et al. successfully stained the chromosomes from animal cells, plant cells as well as polytene chromosomes from *Drosophila*

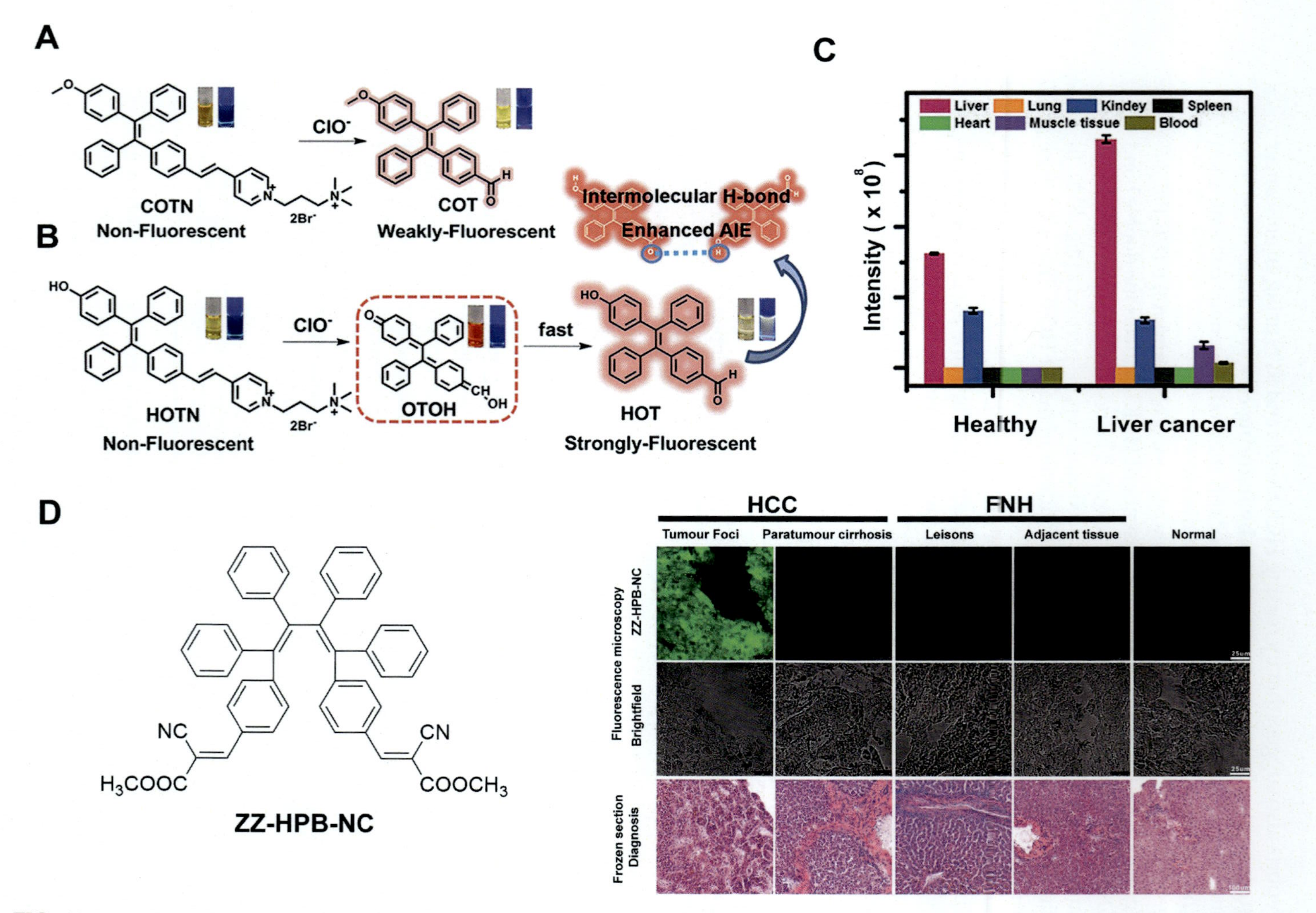

FIG. 10 AIE **HClO** sensors for monitoring of inflammation and cancers. (A) An illustration of **COTN** fluorescence turn-on. (B) **HOTN** transition to **HOT** with enhanced AIE via intermolecular hydrogen bonding and photographs of their visible color changes. (C) Color intensity of fluorescence emission of various organs, tissues, and blood (from left to right: liver, lung, kidney, spleen, heart, muscle tissue, and blood) in the **HOTN**-injected WT mouse compared with another mouse with hepatocarcinoma (liver cancer). (D) Chemical structure of **ZZ-HPB-NC** (left) and fluorescent images of **ZZ-HPB-NC** stained HCC, FNH, and normal liver specimen compared with H&E stained frozen sections. *Panels (A–C): Reprinted (adapted) with permission from X. Han, Y. Ma, Y. Chen, X. Wang, Z. Wang, Enhancement of the aggregation-induced emission by hydrogen bond for visualizing hypochlorous acid in an inflammation model and a hepatocellular carcinoma model, Anal. Chem. 92 (3) (2020) 2830–2838. Copyright (2020) American Chemical Society. Panel (D): Reprinted (adapted) with permission from D. Chen, H. Mao, Y. Hong, Y. Tang, Y. Zhang, M. Li, Y. Dong, Hexaphenyl-1,3-butadiene derivative: a novel "Turn-on" Rapid fluorescent probe for intraoperative pathological diagnosis of hepatocellular carcinoma, Mater. Chem. Front. 4 (9) (2020) 2716–2722. Copyright (2020) Royal Society of Chemistry.*

SCHEME 3 Chemical structures of TTAPE-Me (left) and TTAPE-Et (right).

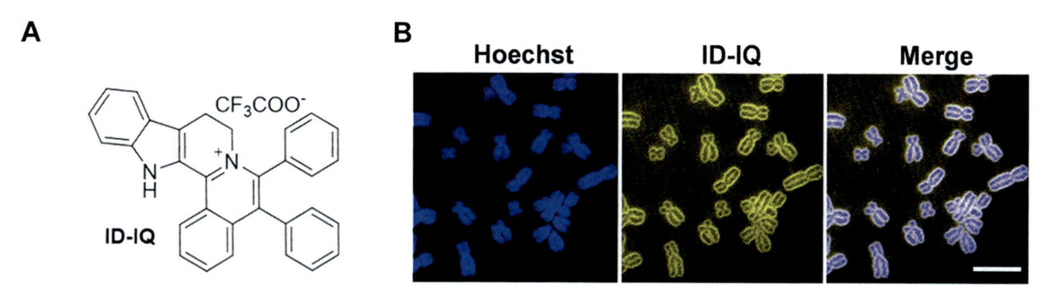

FIG. 11 (A) The chemical structure of the AIE probe: **ID-IQ**. (B) Confocal images of chromosomes stained with Hoechst 33342 and **ID-IQ** in hES2 cells. Scale bar: 10 μm. *Reprinted (adapted) with permission from M.Y. Wu, J.K. Leung, L. Liu, C. Kam, K.Y.K. Chan, R.A. Li, S. Feng, S. Chen, A small-molecule AIE chromosome periphery probe for cytogenetic studies, Angew. Chem. Int. Ed. 59 (2020) 10327–10331. Copyright (2020) John Wiley and Sons.*

salivary glands using water-soluble cationic TPE derivatives **TTAPE-Me** and **TTAPE-Et** (Scheme 3) [65].

Recently, Wu et al. have reported the first small-molecule organic probe for the chromosome periphery (CP), which is an AIE-active molecule. The CP consists of a complex network of proteins and RNAs which has only been identified recently [66]. The current techniques available for CP labeling include the immunofluorescence and expression of genetically encoded fluorescent proteins using Ki-67, Ku70/80 complex, and Bcl2 as targets. No organic fluorescent dyes have so far been reported to selectively label the CP until the development of **ID-IQ** (Fig. 11A) [67]. Labeling the CP in fixed chromosomes with **ID-IQ** showed the chromosome boundaries (Fig. 11B). It enables the rapid segmentation of touching and overlapping chromosomes, accurate localization of the centromere, and rapid identification of chromosome morphology. This staining method is also compatible with fluorescence in situ hybridization (FISH). It is believed that **ID-IQ** will greatly benefit clinical diagnostic testing and genomic research.

8 Microbiology study

The detection and discrimination of microorganisms are of great significance to biomedical applications such as clinical diagnosis and therapeutics. The classic Gram staining method is the most widely used approach for bacterial detection while bearing several shortcomings including long procedures, complex manipulation, and low sensitivity. Alternatively, fluorescence staining is a better choice for fast and sensitive detection of microorganisms. To date, many efforts have been made for the development of AIE probes for microorganism detection, discrimination, and elimination. In this part, we will summarize the previous works of AIE probes for microbiology studies, discuss the pros and cons as well as the strategies of probe development. A brief outlook of the future directions for AIE-based microbiology studies will also be given.

8.1 AIE probes for bacteria research

8.1.1 AIE probes for bacterial detection

The small-molecule AIE probe compound 4 is designed by decorating TPE with an electron-donating ether group and an electron-withdrawing pyridine salt, resulting in a bacteria-targeting fluorescent probe with a red-shift (*ca.* 60 nm) emission (Fig. 12) [68]. Since there are two positively charged amino salt groups in compound 4 and the bacterial cell envelopes of both Gram-positive and Gram-negative bacteria are negatively charged, this probe stains all kinds of bacteria. Owing to its AIE characteristics and good solubility in aqueous solution, compound 4 can detect bacteria in solution with low background signals and the bacterial concentration range of 5×10^6 to 2×10^8 CFU/mL in a linear manner. Besides bacterial imaging, this probe is also used for high-throughput antibiotic screening by measuring the bacterial concentration. The inhibition of bacterial growth by effective antibiotics results in a reduced bacteria number, thus leading to a decreased fluorescence signal.

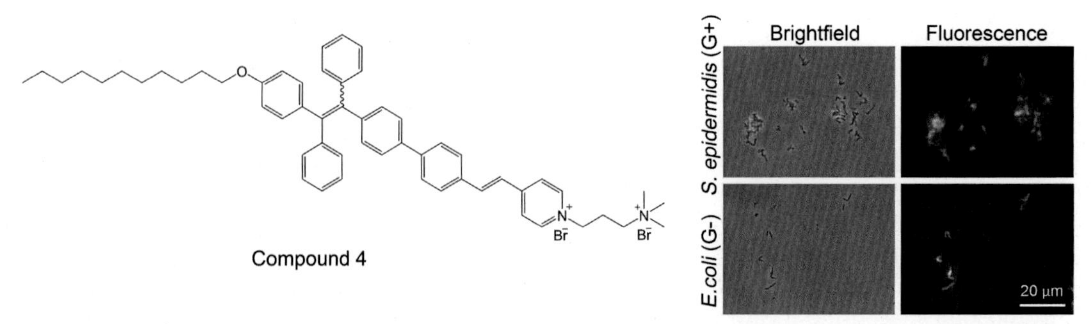

FIG. 12 AIE probes for bacterial detection. The chemical structure of compound 4 (left) and fluorescence detection of Gram-positive (*S. epidermidis*) and Gram-negative (*E. coli*) bacteria with the small-molecule AIE probe (compound 4). *Reprinted (adapted) with permission from E. Zhao, Y. Chen, S. Chen, H. Deng, C. Gui, C.W.T. Leung, Y. Hong, J.W.Y. Lam, B.Z. Tang, A luminogen with aggregation-induced emission characteristics for wash-free bacterial imaging, high-throughput antibiotics screening and bacterial susceptibility evaluation, Adv. Mater. 27 (33) (2015) 4931–4937. Copyright (2015) John Wiley and Sons.*

The probe mentioned above utilized the electrostatic interaction between the positive charges in the probe and negative charges in the bacterial cell envelope. Alternatively, another efficient strategy is to incorporate the probe into bacteria through metabolic labeling. The AIE probe **TPEPy-D-Ala** (Scheme 4) is synthesized through the click reaction between the AIE fluorophore **TPEPy-Butyne** (Scheme 4) and a derivative of unnatural amino acid (3-azido-D-alanine) [69]. When dissolved in solutions, the probe is almost nonfluorescent due to its intramolecular rotation. However, once incorporated into the cell walls of bacteria through metabolic pathways, the probe can emit strong fluorescence due to RIR. Provided that the metabolism of unnatural amino acid D-alanine in the bacterial cell wall biosynthesis does not take place in mammalian cells, the probe can detect both Gram-positive and Gram-negative bacteria without affecting mammalian cells. It is worth noting that **TPEPy-D-Ala** needs to be incorporated into peptidoglycan during metabolic labeling. This process takes 20 min, which limits the immediate bacterial detection by the probe.

To provide scientists with more powerful tools for biological and pharmaceutical research, such as high-throughput screening of antibiotics, efficient evaluation of antibiotics, and classifying bacteria, selective bacterial probes are urgently needed. For different purposes, many AIE probes for selective bacteria detection have been developed according to their corresponding criterion. A cell-impermeable DNA dye, namely **TPE-2BA**, is generated to detect dead bacteria (Fig. 13A) [70]. Owing to its AIE properties and good solubility, the probe is nonfluorescent when dissolved or incubated with live bacteria. It only emits strong fluorescence when incubated with dead bacteria with low background signals. Unlike propidium iodide (PI) which stains both DNAs and RNAs, **TPE-2BA** stains only DNAs. The staining effect of dead bacteria by **TPE-2BA** can last for as long as 3 days, which is suitable for the long-term tracking of bacterial viability.

The replacement of the Gram staining method of bacteria with fluorescence staining is of great interest to researchers for a long time. Through conjugating the triphenylamine (**TPA**) to a functional cationic pyridinium moiety with a carbon–carbon double bond, a benzene ring, or a thiophene unit, a series of AIE probes (**TPy**, **TPPy**, **TTPy**, and **MeOTTPy**) are generated for selective detection of Gram-positive bacteria [71]. One of the examples is shown in Fig. 13B. All these probes exhibit excellent selectivity to Gram-positive bacteria. It is speculated that the probes are more extended to interact with the negatively charged single-layer membrane of Gram-positive bacteria rather than the multilayer membrane of Gram-negative bacteria. The hydrophobicity of the probes affects the intensity of fluorescence signals.

The DPAN-based probe **M1-DPAN** is also an AIE probe that is selective to Gram-positive bacteria (Fig. 13C) [72]. **M1-DPAN** was synthesized from the modification of a lipid probe **DPAS** (which is a nonionized small molecule). The phenyl group of **DPAS** is replaced by a naphthyl group, and two morpholine are added to **DPAS** through a hexyloxy group. The electrostatic interaction between the alkaline morpholine group and the acidic cell wall of Gram-positive bacteria, which contains a large portion of acidic macromolecules, contributes to the anchoring of the probe to the bacterial cell envelope. Subsequently, the insertion of the DPAN moiety to thicken and loosen the cell walls of Gram-positive bacteria through strong hydrophobic interactions results in the AIE effect of the probe. **M1-DPAN** selectively recognized Gram-positive bacteria from Gram-negative bacteria and fungi with the advantages of a strong signal, low background,

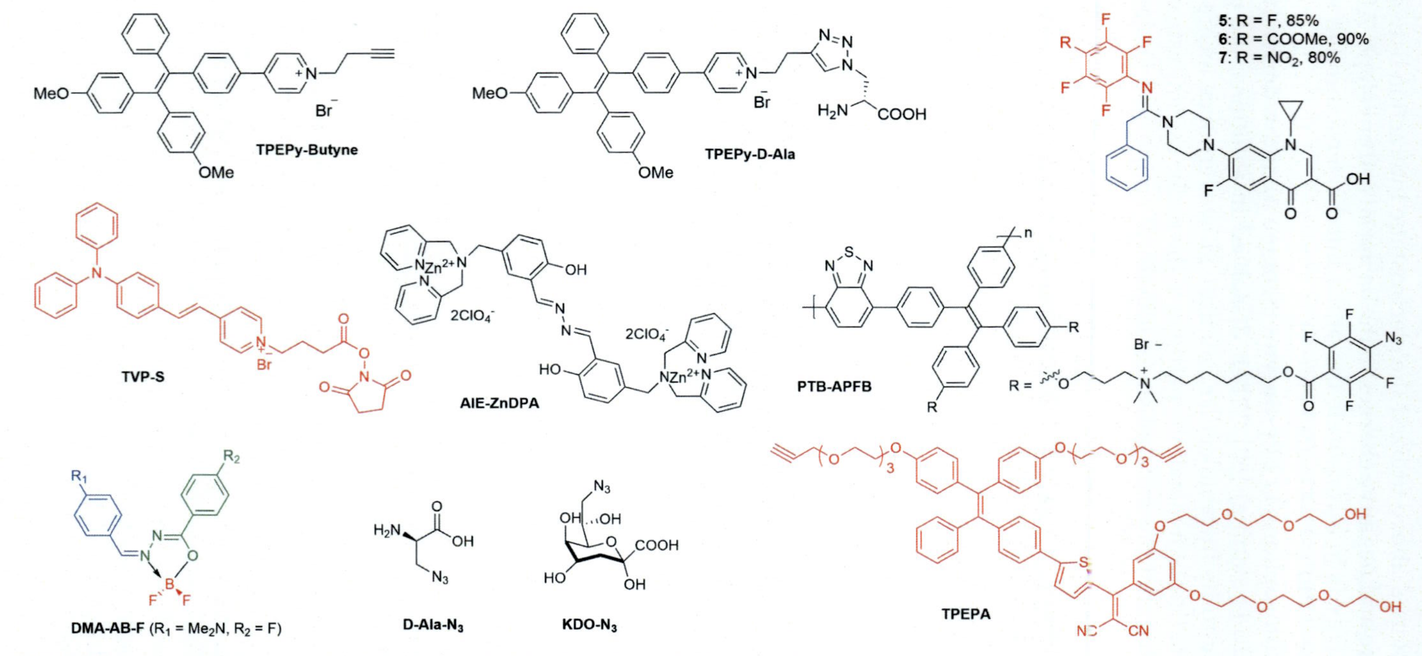

SCHEME 4 Chemical structures of the AIE probes for the killing of microorganisms.

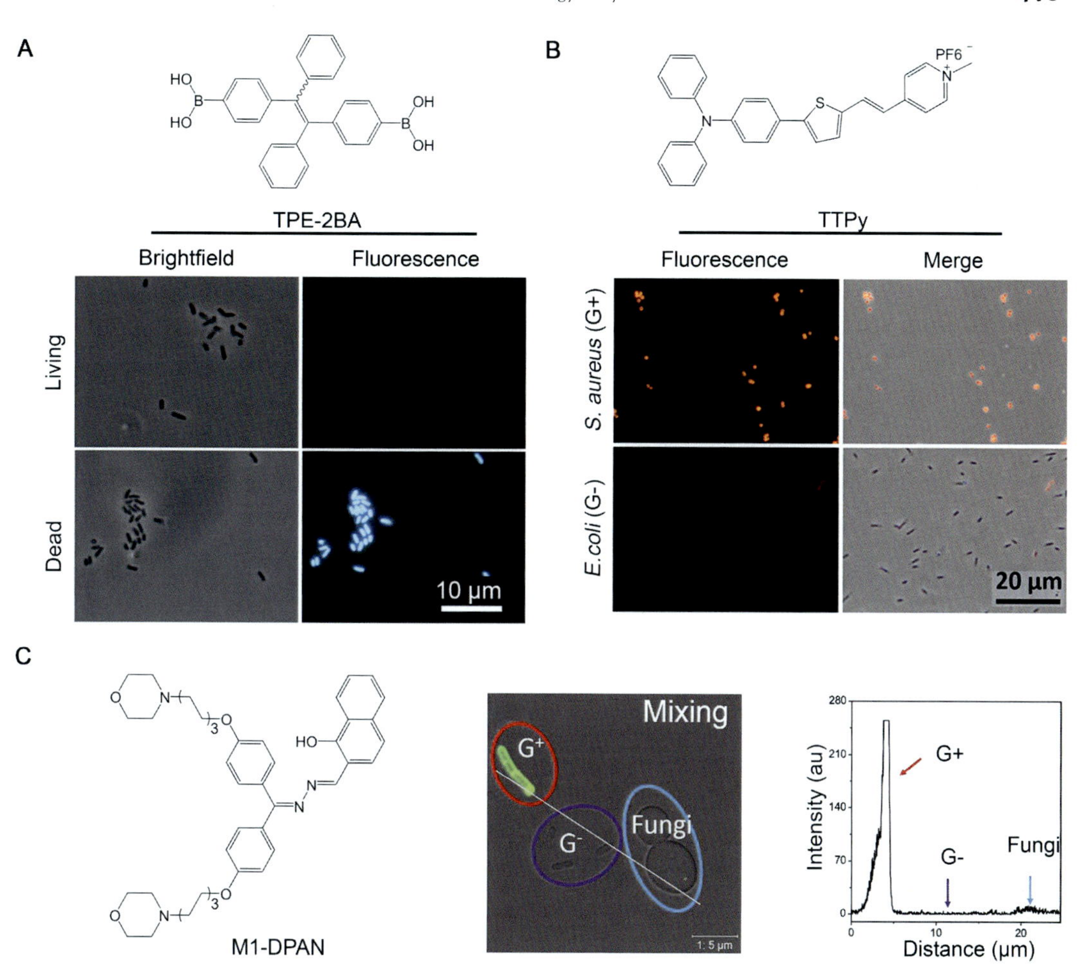

FIG. 13 AIE probes for selective bacterial detection. (A) Selective detection of dead *E. coli* stained with **TPE-2BA**. (B) Selective detection of Gram-positive bacteria stained with **TTPy**. (C) Selective detection of Gram-positive bacteria (*B. subtilis*, the green circle) from Gram-negative bacteria (*P. aeruginosa*, the purple circle) and fungi (*S. cerevisiae*, the cyan circle) by **M1-DPAN** (left). Fluorescence intensity along the white line as shown in the left panel (right). *Panel (A): Reprinted (adapted) with permission from Y. Hong, S. Chen, C.W.T. Leung, J.W.Y. Lam, B.Z. Tang, Water-soluble tetraphenylethene derivatives as fluorescent "light-up" probes for nucleic acid detection and their applications in cell imaging, Chem. Asian J. 8 (8) (2013) 1806–1812. Copyright (2014) John Wiley and Sons. Panel (B): Reprinted (adapted) with permission from D.G. Booth, W.C. Earnshaw, Ki-67 and the chromosome periphery compartment in mitosis, Trends Cell Biol. 27 (12) (2017) 906–916. Copyright (2019) American Chemical Society. Panel (C): Reprinted (adapted) with permission from R. Hu, F. Zhou, T. Zhou, J. Shen, Z. Wang, Z. Zhao, A. Qin, B.Z. Tang, Specific discrimination of gram-positive bacteria and direct visualization of its infection towards mammalian cells by a DPAN-based AIEgen, Biomaterials 187 (2018) 47–54. Copyright (2018) Elsevier.*

and excellent photostability, providing a powerful tool for diagnosis and treatment of infections due to Gram-positive bacteria. [73]

Bacterial detection by imaging-based probes relies on the use of fluorescence microscopy, of which the screening speed and throughput are limited by the methodology. To address these issues, nonimaging-based methods are developed for fast and high-throughput bacterial detection and identification by combining the advantages of AIE probes (such as a strong signal, low background, easy to use, and excellent photostability) and high-throughput methods. Decorating the core AIE fluorophore TPE with a positively charged ammonium group leads to the capability of bacterial binding. Meanwhile, altering its hydrophobicity results in a series of probes with the distinctive signal intensity in response to different bacterial strains (Fig. 14A) [74].

By ligating the TPE core to an alkyl group or a phenyl group through an alkoxy chain, seven different probes are synthesized, namely, **TPE-AMe**, **TPE-AEt**, **TPE-APrA**, **TPE-ABu**, **TPE-ACH**, **TPE-ABn**, and **TPE-AHex**. The calculated log P (C log P; n-octanol/water partition coefficient) values of these probes range from 3.426 to 6.071, which corresponds to different hydrophobicity and critical aggregation concentrations (CACs) in aqueous solutions. Further exploration reveals that the self-assembly behavior of these probes is related to CACs, which are crucial for bacterial selectivity and fluorescence emission. According to the C log P values, these seven probes are classified into three groups: Group A with C log P values ranging from 3 to 5, including **TPE-AMe**, **TPE-Aet**, and **TPE-APrA**, which are more sensitive to Gram-positive bacteria and fungi; Group B with C log P values ranging from 5 to 6, including **TPE-ABu**, **TPE-ACH**, and **TPE-ABn**, which have similar affinity to all three kinds of pathogens; Group C with Clog$P$ values >6, including **TPE-AHex**, which is more selective to Gram-negative bacteria. The combination of these probes gives 17 sensor arrays for pathogen detection and discrimination. The fluorescence intensity pattern of the arrays in response to different pathogens is recorded and analyzed with linear discriminant analysis (LDA), resulting in two-dimensional canonical score plots for pathogen categorization. Of the 17 created arrays, 14 have nearly 100% identification accuracy for seven different pathogen strains. These arrays can detect the mixture of pathogens with 100% identification accuracy.

For a more intuitive observation of the bacterial detection process, fluorescence test strips are made up with electrospun fibrous mats and AIE fluorophores [75,76]. The weak interaction between mannose and bacteria fimbria protein FimH is utilized for the design of Gram-negative bacteria detection test strips [76]. The fluorophore TPE is firstly conjugated to electrospun polystyrene-*co*-maleic anhydride (PSMA) fibers with poly(ethylene glycol) (PEG) diamine (Mw: 2 kDa), followed by decorating with mannose to obtain **PSMA-PEG-TPEC-Man** fibers. Owing to the enrichment of fluorophores in the fiber, the test strips are highly sensitive to Gram-negative bacteria since the detection limit can be as low as 10^2 CFU/mL. This is useful for fast clinical diagnosis without requiring long-term bacterial enrichment processes.

Combining the advantage of electrospun fibrous mats and the enzymatic activity strategy, the interaction between cephalosporin and β-lactamase is used to develop highly selective and sensitive test strips against β-lactamase-expressing pathogens which are often resistant to antibiotics [75]. The AIE fluorophore TPE is firstly decorated with cephalosporin to form a conjugate **TPE-Cep** which is then conjugated to electrospun fibrous mats to form test strips. When the cephalosporin is decomposed by β-lactamase, the aggregation of the TPE

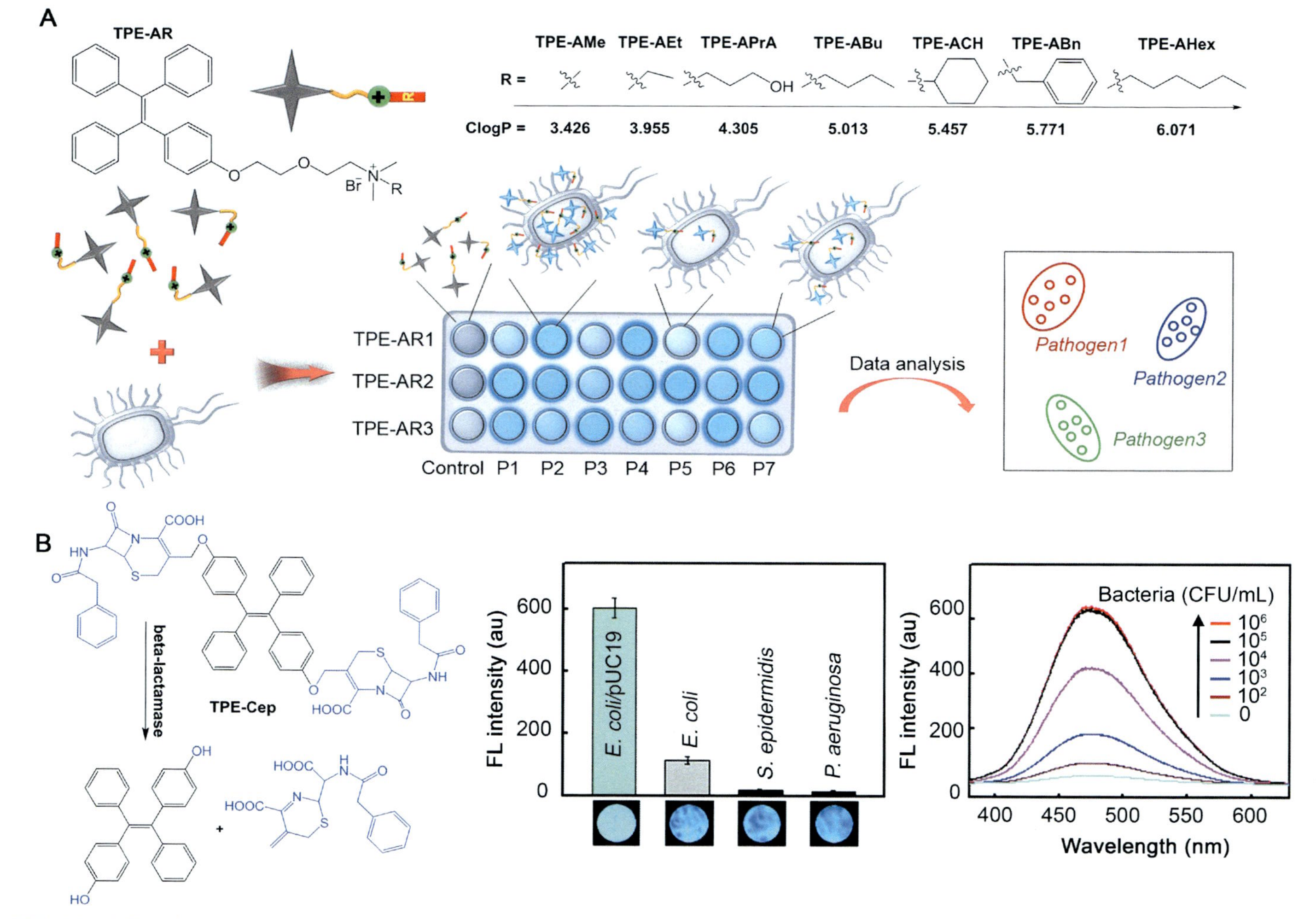

FIG. 14 AIE probes for nonimaging-based bacterial detection. (A) Schematic illustration of the AIE-sensor array that can discriminate Gram-positive bacteria, Gram-negative bacteria, and fungi. (B) Fluorescence intensities of β-lactamase-expressing *E. coli*/pUC19 and other strains incubated with **TPE-Cep** and captured on fibers. *Panel (A): Reprinted (adapted) with permission from C. Zhou, W. Xu, P. Zhang, M. Jiang, Y. Chen, R.T.K. Kwok, M.M.S. Lee, G. Shan, R. Qi, X. Zhou, J.W.Y. Lam, S. Wang, B.Z. Tang, Engineering sensor arrays using aggregation-induced emission Luminogens for pathogen identification, Adv. Funct. Mater. 29 (4) (2019) 1–10. Copyright (2019) American Chemical Society. Panel (B): Reprinted (adapted) with permission from L. Zhao, Y. Liu, Z. Zhang, J. Wei, S. Xie, X. Li, Fibrous testing papers for fluorescence trace sensing and photodynamic destruction of antibiotic-resistant bacteria, J. Mater. Chem. B 8 (13) (2020) 2709–2718. Copyright (2020) Royal Society of Chemistry.*

fluorophore will "light up" the test strips. Following this strategy, the test strips can discriminate β-lactamase-expressing *Escherichia coli* transformed with the pUC19 plasmid DNA (*E. coli*/pUC19) from those not expressing β-lactamase with the detection limit as low as 10^2 CFU/mL (Fig. 14B). Additionally, these test strips can be reused by rinsing with saline which is easy to use and helps to reduce the cost.

8.1.2 AIE probes for bacterial killing

Antibiotics have increased the average life expectancy of humans since discovered. However, the emergence and spread of antibiotic-resistant pathogens threaten public health, driving the development of new and more effective antibiotics. Tracing and observing the interaction between antibiotics and bacteria can help to classify the mechanisms and accelerate the drug-developing process. Some drugs with poor water solubility form nanoaggregates in aqueous solutions after simple processes such as precipitation [77], which possesses better therapeutical effect than carrier-mediated nanoparticles. Under these circumstances, a strategy that combines the technique of pure nano-drug and AIE fluorescence characteristic is designed [78]. The classic antibiotic ciprofloxacin, which inhibits bacterial DNA replication, is used as an initial material. Through rational design and simple synthesis, a perfluoroaryl and a phenyl group are ligated to ciprofloxacin to form a set of propeller-shaped structures (compound **5–7**) (Scheme 4). After crystallization, the intramolecular motion of compound 5 is restricted, and the radiative pathway is activated for fluorescence emission. The diameters of the nanoaggregates of compound 5 formed in DMF, MeCN, acetone, and DMSO range from 40 to 60nm, and they are highly stable under the temperature of 4°C with strong AIE emission. This facilitates its application in bacterial killing and imaging [79].

PDT has been applied to clinical therapeutics for different kinds of tumors since the 1990s [80]. It also shows great potential in clinical diagnosis and therapeutics for microbial infection, especially for antibiotic-resistant pathogens [80]. PS plays a central role in the PDT process which absorbs energy from light illumination followed by releasing the energy, generating either photon emission or cytotoxic radical molecules such as singlet oxygen ($^1O_2$) and reactive oxygen species (ROS) [81]. $^1O_2$ is generated by the direct transfer of energy from the excited PS to a common ground state oxygen molecule (triplet oxygen), and other ROS are generated by transferring protons or electrons from the excited PS to the nearby substrates [82]. Fluorescent materials that can be excited by light illumination for imaging can often be used as PS for PDT too. Different from the conventional fluorescent PS, the AIE PS is able to show both strong fluorescence and good singlet oxygen generation ability in the aggregate state.

Many AIE probes for bacterial imaging are developed to be PSs for bacterial elimination (Scheme 4), of which the selectivity is mainly determined by the ionic charges, hydrophobicity, and degree of asymmetry. A bacteria-targeting PS, namely **AIE-ZnDPA** (structure shown in Scheme 4), is synthesized from salicylaldazine fluorophore and zinc(II)-dipicolylamine (ZnDPA) (Fig. 15A) [83]. The selective electrostatic interaction between the positively charged Zn^{2+} in the probe and the negatively charged bacterial envelope restricts the intramolecular motion of the fluorophore, resulting in selective imaging as well as ablation of both Gram-positive and Gram-negative bacteria over mammalian cells. Binding of **AIE-ZnDPA** to bacteria gives the probe capacity of ROS generation in the presence of white light which facilitates the elimination of bacteria. A similar strategy of decorating AIE fluorophores with a

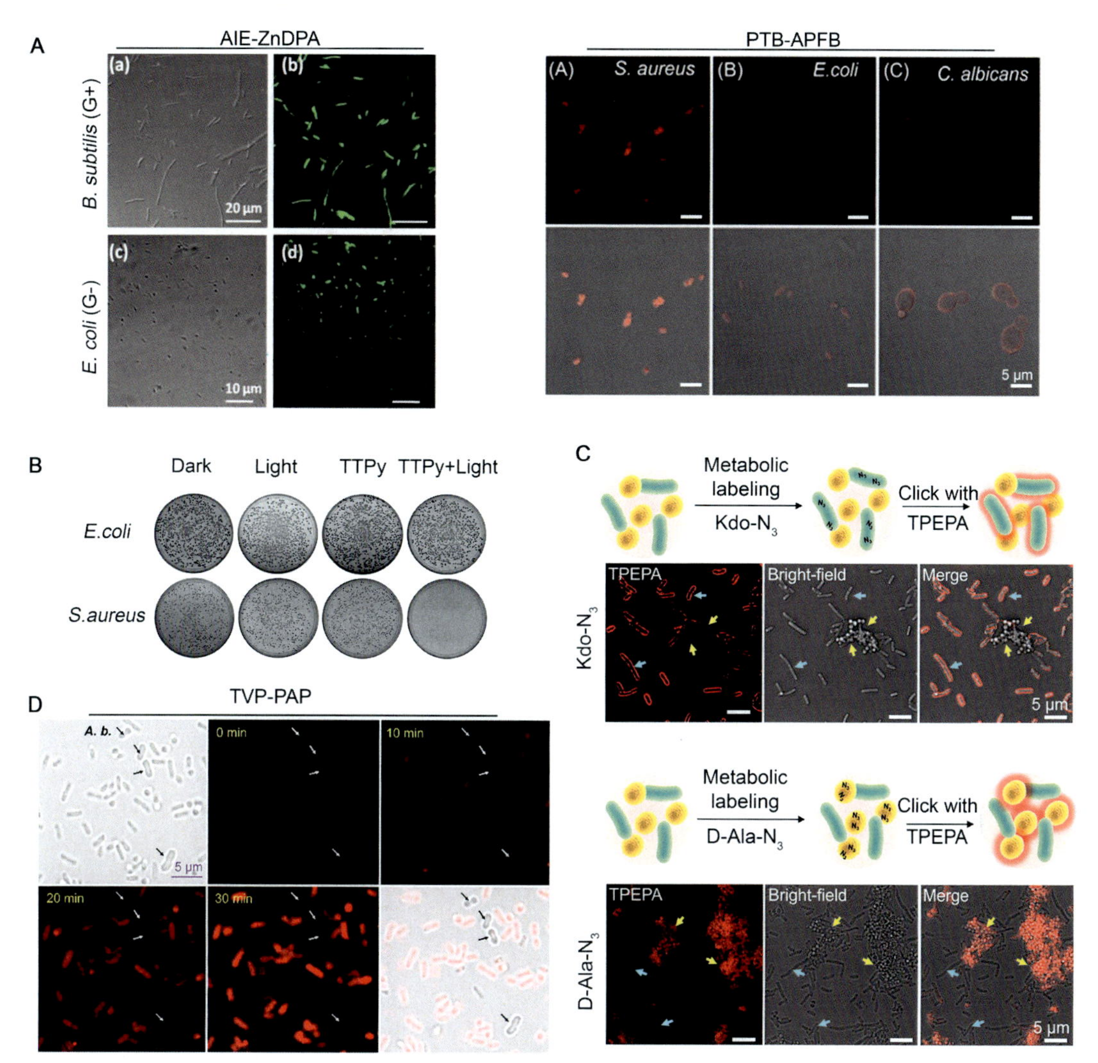

FIG. 15 AIE probes for the PDT of microbial infection. (A) Left: nonselective PDT of both Gram-positive (*B. Subtilis*, a,b) and Gram-negative bacteria (*E. coli*, c,d) with **AIE-ZnDPA**. Right: nonselective PDT of both G+ and G− bacteria with a TPE-based conjugated polymer **PTB-APFB**. (B) Selective PDT of Gram-positive *S. aureus* but not Gram-negative *E. coli* with **TTPy** in the presence of white light ($60\,mW/cm^2$). (C) Selective PDT of Gram-negative *E. coli* (upper row) or Gram-positive *S. aureus* (lower row) with the metabolic labeling strategy of **TPEPA**. (D) Selective labeling of *P. aeruginosa* with **TVP-PAP** which is a phage bioconjugate of **TVP-S**. *Panel A: (i) Reprinted (adapted) with permission from M. Gao, Q. Hu, G. Feng, N. Tomczak, R. Liu, B. Xing, B.Z. Tang, B. Liu, A multifunctional probe with aggregation-induced emission characteristics for selective fluorescence imaging and photodynamic killing of bacteria over mammalian cells, Adv. Healthc. Mater. 4 (5) (2015) 659–663. Copyright (2015) John Wiley and Sons; (ii) Reprinted (adapted) with permission from T. Zhou, R. Hu, L. Wang, Y. Qiu, G. Zhang, Q. Deng, H. Zhang, P. Yin, B. Situ, C. Zhan, A. Qin, B.Z. Tang, An AIE-active conjugated polymer with high ROS-generation ability and biocompatibility for efficient photodynamic therapy of bacterial infections, Angew. Chem. Int. Ed. Engl. 59 (25) (2020) 9952–9956. Copyright (2020) American Chemical Society. Panel B: Reprinted (adapted) with permission from M. Kang, C. Zhou, S. Wu, B. Yu, Z. Zhang, N. Song, M.M.S. Lee, W. Xu, F.J. Xu, D. Wang, L. Wang, B.Z. Tang, Evaluation of structure-function relationships of aggregation-induced emission Luminogens for simultaneous dual applications of specific discrimination and efficient Photodynamic killing of gram-positive bacteria, J. Am. Chem. Soc. 141 (42) (2019) 16781–16789. Copyright (2019) American Chemical Society. Panel C: Reprinted (adapted) with permission from M. Wu, G. Qi, X. Liu, Y. Duan, J. Liu, B. Liu, Bio-orthogonal AIEgen for specific discrimination and elimination of bacterial pathogens via metabolic engineering, Chem. Mater. 32 (2) (2020) 858–865. Copyright (2020) American Chemical Society. Panel D: Reprinted (adapted) with permission from X.W. He, Y.J. Yang, Y.C. Guo, S.G. Lu, Y. Du, J.J. Li, X.P. Zhang, N.L.C. Leung, Z. Zhao, G.L. Niu, S.S. Yang, Z. Weng, R.T.K. Kwok, J.W.Y. Lam, G.M. Xie, B.Z. Tang, Phage-guided targeting, discriminative imaging, and synergistic killing of bacteria by AIE bioconjugates, J. Am. Chem. Soc. 142 (8) (2020) 3959–3969. Copyright (2020) American Chemical Society.*

zinc-containing bacteria-binding group is also used to develop a new PS, namely **TPETH-2Zn**, with improved light absorption of longer wavelength, stronger interaction between the PS and the bacterial cell envelope, and better recognition and phototoxicity to both Gram-positive and Gram-negative bacteria [84]. Different from the decorating strategy, a set of small-molecule planar AIE probes are designed and synthesized, of which the intramolecular motion is restricted with strong intramolecular hydrogen bonds [85]. It is noteworthy that fluoro-substitution in the phenyl group significantly enhances the intramolecular hydrogen bonds, leading to the increased quantum yield of both fluorescence emission and ROS generation, of which **DMA-AB-F** (structure shown in Scheme 4) shows antibacterial capacity through PDT.

A TPE-based conjugated polymer named **PTB-APFB** (structure shown in Scheme 4), however, provides new directions for the development of bacteria-targeting AIE PS (Fig. 15A) [86]. The interaction between **PTB-APFB** and the bacterial envelope over the mammalian cell membrane is achieved by including a positively charged quaternary ammonium group in the middle of the side chain. **PTB-APFB** can be used for the staining and imaging of Gram-positive bacteria, Gram-negative bacteria as well as fungi. Furthermore, **PTB-APFB** shows a good quantum yield of ROS in the presence of white light, which facilitates its application in the PDT of bacterial infection.

Selective bacterial recognition by AIE probes raised the possibility of selective PDT. The abovementioned small-molecule AIE probe for Gram-positive bacteria imaging, namely **TTPy**, is used for selective ablation of Gram-positive bacteria through PDT (Fig. 15B) [71]. Owing to its AIE properties, the ROS generation ability of **TTPy** is higher than those widely used conventional PSs such as hexidium iodide and chlorin e6. The application of **TTPy** in the PDT of *S. aureus* infection also significantly accelerates the wound healing process in rats. By adjusting the steps of the staining process, the metabolic labeling strategy is also successfully applied for the selective PDT of bacterial infection. Two azide-containing metabolic precursors, namely D-Ala-N_3 and 3-deoxy-D-manno-octulosonic acid-N3 (Kdo—N_3), are used to label the envelope of Gram-positive bacteria and Gram-negative bacteria, respectively (Fig. 15C) [87]. The alkyne-containing AIE fluorophore **TPEPA** (structure shown in Scheme 4) is then used to react with the labeled bacteria through a click reaction. The intramolecular motion of **TPEPA** is restricted after the reaction, resulting in the turn-on of fluorescence emission and ROS generation.

Another attempt to construct selective PS for bacterial PDT is the combining of AIE fluorophores and bacteriophages (Fig. 15D) [88]. The AIE fluorophore **TVP-S** (structure shown in Scheme 4) is firstly synthesized as the nonselective AIE core group for imaging and PDT, of which the carboxyl group can react facilely with the amino group of phage proteins on the surface to form the phage bioconjugate (**TVP-PAP**). After modification, the infection ability of the bacteriophages is retained and the AIE fluorophore is largely enriched (approximately 8,200 molecules of **TVP-S** per phage entity), which ensures the good performance in selective bacterial targeting, imaging, and elimination. Through the changing bacteriophage types, selective imaging and elimination of different bacterial species can be easily achieved.

An interesting strategy of generating bacteria-specific AIE PSs involves the production of polymers utilizing bacteria as a "template" [89]. Firstly, three substrate monomers are designed and synthesized as "building blocks," of which the [2-(methacryloyloxy)ethyl] trimethylammonium chloride (TMAEMC) is used as a cation for bacterial binding; the TMAEMC-TPAPy is used as an AIE fluorophore; the [2-(methacryloyloxy)ethyl]dimethyl-(3-sulfopropyl)ammonium hydroxide (DMAPS) is used to increase the solubility of the product polymers. Through a copper-catalyzed atom transfer radical polymerization (ATRP) process, polymers are synthesized on the envelope of the template bacteria which are then separated for the specific recognition of bacteria with the same species as the template. The separated bacterium-templated polymer shows extremely high specificity for the recognition of the corresponding bacterial strains, which facilitates their application in clinical diagnosis and therapeutics. Because of the incorporation of the AIE fluorophores in the polymer, the polymer exhibits low background fluorescence as well as strong ability of ROS generation and fluorescence emission.

8.2 AIE probes for virus detection

Despite not being as well studied in comparison to bacterial detection, AIE probes for the detection of nonbacterial pathogens such as viruses and fungi are also versatile. Through the design and synthesis of a functional AIE substrate **TPE-APP**, a dual-modality system of both fluorometric and colorimetric sensors for the detection of viruses is created by combining the immunoassay and the AIE probe (Fig. 16A) [90]. Viruses are specifically recognized by magnetic bead (MN)-conjugated antibodies, which are then enriched and washed to avoid interference. Afterward, the virus enriched by beads is recognized by another antibody, which is then targeted by biotin-labeled secondary antibodies to form a sandwich-shaped structure, in which the virus is captured in the middle as a bridge to connect the bead-labeled and biotin-labeled antibodies. After washing, the biotin-containing structure is incubated with a streptavidin-labeled ALP (SA-ALP) which can cleave **TPE-APP** to generate a water-insoluble fluorescent **TPE-DMA** and a highly reactive species to produce Ag precipitates on the surface of Au nanoparticles. This results in strong fluorescence emission and an observable color change. This dual-modality system can sensitively and efficiently detect viruses with a detection limit of 1.4 copies/μL under fluorometric modality and semiquantitative detection of viruses in the range of 1.3×10^3 to 2.5×10^6 copies/μL under colorimetric modality. Additionally, the ease of substituting different antibodies in this system makes it a useful platform for both the domestic and clinical diagnosis of corresponding viral infections.

8.3 AIE probes for fungus imaging

For effective discrimination of pathogenic bacteria and fungi, a multifunctional donor-π-acceptor structure AIE probe **IQ-Cm** is designed and synthesized through linking a positively charged AIE fluorophore diphenyl isoquinolinium (IQ) group and a coumarin-derived (Cm) group with a phenyl group (Fig. 16B) [91]. This probe exhibits both AIE and twisted intramolecular charge transfer (TICT) characteristics, and thus it is sensitive to the polarity of the microenvironment. It localizes differently in Gram-negative bacteria (in the cell membrane of compromised cells and the cytoplasm of dead cells), Gram-positive bacteria (in the cytoplasm), and fungi (in mitochondria) because of the different composition of the membrane lipids. This results in different fluorescence signals distinguishable by the naked eye (weak pink fluorescence for Gram-negative bacteria, bright orange fluorescence for Gram-positive bacteria, and strong yellow fluorescence for fungi), which makes it a useful probe for fast and sensitive diagnosis for clinical applications.

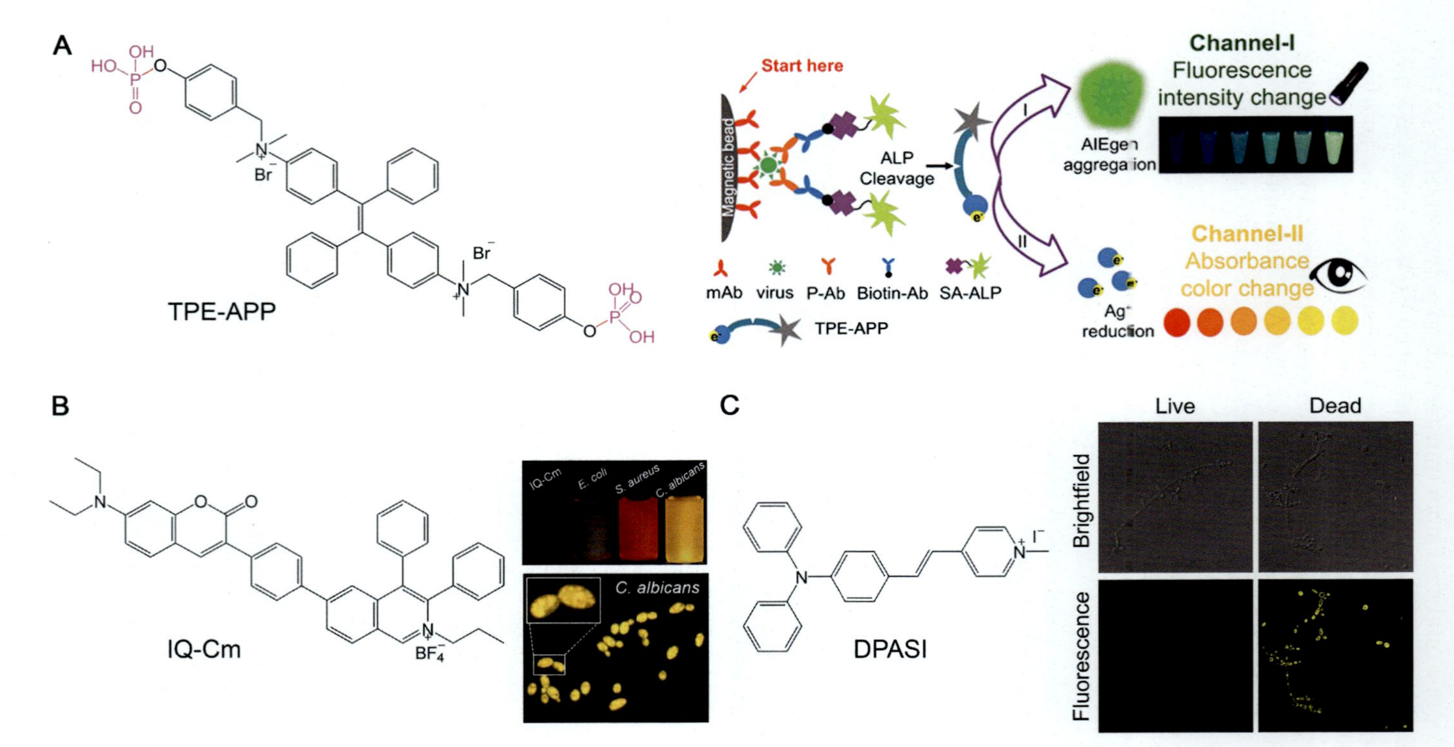

FIG. 16 AIE probes for imaging of viruses and fungi. (A) Schematic illustration of dual-module detection of specific viral species with **TPE-APP**. (B) Discrimination of Gram-positive bacteria (*S. aureus*), Gram-negative bacteria (*E. coli*), and fungi (*C. albicans*) with **IQ-Cm** at 365-nm UV irradiation. (C) Selective detection of dead fungi (*C. albicans*) stained with **DPASI**. *Panel A: Reprinted (adapted) with permission from L.H. Xiong, X. He, Z. Zhao, R.T.K. Kwok, Y. Xiong, P.F. Gao, F. Yang, Y. Huang, H.H.Y. Sung, I.D. Williams, J.W.Y. Lam, J. Cheng, R. Zhang, B.Z. Tang, Ultrasensitive Virion immunoassay platform with dual-modality based on a multifunctional aggregation-induced emission Luminogen, ACS Nano 12 (9) (2018) 9549–9557. Copyright (2020) American Chemical Society. Panel B: Reprinted (adapted) with permission from C.C. Zhou, M.J. Jiang, J. Du, H.T. Bai, G.G. Shan, R.T.K. Kwok, J.H.C. Chau, J. Zhang, J.W.Y. Lam, P. Huang, B. Tang, One stone, three birds: one AIEgen with three colors for fast differentiation of three pathogens, Chem. Sci. 11 (18) (2020) 4730–4740. Copyright (2020) Royal Society of Chemistry.. Panel C: Reprinted (adapted) with permission from X. Ge, M. Gao, B. Situ, W. Feng, B. He, X. He, S. Li, Z. Ou, Y. Zhong, Y. Lin, X. Ye, X. Hu, B.Z. Tang, L. Zheng, One-step, rapid fluorescence sensing of fungal viability based on a bioprobe with aggregation-induced emission characteristics, Mater. Chem. Front. 4 (3) (2020) 957–964. Copyright (2020) Royal Society of Chemistry.*

To address the issue of detecting fungal viability in a fast and effective manner, a water dispersible AIE probe **DPASI** is designed and synthesized by linking a TPA fluorophore and a positively charged pyridinium moiety with a carbon–carbon double bond (Fig. 16C) [92]. **DPASI** possesses a donor-acceptor structure and exhibits both TICT and AIE properties, leading to the efficient labeling of dead fungi and dead *E. coli*. When accumulated in negatively charged mitochondria in the dead fungi, the probe becomes emissive, making it a facile tool for fast detection of fungal viability without the washing steps.

9 Conclusions and perspectives

AIE-based fluorescent materials are usually featured with high brightness, good photostability, superior sensitivity, and nice biocompatibility. By decorating the AIE core fluorophores with different functional groups or moieties, probes with different targetability or tailored functionality can be obtained. In this chapter, we introduced some applications of the AIE molecular probe in biomedicine. These applications involved (but are not limited to) the detection of biomarkers, study of biomacromolecular conformations, visualization of organelles, sensing of endogenous chemicals, efficient imaging of microorganisms including bacteria, viruses, and fungi, and PDT. These examples demonstrate its broad applicability in different areas in biomedical sciences. The smart designs allow for the creation of different sensing mechanisms including simple "turn-on" binding probes. There are some new applications of AIE probes emerging in areas such as CP staining for cytogenetic studies.

The application of AIE probes in microbiology is attractive. Among the pathogenic microorganisms, bacteria have drawn much attention due to the substantial demand for novel clinical therapeutics for antibiotic-resistant bacterial infection. By decorating the commonly used AIE fluorophores (such as TPE and TPA), numerous probes as well as probe arrays are developed for specific detection of bacteria, resulting in the efficient discrimination of live and dead bacteria, Gram-positive and Gram-negative bacteria, special enzyme-expressing bacterial strains, and different bacterial species. Researchers have found or developed many of the AIE PS for PDT, which provides a new direction for developing therapeutics for antibiotic-resistant bacteria. By contrast, the variety and diversity of virus- or fungus-targeting AIE probes are much less available, while the infection and pandemic of these pathogens (for example, COVID-19) can be disastrous. In this sense, more effective AIE probes for direct viral binding and recognition are urgently demanded. An increasing number of pathogenic recognition strategies have been developed in recent years, such as nonionic **M1-DAPN**, bacteriophage-based **TVP-PAP** and bacteria-templated polymer for bacterial detection, and antibody-based viral detection. Meanwhile, several kinds of novel AIE fluorophore cores have also been developed recently to expand the development of AIE strategy, such as planar **DMA-AB-F**, triphenylethylenes linked with a varying number of perylenediimide (PDI) units, (**TriPE-*n*PDIs**, $n = 1$–3) [93], 1,2,3,4-tetraphenyloxazolium (**TPO-P**) and 2,3,5-triphenyloxazolium (**TriPO-PN**) with anion–π^+ interactions [94], and isoquinolinium (**IQ**) [95]. These AIE fluorophores have superior performance in quantum yield, Stokes shift, and NIR emission. Developing new recognition strategies as well as application of the newly developed fluorophores can further improve the development of pathogen-targeting AIE probes for biomedical research and clinical therapeutics in the future.

References

[1] G.P.C. Drummen, Fluorescent probes and fluorescence (microscopy) techniques-illuminating biological and biomedical research, Molecules 17 (12) (2012) 14067–14090.
[2] J. Mei, N.L.C. Leung, R.T.K. Kwok, J.W.Y. Lam, B.Z. Tang, Aggregation-induced emission: together we shine, united we soar! Chem. Rev. 115 (21) (2015) 11718–11940.
[3] S. Nikazar, V.S. Sivasankarapillai, A. Rahdar, S. Gasmi, P.S. Anumol, M.S. Shanavas, Revisiting the cytotoxicity of quantum dots: an in-depth overview, Biophys. Rev. 12 (3) (2020) 703–718.
[4] Y. Huang, J. Xing, Q. Gong, L.-C. Chen, G. Liu, C. Yao, Z. Wang, H.-L. Zhang, Z. Chen, Q. Zhang, Reducing aggregation caused quenching effect through co-assembly of PAH chromophores and molecular barriers, Nat. Commun. 10 (1) (2019) 169.
[5] G.L. Hortin, S.A. Carr, N.L. Anderson, Introduction: advances in protein analysis for the clinical laboratory, Clin. Chem. 56 (2) (2010) 149–151.
[6] H. Tong, Y. Hong, Y. Dong, M. Häußler, J.W.Y. Lam, Z. Li, Z. Guo, Z. Guo, B.Z. Tang, Fluorescent "Light-up" bioprobes based on tetraphenylethylene derivatives with aggregation-induced emission characteristics, Chem. Commun. (2006) 3705–3707.
[7] S. Xie, A.Y.H. Wong, S. Chen, B.Z. Tang, Fluorogenic detection and characterization of proteins by aggregation-induced emission methods, Chem. A Eur. J. 25 (23) (2019) 5824–5847.
[8] Y. Yu, A. Qin, C. Feng, P. Lu, K.M. Ng, K.Q. Luo, B.Z. Tang, An amine-reactive tetraphenylethylene derivative for protein detection in SDS-PAGE, Analyst 137 (23) (2012) 5592.
[9] Y. Yu, J. Li, S. Chen, Y. Hong, K.M. Ng, K.Q. Luo, B.Z. Tang, Thiol-reactive molecule with dual-emission-enhancement property for specific prestaining of cysteine containing proteins in SDS-PAGE, ACS Appl. Mater. Interfaces 5 (11) (2013) 4613–4616.
[10] M. Chevallet, S. Luche, T. Rabilloud, Silver staining of proteins in polyacrylamide gels, Nat. Protoc. 1 (4) (2006) 1852–1858.
[11] S. Xie, A.Y.H. Wong, R.T.K. Kwok, Y. Li, H. Su, J.W.Y. Lam, S. Chen, B.Z. Tang, Fluorogenic Ag+–tetrazolate aggregation enables efficient fluorescent biological silver staining, Angew. Chem. Int. Ed. (2018).
[12] A.Y.H. Wong, S. Xie, B.Z. Tang, S. Chen, Fluorescent silver staining of proteins in polyacrylamide gels, J. Vis. Exp. (146) (2019).
[13] L.-T. Jin, S.-Y. Hwang, G.-S. Yoo, J.-K. Choi, Sensitive silver staining of protein in sodium dodecyl sulfate-polyacrylamide gels using an azo dye, calconcarboxylic acid, as a silver-ion sensitizer, Electrophoresis 25 (15) (2004) 2494–2500.
[14] Y. Hong, L. Meng, S. Chen, C.W.T. Leung, L.T. Da, M. Faisal, D.A. Silva, J. Liu, J.W.Y. Lam, X. Huang, B.Z. Tang, Monitoring and inhibition of insulin fibrillation by a small organic fluorogen with aggregation-induced emission characteristics, J. Am. Chem. Soc. 134 (3) (2012) 1680–1689.
[15] K.J. Robbins, G. Liu, V. Selmani, N.D. Lazo, Conformational analysis of thioflavin T bound to the surface of amyloid fibrils, Langmuir 28 (48) (2012) 16490–16495.
[16] M.I. Ivanova, S.A. Sievers, M.R. Sawaya, J.S. Wall, D. Eisenberg, Molecular basis for insulin fibril assembly, Proc. Natl. Acad. Sci. 106 (45) (2009) 18990–18995.
[17] Q. Li, Z. Li, The strong light-emission materials in the aggregated state: what happens from a single molecule to the collective group, Adv. Sci. 4 (7) (2017) 1600484.
[18] W. Fu, C. Yan, Z. Guo, J. Zhang, H. Zhang, H. Tian, W.H. Zhu, Rational design of near-infrared aggregation-induced-emission-active probes: in situ mapping of amyloid-β plaques with ultrasensitivity and high-fidelity, J. Am. Chem. Soc. 141 (7) (2019) 3171–3177.
[19] G. Hasler, Pathophysiology of depression: do we have any solid evidence of interest to clinicians? World Psychiatry 9 (3) (2010) 155–161.
[20] S.G. Snowden, A.A. Ebshiana, A. Hye, O. Pletnikova, R. O'Brien, A. Yang, J. Troncoso, C. Legido-Quigley, M. Thambisetty, Neurotransmitter imbalance in the brain and Alzheimer's disease pathology, J. Alzheimers Dis. 72 (1) (2019) 35–43.
[21] P. Whitehouse, D. Price, R. Struble, A. Clark, J. Coyle, M. Delon, Alzheimer's disease and senile dementia: loss of neurons in the basal forebrain, Science (80-.) 215 (4537) (1982) 1237–1239.
[22] T. Lazarevic-Pasti, A. Leskovac, T. Momic, S. Petrovic, V. Vasic, Modulators of acetylcholinesterase activity: from Alzheimer's disease to anti-cancer drugs, Curr. Med. Chem. 24 (30) (2017).
[23] M. Pohanka, Inhibitors of acetylcholinesterase and butyrylcholinesterase meet immunity, Int. J. Mol. Sci. 15 (6) (2014) 9809–9825.

[24] L. Peng, G. Zhang, D. Zhang, J. Xiang, R. Zhao, Y. Wang, D. Zhu, A fluorescence "Turn-on" ensemble for acetylcholinesterase activity assay and inhibitor screening, Org. Lett. 11 (17) (2009) 4014–4017.
[25] L. Hartwell, L.E. Hood, M.L. Goldberg, A.E. Reynolds, L.M. Silver, Genetics: From Genes to Genomes, fourth ed., McGraw-Hill Companies, Inc., New York, 2011, pp. 246–287.
[26] Y. Gao, Z. He, X. He, H. Zhang, J. Weng, X. Yang, F. Meng, L. Luo, B.Z. Tang, Dual-color emissive AIEgen for specific and label-free double-stranded DNA recognition and single-nucleotide polymorphisms detection, J. Am. Chem. Soc. 141 (51) (2019) 20097–20106.
[27] G. Biffi, D. Tannahill, J. McCafferty, S. Balasubramanian, Quantitative visualization of DNA G-Quadruplex structures in human cells, Nat. Chem. 5 (3) (2013) 182–186.
[28] I. Cheung, M. Schertzer, A. Rose, P.M. Lansdorp, Disruption of Dog-1 in Caenorhabditis elegans triggers deletions upstream of guanine-rich DNA, Nat. Genet. 31 (4) (2002) 405–409.
[29] R. Rodriguez, K.M. Miller, J.V. Forment, C.R. Bradshaw, M. Nikan, S. Britton, T. Oelschlaegel, B. Xhemalce, S. Balasubramanian, S.P. Jackson, Small-molecule–induced DNA damage identifies alternative DNA structures in human genes, Nat. Chem. Biol. 8 (3) (2012) 301–310.
[30] A. Siddiqui-Jain, C.L. Grand, D.J. Bearss, L.H. Hurley, Direct evidence for a G-Quadruplex in a promoter region and its targeting with a small molecule to repress c-MYC transcription, Proc. Natl. Acad. Sci. 99 (18) (2002) 11593–11598.
[31] S. Müller, S. Kumari, R. Rodriguez, S. Balasubramanian, Small-molecule-mediated G-Quadruplex isolation from human cells, Nat. Chem. 2 (12) (2010) 1095–1098.
[32] Y. Hong, M. Häußler, J.W.Y. Lam, Z. Li, K.K. Sin, Y. Dong, H. Tong, J. Liu, A. Qin, R. Renneberg, B.Z. Tang, Label-free fluorescent probing of G-Quadruplex formation and real-time monitoring of DNA folding by a quaternized tetraphenylethene salt with aggregation-induced emission characteristics, Chem. A Eur. J. 14 (21) (2008) 6428–6437.
[33] X. Wang, A.R. Morales, T. Urakami, L. Zhang, M.V. Bondar, M. Komatsu, K.D. Belfield, Folate receptor-targeted aggregation-enhanced near-IR emitting silica nanoprobe for one-photon in vivo and two-photon ex vivo fluorescence bioimaging, Bioconjug. Chem. 22 (7) (2011) 1438–1450.
[34] C.-K. Lim, S. Kim, I.C. Kwon, C.-H. Ahn, S.Y. Park, Dye-condensed biopolymeric hybrids: chromophoric aggregation and self-assembly toward fluorescent bionanoparticles for near infrared bioimaging, Chem. Mater. 21 (24) (2009) 5819–5825.
[35] F. Hu, B. Liu, Organelle-specific bioprobes based on fluorogens with aggregation-induced emission (AIE) characteristics, Org. Biomol. Chem. 14 (42) (2016) 9931–9944.
[36] J. Smeitink, L. Van Den Heuvel, S. DiMauro, The genetics and pathology of oxidative phosphorylation, Nat. Rev. Genet. 2 (5) (2001) 342–352.
[37] E. Ruiz-Pesini, C. Díez-Sánchez, M.J. López-Pérez, J.A. Enríquez, The role of the mitochondrion in sperm function: is there a place for oxidative phosphorylation or is this a purely glycolytic process? Curr. Top. Dev. Biol. 77 (2007) 3–19.
[38] L. Troiano, A.R.M. Granata, A. Cossarizza, G. Kalashnikova, R. Bianchi, G. Pini, F. Tropea, C. Carani, C. Franceschi, Mitochondrial membrane potential and DNA stainability in human sperm cells: a flow cytometry analysis with implications for male infertility, Exp. Cell Res. 241 (2) (1998) 384–393.
[39] N. Zhao, S. Chen, Y. Hong, B.Z. Tang, A red emitting mitochondria-targeted AIE probe as an indicator for membrane potential and mouse sperm activity, Chem. Commun. 51 (71) (2015) 13599–13602.
[40] B. Situ, X. Ye, Q. Zhao, L. Mai, Y. Huang, S. Wang, J. Chen, B. Li, B. He, Y. Zhang, J. Zou, B.Z. Tang, X. Pan, L. Zheng, Identification and single-cell analysis of viable circulating tumor cells by a mitochondrion-specific AIE bioprobe, Adv. Sci. 7 (4) (2020) 1902760.
[41] C. Gui, E. Zhao, R.T.K. Kwok, A.C.S. Leung, J.W.Y. Lam, M. Jiang, H. Deng, Y. Cai, W. Zhang, H. Su, B.Z. Tang, AIE-active theranostic system: selective staining and killing of cancer cells, Chem. Sci. 8 (3) (2017) 1822–1830.
[42] M.D. Daniell, J.S. Hill, A history of photodynamic therapy, ANZ J. Surg. 61 (5) (1991) 340–348.
[43] R. Ackroyd, C. Kelty, N. Brown, M. Reed, The History of photodetection and photodynamic therapy, Photochem. Photobiol. 74 (5) (2001) 656.
[44] D.E.J.G.J. Dolmans, D. Fukumura, R.K. Jain, Photodynamic therapy for cancer, Nat. Rev. Cancer 3 (5) (2003) 380–387.
[45] R.E. Lawrence, R. Zoncu, The lysosome as a cellular centre for signalling, metabolism and quality control, Nat. Cell Biol. 21 (2) (2019) 133–142.
[46] C. de Duve, R. Wattiaux, Functions of lysosomes, Annu. Rev. Physiol. 28 (1) (1966) 435–492.

[47] T. Sasaki, S. Lian, A. Khan, J.R. Llop, A.V. Samuelson, W. Chen, D.J. Klionsky, S. Kishi, Autolysosome biogenesis and developmental senescence are regulated by both Spns1 and V-ATPase, Autophagy 13 (2) (2017) 386–403.

[48] X. Shi, N. Yan, G. Niu, S.H.P. Sung, Z. Liu, J. Liu, R.T.K. Kwok, J.W.Y. Lam, W.X. Wang, H.H.Y. Sung, I.D. Williams, B.Z. Tang, In vivo monitoring of tissue regeneration using a ratiometric lysosomal AIE probe, Chem. Sci. 11 (12) (2020) 3152–3163.

[49] X. Wang, L. Fan, Y. Wang, C. Zhang, W. Liang, S. Shuang, C. Dong, Visual monitoring of the lysosomal PH changes during autophagy with a red-emission fluorescent probe, J. Mater. Chem. B 8 (7) (2020) 1466–1471.

[50] M. Maekawa, G.D. Fairn, Molecular probes to visualize the location, organization and dynamics of lipids, J. Cell Sci. 127 (22) (2014) 4801–4812.

[51] G. van Meer, D.R. Voelker, G.W. Feigenson, Membrane lipids: where they are and how they behave, Nat. Rev. Mol. Cell Biol. 9 (2) (2008) 112–124.

[52] C.D. Byrne, G. Targher, NAFLD: a multisystem disease, J. Hepatol. 62 (1) (2015) S47–S64.

[53] G.R.V. Hammond, G. Schiavo, R.F. Irvine, Immunocytochemical techniques reveal multiple, distinct cellular pools of PtdIns4P and PtdIns(4,5)P2, Biochem. J. 422 (1) (2009) 23–35.

[54] Z. Wang, C. Gui, E. Zhao, J. Wang, X. Li, A. Qin, Z. Zhao, Z. Yu, B.Z. Tang, Specific fluorescence probes for lipid droplets based on simple AIEgens, ACS Appl. Mater. Interfaces 8 (16) (2016) 10193–10200.

[55] Y. Zhou, J. Hua, G. Barritt, Y. Liu, B.Z. Tang, Y. Tang, Live imaging and quantitation of lipid droplets and mitochondrial membrane potential changes with aggregation-induced emission Luminogens in an in vitro model of liver steatosis, Chembiochem 20 (10) (2019) 1256–1259.

[56] J. Rumin, H. Bonnefond, B. Saint-Jean, C. Rouxel, A. Sciandra, O. Bernard, J.-P. Cadoret, G. Bougaran, The use of fluorescent Nile red and BODIPY for lipid measurement in microalgae, Biotechnol. Biofuels 8 (1) (2015) 42.

[57] T. Fujimoto, Y. Ohsaki, J. Cheng, M. Suzuki, Y. Shinohara, Lipid droplets: a classic organelle with new outfits, Histochem. Cell Biol. 130 (2) (2008) 263–279.

[58] H. Park, S. Li, G. Niu, H. Zhang, Z. Song, Q. Lu, J. Zhang, C. Ma, R.T.K. Kwok, J.W.Y. Lam, K.S. Wong, X. Yu, Q. Xiong, B.Z. Tang, Diagnosis of fatty liver disease by a multiphoton-active and lipid-droplet-specific AIEgen with nonaromatic rotors, Mater. Chem. Front. 5 (2021) 1853–1862.

[59] G.A. Roth, C. Johnson, A. Abajobir, F. Abd-Allah, S.F. Abera, G. Abyu, M. Ahmed, B. Aksut, T. Alam, K. Alam, F. Alla, N. Alvis-Guzman, S. Amrock, H. Ansari, J. Ärnlöv, H. Asayesh, T.M. Atey, L. Avila-Burgos, A. Awasthi, A. Banerjee, A. Barac, T. Bärnighausen, L. Barregard, N. Bedi, E. Belay Ketema, D. Bennett, G. Berhe, Z. Bhutta, S. Bitew, J. Carapetis, J.J. Carrero, D.C. Malta, C.A. Castañeda-Orjuela, J. Castillo-Rivas, F. Catalá-López, J.-Y. Choi, H. Christensen, M. Cirillo, L. Cooper, M. Criqui, D. Cundiff, A. Damasceno, L. Dandona, R. Dandona, K. Davletov, S. Dharmaratne, P. Dorairaj, M. Dubey, R. Ehrenkranz, M. El Sayed Zaki, E.J.A. Faraon, A. Esteghamati, T. Farid, M. Farvid, V. Feigin, E.L. Ding, G. Fowkes, T. Gebrehiwot, R. Gillum, A. Gold, P. Gona, R. Gupta, T.D. Habtewold, N. Hafezi-Nejad, T. Hailu, G.B. Hailu, G. Hankey, H.Y. Hassen, K.H. Abate, R. Havmoeller, S.I. Hay, M. Horino, P.J. Hotez, K. Jacobsen, S. James, M. Javanbakht, P. Jeemon, D. John, J. Jonas, Y. Kalkonde, C. Karimkhani, A. Kasaeian, Y. Khader, A. Khan, Y.-H. Khang, S. Khera, A.T. Khoja, J. Khubchandani, D. Kim, D. Kolte, S. Kosen, K.J. Krohn, G.A. Kumar, G.F. Kwan, D.K. Lal, A. Larsson, S. Linn, A. Lopez, P.A. Lotufo, H.M.A. El Razek, R. Malekzadeh, M. Mazidi, T. Meier, K.G. Meles, G. Mensah, A. Meretoja, H. Mezgebe, T. Miller, E. Mirrakhimov, S. Mohammed, A.E. Moran, K.I. Musa, J. Narula, B. Neal, F. Ngalesoni, G. Nguyen, C.M. Obermeyer, M. Owolabi, G. Patton, J. Pedro, D. Qato, M. Qorbani, K. Rahimi, R.K. Rai, S. Rawaf, A. Ribeiro, S. Safiri, J.A. Salomon, I. Santos, M. Santric Milicevic, B. Sartorius, A. Schutte, S. Sepanlou, M.A. Shaikh, M.-J. Shin, M. Shishehbor, H. Shore, D.A.S. Silva, E. Sobngwi, S. Stranges, S. Swaminathan, R. Tabarés-Seisdedos, N. Tadele Atnafu, F. Tesfay, J.S. Thakur, A. Thrift, R. Topor-Madry, T. Truelsen, S. Tyrovolas, K.N. Ukwaja, O. Uthman, T. Vasankari, V. Vlassov, S.E. Vollset, T. Wakayo, D. Watkins, R. Weintraub, A. Werdecker, R. Westerman, C.S. Wiysonge, C. Wolfe, A. Workicho, G. Xu, Y. Yano, P. Yip, N. Yonemoto, M. Younis, C. Yu, T. Vos, M. Naghavi, C. Murray, Global, regional, and national burden of cardiovascular diseases for 10 causes, 1990 to 2015, J. Am. Coll. Cardiol. 70 (1) (2017) 1–25.

[60] WHO, About Cardiovascular Diseases. https://www.who.int/cardiovascular_diseases/about_cvd/en/. (Accessed 6 January 2021).

[61] B. Situ, M. Gao, X. He, S. Li, B. He, F. Guo, C. Kang, S. Liu, L. Yang, M. Jiang, Y. Hu, B.Z. Tang, L. Zheng, A two-photon AIEgen for simultaneous dual-color imaging of atherosclerotic plaques, Mater. Horiz. 6 (3) (2019) 546–553.

[62] L.M. Coussens, Z. Werb, Inflammation and cancer, Nature 420 (6917) (2002) 860–867.

[63] D. Shi, S. Chen, B. Dong, Y. Zhang, C. Sheng, T.D. James, Y. Guo, Evaluation of HOCl-generating anticancer agents by an ultrasensitive dual-mode fluorescent probe, Chem. Sci. 10 (13) (2019) 3715–3722.

[64] X. Han, Y. Ma, Y. Chen, X. Wang, Z. Wang, Enhancement of the aggregation-induced emission by hydrogen bond for visualizing hypochlorous acid in an inflammation model and a hepatocellular carcinoma model, Anal. Chem. 92 (3) (2020) 2830–2838.

[65] Y. Hong, S. Chen, C.W.T. Leung, J.W.Y. Lam, B.Z. Tang, Water-soluble tetraphenylethene derivatives as fluorescent "light-up" probes for nucleic acid detection and their applications in cell imaging, Chem. Asian J. 8 (8) (2013) 1806–1812.

[66] D.G. Booth, W.C. Earnshaw, Ki-67 and the chromosome periphery compartment in mitosis, Trends Cell Biol. 27 (12) (2017) 906–916.

[67] M.Y. Wu, J.K. Leung, L. Liu, C. Kam, K.Y.K. Chan, R.A. Li, S. Feng, S. Chen, A small-molecule AIE chromosome periphery probe for cytogenetic studies, Angew. Chem. Int. Ed. (2020) 1–6.

[68] E. Zhao, Y. Chen, S. Chen, H. Deng, C. Gui, C.W.T. Leung, Y. Hong, J.W.Y. Lam, B.Z. Tang, A luminogen with aggregation-induced emission characteristics for wash-free bacterial imaging, high-throughput antibiotics screening and bacterial susceptibility evaluation, Adv. Mater. 27 (33) (2015) 4931–4937.

[69] F. Hu, G. Qi, Kenry, D. Mao, S. Zhou, M. Wu, W. Wu, B. Liu, Visualization and in Situ ablation of intracellular bacterial pathogens through metabolic labeling, Angew. Chem. Int. Ed. Engl. 59 (24) (2020) 9288–9292.

[70] E. Zhao, Y. Hong, S. Chen, C.W. Leung, C.Y. Chan, R.T. Kwok, J.W. Lam, B.Z. Tang, Highly fluorescent and photostable probe for long-term bacterial viability assay based on aggregation-induced emission, Adv. Healthc. Mater. 3 (1) (2014) 88–96.

[71] M. Kang, C. Zhou, S. Wu, B. Yu, Z. Zhang, N. Song, M.M.S. Lee, W. Xu, F.J. Xu, D. Wang, L. Wang, B.Z. Tang, Evaluation of structure-function relationships of aggregation-induced emission Luminogens for simultaneous dual applications of specific discrimination and efficient Photodynamic killing of gram-positive bacteria, J. Am. Chem. Soc. 141 (42) (2019) 16781–16789.

[72] R. Hu, F. Zhou, T. Zhou, J. Shen, Z. Wang, Z. Zhao, A. Qin, B.Z. Tang, Specific discrimination of gram-positive bacteria and direct visualization of its infection towards mammalian cells by a DPAN-based AIEgen, Biomaterials 187 (2018) 47–54.

[73] X. Zhang, C. Ren, F. Hu, Y. Gao, Z. Wang, H. Li, J. Liu, B. Liu, C. Yang, Detection of bacterial alkaline phosphatase activity by enzymatic in Situ self-assembly of the AIEgen-peptide conjugate, Anal. Chem. 92 (7) (2020) 5185–5190.

[74] C. Zhou, W. Xu, P. Zhang, M. Jiang, Y. Chen, R.T.K. Kwok, M.M.S. Lee, G. Shan, R. Qi, X. Zhou, J.W.Y. Lam, S. Wang, B.Z. Tang, Engineering sensor arrays using aggregation-induced emission Luminogens for pathogen identification, Adv. Funct. Mater. 29 (4) (2019) 1–10.

[75] L. Zhao, Y. Liu, Z. Zhang, J. Wei, S. Xie, X. Li, Fibrous testing papers for fluorescence trace sensing and photodynamic destruction of antibiotic-resistant bacteria, J. Mater. Chem. B 8 (13) (2020) 2709–2718.

[76] L. Zhao, Y. Chen, J. Yuan, M. Chen, H. Zhang, X. Li, Electrospun fibrous mats with conjugated tetraphenylethylene and mannose for sensitive turn-on fluorescent sensing of Escherichia coli, ACS Appl. Mater. Interfaces 7 (9) (2015) 5177–5186.

[77] H.K. Chan, P.C. Kwok, Production methods for nanodrug particles using the bottom-up approach, Adv. Drug Deliv. Rev. 63 (6) (2011) 406–416.

[78] S. Xie, S. Manuguri, G. Proietti, J. Romson, Y. Fu, A.K. Inge, B. Wu, Y. Zhang, D. Hall, O. Ramstrom, M. Yan, Design and synthesis of theranostic antibiotic nanodrugs that display enhanced antibacterial activity and luminescence, Proc. Natl. Acad. Sci. U. S. A. 114 (32) (2017) 8464–8469.

[79] J. Zhao, Z. Dong, H. Cui, H. Jin, C. Wang, Nanoengineered peptide-grafted hyperbranched polymers for killing of bacteria monitored in real time via intrinsic aggregation-induced emission, ACS Appl. Mater. Interfaces 10 (49) (2018) 42058–42067.

[80] S.S. Lucky, K.C. Soo, Y. Zhang, Nanoparticles in photodynamic therapy, Chem. Rev. 115 (4) (2015) 1990–2042.

[81] B. Zeina, J. Greenman, W.M. Purcell, B. Das, Killing of cutaneous microbial species by photodynamic therapy, Br. J. Dermatol. 144 (2) (2001) 274–278.

[82] J.P. Celli, B.Q. Spring, I. Rizvi, C.L. Evans, K.S. Samkoe, S. Verma, B.W. Pogue, T. Hasan, Imaging and photodynamic therapy: mechanisms, monitoring, and optimization, Chem. Rev. 110 (5) (2010) 2795–2838.

[83] M. Gao, Q. Hu, G. Feng, N. Tomczak, R. Liu, B. Xing, B.Z. Tang, B. Liu, A multifunctional probe with aggregation-induced emission characteristics for selective fluorescence imaging and photodynamic killing of bacteria over mammalian cells, Adv. Healthc. Mater. 4 (5) (2015) 659–663.

[84] G. Feng, C.J. Zhang, X. Lu, B. Liu, Zinc(II)-Tetradentate-coordinated probe with aggregation-induced emission characteristics for selective imaging and photoinactivation of bacteria, ACS Omega 2 (2) (2017) 546–553.

[85] J.S. Ni, T. Min, Y. Li, M. Zha, P. Zhang, C.L. Ho, K. Li, Planar AIEgens with enhanced solid-state luminescence and ROS generation for multidrug-resistant bacteria treatment, Angew. Chem. Int. Ed. Engl. 59 (25) (2020) 10179–10185.

[86] T. Zhou, R. Hu, L. Wang, Y. Qiu, G. Zhang, Q. Deng, H. Zhang, P. Yin, B. Situ, C. Zhan, A. Qin, B.Z. Tang, An AIE-active conjugated polymer with high ROS-generation ability and biocompatibility for efficient photodynamic therapy of bacterial infections, Angew. Chem. Int. Ed. Engl. 59 (25) (2020) 9952–9956.

[87] M. Wu, G. Qi, X. Liu, Y. Duan, J. Liu, B. Liu, Bio-orthogonal AIEgen for specific discrimination and elimination of bacterial pathogens via metabolic engineering, Chem. Mater. 32 (2) (2020) 858–865.

[88] X.W. He, Y.J. Yang, Y.C. Guo, S.G. Lu, Y. Du, J.J. Li, X.P. Zhang, N.L.C. Leung, Z. Zhao, G.L. Niu, S.S. Yang, Z. Weng, R.T.K. Kwok, J.W.Y. Lam, G.M. Xie, B.Z. Tang, Phage-guided targeting, discriminative imaging, and synergistic killing of bacteria by AIE bioconjugates, J. Am. Chem. Soc. 142 (8) (2020) 3959–3969.

[89] G.B. Qi, F. Hu, Kenry, K.C. Chong, M. Wu, Y.H. Gan, B. Liu, Bacterium-templated polymer for self-selective ablation of multidrug-resistant bacteria, Adv. Funct. Mater. 30 (31) (2020) 2001338.

[90] L.H. Xiong, X. He, Z. Zhao, R.T.K. Kwok, Y. Xiong, P.F. Gao, F. Yang, Y. Huang, H.H.Y. Sung, I.D. Williams, J.W.Y. Lam, J. Cheng, R. Zhang, B.Z. Tang, Ultrasensitive Virion immunoassay platform with dual-modality based on a multifunctional aggregation-induced emission Luminogen, ACS Nano 12 (9) (2018) 9549–9557.

[91] C.C. Zhou, M.J. Jiang, J. Du, H.T. Bai, G.G. Shan, R.T.K. Kwok, J.H.C. Chau, J. Zhang, J.W.Y. Lam, P. Huang, B. Tang, One stone, three birds: one AIEgen with three colors for fast differentiation of three pathogens, Chem. Sci. 11 (18) (2020) 4730–4740.

[92] X. Ge, M. Gao, B. Situ, W. Feng, B. He, X. He, S. Li, Z. Ou, Y. Zhong, Y. Lin, X. Ye, X. Hu, B.Z. Tang, L. Zheng, One-step, rapid fluorescence sensing of fungal viability based on a bioprobe with aggregation-induced emission characteristics, Mater. Chem. Front. 4 (3) (2020) 957–964.

[93] Z. Zhao, S. Gao, X. Zheng, P. Zhang, W. Wu, R.T.K. Kwok, Y. Xiong, N.L.C. Leung, Y. Chen, X. Gao, J.W.Y. Lam, B.Z. Tang, Rational design of perylenediimide-substituted triphenylethylene to electron transporting aggregation-induced emission luminogens (AIEgens) with high mobility and near-infrared emission, Adv. Funct. Mater. 28 (11) (2018) 1705609.

[94] J. Wang, X. Gu, P. Zhang, X. Huang, X. Zheng, M. Chen, H. Feng, R.T.K. Kwok, J.W.Y. Lam, B.Z. Tang, Ionization and anion-Π^+ interaction: a new strategy for structural design of aggregation-induced emission luminogens, J. Am. Chem. Soc. 139 (46) (2017) 16974–16979.

[95] M. Jiang, X. Gu, R.T.K. Kwok, Y. Li, H.H.Y. Sung, X. Zheng, Y. Zhang, J.W.Y. Lam, I.D. Williams, X. Huang, K.S. Wong, B.Z. Tang, Multifunctional AIEgens: ready synthesis, tunable emission, mechanochromism, mitochondrial, and bacterial imaging, Adv. Funct. Mater. 28 (1) (2018) 1704589.

CHAPTER

15

Recent advances of aggregation-induced emission nanoparticles (AIE-NPs) in biomedical applications

Soheila Sabouri, Bicheng Yao, and Yuning Hong

Department of Chemistry and Physics, La Trobe Institute for Molecular Science, La Trobe University, Melbourne, VIC, Australia

1 Introduction

Nanotechnology has greatly pushed forward the development of biomedicine. As an essential component of nanotechnology, nanoparticles (NPs) enjoy unique physical and chemical properties, such as stability, enhanced permeability and retention effect (EPR), optical characteristics, and easy surface modification capabilities, which make them a powerful tool for biomedical imaging and therapeutics [1,2]. In recent years, fluorescent nanoparticles (FNPs) have gained much attention due to their prominent photoluminescent features which may visualize related biological processes for analysis or treatment [3]. According to the categories of fluorescent materials, FNPs are mainly divided into two major groups: organic fluorescent nanoparticles (OFNPs) and inorganic fluorescent nanoparticles (IFNPs). In terms of biomedical applications, the OFNPs are more popular with researchers since they show much better biocompatibility and less toxicity for live cells compared to the inorganic ones such as quantum dots (QDs) [4]. Up to now, a variety of OFNPs has been developed for biomedical applications including fluorescent bioimaging, disease diagnostics, drug delivery, therapy, etc. [5,6].

For fabricating OFNPs with high brightness and strong photostability, the choice of fluorescent material is very important. Currently, many organic small molecule fluorescent materials are available with bright emission and plentiful colors ranging from blue to red. However, most of these organic fluorophores show strong fluorescence only in dilute solutions, but they are weakly emissive or nonemissive at high concentrations or in their aggregation state [7].

Aggregation-Induced Emission (AIE)
https://doi.org/10.1016/B978-0-12-824335-0.00007-6

This phenomenon is mainly caused by the strong π-π interactions between the rigid planar fluorescent molecules and is named the aggregation-caused quenching (ACQ) effect. Since OFNPs need to integrate numerous fluorescent molecules into a limited space, the ACQ effect will inevitably limit the application of organic fluorophores in fabricating OFNPs [8]. Fortunately, the emerging AIE fluorogens (AIEgens) in 2001, showing photophysical features completely opposite to ACQ fluorogens, have efficiently addressed this issue. Despite the valuable improvements of AIEgens in various aspects during the past two decades of research, there are still some limitations in utilizing them in the biological application, such as high hydrophobicity, poor biocompatibility, low penetration ability, etc. Developing AIE-nanoparticles (AIE-NPs), on one hand, results in brightly emissive NPs owing to the AIE effect of AIEgens embedded in the nanostructures, on the other hand, promotes biomedical applications of AIEgens due to the improved water solubility and biocompatibility, and easy surface functionalization for specific targeting purposes [9–15]. So far, an ample number of AIE-NPs have been successfully developed and used in biological research. In this chapter, we focus on the latest advances in the design and fabrication of AIE-NPs and their related biomedical applications extending from bioimaging, biosensing, and diagnosis to drug delivery, therapy, and image-guided surgeries.

2 Strategies for the fabrication of AIE-NPs

2.1 Commonly used AIEgens for AIE-NPs fabrication

As the important component of AIE-NPs, AIEgens have a crucial influence on the optical properties and functions of the yielded NPs. Before introducing design strategies of AIE-NPs, some typical AIEgens that have been used for AIE-NP fabrication will be summarized here. The chemical structures of these AIEgens together with their corresponding optical properties are shown in Table 1. Obviously, due to the difference in skeleton structures and functional groups, these AIEgens show various emission wavelengths and fluorescence quantum efficiency. The introduction of electron donating and accepting groups into the conjugation system of AIE cores is usually applied to adjust emission colors. As a result, the emission wavelengths can be tuned to cover the whole visible and near-infrared (NIR) region. Another notable advance of AIEgens is the large Stokes shift (>100nm), which can avoid self-absorption during practical application. Moreover, the AIE molecules can be easily modified with functional groups for further NP preparation. The flourishing development of AIEgens in the past decade, therefore, provides a solid foundation for the design and synthesis of AIE-NPs.

2.2 Design strategies of AIE-NPs

Thanks to their superior performances over conventional organic fluorescent dyes, AIEgens have been widely applied in the preparation of FNPs. Up to now, several strategies have been adopted to encapsulate AIEgens into nanocarriers, producing AIE-NPs with monodispersed sizes, well-controlled shapes, and desired peripheral functional

TABLE 1 Optical properties of the most commonly used AIEgens to fabricate AIE-NPs.

AIEgen	λ_{Em} (nm)	λ_{Ex} (nm)	Quantum yield	Chemical structure
TPE-ITC **(6)** [16]	340	485 (THF/water mixture)	85% in solid state	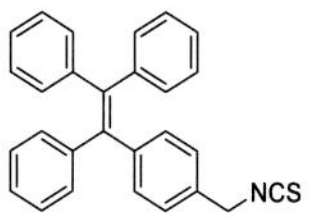 NCS
TPE-FN **(7)** [17]	420	590 (THF/water mixture)	70% in solid state	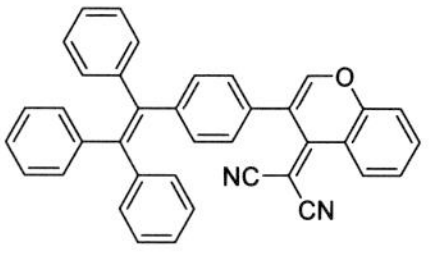 O NC CN
TTF **(8)** [18]	500	660 (THF/water mixture)	52.5% in solid state	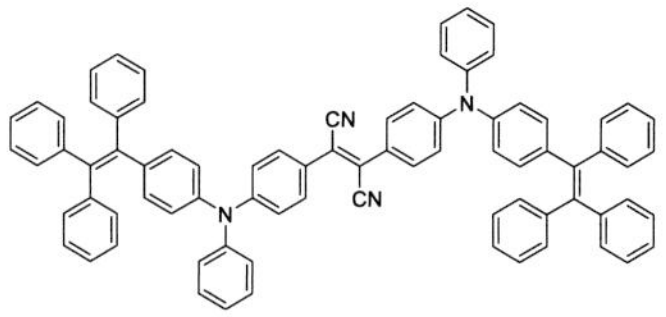 N CN CN N
BTPETD **(9)** [19]	350	539	89% in film state	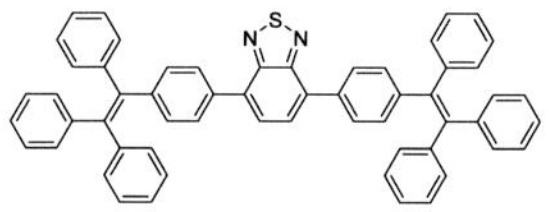 N S N
CAPP **(10)** [20]	430	630 (THF/water mixture)	%	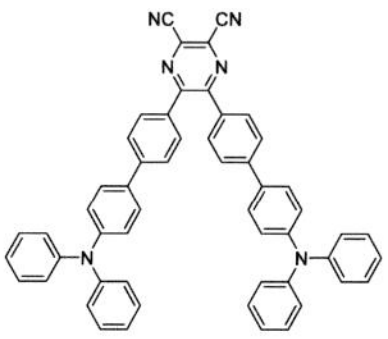 NC CN N N N N
TPE-OXE **(11)** [21]	365	620	57% in solid state	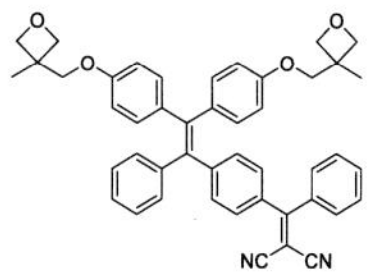 O O O O NC CN
HPS **(12)** [22]	363	497	78% in film state	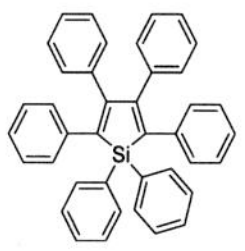 Si

Continued

TABLE 1 Optical properties of the most commonly used AIEgens to fabricate AIE-NPs—cont'd

AIEgen	λ_{Em} (nm)	λ_{Ex} (nm)	Quantum yield	Chemical structure
TPE-EPA-DCM **(13)** [23]	480	633 (THF/water mixture)	12% in BSA matrix with loading ratio of 3.07% TPE-EPA-DCM	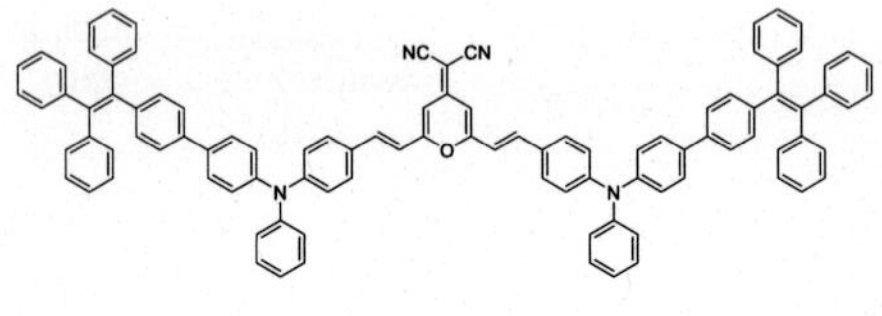
TPE-TETRAD **(14)** [24]	500	680 (THF/water mixture)	23.5% in solid state	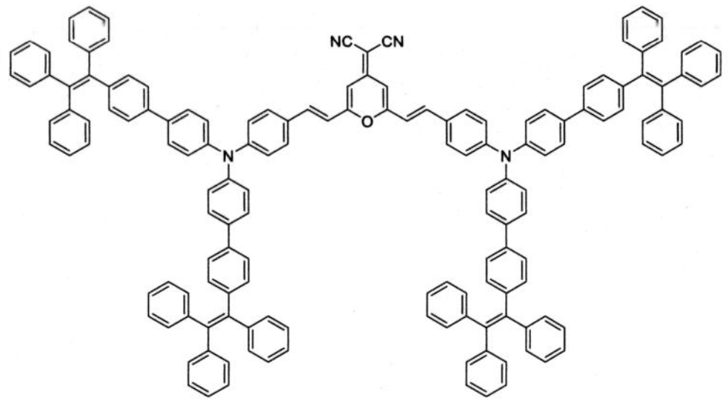

groups. Based on whether chemical bonds are formed between AIEgens and nanocarriers, those strategies can be divided into two major classes: covalent binding method and noncovalent binding method [25]. Since they are closely related to the properties and functions of AIE-NPs, fabrication methods of AIE-NPs will be discussed comprehensively in this section.

2.2.1 Covalent binding method

In a covalent binding method, AIEgens are tightly bound in nanomaterials through chemical bonds, which will prevent the outward diffuse of dye molecules. Thus, the first step for fabricating AIE-NPs via the covalent binding method is the chemical modification of AIEgens, usually with amphiphilic molecules, biomolecules, polymers, or monomers. With these AIE-active building blocks, AIE-NPs can be fabricated through self-assemble, chemical/physical cross-linking, or even polymerization reactions which can form polymer scaffolds of NPs [26].

Bile acids are a class of natural amphiphilic compounds containing both hydrophobic and hydrophilic regions. In 2015, Zhang et al. applied bile acids to conjugate with tetraphenylethylene (TPE) **(1)**, a typical AIE molecule, obtaining six amphiphilic building blocks for NPs fabrication (Fig. 1A) [27]. All the TPE-bile acid conjugates were proved to enjoy AIE features with the blue fluorescence emission. Moreover, AIE-NPs with diameters around 200 nm and narrow size distributions were formed when the TPE-bile acid conjugates **2** or **3** were placed in water-acetone or water–methanol cosolvent systems. It is noteworthy that these AIE-NPs contain amphiphilic binding pockets in their membrane surface, which enables the facile construction of energy-transfer systems through host-guest interactions (Fig. 1A

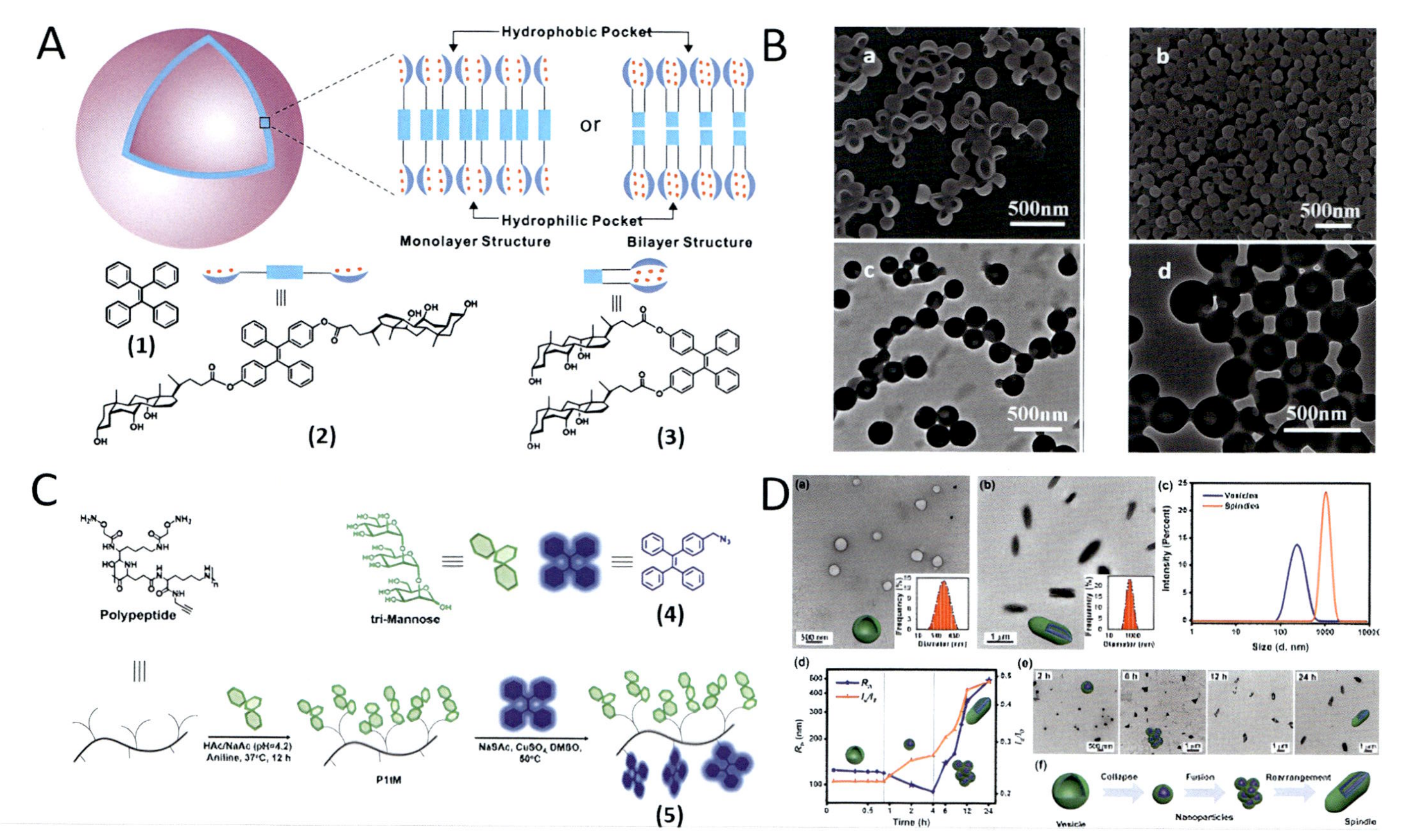

FIG. 1 Chemical structure and schematic illustration of the formed vesicles from **(2) and (3)** (A). SEM and TEM images of the vesicles derived from compound 2 at $f_W = 40\%$, compound 5 at $f_W = 50\%$ in water-acetone cosolvent system (B). Chemical structure and graphical representation of the Synthesis of TPE-Based Glycopolypeptide (P1tM-TPE) **(5)** (C). TEM images and hydrodynamic diameter of P1tM-TPE **(5)** assembly in DMSO-water mixture in at $f_W = 80\%$ and 90%. Relative scattered light intensity (I_S/I_0) and hydrodynamic radius Rh changes, TEM images and schematic illustration of P1tM-TPE **(5)** assembly at different times during the process of f_W from 80% to 90% (D). *Panel (B): Reproduced with permission from M. Zhang, X. Yin, T. Tian, Y. Liang, W. Li, Y. Lan, et al., AIE-induced fluorescent vesicles containing amphiphilic binding pockets and the FRET triggered by host–guest chemistry, Chem. Commun. 51 (50) (2015), 10210–10213. Copyright 2015, Royal Society of Chemistry; Panel (D): Reproduced with permission from H. Chen, E. Zhang, G. Yang, L. Li, L. Wu, Y. Zhang, et al., Aggregation-induced emission luminogen assisted self-assembly and morphology transition of amphiphilic glycopolypeptide with bioimaging application, ACS Macro Lett. 8 (8) (2019), 893–898. Copyright 2019, American Chemistry Society.*

and B). Recently, a kind of biomimetic glycoproteins, termed glycopolypeptides, were linked with TPE-N_3 **(4)** through azide-alkyne reaction to produce an amphiphilic fluorescent macromolecule P1tM-TPE **(5)** (Fig. 1C) [28]. Due to its amphiphilic properties, macromolecule **5** can form assembles in the DMSO-water mixture spontaneously. With the water fraction as 80% and P1tM-TPE **(5)** concentration as 1 mg/mL, fluorescent nanovesicles with a diameter around 240 nm were found under transmission electron microscopy (TEM) (Fig. 1D). Besides vesicles, other self-assemble morphologies including spindles and porous nanosheets can be obtained by adjusting the water fraction of the mixed solvent (Fig. 1D).

Recently, nonspherical NPs generated from a nanocovalent triazine polymer (NCTP) **(15)** were reported by Cui, Liu, and coworkers (Fig. 2A) [29]. In this work, (*E*)-1,2-bis(4-bromophenyl)-1,2-stilbene (TPE-2Br) **(16)** was first synthesized as the AIEgen, and then modified with cyanuric chloride via Friedel-Crafts alkylation to give the final NP building blocks. Due to the thermal polymerization reaction between the triazine units and TPE-2Br **(16)**, a network polymer was synthesized, which can form NPs when aggregated in the aqueous medium. According to the TEM images of those NCTP **(15)** nanoparticles, nonspherical morphology was observed with an average size ranging from 120 to 160 nm (Fig. 2A). Another example using AIEgens as the monomer to construct AIE-NPs through stabilized precipitation polymerization was reported by Wang et al. [30]. In this work, the authors designed a comonomer system containing AIEgen TPE-1VBC **(17)**, maleic anhydride (MAH), and styrene (St) (Fig. 2B). The polymerization was initiated by azobisisobutyronitrile (AIBN) in ethyl butanoate (EB) under nitrogen and 60°C for 240 min, producing fluorescence polymeric nanoparticles with bright yellow emission @ 565 nm, hydrodynamic diameter ranging from 106 to 420 nm, and regular sphere morphology with a narrow polydispersity ($PDI < 0.214$). During the polymerization progress, TPE-1VBC **(17)** also acted as fluorescent reporter to monitor the microenvironmental transformation during the formation of nanostructures (Fig. 2B).

2.2.2 Noncovalent binding method

Even though the covalent bonded AIE-NPs enjoy the advantages of good stability, as well as adjustable dye loading ratio, the chemical modification of AIEgens, is usually difficult and time consuming. In contrast, the noncovalent binding method, which encapsulates AIEgens into nanomaterials through physical interactions, is easier to be implemented [31]. Therefore, the noncovalent binding method is more popular and various standardized operations have been developed for preparing AIE-NPs with fine-controlled size and morphology, high fluorescence efficiency, and diverse functions.

Most of the reported AIEgens, e.g., hexaphenylsilole (HPS) **(12)**, tetraphenylethylene (TPE) **(1)**, tetraphenylpyrazine (TPP) **(18)**, etc., are mainly composed of benzene rings that are highly hydrophobic. Once dispersed in an aqueous solution, these molecules can only form inhomogeneous large particles or precipitates. For some special AIEgens that have moderate hydrophobicity or good crystallization capacity, nanoaggregates or nanocrystals can be formed spontaneously without adding any assisting agents. For example, the silole derivative S-1a **(19)** (Fig. 3A) is an AIEgen that aggregated into NPs with an average diameter of around 220 nm in aqueous media [18]. This is the simplest noncovalent binding method for AIE-NPs fabrication. Besides, Fateminia et al. reported a new AIEgen TPE-FN **(7)** synthesized from TPE and dicyanomethylene-benzopyran [17] (Fig. 3C). Through

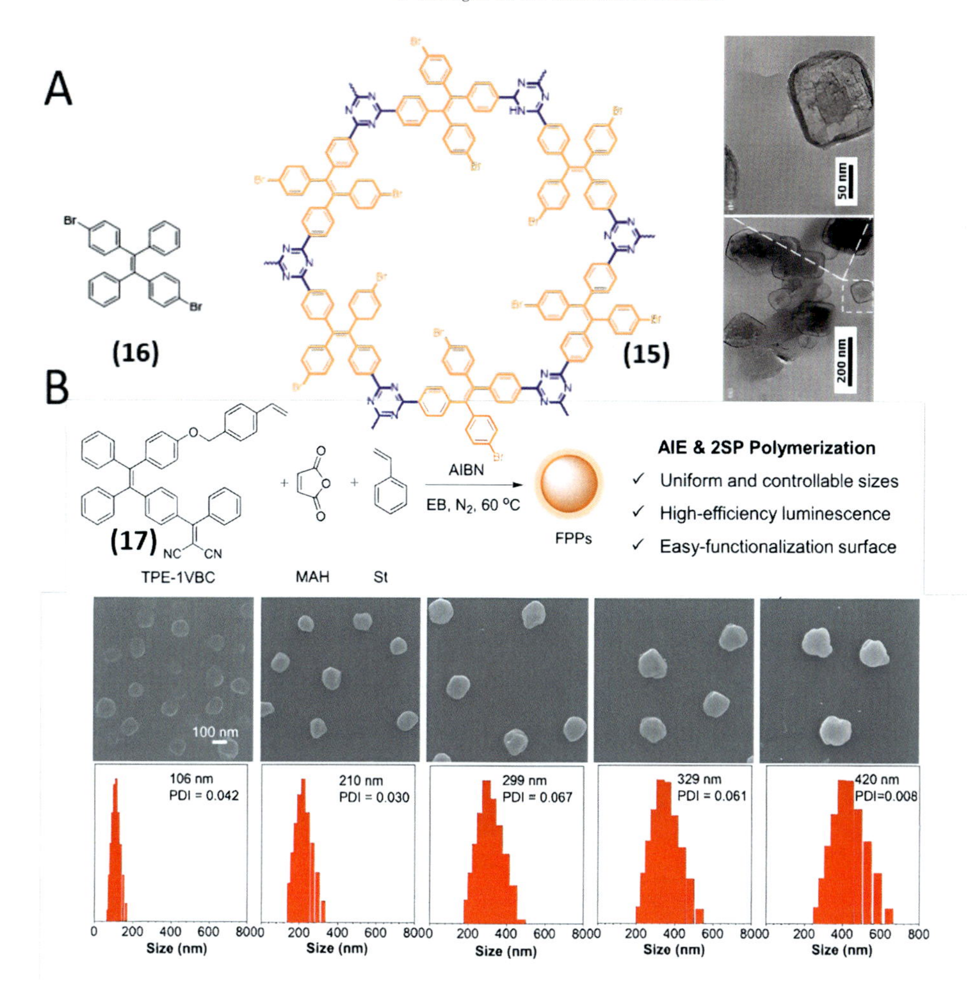

FIG. 2 Chemical structure of the pH-responsive triazine skeleton nanopolymer (NCTP) **(15)** composite and TEM images of NCTP **(15)** particles and NCTP-DOX micelles in PBS solution (A). Chemical structure of TPE-1VBC **(17)** and self-stabilized precipitation (2SP) polymerization of TPE-1VBC **(17)**, maleic anhydride (MAH), and styrene (St) as monomers in ethyl butanoate (EB) solution under the initiation of azobisisobutyronitrile (AIBN) in a nitrogen atmosphere at 60°C. SEM images and DLS analysis of fluorescent polymeric nanoparticles with different sizes (B). *Panel (A): Reproduced with permission from Y. Zhang, X. Peng, X. Jing, L. Cui, S. Yang, J. Wu, et al., Synthesis of pH-responsive triazine skeleton nano-polymer composite containing AIE group for drug delivery, Front. Mater. Sci. 15 (1) (2021), 113–123. Copyright 2021, Springer Nature. Panel (B): Reproduced with permission from G. Wang, L. Yang, C. Li, H. Yu, Z. He, C. Yang, et al., Novel strategy to prepare fluorescent polymeric nanoparticles based on aggregation-induced emission via precipitation polymerization for fluorescent lateral flow assay, Mater. Chem. Front. 5 (5) (2021), 2452–2458. Copyright 2021, Royal Society of Chemistry.*

a stress-induced seed-assisted crystallization method, TPE-FN can be fabricated into uniform nanocrystals with a size of 110 ± 10 nm (Fig. 3C and D). Another commonly used method is using amphiphilic molecules to encapsulate AIEgens physically. Saponin is a class of naturally produced surfactants that can form micelles easily in an aqueous solution. Nicol et al. applied saponin as amphiphilic units to construct two novel AIE-NPs [32].

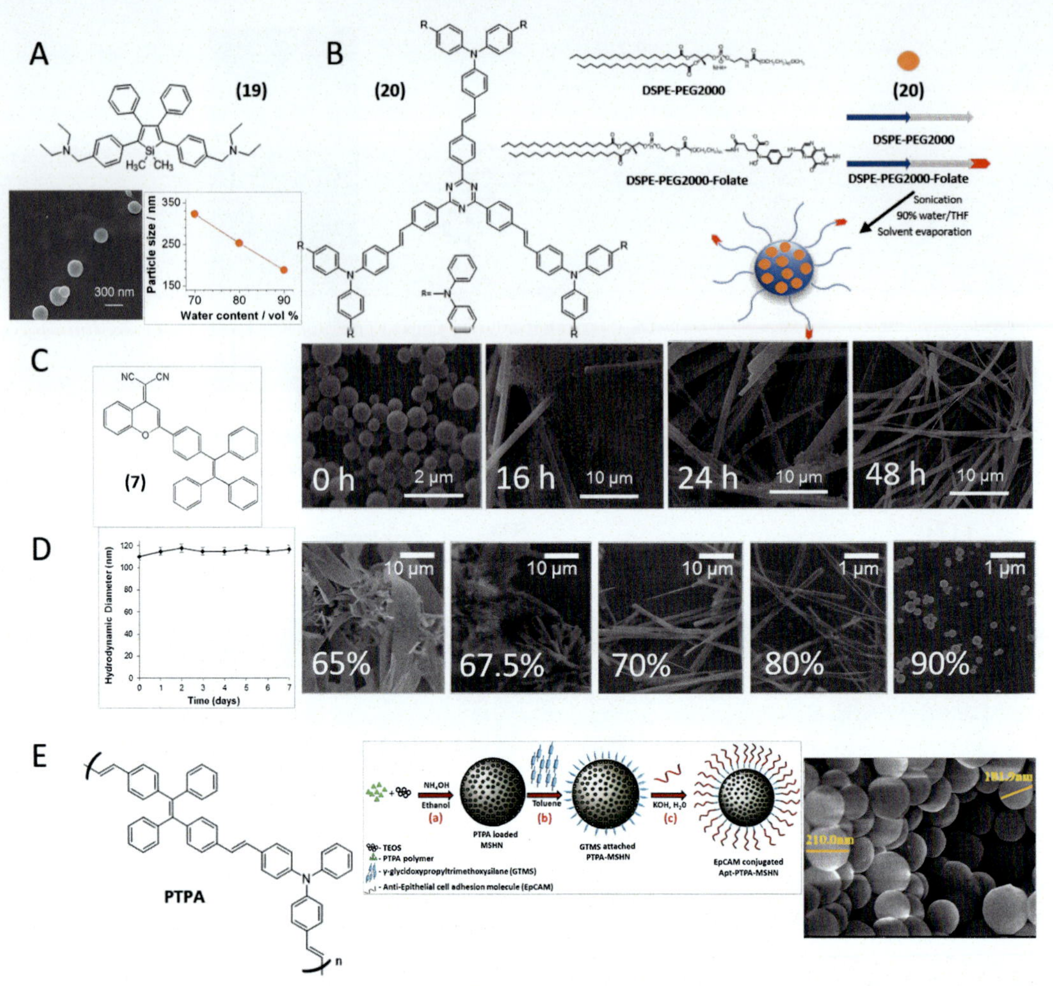

FIG. 3 Chemical structure and the SEM image of nanoaggregates of S-1a **(19)** formed in THF-water mixture (f_w 70%). Variation of the aggregate size of NPs in THF-water mixture (f_w 70%–90%) (A). Chemical structure of T1 **(20)**, DSPE-PEG2000, and DSPE-PEG2000-Folate and schematic illustration of FTNP synthesis (B). Chemical structure and SEM images of TPE-FN **(7)** in THF-water mixture (f_w 70%) during different storage times from 0 to 48 h (C). SEM images of TPE-FN **(7)** NPs in THF-water mixture with different f_w (65%–90%) and hydrodynamic mean particle size of NPs vs. storage time (D). Schematic illustration of PTPA-loaded MSHNs preparation, modification with GTMS, and conjugation of anti-EpCAM aptamer on the GTMS-attached PTPA-MSHNs and SEM image of PTPA MSHNs (E). *Panel (A): Reproduced with permission from Y. Yu, C. Feng, Y. Hong, J. Liu, S. Chen, K.M. Ng, et al., Cytophilic fluorescent bioprobes for long-term cell tracking, Adv. Mater. 23 (29) (2011), 3298–3302. Copyright 2011, WILEY-VCH Verlag GmbH & Co. KGaA, Weinheim; Panel (B): Reproduced from K. Li, Y. Jiang, D. Ding, X. Zhang, Y. Liu, J. Hua, S.-S. Feng, B. Liu, Folic acid-functionalized two-photon absorbing nanoparticles for targeted MCF-7 cancer cell imaging, Chem. Commun. 47 (26) (2011), 7323–7325. Copyright 2011, Royal Society of Chemistry; Panel (D): Reproduced with permission from S.A. Fateminia, Z. Wang, C.C. Goh, P.N. Manghnani, W. Wu, D. Mao, et al., Nanocrystallization: a unique approach to yield bright organic nanocrystals for biological applications, Adv. Mater. 29 (1) (2017), 1604100. Copyright 2016, WILEY-VCH Verlag GmbH & Co. KGaA, Weinheim; Panel (E): Reproduced with permission from S. Dineshkumar, A. Raj, A. Srivastava, S. Mukherjee, S.S. Pasha, V. Kachwal, L. Fageria, R. Chowdhury, I.R. Laskar, Facile incorporation of "aggregation-induced emission"-active conjugated polymer into mesoporous silica hollow nanospheres: synthesis, characterization, photophysical studies, and application in bioimaging, ACS Appl. Mater. Interfaces 11 (34) (2019), 31270–31282. Copyright 2019, American Chemistry Society.*

To embed CAPP **(10)** and TPE-TETRAD **(14)** (Table 1) in the nanostructure, firstly uniform AIE aggregates are formed in the water solution through a nanoprecipitation method. Then saponin was added above the critical micelle concentration (CMC) to encapsulate and stabilize the AIE aggregates. The main driving force for NPs formation is the hydrophobic property of the AIEgens. With the assistance of the DLS technique, the hydrodynamic diameters and PDI values were determined to be 152nm and 0.34 for saponin@CAPP **(10)** NPs and 222nm and 0.55 for saponin@TPE-TETRAD **(14)** NPs. Finally, the as-prepared NPs were used for cell imaging with high permeability and ultrafast delivery into live cancer cells. Besides surfactants, amphiphilic polymer matrices are frequently used for AIEgens encapsulation. For instance, 1,2-distearoyl-sn-glycero-3-phosphoethanolamine-N-[methoxy(polyethylene glycol)-2000] (DSPE-PEG2000), a PEG chain containing lipid molecule DSPE, is extensively used for fabricating AIE-NPs. Li et al. applied folate modified DSPE-PEG5000 for the physical encapsulation of compound T1 **(20)**, which is a hydrophobic AIE and two-photon absorbing chromophore [33]. The modified nanoprecipitation method was applied [34]. Briefly, THF solution of T1, DSPE-PEG2000, and DSPE-PEG5000-Folate was poured into water/THF mixture (9,1, *v*/v), followed by sonicating with a microtip probe sonicator at 12W for 60s. Subsequently, AIE-NPs with various folate contents were formed after stirring the above emulsion at room temperature overnight to evaporate THF. The T1**(20)** molecules were entangled by the hydrophobic DSPE segments forming a core with hydrophilic PEG and PEG-folate chains extended outside in the water. Due to the application of DSPE-PEG amphiphilic polymers, the synthesized AIE-NPs showed smaller hydrodynamic diameters (83–88nm), excellent biocompatibility, and well-controlled surface functional groups (Fig. 3B).

Mesoporous silica nanoparticles (MSNPs) exhibit many advantages in biomedical applications, such as good water dispersity, easy surface modification, tunable pore size and structure, excellent biocompatibility, and cytomembrane penetration capability, etc. Recently, Dineshkumar et al. developed a novel AIE conjugated polymer (CP) PTPA and encapsulated it into mesoporous silica hollow nanospheres (MSHNs) by a facile noncovalent approach [35]. The PTPA-MSHNs were synthesized in situ by polymerizing tetraethoxysilane (TEOS)monomer under vigorous mechanical stirring in an ethanol solution containing PTPA AIEgen. Moreover, surface modification with 3′-glycidoxypropyltrimethoxysilane (GTMS) and anti-EpCAM aptamer was conducted on the PTPA-MSHNs successively to increase their biocompatibility and specificity toward cancer cells (Fig. 3E). Both the PTPA-MSHNs and the modified ones show bright green fluorescence, uniform spherical shape, and an average size ranging from 266 to 342nm. In brief, this work provides a facile noncovalent approach for fabricating AIE-NPs for biomedical applications.

3 Biomedical applications of AIE-NPs

The advantages of bright fluorescence, high stability, and good biocompatibility make AIE-NPs outstanding materials for biomedical applications. One of the most common applications of AIE-NPs is fluorescence bioimaging and tracing which can provide images of living organisms or tissue (e.g., cells, blood vessels, and tumor) with high fluorescent contrast, due to the fantastic optical features of AIEgens. Besides, through surface modification with specific ligands, AIE-NPs are capable to detect biomolecules from body fluids for disease diagnosis purposes. In addition, by integrating photosensing and photothermal functions to the AIE core within NPs, theranostic applications of AIE-NPs can be developed including photodynamic therapy (PDT), photoacoustic imaging (PAI), and photothermal therapy (PTT). In the past decade, much research has been conducted studying the biomedical applications of AIE-NPs and related works have been summarized in several review articles as well [36]. Hence, in this section, recent advances on AIE-NP biomedical applications in the past 5 years will be introduced.

3.1 Fluorescence bioimaging

Biological imaging technologies including optical imaging, magnetic resonance imaging (MRI), computed tomography (CT) [37], ultrasound imaging [38], radio-nuclear imaging, etc., have been widely applied in modern medicine for disease detection, screening, and diagnosis. In the past decades, fluorescence bioimaging as an emerging technology has drawn much research interest. Compared with other technologies, fluorescence bioimaging enjoys the advantages of strong signal intensity, high resolution, cost-efficiency, and simple operation, thus was broadly used to create fluorescent images of cells, tissues, or bodies for biological study or clinical analysis. As mentioned in the introduction section, FNPs have superior photostability and tunable sizes, which make them better fluorescent materials for bioimaging than organic small molecular dyes. Here, we mainly discussed the application of AIE-NPs in cell imaging and tracking, tumor imaging, and vascular imaging.

3.1.1 *Cell imaging*

Profiling and tracing the structural detail and variation of cells is crucial for exploring biological processes of life. In the past years, many AIE-NPs have been developed for staining subcellular structures including cytoplasm, mitochondria, lysosome, etc. In 2018, Tang and coworkers developed a novel deep-red emissive AIEgen TPE-TETRAD (**14**, Table 1), which has both high two-photon absorption cross-section (313 MG at 830 nm) and aggregation-induced three-photon luminescence [24]. These unique optical properties are beneficial for increasing the penetration depth for in vivo fluorescence bioimaging techniques. Thus, AIE-NPs were fabricated using **14** and DSPE–PEG2000–biotin through a modified nanoprecipitation method. The yielded TPE-TETRAD@biotin NPs show a spherical morphology with an average hydrodynamic diameter of 155 nm and a low polydispersity index (PDI) of 0.05, which is ideal for cell permeation and retention (Fig. 4A). After incubated with HeLa cervical cancer cells, TPE-TETRAD@biotin NPs specifically lit up the tubular mitochondrial structure and high-quality fluorescence images were captured (Fig. 4A). The mitochondrial specificity can be ascribed to the peripheral biotin groups which can target the mitochondria of cancer cells via unique avidin-biotin interaction. The results of photostability measurement showed that TPE-TETRAD@biotin NPs have high photostability compared to commercial photostable MitoTracker Green (Fig. 4B), which plays a critical role for in vivo imaging. In addition to the visible tubular mitochondrial structure in HeLa cells, a high rate of mitochondrial colocalization (94.5%) has been recorded (Fig. 4C). In 2017, Xie et al. reported two novel zwitterionic

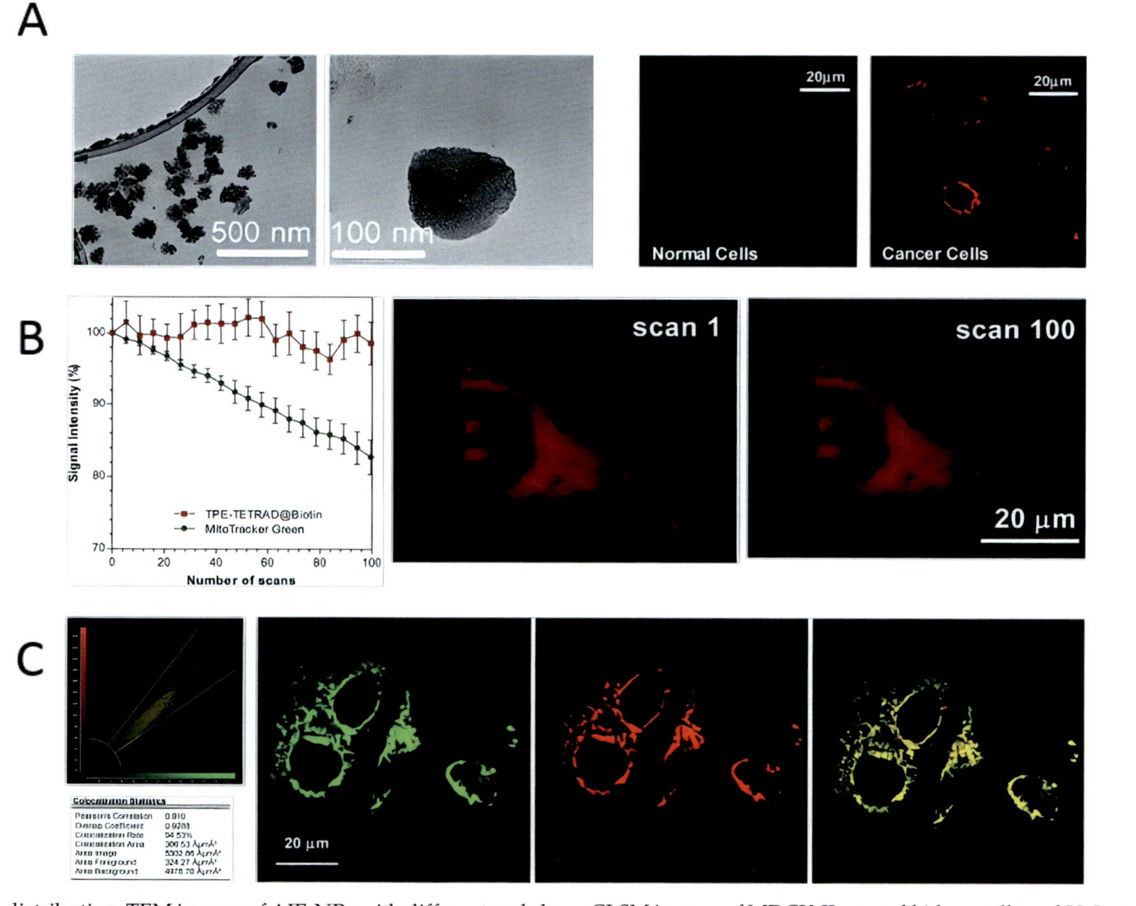

FIG. 4 NP size distribution. TEM images of AIE-NPs with different scale bars. CLSM images of MDCK-II normal kidney cells and HeLa cervical cancer cells after incubation with TPE-TETRAD **(14)**@biotin NPs (A). Photostability comparison between TPE-TETRAD **(14)**@biotin NPs (*red*) and MitoTracker Green (*green*) in HeLa cells with an increasing number of laser scans and fluorescence images at scan 1 and 100 for AIE-NPs in HeLa cells. Scale bar: 20 μm (B). CLSM localization and images of HeLa cells incubated with MitoTracker Green (*green*) and TPE-TETRAD**(14)**@biotin NPs (*red*). The *yellow* signals show merged images (C). *Reproduced with permission from A. Nicol, W. Qin, R.T. Kwok, J.M. Burkhartsmeyer, Z. Zhu, H. Su, et al., Functionalized AIE nanoparticles with efficient deep-red emission, mitochondrial specificity, cancer cell selectivity and multiphoton susceptibility, Chem. Sci. 8 (6) (2017), 4634–4643. Copyright 2017, Royal Society of Chemistry.*

phosphorylcholine copolymers MTP1 and MTP2 with both AIE feature and self-assemble ability [39]. As shown in Fig. 5D, the copolymers were synthesized through RAFT polymerization between a vinyl functionalized AIEgen, NSP2E **(21)**, and a zwitterionic phosphorylcholine monomer MTP as comonomers. Thanks to the hydrophilicity acquired from the zwitterionic component, polymers MTP1 and MTP2 show amphiphilic properties and can form AIE-NPs with hydrophobic NSP2E **(21)** core and zwitterionic phosphorylcholine shell. The intensity-mean hydrodynamic diameters of NPs prepared from MTP1 and MTP2 are 345 ± 22 nm and 147 ± 36 nm, respectively. Due to the strong fluorescence intensity and good particle stability, these AIE-NPs were further applied in cell imaging. Firstly, CCK-8 assay indicated that the AIE-NPs have little influence on the cell viability of A549 cells, manifesting their good cytocompatibility. Afterward, A549 cells were incubated with the MTP1 and MTP2 NPs and observed using a confocal laser scanning microscope (CLSM) under the excitation of a 405 nm laser. The results indicated that the AIE-NPs mainly entered the cytoplasmic regions of A549 cells through endocytosis and solely stained the cytoplasm with strong green fluorescence. Compared with previously reported zwitterionic AIE polymers, MTP1 and MTP2 show ultralow CMC values (0.008 and 0.007 mg/mL for MTP1 and MTP2, respectively) which make them applicable in dilute solutions. Recently, Jana, Ghorai, and coworkers reported an AIEgen-conjugated magnetic NPs for bioimaging applications [40]. Firstly, the γ-Fe_2O_3 NPs were fabricated via a high-temperature colloidal method followed by crosslinking polymerization of acryl monomers on the surface to endow hydrophilicity, ionic charges, and primary amino groups to the surface of NPs. Then, an aldehyde-modified TPE **(1)** derivative reacted with primary amines from the polyacrylate coated NPs, producing the covalently conjugated AIE-NPs MN-TPE-I, -II, and -III. These NPs have small hydrodynamic sizes between 25 and 50 nm, which are much smaller than other reported AIEgen-based NPs. After performing the fluorescence imaging experiments on HeLa cells, nanoparticle MN-TPE-I and -II were found to have good labeling efficiency, mainly localized in cell cytoplasm. As shown in Fig. 5A, the colocalization study with LysoTracker Red, a commercial lysosome dye, further suggested that the AIE-NPs were up taken into the cell via endocytosis which transferred them to the lysosomes (Fig. 5B and C).

3.1.2 Cell tracking

Cell tracking, an important area in biological, medical, and pharmaceutical research gives us the opportunity to evaluate real-time dynamic data of cell evolution, migration, differentiation, and behavior for effective regeneration, and monitor various processes in live cells. In 2018, Li, Song, and coworkers successfully labeled live human embryonic stem (hES) cells and tracked their differentiation process by AIE-NPs prepared from TPE-11 **(22)** (Fig. 6A) [41]. Due to the bola amphiphilic structure composed of one TPEunit and two ionic heads (pyridinium salt-terminated alkyl groups), TPE-11 **(22)** self-assembled [42] in physiological conditions forming flake-like NPs with an average hydrodynamic diameter of about 50 nm. In addition, the zeta potential of TPE-11 **(22)** NPs was measured to be 41.6 mV, which plays a critical role in the endocytosis by cells. When working under a low concentration (8.0 μg/mL), TPE-11 **(22)** NPs showed good biocompatibility, which made them suitable for hES cells labeling and differentiation tracking. The experimental results indicated that hES cells can be labeled with TPE-11 **(22)** NPs showing bright green fluorescence, which

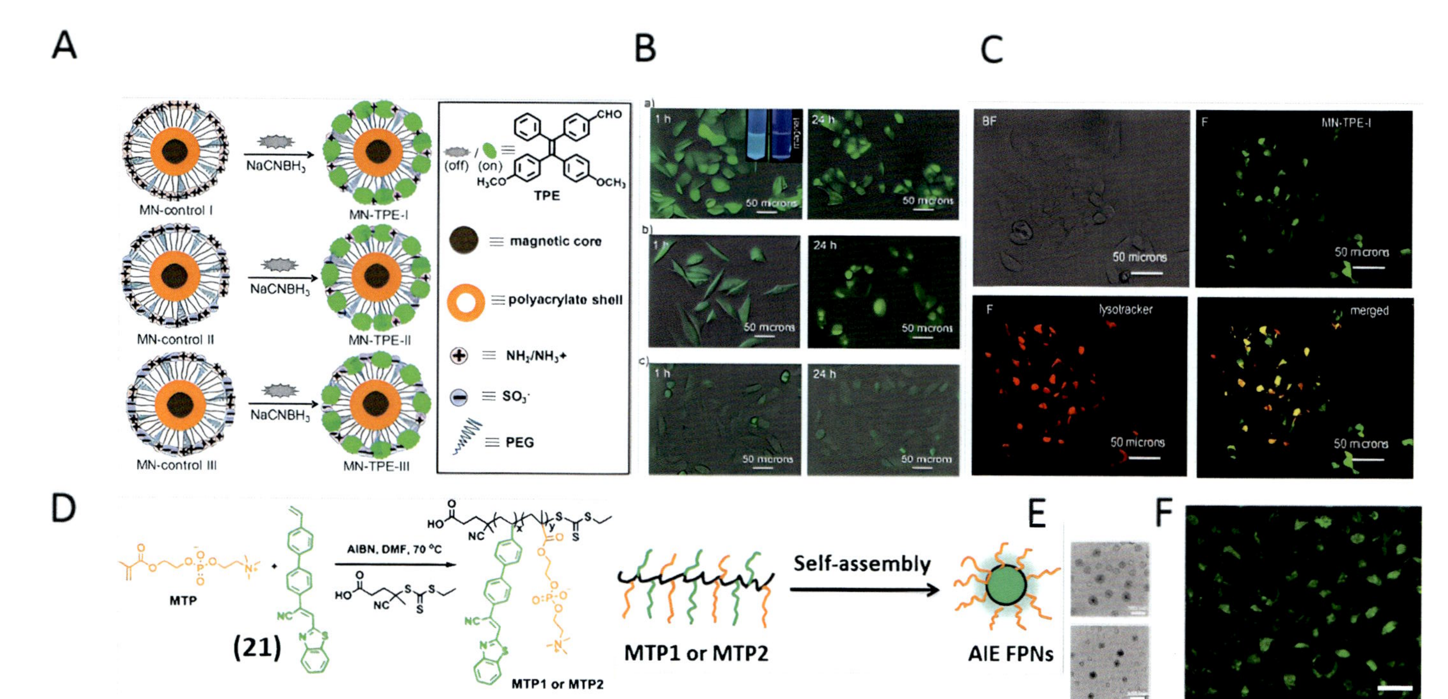

FIG. 5 Schematic illustration of synthesis strategy for AIEgen-Conjugated Magnetic-Fluorescent Nanoparticle (A). Application of MN-TPE-I, MN-TPE-II, and MN-TPE-III as a fluorescent cell imaging probe. Inset shows the magnetic separation of MN-TPE-I labeled cells. Results show that MN-TPE-I and MN-TPE-II can label cells very efficiently as compared to MN-TPE-III and stay/localize inside the cytoplasm (B). Colocalization study of MN-TPE-I with lysotracker red in HeLa cells (C). Schematic illustration of synthesis route of MTP1 and MTP2. RAFT polymerization of MTP with an AIE monomer [NSP2E **(21)**] with a vinyl group to afford the zwitterionic phosphorylcholine polymers, the feed ratio of MTP to NSP2E **(21)** (y/x) was 10:1, and the designed degrees of polymerization were 5 for MTP1 and 10 for MTP2, respectively. The self-assembly of MTP1 or MTP2 into AIE FPNs for cell imaging applications has been shown (D). TEM images of MTP1-NPs (up) and MTP2-NPs (down) dispersed in water (spherical NPs with sizes ranging between 150 and 250 nm and 100–150 nm, respectively) (E). CLSM image of A549 cells incubated with MTP1-NPs (F). *Panel (C): Reproduced with permission from K. Mandal, D. Jana, B.K. Ghorai, N.R. Jana, AIEgen-conjugated magnetic nanoparticles as magnetic–fluorescent bioimaging probes, ACS Appl. Nano Mater. 2 (5) (2019), 3292–3299. Copyright 2019, American Chemistry Society; Panel (E): Reproduced with permission from G. Xie, C. Ma, X. Zhang, H. Liu, X. Guo, L. Yang, et al., Biocompatible zwitterionic phosphorylcholine polymers with aggregation-induced emission feature, Colloids Surf. B: Biointerfaces 157 (2017), 166–173. Copyright 2017, Elsevier.*

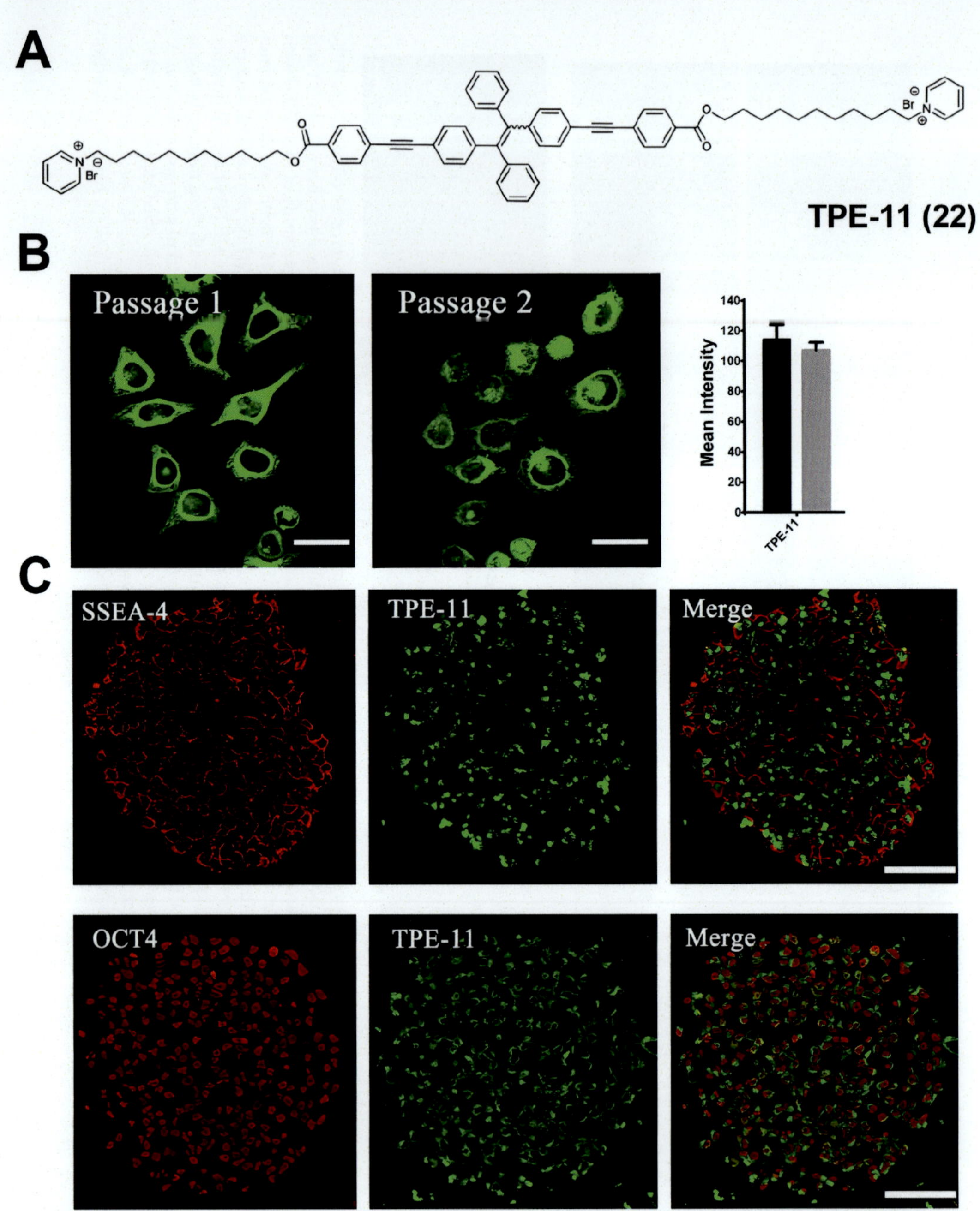

FIG. 6 Chemical structure of TPE-11 **(22)** (A). Fluorescence intensity analysis and persistence on TPE-11 **(22)** resulted from confocal images of HeLa cells before and after removal of the dye and continuously cultured for 24 h. Scale bar: 25 μm. Mean fluorescence intensities of the HeLa cells stained with TPE-11 **(22)** (B). The surface marker (SSEA-4, *red*) and transcription factor (OCT4, *red*) were expressed after TPE-11 **(22)** (*green*) treatment. Scale bar: 100 μm (C). *Reproduced with permission from S. Zhou, H. Zhao, R. Feng, L. Ding, Z. Li, C. Deng, et al., Application of amphiphilic fluorophore-derived nanoparticles to provide contrast to human embryonic stem cells without affecting their pluripotency and to monitor their differentiation into neuron-like cells, Acta Biomater. 78 (2018), 274–284. Copyright 2018, Elsevier.*

allowed long-term monitoring of their differentiation into neuron-like cells over a period of 40 days (Fig. 6). Most recently, Mukadam et al. conducted research on applying an AIE-active medication rilpivirine (RPV) **(23)** for tracking HIV (human immunodeficiency virus) infected cells (**Fig.** 7) [43]. As a fluorescent nonnucleoside reverse transcriptase inhibitor (NNRTI), RPV **(23)** can be used as both a drug and subcellular tracking agent. The molecular RPV **(23)** crystal was simply prepared via an antisolvent precipitation method, followed by coating with DSPE-PEG2000 and phosphatidyl choline (α-PC) to yield RPV **(23)**-NPs (~200 nm) with good stability, lipophilicity, and biocompatibility. Since $CD4^+$ T cell and the monocyte–macrophage (MDM) are the major target cells of HIV infection, the uptake of RPV **(23)**-NPs by the target cells and other HIV-infection resistant lymphocyte subtypes were investigated using flow cytometry and ultraperformance liquid chromatography tandem mass spectrometry (UPLC-MS). The results suggested that RPV **(23)**-NPs can selectively enter both MDM and CD4+ T cells, but with a higher tendency to be uptaken by MDM (Fig. 7). Furthermore, the authors studied subcellular localization of RPV **(23)**-NPs in monocyte-macrophages and $CD4^+$ T cells, confirming their presence in subcellular domains where viral replication occurred. Finally, RPV **(23)**-NPs were successfully applied for the tracking of pharmacokinetics and biodistribution in humanized mouse models.

3.1.3 Tumor imaging

In the cell imaging part, some examples of cancer cell imaging have already been discussed. Here, we will briefly introduce in vivo tumor tissue imaging, which may require better spatial–temporal resolution and tissue penetrability on the AIE-NPs. In 2018, Liu and coworkers reported a polymeric AIE nanorod with excellent tumor penetration and accumulation capabilities [44]. Through the noncovalent binding strategy, three kinds of AIE nanorods were fabricated with three AIEgens, TPE **(1)**, BTPEBT **(24)**, and TPETPAFN **(25)**, showing different emission colors encapsulated in lipid-PEG matrix, respectively (Fig. 8A). For comparison, AIE nanodots with the same chemical components were synthesized as well. Confocal laser scanning microscopy (CLSM) imaging on various cell lines, including NIH-3T3 and HEK293T normal cells, and SKBR-3, MDA-MB-231, HeLa, and MCF-7 cancer cells, demonstrated that the AIE nanorods have higher cancer cell internalization efficiency than nanodots, which can be attributed to the caveolae-mediated endocytosis mechanism (Fig. 8). Subsequently, in vivo tumor imaging in a SKBR-3 tumor-bearing mouse model was carried out using the TPETPAFN **(25)** nanorods for their far-red and near-infrared (FR/NIR) emission. The nanorods demonstrated to have over 1.6-fold higher tumor accumulation ability and a higher tumor penetrability than corresponding nanodots. In summary, the advantages of high tumor accumulation, deep tumor penetration, and stable fluorescence signal make AIE nanorods a powerful tool for tumor targeting and imaging. Another example of AIE-NPs with an accurate diagnosis, high sensitivity, and multimodality imaging ability was designed and realized by Tang and coworkers [45]. In this work, Ag^+ was first reduced in an alkaline condition by the phenol groups in TPE-M2OH **(26)** to form the core of silver NPs (Fig. 9A). Meanwhile, due to the electron donor-acceptor structure, TPE-M2OH **(26)** self-assembled on the surface of silver NPs as a shell layer, producing the final silver@AIEgen core-shell NPs (AACSN) with a mean diameter of 85 nm. It is noteworthy that the fluorescence intensity of AACSN can be enhanced via increasing the TPE-M2OH **(26)** feeding concentration which extended the thickness of the shell layer to restrain the fluorescence

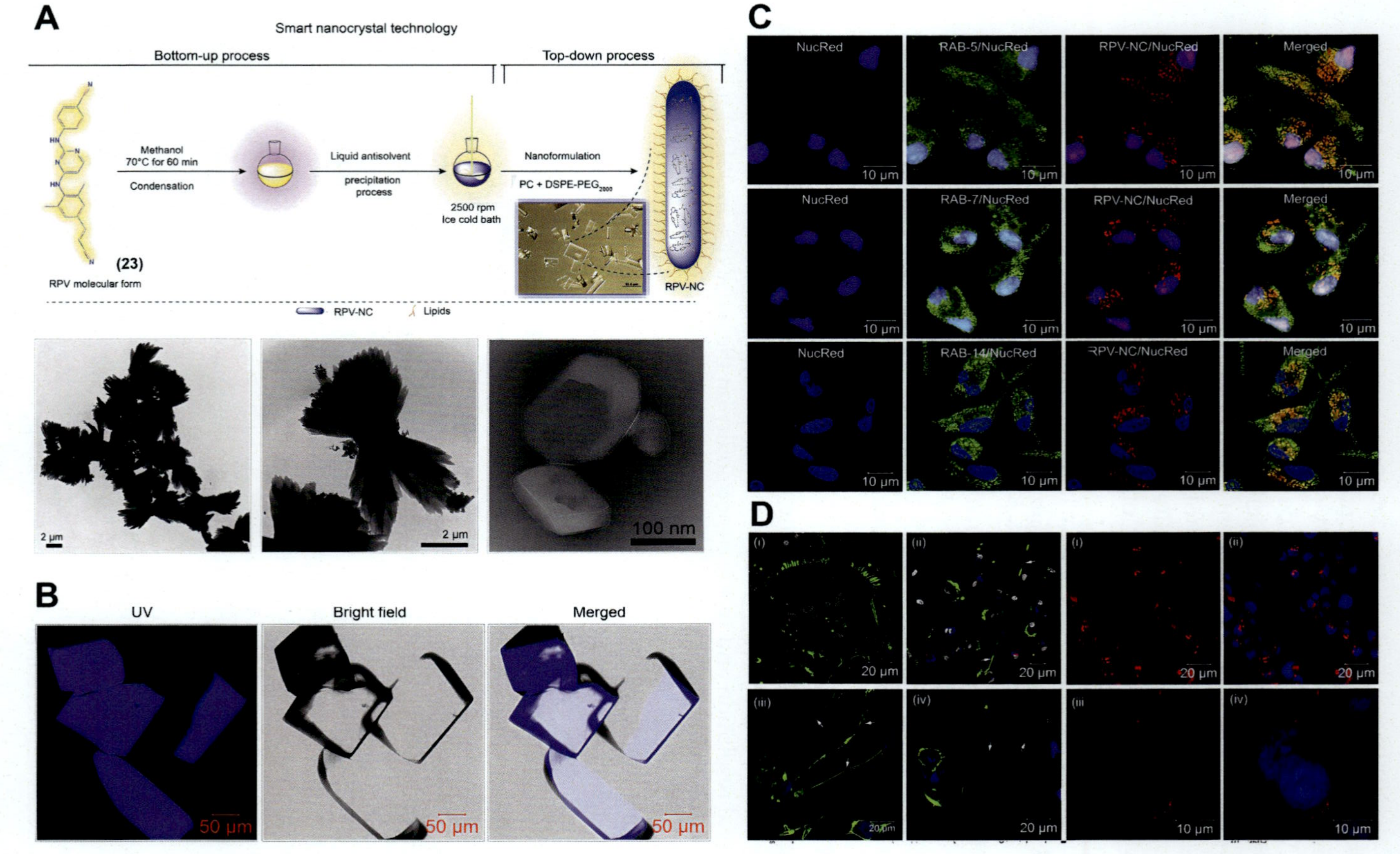

FIG. 7 Schematic representation of the bottom-up and top-down approaches of RPV **(23)** NC synthesis. Low and high magnifications of TEM images of RPV **(23)** NC made by antisolvent precipitation. After homogenization, the surface morphology was uniformed (A). Fluorescence emission of RPV **(23)** NC has been shown by confocal and bright-field images (B). Cellular and subcellular RPV **(23)**-NC localization (C), (D) Confocal images of macrophages treated with RPV **(23)** NC (pseudo-colored red) and Alexa-Flour 488 secondary antibody against primary Rab5, Rab7 or Rab14 antibodies (*green*). *Yellow* areas indicate colocalization of RPV **(23)** NC with endosomal components (C). Cellular distribution of RPV **(23)** NC (*blue*) is visualized in infected macrophages. Virus particles (*red*), cell cytoskeleton, and tunneling nanotubes (*green*) in control and infected cells showed the giant cell form in control macrophages upon infection. Infected macrophages show the formation of tunneling nanotubes (indicated by the *white arrows*) and the presence of RPV **(23)** NC in the cell cytoplasm (D). *Reproduced with permission from I.Z. Mukadam, J. Machhi, J. Herskovitz, M. Hasan, M.D. Oleynikov, W.R. Blomberg, et al., Rilpivirine-associated aggregation-induced emission enables cell-based nanoparticle tracking, Biomaterials 231 (2020), 119669. Copyright 2020, Elsevier.*

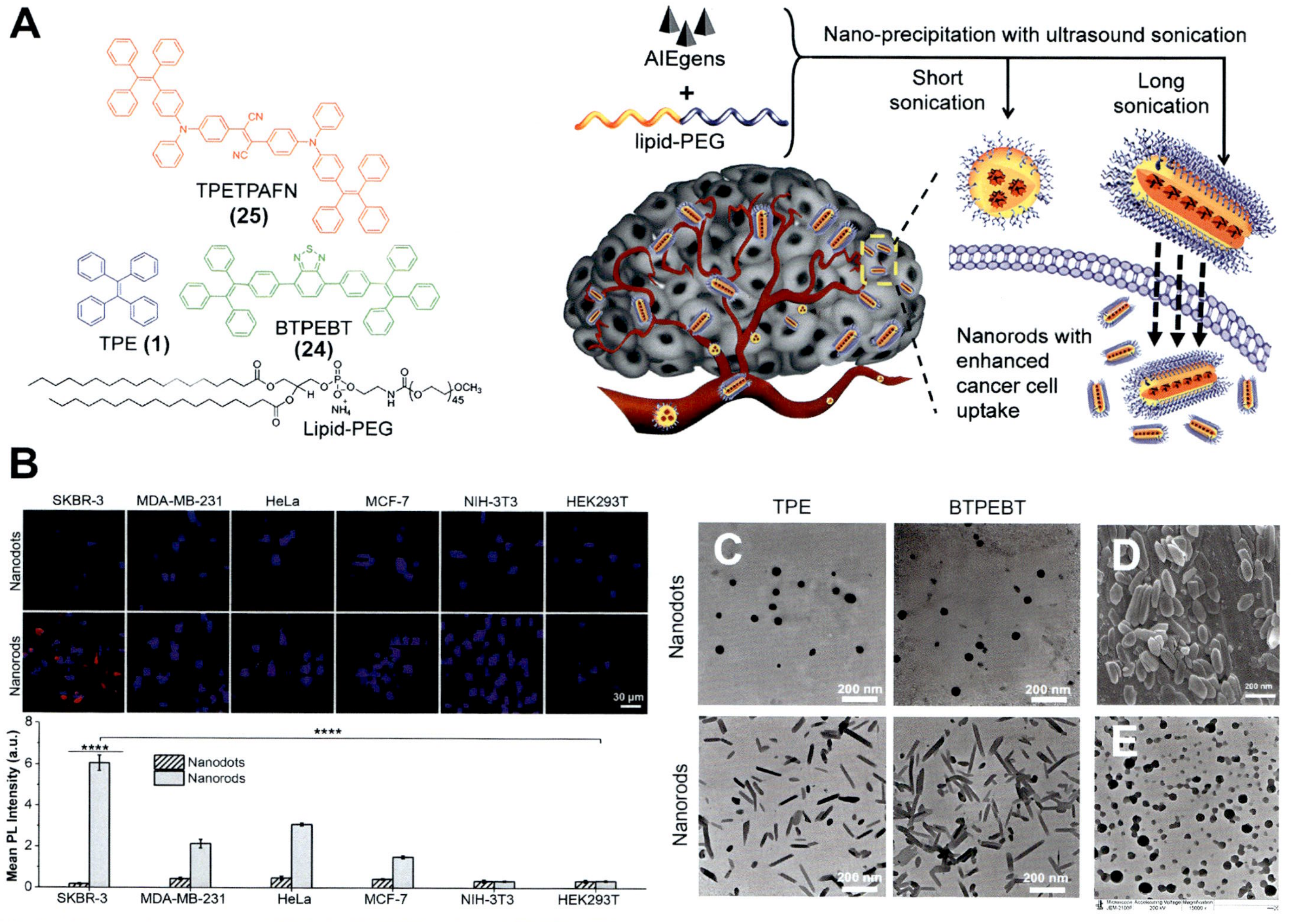

FIG. 8 Chemical structures of TPE (**1**), BTPEBT (**24**), and TPETPAFN (**25**), and schematic illustration of ultrasound sonication induced nanodot-to-nanorod transition and the enhanced tumor accumulation and cancer cell uptake of nanorods (A). Confocal images of different cells incubated with TPETPAFN (**25**) nanodots and nanorods, and mean fluorescence intensities of TPETPAFN (**25**) nanodots and nanorods inside different cells (B). TEM images of TPE and BTPEBT (**24**) nanodots and nanorods (C). SEM image of TPETPAFN (**25**) nanorods (D). TEM image of TPETPAFN (**25**) nanodots fabricated at 80°C with 2 min sonication (E). *Reproduced with permission from G. Feng, D. Mao, J. Liu, C.C. Goh, L.G. Ng, D. Kong, et al., Polymeric nanorods with aggregation-induced emission characteristics for enhanced cancer targeting and imaging, Nanoscale 10 (13) (2018), 5869–5874. Copyright 2018, Royal Society of Chemistry.*

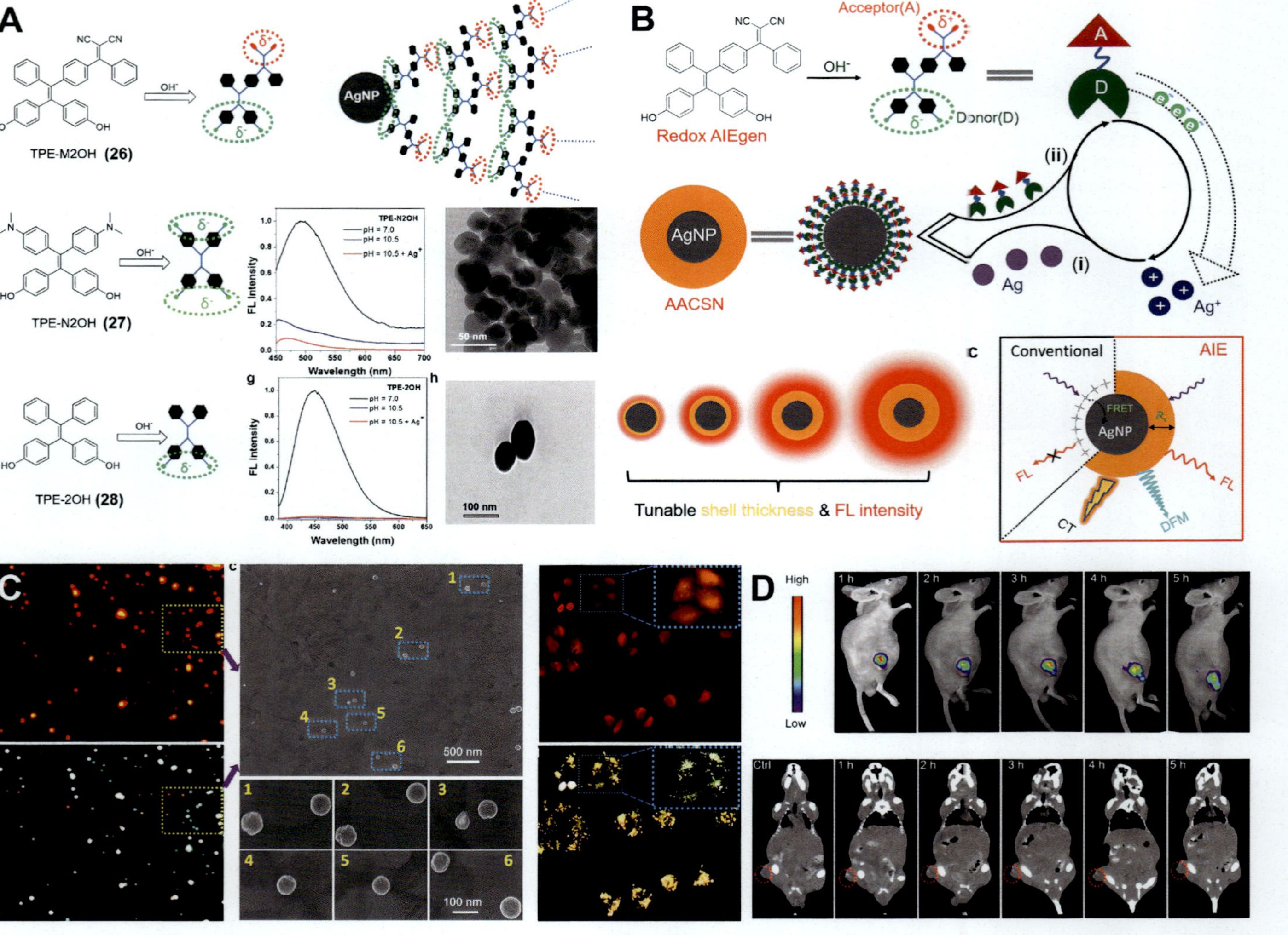

FIG. 9 Chemical structures and self-assembly mechanism of TPE-M2OH **(26)**, TPE-N2OH **(27)**, TPE-2OH **(28)** on the surface of AgNPs. Schematic illustrations of electron-rich and -poor domains of AIEgens under alkaline conditions and attraction of TPE-M2OH **(26)** on the AgNP surface through electron-rich and -poor domains and their fluorescence spectra (A). Schematic representation of the synthesis, tunable properties, and multimodality imaging application of AACSN (B). Single nanoparticle analysis of AACSNs (~20-nm sell thickness), FL and DFM image of AACSNs of the same region. SEM image of AACSNs in the *yellow* box area (C). AACSNs for in vitro and in vivo multimodality imaging. FL and DFM images of the same region of HeLa cells (the left part). FL and CT images of 4T1 tumor-bearing nude mice at different time points after intratumoral injection of AACSNs The tumor site is highlighted with a *red-dotted circle* (D). *Reproduced with permission from X. He, Z. Zhao, L.-H. Xiong, P.F. Gao, C. Peng, R.S. Li, et al., Redox-active AIEgen-derived plasmonic and fluorescent core@ shell nanoparticles for multimodality bioimaging. J. Am. Chem. Soc. 140 (22) (2018), 6904–6911. Copyright 2018, American Chemical Society.*

quenching factors, including fluorescence resonance energy transfer (FRET) and electronic transfer (ET). Thanks to the unique combination of AIEgen and silver NPs that can bring both dark-field microscopy (DFM) and computed tomography (CT) functions, the AACSN were supposed to have multimodality bioimaging capability. Hence, multimodality imaging on HeLa cells (in vitro) and in 4T1 tumor-bearing nude mice (in vivo) were performed with AACSN, showing superior AIE fluorescence signal together with DFM signal with high signal-to-noise ratio and CT signal with large penetration depth (Fig. 9).

3.1.4 Vascular imaging

Vascular and intravascular imaging is a challenging but clinically useful technique for disease diagnosis and monitoring. Due to the excellent biocompatibility and photostability, and high quantum efficiency, deep-red emissive AIE-NPs have been widely utilized in vascular imaging. In 2020, AIE-NPs with intense three-photon fluorescence (3PF) in the deep-red region were designed by Qian, Dong, Han, and coworkers for in vivo brain vascular imaging [46]. As shown in Fig. 10, AIEgen DCPE-TPA **(29)** was first designed and synthesized based on the electron Donor-π-Acceptor (D-π-A) strategy with triphenylamine as the electron donor and two cyan groups as the electron acceptor. Interestingly, DCPE-TPA **(29)** exhibited three different molecular packing models including orange crystals, red crystals, and deep-red emissive amorphous solids. Secondly, a polymeric matrix of Pluronic F-127 was used to encapsulate DCPE-TPA **(29)** physically, resulting in the formation of AIE-NPs with 3PF in the deep-red region (QY = 29.3%) and an average diameter of 185.2 nm. Biocompatible DCPE-TPA **(29)** NPs were further applied for in vivo 3PF imaging of the mouse brain blood vessels. Under the excitation of a 1560 nm fs laser, high-resolution fluorescence images can be

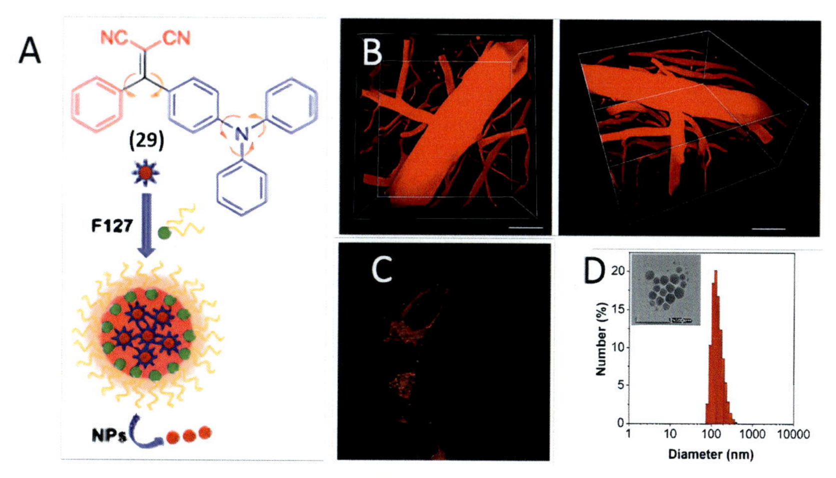

FIG. 10 Chemical structure of DCPE-TPA **(29)** and schematic illustration of NPs formation (A). 3D reconstruction of the vascular, seen from above and the side (B). CLSM imaging of HeLa cells incubated with DCPE-TPA **(29)** NPs (C). DLS and TEM results of DCPE-TPA **(29)** NPs (D). *Reproduced with permission from H. Tian, D. Li, X. Tang, Y. Zhang, Z. Yang, J. Qian, et al., Efficient red luminogen with aggregation-induced emission for in vivo three-photon brain vascular imaging, Mater. Chem. Front. 4 (6) (2020), 1634–1642. Copyright 2020, Royal Society of Chemistry.*

captured, even at a depth of 300 μm. Besides, the authors also got a 3D image of brain blood vessels with an excellent signal-to-background ratio, which revealed their complex vasculature.

3.2 Theranostics

The term theranostics was first proposed by John Funkhouser in 1998 [47]. Through the combination of diagnostics and therapy, theranostics can realize nidus targeting, drug distribution monitoring, and evaluation of therapeutic effect in real-time, which greatly promotes the development of personalized and precise medicine. Nanotechnology allows the combination of both diagnostic and therapeutic functions into single particles and thus has been widely applied for disease theranostics [30,48–50]. In the past few years, as an emerging class of fluorescent nanotechnology, various AIE-NPs have been developed for theranostic applications including photodynamic therapy (PDT), photothermal therapy (PTT), photoacoustic imaging (PAI), drug delivery monitoring, etc. [30,51,52].

3.2.1 Photodynamic therapy

Photodynamic therapy (PDT) refers to the therapeutic technique that applies a photosensitizer (PS) which can produce reactive oxygen species (ROS) under light irradiation to kill cancer or bacterial cells [53]. Due to the merits of being noninvasive, low side effects, precise targeting, and low-cost, PDT has become a promising therapeutic method in several medical fields, e.g., dermatology, urology, ophthalmology, pneumology, cardiology, dentistry, and immunology [42]. The general working procedure of PDT can be divided into four steps: (1) ground-state PS molecule is stimulated to the excited singlet state upon light irradiation; (2) the excited electrons move from excited singlet state to triplet state via the intersystem crossing (ISC) process; (3) the excited triplet can either react with surrounding substrates to generate active oxygenated radicals or transfer energy with oxygen ($^3O_2$) to produce singlet oxygen ($^1O_2$); (4) the generated ROS are highly cytotoxic and have destructive effects on surrounding cells [54,55]. As can be seen, PS plays the most critical role in the process of PDT. According to literature, the commonly used PSs include porphyrin, chlorin, cyanine, and other dyes such as methylene blue, toluidine blue, rose bengal, and hypericin [56]. Recently, AIE-NPs have been reported to act as PSs producing ROS efficiently for PDT. Due to their high brightness and excellent fluorescence imaging ability, these AIE-NPs are usually applied for image-guided PDT (Fig. 11). In this part, recent advances on the image-guided PDT using AIE-NPs will be introduced.

The first example of AIE-NP PS was reported in 2014 by Liu and coworkers [57]. In this work, T-TTD NPs were fabricated by encapsulating the AIE PS TTD in 1,2-distearoyl-sn-glycero-3-phosphoethanolamine-N-[maleimide(polyethylene glycol)] (DSPE-PEG-Mal) matrix and subsequent surface modification with cyclic arginine-glycine-aspartic acid (cRGD) tripeptide as the target moiety for a tumor cell. The T-TTD NPs were proved to specifically enter and light up MDA-MB-231 tumor cells and generate ROS efficiently upon light irradiation to kill the targeted cells. Thanks to the continuous efforts devoted to the development of novel AIE PSs, the image-guided PDT efficiency of AIE-NPs has been gradually improved. Recently, Dai et al. developed a NIR emissive AIE-NP PS which is beneficial for deep tissue

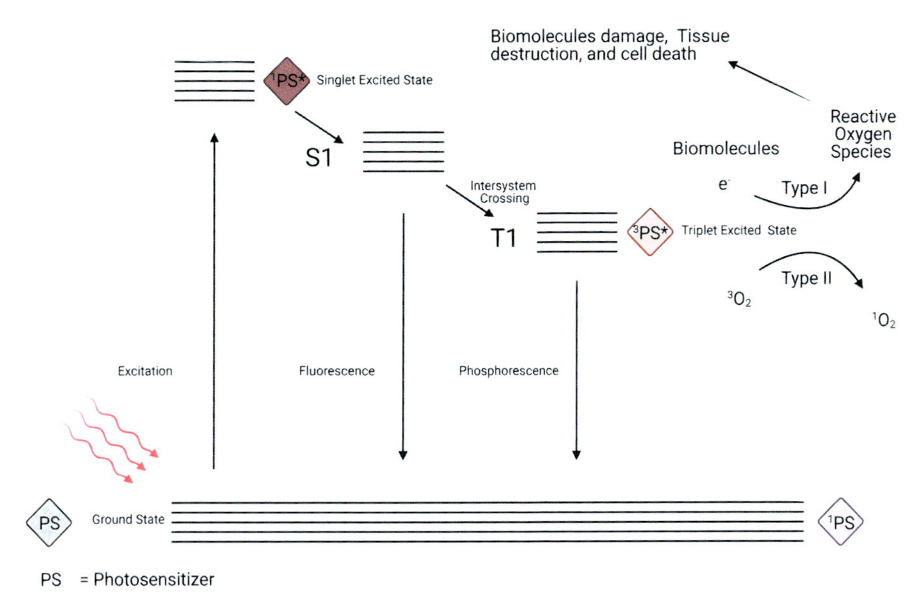

FIG. 11 Schematic diagram of electron transitions of photosensitizers (PS) in their aggregation state.

imaging and in vivo tumor monitoring [58]. Similar to T-TTD NPs, the AIE-NPs (RGD-4R-MPD/TTB) were prepared through a coprecipitation method, imbedding an efficient NIR AIEgen TTB **(30)** into an amphiphilic polymer matrix MPD (Fig. 12A). The synthesized NPs have sphere morphology, with a maximum absorption wavelength at about 550 nm, fluorescence emission peak at approximately 730 nm, solid quantum yields around 3%, and hydrodynamic diameter of 79.1 nm. Thanks to the surface-modified peptide Arg-Gly-Asp-Phe-Gly-Gly-Arg-Arg-Arg-Arg-Cys (RGD-4R), RGD-4R-MPD/TTB NPs acquired enhanced cell membrane penetration ability and can specifically target integrin $\alpha\nu\beta3$ overexpressed cancer cells, which was proven by in vitro cell imaging on SKOV-3, HeLa, PC3, and MCF7 cells using RGD-free, RGD-modified, and RGD-4R modified NPs, respectively. PDT effect of RGD-4R-MPD/TTB**(30)** NPs on integrin $\alpha\nu\beta3$ overexpressed cancer cells was studied, showing that almost all cancer cells were killed within 10 min upon white light irradiation (200 mW/cm^2). Moreover, pharmacokinetics, biodistribution, and long-term tracing of RGD-4R-MPD/TTB**(30)** NPs in vivo were investigated using a SKOV-3 tumor-bearing mouse model. The results revealed that RGD-4R-MPD/TTB**(30)** NPs have excellent in vivo tumor targeting capability, which facilitates their application in image-guided PDT. Finally, the cancer ablation ability of RGD-4R-MPD/TTB**(30)** NPs was further validated in multiple xenograft tumor models (Fig. 12).

Despite the high ROS generation efficiency and ideal tumor cytotoxicity of many reported AIE NP PSs, the relatively short absorption wavelength remains a challenge for application scenarios where deep tissue penetration is required. To solve this problem, one strategy is to apply two-photon excitation with NIR light. Recently, a few two-photon absorption AIE PSs have been synthesized through elaborate molecular design. In 2019, Liu and coworkers reported a polymerization-enhanced strategy to construct two-photon

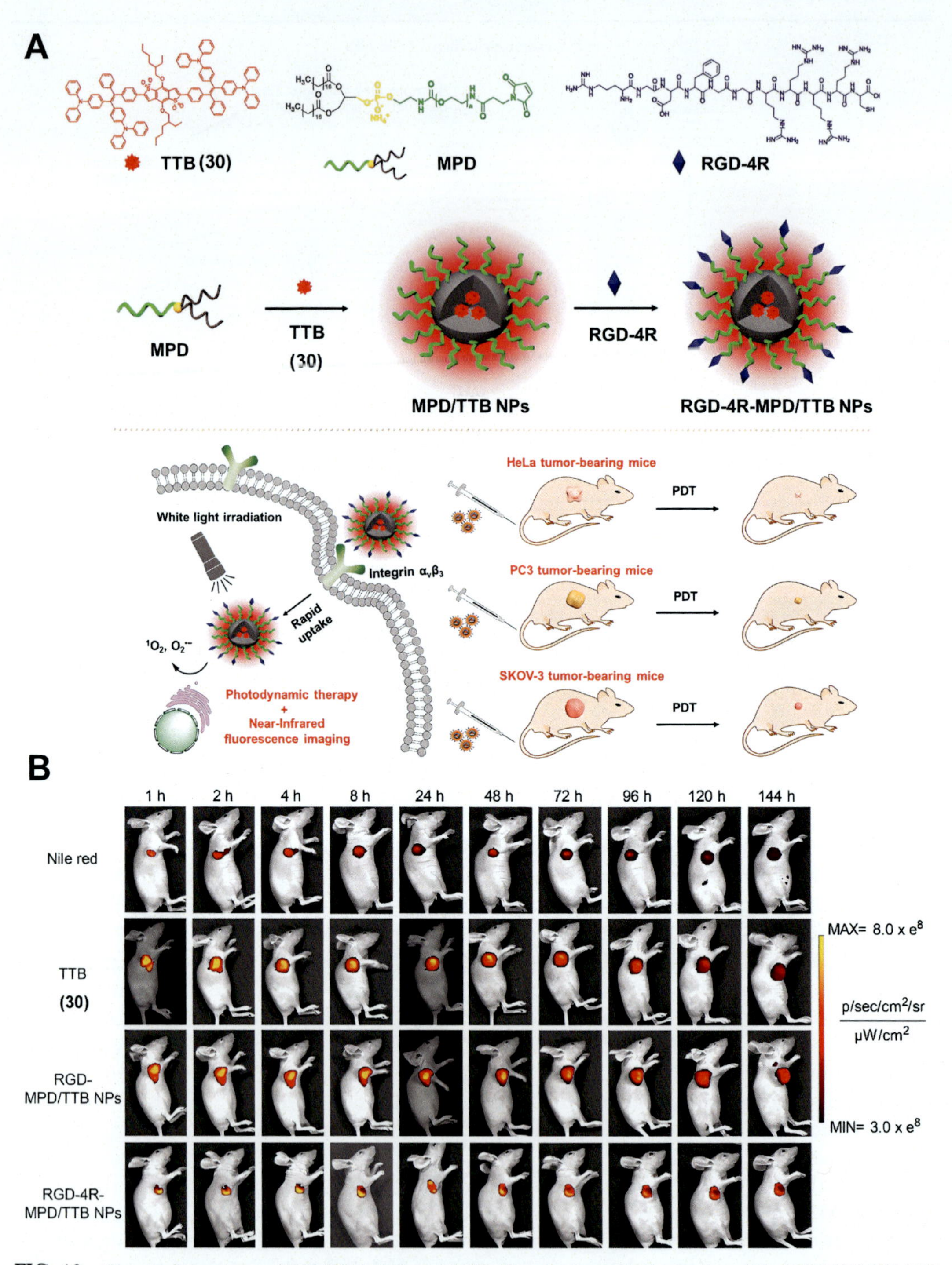

FIG. 12 Chemical structures of TTB **(30)**, MPD, and RGD-4R and schematic illustration of the RGD-4R-MPD-TTB NPs formation, cellular uptake, and PDT process of the RGD-4R-MPD-TTB NPs, and xenograft tumor models (A). In vivo fluorescent images of PC3 tumor-bearing mice after intratumoral administration of Nile red, TTB, RGD-4R-MPD-TTB NPs, and RGD-MDP/TTB NPs for designated time intervals, respectively (B). *Reproduced with permission from J. Dai, Y. Li, Z. Long, R. Jiang, Z. Zhuang, Z. Wang, et al., Efficient near-infrared photosensitizer with aggregation-induced emission for imaging-guided photodynamic therapy in multiple xenograft tumor models, ACS Nano 14 (1) (2019), 854–866. Copyright 2019, American Chemical Society.*

absorption AIE PSs [59]. As we know, conjugated polymers (CPs) have large π-conjugated backbone structures which are beneficial for two-photon absorption [60,61]. Meanwhile, the singlet and triplet state energy gaps of CPs are usually smaller than those of their small molecular analogues, inducing much easier ISC for ROS generation [62,63]. Based on this understanding, the authors designed and synthesized two CPs PTPEDC1 and PTPEDC2 from a small molecule AIE PS (TPEDC). After being encapsulated by an amphiphilic polymer DSPE-PEG-Mal, AIE-NPs exhibited bright emissions in the far-red/NIR region (600–800 nm), good water dispersibility, spherical morphology, and uniform diameters around 30 nm. Subsequently, the $^1O_2$ generation efficiency of these NPs was evaluated, showing that PTPEDC2 NPs have the highest $^1O_2$ generation capability, which is 5.48-fold higher than that of TPEDC NPs. Moreover, the two-photon absorption cross-sections of PTPEDC1 and PTPEDC2 NPs excited at 840 nm were calculated as 3.56×10^5 and 7.36×10^5 GM, respectively, which are dramatically enhanced than the TPEDC NPs. These results manifested the effectiveness of the proposed polymerization-enhanced PS and two-photon absorption strategy. Finally, the TAT peptide functionalized PTPEDC2 NPs (PTPEDC2-TAT NPs) with good cell penetrating ability were applied to in vitro HeLa cell ablation and in vivo zebrafish liver tumor treatment, exhibiting excellent two-photon PDT performance.

3.2.2 *Photoacoustic imaging and photothermal therapy*

Photothermal therapy (PTT) is another emerging therapeutic method for cancers. It applies local heat generated from photothermal agents (PTAs) under visible or NIR light to kill cancer cells. Thanks to the advantages of noninvasiveness, precise targeting, and minor damage to healthy tissues, PTT has been deemed as one of the promising therapies for cancer, including immunotherapy, gene therapy, PDT, etc. During the production of heat from PTAs, ultrasonic waves will be emitted incidentally from the transient thermoelastic expansion of biological molecules. Compared with optical signals, these generated acoustic waves have much lower scattering in biological tissues, thus can be transmitted out from depth without any intensity loss, which is much favorable in deep-tissue imaging [64]. Applying ultrasonic transducers to convert the photoacoustic signals into visual images has been successfully developed into a deep-tissue imaging technique, namely, photoacoustic imaging (PAI) [65–67]. Generally, both PTT and PAI can be achieved simultaneously with one PTA, which provides a powerful intrinsic theranostic platform.

In order to realize the dual modality of PAI and PTT, the PTAs should have both high photothermal conversion efficiency (PCE) and strong NIR absorption [67–69]. According to the Jablonski diagram, the photothermal effect of PTAs comes from the nonradiative decay of the lowest singlet excited state (S_1) toward the ground state (S_0) [12]. Considering that AIEgens usually have propeller structures that are beneficial for the occurrence of nonradiative decay, they have great potential to become PTAs with high PCE. In 2015, through the conjugation between TPE (**1**) and a conventional fluorophore 2,3-bis[4-(diphenylamino)phenyl]fumaronitrile (TPAFN), Liu and coworkers proved that the yielded adduct TPETPAFN (**25**) shows not only AIE characteristics but also 170% higher photoacoustic (PA) signals than TPAFN [70]. This can be explained by the enhanced molecular rotation from the TPE units, which is beneficial for improving the PCE. Since AIE-NPs have good biocompatibility, excellent photostability, readily functionalized surfaces, and enhanced permeability and retention in tumor issues, the authors have further studied the PA behaviors of NPs fabricated from TPETPAFN. The results suggested that the PA signals of TPETPAFN NPs reduced significantly, which is reasonable as the molecular rotation was greatly restricted in the solid NPs. Due to this limitation, AIE-NPs were rarely applied in PAI and PTT research and application.

In 2017, Liu's group promoted the application of AIE-NPs for PAI and PTT on cancer cells through the design and synthesis of novel BTPETTQ **(32)** NPs (Fig. 13A) [71]. Even though the PA signal of the BTPETTQ **(32)** NPs is much lower than that of BTPETTQ **(32)** molecules in DMSO, it is still 15% higher than that of TTQ **(31)** NPs, where TTQ **(31)** is the precursor of BTPETTQ **(32)**, and 50% higher than that of gold nanorods, which is one of the widely used PTAs. This result proved that the nonradiative decay of BTPETTQ **(32)** is not completely suppressed in the NPs due to the loose packing of AIE molecules (Fig. 13). Recently, Tang, Ding, and coworkers further investigated the rotation restriction effects and optimized nonradiative decay to endow AIE-NPs with ideal PA performance [72]. Their strategy was to use long alkyl chains to isolate AIEgens and prevent them from forming close stacking in NPs. As shown in Fig. 14, two novel AIEgens 2TPE-NDTA **(34)** and 2TPE-2NDTA **(35)** were first synthesized through conjugating TPE **(1)** units to naphthalene diimide-fused 2-(1,3-dithiol-2-ylidene)acetonitriles (NDTA **(33)**) with long alkyl chains. Due to the large π-conjugation of **32** core and special electron pushing-pulling structure, both AIEgens exhibited NIR absorption, high molar extinction coefficient, and strong twisted intramolecular charge transfer (TICT) effect. After encapsulating **33** and **34** with DSPE-PEG_{2000}, AIE-NPs with average diameters of about 125 and 152 nm were fabricated. Upon irradiation with NIR laser, both kinds of NPs showed high PCE in the aqueous medium (43.0% for **34** NPs and 54.9% for **35** NPs), which are superior to the PCE value of a previously reported high-performing PTA, poly(cyclopentadithiophene-alt-benzothiadiazole) NPs (PCE = 27.5%). This result can be explained by the high light absorptivity and effective intramolecular rotation maintained by the bulky alkyl chains. In terms of the PAI performance, **35** NPs is slightly better than **34** NPs due to the existence of more alkyl chains. Moreover, the PA intensity of **34** NPs is about 7.6-fold higher than that of **33** NPs. When compared with well-known excellent PA contrast agents, e.g., methylene blue, semiconducting polymer NPs, **35** NPs exhibited several fold enhancements of PA signals as well. Finally, the **35** NPs were applied to a tumor-bearing mice model, giving high-contrast PA imaging of tumor, which further demonstrates the potential ability of these AIE-NPs working as PAI agents for in vivo tumor imaging (Fig. 14).

3.2.3 *Drug delivery/release monitoring*

Drug delivery is a rapidly developing technology that can deliver therapeutic drugs to targeted sites for disease treatment. Using nanotechnology, the solubility of hydrophobic drugs can be enhanced, resulting in increased loading and bioavailability. Thus, nanoparticles are widely used as the carrier for drug loading. In addition to PDT and PTT, AIE-NPs have also been applied in drug delivery, mainly focusing on the distribution and release process study. For instance, light-responsive AIE-NPs were reported to deliver and release doxorubicin (DOX) for cancer treating [73]. In this work, the amphiphilic NP building block TPETP-TK-PEG was prepared by PEGylation of an AIE PS through a ROS

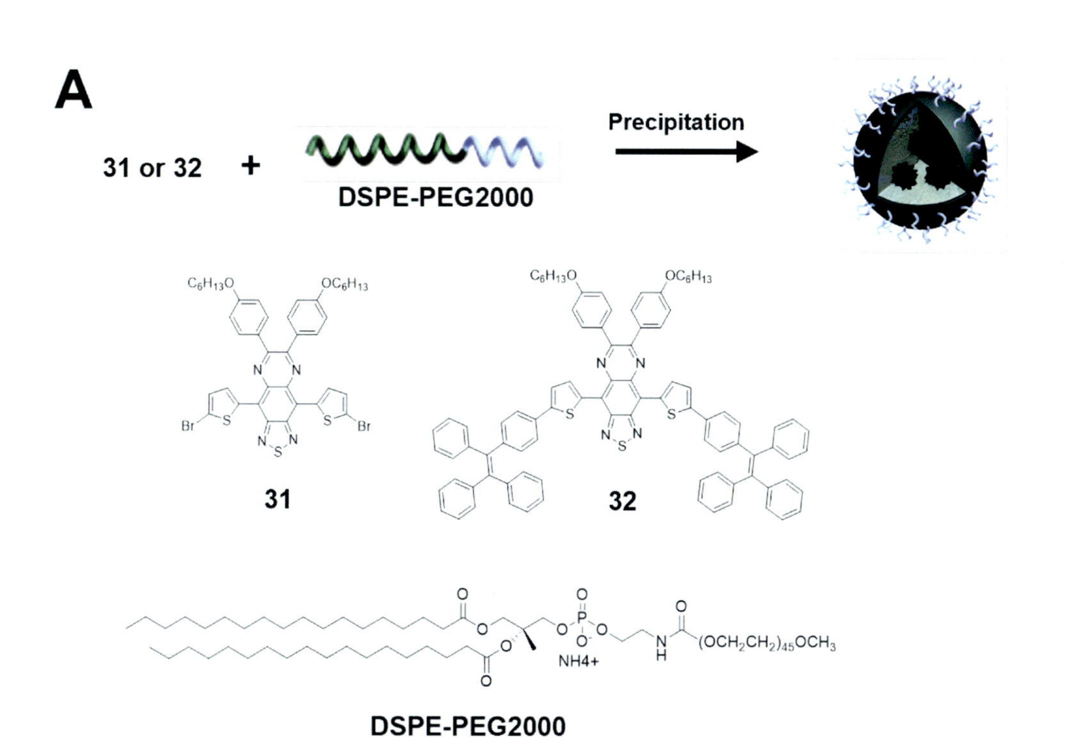

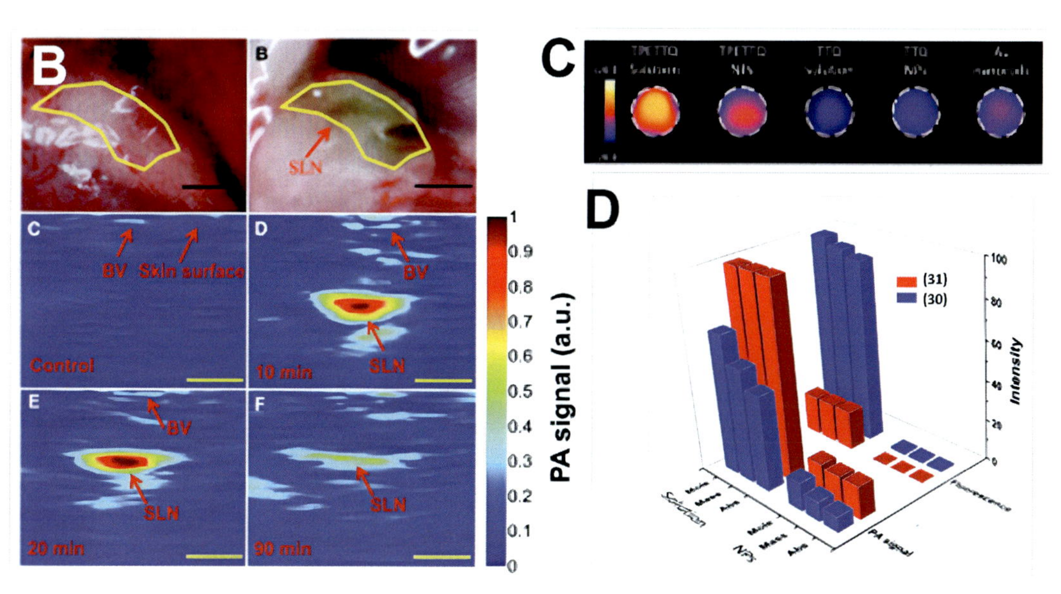

FIG. 13 Chemical structures of DSPE-PEG_{2000} and schematic illustration of the TTQ **(31)** NPs and BTPETTQ **(32)** NPs (A). Hand microscopy images of SLN before and after injection of BTPETTQ **(33)** NPs (upper section). Real-time PA imaging of SLN before and after BTPETTQ **(32)** NPs injection for 10, 20, and 90 min (lower section). Excitation wavelength 800 nm. Scale bar: 1 mm (B). Infrared images of different samples after 808-nm laser irradiation at a laser power of $0.8\,W\,cm^{-2}$ for 10 min. (C). Comparison of PA signal and fluorescence intensity of TTQ **(31)** and BTPETTQ **(32)** solution and NPs upon laser excitation at 700 nm at same mass concentration, molar concentration, and absorbance at 700 nm (D). *Reproduced with permission from X. Cai, J. Liu, W.H. Liew, Y. Duan, J. Geng, N. Thakor, et al., Organic molecules with propeller structures for efficient photoacoustic imaging and photothermal ablation of cancer cells, Mater. Chem. Front. 1 (8) (2017), 1556–1562. Copyright 2017, Royal Society of Chemistry.*

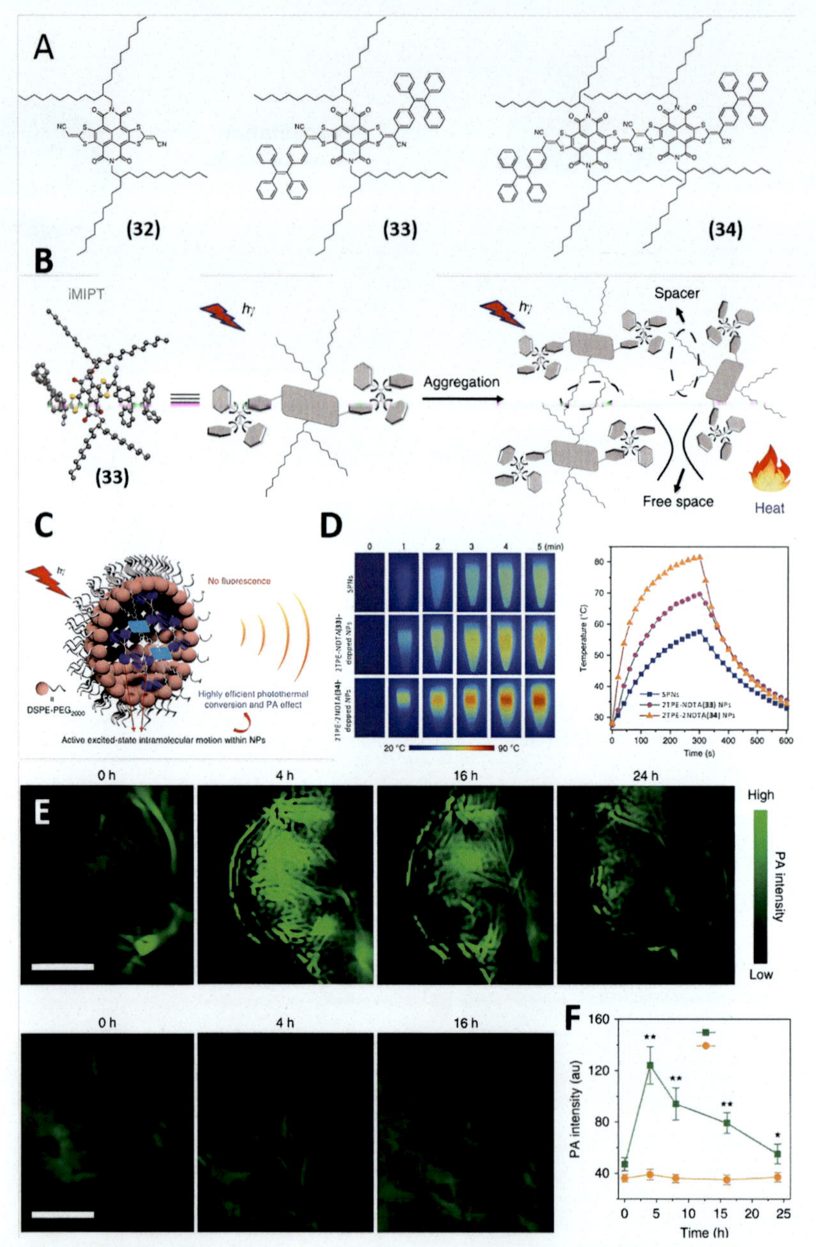

FIG. 14 Chemical structures of NDTA **(33)**, 2TPE-NDTA **(34)**, and 2TPE-2NDTA **(35)** (A). The sketch map of working mechanisms of iMIPT (B). Schematic illustration of 2TPE-NDTA **(34)**-doped NP (C). IR thermal images of various NPs in aqueous solution upon exposure to 808-nm laser irradiation for different times and the temperature changes of solution of various NPs as a function of time. Solutions were irradiated with 808 nm laser for 300 s followed by naturally cooling for another 300 s. (D). PA images of tumors (up) and muscles (down) from 2TPE-2NDTA **(35)**-doped NP-administrate mice. Before (0 h) and after 2TPE-2NDTA **(35)**-doped NPs intravenously injection into xenograft 4T1 tumor-bearing mice for designed time intervals, PA images were taken upon 730-nm pulsed laser irradiation with a laser fluence of 17.5 mJ cm^{-2} and a repetition rate of 10 Hz. The ultrasound frequency is 5 MHz. Scale bars: 3 mm (E). PA intensities of tumor and muscle tissues as a function of time before (0 h) and after 2TPE-2NDTA **(35)**-doped NPs intravenously injection (F). *Reproduced with permission from Z. Zhao, C. Chen, W. Wu, F. Wang, L. Du, X. Zhang, et al., Highly efficient photothermal nanoagent achieved by harvesting energy via excited-state intramolecular motion within nanoparticles, Nat. Commun. 10 (1) (2019), 1–11. Copyright 2019, Springer Nature.*

cleavable thioketal (TK) linker. Through a dialysis method, spherical AIE-NPs with an average hydrodynamic diameter of 110 ± 11 nm and a high DOX loading capacity (9.4%) were obtained. Thanks to the bright fluorescence from the AIEgen, high-quality imaging of the NPs can be achieved for monitoring the drug delivery process. Besides, upon white light ($\lambda = 400$–700 nm) irradiation, ROS were generated from AIE PS to break up both AIE-NPs and *endo*-lysosomal membrane, which significantly increases the intracellular accumulation of DOX in cancer cells. The DOX doping TPETP-TK-PEG NPs were further proven to have improved cytotoxicity toward drug resistance MDA-MB-231 breast cancer cells than free DOX, which is meaningful toward solving the drug resistance problems in cancer treatment.

Since the microenvironment of cancer cells is acidic (pH 6.5–7.2) due to the high rate of fermentative metabolism and less perfusion, plenty of drug delivery designs is based on pH-responsive NPs [37,74]. In 2016, a smart pH-sensitive drug delivery system was reported by Wang et al. [75]. This tadpole-shaped PEG-POSS-$(TPE)_7$ polymer consisted of a rigid cage-shaped POSS unit, a hydrophilic PEG chain, and seven TPE units linked through Schiff base bonds and can self-assemble into NPs in aqueous solutions for DOX loading. Due to the breakage of the Schiff base bonds in acidic conditions, the DOX containing PEG-POSS-$(TPE)_7$ NPs disassembled and rapidly released DOX in tumor cells. Meanwhile, the application of TPE and fluorescence resonance energy transfer (FRET) effect to DOX acceptors allows to monitor the real-time drug localization and release. Recently, Li, Yang, and coworkers reported two-photon AIE-NPs for tumor imaging and pH-sensitive drug delivery [76]. As shown in Fig. 15, a novel two-photon AIEgen (TP) was first synthesized and copolymerized with other functional monomers to yield a pH-responsive triblock copolymer P(TPMA-co-AEMA)-PEI(DA)-Blink-PEG ($PAEE_{Blink}$-DA). Then DOX-loaded $PAEE_{Blink}$-DA micelles were fabricated showing a core-shell nanostructure with a particle size of 75.6 nm and a PDI of 0.112. After the DOX-loaded NPs targeted tumor tissues through the EPR effect, the acidic microenvironment would break the benzoyl imine and amide bonds to cleave the PEG chain and dimethylmaleic anhydride units, resulting in the negative-to-positive charge conversion and hydrophobic-to-hydrophilic conversion on the PEI chain and PolyAEMA segment, respectively. Such a pH response process made the NPs swollen, which is beneficial for maintaining the endocytosis of DOX-loaded $PAEE_{Blink}$-DA NPs by tumor cells and can trigger the release of DOX efficiently. Results indicated that about 90% of the loaded drug released from the $PAEE_{Blink}$-DA NPs at pH 6.8 after 48 h, whereas only 26.5% released under normal physiological pH for the same time. Due to the large two-photon absorption cross-section (up to 265 MG) of the TP unit, two-photon tumor tissue imaging with penetration depth up to 150 μm was proved by blank $PAEE_{Blink}$-DA NPs. In conclusion, AIE-NPs with smart stimuli–response and multiphoton or NIR absorption/emission are favorable for application in fluorescent imaging-guided cancer theranostics (Fig. 15).

3.3 Biosensing

As a new generation of fluorescent labels, FNPs have been employed for developing novel biomolecule detection methods with excellent sensitivity, fast response, high sample

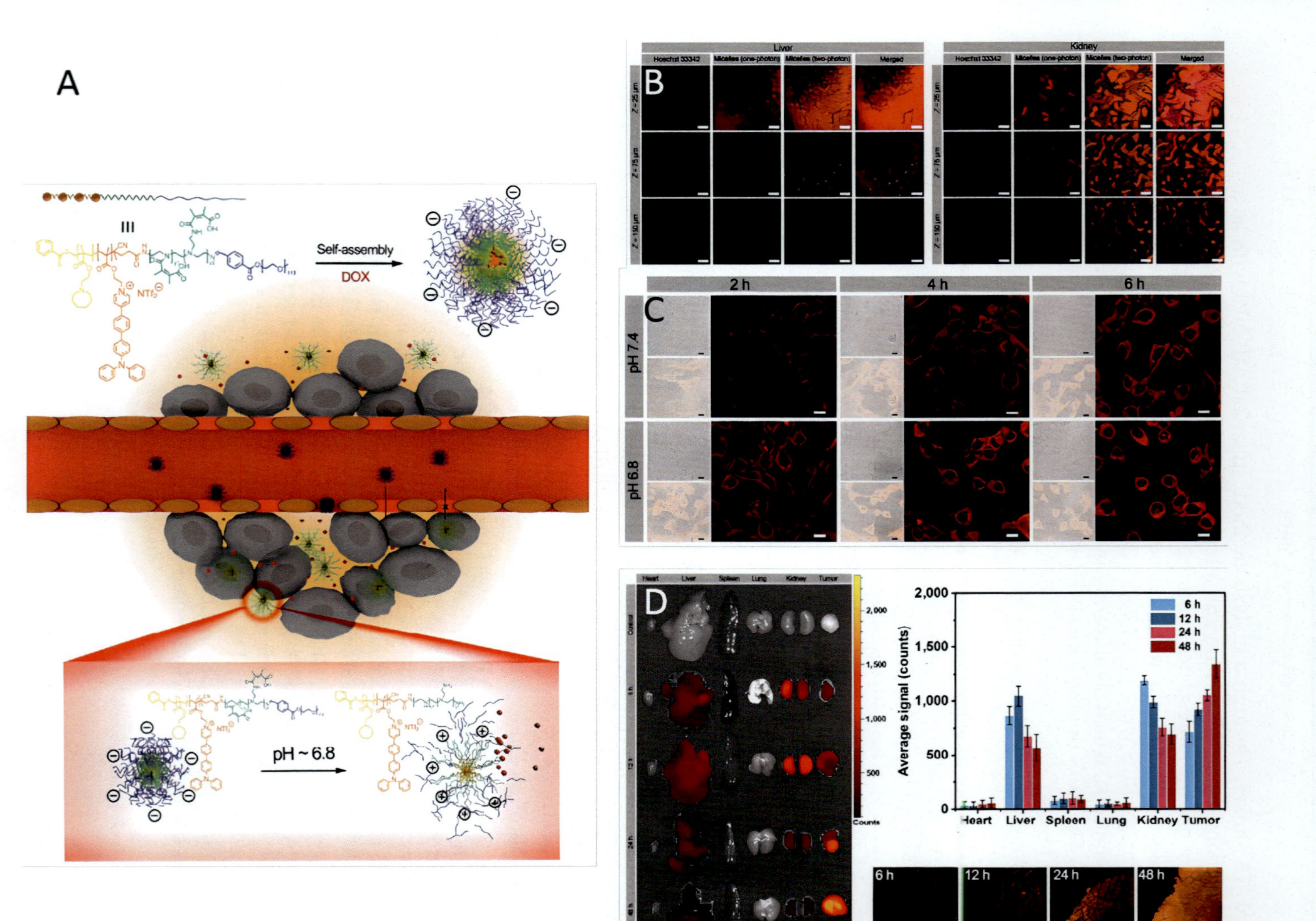

FIG. 15 Illustration of P(TPMA-co-AEMA)-PEI(DA)-Blink-PEG micelles with their pH triggered charge-conversion and drug delivery, along with the two-photon bioimaging (A). Two-photon CLSM images of the liver and the kidney tissues after blank micelles injection for 12h. The scale bar was 100 μm (B). Two-photon CLSM images of 4T1 cells cultured with blank $PAEE_{Blink}$-DA micelles for different times excited at 800 nm. The scale bars were 10 μm (C). Ex vivo fluorescent images of tumors and mean organs from the tumor-bearing mice administrated with blank micelles. Fluorescent intensity of blank $PAEE_{Blink}$-DA micelles in major organs and tumors at different times. Two-photon CLSM images of tumor tissue sections at different times were treated with blank $PAEE_{Blink}$-DA micelles. The scale bars were 100 μm (D). *Reproduced with permission from B. Ma, W. Zhuang, H. He, X. Su, T. Yu, J. Hu, et al., Two-photon AIE probe conjugated theranostic nanoparticles for tumor bioimaging and pH-sensitive drug delivery, Nano Res. 12 (7) (2019), 1703–1712. Copyright 2019, Springer Nature.*

throughput, low cost, easy operation, etc. [77]. Compared with traditional small molecular organic fluorophores, AIEgens have been proved to have better performances in biosensing, including low background noise, good photostability, large Stokes shift, and broad application concentration range. In the aspect of FNPs fabrication, AIEgens are better choices as well, due to their unique AIE characteristics which make them even brighter under aggregated states, rather than becoming dim and like ACQ dyes. In this section, some typical examples of AIE-NPs applied in biosensing will be briefly introduced.

Metal ions play crucial roles in biological processes such as osmotic maintenance, signal transduction, catalysis, proliferation, metabolism, and act as cofactors of biomacromolecules (Fe^{2+} in hemoglobin, Zn^{2+} in zinc finger, and Mn^{2+} in photosystems) [78–81]. It has been proved that metal ion distribution and homeostasis are directly related to cancers, neurodegenerative diseases, and diabetes [82–85]. Attracted by the importance of metal ion detection in biological samples, various AIE-NPs have been designed and synthesized for metal ion sensing. As an important transition metal widely distributed in the human body, the abnormal level of Fe^{3+} can be used to indicate some diseases, e.g., anemia, Parkinson's syndrome, and cancer [86–88]. Therefore, Yang et al. reported an AIE-NPs for intracellular Fe^{3+} sensing in 2016 [89]. A conjugated polymer P2 **(36)** with AIE characteristics was first synthesized containing TPE units and zwitterionic alkyl chains. In order to achieve good biocompatibility, polymer P2 was encapsulated into the DSPE-PEG2000 matrix through a nanoprecipitation method, producing lipid-P2NPs with spherical morphology and an average hydrodynamic diameter approximately 23 nm. In the presence of Fe^{3+}, the intrinsic emission of lipid-P2 NPs at 500 nm gradually decreased with the increase of Fe^{3+} concentration. A good linear correlation between the fluorescence intensity and Fe^{3+} concentration was obtained in the range of 0.1–8 μM. Moreover, the limit of detection (LOD) of this NP-based assay was further identified as 0.22 μM. Finally, the lipid-P2NPs were evaluated as having no obvious cytotoxicity toward A549 cells and applied in the cytoplasm imaging and intracellular Fe^{3+} concentration sensing of A549 cells (Fig. 16A). As we know, fluorescence turn-off probes usually have poor sensitivity due to their high signal-to-noise ratio. Bioprobes with fluorescence turn-on response are more popular in biosensing. Since AIEgens are usually packed tightly in solid particles, AIE-NPs always show bright fluorescence, which makes it a tough job to develop fluorescence turn-on probes based on AIE-NPs. Recently, Li and coworkers reported a fluorescence turn-on strategy for highly sensitive detection of acetylcholinesterase (AChE), whose activity is closely related to neurodegenerative diseases [90]. In this work, a TPE derivative QAU-1 **(37)** with two quaternary ammonium salt moieties was first self-assembled into AIE-NPs in water. These NPs were well characterized showing a uniform size of 5 nm and a zeta value of + 40.87 mV, which indicated their good water stability and strong positive charge. Secondly, highly negative charged gold nanoparticles (AuNPs) with a diameter of 13 nm were synthesized. The AuNPs can bind with QAU-1 NPs via electrostatic interaction and quench the fluorescence of QAU-1 NPs through FRET effect. As a result, the AIE-Au nanoconjugates showed ultralow background fluorescence when applied as a biosensor to evaluate the activity of AChE. The biosensing process was finally realized through the hydrolyzation of acetylthiocholine (ATCh) by AChE, producing thiocholine which can release QAU-1 NPs from the AuNP surface via a competitive displacement process and turn-on the fluorescence of QAU-1 NPs. Thanks to the high quenching efficiency of AuNPs, an ultralow LOD of 0.015 mU/mL was obtained showing the high sensitivity of this assay (Fig. 16B and C).

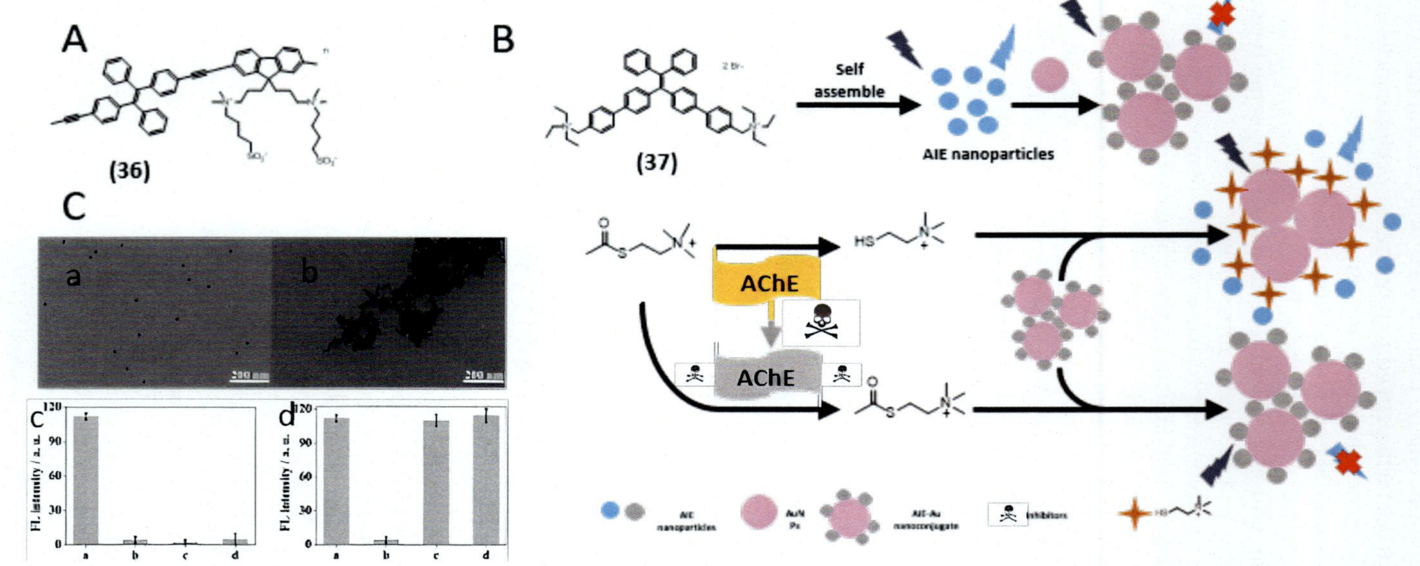

FIG. 16 Chemical structure of polymer P2 **(36)** (A). Chemical structure of QAU-1 **(37)** and the fabrication procedure of AIE-Au nanoconjugates and schematic representation of AIE-Au nanoconjugates in the analysis of AChE activity and inhibitor. (B) TEM characterization of AuNPs (a) and TEM images of AIE-nanoconjugates: AIE-nanoparticle 5.0 μM; AuNPs, 2 nM (b). FL intensity of the sensing system under various circumstances: AIE nanoparticles; AIE-Au nanoconjugates; AIE-Au nanoconjugates in a solution containing ascorbic acid, glycin, alanine, glutamic acid, Cl^-, NO_3^- and SO_4^{2-} with the concentration of 100 nM; AIE-Au nanoconjugates in cabbage extract (c) and FL intensity of AIE nanoparticles toward different conditions: AIE-nanoparticles, AIE-nanoparticles + negative AuNPs (2.0 nM), AIE-nanoparticles + positive AuNPs (2.0 nM), AIE-nanoparticles + positive AuNPs (6.0 nm). *Reproduced with permission from C. Wang, X. Wang, P. Gai, H. Li, F. Li, Target-responsive AIE-Au nanoconjugate for acetylcholinesterase activity and inhibitor assay with ultralow background noise, Sensors Actuators B: Chemical 284 (2019), 118–124. Copyright 2019, Elsevier.*

Instead of interacting with biomolecules to induce fluorescence variation, FNPs are usually applied for fluorescence labeling due to their stable fluorescence signal, excellent colloidal stability, and easy surface modification. Many examples of conventional FNPs have been reported to work as fluorescent indicators in the fluorescent lateral flow assay (FLFA). Most recently, Tang, Gu, Zhang, and coworkers reported a novel AIE-NP FPPs@210 for FLFA application [30]. The detailed fabrication method of FPPs@210 has been introduced in Section 2.1. Since the COVID-19 pandemic has brought an unprecedented influence on our health, economy, and life, the authors employed the antibodies of COVID-19 to verify the feasibility of FPPs@210 NPs for FLFA. A sandwich immunochromatographic method was adopted in this work. Firstly, SARS-COV-2 nucleocapsid protein (N protein) was labeled with FPPs@210 through the click reaction between the anhydride groups and amine groups, followed by a blocking treatment with human serum albumin (HSA) to eliminate nonspecific binding sites on the protein. Afterward, a porous nitrocellulose membrane was drawn with two test lines (M line and G line) which can specifically capture IgM and IgG of SARS-COV-2, respectively, and one control line which can bind with SARS-COV-2N protein and conglutinated with an absorbent pad on the top to complete the test strip fabrication. By dropping the mixture solution of FPPs@210 NPs and body-fluid samples on the bottom of the FLFA strip, the solution flows over the three lines to the water-absorbent pad. Once antigens IgM or IgG of SARS-COV-2 exist in the samples, immune complexes of FPPs@210 NPs and antigen will be formed and retained on the corresponding test line, showing a positive test result of COVID-19. Finally, the results suggested that the AIE-NPs-based FLFA strip is applicable for COVID-19 detection.

4 Metal nanoclusters with AIE characteristics

Metal NPs can be divided into two categories including plasmonic metal NPs and metal nanoclusters (MNCs). The latter ones usually comprise tens or hundreds of atoms and have ultrasmall sizes (<2 nm) which are comparable to the Fermi wavelength of electrons. As a result, MNCs are very different from the plasmonic NPs, showing some molecule-like properties, such as discrete electronic states, photoluminescence (PL), molecular chirality, etc. [91]. Thanks to the efficient electronic transitions between the discrete energy levels, MNCs usually have unique PL properties, which makes them an attractive class of fluorescent nanomaterial. In recent years, many researches have been conducted to understand the origin of fluorescence emission from various MNCs [92]. It was found that both particle size and protecting ligands of MNCs play important roles in their photophysical properties. Considering the advantage that thiolates can well modulate the atomic precision of MNCs in size and structure, thiol-containing compounds including polymers, amino acids, proteins, DNAs, and RNAs are widely utilized as ligands for MNC fabrication. Interestingly, even though the quantum yields of thiolate-protected MNCs rarely exceed 0.1% in solution [93], they become highly emissive in their aggregate states with QY around 5%–20% [91], showing a typical phenomenon of AIE. According to our best knowledge, the connection between MNCs and AIE was first identified by Xie and coworkers in 2012 [94]. After that, much more attention has been drawn to the research of MNCs with AIE features (AIE-MNCs), due to their strong fluorescence in the aggregate state. Most recently, AIE-MNCs were further applied for biomedical applications due to their fine subnanometer size, excellent biocompatibility, low toxicity, strong photostability, and simple synthesis. In this section, we will briefly discuss several biomedical applications of AIE-MNCs to make this chapter all-encompassing. As space is limited, fabrication strategies and mechanism research of AIE-MNCs will not be introduced separately here [95].

One of the fluorescence imaging applications of AIE-MNCs was reported by Yahia-Ammar et al. in 2016. The authors first synthesized glutathione (GSH) protected gold NCs (Au-GSH) through a reducing method [96]. Afterward, a cationic polymer poly(allyl amine hydrochloride) (PAH) was mixed with Au-GSH NCs and formed self-assembly NPs with a spherical shape and average hydrodynamic size of approximately 120 nm. The driven force for the self-assemble process can be ascribed to the electrostatic interaction between the carboxyl group of GSH and the protonated amine group of PAH. After a series of characterizations, the synthesized Au-GSH-PAH NPs showed pH-dependent swelling properties and fourfold fluorescence enhancement compared to Au-GSH NCs, which manifested the AIE characteristics of NCs. Therefore, the Au-GSH-PAH NPs were further applied to fluorescence imaging and drug delivery research. As shown in Fig. 17, the accumulation of biomolecule (peptides and antibodies) loaded Au-GSH-PAH NPs can be clearly observed in the cytoplasm of THP1 cells through CLSM.

The other commonly reported bioapplication of AIE-MNCs is biosensing. Here we would like to introduce a DNA templated silver NCs (DNA-AgNCs) with AIE activity for sensing adenosine triphosphate (ATP) and cytochrome *c* (Cyt c) [97]. As shown in Fig. 18, the DNA-AgNCs were synthesized via reacting Ag^+ ions with cytosine-rich oligonucleotides, followed by reducing Ag^+ ions into Ag(0) using $NaBH_4$ as the reducing agent. By introducing ATP or Cyt c aptamer segment to the cytosine-rich oligonucleotides, DNA-AgNCs with excellent targeting capacity toward ATP and Cyt c were prepared. These NCs show high photostability, good quantum yields ranging from 4.19% to 9.85%, and average diameters around 3.1 nm. After graphene oxide (GO) sheets were uniformly dispersed in an aqueous solution, the DNA-AgNCs can be adsorbed onto the GO surface through hydrophobic and π-π stacking interaction between single-strand DNA chain and GO. It was found that the fluorescence intensity of DNA-AgNCs which have a proper base sequence in the DNA strand was remarkably enhanced once immobilized to the surface of GO. The fluorescence enhancement mechanism of DNA-AgNCs/GO nanohybrids was confirmed by fluorescence spectrometer and transmission electron microscopy to be the formation of compact aggregates of Ag(I)–DNA complex generated from the oxidation of Ag(0) with GO. Finally, aptamer containing DNA-AgNCs/GO nanohybrids were applied for the detection of ATP in lysed *Escherichia coli* DH5 α cells and Cyt c in mouse embryonic stem cells. This label-free biosensing method exhibited high selectivity and low sensitivity for the detection of ATP (LOD = 0.42 nM) and Cyt c (LOD = 2.3 nM), which has significant meanings on designing new AIE-NCs for biomedical applications.

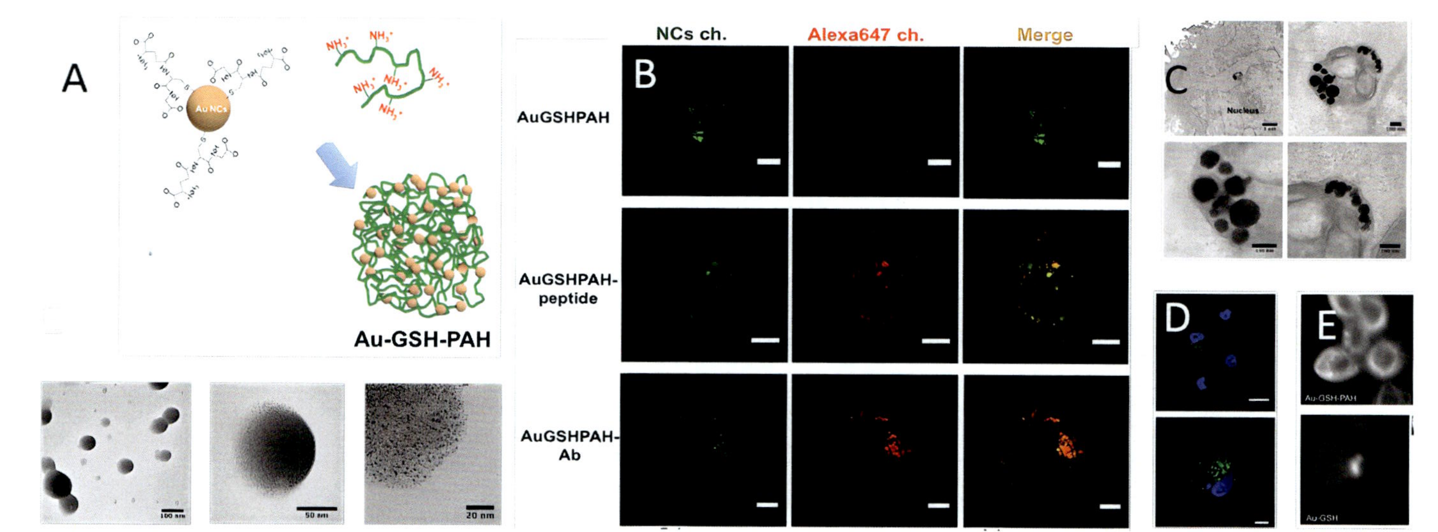

FIG. 17 Schematic illustration of the synthesis of self-assembled AUNCs using a cationic polymer (PAH) as a cross-liking agent and AuNCs stabilized by GSH and TEM images of Au-GSH-PAH AuNCs at different magnifications (A). CLSM images of THP1 cells incubated with Au-GSH-PAH and with the loaded particles Au-GSH-PAH-peptide and Au-GSH-PAH-Ab. Particles are visible with the NC channel (*green*), peptide and antibody with the Alexa647 channel (*red*), and merge both channels (B). Electron microscopy images of Au-GSH-PAH incubated in THP1 cells. The accumulation of the self-assembled particles keeping their morphology into two fused vesicles has been shown (C). CLSM images of THP1 cells incubated with Au-GSH-PAH (*green*). Nucleus was stained with Hoechst (*blue*). Scale bar: 20 μm (up) and 5 μm (down). Au-GSH-PAH accumulated in the cytoplasm (D). Fluorescence intensity of cells incubated with Au-GSH-PAH and Au-GSH was obtained using a FLIM microscope (E).). *Reproduced with permission from A. Yahia-Ammar, D. Sierra, F. Mérola, N. Hildebrandt, X. Le Guével, Self-assembled gold nanoclusters for bright fluorescence imaging and enhanced drug delivery, ACS Nano 10 (2) (2016), 2591–2599. Copyright 2016, American Chemical Society.*

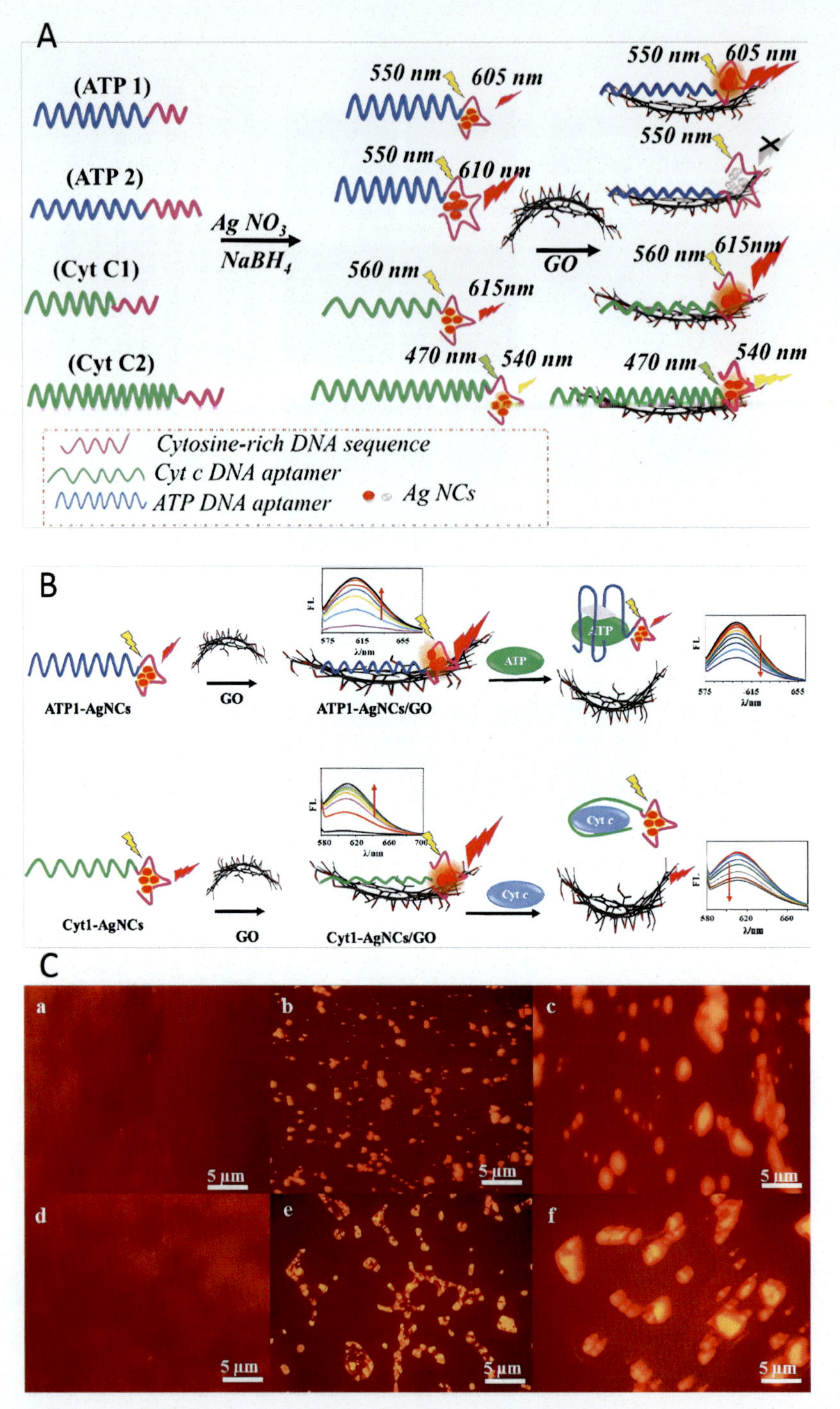

FIG. 18 Schematic representation of the synthesis of DNA-stabilized AgNCs and their photoluminescence after contact with GO (A). Working principles of the label-free ATP and Cyt c sensing based on DNA-stabilized AgNCs/GO nanohybrids (B). (Top) Fluorescence microscopy image of ATP1-AgNCs at 0 min (C, a), 5 min (C, b), and 15 min (C, c) after GO contact. (Bottom) Fluorescence microscopy image of Cyt1–AgNCs at 0 min (C, d), 5 min (C, e), and 15 min (C, f) after GO contact. *Reproduced with permission from M. Shamsipur, K. Molaei, F. Molaabasi, S. Hosseinkhani, A. Taherpour, M. Sarparast, et al., Aptamer-based fluorescent biosensing of adenosine triphosphate and cytochrome c via aggregation-induced emission enhancement on novel label-free DNA-capped silver nanoclusters/graphene oxide nanohybrids, ACS Appl. Mater. Interfaces 11 (49) (2019), 46077–46089. Copyright 2019, American Chemical Society.*

5 Summary and future perspectives

The emergence of AIEgens has opened new areas for researchers in various science disciplines from photophysics to chemistry and biomedicine. During the development process, AIEgens with different structures have been designed and developed for a variety of purposes, along with the understanding of the fundamental mechanisms of the phenomenon. In this chapter, we have summarized the fabrication of AIE-NPs based on organic fluorophores and their applications in biosensing, bioimaging, and theranostics. We have also briefly introduced AIE-active metal nanoclusters and their applications. These AIE-NPs show great capability including good biocompatibility and photostability when compared to traditional luminescent NPs. Through different strategies to decorate the NPs such as taking advantage of biomolecules (nucleic acids, peptides, ligands), they also achieve good cell permeability and selectivity such as targeting cancer cells and tumor regions. However, there are still challenges in the field that could be considered in the future development of AIE-NPs. Designing AIE-NPs with FR/NIR narrow emission to enhance the light penetration depth and weaken biological harmful effect remains one of the challenging parts for in vivo experiments. Ultrasmall AIE-NPs formation is another important area that can attract more attention in the future. Small and uniform functionalized NPs show better cell permeability and biocompatibility. Moreover, the pharmacokinetics and pharmacodynamics studies of AIE-NPs, especially on the human body are yet to be explored. On the other hand, with all the abovementioned advantages, AIE nanomaterials would be a good platform for designing nanosensors for sensing biomarkers in body fluids with high selectivity and sensitivity and thus can pave the way for wider biomedical applications in the near future.

References

[1] S. Das, S. Mitra, S.P. Khurana, N. Debnath, Nanomaterials for biomedical applications, Front. Life Sci. 7 (3–4) (2013) 90–98.

[2] S. Vigneshvar, C. Sudhakumari, B. Senthilkumaran, H. Prakash, Recent advances in biosensor technology for potential applications—an overview, Front. Bioeng. Biotechnol. 4 (2016) 11.

[3] N.S. Vallabani, S. Singh, Recent advances and future prospects of iron oxide nanoparticles in biomedicine and diagnostics, 3 Biotech 8 (6) (2018) 1–23.

[4] F. Fang, M. Li, J. Zhang, C.-S. Lee, Different strategies for organic nanoparticle preparation in biomedicine, ACS Mater. Lett. 2 (5) (2020) 531–549.

[5] V. Vaijayanthimala, H. Chang, Functionalized fluorescent nanodiamonds for biomedical applications, Nanomedicine (London) 4 (1) (2009) 47–55.

[6] R. Serrano García, S. Stafford, Y.K. Gun'ko, Recent progress in synthesis and functionalization of multimodal fluorescent-magnetic nanoparticles for biological applications, Appl. Sci. 8 (2) (2018) 172.

[7] J.V. Jun, D.M. Chenoweth, E.J. Petersson, Rational design of small molecule fluorescent probes for biological applications, Org. Biomol. Chem. 18 (30) (2020) 5747–5763.

[8] M. Ahmed, M. Faisal, A. Ihsan, M.M. Naseer, Fluorescent organic nanoparticles (FONs) as convenient probes for metal ion detection in aqueous medium, Analyst 144 (8) (2019) 2480–2497.

[9] H. Wang, G. Liu, Advances in luminescent materials with aggregation-induced emission (AIE) properties for biomedical applications, J. Mater. Chem. B 6 (24) (2018) 4029–4042.

[10] X. Cai, B. Liu, Aggregation-induced emission: recent advances in materials and biomedical applications, Angew. Chem. 132 (25) (2020) 9952–9970.

[11] X. Yi, J. Li, Z. Zhu, Q. Liu, Q. Xue, D. Ding, In vivo cancer research using aggregation-induced emission organic nanoparticles, Drug Discov. Today 22 (9) (2017) 1412–1420.

[12] W. Wu, Z. Li, Nanoprobes with aggregation-induced emission for theranostics, Mater. Chem. Front. 5 (2) (2021) 603–626.

[13] C. Zhu, R.T. Kwok, J.W. Lam, B.Z. Tang, Aggregation-induced emission: a trailblazing journey to the field of biomedicine, ACS Appl. Bio Mater. 1 (6) (2018) 1768–1786.

[14] X. Zhang, X. Zhang, L. Tao, Z. Chi, J. Xu, Y. Wei, Aggregation induced emission-based fluorescent nanoparticles: fabrication methodologies and biomedical applications, J. Mater. Chem. B 2 (28) (2014) 4398–4414.

[15] L. Yan, Y. Zhang, B. Xu, W. Tian, Fluorescent nanoparticles based on AIE fluorogens for bioimaging, Nanoscale 8 (5) (2016) 2471–2487.

[16] Z. Wang, L. Yang, Y. Liu, X. Huang, F. Qiao, W. Qin, Q. Hu, B.Z. Tang, Ultra long-term cellular tracing by a fluorescent AIE bioconjugate with good water solubility over a wide pH range, J. Mater. Chem. B 5 (25) (2017) 4981–4987.

[17] S.A. Fateminia, Z. Wang, C.C. Goh, P.N. Manghnani, W. Wu, D. Mao, L.G. Ng, Z. Zhao, B.Z. Tang, B. Liu, Nanocrystallization: a unique approach to yield bright organic nanocrystals for biological applications, Adv. Mater. 29 (1) (2017) 1604100.

[18] Y. Yu, C. Feng, Y. Hong, J. Liu, S. Chen, K.M. Ng, K.Q. Luo, B.Z. Tang, Cytophilic fluorescent bioprobes for long-term cell tracking, Adv. Mater. 23 (29) (2011) 3298–3302.

[19] Z. Zhao, C. Deng, S. Chen, J.W. Lam, W. Qin, P. Lu, Z. Wang, H.S. Kwok, Y. Ma, H. Qiu, Full emission color tuning in luminogens constructed from tetraphenylethene, benzo-2, 1, 3-thiadiazole and thiophene building blocks, Chem. Commun. 47 (31) (2011) 8847–8849.

[20] S. Chen, X. Xu, Y. Liu, G. Yu, X. Sun, W. Qiu, Y. Ma, D. Zhu, Synthesis and characterization of n-type materials for non-doped organic red-light-emitting diodes, Adv. Funct. Mater. 15 (9) (2005) 1541–1546.

[21] X. Fang, X. Chen, R. Li, Z. Liu, H. Chen, Z. Sun, B. Ju, Y. Liu, S.X.A. Zhang, D. Ding, Multicolor photo-crosslinkable AIEgens toward compact nanodots for subcellular imaging and STED nanoscopy, Small 13 (41) (2017) 1702128.

[22] G. Yu, S. Yin, Y. Liu, J. Chen, X. Xu, X. Sun, D. Ma, X. Zhan, Q. Peng, Z. Shuai, Structures, electronic states, photoluminescence, and carrier transport properties of 1, 1-disubstituted 2, 3, 4, 5-tetraphenylsiloles, J. Am. Chem. Soc. 127 (17) (2005) 6335–6346.

[23] W. Qin, D. Ding, J. Liu, W.Z. Yuan, Y. Hu, B. Liu, B.Z. Tang, Biocompatible nanoparticles with aggregation-induced emission characteristics as far-red/near-infrared fluorescent bioprobes for in vitro and in vivo imaging applications, Adv. Funct. Mater. 22 (4) (2012) 771–779.

[24] A. Nicol, W. Qin, R.T. Kwok, J.M. Burkhartsmeyer, Z. Zhu, H. Su, W. Luo, J.W. Lam, J. Qian, K.S. Wong, Functionalized AIE nanoparticles with efficient deep-red emission, mitochondrial specificity, cancer cell selectivity and multiphoton susceptibility, Chem. Sci. 8 (6) (2017) 4634–4643.

[25] B. Yang, X. Zhang, X. Zhang, Z. Huang, Y. Wei, L. Tao, Fabrication of aggregation-induced emission based fluorescent nanoparticles and their biological imaging application: recent progress and perspectives, Mater. Today 19 (5) (2016) 284–291.

[26] H. Gao, X. Zhao, S. Chen, AIEgen-based fluorescent nanomaterials: fabrication and biological applications, Molecules 23 (2) (2018) 419.

[27] M. Zhang, X. Yin, T. Tian, Y. Liang, W. Li, Y. Lan, J. Li, M. Zhou, Y. Ju, G. Li, AIE-induced fluorescent vesicles containing amphiphilic binding pockets and the FRET triggered by host–guest chemistry, Chem. Commun. 51 (50) (2015) 10210–10213.

[28] H. Chen, E. Zhang, G. Yang, L. Li, L. Wu, Y. Zhang, Y. Liu, G. Chen, M. Jiang, Aggregation-induced emission luminogen assisted self-assembly and morphology transition of amphiphilic glycopolypeptide with bioimaging application, ACS Macro Lett. 8 (8) (2019) 893–898.

[29] Y. Zhang, X. Peng, X. Jing, L. Cui, S. Yang, J. Wu, G. Meng, Z. Liu, X. Guo, Synthesis of pH-responsive triazine skeleton nano-polymer composite containing AIE group for drug delivery, Front. Mater. Sci. 15 (1) (2021) 113–123.

[30] G. Wang, L. Yang, C. Li, H. Yu, Z. He, C. Yang, J. Sun, P. Zhang, X. Gu, B.Z. Tang, Novel strategy to prepare fluorescent polymeric nanoparticles based on aggregation-induced emission via precipitation polymerization for fluorescent lateral flow assay, Mater. Chem. Front. 5 (5) (2021) 2452–2458.

[31] X. Cai, D. Mao, C. Wang, D. Kong, X. Cheng, B. Liu, Multifunctional liposome: a bright AIEgen–lipid conjugate with strong photosensitization, Angew. Chem. 130 (50) (2018) 16634–16638.

[32] A. Nicol, R.T. Kwok, C. Chen, W. Zhao, M. Chen, J. Qu, B.Z. Tang, Ultrafast delivery of aggregation-induced emission nanoparticles and pure organic phosphorescent nanocrystals by saponin encapsulation, J. Am. Chem. Soc. 139 (41) (2017) 14792–14799.

[33] K. Li, Y. Jiang, D. Ding, X. Zhang, Y. Liu, J. Hua, S.-S. Feng, B. Liu, Folic acid-functionalized two-photon absorbing nanoparticles for targeted MCF-7 cancer cell imaging, Chem. Commun. 47 (26) (2011) 7323–7325.
[34] C. Prashant, M. Dipak, C.-T. Yang, K.-H. Chuang, D. Jun, S.-S. Feng, Superparamagnetic iron oxide–loaded poly (lactic acid)-d-α-tocopherol polyethylene glycol 1000 succinate copolymer nanoparticles as MRI contrast agent, Biomaterials 31 (21) (2010) 5588–5597.
[35] S. Dineshkumar, A. Raj, A. Srivastava, S. Mukherjee, S.S. Pasha, V. Kachwal, L. Fageria, R. Chowdhury, I.R. Laskar, Facile incorporation of "aggregation-induced emission"-active conjugated polymer into mesoporous silica hollow nanospheres: synthesis, characterization, photophysical studies, and application in bioimaging, ACS Appl. Mater. Interfaces 11 (34) (2019) 31270–31282.
[36] M. Yang, J. Deng, H. Su, S. Gu, J. Zhang, A. Zhong, F. Wu, Small organic molecule-based nanoparticles with red/near-infrared aggregation-induced emission for bioimaging and PDT/PTT synergistic therapy, Mater. Chem. Front. 5 (1) (2021) 406–417.
[37] H.B. Ruttala, N. Chitrapriya, K. Kaliraj, T. Ramasamy, W.H. Shin, J.-H. Jeong, J.R. Kim, S.K. Ku, H.-G. Choi, C.S. Yong, Facile construction of bioreducible crosslinked polypeptide micelles for enhanced cancer combination therapy, Acta Biomater. 63 (2017) 135–149.
[38] R. Jiang, M. Liu, H. Huang, L. Mao, Q. Huang, Y. Wen, Q.-Y. Cao, J. Tian, X. Zhang, Y. Wei, Facile fabrication of organic dyed polymer nanoparticles with aggregation-induced emission using an ultrasound-assisted multicomponent reaction and their biological imaging, J. Colloid Interface Sci. 519 (2018) 137–144.
[39] G. Xie, C. Ma, X. Zhang, H. Liu, X. Guo, L. Yang, Y. Li, K. Wang, Y. Wei, Biocompatible zwitterionic phosphorylcholine polymers with aggregation-induced emission feature, Colloids Surf. B: Biointerfaces 157 (2017) 166–173.
[40] K. Mandal, D. Jana, B.K. Ghorai, N.R. Jana, AIEgen-conjugated magnetic nanoparticles as magnetic–fluorescent bioimaging probes, ACS Appl. Nano Mater. 2 (5) (2019) 3292–3299.
[41] S. Zhou, H. Zhao, R. Feng, L. Ding, Z. Li, C. Deng, Q. He, Y. Liu, B. Song, Y. Li, Application of amphiphilic fluorophore-derived nanoparticles to provide contrast to human embryonic stem cells without affecting their pluripotency and to monitor their differentiation into neuron-like cells, Acta Biomater. 78 (2018) 274–284.
[42] M.-I. Sarbu, C. Matei, C.-I. Mitran, M.-I. Mitran, C. Caruntu, C. Constantin, M. Neagu, S.-R. Georgescu, Photodynamic therapy: a hot topic in dermato-oncology, Oncol. Lett. 17 (5) (2019) 4085–4093.
[43] I.Z. Mukadam, J. Machhi, J. Herskovitz, M. Hasan, M.D. Oleynikov, W.R. Blomberg, D. Svechkarev, A.M. Mohs, Y. Zhou, P. Dash, Rilpivirine-associated aggregation-induced emission enables cell-based nanoparticle tracking, Biomaterials 231 (2020), 119669.
[44] G. Feng, D. Mao, J. Liu, C.C. Goh, L.G. Ng, D. Kong, B.Z. Tang, B. Liu, Polymeric nanorods with aggregation-induced emission characteristics for enhanced cancer targeting and imaging, Nanoscale 10 (13) (2018) 5869–5874.
[45] X. He, Z. Zhao, L.-H. Xiong, P.F. Gao, C. Peng, R.S. Li, Y. Xiong, Z. Li, H.H.-Y. Sung, I.D. Williams, Redox-active AIEgen-derived plasmonic and fluorescent core@ shell nanoparticles for multimodality bioimaging, J. Am. Chem. Soc. 140 (22) (2018) 6904–6911.
[46] H. Tian, D. Li, X. Tang, Y. Zhang, Z. Yang, J. Qian, Y.Q. Dong, M. Han, Efficient red luminogen with aggregation-induced emission for in vivo three-photon brain vascular imaging, Mater. Chem. Front. 4 (6) (2020) 1634–1642.
[47] S.S. Kelkar, T.M. Reineke, Theranostics: combining imaging and therapy, Bioconjug. Chem. 22 (10) (2011) 1879–1903.
[48] M.S. Muthu, D.T. Leong, L. Mei, S.-S. Feng, Nanotheranostics- application and further development of nanomedicine strategies for advanced theranostics, Theranostics 4 (6) (2014) 660.
[49] T. Lammers, S. Aime, W.E. Hennink, G. Storm, F. Kiessling, Theranostic nanomedicine, Acc. Chem. Res. 44 (10) (2011) 1029–1038.
[50] L. Ganau, L. Prisco, G.K. Ligarotti, R. Ambu, M. Ganau, Understanding the pathological basis of neurological diseases through diagnostic platforms based on innovations in biomedical engineering: new concepts and theranostics perspectives, Medicines 5 (1) (2018) 22.
[51] J. Qian, B.Z. Tang, AIE luminogens for bioimaging and theranostics: from organelles to animals, Chem 3 (1) (2017) 56–91.
[52] M. Gao, B.Z. Tang, AIE-based cancer theranostics, Coord. Chem. Rev. 402 (2020), 213076.
[53] M.T. Yaraki, Y. Pan, F. Hu, Y. Yu, B. Liu, Y.N. Tan, Nanosilver-enhanced AIE photosensitizer for simultaneous bioimaging and photodynamic therapy, Mater. Chem. Front. 4 (10) (2020) 3074–3085.
[54] M. Li, M. Gao, X. Xie, Y. Zhang, J. Ning, P. Liu, K. Gu, MicroRNA-200c reverses drug resistance of human gastric cancer cells by targeting regulation of the NER-ERCC3/4 pathway, Oncol. Lett. 18 (1) (2019) 145–152.
[55] D.E. Dolmans, D. Fukumura, R.K. Jain, Photodynamic therapy for cancer, Nat. Rev. Cancer 3 (5) (2003) 380–387.

[56] R. Baskaran, J. Lee, S.-G. Yang, Clinical development of photodynamic agents and therapeutic applications, Biomater. Res. 22 (1) (2018) 1–8.
[57] Y. Yuan, G. Feng, W. Qin, B.Z. Tang, B. Liu, Targeted and image-guided photodynamic cancer therapy based on organic nanoparticles with aggregation-induced emission characteristics, Chem. Commun. 50 (63) (2014) 8757–8760.
[58] J. Dai, Y. Li, Z. Long, R. Jiang, Z. Zhuang, Z. Wang, Z. Zhao, X. Lou, F. Xia, B.Z. Tang, Efficient near-infrared photosensitizer with aggregation-induced emission for imaging-guided photodynamic therapy in multiple xenograft tumor models, ACS Nano 14 (1) (2019) 854–866.
[59] S. Wang, W. Wu, P. Manghnani, S. Xu, Y. Wang, C.C. Goh, L.G. Ng, B. Liu, Polymerization-enhanced two-photon photosensitization for precise photodynamic therapy, ACS Nano 13 (3) (2019) 3095–3105.
[60] X. Shen, L. Li, A.C. Min Chan, N. Gao, S.Q. Yao, Q.H. Xu, Water-soluble conjugated polymers for simultaneous two-photon cell imaging and two-photon photodynamic therapy, Adv. Opt. Mater. 1 (1) (2013) 92–99.
[61] W. Wu, R. Tang, Q. Li, Z. Li, Functional hyperbranched polymers with advanced optical, electrical and magnetic properties, Chem. Soc. Rev. 44 (12) (2015) 3997–4022.
[62] W. Wu, D. Mao, S. Xu, F. Hu, X. Li, D. Kong, B. Liu, Polymerization-enhanced photosensitization, Chem 4 (8) (2018) 1937–1951.
[63] S. Liu, H. Zhang, Y. Li, J. Liu, L. Du, M. Chen, R.T. Kwok, J.W. Lam, D.L. Phillips, B.Z. Tang, Strategies to enhance the photosensitization: polymerization and the donor–acceptor even–odd effect, Angew. Chem. Int. Ed. 57 (46) (2018) 15189–15193.
[64] N. Ji, Adaptive optical fluorescence microscopy, Nat. Methods 14 (4) (2017) 374–380.
[65] Y. Liu, P. Bhattarai, Z. Dai, X. Chen, Photothermal therapy and photoacoustic imaging via nanotheranostics in fighting cancer, Chem. Soc. Rev. 48 (7) (2019) 2053–2108.
[66] W.-W. Liu, P.-C. Li, Photoacoustic imaging of cells in a three-dimensional microenvironment, J. Biomed. Sci. 27 (1) (2020) 1–9.
[67] E. Middha, B. Liu, Nanoparticles of organic electronic materials for biomedical applications, ACS Nano 14 (8) (2020) 9228–9242.
[68] J.S. Ni, T. Min, Y. Li, M. Zha, P. Zhang, C.L. Ho, K. Li, Planar AIEgens with enhanced solid-state luminescence and ROS generation for multidrug-resistant bacteria treatment, Angew. Chem. Int. Ed. 59 (25) (2020) 10179–10185.
[69] J.-S. Ni, Y. Li, W. Yue, B. Liu, K. Li, Nanoparticle-based cell trackers for biomedical applications, Theranostics 10 (4) (2020) 1923.
[70] J. Geng, L.-D. Liao, W. Qin, B.Z. Tang, N. Thakor, B. Liu, Fluorogens with aggregation induced emission: ideal photoacoustic contrast reagents due to intramolecular rotation, J. Nanosci. Nanotechnol. 15 (2) (2015) 1864–1868.
[71] X. Cai, J. Liu, W.H. Liew, Y. Duan, J. Geng, N. Thakor, K. Yao, L.-D. Liao, B. Liu, Organic molecules with propeller structures for efficient photoacoustic imaging and photothermal ablation of cancer cells, Mater. Chem. Front. 1 (8) (2017) 1556–1562.
[72] Z. Zhao, C. Chen, W. Wu, F. Wang, L. Du, X. Zhang, Y. Xiong, X. He, Y. Cai, R.T. Kwok, Highly efficient photothermal nanoagent achieved by harvesting energy via excited-state intramolecular motion within nanoparticles, Nat. Commun. 10 (1) (2019) 1–11.
[73] Y. Yuan, S. Xu, C.-J. Zhang, B. Liu, Light-responsive AIE nanoparticles with cytosolic drug release to overcome drug resistance in cancer cells, Polym. Chem. 7 (21) (2016) 3530–3539.
[74] M. Kanamala, W.R. Wilson, M. Yang, B.D. Palmer, Z. Wu, Mechanisms and biomaterials in pH-responsive tumour targeted drug delivery: a review, Biomaterials 85 (2016) 152–167.
[75] X. Wang, Y. Yang, Y. Zhuang, P. Gao, F. Yang, H. Shen, H. Guo, D. Wu, Fabrication of pH-responsive nanoparticles with an AIE feature for imaging intracellular drug delivery, Biomacromolecules 17 (9) (2016) 2920–2929.
[76] B. Ma, W. Zhuang, H. He, X. Su, T. Yu, J. Hu, L. Yang, G. Li, Y. Wang, Two-photon AIE probe conjugated theranostic nanoparticles for tumor bioimaging and pH-sensitive drug delivery, Nano Res. 12 (7) (2019) 1703–1712.
[77] J. Yao, M. Yang, Y. Duan, Chemistry, biology, and medicine of fluorescent nanomaterials and related systems: new insights into biosensing, bioimaging, genomics, diagnostics, and therapy, Chem. Rev. 114 (12) (2014) 6130–6178.
[78] K.P. Carter, A.M. Young, A.E. Palmer, Fluorescent sensors for measuring metal ions in living systems, Chem. Rev. 114 (8) (2014) 4564–4601.
[79] C.P. Sobel, P. Li, The Cognitive Sciences: An Interdisciplinary Approach, Sage Publications, 2013.

[80] H. Bischof, S. Burgstaller, M. Waldeck-Weiermair, T. Rauter, M. Schinagl, J. Ramadani-Muja, W.F. Graier, R. Malli, Live-cell imaging of physiologically relevant metal ions using genetically encoded FRET-based probes, Cell 8 (5) (2019) 492.

[81] D.P. Clark, N.J. Pazdernik, M.R. McGehee, Molecular Biology, Elsevier, 2019.

[82] A. Rauk, The chemistry of Alzheimer's disease, Chem. Soc. Rev. 38 (9) (2009) 2698–2715.

[83] M. Serra, A. Columbano, U. Ammarah, M. Mazzone, A. Menga, Understanding metal dynamics between cancer cells and macrophages: competition or synergism? Front. Oncol. 10 (2020) 646.

[84] M. Kawahara, K.I. Tanaka, M. Kato-Negishi, Zinc, carnosine, and neurodegenerative diseases, Nutrients 10 (2) (2018) 147–167.

[85] G. Bjørklund, M. Dadar, L. Pivina, M.D. Doşa, Y. Semenova, J. Aaseth, The role of zinc and copper in insulin resistance and diabetes mellitus, Curr. Med. Chem. 27 (39) (2020) 6643–6657.

[86] T.A. Rouault, The role of iron regulatory proteins in mammalian iron homeostasis and disease, Nat. Chem. Biol. 2 (8) (2006) 406–414.

[87] R. Gozzelino, P. Arosio, Iron homeostasis in health and disease, Int. J. Mol. Sci. 17 (1) (2016) 130–144.

[88] Y. Li, H. Zhong, Y. Huang, R. Zhao, Recent advances in AIEgens for metal ion biosensing and bioimaging, Molecules 24 (24) (2019) 4593.

[89] D. Yang, F. Li, Z. Luo, B. Bao, Y. Hu, L. Weng, Y. Cheng, L. Wang, Conjugated polymer nanoparticles with aggregation induced emission characteristics for intracellular F e3+ sensing, J. Polym. Sci. A Polym. Chem. 54 (12) (2016) 1686–1693.

[90] C. Wang, X. Wang, P. Gai, H. Li, F. Li, Target-responsive AIE-Au nanoconjugate for acetylcholinesterase activity and inhibitor assay with ultralow background noise, Sensors Actuators B Chem. 284 (2019) 118–124.

[91] N. Goswami, Q. Yao, Z. Luo, J. Li, T. Chen, J. Xie, Luminescent metal nanoclusters with aggregation-induced emission, J. Phys. Chem. Lett. 7 (6) (2016) 962–975.

[92] L. Mao, Y. Wu, C.C. Stoumpos, B. Traore, C. Katan, J. Even, M.R. Wasielewski, M.G. Kanatzidis, Tunable white-light emission in single-cation-templated three-layered 2D perovskites (CH(3)CH(2)NH(3))(4)Pb(3)Br(10-x)Cl (x), J. Am. Chem. Soc. 139 (34) (2017) 11956–11963.

[93] H. Qian, M. Zhu, Z. Wu, R. Jin, Quantum sized gold nanoclusters with atomic precision, Acc. Chem. Res. 45 (9) (2012) 1470–1479.

[94] Z. Luo, X. Yuan, Y. Yu, Q. Zhang, D.T. Leong, J.Y. Lee, J. Xie, From aggregation-induced emission of Au(I)-thiolate complexes to ultrabright Au(0)@Au(I)-thiolate core-shell nanoclusters, J. Am. Chem. Soc. 134 (40) (2012) 16662–16670.

[95] Y. Xiao, Z. Wu, Q. Yao, J. Xie, Luminescent metal nanoclusters: biosensing strategies and bioimaging applications, Aggregate 2 (1) (2021) 114–132.

[96] A. Yahia-Ammar, D. Sierra, F. Mérola, N. Hildebrandt, X. Le Guével, Self-assembled gold nanoclusters for bright fluorescence imaging and enhanced drug delivery, ACS Nano 10 (2) (2016) 2591–2599.

[97] M. Shamsipur, K. Molaei, F. Molaabasi, S. Hosseinkhani, A. Taherpour, M. Sarparast, S.E. Moosavifard, A. Barati, Aptamer-based fluorescent biosensing of adenosine triphosphate and cytochrome c via aggregation-induced emission enhancement on novel label-free DNA-capped silver nanoclusters/graphene oxide nanohybrids, ACS Appl. Mater. Interfaces 11 (49) (2019) 46077–46089.

CHAPTER

16

AIE bio-conjugates for biomedical applications

Zhiyuan Gao and Dan Ding

State Key Laboratory of Medicinal Chemical Biology, Key Laboratory of Bioactive Materials, Ministry of Education, and College of Life Sciences, Nankai University, Tianjin, China

1 Introduction

Bio-conjugates refer to the complex of two biologically related molecules that are linked together by covalent bonds [1–3]. The two parts of the bio-conjugates must be compatible with each other, so that not only can they retain their own functions, such as commonly used targeting functions, detection functions, and toxicity functions, but also complement each other and play a synergistic role [3,4].

Fluorescence-based technology has developed potential biomedical applications due to its high sensitivity, rapid response, and in situ characteristics [5,6]. Traditional fluorescent probes have an aggregation-caused quenching (ACQ) effect, which limits their use in high concentrations or in nanoparticles [7]. The discovery of aggregation-induced emission (AIE) had since solved this problem. As a photosensitizer with excellent optical properties, aggregation-induced emission luminogens (AIEgens) and AIE dots (noncovalent bonding of polymer matrix formed by physical encapsulation) have been widely used in biomedicine [8,9].

AIE bio-conjugates are usually formed by conjugating AIEgens or AIE dots to biomolecules such as sugars, peptides, proteins, enzymes, and DNA [10,11]. The resulting hybrid product usually retains the properties of the two materials, such as the fluorescence properties of AIEgens and the biological functions of biomolecules [12]. In addition, the two parts have complementary advantages. The biomolecules significantly improved the water solubility of AIEgens, which not only increases the biocompatibility of the probe, but also bring about a low emission background that is conducive for high-sensitivity bioassays or high-contrast imaging [13,14]. Besides, AIE dots can also be used as a scaffold for the attachment of a variety of biomolecules, creating a multifunctional AIE biological coupling platform [15,16].

Aggregation-Induced Emission (AIE)
https://doi.org/10.1016/B978-0-12-824335-0.00009-X

2 Reaction for AIE bio-conjugation

Because the premise is to maintain the inherent functions of fragile biomolecules, the conditions for the coupling reaction are relatively strict [17,18]. The following conditions must be met, such as room temperature, aqueous solution, and neutral pH value [19,20]. According to the functional groups possessed or modified by biomolecules, AIEgens or AIE dots that can be conjugated with these functional groups are designed, and AIE bio-conjugates can then be formed through these coupling reactions under mild conditions [21]. Commonly used synthetic methods of AIE bio-conjugates can be roughly divided into two categories: one is conventional coupling reactions used in vitro and the other is click chemical reaction used in vivo.

As shown in Fig. 1A, the synthesis of the first category of AIE bio-conjugates include fourtypes of reactions: (1) condensation reactions of the carboxyl groups (or amines) of AIE probes with the amines or carboxyl groups of the biomolecules using EDC (1-ethyl-3 (3-dimethylaminopropyl) carbodiimide) or NHS (*N*-hydroxy succinimide) as reagents, (2) the condensation reaction between the AIE probe with boric acid and the biomolecule with glycol [22], (3) direct Michael addition reaction between maleimide of the AIE probe and the thiol groups of the biomolecules, and (4) Cu(I) catalyzed click chemical reaction between the AIE probe with azide and the biomolecule with a terminal alkynyl group [23–25].

As shown in Fig. 1B, the synthesis of the second category of AIE bio-conjugates mainly involves the active alkyne-triggered catalyst free click reactions between the reactive triple bonds of AIEgens with the premodified azido groups of the biomolecules [26–28].

In the conventional coupling reactions: (I) The direct reaction between carboxylic acid and amine was difficult. Therefore, in order to improve the coupling efficiency, activators such as NHS and EDC were usually added, which were not suitable for use in vivo. (II) Although the reaction between boric acid and glycol could be completed without a catalyst, the adduct was easily hydrolyzed and the covalent connection method was less stable. (III) The reaction between maleimide and the thiol groups was a highly efficient reaction without a catalyst. However, it was limited to the conjugation of peptides or proteins with thiol-containing amino acids (such as cysteine) [21]. (IV) The reaction between azide and alkyne was highly efficient, stable and highly specific, and could even occur in living cells. However, the participation of Cu(I) could induce the production of toxic reactive oxygen species (ROS), which largely limited its further development in vivo [29]. (V) Ring strain as a means to activate alkynes for catalyst-free click reactions with azides. Therefore, the reaction was a catalyst-free, efficient, stable and highly specific reaction, and it could even be carried out in vivo very well [30].

3 Biosensing

Biosensing is an important means of obtaining biochemical information, which is of great significance to biomedical research, disease diagnosis, and therapy [31,32]. Biosensing is also an effective means to obtain some key information in life processes by detecting various biochemical molecules (such as small molecules, nucleic acids, and proteins) [33].

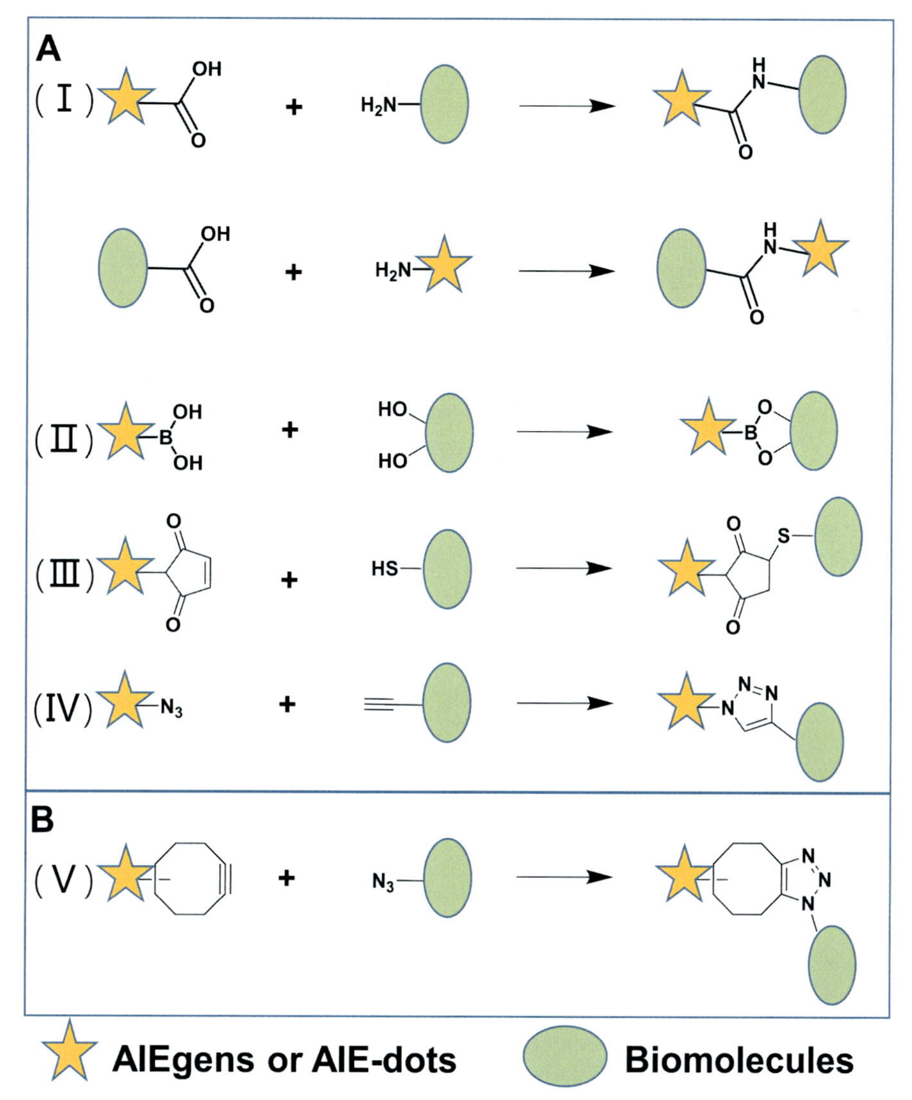

FIG. 1 (A) Schematic diagram of conjugating AIEgens/AIE dots and biomolecules by conventional coupling reactions. (B) Schematic diagram of the conjugation of AIEgens/AIE dots and biomolecules via a novel copper-free click chemistry reaction.

3.1 Small molecules

In biological or medical research, it is often necessary to detect small molecules that play an important role in biological activity, including reactive oxygen species, metal ions, and so on [34–36]. For example, heavy metal mercury ions (Hg^{2+}) are extremely toxic at low concentrations; so, it is very important to develop Hg^{2+} detection methods with high sensitivity and selectivity [37,38]. Xia et al. used the specific binding of thymus (T) and Hg^{2+} to develop

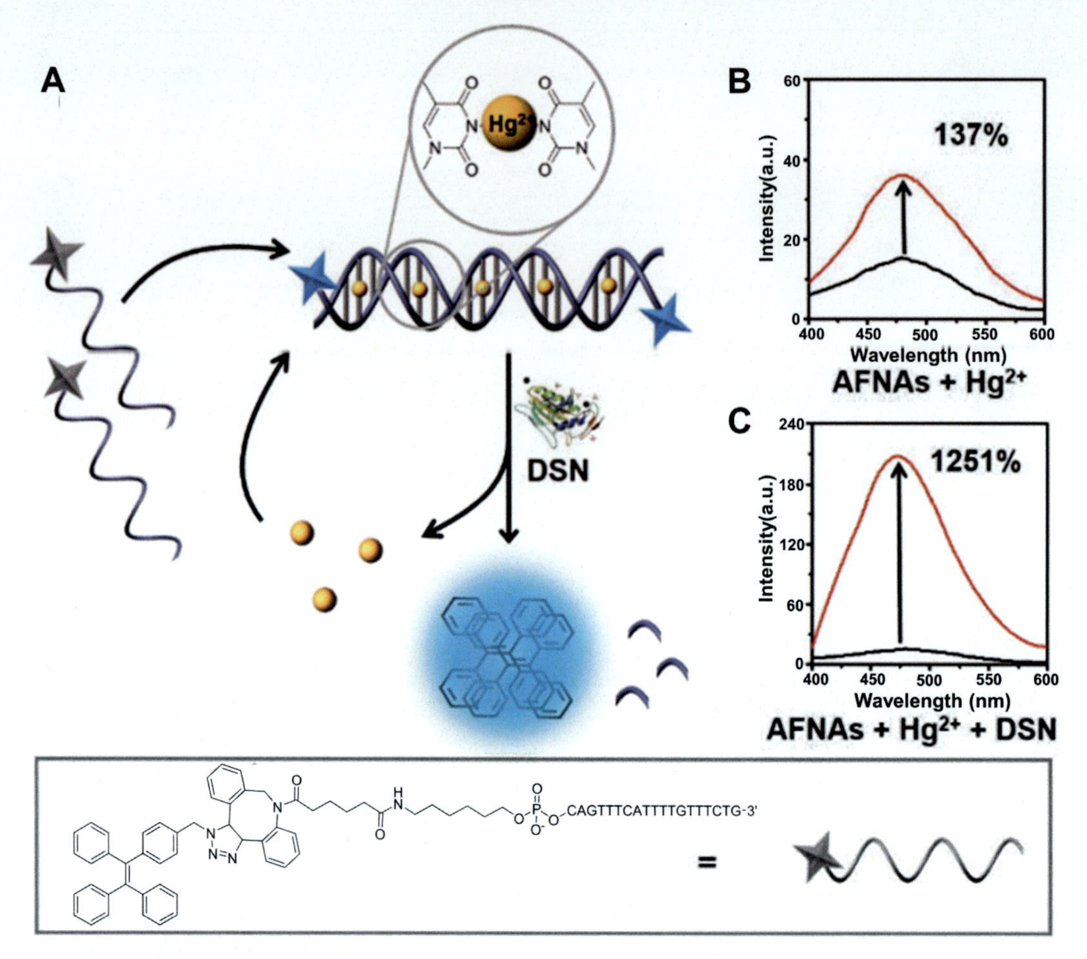

FIG. 2 (A) Schematic diagram of the process of catalytic amplification analysis of Hg^{2+} ions based on the DSN enzyme; (B) after the flexible single-stranded AFNAs is transformed into a rigid double-stranded DNA structure under the action of Hg^{2+}, the maximum emission fluorescence intensity increases by 137%; and (C) after catalytic amplification with the participation of the DSN enzyme, the maximum emission fluorescence intensity increased by 1251%. *Reproduced with permission from reference X. Ou, X. Lou, F. Xia, A highly sensitive DNA-AIEgen-based "turn-on" fluorescence chemosensor for amplification analysis of* Hg^{2+} *ions in real samples and living cells, Sci. China Chem. 60 (2017) 663–669. Copyright 2017, Science China Press and Springer-Verlag Berlin Heidelberg.*

bio-conjugated AFNAs composed of TPE and thymus-rich DNA [39]. As shown in Fig. 2A, the principle of detection involves the specific binding of thymus and Hg^{2+}, which made the single-stranded AFNAs probe became double-stranded, thus increasing the hardness of the probe and inhibited the rotational movement of the benzene ring, thereby turning on the fluorescence of the TPE (Fig. 2B). In addition, duplex-specific nuclease (DSN) enzyme-based target recycling could release more TPE residues and form emissive aggregates in the test medium, which greatly improved the sensitivity of Hg^{2+} detection (Fig. 2C).

3.2 Nucleic acids

The nucleic acid molecule hybridization fluorescent probe is a kind of probe obtained by conjugating a nucleotide sequence complementary to the target DNA/RNA to a fluorophore, and thus using the principle of base complementation to identify the target DNA/RNA [40,41]. Because the complementarity between nucleic acid sequences is very specific, the accuracy of nucleic acid molecular hybridization probes for disease detection is often unparalleled. Zhang et al. designed and synthesized a double-arm "turn-on" probe for the specific detection of nucleic acid sequences [42]. The probe was an AIE fluorescent molecule with a double azide group conjugated to single-stranded DNA (Fig. 3A). Relying on the hydrophilicity and flexibility of single-stranded DNA, the AIE fluorophore could be well dispersed in an aqueous medium. Due to the flexible movement of the benzene ring in the probe, the probe exhibited very weak fluorescence. When hybridized with the complementary strand that needs to be recognized, the double helix structure formed could be used as a semiflexible rod-shaped damper, restricting the free rotation of the benzene ring, and emitting fluorescence that was 6.1 times more intense than the background fluorescence intensity (Fig. 3B and C). In addition, as shown in Fig. 3A, the probe TPE-2DNA_2 with complementary oligonucleotide DNA_2 could hybridize with TPE-2DNA_1 to form a longer strand with tandem repeated units. This self-assembly process further restricted the intramolecular motion of TPE. As a result, the fluorescence intensity was greatly increased compared with nonself-assembled probes (Fig. 3D).

3.3 Proteins

Proteins are the basic components and main functional units of life. Protein detection therefore plays an important role in many aspects [43]. For example, the detection of the most abundant protein in the blood (serum albumin) played an important role in the investigation of bloodstains at crime scenes. Wang et al. reported an AIEgens called TPE-MI that could react with thiol groups on serum albumin and had the ability to visualize bloodstains [44]. Due to the PET (photoinduced electron transfer) effect of TPE-MI with maleimide itself, the fluorescence appeared to be quenched, even in the aggregated state (Fig. 4A). As shown in Fig. 4B, when thiol and maleimide were reacting with no catalyst and mild click chemical reaction, the PET effect of TPE-MI was destroyed; so, the product TPE-SI could form aggregates in the hydrophobic cavity of albumin and restore its strong AIE fluorescence. In real crimes, criminals usually wiped bloodstains to make it difficult to find. Therefore, it is very useful to visualize the transfer of bloodstains at the crime scene. These transferred bloodstains were usually shoe prints and fingerprints left by the criminal. As shown in Fig. 4C, fingerprints were invisible to the naked eye left on the glass slide after dipping the diluted blood with a finger. After spraying TPE-MI, not only the spiral pattern of the Level 1 fingerprint feature could be seen under the microscope, but also the bifurcation and cross structure of the level 2 fingerprint feature could be seen after increasing the magnification (Fig. 4D). What's more striking was that researchers could also observe the fine level 3 features of fingerprints, which referred to the pore structure on the ridge (Fig. 4E). These types of fingerprint information provided powerful help for identifying suspects.

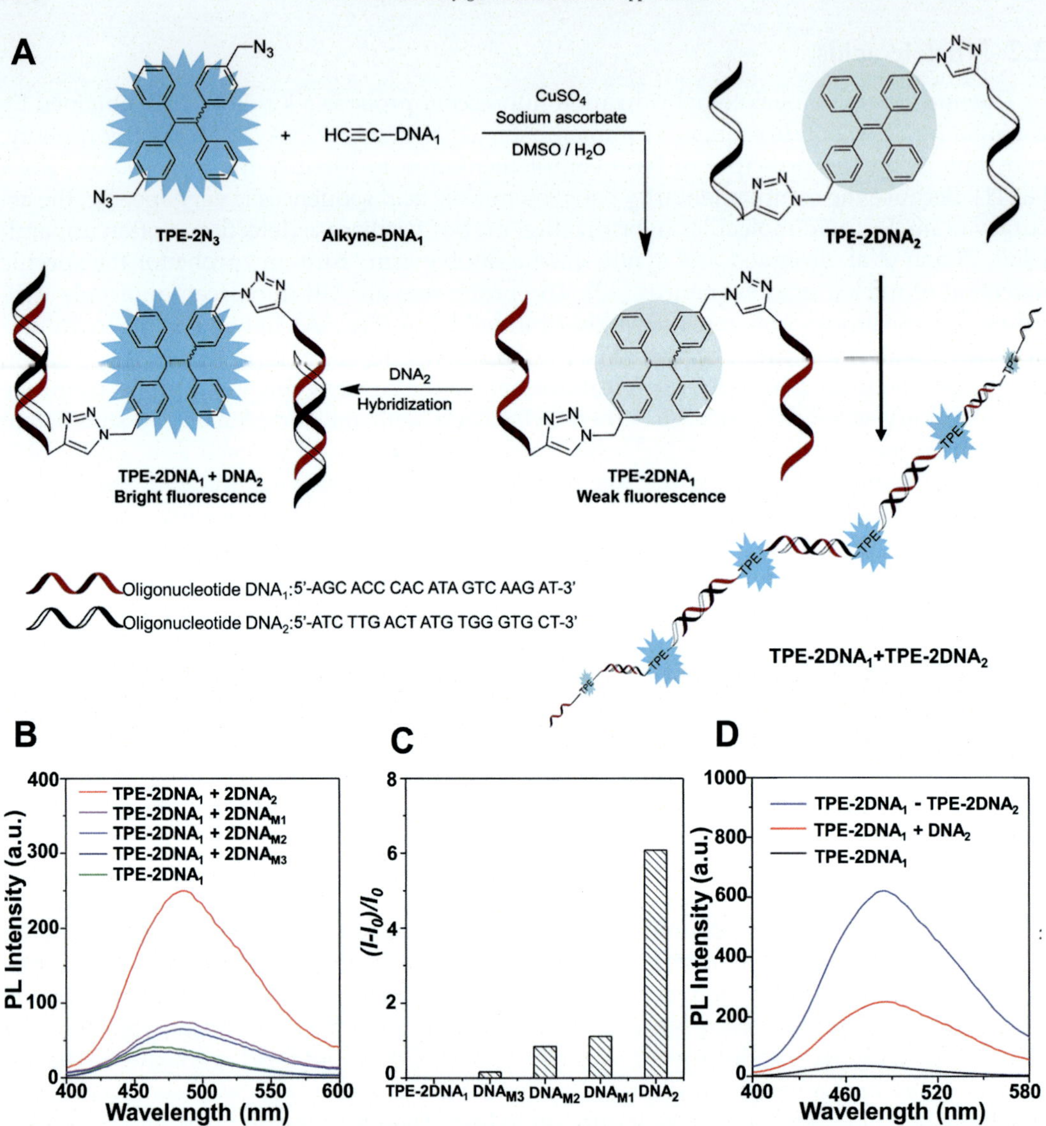

FIG. 3 (A) The route to synthesize TPE-2DNA$_1$ through the click chemistry reaction catalyzed by Cu(I) and its hybridization in nucleic acid hybrid self-assembled duplex (TPE-2DNA$_1$ + TPE-2DNA$_2$); schematic diagram of the application in the test. Oligonucleotides DNA$_1$ and DNA$_2$ were complementary, and their sequences were shown in the Scheme. (B) Photoluminescence (PL) spectra of TPE-2DNA$_1$ alone, and the presence of complementary DNA$_2$, single-base mismatch single-stranded DNA$_{M1}$, double-base mismatched single-stranded DNA$_{M2}$, and random single-stranded DNA$_{M3}$. (C) The fluorescence turn-on value of the probe in the presence of different DNA strands. It was defined as the maximum fluorescence intensity of each sample, and I_0 alone was the background fluorescence of TPE-2DNA$_1$. (D) PL spectra of TPE-2DNA$_1$, duplex with complementary strand (TPE-2DNA$_1$ + DNA$_2$) and self-assembled duplex (TPE-2DNA$_1$ + TPE-2DNA$_2$). *Reproduced with permission from reference R. Zhang, R.T. Kwok, B.Z. Tang, B. Liu, Hybridization induced fluorescence turn-on of AIEgen-oligonucleotide conjugates for specific DNA detection, RSC Adv. 5 (2015) 28332–28337. Copyright 2015, The Royal Society of Chemistry.*

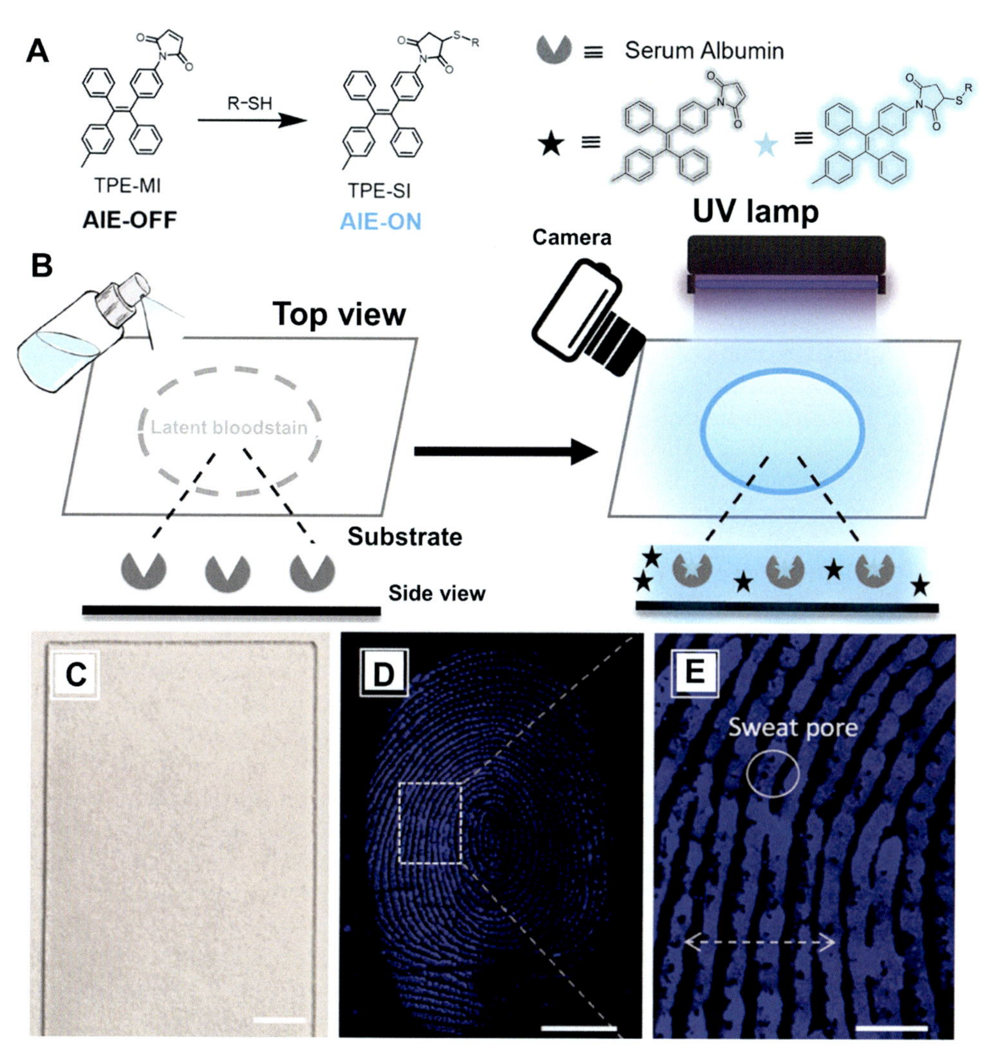

FIG. 4 (A) The structural formula of AIEgen TPE-MI with fluorescence turned off and AIEgen TPE-SI with fluorescence turned on after coupling with serum albumin. (B) TPE-MI was used to spray the surface with potential bloodstains. The schematic diagram shows that the position with potential bloodstains emitted blue fluorescence after being irradiated by the UV lamp. (C) The glass slide with latent blood fingerprint, the photo under ambient light. (D) After spraying the TPE-MI solution, the confocal laser scanning microscope (CLSM) image of latent blood. (E) An enlarged view of the white dashed area in (D). *Reproduced with permission from reference Z. Wang, P. Zhang, H. Liu, Z. Zhao, L. Xiong, W. He, et al., Robust serum albumin-responsive AIEgen enables latent bloodstain visualization in high resolution and reliability for crime scene investigation, ACS Appl. Mater. Interfaces 11 (2019) 17306–17312. Copyright 2019, American Chemical Society.*

4 Cell imaging

Cells are the basic units of life, and research on cell surface markers, organelles [45,46], and intracellular molecules of live-cell imaging is of great significance for understanding life processes, including growth, death, and the development of cancer [47,48].

4.1 Cell surface markers

The cell surface is the outer membrane of bacteria or mammalian cells. Each cell has its own unique membrane structure, especially bacteria and cancer cells [49]. Certain kinds of sugars and proteins are highly expressed on the membrane surface of certain bacterial or cancer cells [50,51]. Therefore, designing probes that specifically target these cell surface markers has become a powerful tool for specific imaging of bacterial and cancer cells, which is of great significance for the diagnosis of bacterial infectious diseases and cancer [52,53].

A unique component of bacterial cell wall is peptidoglycan, which is also the main component of bacterial cell walls. In the cell walls of Gram-positive bacteria, peptidoglycan accounts for 50%–80% of the dry weight of the cell wall, and the content of peptidoglycan in the cell wall of Gram-negative bacteria accounts for 5%–20% of the dry weight of the cell wall [54,55]. In terms of chemical structure, there were many *cis*-diols in the sugar chain part of peptidoglycan. As shown in Fig. 5A, the molecule with boric acid could undergo a dehydration condensation reaction with *cis*-diol under mild conditions without catalysis, and finally conjugated to peptidoglycan. Based on the above reaction, Tang et al. endowed TPE with only two boric acids (TPE-2BA) [56]. To their disappointment, TPE-2BA did not stain live bacteria. Based on the AIE mechanism, they hypothesized that the reason for this phenomenon was that TPE-2BA still contains two benzene rings without boric acid that could rotate freely on the surface of the bacteria. Therefore, Tang et al. further developed probes that could be used for live bacterial imaging [57]. They modified all the rotatable benzene rings on the AIE rotor with boric acid motifs to form TriPE-3BA and TPE-4BA. When they were coupled with the *cis*-diol on the surface of the bacteria, the free movement of all rotors in AIEgen could be inhibited, causing the fluorescence to be turned on. The researchers used the two probes to verify the above hypothesis on Gram-positive *Staphylococcus epidermidis* and Gram-negative *Escherichia coli*, and successfully developed a type of AIE probe for live bacteria imaging (Fig. 5B and C).

Most cancer cells overexpress a protein called LAPTM4B, and the hydrophilic extracellular loop of this protein could specifically bind to peptide IHGHHIISVG (referred to as AP2H) [58]. Based on this, Huang et al. designed an AIE probe that could target a broad-spectrum cancer-related protein LAPTM4B for highly specific detection and localization of tumor markers on the membrane of live cancer cells [59]. As shown in Fig. 6A, the probe (TPE-AP2H) was formed by conjugating TPE and the peptide AP2H through a copper-catalyzed click chemical reaction. The combination of TPE-AP2H and LAPTM4B limited the intramolecular movement of TPE, thereby turning on its fluorescence emission. Next, the probe TPE-AP2H was used to locate the LAPTM4B protein on the cell membrane surface, and the membrane tracker was used for colocalization staining. The results showed that TPE-AP2H was indeed anchored on the cell membrane target (Fig. 6B). The researchers also cocultured TPE-AP2H with cancer cells and normal cells (HepG2, BEL 7402, and HeLa) containing different

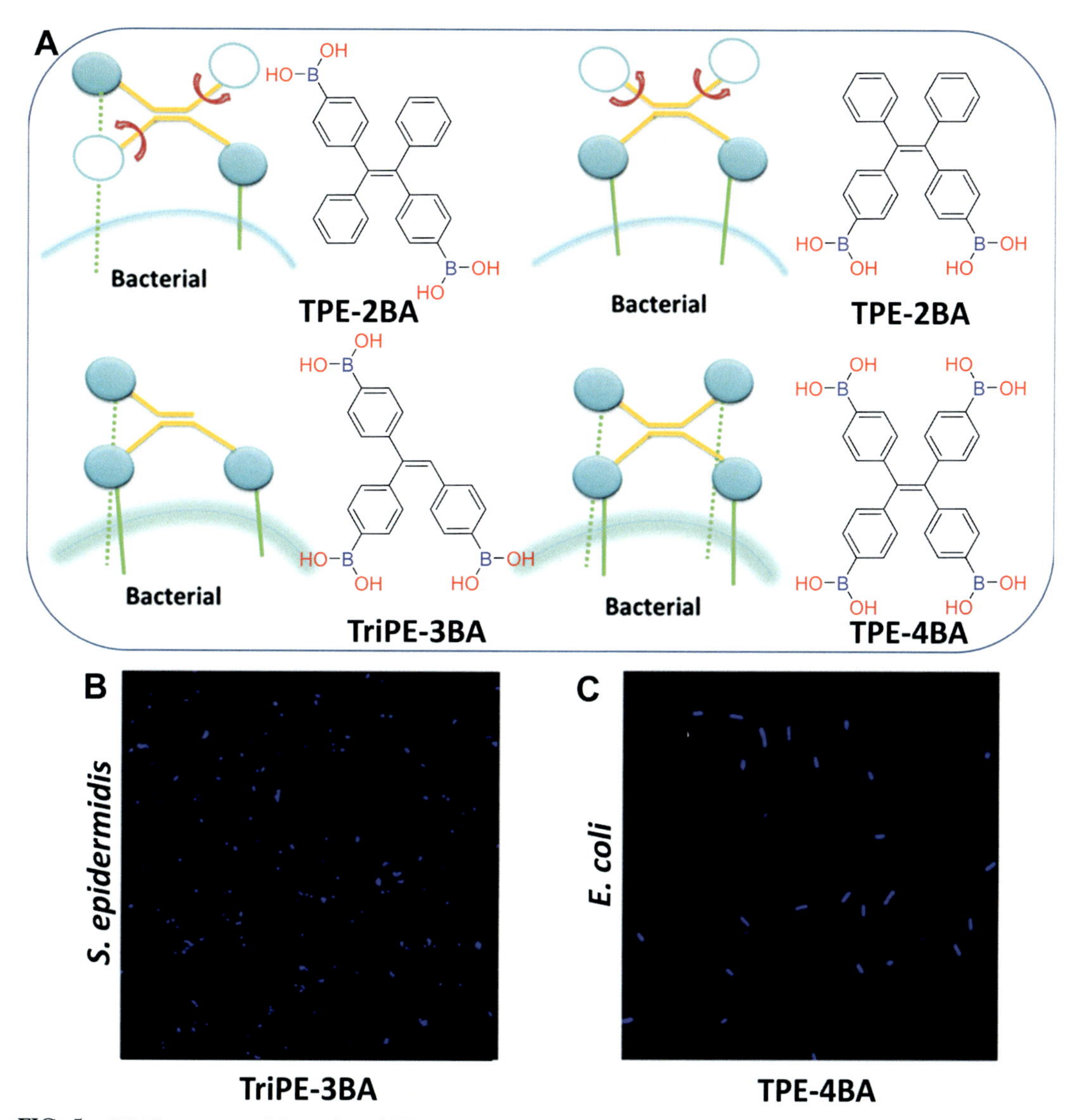

FIG. 5 (A) The structural formulas of TPE-2BA, TriPE-3BA, and TPE-4BA, and the binding status of TPE-2BA, TriPE-3BA, and TPE-4BA after coincubation with bacteria. (B) After incubating with TriPE-3BA, the image of *Staphylococcus epidermidis* as seen under a laser confocal scanning microscope. (C) After incubating with TriPE-4BA, the image of *E. coli* as seen under a laser confocal scanning microscope. *Reproduced with permission from reference T.T. Kong, Z. Zhao, Y. Li, F. Wu, T. Jin, B.Z. Tang, Detecting live bacteria instantly utilizing AIE strategies, J. Mater. Chem. B 6 (2018) 5986–5991. Copyright 2018, The Royal Society of Chemistry.*

amounts of LAPTM4B protein, and they showed different fluorescence intensities (Fig. 6C). In summary, the probe could not only specifically target the overexpressed LAPTM4B protein in most solid tumor cells, but could also distinguish the level of LAPTM4B protein expression in different cancer cells.

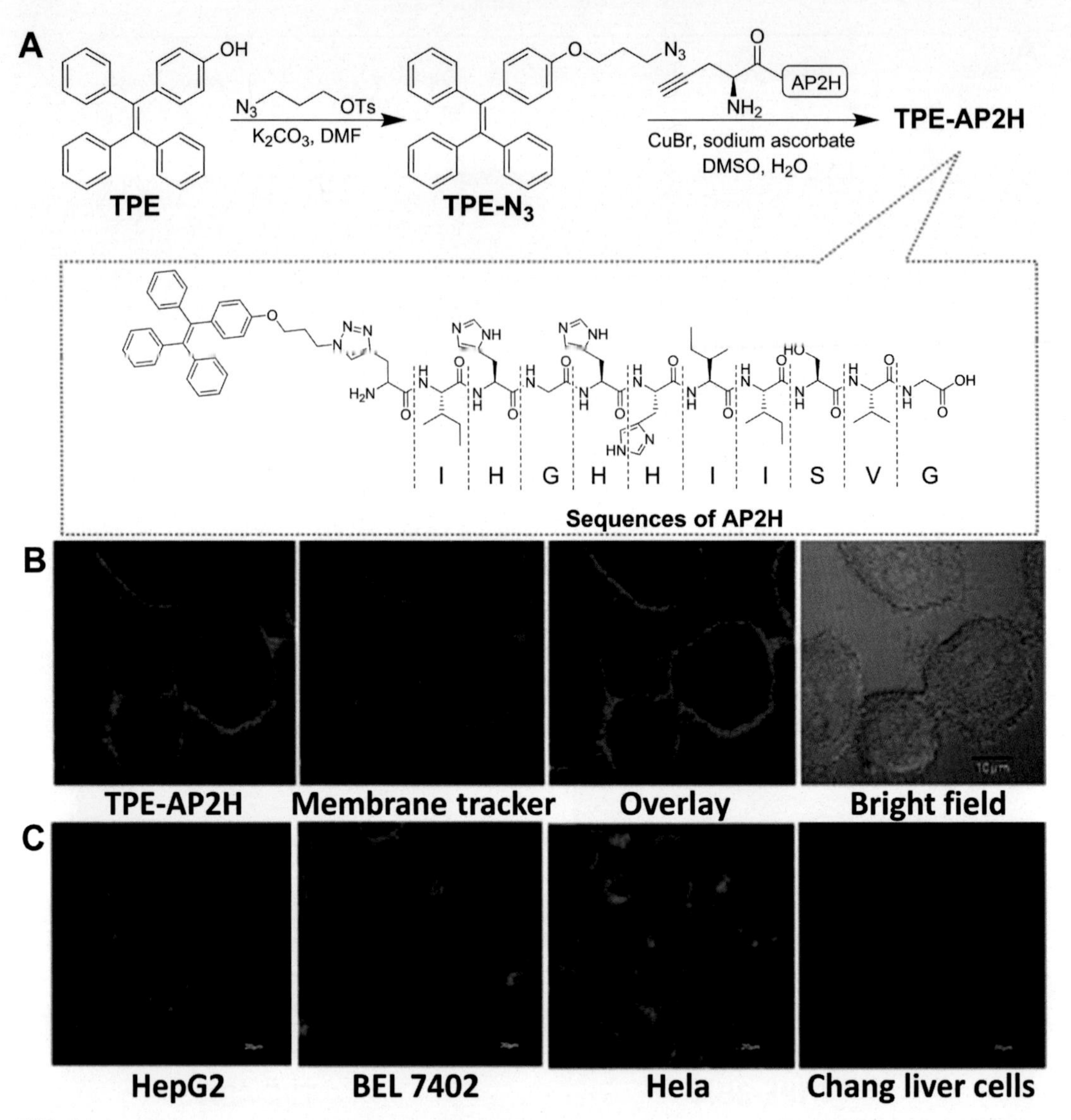

FIG. 6 (A) The chemical structure of probe TPE-AP2H and its synthesis process; (B) CLSM images of BEL 7402 stained by TPE-AP2H and membrane tracker; (C) CLSM images of different types of cancer cells stained by TPE-AP2H (HepG2, BEL 7402, and HeLa). *Reproduced with permission from reference Y. Huang, F. Hu, R. Zhao, G. Zhang, H. Yang, D. Zhang, Tetraphenylethylene conjugated with a specific peptide as a fluorescence turn-on bioprobe for the highly specific detection and tracing of tumor markers in live cancer cells, Chem. Eur. J. 20 (2014) 158–164. Copyright 2014, Wiley-VCH.*

4.2 Organelles

Eukaryotic organelles mainly include cell nucleus, mitochondria, lysosome, endoplasmic reticulum, and Golgi apparatus [60]. They are relatively independent and connected with each other, playing an irreplaceable role in collaboration in cellular processes [60,61]. The imaging of organelles of specific cells was of great significance for studying the functions of organelles in specific cells [62,63]. Shi et al. designed a monoclonal antibody (cetuximab)-conjugated AIEgens (mAb-AIEgens) that could image the mitochondria of cancer cells [64]. As shown in Fig. 7A, AIEgens had strong hydrophilicity after being conjugated with monoclonal antibodies and was dispersed in the cell culture medium; so, it had very low

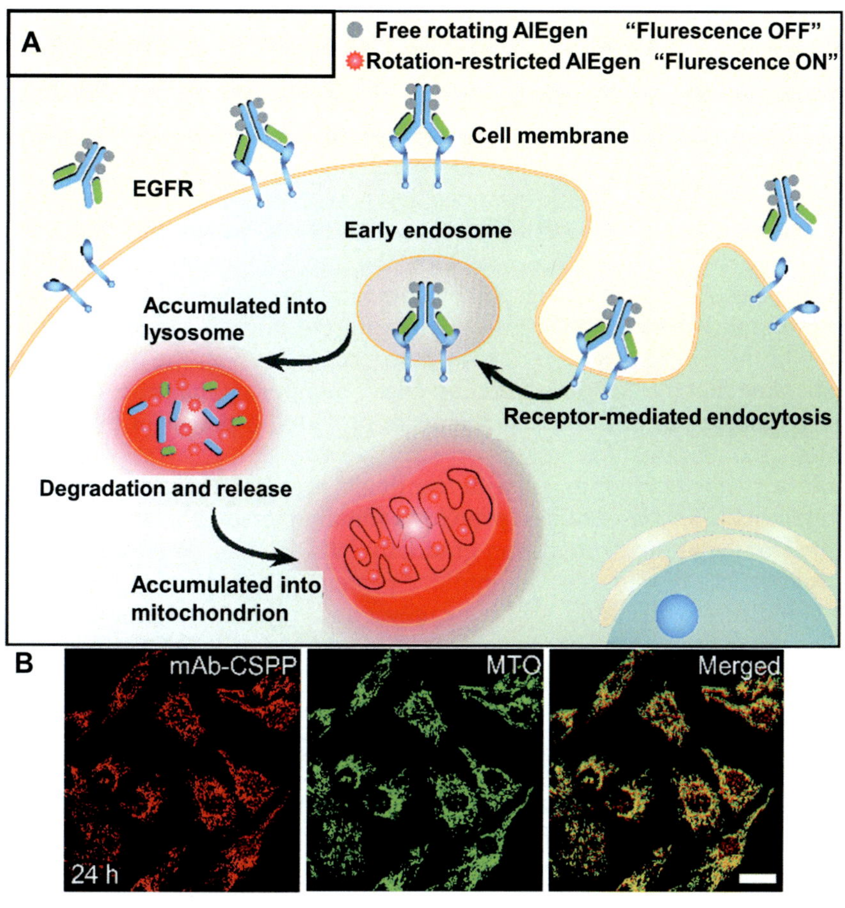

FIG. 7 (A) Schematic diagram of the process of mAb-AIEgens targeting the mitochondria of cancer cells; (B) CLSM images for mAb-AIEgen conjugate staining of HCC827 cells with Mito-Tracker Orange (MTO) counterstain. *Reproduced with permission from reference X. Shi, Y. Chris, H. Su, R.T. Kwok, M. Jiang, Z. He, et al., A red-emissive antibody–AIEgen conjugate for turn-on and wash-free imaging of specific cancer cells, Chem. Sci. 8 (2017) 7014–7024. Copyright 2017, The Royal Society of Chemistry.*

fluorescence background noise. After cetuximab targeted the epidermal growth factor receptor (EGFR) overexpressed by most cancer cells, mAb-AIEgens could first accumulate in the lysosome after entering the cell through receptor-mediated endocytosis. Then, the probe could be hydrolyzed by the hydrolase in the lysosome, and the released AIEgens then selectively accumulated in the mitochondria and emitted strong fluorescence (Fig. 7B). In summary, the probe could not only selectively illuminate the mitochondria of cancer cells, but also exhibited small background noise and strong target signals. Therefore, high-contrast fluorescent images could be obtained on cancer cells.

4.3 Intracellular molecules

As mentioned previously, although AIE bio-conjugates had the ability to verify the existence of bio-targets and monitor their concentration, such probes often lack real-time spatiotemporal information. Therefore, novel probes are needed to detect the positioning and dynamic change of biological macromolecules in living cells [65].

Zhao et al. used AIEgens to develop an intracellular bio-conjugation model that could monitor DNA synthesis and cell proliferation in living cells [66]. TPE-Py-N_3 with an azide group was conjugated with double-stranded DNA with terminal alkynes through a click chemical reaction catalyzed by the Cu(I) complex. As TPE-Py-N_3 was inserted into the hydrophobic core inside the newly synthesized DNA, the fluorescence of AIEgens was turned on gradually (Fig. 8A). As shown in Fig. 8B, by culturing the cells with the essential medium containing EdU for 3h, the fluorescence opening in the nucleus could be observed. With the extension of the incubation time, the fluorescence intensity in the nucleus gradually increased, indicating that the AIEgens were capable of monitoring the process of DNA synthesis and cell proliferation. This AIE bio-conjugate provided a powerful tool for EdU-based DNA synthesis monitoring.

Nondisulfide bonded cysteine (i.e., with free thiols) is a type of residue mainly located inside the protein [67]. As the unfolded protein gradually increases in amount, the thiols exposed on the surface of the protein also gradually increase [68]. Therefore, thiols could be used as a proxy reporter for the level of unfolded proteins. As mentioned above, TPE-MI with maleimide could react with thiols without catalyst and mild click chemical reaction. When TPE-MI was combined with thiols on soluble glutathione, it did not produce fluorescence. However, after binding to the thiols on the unfolded protein, its strong AIE fluorescence was restored due to the rigid local molecular environment provided by the protein [69] (Fig. 9A). As shown in Fig. 9B, the researchers assessed the behavior of TPE-MI in live HeLa cells, and the results show that fluorescence was mainly concentrated in the endoplasmic reticulum region. This is because the endoplasmic reticulum is an important place for protein folding and contains a large amount of unfolded protein. Then, the researchers evaluated the ability of TPE-MI to monitor the unfolded protein content in living cells. When a 42°C heat shock pulse was used to unfold the folded protein in the cell, the heat shock protein that helped restore the folding of the protein could be upregulated. This protective mechanism against heat damage was widespread in bacteria and mammalian cells. As shown in

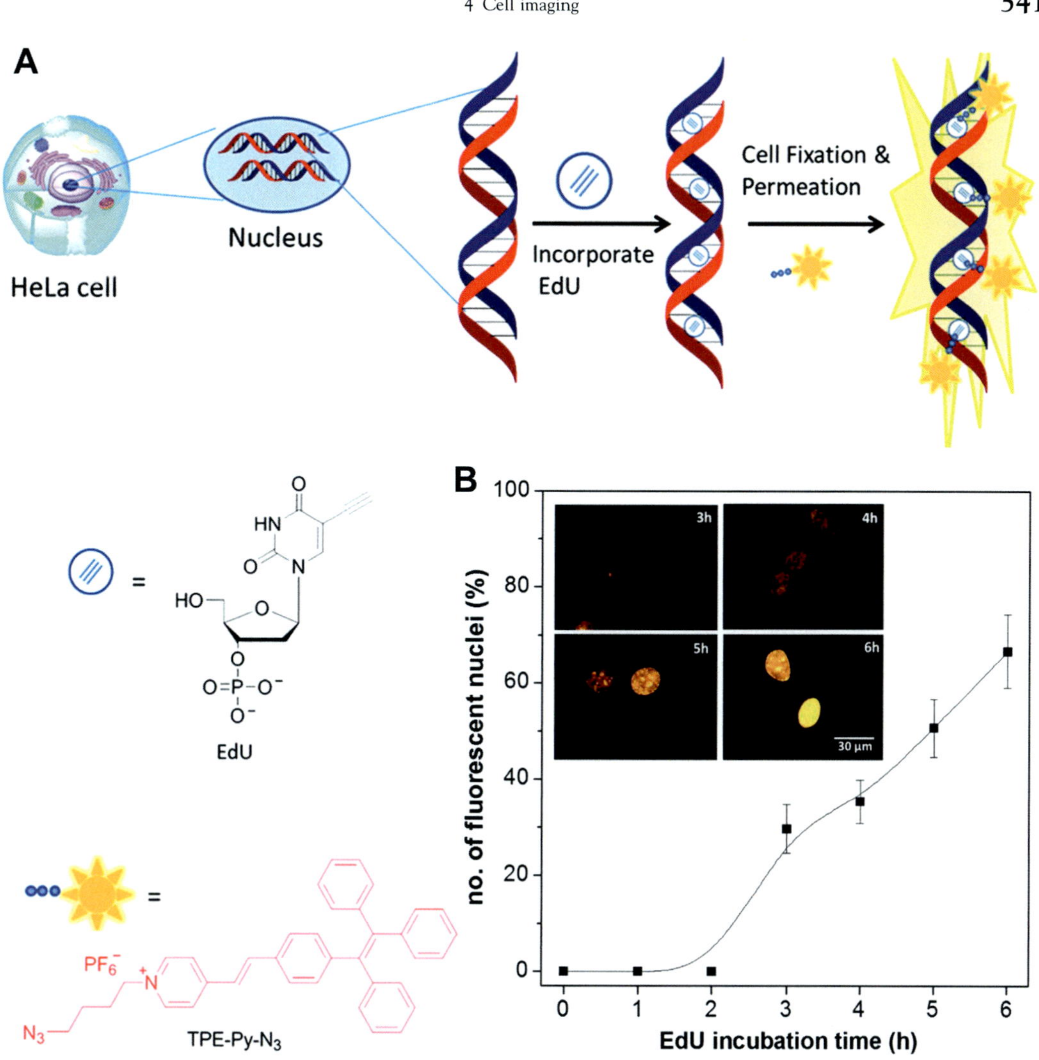

FIG. 8 (A) The chemical structure of EdU and TPE-Py-N3 and the schematic diagram of EdU-based AIE bioconjugates for monitoring DNA synthesis in living cells. (B) Quantitative fluorescence analysis of HeLa cells with increasing incubation time after adding EdU. Inset: fluorescence images of nuclei stained by TPE–Py–N_3. *Reproduced with permission from reference Y. Zhao, Y. Chris, R.T. Kwok, Y. Chen, S. Chen, J.W. Lam, et al., Photostable AIE fluorogens for accurate and sensitive detection of S-phase DNA synthesis and cell proliferation, J. Mater. Chem. B 3 (2015) 4993–4996. Copyright 2015, The Royal Society of Chemistry.*

Fig. 9C, the researchers monitored an increase in the amount of unfolded protein caused by the heat shock in Hela cells within 30 min through a luciferase activity reporter assay. Then, the researchers used the TPE-MI probe to monitor the gradual recovery of HeLa cells from heat shock (Fig. 9D). In summary, TPE-MI is a promising tool to measure the dynamics of protein folding in living cells.

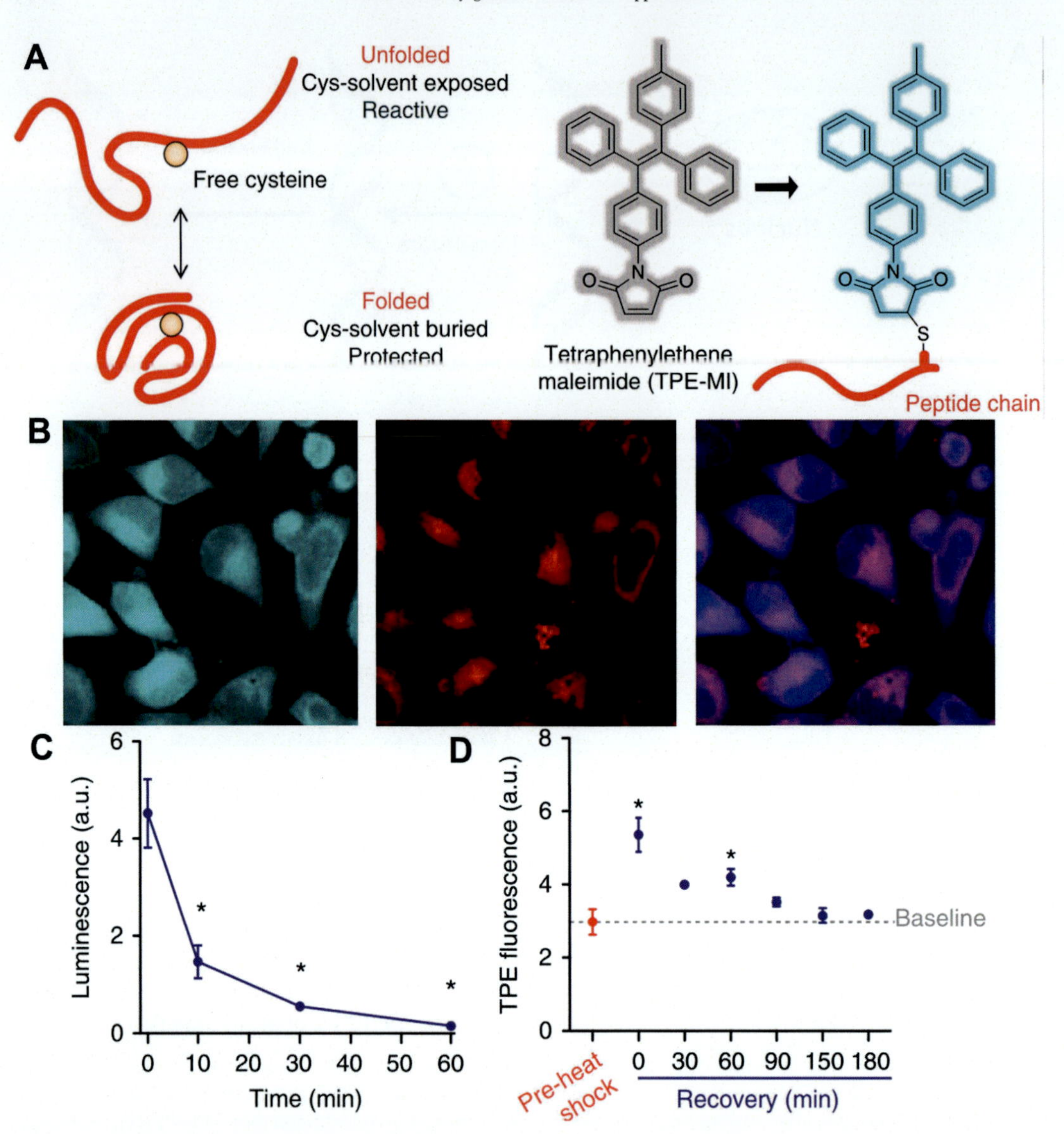

FIG. 9 (A) The chemical structure of TPE-MI of maleimide (MI) conjugated tetraphenylethylene (TPE) and the schematic diagram of monitoring protein folding process through its reaction with thiol on cysteine. After TPE-MI binds to the unfolded protein, it restricts the rotation of the benzene ring, thereby turning on the fluorescence of AIEgens. (B) CLSM images of HeLa cells stained with TPE-MI、CLSM images counterstained with ER Tracker and Overlay of two images colocalized. (C) Unfolding of the proteome by heat shock as assessed by denaturation of *Renilla* luciferase and luciferase activity. (D) Heat shock therapy of HeLa cells at 42 °C for 45 min, and then the fluorescence intensity of HeLa cells is monitored for a subsequent time course of recovery by flow cytometry at 37°C. *Reproduced with permission from reference M.Z. Chen, N.S. Moily, J.L. Bridgford, R.J. Wood, M. Radwan, T.A. Smith, et al., A thiol probe for measuring unfolded protein load and proteostasis in cells, Nat. Commun. 8 (2017) 1–11. Copyright 2017, Springer Nature.*

5 In vivo imaging and image-guided therapy

Compared with cell imaging, in vivo imaging is much more complicated [70,71]. There are often further requirements for AIE bio-conjugates, such as near-infrared luminescent AIE probes with strong tissue penetration [71], negligible toxicity and good biocompatibility, and more specific targeting capabilities [72]. In addition, because AIEgens could be used not only as an imaging tool but also as a phototherapeutic agent, it has been successfully used in imaging-guided therapy [73].

5.1 In vivo imaging

AIE bio-conjugates used in vivo could be divided into two types. One type is bio-conjugates formed in vitro based on traditional coupling reaction, which rely on targeting molecules to attach to the lesion site to complete imaging or therapy in vivo. The second type is based on metabolic labeling and noncopper-catalyzed bioorthogonal reaction methods to form bio-conjugates in vivo, using specific chemical reactions to covalently bind to the lesion site to complete imaging or therapy in vivo.

The first type of bio-conjugates rely on molecules with targeting functions (including antibodies, aptamers, and peptides) to target and specifically attach to cell surface receptors [74,75], which are widely used in in vivo imaging and therapy. However, the effectiveness of these strategies is severely limited due to the insufficient number of target biomolecules of the targeting probes on the cell surface and the nonspecific binding and uptake of imaging probes [76]. These may reduce the specificity and sensitivity of this bio-conjugate. In order to break the inherent limitations of this natural receptor molecular labeling strategy, a molecular labeling strategy combining metabolic labeling technology and bioorthogonal technology has been developed [77]. Metabolic labeling is a labeling method that uses the endogenous synthesis and modification mechanism of living cells to artificially introduce specific affinity tags into biomolecules. A bioorthogonal reaction is a coupling reaction that occurs in a biological system without affecting its inherent biological functions and processes [78]. The combination of bioorthogonal click reaction and metabolic labeling has become a highly specific and sensitive molecular labeling strategy.

The first active targeting strategy relies on natural receptors on the cell surface. However, in the second strategy, regardless of the expression of receptors on the cell surface, artificial receptors based on metabolic markers can be expressed in large numbers on cells, such as the absorption of certain sugars or amino acids by bacterial and cancer cells compared to normal cells with greatly increased utilization, functional groups modified on these sugars or amino acids could be specifically expressed on the surface of bacterial and cancer cells as an artificial receptor [79]. The noncopper-catalyzed bioorthogonal click reaction just provided an opportunity for a probe or drug to covalently link these receptors, thereby realizing high-contrast imaging in vivo and more precise therapy.

Bacterial infection is a serious threat to human health, and it is often difficult to quickly and effectively detect bacteria in vivo. Most clinical tests of bacteria are still based on invasive diagnostic methods such as conventional tissue biopsies and cultures, which are not only tedious and time consuming, but also lack real-time spatiotemporal information.

Therefore, it is of great significance to diagnose live bacterial infectious diseases through noninvasive examination. Based on the in vivo bioorthogonal click chemical system, Liu et al. successfully completed metabolic labeling and targeted imaging of bacteria in vivo [80]. To label the bacteria in vivo, the researchers chose a metabolic label that could be used by bacteria, 3-azido-D-alanine (D-AzAla), which could be selectively integrated into the cell wall of the bacteria and expressed N_3 at the same time. Considering the process of imaging using in vivo bioorthogonal reactions, the most critical and difficult thing was to carry out specific metabolic labeling at the lesion site. The researchers used two carriers to complete the targeted delivery of the metabolic label at the bacterial infection site. One was to use nanoscale metal organic framework (MOF) MIL-100 to load a large amount of hydrophilic D-AzAla to obtain D-AzAla@MIL-100, so that the hydrophilic D-AzAla was not easily removed from the body quickly, and MIL-100 has the H_2O_2-responsive ability to enhance the release of metabolic label at the infection site of bacteria with overexpression of H_2O_2; the second was to use polymers to wrap D-AzAla@MIL-100 to obtain nanoparticles with good biocompatibility (called D-AzAla@MIL-100 NPs), and used its passive targeting ability toward high vascular permeability inflammation sites to complete the enrichment of the metabolic label (Fig. 10A). As a verification of the proposed strategy of bacterial diagnosis, the researchers used methicillin-resistant *Staphylococcus aureus* (MRSA) to construct a mouse bacterial infection model. The fluorescent signal of DBCO-Cy5 confirmed that D-AzAla@MIL-100 NPs could be found in the MRSA infection area (Fig. 10B). Specific aggregation and release of D-AzAla complete the metabolic labeling in the bacteria. As shown in Fig. 10C, AIE NPs-DBCO exhibited more intense fluorescence signals with longer half-life as compared to those obtained from the mice treated with Nano-Only, confirming that this strategy of relying on in vivo bioorthogonal reactions has the ability to diagnose bacterial infection in vivo.

In addition to exogenous pathogenic bacteria, endogenous cancer cells can also be metabolically labeled. Unlike bacterial metabolism label, the cancer cell metabolism label used *N*-acyl-D-mannosamines, which are specific precursors in the synthesis of *N*-acetylneuraminic acid that were highly expressed on the cancer cell membrane. It could be taken up more by cancer cells, efficiently metabolized to the respective *N*-acyl-modified neuraminic acids in vitro and in vivo. When the azide group was modified, ManNAz was obtained, and researchers could use the neuraminic acid biosynthetic pathway to mark the surface of cancer cells with unnatural azide groups. Based on this, Tang et al. developed an NIR emissive AIE dots and modified the cyclooctyne on its surface (DBCO-AIE dots), which could be orthogonal to azide in vivo, to complete tumor metabolic imaging [81] (Fig. 11A). First, the researchers used tetraacetylated ManNAz (Ac4ManNAz) pretreated MCF-7 human breast cancer cells to label MCF-7 cancer cells and completed the in vitro metabolic label visualization through the bioorthogonalization of DBCO-AIE dots on the cell membrane surface (Fig. 11B). As shown in Fig. 11C, compared with mice that were not pretreated with Ac4ManNAz, the fluorescence of the tumor area in the therapy group was significantly increased after 12 h, indicating that Ac4ManNAz had completed the specific tumor labeling in vivo, and the accumulation of AIE-dots at tumor sites could be improved by the in vivo bioorthogonal copper-free click chemistry; so, this strategy could achieve tumor-specific rapid imaging and real-time monitoring.

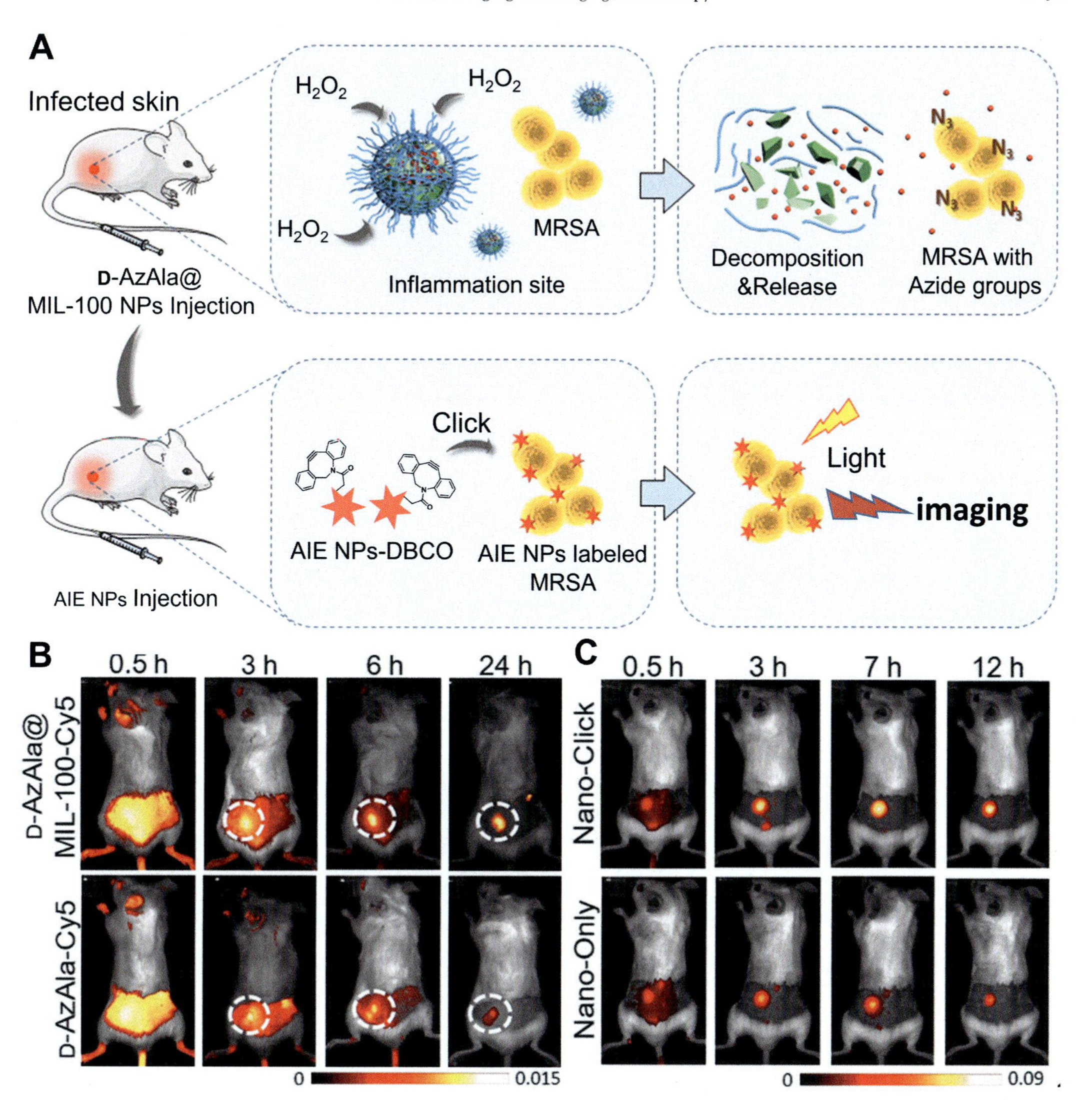

FIG. 10 (A) Schematic illustration of bacterial diagnosis by the H_2O_2-responsive MOFs assisted in vivo metabolic labeling of bacteria. (B) Sequential fluorescence images of bacteria-bearing mice pretreated with D-AzAla@MIL-100 (Fe) NPs and free D-AzAla. (C) Sequential fluorescence images of bacteria-bearing mice pretreated with D-AzAla@MIL-100 (Fe) NPs (Nano-Click group) or saline (Nano-Only group), respectively, followed by intravenous injection of AIE NPs. *Reproduced with permission from reference D. Mao, F. Hu, S. Ji, W. Wu, D. Ding, D. Kong, et al., Metal–organic-framework-assisted in vivo bacterial metabolic labeling and precise antibacterial therapy, Adv. Mater. 30 (2018) 1706831. Copyright 2018, Wiley-VCH.*

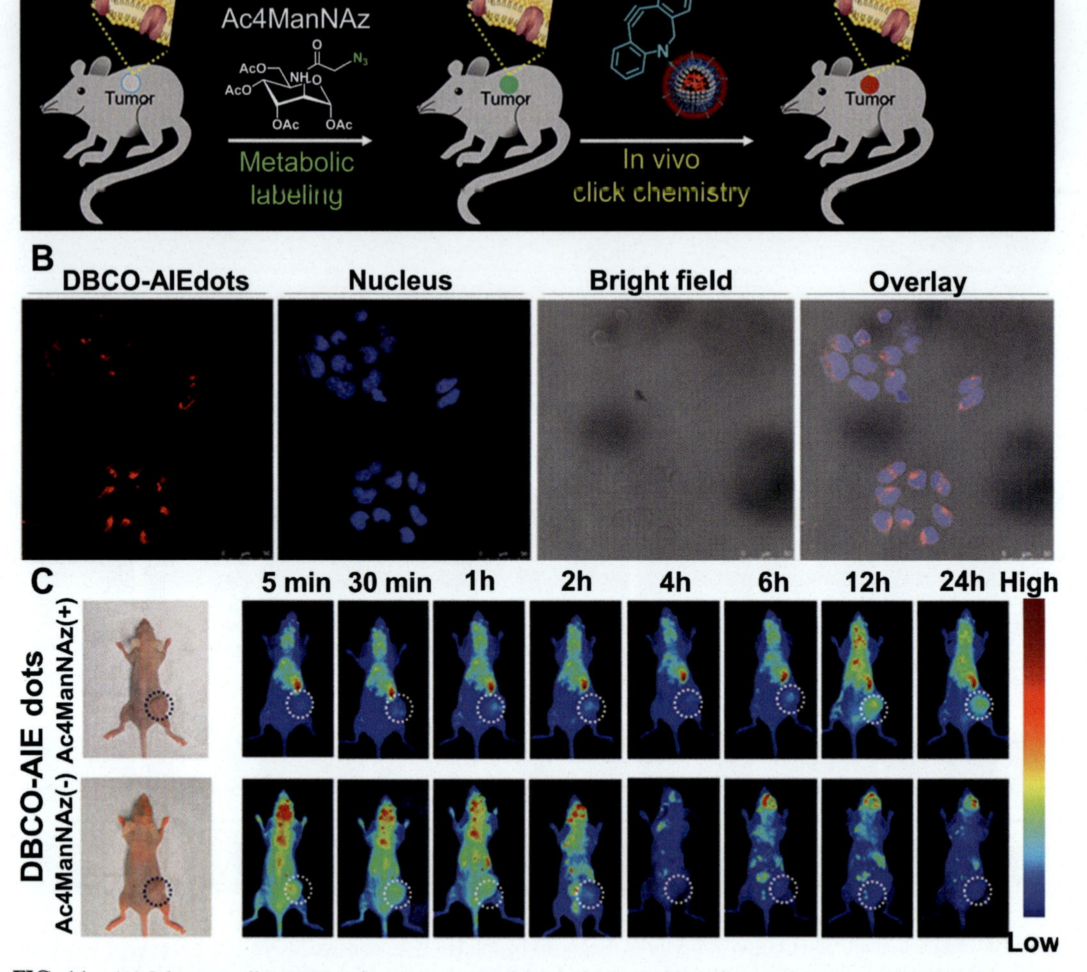

FIG. 11 (A) Schematic illustration of tumor imaging based on in vivo bioorthogonal reaction. (B) CLSM images of AzAcSA pretreated MCF-7 cells upon incubation with DBCO-AIE dots. (C) Time-dependent fluorescence images of in vivo tumor imaging with DBCO-AIE dots after therapy of Ac4ManNAz or not. *Reproduced with permission from reference J.S. Ni, P. Zhang, T. Jiang, Y. Chen, H. Su, D. Wang, et al., Red/NIR-emissive benzo [d] imidazole-cored AIEgens: facile molecular design for wavelength extending and in vivo tumor metabolic imaging, Adv. Mater. 30 (2018) 1805220. Copyright 2018, Wiley-VCH.*

5.2 Image-guided therapy

Facing the global crisis of antibiotic resistance, there is an urgent need for a nonantibiotic tool for image-guided antibacterial therapy. Fortunately, phages have the ability to specifically target and kill their hosts, and can evolve simultaneously to adapt to infection with resistant bacteria. Due to the insufficient ability of phages to treat bacterial infectious diseases by killing pathogenic bacteria in vivo, He et al. used a simple amino-carboxyl reaction to conjugate photodynamic inactivation (PDI)-active AIEgens to phage (PAP) to generate TVP-PAP. TVP-PAP not only retained the ability of phage targeting but also combined the AIEgen's capabilities in imaging and therapy [4] (Fig. 12A). It has thus been successfully used in discriminative imaging and synergistic killing of bacteria. First, the researchers evaluated the targeting and killing ability of AIE bio-conjugate TVP-PAP against specific bacteria in vitro. As shown in Fig. 12B, all *Pseudomonas aeruginosa* were successfully stained within 30min, and the cocultured *Acinetobacter baumannii* (the bacterium indicated by the white arrow) was not stained. After light irradiation, *P. aeruginosa* was effectively killed, but *A. baumannii* was not affected (Fig. 12C). Finally, the researchers evaluated the effect and efficiency of AIE bio-conjugate TVP-PAP in the in vivo synergy against multidrug-resistant bacterial infections. As shown in Fig. 12D and E, after therapy for 8 consecutive days, AIE bio-conjugates TVP-PAP and MDR *P. aeruginosa*-infected wounds in the light therapy group demonstrated significantly faster wound healing, and AIE bio-conjugates successfully achieved the synergistic killing of multidrug resistant bacteria in vivo.

Liu et al. designed a bioorthogonal probe based on AIE bio-conjugates, which can complete the in situ imaging of the tumor by turning on the specific fluorescence of the tumor site [82]. Because this method has excellent targeting properties, it also has excellent effects in image-guided photodynamic therapy. As shown in Fig. 13A, AzAcSA was a metabolic label of cancer cells. The azide group was modified on the cell membrane surface through metabolic engineering, and then BCN-TPET-TEG was bioorthogonally labeled on cancer cells with azide. To verify the entire design process in vitro, BCN-TPET-TEG, which exhibited weak fluorescence in aqueous medium, was added to breast cancer cells modified by metabolic engineering methods, and its fluorescence was turned on at the surface of cancer cells through a bioorthogonal reaction (Fig. 13B). The entire design process was verified in vivo. Compared with mice that were not pretreated with AzAcSA, the fluorescence of the tumor area in the therapy group was significantly enhanced over time, indicating that AzAcSA has completed the specific tumor labeling in vivo, and the BCN-TPET-TEG could be selectively turned on at the tumor site (Fig. 13C). Finally, the researchers proved that this system can inhibit tumor growth in photodynamic therapy (Fig. 13D).

6 Conclusions and perspectives

In this chapter, the biomedical applications of AIE bio-conjugates in biosensing, in vitro and in vivo imaging, and image-guided therapy have been summarized and discussed. The AIE bio-conjugate not only retained the turn-on fluorescence of the AIE probe at high concentrations, but also retained the targeting function of the biomolecule. At the same time, the two parts had complementary advantages. The biomolecule increased the dispersion of

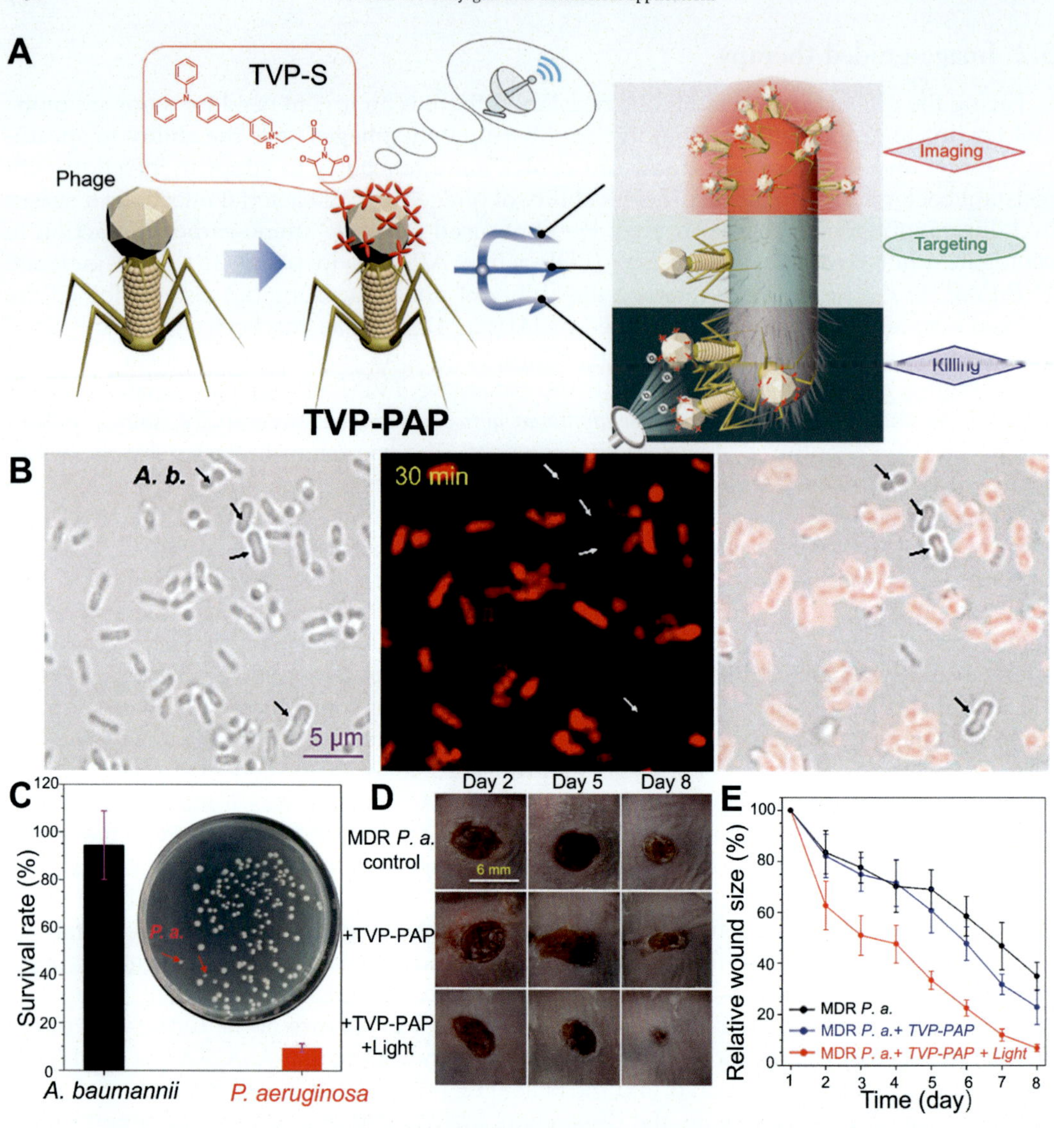

FIG. 12 (A) Schematic diagram of the bacterial identification, real-time fluorescence tracking and collaborative bacterial killing process involved in AIE bio-conjugate TVP-PAP. (B) Specificity test of TVP-PAP by fluorescence imaging of *P. aeruginosa* and *A. baumannii* coincubated with TVP-PAP. Arrows indicate *A. baumannii* without staining by TVP-PAP. (C) The survival rate of *P. aeruginosa* and *A. baumannii* incubated together with TVP-PAP. (D) Photograph of wound healing caused by multidrug resistant (MDR) bacteria after TVP-PAP and white light irradiation. (E) After 8 days of therapy, the effect of synergistic therapy was evaluated by analyzing the relative wound size of MDR bacterial infections. *Reproduced with permission from reference X. He, Y. Yang, Y. Guo, S. Lu, Y. Du, J.J. Li, et al., Phage-guided targeting, discriminative imaging, and synergistic killing of bacteria by AIE bioconjugates, J. Am. Chem. Soc. 142 (2020) 3959–3969. Copyright 2020, American Chemical Society.*

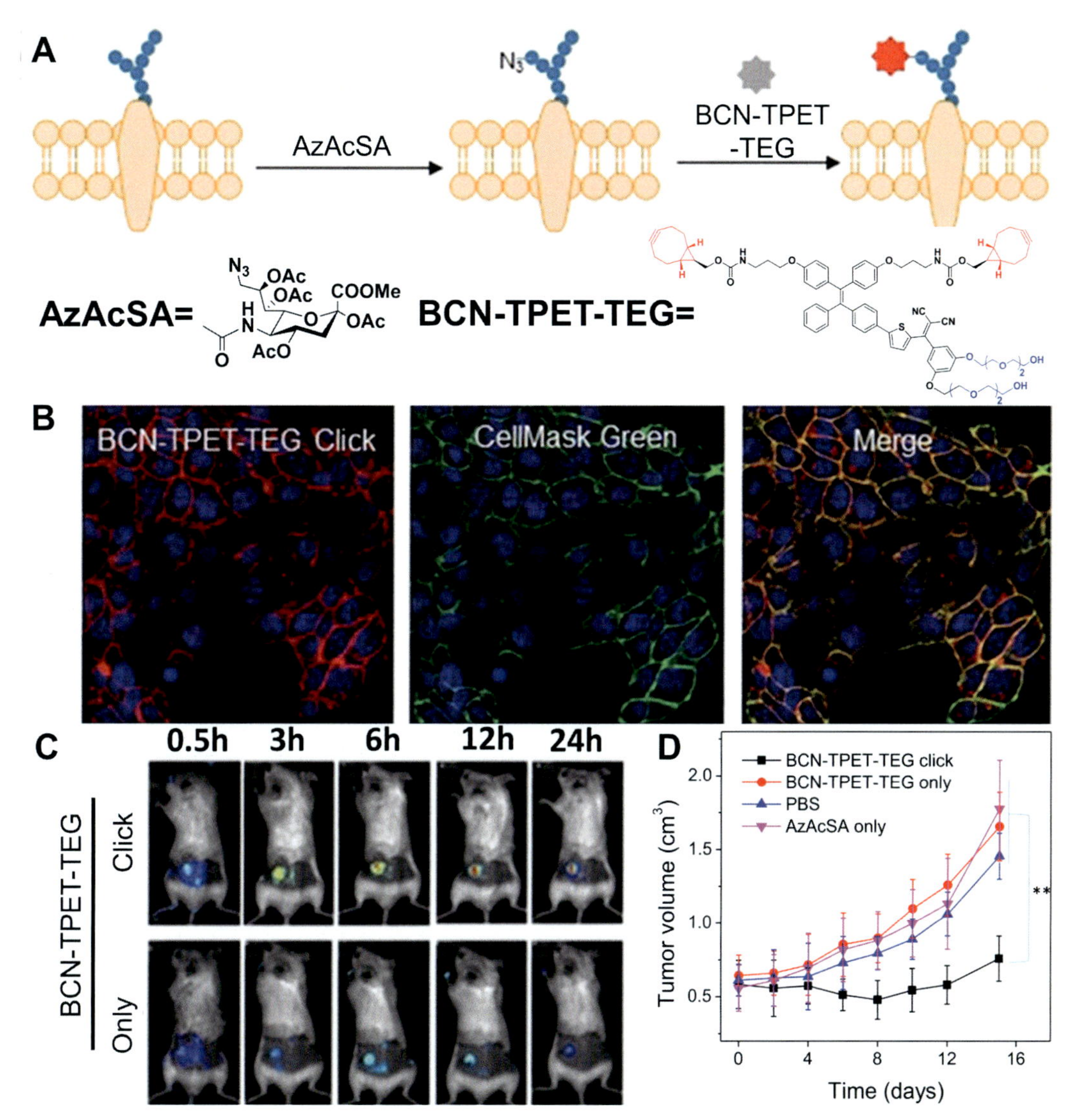

FIG. 13 (A) A schematic diagram of the expression of azide in the membrane of cancer cells and the turn-on of fluorescence at the tumor site. The chemical structure of the metabolic label AzAcSA and the AIE probe BCN-TPET-TEG. (B) CLSM images of in vitro metabolic labeling for the 4T1 cells by copper-free click chemistry after pretherapy of AzAcSA. (C) Time-dependent fluorescence images of in vivo tumor targeting imaging for metabolically labeled mouse injected with BCN-TPET-TEG, whereas the mouse without AzAcSA therapy was used as control. (D) Changes in tumor volume over time in different therapy groups. *Reproduced with permission from reference F. Hu, D. Mao, X. Cai, W. Wu, D. Kong, B. Liu, A light-up probe with aggregation-induced emission for real-time bio-orthogonal tumor labeling and image-guided photodynamic therapy, Angew. Chem. 130 (2018) 10339–10343. Copyright 2018, Wiley-VCH.*

the AIE molecule in water. To the extent that there was almost no background signal, it was conducive to high-contrast imaging. In addition, an emerging metabolic engineering combined with bioorthogonal reaction method introduced artificial receptors through metabolic engineering to solve the shortcomings of insufficient and uneven expression of natural receptors on the cell surface. The bioorthogonal reaction only recognized artificial receptors and did not chemically react with other endogenous substances, which was conducive to accurate imaging and therapy. Future work of AIE bio-conjugates for biomedical applications will mainly focus on the development of more stimulus-responsive AIEgens and AIE dots to result in controllable and targeted conjugation with biomolecules at the local site. In addition, we could also focus on designing AIE bio-conjugates with more multiple modality/functionality. We believe that AIE bio-conjugates will become a powerful tool for solving major problems in life sciences.

References

[1] M. Kwak, A. Herrmann, Nucleic acid amphiphiles: synthesis and self-assembled nanostructures, Chem. Soc. Rev. 40 (2011) 5745–5755.

[2] F.A. Mann, N. Herrmann, F. Opazo, S. Kruss, Quantum defects as a toolbox for the covalent functionalization of carbon nanotubes with peptides and proteins, Angew. Chem. Int. Ed. 59 (2020) 17732–17738.

[3] U. Sb, Growing prospects of dynamic covalent chemistry in delivery applications, Acc. Chem. Res. 52 (2019) 510–519.

[4] X. He, Y. Yang, Y. Guo, S. Lu, Y. Du, J.-J. Li, et al., Phage-guided targeting, discriminative imaging, and synergistic killing of bacteria by AIE bioconjugates, J. Am. Chem. Soc. 142 (2020) 3959–3969.

[5] C. Li, G. Chen, Y. Zhang, F. Wu, Q. Wang, Advanced fluorescence imaging technology in the near-infrared-ii window for biomedical applications, J. Am. Chem. Soc. 142 (2020) 14789–14804.

[6] V.-N. Nguyen, A. Kumar, M.H. Lee, J. Yoon, Recent advances in biomedical applications of organic fluorescence materials with reduced singlet–triplet energy gaps, Coord. Chem. Rev. 425 (2020), 213545.

[7] J. Qi, X. Hu, X. Dong, Y. Lu, H. Lu, W. Zhao, et al., Towards more accurate bioimaging of drug nanocarriers: turning aggregation-caused quenching into a useful tool, Adv. Drug Deliv. Rev. 143 (2019) 206–225.

[8] J. Li, Y. Zhang, P. Wang, L. Yu, J. An, G. Deng, et al., Reactive oxygen species, thiols and enzymes activable AIEgens from single fluorescence imaging to multifunctional theranostics, Coord. Chem. Rev. 427 (2021), 213559.

[9] Y. Zhao, X. Zhao, M.D. Li, L. Za, H. Peng, X. Xie, Crosstalk-free patterning of cooperative-thermoresponse images by the synergy of the AIEgen with the liquid crystal, Angew. Chem. Int. Ed. 59 (2020) 10066–10072.

[10] H. Liu, L.H. Xiong, R.T.K. Kwok, X. He, J.W.Y. Lam, B.Z. Tang, AIE bioconjugates for biomedical applications, Adv. Opt. Mater. 8 (2020) 2000162.

[11] X. Min, M. Zhang, F. Huang, X. Lou, F. Xia, Live cell microRNA imaging using exonuclease III-aided recycling amplification based on aggregation-induced emission luminogens, ACS Appl. Mater. Interfaces 8 (2016) 8998–9003.

[12] X. Lou, Y. Song, R. Liu, Y. Cheng, J. Dai, Q. Chen, et al., Enzyme and AIEgens modulated solid-state nanochannels: in situ and noninvasive monitoring of H2O2 released from living cells, Small Methods 4 (2020) 1900432.

[13] B. He, B. Situ, Z. Zhao, L. Zheng, Promising applications of AIEgens in animal models, Small Methods 4 (2020) 1900583.

[14] F. Wu, X. Wu, Z. Duan, Y. Huang, X. Lou, F. Xia, Biomacromolecule-functionalized AIEgens for advanced biomedical studies, Small 15 (2019) 1804839.

[15] C. Liu, X. Wang, J. Liu, Q. Yue, S. Chen, J.W. Lam, et al., Near-infrared AIE dots with chemiluminescence for deep-tissue imaging, Adv. Mater. 32 (2020) 2004685.

[16] X. Chen, H. Gao, Y. Deng, Q. Jin, J. Ji, D. Ding, Supramolecular aggregation-induced emission nanodots with programmed tumor microenvironment responsiveness for image-guided orthotopic pancreatic cancer therapy, ACS Nano 14 (2020) 5121–5134.
[17] Z. Yang, D. Tang, J. Hu, M. Tang, M. Zhang, H.L. Cui, et al., Near-field nanoscopic terahertz imaging of single proteins, Small (2020) 2005814.
[18] J. Liu, T. Liu, P. Du, L. Zhang, J. Lei, Metal–organic framework (MOF) hybrid as a tandem catalyst for enhanced therapy against hypoxic tumor cells, Angew. Chem. 131 (2019) 7890–7894.
[19] F. Cleeren, J. Lecina, J. Bridoux, N. Devoogdt, T. Tshibangu, C. Xavier, et al., Direct fluorine-18 labeling of heat-sensitive biomolecules for positron emission tomography imaging using the Al 18 F-RESCA method, Nat. Protoc. 13 (2018) 2330–2347.
[20] H. Huang, K. Jia, Y. Chen, Hypervalent iodine reagents enable chemoselective deboronative/decarboxylative alkenylation by photoredox catalysis, Angew. Chem. 127 (2015) 1901–1904.
[21] Y. Yu, J. Liu, Z. Zhao, K.M. Ng, K.Q. Luo, B.Z. Tang, Facile preparation of non-self-quenching fluorescent DNA strands with the degree of labeling up to the theoretic limit, Chem. Commun. 48 (2012) 6360–6362.
[22] J.P. António, R. Russo, C.P. Carvalho, P.M. Cal, P.M. Gois, Boronic acids as building blocks for the construction of therapeutically useful bioconjugates, Chem. Soc. Rev. 48 (2019) 3513–3536.
[23] G. Qi, F. Hu, L. Shi, M. Wu, B. Liu, An AIEgen-peptide conjugate as a phototheranostic agent for phagosome-entrapped bacteria, Angew. Chem. 131 (2019) 16375–16381.
[24] Y. Yuan, S. Xu, X. Cheng, X. Cai, B. Liu, Bioorthogonal turn-on probe based on aggregation-induced emission characteristics for cancer cell imaging and ablation, Angew. Chem. Int. Ed. 55 (2016) 6457–6461.
[25] H. Li, H. Lin, W. Lv, P. Gai, F. Li, Equipment-free and visual detection of multiple biomarkers via an aggregation induced emission luminogen-based paper biosensor, Biosens. Bioelectron. 165 (2020), 112336.
[26] X. Chen, R. Hu, C. Qi, X. Fu, J. Wang, B. He, et al., Ethynylsulfone-based spontaneous amino-yne click polymerization: a facile tool toward regio-and stereoregular dynamic polymers, Macromolecules 52 (2019) 4526–4533.
[27] Y. Shi, T. Bai, W. Bai, Z. Wang, M. Chen, B. Yao, et al., Phenolyne click polymerization: an efficient technique to facilely access regio-and stereoregular poly (vinylene ether ketone) s, Chem Eur J 23 (2017) 10725–10731.
[28] X. Hu, X. Zhao, B. He, Z. Zhao, Z. Zheng, P. Zhang, et al., A simple approach to bioconjugation at diverse levels: metal-free click reactions of activated alkynes with native groups of biotargets without prefunctionalization, Research 2018 (2018).
[29] Z. Yang, W. Yin, S. Zhang, I. Shah, B. Zhang, S. Zhang, et al., Synthesis of AIE-active materials with their applications for antibacterial activity, specific imaging of mitochondrion and image-guided photodynamic therapy, ACS Appl. Bio Mater. 3 (2020) 1187–1196.
[30] C. Boyer, N.A. Corrigan, K. Jung, D. Nguyen, T.-K. Nguyen, N.N.M. Adnan, et al., Copper-mediated living radical polymerization (atom transfer radical polymerization and copper (0) mediated polymerization): from fundamentals to bioapplications, Chem. Rev. 116 (2016) 1803–1949.
[31] F. Li, D. Lyu, S. Liu, W. Guo, DNA hydrogels and microgels for biosensing and biomedical applications, Adv. Mater. 32 (2020) 1806538.
[32] R.M. Kong, X.B. Zhang, Z. Chen, W. Tan, Aptamer-assembled nanomaterials for biosensing and biomedical applications, Small 7 (2011) 2428–2436.
[33] Y. Lu, Y. Yao, Q. Zhang, D. Zhang, S. Zhuang, H. Li, et al., Olfactory biosensor for insect semiochemicals analysis by impedance sensing of odorant-binding proteins on interdigitated electrodes, Biosens. Bioelectron. 67 (2015) 662–669.
[34] J. Cao, W. Wang, B. Bo, X. Mao, K. Wang, X. Zhu, A dual-signal strategy for the solid detection of both small molecules and proteins based on magnetic separation and highly fluorescent copper nanoclusters, Biosens. Bioelectron. 90 (2017) 534–541.
[35] R. Weissleder, K. Kelly, E.Y. Sun, T. Shtatland, L. Josephson, Cell-specific targeting of nanoparticles by multivalent attachment of small molecules, Nat. Biotechnol. 23 (2005) 1418–1423.
[36] P. Zweigenbaum, D. Demner-Fushman, H. Yu, K.B. Cohen, Frontiers of biomedical text mining: current progress, Brief. Bioinform. 8 (2007) 358–375.
[37] N. Dave, M.Y. Chan, P.-J.J. Huang, B.D. Smith, J. Liu, Regenerable DNA-functionalized hydrogels for ultrasensitive, instrument-free mercury (II) detection and removal in water, J. Am. Chem. Soc. 132 (2010) 12668–12673.

[38] R.S. Wijesurendra, M. Ginks, K. Rajappan, Irregular pulse in a patient with a cardiac resynchronization defibrillator, Circulation 138 (2018) 2434–2436.

[39] X. Ou, X. Lou, F. Xia, A highly sensitive DNA-AIEgen-based "turn-on" fluorescence chemosensor for amplification analysis of Hg^{2+} ions in real samples and living cells, Sci. China Chem. 60 (2017) 663–669.

[40] J. Das, I. Ivanov, T.S. Safaei, E.H. Sargent, S.O. Kelley, Combinatorial probes for high-throughput electrochemical analysis of circulating nucleic acids in clinical samples, Angew. Chem. 130 (2018) 3773–3778.

[41] M. Zheng, C. Wiraja, D.C. Yeo, H. Chang, D.C.S. Lio, W. Shi, et al., Oligonucleotide molecular sprinkler for intracellular detection and spontaneous regulation of mRNA for theranostics of scar fibroblasts, Small 14 (2018) 1802546.

[42] R. Zhang, R.T. Kwok, B.Z. Tang, B. Liu, Hybridization induced fluorescence turn-on of AIEgen–oligonucleotide conjugates for specific DNA detection, RSC Adv. 5 (2015) 28332–28337.

[43] O. Novikova, P. Jayachandran, D.S. Kelley, Z. Morton, S. Merwin, N.I. Topilina, et al., Intein clustering suggests functional importance in different domains of life, Mol. Biol. Evol. 33 (2016) 783–799.

[44] Z. Wang, P. Zhang, H. Liu, Z. Zhao, L. Xiong, W. He, et al., Robust serum albumin-responsive AIEgen enables latent bloodstain visualization in high resolution and reliability for crime scene investigation, ACS Appl. Mater. Interfaces 11 (2019) 17306–17312.

[45] M. Hao, S.X. Lin, O.J. Karylowski, D. Wüstner, T.E. McGraw, F.R. Maxfield, Vesicular and non-vesicular sterol transport in living cells: the endocytic recycling compartment is a major sterol storage organelle, J. Biol. Chem. 277 (2002) 609–617.

[46] J. Lippincott-Schwartz, T.H. Roberts, K. Hirschberg, Secretory protein trafficking and organelle dynamics in living cells, Annu. Rev. Cell Dev. Biol. 16 (2000) 557–589.

[47] J. Gala de Pablo, M. Lindley, K. Hiramatsu, K. Goda, High-throughput raman flow cytometry and beyond, Acc. Chem. Res. (2021). eaau0241-15848.

[48] M. Fujimoto, T. Naka, Regulation of cytokine signaling by SOCS family molecules, Trends Immunol. 24 (2003) 659–666.

[49] K.L. Smitten, H.M. Southam, J.B. de la Serna, M.R. Gill, P.J. Jarman, C.G. Smythe, et al., Using nanoscopy to probe the biological activity of antimicrobial leads that display potent activity against pathogenic, multidrug resistant, gram-negative bacteria, ACS Nano 13 (2019) 5133–5146.

[50] W. Kukulski, A glycoprotein in urine binds bacteria and blocks infections, Science 369 (2020) 917–918.

[51] J. Deutscher, C. Francke, P.W. Postma, How phosphotransferase system-related protein phosphorylation regulates carbohydrate metabolism in bacteria, Microbiol. Mol. Biol. Rev. 70 (2006) 939–1031.

[52] C. Yu, H. Nakshatri, J. Irudayaraj, Identity profiling of cell surface markers by multiplex gold nanorod probes, Nano Lett. 7 (2007) 2300–2306.

[53] E. Richie, M. Sullivan, J. Van Eys, A unique surface marker profile in T-cell acute lymphocytic leukemia, Blood 55 (4) (1980) 702–705.

[54] L. Pasquina-Lemonche, J. Burns, R. Turner, S. Kumar, R. Tank, N. Mullin, et al., The architecture of the Gram-positive bacterial cell wall, Nature 582 (2020) 294–297.

[55] K.M. Sandoz, R.A. Moore, P.A. Beare, A.V. Patel, R.E. Smith, M. Bern, et al., β-Barrel proteins tether the outer membrane in many Gram-negative bacteria, Nat. Microbiol. 6 (2021) 19–26.

[56] E. Zhao, Y. Hong, S. Chen, C.W. Leung, C.Y. Chan, R.T. Kwok, et al., Highly fluorescent and photostable probe for long-term bacterial viability assay based on aggregation-induced emission, Adv. Healthc. Mater. 3 (2014) 88–96.

[57] T.T. Kong, Z. Zhao, Y. Li, F. Wu, T. Jin, B.Z. Tang, Detecting live bacteria instantly utilizing AIE strategies, J. Mater. Chem. B 6 (2018) 5986–5991.

[58] K. Zhou, A. Dichlberger, H. Martinez-Seara, T.K. Nyholm, S. Li, Y.A. Kim, et al., A ceramide-regulated element in the late endosomal protein LAPTM4B controls amino acid transporter interaction, ACS Cent. Sci. 4 (2018) 548–558.

[59] Y. Huang, F. Hu, R. Zhao, G. Zhang, H. Yang, D. Zhang, Tetraphenylethylene conjugated with a specific peptide as a fluorescence turn-on bioprobe for the highly specific detection and tracing of tumor markers in live cancer cells, Chem. Eur. J. 20 (2014) 158–164.

[60] C. Greening, T. Lithgow, Formation and function of bacterial organelles, Nat. Rev. Microbiol. 18 (2020) 677–689.

[61] S. Redhai, M. Boutros, The role of organelles in intestinal function, physiology, and disease, Trends Cell Biol. 31 (2021) 485–499.

[62] A.M. Valm, S. Cohen, W.R. Legant, J. Melunis, U. Hershberg, E. Wait, et al., Applying systems-level spectral imaging and analysis to reveal the organelle interactome, Nature 546 (2017) 162–167.
[63] T. Al, N. Najafinobar, F. Penen, E. Kay, P.P. Upadhyay, X. Li, et al., Subcellular mass spectrometry imaging and absolute quantitative analysis across organelles, ACS Nano 14 (2020) 4316–4325.
[64] X. Shi, Y. Chris, H. Su, R.T. Kwok, M. Jiang, Z. He, et al., A red-emissive antibody–AIEgen conjugate for turn-on and wash-free imaging of specific cancer cells, Chem. Sci. 8 (2017) 7014–7024.
[65] I. Macwan, M.D.H. Khan, A. Aphale, S. Singh, J. Liu, M. Hingorani, et al., Interactions between avidin and graphene for development of a biosensing platform, Biosens. Bioelectron. 89 (2017) 326–333.
[66] Y. Zhao, Y. Chris, R.T. Kwok, Y. Chen, S. Chen, J.W. Lam, et al., Photostable AIE fluorogens for accurate and sensitive detection of S-phase DNA synthesis and cell proliferation, J. Mater. Chem. B 3 (2015) 4993–4996.
[67] A.M. Embaby, S. Schoffelen, C. Kofoed, M. Meldal, F. Diness, Rational tuning of fluorobenzene probes for cysteine-selective protein modification, Angew. Chem. Int. Ed. 57 (2018) 8022–8026.
[68] E. Branigan, C. Pliotas, G. Hagelueken, J.H. Naismith, Quantification of free cysteines in membrane and soluble proteins using a fluorescent dye and thermal unfolding, Nat. Protoc. 8 (2013) 2090–2097.
[69] M.Z. Chen, N.S. Moily, J.L. Bridgford, R.J. Wood, M. Radwan, T.A. Smith, et al., A thiol probe for measuring unfolded protein load and proteostasis in cells, Nat. Commun. 8 (2017) 1–11.
[70] B.R. Smith, S.S. Gambhir, Nanomaterials for in vivo imaging, Chem. Rev. 117 (2017) 901–986.
[71] A. Dilipkumar, A. Al-Shemmary, L. Kreiß, K. Cvecek, B. Carlé, F. Knieling, et al., Label-free multiphoton endomicroscopy for minimally invasive in vivo imaging, Adv. Sci. 6 (2019) 1801735.
[72] W. Qin, N. Alifu, J.W. Lam, Y. Cui, H. Su, G. Liang, et al., Facile synthesis of efficient luminogens with AIE features for three-photon fluorescence imaging of the brain through the intact skull, Adv. Mater. 32 (2020) 2000364.
[73] Y.F. Xiao, F.F. An, J.X. Chen, J. Yu, W.W. Tao, Z. Yu, et al., The nanoassembly of an intrinsically cytotoxic near-infrared dye for multifunctionally synergistic theranostics, Small 15 (2019) 1903121.
[74] B. Li, C. Liu, W. Pan, J. Shen, J. Guo, T. Luo, et al., Facile fluorescent aptasensor using aggregation-induced emission luminogens for exosomal proteins profiling towards liquid biopsy, Biosens. Bioelectron. 168 (2020), 112520.
[75] A.S. Klymchenko, Solvatochromic and fluorogenic dyes as environment-sensitive probes: design and biological applications, Acc. Chem. Res. 50 (2017) 366–375.
[76] S. Luo, E. Zhang, Y. Su, T. Cheng, C. Shi, A review of NIR dyes in cancer targeting and imaging, Biomaterials 32 (2011) 7127–7138.
[77] M. Duo, F. Hu, G. Qi, S. Ji, W. Wu, D. Kong, et al., One-step: in vivo metabolic labeling as a theranostic approach for overcoming drug-resistant bacterial infections, Mater. Horiz. 7 (2020) 1138–1143.
[78] M. Sundhoro, S. Jeon, J. Park, O. Ramström, M. Yan, Perfluoroaryl azide Staudinger reaction: a fast and bioorthogonal reaction, Angew. Chem. Int. Ed. 56 (2017) 12117–12121.
[79] L. Zhang, F. Wang, Q. Li, L. Wang, C. Fan, J. Li, et al., Classifying cell types with DNA-encoded ligand-receptor interactions on the cell membrane, Nano Lett. 20 (2020) 3521–3527.
[80] D. Mao, F. Hu, S. Ji, W. Wu, D. Ding, D. Kong, et al., Metal–organic-framework-assisted in vivo bacterial metabolic labeling and precise antibacterial therapy, Adv. Mater. 30 (2018) 1706831.
[81] J.S. Ni, P. Zhang, T. Jiang, Y. Chen, H. Su, D. Wang, et al., Red/NIR-emissive benzo [d] imidazole-cored AIEgens: facile molecular design for wavelength extending and in vivo tumor metabolic imaging, Adv. Mater. 30 (2018) 1805220.
[82] F. Hu, D. Mao, X. Cai, W. Wu, D. Kong, B. Liu, A light-up probe with aggregation-induced emission for real-time bio-orthogonal tumor labeling and image-guided photodynamic therapy, Angew. Chem. 130 (2018) 10339–10343.

CHAPTER

17

AIE-active polymers for explosive detection

Hui Zhou, Ming Hui Chua, Qiang Zhu, and Jianwei Xu

Institute of Materials Research and Engineering, A*STAR (Agency for Science, Technology and Research), Singapore, Singapore

1 Introduction of explosive detection

Explosive detection is of paramount importance for antiterrorism, homeland security, landmine detection, forensic research, and explosives monitoring caused environmental pollution [1]. Notably, most explosives are toxic, which can cause some health problems even on exposure to explosives for a short time, e.g., skin irritation, abnormal liver function, and carcinogenicity. For instance, the United States Environmental Protection Agency (US EPA) has categorized explosive 2,4,6-trinitrotoluene (TNT) as a toxic pollutant in a high concentration level (>2 ppm) [2]. Therefore, fast and accurate identification of explosives is necessary for both health and safety reasons.

Conventionally, explosive detection is performed via instrumental analysis. These include chromatography [3], mass spectrometry [4], Raman spectroscopy [5], X-ray imaging [6], thermal neutron analysis [7], electrochemical assay [8], and ion mobility spectroscopy [9]. However, these analytical methods are costly to operate and inconvenient for on-site detection. Thus, optical sensors, including fluorescence-sensing technology, may provide an alternative solution because of their high sensitivity, fast response time, portability, and low-cost [10]. Several review papers have documented explosive chemosensors based on traditional fluorescence materials from different aspects, e.g., polymer sensors [11], optical chemosensors [12], fluorescent materials-containing polymers [13], frameworks [14], and nanomaterials [15].

There are several classes of explosives. One of the most prevalent classes of explosives is organic nitro-aromatics including the infamous 2,4,6-trinitrotoluene (TNT). Organic nitro-aromatics may undergo highly exothermic redox decomposition to produce nitrogen gas, carbon dioxide, and water. Furthermore, organic nitro-aromatics often have a certain vapor

Aggregation-Induced Emission (AIE)
https://doi.org/10.1016/B978-0-12-824335-0.00013-1

TABLE 1 Chemical structures, names, acronyms, and vapor pressure of nitro-aromatics cited in the chapter.

Class	Name	Acronym	Structure	P_{vap} (Torr)[a]
1.	2-Nitrotoluene; 3-nitrotoluene; 4-nitrotoluene	2-NT; 3-NT; 4-NT		1.44×10^{-2}; 0.107; 4.89×10^{-2}
2.	2,4-Dinitrotoluene; 2,6-dinitrotoluene	2,4-DNT; 2,6-DNT		2.63×10^{-4}; 6.20×10^{-4}
3.	2,4,6-Trinitrotoluene	TNT		5.50×10^{-6}
4.	2,4,6-Trinitrophenol (picric acid)	TNP (PA)		5.80×10^{-9}
5.	*o*-, *m*-, *p*-Dinitrobenzene; 1,3,5-trinitrobenzene	*o*-DNB; *m*-DNB; *p*-DNB; TNB		4.55×10^{-5}; 2.00×10^{-4}; 5.20×10^{-4}; 9.00×10^{-4}

[a] P_{vap}: *vapor pressure at 25°C.*

pressure, which enables them to exist in the gaseous state at mild temperature. The volatility of nitro-aromatics makes them even more dangerous, and harder to be detected by conventional instrumental analysis methods. Table 1 lists chemical structures and vapor pressures of common organic nitroaromatics [16].

Traditional fluorophores suffer from emission quenching in the concentrated or solid state, which is named as aggregation caused quenching (ACQ) and was reported by Förster in 1954 [17]. In 2001, Tang et al. discovered an opposite phenomenon of ACQ, in which fluorophores that are nonemissive in a molecularly dissolved solution appear emissive in the aggregated state. This phenomenon is then termed as aggregation-induced emission (AIE) [18]. Over past 20 years, a library of AIE fluorophores whose fluorescence emission covers the whole electromagnetic region from visible to near-infrared light have been developed, showing valuable potentials for optoelectronics application, chemo-sensing, and bioimaging [19]. For chemical sensing, AIE has been adopted in the development of various fluorescence chemosensors to detect a wide range of chemical analytes, one of which includes explosive compounds. Recently, reports of traditional fluorescence sensors for explosive detection have been summarized in several review articles [12–16], but there are limited reviews particularly focused on AIE materials for explosive detection. Herein, we aim to summarize the fundamentals of explosive detection including the AIE quenching mechanism and the latest progress on AIE-active polymers for explosive detection, especially with the emphasis on the structure-performance relationship of AIE-active polymers. This chapter covers a broad range of AIE-active polymers, including not only conjugated polymers, but also nonconjugated linear and hyper-branched polymers. At the end of the chapter, we also comment on the challenges and outlook of AIE-active polymers for explosive detection.

2 Mechanisms of explosive detection

2.1 Fluorescence quenching theory

Fluorescence quenching is generally classified into two quenching models, static and dynamic quenching, in terms of whether fluorophores and quenchers form non-fluorescence complexes in the ground state or induce energy exchange between them after fluorophore is excited at the molecular level. The quenching model can be identified via measuring time-resolved fluorescence decays of the fluorophore with or without quencher, in which the ratios of fluorescence lifetime, τ_0/τ (y-axis), are plotted versus the quencher concentration (x-axis). In a static quenching model, partial fluorophores and quenchers form nonfluorescence complexes, while the other free fluorophores are still emissive with their native fluorescence decay lifetime, and thus the fluorescence decay lifetime of fluorophore will remain unchanged even though more quenchers are added.

In a dynamic quenching model, quenching occurs when quenchers collide with exciting fluorophores without binding to each other and also reduces the average fluorescence lifetime of all fluorophores. Fluorescence dynamic quenching is illustrated by the Stern-Volmer equation (Eq. 1) using a correlation of lifetime versus quencher concentration ([Q]). Linear Stern-Volmer plots can indicate both static and dynamic quenching with different proportional constants, e.g., binding constant and kinetic constant.

$$\frac{F_0}{F} = 1 + k_q \tau_0 [Q] = 1 + K_D [Q] \tag{1}$$

$$\frac{F_0}{F} = \frac{\tau_0}{\tau} \tag{2}$$

τ_0 = fluorophore lifetimes without quencher.
τ = fluorophore lifetimes with quencher.
F_0 = fluorescence intensities without quenchers.
F = fluorescence intensities with quenchers.
k_q = bimolecular quenching constant.
K_D = Stern-Volmer quenching constant.
$[Q]$ = quencher concentration.

In a static quenching model, fluorescence intensity can be related to quencher concentration through the Stern-Volmer constant (K_{SV}) of the static quenching process as shown in Eq. (3).

$$\frac{F_0}{F} = 1 + K_{SV}[Q] \tag{3}$$

$$\frac{F_0}{F} = (1 + K_D[Q])(1 + K_S[Q]) = 1 + (K_D + K_S)[Q] + K_D K_S [Q]^2 \tag{4}$$

If the fluorescence quenching process contains both dynamic and static models, Eq. (3) will transform into Eq. (4). Eq. (4) will show a linear plot when the quencher concentration is very low due to the less prominent contribution of $[Q]^2$. While in the case of a high quencher concentration, the plot of Eq. (4) will bend upwardly. Generally, when the fluorescence is quenched by both static and dynamic processes, the quencher will bind with many fluorophores. The static quenching process always shows higher sensitivity than the dynamic quenching process because of its larger K_{SV}. In contrast, the dynamic process has a much smaller K_{SV} and exhibits a lower sensitivity, but the dynamic process generally responds faster.

In addition, static and dynamic fluorescence quenching would be affected by temperature and viscosity. Higher temperatures can accelerate the diffusion of quenchers, facilitating dynamic fluorescence quenching. In contrast, higher temperature promotes the dissociation of weakly bound complexes, leading to fewer nonfluorescent fluorophore-quencher complexes in the static quenching process. Ultraviolet-visible (UV-vis) absorption of fluorophores can be used to identify the static and dynamic quenching since dynamic quenching only occurs when the fluorophores are in excited states. Thus, UV-vis absorption spectrum of fluorophores remains unchanged in the dynamic quenching process. In contrast, for the static quenching process, the formed fluorophore-quencher complexes will cause obvious changes in the UV-vis absorption spectra. In addition, the magnitude of constant k_q, which derives from the ratio of K_{SV} and τ_0, can be also used to distinguish the static and dynamic processes. For a dynamic quenching, k_q is around $10^{10} M^{-1} s^{-1}$. In contrast, for the static quenching, the k_q will be always several orders of magnitude larger than that of the dynamic quenching process [20].

2.2 Proposed detection mechanisms

Explosive detection based on fluorescence sensors can identify the presence of explosives through variation in fluorescence spectra or intensity of fluorophores. Due to the high sensitivity of the fluorescence spectroscopy, any slight interference or interaction with fluorophores can lead to a noticeable change in fluorescence properties, such as intensity, wavelength, and fluorescence lifetime. Fig. 1 shows the mechanisms of fluorescence quenching, which are summarized in terms of photoinduced electron transfer (PET, Fig. 1B), Förster resonance energy transfer (FRET, Fig. 1C), and electron exchange (EE, Fig. 1D).

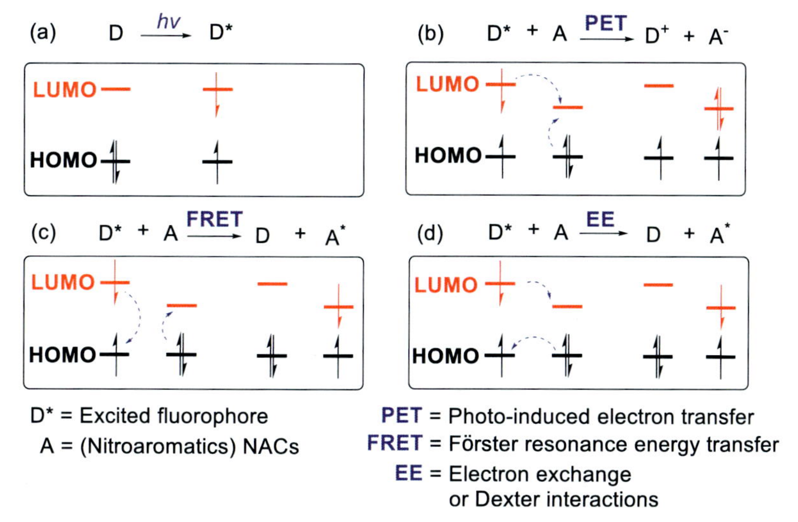

FIG. 1 Schematic illustration of molecular orbitals for three fluorescence quenching mechanisms, including PET, FRET, and EE. *Reproduced with permission from reference H. Zhou, M. H. Chua, J. Xu, Polym. Chem. 10 (2019) 3822–3840. Copyright 2019, the Royal Society of Chemistry.*

2.2.1 PET

Nitroaromatic-type explosive molecules are able to bind with electron-rich species via donor-acceptor interactions, due to their electron-deficient nature [21,22]. In the PET process (Fig. 1B), the photoexcited fluorophore donor donates an electron from its LUMO to the LUMO of the explosive acceptor molecule. The energy gap between LUMO of donor and acceptor molecules is considered the driving force for the PET process. Theoretically, the fluorescence quenching efficiency is determined by the electron transfer rate (K_{et}).

$$K_{et} = A\exp\left(-\frac{\Delta G^*}{kT}\right) = \frac{2\pi^{\frac{3}{2}}}{h\sqrt{\lambda kT}}V^2\exp\left[-\frac{(\Delta G^o + \lambda)^2}{4\lambda kT}\right] \tag{5}$$

$$\lambda = \lambda_i + \lambda_e$$

K_{et} = electron transfer rate
h = Planck constant
k = Boltzmann constant
T = temperature in Kelvin
ΔG^* = Gibbs free energy in PET
ΔG^o = standard Gibbs free energy in PET
V = electron coupling constant between the initial state (D^*A) and the final state (D^+A^-)
λ = energy between different stable states
λ_i = internal part of λ
λ_e = external part of λ

In Marcus theory, the K_{et} is defined by Eq. (5). λ is the energy to distort the product state to adopt the geometry of equilibrium ground state, which is composited by two parts: (i) the internal part (λ_i) is for the geometry changes of donor and acceptor; (ii) the external part (λ_e) is for the changes of surroundings. For the single electron approximation, ΔG^o is originated from the driving force of the electron transfer process through a quantum-chemical calculation. V is the LUMO coupling constant of donor and acceptor and is relevant to their coupling distance and orientation. ΔG^o is the thermodynamic driving force of PET, which is the major mechanism for the fluorescence quenching by nitro-aromatics.

2.2.2 FRET

In FRET, the ground state A is simultaneously excited to its excited state by the energy released from the decay of excited D* to its ground state (Fig. 1C). FRET is not affected by steric factors and electrostatic interactions, because it is originated from long-range dipolar interactions between D* and A.

Typically, energy transfer in FRET is affected by three factors: (i) the distance between D and A; (ii) the orientation of D – A dipoles; and (iii) the extent of spectral overlap between D's fluorescence emission and A's absorption [23]. The efficiency of energy transfer can be defined by Eq. (6). R_0 is the distance between D* and A when the efficiency of energy transfer is 50% (R_0 can be calculated by Eq. 7). J is calculated by Eq. (8), where $F(\lambda)$ is the integrated fluorescence intensity of D in the wavelength range between λ and $\lambda+\Delta\lambda$, and $\varepsilon(\lambda)$ is the molar absorption coefficient of A at wavelength λ [24].

$$E = 1 - \frac{F_0}{F} = \frac{{R_0}^6}{{R_0}^6 + {r_0}^6} \tag{6}$$

$$ {R_0}^6 = 8.79 \times 10^{-5} K^2 K n^{-4} \varnothing J \tag{7}$$

$$J = \frac{\int_0^\infty F(\lambda)\varepsilon(\lambda)\lambda^4 d\lambda}{\int_0^\infty F(\lambda) d\lambda} \tag{8}$$

r_0 = distance between D and A
R_0 = critical distance of D* and A
K^2 = orientation factor of D – A dipoles
n = refractive index of medium
$\varnothing$ = fluorescence quantum yield of D
J = degree of spectral overlap of D's emission and A's absorption

FRET can be used to improve the performance of the explosive detection, mainly through two strategies: (i) designing AIE-active polymers to increase the extent of spectral overlap between the fluorescence of AIE-active polymers and absorption of explosive molecules, thus achieving higher efficiency of FRET and (ii) utilizing electron-deficient property of nitro-aromatics to form complexes with AIE-active polymers, consequently amplifying efficiency of fluorescence quenching.

2.2.3 EE (or Dexter interactions)

In EE, an excited D transfers an electron in the LUMO to an A's LUMO, while at the same time, A transfers an electron from its HOMO to D's HOMO (Fig. 1D) [25]. Thus, the excited state of A is kept. EE only occurs in a special condition with a short distance between D and A, in which spatial overlapping of molecular orbitals between D and A is required. In contrast, FRET can take place over a relatively long distance. The FRET mechanism will be predominant over the EE mechanism if a large spectral overlap exists. If the spectral overlap is small, EE will occur as the predominant mechanism over FRET. In addition, EE needs high concentrations of D and A, while FRET always happens at relatively lower concentrations.

3 AIE conjugated polymers for explosive detection

Many types of AIE-active polymers have been prepared, including linear and hyperbranched, conjugated and nonconjugated polymers. In this chapter, the latest development for synthetic strategies of AIE-active polymers and their performance in explosive detection will be summarized. Generally, the fluorescence quenching of AIE polymers is caused by the electron-transfer between electron-rich AIE-active polymer and electron-deficient nitro-aromatics. Nitro-aromatics with a relatively low LUMO energy level would accept an electron from a photoexcited polymer, resulting in fluorescence quenching of AIE-active polymer (Fig. 2A). The energy gap of LUMOs between polymer and nitro-aromatic is regarded as the "driving force" for fluorescence quenching. PA, TNT, and DNT exhibit LUMOs with energies of −3.89, −3.48, and −2.97 eV respectively. PA exhibits LUMO with the lowest energy, which is widely employed as an analyte in the fluorescence quenching experiment.

Compared to small molecular fluorophores, fluorescence conjugated polymers offer more efficient electronic communication and exciton migration effects between polymer backbones and quenchers. For example, Swager et al. demonstrated that a quencher molecule would cause fluorescence quenching of one whole polymer [26a]. This amplification phenomenon is termed as "the one-point contact, multipoint response effect," or "the molecular wire effect" (Fig. 3B) [26b]. However, unlike conjugated polymers with unique "the molecular wire effect," a quencher can only quench the fluorescence of a small molecule, which does not show a similar amplification effect.

3.1 AIE linear conjugated polymers

AIE linear conjugated polymers generally exhibit high fluorescence efficiencies, which can be used for fluorescence sensing. They can be prepared through the C—C bond coupling reactions, e.g., Stille coupling, Sonogashira coupling, and Suzuki coupling. The chemical structures of AIE-active linear conjugated polymers are shown in Scheme 1, and the performance for explosive detections is summarized in Table 2.

Chan et al. reported the preparation of **P1** and **P2** from disubstituted acetylenes with TPE pendants in the presence of the catalyst of WCl_6–Ph_4Sn. **P1–2** exhibit good solubility in

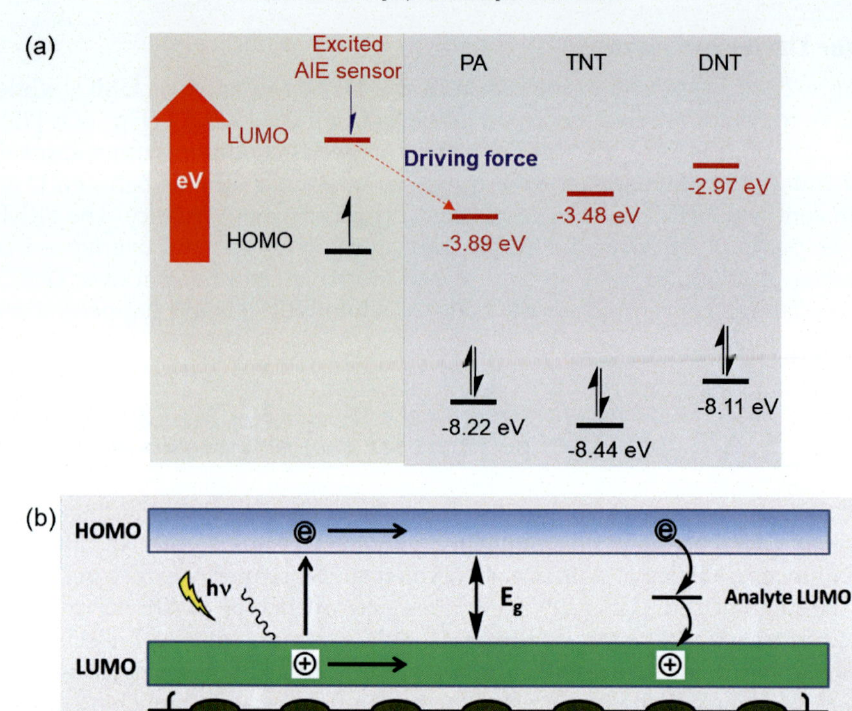

FIG. 2 (A) Schematic illustration for molecular orbitals of the fluorescence quenching process of AIE polymers with various nitro-aromatics. (B) Band diagram illustration of exciton transport and electron transfer fluorescence quenching for fluorescence sensor based on conjugated polymer. *Reproduced with permission from reference H. Zhou, M.-H. Chua, J. Xu, Polym. Chem. 10 (2019) 3822–3840. Copyright 2019, the Royal Society of Chemistry.*

common organic solvents and high thermal stability ($T_d > 400°C$ under N_2). While both polymers display weak fluorescence in solution, **P1** shows AIE characteristics, and **P2** exhibits ACQ properties. The fluorescence of **P1** and **P2** can be quenched by PA in solution. A conjugated copolymer **P3** containing TPE and phenylene-ethynylene (PE) blocks, was prepared through Sonogashira coupling (yield of 96.8%) [30]. Through modification of the ester group on the PE blocks, new polymers **P4** and **P5** were easily prepared from **P3**. **P3–5** show typical aggregation-induced enhanced emission (AIEE) properties, and they could be used for PA detection in solution. Among **P3–5**, **P4** shows the largest K_{sv} as $5.61 \times 10^4\,M^{-1}$ in THF/H_2O (V:V = 1:9) mixture. **P6** [31] was successfully prepared starting from 9-(dibromomethylene)-9H-fluorene with 9-octyl-2,7-bis(4,4,5,5-tetramethyl-1,3,2-dioxaborolan-2-yl)-9H-carbazole, and **P6** can form highly fluorescent AIE nanoparticles in water. Zhou et al. prepared two AIE conjugated polymers poly(triphenyl ethene) **P7** and poly(tetraphenyl ethene) **P8** consisting of TPE and triphenylethene repeating units, respectively, through Suzuki coupling reaction of (2,2-dibromoethene-1,1-diyl)dibenzene and suitable di-boronic compounds [28]. **P7** and **P8** show relatively high molecular weights (M_n: $1.01–1.74 \times 10^4$) with reasonable

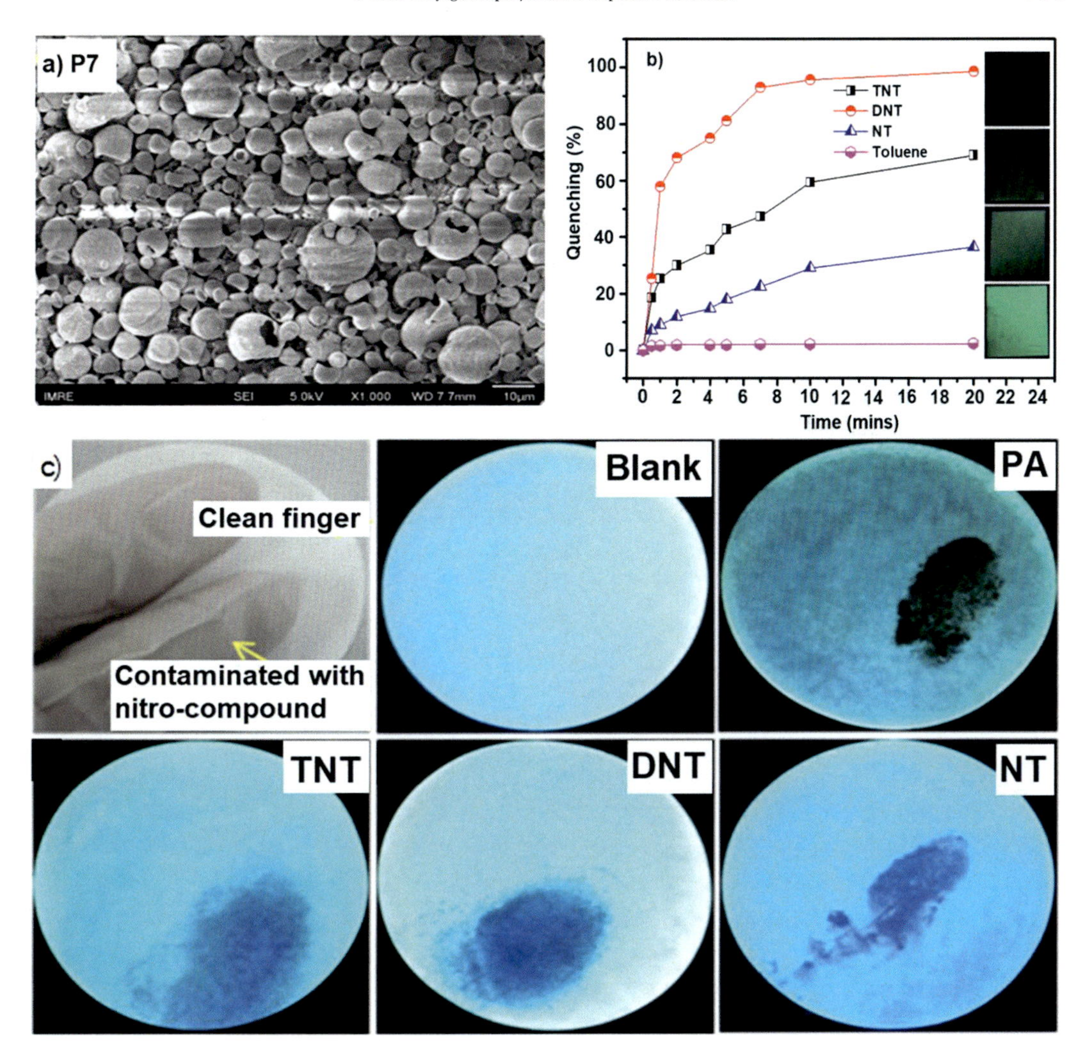

FIG. 3 (A) SEM images of **P7** film fabricated by electrospinning. (B) Fluorescence quenching of porous film **P7** on exposure to saturated vapor of analytes. The inserted photographs are porous film **P7** on exposure to saturated vapor of analytes. (C) Images of paper sensors with a clean finger and contaminated finger printing. Image of paper sensors under 365 nm irradiation with and without analytes. *Reprinted with permission from reference H. Zhou, X. Wang, T.T. Lin, J. Song, B.Z. Tang, J. Xu, Polym. Chem. 7 (2016) 6309–6317. Copyright 2016 the Royal Society of Chemistry.*

polydispersity indices (PDI: 1.5–1.7). Those two polymers show good solubility in common organic solvents and can form very stable nanoparticles in THF/H_2O mixtures without agglomeration when stored at room temperature for several months. Compared to related small AIE-active molecules or monomeric building blocks, **P7** and **P8** show significantly enhanced AIE properties. Many nitro-aromatics could effectively quench the fluorescence of **P7** and **P8** nanoparticles in solution. Taking PA as an example, **P7** displays the K_{sv} as high as

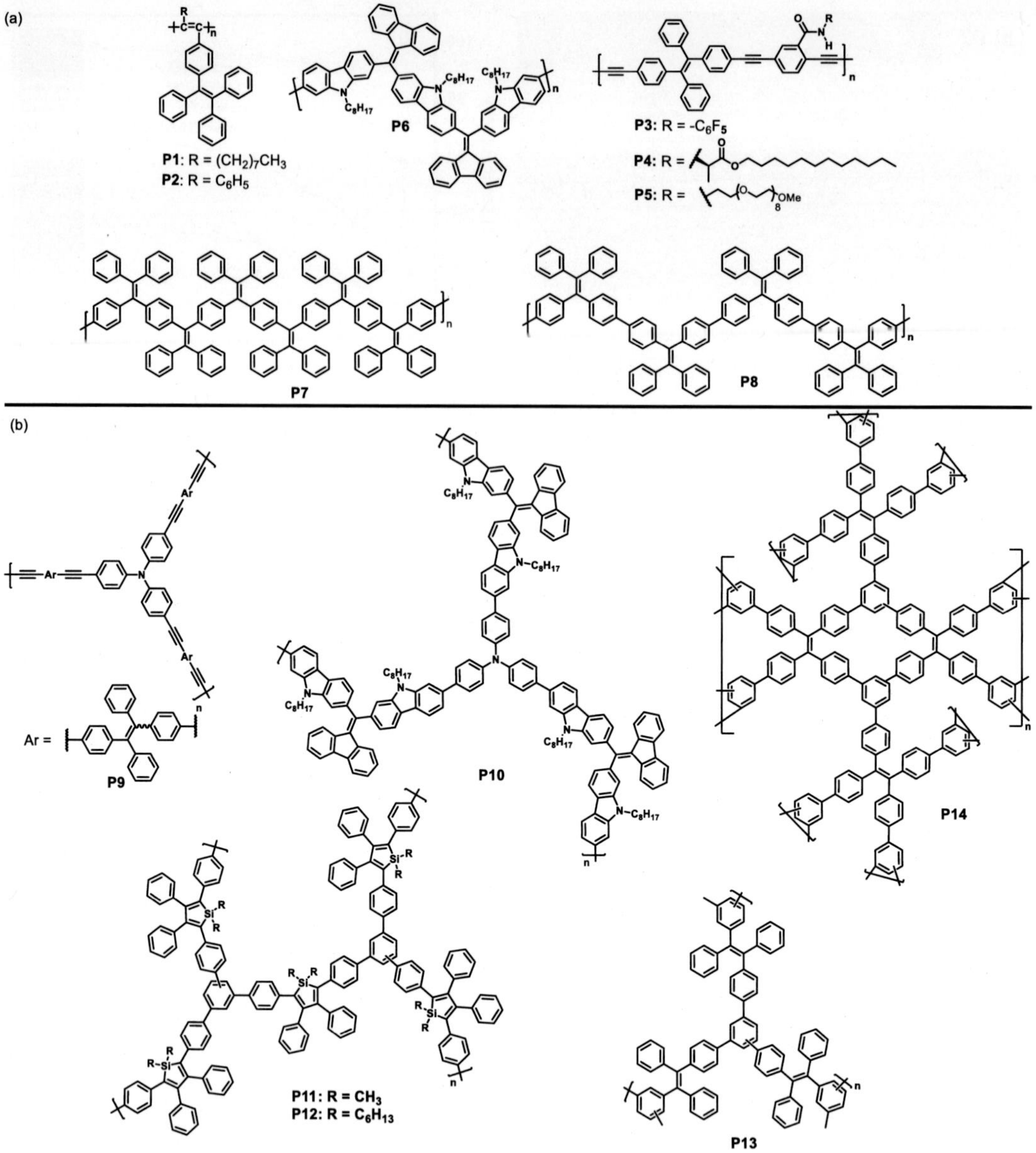

SCHEME 1 Chemical structures of (A) AIE-active linear conjugated polymers and (B) AIE-active hyperbranched conjugated polymers.

$1.80 \times 10^5 M^{-1}$, which is almost 10-fold higher than that of **P8** ($1.22 \times 10^4 M^{-1}$). **P7** and **P8** films with several micrometers thickness could be fabricated by drop-casting and electro-spinning methods. Fig. 3A shows the SEM images of **P7** porous film formed by electrospinning, which display higher fluorescence quenching efficiency to nitro-aromatics vapor when compared with its dense films formed by drop casting (Fig. 3B). The porous **P7** films show an improved

TABLE 2 The performance of AIE-active polymers for explosive detection.

Polymer	FL emission (nm) and Φ_F (%)	State of analyte	Explosives/nitro-compounds	K_{sv} (M^{-1})	Sensitivity
P1	496 (7.0)	Solution	PA	3.09×10^4	1 ppm
P2	529 (14.1)	Solution	PA	2.86×10^4	1 ppm
P3	520 (14.0)	Solution	PA	3.71×10^4	–
P4	520	Solution	PA	5.61×10^4	–
P5	520	Solution	PA	5.37×10^4	–
P7	493 (14.39)	Solution	PA, TNT, DNT, NT	1.80×10^5–3.65×10^3	5 ppb
		Vapor	TNT, DNT, NT	–	5 ppb
P8	480 (18.07)	Solution	PA, TNT, DNT, NT	1.22×10^4	–
P9	526 (>1.89)	Solution	PA	1.67×10^4	0.5 ppm
P10	608	Solution	PA	2.03×10^4	1.0 ppm
P12	510 (18.4)	Solution	PA	2.40×10^4	1.0 ppm
P13	501 (45.4)	Solution	PA	4.29×10^5	1.0 ppm
P14	523	Solution	PA	7.58×10^5	1.0 ppm
P15	468 (9.90)	Solution	PA, TNT, NT	9.72×10^4	2.5, 2.7, 4.0 ppm
P16	468 (8.60)	Solution	PA	6.98×10^4	–
P17	468 (7.80)	Solution	PA	4.27×10^4	–
P18	477 (10.0)	Solution	PA, TNT, NT	2630, 1440, 2320	–
P19	479 (11.0)	Solution	PA, TNT, NT	3040, 2560, 2950	22.9, 22.7, 18.2 ppm
P20	492 (15.0)	Solution	PA, TNT, NT	10,220, 4450, 2200	–
P21	499 (16.0)	Solution	PA, TNT, NT	1.57×10^4, 1.29×10^4, 3410	22.9, 22.7, 18.2 ppm
P23	558	Solution	PA	3.50×10^5	–
P26	469	Vapor	TNT, DNT	–	5 ppb, 100 ppb
P27	452 (31.0)	Solution	PA	8.48×10^5	1 ppm
P31	512 (7.7)	Solution	PA	6.22×10^3	–
P32	512 (22.0)	Solution	PA	2.74×10^4	–
P33	511 (10.6)	Solution	PA	2.01×10^4	–
P34	506 (11.6)	Solution	PA	2.79×10^4	1 ppm

Continued

TABLE 2 The performance of AIE-active polymers for explosive detection—Cont'd

Polymer	FL emission (nm) and Φ_F (%)	State of analyte	Explosives/nitro-compounds	K_{sv} (M^{-1})	Sensitivity
P37	471	Solution	PA	1.62×10^4	1 ppm
P38	471	Solution	PA	5.14×10^4	0.5 ppm
P40	476 (30.1)	Solution	PA	1.42×10^4	1 ppm
P41	548	Solution	PA	1.60×10^5	0.02 ppm
P42	532	Solution	PA	–	–
P44	480 (29.1)	Solution	PA	2.80×10^4	0.1 ppm
P45	480 (12.5)	Solution	PA	3.70×10^4	–
P46	451	Solution	PA	2.07×10^5	–
P47	489	Solution	PA	2.33×10^5	1 ppm
P48	516 (5.2)	Solution	PA	3.39×10^4	5 ppm
P49	485	Solution	PA	1.26×10^4	–
P50	501	Solution	PA, DNT, NBC	2.69×10^5, 1.01×10^5, 9.30×10^3	–
P54	492	Solution	PA	1.64×10^5	1 ppm
P55	480	Solution	PA	3.80×10^5	–
P56	492 (38.31)	Solution	PA, TNT	5.65×10^4, 704	–
P58	470 (16.51)	Solution	PA	1.95×10^4	1 ppm
P59	470	Solution	PA	2.67×10^4	1 ppm
P60	530 (11.49)	Solution	PA	2.24×10^4	–
P61	380 (22.70)	Solution	PA	1.35×10^5	–
P62	490 (34.70)	Solution	PA	6.36×10^4	1 ppm
P63	462	Solution	PA, TNT	–	0.5, 1 ppm
P63	485	Solid	PA, TNT	–	50, 100 ppm

response to nitro-aromatics in an increasing order of NT, TNT, and DNT vapors even its thickness approached a micrometer level, demonstrating less dependence of film thickness on the quenching efficiency in comparison with traditional conjugated fluorescence polymers. A paper sensor can be easily prepared through air-spraying **P7** nanoparticles suspension onto a piece of common filter paper (Fig. 3C). The paper sensor with a surface concentration of **P7** nanoparticles as low as $1.0\,\mu g\,cm^{-2}$ can detect various nitro-aromatics on the contaminated finger at 1 ng scale.

3.2 AIE hyperbranched conjugated polymers

AIE-active hyperbranched conjugated polymers with various topological structures, e.g., star-shape, dendritic patterns, and micro-porosity, can be successfully prepared from corresponding multifunctionalized monomers. For example, Wang et al. prepared triphenylamine-based hyperbranched conjugated poly(aryl-eneethynylene)s **P9** through Sonogashira coupling, which exhibits high molecular weights and good solubility in organic solvents [32]. **P9** shows typical AIE property. **P9** nanoparticles are applied for PA detection in solution with a detection limit of 0.5 ppm, and its Stern-Volmer curve exponentially bends upwards, indicating the super amplification emission quenching effect. Chen et al. prepared hyperbranched conjugated polymer **P10** through dispersion polymerization via germinal Suzuki cross-coupling of 1,1-dibromoolefins [31]. The nanoparticle sizes of **P10** can be well-controlled by adjusting the monomer concentrations, and its well-defined nanoparticles are identified by the SEM images. **P10** with tetra-arylethene building blocks displays enhanced fluorescence emission in THF/H_2O mixture compared to in THF solution, revealing archetypal AIE characteristics. The fluorescent suspension of **P10** nanoparticles exhibits a reasonable quenching efficiency and amplified quenching effect to PA in solution. Liu et al. synthesized two hyperbranched conjugated polymers **P11** and **P12** from 1,1-dialkyl-2,5-bis(4-ethynylphenyl)-3,4-diphenylsiloles in the presence of $TaBr_5$, which show high molecular weights (M_n: 2.50×10^5) and high yield (98%) [33]. These hyperbranched polymers exhibit a high branching degree of 0.55, and display good thermal stability and good solubility in organic solvents.

P11 and **P12** aggregate to form highly emissive nanoparticles that are stable at room temperature for 2 years. Amorphous plates of **P12** emit cyan fluorescence under UV light irradiation (Fig. 4A and B). PA could cause the fluorescence quenching of **P12** in nanoaggregates state (in THF/H_2O mixture) and solution state (in THF) (Fig. 4C and D), and **P12** in THF/H_2O mixture (1:9 *v*/v) shows a detection limit of 1 ppm. Furthermore, the letters "HKUST" written by dropping aliquots of a PA solution onto the filter paper deposited with **P12** can be clearly observed. Hu et al. synthesized a hyperbranched polymer **P13** via polymerization of 1,2-bis(4-ethynylphenyl)-1,2-diphenylethene in the presence of $TaBr_5$ (Scheme 2) with high molecular weight (M_n: 1.58×10^5) [34]. **P13** displays highly thermal stability in air ($T_d > 443$ °C) and typical AIE characteristics with fluorescence efficiency (Φ_F) as high as 81% in solid state. PA could quench the fluorescence of **P13** efficiently with K_{sv} of $4.29 \times 10^5 M^{-1}$. Hu et al. also prepared a hyperbranched poly(tetraphenylethene) **P14** via homopolycyclotrimerization of 1,1,2,2-tetrakis(4-ethynylphenyl)ethane with molecular weight (M_n: 5.28×10^6) and good yield (97%) [35]. **P14** shows good solubility in organic solvents and exhibits extremely high thermal stability ($T_d > 525$°C) in air. **P14** powder emits different colors when exposed to light with different wavelengths, which shows yellow and green emissions when irradiated by daylight and UV light, respectively. **P14** also shows mechano-chromic luminescent, and its sky blue emission at 477 nm could be red-shifted to 505 nm by simple grinding with a glass rod. **P14** displays typical AIEE properties and its fluorescence can be quenched by PA with amplification effect (K_{sv} up to $7.58 \times 10^5 M^{-1}$).

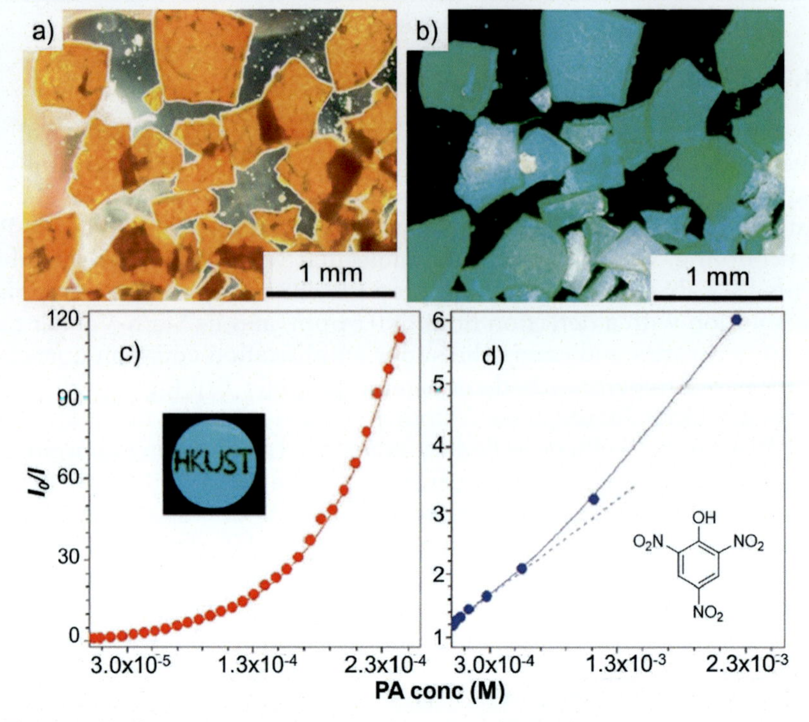

FIG. 4 (A) Photographs of **P12** amorphous plates under daylight. (B) Photographs of **P12** amorphous plates under UV light of 330–380 nm (inserted). Plots of I_0/I vs [PA] for **P12** in (C) THF/H_2O mixture (1:9 v/v) and (D) THF. Inset of (C) fluorescence image of **P12** deposited on a filter paper. The letters "HKUST" were written by dropping aliquots of a PA solution on the filter paper using a capillary tube. *Reprinted with permission from reference J. Liu, Y. Zhong, J.W.Y. Lam, P. Lu, Y. Hong, Y. Yu, Y. Yue, M. Faisal, H.H.Y. Sung, I.D. Williams, K.S. Wong, B.Z. Tang, Macromolecules 43 (2010) 4921–4936. Copyright 2010 the American Chemical Society.*

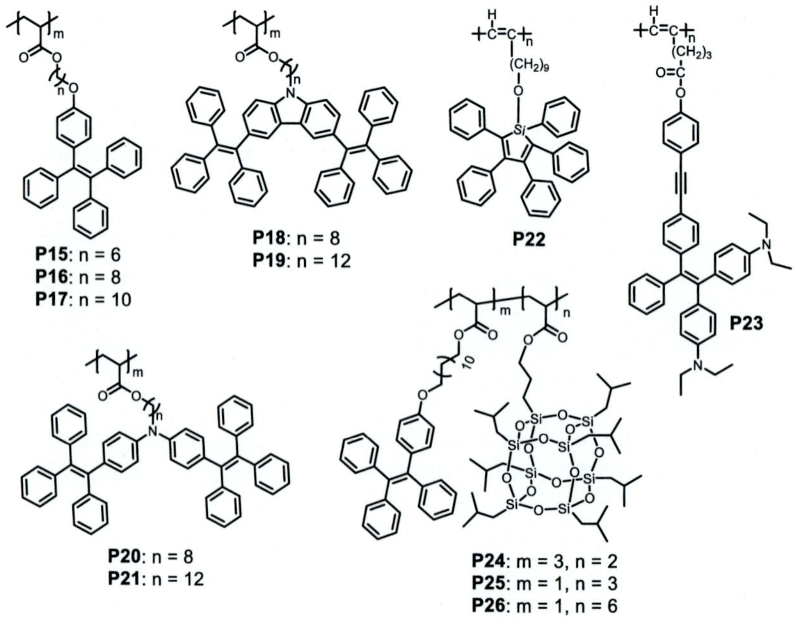

P15: n = 6
P16: n = 8
P17: n = 10

P18: n = 8
P19: n = 12

P22

P23

P20: n = 8
P21: n = 12

P24: m = 3, n = 2
P25: m = 1, n = 3
P26: m = 1, n = 6

SCHEME 2 Chemical structures of nonconjugated linear polymers with AIEgens as pendant groups.

4 AIE nonconjugated polymers and their explosive detection

AIE-active nonconjugated polymers can be synthesized by classic polymerization reactions starting from AIEgen-functionalized monomers, e.g., radical polymerization and atom transfer radical polymerization. In this part, nonconjugated polymers and their performance for explosive detection are summarized, including linear polymers with AIEgens as pendant groups, linear polymers with AIEgens as building blocks in the polymer backbone, and hyperbranched AIE nonconjugated polymers.

4.1 Linear polymers with AIEgens as pendant groups

Generally, TPE-based AIEgens could be synthesized via the McMurry coupling reaction. However, if two different reactants species are involved in the reaction, a mixture of homo- and cross-coupling products will be produced. In order to avoid the side reaction, Zhou et al. have innovated a useful and efficient method to prepare mono-functionalized TPE molecules via the Suzuki coupling reaction starting from 4-(1,2,2-triphenylvinyl)phenol (TPE-OH). This method can generate pure mono-functionalized TPE compounds without tedious separation procedures. Several TPE-containing acrylate monomers [27] are prepared by this method, which are then employed to synthesize nonconjugated polymers **P15–17** with high-molecular weights (M_n: 4.00–6.50 $\times 10^5$) via free radical polymerization. **P15–17** show similar thermal stability with poly(methyl methacrylate) (PMMA) ($T_d > 356$°C), but lower glass transition temperatures ($T_g = 57$–76°C) than PMMA (126°C). Those polymers show excellent solubility in organic solvents and good film-forming ability. Extremely stable **P15–17** nanoparticles could be formed in THF/H_2O mixtures, which appear stable even when stored at 4°C for 9 months. The fluorescence of **P15** nanoparticles and polymer films could be quenched efficiently by almost all nitro-aromatics, e.g., PA, TNT, DNT, and NT, with detection limits of 2.5, 2.7, 4.0, 5.0, and 6.0 ppm, respectively (Fig. 5A–D). Following the similar approach, Chua et al. synthesized four poly(acrylates) **P18–21** with different pendant groups, e.g., 3,6-bis(1,2,2-triphenylvinyl) carbazole and bis(4-(1,2,2-triphenylvinyl)phenyl)-amine [36]. Likewise, those polymers show typical AIE characteristics and the fluorescence of polymer nanoparticles could be quenched by nitro-aromatics PA, TNT, and NT. Paper sensors coated with **P19** and **P21** display reasonable responses to both nitro-aromatics solutions and solid particles. Tang et al. reported silole-containing polyacetylenes **P22** with 1,2,3,4,5-pentaphenylsilolyl pendants as an AIEgen [29]. The **P26** aggregate shows an enhanced emission efficiency up to 46 folds compared to that in dilute solution. Yuan et al. synthesized an *E*-stereoregularity polymer **P23** from 1-pentyne derivatives with aminated TPE monomer through polymerization using organorhodium catalysts [37]. The fluorescence of **P23** nanoaggregates in THF/H_2O mixture (1:9 v/v) is about 57-fold higher than that of its solution in THF, which can be quenched by PA with K_{sv} up to $3.50 \times 10^5\,M^{-1}$. Zhou et al. synthesized AIE-active polymers **P24–26** through free radical polymerization of TPE-containing acrylate monomers and acryloisobutyl polyhedral oligomeric silsesquioxanes (POSS) monomer [38]. Porous films were successfully fabricated via electrospinning **P26** solution in acetone/chloroform mixture (Fig. 5C). Unlike the traditional fluorescence polymers, this porous film displays less dependence of fluorescence

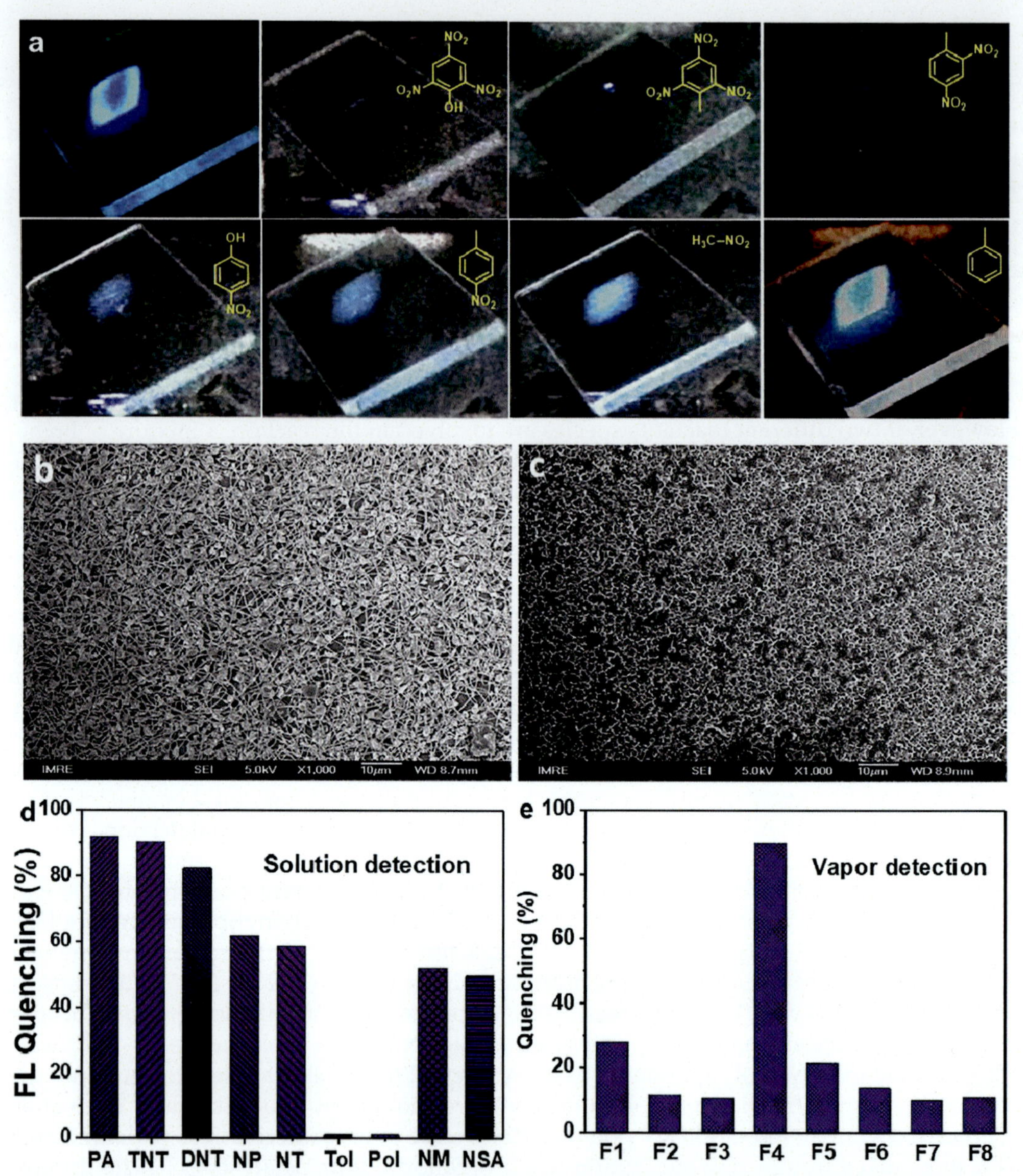

FIG. 5 (A) Photographs of **P15** dense films in the absence and presence of analytes. All photos were under 365 nm radiation. (B) SEM images of films fabricated through electrospun from **P15** and **P26**. (D) Fluorescence quenching of **P15** nanoparticles in THF/H_2O (1:9 v/v) mixtures by different analytes. (E) Fluorescence quenching of **P26** porous films exposed to DNT saturated vapor for 4.0 min at 25°C. *Reprinted with permission from references H. Zhou, J. Li, M.H. Chua, H. Yan, B.Z. Tang, J. Xu, Polym. Chem. 5 (2014) 5628–5637; H. Zhou, Q. Ye, W.T. Neo, J. Song, H. Yan, Y. Zong, B.Z. Tang, T.S. Andy Hor, J. Xu, Chem. Commun. 50 (2014) 13785–13788. Copyright 2014 Royal Society of Chemistry.*

quenching to nitro-aromatics' vapor on film thickness (Fig. 5E). The incorporation of POSS units into polymer enhances the porosity of polymer film, allowing more vapor molecules to penetrate film and subsequently resulting in better sensitivity to nitro-aromatics' vapor. The **P26** porous films fabricated by electrospinning show much higher sensitivity to TNT and DNT vapor, when compared to **P26** dense films fabricated by drop-casting or spin-coating and even **P15** porous film (Fig. 5B).

4.2 Linear nonconjugated polymers with AIEgens as building blocks in the polymer backbone

This section will discuss the linear nonconjugated polymers with AIEgens as building blocks in the polymer backbone, e.g., poly(silylenevinylene)s, polytriazoles, and other types of polymers. The chemical structures of polymers are shown in Schemes 3 and 4, and their performance for explosive detection is summarized in Table 2.

Alkynes and silanes are hydrosilylated to prepare poly(silylenevinylene)s. Poly(silylenevinylene)s may exhibit unique electronic and optical properties, because of the σ^*-π^* hyperconjugation between σ orbitals of silicon atoms and π orbitals of double bonds along the polymer backbones. Lu et al. synthesized *E*-stereoregularity polymers **P27–30** with high molecular weights ($M_n > 3.65 \times 10^4$) through alkyne polyhydrosilylations of 1,2-bis(4-dimethylsilanylphenyl)-1,2-diphenylethene with several diakynes using catalyst $Rh(PPh_3)_3Cl$ [39]. All polymers display good thermal stability ($T_d > 330°C$), and typical AIE characteristics. **P27** shows strong fluorescence emission in both the aggregate state and thin film state, and the fluorescence of regions in the film can be quenched by UV irradiation (Fig. 6A and B). PA would exponentially quench the fluorescence of polymer nanoparticles in solution with a K_{sv} of $8.48 \times 10^5 M^{-1}$ (Fig. 6C and D). Though employing the same synthetic approach, Zhao et al. prepared a series of poly(silylenevinylene)s **P31–34** with a good yield (up to 92%) from 1,1-dimethyl-2,5-bis(4-ethynylphenyl)-3,4-diphenylsilole with aromatic silylhydrides including 1,4-bis(dimethylsilyl)benzene, 4,4-bis(dimethylsilyl) biphenyl, 2,5-bis(dimethylsilyl)thiophene, and 2,7-bis(dimethylsilyl)-9,9-dihexylfluorene [40]. **P31–34** could be prepared with *E*-stereoregularity as high as 99% and good molecular weights (M_n: 9.53×10^4). These polymers also display good thermal stability ($T_d > 420°C$). The fluoresce of **P31–34** is weak in solution and their emissions are significantly enhanced in the aggregated state. PA can quench the fluorescence emission of **P34** in aqueous mixtures with a K_{sv} of up to $2.79 \times 10^4 M^{-1}$.

Yuan et al. synthesized various polytriazoles **P35–38** with good yields (up to 93%) in dimethylformamide at 150°C through metal-free 1,3-dipolar polycycloadditions of TPE-containing diazides and 4,4′-isopropylidenediphenyl diphenylpropiolate [41]. These polytriazoles show good molecular weights ($M_n > 2.68 \times 10^4$) and good thermal stability ($T_d > 375°C$). Polymers **P35–38** are soluble in organic solvents and display typical AIE characteristics. For the PA detection, **P38** shows fluorescence quenching efficiency of $5.14 \times 10^4 M^{-1}$, which is almost 4-folds higher than that of **P37**. By applying the same synthetic method, Li et al. prepared poly(aroxycarbonyltriazole)s **P39–40** from diazide monomers and dipropiolates with a high yield (up to 99%), high molecular weights ($M_n > 2.35 \times 10^4$), and high regioregularity (up to 90%) [42]. More importantly, this reaction

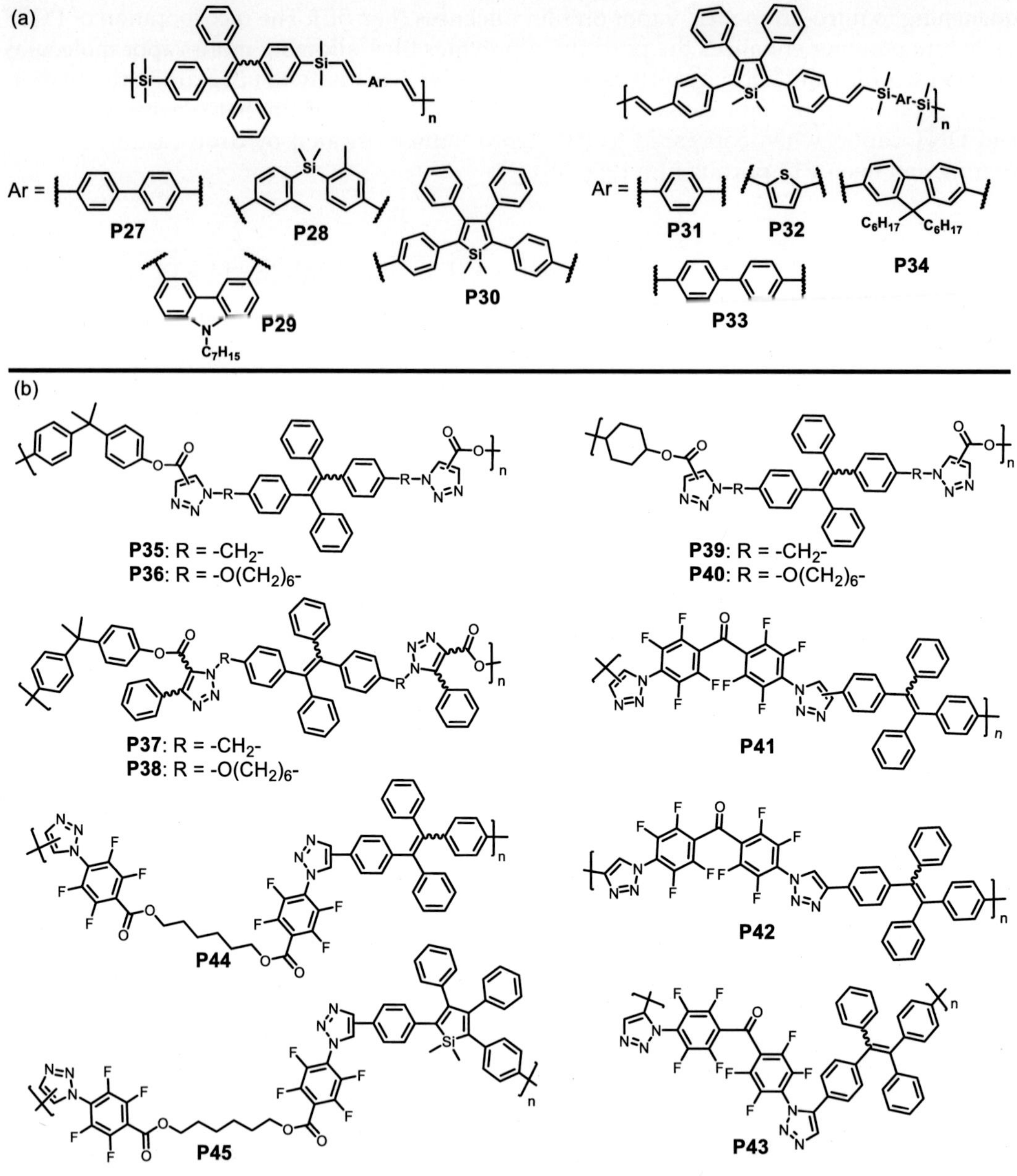

SCHEME 3 Chemical structures of (A) linear AIE-active polymers prepared from hydrosilylations of alkynes and silanes and (B) linearAIE-active polytriazoles.

is tolerant to oxygen and moisture. Wang et al. also synthesized AIEE polytriazole **P41** with a good yield (97.9%) from azide of 4,4′-diazidoperfluorobenzophenone and TPE-containing diyne [43]. **P41** is composed of a mixture of 1,4-regioregular **P42** component (fraction of 42.4%) and 1,5-regioregular **P43** (fraction of 57.6%) component. **P42** and **P43** can be selectively synthesized via Cu(I)- and Ru(II)-catalyzed click polymerizations using the same

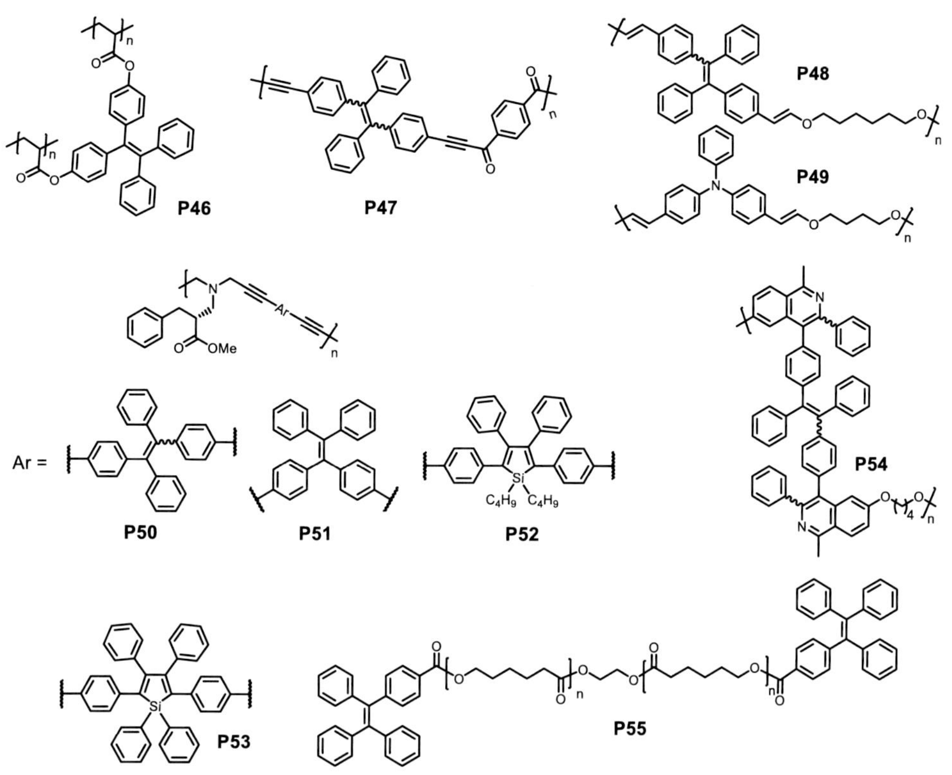

SCHEME 4 Chemical structures of AIE-active linear polymers with AIEgens as building blocks in the polymer backbone.

monomers, respectively. **P41** exhibits high thermal stability ($T_d > 440°C$). Both **P42** and **P43** show typical AIEE property, and their fluorescence of nanoaggregates can be effectively quenched by PA in solution with superamplification quenching effect, and the detection limit is as low as 0.02 ppm. Wu et al. also prepared two AIE-active polytriazoles **P44** and **P45** using the same metal-free polycycloaddition of azides and alkynes, with a high molecular weight (M_n: 3.20×10^4) [44]. Hu et al. synthesized hyperbranched polyacrylates **P46** with closed-loops and glycogen-like structures with M_n of $>8.71 \times 10^4$) in almost quantitative yields [45]. All polymers show good solubility in organic solvents and high thermal stability ($T_d > 357°C$). The fluorescence of **P46** nanoaggregates could be quenched by PA in solution with a K_{sv} of $2.07 \times 10^5 M^{-1}$. Li et al. prepared poly(arylene ynonylene) **P47** from TPE-containing diyne and terephthaloyl dichloride with a reasonable yield (70%) [46]. **P47** is soluble in organic solvents and shows the high molecular weight (M_n: 3.91×10^4). It also shows typical AIEE characteristics, and its fluorescence can be turned "off" by PA and then turned "on" again by hydrazine. It is the first report of a dual fluorescence sensor that is sensitive for both hydrazine and PA. More interestingly, **P47** could further react with hydrazine via click reactions, and the generated polymer shows a much higher K_{sv} (up to $1.09 \times 10^6 M^{-1}$) for PA detection than **P47**, because of its enhanced fluorescence emission and large spectral overlap between its fluorescence emission and UV absorption of PA.

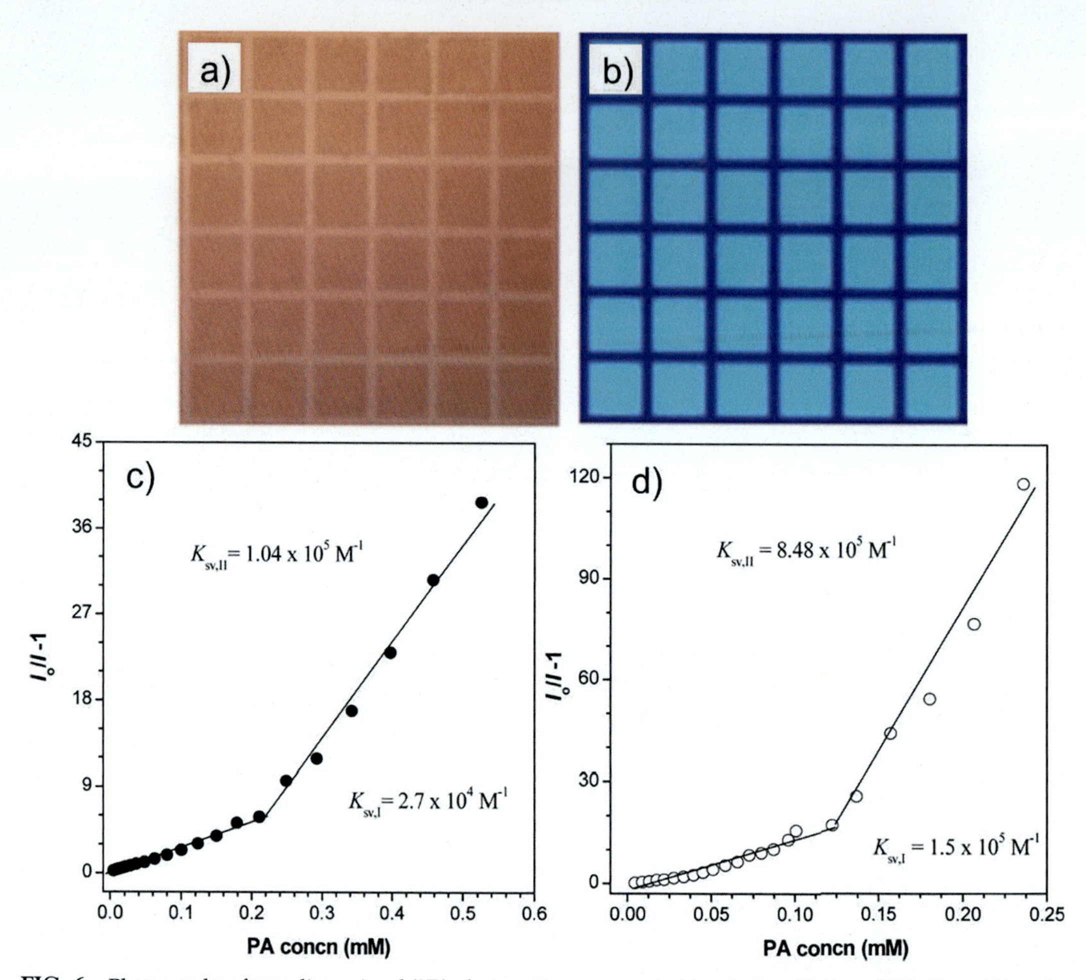

FIG. 6 Photographs of two-dimensional (2D) photo patterns generated by photooxidation of **P27** films: (A) under daylight, (B) under UV radiation. Plots of $(I_0/I-1)$ vs [PA] for **P27** in (C) THF and (D) THF/H_2O mixture (1:9 v/v). *Reprinted with permission from reference P. Lu, J.W.Y. Lam, J. Liu, C.K.W. Jim, W. Yuan, C.Y.K. Chan, N. Xie, Q. Hu, K.K.L. Cheuk, B.Z. Tang, Macromolecules 44 (2011) 5977–5986. Copyright 2011 American Chemical Society.*

Wang et al. synthesized polyhydroalkoxylation from aromatic diynes using phosphazene base, in which region-regular anti-Markovnikov addition polymers poly(vinyl ether)s **P48** and **P49** could be synthesized with good molecular weight (M_n 4.06×10^4) and high yield (99%) [47]. All polymers are soluble in organic solvents and thermally stable. **P48** bearing TPE units is AIE active, while **P49** bearing triphenylamine units display ACQ characteristics. PA could quench the fluorescence of all polymers in the solution, where **P48** shows a higher K_{sv} than **P49**. In addition, the fluorescence of paper strips fabricated from **P48** could be quenched by PA in solution with a detection limit of 5 ppm.

Liu et al. synthesized functional polymers **P50–53** with good molecular weight (M_n 4.38×10^4) and high yield (89%) from primary amines, alkynes, and formaldehyde [48], in which one-pot A3-coupling polymerization catalyzed by copper(I) chloride is tolerant to moisture. **P50–51** with TPE units and **P52–53** with tetraphenyl silole units show typical AIE characteristics. Among them, the fluorescence of **P50** can be quenched by various nitro-aromatics, e.g., PA, DNT, and 4-nitrobenzoyl chloride (NBC), with a K_{sv} of 2.69×10^5, 1.01×10^5, and $9.30\times10^3\,M^{-1}$, respectively. Liu et al. synthesized multisubstituted poly(isoquinoline)s **P54** from internal diynes and *O*-acyloxime derivatives using $[Cp^*RhCl_2]_2$ as catalyst and $Cu(OAc)_2{\cdot}H_2O$ as oxidant, which shows good molecular weight (M_n: 1.83×10^4) and high thermal stability [49]. PA could quench the fluorescence of **P54** aggregate in water/THF with a K_{sv} of $1.64\times10^5\,M^{-1}$. Liang et al. fabricated an AIE organic hybrid nano-sheets through crystalizing **P55**, in which the polyester chain of **P55** is capped by two TPE units [50]. During crystallization, TPE units of **P55** are driven out of the lamellar crystals and finally arrayed on its surface to generate nano-sheets with a thickness of 7.0 nm (Fig. 7A–C). PA would quench the fluorescence of **P55** nano-sheets with a K_{sv} of $3.80\times10^5\,M^{-1}$

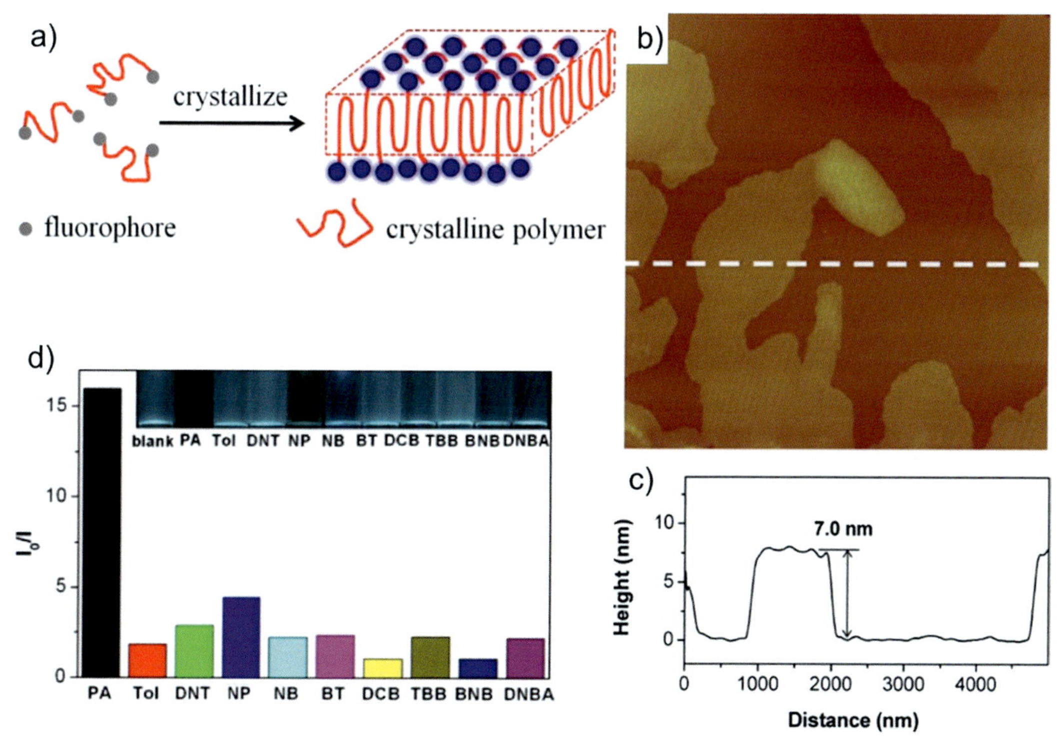

FIG. 7 (A) Schematic illustrations of crystallization-induced nano-sheets of **P55**. (B) Height image of tapping mode AFM to show the thickness of **P55** sheets, image size of $5\times5\,\mu m^2$. (C) Cross-sectional profile along the white dash line in B. (D) Variation of I_0/I of **P55** with addition of PA, Tol, DNT, NP, NB, 2-bromotoluene (BT), 1,2-dichlorobenzene (DCB), 1,2,4-tribromobenzene (TBB), bromo-2-nitrobenzene (BNB), 3,5-dinitrobenzyl alcohol (DNBA). Inset showed the photographs of **P55** suspension with analytes. *Reprinted with permission from reference G. Liang, L.-T. Weng, J.W.Y. Lam, W. Qin, B.Z. Tang, ACS Macro Lett. 3 (2014) 21 – 25. Copyright 2014 American Chemical Society.*

(Fig. 7D). **P55** provides an example of crystallization-induced fluorescence nanomaterials, offering a method to apply AIE materials for nano-devices, biological engineering, etc.

4.3 Hyperbranched AIE nonconjugated polymers

Hyperbranched polymers have received increasing attention because of their unique nature, such as easy preparation process, good solubility, low viscosity, and three-dimensional (3D) structure with multiple end groups. Examples of AIE hyperbranched nonconjugated polymers and their performance for explosive detection are summarized in Scheme 5 and Table 2, respectively.

Wang et al. prepared AIE hyperbranched polytriazoles **P56–57** by triazide-diyne click polymerization using $Cu(PPh_3)_3Br$ as the catalyst [51]. The α_{AIE} values (the ratio of Φ_F or emission intensities for an AIE material in aggregate and solution states) of **P56–57** (348.3) are

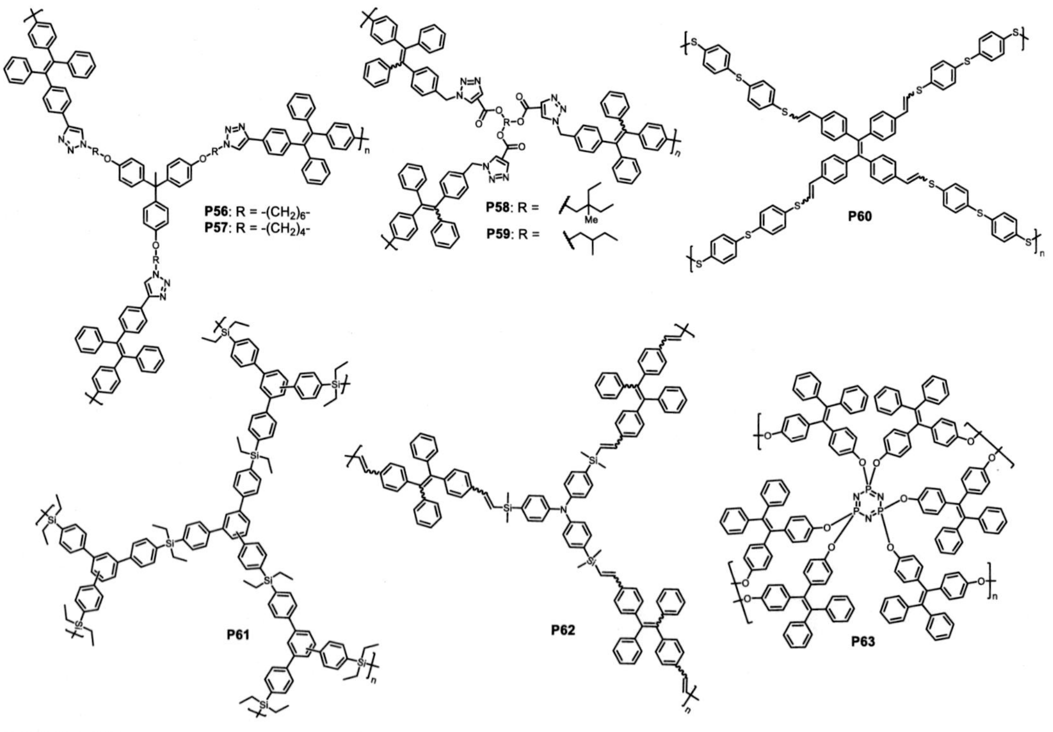

P56: R = $-(CH_2)_6-$
P57: R = $-(CH_2)_4-$
P58: R =
P59: R =
P60
P61
P62
P63

SCHEME 5 Chemical structures of hyperbranched polymers with AIEgens as building blocks in the polymer backbone.

higher than that of small AIE-active molecules, e.g., HPS and TPE. The rotation of phenyl rings in **P56–57** is strongly restricted because of the high compression ratio from solution state to solid state, leading to an increase in Φ_F by 33.31% and 100%, in the aggregate state and in the solid state, respectively. **P56–57** are very soluble in organic solvents and show very good film-forming ability. The fluorescence of **P56–57** can be quenched by many nitro-aromatics with a super amplification effect, but they display different quenching mechanisms for different nitro-aromatics. For PA, the fluorescence of **P56–57** is quenched through an energy transfer mechanism; however, when TNT is used, the mechanism involves the charge transfer between polymers and TNT. Li et al. synthesized hyperbranched polymers **P58–59** by the metal-free click polymerization of tripropiolates and diazide with TPE units [52]. **P58–59** exhibit typical AIE properties and their fluorescence could be quenched by PA with a detection limit of 1 ppm in solution. In addition, **P58–59** is sensitive to UV irradiation, and the high-resolution 2D fluorescent patterns can be generated by photo-cross-linking method using UV irradiation. Yao et al. synthesized hyperbranched poly(vinylene sulfide)s **P60** via thiol-yne click polymerization with a yield of 86%, which exhibits good molecular weights (M_n 6.31×10^4) [53]. **P60** shows typical AIEE characteristics and the fluorescence of its aggregates in THF/H_2O could be efficiently quenched by PA. In addition, **P60** could be further postmodified through thiol-yne click reaction via using the dangling ethynyl groups on its peripheries. Liu et al. synthesized **P61** and applied it for the detection of nitro-aromatics, in which sensing is based on the static quenching mechanism [54]. Compared to linear polymer with the same building blocks (Fig. 8A), **P61** shows faster fluorescence quenching speed with PA, because the 3D globular structure of **P61** may provide more channels for the exciton migration. At the same time, the loose packing structure of **P61** branches may also provide more cavities and additional diffusional pathways for better contacting between analyte molecules and sensing sites, offering a higher K_{sv} of $1.50 \times 10^5 M^{-1}$ (Fig. 8B and C). Lu et al. synthesized **P62** with good molecular weight (M_n: 7.51×10^4) and reasonable yield (71%), through alkyne polyhydrosilylation of 1,2-bis(4-ethynylphenyl)-1,2-diphenylethene and tris(4-dimethylsilylphenyl)amine using rhodium catalyst [55]. **P62** shows high thermal stability ($T_d > 445°C$), and typical AIE characteristics. The fluorescence of **P62** nanoaggregates in the solution can be quenched by PA with K_{sv} of $6.36 \times 10^4 M^{-1}$, which is 7.5-folds larger than K_{sv} of **P62** in the dissolved state, and also threefolds larger than K_{sv} of other polysiloles in aggregate state (up to $2.00 \times 10^4 M^{-1}$). The superamplification effect is observed in fluorescence quenching by PA, which may be caused by the 3D σ^*–π^* conjugation in **P62**. Hu et al. synthesized a cross-linked poly(tetraphenylethylene-cocyclotriphosphazene) **P63** through one-step poly-condensation, which emits strong fluorescence in both suspension and solid states [56]. **P63** also shows good thermal stability because of its cross-linked structures and special building blocks (Fig. 9A). The suspension and powder forms of **P63** shows strong blue fluorescence (Fig. 9B), which could be quenched by PA and TNT. **P63** shows a higher detection limit for PA (0.5 ppm) than that for TNT (1 ppm), which may be caused by an additional energy transfer between **P63** and PA. In addition, the fluorescence of **P63** powder with 25 mg weight could be quenched by 100 ppm TNT and 50 ppm PA respectively, suggesting **P63**'s high sensitivity to nitro-aromatics (Fig. 9C–E).

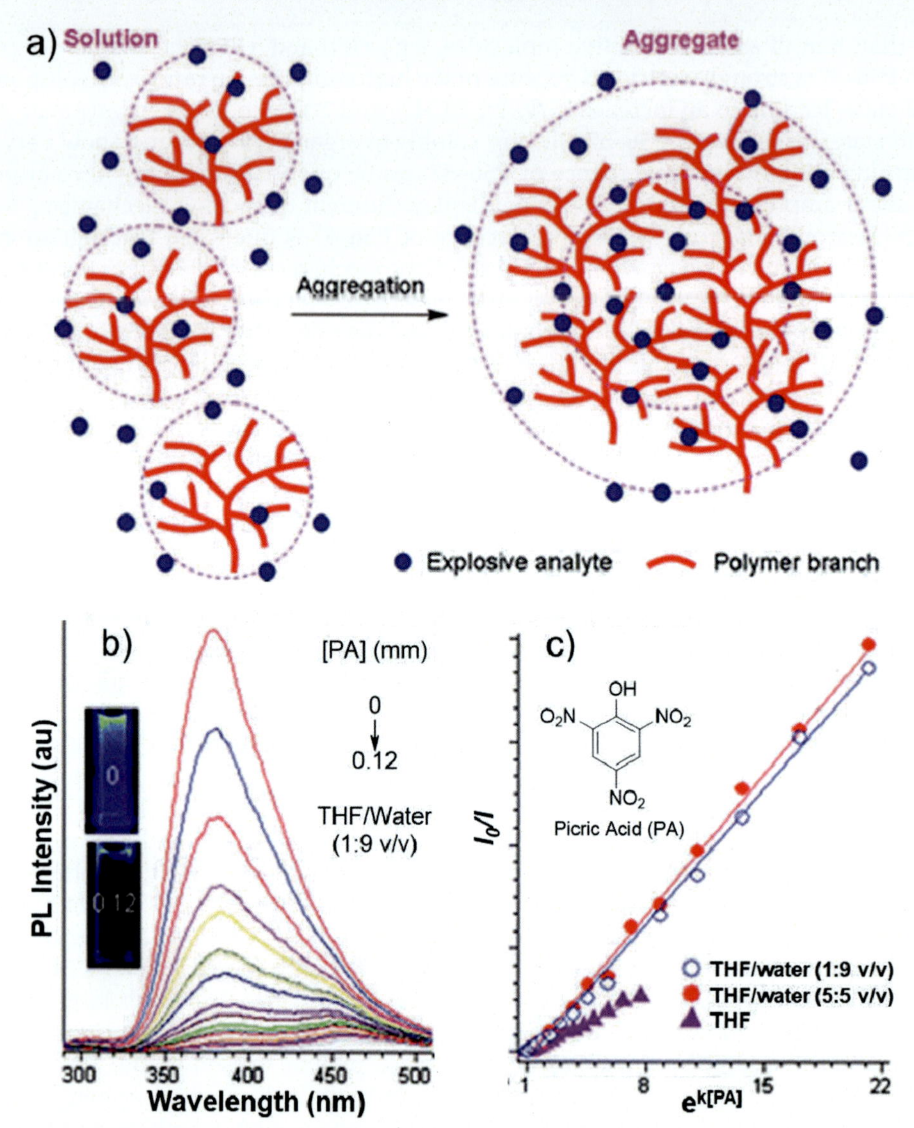

FIG. 8 (A) Illustrations of explosive detection using **P61** in solution and aggregate state. (B) Fluorescence changes of **P61** with the addition of PA in THF/H_2O mixture (1:9v/v). Inset in (B) fluorescence images of **P61** in THF/H_2O mixture (1:9v/v) with [PA] of 0 and 0.12mM, respectively. (C) Plots of I_0/I vs $e^{k[PA]}$ in THF and THF/ H_2O mixtures. *Reprinted with permission from reference J. Liu, Y. Zhong, P. Lu, Y. Hong, J.W.Y. Lam, M. Faisal, Y. Yu, K.S. Wong, B.Z. Tang, Polym. Chem. 1 (2010) 426–429. Copyright 2010 Royal Society of Chemistry.*

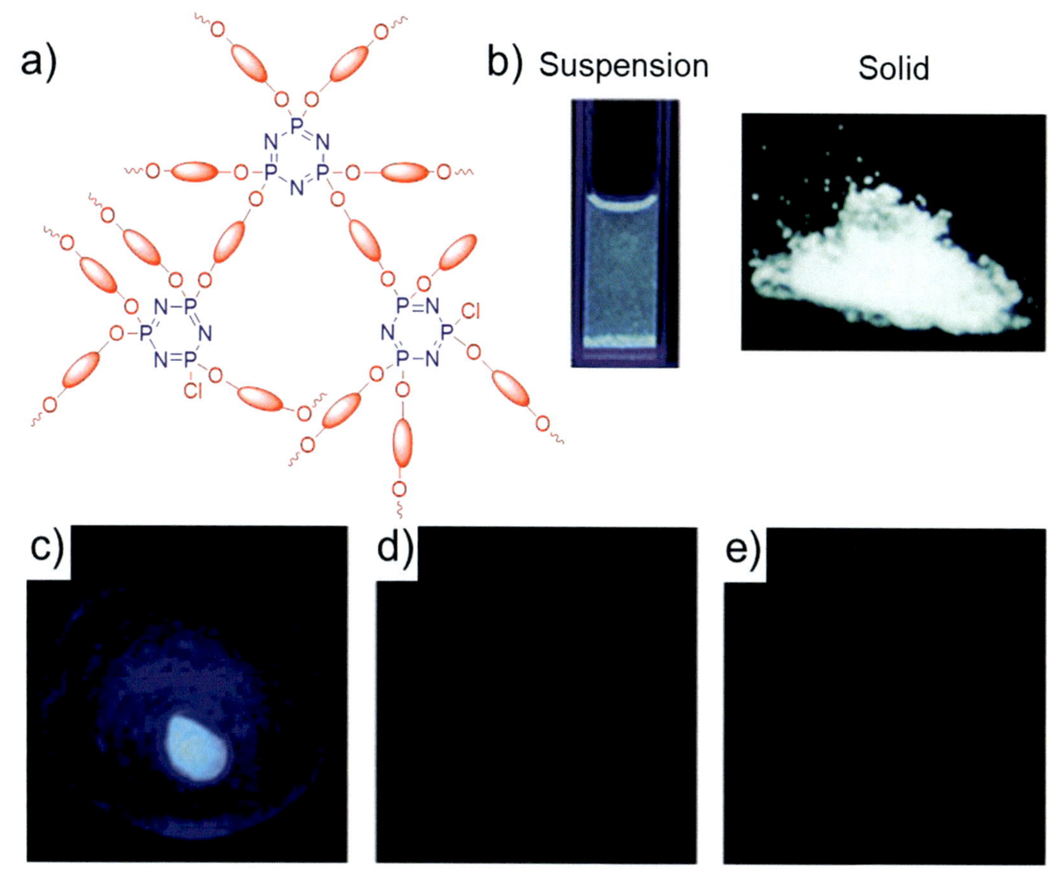

FIG. 9 (A) Structure illustration of **P63**. (B) Fluorescence images of **P63** suspension and powder under 365 nm radiation. Fluorescence images of 25 mg **P63** solid in the absence of analyte (C), in the presence of 100 ppm TNT (D), and 50 ppm PA (E) under 365 nm radiation. *Reprinted with permission from reference X.-M. Hu, Q. Chen, D. Zhou, J. Cao, Y.-J. He, B.-H. Han, Polym. Chem. 2 (2011) 1124–1128. Copyright 2011 Royal Society of Chemistry.*

5 Perspectives and outlooks

In this chapter, we summarized the latest development of a variety of AIE-active polymers-based explosive detection. AIE-active conjugated polymers in general show higher sensitivity than nonconjugated polymers owing to the presence of the "molecular wire effect" in conjugated polymers. For example, a paper sensor fabricated from **P7** nanoparticles shows a detection limit of 5 ppb, which is much lower than these of AIE chemosensors derived from most nonconjugated polymers (ppm level). On the whole, AIE-active polymers exhibit high sensitivity in solution than in solid state because of the low vapor pressure of explosives in solid state, therefore, making it difficult to detect solid-state explosives [27,28]. The detection sensitivity, however, could be improved by modulating the porous morphology of AIE-active polymers. For instance, by incorporation of POSS units as pendant groups into linear polymers **P15** and **P26**, they can form highly porous films via the electrospinning method, offering a better detection sensitivity to nitro-aromatics vapors even when the film thickness is more than 500 nm. To enhance the detection sensitivity to nitro-aromatics, appropriately regulating the LUMO level of AIE-active polymers to match the LUMO level of nitro-aromatics is an alternative method.

Although it has been demonstrated that AIE-active polymers can effectively detect nitro-aromatics type explosives, there are no AIE-active chemosensors that can be used to detect other classes of explosives, particularly acetone peroxides, which has been identified to be responsible for many industrial incidents and are employed in recent terrorist's bomb attacks as well. The lack of chromophores in acetone peroxide makes it hard to be detected. Currently, one indirect detection means is to detect the acetone vapor released during the gradual decomposition of acetone peroxides. Therefore, further AIE research may be channeled to study AIE-based chemosensors for the selective detection of this class of explosives, and thus require the AIE community to pay more attention to the design and synthesis of AIE-active polymers with high sensitivity and selectivity.

Acknowledgment

We acknowledge the fund support from IMRE A*STAR.

References

[1] (a) M.E. Germain, M.J. Knapp, Chem. Soc. Rev. 38 (2009) 2543–2555. (b) Y. Salinas, R. Martinez-Manez, M.D. Marcos, F. Sancenon, A.M. Costero, M. Parra, S. Gil, Chem. Soc. Rev. 41 (2012) 1261–1296.
[2] U. S. E. P. Agency, U. S. Environmental Protection Agency, Washington, DC, 2011.
[3] R. Hodyss, J.L. Beauchamp, Anal. Chem. 77 (2005) 3607–3610.
[4] A. Popov, H. Chen, O.N. Kharybin, E.N. Nikolaev, R.G. Cooks, Chem. Commun. (2005) 1953–1955.
[5] J.M. Sylvia, J.A. Janni, J.D. Klein, K.M. Spencer, Anal. Chem. 72 (2000) 5834–5840.
[6] S.F. Hallowell, Talanta 54 (2001) 447–458.
[7] C. Vourvopoulos, P.C. Womble, Talanta 54 (2001) 459–468.
[8] M. Krausa, K. Schorb, J. Electroanal. Chem. 461 (1999) 10–13.
[9] (a) E. Wallis, T.M. Griffin, N. Popkie Jr., M.A. Eagan, R.F. McAtee, D. Vrazel, J. McKinly, Proc. SPIE Int. Soc. Opt. Eng. 5795 (2005) 54–64. (b) G.A. Eiceman, J.A. Stone, Anal. Chem. 76 (2004) 390A–397A.
[10] (a) S. Yamaguchi, T.M. Swager, J. Am. Chem. Soc. 123 (2001) 12087–12088. (b) S. Zahn, T.M. Swager, Angew. Chem. Int. Ed. 41 (2002) 4226–4230. (c) W. Thomas III, J.P. Amara, R.E. Bjork, T.M. Swager, Chem. Commun. (2005) 4572–4574. (d) A. Narayanan, O.P. Varnavsky, T.M. Swager, T. Goodson III, J. Phys. Chem. C 112 (2008) 881–884. (e) S. Chen, Q. Zhang, J. Zhang, J. Gu, L. Zhang, Sensors Actuators B Chem. 149 (2010) 155–160. (f) A. Rose, Z. Zhu, C.F. Madigan, T.M. Swager, V. Bulovic, Nature 434 (2005) 876–879. (g) I.A. Levitsky, W.B. Euler, N. Tokranova, A. Rose, Appl. Phys. Lett. 90 (2007), 041904. (h) Y. Long, H. Chen, Y. Yang, H. Wang, Y. Yang, N. Li, K. Li, J. Pei, F. Liu, Macromolecules 42 (2009) 6501–6509. (i) J.T. Sarah, C.T. Willianm, J. Mater. Chem. (2006) 2871–2883.
[11] S.J. Toal, W.C. Trogler, J. Mater. Chem. 16 (2006) 2871–2883.
[12] I.A. Buryakov, T.I. Buryakov, V.T. Matsaev, J. Anal. Chem. 69 (2014) 616–631.
[13] (a) S.W. Thomas, G.D. Joly, T.M. Swager, Chem. Rev. 107 (2007) 1339–1386. (b) A. Alvarez, J.M. Costa-Fernández, R. Pereiro, A. Sanz-Medel, A. Salinas-Castillo, Trends Anal. Chem. 30 (2011) 1513–1525. (c) S. Rochat, T.M. Swager, ACS Appl. Mater. Interfaces 5 (2013) 4488–4502.
[14] Z. Hu, B.J. Deibert, J. Li, Chem. Soc. Rev. 43 (2014) 5815–5840.
[15] Y. Ma, S. Wang, L. Wang, Trends Anal. Chem. 65 (2015) 13–21.

[16] (a) ATF, Fed. Regist. 75 (2010) 70291–70293. (b) J. Altschuh, Chemosphere 39 (1999) 1871–1877. (c) NIOSH/OSHA, in: F.W. Mackison, R.S. Stricoff, L.J. Partridge Jr. (Eds.), Occupational Health Guidelines for Chemical Hazards, U.S. Government Printing Office, Washington, DC, 1981. DHHS(NIOSH) Publication No. 81-123 (3 VOLS). (d) C.L. Yaws (Ed.), Handbook of Vapor Pressure, Gulf. Publ. Co, TX, Houston, 1994. (e) T.E. Daubert, R.P. Danner (Eds.), Physical and Thermodynamic Properties of Pure Chemicals Data Compilation, Taylor and Francis, Washington, DC, 1996. (f) M.J. Aernecke, T. Mendum, G. Geurtsen, A. Ostrinskaya, R.R. Kunz, J. Phys. Chem. A 119 (2015) 11514–11522.
[17] T. Forster, K. Kasper, Ein Konzentrationsumschlag der Fluoreszenz, Z. Phys. Chem. 1 (1954) 275–277.
[18] J. Luo, Z. Xie, J.W.Y. Lam, L. Cheng, H. Chen, C. Qiu, H.S. Kwok, X. Zhan, Y. Liu, D. Zhu, B.Z. Tang, Chem. Commun. (2001) 1740–1741.
[19] (a) V.S. Vyas, R. Rathore, Chem. Commun. 46 (2010) 1065–1067. (b) T. Sanji, K. Shiraishi, M. Nakamura, M. Tanaka, Chem. Asian J. 5 (2010) 817–824. (c) S.J. Toal, K.A. Jones, D. Magde, W.C. Trogler, J. Am. Chem. Soc. 127 (2005) 11661–11665. (d) T.L. Andrew, T.M. Swager, J. Am. Chem. Soc. 129 (2007) 7254–7255. (e) S. Kim, H.E. Pudavar, A. Bonoiu, P.N. Prasad, Adv. Mater. 19 (2007) 3791–3795. (f) W.C. Wu, C.Y. Chen, Y. Tian, S.H. Jang, Y. Hong, Y. Liu, R. Hu, B.Z. Tang, Y.T. Lee, C.T. Chen, W.C. Chen, A.K.Y. Jen, Adv. Funct. Mater. 20 (2010) 1413–1423.
[20] X. Sun, Y. Liu, G. Shaw, A. Carrier, S. Dey, J. Zhao, Y. Lei, ACS Appl. Mater. Interfaces 7 (2015) 13189–13197.
[21] (a) M.S. Meaney, V.L. McGuffin, Anal. Chim. Acta 610 (2008) 57–67. (b) B. Valeur, Molecular Fluorescence: Principles and Applications, Wiley-VCH, Weinheim, 2002.
[22] (a) G. He, N. Yan, J. Yang, H. Wang, L. Ding, S. Yin, Y. Fang, Macromolecules 44 (2011) 4759–4766. (b) H. Nie, Y. Lv, L. Yao, Y. Pan, Y. Zhao, P. Li, G. Sun, Y. Ma, M. Zhang, J. Hazard. Mater. 264 (2014) 474–480.
[23] Y.-Z. Zhang, X. Xiang, P. Mei, J. Dai, L.-L. Zhang, Y. Liu, Spectrochim. Acta A 72 (2009) 907–914.
[24] J.R. Lakowicz, Principles of Fluorescence Spectroscopy, Springer, Singapore, 2006.
[25] J.R. Cox, P. Müller, T.M. Swager, J. Am. Chem. Soc. 133 (2011) 12910–12913.
[26] (a) J.-S. Yang, T.M. Swager, J. Am. Chem. Soc. 120 (1998) 11864–11873. (b) H. Zhou, M.H. Chua, J. Xu, Polym. Chem. 10 (2019) 3822–3840.
[27] H. Zhou, J. Li, M.H. Chua, H. Yan, B.Z. Tang, J. Xu, Polym. Chem. 5 (2014) 5628–5637.
[28] H. Zhou, X. Wang, T.T. Lin, J. Song, B.Z. Tang, J. Xu, Polym. Chem. 7 (2016) 6309–6317.
[29] J. Chen, Z. Xie, J.W.Y. Lam, C.C.W. Law, B.Z. Tang, Macromolecules 36 (2003) 1108–1117.
[30] X. Wang, W. Wang, Y. Wang, J.Z. Sun, B.Z. Tang, Polym. Chem. 8 (2017) 2353–2362.
[31] T. Chen, H. Yin, Z.-Q. Chen, G.-F. Zhang, N.-H. Xie, C. Li, W.-L. Gong, B.Z. Tang, M.-Q. Zhu, Small 12 (2016) 6547–6552.
[32] W.Z. Yuan, R. Hu, J.W.Y. Lam, N. Xie, C.K.W. Jim, B.Z. Tang, Chem. Eur. J. 18 (2012) 2847–2856.
[33] J. Liu, Y. Zhong, J.W.Y. Lam, P. Lu, Y. Hong, Y. Yu, Y. Yue, M. Faisal, H.H.Y. Sung, I.D. Williams, K.S. Wong, B.Z. Tang, Macromolecules 43 (2010) 4921–4936.
[34] R. Hu, J.W.Y. Lam, J. Liu, H.H.Y. Sung, I.D. Williams, Z. Yue, K.S. Wong, M.M.F. Yuen, B.Z. Tang, Polym. Chem. 3 (2012) 1481–1489.
[35] R. Hu, J.W.Y. Lam, M. Li, H. Deng, J. Li, B.Z. Tang, J. Polym. Sci. A Polym. Chem. 51 (2013) 4752–4764.
[36] M.H. Chua, H. Zhou, T.T. Lin, J. Wu, J.W. Xu, J. Polym. Sci. A Polym. Chem. 55 (2017) 672–681.
[37] W.Z. Yuan, H. Zhao, X.Y. Shen, F. Mahtab, J.W.Y. Lam, J.Z. Sun, B.Z. Tang, Macromolecules 42 (2009) 9400–9411.
[38] H. Zhou, Q. Ye, W.T. Neo, J. Song, H. Yan, Y. Zong, B.Z. Tang, T.S. Andy Hor, J. Xu, Chem. Commun. 50 (2014) 13785–13788.
[39] P. Lu, J.W.Y. Lam, J. Liu, C.K.W. Jim, W. Yuan, C.Y.K. Chan, N. Xie, Q. Hu, K.K.L. Cheuk, B.Z. Tang, Macromolecules 44 (2011) 5977–5986.
[40] Z. Zhao, T. Jiang, Y. Guo, L. Ding, B. He, Z. Chang, J.W.Y. Lam, J. Liu, C.Y.K. Chan, P. Lu, L. Xu, H. Qiu, B.Z. Tang, J. Polym. Sci. A Polym. Chem. 50 (2012) 2265–2274.
[41] W. Yuan, W. Chi, R. Liu, H. Li, Y. Li, B.Z. Tang, Macromol. Rapid Commun. 38 (2017) 1600745.
[42] H. Li, J. Wang, J.Z. Sun, R. Hu, A. Qin, B.Z. Tang, Polym. Chem. 3 (2012) 1075–1083.
[43] Q. Wang, M. Chen, B. Yao, J. Wang, J. Mei, J.Z. Sun, A. Qin, B.Z. Tang, Macromol. Rapid Commun. 34 (2013) 796–802.
[44] Y. Wu, B. He, C. Quan, C. Zheng, H. Deng, R. Hu, Z. Zhao, F. Huang, A. Qin, B.Z. Tang, Macromol. Rapid Commun. 38 (2017) 1700070.
[45] R. Hu, J.W.Y. Lam, Y. Yu, H.H.Y. Sung, I.D. Williams, M.M.F. Yuenc, B.Z. Tang, Polym. Chem. 4 (2013) 95–105.

[46] J. Li, J. Liu, J.W.Y. Lam, B.Z. Tang, RSC Adv. 3 (2013) 8193–8196.
[47] J. Wang, B. Li, D. Xin, R. Hu, Z. Zhao, A. Qin, B.Z. Tang, Polym. Chem. 8 (2017) 2713–2722.
[48] Y. Liu, M. Gao, J.W.Y. Lam, R. Hu, B.Z. Tang, Macromolecules 47 (2014) 4908–4919.
[49] Y. Liu, M. Gao, Z. Zhao, J.W.Y. Lam, B.Z. Tang, Polym. Chem. 7 (2016) 5436–5444.
[50] G. Liang, L.-T. Weng, J.W.Y. Lam, W. Qin, B.Z. Tang, ACS Macro Lett. 3 (2014) 21–25.
[51] J. Wang, J. Mei, W. Yuan, P. Lu, A. Qin, J. Sun, Y. Ma, B.Z. Tang, J. Mater. Chem. 21 (2011) 4056–4059.
[52] H. Li, H. Wu, E. Zhao, J. Li, J.Z. Sun, A. Qin, B.Z. Tang, Macromolecules 46 (2013) 3907–3914.
[53] B. Yao, T. Hu, H. Zhang, J. Li, J.Z. Sun, A. Qin, B.Z. Tang, Macromolecules 48 (2015) 7782–7791.
[54] J. Liu, Y. Zhong, P. Lu, Y. Hong, J.W.Y. Lam, M. Faisal, Y. Yu, K.S. Wong, B.Z. Tang, Polym. Chem. 1 (2010) 426–429.
[55] P. Lu, J.W.Y. Lam, J. Liu, C.K.W. Jim, W. Yuan, N. Xie, Y. Zhong, Q. Hu, K.S. Wong, K.K.L. Cheuk, B.Z. Tang, Macromol. Rapid Commun. 31 (2010) 834–839.
[56] X.-M. Hu, Q. Chen, D. Zhou, J. Cao, Y.-J. He, B.-H. Han, Polym. Chem. 2 (2011) 1124–1128.

CHAPTER

18

AIE-based chemosensors for vapor sensing

Meng Li[a,b], *Dong Wang*[a], *and Ben Zhong Tang*[c]

[a]Center for AIE Research, Shenzhen University, Shenzhen, China [b]Shenzhen Institute of Aggregate Science and Technology, School of Science and Engineering, The Chinese University of Hong Kong, Shenzhen, Guangdong, China [c]School of Science and Engineering, Shenzhen Key Laboratory of Functional Aggregate Materials, The Chinese University of Hong Kong, Shenzhen, Guangdong, China

1 Introduction

Vapor sensing is critical in numerous areas, including environmental monitoring, food safety management, and protection of homeland security. Traditional vapor sensing techniques including gas chromatography (GC), mass spectrometry (MS), ion mobility spectrometry (IMS), and tunable diode laser absorption spectroscopy (TDLAS) are often inconvenient, expensive, and time consuming, making them less suitable for real-time and on-site detection even though they are highly accurate and sensitive. Therefore, fluorescent chemosensors have aroused increasing attention because of their advantages such as low cost, real-time monitoring, easy operation, and structural diversity for the sensing of various analytes. In spite of these advantages, the usage of fluorescent chemosensors for vapors is still challenging because vapor detection usually proceeds on the interface between the gaseous analytes and fluorophore-loaded thin films. Conventional fluorophores featuring large π-planar structures are highly emissive in solution but faintly emissive in solid due to the aggregation-caused quenching (ACQ) drawbacks. Besides, the compact π-π stacking of planar backbones hampers the diffusion of the analytes in the thin film, resulting in a reduced contact surface and poor responsiveness.

The emergence of aggregation-induced emission (AIE), an innovative concept first coined in 2001 by Tang's group, has paved a new avenue to conquer the ACQ drawbacks. The working principle of the AIE phenomenon is generally believed to be restriction of intramolecular motion (RIM) including both restriction of intramolecular rotation (RIR) and restriction of

Aggregation-Induced Emission (AIE)
https://doi.org/10.1016/B978-0-12-824335-0.00003-9

intramolecular vibration (RIM) [1–3]. On the basis of the RIM mechanism, AIE luminogens (AIEgens) typically have twisted nonplanar conformation with multiple molecular rotors (or vibrators). This specific configuration could inhibit the condensed packing of AIEgens in solid or thin film state, which is favorable for the adsorption and diffusion of the analytes in the porous structure. In addition, the porosity provides the thin film with additional selectivity benefiting from the size-selective effect [4,5].

Based on the RIM theory, a variety of AIE-active scaffolds such as hexaphenylsilole (HPS), tetraphenylethene (TPE), 9,10-distrylanthracene (DSA), tetraphenylpyrazine (TPP), cyanostilbene, and annulene have been developed in the past two decades [6–9]. Taking advantage of the unique luminescent properties including large Stokes' shift, good photostability, and stimulus responsive emission, these AIEgens offer an emerging platform as bio- or chemosensors for the detection of metal ions [10], anions [11], aromatic explosives [12,13], pH values [14], amines [15], and a variety of biomolecules [16–18]. In addition, the intensive emission of AIEgens in thin films also favors vapor/gas sensing with high sensitivity.

In this chapter, we have overviewed recent advances in AIE-based chemosensors for the detection of various vapors mainly including nitroaromatic explosives, acid and amine vapors, and other volatile organic compounds (VOCs). In the first section, we have summarized the sensing mechanisms that are generally involved in vapor sensing events. In the second section, we have talked about the recent development of AIE-based chemosensors for explosive vapors. Since this is the only chapter in the book on AIE-based chemosensors for the detection of explosives, only one figure is presented to give several examples of nitroaromatic explosive vapor sensing. In the third and fourth sections, vapor sensing of volatile acid, amine, and other VOCs (e.g., methanol, phenolic compounds, benzene, toluene) has been introduced. The discussion is organized according to the underlying sensory mechanisms and AIE systems with different structural features including small molecules, polymers, and hybrid materials are covered. Due to the limited space, only some representative examples are introduced and more examples are provided in Table 1. In the last section, a brief summary and further perspectives are listed. We expect that this chapter can provide valuable inspirations for the further development of the AIE-based chemosensors for vapor sensing.

2 Sensing mechanisms of AIE-based chemosensors

There are several mechanisms involved in the sensing event using AIE-based chemosensors to give a distinct optical response such as the color change, fluorescence intensity variation, and spectral shift. The sensing mechanisms are summarized in Fig. 1 including photoinduced electron transfer (PET), excited-state intramolecular charge transfer (ICT), excited-state intramolecular proton transfer (ESIPT), and restriction of intramolecular motion (RIM).

2.1 Photoinduced electron transfer (PET) mechanism

PET refers to an excited-state electron transfer process that can occur in both intramolecular and intermolecular behaviors [19–21]. There are two kinds of PET processes depending on the energy level differences of the highest occupied molecular orbital (HOMO) and lowest unoccupied molecular orbital (LUMO) between the fluorophore and the analyte (intermolecular

TABLE 1 Summary of photophysical properties and sensing performance of AIE-based chemosensors for different vapors.

AIEgens	Luminescent property	Analytes and L.O.D.	Mechanism of detection	Ref.
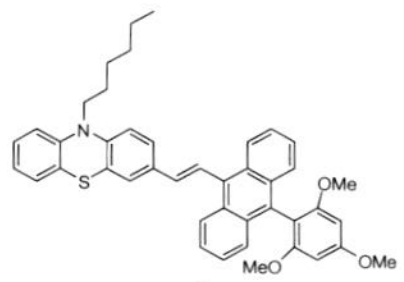 1	λ_{abs} = 400 nm (in acetonitrile) λ_{em} = 624 nm (in acetonitrile)	TNT: saturated vapor	PET	[27]
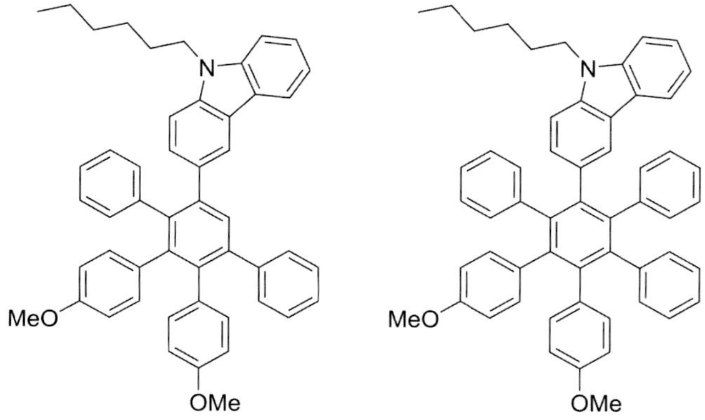	λ_{abs} = 270 nm (in THF) λ_{em} = 360 nm, 380 nm (in THF)	TNT: saturated vapor	PET	[28]
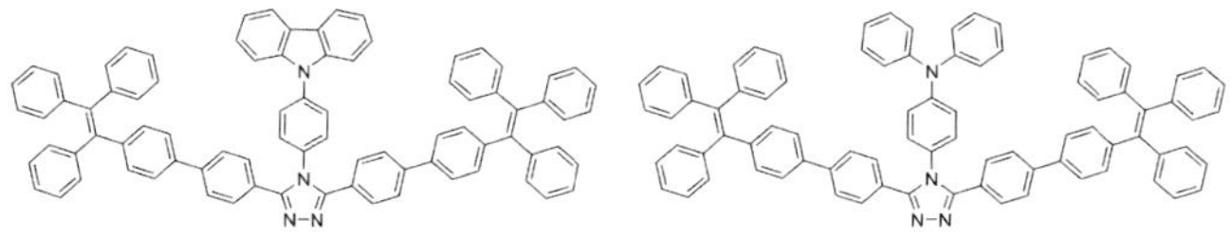	**3a**: λ_{abs} = 300 nm (solid) λ_{em} = 480 nm (solid) **3b**: λ_{abs} = 360 nm (solid) λ_{em} = 475 nm (solid)	DNT: saturated vapor	PET	[29]

Continued

TABLE 1 Summary of photophysical properties and sensing performance of AIE-based chemosensors for different vapors—cont'd

AIEgens	Luminescent property	Analytes and L.O.D.	Mechanism of detection	Ref.
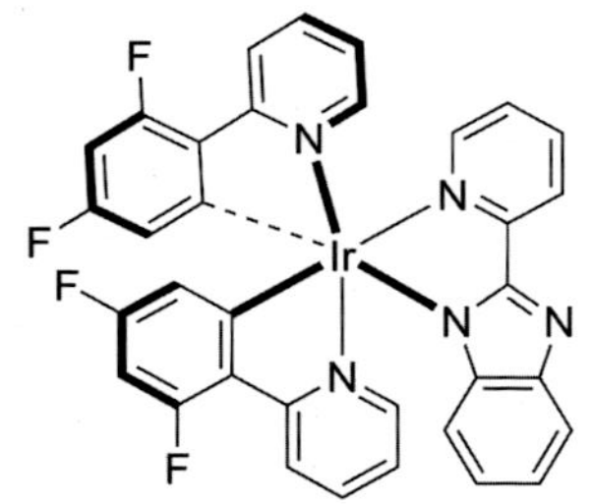 F, F, F, F, N, N, N, N, N, Ir	$\lambda_{abs}=245$ nm, 351 nm (in dichloromethane) $\lambda_{em}=550$ nm (solid, QY: 7.4%)	TNT: saturated vapor	PET	[30]
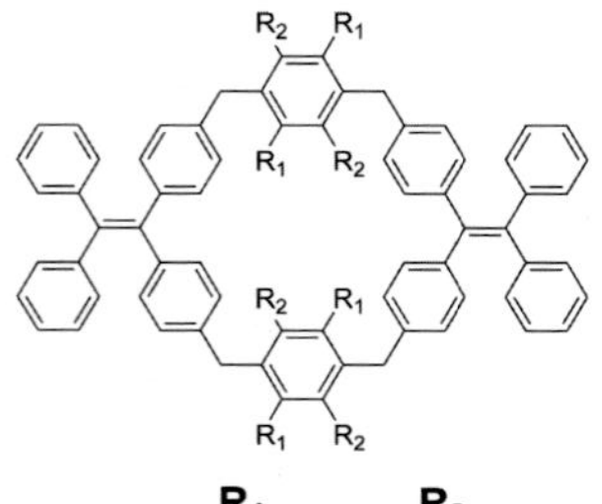 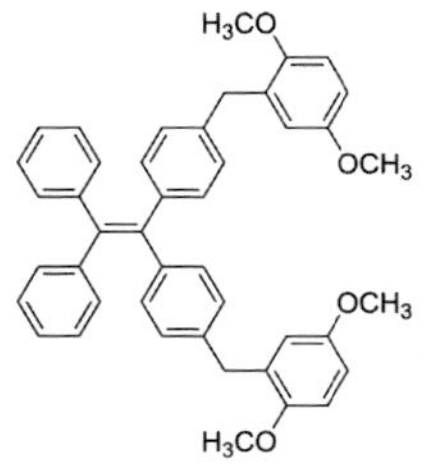 **5a**: R_1 = OCH_3, R_2 = H; **5b**: R_1 = OC_5H_{11}-n, R_2 = H; **5c**: R_1 = OCH_3, R_2 = OCH_3; **6** (H_3CO, OCH_3)	$\lambda_{abs}=320$–350 nm (in THF/H_2O) $\lambda_{em}=450$–480 nm (in THF/H_2O)	TNT: saturated vapor	PET	[31]
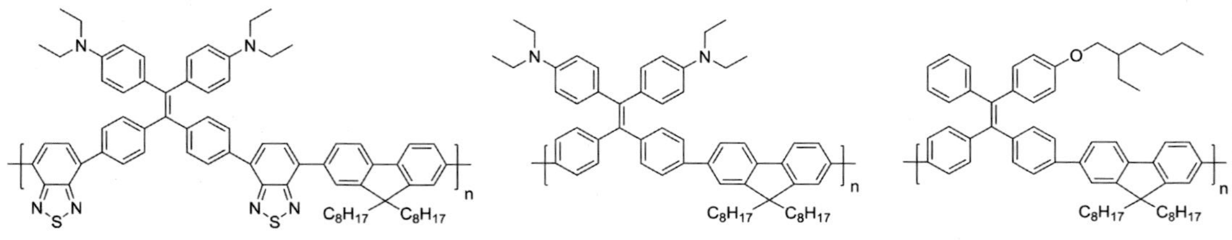 N, N, N, S, N, N, S, N, C_8H_{17}, C_8H_{17}, n; N, N, C_8H_{17}, C_8H_{17}, n; O, C_8H_{17}, C_8H_{17}, n	**P1**: $\lambda_{abs}=327$ nm, 493 nm (solid) $\lambda_{em}=665$ nm (solid) **P2**: $\lambda_{abs}=338$ nm, 390 nm (solid) $\lambda_{em}=550$ nm (solid)	TNT, DNT: saturated vapor	PET	[32]

Structure	Optical properties	Analyte	Mechanism	Ref.
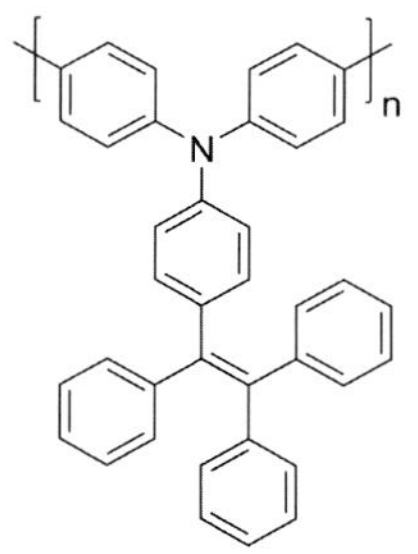 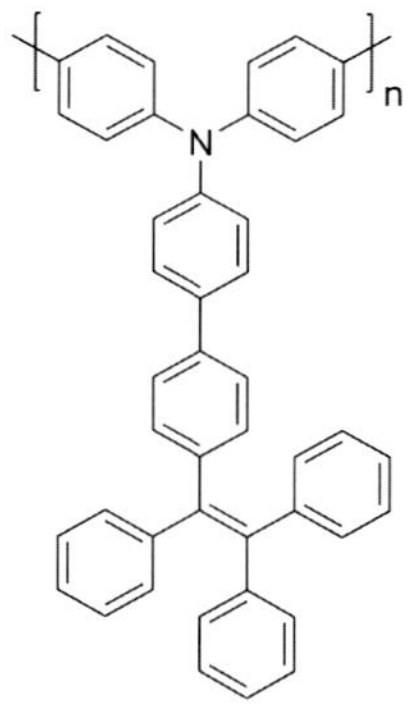	**P3**: $\lambda_{abs} = 362$ nm (solid) $\lambda_{em} = 501$ nm (solid) **P4**: $\lambda_{abs} = 386$ nm (solid) $\lambda_{em} = 511$ nm (solid) **P5**: $\lambda_{abs} = 380$ nm (solid) $\lambda_{em} = 495$ nm (solid)	TNB, DNB, NB: saturated vapor	PET	[33]
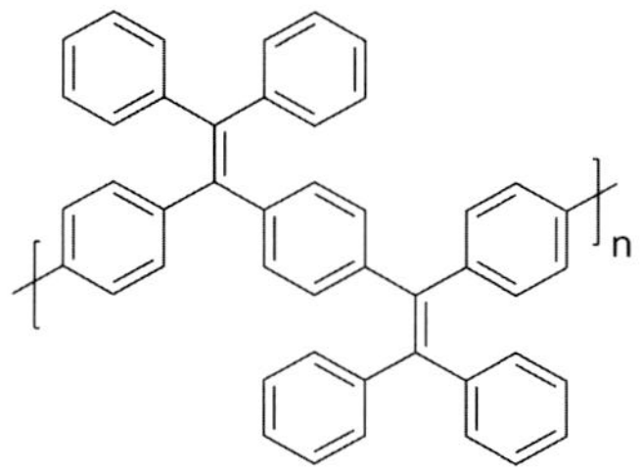 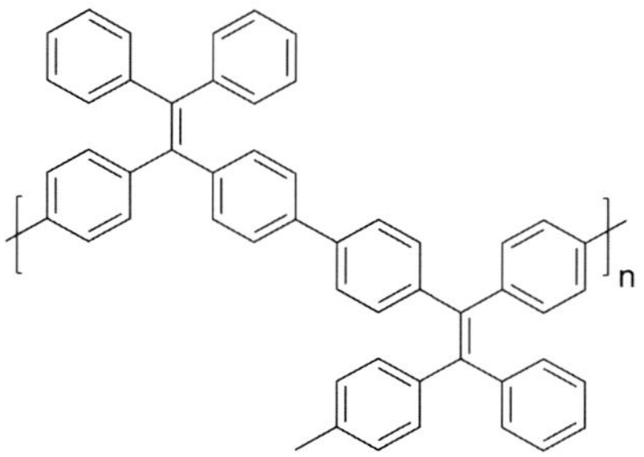	$\lambda_{abs} = 320–340$ nm (THF) $\lambda_{em} = 460–490$ nm (THF:H_2O, 1:9 v/v)	TNT, DNT, NT: saturated vapors	PET	[34]

Continued

TABLE 1 Summary of photophysical properties and sensing performance of AIE-based chemosensors for different vapors—cont'd

AIEgens	Luminescent property	Analytes and L.O.D.	Mechanism of detection	Ref.
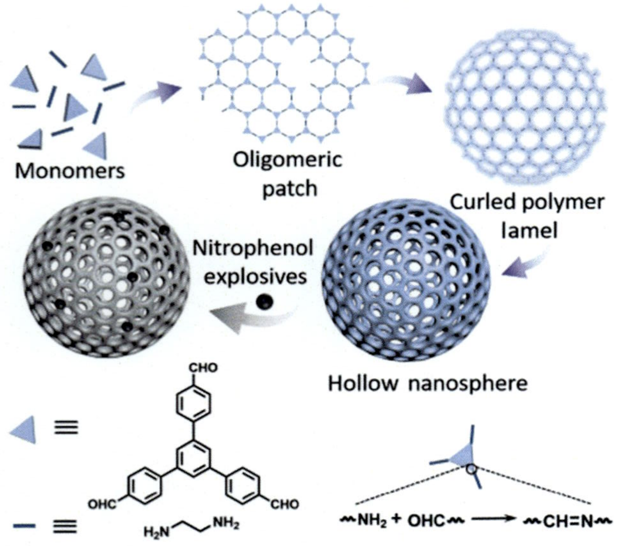	$\lambda_{abs} = 300$ nm (solid) $\lambda_{em} = 460$ nm (solid)	2,4,6-trinitrophenol (TNP): saturated vapor	PET	[35]
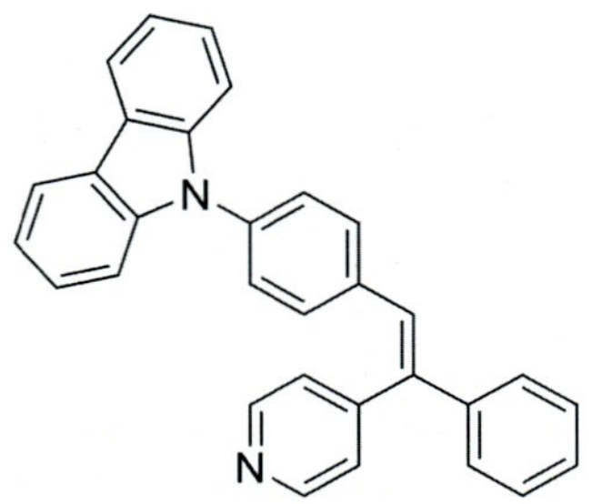	**7a** $\lambda_{abs} = 340$ nm (THF/H_2O) $\lambda_{em} = 450$ nm (THF/H_2O)	TFA: N.A. TEA: N.A.	ICT	[36]

Structure	Photophysical properties	Analyte: LOD	Mechanism	Ref.
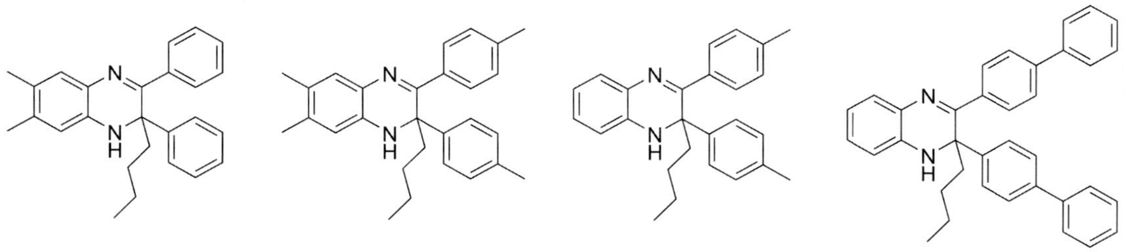	λ_{abs} = 390–420 nm (in THF) λ_{em} = 500–540 nm (in THF)	**8b** NH_3: 690 ppb	ICT	[37]
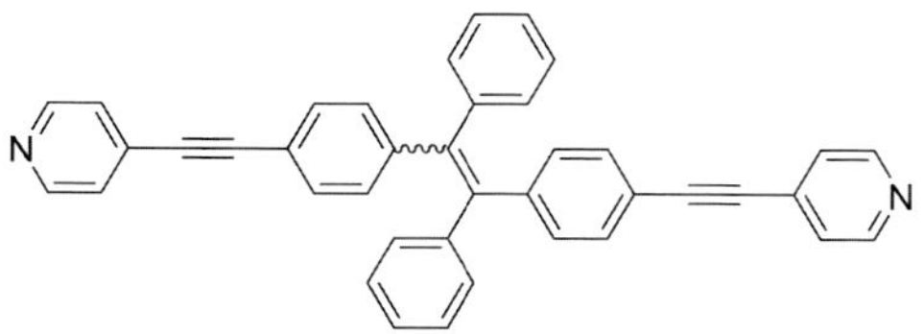	**(Z)-9**: λ_{abs} = 205, 281, 347 nm (in ethanol) λ_{em} = 470 nm (solid, QY = 37.6%) **(E)-9**: λ_{abs} = 204, 289, 345 nm (in ethanol) λ_{em} = 464 nm (solid, QY = 43.8%)	HCl: N.A.	ICT	[39]
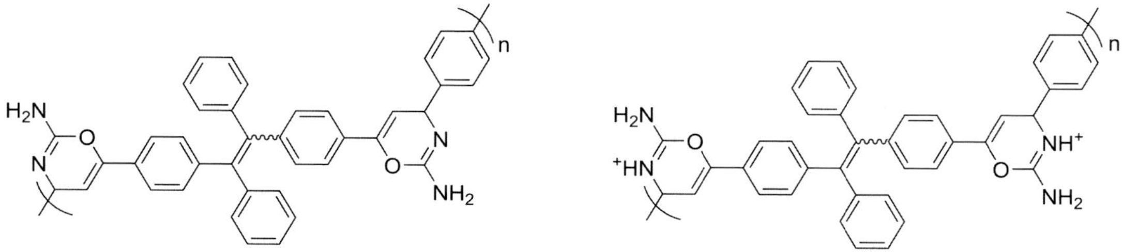	**P9a**: λ_{abs} = 440 nm (solid) λ_{em} = 538 nm (solid) **P9b**: λ_{abs} = 550 nm (solid) λ_{em} = 615 nm (solid)	NH_3: 960 ppb	ICT	[40]

Continued

TABLE 1 Summary of photophysical properties and sensing performance of AIE-based chemosensors for different vapors—cont'd

AIEgens	Luminescent property	Analytes and L.O.D.	Mechanism of detection	Ref.
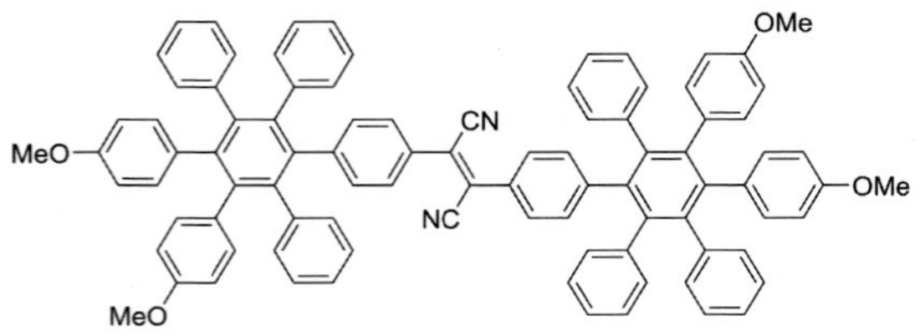	$\lambda_{abs}=282, 372$ nm (in THF) $\lambda_{em}=455$ nm (in THF)	Different amines: N.A.	Aliphatic amines: ICT Aromatic amine: PET	[41]
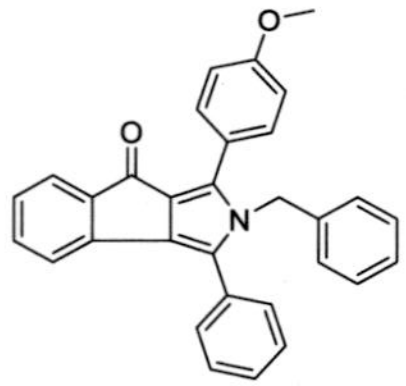 11	$\lambda_{abs}=365$ nm (solid) $\lambda_{em}=527$ nm (solid, QY=8.44%)	TFA: 0.77 ppm HCl: 0.088 ppm HNO_3: 0.38 ppm AcOH: 0.48 ppm Aniline: 6.04 ppb *N*, *N*-dimethylaniline: 8.82 ppb *N*-methylaniline: 3.24 ppb Phenylhydrazine: 0.16 ppb	PET + ICT	[42]
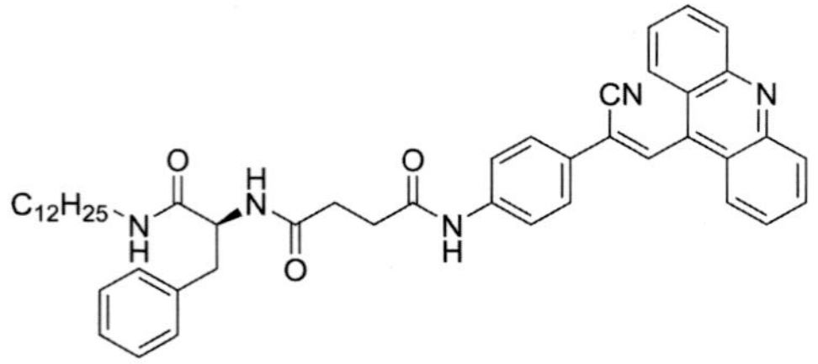	$\lambda_{abs}=365$ nm (solid) $\lambda_{em}=508$ nm (solid)	TFA: 0.5 ppm HCl: 1.26 ppm Acetic acid: 0.12 ppm Aniline: 0.33 ppb *N*-methylaniline: 0.48 ppb *N*, *N*-dimethylaniline: 1.38 ppb Phenylhydrazine: 3.8 ppb	ICT + PET	[43]

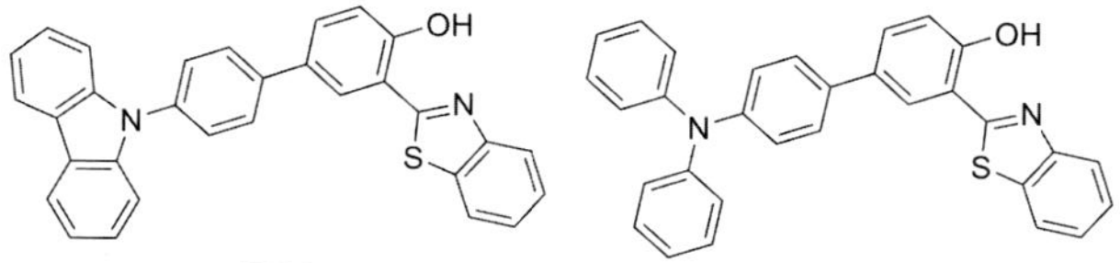 HCP HTPA	**OHBT**: $\lambda_{abs} = 377$ nm (solid) $\lambda_{em} = 502$ nm (solid, 25%) **CHBT**: $\lambda_{abs} = 340$ nm (solid) $\lambda_{em} = 526$ nm (solid, 82%) **DHBT**: $\lambda_{abs} = 378$ nm (solid) $\lambda_{em} = 536$ nm (solid, 31%) **HCP**: $\lambda_{abs} = 396$ nm (solid) $\lambda_{em} = 560$ nm (solid, 70%) **HTPA**: $\lambda_{abs} = 418$ nm (solid) $\lambda_{em} = 588$ nm (solid, 21%)	Different acid and amine vapors: N.A.	ESIPT	[49]
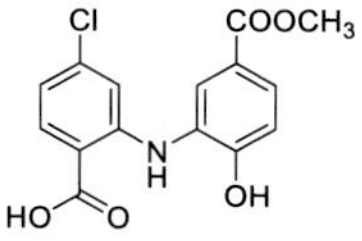 13	$\lambda_{abs} = 330$ nm (in ethanol) $\lambda_{em} = 480$ nm (solid)	NH_3: 20 ppm	RIM	[50]
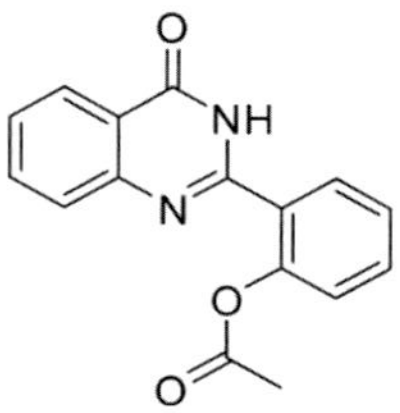 14a	$\lambda_{abs} = 285$ nm (THF/H_2O, 1:99 v/v) **14a** is nonemissve After aminolysis reaction, emissive at 492 nm	NH_3: 8.4 ppm	Aminolysis reaction	[51]

Continued

TABLE 1 Summary of photophysical properties and sensing performance of AIE-based chemosensors for different vapors—cont'd

AIEgens	Luminescent property	Analytes and L.O.D.	Mechanism of detection	Ref.
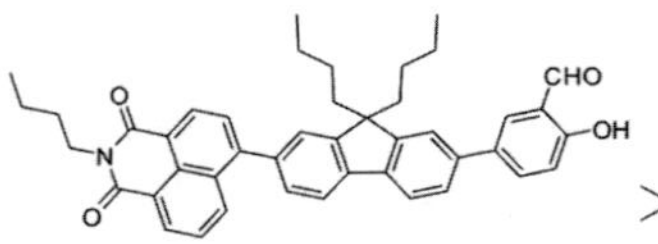 15a 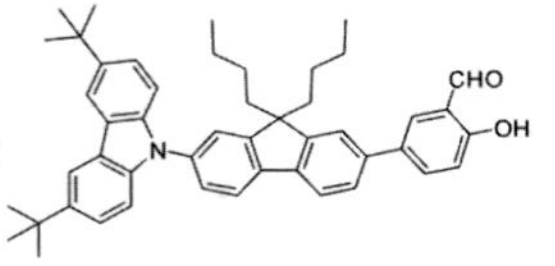 15c	**7a**: λ_{abs} = 299, 322, 374 nm (solid) λ_{em} = 502 nm (solid) **7c**: λ_{abs} = 298, 341 nm λ_{em} = 553 nm (solid)	**7a**: 1.306 ppm for aniline 0.064 ppm for n-propylamine 0.145 ppm for TEA **7c**: 5.32 ppm for aniline 0.79 ppm for n-propylamine 2.714 ppm for TEA	Schiff base reaction	[52]
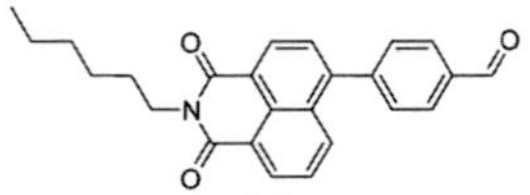 16	λ_{abs} = 352 nm (in DMF) λ_{em} = 426 nm (in DMF)	Hydrazine vapors: 0.003%	Schiff base reaction	[53]
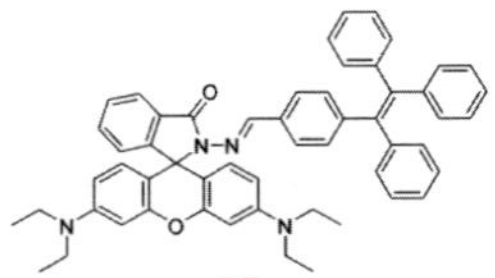 17a	λ_{abs} = 365 nm (THF/H_2O, 9:1 v/v) λ_{em} = 467–500 nm (THF/H_2O, 9:1 v/v)	HCl: 20 ppm	Ring open and close	[54]
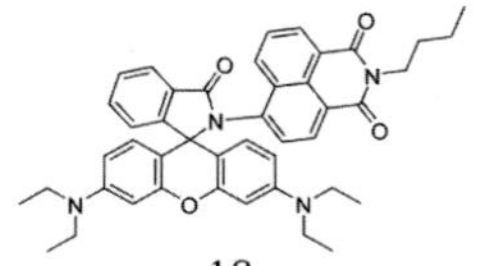 18a	λ_{abs} = 380 nm (pristine); 445 nm (after grinding) λ_{em} = 517 nm (pristine); 582 nm (after grinding)	HCl vapor: N.A.	Ring open and close	[55]

Structure	Photophysical properties	Analyte: LOD	Mechanism	Ref.
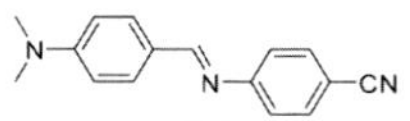 19	$\lambda_{abs} = 356$ nm (in hexane) $\lambda_{em} = 498$ nm (solid)	Methanol: 9.65 ppm	ICT + RIM	[57]
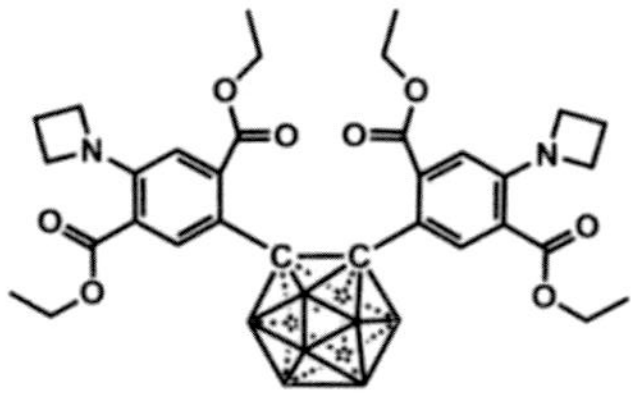 20	$\lambda_{abs} = 310$, 360 nm (Film on smooth glass plate made by spin-coating) $\lambda_{em} = 610$ nm (Film on smooth glass plate made by spin-coating, 73%)	Phenol: 0.4 ppt o-cresol: 0.3 ppt m-cresol: 10 ppt p-cresol: 0.8 ppt	ICT + RIM	[58]

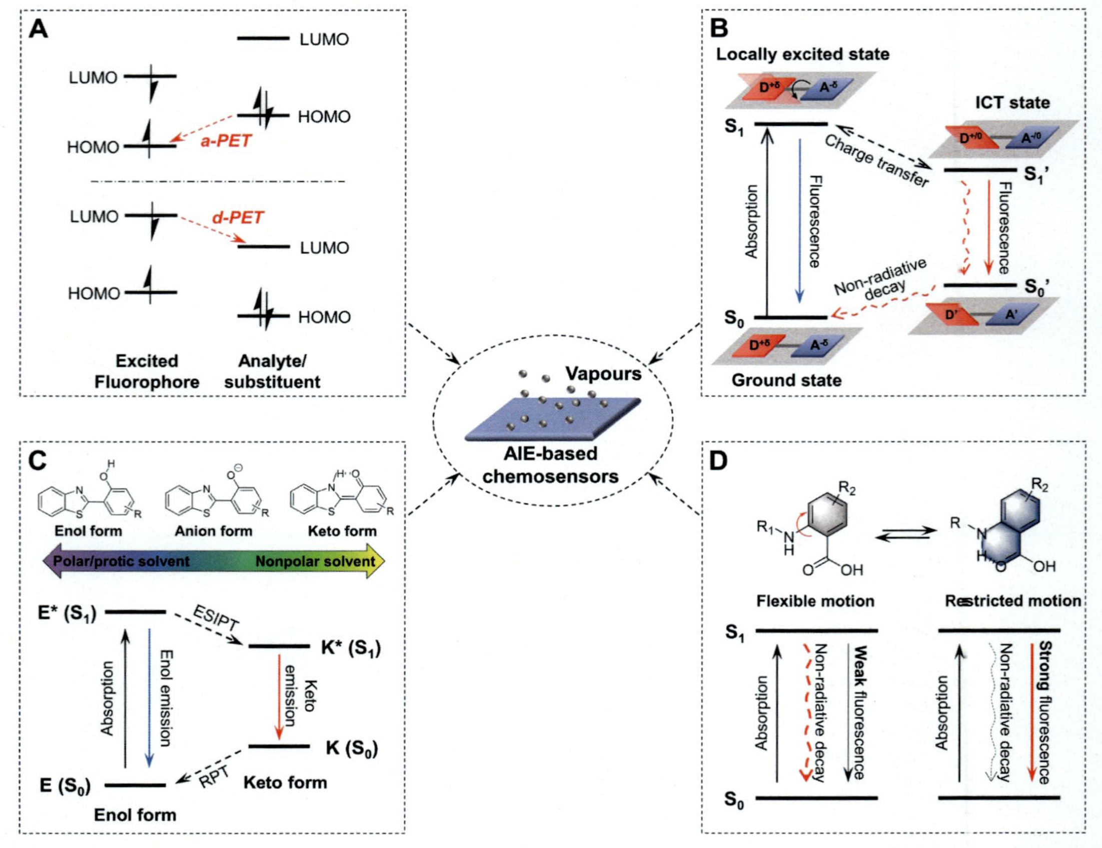

FIG. 1 The sensing mechanisms of AIE-based chemosensors for vapor sensing: (A) photoinduced electron transfer (PET), (B) intramolecular charge transfer (ICT), (C) excited-state intramolecular proton transfer (ESIPT), and (D) restriction of intramolecular motion (RIM).

PET)/substituent (intramolecular PET) (Fig. 1A). When the HOMO energy of the fluorophore is less than that of the analyte/substituent, electron transfer occurs from the analyte/substituent to the fluorophore and this process is called "*a*-PET" as the fluorophore acts as an electron acceptor. On the other hand, if the fluorophore donates one electron from its excited state to the analyte/substituent LUMO, the process is called "*d*-PET" instead. In both situations, the PET process causes fluorescence quenching (reductive quenching and oxidative quenching); therefore, the sensing event based on PET is usually accompanied by fluorescence "off-on" or "on-off" switching. Such an effect has been widely used for the detection of aromatic explosive vapors such as trinitrotoluene (TNT) because the aromatic explosives are electron-deficient with a low LUMO energy. Besides, there are also several examples of the detection of aromatic amines with high HOMO energy, which could donate one electron to fluorophore HOMO in the excited state.

2.2 Intramolecular charge transfer (ICT) mechanism

ICT is a general phenomenon existing in fluorophores containing electron donor-acceptor (D-A) systems [22,23]. Upon photoexcitation, the locally excited (LE) state undergoes a rapid charge transfer transition to the ICT state with a lower energy and subsequently returns to the ground state with a red-shifted emission (Fig. 1B). In a polar solvent, the ICT process occurs faster as the polar ICT species could be better stabilized in the polar environment. Additionally, it is also believed that the ICT state favors multiple nonradiative decays (such as solvent relaxation), leading to an impaired fluorescence efficiency. Thus, fluorophores with ICT effect usually show positive solvatochromic property revealed by the red-shifted emission and reduced fluorescence quantum yield with increasing solvent polarity. Different from the PET process that is accompanied by fluorescence quenching with a neglectable spectral shift, the ICT process typically results in decreased fluorescence efficiency as well as red-shifted emission. Based on the mechanism, ICT emission is quite sensitive to the external perturbations (e.g., polarity, temperature, alkalinity, and acidity) allowing it to be exploited for developing high-performance sensors for various analytes, including amine, acid vapors, and protic solvents such as methanol, which could affect the D-A strength in the heteroatom-containing fluorophores.

2.3 Excited-state intramolecular proton transfer (ESIPT) mechanism

ESIPT is a photo-tautomerization process that produces an intramolecular hydrogen bond between a hydrogen bond donor (—OH and —NH_2) and a hydrogen bond acceptor (=N— and C=O) upon photoexcitation (Fig. 1C). A majority of the ESIPT fluorophores reported to date are derivatives of 2-(2′-hydroxyphenyl)benzimidazole (HBI), 2-(2′-hydroxyphenyl) benzoxazole (HBO), and 2-(2′-hydroxyphenyl)benzothiazole (HBT) [24–26]. Take HBT derivative, for example, ESIPT is a four-level photochemical process ($E \rightarrow E^* \rightarrow K^* \rightarrow K$) involving the photoexcitation of the original state (Enol form), the redistribution of the electronic charge to produce the excited state of Keto form followed by a Keto emission, and eventually the reverse proton transfer (RPT) to produce the original Enol form (Fig. 1). According to the ESIPT mechanism, the optical property of fluorophores with ESIPT effect is sensitive to the stability of the intramolecular hydrogen bond, which could be affected by the polarity of the external environment. Typically, fluorophores with the ESIPT effect feature two emission bands: a short-wavelength emission resulting from Enol emission and a longer one related to the Keto emission (or ESIPT emission). Some research reported a third emission band from the phenolic anion form when the phenolic hydroxyl group was deprotonated by basic species or protic solvents such as ethanol. Therefore, the ESIPT molecule might have three different emission patterns (single emission, dual emission, and triple emission) depending on the preferred conformation in a specific condition. Such environment-based fluorescence responses are quite useful for the sensing of various vapors, especially those with heteroatoms that could affect the formation of the intramolecular hydrogen bond.

2.4 Restriction of intramolecular motion (RIM) mechanism

RIM is a widely accepted mechanism to decipher the AIE phenomenon. The active intramolecular motion favors the nonradiative decay to quench the fluorescence emission. Therefore, the fluorescence signal of AIE-active molecules is sensitive to the analytes that could affect their intramolecular motions (Fig. 1D). For instance, the intramolecular hydrogen bonding formation ("restricted motion" state) and deformation ("flexible motion" motion) in response to analytes could give rise to the fluorescence "on" to "off" switch.

Unlike the sensing event in a homogenous solution, vapor sensing only occurs on the interfaces between the gaseous analytes and the thin films. Thus, in most cases, the responsive behavior relies on the physical adsorption. However, vapor sensing based on chemical reactions and a further alternation of the photophysical property of AIEgens have also been reported. This kind of reactive chemosensors are discussed in Section 4.5.

It is worth mentioning that the above mechanisms are not independent of each other. In a practical case, one or more mechanisms could be involved. To implement the real-world applications, one single sensor is usually not sufficient to realize the sensing and discrimination of various vapors. Therefore, the multifunctional sensor array combining a series of sensors to generate a fingerprint signal is an appealing solution. This is further explained in Section 5.2 with the examples reported in literatures.

3 AIE-based chemosensors for nitroaromatic explosive vapor

Reliable detection of nitroaromatic explosive vapors is of critical importance in various fields ranging from civilian safety to global antiterrorism. Because of the electron-deficient feature of nitroaromatics, the *d*-PET process is likely to occur from the excited-state of electron-rich AIEgens to the nitroaromatic explosives. Thus, the sensitivity of the AIE-based chemosensors is directly related to the quenching efficiency upon exposure to nitroaromatics.

Chakravarty group has reported an electron-rich compound **1** for the selective detection of 2,4,6-trinitrotoluene (TNT) vapor among different nitroaromatic explosives (Fig. 2A and B) [27]. Compound **1** showed a typical AIE characteristic with a low fluorescence quantum yield (QY) of 4% in acetonitrile and a high QY (53%) in the thin film state. The LUMO of AIEgen **1** is close to and higher than that of TNT, benefiting the *d*-PET process from the excited state of **1** to TNT (Fig. 2A). To perform the vapor sensing, an AIEgen **1**-impregnated paper strip was put in the cap of a sealed glass vial with saturated TNT vapor (Fig. 2B). As a result, the yellow

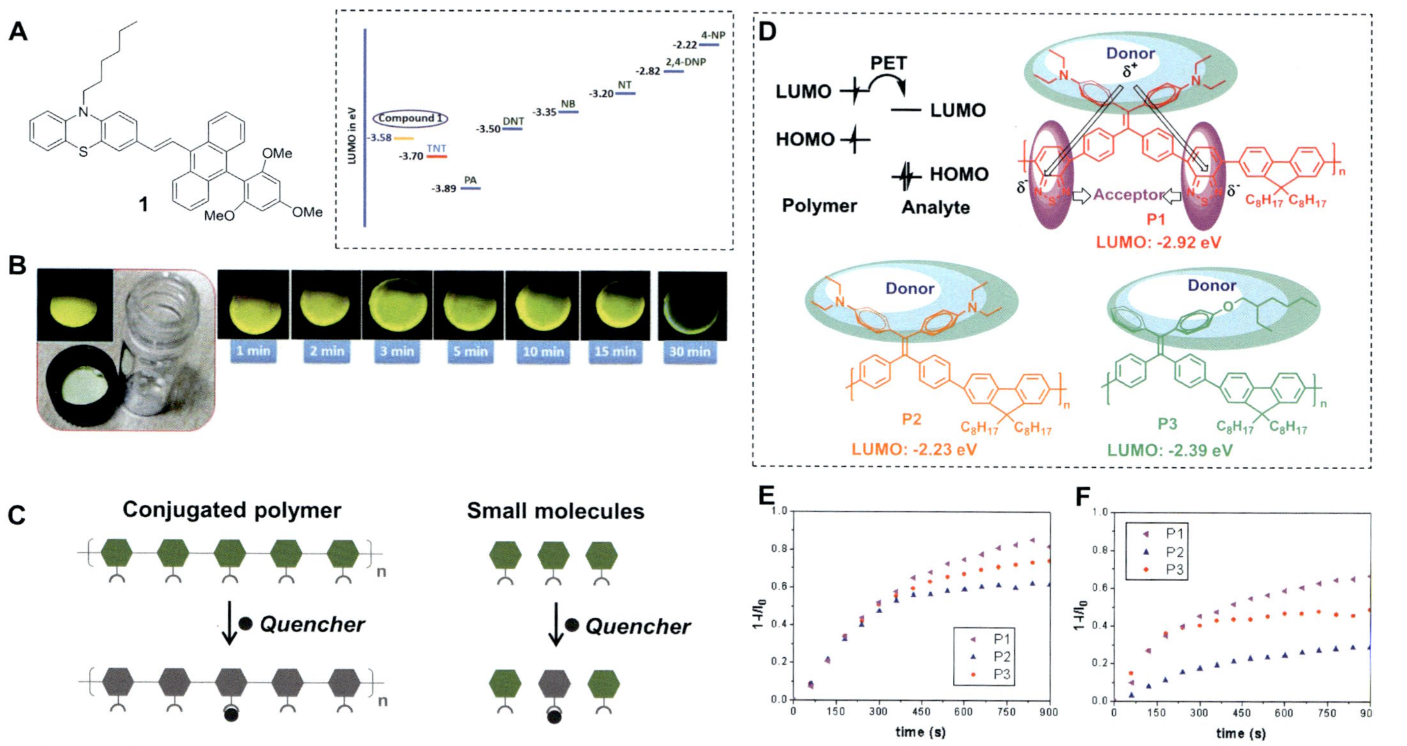

FIG. 2 (A) Chemical structure of AIEgen **1** and the LUMO energy levels of **1** and various nitroaromatics calculated using cyclic voltammetry. (B) Sensing of TNT vapor using the **1**-loaded paper strip at 25°C. 100 mg of solid TNT was placed in a glass vial. Pictures were taken under a 365-nm UV lamp. (C) The schematic illustration of "molecular wire" effect. (D) Chemical structures of **P1**, **P2**, **P3**, and the schematic illustration of PET process from polymer to analyte. (E, F) Time-dependent fluorescence quenching of polymer films upon exposure to saturated (E) DNT and (F) TNT vapors. *(A and B) Reproduced with permission from reference B. Prusti, M. Chakravarty, An electron-rich small AIEgen as a solid platform for the selective and ultrasensitive on-site visual detection of TNT in the solid, solution and vapor states, Analyst 145 (5) (2020) 1687–1694. Copyright 2020 Royal Society of Chemistry. (D–F) Reproduced with permission from reference B. Xu, X. Wu, H. Li, H. Tong, L. Wang, Selective detection of TNT and picric acid by conjugated polymer film sensors with donor–acceptor architecture, Macromolecules 44 (13) (2011) 5089–5092. Copyright 2011 American Chemical Society.*

fluorescence was quenched within 1 min and almost invisible after fuming for 30 min, indicating a fluorescence "turn-off" sensing behavior toward TNT vapor. Other examples of AIE-based chemosensors for aromatic explosive vapor sensing are listed in Table 1. Hetero-oligophenyl-based derivatives **2a** and **2b** with a high AIEE QY (59% and 43%) were reported as a selective fluorescence sensors for the nanomolar detection of TNT vapor [28]. **3a**- and **3b**-based thin films exhibited rapid response (few seconds) to 2,4-dinitrotoluene (DNT) vapor with a high selectivity [29]. Reddy's group reported a cyclometalated iridium(III) complex **4** with aggregation-induced phosphorescent emission (AIPE) (QY = 7.4%) property for the detection of TNT in the vapor phase [30]. Zheng's group demonstrated that the fixation of TPE by a pillar[6]arenes macrocycle (**5a**, **5b**, and **5c**) resulted in an elevated QY compared to the acyclic analogue **6** and the cavity favored the adsorption of TNT vapor for high-performance sensing [31].

Compared with the small molecule fluorophores, conjugated polymers have delocalized π systems that facilitate electron migration along the polymer backbone. Swager et al. have proposed the "molecular wire" effect of conjugated polymers for sensory signal amplification, that is to say, binding to one quencher results in efficient quenching in the entire polymer chain (Fig. 2C). This enables the rapid and sensitive response of fluorescent conjugated polymers as chemosensors. Wang and coworkers have developed three conjugated polymers **P1**, **P2**, and **P3** by Suzuki polymerization [32]. **P1** is a typical D-A polymer with diethylamine-modified TPE as electron-donating group and 2,1,3-benzothiadiazole as electron-accepting group, while **P2** and **P3** were designed with only electron-donating groups (Fig. 2D). The QYs of **P1**, **P2**, and **P3** thin films prepared by spin-coating were measured to be 6.9%, 13.0%, and 43.9%, respectively, which were much higher than the QYs in THF solution (less than 1%), indicating their excellent AIE characteristics. All the polymer-based thin films were able to detect DNT and TNT vapors within 15 min of exposure by a PET-caused quenching process (Fig. 2E and F). Importantly, the fluorescence "turn-off" of the thin films upon explosive treatment and "turn-on" upon methanol washing could be repeated for at least 10 cycles without significant deviations. These results indicate the potential of those AIE-active polymers for reusable film sensors in practical situations. Two polytriphenylamines with TPE side groups (**P4** and **P5**) were prepared with high QY values of 16.9% and 34.3% in solid state. They were applied as film state sensing material for various nitroaromatic explosive vapors including 2,4,6-trinitrobenzene (TNB), 2,4-dinitrobenzene (DNB), and *p*-nitrobenzene (NB) [33]. Tang's group reported the preparation of poly(triphenyl ethene) (**P6**) and poly (tetraphenyl ethene) (**P7**) for DNT and TNT vapor sensing and further demonstrated that the porous films prepared via electrospinning exhibited improved sensitivity than their dense films prepared by drop coating or spin coating [34]. Different from the linear polymer above, Tian's group has prepared conjugated hollow nanospheres (**P8**) constructed by AIE-active building blocks (1,3,5-tris(4-formyl-phenyl)benzene), and the selective response to trinitrophenol (TNP) vapor was proven [35].

4 AIE-based chemosensors for acid and amine vapor

Despite being toxic, irritant, and corrosive to human skin, eyes, and respiratory system organic amines and acid have widespread applications in dyeing, pharmaceutical industry, and materials chemistry,. What's more, biogenic amine vapors that are released during food spoilage is an important indicator for the evaluation of food freshness and edibility. It is thus critical to develop rapid and sensitive methods using chemosensors for the detection of acid and amine vapors.

4.1 Acid and amine vapor sensing based on ICT mechanism

Fluorophores featuring the ICT effect are usually small molecules containing heteroatoms, which enable a reversible protonation and deprotonation transformation in the presence of acid and amine vapors. This process will then lead to the electron redistribution, alternation of electron-donating or accepting ability of the heteroatom-containing group, and ultimately generate a different fluorescence emission.

Tang's group has developed a heteroatom-containing fluorophore **7a** for the sensing of volatile acid and amine vapors (Fig. 3A) [36]. **7a** was nonfluorescent in THF solution but highly fluorescent in the aggregated state (Fig. 3B), indicating the phenomenon of strong AIE effect. **7a** exhibited a weak ICT effect that originated from the D-A interaction between carbazole and pyridine units. The protonated pyridine unit has a much stronger electron-withdrawing ability, hence **7b** showed a strong ICT effect resulting in a red-shifted emission (fluorescent color changed from blue to yellow) with a faint fluorescence (Fig. 3C). **7a** solution was spotted on a thin-layer chromatography (TLC) plate to investigate its vapor sensing behavior. As shown in Fig. 3D, the original sample emitted a greenish-yellow fluorescence due to the weak acidity of the silica gel; the fluorescence turned to red upon fuming with trifluoroacetic acid (TFA) vapor and converted back to blue emission after fuming with trimethylamine (TEA) vapor. This colorimetric response could be repeated for many cycles without destructing the sensor (Fig. 3D). Similarly, a series of dihydropyrazine derivatives (**8a–8d**) with excellent AIE behavior and ICT effect was synthesized. The protonated form of **8b** was developed as a chemosensor for ammonia vapor (detection of limit: 690ppb) and further used to monitor the food spoilage by detecting various biogenic amine vapors [37,38]. Two TPE isomers **(Z)/(E)-9** with unusual negative solvatochromism phenomenon were also reported by Tang's group and both isomers were demonstrated to have a reversible colorimetric response to acid and amine vapors [39].

In another example, a multifunctional heteroatom-containing polymer **P9a** was prepared by metal-free one-pot A^3 polymerization in nearly quantitative yields [40]. The AIE-active moieties (TPE) and nitrogen-rich backbones endowed **P9a** with sufficient solid-state emission and reversible transformation of protonation and deprotonation forms in response to acid-amine treatment (Fig. 3E). Thin film of **P9a** prepared by drop coating exhibited a bright green fluorescence at 538nm (Fig. 3F). The protonation of **P9a** backbone in the presence of acid vapors such as HCl led to strong ICT behavior resulting in a bathochromic-shift to 615nm accompanied by fluorescence quenching and fuming with amine (NH_3) restored the fluorescence emission (Fig. 3F and G). Importantly, the acidified form (**P9b**) had a rapid response to NH_3 (less than 15s) and the detection limit was calculated to be 960ppb. Inspired by these results, **P9b** thin film was explored for in situ biogenic amine detection using scallops as fresh food samples (Fig. 3H). As a result, after standing at room temperature for 12h, the fluorescence of **P9b** thin film was turned on, which was indicative of microbial fermentation. Thus, the acidified thin film (**P9b**) showed the potential as a portable sensor for biogenic amine vapors.

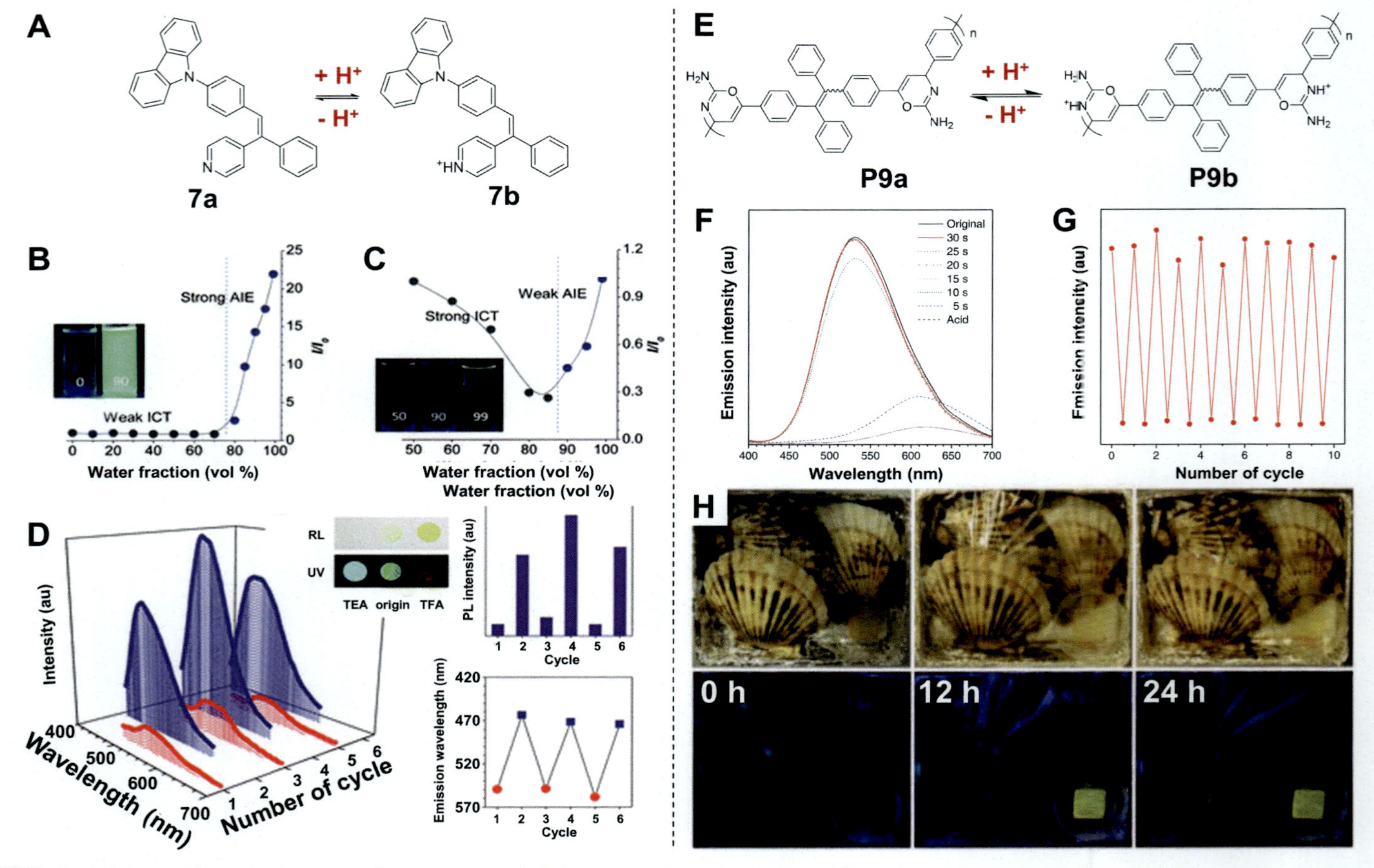

FIG. 3 (A) Reversible transformation between **7a** and **7b** by repeated protonation and deprotonation processes. (B and C) Fluorescence intensity (I/I_0) variation of (B) **7a** and (C) **7b** as a function of water fraction in THF as the solvent. (D) Changes in the fluorescence emission spectrum, intensity, and wavelength of **7a** spot on the TLC plate by repeated fuming with TFA and TEA vapors. Inserts: Photos of **7a** spots on a TLC plate before (central) and after fuming with (left) TEA and (right) TFA vapors taken under normal room lighting (RL) and UV illumination (365 nm). (E) Schematic illustration of the reversible acid and amine detection by protonation of **P9a** and deprotonation of **P9b**. (F) Fluorescence emission spectra of **P9a** thin film (original), **P9b** thin film (acid), and **P9b** thin film after exposure to NH_3 vapor (0.08 M) for different time intervals. (G) Reversible switching of the emission intensity by repeated fuming with NH_3 and HCl vapors. (H) Spoilage detection of scallops in sealed packages for 24 h at room temperature using **P9b** thin film and the photographs taken under normal room illumination (the upper row) and 365-nm UV irradiation (the lower row). *(A–D) Reproduced with permission from references Z. Yang, W. Qin, J.W.Y. Lam, S. Chen, H.H.Y. Sung, I.D. Williams, B.Z. Tang, Fluorescent pH sensor constructed from a heteroatom-containing luminogen with tunable AIE and ICT characteristics, Chem. Sci. 4 (9) (2013) 3725–3730. Copyright (2013) Royal Society of Chemistry. (E–H) Reproduced with permission from reference Y. Hu, T. Han, N. Yan, J. Liu, X. Liu, W.-X. Wang, J.W.Y. Lam, B.Z. Tang, Visualization of biogenic amines and in vivo ratiometric mapping of intestinal pH by AIE-active polyheterocycles synthesized by metal-free multicomponent polymerizations, Adv. Funct. Mater. 29 (31) (2019) 1902240. Copyright (2019) Wiley-VCH.*

4.2 Aromatic amine vapor sensing based on ICT + PET mechanism

In addition to pursuing a high sensitivity for vapor sensing, another challenge to implement the real-world applications is to improve the selectively among different vapors especially those with similar chemical properties such as aliphatic (primary/secondary/tertiary) amines and aromatic amines. Fluorophore with ICT effect could be one promising option to meet the challenge as its optical property can be easily modulated by the D-A strength in response to different analytes. An AIE-active compound **10** featuring a D-A-D system was prepared with methoxy-substituted hexaphenylbenzene (HPB) as the electron donors and fumaronitrile as the electron acceptor (Fig. 4A) [41]. Compound **10** had a maximum emission at 455 nm in THF, with the addition of water, the fluorescence intensity first decreased (ICT effect, 0%–30% water fractions in THF) and then increased (AIE effect, 40%–80% water fractions in THF) showing a new band at 550 nm in the aggregated state (Fig. 4A). Different sensing behaviors of **10** aggregates in H_2O/THF (8:2, v/v) toward aliphatic and aromatic amines are presented in Fig. 4B. Upon addition of aliphatic amines such as TEA to **10** aggregates, the emission at 550 nm decreased along with gradual increase of the band at 450 nm. As for secondary (diethylamine and dimethylamine) and primary aliphatic amines (ethylamine, propylamine, butylamine, hexylamine, dodecylamine, and cyclohexylamine), similar phenomenon was observed, whereas the new blue-shifted emission band appeared at 460 and 500 nm, respectively. This difference was explained by the H-bonding ability of the amines toward methoxy C—H bonds, which decreases from tertiary to secondary and to primary aliphatic amines. Therefore, the sensing mechanism of **10** toward different aliphatic amines was ascribed to the change of D-A strength by H-bonding between amines and methoxy C—H bonds. In the case of aromatic amines such as aniline, a different sensing mechanism was involved. The fluorescence of **10** aggregates was gradually quenched without showing new bands upon the addition of aniline (Fig. 4B). This was explained by the *a*-PET from the HOMO of aniline (−5.63 eV) to that of **10** aggregates (−5.66 eV). ^{1}H NMR study revealed that there was π-π stacking between the donor aniline to the acceptor cyano groups of **10**, whereas no H-bonding interactions between amines and methoxy C—H bonds. Inspired by the excellent selectivity of **10** aggregates toward aliphatic and aromatic amines based on different sensing mechanisms (ICT and *a*-PET), the sensing application of **10** thin films for the detection of amine vapor was investigated. As shown in Fig. 4C, the fluorescence of thin film changed from red color to green after 5 min of exposure to TEA, whereas aniline quenched the fluorescence within 1 min. This suggested a simple method for the discrimination of aliphatic and aromatic amines using a portable sensing device. Likewise, an AIE-active indeno-pyrrole derivative (**11**) was designed with a high sensitivity to aniline (detection limit: 6.04 ppb) and other vapors (Table 1) [42]. The fluorescence of **11** could be reversibly turned off and on upon treatment with acid and aliphatic amine sequentially due to the variation in ICT state. While fuming with aromatic amines such as aniline, the fluorescence of **11** was quenched because of PET effect.

To develop a sensor with high performance, the contact area between the fluorophores and analytes is also an important factor to consider. Hence, Lu and coworkers have devoted their efforts to the preparation of three-dimensional nanofibrous network with porous microarchitecture. To that end, a multifunctional gelator (compound **12**) was synthesized with acridine and cyano groups as electron-withdrawing units to give a low HOMO energy level for

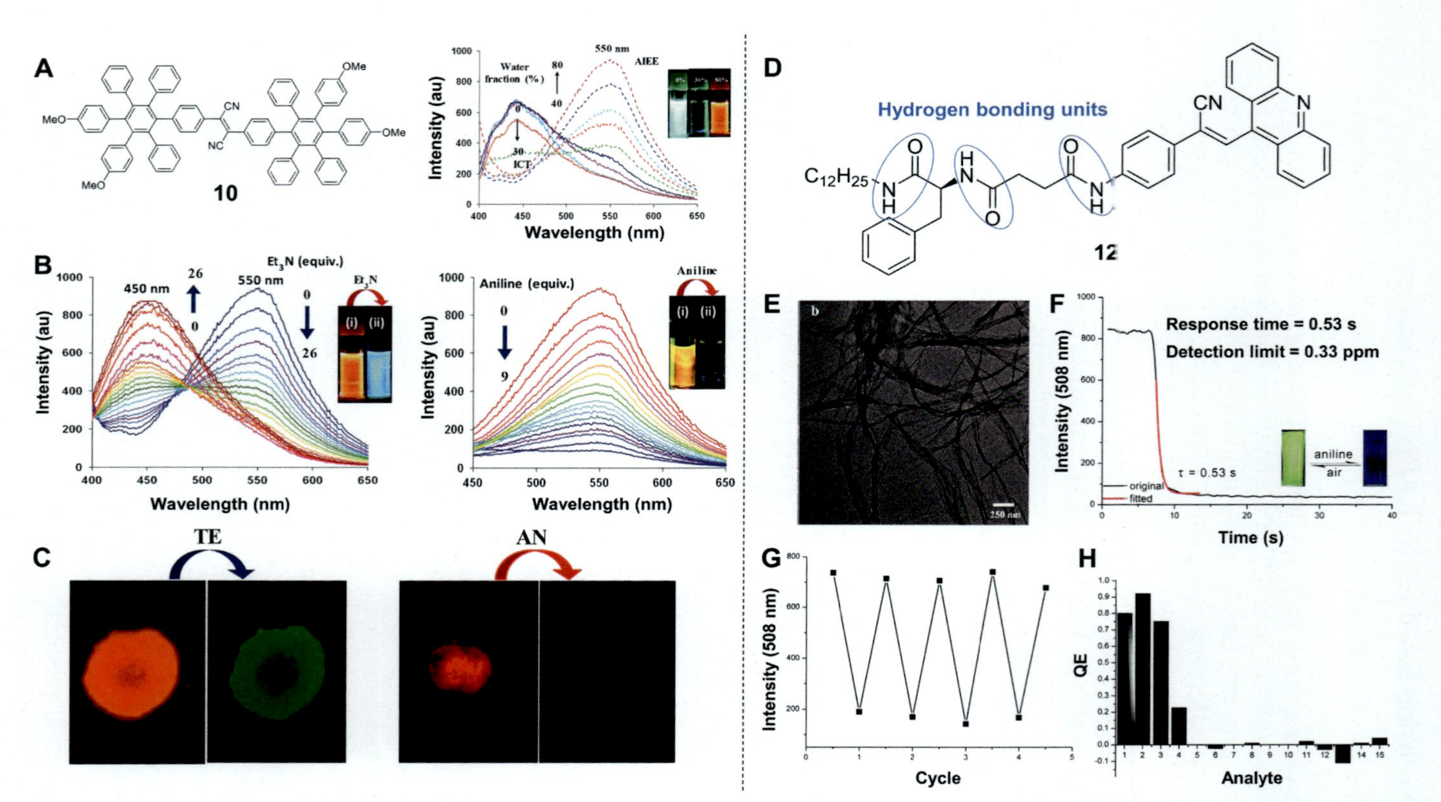

FIG. 4 (A) The chemical structure of AIEgen **10** and fluorescence spectra showing AIEgen **10** (5.0 μM) in various H_2O/THF mixtures (0%–80% volume fraction of water in THF). (B) Fluorescence spectra of AIEgen **10** showing a ratiometric response (left) or quenching (right) in the presence of TEA or aniline in H_2O/THF (8:2, v/v) media (buffered with HEPES, pH 7.05). (C) Photographs of **10** thin films before and after exposure to TEA for 5 min or aniline for 1 min. (D) The chemical structure of AIEgen **12**. (E) TEM image of **12** xerogel film formed in chlorobenzene. (F) Time course of fluorescence quenching of the **12** xerogel film upon exposure to saturated aniline vapor. (G) Cycles of the fluorescence quenching and recovery by exposing the **12** xerogel film to aniline (100 ppm) and air (20 min). (H) Quenching efficiencies (QEs) of the **12** xerogel film in response to vapors of different reagents: (1) aniline, (2) *N*-methylaniline, (3) *N,N*-dimethylaniline, (4) phenylhydrazine, (5) acetone, (6) ethanol, (7) ethyl acetate, (8) CH_2Cl_2, (9) THF, (10) toluene, (11) triethylamine, (12) *n*-butylamine, (13) cyclohexylamine, (14) NH_3, and (15) tributylamine. *(A–C) Reproduced with permission from reference S. Pramanik, H. Deol, V. Bhalla, M. Kumar, AIEE active donor–acceptor–donor-based hexaphenylbenzene probe for recognition of aliphatic and aromatic amines, ACS Appl. Mater. Interfaces 10 (15) (2018) 12112–12123. Copyright (2018) American Chemical Society. (D–H) Reproduced with permission from reference P. Xue, J. Ding, Y. Shen, H. Gao, J. Zhao, J. Sun, R. Lu, Aggregation-induced emission nanofiber as a dual sensor for aromatic amine and acid vapor, J. Mater. Chem. C 5 (44) (2017) 11532–11541. Copyright (2017) Royal Society of Chemistry.*

aromatic amine sensing based on *a*-PET and *N*-dodecyl-L-phenylalaninamide as hydrogen-bonding units to facilitate the gel formation via self-assembly (Fig. 4D) [43]. Compound **12** was nonemissive in THF solution but emitted bright green fluorescence in the aggregated state (water fraction over 60% in THF). Gel phase could be formed when the hot solution of **12** in chlorobenzene was cooled to room temperature and then a uniform xerogel film with strong fluorescence was obtained after removing the solvent. The morphology of the xerogel film was investigated under TEM, and long nanofibers with diameters ranging from 50 to 100 nm were observed, which exhibited favorably enlarged contact surface for analyte interactions (Fig. 4E). When the xerogel film was fumed with aromatic amine vapors such as aniline, its green fluorescence quenched rapidly within 10 s and the film recovered its fluorescence upon air blowing for 20 min to remove aniline (Fig. 4E and F). Furthermore, the HOMO energy of **12** was measured to be −5.73 eV which is lower than that of aniline (−5.63 eV), thus the sensing mechanism of aniline was based on the physical adsorption and a subsequent *a*-PET process. The detection limits of the xerogel film to aniline, *N*-methylaniline, *N*,*N*-dimethylaniline, and phenylhydrazine were as low as 0.33, 0.48, 1.38, and 3.8 ppb, respectively. Moreover, the xerogel film did not show responses to common organic solvents (dichloromethane, THF et al.) or aliphatic amines (TEA, cyclohexylamine et al.) (Fig. 4H). This and several other reports regarding AIE-active xerogel films for the detection of acid and amine vapors provided insight into the nanofibrous films for multiple stimuli responses with high sensitivity [44–48].

4.3 Acid and amine vapor sensing based on ESIPT mechanism

Fluorophores with typical ESIPT effect possess a large Stokes shift, unique emission pattern (Enol emission, Keto emission, and phenolic anion emission), and a special sensitivity to the external environment. This encouraged researchers to explore the applications of ESIPT fluorophores for vapor sensing. A series of fluorescent molecules based on a representative ESIPT scaffold, 2,(2-hydroxyphenyl)benzothiazole (HBT), was designed and synthesized (Fig. 5A) [49]. In solid states, their maximum absorption varied from 340 to 420 nm and the emission color ranged from cyan to orange-red in the order of **OHBT**, **CHBT**, **DHBT**, **HCP**, and **HTPA** (Table 1 and Fig. 5B). They were nonemissive or weakly emissive in THF solution while strongly emissive in solid state, demonstrating their AIE characteristics. Since the ESIPT process is sensitive to the external environment, all the HBT derivatives exhibited solvent-dependent emission and reversible fluorescent responses to amine and acid vapors. Take, for example, **HCP**, a prominent emission band centered at 540 nm (Keto fluorescence, Keto FL) was observed in nonpolar or weakly polar solvents such as hexane, dichloromethane, and toluene due to the formation of intramolecular hydrogen bond; while a strong emission peak at 417 nm (Enol FL) was detected in polar protic solvents such as ethanol because the protic solvent hampers the ESIPT process by forming intermolecular hydrogen bonds; and in polar aprotic solvents such as DMSO, intense dual emission at 417 and 483 nm appeared, ascribed to the Enol and phenolic anion forms, respectively (Fig. 5C). The vapor-responsive property of HCP was examined using **HCP**-loaded filter paper, which initially emitted yellow fluorescence centered at 556 nm resulting from the Keto FL. Upon exposure to isopropylamine vapor for 30 s, the emission color

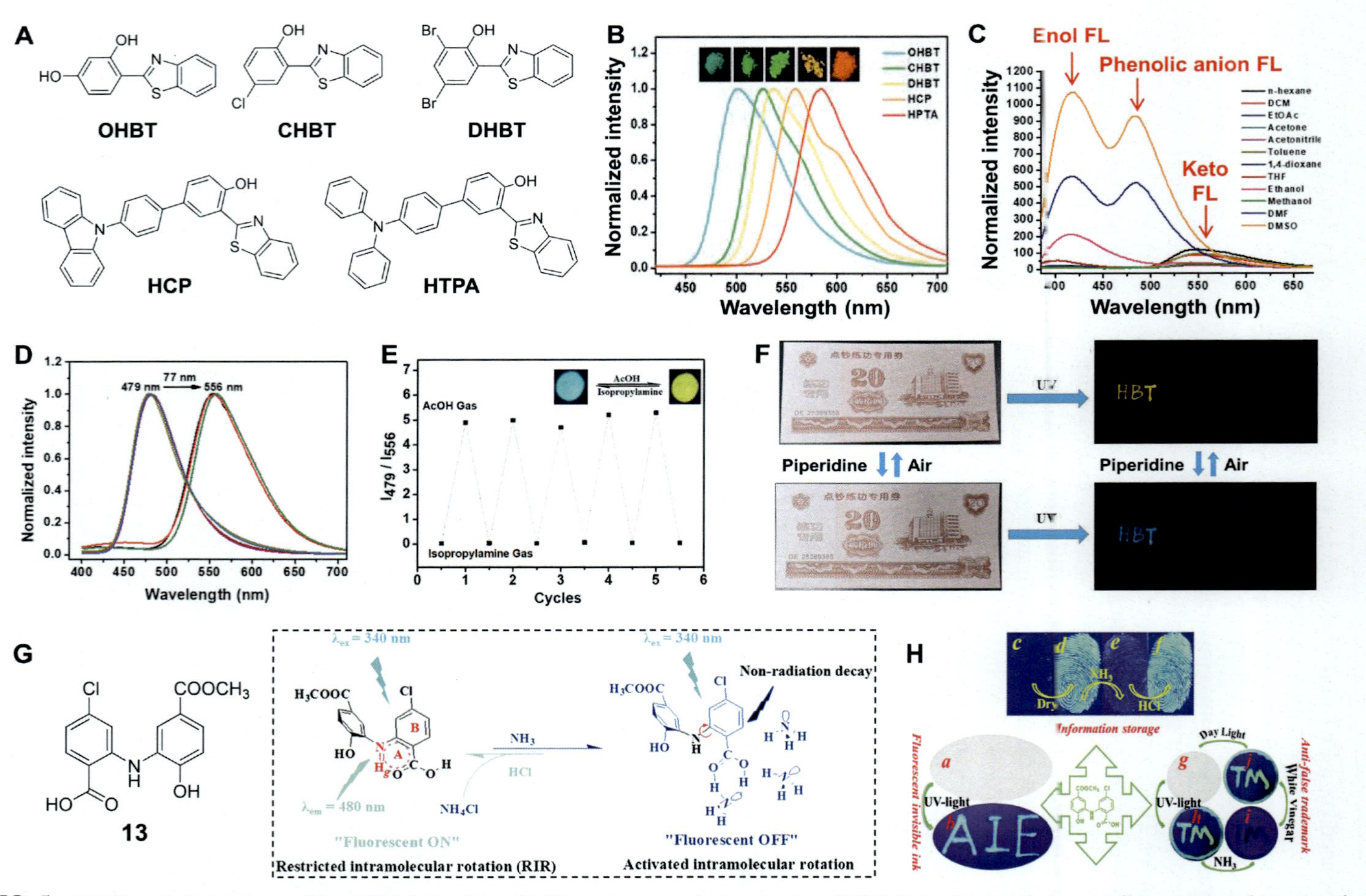

FIG. 5 (A) Chemical structures of five HBT derivatives. (B) Fluorescence-emission spectra of HBT derivatives in their crystalline states and images of the corresponding crystals under 365-nm UV light. (C) Fluorescence spectra of **HCP** (50 μM) in different solvents. (D, E) The normalized fluorescence spectra and fluorescence ratio changes of **HCP**-loaded filter paper by fuming with isopropylamine and AcOH vapors repeatedly. (F) Illustration of **HCP** as an anticounterfeiting ink on a practice banknote with letters of "HBT" after fuming with piperidine and then ambient air. Images were taken under daylight or under 365-nm UV light. (G) Chemical structure of AIEgen **13** and plausible reaction mechanism for the reversibly "Fluorescence ON" and "Fluorescence OFF" phenomena upon treatment with NH_3 and HCl. (H) Practical applications of AIEgen **13** as fluorescent invisible ink, information storage material, and antifalse trademark material. *(A–F) Reproduced with permission from reference B. Li, D. Zhang, Y. Li, X. Wang, H. Gong, Y.-Z. Cui, A reversible vapor-responsive fluorochromic molecular platform based on coupled AIE-ESIPT mechanisms and its applications in anti-counterfeiting measures, Dyes Pigments 181 (2020) 108535. Copyright (2020) Elsevier. (G–H) Reproduced with permission from reference E. Zhang, X. Hou, Z. Zhang, Y. Zhang, J. Wang, H. Yang, J. You, P. Ju, A novel biomass-based reusable AIE material: AIE properties and potential applications in amine/ammonia vapor sensing and information storage, J. Mater. Chem. C 7 (27) (2019) 8404–8411. Copyright (2019) Royal Society of Chemistry.*

changed to blue (phenolic anion emission) centered around 479 nm, which was associated with the deprotonation ability of the alkali vapors (Fig. 5D and E). Note that the fluorescence could return to the original state upon fuming with acetic acid vapor within 30 s or exposed to ambient air for 4–6 h. Such vapor-induced fluorescence switching could be repeated without apparent fatigue (Fig. 5E). To illustrate a practical application as anticounterfeiting labels, the letters "HBT" were written on a banknote using **HCP** solution as fluorescent inks. The letters were invisible under room light and reversible fluorescence-switching under amine vapor (piperidine) and ambient air was distinguished by naked eyes under UV light. The above results demonstrated the opportunities of AIE-active small molecules with ESIPT behavior for solid-state sensing and high-security anticounterfeiting.

4.4 Acid and amine vapor sensing based on RIM mechanism

Intramolecular hydrogen bond could also act as the responsive units by restricting the intramolecular motions. One example is a small molecule fluorophore **13** designed by Ju's group (Fig. 5G) [50]. The fluorescence of **13**-loaded filter paper could be quenched upon NH_3 fuming and then recovered by HCl vapor based on reversible hydrogen bonding destruction and formation process. As illustrated in Fig. 5G, the original state of **13** is highly emissive ("Fluorescence ON") due to the RIM mechanism induced by the formation of intramolecular hydrogen bond between carboxyl O and imino H. After fuming with NH_3, the intramolecular hydrogen bond was destructed resulting in an activated intramolecular motion and promoted nonradiative decay ("Fluorescence OFF"). The detection limit for ammonia vapor was measured to be 20 ppm. The vapor-sensitive fluorescence ON-OFF switching of AIEgen **13** benefited its applications as fingerprint storage material and fluorescent inks (Fig. 5H).

4.5 Acid and amine vapor sensing based on chemical reactions

The above examples of AIE-active chemosensors for the detection of amine and acid vapors are dependent on the photophysical adsorption or simple protonation and deprotonation processes. Other than that, amino and acid vapor sensing based on chemical reactions and a subsequent alternation of photophysical property of AIEgens were also reported in recent 5 years. The involved chemical reactions mainly include aminolysis reaction for amine vapor sensing, Schiff base reaction for hydrazine and primary amine vapor sensing, and ring-opening reaction of rhodamine spirolactam for acid vapor sensing.

4.5.1 Amine vapor sensing based on aminolysis reaction

Tang's group has designed compound **14a** for light-up detection of amine vapors [51]. **14a** containing a phenolic ester could undergo the aminolysis reaction to produce **14b** (Fig. 6A). The fluorescence QY of **14b** increased from 0.3% in THF to 32.4% in the aggregated state (THF: H_2O = 1:99, v/v) and further to 59.4% in the solid state. This AIE property was explained by the RIM mechanism resulting from the formation of intramolecular hydrogen bond (Fig. 6A). A portable sensor was easily prepared by dropping dichloromethane solution of **14a** on a filter paper and then air-dried by evaporation. As shown in Fig. 6A, the **14a**-loaded filter paper

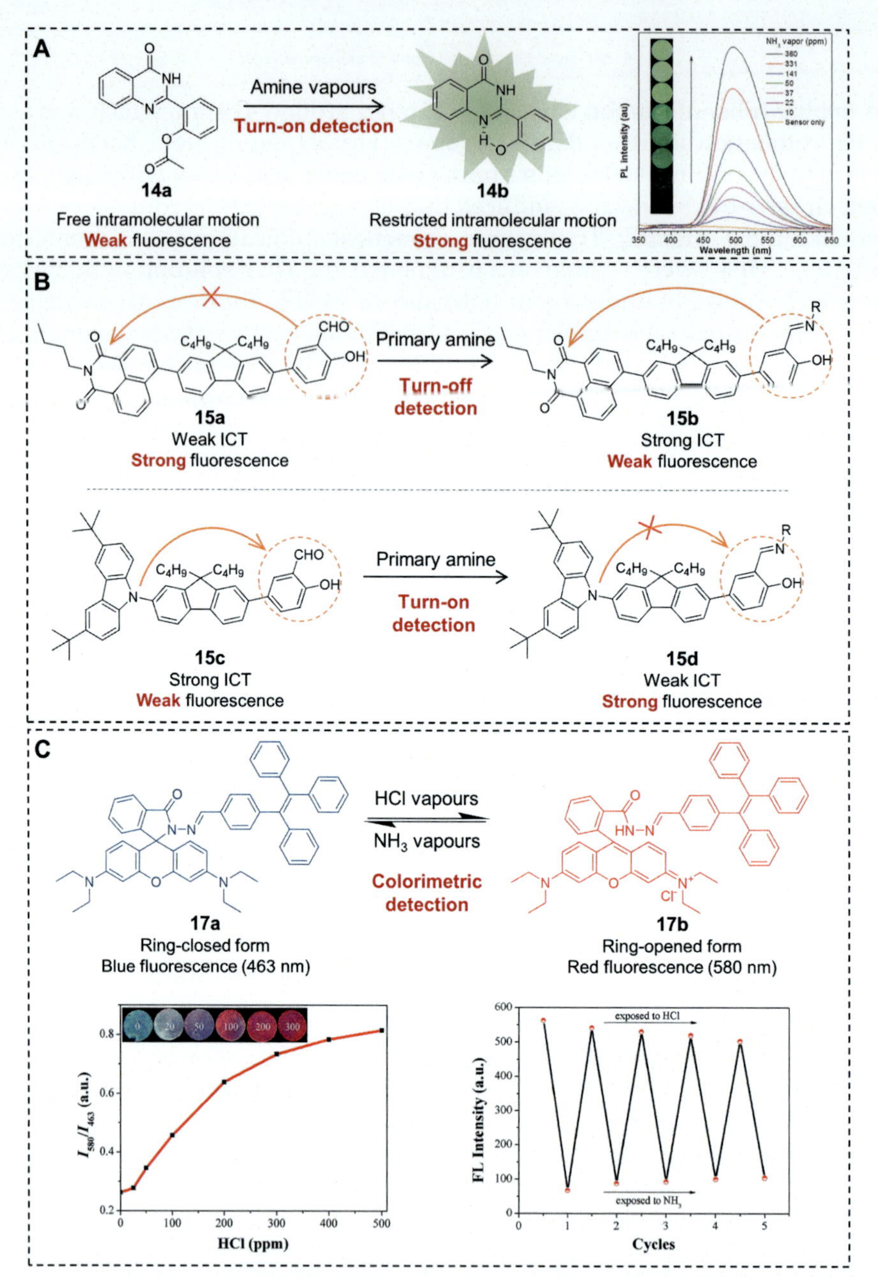

FIG. 6 (A) Design principle of AIEgen **14a** for fluorescence turn-on detection of amine vapors and fluorescent emission spectra of **14a**-loaded filter paper after exposure to ammonia vapor (0, 10, 22, 37, 50, 144, 331, 360 ppm) for 5 min. (B) Design principle of AIEgen **15a** and **15c** for fluorescence turn-off and turn-on detection of amine vapors, respectively. (C) Proposed mechanism of the colorimetric changes of AIEgen **17a** upon addition of HCl and NH_3 vapors; the fluorescence intensity of **17a** to different concentrations of HCl vapor at I_{580}/I_{463} for 3 min at room temperature; fluorescence response and recovery of **17a** thin film (monitoring at 595 nm) from replicate cycling between 500 ppm of HCl and NH_3 vapors. *(A) Reproduced with permission from reference M. Gao, S. Li, Y. Lin, Y. Geng, X. Ling, L. Wang, A. Qin, B.Z. Tang, Fluorescent light-up detection of amine vapors based on aggregation-induced emission, ACS Sens. 1 (2) (2016) 179–184. Copyright (2015) American Chemical Society. (B) Reproduced with permission from reference J. Hu, R. Liu, S. Zhai, Y. Wu, H. Zhang, H. Cheng, H. Zhu, AIE-active molecule-based self-assembled nano-fibrous films for sensitive detection of volatile organic amines, J. Mater. Chem. C 5 (45) (2017) 11781–11789. Copyright (2017) Royal Society of Chemistry. (C) Reproduced with permission from reference Y. Yang, C.-Y. Gao, D. Dong, Tetraphenylethene functionalized rhodamine dye for fluorescence detection of HCl vapor in the solid state, Anal. Methods 8 (44) (2016) 7898–7902. Copyright (2016) Royal Society of Chemistry.*

was almost nonemissive under UV light irradiation, upon fuming with basic vapors such as NH_3, green fluorescence appeared within 5 min. The detection limit for NH_3 was calculated to be 8.4 ppm. This example represents a reactive chemosensor for vapor sensing based on the modulation of intramolecular motion.

4.5.2 Amine vapor sensing based on Schiff base reaction

Another way to obtain a responsive fluorescence signal upon chemical reaction is to modulate the ICT behavior of the chemosensor by changing the electron-donating or electron-accepting ability of the reactive unit. To illustrate that, two D-π-A type fluorophores **15a** and **15c** were developed for "turn-off" and "turn-on" detection of primary amine vapors, respectively (Fig. 6B) [52]. They both consisted of a salicylaldehyde unit as a reactive unit to primary amines, a fluorene ring as π bridge, and a heterocycle structure (naphthalimide derivative for **15a**, and carbazole derivative for **15c**). Both **15a** and **15c** can react with primary amines to produce the Schiff base adduct but exhibited completely opposite responsive behaviors due to the difference in D-A strength. Naphthalimide group in **15a** is an electron-accepting group, carbazole group in **15c** is an electron-donating group, and Schiff base moiety in **15b** and **15d** has a stronger electron-donating ability compared to the salicylaldehyde in **15a** and **15c**. Therefore, **15a** with a weak ICT effect was strongly emissive in solid and suffered from a fluorescence quenching upon reaction with primary amines to produce **15b** with a typical ICT effect. On the contrary, the reaction of **15c** with primary amines produced **15d**, along with impaired ICT effect and significantly enhanced emission. The detection limits of these two reactive AIEgens to primary amines are both ppm level (see Table 1 for details). In another example, AIEE-active aldehyde-modified molecule **16** was developed for the sensing of hydrazine vapor, which could react with **16** to form a hydrazine Schiff base and cause an ICT effect-induced fluorescence quenching [53]. These works represent examples of reaction-induced modulation of ICT behavior for amine vapor sensing.

4.5.3 Acid vapor sensing based on the ring-opening reaction of rhodamine spirolactam

It has been known that rhodamine spirolactam can be easily converted to the ring-opened form by an acid-catalyzed reaction. However, rhodamine spirolactam suffers from the ACQ effect in solid state which is not favorable for the practical applications for vapor sensing. To solve the issue, a TPE-functionalized rhodamine spirolactam derivative **17a** was designed and it was used as a colorimetric fluorescent sensor for HCl vapor sensing [54]. As illustrated in Fig. 6C, **17a**-loaded filter paper emitted blue fluorescence centered at 460 nm, HCl vapor catalyzed the transformation of **17a** to **17b** that emitted red fluorescence centered at 580 nm, and the detection limit was calculated to be 20 ppm. Importantly, the ring-opened form **17b** can be converted back to **17a** after fuming with NH_3, revealing the stability and reusability of the **17a**-based sensor for HCl vapor sensing. In another example, naphthalimide modified rhodamine spirolactam derivative **18a** was reported with a distinct AIE mechanism (elimination of dark states via dimer interactions, EDDI) [55]. The emission color of **18a** powders changed from yellow ($\lambda_{max} = 582$ nm) to red (**18b**, $\lambda_{max} = 617$ nm) upon exposure to HCl vapor and then reversed back after fuming with TEA.

5 AIE-based chemosensors for volatile organic compounds (VOCs)

Volatile organic compounds (VOCs), one of the most common chemicals in industrial production and pesticide residues, are the recognized environmental pollutants. VOC vapor can also cause a variety of severe diseases in people by disrupting nervous system and cellular activity. Therefore, the development of chemosensors with superior sensitivity and high selectivity for various VOC vapors is becoming increasingly important.

5.1 VOC sensing based on RIM+ICT mechanisms

Methanol is regarded as one of the toxic VOCs, as it causes many health problems in humans such as headache, vertigo, nausea, and even visual impairment. Nevertheless, the coexistence of various aliphatic alcohols makes methanol detection challenging due to structural similarity [56]. Therefore, developing reliable sensors for methanol would be particularly important. To that end, a D-A type compound **19** was designed, in which dimethylamino and cyano groups serve as the electron-donating and electron-accepting groups, respectively (Fig. 7A) [57]. The AIE feature of compound **19** was confirmed by QY values, which are 4.7% in THF solution and 14.6% in the solid state. The real-time fluorescent emission spectra of the **19**-loaded paper sensor were recorded upon fuming with saturated methanol vapor. As shown in Fig. 7B, the fluorescence intensity of **19**-loaded paper gradually increased and reached the maximum after fuming for 15 min. The same experiments were also performed with various volatile organic vapors including alcohols, benzene derivatives, and the common organic solvents (Fig. 7C). As a result, methanol vapor evoked the best responsive performance indicating the selective sensing of **19**-based sensor toward methanol. The proposed working mechanism is illustrated in Fig. 7A. The intermolecular hydrogen bonding between imine N of AIEgen **19** and hydroxyl H of methanol could impede the electron transition from donor to acceptor due to the orbital overlap between excited **19** molecule and methanol. On the other hand, with an excessive amount of methanol adsorbed on the film, intramolecular motions (torsion of C=N double bonds and rotation of C—C and C—N single bonds) could be restricted. Thus, the methanol-responsive fluorescence emission was ascribed to the combined effect of weakened ICT and enhanced AIE. As for other alcohols, their diffusion into the film is hindered either by their bulkiness size or by their low volatility, leading only to a faint response.

Phenolic compounds are important chemical widely used in industry and agriculture. But, at the same time, they are generally toxic to organisms. In consideration of that, Fang and coworkers have developed a high-performance fluorescent sensor for phenolic compounds in the air using a crab-shaped o-carborane derivative **20** [58]. **20** is a typical D-A type molecule with azetidine and o-carborane groups serving as electron-donating and electron-accepting units, respectively (Fig. 7D). The fluorescence emission of **20** exhibited a positive solvatochromic effect with a large bathochromic from 586 nm in hexane to 731 nm in DMSO, indicating a typical ICT phenomenon. **20** is highly emissive in solid state with a QY value of 73%, which is much higher than in the solution sate ($\Phi_F < 1\%$), revealing its superior AIE characteristic (Fig. 7D). The sensing behavior of **20**-based film was tested in a homemade sensing platform shown in Fig. 7G. **20**-based sensing film exhibited an enhanced emission upon

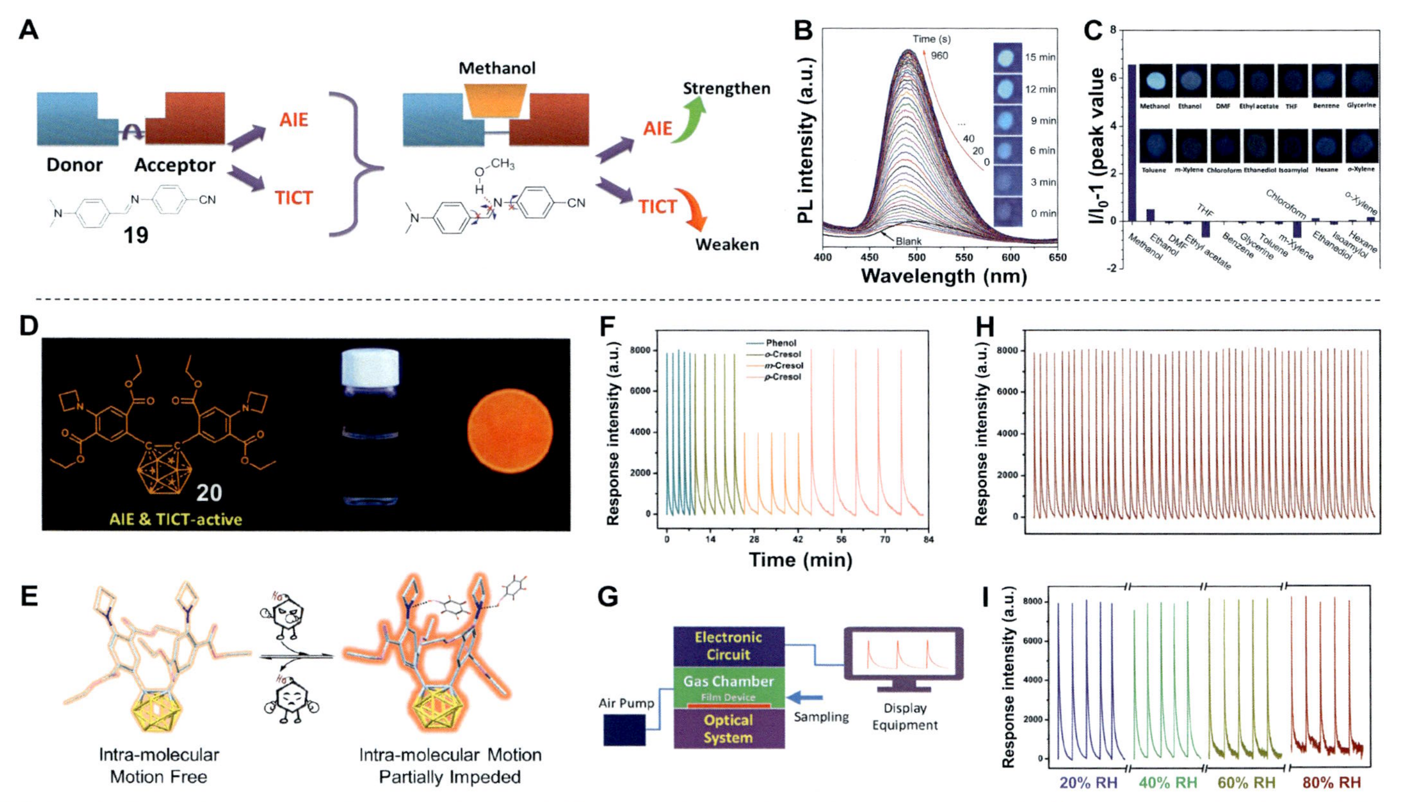

FIG. 7 (A) Chemical structure of AIEgen **19** and schematic illustration showing the working mechanism of the film sensor toward methanol. (B) Time-dependent fluorescence spectra of **19**-loaded cellulose paper exposed to methanol vapor. (C) The selectivity of **19**-loaded cellulose paper to different VOCs. (D) Molecular structure and fluorescent images of the AIEgen **20**. The pictures next to the structure are its fluorescence images in dichloromethane and in film state with glass plate as a substrate. (The photos were taken under 365-nm UV light illumination.) (E) Schematic illustration showing the working mechanism of **20**-based film sensor toward phenol. (F) Typical responses of the **20**-based film device to the saturated vapors of phenol and different kinds of cresol at room temperature ($\lambda_{ex}/\lambda_{em}=350\,nm/620\,nm$). (G) Schematic representation of the homemade conceptual sensor. (H) Performance-stability test of **20**-based film toward the sensing of saturated vapor of phenol (50 repeats). (I) Response performance of **20**-based film sensor toward the saturated vapor of phenol in environment of different humidity with an error of $\pm3\%$ at room temperature. *(A–C) Reproduced with permission from reference R. Zhao, M. Zhang, Y. Liu, X. Zhang, Y. Duan, T. Han, Fabricating D-A type AIE luminogen into film sensor for turn-on detection of methanol vapour, Sensors Actuators B Chem. 319 (2020) 128323. Copyright (2020) Elsevier. (D–I) Reproduced with permission from reference R. Huang, H. Liu, K. Liu, G. Wang, Q. Liu, Z. Wang, T. Liu, R. Miao, H. Peng, Y. Fang, Marriage of aggregation-induced emission and intramolecular charge transfer toward high performance film-based sensing of phenolic compounds in the air, Anal. Chem. 91 (22) (2019) 14451–14457. Copyright (2019) American Chemical Society.*

exposure to phenolic vapor with excellent reproducibility (Fig. 7F–H). The underlying mechanism could be the intermolecular hydrogen bond formation between the azetidine of **20** and hydroxyl of phenolic compounds, which further hinders the intramolecular motion and impairs the electron-donating ability of azetidine (Fig. 7E). Importantly, the fluorescence recovery speed of the sensor upon pumping air to remove phenolic vapor is strongly dependent on the nature of the phenolic compounds, providing an opportunity for discrimination. Furthermore, **20**-based sensor is stable even in a highly humid environment (relative humidity: 80%), which is quite important for practical use (Fig. 7I). The detection limits for phenol, *o*-cresol, *m*-cresol, and *p*-cresol are as low as 0.4, 0.3, 10, and 0.8 ppt, respectively.

5.2 VOCs sensing based on sensor array

Sensor array is a combination of multiple sensors to produce a unique response pattern, allowing the discrimination and identification of various analytes, even those with similar structures such as benzene and toluene.

All the above chemosensors rely on the direct interaction between the AIE-active chemosensor and the analytes. Different from that, Jiang's group has reported an indirect sensing behavior depending on the "polymeric swelling-induced variation of fluorescence intensity" [59]. Firstly, a linear polymer/AIE molecule microwires array was fabricated on a micropillar-structured template with polyethersulfone (PES) as the polymer and **AnPh3** (see Fig. 8D for chemical structure) as the AIE-active molecule (PES:**AnPh3** = 7:3, w/w) (Fig. 8A). The as-prepared PES/$\mathbf{AnPh_3}$ microwires array exhibited uniform composition, smooth surface, and straight boundary as evidenced by SEM, CLSM, and AFM (Fig. 8A). The average size of each microwire is 2 μm in width and 320 nm in height. The fluorescence response of PES/$\mathbf{AnPh_3}$ toward acetone vapor was examined and compared with the PES/$\mathbf{Ph_3A_2}$ as a non-AIE system (Fig. 8C and D). The fluorescence intensity of PES/$\mathbf{Ph_3A_2}$ microwires array showed no difference before and after exposure to acetone vapor (Fig. 8C). In comparison, the fluorescence of PES/$\mathbf{AnPh_3}$ microwires array was quenched by acetone vapor (Fig. 8D). To get insight into the underlying mechanism, AFM images of PES/$\mathbf{AnPh_3}$ microwires arrays were taken before and after fuming with acetone vapor (Fig. 8B). The results demonstrated that PES/$\mathbf{AnPh_3}$ microwires array adsorbed acetone vapor and polymeric swelling occurred and the width increased by $3.85 \pm 0.34\%$ and height increased by $3.31 \pm 0.23\%$. Moreover, the interaction energy between $\mathbf{AnPh_3}$ and PES was calculated to be larger than $\mathbf{AnPh_3}$ dimers, demonstrating the preferred AIE-polymer interaction rather than AIE-AIE interaction. Therefore, the swelling of the PES/$\mathbf{AnPh_3}$ microwires increased the distance between $\mathbf{AnPh_3}$ molecules, leading to the dispersion of $\mathbf{AnPh_3}$ and a decreased fluorescence intensity afterward. To construct the sensor array, eight commercial polymers were used to prepare different polymer/$\mathbf{AnPh_3}$ microwires including poly(styrene-*co*-allyl alcohol), poly(α-methylstyrene), poly(vinyl chloride-*co*-vinyl acetate), poly(vinyl acetate), poly(carbonate bisphenol A), poly(ether sulfone), poly(methyl methacrylate), and poly(vinyl butyral). Principal component analysis (PCA) revealed that polymer/AIE microwire-based sensor array could be used to classify different types of organic vapors (Fig. 8E). Furthermore, the sensor array of different polymer/$\mathbf{AnPh_3}$ based on six synthetic PESs with different side chains was fabricated to identify similar organic vapors, namely, benzene and toluene (Fig. 8F).

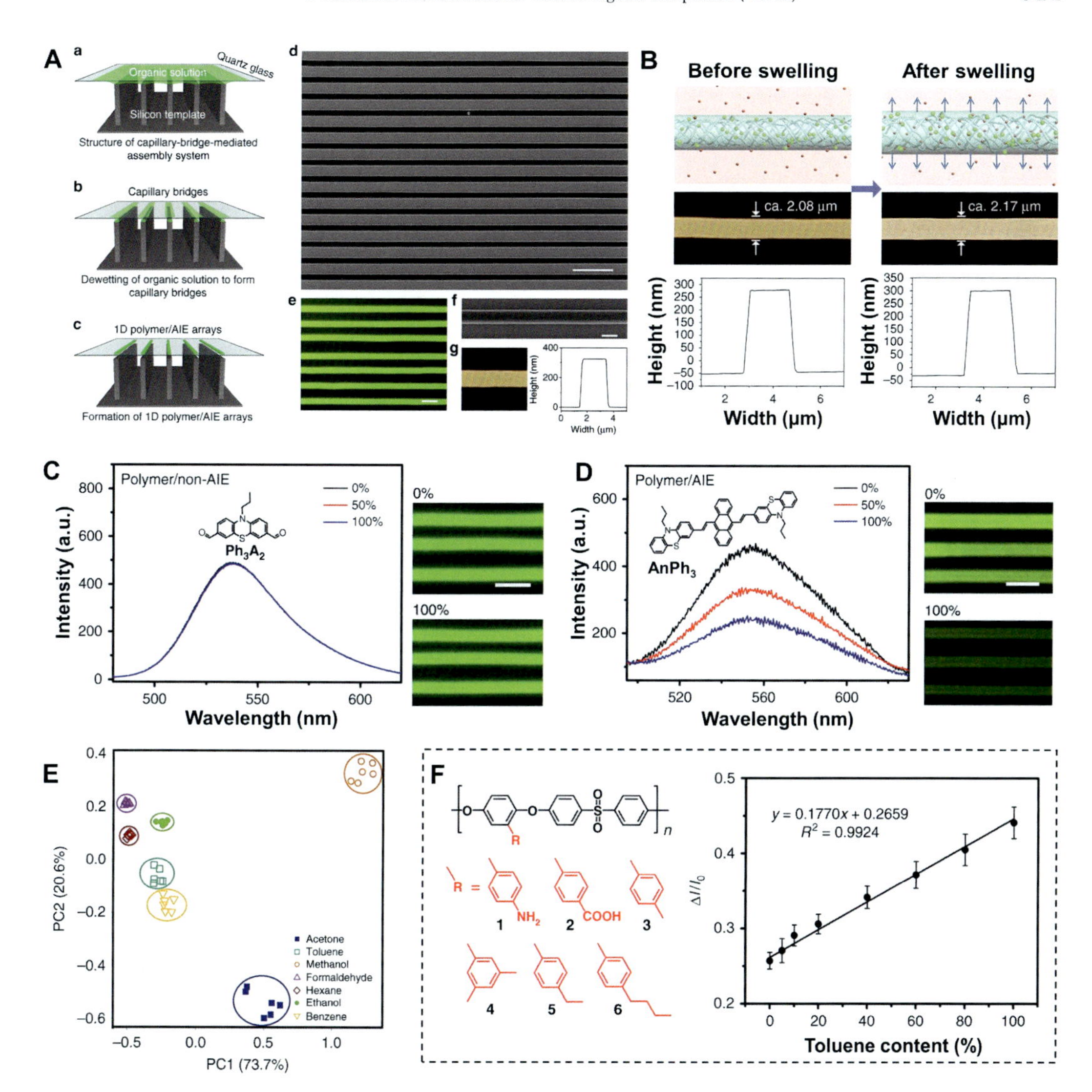

FIG. 8 (A) (a–c) Scheme showing the capillary-bridge-mediated assembly method for the assembly of PES/**AnPh**$_3$ microwires arrays. (d) SEM image and (e) fluorescent image of large-area PES/**AnPh**$_3$ microwire arrays fabricated on quartz glass. The scale bars are 20 and 5μm for (d) and e, respectively. (f) Zoom-in SEM image of (d). The scale bar is 2μm. (g) AFM image. (B) The proposed mechanism of fluorescent intensity variation after exposure of PES/**AnPh**$_3$ microwires to organic vapors. The figures below represent AFM topographies of PES/**AnPh**$_3$ microwires before (left) and after (right) exposure to saturated acetone vapor. (C) PES/non-AIE molecule and (D) PES/AIE molecule microwires before (0%) and after (50% and 100%) exposure to acetone vapor. Inserted photos are fluorescent images of corresponding PES/non-AIE molecule and PES/AIE molecule microwires. All scale bars are 5μm. "100%" represents the saturated acetone vapor in the air, and "50%" represents half of that. (E) Array-based organic vapor sensing and similar organic vapor identification. (F) The molecular structures of six synthetic PESs with different pendant groups and relative variations of the fluorescent intensity ($\Delta I/I_0$) for the polymer/**AnPh**$_3$ microwires derived from 4-methylphenyl-pendant PES in response to the mixed benzene/toluene vapor with different toluene contents. *(A–F) Reproduced with permission from reference X. Jiang, H. Gao, X. Zhang, J. Pang, Y. Li, K. Li, Y. Wu, S. Li, J. Zhu, Y. Wei, L. Jiang, Highly-sensitive optical organic vapor sensor through polymeric swelling induced variation of fluorescent intensity, Nat. Commun. 9 (1) (2018) 3799. Copyright (2018) Springer Nature.*

6 Conclusions and perspectives

The unique features of AIEgens such as good photostability, intensive luminescence in the thin film state, stimuli-responsive emission, structure diversity, and twisted conformation benefit the sensitive and selective response to a variety of vapors. This chapter presents the sensing mechanisms and the recent advances in AIE-based chemosensors for the detection of vapors including aromatic explosives, acid and amine vapors, and organic volatile compounds (VOCs). The sensing of explosive vapors is mostly dependent on the PET-caused quenching due to the electron transfer from the excited state of electron-rich AIEgens to electron-deficient aromatic explosives upon physical adsorption. In the case of other vapors, through molecular engineering of AIEgen structures, the sensing performance could be achieved based on several sensing mechanisms, including PET, ICT, ESIPT, and RIM upon physical adsorption, protonation and deprotonation process, and chemical reactions.

Despite the remarkable progress that has been made on the development of AIEgens for vapor sensing during the past decade, some challenges still remain. (1) The thin films of AIEgens are generally prepared by spin coating, drop coating, and dip coating, which suffer from some drawbacks such as bad reproducibility, uneven thickness over a large area, and limited contact surface with analytes. Even though some solutions have been proposed including electropolymerization and electrospinning, electropolymerization monomers are confined to those with low oxidation potentials and electrospinning is limited to the AIEgens with good solubility and spinnability. Thus, more efforts should be devoted to preparing AIEgen-loaded thin films with high-reproducibility and uniformity. (2) The examples introduced in this chapter mainly utilize AIEgens with UV light as excitation wavelength, which is harmful to human skin and eyes. The development of new AIEgen-based chemosensors with long-wavelength excitation and emission is highly desirable. (3) In a practical case, the temperature and humidity will remarkably affect the detection sensitivity. It is critical to evaluate the performance stability of the AIE-based chemosensors in the humid environment under varying room temperatures. We expect that this chapter would be helpful and suggestive to pave the way for the development of a new generation of AIEgen-based chemosensors for vapor sensing.

References

[1] E.P.J. Parrott, N.Y. Tan, R. Hu, J.A. Zeitler, B.Z. Tang, E. Pickwell-MacPherson, Direct evidence to support the restriction of intramolecular rotation hypothesis for the mechanism of aggregation-induced emission: temperature resolved terahertz spectra of tetraphenylethene, Mater. Horiz. 1 (2) (2014) 251–258.

[2] N.L.C. Leung, N. Xie, W. Yuan, Y. Liu, Q. Wu, Q. Peng, Q. Miao, J.W.Y. Lam, B.Z. Tang, Restriction of intramolecular motions: the general mechanism behind aggregation-induced emission, Chem. Eur. J. 20 (2014) 15349–15353.

[3] Y. Tu, Z. Zhao, J.W.Y. Lam, B.Z. Tang, Mechanistic connotations of restriction of intramolecular motions (RIM), Natl. Sci. Rev. (2020) nwaa260, https://doi.org/10.1093/nsr/nwaa260.

[4] Z. Wang, K. Liu, X. Chang, Y. Qi, C. Shang, T. Liu, J. Liu, L. Ding, Y. Fang, Highly sensitive and discriminative detection of BTEX in the vapor phase: a film-based fluorescent approach, ACS Appl. Mater. Interfaces 10 (41) (2018) 35647–35655.

[5] Z. Zhou, W. Xiong, Y. Zhang, D. Yang, T. Wang, Y. Che, J. Zhao, Internanofiber spacing adjustment in the bundled nanofibers for sensitive fluorescence detection of volatile organic compounds, Anal. Chem. 89 (7) (2017) 3814–3818.

[6] J. Luo, Z. Xie, J.W.Y. Lam, L. Cheng, H. Chen, C. Qiu, H.S. Kwok, X. Zhan, Y. Liu, D. Zhu, B.Z. Tang, Aggregation-induced emission of 1-methyl-1,2,3,4,5-pentaphenylsilole, Chem. Commun. 18 (2001) 1740–1741.
[7] M. Kang, Z. Zhang, N. Song, M. Li, P. Sun, X. Chen, D. Wang, B.Z. Tang, Aggregation-enhanced theranostics: AIE sparkles in biomedical field, Aggregate 1 (2020) 80–106.
[8] J. Mei, N.L.C. Leung, R.T.K. Kwok, J.W.Y. Lam, B.Z. Tang, Aggregation-induced emission: together we shine, united we soar! Chem. Rev. 115 (21) (2015) 11718–11940.
[9] Z. Zhao, X. Zheng, L. Du, Y. Xiong, W. He, X. Gao, C. Li, Y. Liu, B. Xu, J. Zhang, F. Song, Y. Yu, X. Zhao, Y. Cai, X. He, R.T.K. Kwok, J.W.Y. Lam, X. Huang, D. Lee Phillips, H. Wang, B.Z. Tang, Non-aromatic annulene-based aggregation-induced emission system via aromaticity reversal process, Nat. Commun. 10 (1) (2019) 2952–2961.
[10] P. Alam, N.L.C. Leung, J. Zhang, R.T.K. Kwok, J.W.Y. Lam, B.Z. Tang, AIE-based luminescence probes for metal ion detection, Coord. Chem. Rev. 429 (2021), 213693.
[11] M.H. Chua, K.A.-O. Shah, H. Zhou, J. Xu, Recent advances in aggregation-induced emission chemosensors for anion sensing, Molecules 24 (2019) 2711–2752.
[12] Y.-W. Wu, A.-J. Qin, B.Z. Tang, AIE-active polymers for explosive detection, Chin. J. Polym. Sci. 35 (2) (2017) 141–154.
[13] H. Zhou, M.H. Chua, B.Z. Tang, J. Xu, Aggregation-induced emission (AIE)-active polymers for explosive detection, Polym. Chem. 10 (28) (2019) 3822–3840.
[14] S. Chen, Y. Hong, Y. Liu, J. Liu, C.W.T. Leung, M. Li, R.T.K. Kwok, E. Zhao, J.W.Y. Lam, Y. Yu, B.Z. Tang, Full-range intracellular pH sensing by an aggregation-induced emission-active two-channel ratiometric fluorogen, J. Am. Chem. Soc. 135 (13) (2013) 4926–4929.
[15] A. Gupta, Aggregation-induced emission: a tool for sensitive detection of amines, ChemistrySelect 4 (44) (2019) 12848–12860.
[16] D. Wang, B.Z. Tang, Aggregation-induced emission luminogens for activity-based sensing, Acc. Chem. Res. 52 (9) (2019) 2559–2570.
[17] W. Xu, D. Wang, B.Z. Tang, NIR-II AIEgens: a win-win integration towards bioapplications, Angew. Chem. Int. Ed. 59 (2020), https://doi.org/10.1002/anie.202005899.
[18] M. Gao, B.Z. Tang, Fluorescent sensors based on aggregation-induced emission: recent advances and perspectives, ACS Sens. 2 (10) (2017) 1382–1399.
[19] A. Loudet, K. Burgess, BODIPY dyes and their derivatives: syntheses and spectroscopic properties, Chem. Rev. 107 (11) (2007) 4891–4932.
[20] W. Sun, S. Guo, C. Hu, J. Fan, X. Peng, Recent development of chemosensors based on cyanine platforms, Chem. Rev. 116 (14) (2016) 7768–7817.
[21] W. Sun, M. Li, J. Fan, X. Peng, Activity-based sensing and theranostic probes based on photoinduced electron transfer, Acc. Chem. Res. 52 (10) (2019) 2818–2831.
[22] S. Sasaki, G.P.C. Drummen, G.-I. Konishi, Recent advances in twisted intramolecular charge transfer (TICT) fluorescence and related phenomena in materials chemistry, J. Mater. Chem. C 4 (14) (2016) 2731–2743.
[23] J. Wu, W. Liu, J. Ge, H. Zhang, P. Wang, New sensing mechanisms for design of fluorescent chemosensors emerging in recent years, Chem. Soc. Rev. 40 (7) (2011) 3483–3495.
[24] J.E. Kwon, S.Y. Park, Advanced organic optoelectronic materials: harnessing excited-state intramolecular proton transfer (ESIPT) process, Adv. Mater. 23 (32) (2011) 3615–3642.
[25] V.S. Padalkar, S. Seki, Excited-state intramolecular proton-transfer (ESIPT)-inspired solid state emitters, Chem. Soc. Rev. 45 (1) (2016) 169–202.
[26] A.C. Sedgwick, L. Wu, H.-H. Han, S.D. Bull, X.-P. He, T.D. James, J.L. Sessler, B.Z. Tang, H. Tian, J. Yoon, Excited-state intramolecular proton-transfer (ESIPT) based fluorescence sensors and imaging agents, Chem. Soc. Rev. 47 (23) (2018) 8842–8880.
[27] B. Prusti, M. Chakravarty, An electron-rich small AIEgen as a solid platform for the selective and ultrasensitive on-site visual detection of TNT in the solid, solution and vapor states, Analyst 145 (5) (2020) 1687–1694.
[28] M. Kumar, V. Vij, V. Bhalla, Vapor-phase detection of trinitrotoluene by AIEE-active hetero-oligophenylene-based carbazole derivatives, Langmuir 28 (33) (2012) 12417–12421.
[29] Z. Zhang, X. Ai, A. Obolda, A. Abdurahman, F. Li, M. Zhang, A rapid-response fluorescent film probe to DNT based on novel AIE materials, Sensors Actuators B Chem. 281 (2019) 971–976.
[30] K.S. Bejoymohandas, T.M. George, S. Bhattacharya, S. Natarajan, M.L.P. Reddy, AIPE-active green phosphorescent iridium(iii) complex impregnated test strips for the vapor-phase detection of 2,4,6-trinitrotoluene (TNT), J. Mater. Chem. C 2 (3) (2014) 515–523.

[31] J.-H. Wang, H.-T. Feng, Y.-S. Zheng, Synthesis of tetraphenylethylene pillar[6]arenes and the selective fast quenching of their AIE fluorescence by TNT, Chem. Commun. 50 (77) (2014) 11407–11410.
[32] B. Xu, X. Wu, H. Li, H. Tong, L. Wang, Selective detection of TNT and picric acid by conjugated polymer film sensors with donor–acceptor architecture, Macromolecules 44 (13) (2011) 5089–5092.
[33] W. Dong, Y. Pan, M. Fritsch, U. Scherf, High sensitivity sensing of nitroaromatic explosive vapors based on polytriphenylamines with AIE-active tetraphenylethylene side groups, J. Polym. Sci. A Polym. Chem. 53 (15) (2015) 1753–1761.
[34] H. Zhou, X. Wang, T.T. Lin, J. Song, B.Z. Tang, J. Xu, Poly(triphenyl ethene) and poly(tetraphenyl ethene): synthesis, aggregation-induced emission property and application as paper sensors for effective nitro-compounds detection, Polym. Chem. 7 (41) (2016) 6309–6317.
[35] S. Jiang, S. Liu, L. Meng, Q. Qi, L. Wang, B. Xu, J. Liu, W. Tian, Covalent organic hollow nanospheres constructed by using AIE-active units for nitrophenol explosives detection, Sci. China Chem. 63 (4) (2020) 497–503.
[36] Z. Yang, W. Qin, J.W.Y. Lam, S. Chen, H.H.Y. Sung, I.D. Williams, B.Z. Tang, Fluorescent pH sensor constructed from a heteroatom-containing luminogen with tunable AIE and ICT characteristics, Chem. Sci. 4 (9) (2013) 3725–3730.
[37] P. Alam, N.L.C. Leung, H. Su, Z. Qiu, R.T.K. Kwok, J.W.Y. Lam, B.Z. Tang, A highly sensitive bimodal detection of amine vapours based on aggregation induced emission of 1,2-dihydroquinoxaline derivatives, Chem. Eur. J. 23 (59) (2017) 14911–14917.
[38] Y. Jiang, Z. Zhong, W. Ou, H. Shi, P. Alam, B.Z. Tang, J. Qin, Y. Tang, Semi-quantitative evaluation of seafood spoilage using filter-paper strips loaded with an aggregation-induced emission luminogen, Food Chem. 327 (2020), 127056.
[39] Z. Wang, X. Cheng, A. Qin, H. Zhang, J.Z. Sun, B.Z. Tang, Multiple stimuli responses of stereo-isomers of AIE-active ethynylene-bridged and pyridyl-modified tetraphenylethene, J. Phys. Chem. B 122 (7) (2018) 2165–2176.
[40] Y. Hu, T. Han, N. Yan, J. Liu, X. Liu, W.-X. Wang, J.W.Y. Lam, B.Z. Tang, Visualization of biogenic amines and in vivo ratiometric mapping of intestinal pH by AIE-active polyheterocycles synthesized by metal-free multicomponent polymerizations, Adv. Funct. Mater. 29 (31) (2019) 1902240.
[41] S. Pramanik, H. Deol, V. Bhalla, M. Kumar, AIEE active donor–acceptor–donor-based hexaphenylbenzene probe for recognition of aliphatic and aromatic amines, ACS Appl. Mater. Interfaces 10 (15) (2018) 12112–12123.
[42] K. Debsharma, J. Santhi, B. Baire, E. Prasad, Aggregation-induced emission active donor–acceptor fluorophore as a dual sensor for volatile acids and aromatic amines, ACS Appl. Mater. Interfaces 11 (51) (2019) 48249–48260.
[43] P. Xue, J. Ding, Y. Shen, H. Gao, J. Zhao, J. Sun, R. Lu, Aggregation-induced emission nanofiber as a dual sensor for aromatic amine and acid vapor, J. Mater. Chem. C 5 (44) (2017) 11532–11541.
[44] J. Peng, J. Sun, P. Gong, P. Xue, Z. Zhang, G. Zhang, R. Lu, Luminescent nanofibers fabricated from phenanthroimidazole derivatives by organogelation: fluorescence response towards acid with high performance, Chem. Asian J. 10 (8) (2015) 1717–1724.
[45] P. Xue, J. Sun, B. Yao, P. Gong, Z. Zhang, C. Qian, Y. Zhang, R. Lu, Strong emissive nanofibers of organogels for the detection of volatile acid vapors, Chem. Eur. J. 21 (12) (2015) 4712–4720.
[46] P. Xue, B. Yao, Y. Shen, H. Gao, Self-assembly of a fluorescent galunamide derivative and sensing of acid vapor and mechanical force stimuli, J. Mater. Chem. C 5 (44) (2017) 11496–11503.
[47] X. Yang, Y. Liu, J. Li, Q. Wang, M. Yang, C. Li, A novel aggregation-induced-emission-active supramolecular organogel for the detection of volatile acid vapors, New J. Chem. 42 (21) (2018) 17524–17532.
[48] P. Xue, B. Yao, P. Wang, P. Gong, Z. Zhang, R. Lu, Strong fluorescent smart organogel as a dual sensing material for volatile acid and organic amine vapors, Chem. Eur. J. 21 (48) (2015) 17508–17515.
[49] B. Li, D. Zhang, Y. Li, X. Wang, H. Gong, Y.-Z. Cui, A reversible vapor-responsive fluorochromic molecular platform based on coupled AIE-ESIPT mechanisms and its applications in anti-counterfeiting measures, Dyes Pigments 181 (2020), 108535.
[50] E. Zhang, X. Hou, Z. Zhang, Y. Zhang, J. Wang, H. Yang, J. You, P. Ju, A novel biomass-based reusable AIE material: AIE properties and potential applications in amine/ammonia vapor sensing and information storage, J. Mater. Chem. C 7 (27) (2019) 8404–8411.
[51] M. Gao, S. Li, Y. Lin, Y. Geng, X. Ling, L. Wang, A. Qin, B.Z. Tang, Fluorescent light-up detection of amine vapors based on aggregation-induced emission, ACS Sens. 1 (2) (2016) 179–184.
[52] J. Hu, R. Liu, S. Zhai, Y. Wu, H. Zhang, H. Cheng, H. Zhu, AIE-active molecule-based self-assembled nanofibrous films for sensitive detection of volatile organic amines, J. Mater. Chem. C 5 (45) (2017) 11781–11789.

[53] N. Meher, S. Panda, S. Kumar, P.K. Iyer, Aldehyde group driven aggregation-induced enhanced emission in naphthalimides and its application for ultradetection of hydrazine on multiple platforms, Chem. Sci. 9 (16) (2018) 3978–3985.
[54] Y. Yang, C.-Y. Gao, D. Dong, Tetraphenylethene functionalized rhodamine dye for fluorescence detection of HCl vapor in the solid state, Anal. Methods 8 (44) (2016) 7898–7902.
[55] Q. Qi, L. Huang, R. Yang, J. Li, Q. Qiao, B. Xu, W. Tian, X. Liu, Z. Xu, Rhodamine-naphthalimide demonstrated a distinct aggregation-induced emission mechanism: elimination of dark-states via dimer interactions (EDDI), Chem. Commun. 55 (10) (2019) 1446–1449.
[56] L. Hu, J. Sun, J. Han, Y. Duan, T. Han, An AIE luminogen as a multi-channel sensor for ethanol, Sensors Actuators B Chem. 239 (2017) 467–473.
[57] R. Zhao, M. Zhang, Y. Liu, X. Zhang, Y. Duan, T. Han, Fabricating D-A type AIE luminogen into film sensor for turn-on detection of methanol vapour, Sensors Actuators B Chem. 319 (2020), 128323.
[58] R. Huang, H. Liu, K. Liu, G. Wang, Q. Liu, Z. Wang, T. Liu, R. Miao, H. Peng, Y. Fang, Marriage of aggregation-induced emission and intramolecular charge transfer toward high performance film-based sensing of phenolic compounds in the air, Anal. Chem. 91 (22) (2019) 14451–14457.
[59] X. Jiang, H. Gao, X. Zhang, J. Pang, Y. Li, K. Li, Y. Wu, S. Li, J. Zhu, Y. Wei, L. Jiang, Highly-sensitive optical organic vapor sensor through polymeric swelling induced variation of fluorescent intensity, Nat. Commun. 9 (1) (2018) 3799.

CHAPTER

19

AIEgen applications in rapid and portable sensing of foodstuff hazards

Qi Wang, Youheng Zhang, Yanting Lyu, Xiangyu Li, and Wei-Hong Zhu

Institute of Fine Chemicals, East China University of Science and Technology, Shanghai, China

1 Introduction

The nutritional status of the residents in a country represents an important manifestation of the nation's quality of life, which is closely related to social and economic development. Owing to the improvement in the living standard, peoples' demand for food is not only nutrition but also safety, which puts forward higher requirements for food production. In order to ensure food safety, adhering to food production quality standards is of great importance, which can not only reduce the residues of hazardous substances but also avoid the foodborne diseases caused by microorganisms. Consequently, it is necessary to rapidly and portably detect these contaminants, such as pathogens/biotoxin, heavy metals, pesticides, and other illegal additives, in foodstuffs.

Fast detection of these biological and chemical contaminants represents some of the key issues for the assessment of food quality. Up to now, the detection methods mainly include chemical, biological, and multiple combined technologies. Chemical detection methods include chromatography, mass spectrometry, nuclear magnetism, spectroscopy, electrochemistry, and biosensors. Among them, gas chromatography (GC) and high-performance liquid chromatography (HPLC) technologies are relatively well developed, with high separation efficiency and sensitivity for many types of pollutants. The biological detection methods with good detection accuracy mainly include immune, DNA probe, polymerase chain reaction (PCR) technology, and so on. However, most of these strategies involve the use of expensive instruments, long detection procedure, and complicated sample pretreatment, which decrease efficiency and limit their applications.

Aggregation-Induced Emission (AIE)
https://doi.org/10.1016/B978-0-12-824335-0.00021-0

In recent years, fluorescence sensors have been widely used in food safety evaluation, environmental monitoring, and bioimaging due to their simple operation, high sensitivity, fast response speed, and high spatial resolution, especially in realizing qualitative or quantitative detection [1]. However, most of the traditional fluorophores are hydrophobic with large π-conjugated systems, accompanying obvious photobleaching behavior and ACQ effect in high concentrations or aggregation states for low sensitivity with limited applications. Fortunately, aggregation-induced emission (AIE), an emerging photophysical phenomenon, provides the opportunity to address these challenges and fulfills the demands for developing simple, rapid, and accurate detecting methods. This chapter reviews the research progress of AIE materials in the classification and detection of different pollutants, introduces the detection mechanism of AIE materials and the application progress in food safety, and finally prospects the future development of AIE luminogens (AIEgens).

2 AIEgen applications in rapid and portable sensing of foodstuff hazards

2.1 Main contaminants

Food is the first necessity of human survival, and the problem of food safety has become increasingly prominent. Numerous standards are now employed to ensure the quality of food and are adopted in the worldwide food industries [2]. However, global food safety incidents still occur frequently, resulting in serious health and social problems [3]. For example, foodborne pathogens, harmful food additives, pesticides, and heavy metals continue to perplex people's lives. Therefore, it is necessary to carry out food safety inspections, utilize chemical reagents and chemical techniques to detect the edibility of food so as to safeguard consumers' health.

Among them, pathogens are widely distributed in soil and water, decaying fruits and vegetables. They enter the food chain through various ways and then invade the human body through the digestion system, thus threatening human health. Besides, overuse or misuse of food additives during food preparation also poses a threat to health and raises concerns about food safety. In order to extend the shelf life and enhance the taste of food, food factories usually add a certain amount of fluorescent whitening agents, food thickeners, and other food additives. Unfortunately, harmful and illegal chemicals such as melamine and Sudan I have also been found in some foods. Moreover, pesticides are widely used to eliminate harmful organisms that threaten human health. Most pesticides such as organophosphorus pesticides (OPs) and organochlorine pesticides (OCPs) are carcinogens [4], which could accumulate in food. Even at low concentrations, they can block the activity of cholinesterase through neurotoxicity and ultimately endanger human health [5]. In addition, Hg^{2+}, Pb^{2+}, and other heavy metals in food pose a great threat to public health and lead to neurological disorders, cognitive dysfunction, and other diseases. In conclusion, food pollution sources mainly include pesticides, pathogens/toxins, heavy metals, and additives (Fig. 1). Therefore, it is of great importance to develop rapid and accurate detection methods for food pollutants.

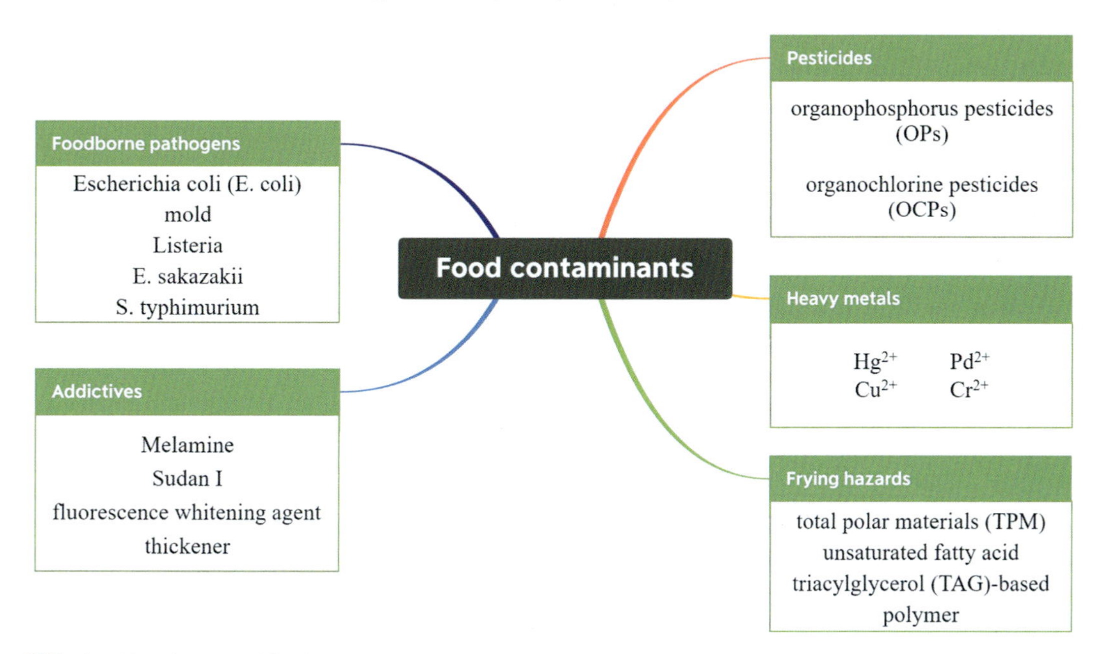

FIG. 1 Classification of food contaminants.

2.2 Conventional detection methods

At present, the detection methods of food pollutants mainly include chemical, biological, and various combination technologies. Among different chemical detection methods, GC, HPLC, gas chromatography–mass spectrometry (GC–MS), and other methods based on large chromatographic instruments have complete detection technology, high separation efficiency, and good detection performance [6,7]. Meanwhile, biological detection methods such as immunoassay [8], DNA technology [9], and polymerase chain reaction (PCR) [10] have been frequently utilized for quantitative analysis and real-time online monitoring. However, they are not suitable for on-site and periodic environmental monitoring due to their high analysis cost, complex instrument maintenance, long analysis time, and dependence on professional technicians.

In order to reduce the detection cost and shorten the analysis time, a large number of rapid detection methods based on spectrometers have been developed, such as near-infrared spectrometers, Raman spectrometers, fluorescence spectrophotometers, and ultraviolet–visible spectrophotometers [11,12]. Although the corresponding portable detector and detection method have been developed, it is difficult to be applied in rapid detection due to the overlapping peaks and poor detection specificity. Notably, fluorescence technology has been widely used in the detection of various analytes due to the advantages of easy operation, high sensitivity, fast response, and high spatial resolution [13–15]. For example, Shen [16] developed a new water-soluble fluorescent probe containing 1,8-naphthalene diformimide dye, quaternary ammonium salt, and borate groups for the detection of organophosphorus pesticides. Yang [17] developed a hemicyanine probe for the detection of benzoyl peroxide in food. Although these "*off–on*" fluorescent probes show good sensing ability, they still have extremely serious photobleaching characteristics and medium signal-to-noise ratio and are limited by pH, temperature, probe concentration, and instrument efficiency due to ACQ [18].

2.3 AIEgens detection methods

The newly developed materials with AIE characteristics are expected to become advanced biosensors for food detection. Due to the restriction of intramolecular movement (RIM), AIE luminogens (AIEgens) show weak emission in solution and the fluorescence increases during the aggregation process [19,20]. Compared with traditional fluorescent dyes such as Rhodamine B, anthocyanin, coumarin, AIEgens have the advantages of wide excitation wavelength range, narrow emission range, high fluorescence intensity, and strong resistance to photobleaching. Subsequently, AIEgens could self-assembles into aggregates with the induction of specific substrates, and the fluorescence significantly enhanced due to the restriction of intramolecular motion (RIM), so as to realize a qualitative analysis of stimulus sources. The use of AIEgens also made it easier to conduct high sensitivity and high-efficiency analysis in vivo imaging and online sensing monitoring [21–25]. Therefore, in recent years, more and more attention has been paid to biosensors based on AIEgens, and they are widely used in the detection of various contaminants and the evaluation of food quality [26–33].

2.3.1 *Foodborne pathogens detection*

Foodborne pathogens are one of the main threats to food safety, and the existence of any pathogenic bacteria in food is a serious public health hazard. They can directly or indirectly contaminate food and water. Oral infection can lead to intestinal infectious diseases, food poisoning, and livestock infectious diseases. Although the well-developed food packaging technology enables food to be stored for a long time, the most common foodborne pathogens such as molds [34], *Escherichia coli* (*E. coli*) [35,36] and Listeria [37] could inevitably break through the safeguards imposed by packing technology and emerge in the process of storage. Therefore, the rapid detection of pathogens is of great importance, which could achieve a real-time evaluation of food contamination and hygiene quality.

Since the surface of bacteria is negatively charged, AIE fluorescent probes with positive charges are generally designed to detect bacteria through electrostatic interaction and molecular motion restriction. Meng [38] reported a multifunctional positive-charged probe, Zinc (II)-lutidine amine (AIE-ZnDPA) (Table 1), for selective imaging and imaging-guided photodynamic killing of bacteria. AIE-ZnDPA first accumulates on the bacterial membrane through electrostatic interaction, and then the AIE effect and excited-state intramolecular proton transfer (ESIPT) are activated by limiting the intramolecular rotation of N—N bonds and forming intramolecular hydrogen bonds in the salicylic acid pyridazine moiety, thus achieving imaging and killing. Although the electrostatic interaction method can selectively recognize bacteria, it cannot distinguish between Gram-positive and Gram-negative bacteria.

The introduction of specific recognition ligands is one of the effective ways to improve the imaging selectivity of AIE sensors. He [50] proposed a strategy to introduce bacteriophages to

TABLE 1 The chemical structures of AIEgens in sensing of foodstuff hazards.

Reference	AIEgens	Structure	Reference	AIEgens	Structure
[27]	OFNs@Au NCs	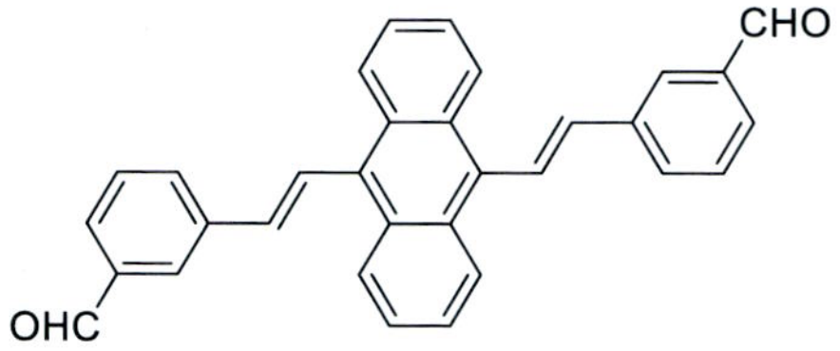	[39]	DSAC2N-C-Apt	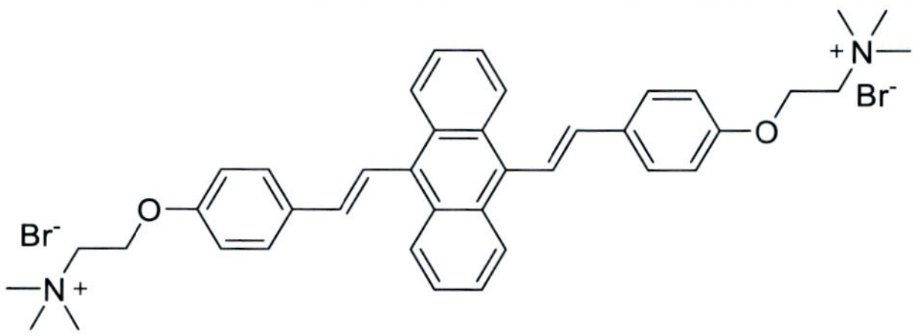
[30]	TPAEQ	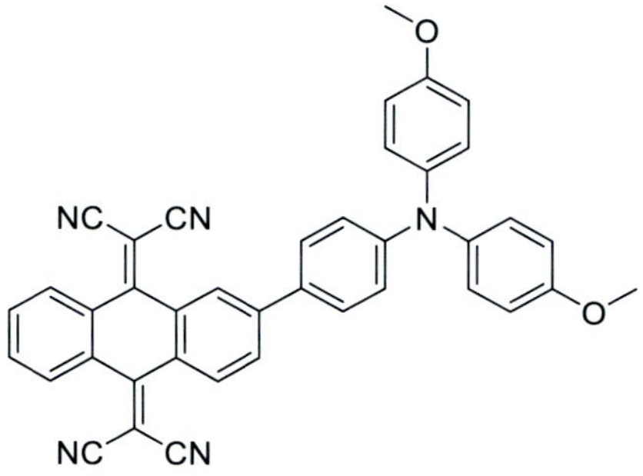	[40]	Pyren-DT	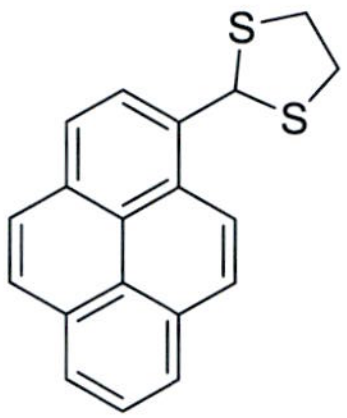
[38]	AIE-ZnDPA	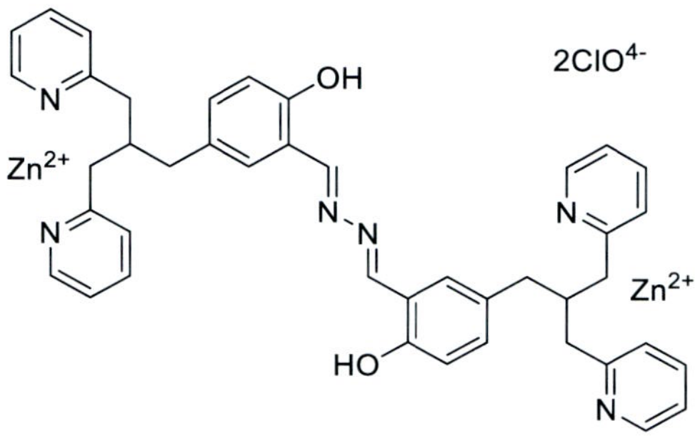	[41]	TPE-triazole-CD	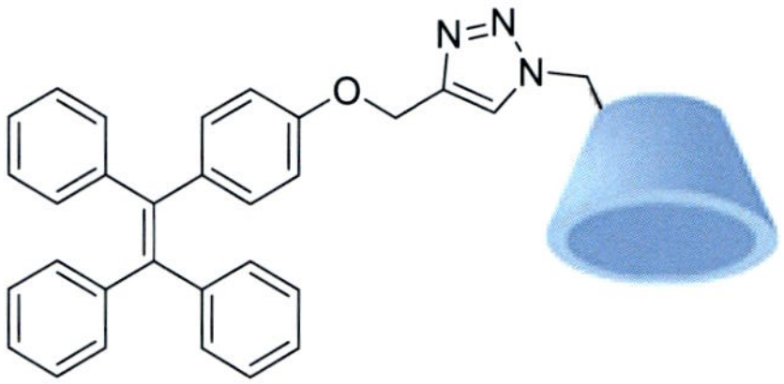

Continued

TABLE 1 The chemical structures of AIEgens in sensing of foodstuff hazards—cont'd

Reference	AIEgens	Structure	Reference	AIEgens	Structure
[42]	BTBP-Gluc	HO_2C, HO, HO, OH, O, N, S, Br	[43]	HMBA-4	COOH, N, OH, H_3C
[44]	AuNPs-PTDNPs	PTD m, n, Cl^-, HN, O, N	[45]	Cu NCs	TFTP SH, F, F, F, F
[46]	BSPOTPE-SiO_2-MnO_2	BSPOTPE $OC_3H_6SO_3^-Na^+$, $^+Na^-O_3SC_3H_6O$	[47]	QM-TPA	NC, CN, N, N
[48]	TPE-LHLHLRL	TPE CHO, HO, OH	[49]	CPA-TPA	N, NC, CN

guide AIEgen to achieve "radar" induction to selectively identify target bacteria, and combining AIEgens with bacteriophages to form a new class of antibacterial biological conjugate TVP-PAP (Fig. 2). By introducing the targeted phage into the probe, AIEgens can specifically recognize the host bacteria and selectively image and kill the target bacteria by photodynamic means through further phage infection and AIE fluorescence emission. Wei [42] reported a novel, sensitive, and low toxic desorbed-based fluorescent probe (BTBP-Gluc). The probe has an insoluble fluorescent group BTBP, which can excite intramolecular proton transfer (ESIPT) and AIE effect. It can be used to detect endogenous-glucuronidase (GUS) activity in living *E. coli*. In addition, the BTBP-Gluc could be further applied in C—F *E. coli* agar due to its low toxicity. The target *E. coli* strains were detected and isolated on the C—F *E. coli* agar plates after being cultivated for 18h. Notably, the recoveries of the inoculated O157:H7 and non-O157:H7 *E. coli* strains were 100% after three repeated tests. It indicated that C—F *E. coli* agar could be a potentially simple sensitive tool for detecting and isolating O157:H7 and non-O157:H7 *E. coli* in real food samples such as milk.

On the other hand, the different structures of pathogens (Gram-negative bacteria, Gram-positive bacteria, and fungi) allow fluorescent probes to penetrate the cell membrane and target different organelles, thereby visually distinguishing these three types [51–53]. Zhou [28] proposed that AIEgens with twisted D-A structure can effectively suppress the nonradiative relaxation of the twisted intramolecular charge transfer (TICT) state through the AIE characteristic and display a variety of fluorescent color visualization responses to the microenvironment (Fig. 3). Therefore, they developed an AIE fluorescent probe IQ-CM, which was composed of diphenyl isoquinolinium (IQ) unit and coumarin (Cm)-derived moiety, with

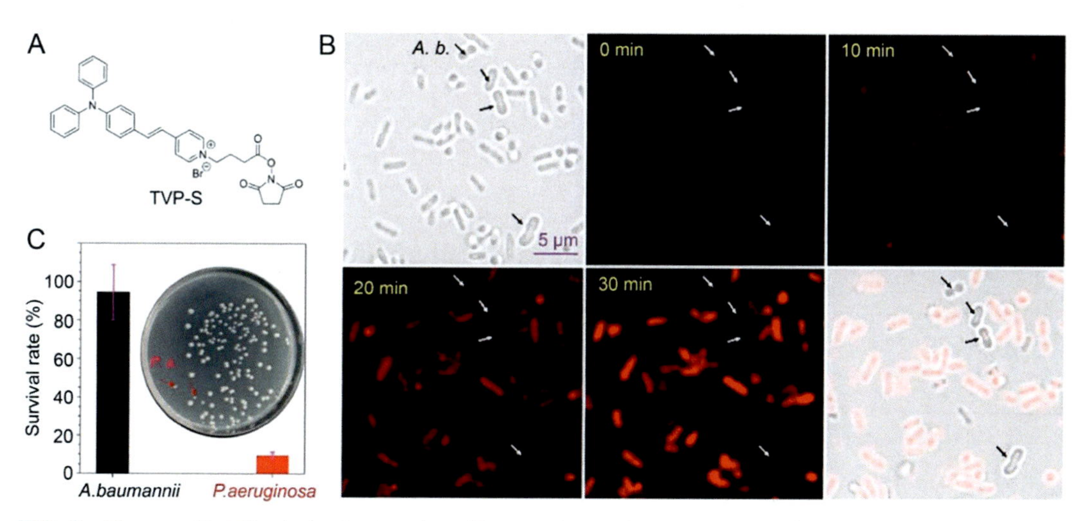

FIG. 2 Phage-guided discriminative imaging of bacteria by AIE bioconjugates. (A) Molecular structure of TVP-S. (B) Specificity fluorescence imaging of *P. aeruginosa* and *A. baumanni* incubated with TVP-PAP. (C) Antibacterial evaluation of *P. aeruginosa* and *A. baumannii* incubated with TVP-PAP [50]. *Reprinted from Journal of the American Chemical Society, Volume 142, Xuewen He, Yujun Yang, Yongcan Guo, Shuguang Lu, Yao Du, Jun-Jie Li, Xuepeng Zhang, Nelson L. C. Leung, Zheng Zhao, Guangle Niu, Shuangshuang Yang, Zhi Weng, Ryan T. K. Kwok, Jacky W. Y. Lam, Guoming Xie, and Ben Zhong Tang, Phage-Guided Targeting, Discriminative Imaging, and Synergistic Killing of Bacteria by AIE Bioconjugates, pp. 3959–3969, Copyright (2020), with permission from ACS.*

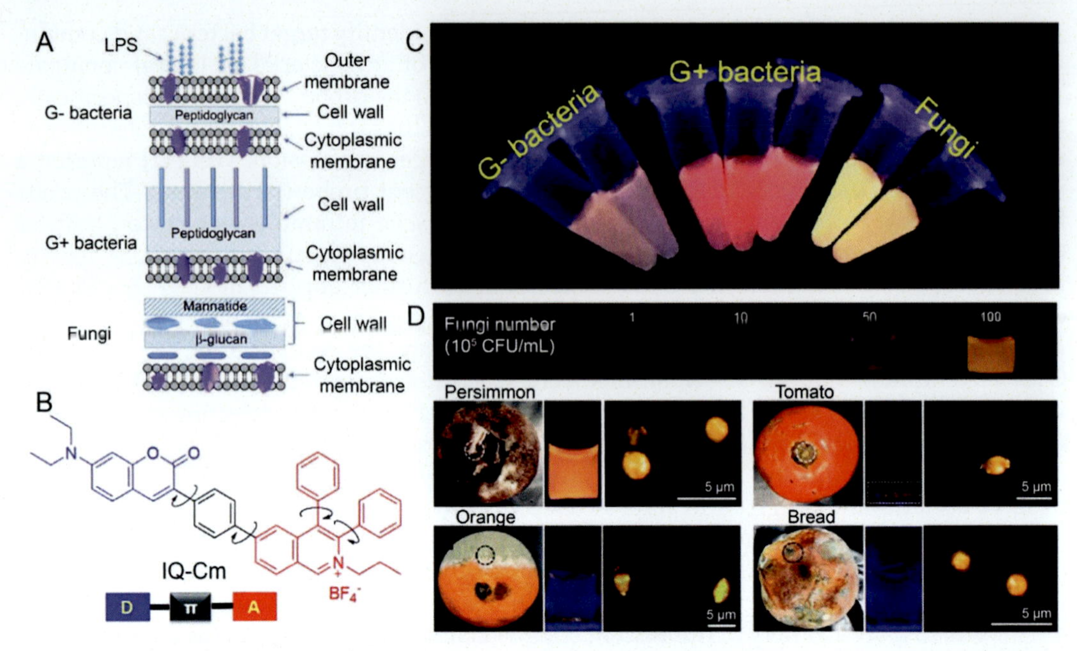

FIG. 3 AIEgen IQ-Cm used three colors for fast differentiation of three pathogens. (A) Schematic representation of cell envelope structures of pathogens. (B) The chemical structure of IQ-Cm. (C) Photographs of IQ-Cm with different pathogens in PBS solutions were obtained under 365nm UV irradiation. (D) Visual detection of mold with IQ-Cm [28]. *Reprinted from Chemical Science, Volume 11, Chengcheng Zhou, Meijuan Jiang, Jian Du, Haotian Bai, Guogang Shan, Ryan T. K. Kwok, Joe H. C. Chau, Jun Zhang, Jacky W. Y. Lam, Peng Huang, and Ben Zhong Tang, One stone, three birds: one AIEgen with three colors for fast differentiation of three pathogens, pp. 4730–4740, Copyright (2020), with permission from RSC.*

TICT and fluorescent color response sensitive to the pathogen's microenvironment. IQ-CM emits three visually distinguishable emission colors in different parts of Gram-negative bacteria, Gram-positive bacteria, and fungi. For example, cationic IQ-CM can target and accumulate in the mitochondria of the fungus, thus producing significant yellow light. Therefore, the bright yellow radiation of IQ-CM labeled fungi greatly promoted the detection, with the naked-eye detection limit of about 10^6 CFU mL $^{-1}$. More importantly, it is easy to determine visually the number of molds growing in food. Therefore, this simple visualization strategy based on a single AIEgen provides a promising platform for rapid detection and diagnosis.

2.3.2 *Pesticide residues detection*

Nowadays, pesticides, especially organophosphorus pesticides (OPs), are widely used in agriculture and forestry planting to prevent diseases and pests. However, OPs are neurotoxins that can reduce the activity of acetylcholine esterase (AChE), cause the accumulation of acetylcholine, and lead to Parkinson's disease, Alzheimer's disease, and other neurological diseases, even death [54,55]. Inappropriate utilization of pesticides leads to excessive pesticide residues in vegetables, fruits, and other agricultural products, which eventually enter the human digestion system and threaten human health. Therefore, rapid and sensitive detection of pesticide residues in food is of great significance for health and environmental protection.

OPs can be phosphorylated by the serine hydroxyl group of AChE to form a stable covalent bond, which prevents the amino group of AChE from binding to the substrate choline, thus reducing the activity of AChE. Therefore, the content of OPs residues can be determined indirectly by detecting the activity of AChE. It is well-known that AChE catalyzes the hydrolysis of acetyl thiocholine iodide (ATCh), which affects the environmental pH value. Therefore, Yue [26] synthesized a tetraphenylethylene (TPE) derivative (TPE-1), whose aldehyde group was pH sensitive (Fig. 4). The irreversible inhibition of AChE activity by OPs inhibited the hydrolysis of ATCh to acetic acid and weakened the protonation of TPE-1, resulting in a linear relationship between fluorescence intensity and OPs concentration. This method has been successfully used for the highly selective detection of OPs.

However, most of the AIE particles used for sensing are unstable in solution, which limits their practical applications [56]. In order to improve the stability and biocompatibility, Chen dispersed amphiphilic polymers in phosphate buffer solution to prepare AIE nanoparticles PTDNPs [44]. Gold nanoparticles (AuNPs) bind with PTDNPs and then quench the fluorescence through fluorescence resonance energy transfer (FRET). ATCh competes with PTDNPs to block

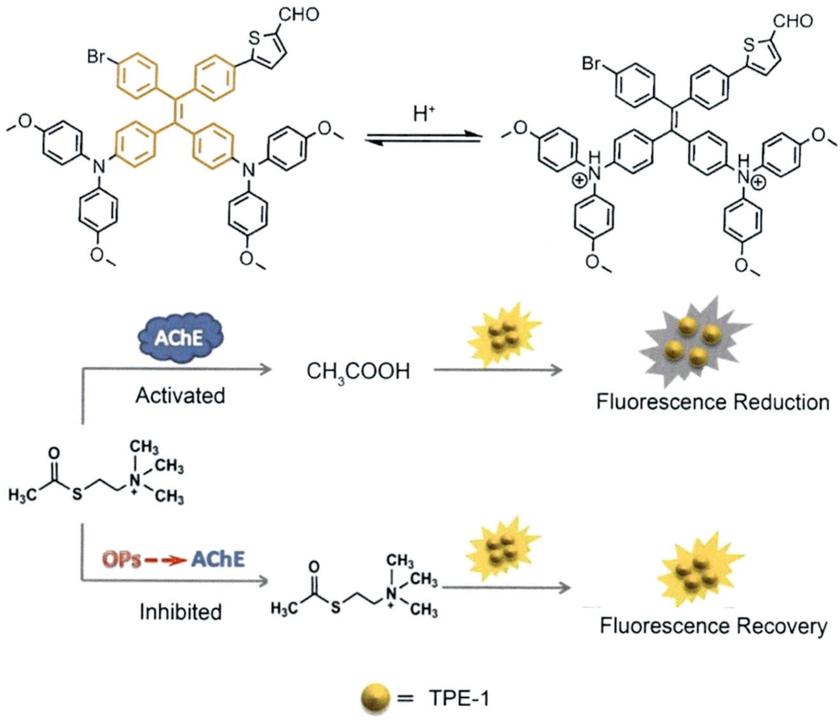

FIG. 4 Schematic illustration of the sensing principles of organophosphorus pesticides (OPs) using TPE-1 [26]. *Reprinted from Sensors and Actuators B: Chemical, Volume 292, Yue Cai, Jingkun Fang, Bingfeng Wang, Fangshuai Zhang, Guang Shao and Yingju Liu, A signal-on detection of organophosphorus pesticides by fluorescent probe based on aggregation-induced emission, pp. 156–163, Copyright (2019), with permission from Elsevier.*

FRET to turn on the fluorescence, enabling PTDNPs to detect successfully OPs with high sensitivity. In addition, Wu [46] designed a sandwich-type nanocomposite AIE fluorescence sensor BSPOTPE-SiO_2-MnO_2 based on 1,2-bis[4-(3-sulfonylpropoxy) phenyl]-1,2-stilbene (BSPOTPE) (Table 1). The fluorescence of the BSPOTPE-SiO_2-MNO_2 sandwich nanocomposite was opened by the AChE-ATCh solution and closed by OPs. It can be seen that BSPOTPE-SiO_2-MnO_2 has a good linear relationship for the detection of paraoxon, which can also achieve a simple and intuitive semiquantitative analysis of paraoxon.

Despite the fact that AIE probes are generally based on the principle of enzyme inhibition, the polypeptide bioprobes also have good applications in the detection of pesticide residues. Wang [48] linked TPE (Table 1) with the peptide sequence LHLHLRL to synthesize the AIE fluorescent probe TPE-peptide for the determination of OPs. OPs can enhance the probe hydrophobicity, and make it easier to accumulate in the peptide fibrils, thus accelerating peptide aggregation of peptides and inducing fluorescence emission. The probe showed a highly sensitive fluorescence response within 0.6–100 μM, and can rapidly detect pesticides within 15 min.

Up to now, there have been many research results based on the AIE method to detect pesticide residues. However, due to the complex composition of food and the wide application of pesticides, it is still difficult to detect a variety of pesticides simultaneously. Therefore, the development of strong antiinterference ability, high selectivity, intuitive, fast field detection method remains the focus of current research.

2.3.3 *Veterinary drug residues detection*

In animal husbandry, veterinary drugs are often used to prevent diseases and promote growth. However, they cannot be completely metabolized, and therefore remained and accumulated in livestock, eventually threatening human health through the food chain. Excessive veterinary drug residues in livestock could cause an allergic reaction, food poisoning, and even death [57]. Therefore, it is crucial to detect and evaluate animal-derived foods (such as meat, milk, and eggs) for the presence of such drug residues.

Amantadine (AMD) is an antiviral drug in livestock, which has been recognized by the international medical community due to having an inhibitory effect on the influenza virus. Enzyme-linked immunosorbent assay (ELISA) is a simple and rapid method for the detection of AMD, but it is not suitable for low concentration detection because of its moderate sensitivity. In order to improve the detection sensitivity, Yu combined AIE and ELISA methods to develop a new immunofluorescence probe (Fig. 5) [58]. In this method, glucose oxidase (GOx) triggered the biological process of GOx/glucose-mediated H_2O_2 production, thereby oxidizing TPE-HPro and lighting up fluorescence. Combined with ELISA, the probe can be used for the quantitative determination of AMD in chicken muscle samples, and its sensitivity was 2.5 times higher than that of traditional immunoassay.

Antibiotic drugs including tetracyclines, penicillin, and chloramphenicol (CAP) are often used to treat bacterial infections in livestock. Among them, CAP has high toxicity and side effects, which not only affects the hematopoietic system of livestock, but also inhibits the hematopoietic function of human bone marrow, resulting in aplastic anemia, hemolytic anemia, and other diseases [59,60]. Zhang [39] has successfully developed a highly sensitive AIE fluorescent probe DSAC_2N-C-Apt for CAP detection. The aptamer C-Apt and the short alkyl chain 9,10-dicyanoanthracene derivative DSAC_2N (Table 1) were used as the recognition groups of CAP and AIE fluorophore, whereas graphene oxide (GO) was used as the

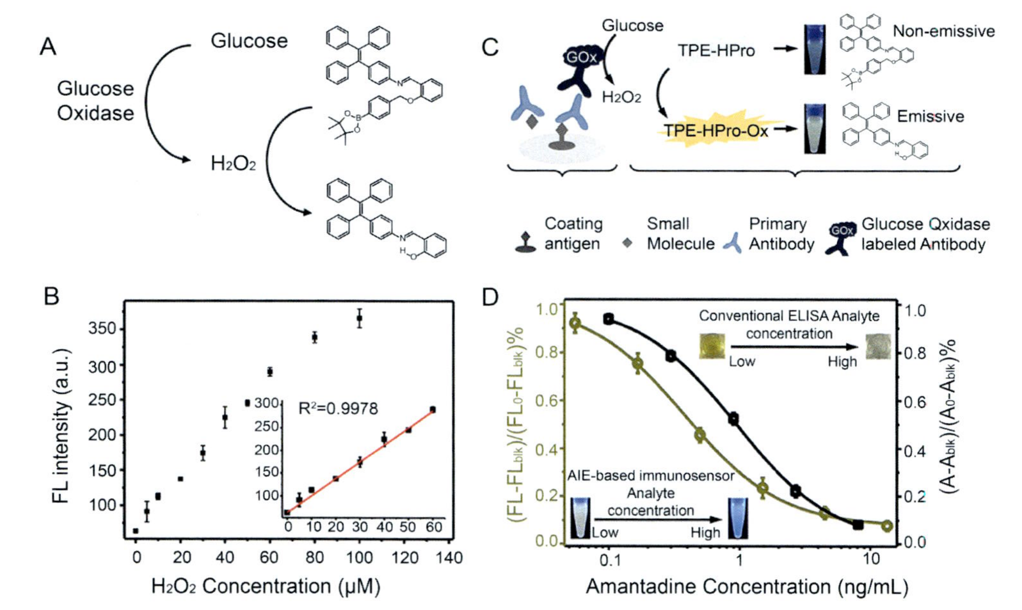

FIG. 5 AIE-based indirect competitive immunoassay for detection of drug residues. (A) Schematic illustration of H_2O_2-triggered "turn-on" fluorescence. (B) Fluorescent intensities plotted against H_2O_2 concentration. (C) Schematic illustration of AIE immunosensor for indirect competitive ELISA assay. (D) Inhibition curve for the quantitative determination of AMD by the AIE immunosensor and the conventional method [58]. *Reprinted from Frontiers in Chemistry, Volume 7, Wenbo Yu, Ying Li, Bing Xie, Mingfang Ma, Chaochao Chen, Chenglong Li, Xuezhi Yu, Zhanhui Wang, Kai Wen, Ben Zhong Tang and Jianzhong Shen, An aggregation-induced emission-based indirect competitive immunoassay for fluorescence "turn-on" detection of drug residues in foodstuffs, p. 228, Copyright (2019), with permission from Frontiers Media S.A.*

fluorescence quencher. C-Apt with $DSAC_2N$ could be adsorbed on GO through π-π stacking and then the fluorescence is quenched by GO. After binding with CAP, the complex (C-Apt-CAP) was formed and desorbed from the surface of GO, then the fluorescence of $DSAC_2N$ was recovered. The detection limit of CAP was $0.36\,ng\,mL^{-1}$, which could be detected by the change of fluorescence intensity. Compared with instrumental analysis and immunoassay, the detection method based on AIE fluorescent sensor has higher sensitivity and better specificity and has broad application prospects in the field of veterinary drug detection.

2.3.4 *Heavy metal detection*

Mercury (Hg), arsenic (As), cadmium (Cd), lead (Pb), and other heavy metals are generally highly toxic, which can cause enzyme and protein inactivation, thereby damaging human organs and causing chronic poisoning and even death [61]. Therefore, monitoring the content of heavy metals in water, food and the environment is a vital task. At present, a variety of fluorescent probes have been successfully applied to detect heavy metal content in the environment [62,63].

The fluorescent probes with thioaldehyde or thioketone structure have high sensitivity and excellent specificity for the detection of Hg^{2+}. Ma [40] synthesized a reactive AIE fluorescent probe Pyren-DT based on Pyrene-1-CHO and ethanedithiol. The probe reacted with Hg^{2+} to generate Pyrene-1-CHO, and the fluorescence intensity was enhanced for semiquantitative detection of Hg^{2+}. Pyren-DT had excellent selectivity and sensitivity to Hg^{2+} with a good linear relationship of 0–6.5 $\mu mol\,L^{-1}$. What's more, the probe showed a better response to Hg^{2+} in samples and was suitable for on-site detection. Further, in order to develop a fluorescent probe for detecting Hg^{2+} in living cells and organisms, Gao [64] developed a water-soluble fluorescence probe MPIPBS based on the polar inversion reaction (Umpolung) induced by Hg^{2+} and the AIE effect (Fig. 6). In the presence of Hg^{2+}, MPIPBS released MPIB and free aldehyde, thus changing the efficiency of intramolecular charge transfer (ICT) and realizing the fluorescence turn-on. MPIPBS has been used to develop portable strips for instant and the quantitative detection of Hg^{2+} in water, urine samples, living cells, and zebrafish.

Pathon [65] developed a fluorescence sensor based on Fe_3O_4 graphene oxide quantum dots (Fe-GQD) to detect As^{3+}.Fe-GQD aggregated with As^{3+} through intermolecular interactions and then turned on the fluorescence. Fe-GQD had good selectivity for As^{3+}, with a detection limit of 5.1 ppb, which was far lower than the allowable detection limit of As^{3+} in drinking water stipulated by the World Health Organization (WHO). Zhang [41] synthesized an AIE fluorescence-responsive sensor TPE-triazole-CD based on cyclodextrin (CD) and TPE (Table 1). The fluorescence was turned on when the complex of Cd^{2+} with triazole bridge and cyclodextrin coordinated was formed. This work provided a simple method to detect Cd^{2+} in a neutral environment based on the AIE effect.

2.3.5 *Food additives detection*

Food additives are nonnutritive substances added to food in low concentrations, which can be used to improve the appearance, flavor, and storage properties of food. However, the abuse of food additives may bring health risks. In recent years, food safety incidents caused by the abuse of food additives also occurred frequently.

Thickener is a type of food additive, which are mainly used to improve and increase the viscosity of food and improve the physical properties. Xu [30] developed the AIE active probe

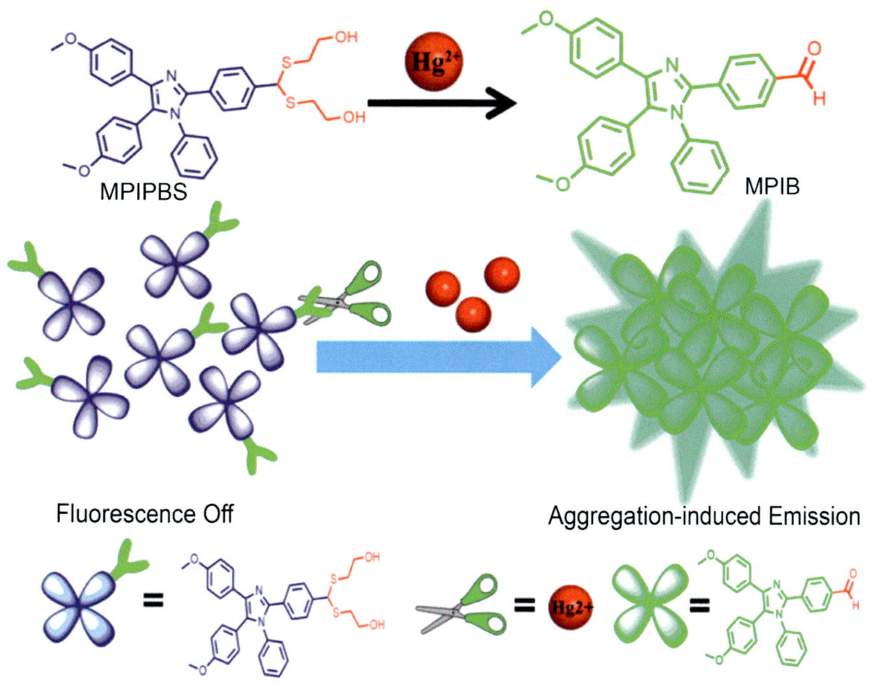

FIG. 6 Chemical structures of probe MPIPBS and design strategy via the Hg^{2+}-induced umpolung reaction and the AIE mechanism [64]. *Reprinted from Journal of Agricultural and Food Chemistry, Volume 67, Tang Gao, Xueyan Huang, Shuai Huang, Jie Dong, Kai Yuan, Xueping Feng, Tingting Liu, Kunqian Yu, and Wenbin Zeng, Sensitive Water-Soluble Fluorescent Probe Based on Umpolung and Aggregation-Induced Emission Strategies for Selective Detection of* Hg^{2+} *in Living Cells and Zebrafish, pp. 2377–2383, Copyright (2019), with permission from ACS.*

TPAEQ (Table 1) with a long emission wavelength and large Stokes shift, which can be used to detect the viscosity of beverages and food thickening effect. With the increase of viscosity, the rotatable partial rotation of trainline and methyl ether in TPAEQ was inhibited, resulting in strong fluorescence emission. In the presence of various inorganic salts, amino acids, and food additives, the fluorescence intensity of TPAEQ was significantly enhanced only in viscous glycerin solutions, suggesting that it can be used in complex environments, such as liquid beverages. Moreover, the fluorescence intensity of TPAEQ enhanced significantly with the increase of food thickener concentration, and there was a linear relationship ranging from 1 to $10\,g\,kg^{-1}$. In addition, since viscosity change is one of the key parameters to reflect the deterioration of fluid food, TPAEQ can be used to detect the viscosity change of liquid beverages through fluorescence response, and subsequently monitor the deterioration of liquid beverages. The method for determining viscosity can be used in food safety inspection applications to facilitate direct field testing.

Melamine is a commonly used raw material in the textile industry and pesticide production. Because of its high nitrogen content and low price, melamine is deliberately and illegally added to milk and infant formula. Ingesting excessive melamine can cause kidney failure and even death in infants. Niu [27] developed a reliable and highly sensitive dual emission proportional fluorescent probe for the detection of Hg^{2+} and melamine in food. OFNs@Au NCs is composed of positively charged AIE organic fluorescent nanoparticles (Table 1) and negatively charged Au nanoclusters, which can be used not only for visual determination but also for the quantitative determination of Hg^{2+} and melamine. The high-affinity interaction between Hg^{2+} and Au resulted in the initial fluorescence quenching of Au NCs. However, due to the higher affinity between melamine and Hg^{2+}, the fluorescence of Au NCs was recovered to detect Hg^{2+}. Consequently, red fluorescence emitted Au NCs was quenched by Hg^{2+} and recovered by melamine, while the green fluorescence Ply-BFSA OFN remained constant, which realized obvious color change and visual detection for Hg^{2+} and melamine. The satisfactory results confirmed that the Ply-BFSA OFNs@Au NCs successfully detected Hg^{2+} and melamine in real samples, such as tap water and milk powder. In addition, the probe broadened the application range of AIE-based organic fluorescent nanoparticles and provided a new method to prepare more sensitive, biocompatible, and visually proportional fluorescent probes.

Rhodamine (Rh) is a potential carcinogen, which can cause abnormal reproductive development and has been banned from the food industry. However, it is still illegally added into food as a pigment due to its low cost. Li [66] used tetraphenylpyrazine (TPP) derivative as the AIE response group to develop a new type of proportional fluorescence sensor AIE-MIPs-1 based on molecular imprinting polymers (MIPs), which can quantitatively detect Rh6G in dried papaya and beverages (Fig. 7). AIE-MIPs-1 exhibited a good linear relationship with Rh6G in the range of 0.0–10.0 $\mu mol\,L^{-1}$, and the detection limit is 0.26 $\mu mol\,L^{-1}$. Further, Li

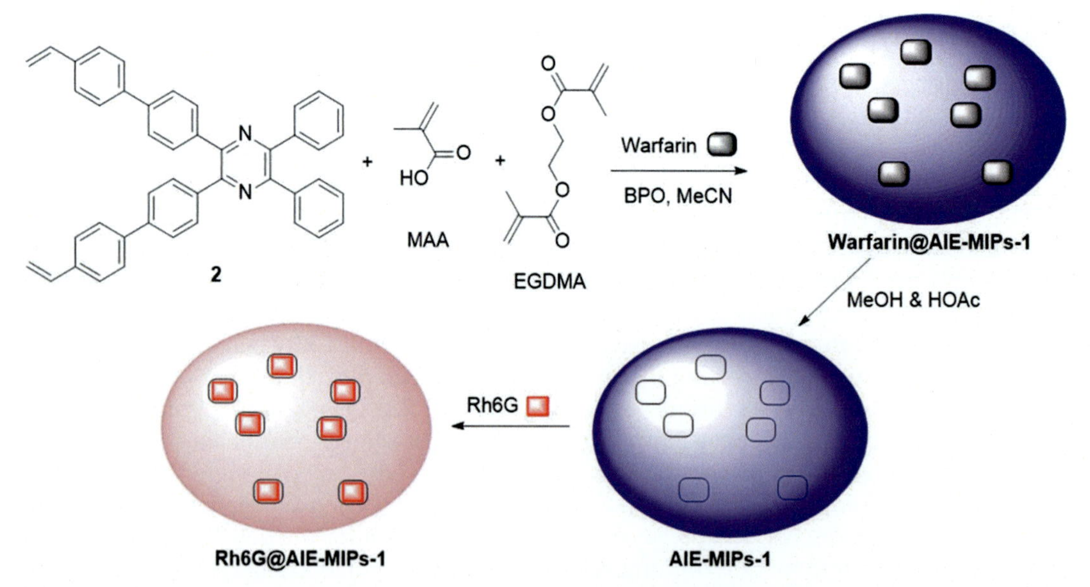

FIG. 7 Schematic illustration for the preparation process of AIE-MIPs-1 and possible detection principle to Rh6G [66]. *Reprinted from Food Chemistry, Volume 287, Yuanyuan Li, Weiye He, Qiuchen Peng, Liyu Hou, Juan He and Kai Li, Aggregation-induced emission luminogen based molecularly imprinted ratiometric fluorescence sensor for the detection of Rhodamine 6G in food samples, pp. 55–60, Copyright (2019), with permission from Elsevier.*

also developed proportional fluorescence sensors TPE-A-MIPs with high sensitivity and selectivity for RhB, with the detection limit of 1.41 $\mu mol\,L^{-1}$ for RhB [67].

2.3.6 Food quality assessment

The original color, taste, and nutritional components of food will change after polluted by external hazards, which will eventually lead to the reduction of food quality. The corruption of fish, meat, and other foods always produces a variety of toxic biogenic amines, such as putrescine, cadaverine, tyramine, and histamine [68].

Han [69] synthesized a class of positional isomers with AIE and intramolecular charge transfer (ICT) effects by attaching a carboxylic group in different sites (para-, meta-, and ortho-position) of an aromatic core (Fig. 8). The strategy of molecular design is to adjust the dipole–dipole direction by changing the position of the carboxylic group, which further controls the self-assembled architecture: The morphology undergoes a transition from 1D nanowire to 2D microsheet, and even to 3D microcube. The prototype of amine sensor based on m-DB self-assemblies was developed accordingly, showing a linear relation with the quantitative determination of amine and a low detection limit of 2.02 Pa. Hou [43], too, developed an amine sensor based on HMBA-4 (Table 1) based on the strategy, which can also be used to quickly detect putrid pork samples. What's more, Han [45] developed a highly photoluminescent (PL) self-assembled copper nanoclusters (Cu NCs) by using 2,3,5,6-tetrafluorothiophenol (TFTP) (Table 1) as both the reducing agent and the protecting ligand, which was capable of rapid, sensitive, and selective detection of histamine. In the presence of histamine, the fluorescence of Cu NCS would be quenched due to the strong interaction between copper atom and histamine, demonstrating a good linear relation between 0.1 and 10 μM and a low detection limit of 60 nm. The sensor based on Cu NCs has been applied in analyzing the amount of histamine in fish, shrimp, and red wine (Fig. 9).

On the other hand, the processing method of food also affects the nutritional composition of food. Fried food is popular because of its unique taste. However, frying is a complex chemical process in which oxidation, hydrolysis, polymerization, and other reactions occur to produce many harmful compounds such as triacylglycerol (TAG)-based polymers and total polar materials (TPM). These compounds are associated with many diseases, such as mutagenesis, cancer, atherosclerosis, and heart disease. Therefore, the quality of frying oil has become one of the most important food safety issues for consumers.

TAG-based polymers are associated with various diseases such as digestive system diseases and cardiovascular diseases, which are important indicators to evaluate food safety. Wu [47] reported an AIEgen probe QM-TPA conjugated with quinoline-malononitrile (QM) and triphenylamine (Table 1), which can detect tag-based polymers in frying oil through a viscosity regulation mechanism. The mechanism involves the increased viscosity of frying oil limiting the intramolecular motion of the AIE probe. This technology is superior to traditional chromatography-based techniques in rapid detection, field analysis, and portable operation in food inspection.

On the other hand, the determination of total polar materials (TPM) is the official standard for the quality determination of frying oil. TPM includes unsaturated fatty acid, monoglycerides, diglycerides, and easily oxidized products (aldehydes or ketones). Many countries have established formulated regulations to restrict the abuse of frying oil, where the TPM level is limited to 24%–30% [70]. Therefore, it is urgently needed to quantify the

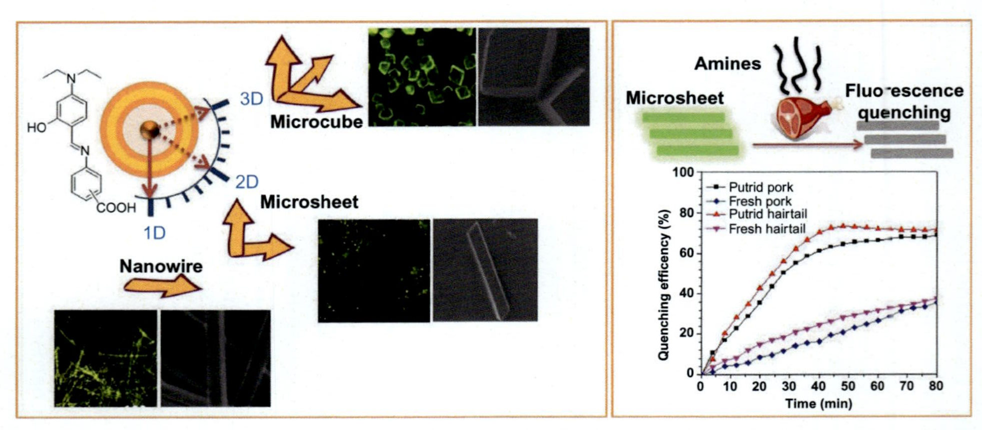

FIG. 8 Schematic illustration of the "rotary knob" with three gears that direct the 1D, 2D, and 3D supramolecular self-assembly and the time-dependent fluorescence quenching profile of the m-DB assemblies under the influence of meat samples [69]. *Reprinted from Sensors and Actuators B: Chemical, Volume 258, Jingqi Han, Yaping Li, Jing Yuan, Zhongfeng Li, Ruixue Zhao, Tianyu Han and Tiandong Han, To direct the self-assembly of AIEgens by three-gear switch: Morphology study, amine sensing and assessment of meat spoilage, pp. 373–380, Copyright (2018), with permission from Elsevier.*

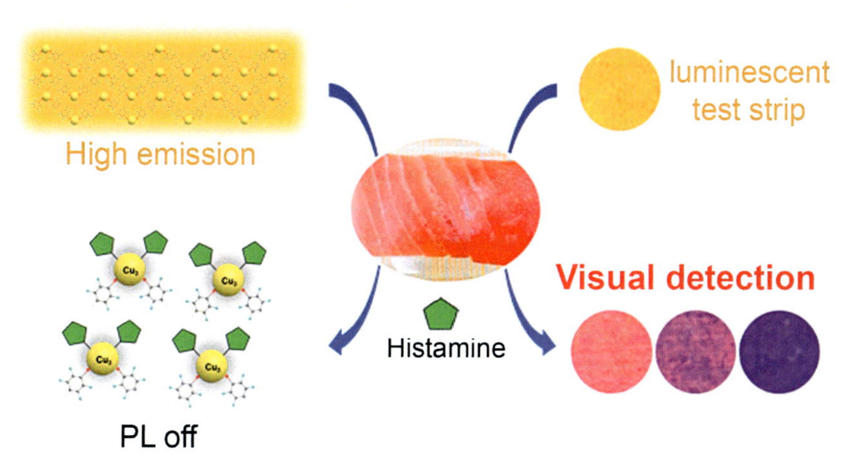

FIG. 9 Chemical structures of highly PL self-assembled copper nanoclusters (Cu NCs) and application of assembled Cu NCs in the detection of histamine [45]. *Reprinted from Analytical Chemistry, Volume 90, Ailing Han, Lin Xiong, Sijia Hao, Yayu Yang, Xia Li, Guozhen Fang, Jifeng Liu, Yong Pei, and Shuo Wang, Sensitive Water-Soluble Fluorescent Probe Based on Umpolung and Aggregation-Induced Emission Strategies for Selective Detection of* Hg^{2+} *in Living Cells and Zebrafish, pp. 9060–9067, Copyright (2018), with permission from ACS.*

TPM in frying oil. Fortunately, viscosity is a reliable parameter for quickly and easily assessing the quality of frying oil in terms of a good relationship between viscosity and TPM%. Therefore, Cui designed a novel D-A type AEE fluorescent probe (CPA-TPA) (Table 1), which responded to the fluctuations of medium viscosity by limiting RIM via the ICT configuration of TPA [49]. The results showed a good linear relationship between fluorescence intensity and TPM, indicating that CPA-TPA is a simple and rapid portable tool to monitor TPM in frying oil.

Since the viscosity of the frying oil system increases with the polymer content or TPM percentage, AIE fluorescent probe with an ultra-sensitive response to viscosity is an ideal candidate for monitoring hazards in frying oil. Compared with chromatographic-based technologies (GC and HPLC), this method solves the problems of large and expensive instruments, complicated preprocessing, and time-consuming procedures, so as to realize rapid responding, real-time sensing, and portable operation.

3 Summary and perspectives

As a new type of fluorescent dye AIEgens have achieved many innovative applications in the field of food safety detection. The fluorescence sensors based on AIEgens have the advantages of fast response, good selectivity, and high efficiency. It can realize the selective identification and detection of pesticides, veterinary drugs, heavy metals, pathogens, and food additives. Moreover, in order to solve the drawbacks of poor biocompatibility and autofluorescence interference, AIEgens are usually further modified to improve the biocompatibility and fluorescence quantum efficiency or connected with phages, peptides, aptamers, enzymes, and other targeted recognition units to improve the sensitivity and accuracy of detection. Furthermore, a combination with other detection technologies (such as PCR technology, sensor arrays, etc.) is also one of the efficient approaches to expand the application range of AIE fluorescence sensors.

Nevertheless, the design of more efficient and sensitive AIE fluorescence sensors still faces many challenges: (i) RIM is the main mechanism of AIE, but many new AIEgens (such as planar AIEgens) cannot be explained by RIM [71], so it is necessary to conduct a more in-depth and systematic study on the AIE mechanism, (ii) the preparation of AIE nanoparticles is an effective method to improve the hydrophilicity and biocompatibility of molecules, which can effectively improve the fluorescence quantum yield. However, there are few researches to analyze the relationship between the characteristics of nanoparticles and the fluorescence quantum yield, (iii) due to the complexity of food composition, improving the selectivity is the key point to accurately identify the target analytes, so it is necessary to combine AIEgens with peptides, aptamers, and molecular imprinting polymer to further improve the targeting ability, (iv) since the signal acquisition of AIEgens mainly depends on some large-scale instruments, the development of portable visual AIE fluorescence sensors for on-site "no instrument" detection has a good application prospect.

Finally, AIEgens have been successfully applied in the field of food safety detection, but related application research is still in the development stage. Although the diversity and complexity of food require high selectivity and sensitivity for AIE fluorescence sensors, we believe that AIE sensors will become an effective tool for food safety evaluation. With the innovative design and in-depth research of AIEgens, it is believed that it would have a broader application in the field of food safety testing.

References

[1] Y. Long, T. Meade, Advances in optical and electrochemical techniques for biomedical imaging, Chem. Sci. 11 (2020) 6940–6941.

[2] M.M. Aung, Y. Chang, Traceability in a food supply chain: safety and quality perspectives, Food Control 39 (2014) 172–184.

[3] Z. Li, N. Su, X. Dong, Y. Yang, Y. Wang, H. Xiao, Edible agro-products quality and safety in China, J. Integr. Agric. 14 (2015) 2166–2175.

[4] M.C.R. Alavanja, J.A. Hoppin, F. Kamel, Health effects of chronic pesticide exposure: cancer and neurotoxicity, Annu. Rev. Public Health 25 (2004) 155–197.

[5] G. Yu, W. Wu, Q. Zhao, X. Wei, Q. Lu, Efficient immobilization of acetylcholinesterase onto amino functionalized carbon nanotubes for the fabrication of high sensitive organophosphorus pesticides biosensors, Biosens. Bioelectron. 68 (2015) 288–294.

[6] A. Malik, C. Blasco, Y. Pico, Liquid chromatography-mass spectrometry in food safety, J. Chromatogr. A 1217 (2010) 4018–4040.

[7] Y. Pico, G. Font, M. Ruiz, M. Fernandez, Control of pesticide residues by liquid chromatography-mass spectrometry to ensure food safety, Mass Spectrom. Rev. 25 (2006) 917–960.

[8] Q. Liao, Q. Gao, J. Wang, Y. Gong, Q. Peng, Y. Tian, Y. Fan, H. Guo, D. Ding, Q. Li, Z. Li, 9,9-Dimethylxanthene derivatives with room-temperature phosphorescence: substituent effects and emissive properties, Angew. Chem. Int. Ed. 59 (2020) 9946–9951.

[9] Y. Li, Y. An, J. Fan, X. Liu, X. Li, F. Hahn, Y. Wang, Y. Han, Strategy for the construction of diverse poly-NHC-derived assemblies and their photoinduced transformations, Angew. Chem. Int. Ed. 59 (2020) 10073–10080.

[10] C. Burtscher, S. Wuertz, Evaluation of the use of PCR and reverse transcriptase PCR for detection of pathogenic bacteria in biosolids from anaerobic digestors and aerobic composters, Appl. Environ. Microbiol. 69 (2003) 4618–4627.
[11] D. McMullin, B. Mizaikoff, R. Krska, Advancements in IR spectroscopic approaches for the determination of fungal derived contaminations in food crops, Anal. Bioanal. Chem. 407 (2015) 653–660.
[12] A. Martins, M. Talhavini, M. Vieira, J. Zacca, J. Braga, Discrimination of whisky brands and counterfeit identification by UV-vis spectroscopy and multivariate data analysis, Food Chem. 229 (2017) 142–151.
[13] H. Zhang, S. Yang, K. De Ruyck, N.V. Beloglazova, S.A. Eremin, S. De Saeger, S. Zhang, J. Shen, Z. Wang, Fluorescence polarization assays for chemical contaminants in food and environmental analyses, TrAC Trends Anal. Chem. 114 (2019) 293–313.
[14] X. Yue, L. Liu, Z. Li, Q. Yang, W. Zhu, W. Zhang, J. Wang, Highly specific and sensitive determination of propyl gallate in food by a novel fluorescence sensor, Food Chem. 256 (2018) 45–52.
[15] M. Heffern, L. Matosziuk, T. Meade, Lanthanide probes for bioresponsive imaging, Chem. Rev. 114 (2014) 4496–4539.
[16] Y. Shen, F. Yan, X. Huang, X. Zhang, Y. Zhang, C. Zhang, J. Jin, H. Li, S. Yao, A new water-soluble and colorimetric fluorescent probe for highly sensitive detection of organophosphorus pesticides, RSC Adv. 6 (2016) 88096–88103.
[17] X. Tian, Z. Li, Y. Pang, D. Li, X. Yang, Benzoyl peroxide detection in real samples and zebrafish imaging by a designed near-infrared fluorescent probe, J. Agric. Food Chem. 65 (2017) 9553–9558.
[18] A. Gandioso, R. Bresoli-Obach, A. Nin-Hill, M. Bosch, M. Palau, A. Galindo, S. Contreras, A. Rovira, C. Rovira, S. Nonell, V. Marchán, Redesigning the coumarin scaffold into small bright fluorophores with far-red to near-infrared emission and large stokes shifts useful for cell imaging, J. Org. Chem. 83 (2018) 1185–1195.
[19] J. Luo, Z. Xie, J. Lam, L. Cheng, H. Chen, C. Qiu, H. Kwok, X. Zhan, Y. Liu, D. Zhu, B.Z. Tang, Aggregation-induced emission of 1-methyl-1,2,3,4,5-pentaphenylsilole, Chem. Commun. (2001) 1740–1741.
[20] J. Mei, Y. Hong, J. Lam, A. Qin, Y. Tang, B.Z. Tang, Aggregation-induced emission: the whole is more brilliant than the parts, Adv. Mater. 26 (2014) 5429–5479.
[21] J. Qian, B.Z. Tang, AIE luminogens for bioimaging and theranostics: from organelles to animals, Chem 3 (2017) 56–91.
[22] W. Fu, C. Yan, Z. Guo, J. Zhang, H. Zhang, H. Tian, W.H. Zhu, Rational design of near-infrared aggregation-induced-emission-active probes: in situ mapping of amyloid-beta plaques with ultrasensitivity and high-fidelity, J. Am. Chem. Soc. 141 (2019) 3171–3177.
[23] M. Kang, C. Zhou, S. Wu, B. Yu, Z. Zhang, N. Song, M. Lee, W. Xu, F. Xu, D. Wang, W. Lei, B.Z. Tang, Evaluation of structure-function relationships of aggregation-induced emission luminogens for simultaneous dual applications of specific discrimination and efficient photodynamic killing of gram-positive bacteria, J. Am. Chem. Soc. 141 (2019) 16781–16789.
[24] X. Gu, X. Zhang, H. Ma, S. Jia, P. Zhang, Y. Zhao, Q. Liu, J. Wang, X. Zheng, J. Lam, D. Dan, B.Z. Tang, Corannulene-incorporated AIE nanodots with highly suppressed nonradiative decay for boosted cancer phototheranostics *in vivo*, Adv. Mater. 30 (2018) 1801065.
[25] R. Wang, K. Dong, G. Xu, B. Shi, T. Zhu, P. Shi, Z. Guo, W. Zhu, C. Zhao, Activatable near-infrared emission-guided on-demand administration of photodynamic anticancer therapy with a theranostic nanoprobe, Chem. Sci. 10 (2019) 2785–2790.
[26] Y. Cai, J. Fang, B. Wang, F. Zhang, G. Shao, Y. Liu, A signal-on detection of organophosphorus pesticides by fluorescent probe based on aggregation-induced emission, Sens. Actuators B 292 (2019) 156–163.
[27] C. Niu, Q. Liu, Z. Shang, L. Zhao, J. Ouyang, Dual-emission fluorescent sensor based on AIE organic nanoparticles and au nanoclusters for the detection of mercury and melamine, Nanoscale 7 (2015) 8457–8565.
[28] C. Zhou, M. Jiang, J. Du, H. Bai, G. Shan, R.T.K. Kwok, J.H.C. Chau, J. Zhang, J.W.Y. Lam, P. Huang, B.Z. Tang, One stone, three birds: one AIEgen with three colors for fast differentiation of three pathogens, Chem. Sci. 11 (2020) 4730–4740.
[29] X. Huang, Q. Guo, R. Zhang, Z. Zhao, Y. Leng, J. Lam, Y. Xiong, B.Z. Tang, AIEgens: an emerging fluorescent sensing tool to aid food safety and quality control, Compr. Rev. Food Sci. Food Saf. 19 (2020) 2297–2329.
[30] L. Xu, L. Ni, F. Zeng, S. Wu, Tetranitrile-anthracene as a probe for fluorescence detection of viscosity in fluid drinks *via* aggregation-induced emission, Analyst 145 (2020) 844–850.

[31] G. Niu, R. Zhang, X. Shi, H. Park, S. Xie, R. Kwok, J. Lam, B.Z. Tang, AIE luminogens as fluorescent bioprobes, TrAC Trends Anal. Chem. 123 (2020), 115769.

[32] D. Wang, B.Z. Tang, Aggregation-induced emission luminogens for activity-based sensing, Acc. Chem. Res. 52 (2019) 2559–2570.

[33] S. Samanta, Y. He, A. Sharma, J. Kim, W. Pan, Z. Yang, J. Li, W. Yan, L. Liu, J. Qu, J.S. Kim, Fluorescent probes for nanoscopic imaging of mitochondria, Chem 5 (2019) 1697–1726.

[34] L. Xu, P. Chen, T. Liu, D. Ren, N. Dong, W. Cui, P. He, Y. Bi, N. Lv, M. Ntakatsane, A novel sensitive visual count card for detection of hygiene bio-indicator-molds and yeasts in contaminated food, LWT- Food Sci. Technol. 117 (2020), 108687.

[35] S. Yang, C. Lin, I.A. Aljuffali, J. Fang, Current pathogenic Escherichia coli foodborne outbreak cases and therapy development, Arch. Microbiol. 199 (2017) 811–825.

[36] T. Saxena, P. Kaushik, M.K. Mohan, Prevalence of *E. coli* O157:H7 in water sources: an overview on associated diseases, outbreaks and detection methods, Diagn. Microbiol. Infect. Dis. 82 (2015) 249–264.

[37] N. Hamidiyan, A. Salehi-Abargouei, Z. Rezaei, R. Dehghani-Tafti, F. Akrami-Mohajeri, The prevalence of Listeria spp. food contamination in Iran: a systematic review and meta-analysis, Food Res. Int. 107 (2018) 437–450.

[38] M. Gao, Q. Hu, G. Feng, N. Tomczak, R. Liu, B. Xing, B.Z. Tang, B. Liu, A multifunctional probe with aggregation-induced emission characteristics for selective fluorescence imaging and photodynamic killing of bacteria over mammalian cells, Adv. Healthc. Mater. 4 (2015) 659–663.

[39] S. Zhang, L. Ma, K. Ma, B. Xu, L. Liu, W. Tian, Label-free aptamer-based biosensor for specific detection of chloramphenicol using AIE probe and graphene oxide, ACS Omega 3 (2018) 12886–12892.

[40] J. Ma, Y. Xiao, C. Zhang, M. Zhang, Q. Wang, W. Zheng, S. Zhang, Preparation a novel pyrene-based AIE-active ratiometric turn-on fluorescent probe for highly selective and sensitive detection of Hg^{2+}, Mater. Sci. Eng., B 259 (2020), 114582.

[41] L. Zhang, W. Hu, L. Yu, Y. Wang, Click synthesis of a novel triazole bridged AIE active cyclodextrin probe for specific detection of Cd^{2+}, Chem. Commun. 51 (2015) 4298–4301.

[42] X. Wei, Off-on fluorogenic substrate harnessing ESIPT and AIE features for *in situ* and long-term tracking of β-glucuronidase in Escherichia coli, Sens. Actuators B 304 (2020), 127242.

[43] J. Hou, J. Du, Y. Hou, P. Shi, Y. Liu, Y. Duan, T. Han, Effect of substituent position on aggregation-induced emission, customized self-assembly, and amine detection of donor-acceptor isomers: implication for meat spoilage monitoring, Spectrochim. Acta A 205 (2018) 1–11.

[44] J. Chen, X. Chen, Q. Huang, W. Li, Q. Yu, L. Zhu, T. Zhu, S. Liu, Z. Chi, Amphiphilic polymer-mediated aggregation-induced emission nanoparticles for highly sensitive organophosphorus pesticide biosensing, ACS Appl. Mater. Interfaces 11 (2019) 32689–32696.

[45] A. Han, L. Xiong, S. Hao, Y. Yang, X. Li, G. Fang, J. Liu, Y. Pei, S. Wang, Highly bright self-assembled copper nanoclusters: a novel photoluminescent probe for sensitive detection of histamine, Anal. Chem. 90 (2018) 9060–9067.

[46] X. Wu, P. Wang, S. Hou, P. Wu, J. Xue, Fluorescence sensor for facile and visual detection of organophosphorus pesticides using AIE fluorogens-SiO_2-MnO_2 sandwich nanocomposites, Talanta 198 (2019) 8–14.

[47] Y. Wu, P. Jin, K. Gu, C. Shi, Z. Guo, Z. Yu, W.-H. Zhu, Broadening AIEgen application: rapid and portable sensing of foodstuff hazards in deep-frying oil, Chem. Commun. 55 (2019) 4087–4090.

[48] J. Wang, J. Zhang, J. Wang, G. Fang, J. Liu, S. Wang, Fluorescent peptide probes for organophosphorus pesticides detection, J. Hazard. Mater. 389 (2020), 122074.

[49] S. Cui, B. Wang, X. Yan, Y. Li, X. Zhou, Y. Wang, L. Chen, A novel emitter: sensing mechanical stimuli and monitoring total polar materials in frying oil, Dyes Pigments 174 (2020), 108020.

[50] X. He, Y. Yang, Y. Guo, S. Lu, Y. Du, J. Li, X. Zhang, N. Leung, Z. Zhao, G. Niu, S. Yang, Z. Weng, R.T.K. Kwok, J.-W.Y. Lam, G. Xie, B.Z. Tang, Phage-guided targeting, discriminative imaging, and synergistic killing of bacteria by AIE bioconjugates, J. Am. Chem. Soc. 142 (2020) 3959–3969.

[51] H. Bai, H. Chen, R. Hu, M. Li, F. Lv, L. Liu, S. Wang, Supramolecular conjugated polymer materials for in situ pathogen detection, ACS Appl. Mater. Interfaces 8 (2016) 31550–31557.

[52] Y. Wang, T. Corbitt, S. Jett, Y. Tang, K. Schanze, E. Chi, D. Whitten, Direct visualization of bactericidal action of cationic conjugated polyelectrolytes and oligomers, Langmuir 28 (2012) 65–70.

[53] H. Yuan, Z. Liu, L. Liu, F. Lv, Y. Wang, S. Wang, Cationic conjugated polymers for discrimination of microbial pathogens, Adv. Mater. 2014 (26) (2017) 4333–4338.

[54] M. Kushwaha, S. Verma, S. Chatterjee, Profenofos, an acetylcholinesterase-inhibiting organophosphorus pesticide: a short review of its usage, toxicity, and biodegradation, J. Environ. Qual. 45 (2016) 1478–1489.
[55] N. Fahimi-Kashani, M. Hormozi-Nezhad, Gold-nanoparticle-based colorimetric sensor array for discrimination of organophosphate pesticides, Anal. Chem. 88 (2016) 8099–8106.
[56] G. Feng, C. Tay, Q. Chui, R. Liu, N. Tomczak, J. Liu, B.Z. Tang, D. Leong, B. Liu, Ultrabright organic dots with aggregation-induced emission characteristics for cell tracking, Biomaterials 35 (2014) 8669–8677.
[57] R.E. Baynes, K. Dedonder, L. Kissell, D. Mzyk, T. Marmulak, G. Smith, L. Tell, R. Gehring, J. Davis, J. Riviere, Health concerns and management of select veterinary drug residues, Food Chem. Toxicol. 88 (2016) 112–122.
[58] W. Yu, Y. Li, B. Xie, M. Ma, C. Chen, C. Li, X. Yu, Z. Wang, K. Wen, B.Z. Tang, J. Shen, An aggregation-induced emission based indirect competitive immunoassay for fluorescence "turn-on" detection of drug residues in foodstuffs, Front. Chem. 7 (2019) 228.
[59] P. Sanaz, M. Jaytry, D. Freddy, R. Johan, B. Ronny, D. Karolien, Aptasensing of chloramphenicol in the presence of its analogues: reaching the maximum residue limit, Anal. Chem. 84 (2012) 6753–6758.
[60] N. Byzova, E. Zvereva, A. Zherdev, S. Eremin, B. Dzantiev, Rapid pretreatment-free immunochromatographic assay of chloramphenicol in milk, Talanta 81 (2010) 843–848.
[61] N. Kumari, S. Jha, S. Misra, S. Bhattacharya, A probe for the selective and parts-per-billion-level detection of copper(II) and mercury(II) using a micellar medium and its utility in cell imaging, Chem. Aust. 79 (2014) 1059–1064.
[62] P. Alam, N. Leung, J. Zhang, R. Kwok, J. Lam, B.Z. Tang, AIE-based luminescence probes for metal ion detection, Coord. Chem. Rev. 429 (2021), 213693.
[63] H. Wan, Q. Xu, P. Gu, H. Li, D. Chen, N. Li, J. He, J. Lu, AIE-based fluorescent sensors for low concentration toxic ion detection in water, J. Hazard. Mater. 403 (2021), 123656.
[64] T. Gao, X. Huang, S. Huang, J. Dong, K. Yuan, X. Feng, T. Liu, K. Yu, W. Zeng, Sensitive water-soluble fluorescent probe based on umpolung and aggregation-induced emission strategies for selective detection of Hg^{2+} in living cells and zebrafish, J. Agric. Food Chem. 67 (2019) 2377–2383.
[65] S. Pathan, M. Jalal, S. Prasad, S. Bose, Aggregation-induced enhanced photoluminescence in magnetic graphene oxide quantum dots as a fluorescence probe for As(III) sensing, J. Mater. Chem. A 7 (2019) 8510–8520.
[66] Y. Li, W. He, Q. Peng, L. Hou, J. He, K. Li, Aggregation-induced emission luminogen based molecularly imprinted ratiometric fluorescence sensor for the detection of rhodamine 6G in food samples, Food Chem. 287 (2019) 55–60.
[67] Y. Li, L. Hou, F. Shan, Z. Zhang, Y. Li, Y. Liu, Q. Peng, J. He, K. Li, A novel aggregation-induced emission luminogen based molecularly imprinted fluorescence sensor for ratiometric determination of Rhodamine B in food samples, Chemistry Select 4 (2019) 11256–11261.
[68] A. Poghossian, H. Geissler, M. Schoning, Rapid methods and sensors for milk quality monitoring and spoilage detection, Biosens. Bioelectron. 140 (2019) 18–31.
[69] J. Han, Y. Li, J. Yuan, Z. Li, R. Zhao, T. Han, T. Han, To direct the self-assembly of AIEgens by three-gear switch: morphology study, amine sensing and assessment of meat spoilage, Sens. Actuators B 258 (2018) 373–380.
[70] R. Stier, Ensuring the health and safety of fried foods, Eur. J. Lipid Sci. Technol. 115 (2013) 956–964.
[71] J. Ni, T. Min, Y. Li, M. Zha, P. Zhang, C. Hou, K. Li, Planar AIEgens with enhanced solid-state luminescence and ROS generation for multidrug-resistant bacteria treatment, Angew. Chem. Int. Ed. 59 (2020) 10179–10185.

CHAPTER

20

Computational modeling of AIE luminogens

Qian Peng[a], *Zhigang Shuai*[b], *and Qi Ou*[b]

[a]School of Chemical Sciences, University of Chinese Academy of Sciences, Beijing, China [b]MOE Key Laboratory of Organic OptoElectronics and Molecular Engineering, Department of Chemistry, Tsinghua University, Beijing, China

1 Introduction

Over the past decades, organic luminescence continuously plays an important role in the development of science and technology, which has been widely applied in promising flexible displays, solid-state lighting, organic lasers, chemical/biological sensors, etc. [1–6]. For practical applications, these organic materials are required to have a high luminescence quantum efficiency in the solid phase. Traditionally, organic molecules with extended conjugation can emit bright light in solution, which may unfortunately suffer from an undesirable concentration- or aggregation-caused quenching of luminescence [7]. Scientists had hence focused on the investigation of organic systems with strong fluorescence in solution for a long time, which strongly limited the development and innovation of organic luminescent materials in solid states. In 2001, Tang et al. coined the term aggregation-induced emission (AIE) to describe a phenomenon of organic molecules showing a strong increase in luminescence upon aggregation [8], which opened up a very broad aperture for scientist to rethink the design strategy of high-efficiency solid-state organic systems. Since then, numerous AIEgens have been synthesized and many applications have been exploited [9–11]. Deep understanding of the inherent mechanism of AIE is strongly desired to pave the way toward designing novel AIEgens and the development of their applications.

One of the earliest and probably the most widely applied descriptions to the luminescence property of aggregates is Kasha's exciton model (Fig. 1B). According to Kasha's picture, molecular dimers with positive exciton coupling due to transition dipole–dipole interaction are recognized as H-aggregates, of which the lowest excited state is a transition dipole-forbidden antisymmetric Frenkel exciton (FE), and hence the emission of such H-aggregate dimers is prohibited. Molecular dimers with negative exciton coupling are recognized as J-aggregates,

Aggregation-Induced Emission (AIE)
https://doi.org/10.1016/B978-0-12-824335-0.00015-5

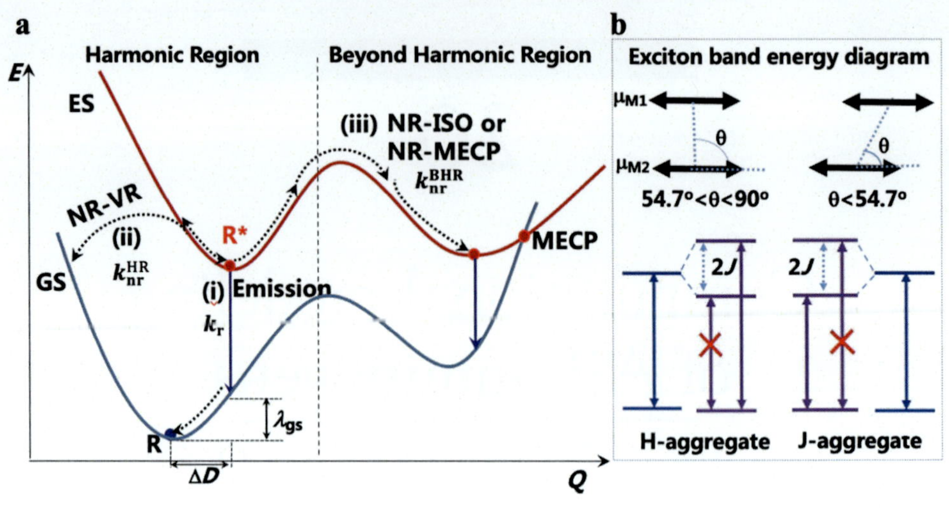

FIG. 1 (A) Schematic graph of the decay pathways from an excited state (ES) to ground state (GS): (i) radiative process, fluorescence for singlet state and phosphorescence for triplet state; (ii) the nonradiative decay process induced by vibration relaxation in a harmonic region (NR-VR); and (iii) the nonradiative decay pathway via isomerization or a MECP (NR-ISO or NR-MECP). (B) Exciton band energy diagram. *Reproduced with permission from Q. Peng, Z. Shuai, Molecular mechanism of aggregation-induced emission, Aggregate 2 (2021) e91, Copyright 2021 John Wiley and Sons.*

of which the lowest excited state becomes the transition dipole-allowed symmetric FE and its transition dipole moment is strengthened by a factor of $\sqrt{2}$ compared to that of the monomer, leading to an enhanced emission, even superradiance upon aggregation.

Such enhancement in J-aggregates was once regarded as the principal cause for the enhancement of solid-state luminescence for a long time [12,13]. The earliest J-aggregation was proposed independently by Scheibe and Jelley, in which the transition dipole moment is enlarged in compact-conjugation planar systems compared to that of the isolated molecules. However, this classical Kasha's model for J-aggregation cannot effectively explain the phenomena of the overwhelming majority of newly synthesized AIEgens because the dipole–dipole interaction of those AIEgens is usually extremely weak owing to their flexible three-dimensional confirmation [9]. As a result, various mechanisms beyond Kasha's aggregate picture have sprung up for AIE phenomena for different organic systems, such as restriction of the intramolecular rotation (RIR) [8], restriction of intermolecular motion (RIM) [14], blocking of nonradiative decay channels [15,16], the restriction of the E/Z isomerization process [17], the excited-state intramolecular proton transfer [18], the blockage of access to the dark state via isomerization [19], restricted access to conical interaction [49,86], crystalline-induced reverse from dark to bright state [20,21], Herzberg-Teller vibronic coupling induced emission [22], etc. Nevertheless, a clear and comprehensive picture has yet to be rectified, for the microscopic AIE mechanism, the corresponding relationship between the photophysical property and mechanism, and general principles of aggregation behavior modulation.

From the perspective of microscopic processes, once the excited state (ES) of a single molecule is formed by photoexcitation, there are commonly three decay pathways from the ES (the first singlet S_1 or triplet T_1 state) to the ground state (GS) S_0 (as seen in Fig. 1A): (i) radiative process (k_r), in which the ES has significant transition dipole or oscillator strength, which is the first prerequisite for luminescence; (ii) the nonradiative decay induced by vibration relaxation (NR-VR) in a harmonic region (k_{nr}^{HR}), which is expected to be slow for high-luminescence quantum efficiency; and (iii) the nonradiative decay beyond the harmonic region (NR-BHR), either via isomerization (NR-ISO) or a minimum energy crossing point (NR-MECP) to the GS beyond the harmonic region (k_{nr}^{BHR}), which is expectedly avoided for excellent luminescent materials. When going to aggregate, the light-emitting state can either become a Frenkel exciton owing to strong excitonic coupling or remain as the ES of a single molecule. The former exhibits remarkably different optical spectra relative to those of single molecules, namely, significantly blue-shifted/enhanced absorption spectrum and red-shifted/weakened emission spectrum in a H-aggregate and red-shifted/strengthened absorption and emission spectra in a J-aggregate (see Fig. 1B), while the changes of emission position and nonradiative decays for the latter case mainly stem from the relaxation of molecular geometrical/electronic structures resulting from intermolecular electrostatic interactions between the light-emitting molecule and the surrounding molecules.

In this chapter, we introduce the computational modeling for AIEgens by first comparatively investigating the effect of excitonic coupling on the optical spectra in conventional H- and J-aggregates and newly reported AIEgens, which evince that the emission property of typical AIEgens is merely influenced by the intermolecular excitonic coupling (J) owing to remarkably strong intramolecular electron-vibration coupling (λ) and relatively small excitonic coupling (with $J/\lambda < 0.17$) at room temperature. Then, we introduce the quantitative evaluation of the luminescence quantum yield (Φ_{LQY}), which is determined by three microscopic process (radiative, NR-VR, and NR-BHR), by combining quantum chemistry calculations and thermal vibration correlation function (TVCF) rate theory developed by our group [32,88,91] in Section 3. The rest of the chapter is then divided into two parts, (i) the prohibition of the nonradiative processes in aggregates and (ii) the enhancement of the radiative process, both of which lead to the AIE phenomenon. In Section 4, we introduce the numerical examples on the blockage of various nonradiative decay channels in aggregates, including vibration relaxation (NR-VR) in a harmonic region or isomerization (NR-ISO) or minimum energy crossing point (NR-MECP) beyond the harmonic region. Furthermore, numerical examples on the generation of a bright light-emitting state with enhanced radiative decay rate upon aggregation are presented in Section 5, in which the microscope mechanisms of the inversion from transition dipole-forbidden "dark" state to dipole-allowed "bright" upon aggregation is disclosed. Altogether, we hope this chapter can provide a comprehensive photophysical picture for readers to understand the emission phenomena of organic systems in different environments in depth.

2 Effect of excitonic coupling and electron-vibration coupling on emission in aggregates

In this section, we systematically introduce the computation modeling of the vibrationally resolved spectra for conventional H-/J-aggregates and newly reported AIEgens, with and without considering intermolecular electrostatic interaction from the surrounding molecules and excitonic coupling. Contrary to the solution phase, organic compounds in aggregates always have different photo-responsive behaviors owing to a variety of intermolecular interactions as well as the exciton coupling. Therefore, the vibrationally resolved optical spectrum is an effective means to probe emissive species and reveal luminescent essence because it can provide detailed information about geometrical relaxation, electronic transition, electrostatic interaction, excitonic coupling, etc. By analyzing the origin of these spectra, this section paves the way for disclosing the luminescence mechanism of organic aggregates [23].

Intramolecular electron-vibration coupling (λ), characterizing the molecular reorganization energy released from geometrical relaxation upon photoexcitation, determines the fine structure of the optical spectrum for a single molecule. The excitonic coupling ($J \propto \mu_1 . \mu_2 / d^3$, $\mu_{1(2)}$ is the transition dipole moment vector for molecule 1(2) and d is the distance between the geometrical center of two molecules), reflecting the degree of the intermolecular resonant conversion, can significantly alter the peak position and intensity of absorption or emission spectrum. Through an exciton-vibronic coupling model with typical parameters derived from organic molecules, it is found that when decreasing the ratio of J/λ, the optical spectra of aggregate are gradually getting closer, in both lineshape and peak position, to those of the monomer, and when λ is increased such that $J/\lambda \leq 0.17$, the spectra are almost identical to monomer [23]. To verify this point in real systems, we calculate J/λ and the optical spectra with and without surrounding molecules and excitonic coupling of a series of fluorophores, including four conventional non-AIEgens (DSB, 6T, anthracene, rubrene) and five AIEgens (DCDPP, CB, HPS, BFTPS, BTPES) shown in Fig. 2. Here, the electrostatic interaction from the surroundings is taken into account for the geometrical optimizations and frequency calculations of the two compounds via the QM/MM method. It is interesting to find that the values of J/λ of conjugated non-AIEgens are all larger than 0.17, while those of AIEgens are all less than 0.17, as shown in Fig. 3A. Furthermore, the spectra of the non-AIEgens undergo significant change after taking excitonic coupling into consideration, while those of AIEgens remain roughly unchanged. These findings fully demonstrate the competition between intermolecular excitonic coupling (energy transfer) and intramolecular vibronic relaxation in aggregates and there exists a critical value of J/λ for the optical spectra. The spectra of DSB and 6T experience the largest modification owing to very large J/λ. Prototypical H-aggregation behaviors as indicated by classical Kasha's picture are observed for these two compounds, namely, sharply blue-shifted and significantly intensity-enhanced absorption spectra and red-shifted and intensity-reduced emission spectra from solution to aggregate. The spectra of anthracene and rubrene exhibit slight variance upon aggregation because of small J/λ. Notably, the spectra of the AIEgens are independent of the excitonic coupling because of the weak intermolecular excitonic coupling compared to strong intramolecular electron-vibration interaction. This might also imply that the enhanced fluorescence in AIEgens would not be attributed to the excitonic coupling effect led by the J-aggregation.

3 The quantitative calculation of luminescence quantum yield

In this section, we will introduce our approach for the quantitative evaluation of the radiative and nonradiative decay rate constants for organic fluorophores. From Fig. 1A, the NR-VR in the harmonic region and NR-MECP beyond the harmonic region contribute and

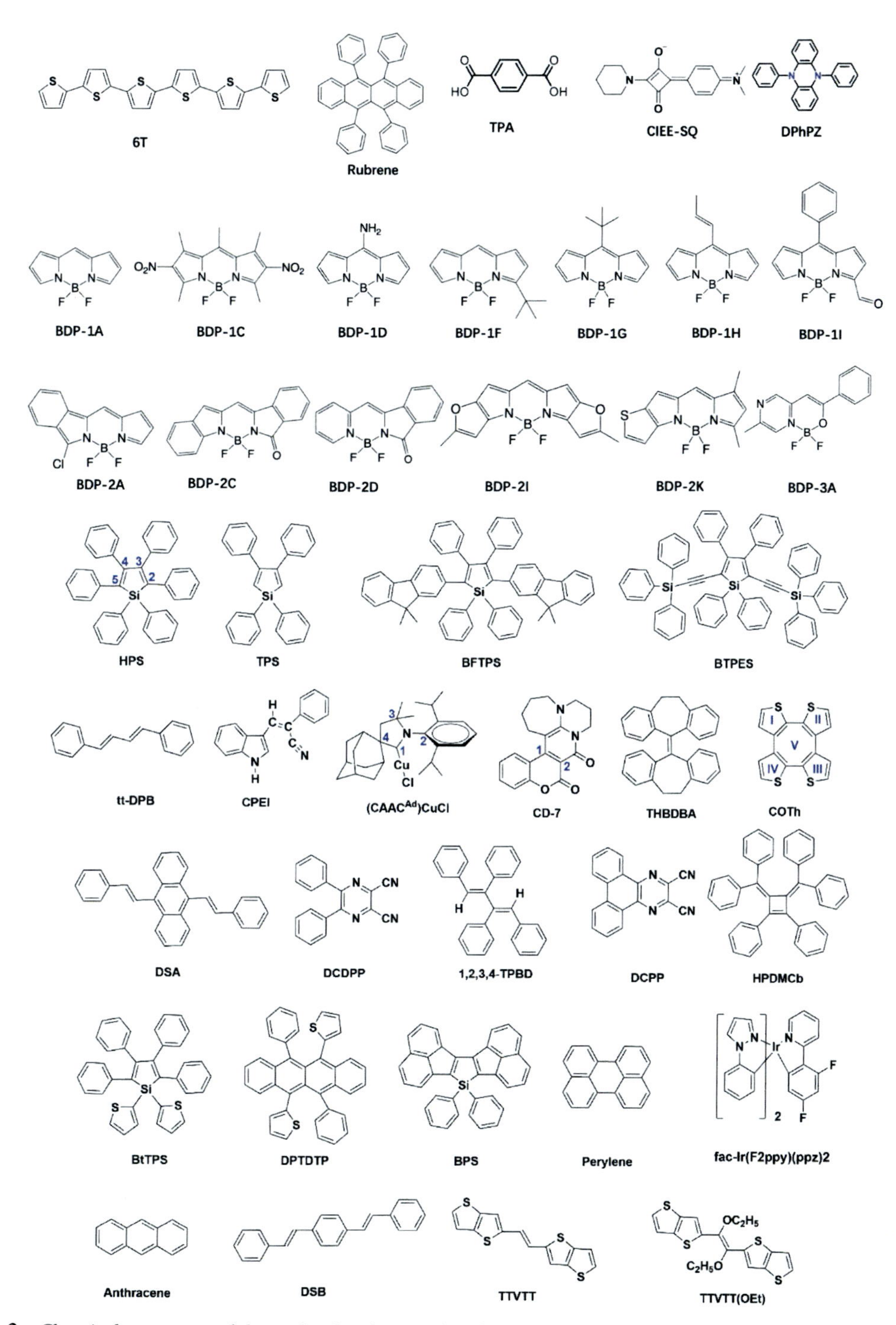

FIG. 2 Chemical structures of the molecules discussed in this chapter.

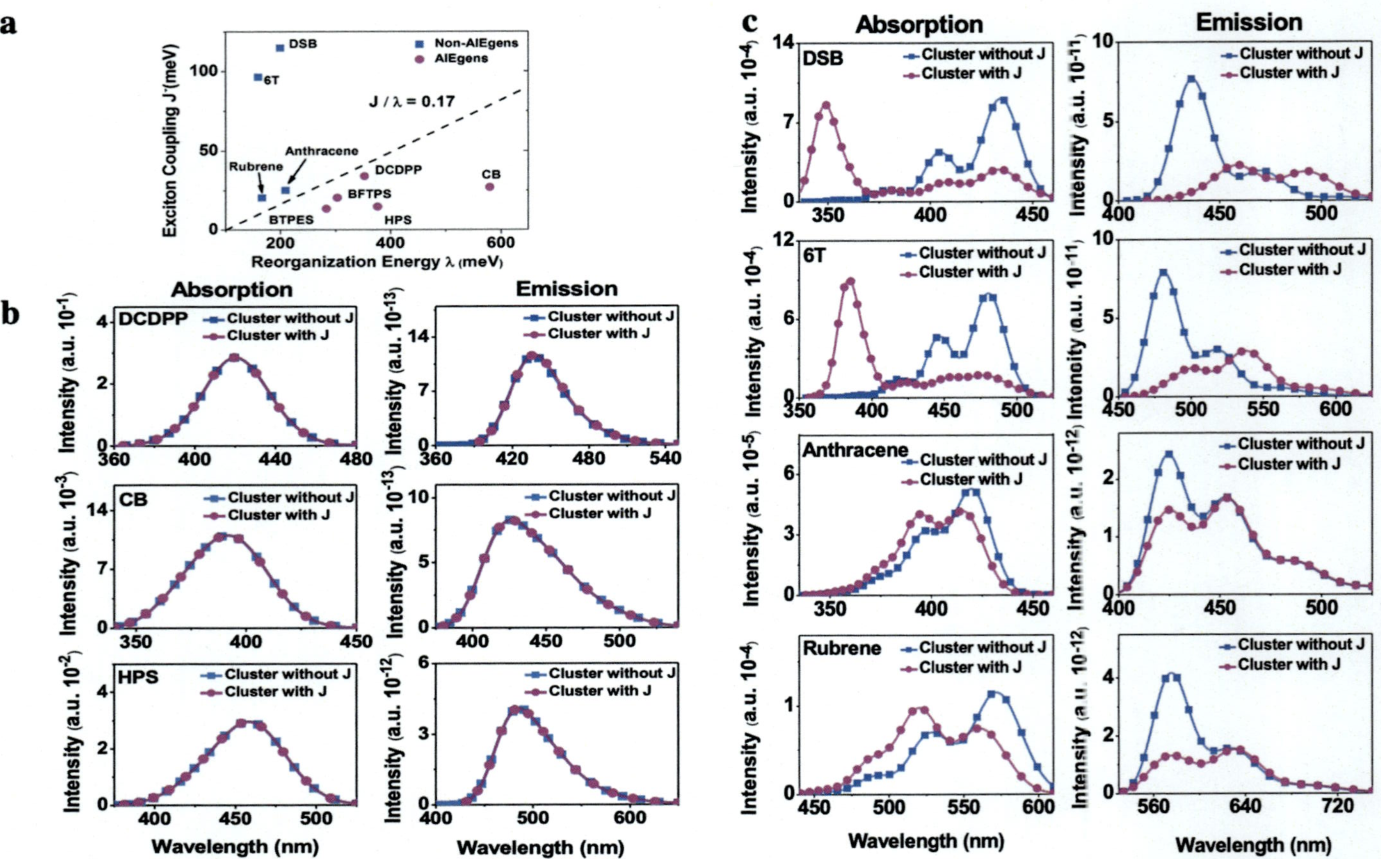

FIG. 3 Illustration of the orientation of the transition dipole moments and exciton band energy diagram for H- and J-aggregates (A) and calculated J/λ (B) and optical spectra of the non-AIEgens (C) and AIEgens (D) in cluster with and without considering J at 298 K. *Reproduced with permission from W. Li, Q. Peng, Y. Xie, T. Zhang, Z. Shuai, Effect of intermolecular excited-state interaction on vibrationally resolved optical spectra in organic molecular aggregates, Acta Chim. Sin. 74 (2016) 902. https://doi.org/10.6023/A16080452. Copyright 2016 Shanghai Institute of Organic Chemistry, Chinese Academy of Sciences.*

compete each other to determine the nonradiative process. As a matter of fact, determination of nonradiative decay channel remains an open question since long. In 1923, the predissociation phenomenon of O_2 observed by Henri was regarded as the earliest nonradiative transition phenomena [81]. In 1932, Zener [24] claimed theoretically that molecules can undergo nonradiative transition through nonadiabatic crossing of energy levels [24]. In 1937, Teller extended the nonradiative transition to polyatomic molecules with conical intersection (CI) between potential surfaces of two electronic states with identical symmetry [25]. In 1950, Huang and Rhys proposed a multiphonon relaxation theory for the nonradiative transitions in F-centerswith displaced harmonic oscillator model [26]. From 1961 to 1965, Robinson, Frosch, and Ross treated the internal conversion (IC) as a tunneling process, in which the crossing of potential surfaces rarely occurred, and the IC rate constant is the product of the square of perturbation matrix with the Franck-Condon (FC) factor [27,28]. In 1966, Lin wrote the general expression of nonradiative transition rate constant based on Fermi's golden rule and displaced harmonic oscillator model [29]. And then Englman and Jortner derived the energy gap law for the nonradiative transition in large molecules by adopting weak-coupling approximation, followed by Fischer and Schneider by adding an anharmonic term to the formula in 1971 [30]. Later, Lin discussed Duschinsky rotation effect of several vibrational modes on the transition rate in 1998 [82]. In 2007, Peng et al. obtained a time-integrated analytical formalism by considering the Duschinsky rotation effect of all vibration normal modes via the thermal vibration correlation function (TVCF) [91]. Islampour et al. also obtained similar formalism by employing a generation function independently [31]. In 2008, applying the TVCF approach, Niu, Peng, and Shuai further deduced a more general analytical formalism, replacing nonadiabatic coupling prefactor of one promoting-mode by a multimode coupling matrix [32]. In 2013, Peng et al. started from second-order perturbation and combined spin-orbit coupling and nonadiabatic coupling to derive nonradiative decay rate between triplet and singlet states [33]. Marian et al. investigated the effect of spin-vibronic coupling on the intersystem crossing rate [34]. The excitonic coupling for nonradiative transition rate was considered by perturbation for molecular aggregate[35]. At the same time, CIs have been widely discussed, but only as a pathway to account for nonemissive issues, instead of light-emitting process [36].

From the perspective of dynamics theory, the molecular quantum nonadiabatic dynamics simulation can, in principle, provide a more general picture regardless of the nonradiative transition induced by VR or MECP. However, the typical time scale of the available nonadiabatic dynamics simulation is in the order of ~fs or ~ps [37], which is about 3–4 orders of magnitude shorter than the radiative/nonradiative decay rate constant of ca. 10 ns for luminescent organic systems. Moreover, favored organic molecules for practical applications in the field of light-emitting devices have dozens or hundreds of atoms, which are far beyond the computational ability of quantum dynamics simulations. Therefore, the kinetic TVCF formalism for k_{nr}^{HR} of a VR-induced transition in the harmonic region [38] (channel II in Fig. 1A) and transition state theory (TST) for k_{nr}^{BHR} of an MECP or isomerization one beyond the harmonic region [39] (channel III in Fig. 1A) seem to be practical approaches for quantitatively predicting luminescence quantum yield (Φ_{LQY}). Key parameters obtained from quantum chemistry calculation for the rate constant evaluation include (i) the transition dipole moment essential to the radiative process, (ii) the nonadiabatic coupling between the light-emitting state and the ground state, essential to the nonradiative process, and (iii) the spin-orbit coupling between singlets and triplets for triplet-involved processes such as intersystem crossing and phosphorescence. The solution and aggregation effects are considered by the polarizable continuum model (PCM) and the quantum mechanics/molecular mechanics approach (QM/MM), respectively (the computational models are shown in Fig. 4.

In order to check the reliability and practicability of the prediction of k_{nr}^{HR}, we firstly compare the experimental Φ_{LQY} and theoretical results via combining TVCF formalism and quantum chemistry calculations for a variety of organic systems (Fig. 2), as shown in Fig. 5, as well as the values of the radiative and nonradiative rate constants listed in Tables 1 and 2. Here, $\Phi_{LQY} = \frac{k_r}{k_r + k_{nr}^{HR}}$. The radiative decay rate constant k_r was evaluated via TVCF formulism as $k_r = \int_0^\infty \sigma_{em}(\omega) d\omega$ [88]. The k_{nr}^{HR} is approximated to be IC rate constant by neglecting the

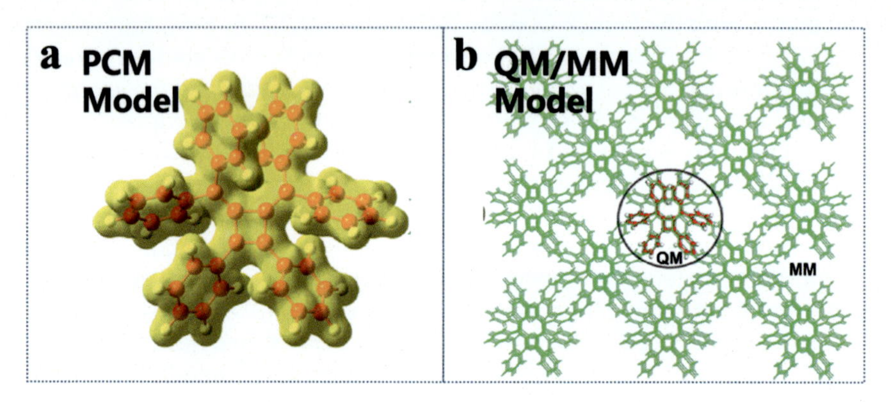

FIG. 4 Setups of the PCM and QM/MM model (Taking HPDMCb as an example). *Reprinted with permission from T. Zhang, Q. Peng, C. Quan, H. Nie, Y. Niu, Y. Xie, Z. Zhao, B. Z. Tang, Z. Shuai, Using the isotope effect to probe an aggregation induced emission mechanism: theoretical prediction and experimental validation, Chem. Sci. 7 (2016) 5573–5580, Copyright 2016 the Royal Society of Chemistry.*

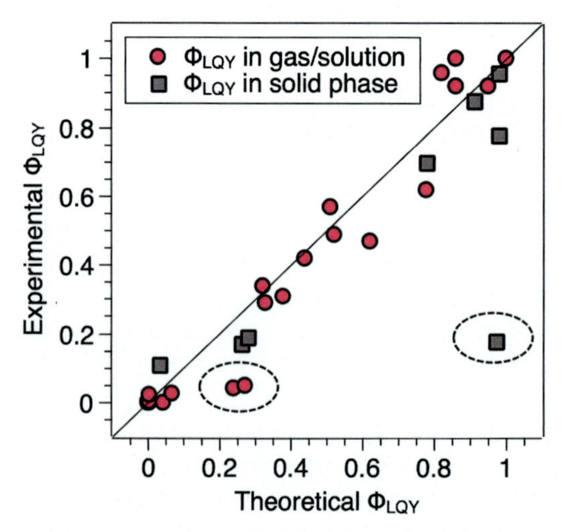

FIG. 5 The comparison between the theoretical and experimental luminescence quantum efficiency for the organic systems in solution of solid phase. *Reproduced with permission from Q. Peng, Z. Shuai, Molecular mechanism of aggregation-induced emission, Aggregate 2 (2021) e91, Copyright 2021 John Wiley and Sons.*

TABLE 1 Summary of the radiative and NR-VR rate constants and luminescence quantum yield calculated by TVCF method, and experimental data for organic compounds in solution.

Molecule	k_r	k_{nr}^{HR}	Φ_{LQY}^{cal}	Φ_{LQY}^{exp}
BDP-1A[a]	1.90×10^8	0.32×10^8	0.86	0.92
BDP-1C[a]	2.67×10^8	4.44×10^8	0.38	0.31
BDP-1D[a]	2.20×10^8	0.11×10^8	0.95	0.92
BDP-1F[a]	2.18×10^8	0.36×10^8	0.86	1.00
BDP-1G[a]	1.17×10^8	3.73×10^8	0.24	0.04
BDP-1H[a]	1.12×10^8	3.09×10^8	0.27	0.05
BDP-1I[a]	1.65×10^8	3.29×10^8	0.33	0.29
BDP-2A[a]	1.87×10^8	0.52×10^8	0.78	0.62
BDP-2C[a]	2.32×10^8	2.12×10^8	0.52	0.49
BDP-2D[a]	2.59×10^8	1.60×10^8	0.62	0.47
BDP-2I[a]	2.52×10^8	0.56×10^8	0.82	0.96
BDP-2K[a]	2.07×10^8	4.35×10^8	0.32	0.34
BDP-3A[a]	2.30×10^8	2.17×10^8	0.51	0.57
tt-DPB[b]	9.58×10^8	11.9×10^8	0.44	0.42
Perylene[c]	0.91×10^8	0.72×10^3	1.00	1.00
DCDPP[d]	0.93×10^7	4.45×10^9	2.09×10^{-3}	1.50×10^{-4}
1,2,3,4-TPBD[b]	4.80×10^8	1.09×10^{10}	4.20×10^{-2}	1.10×10^{-3}
DMTPS[e]	1.20×10^8	1.80×10^{11}	6.66×10^{-4}	2.20×10^{-4}
HPDMCb[f]	0.86×10^8	1.31×10^{11}	6.59×10^{-4}	1.7×10^{-3}
TPS[g]	0.93×10^6	1.62×10^{10}	1.0×10^{-4}	2.0×10^{-4}
HPS[g]	4.98×10^7	2.53×10^8	2.0×10^{-3}	2.2×10^{-3}
BTPES[g]	6.76×10^7	2.66×10^8	2.5×10^{-3}	2.3×10^{-2}
BFTPS[g]	1.22×10^8	1.66×10^9	6.86×10^{-2}	2.6×10^{-2}
APPEF[h]	0.47×10^8	1.27×10^8	3.69×10^{-3}	1.10×10^{-3}

[a] *Ou et al. [40].*
[b] *Leung et al. [14].*
[c] *Peng et al. [92].*
[d] *Wu et al. [41].*
[e] *Wu et al. [42].*
[f] *Zhang et al. [78].*
[g] *Xie et al. [93].*
[h] *Peng et al. [94].*

TABLE 2 Summary of the radiative and NR-VR rate constants and luminescence quantum yield calculated by TVCF method, and experimental data for organic compounds in solid phase.

Molecule	k_r	k_{nr}^{HR}	Φ_{LQY}^{cal}	Φ_{LQY}^{exp}
COTh[a]	6.05×10^5	1.87×10^7	0.031	0.11
HPDMCb[b]	7.95×10^5	2.29×10^7	0.78	0.70
TPS[c]	1.15×10^6	3.32×10^6	0.26	0.17
HPS[c]	7.43×10^7	1.56×10^6	0.98	0.78
BTPES[c]	6.57×10^7	1.93×10^6	0.97	0.18
BFTPS[c]	1.14×10^8	1.07×10^7	0.91	0.88
CPEI[d]	2.17×10^8	5.64×10^8	0.28	0.19
(CAACAd)CuCl[e]	7.83×10^5	1.47×10^4	0.98	0.96

[a] *Zhao et al. [43].*
[b] *Zhang et al. [78].*
[c] *Xie et al. [93].*
[d] *Wu et al. [42].*
[e] *Lin et al. [44].*

insignificant S→T ISC rate constants for these systems, which is evaluated as $k_{nr}^{HR} = \frac{1}{\hbar^2}\sum_{kl} R_{kl} \int_{-\infty}^{\infty} dt Z_i^{-1} e^{i\omega t} \rho_{kl}(t,T)$ with $\rho_{kl}(t,T)$ corresponding to the TCVF and $R_{kl}(t,T)$ referring to nonadiabatic coupling and Z_i being the partition function for the initial state parabola [32]. From Fig. 4, it is obvious that TVCF formulism has achieved a big success in quantitatively predicting Φ_{LQY} for organic systems either in the gas phase, solution, or solid phase according to the good linear relationship between the calculated and experimental Φ_{LQY}.

As seen in Fig. 5, three large deviations are found in two molecules in solution (BDP-1G and BDP-1H), and one in the solid phase (BTPES), in which the calculated Φ_{LQY} values are significantly larger than the experimental counterparts. In terms of the excited state decay processes shown in Fig. 1A and analyzed above, the k_{nr}^{BHR} away from equilibrium point of S_1 to MECP should be also considered in addition to k_{nr}^{HR}, and in this case the Φ_{LQY} becomes $k_r/(k_r + k_{nr}^{HR} + k_{nr}^{BHR})$. We here choose BDP-1A and BDP-1G, which are "good" and "bad" representatives, respectively, to quantitatively compare the contribution from k_{nr}^{HR} and k_{nr}^{BHR} to the whole nonradiative decay rate [40]. We first locate the S_1/S_0 MECPs of both systems by using the penalty function method developed by Levine et al. [45] at the spin-flip TDDFT (SF-TDDFT) level in the Q-Chem program [46]. The obtained MECP geometry of BDP-1G is very close to that determined at the CASPT2 level in Jiao et al. [47], indicating SF-TDDFT is a reliable alternative in describing the MECP for such systems. The k_{nr}^{BHR} is evaluated by the TST approach as $k_{nr}^{MECP} = \frac{k_B T}{h} \exp\left(\frac{-\Delta G^{\neq}}{RT}\right)$ with $\Delta G^{\neq}$ denoting the Gibbs free energy of activation from the FC point to the MECP geometry along the reaction path in the S_1 state. As seen from the calculated results presented in Fig. 6, the S_1/S_0 MECP of BDP-1A lies much higher in energy than the FC point, and the resultant k_{nr}^{BHR} is approximately vanishing due to an extremely large $\Delta G^{\ddagger} = 20.432\,\text{kcal/mol}$. Thus, the calculated Φ_{LQY} that includes only the k_{nr}^{HR} and k_r is in

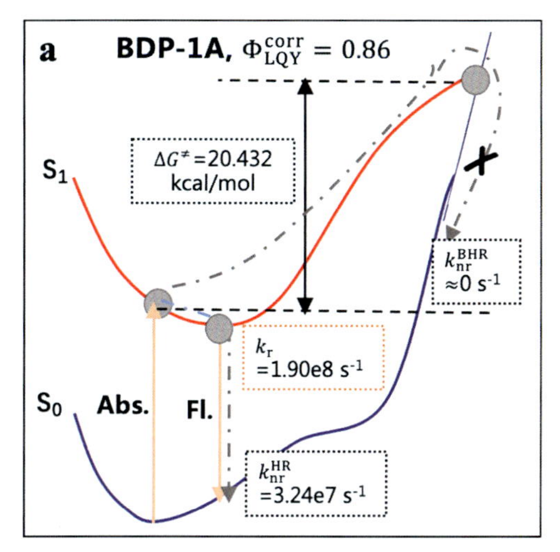

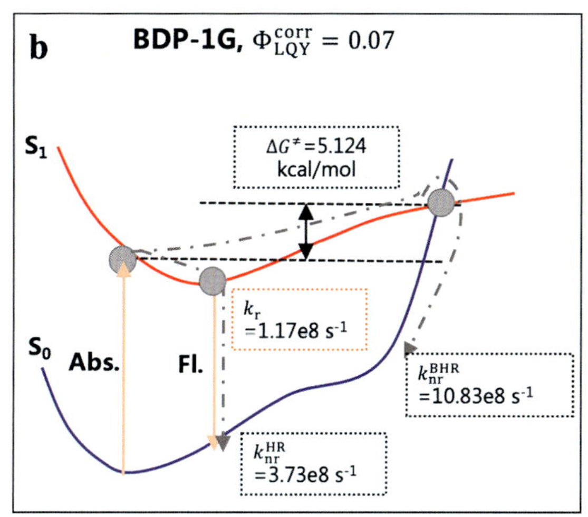

FIG. 6 Schematic graph of the two nonradiative decay channels NR-VR and NR-MECP (A) BDP-1A and (B) BDP-1G with the Gibbs free energy of activation $\Delta G^{\neq}$, the nonradiative decay rate constants, and the corrected quantum efficiency. *Reprinted with permission from Q. Ou, Q. Peng, Z. Shuai, Toward quantitative prediction of fluorescence quantum efficiency by combining direct vibrational conversion and surface crossing: BODIPYs as an example, J. Phys. Chem. Lett. 11 (18) (2020) 7790–7797. https://doi.org/10.1021/acs.jpclett.0c02054. Copyright 2020 American Chemical Society.*

good consistence with experimental values for the BDP-1A compound, as shown in Fig. 5. On the contrary, k_{nr}^{BHR} ($10.83 \times 10^8 s^{-1}$) of BDP-1G is 2–3 times larger than k_{nr}^{HR} ($3.73 \times 10^8 s^{-1}$), which results in a significant decrease of Φ_{LQY} from 0.24 to 0.07, much closer to the experimental value of 0.04. Therefore, in the case of BDP-1G, both k_{nr}^{BHR} and k_{nr}^{HR} should be taken into consideration, so that the theoretically calculated luminescence yield is in line with the experimentally measured values. It should be noted that $\Delta G^{\neq}$ is the energy difference between the S_1 states at the S_0 geometry and MECP geometry as plotted in Fig. 6 because there is no transition state along the reaction path. Considering the fact $k_r \sim 10^{7-9}\ s^{-1}$ for typical organic molecules, channel III can be safely neglected when $\Delta G^{\neq} > 10\,kcal/mol$ ($k_{nr}^{BHR} < 10^6 s^{-1}$). Otherwise, it would play an important role in the nonradiative decay processes, especially when $\Delta G^{\neq} < 6\,kcal/mol$ ($k_{nr}^{BHR} > 10^8 s^{-1}$).

Taking microprocesses into account in the determination of Φ_{LQY} and considering the effect of molecular aggregation within such processes, the mechanism of AIE is naturally revealed for the AIEgens. By quantitatively investigating k_r and k_{nr}^{HR} via the TVCF approach of a series of AIEgens in different environments, including temperatures, gas phase, solution, and solid phase, Shuai's group reported that the nonradiative decay channels are significantly blocked due to the decrease of electron-vibration coupling (reorganization energy) and the vibration-vibration coupling (Duschinsky rotation effect) in rigid environmental conditions [15,41,42,51,78]. Li et al. claimed that strong fluorescence is caused by the removal of MECP from solution to aggregates for some AIEgens [48,49]. The elimination of photoisomerization was also found to introduce the occurrence of the AIE phenomenon from solution to aggregate [59]. Other groups proposed the photoinduced ring-closed nonradiative process in TPE by a nonradiative dynamic method [50]. These indicate that the two kinds of nonradiative decay processes are system-dependent and environment-dependent, and these either coexist or compete with each other, or one of them becomes the dominant rate-controlling step.

Although the nonradiative decay process cannot be visualized via experiments, the related photophysical property and signals are likely to behave selectively, particularly for different mechanisms. In the following sections, we review the quantitative description of these nonradiative decay pathways, disclose different AIE mechanisms of some typical AIEgens, and present unique characteristics of photophyscial behaviors that correspond to different mechanisms.

4 The elimination of nonradiative channels in aggregates

4.1 The restriction of nonradiative decay induced by vibration relaxation in a harmonic region

As can be seen from the comparison between the radiative and nonradiative decay rate constants listed in Tables 1 and 2 for AIEgens in the gas phase/solution and solid phase, one easily finds that for these systems, the radiative decay rate constant is almost independent of the environment, while the nonradiative decay rate constants greatly decrease from solution to aggregate, which leads to the occurrence of strong fluorescence in the aggregate. These findings indicate another AIE mechanism that is different from what we have introduced in Section 3, i.e., the blockage of the nonradiative decay channels upon aggregation. Taking HPS an example, we disclose the origin of the change of the nonradiative decay rate by analyzing nonadiabatic coupling, electron-vibration coupling (vibration relaxation energy, reorganization energy), and vibration-vibration mode mixing (Duschinsky rotation effect, DRE) during the excited state decay process in Fig. 7 [51]. Nonadiabatic coupling provides the driving force of internal conversion from S_1 to S_0, whose coupling matrix elements are calculated via the first-order perturbation theory; the electron-vibration coupling (reorganization energy, relaxation energy) characterizes the ability of molecular vibrations to accept the electronic excited-state energy of the internal conversion, which is computed via $\lambda_i = \frac{1}{2}\omega_i^2 \Delta D_i^2$ for the ith normal mode, and D is the displacement between the equilibrium geometries of the two electronic states; the mode mixing (Duschinsky rotation effect, DRE) corresponds to the multichannel internal conversion of a vibration mode of S_1 to various modes of S_0, in which the Duschinsky rotation matrix (DRM) is obtained via DRM$=\boldsymbol{L}_i^T \boldsymbol{L}_f$ ($\mathbf{L}_{i(f)}$ are the normal modes with mass weighted in the initial (final) molecular electronic states). From Fig. 7, it can be seen that while the nonadiabatic coupling is insensitive to the environment, the electron-vibration coupling and the degree of mode mixing are very susceptible to aggregation: (i) the low-frequency vibration modes are significantly blue shifted with the number of low-frequency modes ($\omega < 100\,\mathrm{cm}^{-1}$) decreased from 15 to 6; (ii) the electron-vibration couplings are considerably weakened, especially those of the low-frequency modes; and (iii) many modes are segregated from each other with many vanishing off-diagonal terms from Fig. 7C to D (when the off-diagonal element of the Duschinsky rotation matrix is not zero, the corresponding two modes mix with each other) when going from solution to aggregate. These changes slow down the nonradiative decay from S_1 to S_0, leading to strong light emitting in aggregate.

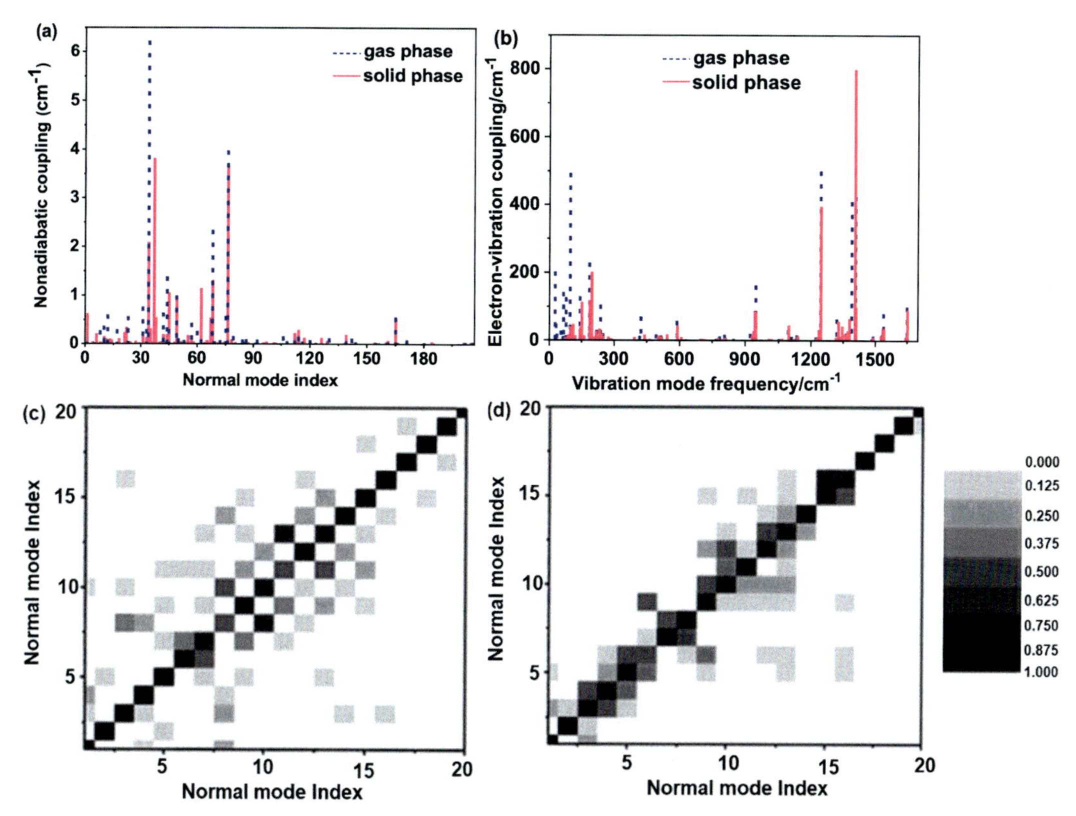

FIG. 7 The calculated nonadiabatic coupling, electron-vibration coupling, and Duschinsky rotation matrix (the degree of mode mixing) of HPS in the gas phase and solid phase. *Reprinted (adapted) with permission from M. Dommett, M. Rivera, R. Crespo-Otero, How inter- and intramolecular processes dictate aggregation-induced emission in crystals undergoing excited-state proton transfer, J. Phys. Chem. Lett. 8 (24) (2017) 6148–6153. https://doi.org/10.1021/acs.jpclett.7b02893. Copyright 2017 American Chemical Society.*

Based on the above analysis, we get the micromechanism of the kind of AIEgen and plot the schematic change of the potential energy surfaces (PESs) from solution to aggregate in Fig. 8. In light of the important contribution of low-frequency modes to the AIE of HPS, we present the PESs of S_0 and S_1 of two low-frequency modes as a representative. In solution, the PESs of S_0 and S_1 are very flat and energetically well separated from each other. Moreover, each mode of S_1 mixes with a variety of modes of S_0. At room temperature, various vibrational modes with high quantum numbers are activated to dissipate the excited state owing to strong electron-vibration coupling, and a great number of nonradiative channels are opened due to the effective mode mixings. Both of these two processes highly speed up the nonradiative decay rate. For aggregates, the PESs become very steep because the intramolecular vibrational motions are restricted by the intermolecular interactions in compact molecular-packing aggregates. Accordingly, the number of activated vibration states and the nonradiative decay channels is decreased because of the weak electron-vibration couplings and subtle mixing among modes. Consequently, the nonradiative decay rate is sharply decreased. Theoretical

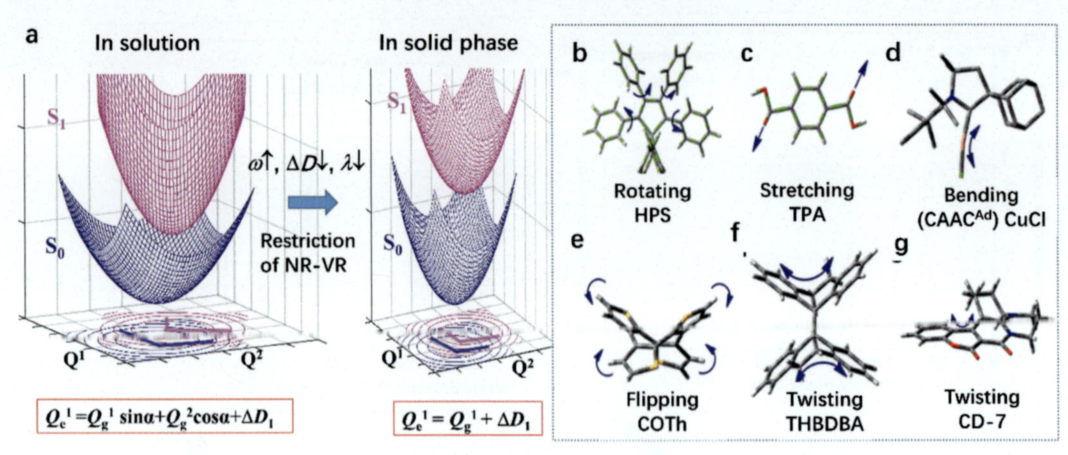

FIG. 8 (A) PES scheme of AIEgens in solution and solid states and (B–G) view of vibrational modes involved in the NR-VR channels with dominant contribution. *Reproduced with permission from Q. Peng, Z. Shuai, Molecular mechanism of aggregation-induced emission, Aggregate 2 (2021) e91, Copyright 2021 John Wiley and Sons.*

investigations of a tremendous amount of systems demonstrate that aggregation can effectively attenuate the couplings between electron and a great variety of vibration modes, including rotating [14,41,51,52], twisting [53], stretching [20], bending [44], and flipping [43] vibrations by virtue of their specific intermolecular interactions (see Fig. 8B–G), which all significantly restrict the molecular geometrical relaxation and slow down the nonradiative decay, thus inducing strong fluorescence upon aggregation. These findings provide sound theoretical evidences for the RIR, RIV, and RIM mechanism proposed by Tang's group from the perspective of molecular conformations [8,9,53]. So far, theoretical calculations have drawn a very clear picture to demonstrate the change of electron–electron, electron-vibration, vibration-vibration coupling, and the effect of them on the nonradiative decay rate from solution to aggregate, which provides deep insights into various experimental phenomena.

4.2 The elimination of nonradiative channels beyond the harmonic region

4.2.1 Restricted access to an MECP

As analyzed above, the nonradiative decay via MECP may become dominant and quench the emission in solution, and its removal in aggregate leads to bright emission, i.e., the AIE phenomenon. Li and Blancafort first proposed the restricted access to a conical intersection (RACI) mechanism to rationalize the AIE phenomenon of diphenyl dibenzofulvene [86]. Since then, the RACI mechanism has been widely applied to rationalize the AIE phenomenon for many AIEgens [49,83,84,89]. Herein, we illustrate the RACI mechanism via an example of 4-diethylamino-2-benzylidene malonic acid dimethyl ester (BIM) with AIE activity. Its fluorescence quenching in solution and fluorescence enhancement in crystals are investigated by the quantum chemistry calculations via combining QM (including TDDFT and CASSCF) and QM:MM approaches [54]. The calculated S_1 potential energy surface along the torsion coordinate of the styrene double bond in solution shows that upon excitation, BIM first initiates a local minimum (S_1-EM) within the FC region, then experiences a barrierless relaxation to a low-energy intermediate with charge-transfer character (S_1-CT), then reaches an S_1/S_0 conical intersection (S_1/S_0-CIb), and finally nonradiatively decay to S_0 via the CIb, as depicted in Fig. 9. The barrier to access S_1/S_0-CIb from S_1-EM is very small, which indicates that S_1/S_0-CIb is responsible for the fluorescence quenching of RIM in methanol. In the crystalline phase, the rigid environment precludes the torsion of the styrene double bond, which restricts the access to a CI with much higher energy. As a result, RIM is trapped in the S_1-EM of BIM, emitting strong fluorescence upon aggregation.

How can we prove the occurrence of nonradiative decay via VR (k_{nr}^{HR}), or nonradiative decay via CI or MECP (k_{nr}^{BHR}), or both of them? We herein demonstrate this by investigating the temperature dependence of the nonradiative decay. Because the nonradiative decay via MECP always needs to cross over an energy barrier, either to a transition state (TS) or to a high-energy MECP, k_{nr}^{BHR} is likely to be very sensitive to temperature and have a sudden growth when energy reaches the mutation point at a certain temperature. While the nonradiative decay via VR is a decay process from a high-energy excited state to a low-energy ground state without an energy barrier, k_{nr}^{HR} is not as sensitive to temperature as k_{nr}^{BHR} and acquires a monotonically slight increase with the increase of temperature. Consequently, the excited-state lifetime and luminescence quantum efficiency combining the two nonradiative processes will exhibit nonmonotonic behaviors with a "knee point" as temperature increases. Keeping this in mind, we exemplify *fac*-Ir(F_2ppy)(ppz)$_2$ (ppy = 2-phenylpyridine and ppz = phenylpyrazole) with comprehensive photophysical data from experiments [55] to quantitatively investigate the k_r, k_{nr}^{HR} and k_{nr}^{BHR} and excited-state lifetime at different temperatures in THF solution [56]. The emissive state of *fac*-Ir(F_2ppy)(ppz)$_2$ is a triplet excited state. We firstly construct the potential energy surface (PES) of T_1 and S_0 by locating the minimum point of S_0 (S_0 geometry) and the local minimum point of T_1 with metal–ligand charge transfer transition (^{3}MLCT) in the harmonic region, as well as the TS and the local minimum with metal-

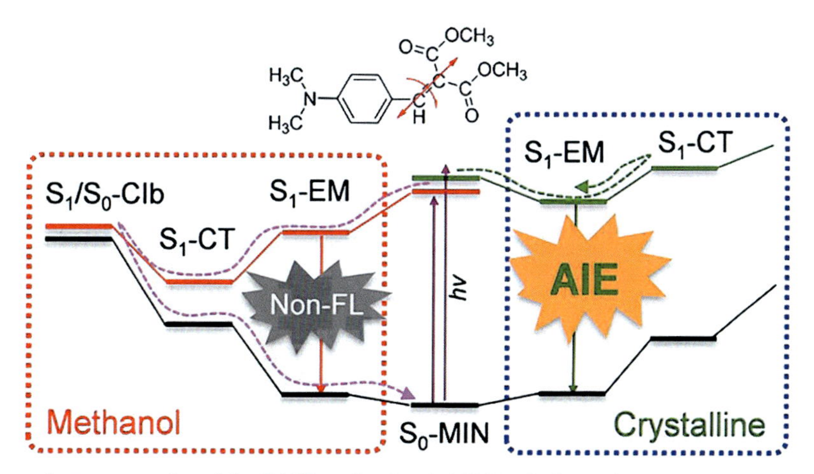

FIG. 9 Schematic representation of the RACI mechanism in BIM, including relative energies in eV. *Reprinted with permission from B. Wang, X. Wang, W. Wang, F. Liu, Exploring the mechanism of fluorescence quenching and aggregation-induced emission of a phenylethylene derivative by QM (CASSCF and TDDFT) and ONIOM (QM:MM) calculations, J. Phys. Chem. C 120 (2016) 21850–21857. Copyright 2016 American Chemical Society.*

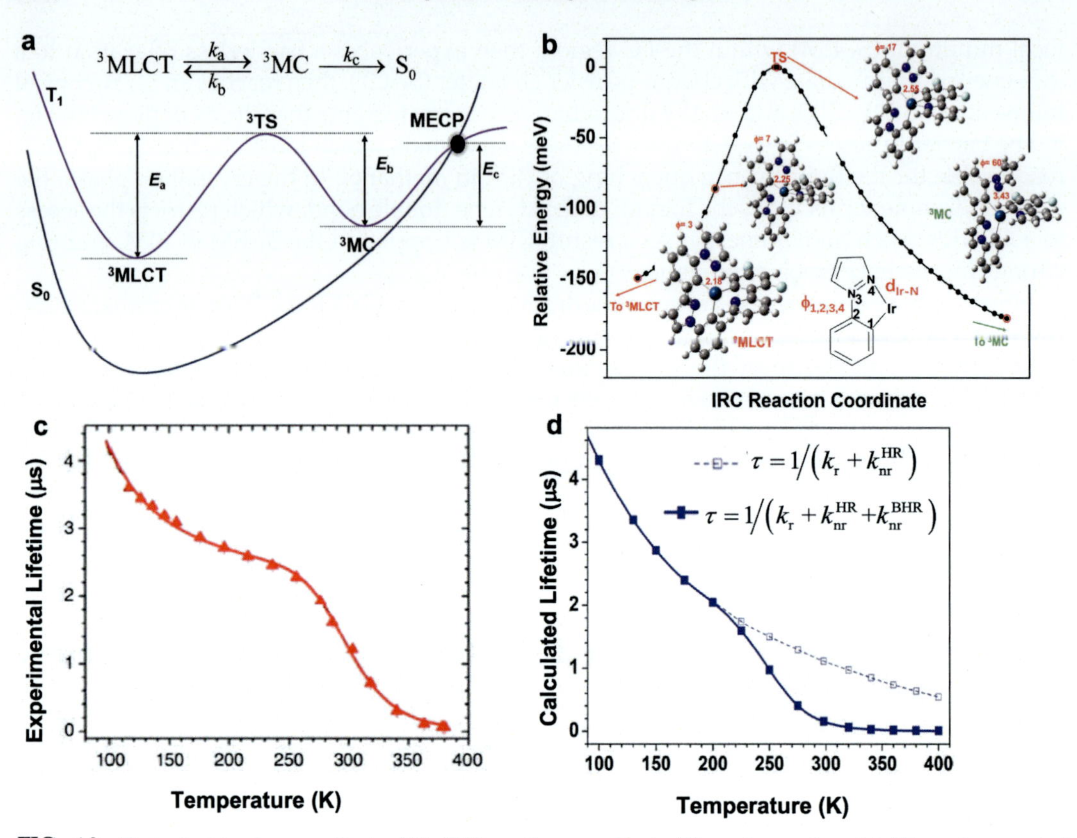

FIG. 10 Excited state decay pathway (A), IRC reaction coordinate (B), and experimental (C), and computed (D) temperature dependence of the excited-state lifetime of *fac*-Ir(F2ppy)(ppz)2. *Reprinted (adapted) with permission from X. Zhang, D. Jacquemin, Q. Peng, Z. Shuai, D. Escudero, General approach to compute phosphorescent OLED efficiency, J. Phys. Chem. C 122 (11) (2018) 6340–6347. https://doi.org/10.1021/acs.jpcc.8b00831. Copyright 2018 American Chemical Society.*

center transition (MC) and MECP between T_1 and S_0 beyond the harmonic region, as seen in Fig. 10A. Herein, the optimizations except for the MECP are performed at the level of B3LYP with LANL2DZ for the iridium atom and 6-31G (d,p) basis set for the others in the Gaussian 09 package. The MECP is done using Harvey's algorithm at the B3LYP/ECP-60-mwb/def2-SVP level and its activation energy is corrected by PWPB95-D3/ECP-60-mwb/def2-SVP in the ORCA package [57]. From Fig. 10A, it is easily seen that the emissive ^{3}MLCT can undergo radiative decay directly to S_0 as light (k_r), nonradiative decay to S_0 via vibration relaxation (k_{nr}^{HR}), and undergo nonradiative decay indirectly to S_0 by a two-step reaction of $^3\text{MLCT} \underset{}{\overset{\text{TS}}{\rightleftharpoons}} {}^3\text{MC} \xrightarrow{\text{MECP}} S_0$. In the two-step reaction, the barrier to populate the TS plays a decisive role because of the largest geometrical relaxations taking place from the ^{3}MLCT (pseudo-octahedral) to the ^{3}MC (trigonal bipyramid) point via the breakage of the Ir—N bond and the rotation of the relative ligand [58].

Based on the obtained geometrical and electronic structure information, k_r is calculated with the Einstein relationship by considering Boltzmann population of three substates of the triplet state; the k_{nr}^{HR} is computed with the TVCF approach in the MOMAP program; and adapting the kinetic model in Sajoto *et al.* [55], in which ^{3}MLCT →^{3}MC is the rate-limiting step and the equilibration between ^{3}MLCT and ^{3}MC is considered. With the steady-state approximation, the rate constant can be simply calculated as $k_{nr}^{BHR} = A_0 A\exp(-E_a/k_B T)$. Here, A_0 is a temperature-dependent prefactor and $A_0 = [1+\exp((E_b - E_c)/k_B T)]^{-1}$; A stands for the preexponential factor obtained by canonical variational transition state theory (CVT) implemented in the POLYRATE program; and E_a, E_b, and E_c are the activations marked in Fig. 10A. Fig. 10C and D show the experimental and calculated plot of the excited state lifetime of $\tau = 1/(k_r + k_{nr}^{HR} + k_{nr}^{BHR})$ versus temperature, as well as the calculated $\tau = 1/(k_r + k_{nr}^{HR})$ for comparison. Explicit numbers are collected in Table 3. It can be seen in Fig. 10D that the sigmoid-like dependency is well reproduced by the calculated results. Three distinct regimes are depicted in Fig. 10D, i.e., a slight lifetime decrease at 77–220 K and a second more significant drop at 220–320 K, and a third almost unchanged tendency at temperature higher than 320 K. The two key inflection points are at ca. 220 and 320 K, which is in good agreement with experimentally observed values, i.e., 250 and 340 K. It is easily understood from the data in Table 3 that k_r slightly grows up owing to the increasing population of the substates with the largest k_r value among the three substates of ^{3}MLCT observed at 77–130 K, and then it is almost unchanged when $T > 130$ K because the thermally activated vibrations always largely the spectrum lineshape other than the integral area with significant redshift. k_{nr}^{HR} is expected to increase slightly because of small relaxation energy that exclusively comes from

TABLE 3 Computed k_r, k_{nr}^{HR} and k_{nr}^{BHR}, global lifetimes (τ) and Φ_{LQY} values at different temperatures for *fac*-Ir(F_2ppy)(ppz)$_2$.

Temperature (K)	k_r (s^{-1})	k_{nr}^{HR} (s^{-1})	k_{nr}^{BHR} (s^{-1})	τ (μs)	Φ_{LQY}
77	1.60×10^4	1.78×10^5	4.43×10^{-12}	5.16	0.08
100	3.99×10^4	1.93×10^5	1.92×10^{-6}	4.29	0.17
130	8.23×10^4	2.16×10^5	4.31×10^{-2}	3.35	0.28
150	1.13×10^5	2.36×10^5	3.67×10^0	2.86	0.32
175	1.52×10^5	2.66×10^5	2.26×10^2	2.39	0.36
196	1.82×10^5	2.97×10^5	4.93×10^3	2.04	0.38
225	2.22×10^5	3.53×10^5	5.37×10^4	1.59	0.35
250	2.52×10^5	4.16×10^5	3.62×10^5	0.97	0.25
275	2.80×10^5	4.96×10^5	1.72×10^6	0.40	0.11
298	3.03×10^5	6.00×10^5	5.69×10^6	0.15	0.05
320	3.23×10^5	7.07×10^5	1.52×10^7	0.06	0.02
340	3.40×10^5	8.38×10^5	3.33×10^7	0.03	0.01
360	3.55×10^5	1.00×10^6	6.67×10^7	0.02	0.005
380	$\mathbf{3.70\times10^5}$	$\mathbf{1.20\times10^6}$	$\mathbf{1.24\times10^8}$	**0.008**	**0.003**
400	$\mathbf{3.83\times10^5}$	$\mathbf{1.46\times10^6}$	$\mathbf{2.17\times10^8}$	**0.005**	**0.002**

high-frequency stretching and in-plane deformation vibrations [56]. At the low-temperature regime, only the k_r and k_{nr}^{HR} components contribute to the lifetime because the temperature cannot provide enough energy to reach the barrier for proceeding significant k_{nr}^{BHR}. Therefore, the temperature effect is not obvious at low temperature. However, when temperature reaches a certain point, the molecule has enough energy to cross over the barrier to produce the transform $^3MLCT \rightarrow TS \rightarrow ^3MC$. k_{nr}^{BHR} becomes significant only when the temperature goes above ca. 230 K and largely influences the quantum efficiency and lifetime by directly competing with radiative decay at room temperature, and then evolves into the most prominent deactivation channel that significantly quenches the luminescence at high temperatures. The relevance of the k_{nr}^{BHR} is further illustrated in Fig. 10D by switching off its contribution to a lifetime of $\tau = 1/(k_r + k_{nr}^{HR})$ (dashed line in Fig. 10D). Without taking k_{nr}^{BHR} into consideration, the sigmoid-like temperature-dependence will disappear, which will lead to the inconsistency with the experimental observation. At the same time, this work provides a criterion that justifies whether the MECP occurs in the excited-state decay processes both in solution and in aggregates.

4.2.2 *Aggregation-dispelled isomerization*

Apart from the MECP between the light-emitting state and the ground state, another nonradiative decay channel that may occur beyond the harmonic region is via isomerization. Generally, most AIEgens have twist/rotor structures while aggregation-caused quenching (ACQ) molecules possess planar conjugated structures. However, such a twist-rotor structure may give rise to the occurrence of isomerization, which may significantly alter the light-emitting properties. Recently, Huang's group reported six planar molecules with analogous structure that show completely different emission behaviors in experiment. 1,2-Di(2-selenophenyl)-ethene (SVS), 1,2-di(2-thienyl)-ethene (TVT), and (E)-1,2-bis(thieno[3,2-*b*]thiophen-2-yl)ethane (TTVTT) molecules exhibit typical ACQ phenomena, while SVS(OEt), TVT(OEt), and TTVTT(OEt) show interesting AIE behavior [59]. Therefore, it is very meaningful to investigate the mechanism of their completely different emission behaviors that arise from their similar planar structures.

We choose TTVTT and TTVTT(OEt) as representatives and firstly construct the potential energy surfaces (PESs) of S_0 and S_1 by a relaxed scan at the level of (TD)B3LYP/6-31G(d, p) in the Gaussian 09 package, as seen in Fig. 11A and B. It is obvious that the PESs of the two compounds have double wells, corresponding to cis- and trans-geometries, respectively. The trans-geometry is more stable since it is lower in energy than the cis-geometry, which is in line with the single-crystal X-ray structure. The isomerization barriers from trans to cis are very high for TTVTT(OEt) (1.26 eV) and TTVTT (1.23 eV) in the S_0 state. However, the barriers sharply reduce in the S_1 state up to 0.45 eV of TTVTT(OEt) and 0.67 eV of TTVTT. Thus, the introduction of -OCH_3 destabilizes the planar conformation and facilitates the occurrence of isomerization from trans- to cis-TTVTT(OEt) in the S_1 state [60].

We further compare the luminescence property of trans- and cis-TTVTT(OEt) by computing their k_r and k_{nr}^{HR} in the harmonic range of locally optical transition using the TVCF approach in the MOMAP program. It is found that k_r of trans-TTVTT(OEt) ($1.1 \times 10^8 s^{-1}$) is two orders of magnitude larger than that of cis-TTVTT(OEt) ($6.2 \times 10^6 s^{-1}$), which mainly stems from larger transition dipole moment of trans-TTVTT(OEt) (18.56 Debye) with a more planar conjugated structure than that of cis-TTVTT(OEt) (6.42 Debye) with twisted

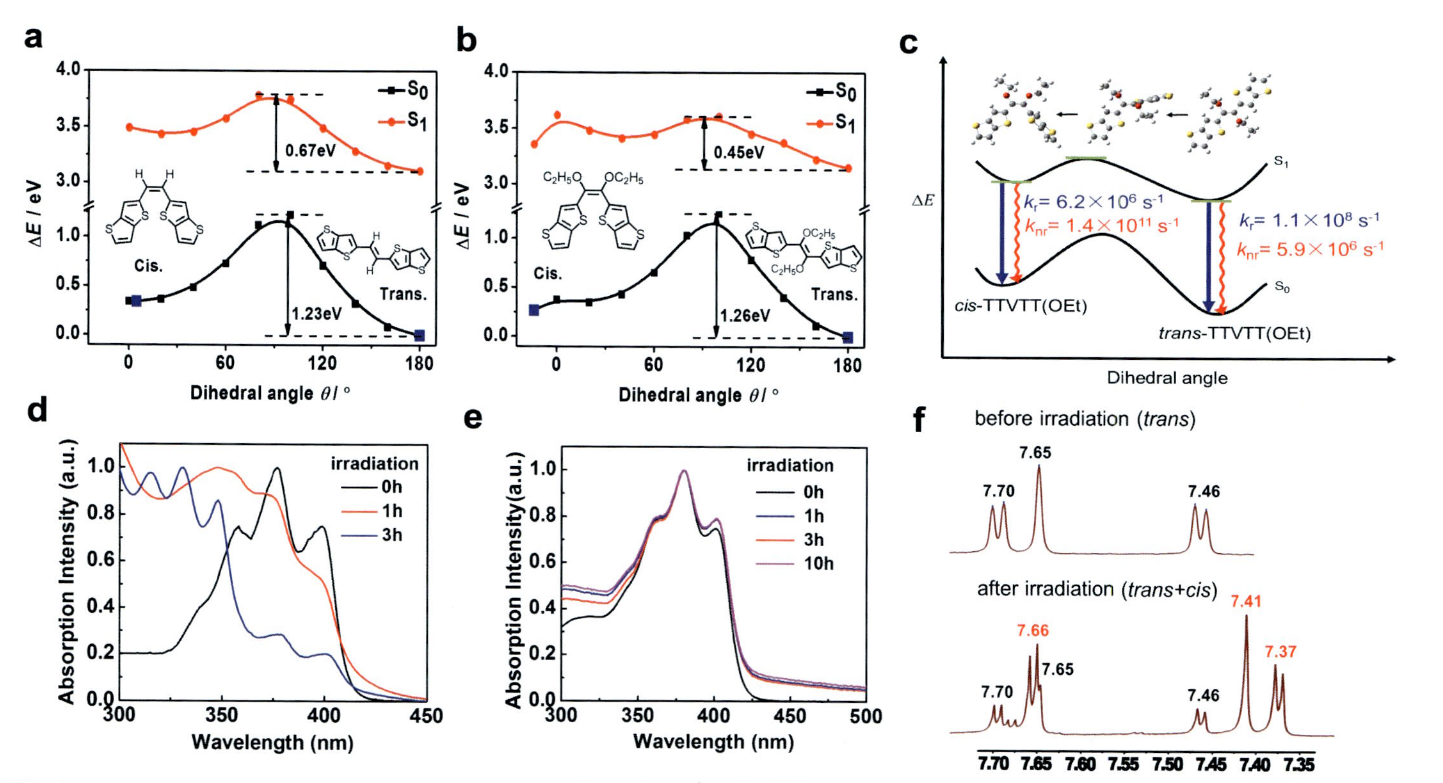

FIG. 11 Potential energy surfaces along the central vinyl dihedral angle of TTVTT (A) and TTVTT(OEt) (B); the excited-state decay rates of isolated trans- and cis-TTVTT(OEt) (C); UV–vis (D) and H NMR spectra (F) of TTVTT(OEt), and UV–vis (E) of TTVTT before and after irradiation in THF solution. *Reproduced with permission from L. Yang, P. Ye, W. Li, W. Zhang, Q. Guan, C. Ye, T. Dong, X. Wu, W. Zhao, X. Gu, Q. Peng, B. Tang, H. Huang, Uncommon aggregation-induced emission molecular materials with highly planar conformations, Adv. Opt. Mater. 6 (2018) 1701394. https://doi.org/10.1002/adom.201701394. Copyright 2018 John Wiley and Sons.*

conformation. On the contrary, k_{nr}^{HR} of trans-TTVTT(OEt) ($5.9 \times 10^6 s^{-1}$) is five orders of magnitude smaller than that of cis-TTVTT(OEt) ($1.4 \times 10^{11} s^{-1}$), which is caused by the rigidity of trans-TTVTT(OEt) and flexibility of cis-TTVTT(OEt). Consequently, trans-TTVTT(OEt) emits very bright fluorescence with $k_r \gg k_{nr}^{HR}$ whereas cis-TTVTT(OEt) shows dark fluorescence with $k_r \ll k_{nr}^{HR}$ in the gas phase. Overall, the whole photophyscial process in solution is described as the following: after absorbing a photon, the stable trans-TTVTT(OEt) firstly jumps from S_0 to S_1, and then transforms to cis-TTVTT(OEt) along the S_1-PES, finally nonradiatively relaxing to S_0 of cis-TTVTT(OEt) without generating visible fluorescence. The lifetime of cis-TTVTT(OEt) is calculated to be 7.1 ps, which is consistent with the experimental value (<30 ps). Upon aggregation, the isomerization from trans-TTVTT(OEt) to cis-TTVTT(OEt) are inhibited, resulting in a strong fluorescence from trans-TTVTT(OEt). Therefore, the aggregation-dispelled photoisomerization from the bright state (trans-conformation) to the dark state (cis-conformation) is responsible for the unusual AIE activity of these highly planar molecules.

To confirm the photoisomerization mechanism proposed above, a series of measurements was performed for TTVTT(OEt) in THF, including UV–Vis and ^1H NMR, before and after irradiation by an ultraviolet lamp. These results are shown in Fig. 10D–F. As expected, after a period of irradiation, a new set of absorption bands starts to appear and the intensity is gradually increased, while the intensity of original bands is contrarily decreased (Fig. 11D), indicating the generation of a new compound. A new set of peaks (7.66, 7.41, 7.37 ppm) in 1H NMR also arises after 3 h irradiation, which underlines that the new compound is *cis*-TTVTT(OEt). By this time, the mixture of *cis*- and *trans*-TTVTT(OEt) exhibit nonemissive characteristics in solution, which verify the dark S_1 state of *cis*-TTVTT(OEt) with $k_r \ll k_{nr}^{HR}$. However, the UV–vis spectra of TTVTT (Fig. 11E) are almost identical before and after irradiation for 10 h, suggesting photoisomerization barely happened in TTVTT solution. The solid-phase TTVTT(OEt), upon ultraviolet light irradiations, also displays unchanged UV–vis spectrum before and after irradiation, indicating the hindrance of photoisomerization. These perfectly confirm the mechanism of aggregation-dispelled isomerization in TTVTT derivatives, which can be extended to explain the AIE phenomena of a series of systems.

5 Enhancement of the radiative processes in aggregates

The radiation process involves the interaction between light (electromagnetic field) and a molecule (transition moment), which requires a significantly large transition dipole moment between the excited and ground states of a molecule for strong emission. The direct electric transition dipole moment between two electronic states is proportional to the overlap integral of the wave functions of the involved electronic states under the Franck-Condon (FC) principle, $\mu \propto \int \phi_i^*(R,r)\phi_f(R,r)dr$.Thus, the properties of wave functions are essential to transition dipole moments, including shape, symmetry, and spatial distribution, and thereupon several selection rules are generated: (i) spin-allowed if the spin multiplicities of the two states are identical, (ii) symmetry-allowed if the transition happens between a gerade state and a ungerade state, and (iii) overlap-allowed if the wave functions of two states are not completely separated in space [7]. When μ of an excited state is large enough to generate considerably fast radiative decay according to Einstein's spontaneous emission theory,

$k_r = \frac{8\pi^2 \Delta E^3 \mu^2}{3\varepsilon_0 \hbar c^3} = \frac{f \Delta E^2}{1.5}$ (f is the dimensionless oscillator strength and ΔE is the transition energy in the unit of wavenumber), the excited state is called as the "bright" state, otherwise the "dark" state. Scientists are always prone to investigate the organic systems with $\pi \rightarrow \pi^*$ bright transition. Recently, it is found that the "dark" state with $n \rightarrow \pi^*$ or $\sigma \rightarrow \pi^*$ or fully charge-transfer (CT) or symmetry-forbidden transitions can be converted to the "bright" state owing to electrostatic interactions in some organic compounds when going from solution to aggregates [20,21,61]. Besides, a "dark" state with symmetry-forbidden transition is also found to emit bright light owing to the Herzberg-Teller vibronic coupling in 5,10-diphenylphenazine (DPhPZ) aggregate [22]. The aforementioned inversion of the light-emitting state promotes the enhanced fluorescence in aggregates, and the reported novel cores/backbones of AIEgens based on such inversion of state largely broaden the scope of excellent AIEgens. Several examples are presented in the following.

5.1 Aggregation-induced reverse from dark n-π* to bright π-π* state

A prototypical aromatic diacid, terephthalic acid (TPA, Fig. 2), was reported to exhibit a unique phenomenon of strong fluorescence, delayed fluorescence, and phosphorescence upon crystallization, in contrast to extremely weak visible light in solution and amorphous phases [62]. To find out the morphology dependence of luminescence property, comparatively analysis has been proposed to explore the geometrical and electronic structures of the low-lying excited states of the isolated and crystalline-state TPA at hybrid CASPT2 (8e, 8o)/ANO-RCC-VDZP/AMBER level by interfacing MOLCAS [63] and TINKER [64] packages. It was found that electrostatic interaction from the surrounding molecules in solid phase can significantly lift the excited state with $n \rightarrow \pi^*$ transition and slightly lower the excited state with $\pi \rightarrow \pi^*$ transition in energy, as seen in Fig. 12. Upon crystallization, the $^1(\pi,\pi^*)$ state shifts to the lowest excited state S_1 (4.76 eV) from the second lowest excited state S_2 (4.99 eV) in the

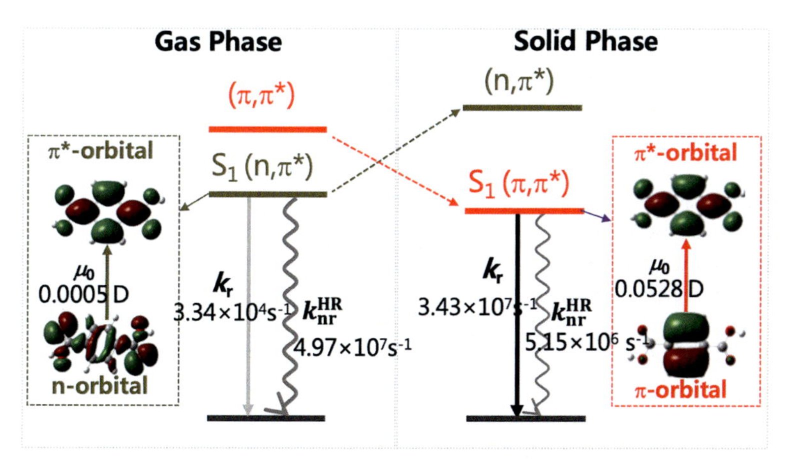

FIG. 12 Energy level, natural transition orbitals and transition property of the excited states, and the radiative and nonradiative transition rate constants for TPA in the gas and solid phases. *Reprinted with permission from H. Ma, W. Shi, J. Ren, W. Li, Q. Peng, Z. Shuai, Electrostatic interaction-induced room-temperature phosphorescence in pure organic molecules from QM/MM calculations, J. Phys. Chem. Lett. 7 (15) (2016) 2893–2898. https://doi.org/10.1021/acs.jpclett.6b01156. Copyright 2016 American Chemical Society.*

gas phase, whereas the $^1(n,\pi^*)$ state is lifted to the S_2 state (5.05 eV) from the S_1 state (4.81 eV) in the gas phase. Such a change in the energetic order is because the n-orbital is sensitive to the electrostatic forces owing to the dense charge on the oxygen atom, while the π-orbital is insensitive due to uniform distribution of the charge. These findings highlight that crystallization induces the conversion from dark $^1(n, \pi^*)$ to bright $^1(\pi,\pi^*)$ for the S_1 state in TPA, recovering the strong emission on the basis of the Kasha rule. The resultant transition dipole moment of S_1 is largely increased from 4.97×10^{-4} to 5.28×10^{-2} upon aggregation, which meets the necessary requirement for fluorescence ($k_F \propto \Delta E_{S1}^3 \mu_{S1}^2$) or phosphorescence $\left(k_P \propto \Delta E_{T1}^3 \frac{\left|\xi_{(S_1,T_1)}\right|^2 \cdot \mu^2{}_{S1}}{\Delta E_{S1T1}^2} \right)$ [20].

The electrostatic interaction not only gives rise to a faster radiative decay by altering the energetic order of the excited states, but also slows down the nonradiative decay of $S_1 \rightarrow S_0$ upon crystallization. The rate constants of the radiative and nonradiative decay from S_1 to S_0 are calculated via the TVCF method in our home-built MOMAP program [16,32,33,65,87,91]. The resultant k_F is greatly fastened by three orders of magnitudes from $3.34 \times 10^4\,s^{-1}$ to $3.43 \times 10^7\,s^{-1}$, while k_{nr}^{HR} is decreased by one order of magnitude from $4.97 \times 10^7\,s^{-1}$ to $5.15 \times 10^6\,s^{-1}$, as TPA goes from the gas phase to the crystalline phase. The enhancement of radiative decay rate and the reduction of nonradiative decay rate concertedly lead to the occurrence of strong fluorescence and phosphorescence upon crystallization, which rationalize the mechanism of the crystallization-induced double emission phenomenon as observed in experiments.

5.2 Crystallization-induced reversal from the dark (n+σ)-π* to the bright π-π* state

A unique squaraine (SQ) derivative (CIEE-SQ, Fig. 2) is found to emit continuously enhanced light from crystal to cocrystal with CIEE-SQ:$CHCl_3$=1 while weakly fluorescing in solution, which have rarely been observed in previously reported SQ derivatives [21]. To unveil the inherent micro-mechanism, the geometrical and electronic structures of the CIEE-SQ in isolated, crystalline, and cocrystalline phases are carried out, using the ONIOM approach with two layers TD/M062X/6-31G*/UFF in the Gaussian 16 package [66]. It should be noted that there are three conformations (**C1**, **C2** and **C3**) in the CIEE-SQ crystal, which are calculated and discussed in details in [21]. Here, we take **C2** as a representative to illustrate the changes of the geometrical and electronic property from isolated to aggregated CIEE-SQ. It is firstly found that the geometry of the S_0 state (S_0-geometry) of the isolated CIEE-SQ is similar to that of the aggregated CIEE-SQ, which is confirmed by the analogous absorption spectra in dilute solution and crystalline phase as observed in experiments. Secondly, there is a dramatic geometric change to the S_1 geometry upon excitation. The S_1 geometry of the isolated molecule tends to maximally extend with the decrease of the θ angle from 54.80° of the S_0 state to 18.48° of the S_1 state, while that of the crystalline structure becomes more distorted with the increase of the θ angle from 58.20° to 71.52° and the ϕ angle from 3.70° to 15.66° (see Fig. 13A). Such geometric discrepancy would lead to the distinct electronic transition properties in diluted solution and in aggregates. Thirdly, as seen in Fig. 13B, the calculated oscillator strength ($f = \frac{2}{3}\Delta E_{S1}\mu_{S1}^2$) of the S_1 state is zero because the transition is overlap-

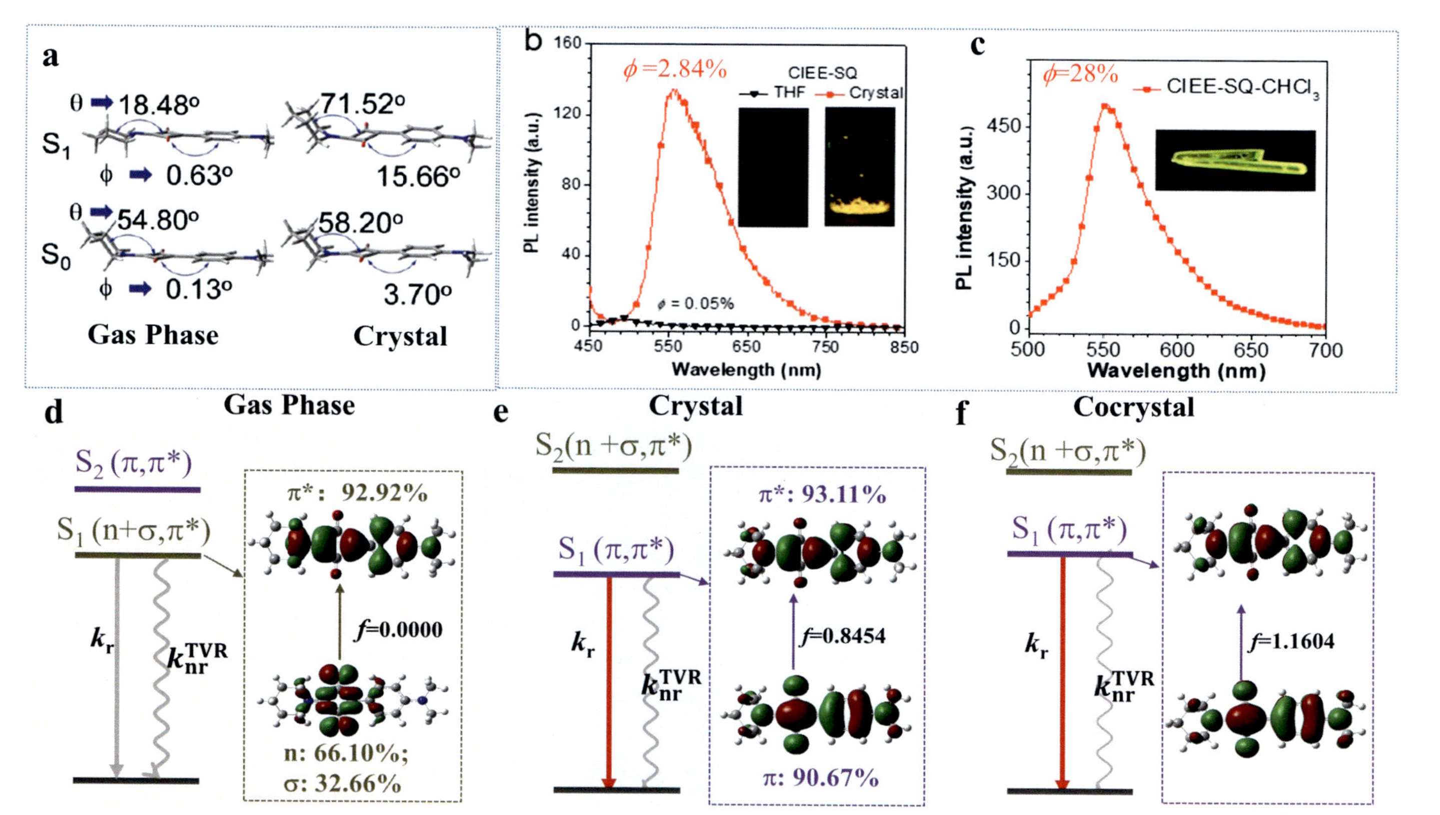

FIG. 13 The equilibrium geometries of S0 and S1 states in gas and solid phases (A); PL spectrum and efficiency in THF/crystal (B) and cocrystal (C), and energy levels, transition orbitals, and oscillator strengths of the low-lying excited states based on the S_1 geometry in the gas phase (D), crystal (E) and cocrystal (F). *Reproduced with permission from S. Yang, P.A. Yin, L. Li, Q. Peng, X. Gu, G. Gao, J. You, B.Z. Tang, Crystallization-induced reversal from dark to bright excited states for construction of solid-emission-tunable squaraines, Angew. Chem. Int. Ed. 59 (25) (2020) 10136–10142. https://doi.org/10.1002/anie.201914437. Copyright 2020 John Wiley and Sons.*

forbidden from mixed n- and σ-orbital to π*-orbital, while that of the S_2 state is 0.8900 owing to its overlap-allowed transition from π- to π*-orbitals for the isolated CIEE-SQ. On the basis of the Kasha rule, these results lead to extremely weak fluorescence in dilute solution, as observed in experimental phenomena. On the contrary, the energetic order of the low-lying excited states upon aggregation (**C2** confirmation) is reversed, that is, the S_1 state turns to be bright $^1(\pi, \pi^*)$ with $f = 0.8454$ while the S_2 state coverts to a dark state, which rationalizes the enhancement of the luminescence from in solution to aggregate in experiments.

Except for the radiative decay, the nonradiative decay simultaneously takes place to dissipate the excited state energy. As is known from previous investigations, the geometrical relaxation energy (λ), also called the reorganization energy or electron-vibration coupling, can be applied to characterize the rate of the nonradiative decay process. The calculated λ of isolated molecule and **C2** in crystal are $4928.78\,\mathrm{cm}^{-1}$ and $3273.53\,\mathrm{cm}^{-1}$, respectively, which indicates that the rigid environment in crystal significantly depress λ to restrict the nonradiative decay to some extent and facilitate the radiative decay. Considering the fact that the electrostatic interaction is able to stabilize $^1(\pi, \pi^*)$ and restrict the intramolecular motions to a greater extent, we introduce chloroform ($CHCl_3$) into the modeling of the CIEE-SQ crystal to form a more compact cocrystal conformation. As expected, the participation of chloroform further stabilizes the $^1(\pi, \pi^*)$ state with f increased from 0.8454 to 1.1604 and λ reduced from 3273.53 to $870.80\,\mathrm{cm}^{-1}$ when going from crystal to cocrystal, which significantly boosts the fluorescence in the cocrystal and agrees with the experimental measurements quite well, i.e., quantum efficiency improved from 2.84% to 28.0% upon cocrystallization. Overall, crystallization and/or cocrystallization induce the conversion from the dark $^1(n + \sigma, \pi^*)$ state to bright $^1(\pi, \pi^*)$ state that proceeds with significantly fast radiative decay rate and simultaneously suppresses the nonradiative decay rate, boosting strong fluorescence. More importantly, a novel type of AIEgens without typical π-conjugation feature has been exploited for high-performance solid-state luminescence, which inspires researchers to produce excellent solid-state fluorophores by excited-state manipulation.

6 Conclusions and outlook

In this chapter, we first demonstrate that the AIE phenomena among various reported AIEgens could not be described via Kasha's J-aggregation picture, due to the fact that most of the AIEgens have significant intramolecular electron-vibration couplings, which considerably hinders the intermolecular excitonic coupling, being J- or H-type. To obtain a comprehensive and profound theoretical picture of the AIE mechanism, we introduce the quantitative evaluation of the luminescence quantum yield, followed by two processes that lead to the AIE phenomenon with various numerical examples: (i) the blockage of the nonradiative decay in aggregates within and beyond the harmonic region, and (ii) the enhancement of radiative decay via the generation of bright state upon aggregation, which might be used to explain the recovery of fluorescence in aggregate for AIEgens with charge transfer, high symmetry and other transition dipole-forbidden properties.

From the microscopic perspective, three common decay processes, i.e., radiative decay (with rate constant k_r), nonradiative caused by vibration relaxation in the harmonic region (NR-*VR*) (with rate constant k_{nr}^{HR}), are quantitatively and comprehensively explored by combining quantum chemistry calculations and TVCF rate theory. The radiative, NR-VR, and

NR-MECP processes exhibit different sensitivity toward temperature, aggregation, etc. for distinct systems. In NR-VR-dominated cases of AIEgens, the restriction of NR-VR is realized by the decoupling of electron-vibration and vibration-vibration among various kinds of normal modes, such as rotating, stretching, bending, flipping, and twisting vibrations, which are experimentally confirmed visualizable signals. When the NR-VR is superior, the lifetime of excited state exhibits a monotonic decrease with the increase of temperature. Once the NR-MECP appears, it becomes sigmoid-like dependency of lifetime on the temperature with two key inflection points, as observed in experiment. This is because the NR-MECP process is more sensitive to temperature than NR-VR. When going from solution to aggregate, the barrier of MECP grows higher owing to intermolecular interaction in rigid aggregate, which significantly inhibits NR-MECP and hence largely enhances the luminescence. Altogether, a clear physical image is presented about the microcosmic mechanism of aggregation-induced emission of organic compounds.

Looking forward, accurate description of the electronic structure and decay processes of an excited state molecule is still a long-term challenge owing to complicated electron–electron correlation and electron-*v*ibration coupling which need to be considered, let alone the excited state of an aggregate [67]. There are a lot of processes and effects that ha*v*e not been covered in the calculations of the investigated systems: (i) The intersystem crossing through *V*R or MECP between S_1 and $T_{n(n1)}$ are not considered because of insignificant spin-orbit coupling and large singlet-triplet splitting energy [68]; (ii) the anharmonic effect is not involved in TVCF method. Because tremendous amount of normal modes in the systems with large number of atoms share the excess excitation energy which sharply decrease the anharmonic effect of potential energy surface [69]; (iii) only one molecule is dealt with QM calculation without taking intermolecular excitonic coupling and charge transfer into consideration. It was found that the excitonic coupling has very minor effect on the optical spectra and NR-VR rate constant in the typical AIEgens [23,35]; (iv) the polarizable force field is not used in QM/MM protocol for these systems are not charge-transfer ones [19]. However, these aspects might play an outstanding role in the excited-state decay processes for some systems, which should be carefully checked and handled. These issues have been actively pursued in our group.

In addition, from the perspective of experimental phenomena, we have not covered the room temperature phosphorescence of purely organic systems [71], the aggregation-induced delayed fluorescence [85], clusterization-triggered emission of nonaromatic molecules [72], mechanochromic luminescence of organic aggregates [73,74], aggregation-induced circularly polarized luminescence [75], multicolor of a single molecule in aggregate [76], and so on [9,10], which are developed recently research interests in the field of organic luminescence. It is eagerly demand to demystify the microcosmic mechanism behind the newly interesting experimental phenomena with different categories.

References

[1] Y. Jiang, Y.Y. Liu, X. Liu, H. Lin, K. Gao, W.Y. Lai, W. Huang, Organic solid-state lasers: a materials view and future development, Chem. Soc. Rev. 49 (16) (2020) 5885–5944, https://doi.org/10.1039/d0cs00037j.

[2] J. Kido, M. Kimura, K. Nagai, Multilayer white light-emitting organic electroluminescent device, Science 267 (5202) (1995) 1332–1334, https://doi.org/10.1126/science.267.5202.1332.

[3] S. Reineke, F. Lindner, G. Schwartz, N. Seidler, K. Walzer, B. Lüssem, K. Leo, White organic light-emitting diodes with fluorescent tube efficiency, Nature 459 (7244) (2009) 234–238, https://doi.org/10.1038/nature08003.

[4] I.D.W. Samuel, G.A. Turnbull, Organic semiconductor lasers, Chem. Rev. 107 (4) (2007) 1272–1295, https://doi.org/10.1021/cr050152i.

[5] C.W. Tang, S.A. Vanslyke, Organic electroluminescent diodes, Appl. Phys. Lett. 51 (12) (1987) 913–915, https://doi.org/10.1063/1.98799.

[6] G. Zhang, G.M. Palmer, M.W. Dewhirst, C.L. Fraser, A dual-emissive-materials design concept enables tumour hypoxia imaging, Nat. Mater. 8 (9) (2009) 747–751, https://doi.org/10.1038/nmat2509.

[7] N.J. Turro, V. Ramamurthy, J.C. Scaiano, Modern Molecular Photochemistry of Organic Molecules, University Science Books, Sausalito, CA, 2010.

[8] J. Luo, Z. Xie, Z. Xie, J.W.Y. Lam, L. Cheng, H. Chen, C. Qiu, H.S. Kwok, X. Zhan, Y. Liu, D. Zhu, B.Z. Tang, Aggregation-induced emission of 1-methyl-1,2,3,4,5-pentaphenylsilole, Chem. Commun. 18 (2001) 1740–1741, https://doi.org/10.1039/b105159h.

[9] J. Mei, N.L.C. Leung, R.T.K. Kwok, J.W.Y. Lam, B.Z. Tang, Aggregation-induced emission: together we shine, united we soar! Chem. Rev. 115 (21) (2015) 11718–11940, https://doi.org/10.1021/acs.chemrev.5b00263.

[10] W. Zhao, Z. He, B.Z. Tang, Room-temperature phosphorescence from organic aggregates, Nat. Rev. Mater. 5 (12) (2020) 869–885, https://doi.org/10.1038/s41578-020-0223-z.

[11] Z. Zhao, H. Zhang, J.W.Y. Lam, B.Z. Tang, Aggregation-induced emission: new vistas at the aggregate level, Angew. Chem. Int. Ed. 59 (25) (2020) 9888–9907, https://doi.org/10.1002/anie.201916729.

[12] E.E. Jelley, Spectral absorption and fluorescence of dyes in the molecular state [1], Nature 138 (3502) (1936) 1009–1010, https://doi.org/10.1038/1381009a0.

[13] G. Scheibe, Über die Veränderlichkeit der Absorptionsspektren in Lösungen und die Nebenvalenzen als ihre Ursache, Angew. Chem. 50 (1937) 212–219.

[14] N.L.C. Leung, N. Xie, W. Yuan, Y. Liu, Q. Wu, Q. Peng, Q. Miao, J.W.Y. Lam, B.Z. Tang, Restriction of intramolecular motions: the general mechanism behind aggregation-induced emission, Chem. A Eur. J. 20 (47) (2014) 15349–15353, https://doi.org/10.1002/chem.201403811.

[15] Q. Peng, Z. Shuai, Molecular mechanism of aggregation-induced emission, Aggregate 2 (2021) e91, https://doi.org/10.1002/agt2.91.

[16] Q. Peng, Y. Yi, Z. Shuai, J. Shao, Toward quantitative prediction of molecular fluorescence quantum efficiency: role of Duschinsky rotation, J. Am. Chem. Soc. 129 (30) (2007) 9333–9339, https://doi.org/10.1021/ja067946e.

[17] J.W. Chung, S.J. Yoon, B.K. An, S.Y. Park, High-contrast on/off fluorescence switching via reversible e–Z isomerization of diphenylstilbene containing the α-cyanostilbenic moiety, J. Phys. Chem. C 117 (21) (2013) 11285–11291, https://doi.org/10.1021/jp401440s.

[18] T. Mutai, H. Sawatani, T. Shida, H. Shono, K. Araki, Tuning of excited-state intramolecular proton transfer (ESIPT) fluorescence of imidazo[1,2-a]pyridine in rigid matrices by substitution effect, J. Org. Chem. 78 (6) (2013) 2482–2489, https://doi.org/10.1021/jo302711t.

[19] Z. Tu, G. Han, T. Hu, R. Duan, Y. Yi, Nature of the lowest singlet and triplet excited states of organic thermally activated delayed fluorescence emitters: a self-consistent quantum mechanics/embedded charge study, Chem. Mater. 31 (17) (2019) 6665–6671, https://doi.org/10.1021/acs.chemmater.9b00824.

[20] H. Ma, W. Shi, J. Ren, W. Li, Q. Peng, Z. Shuai, Electrostatic interaction-induced room-temperature phosphorescence in pure organic molecules from QM/MM calculations, J. Phys. Chem. Lett. 7 (15) (2016) 2893–2898, https://doi.org/10.1021/acs.jpclett.6b01156.

[21] S. Yang, P.A. Yin, L. Li, Q. Peng, X. Gu, G. Gao, J. You, B.Z. Tang, Crystallization-induced reversal from dark to bright excited states for construction of solid-emission-tunable squaraines, Angew. Chem. Int. Ed. 59 (25) (2020) 10136–10142, https://doi.org/10.1002/anie.201914437.

[22] P. Yin, Q. Wan, Y. Niu, Q. Peng, Z. Wang, Y. Li, A. Qin, Z. Shuai, B.Z. Tang, Theoretical and experimental investigations on the aggregation-enhanced emission from dark state: vibronic coupling effect, Adv. Electron. Mater. 6 (2020), https://doi.org/10.1002/aelm.202000255.

[23] W. Li, Q. Peng, Y. Xie, T. Zhang, Z. Shuai, Effect of intermolecular excited-state interaction on vibrationally resolved optical spectra in organic molecular aggregates, Acta Chim. Sin. 74 (2016) 902, https://doi.org/10.6023/A16080452.

[24] C. Zener, Non-adiabatic crossing of energy levels, Proc. R. Soc. Lond. A 137 (1932) 696–702.

[25] E. Teller, The crossing of potential surfaces, J. Phys. Chem. 41 (1) (1937) 109–116, https://doi.org/10.1021/j150379a010.

[26] K. Huang, A. Rhys, Theory of light absorption and non-radiative transitions in F-centres, Proc. R. Soc. Lond. A 204 (1950) 406–423, https://doi.org/10.1098/rspa.1950.0184.

[27] J. Byrne, E. McCoy, I. Ross, Internal conversion in aromatic and N-heteroaromatic molecules, Aust. J. Chem. 18 (1965) 1589–1603, https://doi.org/10.1071/CH9651589.

[28] G.W. Robinson, R.P. Frosch, Electronic excitation transfer and relaxation, J. Chem. Phys. 38 (5) (1963) 1187–1203, https://doi.org/10.1063/1.1733823.

[29] S.H. Lin, Rate of interconversion of electronic and vibrational energy, J. Chem. Phys. 44 (1966) 3759–3767, https://doi.org/10.1063/1.1726531.

[30] R. Englman, J. Jortner, The energy gap law for radiationless transitions in large molecules, Mol. Phys. 18 (1970) 145–164, https://doi.org/10.1080/00268977000100171.

[31] R. Islampour, M. Miralinaghi, Dynamics of radiationless transitions: effects of displacement-distortion- rotation of potential energy surfaces on internal conversion decay rate constants, J. Phys. Chem. A 111 (38) (2007) 9454–9462, https://doi.org/10.1021/jp073280e.

[32] Y. Niu, Q. Peng, Z. Shuai, Promoting-mode free formalism for excited state radiationless decay process with Duschinsky rotation effect, Sci. China, Ser. B: Chem. 51 (12) (2008) 1153–1158, https://doi.org/10.1007/s11426-008-0130-4.

[33] Q. Peng, Y. Niu, Q. Shi, X. Gao, Z. Shuai, Correlation function formalism for triplet excited state decay: combined spin-orbit and nonadiabatic couplings, J. Chem. Theory Comput. 9 (2) (2013) 1132–1143, https://doi.org/10.1021/ct300798t.

[34] T.J. Penfold, E. Gindensperger, C. Daniel, C.M. Marian, Spin-vibronic mechanism for intersystem crossing, Chem. Rev. 118 (15) (2018) 6975–7025, https://doi.org/10.1021/acs.chemrev.7b00617.

[35] W. Li, L. Zhu, Q. Shi, J. Ren, Q. Peng, Z. Shuai, Excitonic coupling effect on the nonradiative decay rate in molecular aggregates: formalism and application, Chem. Phys. Lett. 683 (2017) 507–514, https://doi.org/10.1016/j.cplett.2017.03.077.

[36] W. Domcke, D. Yarkony, H. Köppel, Conical Intersections: Electronic Structure, Dynamics & Spectroscopy, World Scientific, 2004.

[37] T.R. Nelson, A.J. White, J.A. Bjorgaard, A.E. Sifain, Y. Zhang, B. Nebgen, S. Fernandez-Alberti, D. Mozyrsky, A.E. Roitberg, S. Tretiak, Non-adiabatic excited-state molecular dynamics: theory and applications for modeling photophysics in extended molecular materials, Chem. Rev. 120 (4) (2020) 2215–2287, https://doi.org/10.1021/acs.chemrev.9b00447.

[38] Z. Shuai, Q. Peng, Excited states structure and processes: understanding organic light-emitting diodes at the molecular level, Phys. Rep. 537 (4) (2014) 123–156, https://doi.org/10.1016/j.physrep.2013.12.002.

[39] K.J. Laidler, J.H. Meiser, Physical Chemistry, Benjamin/Cummings Pub. Co, Menlo Park, Calif, 1982.

[40] Q. Ou, Q. Peng, Z. Shuai, Toward quantitative prediction of fluorescence quantum efficiency by combining direct vibrational conversion and surface crossing: BODIPYs as an example, J. Phys. Chem. Lett. 11 (18) (2020) 7790–7797, https://doi.org/10.1021/acs.jpclett.0c02054.

[41] Q. Wu, C. Deng, Q. Peng, Y. Niu, Z. Shuai, Quantum chemical insights into the aggregation induced emission phenomena: a QM/MM study for pyrazine derivatives, J. Comput. Chem. 33 (23) (2012) 1862–1869, https://doi.org/10.1002/jcc.23019.

[42] Q. Wu, Q. Peng, T. Zhang, Z. Shuai, Theoretical study on the aggregation induced emission, Sci. Sin. Chim. 43 (9) (2013) 1078–1089, https://doi.org/10.1360/032013-201.

[43] Z. Zhao, X. Zheng, L. Du, Y. Xiong, W. He, X. Gao, C. Li, Y. Liu, B. Xu, J. Zhang, F. Song, Y. Yu, X. Zhao, Y. Cai, X. He, R.T.K. Kwok, J.W.Y. Lam, X. Huang, D.L. Phillips, et al., Non-aromatic annulene-based aggregation-induced emission system via aromaticity reversal process, Nat. Commun. 10 (2019) 2952, https://doi.org/10.1038/s41467-019-10818-5.

[44] S. Lin, Q. Peng, Q. Ou, Z. Shuai, Strong solid-state fluorescence induced by restriction of the coordinate bond bending in two-coordinate copper(I)-carbene complexes, Inorg. Chem. 58 (21) (2019) 14403–14409, https://doi.org/10.1021/acs.inorgchem.9b01705.

[45] B.G. Levine, J.D. Coe, T.J. Martínez, Optimizing conical intersections without derivative coupling vectors: application to multistate multireference second-order perturbation theory (MS-CASPT2), J. Phys. Chem. B 112 (2) (2008) 405–413, https://doi.org/10.1021/jp0761618.

[46] Y. Shao, Z. Gan, E. Epifanovsky, A.T.B. Gilbert, M. Wormit, J. Kussmann, A.W. Lange, A. Behn, J. Deng, X. Feng, D. Ghosh, M. Goldey, P.R. Horn, L.D. Jacobson, I. Kaliman, R.Z. Khaliullin, T. Kuś, A. Landau, J. Liu, et al., Advances in molecular quantum chemistry contained in the Q-Chem 4 program package, Mol. Phys. 113 (2) (2015) 184–215, https://doi.org/10.1080/00268976.2014.952696.

[47] L. Jiao, C. Yu, J. Wang, E.A. Briggs, N.A. Besley, D. Robinson, M.J. Ruedas-Rama, A. Orte, L. Crovetto, E.M. Talavera, J.M. Alvarez-Pez, M. Van Der Auweraer, N. Boens, Unusual spectroscopic and photophysical

properties of meso-tert-butylBODIPY in comparison to related alkylated BODIPY dyes, RSC Adv. 5 (109) (2015) 89375–89388, https://doi.org/10.1039/c5ra17419h.

[48] R. Crespo-Otero, Q. Li, L. Blancafort, Exploring potential energy surfaces for aggregation-induced emission—from solution to crystal, Chem. Asian J. 14 (6) (2019) 700–714, https://doi.org/10.1002/asia.201801649.

[49] X.L. Peng, S. Ruiz-Barragan, Z.S. Li, Q.S. Li, L. Blancafort, Restricted access to a conical intersection to explain aggregation induced emission in dimethyl tetraphenylsilole, J. Mater. Chem. C 4 (14) (2016) 2802–2810, https://doi.org/10.1039/c5tc03322e.

[50] J. Guan, R. Wei, A. Prlj, J. Peng, K.H. Lin, J. Liu, H. Han, C. Corminboeuf, D. Zhao, Z. Yu, J. Zheng, Direct observation of aggregation-induced emission mechanism, Angew. Chem. Int. Ed. 59 (35) (2020) 14903–14909, https://doi.org/10.1002/anie.202004318.

[51] T. Zhang, Y. Jiang, Y. Niu, D. Wang, Q. Peng, Z. Shuai, Aggregation effects on the optical emission of 1,1,2,3,4,5-hexaphenylsilole (HPS): a QM/MM study, J. Phys. Chem. A 118 (39) (2014) 9094–9104, https://doi.org/10.1021/jp5021017.

[52] Q. Wu, T. Zhang, Q. Peng, D. Wang, Z. Shuai, Aggregation induced blue-shifted emission-the molecular picture from a QM/MM study, Phys. Chem. Chem. Phys. 16 (12) (2014) 5545–5552, https://doi.org/10.1039/c3cp54910k.

[53] F. Bu, R. Duan, Y. Xie, Y. Yi, Q. Peng, R. Hu, A. Qin, Z. Zhao, B.Z. Tang, Unusual aggregation-induced emission of a Coumarin derivative as a result of the restriction of an intramolecular twisting motion, Angew. Chem. Int. Ed. 54 (48) (2015) 14492–14497, https://doi.org/10.1002/anie.201506782.

[54] B. Wang, X. Wang, W. Wang, F. Liu, Exploring the mechanism of fluorescence quenching and aggregation-induced emission of a phenylethylene derivative by QM (CASSCF and TDDFT) and ONIOM (QM:MM) calculations, J. Phys. Chem. C 120 (2016) 21850–21857.

[55] T. Sajoto, P.I. Djurovich, A.B. Tamayo, J. Oxgaard, W.A. Goddard, M.E. Thompson, Temperature dependence of blue phosphorescent cyclometalated Ir(III) complexes, J. Am. Chem. Soc. 131 (28) (2009) 9813–9822, https://doi.org/10.1021/ja903317w.

[56] X. Zhang, D. Jacquemin, Q. Peng, Z. Shuai, D. Escudero, General approach to compute phosphorescent OLED efficiency, J. Phys. Chem. C 122 (11) (2018) 6340–6347, https://doi.org/10.1021/acs.jpcc.8b00831.

[57] F. Neese, Orca, an Ab Initio, DFT and Semiempirical SCF-MO Package 2.8.0 R2327, University of Bonn, Bonn, Germany (2011).

[58] D. Escudero, Quantitative prediction of photoluminescence quantum yields of phosphors from first principles, Chem. Sci. 7 (2) (2016) 1262–1267, https://doi.org/10.1039/c5sc03153b.

[59] L. Yang, P. Ye, W. Li, W. Zhang, Q. Guan, C. Ye, T. Dong, X. Wu, W. Zhao, X. Gu, Q. Peng, B. Tang, H. Huang, Uncommon aggregation-induced emission molecular materials with highly planar conformations, Adv. Opt. Mater. 6 (2018) 1701394, https://doi.org/10.1002/adom.201701394.

[60] L. Greb, J.M. Lehn, Light-driven molecular motors: imines as four-step or two-step unidirectional rotors, J. Am. Chem. Soc. 136 (38) (2014) 13114–13117, https://doi.org/10.1021/ja506034n.

[61] B.H. Jhun, S.Y. Yi, D. Jeong, J. Cho, S.Y. Park, Y. You, Aggregation of an n-π^* molecule induces fluorescence turn-on, J. Phys. Chem. C 121 (21) (2017) 11907–11914, https://doi.org/10.1021/acs.jpcc.7b02797.

[62] Y. Gong, L. Zhao, Q. Peng, D. Fan, W.Z. Yuan, Y. Zhang, B.Z. Tang, Crystallization-induced dual emission from metal- and heavy atom-free aromatic acids and esters, Chem. Sci. 6 (8) (2015) 4438–4444, https://doi.org/10.1039/c5sc00253b.

[63] F. Aquilante, L. De Vico, N. Ferré, G. Ghigo, P.A. Malmqvist, P. Neogrády, T.B. Pedersen, M. Pitoňák, M. Reiher, B.O. Roos, L. Serrano-Andrés, M. Urban, V. Veryazov, R. Lindh, Software news and update MOLCAS 7: the next generation, J. Comput. Chem. 31 (1) (2010) 224–247, https://doi.org/10.1002/jcc.21318.

[64] G. Moyna, H.J. Williams, R.J. Nachman, A.I. Scott, Conformation in solution and dynamics of a structurally constrained linear insect kinin pentapeptide analogue, Biopolymers 49 (5) (1999) 403–413, https://doi.org/10.1002/(SICI)1097-0282(19990415)49:5<403::AID-BIP6>3.0.CO;2-T.

[65] Z. Shuai, Thermal vibration correlation function formalism for molecular excited state decay rates, Chin. J. Chem. (2020), https://doi.org/10.1002/cjoc.202000226.

[66] M.J. Frisch, G.W. Trucks, H.B. Schlegel, G.E. Scuseria, M.A. Robb, J.R. Cheeseman, G. Scalmani, V. Barone, G.A. Petersson, H. Nakatsuji, X. Li, M. Caricato, A.V. Marenich, J. Bloino, B.G. Janesko, R. Gomperts, B. Mennucci, H.P. Hratchian, J.V. Ortiz, et al., Gaussian 16 Rev. C.01, 2016. Wallingford, CT.

[67] Z. Shuai, W. Xu, Q. Peng, H. Geng, From electronic excited state theory to the property predictions of organic optoelectronic materials, Science China Chem. 56 (9) (2013) 1277–1284, https://doi.org/10.1007/s11426-013-4916-7.

[68] Z. Shuai, Q. Peng, Organic light-emitting diodes: theoretical understanding of highly efficient materials and development of computational methodology, Natl. Sci. Rev. 4 (2) (2017) 224–239, https://doi.org/10.1093/nsr/nww024.
[69] Y. Jiang, Q. Peng, X. Gao, Z. Shuai, Y. Niu, S.H. Lin, Theoretical design of polythienylenevinylene derivatives for improvements of light-emitting and photovoltaic performances, J. Mater. Chem. 22 (2012) 4491, https://doi.org/10.1039/c1jm14956c.
[71] Q. Peng, H. Ma, Z. Shuai, Theory of long-lived room-temperature phosphorescence in organic aggregates, Acc. Chem. Res. (2020), https://doi.org/10.1021/acs.accounts.0c00556.
[72] H. Zhang, Z. Zhao, P.R. McGonigal, R. Ye, S. Liu, J.W.Y. Lam, R.T.K. Kwok, W.Z. Yuan, J. Xie, A.L. Rogach, B.Z. Tang, Clusterization-triggered emission: uncommon luminescence from common materials, Mater. Today 32 (2020) 275–292, https://doi.org/10.1016/j.mattod.2019.08.010.
[73] Q. Li, Z. Li, Molecular packing: another key point for the performance of organic and polymeric optoelectronic materials, Acc. Chem. Res. 53 (4) (2020) 962–973, https://doi.org/10.1021/acs.accounts.0c00060.
[74] S. Mukherjee, P. Thilagar, Renaissance of organic triboluminescent materials, Angew. Chem. Int. Ed. 58 (24) (2019) 7922–7932, https://doi.org/10.1002/anie.201811542.
[75] M. Hu, H.-T. Feng, Y.-X. Yuan, Y.-S. Zheng, B.Z. Tang, Chiral AIEgens—chiral recognition, CPL materials and other chiral applications, Coord. Chem. Rev. 416 (2020), https://doi.org/10.1016/j.ccr.2020.213329, 213329.
[76] Z. He, W. Zhao, J.W.Y. Lam, Q. Peng, H. Ma, G. Liang, Z. Shuai, B.Z. Tang, White light emission from a single organic molecule with dual phosphorescence at room temperature, Nat. Commun. 8 (2017) 416, https://doi.org/10.1038/s41467-017-00362-5.
[78] T. Zhang, H. Ma, Y. Niu, W. Li, D. Wang, Q. Peng, Z. Shuai, W. Liang, Spectroscopic signature of the aggregation-induced emission phenomena caused by restricted nonradiative decay: a theoretical proposal, J. Phys. Chem. C 119 (2015) 5040–5047, https://doi.org/10.1021/acs.jpcc.5b01323.
[81] V. Henri, The structure of molecules and the absorption spectra of gaseous substances, Compt. Rend. (177) (1923) 1037.
[82] M. Hayashi, A.M. Mebel, K.K. Liang, S.H. Lin, *Ab initio* calculations of radiationless transitions between excited and ground singlet electronic states of ethylene, J. Chem. Phys. 108 (5) (1998) 2044–2055, https://doi.org/10.1063/1.475584.
[83] M. Dommett, M. Rivera, R. Crespo-Otero, How inter- and intramolecular processes dictate aggregation-induced emission in crystals undergoing excited-state proton transfer, J. Phys. Chem. Lett. 8 (24) (2017) 6148–6153, https://doi.org/10.1021/acs.jpclett.7b02893.
[84] Y.-J. Gao, X.-P. Chang, X.-Y. Liu, Q.-S. Li, G. Cui, W. Thiel, Excited-state decay paths in tetraphenylethene derivatives, J. Phys. Chem. A 121 (13) (2017) 2572–2579, https://doi.org/10.1021/acs.jpca.7b00197.
[85] J. Huang, H. Nie, J. Zeng, Z. Zhuang, S. Gan, Y. Cai, J. Guo, S.-J. Su, Z. Zhao, B.Z. Tang, Highly efficient nondoped OLEDs with negligible efficiency roll-off fabricated from aggregation-induced delayed fluorescence luminogens, Angew. Chem. Int. Ed. Engl. 56 (42) (2017) 12971–12976, https://doi.org/10.1002/anie.201706752.
[86] Q. Li, L. Blancafort, A conical intersection model to explain aggregation induced emission in diphenyl dibenzofulvene, Chem. Commun. (Camb.) 49 (53) (2013) 5966–5968, https://doi.org/10.1039/c3cc41730a.
[87] Y. Niu, Q. Peng, H. Geng, Y. Yi, L. Wang, G. Nan, D. Wang, Z. Shuai, MOlecular MAterials Property Prediction Package (MOMAP) 1.0: a software package for predicting the luminescent properties and mobility of organic functional materials, Mol. Phys. 116 (7–8) (2018) 1078–1090, https://doi.org/10.1080/00268976.2017.1402966.
[88] Y. Niu, Q. Peng, C. Deng, X. Gao, Z. Shuai, Theory of excited state decays and optical spectra: application to polyatomic molecules, J. Phys. Chem. A 114 (30) (2010) 7817–7831, https://doi.org/10.1021/jp101568f.
[89] S. Sasaki, S. Suzuki, W.M.C. Sameera, K. Igawa, K. Morokuma, G.-I. Konishi, Highly twisted *N,N*-dialkylamines as a design strategy to tune simple aromatic hydrocarbons as steric environment-sensitive fluorophores, J. Am. Chem. Soc. 138 (26) (2016) 8194–8206, https://doi.org/10.1021/jacs.6b03749.
[91] Q. Peng, Y. Yi, Z. Shuai, J. Shao, Excited state radiationless decay process with Duschinsky rotation effect: formalism and implementation, J. Chem. Phys. 126 (2007) 114302, https://doi.org/10.1063/1.2710274.
[92] Q. Peng, Y. Niu, Z. Wang, Y. Jiang, Y. Li, Y. Liu, Z. Shuai, Theoretical predictions of red and near-infrared strongly emitting X-annulated rylenes, J. Chem. Phys. 134 (7) (2011), https://doi.org/10.1063/1.3549143, 074510.
[93] Y. Xie, T. Zhang, Z. Li, Q. Peng, Y. Yi, Z. Shuai, Influences of conjugation extent on the aggregation-induced emission quantum efficiency in Silole derivatives: a computational study, Chem. Asian J. 10 (10) (2015) 2154–2161, https://doi.org/10.1002/asia.201500303.
[94] Q. Peng, Y. Niu, C. Deng, Z. Shuai, Vibration correlation function formalism of radiative and non-radiative rates for complex molecules, Chem. Phys. 370 (1–3) (2010) 215–222, https://doi.org/10.1016/j.chemphys.2010.03.004.

Index

Note: Page numbers followed by *f* indicate figures, *t* indicate tables, and *s* indicate schemes.

F

G

H

I

V

Printed in the United States
by Baker & Taylor Publisher Services